ELECTRON DYNAMICS IN MOLECULAR INTERACTIONS

Principles and Applications

ELECTRON DYNAMICS IN MOLECULAR INTERACTIONS

Principles and Applications

Frank Hagelberg

East Tennessee State University, USA

ICP Imperial College Press

Published by

Imperial College Press
57 Shelton Street
Covent Garden
London WC2H 9HE

Distributed by

World Scientific Publishing Co. Pte. Ltd.
5 Toh Tuck Link, Singapore 596224
USA office: 27 Warren Street, Suite 401-402, Hackensack, NJ 07601
UK office: 57 Shelton Street, Covent Garden, London WC2H 9HE

Library of Congress Control Number: **2013956293**

British Library Cataloguing-in-Publication Data
A catalogue record for this book is available from the British Library.

ELECTRON DYNAMICS IN MOLECULAR INTERACTIONS
Principles and Applications

ISBN 978-1-84816-487-1

Typeset by Stallion Press
Email: enquiries@stallionpress.com

Printed in Singapore

Dedication

When rays of fire with boist′rous fury stream
From sky to earth in bold trajectories,
Or rushing air descends from heights supreme,
If devastating gale, if gentle breeze,

Then think, my Love, that tiny moleckules
Form all there is, from lowly to sublime,
And that thy wondrous mind commands the rules
To trace with ease their tumbling dance through time.

You be their mistress, you their humble maid –
Their laws lay down and patiently record
What in the stillness of thy thought you braid.
A book on nature′s course be thy reward.

Appease and call to order nature′s rage
With numbers, figures, symbols, page by page.

Erasmus Owen [1]

Dedication

When rays of fire with boisterous fury stream
From sky to earth in bold trajectories,
Or rushing air descends from heights supreme,
If devastating gale, if gentle breeze,

Then think, my Love, that tiny molecules
Form all there is, from lowly to sublime,
And that thy wondrous mind commands the rules
To trace with care their tumbling dance through time.

You be their mistress, you their humble maid –
Their laws lay down and patiently record
What in the fullness of thy thought you braid.
A book on nature's course be thy reward.

Appear, and call in order nature's rage
With numbers, figures, symbols, page by page.

Erasmus Owen [1]

Contents

About This Book xix

Introduction: Electron Multistate Molecular Dynamics xxv

Part I: Preparations 1

1 *Ab Initio* Theory of Electronic Structure 3

1.1 Molecular Orbitals 4
1.1.1 Molecular and atomic orbitals 9
1.1.2 Expectation values from molecular orbitals: The example of spin-orbit coupling 11
1.2 The Minimal LCAO Model 12
1.3 Hartree–Fock Theory 16
1.3.1 The Hartree–Fock equations 17
1.3.2 Koopmans' theorem 21
1.3.3 The Hartree–Fock Hamiltonian 22
1.4 The Restricted and the Unrestricted Hartree–Fock Formalism 23
1.4.1 The restricted Hartree–Fock method 24
1.4.2 The unrestricted Hartree–Fock method 25
1.4.3 The Roothaan formalism 27
1.5 Post-Hartree–Fock Methods 30
1.5.1 Configuration interaction 31
1.5.2 Many-body perturbation theory 34
1.6 Excited Electronic States 36
1.7 Appendix: The Functional Derivative 41

2 The Adiabatic and the Diabatic Representation **45**

2.1 The Born–Oppenheimer Approximation 46
2.2 Harmonic Vibrational Modes 50
2.3 Adiabatic and Diabatic Frames 54
2.3.1 The diabatic approximation 58
2.4 Gauge Theoretical Form of the Nuclear Equation 59
2.5 Avoided Crossings, Degeneracies, Conical Intersections 63
2.5.1 Avoided crossings 63
2.5.2 Conical intersections and the Jahn–Teller effect 65
2.5.3 Jahn–Teller distortion 68
2.5.4 The molecular Aharonov–Bohm effect 71
2.5.5 The geometric phase in molecular pseudorotation 75
2.6 Locating the Seam 78
2.6.1 Intersection-adapted coordinates 79
2.6.2 Determination of the seam space 82
2.6.3 Seam subspaces by Lagrangian minimization 84

3 Basic Concepts of Scattering Theory **87**

3.1 The Time-Dependent and the Time-Independent View of Scattering Processes 88
3.2 Quantum Mechanical Equations of Motion 89
3.3 The Scattering Matrix 94
3.3.1 The Møller operators 96
3.3.2 The Lippmann–Schwinger equations 98
3.3.3 Unitarity of the S-matrix 100
3.4 Elastic Scattering by a Spherical Potential 104
3.4.1 The asymptotic scattering solution 104
3.4.2 T-, S-, and K-matrix boundary conditions . . . 107
3.4.3 The elastic cross section 111
3.5 Resonances 113

4 **Semiclassical Notions** 121
4.1 Path Integrals and the Quantum Propagator 122
4.1.1 The quantum and the semiclassical propagator 123
4.1.2 The Van Vleck propagator 128
4.1.3 The monodromy matrix 133
4.2 The WKB Approximation 136
4.2.1 The WKB wave function 136
4.2.2 The Bohr–Sommerfeld quantization rules for bound WKB states 144
4.3 The Wigner Function: A Quantum Mechanical Phase Space Distribution 146
4.3.1 Defining properties of the Wigner function . . . 146
4.3.2 Time dependence of the Wigner function 149
4.3.3 The Moyal formalism 153
4.4 Coherent States 154
4.4.1 Coherent and particle number states 155
4.4.2 Coherent states as minimal uncertainty solutions 158
4.4.3 The nuclear coherent state 159

5 **Open Systems: Elements of Rate Theory** 161
5.1 Classical Rate Theory 163
5.2 Quantum Transition State Theory 169
5.2.1 The quantum transition state approximation 171
5.3 The Euclidean Path Integral 174
5.3.1 Classical polymer isomorphism 178
5.4 Centroid Dynamics 181
5.5 The Path Integral Form of the Golden Rule Rate Constant 187
5.6 Beyond the Golden Rule: Reduced Density Matrix Theory 190
5.6.1 A two-state problem 195

Part II: Methods **199**

6 Time-Independent Theory of Molecular Collisions I: Multichannel Scattering **201**

6.1 The Multichannel Problem . . . 201
6.2 The Lippmann–Schwinger Equation for Inelastic Scattering . . . 206
6.3 The Born Approximation . . . 210
6.3.1 The distorted-wave Born approximation (DWBA) . . . 212
6.4 Microreversibility . . . 213
6.5 R-matrix and Log Derivative Propagation . . . 217
6.5.1 The log derivative method . . . 220
6.6 Reactive Scattering I: The Differential Equation Approach . . . 223
6.6.1 Jacobi coordinates . . . 224
6.6.2 Hyperspherical coordinates . . . 227
6.7 Space-Fixed and Body-Fixed Frames of Reference . . . 229
6.7.1 Space-fixed representation . . . 230
6.7.2 Body-fixed representation . . . 232
6.8 Reactive Scattering II: The Integral Equation Approach . . . 238

7 Time-Independent Theory of Molecular Collisions II: The Electronic Problem **245**

7.1 Inclusion of the Electronic System . . . 246
7.1.1 The triatomic case . . . 247
7.1.2 The adiabatic case . . . 251
7.1.3 The diabatic case . . . 253
7.2 Case Study: The $F + H_2$ Reaction . . . 255
7.3 Variational Procedures . . . 263
7.3.1 The Kohn variational principle . . . 267
7.3.2 Kohn anomalies . . . 269

7.4 Case Study: Quenching of the Sodium Atom 3*p* State by Interaction with Hydrogen Molecules 271
7.4.1 Basis sets 277
7.4.2 Algebraic realization of the outgoing wave variational principle 279
7.4.3 Exciplex funnel dynamics 280
7.5 The Landau–Zener–Stückelberg Model of Nonadiabatic Transitions 282

8 The Time-Dependent Self-Consistent Field Theory 291

8.1 Time-Dependent Variational Principles 292
8.1.1 Time-dependent perturbations 295
8.1.2 Free and forced oscillations 299
8.2 The Time-Dependent Hartree–Fock Theory: Application to Molecules 300
8.3 Wave-Function-Based *Ab Initio* Molecular Dynamics 305
8.3.1 Direct molecular dynamics in the time-dependent Hartree–Fock framework 305
8.3.2 Classical trajectories within TDHF dynamics 307
8.3.3 The Hellmann–Feynman theorem 312
8.3.4 Ehrenfest dynamics 315
8.3.5 Car–Parrinello dynamics 318
8.4 Time-Dependent Hartree–Fock Dynamics in the Eikonal Approximation 322
8.4.1 The eikonal approximation 324
8.4.2 TDHF approach to the electronic problem within the eikonal approximation 327
8.4.3 The Liouville–von Neumann equation in a traveling orbital basis 329
8.5 Case Study: Light Emission in Slow Proton–Hydrogen Collisions 333

9 Evolution of Coherent Molecular States: Electron Nuclear Dynamics Theory 341

9.1 The Thouless Representation 343
9.2 The END Equations 348
9.2.1 Derivation of the END equations 349
9.2.2 Interpretation of the END equations 354
9.3 Two Special Cases: The Boosted Self-Consistent Field and the Linearized END Equations 357
9.3.1 The boosted electronic system 357
9.3.2 The linear version of the electronic END equations 360
9.4 Inclusion of Nuclear Quantum Effects 361
9.4.1 Trajectory interference 362
9.4.2 Case study: H impact on molecular and atomic targets by END theory 366
9.4.3 Rovibrational analysis of the nuclear system . . . 372
9.5 Nonadiabatic Effects in Bound Systems by END Theory: The Pseudorotation of H_3^+ 376

10 The Classical Electron Analog 383

10.1 Critique of the Ehrenfest Representation 384
10.2 The Classical Electron Analog 386
10.2.1 The CEA equations of motion 388
10.2.2 Adiabatic representation of the Hamilton function 389
10.2.3 The classical analog of the electronic two-state problem 391
10.3 CEA Theory Applied to a Conical Intersection Problem 395

11 Hopping and Spawning 403

11.1 The Trajectory Surface Hopping Method 404
11.2 The Fewest Switches Algorithm 408
11.2.1 Three test cases 413
11.2.2 Complex-valued trajectories 418

11.3 Spawning 419
11.3.1 Applications to model problems 425
11.4 Case Study: The Dynamics of Na*-Quenching by Collision with Hydrogen Molecules 429
11.5 Comparison with Other Methods 433

12 Semiclassical Propagator Techniques 437

12.1 The Path Integral Approach to Molecular Dynamics 438
12.2 Semiclassical Propagation and Surface Hopping 442
12.3 The Initial Value Representation 447
12.4 The Mapping Approach to Electronic Degrees of Freedom 451
12.4.1 The Schwinger mapping formalism 452
12.4.2 Extension to general N-level systems 454
12.5 The Mapping Technique Applied to Nonadiabatic Dynamics 457
12.5.1 The SC-IVR approach applied to nonadiabatic model cases 460
12.5.2 Comparison with the Ehrenfest model 464
12.6 Case Study: The S_1–S_2 Transition in Pyrazine: SC-IVR Treatment of a Conical Intersection Problem 466
12.7 Numerical Procedures for Semiclassical Propagation Methods 471
12.7.1 Monte Carlo integration 473
12.7.2 Filinov filtering 476
12.7.3 The forward-backward initial value representation 479
12.8 Cellular Dynamics 482

13 Quantum Hydrodynamics I: Coupled Trajectories in Bohmian Mechanics 491

13.1 Elements of the Quantum Theory of Motion 492
13.1.1 Quantum trajectories 495
13.1.2 The pilot wave and the guided particle 498

13.2 Lagrangian Quantum Hydrodynamics 499
13.2.1 Assembling the wave function 502
13.2.2 Technical challenges for quantum trajectory propagation 503
13.3 Nonadiabatic Lagrangian Quantum Hydrodynamics 505
13.4 The Classical Limit of the Quantum Theory of Motion 512

14 Quantum Hydrodynamics II: The Semiclassical Liouville–Von Neumann Equation 517
14.1 The Semiclassical Liouville Formalism for Multistate Problems 518
14.1.1 Two coupled states: A model problem 524
14.2 Phase Space Trajectory Implementation 528
14.3 Generalized Quantum Hydrodynamics: Mixed States 534
14.3.1 Pure states 538
14.3.2 Mixed states 540
14.4 Coupled Electronic States 542

15 Wave Packet Propagation Methods 547
15.1 The Grid Representation 548
15.1.1 The discrete variable representation (DVR) . . . 553
15.1.2 The fast Fourier transform (FFT) 555
15.2 Numerical Wave Packet Propagation Techniques 557
15.2.1 The Crank–Nicolson scheme 557
15.2.2 Split operator propagation 559
15.2.3 Propagator expansion techniques 560
15.3 The Multiconfiguration Time-Dependent Hartree (MCTDH) Method 564
15.3.1 The time-dependent Hartree (TDH) approach . 565
15.3.2 The multiconfiguration time-dependent Hartree (MCTDH) approach 567
15.3.3 The MCTDH equations 569

15.4 Case Study: Photostability of Biologically Relevant Molecules 574
15.4.1 Ultrafast deexcitation by passage through conical intersections in nucleic acid bases and base pairs 575
15.4.2 Dynamics at the $^1\pi\sigma^*$-S_0 conical intersection of pyrrole 579

16 Density Functional Dynamics 587

16.1 Fundamentals of Density Functional Theory 588
16.1.1 Exchange-correlation potentials 593
16.2 Excited Electronic States in DFT 595
16.3 Time-Dependent Density Functional Theory 598
16.3.1 TDDFT in the linear response domain 601
16.3.2 Time-dependent current density functional theory 603
16.4 Direct Molecular Dynamics Based on DFT 605
16.4.1 Calculating molecular photoabsorption spectra 606
16.4.2 Molecular bonding properties analyzed by the electron localization function 608
16.4.3 Combining TDDFT with standard methods of nonadiabatic dynamics 609

17 Decoherence 613

17.1 The Dissipative Liouville–von Neumann Equation . . . 615
17.2 Evaluating Decoherence Times in a Semiclassical Framework 626
17.2.1 Ensemble average of the decoherence function 632
17.3 Case Study: The Dynamics of Electron Hydration . . . 636
17.3.1 Isotope effects in hydrated electron relaxation 638
17.4 Continuous Surface Switching: A Compromise between Mean-Field and Individual Surface Propagation 644

17.5 Decay of Mixing 648
17.5.1 Decoherence time 657
17.5.2 Determining the decoherent state 658

Part III: Special Topics **661**

18 Ultrafast Optical Spectroscopy **663**

18.1 Linear and Nonlinear Polarization 664
18.1.1 Deriving the pump–probe signal 666
18.2 Theory of Nonlinear Polarization in Femtosecond Molecular Spectroscopy 669
18.2.1 The perturbative approach 672
18.2.2 The non-perturbative approach 681
18.3 Polarization Studies of *cis-trans* Isomerization 682
18.3.1 Adiabatic formulation 688
18.4 The Density Matrix Approach to Simulating Pump–Probe Signals 690
18.4.1 The pump–probe signal 698
18.5 Case Study: Ultrafast Spectroscopy on Non-Stoichiometric Alkali-Halide Clusters 700
18.5.1 Effective single-electron systems of the form Na_nF_{n-1} 701
18.5.2 Extension to nonadiabatic dynamics 706
18.6 Appendix: Derivation of the Pump–Probe Signal $S(t_\mathrm{d})$ 711

19 Optical Control of Electron Multistate Molecular Dynamics **715**

19.1 Interaction of a Molecule with a Pulse of Light 716
19.2 The Tannor–Rice Scheme: Optimal Control 719
19.3 The Brumer–Shapiro Scheme: Coherent Control 725
19.4 Case Study: Coherent Control of ICN Photodissociation 730
19.5 Optimal Control in Pump–Probe Spectroscopy 736
19.5.1 Case study: Application to Na_3F_2 743

20 **Electron Transfer in Condensed Media** 749

20.1 The Electronic Hamiltonian 753
20.2 Electronic–Vibronic Coupling: The Spin-Boson Hamiltonian 756
20.3 Adiabatic versus Nonadiabatic Electron Transfer 760
20.4 Thermally Activated Transfer 763
20.5 Inclusion of Nuclear Tunneling 767
20.5.1 The continuous limit of nuclear frequencies . . . 771
20.6 Effects of Polar Solvents on Electron Transfer 774
20.6.1 The dielectric displacement field 776
20.6.2 Polarization and polarizability 778
20.6.3 The free energy functional 783
20.6.4 The electron transfer rate in a polar environment 786
20.7 Ultrafast Electron Transfer 790
20.8 Case Study: Aqueous Ferrous–Ferric Exchange 793
20.8.1 Monte Carlo modeling 793
20.8.2 Euclidean path integral simulations 798
20.8.3 Recent quantum dynamical extensions 804
20.9 Appendix: Formulae Relevant for Electron Transfer Theory within the Marcus Model 805
20.9.1 Electron transfer in a vibrational bath: Formal procedures used in the derivation of the rate constant 806
20.9.2 Derivation of the effective free energy functional Eq. (20.109) 810
20.9.3 The density of states for electron transfer in a solvent: Calculating the trace Eq. (20.118) 813

21 **Electronic Friction in Molecule–Surface Interactions** 817

21.1 Langevin Formulation of Ehrenfest Dynamics 820
21.2 An *Ab Initio* Model for Electronic Friction 824
21.3 Case Study: Nonadiabatic Effects in the Interaction between the Cu(100) Surface and a CO Molecule 828

21.3.1 Vibrational relaxation of CO on the Cu(100) surface: The impact of electronic friction 828
21.3.2 Vibrational excitation and hot diffusion . . . 835
21.4 Beyond Langevin Theory 838

Epilogue **841**

Bibliography **847**

Index **875**

About This Book

The quantum molecular dynamics of electron multistate problems is a young and still emergent field. Traditional schemes of molecular dynamics are confined to a single electronic state, usually the ground state of the considered system, and most processes of current chemical interest can be well described within this constraint. As, however, the perspective widens to include multiple electronic states, a broad range of complex novel phenomena of both systematic and practical interest comes into view. These comprise photochemical and photobiological processes involving non-radiative deexcitation, photoinduced unimolecular decay, dissociative electron attachment, chemiluminescent reactions, atom–molecule or intermolecular collisions in the physical energy regime, molecule–surface interactions or electron transfer processes, and much more. Theoretical understanding of these matters has acquired urgency as femtosecond laser pulses with power densities exceeding 10^{16} W/cm^2 have become available, providing tools that make it possible to trace the evolution of electronically excited molecular species. The ability of experimentalists to resolve ultrafast processes associated with dramatic changes of the electronic system of a molecule calls for an adequate response from the side of theorists. A great variety of models have been devised to address electron multistate situations. While many describe nuclear dynamics in a landscape of coupled potential energy surfaces, making allowance for electronic excitation and deexcitation, some sacrifice this approach in favor of direct or *ab initio* dynamics where both electronic and nuclear degrees of freedom are treated as dynamic variables. While some approximate the full solution of the time-dependent Schrödinger equation or the Liouville–von Neumann equation, others take recourse to semiclassical or even classical models, utilizing trajectories rather than wave functions or density matrices advancing in time. Further, procedures that use time as the independent variable are to be contrasted with stationary approaches that operate within the energy representation. The purpose of this book is to provide an introduction to the rich diversity of theories of electron dynamics in molecular processes, including related computational procedures.

Substantial progress has been made recently in addressing the problem of several coupled potential energy surfaces in quantum molecular dynamics. Not only has the great complexity of this problem been widely acknowledged, but also its high degree of practical relevance. The search for more general and refined strategies for the quantitative analysis of nonadiabatic interactions in polyatomic molecules is at the forefront of theoretical physical chemistry.

Within chemistry, the coupling between nuclear and electronic degrees of freedom was for a long time viewed as an effect of higher order, a largely negligible correction to molecular dynamics proceeding on a single potential energy surface. Nonadiabatic effects, involving the breakdown of the Born–Oppenheimer approximation that separates electronic and nuclear motion, were studied mostly by applying quantum mechanical or semiclassical theory to model systems. With the increase of computer capacity during the last three decades, however, realistic studies of excited electronic states and, concomitantly, the mapping of excited potential energy surfaces at great accuracy and for a broad variety of molecules became feasible. New insight was gained into the key significance of certain critical configurations within the potential energy landscape, namely *avoided crossings* and *conical intersections* (formerly referred to as *electronic funnels*) for electron dynamics in general and for photochemistry in particular, rationalizing the subpicosecond time scale of internal conversion events in a large class of organic molecules. In the wake of this development, nonadiabatic phenomena were characterized in terms of geometric, topological, and energetic properties of these structures, and powerful algorithms were designed to trace their locations.

Excellent survey articles dealing with these subjects and with the theories designed to address them have appeared in various journals (e.g. [2]–[6]). Also, monographs on quantum molecular dynamics are available that devote some sections to the theme of this book or deal with subtopics of the book in greater detail than presented here (e.g. [7]–[12]). A unified presentation and comparative discussion of molecular dynamics theories involving multiple electronic states, however, seems to be lacking so far. To fill this gap was the main motivation for writing this book.

The term *electron dynamics* is understood here in the twofold sense of direct electron dynamics, as described by approaches that assign time-dependent variables to the electronic degrees of freedom, and of nuclear dynamics proceeding on coupled potential energy surfaces. Doing full justice to this wide field or referring to all of the theories proposed in this discipline is beyond both the ability and the ambition of the author. A factor that further complicates the task is that the subject is in rapid motion. Posing formidable conceptual as well as technical challenges, electron multistate molecular dynamics is a constantly expanding area of quantum chemistry and molecular physics with swiftly shifting frontiers. Therefore, this effort aims at a selection of theoretical and computational methods prevalent in this area during the last two or three decades rather than a complete and seamless account.

At the core of this text is a synoptic presentation of partially complementary and partially competing theories that address the problem of electronically nonadiabatic interactions in molecular dynamics. In the recent past, this challenge has been confronted by novel computational approaches, such as the multiconfiguration time-dependent Hartree (MCTDH) or the electron nuclear dynamics (END) theory. Their explanatory power has been demonstrated extensively by comparison with experiment. The main objective of the book is to introduce these theoretical efforts and to contrast them with more traditional schemes of electron dynamics in molecules. The presented models are derived from their quantum theoretical roots, their interrelations are discussed, and characteristic applications of these models to concrete chemical systems are outlined.

The first five chapters are designed to develop the theme of this book in an *ab initio* fashion, providing the quantum chemical, semiclassical, as well as quantum dynamical tools required for a full appreciation of the later chapters that deal with electron dynamics. In this way, this volume may be used as a textbook, supplementing the reading list for graduate or advanced undergraduate courses in molecular physics or physical chemistry. Thus, the first part of this text sets the frame in which the various approaches to nonadiabatic molecular dynamics appear, containing basic tenets of electronic structure theory, scattering theory, some essential ideas of semiclassical molecular physics, the treatment of open systems, as well as the fundamental distinction between the adiabatic and the diabatic representation of the electronic system which pervades the whole text.

The central part of this book, comprising Chapters 6–17, presents a comparative survey of theoretical and computational methods dealing with quantum dynamics in an electron multistate scenario. A guiding scheme for discussing these methods is given by three basic distinctions. First, time-dependent and time-independent procedures are compared, where emphasis is placed on the former. Further, methods that invoke the classical trajectory concept, implementing a Newtonian approach to nuclear dynamics, are contrasted with quantal models based on nuclear wave packet propagation. Lastly, dynamics proceeding in a landscape of coupled potential energy surfaces is distinguished from *ab initio* dynamics. These oppositions provide natural categories for characterizing the methods of electron multistate molecular dynamics included in this volume.

The last segment of this book deals with topical applications of the described theoretical models to areas as diverse as optical spectroscopy, electron transfer within or between molecular centers embedded in condensed media, or the interaction of molecules with surfaces.

To show the surveyed theoretical and computational methods in operation, case studies are included, involving, wherever possible, their application to physical molecular systems. In most cases, these examples comprise an assessment of the respective results by comparison with experimental findings. For appraising mixed classical-quantal or semiclassical methods, validation by the standard of full quantum mechanical solutions is of almost equal importance. Not rarely, this type of examination requires making reference to unrealistic (e.g. one-dimensional) test situations, but it often reveals salient strengths or weaknesses of the tested scheme with great clarity and thus is of high practical value. Rather than using superlatives as criteria for selecting applications of a given method for more detailed presentation in this volume, and thus including preferably the most recent or most extensive examples for the use of the method, I attempted to choose cases best suited to illustrate its characteristic traits.

The text is directed at researchers in the field of physical chemistry as well as graduate or advanced undergraduate students. The essential prerequisites are regular introductory courses in quantum mechanics and thermodynamics as well as some basic knowledge of atomic and molecular architecture.

The reader will find exercises scattered through each chapter. They are intended to provide some basic experience in handling the concepts introduced in this text, and also to close some unavoidable gaps in its narrative. Most exercises are at a moderate level of difficulty, a few of them test not only the comprehension but also the perseverance of the reader. About the notation used throughout the text, symbols for arrays are set in bold print and operators are distinguished by hat symbols; electronic coordinates are denoted by lower case and nuclear coordinates by upper case letters.

I hope not to have failed too badly in my intention to write a book that suffers neither from appendicitis nor acronymitis. While proliferation of acronyms has been avoided, persistently recurring terms, such as time-dependent Schrödinger equation, appear in abbreviated form throughout the text. Readers particularly interested in time-dependent methods of electron dynamics can focus on Chapters 8–17 without going through all of the preceding text. Frequent specific references to earlier book sections aid the reader in looking up materials needed for appreciating the second part of this book. Some sections intended to provide a deeper or more detailed understanding of the topics discussed in the respective chapter were not included in this volume in order to keep it at a manageable size. These portions are labeled by section numbers with asterisks, and they are posted on my website (www.etsu.edu/physics/comp_chem.html). It is probable that this text contains mistakes, in spite of all my efforts to avoid and eliminate them. Corrections of any errors that come to my attention after publication of the book will also be found on my website.

My gratitude goes to all those who have contributed to completing this book through advice, criticism, encouragement, through alerting me to relevant publications and ongoing projects, or even letting me have a look at their own manuscripts prior to publication. For very helpful and stimulating conversations on and around the theme of the book I am particularly thankful to Vlasta Bonacic-Koutecky, Remigio Cabrera-Trujillo, Wolfgang Domcke, Raymond Kapral, William A. Lester, Todd J. Martinez, David Micha, William A. Miller, Jorge A. Morales, Yngve Öhrn, Michael Thoss, and Donald Truhlar. I have benefited tremendously from the constant support I received from Eric Deumens who read the whole text and made invaluable comments. For technical help with producing the manuscript I thank Hao Hu, David Dixon, Brooke Wedding, Leo Memolo, and, in the very first place, Jianhua Wu who was always available to assist me with great efficiency and unfailing patience. I want to thank Lizzie Bennet, Katie Lydon, and Jacqueline Downs at Imperial College Press, for accompanying my progress with gentle firmness. Lastly, I thank my wife Betsy and my children Ellen, Julia, and Philip for keeping me steadfast and cheerful on the whole long journey of writing this book.

Introduction: Electron Multistate Molecular Dynamics

Most applications of molecular dynamics rest on the premise that the electronic state of lowest energy is well separated from excited electronic states. Molecular motion, accordingly, is entirely determined by the ground state potential energy surface of the considered molecular system. The theoretical framework for this assumption is given by the *adiabatic* or *Born–Oppenheimer approximation* [13] that applies wherever the nuclear dynamics of a system is confined to a single potential energy surface. The advent of femtosecond laser spectroscopy that allows us to resolve ultrafast processes associated with dramatic changes of the electronic system of a molecule, however, challenged this notion. In the wake of this development, the attention of physical chemists during the last two or three decades turned increasingly to phenomena that transcend the frame of the Born–Oppenheimer approximation, as prototypically realized by photochemical and photobiological reactions involving non-radiative deexcitation. These are specified by transitions between at least two coupled potential energy surfaces, accentuating the need for theories of *electronic multistate dynamics*, the theme of this monograph.

While transitions between electronic states will generally differ in energy, the topic of this text may equivalently be characterized in terms of time. Thus, the Born–Oppenheimer approximation relies on the separability of two time scales that govern the temporal behavior of the studied system. These are defined by the rapid motion of the electrons as opposed to the comparatively slow motion of the nuclei, with the concepts of the *adiabatic electronic state* and, in consequence, the *adiabatic potential energy surface* that determines the nuclear dynamics as the mathematical reflections of this difference. As these two time scales become comparable, the nuclear and the electronic degrees of freedom cease to ignore each other and interact, giving rise to *nonadiabatic effects*.

In principle, these situations can be captured by allowing for multiple adiabatic states instead of just a single one. This strategy relies on the *adiabatic representation*, which, in actual calculations, is realized by expanding the overall molecular wave function that describes both the electronic and the nuclear subsystem into adiabatic basis functions. The basis states are coupled by the operator of the nuclear kinetic energy. If the Hamiltonian matrix of the molecular system is evaluated in the adiabatic representation, this coupling introduces non-diagonal elements that measure the interaction between the electronic and the nuclear degrees of freedom in the studied problem. As the numerical treatment of these terms is usually difficult, deviating from the adiabatic frame is often advisable when they gain importance. In these cases the *diabatic representation* of the molecular wave function may be preferable, which amounts to a shift of the agent that couples the electronic basis states from the nuclear kinetic to the nuclear portion of the potential energy operator. Like its adiabatic counterpart, the diabatic representation allows for an approximation that holds in the limit of negligible coupling. Its conditions of validity may be viewed as the reverse of those defining the adiabatic approximation, involving nuclear motion too fast for the active electrons to adjust to it. This extreme case is of much lower chemical relevance than the situation of rapid electrons and slow nuclei but may be realized in the physical regime of high kinetic nuclear energies or in electron transfer processes.

Both the adiabatic and the diabatic approximation require a large enough difference between the nuclear and the electronic time regime, and thus the characteristic velocities of the two particle types. The relative size of these scales depends on the nuclear dynamics or the *topographical features of the potential energy landscape* in which a given molecular system evolves, or a combination of these effects. Not only the approximate matching between the velocities of the nuclei and the electrons provides an intuitive criterion for the presence of appreciable interaction between the nuclear and the electronic degrees of freedom, but also the separation between adjacent potential energy surfaces. Sufficiently small energy differences put the Born–Oppenheimer approximation at risk, and it breaks down around points of degeneracy between two potential energy surfaces of the same spatial and spin symmetry. In the case of polyatomic systems, the nuclear geometry adopted at such a point is referred to as *conical intersection*. These configurations, once believed to be pathological rarities, are now acknowledged as ubiquitous features of the potential energy landscape.

The challenge of molecular interactions involving a multiplicity of electronic states is the ancestral dominion of quantum theory. Much of this text will deal with attempts to address these problems by approximating solutions to the time-dependent Schrödinger equation (TDSE), or by solving equations that approximate the TDSE. The realization that ideally both the nuclear and the electronic degrees of freedom should receive a fully quantal treatment seems to imply a hierarchy of approaches which leads up to a consistent quantum mechanical formulation of the dynamics of nuclei and electrons. Methods satisfying this rigid standard to varying degrees have been developed, are in use, and will be covered in this monograph. They are, however, often prohibitively expensive in terms of computational effort, confining the set of tractable physical systems to a few particles only. Shifting the frontier of electronic multistate dynamics means in these cases improving computation routines or computer technology. Much of this monograph is devoted to alternative schemes that tend to be less costly and more imaginative, that do not furnish an ultimate solution to the problem of electron dynamics in molecular interactions but are, in many cases, well adjusted to capture the essential features of a given dynamical situation. The majority of these procedures borrow the trajectory concept from classical physics and combine a classical representation of the nuclei with a quantum mechanical description of the electrons. In the physical regime, characterized by high nuclear kinetic energies, this simplification is often legitimate as typical nuclear quantum effects, involving zero-point motion, tunneling, or separation of the nuclear wave packet, are not likely to be prevalent in this case. Semiclassical or mixed quantum-classical answers to the question for the dynamics of electrons and nuclei in molecules will be introduced and discussed throughout this book.

The chapters of this volume are grouped into three parts. To make the presentation as self-contained as possible, the first five chapters provide introductory materials that are crucial for understanding quantum dynamics in general and electron dynamics in particular, encompassing basic concepts of quantum chemistry, scattering theory, semiclassical physics, approaches to the treatment of open systems, and highlighting in some detail the contrasting features of the adiabatic and the diabatic representation of the molecular wave functions. The core of the book is comprised in the following 12 chapters (6–17) that attempt to provide a survey of the currently most prevalent theoretical and computational approaches to the challenge of electron dynamics. Reflecting the present-day trend in quantum dynamic modeling, the presentation emphasizes the time-dependent perspective while also giving room to stationary models, as described in Chapters 6 and 7. The final segment of the text (Chapters 18–21) offers selected topical applications of the theories introduced, dealing in particular with optical spectroscopy of ultrafast processes and control of reaction dynamics as well as electron transfer phenomena in the condensed phase and molecule–surface interactions.

The first two chapters of this monograph center on stationary features of basic relevance to molecular quantum dynamics, involving the definition of adiabatic states as well as alternative representations of the molecular wave function and the geometric properties of potential energy surfaces. Time makes its first appearance in Chapter 3 which accentuates the double-sidedness of the term *dynamics*, in the sense that the full solution of the time-dependent Schrödinger equation may be achieved by solving the corresponding stationary problem rather than propagating an initial wave packet. While the two preceding chapters deal with the discrete spectrum of the molecular Hamiltonian, this chapter explores the continuous part, adding fundamentals of scattering theory.

As complete sets of basis functions are unavailable for the case of molecules, both stationary and dynamic molecular problems allow only approximate solutions, restricted to strategically chosen segments of the full Hilbert space. In particular, the essential challenge of quantum molecular dynamics, namely the interaction between the electronic and nuclear degrees of freedom, cannot be met in terms of a complete and accurate quantum mechanical representation of both sets of parameters. While some models succeed in treating the electrons as well as the nuclei in a consistent quantum theoretical framework, these approaches are computationally expensive and thus largely confined to low-dimensional situations. Most procedures currently in use to simulate physical systems revert instead to *semiclassical theory*, combining the use of classical theory to describe the nuclei with a quantum treatment of the electrons. For a large part, this monograph is about the various realizations of this scheme.

Semiclassical notions that pervade much of the text are summarily introduced in Chapter 4, including chiefly the Van Vleck propagator as a semiclassical version of the quantum propagator in path integral theory, the WKB approximation, the Wigner function, and the concept of the coherent state. These entities inhabit a space intermediate between the classical and the quantum sphere, retaining critical quantum features while affording the relative ease of calculating with classical quantities, such as well-defined trajectories in the place of wave packets or observables adopting sharp values instead of Hermitian operators.

Energy or particle transfer between molecules and their environment play a crucial role in many molecular processes of practical interest, such as electron transfer reactions in biological media. These tend to be slow by the measure of the environmental degrees of freedom. Therefore, direct simulation of the latter is often not viable, neither on the grounds of quantum nor classical mechanics. Problems of this type are adequately addressed by a statistical approach that allows us to infer rate constants for reactions between equilibrium states. Chapter 5 places these ideas into the context of quantum molecular dynamics, emphasizing the task of obtaining the relevant partition functions which may be framed as a problem of dynamics in imaginary time. Rate theory on the basis of Fermi's golden rule is contrasted with Redfield theory which employs a reduced matrix formalism to incorporate the impact of the environment on the studied molecular process.

Quantum molecular dynamics in the framework of time-independent theory is the topic of the two following chapters, 6 and 7, where the focus is on reactive scattering between atoms and molecules. This segment of the monograph is divided into a general part that introduces the multichannel formalism of scattering theory without making explicit reference to electrons and nuclei (Chapter 6), and a more specific part that addresses electronic interactions in molecular collisions (Chapter 7). In the former chapter, fundamental concepts and methods are developed, such as the notion of microreversibility, the basic tenets of inelastic scattering as described by the Lippmann–Schwinger equations, and the question of adequate systems of reference to represent molecular scattering events in stationary models.

While abstract variables are utilized in Chapter 6, the following chapter assigns to them the meaning of nuclear and electronic coordinates and illustrates the previously introduced notions of time-independent molecular scattering by the example of reactions between H_2 and halogen or alkali atoms. Further, the variational principle as a tool to derive electronic solutions for collision problems and to generate numerical approximations to wave functions and transition matrix elements is highlighted at this place. An analytic model of great historical consequence, namely the treatment of the transition dynamics in a diabatic two-level situation by Landau, Zener, and Stückelberg ([14]–[16]), concludes our account of stationary approaches to the electronic problem in reactive scattering.

The middle section of this text, comprising Chapters 8–16, surveys the prevailing time-dependent theories of quantum molecular dynamics with emphasis on electronic transitions in general, and specifically on electronically nonadiabatic processes. The primacy enjoyed originally by the stationary approach to analyzing physical quantum systems was hardly challenged until the mid-seventies, when the propagation of Gaussians was established as a powerful and practicable tool of dynamic theory [17]. Making the step from the stationary Schrödinger equation to the TDSE means asking for wave packets instead of energy eigenfunctions as solutions of the problem. The development of efficient and elegant algorithms of wave packet propagation in the seventies and eighties, and their subsequent extensions and refinements, helped fortify the appreciation of time as the fundamental independent variable of quantum dynamics. A major technical advantage of the time-dependent as compared with the stationary frame is that the often complicating distinction between normalizable bound and periodic continuum states is replaced by the unifying concept of the square-integrable wave packet. Further, physical intuition may be partial to the latter, as it can claim to be the rightful quantum heir to the classical trajectory and often allows interpretation of a given process in concrete visual terms more readily than the stationary wave function.

Using either approach to treat physical systems, one may, with few exceptions, only hope to arrive at *approximate solutions to approximated problems*. Chapter 8 explores various aspects of the time-dependent self-consistent field theory, the temporal counterpart of stationary Hartree–Fock theory, as introduced in the first chapter. Employing a variational principle, we will derive the time-dependent Hartree–Fock equation as an approximation to the TDSE. Having arrived at a formalism that succeeds, within certain limits, at describing the temporal variation of the electronic system, we are prepared to address the larger problem of molecular dynamics in the time-dependent frame. In this context, the most essential question is how to integrate nuclear motion, and thus the interaction between electronic and nuclear degrees of freedom, into the model. There is no unique way to tackle this task. The implementations of the nuclear subsystem in quantum molecular dynamics form a hierarchy reaching from a single trajectory described by means of pure Newtonian mechanics to nuclear wave packets evolving subject to quantum theory. Introducing initially classical representations of the nuclei, we explore the methodological notion of electronic wave packets developing along nuclear trajectories and contrast popular adiabatic with nonadiabatic schemes. Among the first of these two classes, we emphasize *Born–Oppenheimer* and *Car–Parrinello* dynamics [18]. While the former necessitates optimizing the electronic system at every nuclear time step, the latter is free of this constraint since it operates with electron orbitals as dynamic quantities, tying the evolution of the system as a whole to a single potential energy surface through a Lagrangian minimization procedure. Eliminating this restriction yields the *Ehrenfest representation* of quantum dynamics. The nuclear trajectory results here from an interplay of various electronic states, related to a multiplicity of potential energy surfaces. In consequence, the nuclei follow an effective path defined by an average surface. The extreme case of vanishing *back action* of the electronic on the nuclear degrees of freedom is given by the *classical path method.* This is applicable in cases where electronic transitions are expected to have little impact on the nuclear motion, as might be realized by the passage of fast projectiles through a gaseous medium [19].

The coupling between quantum mechanical electrons and the classical nuclei may be realized at different levels of theoretical rigor. This is illustrated by the *eikonal representation* (Section 8.4). The potential felt by the nuclei is here specified as a series of additive terms whose truncation at the lowest order, corresponding to the least detailed treatment of nonadiabatic effects, yields the *eikonal approximation* [20]. Consistent inclusion of the interactions between the classical nuclear and the quantum mechanical electronic degrees of freedom is achieved by *electron nuclear dynamics*

(END) theory [2], as discussed in Chapter 9. Rather than relying on the time-dependent Hartree–Fock formalism, END operates in a generalized phase space inhabited by coherent states for both the electronic and the nuclear subsystem. The equations of motion that govern molecular evolution in this space are derived from the time-dependent variational principle. As this approach is adopted, their coupling acquires a geometric interpretation, emerging from the curved metric of the generalized phase space. In physical terms, the back action of the electrons on the nuclei includes momentum and energy transfer and is not restricted to defining the nuclear potential energy.

Since coherent states may be viewed as hybrids between the quantum and the classical world, END provides an example of a uniformly nonlinear semiclassical model approximating the linear quantum theory. Treating nonadiabatic effects with greater accuracy, and tracing their origin with greater depth than semiclassical models of quantum dynamics that are rooted in the time-dependent Hartree–Fock theory, END does not yet transcend the limits of the Ehrenfest representation which is predicated on the assumption that an averaged potential energy surface is a physically meaningful construct. Also, while END eases the disparity between the description of the electrons and the nuclei by assigning coherent states to both of them, the classical limit of the nuclear coherent state is taken in present applications of the theory. The asymmetry between electronic and nuclear degrees of freedom, as a typical shortcoming of semiclassical models, may be remedied by a consistent quantal approach for both subsystems. The surprising alternative to this, for most applications, prohibitively expensive strategy, is a uniformly classical treatment of both parameter sets. This is the stance of the classical electron analog (CEA) model [21] introduced in Chapter 10. Nuclear and suitably defined electronic degrees of freedom are here both subjected to a Hamilton–Jacobi formalism whose exact equivalence to the TDSE is readily demonstrated. Further, by borrowing from classical S-matrix theory, the predicament of the Ehrenfest representation, involving an average over the physically real potential energy surfaces, can be overcome in the framework of the CEA.

The most intuitive and direct semiclassical scheme that goes beyond the Ehrenfest representation involves constraining the nuclei to move on one potential energy surface at any given time, and simultaneously allowing for jumps between discrete electronic levels. The latter occur in potential energy landscape segments with sufficiently strong nonadiabatic interaction, as specified by a preset threshold. The concept of *trajectory surface hopping* (TSH) was originally proposed in 1967 [22] and has since diversified into a broad spectrum of differing implementations (e.g. [23]). The main ideas governing the principles and the practice of this approach are presented in Chapter 11 in combination with a derivative of the hopping method that supplements the hopping methodology by a quantum description of the nuclear system. The *spawning technique* operates with classical trajectories dressed with Gaussian wave packets that provide an approximate description of dynamic nuclear quantum states. A *spawning criterion*, associated with generating nuclear states advancing along a new trajectory, here takes the place of the condition for hopping events between potential energy surfaces. Spawning thus makes allowance for the splitting of the nuclear wave packet, associated with partial transfer of nuclear population between potential energy surfaces. Both interstate and intrastate interference effects can, in principle, be described with arbitrary accuracy, allowing for a definition of the full TDSE solution as limiting case in terms of the spawning threshold as well as the number of Gaussian wave packets invested.

While Chapter 11 introduces the hopping model in an *ad hoc* manner on the basis of plausibility arguments, the following chapter derives it systematically within the Feynman formulation of quantum mechanics. In particular, trajectory hopping is deduced from *semiclassical propagator* theory. The semiclassical, or Van Vleck–Gutzwiller propagator, while capturing a wide range of quantum phenomena, is a classical object by design, consisting of quantities defined within the framework of classical mechanics. As the semiclassical propagator formalism also provides a theoretical basis for the classical electron analog, it turns out to be the common root from which the two seemingly diverse approaches to the nonadiabatic problem in quantum dynamics presented in Chapters 10 and 11 originate. With respect to the utilization of this instrument for the purposes of quantum dynamics, the *initial value representation* (IVR) of the propagator [24] circumvents the need to engage in time-consuming root search and thus makes it an efficient tool for simulating molecular processes. Extending and refining numerical procedures that help brave the mathematical challenges of the semiclassical propagator is a field of lively current experimentation, as highlighted in the last two sections of Chapter 12.

The next two chapters lead from quantum dynamics to quantum hydrodynamics, or from particle to fluid mechanics. The starting point is here defined by the nuclear wave packet at the initial time of the evolution, and thus by a full quantum approach. This wave packet, however, is broken up into a number of discrete elements which are understood as fluid cells. The *quantum trajectory model* provides a set of rules that specify how to predict the future course of the *quantum fluid*, and how to interpret molecular interaction in terms of its spatio-temporal features. To stay within the imagery of hydrodynamics, and to exploit propagation schemes designed for fluid analysis while implementing at the same time quantum dynamics, one employs the *quantum theory of motion* [25], as proposed originally by Louis de Broglie and David Bohm. Each trajectory carries both density and phase information, as extracted directly from the polar representation of the wave packet, and equations of motion are readily formulated for the respective quantities, the wave packet modulus and the action integral. These are advanced through a potential energy landscape augmented by a *quantum potential* which, providing a measure for the curvature of the modulus, depends on the wave packet and is thus non-local and also the source of a highly nonlinear ingredient of the theory. The quantum potential and the divergence of the fluid velocity field function as agents that couple the trajectories, serving as media through which the simultaneously advancing fluid cells communicate. This formalism is equivalent to the exact quantum description in the sense that it yields, through assembling the fluid elements at any given time, the full solution of the TDSE in the continuous limit. While originally acknowledged as realization of the causal interpretation of quantum mechanics, the quantum theory of motion has more recently proved to be an expedient device for solving numerical problems, as emphasized throughout Chapter 13. In the last two sections of the chapter, nonadiabatic generalizations of quantum trajectory schemes are discussed.

The hydrodynamic model may be used to simulate the time behavior of the nuclear wave packet or, alternatively, the nuclear density matrix. In the first case, the hydrodynamic equations of motion emulate the TDSE, in the second the Liouville–von Neumann equation. With reference to molecular dynamics, the latter strategy allows for direct modeling of populations and coherences characteristic of molecular evolution in a multistate scenario. It is thus particularly well adjusted to addressing nonadiabatic situations. Chapter 14 will draw the contours of density-matrix-based hydrodynamic theory and also outline extensions of this theory to statistical mixtures as opposed to pure states [26], accentuating the fact that the hydrodynamic approach in its Liouville–von Neumann variant is well adjusted to accommodate problems involving open systems.

Several methods based on trajectories, such as the Gaussian-dressed paths of the spawning technique or the interacting trajectories of Bohmian mechanics, are rich enough to generate exact quantum solutions in a well-defined asymptotic limit, which, however, may be difficult or impossible to attain in applications to physical systems. The ultimate measure of success for any of these methods is the extent to which they reproduce the results of a *full quantal treatment* of the nuclear degrees of freedom. Evidently, any uniformly quantal instead of semiclassical approach for both the electrons and the nuclei is unburdened by the need to demonstrate that the quantum limit can in principle be reached, which eliminates a potentially serious justification issue. Further, placing both subsystems on the same footing avoids the inevitable lopsidedness of semiclassical theory. While this can be achieved in the framework of a homogeneously classical treatment, as exemplified by the classical electron analog, a model of this type has in general to import signature quantum features, such as spin or zero-point oscillations, in an *a posteriori* manner. Chapter 15 deals with wave packet representations of the nuclear system, without any borrowing from classical terminology. This program raises, in the first place, the technical question of how to use the TDSE as a device for wave packet propagation. Addressing this item, Chapter 15 focuses on grid solutions of the TDSE as well as procedures to advance wave packets in time. Since solving the time-dependent problem exactly is not feasible for cases of physical or chemical relevance, molecular dynamics simulations have to employ models that approximate the TDSE in a precisely specifiable sense. Such a model is available with the multiconfiguration time-dependent Hartree (MCTDH) theory as a viable and popular compromise between rigor and practicability [5]. We will trace the basic tenets of this method and outline its extension to problems involving a multiplicity of potential energy surfaces.

Among the opposing notions that pervade this text, such as classical versus quantal or stationary versus time-dependent, is the alternative between direct dynamics and propagation on individual potential energy surfaces, implying that the former label refers to procedures that do not require potential energy surface construction. This approach is represented chiefly by Ehrenfest dynamics in its TDHF or END realization. Both theories are anchored in an *ab initio* description of the electronic system. The same methodological idea, however, can be implemented by use of *density functional theory* (DFT) [27] as the underlying electronic structure formalism. Here the center stage, formerly held by the ground state wave function Ψ_0, is occupied by the ground state density ρ_0 as a simpler quantity of lower redundancy and higher computational accessibility. The price to be paid for this virtue is that the exact form of the DFT exchange and correlation functionals is unknown, and no unique prescription for how to construct these potentials can be given to this date. Whereas DFT has enjoyed enormous success as a tool for electronic structure calculations in quantum chemistry and materials science, extending its ground state formalism to excited states is less evident than in the case of *ab initio* techniques. Nevertheless, DFT-based methods of multistate dynamics are presently developed, reaching an ever-growing community of users. In Chapter 16, a condensed presentation of DFT fundamentals is followed by an outline of the *time-dependent density functional theory* (TDDFT) and its emergent usage for the purposes of nonadiabatic molecular dynamics.

Figure 1 subdivides all time-dependent methods mentioned before into two categories, namely *ab initio* electron dynamics and nuclear dynamics on coupled electronic potential energy surfaces. In both cases, adiabatic and nonadiabatic evolution are contrasted, where Car–Parrinello and Born–Oppenheimer dynamics represent the adiabatic limits of the *ab initio* and potential energy surface propagation approach, respectively. The models listed under "no back action", the classical path method and Redfield theory, are of a very different nature, as the latter is a density matrix procedure that does make reference to classical trajectories. A common element of both schemes is, however, that the electronic system is determined by a group of variables on which it does not exert any influence, namely the nuclear coordinates of the classical path method, and the environmental parameters of Redfield theory. The arrows pointing from the field labeled semiclassical

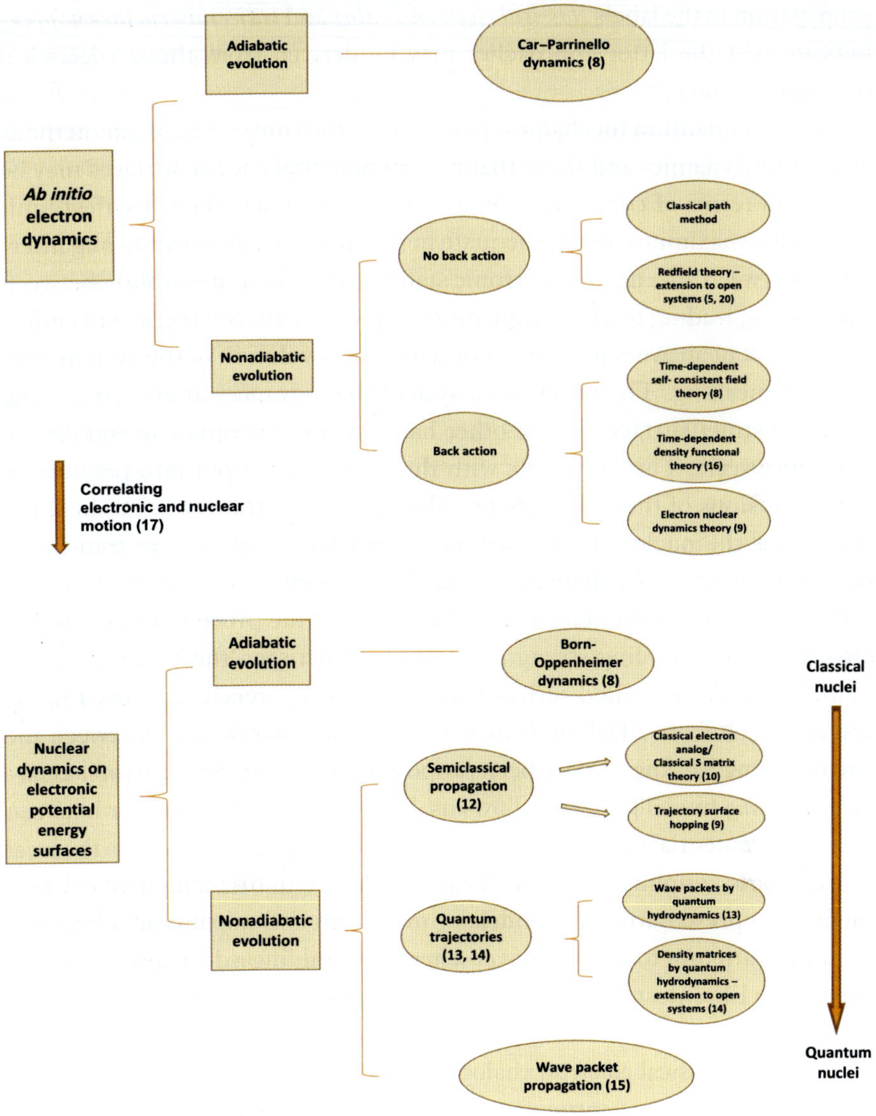

Figure 1: Overview of the time-dependent methods treated in Chapters 8–17, related to *ab initio* electron dynamics and nuclear dynamics on coupled electronic potential energy surfaces. The numbers refer to the chapters that introduce the respective method.

propagation to the labels *classical electron analog* and *trajectory surface hopping* indicate that the latter approaches may be derived from the semiclassical propagator concept.

From a quantum mechanical perspective, the contrast between methods of *ab initio* dynamics and those that rely on potential energy surfaces may be framed in terms of coherence. On a scale of coherence, the Ehrenfest limit defines the maximum, being adequate in regimes of sufficiently strong interaction between alternative electronic states. Large enough coupling between these states, leading to a high transition frequency between them, will endow the concept of an average surface governing the evolution of the system with physical meaning. The notion of a system that advances at any given time on one distinct surface, on the other hand, marks the opposite end on the coherence scale. This conforms with the methodology put into practice by earlier versions of the TSH scheme. Wherever this strategy is justified, the passage of the nuclear framework is punctuated by electronic transitions occurring at sites of enhanced nonadiabatic coupling but takes place in a well-defined electronic state at all other stages of the process. In intermediate cases, with coupling strengths somewhere between these extremes, the unmodified TSH method suffers from a lack of coherence, the pure Ehrenfest approach from a lack of *decoherence*. Hybrids that mediate between the two opposing paradigms have been proposed, among them the *continuous surface switching* model [29] that allows us, by varying degrees, to alter between individual potential energy surfaces and their average, in response to the local nonadiabatic coupling strength. As an alternative to balancing two existing methods representative for models of minimum and maximum coherence, one might attempt to devise a novel scheme by building a decoherence mechanism into Ehrenfest theory. The *decay-of-mixing algorithm* [30] realizes this idea, accomplishing decoherent molecular evolution by a suitable reformulation of the classical electron analog. Chapter 17 places these topics into the wider context of the quantum mechanical concept of decoherence. In particular, augmentation of the Liouville–von Neumann equation by a dissipative term to incorporate symmetric interaction with environmental variables is discussed.

The last part of this monograph, encompassing Chapters 18–21, deals with applications of the presented notions and methods in topical fields of chemistry and molecular physics that have not been touched on in the earlier chapters. Chapters 18 and 19 highlight two aspects of optical spectroscopy involving electronically nonadiabatic situations. Basic features of the *pump–probe technique* as a means to monitor ultrafast optical transitions in a time-resolved mode are summarized in the first two sections of Chapter 18, along with perturbative as well as non-perturbative computational methods [3] to predict and interpret the results of these measurements. The use of non-perturbative theory to simulate pump–probe spectra is illustrated by the phenomenon of *cis-trans isomerization*, involving deexcitation by passage through a conical intersection. This reaction is of great relevance for understanding both the mechanisms underlying complex nonadiabatic processes and the role played by these processes in nature. In particular, cis-trans isomerization of the retinal chromophore in the visual pigment rhodopsin has been shown to initialize vision by triggering excitation of the retinal rod cell. The nonadiabatic character of this reaction confines its time scale to the hundred-femtosecond range.

A further focus of Chapter 18 is modeling the pump–probe signal by perturbation theory in the framework of the semiclassical Liouville–von Neumann formalism. The interaction between the molecule and the radiation field is here described by a perturbing Liouville operator, and the underlying molecular dynamics is captured by the time dependence of nuclear density matrices in a landscape of potential energy surfaces. We review the application of this approach to small inorganic units, non-stoichiometric alkali-halide clusters of the form Na_nF_m [28]. The rationale for the interest in these species is that the indices n and m may be adjusted to yield effective one-electron or multi-electron systems (e.g. Na_2F versus Na_3F), or to display adiabatic or nonadiabatic dynamics following population of their first excited states by a pump laser pulse (e.g. Na_4F_3 versus Na_3F_2), resulting in a considerable diversity of configurations of interest for assessing models of ultrafast molecular processes.

In the following chapter, the principles of optical femtosecond spectroscopy involving multistate molecular dynamics are applied to the practical discipline of quantum control. This effort is directed at designing an electromagnetic field pulse optimally suited to guide a given set of reactants towards a targeted product. Two ways to pursue this goal are outlined: a Lagrangian optimization procedure due to Tannor and Rice [31] and the *coherent control* scheme of Brumer and Shapiro [32] that yields the desired target state through manipulation of the interference between two alternative excitation routes. In some detail, we discuss the fundamentals of optimal control in pump–probe spectroscopy and illustrate the viability of this notion once more by the example of multistate dynamics involving non-stoichiometric alkali-halide clusters.

The last two chapters of this survey go beyond intermolecular problems, commenting on molecules in contact with systems of higher dimension, namely those surrounded by bulk phases (Chapter 20) and attached to surfaces (Chapter 21). We will limit our discussion of molecules in condensed matter to one specific, intensely investigated type of electronic interactions. This is intramolecular electron transfer, involving the transition of an electron from a donor to an acceptor site within the embedded molecular unit. These processes can often be specified as slow in comparison with the nuclear motion of the molecule. They also occur at times substantially larger than those typical for geometric rearrangement of the embedding medium which may be a polar solvent. The electronic motion of the solvent molecules, being still faster than the relaxation of their nuclear framework, completes the hierarchy of time scales relevant for electron transfer in the condensed phase. Electron transfer within molecules in condensed matter thus provides a case where motion of the electrons can generally be considered to proceed at a slower pace than that of the nuclei. Accordingly, a diabatic model frequently sets an adequate frame for analyzing the transfer mechanism, and rate theory based on Fermi's golden rule is applicable. Along these lines, we will consider the impact of the *vibrational bath* (Sections 20.4 and 20.5) as well as the *polar environment* (Section 20.6) on the transfer process. In both cases, we will make reference to *Marcus theory* [33] as a formal scheme of interpretation that allows for a strictly analogous description of both interaction types in terms of the transfer rate constant. The final section of Chapter 20 is devoted to the challenge of simulating electron transfer in a solvent by removing the premise of a dielectric continuum model that underlies the preceding account on molecule–solvent processes. Specifically, we will outline techniques based on classical statistics as well as quantum mechanical Euclidean path integral theory to investigate the ferrous–ferric exchange as a prototypical example of electron transfer between metal atom centers.

While Chapter 20 invites readers for an excursion in the condensed phase, Chapter 21 concentrates on molecules in contact with surfaces. As a paradigmatic case for the coupling between electronic and nuclear degrees of freedom in molecule-surface interactions, we focus on vibronic deexcitation of a molecular adsorbate attaching to a metal substrate. The life times of the excited vibrational state have been found to be several orders of magnitude lower for metal than for semiconductor or insulator substrates, implying excitation transfer from the vibronic subsystem of the adsorbate to the conduction electrons of the surface as the leading cause of the observed transition. The particular challenge posed by this problem, associated with a continuum of electronic excitations, is met, to a certain approximation, by the *Langevin formulation of Ehrenfest dynamics* which captures *molecular dynamics with electronic friction*. Focusing on the intensely studied vibronic relaxation of CO in contact with a copper surface, we discuss the Langevin model as an explanatory device that uses the terminology of stochastic processes, involving friction kernels and fluctuating forces, to address nonadiabatic processes, and we introduce techniques for its quantitative analysis. The final section of Chapter 21 comments on the limitations of this approach.

The list of topics presented in the last four chapters is in no way exhaustive but still may give an impression of the wide thematic range as well as the efficiency and flexibility of the methods introduced in the second part of this book, as they are adjusted to research areas of high current interest. The emerging applications of electron dynamics theories in areas as diverse as nanoscience and benchmark quantum chemistry of small molecules warrant a separate monograph. In the meantime, basic research on the origin of nonadiabatic electron dynamics in molecules continues [34], and ever more powerful algorithms to simulate the respective phenomena are designed. The concepts and techniques described in this book should familiarize readers with the state of the art of molecular electron dynamics and prepare them for appreciating the present developments in this field.

paradigmatic case for the coupling between electronic and nuclear degrees of freedom in molecule-surface interactions, we focus on vibronic deexcitation of a molecular adsorbate attaching to a metal substrate. The lifetimes of the excited vibrational state have been found to be several orders of magnitude lower for metal than for semiconductor or insulator substrates, implying excitation transfer from the vibronic subsystem of the adsorbate to the conduction electrons of the surface as the leading cause of the observed transition. The particular challenge posed by this problem, associated with a continuum of electronic excitations, is met, to a certain approximation, by the *Langevin formulation of Ehrenfest dynamics* which captures *molecular dynamics with electronic friction*. Focusing on the intensely studied vibronic relaxation of CO in contact with a copper surface, we discuss the Langevin model as an explanatory device that uses the terminology of stochastic processes, involving friction kernels and fluctuating forces, to address nonadiabatic processes, and we introduce techniques for its quantitative analysis. The final section of Chapter 21 comments on the limitations of this approach.

The list of topics presented in the last four chapters is in no way exhaustive but still may give an impression of the wide thematic range as well as the efficiency and flexibility of the methods introduced in the second part of this book, as they are adjusted to research areas of high current interest. The emerging applications of electron dynamics theories in areas as diverse as nanoscience and benchmark quantum chemistry of small molecules warrant a separate monograph. In the meantime, basic research on the origin of nonadiabatic electron dynamics in molecules continues [3-4], and ever more powerful algorithms to simulate the respective phenomena are designed. The concepts and techniques described in this book should familiarize readers with the state of the art of molecular electron dynamics and prepare them for appreciating the present developments in this field.

Part I: Preparations

Chapter 1

Ab Initio Theory of Electronic Structure

The first section of this text deals with fundamental quantum mechanical models of the molecular electronic system. Thus, everything that follows is based on the material presented in this chapter. While later chapters will address the time-dependent Schrödinger equation (TDSE) and its approximate solution for cases of molecular interaction, we will emphasize here its stationary counterpart and summarize some of the historically most influential methods of solving the Schrödinger equation for electrons in molecules. Among these methods, the Hartree–Fock procedure is of primary importance and will therefore be covered in greater detail. The Hartree–Fock formalism has opened up an enormous range of chemical systems to computational treatment since it accomplishes a mean-field approximation, reducing a complex many-electron situation to a set of single-electron problems. This great virtue can also be framed as a great shortcoming, as it implies that the Hartree–Fock theory does not include many-body effects. With respect to the electronic system of a molecule, this means that it fails to account for electron–electron correlation. We will discuss several models that incorporate these effects by building on the original Hartree–Fock scheme in a variational or perturbative way.

As far as dynamics is concerned, this chapter marks the starting point for the main avenues that we will follow later in this text. First, the time-dependent Hartree–Fock theory, to be introduced in Chapter 8, is paradigmatic for models involving direct electron dynamics. Second, nuclear dynamics in a landscape of potential energy surfaces is also rooted in the electronic structure methods that are the subject of this chapter as they enable us to derive these surfaces, or patches of these surfaces if an *on-the-fly* procedure is used. Electronic structure methods are applied here to derive local forces, and thus the construction of global potential energy surfaces is bypassed.

To appreciate the quantum dynamics of molecular electronic systems in all its facets, one needs familiarity with electronic states in molecules. The following sections thus address the static problem of describing the electronic system of a molecule at equilibrium. While the basic concepts will be developed in Sections 1.1–1.5 by considering the electronic ground state, problems posed by excited states will be covered in Section 1.6.

1.1 Molecular Orbitals

For everything that follows in the present chapter, we will assume that separation of nuclear and electronic degrees of freedom has been accomplished. This *Born–Oppenheimer separation* [13] makes it possible to focus on the electronic molecular states alone, which is our objective in this section. The theoretical background for this approach will be supplied in the next chapter.

In the framework of this approximation, the state of the molecule is described by an overall wave function of the form

$$\Psi(\mathbf{x},\mathbf{X}) = \chi(\mathbf{X})\psi(\mathbf{x},\mathbf{X}), \tag{1.1}$$

with χ as nuclear and ψ as electronic contribution. The latter is explicitly dependent on the electronic space coordinate $\mathbf{x}$ and parametrically on the nuclear coordinate $\mathbf{X}$. The function ψ is an eigenstate of the electronic Hamiltonian $\hat{H}_{\text{el}}$ according to

$$\hat{H}_{\text{el}}\psi(\mathbf{x},\mathbf{X}) = E_{\text{el}}\psi(\mathbf{x},\mathbf{X}), \tag{1.2}$$

where $\hat{H}_{\text{el}}$ denotes the non-relativistic operator for a system with N electrons and M nuclei,

$$\hat{H}_{\text{el}} = -\frac{\hbar^2}{2m}\sum_{i=1}^{N}\frac{\partial^2}{\partial\mathbf{x}_i^2} - \sum_{i=1}^{N}\sum_{k=1}^{M}\frac{e^2}{4\pi\varepsilon_0}\frac{Z_k}{|\mathbf{x}_i-\mathbf{X}_k|} + \sum_{i=1}^{N}\sum_{j>i}^{N}\frac{e^2}{4\pi\varepsilon_0}\frac{1}{|\mathbf{x}_i-\mathbf{x}_j|}, \tag{1.3}$$

with m and e as the electron mass and the elementary charge, respectively. We observe that the first two terms contain just one electron index while the third involves two sets of electronic coordinates. Employing the nomenclature

$$\hat{h}_i = -\frac{\hbar^2}{2m}\frac{\partial^2}{\partial\mathbf{x}_i^2} - \sum_{k=1}^{N}\frac{e^2}{4\pi\varepsilon_0}\frac{Z_k}{|\mathbf{x}_i-\mathbf{X}_k|}, \tag{1.4}$$

we express $\hat{H}_{el}$ as the sum of one-body and many-body components:

$$\hat{H}_{el} = \sum_{i=1}^{N} h_i + \sum_{i=1}^{N}\sum_{j>i}^{N} \frac{e^2}{4\pi\varepsilon_0} \frac{1}{|\mathbf{x}_i - \mathbf{x}_j|}. \tag{1.5}$$

By construction, the eigenvalue E_{el} depends on the nuclear coordinate: $E_{el} \equiv E_{el}(\mathbf{X})$. Within the *Born–Oppenheimer approximation*, E_{el} in conjunction with the internuclear repulsion U_N yields the potential energy function that determines the nuclear motion. With $E_{pot} = E_{el} + U_N$, the nuclear Hamiltonian for a system with N nuclei whose masses are $M_1, M_2, \ldots, M_N$ may be written as

$$\hat{H}_N = -\frac{\hbar^2}{2M_i} \sum_{i=1}^{N} \frac{\partial^2}{\partial \mathbf{X}_i^2} + E_{pot}(\mathbf{X}). \tag{1.6}$$

Therefore, it holds that

$$\hat{H}_N \chi(\mathbf{X}) = E\chi(\mathbf{X}). \tag{1.7}$$

Here, E stands for the total molecular energy within the Born–Oppenheimer approximation. Dealing with electronic structure theory, the present chapter emphasizes Eq. (1.2) which poses, in the general case, a highly non-trivial many-body problem. As mentioned earlier, we will focus on strategies of replacing this many-body problem with many one-body problems. The natural starting point of this undertaking is the one-particle operator $\hat{h}_i$. More specifically, we introduce a *molecular orbital* (MO) function $\Phi_i(\mathbf{x})$ by the eigenvalue equation

$$\hat{h}\,\Phi_i(\mathbf{x}) = \varepsilon_i \Phi_i(\mathbf{x}), \tag{1.8}$$

where $\hat{h}$ is of the form (1.4) with $\mathbf{x}_i = \mathbf{x}$, and ε_i signifies the *molecular orbital energy* associated with the MO Φ_i. We point out that the one-particle operator $\hat{h}$ which defines the eigenstate Φ_i as well as the eigenenergy ε_i does not need to conform with Eq. (1.4), as it may involve an effective interaction, namely the averaged interelectronic repulsion term. We will return to this choice in further detail later. Assuming that a spectrum $\varepsilon_j (j = 1, ..., N)$, along with the corresponding eigenvectors Φ_j, has been identified, it is easy to specify an eigenfunction of the one-body part $\sum_{i=1}^{N} \hat{h}_i$ of the electronic

Hamiltonian $\hat{H}_{\rm el}$, namely

$$\sum_{i=1}^{N} \hat{h}_i \psi^{\rm H}(\mathbf{x}_1, \ldots, \mathbf{x}_{\rm N}) = \varepsilon^{\rm T} \psi^{\rm H}(\mathbf{x}_1, \ldots, \mathbf{x}_{\rm N}), \tag{1.9}$$

with $\varepsilon^{\rm T} = \sum_{j=1}^{N} \varepsilon_j$, and $\psi^{\rm H}(\mathbf{x}_1, \ldots, \mathbf{x}_{\rm N})$ as a *Hartree product* [35], defined by

$$\psi^{\rm H}(\mathbf{x}_1, \ldots, \mathbf{x}_{\rm N}) \equiv \Phi_{k_1}(\mathbf{x}_1)\Phi_{k_2}(\mathbf{x}_2) \ldots \Phi_{k_N}(\mathbf{x}_{\rm N}), \quad k_1, \ldots, k_N = 1, \ldots, N. \tag{1.10}$$

The MOs $\{\Phi_i\}$ define a set of orthonormal functions, satisfying $\int d\mathbf{x}\, \Phi_i^*(\mathbf{x})\Phi_j(\mathbf{x}) = \langle \Phi_i, \Phi_j \rangle = \delta_{ij}$. The index k_i may be interpreted as a function that maps the set $\{1, 2, \ldots, N\}$ onto itself. It is evident that the Hartree product does not do justice to the status of electrons as identical particles. Thus, if one adopts an electronic wave function of the form (1.10), individual electrons may be distinguished by their different probability densities, assuming that they occupy different MOs. With reference to definition (1.10), this probability density is $|\Phi_{k_1}|^2$ for the first electron, $|\Phi_{k_2}|^2$ for the second etc. This deficiency can be remedied by proper symmetrization of the Hartree product. The fermionic character of the electron, however, requires antisymmetrization rather than symmetrization. The overall wave function has to obey the *Pauli exclusion principle*:

$$\psi^{\rm H}(\mathbf{x}_1, \ldots, \mathbf{x}_i, \ldots, \mathbf{x}_j, \ldots, \mathbf{x}_{\rm N}) = -\psi^{\rm H}(\mathbf{x}_1, \ldots, \mathbf{x}_j, \ldots, \mathbf{x}_i, \ldots, \mathbf{x}_{\rm N}), \quad 1 < i, j < N. \tag{1.11}$$

For a two-electron system involving the MOs Φ_1 and Φ_2, the linear combination of Hartree products that satisfies Eq. (1.11) is given by

$$\psi(\mathbf{x}_1, \mathbf{x}_2) = \frac{1}{\sqrt{2}}[\Phi_1(\mathbf{x}_1)\Phi_2(\mathbf{x}_2) - \Phi_1(\mathbf{x}_2)\Phi_2(\mathbf{x}_1)], \tag{1.12}$$

where the prefactor ensures the normalization of the overall wave function. We abbreviate the result (1.12) by writing

$$\psi(\mathbf{x}_1, \mathbf{x}_2) = \frac{1}{\sqrt{2}} \det\,[\Phi_i(\mathbf{x}_1), \Phi_j(\mathbf{x}_2)], \quad i, j = 1, 2. \tag{1.13}$$

It is easily demonstrated that the extension of this solution to the case of N electrons is given by the determinant

$$\psi^{\mathrm{H}}(\mathbf{x}_1, \mathbf{x}_2, \ldots, \mathbf{x}_\mathrm{N}) = \frac{1}{\sqrt{N!}} \det\left[\Phi_{k_1}(\mathbf{x}_1), \Phi_{k_2}(\mathbf{x}_2), \ldots, \Phi_{k_N}(\mathbf{x}_\mathrm{N})\right], \quad k_1, \ldots, k_N = 1, N. \tag{1.14}$$

So far, we have described the MOs as functions of spatial coordinate sets $\mathbf{x}_i (i = 1, \ldots, N)$. The notion of the MO may be conveniently extended from that of a *spatial orbital* to a *spin orbital* by including an additional electron spin variable. Formally, the spin functions α (*spin up*) and β (*spin down*) depend on a coordinate ω such that

$$\int d\omega\, \alpha^*(\omega)\alpha(\omega) \equiv \langle\alpha|\alpha\rangle = 1, \tag{1.15a}$$

$$\int d\omega\, \beta^*(\omega)\beta(\omega) \equiv \langle\beta|\beta\rangle = 1, \tag{1.15b}$$

$$\int d\omega\, \alpha^*(\omega)\beta(\omega) \equiv \langle\alpha|\beta\rangle = 0 = \langle\beta|\alpha\rangle = \int d\omega\, \beta^*(\omega)\alpha(\omega). \tag{1.15c}$$

We thus make the transition from spatial to spin orbitals by introducing the generalized coordinate set $\mathbf{x}' \equiv (\mathbf{x}, \omega)$ as well as the spin orbital function ζ, stipulating

$$\zeta(\mathbf{x}') = \Phi(\mathbf{x})\sigma(\omega), \tag{1.16}$$

where the spin function $\sigma(\omega)$ may be realized by $\alpha(\omega)$ or $\beta(\omega)$, denoting an alpha or a beta spin orbital, respectively. The counterpart of Eq. (1.14) for spin instead of spatial orbitals, i.e.

$$\psi^{\mathrm{H}}(\mathbf{x}'_1, \mathbf{x}'_2, \ldots, \mathbf{x}'_N) = \frac{1}{\sqrt{N!}} \det\left[\zeta_{k_1}(\mathbf{x}'_1), \zeta_{k_2}(\mathbf{x}'_2), \ldots, \zeta_{k_N}(\mathbf{x}'_N)\right], \quad k_1, \ldots, k_N = 1, N, \tag{1.17}$$

is referred to as a *Slater determinant*. We find the analog of Eq. (1.12) by expanding this determinant for the special case of two electrons where opposite spins are assumed:

$$\begin{aligned}\psi(\mathbf{x}_1',\mathbf{x}_2') &= \frac{1}{\sqrt{2}}[\zeta_1(\mathbf{x}_1')\zeta_2(\mathbf{x}_2') - \zeta_2(\mathbf{x}_1')\zeta_1(\mathbf{x}_2')] \\ &= \frac{1}{\sqrt{2}}[\Phi_1(\mathbf{x}_1)\alpha(\omega_1)\Phi_2(\mathbf{x}_2)\beta(\omega_2) - \Phi_2(\mathbf{x}_1)\beta(\omega_1)\Phi_1(\mathbf{x}_2)\alpha(\omega_2)].\end{aligned} \tag{1.18}$$

Using Eqs. (1.15a–1.15c), we derive from this expression the probability density $\rho(\mathbf{x}_1,\mathbf{x}_2)$, which results in

$$\rho(\mathbf{x}_1,\mathbf{x}_2) = \frac{1}{2}[|\Phi_1(\mathbf{x}_1)|^2|\Phi_2(\mathbf{x}_2)|^2 + |\Phi_2(\mathbf{x}_1)|^2|\Phi_1(\mathbf{x}_2)|^2]. \tag{1.19}$$

From the form of the wave function (1.18) one concludes that the two electrons are entirely uncorrelated. This is readily demonstrated by considering the extreme case $\Phi_1 = \Phi_2$ which leads to the non-vanishing probability density product $\rho(\mathbf{x}_1,\mathbf{x}_2) = |\Phi_1(\mathbf{x}_1)|^2|\Phi_1(\mathbf{x}_2)|^2$. Thus, the single-determinantal approach does permit the simultaneous population of one and the same spatial MO by both electrons. Following the logic of the Pauli exclusion principle, one expects that this becomes invalid as soon as both electrons have the same spin. Replacing all beta by alpha orbitals in Eq. (1.18), and calculating again the probability density, we arrive at

$$\begin{aligned}\rho(\mathbf{x}_1,\mathbf{x}_2) = \frac{1}{2}[&|\Phi_1(\mathbf{x}_1)|^2|\Phi_2(\mathbf{x}_2)|^2 + |\Phi_2(\mathbf{x}_1)|^2|\Phi_1(\mathbf{x}_2)|^2 \\ &- \Phi_1^*(\mathbf{x}_1)\Phi_2^*(\mathbf{x}_2)\Phi_2(\mathbf{x}_1)\Phi_1(\mathbf{x}_2) \\ &- \Phi_1(\mathbf{x}_1)\Phi_2(\mathbf{x}_2)\Phi_2^*(\mathbf{x}_1)\Phi_1^*(\mathbf{x}_2)].\end{aligned} \tag{1.20}$$

As one confirms immediately, the interference term that appears in Eq. (1.20) but not in Eq. (1.19) ensures for the case of identical spins that $\rho(\mathbf{x}_1,\mathbf{x}_2) = 0$ if $\Phi_1 = \Phi_2$. Two electrons with the same spin functions will therefore never occupy the same spatial orbitals. In particular, Eq. (1.20) implies that $\rho(\mathbf{x}_1,\mathbf{x}_1) = 0$, while this is generally not true for two electrons with opposite spins. In the parallel spin case, each electron is surrounded by a so-called *Fermi hole*, reflecting a certain degree of electron–electron correlation, which is completely absent from the single-determinantal description of two electrons with opposite spins.

1.1.1 *Molecular and atomic orbitals*

Usually, MOs are defined as linear combinations of atomic orbitals (LCAOs), according to

$$\Phi = \sum_{k=1}^{K} a_k \phi_k, \tag{1.21}$$

where the atomic orbitals (AOs) are centered on individual nuclei. They are generally non-orthogonal, albeit linearly independent functions. The AOs, in turn, may be composed of *Gaussian primitives* $\xi(\mathbf{x})$:

$$\xi(\mathbf{x}) \sim (x - X)^{N_x}(y - Y)^{N_y}(z - Z)^{N_z} \exp(-\alpha|\mathbf{x} - \mathbf{X}|^2), \tag{1.22}$$

where the exponents N_x, N_y, and N_z reflect the angular character of the considered AO. Other functions have been used for the primitives. Thus, plane waves are a particularly adequate choice for problems with periodic boundary conditions. The localized nature of Gaussian functions, combined with the advantage that they allow all matrix elements of the interaction terms contained in the electronic Hamiltonian to be evaluated analytically, accounts for their frequent use in gas phase situations. This holds also, on the most basic level, for the overlap matrix element S_{ij}:

$$S_{ij} = \int d\mathbf{x}\, \phi_i(\mathbf{x})^* \phi_j(\mathbf{x}). \tag{1.23}$$

From the explicit form of the AOs (1.22), it is plausible that they are non-orthogonal since different ϕ_i may be centered on different nuclei within a molecule, overlapping each other to a smaller or larger degree. In terms of the overlap matrix, the scalar product of two MOs Φ_i and Φ_j is written as

$$\langle \Phi_i, \Phi_j \rangle = \sum_m \sum_k a_{mi}^* S_{mk} a_{kj}, \tag{1.24}$$

which reduces to δ_{ij} if the MOs are orthonormal, as is customarily the case. Since the AOs are normalized, the diagonal elements of $\mathbf{S}$ are equal to one while it holds for the non-diagonal elements that $|S_{mk}| < 1$, $m \neq k$. Evidently, $\mathbf{S}$ is Hermitian. The AOs are generally chosen as real functions, making $\mathbf{S}$ real and symmetric. In addition, it has the positive-definite property, i.e. all of its eigenvalues are positive. The LCAO approach is often used to express the charge density $\rho(\mathbf{x})$ of a molecular system. For a one-determinantal state ψ, this quantity is the probability density projected on a spatial location $\mathbf{x}$. Thus

$$\rho(\mathbf{x}) = \langle \psi(\mathbf{x}) | \hat{\rho} | \psi(\mathbf{x}) \rangle, \tag{1.25}$$

which involves the density operator $\hat{\rho} = \sum_{i=1}^{N} \delta(\mathbf{x}_i - \mathbf{x})$ projecting each set of electron coordinates on the location $\mathbf{x}$. By use of Eq. (1.17) for the wave function ψ, and further by converting from spin to spatial orbitals as well as exploiting the orthonormality relations valid for both the spin functions and the spatial MOs, we go from Eq. (1.25) to

$$\rho = \sum_{i=1}^{N} |\Phi_i|^2. \tag{1.26}$$

Note that integrating $\rho(\mathbf{x})$ with respect to the coordinate $\mathbf{x}$ yields just the number of electrons, $\int d\mathbf{x}\, \rho(\mathbf{x}) = N$. Inserting the LCAO expansion (1.21) into Eq. (1.26), we obtain the charge density in terms of the AOs:

$$\begin{aligned}
\rho(\mathbf{x}) &= \sum_{i=1}^{N} \Phi^*(\mathbf{x})\Phi(\mathbf{x}), \\
&= \sum_{i=1}^{N} \sum_{k=1}^{K} a_{ki}^* \phi_k^*(\mathbf{x}) \sum_{j=1}^{K} a_{ji}\phi_j(\mathbf{x}), \\
&= \sum_{j,k=1}^{K} \left[\sum_{i=1}^{N} a_{ji} a_{ki}^* \right] \phi_j(\mathbf{x})\phi_k^*(\mathbf{x}), \\
&= \sum_{j,k=1}^{K} P_{jk}\phi_j(\mathbf{x})\phi_k^*(\mathbf{x}).
\end{aligned} \tag{1.27}$$

Equation (1.27) expresses the charge density through the *charge density matrix* or *charge density bond order matrix* $\mathbf{P}$, given by a summation over all indices of occupied MOs, namely by $P_{jk} = \sum_{i=1}^{N} a_{ji}a_{ki}^*$. In the case of a closed-shell system, which implies complete pairing of alpha and beta electrons, such that each spatial orbital is populated by two electrons with opposite spins, the sum can be truncated to yield $P_{jk} = 2\sum_{i=1}^{N/2} a_{ji}a_{ki}^*$.

1.1.2 *Expectation values from molecular orbitals: The example of spin-orbit coupling*

The representation (1.21) of a molecular orbital in an atomic basis leads to a simple scheme for calculating expectation values of an electron described by an MO Φ in terms of atomic basis function $\{\phi_k\}$. Specifically, the expectation value $\langle \hat{O} \rangle$ of an observable $\hat{O}$ is obtained from

$$\langle \hat{O} \rangle = \sum_{k,k'}^{K} a_k^* a_{k'} \langle \phi_k | \hat{O} | \phi_{k'} \rangle. \tag{1.28}$$

While this scheme is elementary, its implementation may be challenging if the structure of the operator $\hat{O}$ is sufficiently complex. As an example, we discuss *spin-orbit coupling*, a contribution absent from the non-relativistic approximation (1.3) to the full molecular Hamiltonian. Extension to the relativistic case [36,37] up to the order of $(\frac{v}{c})^4$ in the electron's velocity yields various additional terms, among them corrections to the kinetic energy, the *Darwin term* that affects only atomic states of *s* symmetry, and spin-orbit coupling. The latter may be understood with classical arguments as an effect due to the electron motion in an electric field **E**, giving rise to a magnetic field at the site of the electron that interacts with the magnetic moment associated with the electron's spin. This component may be written as

$$\hat{H}_{\mathrm{SO}} \equiv \frac{1}{2mc^2} \hat{\mathbf{S}} \cdot \hat{\mathbf{p}} \times \mathbf{E}, \tag{1.29}$$

where $\hat{\mathbf{S}}$ and $\hat{\mathbf{p}}$ stand for the spin and the linear momentum operator. In a system where the z-component is chosen to be diagonal, the matrix representation of the electron spin operator $\hat{\mathbf{S}} = \frac{\hbar}{2}\boldsymbol{\sigma}$ is defined by the *Pauli matrices*

$$\boldsymbol{\sigma}_x = \begin{pmatrix} 0 & 1 \\ 1 & 0 \end{pmatrix}, \quad \boldsymbol{\sigma}_y = \begin{pmatrix} 0 & -i \\ i & 0 \end{pmatrix}, \quad \boldsymbol{\sigma}_z = \begin{pmatrix} 1 & 0 \\ 0 & -1 \end{pmatrix}. \tag{1.30}$$

As the electric field **E** is derived from the overall Coulomb potential $V_{\mathrm{coul}}(\mathbf{x})$ at the electron's site by $\mathbf{E} = -\frac{\delta V_{\mathrm{coul}}}{\delta \mathbf{x}}$, the special case of a central Coulomb potential $V_{\mathrm{coul}}(r)$ simplifies formula (1.29) to yield

$$\hat{H}_{\mathrm{SO}} = \frac{1}{2mc^2} \hat{\mathbf{S}} \cdot \hat{\mathbf{L}} \frac{1}{r} \frac{dV_{\mathrm{coul}}(r)}{dr}. \tag{1.31}$$

Combining these premises with the spatial MO expansion (1.21), one finds the expectation value for the spin-orbit coupling of an electron in the spin orbital $\zeta = \Phi\sigma$ by calculating

$$\langle \hat{H}_{\rm SO} \rangle = \sum_{k,k'}^{K} a_k^* a_{k'} \langle \phi_k(\mathbf{x})\sigma(\omega) | \hat{H}_{\rm SO} | \phi_{k'}(\mathbf{x})\sigma(\omega) \rangle. \tag{1.32}$$

Reformulating the operator (1.29) slightly, this becomes [38]

$$\langle \hat{H}_{\rm SO} \rangle = -\frac{1}{2mc^2 i} \sum_{k,k'}^{K} a_k^* a_{k'} \langle \phi_k | [\nabla \times \nabla V_{\rm coul}]_z | \phi_{k'} \rangle \langle \sigma | \hat{S}_z | \sigma \rangle. \tag{1.33}$$

Note that the *non-collinear* case, where admission is made for different spin functions in the MO expansion, requires summation over all three components of the spin-orbit coupling operator. The first of the two scalar products combined in (1.33) can be conveniently expressed in the following way:

$$\langle |\phi_k | [\nabla \times \nabla V_{\rm coul}]_z | \phi_{k'} \rangle = \left\langle \frac{d\phi_k}{dy} | V_{\rm coul} | \frac{d\phi_{k'}}{dx} \right\rangle - \left\langle \frac{d\phi_k}{dx} | V_{\rm coul} | \frac{d\phi_{k'}}{dy} \right\rangle. \tag{1.34}$$

Relations (1.33) and (1.34) provide a basic prescription for an approximate computational treatment of the spin-orbit effect for an electron described by a well-defined MO. In Exercise 1.1 this scheme is extended to the case of a two-electron system represented by the determinant (1.13). The result is readily generalized to N electrons and arbitrary single-particle operators.

Exercise 1.1. *Spin-orbit expectation value for a two-electron system.*

Assuming an ensemble of two electrons described by the determinant (1.13), show that

$$\langle \hat{H}_{\rm SO} \rangle = \sum_{k=1}^{2} \langle \phi_k | \hat{H}_{\rm SO} | \phi_k \rangle. \tag{1.35}$$

1.2 The Minimal LCAO Model

We illustrate the concepts of the molecular orbital and the Slater determinant by the example of the minimal LCAO theory of H_2. On the most elementary level, the quantum chemical description of the H_2 molecule consists of merely two H(1*s*) basis functions, each of them centered on one H nucleus.

This choice allows a symmetrized and an antisymmetrized MO to be constructed. The *gerade* alternative is invariant with respect to the exchange of the two centers while the *ungerade* MO switches its sign upon this operation. Explicitly, the two normalized orbitals have the form

$$\Phi_g = \frac{1}{\sqrt{2(1+S_{12})}}(\phi_1 + \phi_2), \tag{1.36a}$$

$$\Phi_u = \frac{1}{\sqrt{2(1+S_{12})}}(\phi_1 - \phi_2), \tag{1.36b}$$

where the subscripts g and u stand for the labels gerade and ungerade, respectively. Figure 1.1 illustrates these two states in an elementary molecular orbital diagram. The H_2 ground state involves an electron pair occupying the state Φ_g. The electron density adopts a local maximum at the midpoint between the two centers while it vanishes at this place for the Φ_u alternative. Thus, the MO Φ_g describes a bonding situation and exhibits the salient features of the H_2 ground state potential energy as a function of the internuclear distance. The MO Φ_u, in contrast, leads to repulsion between the two H atoms, realizing an *antibonding* configuration. Since each MO may be occupied by two electrons with opposite spins, the H_2 ground state determinant in the minimal model is written as

$$\psi(\mathbf{x}_1', \mathbf{x}_2') = \frac{1}{\sqrt{2}} \det\left[\Phi_g(\mathbf{x}_1)\alpha(\omega_1), \Phi_g(\mathbf{x}_2)\beta(\omega_2)\right]. \tag{1.37}$$

This basic approach yields the bonding H_2 ground state potential energy function sketched in Figure 1.2. The corresponding curve derived from the

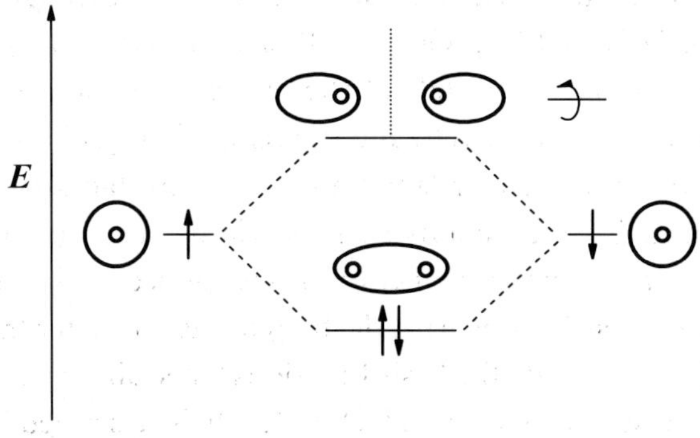

Figure 1.1: Molecular orbital diagram, showing the emergence of a bonding and an antibonding electron configuration as an H_2 molecule is formed from two isolated H atoms.

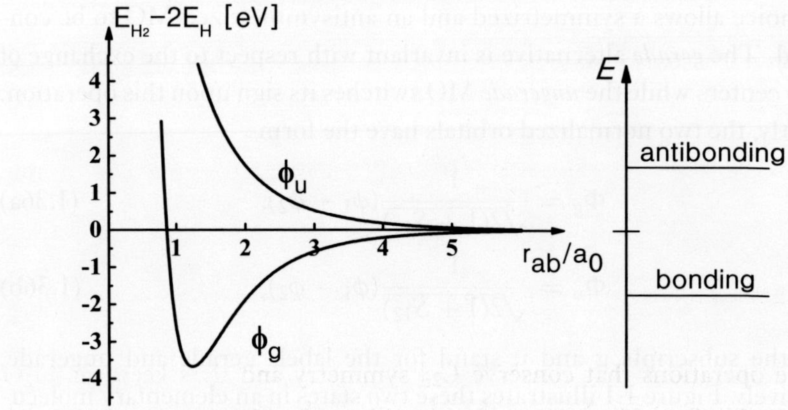

Figure 1.2: The bonding H_2 ground state potential energy curve, based on a symmetric (gerade) MO, and the repulsive antibonding curve, resulting from an antisymmetric (ungerade) MO (see formulae (1.36a) and (1.36b)).

ungerade MO Φ_u describes a repulsive situation. Although not quantitatively reliable, the potential energy curves obtained within a minimal model of the H_2 molecule make a valid qualitative statement.

Molecular symmetry provides a powerful guiding principle for constructing molecular orbitals, which is indispensable especially in the case of polyatomic systems. If the symmetry of the studied molecule has been identified, the molecular orbitals can be generated as *irreducible representations* of the respective point group. The composition of these orbitals from atomic basis functions mirrors the symmetry operations contained in this group. Among the popular examples for this correspondence is the water molecule. In the assumed minimal model, its molecular orbitals combine from basis functions of types O(1s), O(2s), O($2p_x$), O($2p_y$), O($2p_z$), $H_1(1s)$, $H_2(1s)$, where O(1s) and $H_1(1s)$ denote the 1s orbitals of the oxygen atom and the first of the two hydrogen atoms, respectively, and the remaining symbols are defined in an analogous way. Symmetry-adapted molecular orbitals are designed from this set of atomic basis functions by subdividing them according to their behavior under the elementary operations comprised in the point group of water, C_{2v}. This is easily done with the help of the character table of the C_{2v} group [39,40]. To associate the basis functions listed above with the C_{2v} representations, the structure of the H_2O molecule is described in a Cartesian coordinate system, as shown in Figure 1.3.

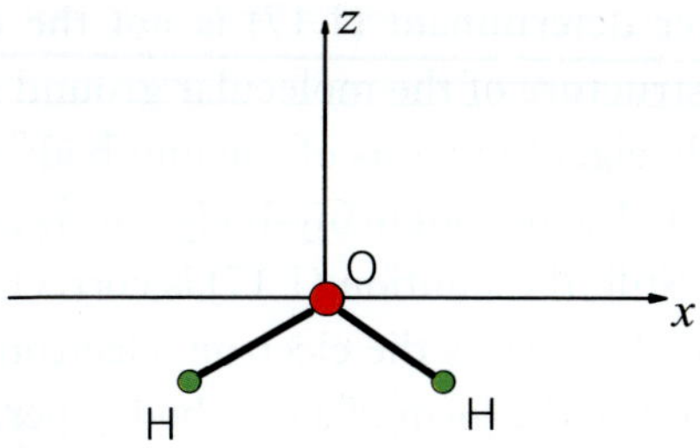

Figure 1.3: Orientation of the H_2O molecule in a Cartesian coordinate system.

The operations that conserve C_{2v} symmetry and thus keep the H_2O geometry invariant are, besides the identity $\hat{I}$, the principal rotation $\hat{C}_2$ as well as the reflections $\hat{\sigma}(xz)$ and $\hat{\sigma}(yz)$ with respect to the $xz-$ and the $yz-$ plane, respectively. Classifying the basis functions by their transformation properties under $\hat{I}$, $\hat{C}_2$, $\hat{\sigma}(xz)$, and $\hat{\sigma}(yz)$, we arrive at the following table which assigns to each function the corresponding representation of C_{2v} along with the defining set of characters.

C_{2v}	I	C_2	$\sigma(xz)$	$\sigma(yz)$	Atomic basis functions
A_1	1	1	1	1	O(1s),O(2s),O(2p_z),H_1(1s)+H_2(1s)
A_2	1	1	−1	−1	
B_1	1	−1	1	−1	O(2p_x),H_1(1s)−H_2(1s)
B_2	1	−1	−1	1	O(2p_y)

The labels A, B stand for a one-dimensional representation of the group in question, while E and T are used to denote two- and three-dimensional representations, respectively. From the table, the five molecular orbitals occupied by the closed-shell electronic system of H_2O are of the types A_1, B_1, and B_2. Combining the LCAO procedure with self-consistent field theory, as outlined in the following section, one obtains optimized molecular orbitals as well as orbital energies $\varepsilon_j, j = 1,\ldots,5$. The resulting ground state electron configuration is $(1A_1)^2(2A_1)^2(3A_1)^2(1B_1)^2(1B_2)^2$ where the symbol $(1A_1)^2$ denotes the first molecular orbital of A_1 character and occupied by an electron pair, and correspondingly for the following symbols. Quantum chemistry generalizes these minimal LCAO results in two directions: the size of the basis set and the extension of the physical system as given by the number of its atomic constituents.

Clearly, the Slater determinant (1.17) is not the definitive answer to the question for the structure of the molecular ground state wave function, since it contains solely eigenfunctions of the one-body contributions to the electronic Hamiltonian, leaving out many-body effects associated with interelectronic repulsion. Still, the solution (1.17) is correct in the framework of mean-field theory which replaces the electron–electron interaction with an effective potential that has the form of a one-body operator.

Such a theory is available with the Hartree–Fock approach. In the following, we summarize the most fundamental tenets of this procedure which have had an enormous historical impact.

1.3 Hartree–Fock Theory

The Hartree–Fock method is variational. It consists in a prescription for obtaining the many-electron determinant ψ_0 that minimizes the expectation value of the total electronic energy E_0, given by

$$E_0 \equiv \langle \psi_0 | \hat{H}_{\mathrm{el}} | \psi_0 \rangle. \tag{1.38}$$

Using the definition of the electronic Hamiltonian and the Slater determinant, we arrive at an expression for E_0 that contains the spin orbitals explicitly:

$$E_0 = \sum_{i=1}^{N} \langle \zeta_i | \hat{h} | \zeta_i \rangle + \frac{1}{2} \frac{e^2}{4\pi\varepsilon_0} \sum_{i,j=1}^{N} \left\langle \zeta_i \zeta_j \left| \frac{(1-\hat{p}_{ij})}{r_{12}} \right| \zeta_i \zeta_j \right\rangle, \tag{1.39}$$

where we identify $\hat{p}_{ij}$ with an exchange operator which swaps the indices i and j, and r_{12} stands for $|x_1 - x_2|$.

Exercise 1.2. Prove Eq. (1.39) by use of the definitions (1.16) and (1.38).

We simplify Eq. (1.39) by introducing a special nomenclature for the two electron terms, rewriting E_0 as

$$E_0 = \sum_{i} \langle \zeta_i | \hat{h} | \zeta_i \rangle . + \frac{1}{2} \sum_{i,j=1}^{N} \mathcal{J}_{ij} - K_{ij}. \tag{1.40}$$

The latter formula involves the *Coulomb integral* $\mathcal{J}_{ij}$, defined as

$$\mathcal{J}_{ij} \equiv \left\langle \zeta_i \zeta_j \left| \frac{e^2}{4\pi\varepsilon_0} \frac{1}{r_{12}} \right| \zeta_i \zeta_j \right\rangle = \frac{e^2}{4\pi\varepsilon_0} \int d\mathbf{x}'_1 d\mathbf{x}'_2 |\zeta_i(\mathbf{x}'_1)|^2 \frac{1}{r_{12}} |\zeta_j(\mathbf{x}'_2)|^2 \tag{1.41}$$

and the *exchange integral* K_{ij}, defined as

$$K_{ij} \equiv \left\langle \zeta_i \zeta_j \left| \frac{e^2}{4\pi\varepsilon_0} \frac{1}{r_{12}} \right| \zeta_j \zeta_i \right\rangle$$

$$= \frac{e^2}{4\pi\varepsilon_0} \int d\mathbf{x}'_1 d\mathbf{x}'_2 \zeta_i^*(\mathbf{x}'_1)\zeta_j^*(\mathbf{x}'_2)\frac{1}{r_{12}}\zeta_j(\mathbf{x}'_1)\zeta_i(\mathbf{x}'_2). \tag{1.42}$$

Obviously, Eq. (1.41) is identical with the classical potential energy due to the electrostatic interaction of two charge distributions, $\rho_i(\mathbf{x}') = |\Phi_i(\mathbf{x}')|^2$ and $\rho_j(\mathbf{x}') = |\Phi_j(\mathbf{x}')|^2$, while the appearance of the exchange contribution is, as a consequence of the Pauli exclusion principle, a pure quantum effect. By inspection of Eqs. (1.39)–(1.42), K_{ii} reduces to $\mathcal{J}_{ii}$, so that there is no need to exclude the case $i = j$ explicitly.

1.3.1 *The Hartree–Fock equations*

With the statements (1.40)–(1.42), all tools are in place to tackle the crucial task of minimizing the ground state energy (1.38). This proceeds by introducing the functional $L \equiv L[\zeta_i]$, i.e. a function with the spin orbitals ζ_i as arguments. This entity is required to be stationary for the optimal set of spin orbitals. It is, however, not simply identical with the ground state energy. This is related to the fact that the ζ_i cannot be chosen freely but are constrained by the orthonormality condition

$$\langle \zeta_i, \zeta_j \rangle = \delta_{ij}. \tag{1.43}$$

Taking into account Eq. (1.43), the functional to be optimized becomes

$$L[\zeta_i] = E_0[\zeta_i] - \sum_{i=1}^{N}\sum_{j=1}^{N} \varepsilon_{ij}(\langle \zeta_i, \zeta_j \rangle - \delta_{ij}) \tag{1.44}$$

where the $N \times N$ constraints (1.43) correspond to an equal number of *Lagrange multipliers* ε_{ij}. Since the functional L is real, and $\langle \zeta_i, \zeta_j \rangle = \langle \zeta_j, \zeta_i \rangle^*$, we conclude that the ε_{ij} form a Hermitian matrix:

$$\varepsilon_{ij} = \varepsilon_{ji}^*. \tag{1.45}$$

Stationary behavior of L is achieved if

$$\delta L = \delta E_0 - \sum_{i=1}^{N}\sum_{j=1}^{N} \varepsilon_{ij}\delta(\langle \zeta_i, \zeta_j \rangle) = 0 \tag{1.46}$$

under an infinitesimal variation of the spin orbitals:

$$\zeta_i \to \zeta_i + \delta\zeta_i. \tag{1.47}$$

Inserting Eq. (1.40) into Eq. (1.46), we obtain after some algebra (see for instance [35])

$$\delta L = \sum_{i=1}^{N} \langle \delta\zeta_i | \hat{h} | \zeta_i \rangle + \frac{e^2}{4\pi\varepsilon_0} \times \sum_{i,j=1}^{N} \left(\left\langle \delta\zeta_i \zeta_i \left| \frac{1}{r_{12}} \right| \zeta_j \zeta_j \right\rangle - \left\langle \delta\zeta_i \zeta_j \left| \frac{1}{r_{12}} \right| \zeta_j \zeta_i \right\rangle - \varepsilon_{ij} \left\langle \delta\zeta_i | \zeta_j \right\rangle \right) + c.c. \tag{1.48}$$

This relation translates into a set of integro-differential equations whose solutions are the desired optimal spin orbitals. To derive these equations, we reformulate Eq. (1.48):

$$\sum_{i=1}^{N} \int d\mathbf{x}'_1 \delta\zeta_i^*(\mathbf{x}'_1) \times \left[h(\mathbf{x}'_1)\zeta_i(\mathbf{x}'_1) + \sum_{j=1}^{N} [\hat{\mathcal{J}}_j(\mathbf{x}'_1) - \hat{K}_j(\mathbf{x}'_1)]\zeta_i(\mathbf{x}'_1) - \sum_{j=1}^{N} \varepsilon_{ji}\zeta_j(\mathbf{x}'_1) \right] + c.c. = 0. \tag{1.49}$$

For convenience, we have introduced the Coulomb and exchange operators $\hat{\mathcal{J}}_j(\mathbf{x}')$ and $\hat{K}_j(\mathbf{x}')$, respectively, which act on spin orbitals in the following way:

$$\hat{\mathcal{J}}_j(\mathbf{x}'_1)\zeta_i(\mathbf{x}'_1) = \frac{e^2}{4\pi\varepsilon_0} \left[\int d\mathbf{x}'_2 \zeta_j^*(\mathbf{x}'_2) \frac{1}{r_{12}} \zeta_j(\mathbf{x}'_2) \right] \zeta_i(\mathbf{x}'_1), \tag{1.50a}$$

and

$$\hat{K}_j(\mathbf{x}'_1)\zeta_i(\mathbf{x}'_1) = \frac{e^2}{4\pi\varepsilon_0} \left[\int d\mathbf{x}'_2 \zeta_j^*(\mathbf{x}'_2) \frac{1}{r_{12}} \zeta_i(\mathbf{x}'_2) \right] \zeta_i(\mathbf{x}'_1). \tag{1.50b}$$

The variation $\delta\zeta_i^*(\mathbf{x}'_1)$ is arbitrary, implying that the second factor of the integrand in Eq. (1.49) vanishes identically. Thus,

$$\hat{h}(\mathbf{x}'_1)\zeta_i(\mathbf{x}'_1) + \sum_{j=1}^{N} [\hat{\mathcal{J}}_j(\mathbf{x}'_1) - \hat{K}_j(\mathbf{x}'_1)]\zeta_i(\mathbf{x}'_1) = \sum_{j=1}^{N} \varepsilon_{ji}\zeta_j(\mathbf{x}'_1). \tag{1.51}$$

By use of a variational principle, we have derived N coupled equations of the form

$$\hat{F}(\mathbf{x}'_1)\zeta_i(\mathbf{x}'_1) = \sum_{j=1}^{N} \varepsilon_{ji}\zeta_j(\mathbf{x}'_1) \tag{1.52}$$

where $\hat{F} \equiv \hat{h} + \sum_{j=1}^{N} \hat{\mathcal{J}}_j - \hat{K}_j$ is the *Fock operator*. We note that the variational principle that led to Eq. (1.52) could be implemented more directly by requiring the functional derivatives of L with respect to ζ_i and ζ_i^* to vanish: $\frac{\delta L}{\delta \zeta_i} = \frac{\delta L}{\delta \zeta_i^*} = 0$. For a condensed introduction to the concept of the functional derivative, consult the appendix of this chapter.

From the definition (1.52) in conjunction with Eqs. (1.43) and (1.45), we infer immediately the Hermitian property of the Fock operator. The identity (1.52) does not yet represent the ultimate form of the equation that leads to the optimal set $\{\zeta_i\}$ since the latter is not unique. More specifically, the spin orbitals $\{\zeta_i\}$ obtained from Eq. (1.52) may be transformed among themselves by the action of a unitary matrix $\mathbf{U}$ without any change of the expectation value $E_0 = \langle \psi_0 | \hat{H}_{\text{el}} | \psi_0 \rangle$. In the following step we will examine the effect of such a transformation on all ingredients of Eq. (1.52), starting with the spin orbitals which change under the action of $\mathbf{U}$ as

$$\tilde{\zeta}_i = \sum_{j=1}^{N} \zeta_j U_{ji}. \tag{1.53}$$

How does $\mathbf{U}$ affect the Fock operator? Not at all. This is demonstrated by a simple calculation:

$$\begin{aligned}\sum_{j=1}^{N} \tilde{\hat{\mathcal{J}}}_j(\mathbf{x}'_1) &\equiv \sum_{j=1}^{N} \frac{e^2}{4\pi\varepsilon_0} \int d\mathbf{x}'_2\, \tilde{\zeta}_j^*(\mathbf{x}'_2) \frac{1}{r_{12}} \tilde{\zeta}_j(\mathbf{x}'_2) \\ &= \sum_{k,m=1}^{N} \left[\sum_{j=1}^{N} U_{kj}^* U_{mj} \right] \frac{e^2}{4\pi\varepsilon_0} \int d\mathbf{x}'_2 \zeta_k^*(\mathbf{x}'_2) \frac{1}{r_{12}} \zeta_m(\mathbf{x}'_2).\end{aligned} \tag{1.54}$$

From $\sum_{j=1}^{N} U_{kj}^* U_{mj} = \delta_{km}$, we conclude that

$$\sum_{j=1}^{N} \tilde{\hat{\mathcal{J}}}_j(\mathbf{x}'_1) = \sum_{j=1}^{N} \hat{\mathcal{J}}_j(\mathbf{x}'_1). \tag{1.55}$$

An analogous computation establishes the invariance of the exchange operator under application of **U**. Therefore

$$\tilde{\hat{F}}(\mathbf{x}'_1) = \hat{F}(\mathbf{x}'_1). \tag{1.56}$$

The transformation law for the Lagrange multiplier matrix follows immediately from Eq. (1.56):

$$\tilde{\boldsymbol{\varepsilon}} = \mathbf{U}^\dagger \boldsymbol{\varepsilon} \mathbf{U}. \tag{1.57}$$

We can always accomplish diagonalization of $\boldsymbol{\varepsilon}$ by an appropriate choice of **U**. Assuming thus a diagonal matrix $\boldsymbol{\varepsilon}'$, we may write down the transformed Eq. (1.52). It is

$$\tilde{\hat{F}}(\mathbf{x}'_1)\tilde{\zeta}_i(\mathbf{x}'_1) = \tilde{\varepsilon}_i \tilde{\zeta}_i(\mathbf{x}'_1). \tag{1.58}$$

By dropping the tilde symbols, we arrive at the *Hartree–Fock equation* in its *canonical form*:

$$\hat{F}(\mathbf{x}'_1)\zeta_i(\mathbf{x}'_1) = \varepsilon_i \zeta_i(\mathbf{x}'_1). \tag{1.59}$$

The Fock operator contains all occupied spin orbitals except ζ_i. Therefore, the system (1.59) consists of N coupled, highly nonlinear equations.

We turn now to the interpretation of the spin orbital eigenenergies, or Fock operator eigenvalues ε_i. In particular, we want to explore their relation to the ground state energy expectation value E_0. With the help of Eqs. (1.50a), (1.50b), and (1.59), one readily derives the connection between the ε_i and the Coulomb and exchange terms:

$$\begin{aligned}\varepsilon_i &= \langle \zeta_i | \hat{F} | \zeta_i \rangle = \left\langle \zeta_i \middle| \hat{h} + \sum_{j=1}^{N} \hat{\mathcal{J}}_j - \hat{K}_j \middle| \zeta_i \right\rangle \\ &= \langle \zeta_i | \hat{h} | \zeta_i \rangle + \left\langle \zeta_i \middle| \sum_{j=1}^{N} \hat{\mathcal{J}}_j - \hat{K}_j \middle| \zeta_i \right\rangle = h_{ii} + \sum_{j=1}^{N} \mathcal{J}_{ij} - K_{ij}. \end{aligned} \tag{1.60}$$

By comparison with Eq. (1.40) we realize that adding up the ε_i by summing over all occupied spin orbitals does not yield the ground state energy E_0. It holds that

$$\sum_{j=1}^{N} \varepsilon_i = \sum_{j=1}^{N} h_{ii} + \sum_{i,j=1}^{N} \mathcal{J}_{ij} - K_{ij} \neq E_0. \tag{1.61}$$

In particular, we find that $\sum_{j=1}^{N} \varepsilon_i - E_0 = \frac{1}{2} \sum_{i,j=1}^{N} \mathcal{J}_{ij} - K_{ij}$. The reason for this discrepancy lies in the structure of the Fock operator. The eigenvalue ε_i contains the Coulomb and exchange interaction of an electron in the spin orbital ζ_i with some other electron in the spin orbital ζ_j. The same interaction, however, is included in the eigenvalue ε_j. Thus, the Coulomb and exchange contributions are counted twice in the sum over all spin orbital eigenenergies, which consequently deviates from the ground state energy.

1.3.2 *Koopmans' theorem*

The meaning of the Fock operator eigenvalues by themselves is most easily explored by considering an $(N - 1)$-electron system that emerges from the Hartree–Fock ground state as represented by a determinant built from N occupied orbitals $\{\zeta_i\}$ by removing an electron from the spin orbital with label k. If we further assume that the other $N - 1$ spin orbitals remain unaffected by this change, the difference between the N-particle energy ${}^{N}E_0$ and the $(N - 1)$-particle energy ${}^{N-1}E_k$ is found to be

$$ {}^{N-1}E_k - {}^{N}E_0 = -\varepsilon_k. \tag{1.62} $$

Thus the eigenvalue ε_k, with reversed sign, provides an approximation to the energy required to ionize the N-electron system by removing one electron from the spin orbital ζ_k. The hypothesis underlying Eq. (1.62), i.e. the assumption that all spin orbitals with labels $i \neq k$ remain invariant as ζ_k is emptied, is, however, generally unrealistic. A refined model of ionization will include the effect of electronic relaxation on the quantity ${}^{N-1}E_k$.

The argument outlined in the foregoing paragraph may be extended from electron subtraction to electron addition. While in the former case an occupied orbital of the N-particle determinant is eliminated, an unoccupied, or virtual, orbital is populated in the latter. In this way, an $(N + 1)$-electron system is generated where it is understood that the previously occupied orbitals of the N-electron system are unaltered as the particle number increases by one. Assigning a label s, with $s = N + 1, N + 2, \ldots$ to the virtual orbital to be filled with an additional electron, we arrive at a result analogous to Eq. (1.62) by comparing the expectation values ${}^{N}E_0$ and ${}^{N+1}E_s$.

More specifically,

$$^{N}E_0 - {}^{N+1}E_s = -\varepsilon_s. \tag{1.63}$$

This relation suggests an interpretation of the virtual spin orbital eigenvalues. They may be viewed as approximations to the energy change undergone by an ensemble of N electrons as one more electron is added, or the *electron affinities* associated with the virtual spin orbitals $\{\zeta_i\}$. The relations (1.62) and (1.63) embody *Koopmans' Theorem* which is an often useful tool for estimating the ionization energies and, to a lesser degree, the electron affinities of a molecular system. It relates the spin orbital eigenvalues of occupied (virtual) states to the energy change of an N-electron system as one electron is removed from (filled into) the orbital with label $k(s)$.

Exercise 1.3. Establish the validity of relations (1.62) and (1.63).

The precision of quantities evaluated by means of Koopmans' theorem is limited by the approximations inherent in its derivation, i.e. the neglect of relaxation upon addition or subtraction of one electron and the use of a single determinant for the wave function of the systems. While the relaxation effect will decrease the energy $^{N-1}E_k$ in Eq. (1.62), and therefore diminish the magnitude of the ionization energy, inclusion of electron–electron correlation is predicted to have the opposite impact. This is due to the fact that many-body phenomena lower the energy of an N-electron ensemble more than that of $N - 1$ electrons. Consequently, the two simplifications cooperate to distort the correct result in opposing directions, and partial cancellation of errors ensues. This argument does not hold for the electron affinities, where both the relaxation and the correlation error act in the same direction. Accordingly, ionization energies based on Koopmans' theorem tend to be of better accuracy than electron affinities.

1.3.3 *The Hartree–Fock Hamiltonian*

Hartree–Fock theory establishes the many-electron Slater determinant as an approximate solution to the multi-electron problem, arising from a variational approach that involves minimization of the energy expectation value E_0. From the Fock operator eigenvalue in Eq. (1.59), it follows immediately that

$$\sum_{i=1}^{N} \hat{F}(\mathbf{x}'_i)\psi_0(\mathbf{x}'_1, \mathbf{x}'_2, \ldots, \mathbf{x}'_N) = \sum_{i=1}^{N} \varepsilon_i \psi_0(\mathbf{x}'_1, \mathbf{x}'_2, \ldots, \mathbf{x}'_N) \tag{1.64}$$

with ψ_0 as the variationally optimized Slater determinant. We refer to the sum of the Fock operators, $\sum_{i=1}^{N} \hat{F}(\mathbf{x}'_\mathrm{i})$, as the *Hartree–Fock operator*, $\hat{H}_0$. The relation of $\hat{H}_0$ to the true electronic Hamiltonian is evidently

$$\hat{H}_{\mathrm{el}} = \hat{H}_0 + (\hat{H}_{\mathrm{el}} - \hat{H}_0) \equiv \hat{H}_0 + \hat{V}. \tag{1.65}$$

The expectation value of the operator $\hat{V} = \frac{1}{4\pi\varepsilon_0} \sum_{i=1}^{N} \sum_{j>i}^{N} \frac{e^2}{|\mathbf{x}'_\mathrm{i} - \mathbf{x}'_\mathrm{j}|} - \sum_{i,j=1}^{N} [\hat{\mathcal{J}}_j(\mathbf{x}'_i) - \hat{K}_j(\mathbf{x}'_i)]$ with respect to the Hartree–Fock determinant results as

$$\langle \psi_0 | \hat{V} | \psi_0 \rangle = -\frac{1}{2} \sum_{i,j=1}^{N} \left[\mathcal{J}_{ij} - K_{ij} \right]. \tag{1.66}$$

This clarifies that the Hartree–Fock energy E_0 may be understood as the first approximation to the true eigenvalue E_{el} of the electronic Hamiltonian $\hat{H}_{\mathrm{el}}$ in the sense of perturbation theory with $\hat{V}$ as perturbation operator:

$$E_{\mathrm{el}} = E_{\mathrm{el}}^{(0)} + E_{\mathrm{el}}^{(1)} + E_{\mathrm{el}}^{(2)} + \cdots \tag{1.67}$$

$$E_0 = E_{\mathrm{el}}^{(0)} + E_{\mathrm{el}}^{(1)} \tag{1.68}$$

where the first energy correction $E_{\mathrm{el}}^{(1)}$ is just $\langle \psi_0 | \hat{V} | \psi_0 \rangle$.

1.4 The Restricted and the Unrestricted Hartree–Fock Formalism

So far, the Hartree–Fock procedure has been formulated in terms of spin orbitals. In the following we will make the transition from spin to spatial orbitals, discussing the *restricted* and the *unrestricted* case. The first alternative is realized when alpha and beta electrons are perfectly paired, each occupied spatial orbital being populated by two electrons with opposite spins. The second alternative allows different spatial orbitals for the two spins. *Open-shell* systems, i.e. systems with non-zero total spin, necessitate the further distinction between *restricted* and *unrestricted open-shell* cases. The former imposes identical spatial orbitals upon all paired electrons of the open-shell system, while the latter relaxes this constraint. We start with a description of the restricted Hartree–Fock scheme.

1.4.1 *The restricted Hartree–Fock method*

The complete pairing of alpha and beta electrons in a closed-shell situation implies that we can use Eq. (1.16) to write the spin orbitals as

$$\zeta_i(\mathbf{x}') = \Phi_i(\mathbf{x})\alpha(\omega), \tag{1.69a}$$

$$\text{or} \qquad \zeta_i(\mathbf{x}') = \Phi_i(\mathbf{x})\beta(\omega). \tag{1.69b}$$

The Hartree–Fock equations for the restricted case (RHF equations) can be derived in a straightforward fashion by use of the Fock operator as defined by Eqs. (1.51) and (1.52). Focusing on the alpha subset, we find

$$\hat{F}(\mathbf{x}'_1)\Phi_i(\mathbf{x}_1)\alpha(\omega_1) = \varepsilon_i\Phi_i(\mathbf{x}_1)\alpha(\omega_1). \tag{1.70}$$

We could have settled for the beta electrons just as well, as the meaning of *closed shell* implies identical results for both spins. Multiplication by $\alpha(\omega_1)^*$ from the left and integration over ω_1 eliminates the spin function on the right side of Eq. (1.70), leading to

$$\left[\int d\omega_1\alpha^*(\omega_1)\hat{F}(\mathbf{x}_1)\alpha(\omega_1)\right]\Phi_i(\mathbf{x}_1) = \varepsilon_i\Phi_i(\mathbf{x}_1). \tag{1.71}$$

If one inserts into Eq. (1.71) the explicit form of the Fock operator,

$$\hat{F}(\mathbf{x}_1) = \hat{h}(\mathbf{x}_1) + \frac{e^2}{4\pi\varepsilon_0}\sum_{i=1}^{N}\int d\mathbf{x}'_2\,\zeta_i^*(\mathbf{x}'_2)\frac{1}{r_{12}}(1-\hat{p}_{ij})\zeta_i(\mathbf{x}'_2), \tag{1.72}$$

and applies the substitutions (1.69a) and (1.69b), the Hartree–Fock equations for the restricted case are readily derived. In performing this computation, one has to take into account that the number of electrons occupying alpha and beta spin orbitals ζ_i is both $N/2$, and to use the orthonormality relations (1.15a)–(1.15c). The result is

$$\hat{F}(\mathbf{x}_1)\Phi_i(\mathbf{x}_1) = \left[\hat{h}(\mathbf{x}_1) + \sum_{j=1}^{N/2} 2\hat{\mathcal{J}}_j(\mathbf{x}_1) - \hat{K}_j(\mathbf{x}_1)\right]\Phi_i(\mathbf{x}_1) = \varepsilon_i\Phi_i(\mathbf{x}_1), \tag{1.73}$$

and the Coulomb and exchange operators are expressed entirely in terms of the spatial orbitals:

$$\hat{\mathcal{J}}_j(\mathbf{x}_1)\Phi_i(\mathbf{x}_1) = \frac{e^2}{4\pi\varepsilon_0}\left[\int d\mathbf{x}_2\, \Phi_j^*(\mathbf{x}_2)\frac{1}{r_{12}}\Phi_j^*(\mathbf{x}_2)\right]\Phi_i(\mathbf{x}_1) \tag{1.74a}$$

and

$$\hat{K}_j(\mathbf{x}_1)\Phi_i(\mathbf{x}_1) = \frac{e^2}{4\pi\varepsilon_0}\left[\int d\mathbf{x}_2\, \Phi_j^*(\mathbf{x}_2)\frac{1}{r_{12}}\Phi_i(\mathbf{x}_2)\right]\Phi_j(\mathbf{x}_1). \tag{1.74b}$$

The Hartree–Fock ground state energy for the restricted case follows immediately from Eq. (1.40) as one observes that the summation extends over equal numbers of alpha and beta spins with identical spatial orbitals. Thus,

$$\sum_{i=1}^{N}\zeta_i = \sum_{j=1}^{N/2}\Phi_j\alpha + \sum_{k=1}^{N/2}\Phi_k\beta. \tag{1.75}$$

Employing this approach and using the orthonormality of the spin functions, we arrive at

$$\begin{aligned} E_0 &= 2\sum_{i=1}^{N/2}\langle\Phi_i|\hat{h}|\Phi_i\rangle + \frac{e^2}{4\pi\varepsilon_0}\sum_{j=1}^{N/2}\sum_{k=1}^{N/2}2\left\langle\Phi_j\Phi_k\left|\frac{1}{r_{12}}\right|\Phi_j\Phi_k\right\rangle \\ &\quad - \left\langle\Phi_j\Phi_k\left|\frac{1}{r_{12}}\right|\Phi_k\Phi_j\right\rangle \\ &= 2\sum_{i=1}^{N/2}h_{ii} + \sum_{j=1}^{N/2}\sum_{k=1}^{N/2}2\mathcal{J}_{jk} - K_{jk}. \end{aligned} \tag{1.76}$$

1.4.2 *The unrestricted Hartree–Fock method*

The rigid pairing constraint imposed on the restricted scheme, involving spatial orbitals that are shared by an alpha and a beta electron, is abandoned in the unrestricted case, making allowance for a higher degree of variational freedom and therefore a lowering of the Hartree–Fock energy. This added flexibility, however, comes with a sacrifice: the unrestricted Hartree–Fock (UHF) determinant is no longer an eigenfunction of the total electronic spin operator $\hat{S}^2$. This may give rise to unrealistic spin expectation values for the studied system. This *spin contamination* can, in principle, be remedied by considering a multiplicity of unrestricted determinants. We develop the unrestricted variant in analogy to the restricted formalism, as outlined earlier, replacing Eqs. (1.69a) and (1.69b) with

$$\zeta_{i\alpha}(\mathbf{x}') = \Phi_{i\alpha}(\mathbf{x})\alpha(\omega) \tag{1.77a}$$

$$\text{or} \quad \zeta_{i\beta}(\mathbf{x}') = \Phi_{i\beta}(\mathbf{x})\beta(\omega). \tag{1.77b}$$

Obviously, the unrestricted reduces to the restricted case if one stipulates $\Phi_{i\alpha} = \Phi_{i\beta}$. Assuming a total number of N^{α} (N^{β}) electrons with alpha (beta) spin and setting $N = N^{\alpha} + N^{\beta}$, we derive the explicit form of the unrestricted Fock operator by inserting Eqs. (1.77a), (1.77b) into Eq. (1.73), and obtain

$$\hat{F}^{\alpha}(\mathbf{x}_1) = \hat{h}(\mathbf{x}_1) + \sum_{i=1}^{N^{\alpha}} [\hat{\mathcal{J}}_i^{\alpha}(\mathbf{x}_1) - \hat{K}_i^{\alpha}(\mathbf{x}_1)] + \sum_{i=1}^{N^{\beta}} \hat{\mathcal{J}}_i^{\beta}(\mathbf{x}_1). \tag{1.78}$$

The operator $\hat{F}^{\alpha}(\mathbf{x}_1)$ acts on alpha spin orbitals, while the Coulomb component refers to the interelectronic repulsion between two alpha electrons $\mathcal{J}_i^{\alpha}$. The parallel expression for F^{β} results from exchanging the indices α and β in Eq. (1.78). The unrestricted counterpart of Eq. (1.73) then consists of the equation pair

$$\hat{F}^{\alpha}(\mathbf{x}_1)\Phi_i^{\alpha}(\mathbf{x}_1) = \varepsilon_i^{\alpha}\Phi_i^{\alpha}(\mathbf{x}_1), \tag{1.79a}$$

$$\hat{F}^{\beta}(\mathbf{x}_1)\Phi_i^{\beta}(\mathbf{x}_1) = \varepsilon_i^{\beta}\Phi_i^{\beta}(\mathbf{x}_1). \tag{1.79b}$$

The occurrence of a beta Coulomb potential term in Eq. (1.78) and the respective alpha term in the analogous equation for $\hat{F}^{\beta}$ implies that these two integro-differential equations are coupled. The ground state energy for the unrestricted case is found from a direct generalization of Eq. (1.76), namely

$$E_0 = \sum_{i=1}^{N^{\alpha}} h_{ii}^{\alpha} + \sum_{i=1}^{N\beta} h_{ii}^{\beta} + \frac{1}{2}\sum_{j=1}^{N^{\alpha}}\sum_{k=1}^{N^{\alpha}} (\mathcal{J}_{jk}^{\alpha\alpha} - K_{jk}^{\alpha\alpha})$$
$$+ \frac{1}{2}\sum_{j=1}^{N^{\beta}}\sum_{k=1}^{N^{\beta}} (\mathcal{J}_{jk}^{\beta\beta} - K_{jk}^{\beta\beta}) + \sum_{j=1}^{N^{\alpha}}\sum_{k=1}^{N^{\beta}} \mathcal{J}_{jk}^{\alpha\beta}, \tag{1.80}$$

where $\mathcal{J}_{jk}^{\alpha\alpha} \equiv \langle\Phi_j^{\alpha}|\hat{\mathcal{J}}_k^{\alpha}|\Phi_j^{\alpha}\rangle$, with the corresponding definition for $\mathcal{J}_{jk}^{\beta\beta}$. Accordingly, $\mathcal{J}_{jk}^{\alpha\beta} = \mathcal{J}_{kj}^{\beta\alpha}$ is shorthand for $\langle\Phi_j^{\alpha}|\hat{\mathcal{J}}_k^{\beta}|\Phi_j^{\alpha}\rangle = \langle\Phi_k^{\beta}|\hat{\mathcal{J}}_j^{\alpha}|\Phi_k^{\beta}\rangle$. Analogous assignments are made for $K_{jk}^{\alpha\alpha}$ and $K_{jk}^{\beta\beta}$. It is directly seen from formula (1.80) that the self-interaction vanishes, i.e. the Coulomb and exchange contributions due to the interaction of some electron in an orbital Φ_j^α or Φ_j^β with itself cancel, as $\mathcal{J}_{jj}^{\alpha\alpha} = K_{jj}^{\alpha\alpha}$ and $\mathcal{J}_{jj}^{\beta\beta} = K_{jj}^{\beta\beta}$.

The spin contamination problem, as a major deficiency of the unrestricted Hartree–Fock method, can be alleviated by projecting the wave function that results from the variational procedure on spin operator eigenfunctions. This improvement, however, comes at the cost of admitting a linear combination of determinants as ground state function instead of a single determinant (for a condensed account of this matter see [41]).

1.4.3 *The Roothaan formalism*

The Hartree–Fock equations for the restricted as well as unrestricted case, Eq. (1.73) and Eq. (1.78), respectively, are in general too complex to admit direct numerical solutions. However, the LCAO approach Eq. (1.21) has been used to cast these equations into matrix form and thus to reduce the task to an algebraic procedure, the *self-consistent field* method. Using the restricted Hartree–Fock scheme, we substitute for the MO Φ_i the linear combination $\Phi_i = \sum_{k=1}^{K} a_{ki}\phi_k$ to obtain

$$\hat{F}(\mathbf{x}_1)\sum_{k=1}^{K} a_{ki}\phi_k(\mathbf{x}_1) = \varepsilon_i \sum_{k=1}^{K} a_{ki}\phi_k(\mathbf{x}_1). \tag{1.81}$$

From this approach, a matrix equation is generated as one multiplies from the left by ϕ_m^* and subsequently integrates over $\mathbf{x}_1$. This yields

$$\sum_{k=1}^{K} a_{ki} \int d\mathbf{x}_1 \phi_m^*(\mathbf{x}_1)\hat{F}(\mathbf{x}_1)\phi_k(\mathbf{x}_1) = \varepsilon_i \sum_{k=1}^{K} a_{ki} \int d\mathbf{x}_1\, \phi_m^*(\mathbf{x}_1)\phi_k(\mathbf{x}_1). \tag{1.82}$$

This expression involves the overlap matrix $\mathbf{S}$ of dimension $K \times K$, defined by $S_{mk} = \int d\mathbf{x}_1\, \phi_m^*(\mathbf{x}_1)\phi_k(\mathbf{x}_1)$ (see formula (1.23)). Further, we define the Fock matrix $\mathbf{F}$ by

$$F_{mk} = \int d\mathbf{x}_1 \phi_m^*(\mathbf{x}_1)\hat{F}(\mathbf{x}_1)\phi_k(\mathbf{x}_1) \tag{1.83}$$

which represents the Fock operator in the atomic basis. Reformulating Eq. (1.82) by use of S_{mk} and F_{mk}, we find

$$\sum_{k=1}^{K} F_{mk}a_{ki} = \varepsilon_i \sum_{k=1}^{K} S_{mk}a_{ki} \tag{1.84}$$

or, in compact matrix notation,

$$\mathbf{Fa} = \mathbf{Sa}\boldsymbol{\varepsilon} \tag{1.85}$$

where the coefficient matrix $\mathbf{a}$ with elements a_{ki} and the orbital energy eigenvalue matrix $\boldsymbol{\varepsilon}$ have been introduced. The latter is diagonal with elements $\varepsilon_{ij} = \varepsilon_i \delta_{ij}$. Formula (1.85) embodies the *Roothaan equations* which are obtained by combining the Hartree–Fock theory with an LCAO approach for the molecular orbitals. The Roothaan equations are readily extended to the unrestricted case where they assume the form

$$\mathbf{F}^\alpha \mathbf{a}^\alpha = \mathbf{S}\mathbf{a}^\alpha \boldsymbol{\varepsilon}^\alpha \tag{1.86a}$$

$$\mathbf{F}^\beta \mathbf{a}^\beta = \mathbf{S}\mathbf{a}^\beta \boldsymbol{\varepsilon}^\beta, \tag{1.86b}$$

where $\mathbf{F}^{\alpha,\beta}$ contains the Fock operators $\hat{F}^{\alpha,\beta}$, and the matrices $\boldsymbol{\varepsilon}^{\alpha,\beta}$ contain the orbital energy eigenvalues for the spin alpha and beta subsystems. Both equations are coupled, as Eq. (1.78) and its counterpart for the beta subsystem are. Applying the self-consistent field method requires further expansion of the Fock matrix element (1.83). In the following, we show that F_{mk} is naturally represented in terms of the charge density matrix $\mathbf{P}$ (see the discussion following Eq. (1.27)). In order to see this, we rewrite F_{mk} using the explicit form of $\hat{F}$ as given by Eq. (1.73) for the restricted case. Thus [35]

$$\begin{aligned} F_{mk} &= \int d\mathbf{x}_1\, \phi_m^*(\mathbf{x}_1)\hat{h}(\mathbf{x}_1)\phi_k(\mathbf{x}_1) \\ &\quad + \sum_{i=1}^{N/2} \int d\mathbf{x}_1\, \phi_m^*(\mathbf{x}_1)(2\hat{\mathcal{J}}_i(\mathbf{x}_1) - \hat{K}_i(\mathbf{x}_1))\phi_k(\mathbf{x}_1). \end{aligned} \tag{1.87}$$

The definitions (1.74a) and (1.74b) of the operators $\hat{\mathcal{J}}_i$ and $\hat{K}_i$ in conjunction with the LCAO expansion (1.21) turn Eq. (1.87) into

$$\begin{aligned} F_{mk} &= h_{mk} + \sum_{i=1}^{N/2} \sum_{p,q=1}^{K} a_{pi} a_{qi}^* [2(mk|qp) - (mp|qk)] \\ &= h_{mk} + \sum_{p,q=1}^{K} P_{pq} \left[(mk|qp) - \frac{1}{2}(mp|qk) \right] \\ &= h_{mk} + G_{mk} \end{aligned} \tag{1.88}$$

where the first term, h_{mk}, includes the one-body contributions to the Fock operator. The second term contains the two-electron integral symbol $(ij|mn)$, defined as

$$(ij|mn) \equiv \frac{1}{4\pi\varepsilon_0} \int d\mathbf{x}_1 d\mathbf{x}_2\, \phi_i^*(\mathbf{x}_1)\phi_j(\mathbf{x}_1)\frac{e^2}{r_{12}}\phi_m^*(\mathbf{x}_2)\phi_n(\mathbf{x}_2). \tag{1.89}$$

We note that the frequently utilized symbol

$$\langle ij||mn\rangle \equiv (ij|mn) - \frac{1}{2}(in|mj) \tag{1.90}$$

makes it possible to write the Fock matrix element (1.88) in a still more compact form. Evaluating all integrals of this type as they occur in the course of a Hartree–Fock computation is in general an extensive task and the limiting factor in the actual solution of the Roothaan equations.

Apart from this practical difficulty, the treatment of the matrix equation (1.85) seems to be greatly simplified by diagonalization as well as normalization of the overlap matrix $\mathbf{S}$, reducing the latter to the unit matrix $\mathbf{I}$. If this goal is achieved by a transformation matrix $\mathbf{M}$, the correspondingly transformed Fock and coefficient matrices may be defined as $\mathbf{F}' \equiv \mathbf{M}^\dagger\mathbf{F}\mathbf{M}$ and $\mathbf{a}' = \mathbf{M}^{-1}\mathbf{a}$, respectively, leading to the equation system

$$\mathbf{F}'\mathbf{a}' = \mathbf{a}'\boldsymbol{\varepsilon}. \tag{1.91}$$

The simple appearance of this relation, however, is deceiving since $\mathbf{F}'$ depends on $\mathbf{a}'$, and thus the Roothaan equations remain highly nonlinear in any given atomic basis. Solving them requires the use of an iterative method. The starting point of this self-consistent field procedure is the selection of an appropriate atomic basis set $\{\phi_i\}$, followed by evaluating the one-electron contributions h_{mk} to the Fock matrix as well as the two-electron integrals. The matrix $\mathbf{S}$ is diagonalized and the two-electron term, G_{mk}, of Eq. (1.88) is estimated by guessing the charge density matrix $\mathbf{P}$. Both $\mathbf{a}'$ and $\boldsymbol{\varepsilon}$ are determined from Eq. (1.91) which in turn yields the coefficient matrix as one reverses the above relation. The new MOs thus generated are employed to compute an improved guess at the charge density matrix, and the procedure is repeated. It terminates at step k if the charge density matrix produced in this cycle is sufficiently close to that obtained at step $k-1$. In this way, the exact solution of the Hartree–Fock equations can be accomplished in the ideal case of a complete atomic basis. As the basis $\{\phi_i\}$ increases, the Roothaan approach approximates the Hartree–Fock limit.

1.5 Post-Hartree–Fock Methods

The Hartree–Fock formalism has proven to be successful in the qualitative, and frequently also in the quantitative description of quantum chemical systems. Its fundamental methodological limitation, however, is its mean-field character as reflected by the form of the Hartree–Fock solution, a single determinant composed of molecular orbitals. Therefore it fails to include the correlation between electrons. Conceptually, we distinguish between a dynamic and a static component of the overall correlation effect.

The dynamic part is related to the approximate treatment of the interelectronic repulsion term by Hartree–Fock theory. Each electron feels the average of the Coulomb and exchange potentials due to all remaining electrons. Moving in an effective one-body instead of the accurate many-body interelectronic potential, two electrons may approach each other too closely which results in an overestimation of the repulsion between them.

Static correlation manifests itself prototypically in bond dissociation which is insufficiently captured by any description based on the restricted Hartree–Fock method. The ground state determinant for H_2, for instance, may be expanded into *covalent* and *ionic* contributions, the latter referring to an $H^- - H^+$ or an $H^+ - H^-$ configuration, where two electrons are attached to one of the nuclei, the former to an even distribution of the electrons among the H centers. At large internuclear distances, where H_2 separates into two H atoms, the weights of the ionic and the covalent contributions continue to be equal which is evidently unphysical. This case can be remedied by adding a further determinant to the solution which cancels the unwanted admixtures (see also Exercise 1.6). Dynamic and static correlation do not depend on each other since the former phenomenon refers to the short range, the latter to the long range where the interelectronic Coulomb term is of minor importance.

We point out that the unrestricted Hartree–Fock theory arrives at a qualitatively correct model of molecular dissociation. However, this success is achieved in the spirit of improvisation and *ad hoc* amendment, as may be once more demonstrated by the example of the H_2 dissociation problem: the unrestricted determinant ceases to have spin singlet character in the asymptotic regime and is described as a hybrid of singlet and triplet which both contribute with equal weights. This again fails to do justice to the actual physical situation. Various approaches have been put forward to incorporate the correlation effect. In the following, we mention *configuration interaction* and the *many-body perturbation theory*.

1.5.1 *Configuration interaction*

The systematic solution of the electron correlation problem consists in replacing the single determinant approach that underlies the various realizations of the Hartree–Fock method with a multideterminant picture. Thus, the variational search for the optimal wave function now proceeds in a subspace defined by N_C linearly independent determinants, starting with the Hartree–Fock solution ψ_0:

$$\psi = \sum_{k=0}^{N_C} c_k \psi_k, \tag{1.92}$$

where N_C stands for the number of configurations represented by the included determinants ψ_k. This extension of the Hartree–Fock formalism defines the configuration interaction (CI) method. From the variational principle, one concludes that the linear combination (1.92) which contains the Hartree–Fock wave function as one of the basis vectors corresponds to a lower energy than that of the Hartree–Fock limit, E_0. This energy decrease is ascribed to the inclusion of electron correlation. Taking into account the full correlation effect is tantamount to solving the Schrödinger equation (1.2), and so we may define the correlation energy E_{co} as the energy difference that bridges the gap between the accurate non-relativistic total energy E_{el} of the stationary molecular system and the Hartree–Fock limit:

$$E_{\text{co}} = E_{\text{el}} - E_0. \tag{1.93}$$

If the configuration basis $\{\psi_k\}$ is complete, the ansatz (1.92) leads to the precise correlation energy and thus to the true eigenvalue E_{el} of the Schrödinger equation. Although completeness cannot be achieved in practice, a simple construction principle may be outlined whose extension *ad infinitum* will yield a complete basis. The prescription is this: starting with the Hartree–Fock determinant ψ_0, one replaces an occupied orbital j by a virtual orbital s, and thus creates a sequence of substitutions ψ_j^s with coefficients c_j^s. The ψ_j^s are referred to as singly excited determinants. Following the same scheme, one arrives at double and triple excitations, ψ_{jk}^{st}, ψ_{jkl}^{stu}, etc. The expansion

(1.92) is written as

$$\psi = \psi_0 + \sum c_j^s \psi_j^s + \sum c_{jk}^{st} \psi_{jk}^{st} + \cdots, \tag{1.94}$$

where repeated configurations are excluded. The most important excited determinant is in general the one that corresponds to a transition of an electron from the highest occupied molecular orbital (HOMO) to the lowest unoccupied molecular orbital (LUMO). For organic molecules, this single substitution provides frequently a good approximation to the first excited state of the system [42]. Even if the first excited state is of a more intricate nature, the energy difference between the HOMO and the LUMO level gives a measure of the effort it takes to excite the system, and thus of the stability of its ground state. By construction, the excitations comprised in Eq. (1.94) are orthonormal. The coefficients c_j^s, etc., can thus be determined by means of a variational approach in analogy to Eq. (1.46) where the orthonormality constraint for the spin orbitals is now replaced by that for the configurations. The AO coefficients are not reoptimized in this procedure. It converges to the true correlation energy in the limit of an infinite atomic basis as well as an infinite number of configurations.

The variational scheme implies that unfreezing the MO coefficients, thus treating them as variables to be optimized during the configuration interaction procedure, will yield an increased amount of correlation energy, and therefore a more adequate wave function. The resulting improved method is the *multiconfiguration-self-consistent field* (MCSCF) theory. This improvement, however, comes at the price of substantially enhanced computational effort.

Among the problems of the configuration interaction approach as a tool for quantum chemical calculations are the generally slow convergence of the series (1.94) and the lack of *size consistency*. This term refers to the requirement that the total energy of two non-interacting systems should equal the sum of the energies of the two systems, a condition that is of obvious importance for the correct description of molecular dissociation. Consider the following intuitive special case of size consistency: two H_2 molecules are at a sufficiently large distance from each other to exclude any interaction between them. The total energy of this arrangement must then equal the sum of the subsystem energies. While configuration interaction meets this requirement, truncated configuration interaction does not. Applying the configuration interaction technique to this situation, we might assume that the expansion is truncated after the terms of second order, leading up to doubly excited determinants. As we treat each of the H_2 units in isolation from the other, we thus include the case of all four electrons populating virtual orbitals. However, interpreting the system as H_4 instead of (H_2, H_2), we are unable to realize this configuration, which would involve a fourfold substitution from the Hartree–Fock reference determinant. Viewing the two H_2 molecules as separate entities will therefore lead to a lower total energy than treating them as one supermolecule. These limitations motivate the search for alternative methods. An approximation that is not compromised by the size consistency problem is given with the *many-body perturbation theory*, as discussed in the following subsection.

Before continuing our review of post-Hartree–Fock procedures, we touch on a model that exhibits a formal similarity with the configuration interaction approach, the *valence bond* method [43, 44]. The overall wave function is here expressed as an antisymmetrized product of electron-pair functions. The latter comprise two factors, namely symmetric spatial and antisymmetric spin functions. In general, the spatial part is composed of atomic orbitals. While this prescription does not lead to a single-determinant wave function, it is in accordance with the antisymmetry constraint imposed by the Pauli principle. One concludes that the valence bond solution ψ_{VB} can be represented as a linear combination of determinants. Thus, one may write in analogy to Eq. (1.92):

$$\psi_{\mathrm{VB}} = \sum_{k=0}^{N_{\mathrm{C}}} c_k \tilde{\psi}_k, \tag{1.95}$$

where the $\{\tilde{\psi}_k\}$ are a set of determinants. However, in contrast to the $\{\psi_k\}$ of Eq. (1.92), these are built of non-orthogonal atomic rather than orthogonal molecular orbitals. In this sense, valence bond theory can be regarded as complementary to molecular orbital theory, as much as the concept of a localized electron pair complements that of a canonical orbital which may be spread over the entire molecule. The *generalized valence bond theory* (GVB, [45]), which allows for self-consistent variational determination of the non-orthogonal valence bond orbitals, may be understood as a version of MCSCF theory.

1.5.2 *Many-body perturbation theory*

In numerous systems, electron correlation effects can be understood as minor corrections. In this case, $E_{co} \ll E_0$, and the partition (1.65) may be used which subdivides the Hamiltonian $\hat{H}_{el}$ into a reference Hamiltonian $\hat{H}_0$ as a zeroth-order term and a perturbation operator $\hat{H}_{el} - \hat{H}_0 \equiv \hat{V}$. If this subdivision is made, as pointed out earlier (see Eqs. (1.40), (1.66), and (1.68)), the zeroth approximation to E_{el}, $E_{el}^{(0)}$, is the sum of the orbital energies and thus different from the Hartree–Fock energy. Adopting $\hat{V}$ as perturbation term and making reference to the expansion (1.67), we conclude that

$$E_{co} = E_{el} - E_0 = E_{el} - E_{el}^{(0)} - E_{el}^{(1)}. \tag{1.96}$$

Therefore, the correlation effect is contained in the perturbation terms of second or higher order. In the following, we will focus on $E_{el}^{(2)}$ as the leading order term of the correlation energy. The corrections to both the wave function and its energy eigenvalue may be derived from the customary Rayleigh–Schrödinger perturbation theory where one includes the effect of a small perturbation $\hat{V}$ on a reference system characterized by a Hamiltonian $\hat{H}_0$ by considering a full Hamiltonian of the form $\hat{H}_0 + \lambda\hat{V}$, expanding both the wave functions and the eigenvalues with respect to the real parameter λ. The reference Hamiltonian is here the zeroth-order Hamiltonian $\hat{H}_0$ specified in Section 1.3.3. Its eigenfunctions are used to construct a complete basis for further development. This basis is given by the set containing the Hartree–Fock ground state determinant besides the excitations ψ_j^s, ψ_{jk}^{st} etc., according to

$$\hat{H}_0\psi_i = E_{el,i}^{(0)}\psi_i \tag{1.97}$$

with $\psi_i = \psi_0, \psi_i^s, \psi_{ik}^{st}$, etc., and $E_{el,0}^{(0)} \equiv E_{el}^{(0)}$. The eigenvalues $E_{el,i}^{(0)}$ are readily found. As an example, we consider the double substitution ψ_{ik}^{st},

$$\hat{H}_0\psi_{ik}^{st} = [E_{el,0}^{(0)} - (\varepsilon_i + \varepsilon_k - \varepsilon_s - \varepsilon_t)]\psi_{ik}^{st}, \tag{1.98}$$

which follows immediately from the definition of the zeroth-order energy eigenvalue as the sum over the energies of the occupied orbitals. To derive $E_{el}^{(2)}$, we employ the expression for the second-order energy correction,

$$E_{el}^{(2)} = \sum_{i=1} \frac{|\langle\psi_0|\hat{V}|\psi_i\rangle|^2}{E_{el,0}^{(0)} - E_{el,i}^{(0)}}, \tag{1.99}$$

where the sum runs over all basis vectors i excluding the ground state, ψ_0. Evaluating the scalar product in the numerator of Eq. (1.99) might appear to be a daunting task. It is, however, greatly simplified by the finding that the only contributing ψ_i are the doubly excited states ψ_{jk}^{st}. In fact, it is easily concluded from the two-particle character of $\hat{V}$ that this operator does not couple ψ_0 with threefold or higher substitutions (see Exercise 1.4).

Exercise 1.4. Use the explicit form of $\hat{V}$ and the orthonormality of the basis vectors comprised of the ground state ψ_0 and the excited determinants (ψ_j^s, $\psi_{jk}^{st}, \ldots$) to prove that the scalar product $\langle\psi_0|\hat{V}|\psi_i\rangle$ assumes non-zero values for maximally twofold substitutions.

What about single substitutions? Let us analyze the special case $\psi_i = \psi_j^s$ by examining the scalar product $\langle\psi_0|\hat{V}|\psi_j^s\rangle$. It may be written as

$$\langle\psi_0|\hat{V}|\psi_j^s\rangle = \langle\psi_0|(\hat{H}_{\text{el}} - \hat{H}_0)|\psi_j^s\rangle = \langle\psi_0|\hat{H}_{\text{el}}|\psi_j^s\rangle - \langle\psi_0|\hat{H}_0|\psi_j^s\rangle. \tag{1.100}$$

The second term on the right-hand side of Eq. (1.100) clearly reduces to zero since ψ_j^s is an eigenfunction of the Hartree–Fock Hamiltonian. With respect to the first term, we observe that

$$\langle\psi_0|\hat{H}_{\text{el}}|\psi_j^s\rangle = \langle\zeta_j|\hat{h}|\zeta_s\rangle + \sum_i (js|ii) - (ji|si), \tag{1.101}$$

where we have utilized the nomenclature introduced by Eq. (1.89) for the two particle matrix elements involving the Coulomb repulsion operator $\frac{1}{r_{12}}$. The indices i, j, s label spin orbitals.

Exercise 1.5. Prove identity (1.101).

By Eq. (1.72), this result equals the Fock operator matrix element $\langle\zeta_j|\hat{F}|\zeta_s\rangle$. Since the index s refers to a virtual and j to an occupied orbital, this latter expression vanishes. Thus,

$$\langle\psi_0|\hat{V}|\psi_j^s\rangle = 0, \tag{1.102}$$

which demonstrates that $\hat{V}$ does not mix the Hartree–Fock ground state with first-order substitutions. The observation that the electronic Hamiltonian $\hat{H}_{\text{el}}$ does not couple the Hartree–Fock reference determinant to any single excitation arising from this state is the content of the *Brillouin theorem*. We add a caveat: although Eq. (1.102) excludes any direct mixing of the Hartree–Fock ground state with single substitutions, the latter still can exert indirect influence on the ground state since they couple to double substitutions which in turn interact with the ground state.

Similar considerations as presented in the previous paragraph can be employed to evaluate the scalar product $\langle\psi_0|\hat{V}|\psi_{jk}^{st}\rangle$ and thus to derive an explicit form of $E_{\rm el}^{(2)}$. Using once more the definition $\hat{\hat{V}} = \hat{H}_{\rm el} - \hat{H}_0$, we realize that neither the operator $\hat{H}_0$ nor the one-body components $\hat{h}_i$ of $\hat{H}_{\rm el}$ ($\hat{H}_{\rm el} = \sum_{\rm n} \hat{h}_{\rm n} + \frac{e^2}{4\pi\varepsilon_0}\sum_{m>n}\frac{1}{|\mathbf{x}_n-\mathbf{x}_m|}$) contribute to the scalar product. The remainder is summarized as

$$E_{\rm el}^{(2)} = \frac{\sum_{j<k,s<t} |\langle\psi_0|\sum_{m>n}\frac{e^2}{4\pi\varepsilon_0}\frac{1}{|\mathbf{x}_n-\mathbf{x}_m|}|\psi_{jk}^{st}\rangle|^2}{(\varepsilon_j+\varepsilon_k-\varepsilon_s-\varepsilon_t)}$$

$$= \frac{\sum_{j<k,s<t} |(js|kt)-(jt|ks)|^2}{(\varepsilon_j+\varepsilon_k-\varepsilon_s-\varepsilon_t)} \tag{1.103}$$

where the right-hand side exploits once more the compact two-electron integral notation of Eq. (1.89). The term $E_{\rm el}^{(2)}$ is frequently referred to as the energy correction due to electron correlation within the Møller–Plesset perturbation theory at second order (MP2). For quantum dynamics, the proper inclusion of electron correlation effects is often essential, particularly in the chemical region where the nuclear kinetic energy is small. If the dynamic treatment is based on potential energy surfaces, these are, in general, constructed by use of a correlated procedure. In these cases, a correlated formalism is applied prior to simulating the evolution of the nuclear degrees of freedom. *Ab initio* dynamics procedures, however, which include electronic degrees of freedom as dynamic variables, are challenged to incorporate the electron correlation effect directly into the dynamical scheme.

1.6 Excited Electronic States

The availability of excited electronic states is a prerequisite for many of the dynamic simulations to be presented in this text. This holds in particular for procedures that rely on potential energy surfaces. Including excited potential energy surfaces in the computation requires a sufficiently accurate understanding and numerical representation of the underlying electronic states. *On-the-fly* methods do not involve previously constructed excited surfaces, but the respective state is generated along the trajectory of the moving system. In either case, the adequate treatment of electronic excitations is a crucial element of the theory.

The excited electronic states of relevance for most topics negotiated in this monograph are single-electron excitations of an N-electron system. Although not identical with the single substitutions introduced earlier (ψ_j^s, see Section 1.5.1), they can be expressed in terms of these states. Formally, a basis capable of representing single excitations can be generated by simultaneous application of fermionic annihilation and creation operators, labeled $\hat{a}_j$ and $\hat{a}_s^\dagger$, respectively, on a reference determinant which may be chosen as the Hartree–Fock ground state:

$$\begin{aligned}\hat{a}_s^\dagger \hat{a}_j \,\psi_0 &= \hat{a}_s^\dagger \hat{a}_j \;\det\left[\Phi_1, \ldots, \Phi_j, \ldots, \Phi_N\right] \\ &= \det\left[\Phi_1, \ldots, \Phi_s, \ldots \Phi_N\right] = \psi_j^s. \end{aligned} \tag{1.104}$$

Obviously, configurations involving n excited electrons may be produced by cooperative action of n annihilation and creation operator pairs on the same state of reference.

As far as *ab initio* approaches are concerned, the methodology used to generate ground states is essentially applicable to excited states as well. For instance, the techniques discussed in the previous sections of this chapter will yield an excited state if one imposes on the system a spin multiplicity that differs from that of the ground state.

In principle, excited states may be represented in a configuration interaction basis as described in Section 1.5.1. The task of deriving the energies and wave functions for the ground state as well as excited states of the considered system is then reduced to determining the eigenvalues and eigenfunctions, respectively, of the Hamiltonian matrix given by

$$\begin{pmatrix} \langle\psi_0|\hat{H}|\psi_0\rangle & \langle\psi_0|\hat{H}|\psi_1\rangle & . & . \\ \langle\psi_1|\hat{H}|\psi_0\rangle & . & . & . \\ . & . & . & . \\ . & . & . & . \end{pmatrix}, \tag{1.105}$$

where the index $i = 0, 1, \ldots$ labels a CI basis $\{\psi_i\}$ that includes double or higher substitutions. This treatment yields all states of interest concurrently. It is typically extended to the first few excited states, which can often be obtained with good accuracy, while the problem usually becomes intractable for a larger number of solutions.

As mentioned earlier, single substitutions within a ground state determinant provide often useful approximations to single excitations. Therefore the arguably most common way to obtain information about these states is by utilizing the single-excitation configuration interaction (CIS) method [46]. The excited state is realized by moving one electron from an occupied into a virtual orbital and thus represented by a state of the form ψ_j^s. Adopting this scheme does not compromise the quality of the ground state since, by the Brillouin theorem (1.102), single substitutions do not mix with the ground state determinant. Since, however, the substitutions are composed of molecular orbitals optimized for the ground state, the CIS model yields rather an estimate of the excited state than a systematic approximation. The respective energies, and therefore the energy gap of the studied species, tend to be too high.

Numerous systems involving the ground state as well as excited states require an MCSCF treatment for adequate representation. MCSCF theory poses a twofold optimization problem as both the orbital coefficients and the coefficients of the basis states are subjected to variation (see Section 1.5.1). A famous multiconfiguration case is the H_2 molecule at elongated bond length, where the correct dissociation behavior is obtained by the superposition of a covalent and an ionic solution, as pointed out earlier. For a more detailed motivation, see the following exercise.

Exercise 1.6. Show that in the dissociation limit, the H_2 molecule is adequately described by a superposition of two configurations with equal weights, namely a covalent and an ionic contribution, ψ_{cov} and ψ_{ion}, respectively. Here ψ_{cov} (ψ_{ion}) combines two spin orbitals of the form $(\phi_1(1s) + \phi_2(1s))\sigma$, $((\phi_1(1s) - \phi_2(1s))\sigma)$ with $\sigma = \alpha, \beta$.

An important application of the MCSCF principle to the challenge of determining excited electronic states is the *complete active space self-consistent field* (CASSCF) model [47] that achieves a significant improvement over the level of the CIS method. It is based on subdividing the entire single-electron orbital space into three groups of spatial molecular orbitals. These are the *core*, *active*, and *virtual* orbitals. Each core orbital is occupied by two electrons, each active orbital by two, one, or zero, and each virtual orbital is unoccupied. A CASSCF computation proceeds in the *active space* which consists of all Slater determinants that can be formed by distributing N_{act} electrons in all possible ways among the active orbitals. The number N_{act} is given by $N - N_{core}$ where N is the total number of electrons and N_{core} the number of core electrons. The most elementary information about a CASSCF calculation is thus given by a pair (M_{act}, N_{act}) if M_{act} stands for the number of active molecular orbitals. By the example of formaldehyde (COH_2), if one freezes the 1*s* orbitals of both C and O, the core includes the corresponding two molecular orbitals, both of A_1 symmetry. The remaining twelve electrons may be distributed over the sequence of five A_1, three B_1, and two B_2 orbitals, resulting in a (10, 12) CASSCF approach. This yields an active space comprising 11,148 Slater determinants. To reduce the very respectable size of this space, one imposes constraints on its basis states. These may be chosen as linear combinations of Slater determinants which are constrained by a certain symmetry requirement. It is common practice to define the basis in terms of *configuration state functions* (CSFs) or *spin-adapted configurations* instead of substitutions. The CSFs are defined as eigenfunctions of the $\hat{S}^2$ operator built from Slater determinants. Choosing a target state multiplicity and eliminating all functions that correspond to $\hat{S}^2$ eigenvalues incompatible with this choice can reduce the active space size very efficiently. Containing the set of all substitutions within the considered active space, the CASSCF method is size-consistent.

The MCSCF approach is to be used whenever the electronic state of interest is not definable in terms of a unique dominant configuration. This is the case for bond-breaking problems, as discussed earlier by the example of H_2. Likewise, radicals fall into this category as well as systems involving points of avoided crossing or conical intersections, as will be further detailed in Section 2.5. Excited states often require inclusion of multiple reference configurations since, in contrast to ground states, their energy separation from adjacent states tends to be small. In analogy to SCF theory, corrections to MCSCF solutions can be accomplished, as the reference configurations are augmented by single, double, or higher substitutions. This approach is known as *multireference configuration interaction* (MRCI). As demonstrated by the case of H_2 at elongated bond length, the static correlation problem can be efficiently addressed by MCSCF theory. Dynamic correlation, however, which is prevalent in the regime of small internuclear distances, is not contained in the MCSCF scheme. MRCI, involving a CI expansion for each reference determinant, has been shown to result in a strongly improved description of dynamic correlation [48]. This feature, in turn, leads to a better balance in the treatment of multiple states and in consequence to more accurate excitation energies.

A highly correlated non-variational single-reference procedure is the *coupled-cluster* method [49, 50]. It introduces *cluster operators* $\hat{T}_n, n = 1, 2, \ldots, N$, that act on an N-electron system, generating n-fold excitations from a single-determinant reference state ψ_0. The operator $\hat{T}_1$ is given by

$$\hat{T}_1 \psi_0 = \sum_{j,s} t_j^s \psi_j^s \tag{1.106}$$

with an analogous definition for $\hat{T}_n$. The *cluster amplitudes* t_j^s are obtained from orthogonality relations. This approach leads to the following form of the stationary Schrödinger equation:

$$\hat{H} \exp(\hat{T}) \psi_0 = E \exp(\hat{T}) \psi_0. \tag{1.107}$$

Here $\hat{T}$ stands for $\hat{T}_1 + \hat{T}_2 + \cdots + \hat{T}_N$, and $\exp(\hat{T})\psi_0$ may be shown to be the exact many-electron wave function of the system. The size consistency of the coupled-cluster method depends on the reference state ψ_0. If, for instance, the method is applied to H_2 close to the dissociation limit, size consistency will be found for a UHF, but not for an RHF wave function ψ_0.

For the treatment of excited electronic states (as much as ground states for radicals or systems on the verge of dissociation) the single-reference character of the coupled-cluster theory becomes a severe restriction. Multireference extensions of the formalism [51, 52] address this problem. Less time-consuming procedures are based on the *equation-of-motion* (EOM) variant of coupled-cluster theory [53, 54] which, as the theory itself, originated in the field of nuclear physics rather than quantum chemistry. This technique has been demonstrated to predict excited state energies with accuracies on the order of 0.1 eV (see e.g. [55]).

Various other techniques not discussed in the present context have been successful in describing ground states with high precision and generating excited states by forcing the respective wave functions to be orthogonal to the ground state. This has been done, for instance, by use of various *quantum Monte Carlo* methods [56] which implement statistical integration schemes.

Alternative methods of treating electronic excitations are provided by *density functional theory*, as surveyed in Chapter 16. Some of the related strategies will be summarized in Section 16.2.

1.7 Appendix: The Functional Derivative

Introducing the functional derivative, we ask for a suitable generalization of the familiar prescription for differentiating a function f with respect to a one-dimensional variable, namely

$$f'(x) \equiv \lim_{\varepsilon \to 0} \frac{f(x+\varepsilon) - f(x)}{\varepsilon}. \tag{1.108}$$

Functional differentiation involves deriving a function with respect to another function as opposed to some variable x. As an intermediate step, we consider a function $f(\mathbf{x})$, where $\mathbf{x}$ is a vector with N components. Extending Eq. (1.108) to the case of partial differentiation of a function $F(\mathbf{x})$ with respect to a component x_n, we write

$$\frac{\partial F(\mathbf{x})}{\partial x_n} = \lim_{\varepsilon \to 0} \frac{F(\mathbf{x} + \varepsilon \mathbf{e}_n) - F(\mathbf{x})}{\varepsilon}, \tag{1.109}$$

with $\mathbf{e}_n$ as the nth vector of the orthonormal Cartesian basis $\mathbf{e}_1, \mathbf{e}_2, \ldots, \mathbf{e}_\mathrm{N}$. In the following step we replace the discrete set of variables $\{x_1, x_2, \ldots, x_N\}$ with a continuous set, defined by the function $f(x)$ in the sense that $f(x_i)$ is an argument of the *functional* F for any arbitrary choice of x_i. To accomplish a definition of the functional derivative in analogy to the partial derivative (1.109), a basis is needed for representation of $f(x)$. Such a basis is given by the continuous set $\{.., \delta(x - x'), \ldots\}$. With these premises, the desired extension of Eq. (1.109) for the continuous case becomes

$$\frac{\delta F[f]}{\delta f(x')} \equiv \lim_{\varepsilon \to 0} \frac{F[f(x) + \varepsilon \delta(x - x')] - F[f(x)]}{\varepsilon}. \tag{1.110}$$

Using this definition, the second functional derivative is introduced by the formula

$$\frac{\delta^2 F[f]}{\delta f(x_2)\delta f(x_1)} = \lim_{\varepsilon_{1,2}\to 0} \frac{1}{\varepsilon_1 \varepsilon_2}\{F[f(x) + \varepsilon_1 \delta(x - x_1) + \varepsilon_2 \delta(x - x_2)] - F[f(x) + \varepsilon_1 \delta(x - x_1)] - F[f(x) + \varepsilon_2 \delta(x - x_2)] + F[f(x)]\}. \tag{1.111}$$

Analogous prescriptions yield the higher derivatives. Once it is possible to construct functional derivatives of arbitrary order, a functional $F[f]$ may be Taylor-expanded according to

$$F[f] = F[0] + \int dx_1 \frac{\delta F[f]}{\delta f(x_1)}|_{f=0} f(x_1) + \frac{1}{2}\int dx_1 dx_2 \frac{\delta^2 F[f]}{\delta f(x_1)\delta f(x_2)}|_{f=0} f(x_1) f(x_2) + \cdots. \tag{1.112}$$

Readers who resent the occurrence of integrals in Eq. (1.112) are once more reminded of multidimensional calculus. For instance, the Taylor expansion of the function $f(x_1, x_2)$ contains the scalar product $\frac{\partial f}{\partial \mathbf{x}}|_{\mathbf{x}=0} \cdot \mathbf{x}$, with $\mathbf{x} = (x_1, x_2)$, as first-order term. To generalize this expression for functional derivatives, the generalization of the scalar product for functions is needed, which involves integrating products of functions. This implies also that for functional derivatives, the equivalent of the standard chain differentiation rule is given by the following:

$$\frac{\delta F[f[g]]}{\delta g(x)} = \int dy \frac{\delta F[f]}{\delta f(y)}|_{f=f[g]} \frac{\delta f(y)[g]}{\delta g(x)}. \tag{1.113}$$

To conclude, the reader may perform the following elementary exercises which accentuate the parallels and dissimilarities between regular and functional differentiation.

Exercise 1.7. Use the definition (1.110) of the functional derivative to establish the following relations:

$$\frac{\delta I[\psi]}{\delta \psi(x)} = 2\psi(x), \quad \text{if } I[\psi] = \int dx\, \psi^2(x); \tag{1.114}$$

$$\frac{\delta\psi(x')}{\delta\psi(x)} = 2\delta(x - x'); \tag{1.115}$$

$$\frac{\delta\langle V_{\text{ext}}\rangle[\rho]}{\delta\rho(\mathbf{x})} = \hat{V}_{ext}(\mathbf{x}); \tag{1.116}$$

with $\langle V_{\text{ext}}\rangle = \int d\mathbf{x}\ \rho(\mathbf{x})V_{\text{ext}}(\mathbf{x})$ as the expectation value of the external potential $V_{\text{ext}}(\mathbf{x})$. The last example is of relevance for density functional theory, as treated in Chapter 16.

Chapter 2

The Adiabatic and the Diabatic Representation

The *ab initio* methods outlined in the previous chapter all rest on a common assumption: that of perfect separability of nuclear and electronic motion. The discussed electronic structure theories contain the nuclear degrees of freedom exclusively as classical parameters that define the geometry of the molecule. For a large class of problems, this methodological limitation is justified. Chemical processes are usually confined to a single potential energy surface, and the time scales of nuclear and electronic evolution tend to be widely separated. If this premise holds, the *fast* electrons are unaffected by the motion of the *slow* nuclei which can be considered static at any given moment. This is the intuitive content of the *adiabatic* or *Born–Oppenheimer approximation*. We characterize this approach in Sections 2.1–2.3 and trace out its domain of validity.

In the opposite dynamic extreme, the nuclei move much more rapidly than the electrons. This is the *diabatic* case which is of less relevance to chemistry than the adiabatic alternative, but prevails in various special reactions, such as electron transfer through potential energy barriers where the active electronic degrees of freedom vary at a substantially slower rate than the nuclear subsystem (see Chapter 20). If this condition is realized, a *diabatic approximation* may be employed as the counterpart of the adiabatic approximation.

It is not difficult to identify situations where the neglect of the coupling between electrons and nuclei is not warranted any longer. The mentioned approximations break down if the time scales of electronic and nuclear processes become similar, and therefore the dynamics of the electronic system ceases to be unanimously adiabatic or diabatic. This occurs in cases of sufficiently fast nuclear motion, as is often realized for molecular scattering events in the physical regime. Another threat to the adiabatic model arises from the topography of the potential energy landscape. Traditional chemistry, for the largest part, deals with reactions that are well described as proceeding on the lowest or ground state potential energy surface of the studied chemical system. They typically take place in a regime well separated from the first excited state. This condition, however, cannot be taken for granted. Recent theoretical chemistry has paid much attention to regions where the energy difference between adjacent surfaces is small, and specifically to intersections between these surfaces. These configurations have turned out to be much more common than previously believed. Far from being pathological exceptions, they are regular features of any polyatomic species [57]. It is plausible that close to the crossing or near-crossing point of potential energy surfaces, transitions between these surfaces are likely, and restriction to a single surface fails to capture the dynamics of the system.

Sections 2.4 and 2.5 deal largely with the conceptual and mathematical tools needed to address *nonadiabatic* problems where the cumbersome twofold negation indicates that the considered molecular evolution unfolds on coupled potential energy surfaces. We will discuss both the adiabatic and the diabatic representation of a molecular system. Both are formally equivalent but, due to the unavailability of a complete electronic basis, of different relevance from a practical point of view. The phenomena of conical intersections and avoided crossings will be explored in Section 2.5. In particular, it will be shown that a coincidence between two adiabatic potential energy surfaces affects the dynamics of a system even if its evolution proceeds entirely on one of the surfaces, an effect that is related to the appearance of a *geometric phase* in the presence of degeneracies. We conclude this chapter by summarizing procedures to locate the *seam*, the set of all points that maintain degeneracy between two or more potential energy surfaces up to linear order in the nuclear coordinates.

2.1 The Born–Oppenheimer Approximation

Generalizing the approach (1.2), we write the stationary non-relativistic Schrödinger equation for the molecular problem as

$$\hat{H}\Psi(\mathbf{x}, \mathbf{X}) = E\Psi(\mathbf{x}, \mathbf{X}). \tag{2.1}$$

The eigenvalue E denotes the total molecular energy and the eigenfunction $\Psi(\mathbf{x}, \mathbf{X})$ a bound state solution. The function $\Psi(\mathbf{x}, \mathbf{X})$ involves electronic and nuclear degrees of freedom, as indicated by the position variables $\mathbf{x}$ and $\mathbf{X}$, respectively. Initially, $\mathbf{x}$ and $\mathbf{X}$ are understood as Cartesian. The transition from Cartesian coordinates to *normal modes* will be outlined in the following section.

The Hamiltonian $\hat{H}$ is partitioned into three contributions:

$$\hat{H} = \hat{T}_{\mathrm{N}} + \hat{T}_{\mathrm{el}} + U(\mathbf{x}, \mathbf{X}), \tag{2.2}$$

where the operator $\hat{T}_{\mathrm{N}} = -\sum_i \frac{\hbar^2}{2M_i}\frac{\partial^2}{\partial \mathbf{X}_i^2}$ stands for the kinetic energy of the nuclei, $\hat{T}_{\mathrm{el}} = -\sum_k \frac{\hbar^2}{2m}\frac{\partial^2}{\partial \mathbf{x}_k^2}$ for that of the electrons, and the multiplicative operator U for the potential energy of the nuclei and the electrons. Introducing the electronic Hamiltonian as the sum of the contributions to H that contain the electronic coordinates $\mathbf{x}$, we write

$$\hat{H}_{\mathrm{el}} = \hat{T}_{\mathrm{el}} + U_{\mathrm{el}}(\mathbf{x}, \mathbf{X}). \tag{2.3}$$

The potential energy U_{el} contains two components, the interelectronic repulsion $\hat{U}_{\mathrm{ee}}$ as well as the attraction between the nuclei and the electrons, U_{eN}.

$$U_{\mathrm{el}}(\mathbf{x}, \mathbf{X}) = U_{ee}(\mathbf{x}) + U_{eN}(\mathbf{x}, \mathbf{X}). \tag{2.4}$$

The *adiabatic representation* expresses the overall molecular wave function as a sum over wave function products according to

$$\Psi(\mathbf{x}, \mathbf{X}) = \sum_n \chi_n(\mathbf{X})\psi_n(\mathbf{x}, \mathbf{X}). \tag{2.5}$$

This ansatz may be viewed as an expansion of the molecular wave function $\Psi(\mathbf{x}, \mathbf{X})$ into a basis given by the set $\psi_n(\mathbf{x}, \mathbf{X})$ and with coefficients $\chi_n(\mathbf{X})$. The wave functions $\psi_n(\mathbf{x}, \mathbf{X})$ are parametrically dependent on the nuclear coordinates. They are understood as eigenfunctions of $\hat{H}_{\mathrm{el}}$:

$$\hat{H}_{\mathrm{el}}\psi_n(\mathbf{x}, \mathbf{X}) = E_{el,n}\psi_n(\mathbf{x}, \mathbf{X}). \tag{2.6}$$

They also form a *complete basis of electronic states*[1] for any choice of the parameter $\mathbf{X}$:

$$\sum_n \psi_n^*(\mathbf{x}', \mathbf{X})\psi_n(\mathbf{x}, \mathbf{X}) = \delta(\mathbf{x}' - \mathbf{x}). \tag{2.7}$$

[1]Here and in the following, we assume that the *bound states* that define the discrete spectrum of a molecule form a complete basis. For inclusion of the continuous states, see Eq. (3.1).

In the present section, we will not address the problem of degenerate electronic eigenvalues and assume that the basis $\psi_n(\mathbf{x},\mathbf{X})$ consists of electronic singlet states. Likewise, we define the coefficients $\chi_m(\mathbf{X})$ as nuclear wave functions. No orthonormality constraint is imposed on the set $\{\chi_m\}$. The integral over the square magnitudes of the functions χ_m, $\int d\mathbf{X}\chi_m^*\chi_m$, is interpreted as the population of the mth electronic state.

We proceed by inserting Eq. (2.5) into the stationary Schrödinger equation (2.1). Multiplying the result from the left by $\psi_n^*(\mathbf{x},\mathbf{X})$ and integrating over $\mathbf{x}$ leaves us with the following *stationary Schrödinger equation for the nuclear motion*:

$$(\hat{T}_{\rm N} + E_{{\rm el},n} + U_{\rm N})\chi_m(\mathbf{X}) - \sum_n \sum_i \frac{\hbar^2}{2M_i} \left(2\mathbf{A}_{mn,i} \cdot \frac{\partial}{\partial \mathbf{X}_i} + B_{mn,i}\right) \chi_n(\mathbf{X}) = E\chi_m(\mathbf{X}), \qquad (2.8)$$

where $U_{\rm N}$ stands for the purely nuclear interaction potential

$$U_{\rm N}(\mathbf{X}) = \frac{1}{2}\frac{1}{4\pi\varepsilon_0}\sum_{k\neq m}\frac{Z_k Z_m e^2}{|\mathbf{X}_k - \mathbf{X}_m|}. \qquad (2.9)$$

and the arrays $\mathbf{A}_{mn,i}$ and $B_{mn,i}$ are defined as

$$\mathbf{A}_{mn,i} = \left\langle \psi_m \middle| \frac{\partial}{\partial \mathbf{X}_i} \middle| \psi_n \right\rangle \qquad (2.10)$$

and

$$B_{mn,i} = \left\langle \psi_m \middle| \frac{\partial^2}{\partial \mathbf{X}_i^2} \middle| \psi_n \right\rangle. \qquad (2.11)$$

The scalar product in Eqs. (2.10) and (2.11) involves integration over the electronic coordinates. These expressions determine the interaction between nuclear and electronic degrees of freedom, or the "friction" between electrons and nuclei in a given problem. It should be noted that the *derivative coupling matrix* $\mathbf{A}_{mn,i}$ is anti-Hermitian. This implies that all of its diagonal elements are zero if real-valued wave functions are used. It therefore induces transitions between electronic states ψ_n and ψ_m, with $n \neq m$.

The two matrices (2.10) and (2.11) may be combined into the *nonadiabaticity*, an operator matrix $\hat{\Lambda}_{mn}(\mathbf{X})$ defined as

$$\hat{\Lambda}_{mn}(\mathbf{X}) = -\sum_i \frac{\hbar^2}{M_i}\left(\mathbf{A}_{mn,i}\cdot\frac{\partial}{\partial \mathbf{X}_i} + \frac{1}{2}B_{mn,i}\right). \tag{2.12}$$

The dimension of $\hat{\mathbf{\Lambda}}$ is $N \times N$ if N basis functions ψ_n are included. In the limit of completeness, $N \to \infty$. If the coupling between different electronic states is small enough to be neglected, the nonadiabaticity reduces to the diagonal elements of $B_{mn,i}$ since $\hat{\Lambda}_{mm}(\mathbf{X}) = -\sum_i \frac{\hbar^2}{2M_i}B_{mm,i}$. The *adiabatic approximation* (or *Born–Huang approximation*) consists in the neglect of all non-diagonal elements of $\hat{\mathbf{\Lambda}}$. This approach eliminates any coupling terms between electronic wave functions with different labels n in the stationary Schrödinger equation (2.8) and thus implies a reduction of the sum (2.5) to a product of a nuclear and an electronic contribution [13] according to

$$\Psi_m(\mathbf{x},\mathbf{X}) = \chi_m(\mathbf{X})\psi_m(\mathbf{x},\mathbf{X}) \tag{2.13}$$

for the electronic state with label m. With this ansatz, a separation of the nuclear and the electronic contribution to the wave function is achieved. The wave function (2.13) is adequate in cases where the electrons adjust instantaneously to the nuclear motions, and thus the electronic system behaves at any given time as if the nuclei are fixed in their momentary positions. The adiabatic approximation is frequently called the *Born–Oppenheimer approximation*, and Eq. (2.13) the *Born–Oppenheimer wave function*. Other authors define the Born–Oppenheimer approximation by the complete neglect of $\hat{\mathbf{\Lambda}}$.

We will now use the factorized form (2.13) of the wave function to derive separate statements about the electronic and the nuclear system of the considered molecule. Adopting the product function (2.13) as solution, we write the stationary Schrödinger equation in the following form:

$$(\hat{H}_{\text{el}} + \hat{T}_{\text{N}} + U_{\text{N}}(\mathbf{X}))\chi_m(\mathbf{X})\psi_m(\mathbf{x},\mathbf{X}) = E\chi_m(\mathbf{X})\psi_m(\mathbf{x},\mathbf{X}) \tag{2.14}$$

or

$$\frac{\hat{H}_{\text{el}}\psi_m(\mathbf{x},\mathbf{X})}{\psi_m(\mathbf{x},\mathbf{X})} = E - \frac{(\hat{T}_{\text{N}} + U_{\text{N}}(\mathbf{X}))\chi_m(\mathbf{X})}{\chi_m(\mathbf{X})}. \tag{2.15}$$

Since the right-hand side of this identity is independent of the electronic coordinate $\mathbf{x}$, the same must hold for the left-hand side which therefore depends only on the nuclear coordinate **X**. From this observation and Eq. (2.6), it follows that

$$\frac{\hat{H}_{\rm el}\psi_m(\mathbf{x},\mathbf{X})}{\psi_m(\mathbf{x},\mathbf{X})} = E_{{\rm el},m}(\mathbf{X}) \equiv V_m^{\rm A}(\mathbf{X}). \tag{2.16}$$

We name the quantity $V_m^{\rm A}(\mathbf{X})$ an *adiabatic potential*. One may justify the use of this term by inspecting the right-hand side of Eq. (2.15):

$$V_m^{\rm A}(\mathbf{X}) = E - \frac{(\hat{T}_{\rm N} + U_{\rm N}(\mathbf{X}))\chi_m(\mathbf{X})}{\chi_m(\mathbf{X})} \tag{2.17}$$

and therefore, with $V_m^{\rm APES}(\mathbf{X}) \equiv V_m^{\rm A}(\mathbf{X}) + U_{\rm N}(\mathbf{X})$,

$$(\hat{T}_{\rm N} + V_m^{\rm APES}(\mathbf{X}))\chi_m(\mathbf{X}) = E\chi_m(\mathbf{X}). \tag{2.18}$$

Within this approximation, the nuclear wave functions obey a stationary Schrödinger equation where the sum $V_m^{\rm APES}(\mathbf{X})$ plays the role of an *adiabatic potential energy surface* that determines the nuclear motion. If the nuclear potential is augmented by the diagonal elements of the nonadiabaticity (2.12), the definition of the potential energy surface changes to $V_m^{\rm APES}(\mathbf{X}) \equiv V_m(\mathbf{X}) + U_{\rm N}(\mathbf{X}) + \hat{\Lambda}_{mm}(\mathbf{X})$.

Adopting the adiabatic approximation, one splits the molecular problem into two sequentially solvable subproblems. The first one is posed by Eq. (2.6). It is purely electronic in nature, involving the evaluation of stationary electronic wave functions and the adiabatic potentials $V_m^{\rm A}(\mathbf{X})$, possibly by use of the *ab initio* methods surveyed in Chapter 1. Once this task is completed, the nuclear wave functions may be obtained from Eq. (2.18), as will be discussed in the following section. In Section 2.3, we will examine the conditions for the validity of this equation in greater detail.

2.2 Harmonic Vibrational Modes

The approximation (2.18) provides the basis for defining the vibrational modes of the nuclei. For this purpose, we arrange the Cartesian nuclear coordinates into an array $\{X_{1x}, X_{1y}, X_{1z}, X_{2x}, X_{2y}, X_{2z}, \ldots, X_{Mx}, X_{My}, X_{Mz}\}$, where the origin is identified with the center of mass. We further assume that the studied system adopts an equilibrium structure at the position $\mathbf{X}^0 = \{X_{1x}^0, X_{1y}^0, X_{1z}^0, \ldots, X_{Mx}^0, X_{My}^0, X_{Mz}^0\}$ and utilize the fact that the forces $\frac{\partial V^{\mathrm{APES}}(\mathbf{X})}{\partial \mathbf{X}}|_{\mathbf{X}=\mathbf{X}^0}$ are zero. Expansion of $V^{\mathrm{APES}}(\mathbf{X})$ around $\mathbf{X}^0$ up to second order yields

$$V^{\mathrm{APES}}(\mathbf{X}) \approx V^{\mathrm{APES}}(\mathbf{X}^0) + \frac{1}{2}\sum_{i,j=1}^{M}\sum_{p,q=x,y,z} W_{ip,jq}(X_{ip} - X_{ip}^0)(X_{jq} - X_{jq}^0) \tag{2.19}$$

with

$$W_{ip,jq} = \left.\frac{\partial^2 V^{\mathrm{APES}}(\mathbf{X})}{\partial X_{ip}\partial X_{jq}}\right|_{X_{ip}=X_{ip}^0, X_{jq}=X_{jq}^0} \tag{2.20}$$

as the *Hessian matrix*. The condition that the first derivative of V^{APES} vanishes at $\mathbf{X}^0$ is a necessary but not a sufficient criterion for the occurrence of a potential energy surface minimum at this point. Stability at $\mathbf{X}^0$ is guaranteed, however, if all eigenvalues of the Hessian matrix are non-negative. The harmonic approximation (2.19) is often adequate, although higher-order or *anharmonic* terms need to be included in some cases to represent the vibrational structure of the considered system with sufficient accuracy. Rotations are neglected in the present analysis as well as the possible coupling between vibrational and rotational degrees of freedom which is a source of anharmonicity.

The harmonic approximation to the effective nuclear Hamiltonian $\hat{H}_{\mathrm{N}}$ is obtained by inserting (2.19) into (2.18). To simplify the appearance of this expression, we introduce *mass-weighted coordinates* by making the assignment

$$R_{iq} \equiv \sqrt{2\frac{M_i}{\hbar^2}}(X_{iq} - X_{iq}^0), \tag{2.21}$$

which leads to

$$\hat{H}_{\mathrm{N}} = -\sum_{m=1}^{3M}\frac{\partial^2}{\partial R_m^2} + \frac{1}{2}\sum_{m=1}^{3M}\sum_{n=1}^{3M}\frac{\partial^2 V^{\mathrm{APES}}(\mathbf{R})}{\partial R_m \partial R_n}|_{\mathbf{R}=\mathbf{0}} R_m R_n, \tag{2.22}$$

where we have replaced the index pairs (ip) and (jq) with the collective indices m and n, both ranging from 1 to $3M$. Without restricting the general validity of the following arguments, the constant $V^{\text{APES}}(\mathbf{X}^0)$ in expression (2.19) has been set equal to zero. The complexity of the corresponding eigenvalue equation (2.18) is further reduced by eliminating the non-diagonal terms from the Hessian matrix in the **R** coordinates, $\mathbf{W}'$. This operation generates the *vibrational modes* of the problem. We thus define

$$\frac{1}{2}\frac{\partial^2 V^{\text{APES}}(\mathbf{R})}{\partial R_m \partial R_n}\bigg|_{\mathbf{R}=\mathbf{0}} \equiv \mathbf{W}'_{mn} = (\mathbf{O}^\dagger \widetilde{\mathbf{W}} \mathbf{O})_{mn}. \tag{2.23}$$

The matrix $\widetilde{\mathbf{W}}$ is diagonal, comprising the eigenvalues of $\mathbf{W}'$, and the diagonalizing matrix **O** is orthonormal. The transformation described by **O** creates a new set of coordinates, $\widetilde{\mathbf{R}}$, which are linear combinations of the previous coordinates **R** according to

$$\widetilde{R}_i = \sum_{l=1}^{3M} O_{il} R_l. \tag{2.24}$$

With $\widetilde{\mathbf{R}}$, we have identified a system of *normal* or *internal coordinates* in which the Hessian matrix is diagonal. Relation (2.24) implies $O_{il} = \frac{\partial \widetilde{R}_i}{\partial R_l}$ which can be utilized to transform the operator of the nuclear kinetic energy. Specifically, we find

$$\begin{aligned}
\frac{\partial^2}{\partial R_m^2} &= \frac{\partial}{\partial R_m}\left[\sum_l \left(\frac{\partial \widetilde{R}_l}{\partial R_m}\right)\frac{\partial}{\partial \widetilde{R}_l}\right] \\
&= \frac{\partial}{\partial R_m}\left[\sum_l O_{lm}\frac{\partial}{\partial \widetilde{R}_l}\right] \\
&= \sum_l O_{lm}\sum_k \left(\frac{\partial \widetilde{R}_k}{\partial R_m}\right)\frac{\partial^2}{\partial \widetilde{R}_k \partial \widetilde{R}_l} \\
&= \sum_{l,k} O_{lm} O_{km}\frac{\partial^2}{\partial \widetilde{R}_k \partial \widetilde{R}_l}.
\end{aligned} \tag{2.25}$$

We use this result to evaluate the kinetic operator $-\sum_{m=1}^{3M}\frac{\partial^2}{\partial R_m^2}$, summing the last expression of Eq. (2.25) over the index m, and further exploiting the orthonormal property of the matrix **O**:

$$\sum_m O_{lm} O_{km} = \sum_m O^\dagger_{ml} O^\dagger_{mk} = \delta_{lk}. \tag{2.26}$$

The combination of Eqs. (2.25) and (2.26) yields the nuclear kinetic energy operator in the coordinates $\widetilde{\mathbf{R}}$. We find

$$-\sum_{m=1}^{3M} \frac{\partial^2}{\partial R_m^2} = -\sum_{k=1}^{3M} \frac{\partial^2}{\partial \widetilde{R}_k^2}, \tag{2.27}$$

and conclude that the form of the operator $\hat{T}_\mathrm{N}$ remains unaffected by the transition from the old to the new coordinates. Expressing the Hamiltonian $\hat{H}_\mathrm{N}$ in the system $\widetilde{\mathbf{R}}$, we obtain

$$\hat{H}_\mathrm{N} = \sum_{k=1}^{3M} -\frac{\partial^2}{\partial \widetilde{R}_k^2} + \widetilde{W}_{kk}\widetilde{R}_k^2. \tag{2.28}$$

The respective eigenvalue equation (2.18) is therefore separable in all coordinates $\widetilde{\mathbf{R}} = \{\widetilde{R}_1, \widetilde{R}_2, \ldots, \widetilde{R}_{3M}\}$. This feature is reflected by the vibrational wave function which factorizes into the product

$$\chi_m^j(\widetilde{\mathbf{R}}) = \chi_{m,1}^j(\widetilde{R}_1)\chi_{m,2}^j(\widetilde{R}_2)\ldots\chi_{m,3M}^j(\widetilde{R}_{3M}) \tag{2.29}$$

where each of the one-dimensional factor functions satisfies the equation

$$\left[-\frac{\partial^2}{\partial \widetilde{R}_k^2} + \widetilde{W}_{kk}\widetilde{R}_k^2\right]\chi_{m,k}^j(\widetilde{R}_k) = E_k\chi_{m,k}^j(\widetilde{R}_k). \tag{2.30}$$

We recognize in this identity the stationary Schrödinger equation for a one-dimensional harmonic oscillator (see, for instance, [58]). Therefore, we can infer directly the connection between the eigenvalues E_k and the force constants $\widetilde{W}_{kk}$, namely

$$E_k = (2n_k + 1)\sqrt{\widetilde{W}_{kk}}, \quad n_k = 0, 1, 2, \ldots. \tag{2.31}$$

Here, the integer n_k characterizes the state of the kth vibrational degree of freedom. How far beyond the ground state $n_k = 0$ this index can be carried in a meaningful way depends on the limits of the harmonic approximation in a given problem. The square roots of the Hessian eigenvalues $\widetilde{W}_{kk}$ are proportional to the vibration frequencies. These turn out to be real if the function V^{APES} has been expanded about a true minimum of the potential energy surface, while expansion about a saddle point yields one imaginary frequency. We mention that the mapping from the original Cartesian to the normal coordinates yields six zero-frequency modes for polyatomic molecules and five for diatoms. These refer to three translational and three or two rotational degrees of freedom, respectively.

2.3 Adiabatic and Diabatic Frames

Under what circumstances is the adiabatic approximation applicable? In terms of perturbation theory, one might interpret the nuclear kinetic energy as a perturbing operator. One will then require that the factorized form (2.13) of the molecular wave function emerge as a solution at a well-defined order of approximation with respect to a properly chosen expansion parameter. The latter should be small in the validity regime of the Born–Oppenheimer approximation. An intuitive criterion for this regime is given by the condition

$$\frac{\hbar\omega}{|E^0_{el,m} - E^0_{el,n}|} \ll 1. \tag{2.32}$$

In this expression, ω denotes a characteristic frequency of the nuclear system that emerges from the harmonic approach (2.31). The content of the criterion (2.32) is that the adiabatic approximation is justified whenever the separation of the adiabatic electronic energies at the equilibrium geometry of the molecule by far exceeds the energy associated with the vibrational motion of the nuclei. A premise for the derivation of the inequality (2.32) is that the vibrational spectrum (2.31) is discrete, i.e. the matrix **W** is positive definite.

A rough estimate of ω and of the averaged energy gap $\langle \Delta E_{\mathrm{el}} \rangle$ between two states of the electronic system leads to

$$\frac{\hbar\omega}{\langle \Delta E_{\mathrm{el}} \rangle} \approx \sqrt{\frac{m}{M_{\mathrm{N}}}}, \tag{2.33}$$

where m is the electronic and M_{N} a typical nuclear mass which could be chosen as the average mass of the nuclei present in the system. Relation (2.33) clarifies that the validity of the adiabatic approximation is ultimately rooted in the small magnitude of the ratio between the electronic and the nuclear mass [59]. In the following, we rationalize the estimate (2.33). First, we adopt the harmonic approach (2.28) and equate the nuclear energy eigenvalue E in Eq. (2.18) to $\hbar\sqrt{\frac{K}{M_{\mathrm{N}}}}$ where K stands for a harmonic spring constant. The magnitude of the electronic potential energy is estimated to be $\frac{e^2}{4\pi\varepsilon_0}\frac{1}{r_{\mathrm{el}}}$, where we assume predominantly electrostatic interaction, determined by a length scale r_{el} which characterizes the dimension of the molecule. We use this quantity to assess the order of the average electronic velocity $\langle v_{\mathrm{el}} \rangle$ as $\langle v_{\mathrm{el}} \rangle \approx \frac{\hbar}{m r_{\mathrm{el}}}$.

The latter finding in combination with the virial theorem applied to the Coulomb potential leads to $\frac{e^2}{4\pi\varepsilon_0}\frac{1}{r_{\rm el}} \approx \frac{\hbar^2}{mr_{\rm el}^2}$. As the nuclear displacement becomes comparable to the molecular size, the vibrational potential energy reaches the range of the electronic energy [60]. Therefore, we set $Kr_{\rm el}^2 \approx \frac{\hbar^2}{mr_{\rm el}^2}$, yielding as estimate for the spring constant

$$K \approx \frac{\hbar^2}{mr_{\rm el}^4}. \tag{2.34}$$

Finally, we use $\langle \Delta E_{\rm el} \rangle \Delta T_{\rm el} \approx \hbar$ to obtain an approximate expression for the averaged energy gap $\langle \Delta E_{\rm el} \rangle$ between adjacent electronic states, where $\Delta T_{\rm el}$ denotes the period for bound electronic motion. From $\Delta T_{\rm el} \approx \frac{r_{\rm el}}{\langle v_{\rm el} \rangle} \approx \frac{mr_{\rm el}^2}{\hbar}$, $\langle \Delta E_{\rm el} \rangle \approx \frac{\hbar^2}{mr_{\rm el}^2}$. This result in conjunction with relation (2.34) yields the desired order of the ratio between the nuclear vibrational energy and the averaged electronic energy gap as

$$\frac{\hbar\omega}{\langle \Delta E_{\rm el} \rangle} \approx \frac{\hbar\sqrt{\frac{K}{M_{\rm N}}}}{\langle \Delta E_{\rm el} \rangle} \approx \sqrt{\frac{m}{M_{\rm N}}}. \tag{2.35}$$

The square root of this ratio, $(\frac{m}{M_{\rm N}})^{\frac{1}{4}}$, has been identified with the perturbation parameter κ [13,61], making the term $\frac{\hbar\omega}{\langle \Delta E_{\rm el} \rangle}$ small in the sense of second order perturbation theory.

Exercise 2.1. Making a further order-of-magnitude estimate, we may relate the potential energy of the nuclei in their vibrational ground states to their total energy by

$$\frac{M_N\omega^2\delta^2}{2} \approx \frac{\hbar\omega}{2} \tag{2.36}$$

where δ is a typical displacement from the nuclear equilibrium position. Show: $\delta \approx (\frac{m}{M_N})^{\frac{1}{4}} r_{\rm el}$.

Acknowledging that the root mean square displacement from the equilibrium position $\mathbf{X}^0$ of the harmonic oscillator scales as κ (see Exercise 2.1, and also [60]), we stipulate $\mathbf{X} = \mathbf{X}^0 + \kappa\widetilde{\mathbf{X}}$. This defines a perturbation approach whose zeroth approximation involves a (fictitious) case of motionless nuclei fixed at their equilibrium positions. In this situation, the electronic eigenfunctions $\psi_m(\mathbf{x}, \mathbf{X})$ reduce to $\psi_m(\mathbf{x}, \mathbf{X}^0) \equiv \phi_m(\mathbf{x})$.

The perturbing interaction is given by the nuclear kinetic energy $\hat{T}_\mathrm{N} = \kappa^4 \hat{H}_1$ with $\hat{H}_1 = -\sum_i \frac{M_\mathrm{N}}{M_i} \frac{\hbar^2}{2m} \frac{\partial^2}{\partial \mathbf{X}_i^2}$. Considering the dependence of the coordinate $\mathbf{X}$ on the parameter κ, one realizes that $\hat{T}_\mathrm{N}$ involves an effect of second order, since $\hat{T}_\mathrm{N} = \kappa^4 \hat{H}_1 = \kappa^2 \hat{H}_1^{(2)}$ where $\hat{H}_1^{(2)} = -\sum_i \frac{M_\mathrm{N}}{M_i} \frac{\hbar^2}{2m} \frac{\partial^2}{\partial \widetilde{\mathbf{X}}_i^2}$. The substitution of $\mathbf{X}^0 + \kappa\widetilde{\mathbf{X}}$ for $\mathbf{X}$ implies that the potential energy $U(\mathbf{x}, \mathbf{X}) = U(\mathbf{x}, \mathbf{X}^0 + \kappa\widetilde{\mathbf{X}})$ (see Eq. (2.2)) also contributes to the perturbation.

The reader will find the details of this perturbation treatment in [61]. The zero-order equation for the mth wave function yields

$$\Psi_m^{(0)}(\mathbf{x}, \widetilde{\mathbf{X}}) = \chi_m^{(0)}(\widetilde{\mathbf{X}})\psi_m^{(0)}(\mathbf{x}). \tag{2.37}$$

At this level of approximation, the function $\chi_m^{(0)}(\widetilde{\mathbf{X}})$ is an undetermined factor, multiplying the electronic part which depends solely on the electronic coordinates. This complete separation of nuclear and electronic degrees of freedom is known as the *crude adiabatic approximation*. Carrying the perturbation approach to the second-order level, as demonstrated in [61], one arrives at an equation that involves the nuclear kinetic energy which, from our considerations before, contains κ^2 as a factor. This equation makes a statement about the zeroth-order nuclear wave function which is given by

$$(\hat{T}_\mathrm{N} + V_m^{(2)}(\widetilde{\mathbf{X}}))\chi_m^{(0)}(\widetilde{\mathbf{X}}) = E\chi_m^{(0)}(\widetilde{\mathbf{X}}). \tag{2.38}$$

Exercise 2.2. *Perturbation of the electronic motion by the nuclear kinetic energy.*

Expand the molecular Hamiltonian (2.2) with respect to the perturbation parameter κ. Do the same with the wave functions ψ_m and Ψ and their energy eigenvalues, where $\psi_m^{(0)}$ is the electronic state occupied in the unperturbed case. Use the stationary Schrödinger equation (2.1) and its counterpart for the electronic system (2.6) to derive equations for $\psi_m^{(i)}$ and $\Psi_m^{(i)}$ with $i = 0$, 1, 2. Show that the first-order solution $\Psi_m^{(1)}$ is given by

$$\Psi_m^{(1)}(\mathbf{x}, \widetilde{\mathbf{X}}) = \chi^{(0)}(\widetilde{\mathbf{X}})\psi_m^{(1)}(\mathbf{x}, \widetilde{\mathbf{X}}) + \chi^{(1)}(\widetilde{\mathbf{X}})\psi_m^{(0)}(\mathbf{x}). \tag{2.39}$$

The second-order expression $V_m^{(2)}(\widetilde{\mathbf{X}})$ in Eq. (2.38), is identical with the harmonic potential in Eq. (2.19). Therefore the factor $\chi_m^{(0)}$ turns out to be the vibrational nuclear wave function as obtained in the framework of the harmonic approximation introduced in the previous section. Continuing the perturbation development and neglecting all terms of third or higher order, we arrive at the wave function

$$\Psi_m(\mathbf{x}, \widetilde{\mathbf{X}}) = (\chi_m^{(0)}(\widetilde{\mathbf{X}}) + \kappa\chi_m^{(1)}(\widetilde{\mathbf{X}}) + \kappa^2\chi_m^{(2)}(\widetilde{\mathbf{X}}))\psi_m(\mathbf{x}, \widetilde{\mathbf{X}}), \tag{2.40}$$

which is of the Born–Oppenheimer form (2.13). In this case, the full parametric dependence of the electronic part on the nuclear coordinates is recovered, which was suppressed in the crude adiabatic approximation (2.37). The latter approach, however, can be extended to provide the full solution of the molecular problem, as one expands the molecular wave function according to

$$\Psi(\mathbf{x}, \mathbf{X}) = \sum_n \chi_n(\mathbf{X})\psi_n(\mathbf{x}, \mathbf{X}^0) = \sum_n \chi_n(\mathbf{X})\phi_n(\mathbf{x}), \tag{2.41}$$

equating the set $\{\psi_n(\mathbf{x}, \mathbf{X}^0)\}$ with the set $\{\phi_n(\mathbf{x})\}$. Evidently, Eq. (2.37) may be viewed as a special case of Eq. (2.41) which defines the so-called *diabatic representation*. By virtue of the assumed completeness of both basis sets, $\phi_n(\mathbf{x})$ and $\psi_n(\mathbf{x}, \mathbf{X})$, Eqs. (2.5) and (2.41) are equivalent in the sense that a unitary transformation can be found that maps the one into the other.

Proceeding as in the case of the adiabatic representation (2.5), i.e. inserting Eq. (2.41) into the stationary Schrödinger equation and bracketing the resulting equation with $\psi_m^*(\mathbf{x}, \mathbf{X}^0) = \phi_m^*(\mathbf{x})$ from the left, yields a considerably simpler result than Eq. (2.8) since all derivatives of the electronic eigenfunctions with respect to the nuclear coordinates vanish. However, the potential energy matrix $\mathbf{V}$, which is diagonal in the adiabatic representation, acquires now non-diagonal elements for general values of $\mathbf{X}$. More specifically, we obtain

$$V_{mn}(\mathbf{X}) = \langle\phi_m|\hat{T}_{\mathrm{el}}(\mathbf{x}) + U_{\mathrm{ee}}(\mathbf{x}) + U_{\mathrm{eN}}(\mathbf{x}, \mathbf{X}) + U_{\mathrm{N}}(\mathbf{X})|\phi_n\rangle. \tag{2.42}$$

In zero-order perturbation theory with respect to the parameter κ, however, V_{mn} is diagonal, with the elements V_{nn} given by the potential energy surface values $V_n^{\mathrm{APES}}(\mathbf{X}^0)$.

The full diabatic representation is obtained by sacrificing the constraint that $\mathbf{X}^0$ refers to an equilibrium position of the nuclei, and thus the harmonic approximation for the nuclear motion. The reference configuration chosen for $\mathbf{X}^0$ may be dictated by computational practicability rather than the actual physical situation at hand. Consequently, the full molecular wave function is expressed as

$$\Psi(\mathbf{x}, \mathbf{X}) = \sum_n \chi_{0,n}(\mathbf{X})\psi_n(\mathbf{x}, \mathbf{X}_0). \tag{2.43}$$

The functions $\chi_{0,n}(\mathbf{X})$ denote a set of nuclear solutions that refers to the configuration $\mathbf{X}^0$.

2.3.1 *The diabatic approximation*

On the basis of Eq. (2.43) we will investigate the nature of diabatic coupling in further detail and contrast the adiabatic with the diabatic approximation. By adding and subtracting $U_{\text{eN}}(\mathbf{x}, \mathbf{X}^0) + U_{\text{N}}(\mathbf{X}^0)$ to and from the operators in the matrix element (2.42), we take advantage of the definition of the set $\{\phi_n\}$ as orthonormal eigenfunctions of $\hat{H}_{\text{el}}$ at $\mathbf{X} = \mathbf{X}^0$:

$$\begin{aligned} V_{mn}(\mathbf{X}) = {} & V_m^{\text{APES}}(\mathbf{X}^0)\delta_{mn} \\ & + \langle \phi_m | U_{\text{eN}}(\mathbf{x}, \mathbf{X}) + U_{\text{N}}(\mathbf{X}) - U_{\text{eN}}(\mathbf{x}, \mathbf{X}^0) - U_{\text{N}}(\mathbf{X}^0) | \phi_n \rangle. \end{aligned} \tag{2.44}$$

The second term in this sum couples electronic states, with $m \neq n$. The diagonal elements $V_{nn}(\mathbf{X})$ are termed *diabatic potentials*, the elements $V_{mn}(\mathbf{X}), m \neq n$, *diabatic coupling potentials*. It should be noted that the diabatic coupling terms appear in the Hamiltonian as local functions and are thus mathematically more easily tractable than the corresponding terms (2.10) and (2.11) in the adiabatic representation.

In cases where the diabatic coupling is weak enough to be neglected, the diabatic approximation applies. The nuclear wave functions $\chi_{0,n}(\mathbf{X})$ then obey the equation

$$(\hat{T}_{\text{N}} + V_m^{\text{D}}(\mathbf{X}))\chi_{0,m}(\mathbf{X}) = E\chi_{0,m}(\mathbf{X}), \tag{2.45}$$

where $V_m^{\text{D}}(\mathbf{X}) \equiv V_{mm}(\mathbf{X})$. These potentials are much more easily constructed than the adiabatic potentials $V_m^{\text{A}}(\mathbf{X})$. By definition of the wave functions ϕ_n, after all, evaluating the electronic states at the reference location $\mathbf{X}^0$ is sufficient to generate the $V_m^{\text{D}}(\mathbf{X})$ while they need to be known in an extended segment of the nuclear coordinate space in order to calculate the $V_m^{\text{A}}(\mathbf{X})$. The adiabatic approximation, however, is of greater use in physical chemistry than its diabatic equivalent. The reason for this asymmetry resides in the different coupling mechanisms that underlie the two representations. In the adiabatic case, the nuclear kinetic energy couples two electronic states according to expression (2.12). In the diabatic frame, the potential energy performs this function (see Eq. (2.44)). Consequently, a sufficiently small nuclear kinetic energy is among the validity conditions of the adiabatic approximation. The diabatic approximation, in contrast, requires that the electronic and nuclear subsystems be weakly coupled by a potential energy operator. The latter condition is realized if the nuclear motion is too fast for the electrons to adapt to it, such that they may be assumed to maintain a fixed nuclear reference configuration $\mathbf{X}^0$. In this extreme, the approach $\psi_m(\mathbf{x}, \mathbf{X}) = \psi_m(\mathbf{x}, \mathbf{X}^0) = \phi_m(\mathbf{x})$ for the electronic wave function is physically justified. On the grounds of these arguments, the adiabatic and diabatic approximations are opposed to each other in the sense that they apply to the limiting cases of low and high nuclear kinetic energy, respectively. Evidently, the former alternative is more appropriate for most situations of relevance to chemical dynamics.

2.4 Gauge Theoretical Form of the Nuclear Equation

In the adiabatic representation, the nuclear kinetic energy contribution assumes a highly complex appearance, while the potential energy is given by a diagonal matrix. The diabatic representation reverses these roles and casts the nuclear kinetic energy into diagonal form, at the expense of complicating the potential energy part of the Hamiltonian. This difference can be understood in geometric terms. Thus, the nuclear kinetic energy operator in Eq. (2.8) may be written in a more compact form, defining [62]

$$\mathbf{K}_i = -\mathbf{I}\frac{\partial}{\partial R_i^2} + \hat{\mathbf{\Lambda}}_i \tag{2.46}$$

and

$$\hat{\mathbf{\Lambda}}_i = -2\mathbf{A}_i\frac{\partial}{\partial R_i} - \mathbf{B}_i. \tag{2.47}$$

As in Eq. (2.12), $\mathbf{A}_i$ and $\mathbf{B}_i$ denote $N \times N$ matrices if N basis functions are considered, and $\mathbf{I}$ stands for the unit matrix in the space spanned by this basis. In an effort to reduce the nuclear kinetic component of Eq. (2.8) to its simplest possible structure, we utilize scaled coordinates $R_i = \sqrt{\frac{2M_i}{\hbar^2}}X_i$, as introduced before in the context of vibrational modes (see Eq. (2.21)) where the reference point $\mathbf{X^0}$ has been set equal to zero. The first derivative matrices $\mathbf{A}_i$ are modified accordingly, and the matrices of the second derivatives, $\mathbf{B}_i$, are expressed through the $\mathbf{A}_i$ as follows:

$$\mathbf{B}_i = \frac{\partial \mathbf{A}_i}{\partial R_i} + \mathbf{A}_i\mathbf{A}_i. \tag{2.48}$$

Exercise 2.3. Use the definitions of **A** and **B** to prove relation (2.48).

Summarizing Eqs. (2.46)–(2.48), we arrive at an alternative formulation of Eq. (2.8) that makes explicit the gauge invariance of the stationary Schrödinger equation in the adiabatic representation:

$$\left[-\sum_i \left(\mathbf{I}\frac{\partial}{\partial R_i} + \mathbf{A}_i\right)\left(\mathbf{I}\frac{\partial}{\partial R_i} + \mathbf{A}_i\right) + \mathbf{V} - E\mathbf{I}\right]\boldsymbol{\chi} = 0, \tag{2.49}$$

with $\boldsymbol{\chi}$ as the vector of vibrational wave functions and $\mathbf{V}$ as the diagonal potential energy matrix whose elements are potential energy surface functions $V_n^{\mathrm{APES}}(\mathbf{R})$. The operator $(\mathbf{I}\frac{\partial}{\partial R_i} + \mathbf{A}_i)$ is interpreted as a covariant derivative corresponding to the curvilinear character of the adiabatic electronic basis $\psi_n(\mathbf{x}, \mathbf{R})$ that exhibits parametric dependence on the set of variables $\{\mathbf{R}\}$, in contrast to the diabatic basis $\psi_n(\mathbf{x}, \mathbf{R}_0)$, and thus gives rise to the vector potential term $\mathbf{A}_i$.

For real-valued electronic states ψ_{n} the adiabatic approximation will not involve any covariant derivative since the non-diagonal elements of $\mathbf{A}_i$ are zero by definition, and the diagonal elements by the choice of real ψ_{n}. If, however, complex-valued electronic basis functions are admitted, the derivative coupling of the reference state with itself is, in general, not zero. The adiabatic pseudorotation of a triatomic molecule around a conical intersection, to be discussed later, will give an example of this phenomenon. Thus, if $\Psi_m(\mathbf{x}, \mathbf{R}) = \chi_m(\mathbf{R})\psi_m(\mathbf{x}, \mathbf{R})$ (see Eq. (2.13)), the identity (2.49) changes into

$$\left[-\sum_i \left(\frac{\partial}{\partial R_i} + A_i\right)\left(\frac{\partial}{\partial R_i} + A_i\right) + V_m^{\mathrm{APES}} - E\right]\chi_m = 0. \tag{2.50}$$

The elements A_i are the components of a vector $\mathbf{A} = \langle\psi_m(\mathbf{x}, \mathbf{R})|\frac{\partial}{\partial \mathbf{R}}|\psi_m(\mathbf{x}, \mathbf{R})\rangle$. Assume we have found a solution $\chi_m^{A=0}$ of Eq. (2.50) for vanishing vector potential **A**. The corresponding solution for non-zero vector potential can then be written as

$$\chi_m(\mathbf{R}) = \exp\left[- \int_{\mathbf{R}_0}^{\mathbf{R}} \sum_i A_i(\mathbf{R}')dR'_i \right] \chi_m^{A=0}(\mathbf{R}). \qquad (2.51)$$

This relation exemplifies a general phenomenon associated with gauge symmetries: the presence of a gauge field is manifested by the appearance of a phase factor in the wave function, involving an *adiabatic phase*

$$\varphi(\mathbf{R}_0, \mathbf{R}) \equiv i \int_{\mathbf{R}_0}^{\mathbf{R}} \sum_i A_i(\mathbf{R}')dR'_i, \qquad (2.52)$$

so that $\exp[-\int_{\mathbf{R}_0}^{\mathbf{R}} \sum_i A_i(\mathbf{R}')dR'_i] = \exp[i\varphi(\mathbf{R}_0, \mathbf{R})]$. Definition (2.52) exploits the entirely imaginary nature of the components A_i, thus ensuring that the phase φ is a real quantity. The adiabatic phase is accumulated along a path that leads from some initial location $\mathbf{R}_0$ to the point **R** at which the solution χ_m is evaluated. The implied reference to a path does not give rise to any difficulty as long as the vector potential is integrable in the considered segment of the nuclear coordinate space. This condition, however, is not always satisfied, as will be discussed in further detail in the following sections. Thus, we will see that singularities of the vector potential arise at points where two potential energy surfaces become degenerate. Further, we will comment on the adiabatic phase in the context of closed paths encircling the point of degeneracy in the nuclear coordinate space, the *geometric phase*.

The discussion will now be extended to more states than one, and it will be shown that a phase factor analogous to that defined by Eq. (2.52) determines the transition from the diabatic to the adiabatic frame. In this case, a line integral over the coupling elements of the derivative matrix **A** is to be computed, as we demonstrate in the following.

Although the diabatic and adiabatic *approximations* apply to opposite dynamical extremes, the adiabatic and diabatic *representations* are equivalent due to the completeness of the nuclear and electronic basis sets involved in both, as pointed out earlier. It therefore holds that

$$\Psi(\mathbf{x}, \mathbf{R}) = \sum_n \chi_{0,n}(\mathbf{R})\psi_n(\mathbf{x}, \mathbf{R}_0) = \sum_m \chi_m(\mathbf{R})\psi_m(\mathbf{x}, \mathbf{R}). \qquad (2.53)$$

We can construct the unitary matrix $\mathbf{W}$ that transforms one representation into the other by expressing the basis set $\{\chi_m\}$ in terms of $\{\chi_{0,n}\}$. We obtain

$$\chi_m(\mathbf{R}) = \sum_n W_{mn}(\mathbf{R}, \mathbf{R}_0)\chi_{0,n}(\mathbf{R}), \tag{2.54}$$

where

$$W_{mn}(\mathbf{R}, \mathbf{R}_0) = \langle \psi_m(\mathbf{x}, \mathbf{R}) | \psi_n(\mathbf{x}, \mathbf{R}_0) \rangle. \tag{2.55}$$

This relation, however, is of theoretical rather than practical significance, as it makes full knowledge of the adiabatic eigenfunctions $\psi_m(\mathbf{x}, \mathbf{R})$ a prerequisite for computing the transformation matrix elements. This task is greatly simplified by exploiting the diagonal form of the electronic Hamiltonian in the adiabatic basis. We can conclude that the non-diagonal potential energy matrix $\mathbf{V}$ (see Eq. (2.44)) is diagonalized by the transformation matrix $\mathbf{W}$ according to [7]

$$V_n(\mathbf{R})\delta_{nk} = \sum_{m,l} W^\dagger_{nm}(\mathbf{R}, \mathbf{R}_0) V_{ml}(\mathbf{R}) W_{lk}(\mathbf{R}, \mathbf{R}_0). \tag{2.56}$$

The diagonal elements on the left-hand side of this identity are the adiabatic potential energy surfaces $V_n^{\mathrm{APES}}(\mathbf{R})$, as introduced by Eq. (2.18). Using relation (2.56), one achieves the desired transition between the two frames by diagonalizing the diabatic potential energy matrix, which is more easily generated than its adiabatic counterpart. In this way, the less complex diabatic approach can be utilized to address nonadiabatic situations.

In the special case of two interacting electronic states $\mathbf{W}$ may be written as

$$\mathbf{W} = \begin{pmatrix} \cos\theta(\mathbf{R}) & \sin\theta(\mathbf{R}) \\ -\sin\theta(\mathbf{R}) & \cos\theta(\mathbf{R}) \end{pmatrix}. \tag{2.57}$$

To evaluate the angular coupling parameter $\theta(\mathbf{R})$, we consider the scalar product of the adiabatic basis functions, $\langle \psi_m(\mathbf{x}, \mathbf{R}) | \psi_n(\mathbf{x}, \mathbf{R}) \rangle = \delta_{mn}$. Using the relation $\psi_m(\mathbf{x}, \mathbf{R}) = \sum_n W_{mn}\psi_n(\mathbf{x}, \mathbf{R}_0)$, we observe that

$$\left\langle \psi_2(\mathbf{x}, \mathbf{R}) \left| \frac{\partial \psi_1(\mathbf{x}, \mathbf{R})}{\partial \mathbf{R}} \right.\right\rangle = -\left\langle \psi_1(\mathbf{x}, \mathbf{R}) \left| \frac{\partial \psi_2(\mathbf{x}, \mathbf{R})}{\partial \mathbf{R}} \right.\right\rangle = \frac{\partial \theta(\mathbf{R})}{\partial \mathbf{R}} \tag{2.58}$$

and therefore

$$\theta(\mathbf{R}) = \theta(\mathbf{R}_0) + \int_{\mathbf{R}_0}^{\mathbf{R}} d\mathbf{R}' \left\langle \psi_2(\mathbf{x}, \mathbf{R}') | \frac{\partial \psi_1(\mathbf{x}, \mathbf{R}')}{\partial \mathbf{R}'} \right\rangle, \tag{2.59}$$

stipulating that $\mathbf{W}$ reduces to the 2×2 unit matrix as $\mathbf{R} = \mathbf{R}_0$. Conditions for the existence of a non-trivial diabatic basis, i.e. one that is not strictly independent of the nuclear coordinate $\mathbf{R}$, are derived in Section 2.4*.

2.5 Avoided Crossings, Degeneracies, Conical Intersections

So far we have referred to mild-mannered systems with well-separated adiabatic potentials derived from electronic singlets. By the criterion (2.32), the Born–Oppenheimer approximation becomes endangered in regions characterized by close proximity of the zeroth order energies $E^0_{\mathrm{el},m}$ and $E^0_{\mathrm{el},n}$. The scheme breaks down in cases where two or more adiabatic potential energy surfaces coincide. In the following, we will focus on these situations and raise the question of their relevance to quantum molecular dynamics. Do singularities associated with degeneracies or quasi-degeneracies of adiabatic potentials have the status of isolated accidents, interesting curiosities rather than configurations of practical consequence for molecular dynamics? In the course of the subsequent discussion it will turn out that they are in fact a frequent sight in the landscapes of molecular potential energy surfaces, far from being rare anomalies, and that they are of fundamental importance for both the theory and the application of quantum molecular dynamics.

2.5.1 *Avoided crossings*

To investigate the limits of the adiabatic approximation quantitatively, we analyze the derivative coupling matrix **A** as given by formula (2.10). We will proceed by deriving both sides of eigenvalue equation (2.6) for the electronic system with respect to the nuclear coordinate X_i:

$$\frac{\partial}{\partial X_i}\hat{H}_{\mathrm{el}}\psi_n(\mathbf{x},\mathbf{X}) = \frac{\partial}{\partial X_i}V_n^{\mathrm{A}}(\mathbf{X})\psi_n(\mathbf{x},\mathbf{X}), \tag{2.60}$$

equating $E_{\mathrm{el},m}(\mathbf{X})$ with $V_m^{\mathrm{A}}(\mathbf{X})$ (see Eq. (2.16)). This yields

$$\frac{\partial \hat{H}_{\mathrm{el}}(\mathbf{x},\mathbf{X})}{\partial X_i}|\psi_n(\mathbf{x},\mathbf{X})\rangle + \hat{H}_{\mathrm{el}}(\mathbf{x},\mathbf{X})|\frac{\partial}{\partial X_i}\psi_n(\mathbf{x},\mathbf{X})\rangle$$
$$= \frac{\partial V_n^{\mathrm{A}}(\mathbf{X})}{\partial X_i}|\psi_n(\mathbf{x},\mathbf{X})\rangle + V_n^{\mathrm{A}}(\mathbf{X})|\frac{\partial}{\partial X_i}\psi_n(\mathbf{x},\mathbf{X})\rangle.$$

Multiplying from the left by $\langle\psi_m(\mathbf{x},\mathbf{X})|$, where $m \neq n$, we find

$$\left\langle \psi_m(\mathbf{x},\mathbf{X}) \left| \frac{\partial \hat{H}_{\text{el}}(\mathbf{x},\mathbf{X})}{\partial X_i} \right| \psi_n(\mathbf{x},\mathbf{X}) \right\rangle$$

$$+ \left\langle \psi_m(\mathbf{x},\mathbf{X}) | \hat{H}_{\text{el}}(\mathbf{x},\mathbf{X}) | \frac{\partial}{\partial X_i} \psi_n(\mathbf{x},\mathbf{X}) \right\rangle$$

$$= V_n^{\text{A}}(\mathbf{X}) \left\langle \psi_m(\mathbf{x},\mathbf{X}) \left| \frac{\partial}{\partial X_i} \right| \psi_n(\mathbf{x},\mathbf{X}) \right\rangle$$

or

$$A_{mn,i}(\mathbf{X}) = \left\langle \psi_m(\mathbf{x},\mathbf{X}) \left| \frac{\partial}{\partial X_i} \right| \psi_n(\mathbf{x},\mathbf{X}) \right\rangle$$

$$= \frac{\left\langle \psi_m(\mathbf{x},\mathbf{X}) \left| \frac{\partial \hat{H}_{\text{el}}(\mathbf{x},\mathbf{X})}{\partial X_i} \right| \psi_n(\mathbf{x},\mathbf{X}) \right\rangle}{V_n^{\text{A}}(\mathbf{X}) - V_m^{\text{A}}(\mathbf{X})}. \tag{2.61}$$

This mode of representing the derivative coupling clarifies that the coupling matrix $\mathbf{A}$ and thus the nonadiabaticity $\hat{\mathbf{\Lambda}}$ become singular at points of coincidence between the adiabatic potentials $V_n^{\text{A}}(\mathbf{X})$ and $V_m^{\text{A}}(\mathbf{X})$. The same statement can be made about adiabatic potential energy surfaces $V_n^{\text{APES}}(\mathbf{X})$ and $V_m^{\text{APES}}(\mathbf{X})$ which differ from the adiabatic potentials only by the nuclear repulsion term, and thus $V_n^{\text{APES}}(\mathbf{X}) - V_m^{\text{APES}}(\mathbf{X}) = V_n^{\text{A}}(\mathbf{X}) - V_m^{\text{A}}(\mathbf{X})$.

In the diabatic frame, the degeneracy at a certain nuclear configuration is not associated with a singularity in the coupling terms. The case of coinciding diabatic potentials can be investigated by diagonalization of the potential energy matrix $\mathbf{V}$ (see Eq. (2.44)). This is easily done for the assumed two-state situation. Diagonalizing the matrix

$$\begin{pmatrix} V_{11} & V_{12} \\ V_{21} & V_{22} \end{pmatrix}, \tag{2.62}$$

one arrives at the adiabatic potentials V_+^{A} and V_-^{A} as functions of the diabatic potential energy matrix elements. More specifically

$$V_\pm^{\text{A}} = \frac{1}{2}(V_{11} + V_{22}) \pm \frac{1}{2}[(V_{11} - V_{22})^2 + 4|V_{12}|^2]^{\frac{1}{2}}, \tag{2.63}$$

where the Hermitian property of $\mathbf{V}$ has been used. At the point of degeneracy in the diabatic frame, the respective adiabatic states are separated by an energy difference

$$V_+^{\mathrm{A}} - V_-^{\mathrm{A}} = 2|V_{12}|, \tag{2.64}$$

and so by twice the magnitude of the diabatic coupling potential. For $V_{12} \neq 0$, Eqs. (2.63) and (2.64) give an example of an *avoided crossing* of two adiabatic potentials. The coincidence of two diabatic potentials V_1^{D} and V_2^{D} at a nuclear position $\mathbf{X}$ does not imply that of the adiabatic potentials V_+^{A} and V_-^{A}. More specifically, we conclude from Eq. (2.63) that two independent conditions have to be met to guarantee the crossing of V_+^{A} and V_-^{A}, namely $V_{11} = V_{22}$ and $V_{12} = 0$. Given a problem with M internal nuclear degrees of freedom, these two equations will, in general, define a hypersurface of dimension $M - 2$. This implies for a triatomic molecule that the places of coincidence between the two adiabatic potentials are found along a one-dimensional seam. For a diatomic system, where the internuclear distance R is the only internal degree of freedom, no crossing points will be found, unless both conditions yield accidentally the same value of R, which is unlikely.

2.5.2 *Conical intersections and the Jahn–Teller effect*

In the following we will address the case of an electronic degeneracy involving two electronic states of a triatomic homonuclear molecule that interact with two vibrational modes. From the earlier discussion, we will expect for this case a zero-dimensional manifold, i.e. an isolated point of coincidence in the nuclear configuration space, or possibly several isolated points. As we consider a molecule at a point of D_{3h} symmetry, an electronic doublet $|E_\theta\rangle, |E_\varepsilon\rangle$ couples to a pair of degenerate nuclear coordinates labeled $\widetilde{R}_\theta$ and $\widetilde{R}_\varepsilon$, where we use the notation introduced for vibrational modes in Section 2.2. These internal coordinates are associated with non-totally symmetric modes, related to shape deformations of the molecule, as shown in Figure 2.1. The $\widetilde{R}_\theta$ coordinate conserves and the $\widetilde{R}_\varepsilon$ coordinate breaks C_{2v} symmetry.

We construct now the diabatic potential matrix in the $E \times e$ space, where E stands for the D_{3h} doublet $\{|E_\theta\rangle, |E_\varepsilon\rangle\}$, which are both chosen as real-valued functions, and e for the coordinate pair $\{\widetilde{R}_\theta, \widetilde{R}_\varepsilon\}$. The electronic basis functions $\psi_n(\mathbf{x}, \widetilde{\mathbf{R}}_0)$, $n = 1, 2$, are here given by $|E_\theta\rangle$ and $|E_\varepsilon\rangle$, and the fixed

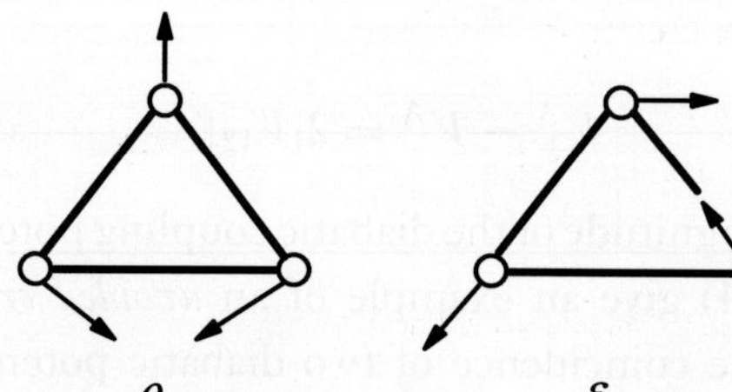

Figure 2.1: Sketch of the normal modes $\widetilde{R}_\theta$ and $\widetilde{R}_\varepsilon$.

location $\widetilde{\mathbf{R}}_0$ in nuclear coordinate space corresponds to the equilateral geometry of the triatomic species. Note that at $\widetilde{\mathbf{R}} = \widetilde{\mathbf{R}}_0$, there is no difference between the electronic wave functions in diabatic and adiabatic representation since $\psi_n(\mathbf{x}, \widetilde{\mathbf{R}})$ and $\psi_n(\mathbf{x}, \widetilde{\mathbf{R}}_0)$ are identical. Expanding the matrix (2.62) around the point of D_{3h} symmetry, we express the totally symmetric term of second order in the nuclear coordinates as

$$\mathbf{V}_0 = \frac{1}{2}\hbar\omega(Q_\theta^2 + Q_\varepsilon^2)\mathbf{1}, \tag{2.65}$$

where, for convenience, dimensionless coordinates

$$Q_i = \sqrt{\frac{\hbar\omega}{2}}\widetilde{R}_i, \quad \text{with } i = \theta, \varepsilon, \tag{2.66}$$

are utilized.[2] Here we take advantage of the degeneracy of the modes $\widetilde{R}_\theta$ and $\widetilde{R}_\varepsilon$ at D_{3h} symmetry, implying $\omega_\theta = \omega_\varepsilon = \omega$.

To construct the first-order interaction between the vibrational modes and the electronic states, we expand the diabatic potential energy matrix (2.44) with respect to Q, setting V_m^{APES} equal to zero at the point of degeneracy. The linear term reads

$$V_{1,nm} = \sum_{i=\theta,\varepsilon} \left\langle E_m \left| \frac{\partial[U_{\mathrm{eN}}(\mathbf{x}, \mathbf{Q}) + U_{\mathrm{N}}(\mathbf{Q})]}{\partial Q_i} \right| E_n \right\rangle \Bigg|_{\mathbf{Q}_0} Q_i, \tag{2.67}$$

with $m, n, i = \theta, \varepsilon$ and $\mathbf{Q}_0 = \mathbf{0}$. Each of the two coordinates engenders a 2×2 matrix. Using the Wigner–Eckart theorem, we decompose

$$\left\langle E_m \left| \frac{\partial[U_{\mathrm{eN}}(\mathbf{x}, \mathbf{Q}) + U_{\mathrm{N}}(\mathbf{Q})]}{\partial Q_i} \right| E_n \right\rangle \tag{2.68}$$

[2]Note that the $\{\widetilde{R}_i\}$ were introduced as mass-weighted coordinates in Section 2.2.

into a reduced matrix element

$$V_E = \left\langle E \left|\left| \frac{\partial[U_{eN}(\mathbf{x},\mathbf{Q}) + U_N(\mathbf{Q})]}{\partial Q_i} \right|\right| E \right\rangle\Bigg|_{\mathbf{Q}_0} \tag{2.69}$$

and a Clebsch–Gordan coefficient matrix $\mathbf{C}_{Ei}$.

$$\mathbf{V}_1 = \sum_i V_E \mathbf{C}_{Ei} Q_i, \quad i = \theta, \varepsilon. \tag{2.70}$$

The matrices $\mathbf{C}_{Ei}$ are given by [63]

$$\mathbf{C}_{E\theta} = \begin{pmatrix} -1 & 0 \\ 0 & 1 \end{pmatrix} = -\boldsymbol{\sigma}_z, \quad \mathbf{C}_{E\varepsilon} = \begin{pmatrix} 0 & 1 \\ 1 & 0 \end{pmatrix} = \boldsymbol{\sigma}_x. \tag{2.71}$$

From this one derives the explicit form of the diabatic potential energy matrix up to first order in the normal modes Q_θ and Q_ε as [64]

$$\mathbf{V} = \frac{1}{2}\hbar\omega(Q_\theta^2 + Q_\varepsilon^2)\mathbf{1} + V_E \begin{pmatrix} -Q_\theta & Q_\varepsilon \\ Q_\varepsilon & Q_\theta \end{pmatrix}. \tag{2.72}$$

In accordance with our expectation of an isolated point of coincidence, we find that the degeneracy is lifted as we move away from the origin, $Q_\theta = Q_\varepsilon = 0$. This effect is linear in the normal modes Q_θ and Q_ε. Whenever the removal of the degeneracy is linear in the displacement coordinates, the respective point of coincidence is referred to as a *conical intersection*. The *branching space* is defined as the subset of nuclear coordinates that lift the degeneracy to linear order, in this case it is the set $\{Q_\theta, Q_\varepsilon\}$. Carrying out the transformation from $\{Q_\theta, Q_\varepsilon\}$ to the complex variable pair $\{Q_\theta + iQ_\varepsilon, Q_\theta - iQ_\varepsilon\}$, the potential is conveniently expressed through the Pauli matrices $\boldsymbol{\sigma}_0$, $\boldsymbol{\sigma}_x$, and $\boldsymbol{\sigma}_y$:

$$\mathbf{V} = \mathbf{V}_0 + \mathbf{V}_1 = \frac{1}{2}\hbar\omega(Q_\theta^2 + Q_\varepsilon^2)\boldsymbol{\sigma}_0 + V_E(Q_\theta\boldsymbol{\sigma}_x + Q_\varepsilon\boldsymbol{\sigma}_y), \tag{2.73}$$

where $\boldsymbol{\sigma}_0$ denotes the 2×2 unit matrix. The transition from Eq. (2.72) to Eq. (2.73) is tantamount to a change of the electronic basis at $\mathbf{Q}_0$ from $\{|E_\theta\rangle, |E_\varepsilon\rangle\}$ to $\{|E_1\rangle, |E_2\rangle\}$ with $|E_1\rangle = 2^{-\frac{1}{2}}(|E_\theta\rangle + i|E_\varepsilon\rangle)$ and $|E_2\rangle = 2^{-\frac{1}{2}}(|E_\theta\rangle - i|E_\varepsilon\rangle)$. Note that the two latter basis vectors are complex conjugates of each other. In polar coordinates

$$Q_\theta + iQ_\varepsilon = \rho \exp(i\alpha), \quad Q_\theta - iQ_\varepsilon = \rho \exp(-i\alpha). \tag{2.74}$$

With these substitutions, the coupling term in Eq. (2.73) becomes

$$V_E \rho \begin{pmatrix} 0 & \exp(-i\alpha) \\ \exp(i\alpha) & 0 \end{pmatrix}. \tag{2.75}$$

The eigenvectors of this matrix are the adiabatic states corresponding to $|E_1\rangle, |E_2\rangle$. We obtain

$$\psi_\pm = \frac{1}{\sqrt{2}}(\exp(-i\alpha/2)|E_1\rangle \pm \exp(i\alpha/2)|E_2\rangle). \tag{2.76}$$

By application of the linear vibronic coupling matrix (2.75) on both ψ_- and ψ_+ we find for these eigenfunctions the eigenvalues $-V_E\rho$ and $+V_E\rho$, respectively. Expressed in the adiabatic basis $\{\psi_+, \psi_-\}$, the matrix $\widetilde{\mathbf{V}}$ adopts a compact diagonal form, namely:

$$\widetilde{\mathbf{V}} = \frac{1}{2}\hbar\omega\rho^2\boldsymbol{\sigma}_0 + V_E\rho\boldsymbol{\sigma}_z, \tag{2.77}$$

from which one finds immediately the desired adiabatic potential energy surfaces, given by

$$V_\mp^{\mathrm{APES}} = \frac{1}{2}\hbar\omega\rho^2 \mp V_E\rho. \tag{2.78}$$

In keeping with the assumed degeneracy of $|E_\theta\rangle$ and $|E_\varepsilon\rangle$ at the place of D_{3h} symmetry, $\mathbf{Q}_0$, the two sheets V_+^{APES} and V_-^{APES} coincide at $\rho = 0$. Figure 2.2 illustrates the resulting potential energy landscape.

2.5.3 *Jahn–Teller distortion*

The most important observation related to the potential energy surface shapes is that the maximally symmetric structure, namely that of an equilateral triangle, located at the point of degeneracy, is unstable and thus subject to distortion. More precisely, the explicit form of the lower potential energy sheet, $V_-^{\mathrm{APES}} = \frac{1}{2}\hbar\omega\rho^2 - V_E\rho$, implies the existence of a gradient $-V_E$ at the point $\rho = 0$.

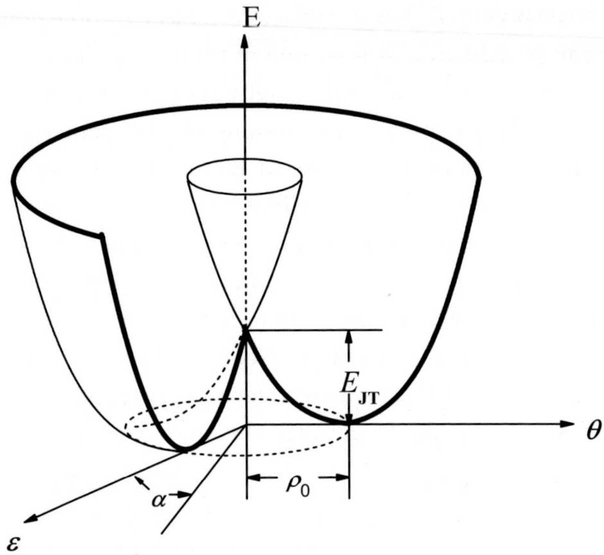

Figure 2.2: Mexican hat profile of the adiabatic potential energy surfaces for the $E \times e$ situation. The adiabatic potentials V_+^{APES} and V_-^{APES} are shown as functions of the deformation modes Q_θ and Q_ε. Only the linear interaction between electronic and vibrational degrees of freedom is included. The V_-^{APES} sheet is marked by a bold contour line.

From the explicit form of V_-^{APES}, a value of $\rho_0 = \frac{V_E}{\hbar\omega}$ is determined as the trough radius. This yields a potential trough depth of magnitude

$$|V_-(\rho = \rho_0)| \equiv E_{\text{JT}} = \frac{V_E^2}{2\hbar\omega} \tag{2.79}$$

with E_{JT} as the *Jahn–Teller stabilization energy*.

This quantity measures the energy decrease associated with the deformation of the system from the point of degeneracy, $\rho = 0$, to the equilibrium structure adopted at $\rho = \rho_0$. This distortion may be described in terms of symmetry lowering which leads in the case considered here from the point group D_{3h} to C_{2v}. Evidently, such distortion of the geometry realized at the point of degeneracy will be observed whenever the derivative of the linear potential matrix element does not vanish at this point. A direct generalization of the results developed in the preceding discussion for the special case of a triatomic species is given by the Jahn–Teller theorem [65] which may be stated as follows [66]:

If N adiabatic potential energy surfaces intersect at some point $\mathbf{Q}_0$ in the space of nuclear coordinates, at least one of them exhibits a non-vanishing derivative at $\mathbf{Q}_0$ with respect to a non-totally symmetric normal mode. This statement is valid for any polyatomic system whose geometry is not linear at $\mathbf{Q}_0$. A further exception is the case of Kramers' degeneracy associated with electronic spin doublets. One concludes that any nonlinear system with an orbitally degenerate electronic ground state is subject to distortion, which is frequently presented as the intuitive content of the Jahn–Teller theorem.

This theorem rests on the observation that the reduced matrix element of first order in the normal coordinates (see Eq. (2.75) and the related discussion) remains finite at the locus of degeneracy, reflecting the presence of an active distortion mode. In general, this matrix element may be written as $V_\Gamma = \langle \psi_{\Gamma'} \| \frac{\partial [U_{\mathrm{eN}} + U_{\mathrm{N}}(Q)]}{\partial Q_\Gamma} \| \psi_{\Gamma'} \rangle$. The symbols Γ and Γ' refer here to irreducible representations of the relevant point group G. In the preceding paragraphs we introduced the electronic doublet $\{E_\theta, E_\varepsilon\}$ of the point group D_{3h} which provides an example of a realization of the states $\psi_{\Gamma'}$. From group theory, one infers that the reduced matrix element V_Γ will be finite if, and only if, the representation Γ is contained in the product $\Gamma' \times \Gamma'$, where the symmetric square $[\Gamma'^2]$ has to be inspected if Γ' is a normal representation and the antisymmetric square $\{\Gamma'^2\}$ in case of a double representation. The proof provided by Jahn and Teller proceeds by listing all point symmetry groups and examining both the possible symmetric or antisymmetric squares and the non-totally symmetric normal modes. As a result, each one of the investigated squares turns out to contain a representation of such a mode, demonstrating the existence of a vibrational degree of freedom that causes distortion, thereby lowering both the symmetry and the energy of the respective system.

As made explicit in our formulation of the Jahn–Teller theorem, the previous arguments do not hold for linear molecules. We qualify this statement by pointing out that linear molecules do not undergo distortion as a result of the first-order coupling $\mathbf{V}_1$ (Eq. (2.70)) but may bend as higher orders of the vibronic interaction are considered. As a general result, first-order coupling by any bending mode does not occur between degenerate electronic states of linear molecules. This is rooted in the finding that these states share the same well-defined orbital angular momentum L, and their symmetry may be of the $\Pi, \Delta, \ldots$, etc., type. Further, their parity is entirely

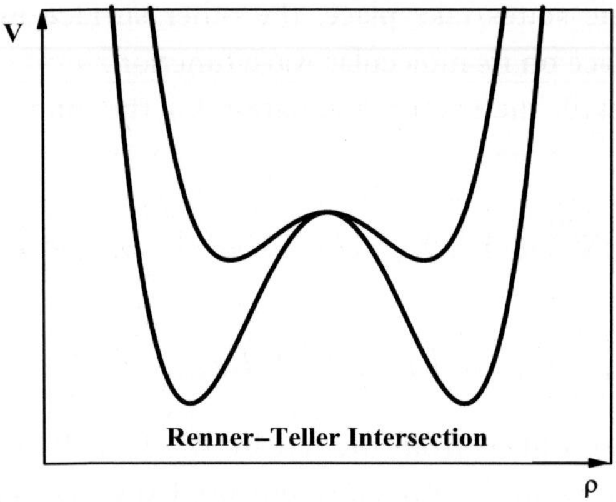

Figure 2.3: Glancing coincidence of two surfaces of revolution, displaying a possible Renner–Teller configuration. The figure might refer to a linear polyatomic molecule that distorts as a result of vibronic interactions at quadratic or higher order.

determined by L, being $(-1)^L$. These two observations combined with the fact that the bending mode for linear molecules is odd with respect to space reflection imply that the first-order coupling element for the distortion of linear molecules vanishes. *Renner–Teller* distortions [67], however, which are a consequence of vibronic coupling at quadratic or higher order, can induce the bending of a linear molecule. Geometrically, the respective coincidence is not a conical, but a glancing intersection, as illustrated by Figure 2.3.

2.5.4 *The molecular Aharonov–Bohm effect*

The nonadiabaticity $\hat{\mathbf{\Lambda}}$ cannot be neglected for Jahn–Teller systems since, according to Eq. (2.61), the derivative coupling matrix is singular at the point of degeneracy. We write down a solution of the stationary Schrödinger equation, choosing the ansatz

$$\Psi_{\pm} = \chi_{\pm}(\rho, \alpha)\psi_{\pm}(\rho, \alpha, \mathbf{x}). \tag{2.80}$$

The product form of these solutions assumes the validity of the Born–Oppenheimer approximation. We thus refer to a system whose evolution proceeds entirely on one of the two potential energy surfaces, the upper or the lower. It will be seen, however, that the coincidence between the surfaces must be taken into account for a correct treatment of the dynamic problem. Although the system is confined to a single surface, and no transitions between electronic states take place, the other surface exerts a defining, if indirect, influence on its molecular wave function.

In greater detail, the resulting equation for the nuclear wave function (2.8) is [68]

$$-\hbar\omega\left[\frac{1}{2}\nabla^2\chi_\pm + \nabla\chi_\pm\cdot\langle\psi_\pm|\nabla\psi_\pm\rangle + \frac{1}{2}\chi_\pm\langle\psi_\pm|\nabla^2\psi_\pm\rangle\right]$$
$$+\left(\frac{1}{2}\hbar\omega\rho^2 \pm V_E\rho\right)\chi_\pm = E\chi_\pm, \tag{2.81}$$

where we have multiplied from the left by $\psi_\pm^*$ as well as integrated with respect to the electronic coordinates and used the gradient symbol ∇ to denote the vector operator $(\boldsymbol{\rho}\frac{\partial}{\partial\rho} + \boldsymbol{\alpha}\frac{1}{\rho}\frac{\partial}{\partial\alpha})$. With these assignments, the nonadiabatic coupling terms in Eq. (2.81) are easily evaluated to yield

$$\langle\psi_\pm|\nabla\psi_\pm\rangle = (0,0) \tag{2.82}$$

and

$$\langle\psi_\pm|\nabla^2\psi_\pm\rangle = -\frac{1}{4\rho^2}. \tag{2.83}$$

Exercise 2.4. Use the diabatic potential energy matrix (2.72) to derive a pair of real-valued adiabatic eigenfunctions as an alternative to the complex-valued set (2.76). Show that the real-valued basis also satisfies the relations (2.82) and (2.83).

Combining Eq. (2.81) with Eqs. (2.82) and (2.83), we obtain the differential equation for the vibrational function χ_- that is associated with the electronic ground state ψ_-:

$$\left\{\frac{\hbar\omega}{2}\left[-\frac{1}{\rho}\left(\frac{\partial}{\partial\rho}\right)\rho\left(\frac{\partial}{\partial\rho}\right) - \frac{1}{\rho^2}\frac{\partial^2}{\partial\alpha^2} + \frac{1}{4\rho^2} + \rho^2\right] - V_E\rho\right\}\chi_- = E\chi_-. \tag{2.84}$$

The second derivative coupling, as given by Eq. (2.83), thus adds a centripetal contribution proportional to $\frac{1}{\rho^2}$ to the remaining potential terms.

We are now prepared to address the periodic motion of a triatomic system in the adiabatic ground state potential shown in Figure 2.2. Obviously, this process does not have the character of a physical rotation in two-dimensional space, but rather that of a periodic structural deformation in the space spanned by the normal modes Q_θ and Q_ε, as illustrated in Figure 2.1, or a *pseudorotation*. To first order in the vibronic coordinates, as considered here, the evolution of the system may thus be viewed as an unhindered cyclic shape variation around a center defined by the equilateral geometry that marks the point of degeneracy. This is shown in Figures 2.4 and 2.5.

In particular, Figure 2.5 clarifies the relation of the molecular pseudorotation to individual atomic rotations. Each atom can be imagined to move at constant angular speed on a circular track, where the three circles have equal radii. The three rotations are coordinated, as each one is shifted in phase by angle of $2\pi/3$ with respect to the following one. Therefore, pseudorotation may be characterized as a wave of distortions traversing the molecule in cyclic order.

Tracing the electronic solution (2.76) through a full pseudorotation cycle by traversing the interval $0 \leq \alpha \leq 2\pi$, we notice that the parametric dependence of this solution on the nuclear coordinates engenders an interesting phenomenon. The function ψ_- turns out to be multivalued, as it switches its sign upon one full pseudorotation and therefore repeats with a period of 4π. This effect does not pose a fundamental problem, since only the wave function (2.80), describing the molecule as a whole, is constrained

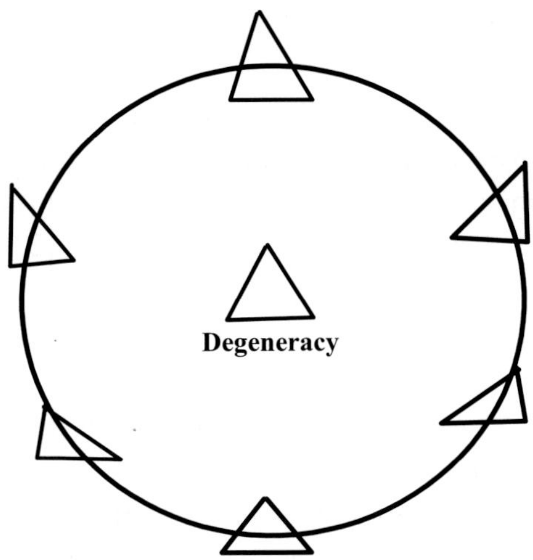

Figure 2.4: Pseudorotation of a triatomic molecule in the circular ground state potential energy trough arising from $E \times e$ Jahn–Teller distortion.

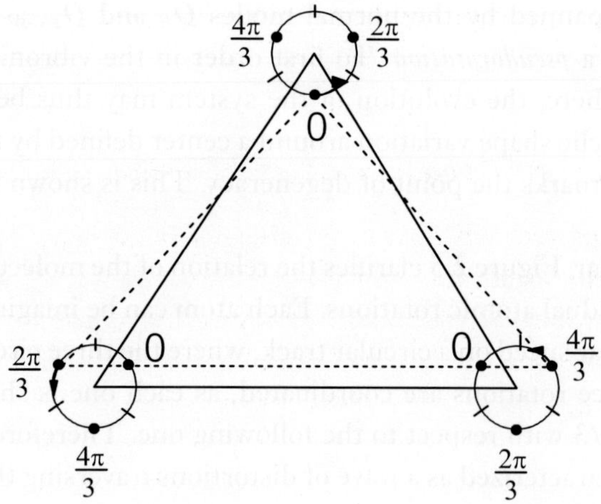

Figure 2.5: Origin of the pseudorotation of a triatomic molecule from three coordinated individual atomic rotations. The circles around the corners of the equilateral triangle represent atomic trajectories. Each rotation is phase shifted with respect to the next by an angle of $2\pi/3$, in cyclic order.

to be single-valued. Consequently, the sign change upon pseudorotation about an angle of 2π must be compensated by an equal change occurring in the vibronic wave function χ_-. Factorizing the vibrational solution according to

$$\chi_-(\rho, \theta) = R(\rho)T(\theta) \tag{2.85}$$

with $R(\rho)$ as the harmonic oscillator ground state wave function, and $T(\theta) = \exp(iq\theta)$, we enforce the 4π periodicity of the angular part of the vibrational component by stipulating that q adopts half-odd-integer values: $q = \pm\frac{1}{2}, \pm\frac{3}{2}, \pm\frac{5}{2}, \ldots$. Using the ansatz (2.85) in Eq. (2.84), we arrive at

$$E = \hbar\omega \left\{ \frac{q^2}{2k^2} + \frac{1}{2}\left(1 - k^2\right) \right\}, \quad q = \frac{1}{2}, \frac{3}{2}, \frac{5}{2}, \ldots, \tag{2.86}$$

with $k \equiv V_E/\hbar\omega$. The requirement of single-valuedness imposed on the molecular wave function thus introduces naturally a quantization rule for q which is interpreted as the eigenvalue of a pseudorotation angular momentum. The latter is to be understood as analogous to the familiar orbital angular momentum in two-dimensional space. The discrete values adopted by q, in turn, yield a quantization rule for the total energy of the pseudorotating triatom.

2.5.5 *The geometric phase in molecular pseudorotation*

In the following we will discuss an alternative strategy for treating the ambiguity of the electronic contribution shown in Eq. (2.76). Instead of using the vibrational part to make the overall wave function unique, one might remedy the electronic part by multiplication with a factor $\exp\left(\frac{i\alpha}{2}\right)$. The electronic wave function (2.76) thus turns into

$$\psi'_{+} = \frac{1}{\sqrt{2}}(|E_1\rangle + \exp(i\alpha)|E_2\rangle), \tag{2.87}$$

$$\psi'_{-} = \frac{1}{\sqrt{2}}(|E_1\rangle - \exp(i\alpha)|E_2\rangle). \tag{2.88}$$

With recourse to Eq. (2.81), we arrive now at nonadiabatic coupling terms that differ from those of Eqs.(2.82) and (2.83), namely

$$\langle\psi'_{\pm}|\,\nabla\,\psi'_{\pm}\rangle = \left(0, \frac{i}{2\rho}\right) \tag{2.89}$$

and

$$\langle\psi'_{\pm}|\,\nabla^2\,\psi'_{\pm}\rangle = -\frac{1}{2\rho^2}. \tag{2.90}$$

In particular, a purely imaginary vector potential results. This vector potential can be eliminated from the stationary Schrödinger equation (2.81) for the vibrational solution by applying a phase factor $\exp\left(-\int_0^{\alpha} \frac{i}{2\rho}\rho\, d\alpha'\right) = \exp(-i\alpha/2)$ to the single-valued basis, in accord with Eq. (2.51). From this manipulation the double-valued basis (2.76) is recovered, and in further consequence the set of half-odd-integer eigenvalues for the pseudorotation angular momentum q. As we retain the basis given in Eqs. (2.87) and (2.88) as the electronic component, however, a new quantization condition for q emerges (see Exercise 2.5).

Exercise 2.5. Use the single-valued electronic basis given in Eqs. (2.87) and (2.88) to obtain from Eq. (2.81) the analog of Eq. (2.84) for the vibrational function. Employ the approach (2.85) and compare with the spectrum (2.86) to show that the same sequence of spectral terms emerges as in the case of a double-valued electronic basis. The order of vibrational energies is thus invariant under a change of gauge, as it should be.

In this case, the uniqueness of the molecular wave function demands 2π periodicity for the vibrational part, which translates into the requirement of integer pseudorotation quantum number q. Both approaches, i.e. the multivalued as well as the single-valued electronic basis, give rise to the same succession (2.86) of total energy levels, as will be shown in Exercise 2.5. As an aside, a single-valued wave function that is adiabatically transported through the nuclear parameter space cannot be chosen consistently as a real function if electronic degeneracies are present. Locally, this choice is legitimate, but globally, a complex geometric phase factor is required to remove the ambiguity of the wave function.

We have seen that the problem of the double-valued electronic wave function associated with the linear $E \times e$ Jahn–Teller problem may be remedied in two alternative ways. The first avenue involves compensating for the sign change upon a full pseudorotation by forcing the same behavior on the vibrational component, and thus restricting the pseudorotation angular momentum q to half-odd-integer values. In this case, the vector potential vanishes and thus the wave function does not contain any phase factor related to the vector potential **A**. In the second case, admission is made for such a phase factor, corresponding to the appearance of a finite vector potential. This procedure turns the electronic wave function from a multivalued into a single-valued quantity.

In view of the simple structure of the vector potential (2.89), the adiabatic phase accumulated along a circular path C_0 that leads around the locus of degeneracy in the linear $E \times e$ Jahn–Teller problem can be easily evaluated. Thus

$$\varphi(C_0) = i \int_0^{2\pi} \frac{i}{2\rho} \rho d\alpha = -\pi. \tag{2.91}$$

It should be noted that this result is independent of the radius ρ. Consequently, any roundtrip enclosing the conical intersection will yield the indicated result. Further, this finding is unaffected by the dynamics of the investigated process as long as the latter remains adiabatic. This may be understood by inspecting the general expression for the phase accumulated along some closed path C_0:

$$\begin{aligned}\varphi(C_0) &= i\oint_{C_0}\sum_j A_j\,dQ'_j \\ &= i\oint_{C_0}\sum_j \langle\psi_m(\mathbf{x},\mathbf{Q}')\left|\frac{\partial}{\partial Q'_j}\right|\psi_m(\mathbf{x},\mathbf{Q}')\rangle dQ'_j.\end{aligned} \tag{2.92}$$

The adiabatic phase acquired on a roundtrip encircling any region of the potential energy surface, therefore, is entirely determined by the geometry of the surface and the path taken by the system as it evolves, rationalizing the term *geometric phase*. This concept was first introduced by Longuet-Higgins in the context of the Jahn–Teller problem [69]. The character of the geometric phase as a general physical phenomenon related to the development of a system that contains a set of "fast variables" with a parametric dependence on a set of "slow variables" was analyzed by M.V. Berry [70], and thus the frequently encountered alternative name *Berry's phase*.

The geometric phase, acting like a memory of the system, reflects a path dependence of the molecular wave function that is adiabatically transported through a potential energy landscape with points of degeneracy. One verifies easily that the geometric phase is -2π as two pseudorotation cycles are completed, which reverses the sign change of the wave function upon a single cycle. Adiabatic propagation around a conical intersection is thus analogous to parallel transport on a Moebius strip where the initial condition is restored upon two roundtrips. The path-dependent phase shift of the molecular wave function due to a point of degeneracy causes wave packet interference phenomena which differ markedly from those recorded in the absence of any intersection. These findings demonstrate the tangible impact of the adiabatic vector potential $\mathbf{A}$, and thus a parallel can be drawn to the Aharonov–Bohm effect [71] which consists in the observable influence of an electromagnetic vector potential on charged particle interference. The non-locality of the geometric phase which is accrued by a nuclear wave packet moving along a path that encloses a point of intersection, irrespective of its distance from this point, represents another analog to the Aharonov–Bohm effect. Consequently, the manifestation of the adiabatic vector potential in molecular interactions is aptly referred to as *molecular Aharonov–Bohm effect* [72].

Conical intersections, giving rise to non-trivial topological effects in the evolution of a molecular system, appear frequently and exert crucial influence on a broad range of electronically nonadiabatic phenomena [57]. In particular, they define a path between electronic states, allowing for fast radiationless deexcitation which is at the core of numerous photochemical processes [73, 74]. Measurable spectroscopic signatures of conical intersections are available from experiments involving detection of vibrational wave packets that move between electronic states in less than 100 fs [75, 76], from measured photoproduct branching ratios [77], or from polarized femtosecond pump–probe observation of electronic wave packets passing through an intersection [78]. Biological implications of electronic transition through conical intersections will be discussed in Chapter 15. Further, as has been known for several decades, scattering systems consisting of alkali and other metal trimers X_3 display a direct reflection of the geometric phase effect [79, 80]. Neglecting a sign change caused by this effect in the electronic wave function upon cyclic permutation of the three atoms leads to the wrong relative sign between the direct and the exchange amplitude for the reaction $X + X_2 \rightarrow X_2 + X$. Both the theoretical predictions related to conical intersections and the experimental examination of these phenomena motivate the need for search strategies designed to locate their places in a potential energy landscape. In Section 2.6 we look for such strategies.

An extension of the $E \times e$-problem to quadratic order, emphasizing that this problem contains far more points of degeneracy than just the origin, $\rho = 0$, is found in Section 2.7*.

2.6 Locating the Seam

From the arguments delineated in Sections 2.5 and 2.6, conical intersections and avoided crossings between potential energy surfaces may exert a substantial impact on the dynamics of a given molecular problem. Locating these regions of pronounced interaction between different electronic states is of major importance for the treatment of nonadiabatic situations. However, the procedure of direct inspection, involving the full determination of the intersecting potential energy surfaces will, in the general case, require prohibitive computational effort. Algorithms that allow us to identify zones of close proximity between these surfaces without their prior computation are thus of major interest for nonadiabatic dynamics. In the present section, we will introduce some of the relevant approaches proposed to tackle this problem.

In Section 2.5 we considered a problem involving two internal nuclear coordinates. Expanding the Hamiltonian up to linear order, we found one isolated point of degeneracy between the two interacting electronic states. We have seen, however, that in general, the places where degeneracy is maintained up to linear order in the nuclear coordinates form regions of dimension $M-2$ in the nuclear coordinate space, with M as the total number of internal coordinates. This region is called the *seam*.

2.6.1 *Intersection-adapted coordinates*

For a sufficient number of nuclear degrees of freedom, the seam consists of a multidimensional subspace of intersection points. For the purpose of identifying strategies that locate the seam, it is useful to cast the Hamiltonian in the neighborhood of a degeneracy into geometric terms which are adjusted to the structural features of a conical intersection. From a purely geometric viewpoint, the characteristic shape of a double cone, as illustrated by Figure 2.2 in the context of the linear $E \times e$ Jahn–Teller problem, is generated in three-dimensional space by a line segment of length ρ that includes an angle θ with the z-axis, where $0 \leq \theta \leq \pi/2$. Rotating this segment about the z-axis and reflecting the resulting cone with respect to the xy plane creates a symmetric double-cone structure. Among the properties of this structure is its steepness or *pitch*, for which $\cos\theta$ provides a measure. One might choose the bounding curves of the three-dimensional double-cone surface to be elliptic instead of circular, thus admitting a certain extent of *asymmetry*. Further, its principal axis does not necessarily have to align with the z-axis. It might be reoriented into the direction $(\theta^{\mathrm{t}}, \varphi^{\mathrm{t}})$ which defines its *tilt*. The salient topographic features of conical intersections are summarized in Figure 2.6.

With reference to the discussion in Section 2.3, we write the diabatic potential energy matrix (2.42) for two interacting electronic states as [81]

$$\mathbf{V}(\mathbf{R}) = \begin{pmatrix} V_{11}(\mathbf{R}) & V_{12}(\mathbf{R}) \\ V_{21}(\mathbf{R}) & V_{22}(\mathbf{R}) \end{pmatrix} = S(\mathbf{R})\mathbf{I} + \begin{pmatrix} -G(\mathbf{R}) & W(\mathbf{R}) \\ W(\mathbf{R}) & G(\mathbf{R}) \end{pmatrix}. \tag{2.93}$$

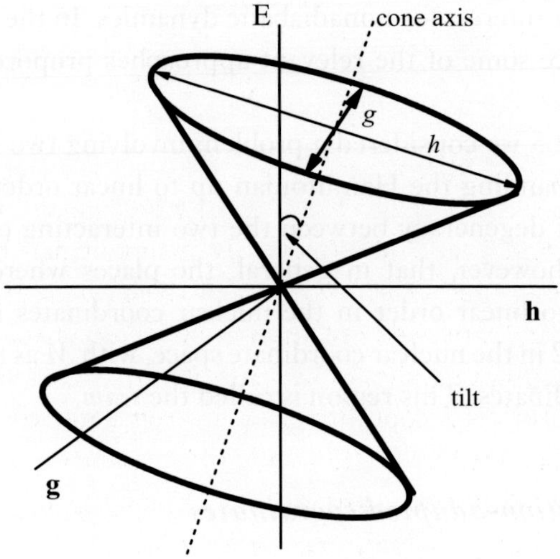

Figure 2.6: Double-cone structure. The basic topographic parameters, pitch, asymmetry, and tilt are indicated.

Here, $S(\mathbf{R}) \equiv \frac{1}{2}[V_{11}(\mathbf{R}) + V_{22}(\mathbf{R})]$, $G(\mathbf{R}) \equiv \frac{1}{2}[-V_{11}(\mathbf{R}) + V_{22}(\mathbf{R})]$, and $W(\mathbf{R}) \equiv V_{12}(\mathbf{R})$. We operate here with real matrix elements, such that $V_{12}(\mathbf{R}) = V_{21}(\mathbf{R})$. With Eq. (2.93), we revert to mass-scaled Cartesian coordinates, as introduced by Eq. (2.21). This change is motivated by the strategy to locate the seam which will be described in the following.

We will clarify that this involves a special set of conical-intersection-adapted coordinates, to be derived from the Cartesian position **R**. The form of **V** as indicated in Eq. (2.93) is, for instance, realized by the potential (2.72) for the linear $E \times e$ Jahn–Teller effect. The eigenvalues of **V** yield the adiabatic potential energy surfaces $V_{\pm}^{\mathrm{APES}}(\mathbf{R})$ as

$$V_{\pm}^{\mathrm{APES}}(\mathbf{R}) = S(\mathbf{R}) \pm \sqrt{G(\mathbf{R})^2 + W(\mathbf{R})^2}. \tag{2.94}$$

From this relation, the criterion for finding an intersection at the position $\mathbf{R}_0$ is $G(\mathbf{R}_0) = W(\mathbf{R}_0) = 0$. Inspecting the gradients $\frac{\partial G(\mathbf{R})}{\partial \mathbf{R}}|_{\mathbf{R}=\mathbf{R}_0} \equiv \mathbf{g}(\mathbf{R}_0)$ and $\frac{\partial W(\mathbf{R})}{\partial \mathbf{R}}|_{\mathbf{R}=\mathbf{R}_0} \equiv \mathbf{h}(\mathbf{R}_0)$, we can examine if the intersection is conical. This holds if $g(\mathbf{R}_0) \neq 0$ and $h(\mathbf{R}_0) \neq 0$. In the vicinity of conical intersections, it is advantageous to operate with intersection-adapted coordinates [81, 82] $x, y, w_i, i = 1, \ldots, M^{\mathrm{int}} - 2$, where M^{int} is the number of internal nuclear coordinates. The x- and y-axes are defined by the unit vectors $\hat{\mathbf{x}}$ and $\hat{\mathbf{y}}$, according to

$$\hat{\mathbf{x}} = \frac{\mathbf{g}(\mathbf{R}_0)}{g}, \quad \hat{\mathbf{y}} = \frac{\mathbf{h}(\mathbf{R}_0)}{h} \tag{2.95}$$

and $g = |\mathbf{g}(\mathbf{R}_0)|$, $h = |\mathbf{h}(\mathbf{R}_0)|$.

The vectors $\hat{\mathbf{x}}$ and $\hat{\mathbf{y}}$ determine the *branching plane* (or *gh-plane*). We mention that $\hat{\mathbf{x}}$ and $\hat{\mathbf{y}}$ are not necessarily orthonormal, but are also not collinear if the condition $G(\mathbf{R}_0) = W(\mathbf{R}_0) = 0$ comprises two independent geometric constraints, reducing the number of internal coordinates by two. An orthogonalization technique for $\hat{\mathbf{x}}$ and $\hat{\mathbf{y}}$ is outlined in reference [57]. The remaining $M^{\text{int}} - 2$ coordinates w_i are consequently associated with the seam space, spanned by the mutually orthonormal unit vectors $\hat{\mathbf{w}}_i, i = 1, \ldots, M^{\text{int}} - 2$. With these assignments, the expansion of the matrix $\mathbf{V}$ around the point $\mathbf{R}_0$ up to linear order is as follows:

$$\mathbf{V}_1(\mathbf{x}, \mathbf{y}, \mathbf{w}) = \left(s_x x + s_y y + \sum_{i=1}^{M^{\text{int}}-2} s_{w_i} w_i \right) \mathbf{I} + \begin{pmatrix} -gx & hy \\ hy & gx \end{pmatrix}, \tag{2.96}$$

employing the projections $s_x = \frac{\partial S(\mathbf{R})}{\partial \mathbf{R}}|_{\mathbf{R}=\mathbf{R}_0} \cdot \hat{\mathbf{x}}$, $s_y = \frac{\partial S(\mathbf{R})}{\partial \mathbf{R}}|_{\mathbf{R}=\mathbf{R}_0} \cdot \hat{\mathbf{y}}$, and $s_{w_i} = \frac{\partial S(\mathbf{R})}{\partial \mathbf{R}}|_{\mathbf{R}=\mathbf{R}_0} \cdot \hat{\mathbf{w}}_i$. Replacing the tuning and coupling coordinates x and y, respectively, by their polar representations $x = \rho \cos\alpha$, $y = \rho \sin\alpha$, one expresses the eigenvalues of $\mathbf{V}_1$, which are identical with those of $\mathbf{V}$ (see Eq. (2.93)) in the vicinity of the conical intersection, as functions of ρ, α, and w_i:

$$\begin{aligned} V_{1,\pm}^{\text{APES}}(\mathbf{R}) &= \rho(s_x \cos\alpha + s_y \sin\alpha) \\ &\quad + \sum_{i=1}^{M^{\text{int}}-2} s_{w_i} w_i \pm \rho\sqrt{(g\cos\alpha)^2 + (h\sin\alpha)^2} \\ &= \rho(s_x \cos\alpha + s_y \sin\alpha) + \sum_{i=1}^{M^{\text{int}}-2} s_{w_i} w_i \\ &\quad \pm \rho\delta_{gh}\sqrt{1 + \Delta_{gh}\cos(2\alpha)}. \end{aligned} \tag{2.97}$$

In the last line, we have cast the eigenvalues of $\mathbf{V}$ into a form that contains the geometric parameters of the double cone explicitly. Thus, s_x and s_y measure the tilt, while the asymmetry and the pitch are given by the quantities $\Delta_{gh} \equiv (g^2 - h^2)/(g^2 + h^2)$ and $\delta_{gh}^2 \equiv \frac{1}{2}(g^2 + h^2)$, respectively.

2.6.2 *Determination of the seam space*

We will now turn to the problem of locating the seam, understood as the task of identifying demarcation strategies for zones of coincidence between potential energy surfaces. We assume the degeneracy of K surfaces at a point $\mathbf{R}_0$, and decompose the electronic wave function $\psi_{\mathrm{n}}(\mathbf{x},\mathbf{R})$ employed in the adiabatic representation Eq. (2.6) as follows:

$$\psi_n(\mathbf{x},\mathbf{R}) = \sum_{i=1}^{NC} c_i^n(\mathbf{R}_0)\zeta_i(\mathbf{x},\mathbf{R}), \tag{2.98}$$

an approach reminiscent of the crude adiabatic representation. The upper limit of this summation, NC, stands for the dimension of the configuration state basis $\{\zeta_i\}$. We express the Hamiltonian $\hat{H}_{\mathrm{el}} + U_{\mathrm{N}}$ in this basis and label the resulting matrix representation $\mathbf{V}_\zeta$. In the next step, we subdivide the set of vectors $\{\mathbf{c}^j\}$ into two classes. In particular, we define the *Q-space* by specifying the K vectors ${}^{Q}\mathbf{c}^n(\mathbf{R}_0), 1 \leq n \leq K$. The ${}^{Q}\mathbf{c}^n(\mathbf{R}_0)$ are understood as eigenvectors of $\mathbf{V}_\zeta$ at $\mathbf{R}_0$, such that $[\mathbf{V}_\zeta(\mathbf{R}_0) - V_n^{\mathrm{APES}}(\mathbf{R}_0)\mathbf{I}]\,{}^{Q}\mathbf{c}^n(\mathbf{R}_0) = 0$. The second class consists of the vectors ${}^{P}\mathbf{c}^k(\mathbf{R}_0), K+1 \leq k \leq NC$, which provide a basis for the orthogonal complement of Q, labeled P-space. We use these vectors to introduce the matrices $\mathbf{V}_\zeta^{QQ}(\mathbf{R}), \mathbf{V}_\zeta^{PP}(\mathbf{R}), \mathbf{V}_\zeta^{QP}(\mathbf{R})$, and $\mathbf{V}_\zeta^{PQ}(\mathbf{R})$ which operate entirely in Q-space (QQ) or entirely in P-space (PP), or communicate between these spaces (QP, PQ). Their elements are

$$\begin{aligned} V_{\zeta,nm}^{QQ}(\mathbf{R}) &= \mathbf{c}^n(\mathbf{R}_0)^\dagger \mathbf{V}_\zeta(\mathbf{R})\mathbf{c}^m(\mathbf{R}_0), \\ V_{\zeta,kn}^{QP}(\mathbf{R}) &= \mathbf{c}^k(\mathbf{R}_0)^\dagger \mathbf{V}_\zeta(\mathbf{R})\mathbf{c}^n(\mathbf{R}_0), \quad n,m \leq K;\ k,l > K, \\ V_{\zeta,lk}^{PP}(\mathbf{R}) &= \mathbf{c}^k(\mathbf{R}_0)^\dagger \mathbf{V}_\zeta(\mathbf{R})\mathbf{c}^l(\mathbf{R}_0), \end{aligned} \tag{2.99}$$

and correspondingly for $V_{\zeta,kn}^{PQ}$. This leads to the eigenvalue equation

$$\begin{pmatrix} \mathbf{V}_\zeta^{QQ}(\mathbf{R}) - V_i^{\mathrm{APES}}(\mathbf{R})\mathbf{I} & \mathbf{V}_\zeta^{QP}(\mathbf{R}) \\ \mathbf{V}_\zeta^{PQ}(\mathbf{R}) & \mathbf{V}_\zeta^{PP}(\mathbf{R}) - V_i^{\mathrm{APES}}(\mathbf{R})\mathbf{I} \end{pmatrix} \begin{pmatrix} {}^{Q}\mathbf{c}(\mathbf{R}) \\ {}^{P}\mathbf{c}(\mathbf{R}) \end{pmatrix} = \begin{pmatrix} \mathbf{0} \\ \mathbf{0} \end{pmatrix}, \tag{2.100}$$

where $1 \leq i \leq NC$. By construction, $V_{\zeta,nm}^{QQ}, n \neq m$, vanishes at the point of degeneracy:

$$V_{\zeta,nm}^{QQ}(\mathbf{R}_0) = 0.$$

At the same time, $V^{QQ}_{\zeta,nn}(\mathbf{R}_0) = V^{\text{APES}}_n(\mathbf{R}_0)$, and

$$V^{QP}_{\zeta,nk}(\mathbf{R}_0) = V^{PQ}_{\zeta,kn}(\mathbf{R}_0) = 0. \tag{2.101}$$

With these premises, we can derive conditions under which the K-fold intersection at $\mathbf{R}_0$ is preserved in the vicinity of this point. Focusing on some place $\mathbf{R}_\varepsilon$ close to $\mathbf{R}_0$, one finds to linear order

$$V_\zeta(\mathbf{R}_\varepsilon) = \mathbf{V}_\zeta(\mathbf{R}_0) + \sum_j \left.\frac{\partial \mathbf{V}_\zeta(\mathbf{R})}{\partial R_j}\right|_{\mathbf{R}_0} \cdot \delta\mathbf{R}_j, \tag{2.102}$$

where $\delta\mathbf{R} = \mathbf{R}_\varepsilon - \mathbf{R}_0$. Specifically, we evaluate $\mathbf{V}^{QQ}_{\zeta,nm}(\mathbf{R}_\varepsilon)$ from

$$V^{QQ}_{\zeta,nn}(\mathbf{R}_\varepsilon) = V_n(\mathbf{R}_0) + \mathbf{g}^n(\mathbf{R}_0)^\dagger \cdot \delta\mathbf{R} \tag{2.103}$$

$$V^{QQ}_{\zeta,nm}(\mathbf{R}_\varepsilon) = \mathbf{h}^{nm}(\mathbf{R}_0)^\dagger \cdot \delta\mathbf{R}, \quad n \neq m, \tag{2.104}$$

with the definitions $g^n_j(\mathbf{R}_0) = \mathbf{c}^n(\mathbf{R}_0)^\dagger \frac{\partial \mathbf{V}_\zeta}{\partial R_j} \mathbf{c}^n(\mathbf{R}_0)$ and $h^{nm}_j(\mathbf{R}_0) = \mathbf{c}^n(\mathbf{R}_0)^\dagger \frac{\partial \mathbf{V}_\zeta}{\partial R_j}$ $\mathbf{c}^m(\mathbf{R}_0)$. From the diagonal matrix element (2.103), we conclude that the degeneracy is preserved at $\mathbf{R}_\varepsilon$ if

$$\mathbf{g}^n(\mathbf{R}_0)^\dagger \cdot \delta\mathbf{R} = \mathbf{g}^M(\mathbf{R}_0)^\dagger \cdot \delta\mathbf{R}, \quad n < M, \tag{2.105}$$

while Eq. (2.104) yields the condition

$$\mathbf{h}^{nm}(\mathbf{R}_0)^\dagger \cdot \delta\mathbf{R} = 0, \quad m < n \leq M. \tag{2.106}$$

In order to cast Eq. (2.105) into the form of an orthogonality condition, we introduce the difference vector $\mathbf{g}^{nK}(\mathbf{R}_0) = \mathbf{g}^n(\mathbf{R}_0) - \mathbf{g}^K(\mathbf{R}_0)$ and observe that $\mathbf{g}^{nK}(\mathbf{R}_0) \cdot \delta\mathbf{R} = 0$. We recognize in $\mathbf{g}^{nm}$ and $\mathbf{h}^{nm}$ direct generalizations of the vectors $\mathbf{g}$ and $\mathbf{h}$ that were introduced in the context of Eq. (2.95) and applied to the special case of two intersecting potential energy surfaces instead of the K-fold degeneracy treated here. Equations (2.105) and (2.106) yield $K-1$ and $K(K-1)/2$ constraints, respectively, adding to a sum of $K' = (K-1)(K+2)/2$ *in toto*, which is the dimension of the space spanned by $\mathbf{g}^{nK}(\mathbf{R}_0)$ and $\mathbf{h}^{nm}(\mathbf{R}_0)$. We note that for $K = 2$, where $\mathbf{g}^{nK}(\mathbf{R}_0)$ and $\mathbf{h}^{nm}(\mathbf{R}_0)$ define the gh-plane, this dimension coincides with the number of intersecting potential energy surfaces. From Eqs. (2.105) and (2.106), one infers that the degeneracy persists up to linear order if the displacement $\delta\mathbf{R}$ is confined to

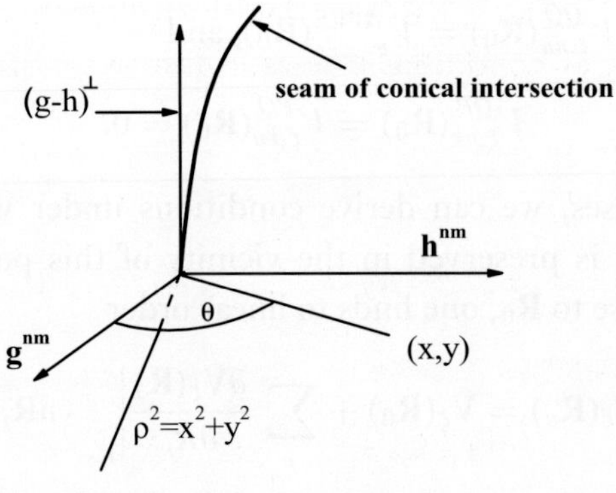

Figure 2.7: Representation of the gh-space as well as the $(gh)^{\perp}$-space for the case $M^{\text{int}} = 3$, $K = 2$. The $(g - h)^{\perp}$-space is spanned by a vector perpendicular to the gh-plane and tangent to the seam line. Note that $\mathbf{g}^{nm}(\mathbf{R}_0)$ and $\mathbf{h}^{nm}(\mathbf{R}_0)$ do not necessarily have to be orthogonal.

the orthogonal complement $(gh)^{\perp}$ of the gh-space. For $M^{\text{int}} = 3$ and $K = 2$, this concept assumes a particularly simple geometric meaning that is illustrated in Figure 2.7. In this case, the seam space is one-dimensional. More specifically, it is a line whose tangent vector at $\mathbf{R}_0$ is perpendicular to the gh-plane. This vector forms the basis of the $(gh)^{\perp}$-space, a case that is realized for the triatomic molecule considered in Section 2.5. If the degeneracy occurs at all loci of D_{3h} symmetry, the gh-plane is given by all deformations of the equilateral shape, as described before by the normal modes Q_{θ} and Q_{ϵ}. The remaining dimension, corresponding to the totally symmetric mode that changes the size but not the shape of the molecule at the point of degeneracy, is the seam line, or the $(gh)^{\perp}$-space.

2.6.3 *Seam subspaces by Lagrangian minimization*

For a polyatomic problem, the number M^{int} of internal nuclear coordinates can be large, which may necessitate the use of different strategies for locating the seam than those appropriate for small systems with $M^{\text{int}} = 2$ or 3. In the case of a high seam dimension, however, it will in general be possible to restrict the search to a certain subspace of the seam. The search then proceeds under two groups of constraints [83]. If two electronic states are considered, ψ_n and ψ_m, the first set of conditions requires them to be degenerate. The second group defines this subspace, for instance by freezing certain bond angles or distances that are not expected to change in the considered process. These two groups of conditions translate into two systems of equations, containing

(1) energy constraints:

$$V_n^{\text{APES}}(\mathbf{R}) = V_m^{\text{APES}}(\mathbf{R}), \quad \text{and} \tag{2.107}$$

(2) geometry constraints:

$$F_i(\mathbf{R}) = 0, \quad i = 1, N^{\text{con}}, \tag{2.108}$$

with N^{con} as the number of geometry constraints. The latter are exemplified by $F_i(\mathbf{R}) = R^2(1,i) - D^2$, where R stands for the distance between the nuclear centers whose labels are indicated in the parenthesis, and D is a constant. To incorporate the intersection requirement into the search formalism without relying on prior evaluation of the potential energy surfaces, one may employ a procedure similar to the one that led to the definition of the seam by relations (2.105) and (2.106). Thus, we refer to an arbitrary point $\mathbf{R}_a$ such that $V_n^{\text{APES}}(\mathbf{R}_a) \neq V_m^{\text{APES}}(\mathbf{R}_a)$. At $\mathbf{R}_0$, however, we require $V_n^{\text{APES}}(\mathbf{R}_0) = V_m^{\text{APES}}(\mathbf{R}_0)$ to be valid up to linear order. Again, the expansion (2.98) is used, but now in the coordinate $\mathbf{R}_a$ instead of $\mathbf{R}_0$. The Q-space is therefore constructed by use of the condition

$$(\mathbf{V}_\zeta(\mathbf{R}_a) - V_i^{\text{APES}}(\mathbf{R}_a)\mathbf{I})\,{}^{Q}\mathbf{c}_i(\mathbf{R}_a) = \mathbf{0}, \quad i = m, n. \tag{2.109}$$

To first order, we find the following relation between the diagonal elements at $\mathbf{R}_0$ and the eigenvalues at $\mathbf{R}_a$:

$$V_{\zeta,nn}^{QQ}(\mathbf{R}_0) - V_{\zeta,mm}^{QQ}(\mathbf{R}_0) = \Delta V_{nm}^{\text{APES}}(\mathbf{R}_a) + \mathbf{g}^{nm}(\mathbf{R}_a)^\dagger \cdot \delta\mathbf{R} = 0, \tag{2.110}$$

where $\Delta V_{nm}^{\text{APES}}(\mathbf{R}_a) \equiv V_n^{\text{APES}}(\mathbf{R}_a) - V_m^{\text{APES}}(\mathbf{R}_a)$ and $\delta\mathbf{R} = \mathbf{R}_0 - \mathbf{R}_a$. The corresponding condition for the nondiagonal matrix elements of $\mathbf{V}_\zeta$ is analogous to Eq. (2.106):

$$V_{\zeta,nm}^{QQ}(\mathbf{R}_0) = \mathbf{h}^{nm}(\mathbf{R}_a)^\dagger \cdot \delta\mathbf{R} = 0, \quad m < n \leq K. \tag{2.111}$$

A possible strategy of locating the seam under the selected geometric constraints and in accord with Eqs. (2.110) and (2.111) is *Lagrangian minimization*. A Lagrangian that incorporates all conditions imposed on the problem is given by the function

$$L(\mathbf{R}, \xi, \lambda) = V_n^{\text{APES}}(\mathbf{R}) + \xi_1 \Delta V_{nm}^{\text{APES}}(\mathbf{R}) + \xi_2 V_{\xi,nm}^{QQ}(\mathbf{R}) + \sum_{i=1}^{N^{\text{con}}} \lambda_i F_i(\mathbf{R}). \tag{2.112}$$

In this expression, the vectors $\boldsymbol{\xi}$ and $\boldsymbol{\lambda}$ contain the Lagrange multipliers related to the energy and geometry constraints, respectively. The variation of L with respect to the variables $\mathbf{R}$, $\boldsymbol{\xi}$, and $\boldsymbol{\lambda}$ is required to be stationary through second order which may be implemented by use of the Newton–Raphson method [82]. Equations (2.110) and (2.111) emerge naturally from this Lagrangian minimization approach. Employing this scheme, surfaces of conical intersection have been determined for a variety of chemical species. A case study on hydroxylamine (NH_2OH) can be found in [82].

Strategies for locating conical intersections are not only relevant in view of their ubiquitous nature or their role as touchstones for any model of nonadiabatic dynamics. Being associated with the geometric phase phenomenon, conical intersections also influence adiabatic processes, requiring a modification of the Born–Oppenheimer approximation for a polyatomic system in their vicinity. Their systematic as well as practical importance for the electron dynamics in molecules will be highlighted in several of the following chapters that explore, for instance, their role in determining the life times of photo-excitations for many biologically active molecules (Section 15.4), or the mechanism of *cis-trans* isomerization processes (Section 18.3).

Chapter 3

Basic Concepts of Scattering Theory

From a physical perspective, reactions between molecules are understood as collisions proceeding with or without energy or particle transfer. The vast majority of the interactions that fall into the scope of this monograph may be viewed as scattering processes. While the previous two chapters dealt with the discrete spectrum, in the present chapter we start exploring the continuous spectrum and develop basic notions of the scattering formalism, as far as they are of relevance for molecular dynamics.

Addressing scattering problems requires us to make a primary decision, namely that between a time-dependent or time-independent mode of description. One might argue that giving preference to one over the other is largely a matter of taste, since both frames are formally equivalent, as is demonstrated in Section 3.1. Practical considerations, nevertheless, speak often in favor of the time-dependent approach, involving L^2-integrable wave packets instead of periodic functions.

To begin, we will adopt the perspective of time-dependent theory and highlight various ways of framing the quantum mechanical equations of motion in Section 3.2. In the following section, we adopt the energy representation as operational framework, confining the time dependence of the problem to the time evolution operators. Applying the *Møller operators* to the eigenstates of the unperturbed Hamiltonian leads to the *Lippmann–Schwinger equations* as the cornerstones of the time-independent theory. Essential concepts needed for extensions of this theory to the multichannel scenario and applications to electronic processes in Chapters 6 and 7, respectively, are developed in Section 3.4 by considering the most elementary case of scattering, involving elastic interactions with potentials. The final section of this chapter highlights resonances and thus the continuation to complex-valued energies.

3.1 The Time-Dependent and the Time-Independent View of Scattering Processes

Assuming that a complete basis is available for the considered physical system, one may derive the wave function in the energy and the time representation as equally valid modes of description. This becomes evident as one inspects the general solution of the time-dependent Schrödinger equation, namely a wave packet that may be written explicitly as

$$\Phi(\mathbf{x},\mathbf{X};t) = \int dE' \sum_{n=0}^{n_{\max}} c(E',n)\Psi(\mathbf{x},\mathbf{X};E',n)\exp\left(-\frac{i}{\hbar}E't\right) \tag{3.1}$$

where $E' = E$ for $E' \geq 0$ and $E' = E_0, E_1, E_2, \ldots$ for $E' < 0$, and the symbol $\int \Sigma$ denotes summation over the discrete as well as integration over the continuous part of the spectrum. The time-dependent solution is thus represented by (3.1) as expansion in a basis defined by the stationary states of the problem. All energies are described as eigenvalues of the relevant Hamiltonian $\hat{H}$, such that

$$\hat{H}\Psi(\mathbf{x},\mathbf{X};E) = E\Psi(\mathbf{x},\mathbf{X};E) \quad \text{for the continuous domain,} \tag{3.2a}$$

$$\hat{H}\Psi_n(\mathbf{x},\mathbf{X}) = E_n\Psi_n(\mathbf{x},\mathbf{X}) \quad \text{for the discrete domain.} \tag{3.2b}$$

The factor $c(E',n)$ is a function $c(E')$ for $E' \geq 0$ and reduces to a set of coefficients $c_0, c_1, c_2, \ldots$ for $E' < 0$. This entity may be constructed from the prescription

$$c(E) = \langle \Psi | \Phi(\mathbf{x},\mathbf{X};t=0) \rangle, \tag{3.3a}$$

$$\text{or} \quad c_n = \langle \Psi_n | \Phi(\mathbf{x},\mathbf{X};t=0) \rangle. \tag{3.3b}$$

Relations (3.1)–(3.3b) imply the complete equivalence of the time-dependent and the stationary view. If the solutions of the four groups of time-independent equations are known, we are able to generate the wave packet $\Phi(\mathbf{x},\mathbf{X};t)$ at any arbitrary time. Intuitively, however, a wave packet as the solution of the scattering problem may seem more appealing than the eigenfunctions of Eqs. (3.2a) and (3.2b), since the former can be interpreted as a quantum mechanical extension of a classical ensemble of trajectories, lending itself more naturally to a representation of the actual physical process than the solution of the stationary Schrödinger equation. While the time-independent approach operates with extended, periodic states, the time-dependent scheme employs wave packets that are localized in configuration space at the initial time and generally maintain a high degree of localization in the course of the interaction. Further, from a technical point of view, the difference between the two frames corresponds to that between an eigenvalue and an initial value problem. Besides being better adjusted to the experimental realization of scattering processes, the latter is often mathematically less challenging than the former. Especially in situations governed by continuum functions satisfying complicated boundary conditions, the wave-packet-based procedure appears preferable. Also, the N^3 scaling behavior of the time-independent procedure, where N is the number of basis functions used to expand the eigenfunctions, provides an argument in favor of the time-dependent method in the case of complex polyatomic problems. The localized character of a wave packet is often of substantial advantage for techniques that employ approximate wave functions, such as the time-dependent Hartree–Fock formalism (Chapter 8). In Chapter 15 we will review some of the prevailing schemes used for the propagation of wave packets and thus for the construction of time-dependent solutions.

3.2 Quantum Mechanical Equations of Motion

Various frames are used to describe the evolution of quantum mechanical systems. The respective representations, known as *Schrödinger*, *Heisenberg* and *interaction pictures*, differ in the classification of the theory elements, i.e. operators and wave functions, as static or dynamic. Introducing the indices S, H and I to designate the three frames, we write the time-dependent Schrödinger equation (TDSE) as

$$i\hbar \frac{\partial \Psi_S(t)}{\partial t} = \hat{H}\Psi_S(t), \tag{3.4}$$

with a Hamiltonian $\hat{H}$ that does not explicitly depend on time. The Schrödinger picture assigns the time dependence of the studied process to the state vector Ψ, while the observables are treated as time-independent unless they contain time explicitly. The opposite extreme, involving static wave functions and dynamic operators, is realized by the Heisenberg picture. Thus, defining Ψ_S^a and Ψ_S^b as two wave functions that satisfy Eq. (3.4), and making use of $\Psi_S^a(t) = \exp(-\frac{i}{\hbar}\hat{H}t)\Psi_S^b(0)$ as formal solution of Eq. (3.4), we write the matrix element of some arbitrary observable $\hat{O}_S$ as

$$\langle \Psi_S^a(t)|\hat{O}_S|\Psi_S^b(t)\rangle = \left\langle \Psi_S^a(0) \left| \exp\left(\frac{i}{\hbar}\hat{H}t\right) \hat{O}_S \exp\left(-\frac{i}{\hbar}\hat{H}t\right) \right| \Psi_S^b(0) \right\rangle. \tag{3.5}$$

Deriving Eq. (3.5) with respect to time yields

$$\begin{aligned} \frac{d}{dt}\langle \Psi_S^a(t)|\hat{O}_S|\Psi_S^b(t)\rangle \\ = \left\langle \Psi_S^a(0) \left| \exp\left(\frac{i}{\hbar}\hat{H}t\right) \frac{\partial \hat{O}_S}{\partial t} \exp\left(-\frac{i}{\hbar}\hat{H}t\right) \right| \Psi_S^b(0) \right\rangle \\ + \frac{1}{i\hbar}\left\langle \Psi_S^a(0) \left| \left[\exp\left(\frac{i}{\hbar}\hat{H}t\right) \hat{O}_S \exp\left(-\frac{i}{\hbar}\hat{H}t\right), \hat{H} \right] \right| \Psi_S^b(0) \right\rangle. \end{aligned} \tag{3.6}$$

In the Heisenberg picture, the state vectors are the constants of motion

$$\Psi_H \equiv \Psi_S(0). \tag{3.7}$$

Correspondingly, all of the time dependence is carried by the operators according to

$$\hat{O}_H \equiv \exp\left(\frac{i}{\hbar}\hat{H}t\right) \hat{O}_S \exp\left(-\frac{i}{\hbar}\hat{H}t\right). \tag{3.8}$$

Differentiating Eq. (3.8), we derive the equation of motion in the Heisenberg picture:

$$\frac{d\hat{O}_H}{dt} = \frac{\partial \hat{O}_S}{\partial t} + \frac{1}{i\hbar}[\hat{O}_H, \hat{H}]. \tag{3.9}$$

From the definition (3.8), it is clear that the Hamiltonian is the same in the Schrödinger and Heisenberg pictures. A compromise between the latter two frames is given with the interaction picture where the load of the time dependence is equally shared by wave functions and observables. Consequently, this frame involves two equations of motion. It is based on the following partition of the Hamiltonian:

$$\hat{H} = \hat{H}_0 + \hat{H}' \tag{3.10}$$

where it is assumed that $\hat{H}_0$ does not contain any explicit time dependence. A variety of realizations of the scheme (3.10) are conceivable. Thus, $\hat{H}_0$

could be the Hamiltonian of a system with well-known solutions and $\hat{H}'$ could play the role of a perturbation. Alternatively, $\hat{H}_0$ and $\hat{H}'$ might be chosen as operators of the kinetic and the potential energy, respectively. Whatever meaning is assigned to the partial Hamiltonians, a state vector and an observable in the interaction picture are defined as

$$\Psi_I(t) \equiv \exp\left[\frac{i}{\hbar}\hat{H}_{0S}t\right]\Psi_S(t) \tag{3.11}$$

and

$$\hat{O}_I(t) \equiv \exp\left[\frac{i}{\hbar}\hat{H}_{0S}t\right]\hat{O}_S \exp\left[-\frac{i}{\hbar}\hat{H}_{0S}t\right] \tag{3.12}$$

at any time.

For simplicity of notation, we will designate the Schrödinger picture as default frame and thus use the index S only when a comparison between the various pictures is emphasized. The index I, however, will continue to label the interaction picture operators and wave functions. The interaction and Heisenberg pictures obviously reduce to each other in the special case $\hat{H}' = 0$. Although it holds that $\hat{H}_{0I} = \hat{H}_{0S}$, the reference Hamiltonians $\hat{H}_0$ are different in the Heisenberg and the Schrödinger picture, and the Schrödinger and interaction Hamiltonians are not identical. By differentiating Eqs. (3.11) and (3.12) with respect to time, we arrive at the equations of motion in the interaction picture:

$$i\hbar\frac{\partial \Psi_I(t)}{\partial t} = \hat{H}_I'\Psi_I(t), \tag{3.13}$$

$$\frac{d\hat{O}_I}{dt} = \frac{\partial \hat{O}_I}{\partial t} + \frac{1}{i\hbar}[\hat{O}_I, \hat{H}_{0I}]. \tag{3.14}$$

Exercise 3.1. Verify relations (3.13) and (3.14).

Thus the evolution of the state vector is governed by $\hat{H}'$, that of the observables by $\hat{H}_0$. The interaction picture is of particular advantage wherever $\hat{H}'$ is small in the sense of perturbation theory. If this is the case, the time dependence of the state vector is weak, as manifested by Eq. (3.13), while the observables are independent of $\hat{H}'$ by construction (see Eqs. (3.12) and (3.14)). The interaction picture has been established as the most convenient frame for developing the fundamental tenets of the time-dependent theory of scattering. In collision problems, the reference Hamiltonian $\hat{H}_0$ in Eq. (3.10) is naturally chosen as the Hamiltonian of the free, or asymptotic problem, and $\hat{H}'$ correspondingly as the scattering potential V that vanishes asymptotically. Therefore,

$$\hat{H}'_{\mathrm{I}}(t) = \exp\left[\frac{i}{\hbar}\hat{H}_0 t\right] V \exp\left[-\frac{i}{\hbar}\hat{H}_0 t\right]. \tag{3.15}$$

Constructing the unitary linear operator $\hat{U}_{\mathrm{I}}(t, t_0)$ that is the interaction picture counterpart of the Schrödinger time evolution operator $\hat{U}(t, t_0) = \exp\left(-\frac{i}{\hbar}\hat{H}(t - t_0)\right)$ is now an easy task. Thus, from (3.11),

$$\begin{aligned} \Psi_{\mathrm{I}}(t) &= \exp\left[\frac{i}{\hbar}\hat{H}_0 t\right] \exp\left[-\frac{i}{\hbar}\hat{H}(t - t_0)\right] \Psi_{\mathrm{S}}(t_0) \\ &= \exp\left[\frac{i}{\hbar}\hat{H}_0 t\right] \exp\left[-\frac{i}{\hbar}\hat{H}(t - t_0)\right] \exp\left[-\frac{i}{\hbar}\hat{H}_0 t_0\right] \Psi_{\mathrm{I}}(t_0) \\ &\equiv \hat{U}_{\mathrm{I}}(t, t_0)\Psi_{\mathrm{I}}(t_0). \end{aligned} \tag{3.16}$$

The explicit form of the time evolution operator in the interaction picture is therefore

$$\hat{U}_{\mathrm{I}}(t, t_0) = \exp\left[\frac{i}{\hbar}\hat{H}_0 t\right] \exp\left[-\frac{i}{\hbar}\hat{H}(t - t_0)\right] \exp\left[-\frac{i}{\hbar}\hat{H}_0 t_0\right]. \tag{3.17}$$

Further, one readily shows that the following decomposition is valid:

$$\hat{U}_{\mathrm{I}}(t, t_0) = \hat{U}_{\mathrm{I}}(t, t')\hat{U}_{\mathrm{I}}(t', t_0), \tag{3.18}$$

and that the adjoint of $\hat{U}_{\mathrm{I}}(t, t_0)$ is given by

$$\hat{U}^{\dagger}_{\mathrm{I}}(t, t_0) = \exp\left[\frac{i}{\hbar}\hat{H}_0 t_0\right] \exp\left[\frac{i}{\hbar}\hat{H}(t - t_0)\right] \exp\left[-\frac{i}{\hbar}\hat{H}_0 t\right] = \hat{U}_{\mathrm{I}}(t_0, t). \tag{3.19}$$

In what follows, we construct a useful alternative representation of the interaction frame propagator $\hat{U}_{\mathrm{I}}(t, t_0)$ by formal integration of Eq. (3.13) which yields for the scattering solution $\Psi_{\rm I}(t)$

$$\Psi_{\rm I}(t) = \Psi_{\rm I}(t_0) - \frac{i}{\hbar}\int_{t_0}^{t} \hat{H}'_{\rm I}(\tau)\Psi_{\rm I}(\tau)d\tau. \tag{3.20}$$

We represent the solution as a perturbation series whose first term is defined by the interaction-free case. Thus, the zeroth approximation to $\Psi_{\rm I}(t)$ is

$$\Psi_{{\rm I},0}(t) = \Psi_{\rm I}(t_0). \tag{3.21}$$

The nth term of the series is given by the recursive formula

$$\Psi_{{\rm I},n}(t) = -\frac{i}{\hbar}\int_{t_0}^{t} \hat{H}'_{\rm I}(\tau)\Psi_{{\rm I},n-1}(\tau)d\tau. \tag{3.22}$$

Expanding this function into all contributing orders of approximation we arrive at an n-fold product of integrals, namely

$$\Psi_{{\rm I},n}(t) = \left(-\frac{i}{\hbar}\right)^n \left[\int_{t_0}^{t} \hat{H}'_{\rm I}(\tau_n)d\tau_n \int_{t_0}^{\tau_n} \hat{H}'_{\rm I}(\tau_{n-1})d\tau_{n-1}\ldots\right.$$
$$\left.\times \int_{t_0}^{\tau_2} \hat{H}'_{\rm I}(\tau_1)d\tau_1\right]\Psi_{\rm I}(t_0). \tag{3.23}$$

We point out that the domain of the time variable $\tau_i, i < n$, is bounded from above by the value of the following variable τ_{i+1}. This condition imposes a time order on the structure of Eq. (3.23). The cumbersome restrictions on the time domains are removed by introducing the time order operator $\hat{T}$ which arranges the partial operators $\hat{H}'_{\rm I}(\tau_i)$, $1 \leq i \leq n$, according to consecutive times, such that those that refer to earlier times are positioned on the right of those referring to later times. If, for instance, $t_1 < t_2$, then $\hat{T}(\hat{H}'_{\rm I}(t_1)\hat{H}'_{\rm I}(t_2)) = \hat{H}'_{\rm I}(t_2)\hat{H}'_{\rm I}(t_1)$. Using the time order operator, the representation (3.23) reads

$$\Psi_{{\rm I},n}(t)$$
$$= \left(-\frac{i}{\hbar}\right)^n \frac{1}{n!}\hat{T}\int_{t_0}^{t} \hat{H}'_{\rm I}(\tau_n)\hat{H}'_{\rm I}(\tau_{n-1})\cdots\hat{H}'_{\rm I}(\tau_1)d\tau_n d\tau_{n-1}\cdots d\tau_1\Psi_{\rm I}(t_0). \tag{3.24}$$

The factor $\frac{1}{n!}$ corrects this product for multiple counting. The solution $\Psi_{\rm I}(t)$ results as the sum over all partial wave functions $\Psi_{{\rm I},n}(t)$ defined by

Eq. (3.24):

$$\Psi_I(t) = \hat{T}\sum_{n=0}^{\infty}\frac{1}{n!}\prod_{m=0}^{n}\left[-\frac{i}{\hbar}\int_{t_0}^{t}\hat{H}'_I(\tau_m)d\tau_m\right]\Psi_I(t_0). \tag{3.25}$$

This finding is cast into a more concise notation by writing

$$\Psi_I(t) = \hat{T}\exp\left[-\frac{i}{\hbar}\int_{t_0}^{t}dt'\,\hat{H}'_I(t')\right]\Psi_I(t_0). \tag{3.26}$$

The meaning of the time-ordering prescription is clear only in the expanded form (3.25). The compact version (3.26), however, is very helpful in formal manipulations. The time evolution operator $\hat{U}_I(t, t_0)$ results as

$$\hat{U}_I(t, t_0) = \hat{T}\exp\left[-\frac{i}{\hbar}\int_{t_0}^{t}dt'\,\hat{H}'_I(t')\right]. \tag{3.27}$$

Combining Eq. (3.27) with Eq. (3.11) and Eq. (3.26) we find

$$\Psi(t) = \exp\left[-\frac{i}{\hbar}\hat{H}_0 t\right]\Psi_I(t) = \hat{U}_0(t, t_0)\tilde{\hat{U}}_I(t, t_0)\Psi(t_0), \tag{3.28}$$

and therefore

$$\hat{U}(t, t_0) = \hat{U}_0(t, t_0)\tilde{\hat{U}}_I(t, t_0), \tag{3.29}$$

where the nomenclature $\hat{U}_0(t, t_0)$ has been introduced for the time evolution operator $\exp[-\frac{i}{\hbar}\hat{H}_0(t - t_0)]$, and $\tilde{\hat{U}}_I(t, t_0)$ for $\exp[-\frac{i}{\hbar}\hat{H}_0 t_0]$ $\hat{U}_I(t, t_0)\exp[\frac{i}{\hbar}\hat{H}_0 t_0]$. Needless to say, the latter differs from expression (3.27) just by a unitary transformation. Both propagators are equal for the choice $t_0 = 0$. The product representation of $\hat{U}$ defined by Eq. (3.29) provides a convenient partition into two separate propagation agents, one related to the reference Hamiltonian, the other to a perturbation. This subdivision proves to be useful for a broad range of dynamical problems involving the response of a system to an external disturbance (see, for instance, Section 5.4.1).

In the following section, we will discuss a limiting case of $\hat{U}_I(t, t_0)$ which consists in the projection of a state in the far past, represented by $t_0 \to -\infty$, onto a state in the far future ($t \to \infty$).

3.3 The Scattering Matrix

Asymptotically, scattering processes as considered in the previous section are determined by the Hamiltonian $\hat{H}_0$ for the free system. It is therefore very suggestive to adopt the energy representation and to describe the region beyond the interaction zone by eigenstates of $\hat{H}_0$. The respective wave functions are $\Psi_{\rm I}(-\infty)(\Psi_{\rm I}(+\infty))$, referring to the infinite past (future). We introduce an operator $\hat{S}$ that maps the asymptotic solutions into each other by

$$\Psi_{\rm I}(+\infty) = \hat{S}\Psi_{\rm I}(-\infty). \tag{3.30}$$

This so-called *scattering operator* may be expressed as a special case of the time evolution operator $\hat{U}_{\rm I}(t, t_0)$. More specifically

$$\hat{S} \equiv \lim_{\substack{t\to+\infty \\ t_0\to-\infty}} \hat{U}_{\rm I}(t, t_0). \tag{3.31}$$

This definition, involving a double limit, assumes the existence of $\hat{U}_{\rm I}(t, t_0)$ in both the remote past and the distant future. However, these conditions are not necessarily satisfied. Since the asymptotic states are chosen as eigenfunctions of $\hat{H}_0$, belonging to the continuous spectrum, our attempt to construct the $\hat{S}$ operator according to prescription (3.31) will encounter serious difficulties. This may be seen in the following way [84]. Let us assume that the solution of the TDSE at $t = t_0$ is

$$\Psi(t_0) = \exp\left[-\frac{i}{\hbar}E_i t_0\right]\varphi_i \tag{3.32}$$

and $\hat{H}_0\,\varphi_i = E_i\,\varphi_i$. We thus specify a certain continuum eigenfunction of H_0, with energy eigenvalue E_i, as the initial state of the scattering process. From (3.11), one concludes that

$$\Psi_{\rm I}(t_0) = \varphi_i \tag{3.33}$$

and in particular $\Psi_{\rm I}(-\infty) = \varphi_i$.

Analogous considerations apply to the asymptotic wave function for large positive times, i.e. $\Psi_{\rm I}(+\infty)$ is a linear combination of outgoing asymptotic eigenstates of $\hat{H}_0$. The amplitude of the eigenstate Ψ_f in the final wave function $\Psi_{\rm I}(+\infty)$ is therefore given by the matrix element

$$\begin{aligned} S_{if} = \langle\varphi_f|\hat{S}|\varphi_i\rangle &= \lim_{\substack{t\to+\infty \\ t_0\to-\infty}} \langle\varphi_f|\hat{U}_{\rm I}(t, t_0)|\varphi_i\rangle \\ &= \lim_{\substack{t\to+\infty \\ t_0\to-\infty}} \langle\hat{U}_{\rm I}(0, t)\varphi_f|\hat{U}_{\rm I}(0, t_0)\varphi_i\rangle, \end{aligned} \tag{3.34}$$

by use of Eqs. (3.18) and (3.19). By Eq. (3.17), the ket part of S_{if} is seen to oscillate as a function of t_0 and thus the limit $t_0 \to -\infty$ is in general undefined. Therefore, in order to employ the scattering operator, a limiting process must be introduced that eliminates the oscillatory behavior in the asymptotic regime. Such a device is available with a procedure introduced by Gell-Mann and Goldberger [85] that provides a damping mechanism for asymptotic oscillations. Utilizing this technique, one finds the limits of large negative and positive times for some wave function $\Phi(t)$ from

$$\lim_{t\to-\infty} \Phi(t) = \lim_{\varepsilon\to 0^+} \frac{\varepsilon}{\hbar} \int_{-\infty}^{0} dt' \exp\left(\frac{\varepsilon}{\hbar} t'\right) \Phi(t'), \tag{3.35}$$

$$\lim_{t\to+\infty} \Phi(t) = \lim_{\varepsilon\to 0^+} \frac{\varepsilon}{\hbar} \int_{0}^{\infty} dt' \exp\left(-\frac{\varepsilon}{\hbar} t'\right) \Phi(t'). \tag{3.36}$$

Through integration by parts, one readily confirms that this procedure reproduces any existing limit of the form $\lim_{t\to-\infty}\Phi(t)$ or $\lim_{t\to\infty}\Phi(t)$. If, however, $\Phi(t)$ exhibits oscillatory behavior for $t \to -\infty, +\infty$, the oscillations will be filtered out as the formulae (3.35) and (3.36) are applied. An elementary example is given in Exercise 3.2.

Exercise 3.2. Consider the function $\Phi(t) = \sin(at)$. Apply prescription (3.35) to construct its limit for large negative times. In particular, show

$$\lim_{\varepsilon\to 0^+} \frac{\varepsilon}{\hbar} \int_{-\infty}^{0} dt' \exp\left(\frac{\varepsilon}{\hbar} t'\right) \Phi(t') = -\lim_{\varepsilon\to 0^+} \frac{a\varepsilon/\hbar}{\left(\frac{\varepsilon}{\hbar}\right)^2 + a^2} = 0. \tag{3.37}$$

The oscillatory character of the asymptotic solution (3.32) is of course due to our choice of an eigenfunction of the free Hamiltonian $\hat{H}_0$ as the initial wave function in the scattering process. Instead, a wave packet may be selected as the initial state whose limit is well defined as $t \to -\infty$.

3.3.1 *The Møller operators*

The limiting scheme shown in (3.35) and (3.36) lays a mathematical foundation for evaluating the scattering matrix elements (3.34). We give separate consideration to the structure of the bra and the ket part of S_{if} by introducing the *Møller operators* $\hat{\Omega}^+$ and $\hat{\Omega}^-$ through

$$\hat{\Omega}^{\pm} = \lim_{t_0 \to \mp\infty} \hat{U}_{\mathrm{I}}(0, t_0). \tag{3.38}$$

Cast into the interaction frame, $\hat{\Omega}^+$ becomes the time evolution operator $\hat{U}_{\mathrm{I}}(t, t_0)$:

$$\begin{aligned} \hat{\Omega}_{\mathrm{I}}^{+}(t) &= \exp\left(\frac{i}{\hbar}\hat{H}_0 t\right)\hat{\Omega}^{+}\exp\left(-\frac{i}{\hbar}\hat{H}_0 t\right) \\ &= \lim_{t_0 \to -\infty} \hat{U}_{\mathrm{I}}(t, t_0) = \hat{T}\exp\left[-\frac{i}{\hbar}\int_{-\infty}^{t} dt'\,\hat{H}_{\mathrm{I}}'(t')\right] \end{aligned} \tag{3.39}$$

where the identity (3.27) has been used in the last step. This relation implies a representation of the scattering operator $\hat{S}$ in terms of the Møller operator $\hat{\Omega}_{\mathrm{I}}^{+}(t)$. More specifically, it holds that

$$\hat{S} = \hat{\Omega}_{\mathrm{I}}^{+}(\infty) = \hat{T}\exp\left[-\frac{i}{\hbar}\int_{-\infty}^{\infty} dt'\,\hat{H}_{\mathrm{I}}'(t')\right]. \tag{3.40}$$

In order to make the transition to the time-independent frame for the description of scattering processes, we clarify the operation of $\hat{\Omega}^{\pm}$ on an arbitrary continuum eigenfunction Ψ_a of $\hat{H}_0$ with eigenvalue E_a. Examining the action of $\hat{\Omega}^+$, we obtain by means of Eq. (3.35),

$$\begin{aligned} \hat{\Omega}^{+}\varphi_a &= \lim_{\varepsilon \to 0^+} \frac{\varepsilon}{\hbar}\int_{-\infty}^{0} dt'\,\exp\left(\frac{\varepsilon}{\hbar}t'\right)\exp\left(\frac{i}{\hbar}\hat{H}t'\right)\exp\left(-\frac{i}{\hbar}E_a t'\right)\varphi_a \\ &= \lim_{\varepsilon \to 0^+} \frac{i\varepsilon}{E_a - H + i\varepsilon}\varphi_a \\ &= \lim_{\varepsilon \to 0^+} \frac{1}{E_a - H + i\varepsilon}(i\varepsilon + E_a - H_0)\varphi_a \\ &= \varphi_a + \lim_{\varepsilon \to 0^+} \frac{1}{E_a - H + i\varepsilon}V\varphi_a \\ &\equiv \Psi_a^{+}, \end{aligned} \tag{3.41}$$

where we have used Eq. (3.10) with the scattering potential V as $\hat{H}'$. The scattering Schrödinger wave function Ψ_a^+ thus defined satisfies $\hat{H}\Psi_a^+ = E_a\Psi_a^+$ and therefore shares the eigenenergy E_a with the asymptotic solution Ψ_a. The symbol + signifies that the function Ψ_a^+ obeys *outgoing boundary conditions*. The analogous treatment for the case of $\hat{\Omega}^-$ yields a scattering state Ψ_a^- that fulfils *incoming boundary conditions*. Summarizing both results in

$$\Psi_a^\pm = \varphi_a + \lim_{\varepsilon\to 0^+} \frac{1}{E_a - H \pm i\varepsilon} V\varphi_a. \tag{3.42}$$

This relation is known as the *Lippmann–Schwinger equation* [86]. It may be shown (see Exercise 3.3) to be equivalent to

$$\Psi_a^\pm = \varphi_a + \lim_{\varepsilon\to 0^+} \frac{1}{E_a - H_0 \pm i\varepsilon} V\Psi_a^\pm. \tag{3.43}$$

Exercise 3.3. Establish relation (3.43). In a first step, prove the *resolvent identity* $(E_a - \hat{H} \pm i\varepsilon)^{-1} = (E_a - \hat{H}_0 \pm i\varepsilon)^{-1} + (E_a - \hat{H}_0 \pm i\varepsilon)^{-1} V (E_a - \hat{H} \pm i\varepsilon)^{-1}$.

3.3.2 *The Lippmann–Schwinger equations*

In the following, we approach the crucial identities (3.42) and (3.43) from a somewhat different angle, namely that of the time-independent Schrödinger equation, written as

$$[E_a - \hat{H}_0]\Psi_a = V\Psi_a. \tag{3.44}$$

The formal solution of this equation is

$$\Psi_a = \varphi_{a,i} + \frac{1}{E_a - \hat{H}_0} V\Psi_a, \tag{3.45}$$

with Ψ_a as solution of the homogeneous problem $[E_a - \hat{H}_0]\varphi_{a,i} = 0$. The inverse of the operator $[E_a - \hat{H}_0]$ may be defined by its spectral resolution:

$$\frac{1}{E_a - \hat{H}_0} = \int dE' \frac{|\varphi_{a,E'}\rangle\langle\varphi_{a,E'}|}{E - E'}. \tag{3.46}$$

We have added subscripts E, E' to the homogeneous solution in order to emphasize that it is part of a continuum of wave functions that depend parametrically on the energy. The representation (3.46) clarifies immediately that (3.45) is an integral equation, namely

$$\Psi_a = \varphi_{a,i,E} + \int dE' \frac{\psi_{a,E'}\langle\psi_{a,E'}|V|\Psi_{a,E}\rangle}{E - E'}. \tag{3.47}$$

The energy integral is not a well-defined Riemann integral due to the singularity at $E = E'$. By allowing E to be complex with a small imaginary part $\pm i\varepsilon$, however, one can understand the integral as the limit of Riemann integrals as ε tends towards zero. Implementing this idea generates the Lippmann–Schwinger equation (3.43). Depending on the sign of $\pm\varepsilon$, the integral (3.46) tends to different limiting values, corresponding to different scattering solutions Ψ_a^+ and Ψ_a^- which are not orthogonal to each other.

The Lippmann–Schwinger equations (3.42) and (3.43) are customarily cast into the more compact Green's function formalism, based on the two *Green's operators* $\hat{G}_0^\pm$ and $\hat{G}^\pm$:

$$\hat{G}_0^\pm = \lim_{\varepsilon\to 0^+} \frac{1}{E_a - \hat{H}_0 \pm i\varepsilon}, \tag{3.48}$$

$$\hat{G}^\pm = \lim_{\varepsilon\to 0^+} \frac{1}{E_a - \hat{H} \pm i\varepsilon}. \tag{3.49}$$

The scattering solutions are thus obtained from the identities

$$\Psi_a^\pm = \varphi_{a,i} + \hat{G}^\pm V \varphi_{a,i}. \tag{3.50}$$

$$\Psi_a^\pm = \varphi_{a,i} + \hat{G}_0^\pm V \Psi_a^\pm, \tag{3.51}$$

in which one recognizes alternative realizations of the Lippmann–Schwinger equations. Specifically, relation (3.51) represents a condensed notation of the Green's function procedure which is customary in the treatment of inhomogeneous differential equations.

Exercise 3.4. Show that the following operator analogs of the Lippmann–Schwinger equations for the scattering wave function are valid: $\hat{G}^\pm = \hat{G}_0^\pm + \hat{G}_0^\pm V \hat{G}^\pm$ and $\hat{G}^\pm = \hat{G}_0^\pm + \hat{G}^\pm V \hat{G}_0^\pm$.

We further introduce the decomposition of the Green's operators $\hat{G}^\pm$ and $\hat{G}_0^\pm$ into a real and an imaginary part, which is instrumental for the definition of real-valued scattering solutions. This is achieved by use of the relation

$$\frac{1}{x - x_0 \pm i\varepsilon} = P\frac{1}{x - x_0} \mp i\pi\delta(x - x_0), \tag{3.52}$$

where P denotes the *Cauchy principal value*. The first term on the right-hand side of Eq. (3.52) is explicitly written as

$$P\frac{1}{x - x_0} = \lim_{\varepsilon\to 0} \frac{1}{2}\left(\frac{1}{x - x_0 + i\varepsilon} + \frac{1}{x - x_0 - i\varepsilon}\right). \tag{3.53}$$

Invoking the delta function representation

$$\delta(y) = \lim_{\varepsilon \to 0} \frac{1}{2i\pi} \left(\frac{1}{y - i\varepsilon} - \frac{1}{y + i\varepsilon} \right) = \lim_{\varepsilon \to 0} \frac{\varepsilon/\pi}{y^2 + \varepsilon^2}, \tag{3.54}$$

and referring to $\hat{G}^{\pm}$ alone, we readily conclude that the latter operator may be partitioned according to

$$\hat{G}^{\pm} = \hat{G}^{\mathrm{p}}(E) \mp i\pi\delta(E - \hat{H}) \tag{3.55}$$

with $\hat{G}^{\mathrm{p}}(E) \equiv P\frac{1}{E-\hat{H}}$. This expansion of $\hat{G}^{\pm}$ may be viewed as subdividing the operator into a Hermitian and an anti-Hermitian part. A relation analogous to Eq. (3.55) is valid for the $\hat{G}_0^{\pm}$ operator, whose principal part is labeled $\hat{G}_0^{\mathrm{p}}$. The notation $\delta(E - \hat{H})$ is explained by [87]

$$\delta(E - \hat{H})\Psi(E_1) = \delta(E - E_1)\Psi(E_1). \tag{3.56}$$

A principal-value scattering state Ψ is constructed from $\hat{G}^{\mathrm{p}}$ by means of the prescription

$$\Psi_a = \varphi_{a,i} + \hat{G}^{\mathrm{p}} V \Psi_{a,i} \tag{3.57}$$

or

$$\Psi_a = \varphi_{a,i} + \hat{G}_0^{\mathrm{p}} V \Psi_a. \tag{3.58}$$

As the asymptotic solution $\varphi_{a,i}$ is chosen as a real-valued function, the state Ψ_a is seen to be real-valued itself. Solutions of the type (3.57) and (3.58) have a standing wave character.

3.3.3 *Unitarity of the S-matrix*

The Møller operators map an eigenfunction of $\hat{H}_0$ onto one of $\hat{H}$. The spectrum of $\hat{H}_0$, however, is entirely continuous while no such restriction applies to $\hat{H}$. Thus, by action of $\hat{\Omega}^{\pm}$, the entire set of the $\hat{H}_0$ eigenstates is projected onto the continuous subset of $\hat{H}$. Mathematically, this characteristic feature of the Møller operators is reflected by their property of *unidirectional unitarity* [7]. Specifically, it holds that

$$(\hat{\Omega}^{\pm})^{\dagger}\hat{\Omega}^{\pm} = \hat{I}. \tag{3.59}$$

The corresponding relation for the adjoint $\hat{\Omega}^{\dagger\pm}$, which would complete the proof of unitarity, is not valid in general. Thus, Eq. (3.59) implies that the adjoint Møller operator maps an eigenvector of $\hat{H}$ into an asymptotic wave function. The product $\hat{\Omega}^{\pm}(\hat{\Omega}^{\pm})^{\dagger}$ can therefore yield the identity only if the former belongs to the continuous spectrum of $\hat{H}$. On the other hand, cases may be found where

$$\hat{\Omega}^{\pm}(\hat{\Omega}^{\pm})^{\dagger} \neq I. \tag{3.60}$$

Such a case is realized if the $(\hat{\Omega}^{\pm})^{\dagger}$ acts on a bound eigenstate of $\hat{H}$. None of these states can be projected by this operator on any eigenstate of $\hat{H}_0$. The quasi-unitarity of the Møller operators may be readily used to prove the unitarity of the scattering operator. The properties (3.18) and (3.19) of $\hat{U}_{\text{I}}$ as well as the definitions (3.31) and (3.38) of $\hat{S}$ and $\hat{\Omega}^{\pm}$, respectively, cooperate to yield

$$\hat{S} = (\hat{\Omega}^{-})^{\dagger}\hat{\Omega}^{+}. \tag{3.61}$$

Since both the scattering operator and its adjoint act exclusively on asymptotic states, only the continuous subset of the eigenfunctions of $\hat{H}$ enter in this case, and the relation $\hat{\Omega}^{\pm}(\hat{\Omega}^{\pm})^{\dagger} = \hat{I}$ is satisfied. The latter formula in conjunction with Eq. (3.59) establishes

$$\hat{S}\hat{S}^{\dagger} = \hat{S}^{\dagger}\hat{S} = I \tag{3.62}$$

and thus the unitarity of $\hat{S}$. By the definition (3.30) of the scattering operator, implying

$$\begin{aligned} \langle\Psi_{\text{I}}(-\infty)|\Psi_{\text{I}}(-\infty)\rangle &= \langle\Psi_{\text{I}}(-\infty)|\hat{S}^{\dagger}\hat{S}|\Psi_{\text{I}}(-\infty)\rangle \\ &= \langle\hat{S}\Psi_{\text{I}}(-\infty)|\hat{S}\Psi_{\text{I}}(-\infty)\rangle = \langle\Psi_{\text{I}}(+\infty)|\Psi_{\text{I}}(+\infty)\rangle, \end{aligned} \tag{3.63}$$

the property (3.62) signifies the conservation of particle number in scattering processes.

If no interaction is involved, the scattering matrix reduces trivially to unity. This is acknowledged by partitioning **S** according to

$$\mathbf{S} = \mathbf{I} - 2i\mathbf{T}, \tag{3.64}$$

where the description of the scattering process is confined to the *transition matrix* **T** which is given by the elements

$$T_{if} = \langle\varphi_f|\hat{T}|\varphi_i\rangle, \tag{3.65}$$

with $\hat{T}$ as *transition operator*. The factor $-2i$ in Eq. (3.64) is conventional. Relation (3.64) implies that the unitarity of the S-matrix is guaranteed as long as the transition matrix obeys:

$$2\mathbf{T}^\dagger\mathbf{T} = i(\mathbf{T} - \mathbf{T}^\dagger). \tag{3.66}$$

The unitary property of the scattering matrix manifests the conservation of flux in scattering processes. The conservation of energy in scattering interactions, as encoded by the commutation relation

$$[\hat{H}_0, \hat{S}] = 0, \tag{3.67}$$

is derived by utilizing the Møller operators. This task is to be confronted in Exercise 3.5.

Exercise 3.5. The vanishing of the commutator $[\hat{H}_0, \hat{S}]$ is a direct consequence of the *intertwining relation*

$$\hat{H}\hat{\Omega}^\pm = \hat{\Omega}^\pm \hat{H}_0.$$

Use the explicit form of the Møller operators to prove this relation. On the basis of this result, demonstrate that Eq. (3.67) is satisfied.

The S-matrix element (3.34) may be reformulated in terms of the scattering functions. Setting the index a equal to i or f, we obtain from Eq. (3.61)

$$S_{if} = \langle \Psi_f^- | \Psi_i^+ \rangle = \langle \Psi_f^+ | \Psi_i^+ \rangle + \langle \Psi_f^- - \Psi_f^+ | \Psi_i^+ \rangle, \tag{3.68}$$

where Ψ_i^+ is generated by advancing the asymptotic wave function Ψ_i that is realized in the "far past" to the time $t = 0$ when the scattering process is taking place. The relation between Ψ_f^- and φ_f is strictly analogous. Both Ψ_i^+ and Ψ_f^- are here understood as eigenfunctions of $\hat{H}$ while φ_i and φ_f are two independent eigenfunctions of the unperturbed Hamiltonian $\hat{H}_0$. Thus, by applying the operator $\hat{G}^+$ ($\hat{G}^-$) on φ_f (φ_i), we may generate the solutions Ψ_f^+ (Ψ_i^-) according to Eq. (3.50). It can be shown that $\Psi_f^\pm$ and $\Psi_i^\pm$ are orthogonal to each other (for a proof of this statement, see, for instance, [88]). In the following, we remodel both terms on the right-hand side of Eq. (3.68). We adopt energy normalization such that $\langle \Psi_f^+ | \Psi_i^+ \rangle = \delta(E_f - E_i)\delta_{if}$, where the indices i, f comprise all discrete quantum numbers that characterize the scattering solution. With the help of Eq. (3.42), one realizes that

$$\Psi_f^- - \Psi_f^+ = \left(\frac{1}{E_f - \hat{H} - i\varepsilon} - \frac{1}{E_f - \hat{H} + i\varepsilon} \right) V\varphi_f, \tag{3.69}$$

where it is understood that ultimately the limit $\varepsilon \to 0^+$ is taken. Inserting Eq. (3.69) into Eq. (3.68) yields

$$\begin{aligned} S_{if} &= \delta(E_f - E_i)\delta_{if} + \left\langle \left(\frac{1}{E_f - \hat{H} - i\varepsilon} - \frac{1}{E_f - \hat{H} + i\varepsilon} \right) V\varphi_f | \Psi_i^+ \right\rangle \\ &= \delta(E_f - E_i)\delta_{if} - \left\langle \varphi_f | V \left(\frac{1}{E_f - \hat{H} - i\varepsilon} - \frac{1}{E_f - \hat{H} + i\varepsilon} \right) \Psi_i^+ \right\rangle \\ &= \delta(E_f - E_i)\delta_{if} - \frac{2i\varepsilon}{(E_f - E_i)^2 + \varepsilon^2} \langle \varphi_f | V\Psi_i^+ \rangle. \end{aligned} \tag{3.70}$$

As we carry out the operation $\varepsilon \to 0^+$ and observe that $\lim_{\varepsilon \to 0^+} \frac{\varepsilon/\pi}{(E_f - E_i)^2 + \varepsilon^2}$ is a representation of the delta function $\delta(E_f - E_i)$ [58], the scattering operator matrix element turns into the compact expression

$$S_{if} = \delta(E_f - E_i)(\delta_{if} - 2\pi i \langle \varphi_f | V\Psi_i^+ \rangle). \tag{3.71}$$

We generalize this result by defining an *on-shell scattering matrix element* $S_{if}(E)$ through the relation

$$\langle \varphi_f(E') | \hat{S} | \varphi_i(E) \rangle = \delta(E - E') S_{if}(E). \tag{3.72}$$

From (3.71), $S_{if}(E)$ may be written explicitly as

$$S_{if}(E) = \delta_{if} - 2\pi i \langle \varphi_f | V\Psi_i^+ \rangle \tag{3.73}$$

where, by comparison with Eq. (3.64), the scalar product $\langle \varphi_f | V\Psi_i^+ \rangle$ is to be interpreted as a transition matrix element. We notice in (3.71) and (3.73) a factor π that is lacking in the definition of the T-matrix given in Eq. (3.64). This factor is due to the energy normalization chosen to represent the on-shell scattering matrix [7]. An important conclusion to be drawn from Eqs. (3.71) and (3.73) is that the $\hat{T}$ operator projects an asymptotic initial solution onto the outgoing scattering wave function according to

$$\hat{T}\varphi_i = V\Psi_i^+. \tag{3.74}$$

Exercise 3.6. Use the properties of the S-matrix to demonstrate that $T_{if} = \langle \varphi_f | V \Psi_i^+ \rangle = \langle \Psi_f^- | V \varphi_i \rangle$.

3.4 Elastic Scattering by a Spherical Potential

This section provides a condensed summary of the most elementary scattering situation, namely elastic scattering by a spherically symmetric potential. While this case gives a basic example of single-channel scattering and is thus quite remote from the main theme of this text, which involves multiple electronic channels, it will allow us to introduce various fundamental concepts of relevance for complex dynamic interactions between molecules in a systematic fashion. We employ the time-independent approach derived in the preceding section from the time-dependent formalism, i.e. we seek to determine scattering states as eigenfunctions of the Hamiltonian that governs the considered problem.

3.4.1 *The asymptotic scattering solution*

The Hamiltonian contains a central potential $V(R)$ such that

$$\hat{H} = -\frac{\hbar^2}{2\mu}\frac{1}{R}\frac{\partial^2}{\partial R^2}R + \frac{\hat{L}^2}{2\mu R^2} + V(R) \tag{3.75}$$

with $\hat{L}$ as the orbital angular momentum operator and μ the effective mass characteristic of the problem at hand. Solving the stationary Schrödinger equation

$$\hat{H}\Psi = E\Psi \tag{3.76}$$

is facilitated by partial wave analysis. More specifically, we express the solution as a sum over partial waves, employing the ansatz

$$\Psi(R) = \frac{1}{R}\sum_{l=0}^{\infty} a_l \psi_l(R) P_l(\cos\theta), \tag{3.77}$$

which includes the Legendre polynomials P_l. Inserting this wave function into the stationary Schrödinger equation and using closure with respect to the Legendre polynomials yields a set of equations that allow us to determine the radial functions $\psi_l(R)$:

$$\left[\frac{d^2}{dR^2} + k^2 - U - \frac{l(l+1)}{R^2}\right]\psi_l(R) = 0, \tag{3.78}$$

where we have substituted k^2 for $\frac{2\mu E}{\hbar^2}$ and U for $\frac{2\mu V}{\hbar^2}$. The term $U - \frac{l(l+1)}{R^2}$ represents an effective potential U_l^{eff} determined by the orbital angular momentum quantum number l. We assume that $U_l^{\mathrm{eff}}(R)$ tends to zero for large R. This implies that $\psi_l(R)$ merges into a periodic function as $R \to \infty$. Consequently, we may write ψ_l as $u(R)\exp(ikR)$, with the understanding that the function u becomes nearly constant in the asymptotic region. As shown for instance in [89], this requirement is only fulfilled if $RU(R) \to 0$ as $R \to \infty$. Thus, at a sufficiently large distance, $U_l^{\mathrm{eff}}(R)$ must fall off faster than the Coulomb potential. We also stipulate that the regime $R \approx 0$ is dominated by the centripetal potential, setting $\lim_{R\to 0} R^2 U_l^{\mathrm{eff}}(R) = l(l+1)$. If this condition is satisfied, Eq. (3.78) implies that the behavior of ψ_l with R close to the origin is as R^{-l} or R^{l+1}. Only the latter of these two alternatives, i.e. only the regular or bounded one among the two mathematically permitted independent solutions, is physically acceptable. This restriction to one solution excludes complex choices for ψ_l, excepting a constant multiplying factor, since for any complex ψ_l, the conjugate ψ_l^* would represent a second, equally admissible independent solution. The general asymptotic form of $\psi_l(R)$ is therefore written as

$$\lim_{R\to\infty} \psi_l(R) = C_l \sin(kR + \eta_l), \tag{3.79}$$

where C_l is a constant factor and η_l is real. Phrasing this expression such that it involves both the radially incoming and outgoing waves, we rewrite Eq. (3.79) as

$$\psi_l(R) \sim \frac{1}{2i}\left[\exp(ikR + i\eta_l) - \exp(-ikR - i\eta_l)\right]. \tag{3.80}$$

Obviously, both components have to be of equal amplitude. Relation (3.80) shows explicitly that elastic scattering manifests itself by a phase shift between the incoming and the outgoing part which is characteristic of the scattering potential $U_l^{\mathrm{eff}}(R)$. Conventionally, one subtracts the effect of the centripetal contribution by introducing the phase δ_l as a quantity that reflects the impact of $U(R)$ alone, setting

$$\delta_l = \eta_l - \eta_l'. \tag{3.81}$$

The purely centripetal phase η_l' is identified as one solves the well-known equation [112]

$$\left[\frac{d^2}{dR^2} + k^2 - \frac{l(l+1)}{R^2}\right]\varphi_l(R) = 0. \tag{3.82}$$

In the absence of the scattering potential $U(R)$, the bounded solution of the stationary Schrödinger equation is given by $\varphi_l(R) = kR\widetilde{j}_l(kR)$, where $\widetilde{j}_l$ denotes the spherical Bessel function of order l. Asymptotically, it behaves as

$$\lim_{R\to\infty} \varphi_l(R) = D_l \sin\left(kR - \frac{l\pi}{2}\right). \tag{3.83}$$

Therefore, the centripetal contribution to the phase is $\eta_l' = l\pi/2$ and $\psi_l(R)$ is written in the limit of large radii as

$$\lim_{R\to\infty} \psi_l(R) = C_l \sin\left(kR - \frac{l\pi}{2} + \delta_l\right), \tag{3.84}$$

where C_l and D_l are constants. The latter solution may be expanded into a linear combination of the *Riccati–Bessel* and *Riccati–Neumann* functions $j_l(kR)$ and $n_l(kR)$, respectively, which are related to the spherical Bessel and Neumann functions $\widetilde{j}_l(kR)$ and $\widetilde{n}_l(kR)$ through $j_l(kR) = kR\widetilde{j}_l(kR)$ and $n_l(kR) = kR\widetilde{n}_l(kR)$. Using these functions, the scattering solution (3.84) becomes

$$\lim_{R\to\infty} \psi_l(R) = A_l j_l(kR) + B_l n_l(kR), \tag{3.85}$$

where

$$C_l^2 = A_l^2 + B_l^2 \tag{3.86}$$

and

$$\tan\delta_l = -\frac{B_l}{A_l}. \tag{3.87}$$

We further introduce the complex-valued Riccati–Hankel functions $h^{\pm}(kR)$ by $h^{\pm}(kR) = n_l(kR) \pm i j_l(kR)$. Their asymptotic behavior is

$$\lim_{R\to\infty} h^{\pm}(kR) = F_l \exp\left[\pm i(kR - \frac{l\pi}{2})\right]. \tag{3.88}$$

In terms of the Riccati–Hankel functions the scattering solution (3.85) assumes the form

$$\lim_{R\to\infty} \psi_l(R) = \frac{1}{2i}[C_l^+ h^+(kR) - C_l^- h^-(kR)]. \tag{3.89}$$

Here, the coefficient C_l^+ denotes $A_l + iB_l$, and $C_l^- = C_l^{+*}$.

3.4.2 *T-, S-, and K-matrix boundary conditions*

Having specified the asymptotic elastic scattering wave functions in the case of a spherical potential by direct inspection of the stationary Schrödinger equation, we will analyze now the same problem from an integral equation viewpoint. For this purpose we employ the Lippmann–Schwinger equation (3.51) in the form

$$\psi_l^+ = \varphi_l + \hat{G}_{0,l}^+ V \psi_l^+, \tag{3.90}$$

with V as a spherically symmetric potential as in Eq. (3.75) and φ_l as the solution of the homogeneous problem which is obtained by neglecting this potential. Such a solution is given by the Riccati–Bessel function $j(kR)$ as defined before. A coordinate representation of the recursion relation (3.90) is given by

$$\psi_l^+(R) = j_l(kR) + \int_0^\infty dR' G_{0,l}^+(R,R')V(R')\psi_l^+(R'), \tag{3.91}$$

with $G_{0,l}^+(R,R')$ as *the radial Green's function*. Inserting the integral form (3.91) of the solution into (3.78), we conclude that $G_{0,l}^+(R,R')$ satisfies the relation

$$\left[\frac{d^2}{dR^2} + k^2 - \frac{l(l+1)}{R^2}\right] G_{0,l}^+(R,R') = \frac{2\mu}{\hbar^2}\delta(R-R'). \tag{3.92}$$

The radial Green's function is obtained by treating the domains $R < R'$ and $R > R'$ in separation from each other. One confirms immediately that Eq. (3.92) is solved by

$$G_{0,l}^+(R,R') = \begin{cases} A_<(R')j_l(kR) & \text{if } R < R', \\ A_>(R')h_l^+(kR) & \text{if } R > R', \end{cases} \tag{3.93}$$

where the regular solution of the homogeneous problem has been chosen for the region close to the origin and the outgoing Hankel function for the zone that merges into the asymptotic regime. To evaluate the factors $A_<(R')$ and $A_>(R')$, we use the continuity of the Green's function as well as the discontinuity of its first derivative at the place $R = R'$. Specifically

$$A_<(R')j_l(kR') = A_>(R')h_l^+(kR'), \tag{3.94}$$

and

$$A_>(R')h_l^{+\prime}(kR') - A_<(R')j_l'(kR') = \frac{2\mu}{\hbar^2}. \tag{3.95}$$

In the latter equation we employ the compact notation $h_l^{+\prime}$ and j_l' to denote the derivatives of the Riccati–Bessel and Riccati–Neumann functions with respect to R. The statement (3.95) comprises the 'jump condition' that has to be satisfied by the first derivative of a Green's function $G(R, R')$ in a one-dimensional situation [90], or in single-channel conditions, as investigated in this chapter.

The Wronskian for the two classes of solutions, regular and irregular, is obtained from

$$W(R) = j_l'(kR)h_l^+(kR) - j_l(kR)h_l^{+\prime}(kR) = k. \tag{3.96}$$

Comparison between the discontinuity relation (3.95) and the Wronskian formula (3.96) yields the coefficients $A_<(R')$ and $A_>(R')$ as

$$A_<(R') = -\frac{2\mu}{\hbar^2 k}h_l^+(kR'), \tag{3.97}$$

$$A_>(R') = -\frac{2\mu}{\hbar^2 k}j_l(kR'). \tag{3.98}$$

We use these results to write the Green's function (3.93) in the following compact form:

$$G_{0,l}^+(R, R') = -\frac{2\mu}{\hbar^2 k}j_l(kR_<)h_l^+(kR_>). \tag{3.99}$$

For each pair R and R', $R_<$ signifies the smaller and $R_>$ the larger of the two. Inserting this expression into (3.91), we rewrite the scattering solution

ψ_l^+ in the following way:

$$\psi_l^+(R) = j_l(kR) - \frac{2\mu}{\hbar^2 k}\int_0^\infty dR'\, j_l(kR_<)h_l^+(kR_>)V(R')\psi_l^+(R'). \qquad (3.100)$$

Taking the limit $R \to \infty$ makes it possible to eliminate the variables $R_<$ and $R_>$ from Eq. (3.100), leading to

$$\lim_{R\to\infty} \psi_l^+(R) = j_l(kR) - \frac{2\mu}{\hbar^2 k}h_l^+(kR)\int_0^\infty dR'\, j_l(kR')V(R')\psi_l^+(R'). \qquad (3.101)$$

In the integral on the right-hand side we recognize the scalar product $\langle j_l, |V|\psi_l^+\rangle$ which, in keeping with Eq. (3.74) and Exercise 3.6, is associated with the transition operator matrix element. We make the identification

$$T_l = \frac{2\mu}{\hbar^2 k}\int_0^\infty dR'\, j_l(kR')V(R')\psi_l^+(R'). \qquad (3.102)$$

This assignment allows the asymptotic solution (3.101) to be cast into the compact form

$$\lim_{R\to\infty} \psi_l^+(R) = j_l(kR) - T_l h_l^+(kR). \qquad (3.103)$$

Recalling the operator equation $\hat{S} = \hat{I} - 2i\hat{T}$ (3.64), we reexpress this result in terms of the scattering matrix as

$$\lim_{R\to\infty} \psi_l^+(R) = \frac{1}{2i}[-h_l^-(kR) + S_l h_l^+(kR)]. \qquad (3.104)$$

Formula (3.103) describes the asymptotic form of $\psi_l^+(R)$ by *T-matrix boundary conditions*, associated with a standing wave as incoming solution, while (3.104) realizes *S-matrix boundary conditions*, involving an incident wave function with a definite momentum. An explicit expression for the factor T_l is derived by comparing Eq. (3.84) with Eq. (3.103). This yields

$$T_l = -\sin(\delta_l)\exp(i\delta_l). \qquad (3.105)$$

Utilizing this result, we establish the explicit form of S_l after a quick calculation to be

$$S_l = \exp(2i\delta_l). \qquad (3.106)$$

This implies that under conditions of elastic scattering, the current density associated with ψ_l^+, $(|S_l|^2 - 1)\hbar k/\mu$ (see Exercise 3.7), always reduces to zero, in keeping with expectation. Both the S-matrix and the current density expression have to be modified as admission is made for complex-valued energies and wave numbers. This extension of the theory becomes relevant in the context of resonances, as considered in further detail in Section 3.5.

Exercise 3.7. Use Eq. (3.104) to show that the one-particle current density resulting from elastic scattering by a potential is given by $(|S_l|^2 - 1)\hbar k/\mu$.

A real-valued homogeneous solution ψ_l of Eq. (3.78) is obtained by reformulating the asymptotic wave function given by Eq. (3.108) in the following way:

$$\lim_{R\to\infty} \psi_l^+(R) = \psi_l(R)(1 - iT_l). \tag{3.107}$$

This equation is satisfied if $\psi_l(R)$ is chosen for large R as

$$\lim_{R\to\infty} \psi_l(R) = j_l(kR) + n_l(kR)K_l. \tag{3.108}$$

Comparison with Eq. (3.103) yields the relation between the factors T_l and K_l. It turns out to be

$$K_l = -\frac{T_l}{1 - iT_l}. \tag{3.109}$$

Exploiting Eqs. (3.105) and (3.106), K_l may as well be expressed in terms of S_l:

$$K_l = i\frac{1 - S_l}{1 + S_l}. \tag{3.110}$$

Either of the preceding identities leads to an explicit formula for the factor K_l, namely

$$K_l = \tan \delta_l, \tag{3.111}$$

which demonstrates that the bounding function as given by the right-hand side of Eq. (3.108) is indeed real-valued. The solution on the right-hand side of Eq. (3.108) defines the *K-matrix boundary condition* as opposed to T- and S-matrix boundary conditions which apply to complex-valued asymptotic wave functions.

From the foregoing discussion it is obvious that the wave function

$$\psi_l(R) = \frac{\psi_l^+(R)}{1 - iT_l} \tag{3.112}$$

satisfies the K matrix boundary condition (3.108). This real function is a principal-value scattering state which may be generated by solving the Lippmann–Schwinger equations (3.57) and (3.58). From Eq. (3.102) in combination with Eqs. (3.109) and (3.112), we conclude that

$$K_l = -\frac{2\mu}{\hbar^2 k}\frac{\langle j_l|V|\psi_l^+\rangle}{1 - iT_l} = -\frac{2\mu}{\hbar^2 k}\langle j_l|V|\psi_l\rangle. \tag{3.113}$$

This result explicitly specifies the elements of the K-matrix, often also referred to as the *reactance matrix*.

3.4.3 *The elastic cross section*

The experimentally accessible quantity that makes it possible to examine theoretical predictions related to scattering processes is the differential or integral cross section of the considered interaction. The differential cross section per solid angle $\frac{d\sigma}{d\Omega}$ is defined as the ratio of the scattered flux emerging from the region of interaction to the incoming flux. Here $d\mathbf{\Omega}$ is the vector solid angle subtended by an area element dS of a sphere surrounding the scattering center. To make contact with experiment, we associate the incoming wave with a monoenergetic beam of particles at a well-defined momentum of magnitude $\hbar k$, and incident along the z-axis. If we denote the number of particles passing per unit of time through dS by dN, this portion of the number of incident particles may be described in terms of the incoming or the scattered current density, $\mathbf{j}_{\text{inc}}$ or $\mathbf{j}_{\text{sc}}$, respectively:

$$dN = R^2\mathbf{j}_{\text{sc}} \cdot d\mathbf{\Omega} = j_{\text{inc}} d\sigma. \tag{3.114}$$

The latter formula may be read as definition of the differential cross section $d\sigma$. The scattering geometry implies that the incoming wave is proportional to $\exp(ikz)$. The scattered state, in contrast, is given by a spherically symmetric outgoing wave modulated by a factor that depends on the polar angle θ. In summary, the asymptotic form of the solution may be written as

$$\lim_{R\to\infty} \Psi = \exp(ikz) + f(\theta)\frac{\exp(ikR)}{R}, \tag{3.115}$$

where the function $f(\theta)$ stands for the *scattering amplitude*. From Eq. (3.114), the differential cross section involves the projection of the current density associated with the second term of the superposition on the orientation of $d\boldsymbol{\Omega}$ which is the radial direction. Therefore

$$\frac{d\sigma}{d\Omega} = \frac{R^2 j_{sc,R}}{j_{inc}}. \tag{3.116}$$

A quick calculation yields $\frac{\hbar k}{\mu R^2}|f(\theta)|^2$ for the radial component of the scattered current density $j_{sc,R}$, and $\frac{\hbar k}{\mu}$ for j_{inc}. One concludes that

$$\frac{d\sigma}{d\Omega} = |f(\theta)|^2. \tag{3.117}$$

It is inviting to equate the ansatz (3.115) to the partial wave expansion of the solution as represented by Eq. (3.77). Using the partial wave representation of $\exp(ikz)$, namely

$$\lim_{R\to\infty} \exp(ikz) = \frac{1}{kR}\sum_{l=0}^{\infty}(2l+1)i^l P_l(\cos\theta)\sin\left(kR - \frac{l\pi}{2}\right), \tag{3.118}$$

and inserting into Eq. (3.77) the result of the phase shift analysis (3.84), where the factors C_l are absorbed by the coefficients a_l, we derive from this comparison explicit expressions for both the coefficients a_l

$$a_l = \frac{1}{k}(2l+1)i^l \exp(i\delta_l), \tag{3.119}$$

and the scattering amplitude $f(\theta)$,

$$f(\theta) = -\frac{1}{k}\sum_{l=0}^{\infty}(2l+1)\sin\delta_l \exp(i\delta_l)P_l(\cos\theta). \tag{3.120}$$

As we substitute for $\sin\delta_l \exp(i\delta_l)$ by use of Eq. (3.105), the scattering amplitude assumes the more compact form

$$f(\theta) = \frac{1}{k}\sum_{l=0}^{\infty}(2l+1)i^l\, T_l P_l(\cos\theta). \tag{3.121}$$

Exercise 3.8. Show by explicit calculation, i.e. by equating expressions (3.77) and (3.115), that formulae (3.119) and (3.120) are valid.

Combining Eq. (3.117) with Eq. (3.120) yields a simple relation between the integral cross section for particle scattering by spherically symmetric potentials and the scattering phases. More specifically

$$\sigma = \int_0^{\pi} d\theta \, 2\pi \sin\theta |f(\theta)|^2 = \frac{4\pi}{k^2} \sum_{l=0}^{\infty} (2l+1) \sin^2 \delta_l, \tag{3.122}$$

where the orthonormality relation for Legendre polynomials

$$\frac{2l+1}{2} \int_0^{\infty} d\theta \sin\theta P_l(\cos\theta) P_l'(\cos\theta) = \delta_{ll'} \tag{3.123}$$

has been utilized.

3.5 Resonances

As long as the energy E is chosen as a real-valued quantity, the scattering solutions reduce at large distances to the form (3.84), allowing for a complex-valued amplitude C_l (see [92]). This finding involves a scattering matrix element of unit magnitude ($S_l = \exp(2i\delta_l)$, see Eq. (3.106)) and therefore vanishing radial current density. This feature of the theory is inadequate for the description of resonance phenomena. These arise, for instance, if a particle undergoing a scattering process is temporarily trapped by a potential barrier from which it may escape through tunneling, a case referred to as *shape resonance*. *Feshbach resonances*, in contrast, require a multidimensional scattering situation. Here the energy associated with the scattering coordinate is transferred for a finite period to other degrees of freedom of the system as a whole, resulting in the formation of a quasi-bound state that decays with a characteristic life time. An example is given by the excitation of a He^+ ion by an incident electron that may combine with the ionic core to form a short-lived metastable composite before being ejected as the ion returns to its ground state. Both resonance types are illustrated in Figure 3.1.

Feshbach resonances lead frequently to *predissociation dynamics* (see also Section 7.4), associated with the coupling between a bound and a continuum state of a molecule. *Electronic predissociation* [93,94] is a typical nonadiabatic

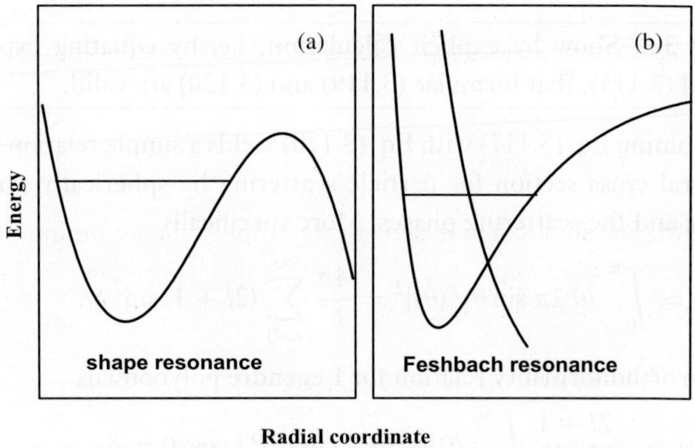

Figure 3.1: Sketches of typical configurations for (a) shape and (b) Feshbach resonances. In both cases, the features of the potential landscape give rise to the temporary formation of bound or quasi-bound states. In the situation illustrated by sketch (b), the resonance effect is associated with the partial coincidence of regions supporting discrete and continuous states in the potential energy landscape.

phenomenon. Adopting the adiabatic representation, one can describe this effect in terms of two potential energy surfaces that are coupled by an avoided crossing or a conical intersection (see Section 2.5). As the state of higher energy is prepared, the vibrational subsystem splits into two components: one that remains confined to the upper surface, and a second one that undergoes a transition to the lower surface where it is not bound any longer. This second process gives rise to the dissociation of the molecule. The fragmentation mechanism is in this case not tunneling, but inelastic electron–phonon scattering. Electronic predissociation involving diatomic systems has been described in great detail for NaI [95] and Ne_2^+ [96, 97]. In the latter case the effect was found to occur after electron-beam ionization of Ne_2.

In contrast to the elastic scattering processes considered so far, resonances do not involve cancellation of the overall flux, since the incoming and outgoing components are time-shifted with respect to each other. A limiting case is realized if no incoming flux is present during the gradual decay of the metastable state. In this situation, the radial current density is purely outgoing.

To capture the essential traits of a resonance, we make admission for a complex-valued total energy and correspondingly introduce a complex wave number $k = k_1 + ik_2$ as well as a complex phase shift $\delta_l = \delta_{1l} + i\delta_{2l}$. From the definition $E = \frac{2\mu}{\hbar^2}k^2$ we infer the real and the imaginary part of the total energy:

$$E = \frac{2\mu}{\hbar^2}(k_1 + ik_2)^2 = \frac{2\mu}{\hbar^2}(k_1^2 - k_2^2 + 2ik_1k_2). \tag{3.124}$$

To calculate the radial current density we employ the asymptotic form of the scattering solution as given by Eq. (3.104) which separates the incoming and the outgoing contributions. This yields

$$\mathcal{J}_R = \frac{2\mu}{\hbar^2}[-k_1 \sinh(2k_2R + 2\delta_{2l}) + k_2 \sin(2k_1R + 2\delta_{1l})]. \tag{3.125}$$

The first term in this equation strongly dominates the second at large distances. The imaginary part of the energy (3.124) translates into a decay factor, as can be seen by inspection of the wave function time dependence at an energy $E = E_{Rl} - \frac{i}{2}\Gamma_{Rl}$:

$$\exp\left[-i\frac{Et}{\hbar}\right] = \exp\left(-i\frac{E_{Rl}t}{\hbar}\right)\exp\left[-\frac{\Gamma_{Rl}t}{2\hbar}\right], \tag{3.126}$$

implying the existence of a decay time constant

$$\tau = \frac{\hbar}{\Gamma_{Rl}}. \tag{3.127}$$

Let us investigate the disintegration of a metastable quasi-bound state in the case where the corresponding asymptotic wave function (3.104) consists entirely of the outward component $S_l h^+(kR)$. This is achieved if the scattering matrix element S_l is infinite under resonance conditions, related to a singularity of S_l at the eigenvalue $E = E_{Rl} - \frac{i}{2}\Gamma_{Rl}$. Analyzing the time dependence of the wave function in further detail reveals that the amplitude S_l carries temporal information about the decay process. To see this, we rewrite the product $S_l(E)\exp(-i\frac{Et}{\hbar})$ in the following manner:

$$S_l(E)\exp\left(-\frac{i}{\hbar}Et\right) = \exp\left\{-i\left[i\ln(S_l(E)) + \frac{Et}{\hbar}\right]\right\}. \tag{3.128}$$

Subsequently, $S_l(E)$ is expanded around E_{Rl} with respect to E. For this purpose, the scattering matrix factor is represented as $S_l = \exp[2i(\delta_l^{(0)} + \delta_l^{(1)})]$, distinguishing a phase $\delta_l^{(0)}$ whose energy dependence is negligibly weak in the vicinity of E_{Rl} from a phase $\delta_l^{(1)}$ which changes rapidly as E passes through E_{Rl}. We will confirm later that $\delta_l^{(1)}(E_{Rl}) = 0$. With these premises, the expansion of $S_l(E)$ up to first order results in

$$S_l = S_l(\delta_l^{(0)}) \exp\left[\frac{\hbar \frac{d}{dE} \ln (S_l(E))|_{E_{Rl}}(E - E_{Rl})}{\hbar}\right] \tag{3.129}$$

with $S_l(\delta_l^{(0)}) = \exp(2i\delta_l^{(0)})$. This expression turns Eq. (3.128) into

$$S_l(E) \exp\left(-i\frac{Et}{\hbar}\right)$$
$$= S_l(\delta_l^{(0)}) \exp\left(-i\frac{E_{Rl}t}{\hbar}\right) \exp\left[-i\frac{(i\hbar \frac{d}{dE} \ln (S_l(E))|_{E_{Rl}} + t)(E - E_{Rl})}{\hbar}\right]. \tag{3.130}$$

The presence of a resonance thus modifies the wave function time dependence on comparison with the regular form $\exp(-i\frac{Et}{\hbar})$ by a time delay

$$T_{\mathrm{d}} = -i\hbar \left[\frac{d}{dE} \ln (S_l(E))\right]\bigg|_{E_{Rl}}. \tag{3.131}$$

A particle with energy E_{Rl} will experience maximum time delay. Consequently, we require the function $T_{\mathrm{d}}^{-1}(E)$ to attain a minimum at $E = E_{Rl}$. Expanding the reciprocal time delay up to second order about this energy yields [88]

$$T_{\mathrm{d}}^{-1}(E) = \alpha + \beta(E - E_{Rl})^2. \tag{3.132}$$

Combining Eqs. (3.131) and (3.132), and integrating $\frac{d}{dE} \ln (S_l(E))$, we find

$$S_l = S_l(\delta_l^{(0)}) \exp\left[2ir \tan^{-1} \frac{v/2}{E_{Rl} - E}\right], \tag{3.133}$$

where $v = 2(\alpha/\beta)^{\frac{1}{2}}$ and $r = \frac{1}{2}(\alpha\beta)^{-\frac{1}{2}}$ [88]. The requirement that S be analytic in the upper half of the complex energy plane restrains the parameter

r to integers: $r = 0, 1, 2, \ldots$.[1] The most relevant case is realized for $r = 1$ which results in the *Breit–Wigner phase relation*, as described later. With this choice, S_l becomes

$$\begin{aligned} S_l &= S_l(\delta_l^{(0)}) \exp[2i\delta_l^{(1)}] \\ &= S_l(\delta_l^{(0)}) \frac{1 + i \tan \delta_l^{(1)}}{1 - i \tan \delta_l^{(1)}} \\ &= S_l(\delta_l^{(0)}) \frac{E_{Rl} - E + i(\alpha/\beta)^{\frac{1}{2}}}{E_{Rl} - E - i(\alpha/\beta)^{\frac{1}{2}}}. \end{aligned} \tag{3.134}$$

Exercise 3.9. Derive the final expression of Eq. (3.134).

Recalling that S_l has a singularity at the eigenvalue $E = E_{Rl} - i\Gamma_{Rl}/2$, one infers the identity

$$(\alpha/\beta)^{\frac{1}{2}} = \frac{\Gamma_{Rl}}{2}, \tag{3.135}$$

and S_l turns into

$$S_l = S_l(\delta_l^{(0)}) \frac{E - E_{Rl} - i\frac{\Gamma_{Rl}}{2}}{E - E_{Rl} + i\frac{\Gamma_{Rl}}{2}}. \tag{3.136}$$

A short calculation verifies that $\delta_l^{(1)}$ may be conveniently cast into the following form:

$$\begin{aligned} \delta_l^{(1)} &= \frac{i}{2}\left[\ln\left(E - E_{Rl} + i\frac{\Gamma_{Rl}}{2}\right) - \ln\left(E - E_{Rl} - i\frac{\Gamma_{Rl}}{2}\right)\right] \\ &= \tan^{-1}\left[\frac{\Gamma_{Rl}}{2(E_{Rl} - E)}\right]. \end{aligned} \tag{3.137}$$

The phase thus reduces to the well-known Breit–Wigner form $\delta_l = \delta_l^{(0)} + \delta_l^{(1)} = \delta_l^{(0)} + \tan^{-1}[\frac{\Gamma_{Rl}}{2(E_{Rl}-E)}]$. As E traverses the point E_{Rl}, moving from smaller to larger energies, $\delta_l^{(1)}$ changes from $+\pi/2$ to $-\pi/2$.

[1] It is easily verified that negative whole numbers as choices for r do not add any new information, as they give rise to the same resonance scattering cross sections as the corresponding integers.

Our derivation of $\delta_l^{(1)}$ allows for a statement about the delay time due to the presence of a resonance. Generalizing formula (3.131), we have

$$T_{\rm d} = 2\hbar \left. \frac{d\delta_l}{dE} \right|_{E_{Rl}} . \tag{3.138}$$

We use the last expression given in (3.137) to compute $\frac{d}{dE}\delta_l^{(1)}|_{E_{Rl}} = \frac{2}{\Gamma_{Rl}}$. This leads to

$$T_{\rm d} = 2\hbar \left. \frac{d\delta_l^{(0)}}{dE} \right|_{E_{Rl}} + \frac{4\hbar}{\Gamma_{Rl}}, \tag{3.139}$$

where the second term on the right-hand side is associated with the time spent by the particle in the metastable state that causes the resonance.

On the basis of the information obtained about the scattering phases characteristic of a resonance situation, we can proceed to examine the behavior of the respective cross section as a function of energy. Utilizing

$$\sigma_l = \frac{4\pi}{k^2}(2l+1)\sin^2\delta_l \tag{3.140}$$

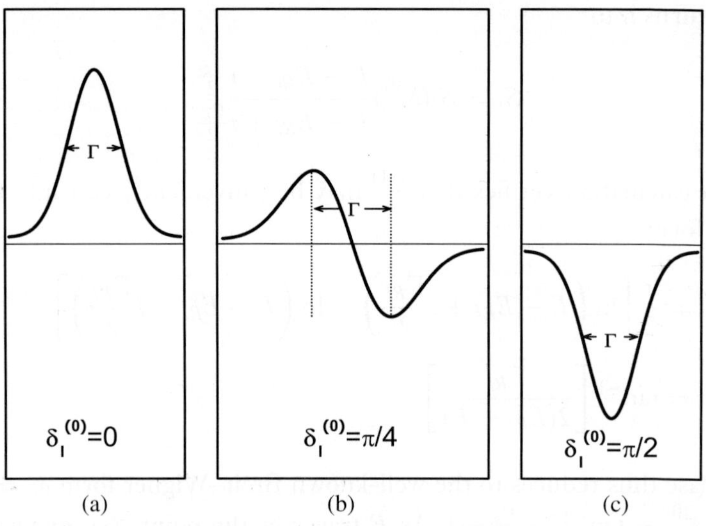

Figure 3.2: Three cases of resonant scattering, distinguished by the values adopted by the parameter $\delta_l^{(0)}$: (a) $\delta_l^{(0)} = 0$, (b) $\delta_l^{(0)} = \pi/4$, (c) $\delta_l^{(0)} = \pi/2$. The diagrams display the cross section σ_l versus the energy difference $E - E_{Rl}$. The abscissa lines are not defined by $\sigma_l = 0$ which, for cases (b) and (c), would be in conflict with the positive-definite character of the cross section, but refer to the offset $\frac{4\pi}{k^2}(2l+1)\sin^2\delta_l^{(0)}$ (see formula (3.141).

(see Eq. 3.122)) we find [92]

$$\sigma_l = \frac{4\pi}{k^2}(2l+1)\left[\sin^2\delta_l^{(0)} + \frac{\Gamma_{Rl}^2\cos(2\delta_l^{(0)}) - 2\Gamma_{Rl}(E-E_{Rl})\sin(2\delta_l^{(0)})}{4(E-E_{Rl})^2+\Gamma_{Rl}^2}\right]. \tag{3.141}$$

Plotting the cross section versus the energy difference $(E-E_{Rl})$ in Figure 3.2, we use the phase $\delta_l^{(0)}$ as a parameter. From the images, this quantity has a strong impact on the resonance line shape. The widths of the curves shown in the figure are defined by the imaginary part of the resonance energy, $\Gamma/2$.

A resonance state is understood as a composite of a scattering and a bound state, as represented by a continuous function with a strongly enhanced amplitude in the interaction region. This hybrid character is captured by the *Feshbach partitioning* [7, 98] which consists in subdividing the overall wave function into a regular continuum component Ψ_{C} and an admixture Ψ_{R} that is nonzero only within a finite range, the innermost scattering zone: $\Psi = \Psi_{\mathrm{C}} + \Psi_{\mathrm{R}}$. The contribution Ψ_{R} may be identified with a quasi-bound state whose time evolution contains information about the emergence of resonances and their decay.

Chapter 4

Semiclassical Notions

In the two preceding chapters, the ground was laid for treating electronic phenomena in dynamic interactions between molecules. In full quantum mechanical rigor, these are described as scattering processes in a landscape of multiple potential energy surfaces, given in an adiabatic or diabatic frame, depending on the demands of the case at hand. Any specific problem may be addressed in the energy or in the time representation. However, as noted in Section 3.1, both viewpoints are equivalent only in the limit of completeness which is practically unattainable. When following both of these avenues, the time-independent and the time-dependent approach in Chapters 6–7 and 15, respectively, we will notice that the complexity of physical systems often necessitates working with a drastically truncated basis which raises the question of the validity and accuracy of the solution.

As an alternative to *approximating the solution* we will, throughout this text, emphasize the strategy of *approximating the problem*. In particular, semiclassical strategies of addressing quantal problems have been seen to provide powerful tools for exploring nonadiabatic situations. While the electronic degrees of freedom in a molecular problem are irreducibly quantum mechanical,[1] the nuclear degrees of freedom allow, to various extents, for classical treatment. The translational motion of a nuclear wave packet can be described largely in terms of a classical trajectory. Not rarely, the rotational coordinates behave classically as well, due to the narrowly spaced, quasi-continuous nature of the rotational spectrum. The vibrational nuclear system, however, requires in general inclusion of quantum elements

[1]The discussion presented in Chapters 10 and 12 qualifies this statement to a certain extent. As outlined there, electron dynamics can be treated by mapping the electronic degrees of freedom onto classical action-angle variables, once stationary states for the electronic system have been identified.

for appropriate representation. Chapters 8–14 of this text will highlight semiclassical approaches and discuss their utility for molecular dynamics involving electronic effects.

The following sections are intended to introduce some of the basic concepts of semiclassical physics of relevance to the electronic problem in molecular processes. We begin with Feynman's path integral formulation of the quantum propagator and its semiclassical realization by use of the classical action functional.

The next segment of this chapter contains a condensed summary of the Wentzel–Kramers–Brillouin (WKB) method which is based on the separation of the quantum effect into a *quantum potential* whose neglect transforms the TDSE into the quantum equivalent of the classical Hamilton–Jacobi equation. This argument lays the basis for the quantum trajectory procedures to be discussed in Chapter 13.

We continue by focusing on a quantum analog of the classical phase space distribution which admits simultaneously sharp position and momentum coordinates. The quantum generalization of this classical concept is accomplished by the *Wigner function*, a typically semiclassical entity which depends on both groups of variables, allowing us to recover the respective quantum probability distributions upon reduction to position or momentum space. Last, we deal with *coherent states*, quantum states on the verge of the classical realm. We develop the crucial features of these objects and underscore their minimal uncertainty character. The nuclear coherent state is introduced in preparation for coherent state dynamics to be treated further in Chapter 9.

4.1 Path Integrals and the Quantum Propagator

The path integral formalism as introduced by Feynman [99] has proven its efficiency and explanatory power in many disciplines of quantum physics, in quantum dynamics, statistics, field theory, and beyond. Numerous monographs on the topic (for a selection, see [100, 102]) testify to the wide use and high versatility of the method. Its popularity is also rooted in its intuitive appeal, as it may be viewed as a direct continuation of the trajectory concept of classical mechanics into quantum theory. It is not the algebraic objects of canonical quantum theory, such as Hilbert space vectors and operators that are at the center of the path integral approach, but a probability amplitude referring to the motion of the considered system along a specific trajectory. The latter may deviate from the path taken by a classical particle, as generated by the principle of minimal action. As described in further detail later, the overall probability for a particle to advance from an initial space-time point (x_i, t_i) to a destination (x_f, t_f) is found from the superposition of the amplitudes associated with all possible trajectories between the beginning and the end. The *quantum propagator* governing this transition will be shown to depend on an analog of the action functional that determines the classical equivalent of the quantum process.

Path integrals allow for a rigorous representation of quantum mechanics. For their practical use, however, approximations have to be made. The most essential modifications of the theory involve absorbing elements of classical mechanics into the quantum formalism. The classical action becomes thus the centerpiece of the quantum propagator. In what follows, we summarize the basic tenets of the path integral theory as far as it is of importance for quantum dynamics.

4.1.1 *The quantum and the semiclassical propagator*

To begin, we introduce operators $\hat{\mathbf{x}}$ and $\hat{\mathbf{p}}$ that stand for generalized position and momentum observables, obeying the commutation relations

$$[\hat{x}_i, \hat{p}_j] = i\hbar\delta_{ij}. \tag{4.1}$$

For the sake of simplicity, we refer here to a one-dimensional situation and identify $\hat{x}$ with the corresponding position operator and $\hat{p}$ with its conjugate momentum. The respective eigenfunctions are expressed in the abstract ket notation $|x\rangle$ and $|p\rangle$. They satisfy the eigenvalue equations

$$\hat{x}|x\rangle = x|x\rangle, \tag{4.2a}$$

$$\hat{p}|p\rangle = p|p\rangle. \tag{4.2b}$$

Adopting the Heisenberg picture (see Section 3.2), we define the time-dependent position operator $x(t)$ in accordance with Eq. (3.8) by

$$\hat{x}(t) = \exp\left(\frac{i}{\hbar}\hat{H}t\right)\hat{x}\exp\left(-\frac{i}{\hbar}\hat{H}t\right), \tag{4.3}$$

where a Hamiltonian $\hat{H}$ has been specified for the physical system of interest. The time-shifted operator $\hat{p}(t)$ is defined in the same manner. Further, we employ time-dependent position and momentum eigenfunctions $|x, t\rangle$ and $|p, t\rangle$, respectively. Thus, a one-dimensional system may adopt the position eigenvalue $x_1(x_2)$ at a time $t_1(t_2)$, in general:

$$\hat{x}(t)|x_i, t_i\rangle = x_i|x_i, t_i\rangle, \quad i = 1, 2, \ldots \tag{4.4}$$

The time dependence of the eigenstate $|x_i, t_i\rangle$ is separated according to

$$|x_i, t_i\rangle = \exp\left(\frac{i}{\hbar}\hat{H}t_i\right)|x_i\rangle. \tag{4.5}$$

We point out that the appearance of a factor $+i$ instead of $-i$ in the exponent (see formulae (3.4) and (3.5)) is due to our use of the Heisenberg instead of the Schrödinger picture in the present context. It is readily shown that the identity (4.5) is in agreement with Eqs. (4.2a) and (4.3).

The probability of detecting a quantum system in the state $|x_2, t_2\rangle$ after it has been found to occupy the location x_1 at the time t_1 is $|\langle x_2, t_2|x_1, t_1\rangle|^2$. Summarizing Eqs. (4.4) and (4.5) we derive for the transition amplitude between the two eigenstates the expression [103]

$$\langle x_2, t_2|x_1, t_1\rangle = \langle x_2|\exp\left[-\frac{i}{\hbar}\hat{H}(t_2 - t_1)\right]|x_1\rangle. \tag{4.6}$$

This quantum propagator contains the probability that a particle in one dimension advances from (x_1, t_1) to (x_2, t_2). Further, we generate $|x_2\rangle$ from $|x_1\rangle$ by a linear translation of the form

$$|x_2\rangle = \exp\left(-\frac{i}{\hbar}\hat{p}(x_2 - x_1)\right)|x_1\rangle, \tag{4.7}$$

involving the momentum operator $\hat{p}$. Combining relations (4.6) and (4.7), we obtain

$$\langle x_2, t_2|x_1, t_1\rangle = \langle x_1|\exp\left(+\frac{i}{\hbar}\hat{p}(x_2 - x_1)\right)\exp\left(-\frac{i}{\hbar}\hat{H}(t_2 - t_1)\right)|x_1\rangle. \tag{4.8}$$

To simplify the propagator further, we introduce the plane wave basis $|p_1\rangle = (\frac{1}{2\pi\hbar})^{\frac{1}{2}} \exp(-\frac{i}{\hbar}p_1 x)$ and insert into Eq. (4.8) the decomposition of unity

$$\int_{-\infty}^{\infty} dp_1 |p_1\rangle\langle p_1| = 1 \tag{4.9}$$

to find, with $\delta t_1 \equiv t_2 - t_1$,

$$\langle x_2, t_2|x_1, t_1\rangle = \int dp_1 \exp\left[\frac{i}{\hbar}p_1(x_2 - x_1)\right] \langle x_1|p_1\rangle\langle p_1| \times \exp\left(-\frac{i}{\hbar}H\delta t_1\right)|x_1\rangle, \tag{4.10}$$

which is derived by letting the operator $\exp[-\frac{i}{\hbar}\hat{p}(x_2 - x_1)]$ act on the functions $|p_1\rangle$. At this place, $\hat{H}$ is identified with the classical Hamiltonian function $H(x,p) = \frac{p^2}{2m} + V(x)$ for a particle of mass m subjected to a potential V. In addition, we realize that $\langle x_1|p_1\rangle = (\frac{1}{2\pi\hbar})^{\frac{1}{2}} \exp(\frac{i}{\hbar}p_1 x_1)$ and conclude that Eq. (4.10) takes the form

$$\langle x_2, t_2|x_1, t_1\rangle = \frac{1}{2\pi\hbar}\int_{-\infty}^{\infty} dp_1 \exp\left[\frac{i}{\hbar}(-H(x_1, p_1) + p_1\dot{x}_1)\delta t_1\right]. \tag{4.11}$$

Exercise 4.1. Show that $\langle p|x\rangle = \sqrt{\frac{1}{2\pi\hbar}} \exp(-\frac{i}{\hbar}px)$. Consider $\langle p|x\rangle$ as an element of a unitary transformation matrix $\mathbf{U}(x,p)$, and make use of the relation $\int dx\, U(x,p)U^*(x,p') = \delta(p - p')$.

Statement (4.11) is based on the assumption of a well-defined velocity $\dot{x}_1 = \frac{x_2 - x_1}{\delta t_1}$ in the time interval δt_1 which is chosen sufficiently small to justify substituting the position derivative. Carrying out the integration over p_1 in Eq. (4.11),[2] we arrive at

$$\langle x_2, t_2|x_1, t_1\rangle = \sqrt{\frac{m}{2\pi i\hbar\delta t_1}} \exp\left[\frac{i}{\hbar}\left(\frac{m\dot{x}_1^2}{2} - V(x_1)\right)\delta t_1\right]. \tag{4.12}$$

In the expression $\frac{m\dot{x}_1^2}{2} - V(x_1)$, we recognize the classical Lagrangian $L \equiv L[x(t_1), \dot{x}(t_1)]$, neglecting possible vector potential contributions, and

[2]The integration formula used in going from (4.11) to (4.12) is

$$\int_{-\infty}^{\infty} dy \exp(-ay^2 + by) = \sqrt{\frac{\pi}{a}} \exp\left(\frac{b^2}{4a}\right).$$

limiting the discussion to time-independent potentials such that L does not contain time as an explicit argument. Substituting L, we interpret the result (4.12) as a relation between the quantum propagator of the considered problem and the Lagrangian of its classical analog. This finding is immediately generalized to the case of a finite time interval consisting of n segments $\delta t : t_f - t_i = n\delta t$. Using the completeness of the position operator eigenstates at an arbitrary time t:

$$\int_{-\infty}^{\infty} dx|x,t\rangle\langle x,t| = 1, \tag{4.13}$$

we insert unity $n-1$ times into the scalar product $\langle x_f, t_f|x_i, t_i\rangle$ and obtain

$$\langle x_f, t_f|x_i, t_i\rangle = \int dx_1 \int dx_2 \ldots \int dx_{n-1}\langle x_f, t_f|x_{n-1}, t_{n-1}\rangle \ldots \\ \times \langle x_2, t_2|x_1, t_1\rangle\langle x_1, t_1|x_i, t_i\rangle. \tag{4.14}$$

In the following move, the scalar product factors in Eq. (4.14) are replaced by their explicit form as given by Eq. (4.12). This leads to

$$\langle x_f, t_f|x_i, t_i\rangle = \left[\sqrt{\frac{m}{2\pi i\hbar\delta t}}\right]^n \prod_j \int_{x_i}^{x_f} dx_j \exp\left(\frac{i}{\hbar}L[x(t_j), \dot{x}(t_j)]\delta t\right), \tag{4.15}$$

where the product extends over the indices $1, 2, \ldots, n-1$ of the intermediate positions. We point out that the "classical turn" made earlier by substituting the Hamilton function for the Hamilton operator is not mandatory for achieving factorization of the propagator. By use of the Trotter formula, to be introduced in Section 5.4, one may decompose the propagator to arrive at the quantum equivalent of Eq. (4.15).

The subsequent step is taking the limit $n \to \infty$, thus letting Eq. (4.15) tend from a grid of finite intervals to an infinitely fine subdivision between the bounding locations x_i and x_f. This defines a propagator K by

$$K(x_i, x_f; t_f - t_i) \equiv \int_{x_i}^{x_f} Dx \exp\left(\frac{i}{\hbar}\int_{t_i}^{t_f} dt\, L[x(t_j), \dot{x}(t_j)]\right) \tag{4.16}$$

with the symbol Dx denoting the measure

$$Dx \equiv \lim_{n\to\infty}\left[\frac{m}{2\pi i\hbar\delta t}\right]^{\frac{n}{2}} dx_1 dx_2 \ldots dx_{n-1}. \tag{4.17}$$

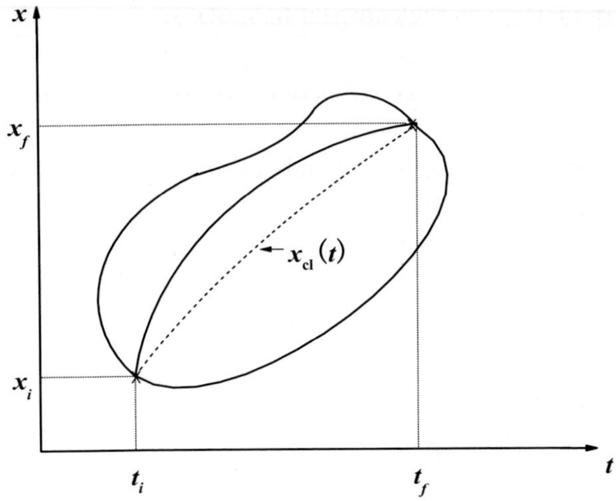

Figure 4.1: Set of trajectories leading from an initial point x_i to a final point x_f. Each path is traversed in a fixed time interval $(t_f - t_i)$. The path followed by a classical particle of mass m and constrained to move from x_i to x_f within this time interval is marked by the dashed line.

The propagator (4.16) is to be understood as the sum, or rather integral, over all possible paths connecting the limits (t_i, x_i) and (t_f, x_f) as shown in Figure 4.1. It has been termed *sum over histories* [99, 100].

The limitation process that leads from Eq. (4.15) to Eq. (4.16) can thus be viewed as a transition from polygonal or *broken* paths, consisting of a finite number of straight segments, to smooth trajectories in the sense of classical mechanics.

By definition, the propagator $K(x_i, x_f; t_f - t_i)$ assembles the wave function $|x_f, t_f\rangle$ that originates from an initial state $|x_i, t_i\rangle$:

$$|x_f; t_f\rangle = \int dx_i K(x_i, x_f; t_f - t_i)|x_i; t_i\rangle. \tag{4.18}$$

This identity may be interpreted as the TDSE cast into integral equation form.

Expression (4.16) assumes its simplest shape as one substitutes the classical action functional $S[x(t); t_f - t_i]$, recalling that (see, for instance, [104])

$$S[x(t); t_f - t_i] = \int_{t_i}^{t_f} dt\, L[x(t), \dot{x}(t)]. \tag{4.19}$$

This turns Eq. (4.16) into the compact formula

$$K(x_f, x_i; t_f - t_i) = \int_{x_i}^{x_f} Dx \exp\left(\frac{i}{\hbar} S[x(t); t_f - t_i]\right). \tag{4.20}$$

Adopting this form, we can immediately evaluate the propagator K^{free} for a free quantum particle moving from x_i to x_f in a time interval $t_f - t_i$. We realize that the classical action for such a particle is given by

$$S_{\text{cl}} = \frac{m}{2}\frac{(x_f - x_i)^2}{t_f - t_i} \tag{4.21a}$$

and use the relation

$$\int_{-\infty}^{\infty} \left[\frac{m}{2\pi i\hbar\delta t}\right]^{\frac{1}{2}} \exp\left(\frac{m}{2i\hbar\delta t}[(x_{m+1} - x_m)^2 - (x_m - x_{m-1})^2]\right) dx_m$$

$$= \left[\frac{m}{2\pi i\hbar\delta t}\right]^{\frac{1}{2}} \exp\left(\frac{m}{2i\hbar\delta t}(x_{m+1} - x_{m-1})^2\right) \tag{4.21b}$$

to find

$$K^{\text{free}}(x_f, x_i; t_f - t_i) = \left[\frac{m}{2\pi i\hbar(t_f - t_i)}\right]^{\frac{1}{2}} \exp\left(\frac{im(x_{m+1} - x_{m-1})^2}{2\hbar(t_f - t_i)}\right). \tag{4.21c}$$

Exercise 4.2. Derive the propagator (4.21c) for a free quantum particle from relations (4.20–4.21b).

4.1.2 *The Van Vleck propagator*

The histories summed over in the path integral formula (4.20) are assigned phases through the exponential factor which facilitate interference between trajectories. The most important observation regarding this structure is that the behavior of the factor is, in general, highly oscillatory in the semiclassical regime where it holds that $S[x(t); t_f - t_i] \gg \hbar$, implying the cancellation of the contributions due to adjacent paths. Exceptions to this rule occur in the vicinity of the classical paths, i.e. the trajectory or trajectories that the considered particle would take according to the laws of classical mechanics. This feature is due to the stationary character of the classical action functional along these paths. Assuming we know the course a classical particle would take from x_i to x_f,[3] we may write the propagator as [105]

$$K_V(x_f, x_i; t_f - t_i) \approx C \exp\left(\frac{i}{\hbar} S[x_{cl}(t); t_f - t_i]\right), \tag{4.22}$$

where the subscript V stands for Van Vleck and $x_{cl}(t)$ denotes a classical trajectory, with t as the continuous time variable. How to determine the factor C?

Let us imagine that a path $x_{cl}(t)$ has been found by solving the Euler–Lagrange equation

$$\frac{d}{dt}\frac{\delta L}{\delta \dot{x}} - \frac{\delta L}{\delta x} = 0. \tag{4.23}$$

The trajectory $x_{cl}(t)$ satisfies the boundary conditions $x_{cl}(t_i) = x_i$ and $x_{cl}(t_f) = x_f$. We further define a small deviation $y(t)$ from the classical path $x_{cl}(t)$ by stipulating

$$y(t) = x(t) - x_{cl}(t) \tag{4.24}$$

with $x(t)$ as a trajectory in the neighborhood of $x_{cl}(t)$. This leads to the following expansion of the action functional:

$$\begin{aligned} S[x(t); t_f - t_i] &= S[x_{cl}(t) + y(t); t_f - t_i] \\ &= S[x_{cl}(t); t_f - t_i] + \delta S[x_{cl}(t); t_f - t_i] \\ &\quad + \delta^2 S[x_{cl}(t); t_f - t_i] + \cdots \end{aligned} \tag{4.25}$$

The δ sign stands here for the variation of $x_{cl}(t)$ by the small deformations $y(t)$. From Hamilton's principle, the action functional is stationary with respect to infinitesimal deviations from the trajectory $x_{cl}(t)$, such that the linear variation of the action functional vanishes [104]. Since we tolerate only small transgressions from the classical path, we are allowed to neglect higher than second variations. Therefore, the exponential in the path integral (4.16) can be factorized in the following manner:

$$K(x_f, x_i; t_f - t_i) = \exp\left(\frac{i}{\hbar} S_{cl}\right) \int Dx \exp\left(\frac{i}{\hbar} \delta^2 S_{cl}\right), \tag{4.26}$$

where $S[x_{cl}(t); t_f - t_i]$ has been replaced by S_{cl} and $\delta^2 S[x_{cl}(t); t_f - t_i]$ by $\delta^2 S_{cl}$. The second variation of S is determined by the definition

[3] Later, we will more realistically include a multiplicity of classical trajectories.

$S = \int_{t_1}^{t_2} dt\, L[x(t_1), \dot{x}(t_1)] = \int_{t_1}^{t_2} dt T[\dot{x}(t)] - V[x(t)] = \int_{t_1}^{t_2} dt\, p(t)\dot{x}(t) - H$ $[x(t), \dot{x}(t)]$ and found from

$$\delta^2 S_{\text{cl}} = \int_{t_i}^{t_f} dt \left. \frac{1}{2}\frac{\partial^2 L}{\partial \dot{x}^2}\right|_{\dot{x}_{\text{cl}}} \dot{y}(t)^2 + \left. \frac{1}{2}\frac{\partial^2 L}{\partial x^2}\right|_{x_{\text{cl}}} y(t)^2$$

$$= \int_{t_i}^{t_f} dt \frac{1}{2} m\dot{y}(t)^2 - \left. \frac{1}{2}\frac{\partial^2}{\partial x^2} V(x)\right|_{x_{\text{cl}}} y(t)^2. \tag{4.27}$$

Using again a polygonal approximation to a path $x(t)$, corresponding to n intermediate instants given by $t_k = t_i + \frac{k}{n}(t_f - t_i)$, we observe that $\dot{y}(t_k) \approx \frac{y(t_{k+1}) - y(t_k)}{\delta t}$, and the second factor of the propagator (4.26) becomes

$$\int Dx \exp\left(\frac{i}{\hbar}\delta^2 S_{\text{cl}}\right)$$

$$= \lim_{n\to\infty} \left[\frac{m}{2\pi i \hbar \delta t}\right]^{\frac{n+1}{2}} \int dy_1 \int dy_2 \ldots \int dy_n$$

$$\times \exp\left(\frac{i}{\hbar}\sum_{k=0}^{n}\left[\frac{m}{2\delta t}(y_{k+1} - y_k)^2 - \frac{1}{2}\delta t V_{xx,k} y_k^2\right]\right), \tag{4.28}$$

where $V_{xx,k}$ is shorthand for the second derivative with respect to the variable x, evaluated along the classical path at the intermediate position with index k. The integral (4.28) can be evaluated analytically by introducing the vector [101]

$$\mathbf{y} = \begin{pmatrix} y_1 \\ \vdots \\ y_n \end{pmatrix}, \tag{4.29}$$

such that the sum in the exponential adopts the compact form $\mathbf{y}^{\mathrm{T}}\mathbf{A}\mathbf{y}$. The tridiagonal matrix $\mathbf{A}$ is easily extracted from formula (4.28). We consider next a unitary transformation that turns $\mathbf{A}$ into the diagonal matrix $\widetilde{\mathbf{A}}$, according to

$$\mathbf{A} = \mathbf{U}^{\dagger}\widetilde{\mathbf{A}}\mathbf{U}, \quad \widetilde{\mathbf{y}} = \mathbf{U}\mathbf{y}. \tag{4.30}$$

The integral in Eq. (4.28) becomes

$$\int d^n\mathbf{y} \exp(\mathbf{y}^{\mathrm{T}}\mathbf{A}\mathbf{y}) = \int d^n\widetilde{\mathbf{y}} \exp(\widetilde{\mathbf{y}}^{\mathrm{T}}\widetilde{\mathbf{A}}\widetilde{\mathbf{y}}). \tag{4.31}$$

As a consequence of $|\det \mathbf{U}| = 1$, the Jacobian that accompanies the coordinate change from $\mathbf{y}$ to $\widetilde{\mathbf{y}}$ is unity. The seemingly complicated integration involved in formula (4.28) is now reduced to the trivial task of carrying out an elementary Gaussian integral n times:

$$\int d^n\widetilde{\mathbf{y}} \exp(\widetilde{\mathbf{y}}^{\mathrm{T}}\widetilde{\mathbf{A}}\widetilde{\mathbf{y}}) = \prod_{j=1}^{n} \sqrt{\frac{\pi}{\widetilde{A}_{jj}}} = \frac{\pi^{\frac{n}{2}}}{\sqrt{\det \mathbf{A}}}, \tag{4.32}$$

turning the path integral (4.26) into

$$K(x_f, x_i; t_f - t_i) = \exp\left(\frac{i}{\hbar} S_{\mathrm{cl}}\right) \lim_{n\to\infty} \frac{\pi^{\frac{n}{2}}}{\sqrt{\det \mathbf{A}}}. \tag{4.33}$$

The remaining challenge of computing $\det \mathbf{A}$ in the limit of infinitely many intermediate points is addressed in [101]. We emphasize that formula (4.33) gives the exact form of the path integral for the case of a quadratic Lagrangian, i.e. for a functional $L[x(t_j), \dot{x}(t_j)]$ that does not contain higher powers than x^2, $\dot{x}^2$, and $\dot{x}x$.

In the following step, we present an explicit evaluation of the factor multiplying $\exp(\frac{i}{\hbar}S_{\mathrm{cl}})$ in the semiclassical limit of $\hbar$ tending towards zero. Specifically, one may make use of the observation that

$$K(x_f, x_i; 0) = \delta(x_f - x_i) \tag{4.34}$$

in conjunction with the propagator group property

$$K(x_f, x_i; t_f - t_i) = \int_{-\infty}^{+\infty} dx\, K(x_f, x; t_f - t)\, K(x, x_i; t - t_i). \tag{4.35}$$

Leaving the factor undetermined, we define the following ansatz for the two propagator factors on the right-hand side of Eq. (4.35) [104]:

$$K(x_f, x; t_f - t) \equiv A(t) \exp\left(\frac{i}{\hbar} S_{\mathrm{cl}}(x_f, x; t_f - t)\right), \tag{4.36}$$

$$\begin{aligned} K(x, x_i; t - t_i) &= K^*(x_i, x; t_i - t) \\ &= A^*(t) \exp\left(-\frac{i}{\hbar} S_{\mathrm{cl}}(x_i, x; t_i - t)\right). \end{aligned} \tag{4.37}$$

Combining this with Eqs. (4.34) and (4.35), one infers

$$\delta(x_f - x_i) = \int_{-\infty}^{+\infty} dx |A(t)|^2$$
$$\times \exp\left(\frac{i}{\hbar}[S_{\text{cl}}(x_f, x; t - t_i) - S_{\text{cl}}(x_i, x; t - t_i)]\right). \qquad (4.38)$$

At this juncture, we implement the semiclassical condition $\hbar \to 0$ by setting $x_f = x_i + \Delta$, letting

$$S_{\text{cl}}(x_i + \Delta, x; t - t_i) - S_{\text{cl}}(x_i, x; t - t_i) = \frac{\partial S_{\text{cl}}(x_i + \Delta, x; t - t_i)}{\partial x_i}\Delta. \qquad (4.39)$$

The justification for this move is found in the characteristic structure of Eq. (4.38) where $\hbar$ appears in the denominator of the exponential, confining the numerator, and thus the increment Δ, likewise to small values. The tendency of the integral towards rapid variation implies that the most significant contributions to the integrand stem from regions infinitesimally close to the point x_i, such that the difference of actions is well approximated by the differential. We now reformulate relation (4.38) by defining

$$f(x) \equiv \frac{\partial S_{\text{cl}}(x_i + \Delta, x; t - t_i)}{\partial x_i} \qquad (4.40)$$

to find

$$\delta(x_f - x_i) = \int_{-\infty}^{+\infty} df \left|\frac{dx}{df}\right| |A(t)|^2 \exp\left(\frac{i}{\hbar} f(x)(x_f - x_i)\right), \qquad (4.41)$$

where f has been substituted as integration variable for x. Multiplying the integral by one yields

$$\delta(x_f - x_i) = \int_{-\infty}^{+\infty} \frac{df}{2\pi\hbar} \exp\left(\frac{i}{\hbar} f(x)(x_f - x_i)\right) \frac{2\pi\hbar |A(t)|^2}{|\frac{df}{dx}|}, \qquad (4.42)$$

The right-hand side of this expression factors into a representation of $\delta(x_f - x_i)$ and a composite of terms that do not depend on the integration variable. We reiterate that the action functional does not contain higher powers of x than two, such that $\frac{df}{dx}$ is constant. The identity (4.42) leads to the conclusion

$$|A(t)|^2 = \frac{1}{2\pi\hbar}\left|\frac{df}{dx}\right| = \frac{1}{2\pi\hbar}\left|\frac{\partial^2 S_{\rm cl}(x_i, x; t-t_i)}{\partial x_i \partial x}\right|. \tag{4.43}$$

This relation specifies the amplitude $A(t)$ except for a phase factor. We choose this factor as $\exp(-\frac{i\pi}{4})$ and define the resulting *Van Vleck propagator* $K_{\rm V}$ as:

$$K_{\rm V}(x_f, x_i; t_f - t_i) = \sqrt{\frac{1}{2\pi i\hbar}\left|\frac{\partial^2 S_{\rm cl}(x_i, x_f; t_f - t_i)}{\partial x_i \partial x_f}\right|}\exp\left(\frac{i}{\hbar}S_{\rm cl}\right). \tag{4.44}$$

Working out the following exercise, the reader will find the explicit form of the Van Vleck propagator for the elementary case of a free particle, and justify the phase of the amplitude included in Eq. (4.44).

Exercise 4.3. Determine the Van Vleck propagator (4.44) for the free particle action (4.21a), and show that the choice of $\exp(-\frac{i\pi}{4})$ for the phase factor of the amplitude $A(t)$ reduces $K_{\rm V}$ to $\delta(x_f - x_i)$ as $t \to t_i$.

4.1.3 *The monodromy matrix*

The amplitude of $K_{\rm V}$ warrants further interpretation. What is the physical meaning of the function $f = \frac{\partial S_{\rm cl}}{\partial x_i}$ that was substituted in Eq. (4.41) for the integration variable x? By inspection of the first variation of $S_{\rm cl}$, we arrive at the quasi-Hamilton equations [104] (see Exercise 4.4):

$$p_i = -\frac{\partial S_{\rm cl}}{\partial x_i}, \tag{4.45a}$$

$$p_f = \frac{\partial S_{\rm cl}}{\partial x_f}. \tag{4.45b}$$

From Eq. (4.45a) we have

$$\frac{\partial^2 S_{\rm cl}}{\partial x_i \partial x_f} = -\frac{\partial p_i}{\partial x_f}. \tag{4.46}$$

Exercise 4.4. Demonstrate the validity of Eqs. (4.45a) and (4.45b) by investigating $\delta S_{\rm cl}$, the first variation of $S_{\rm cl}$.

In view of Eq. (4.46), the Van Vleck propagator (4.44) becomes

$$K_V(x_f, x_i; t_f - t_i) = \sqrt{\frac{1}{2\pi i\hbar}} \frac{\exp\left(\frac{i}{\hbar}S_{cl}\right)}{\sqrt{\left|\frac{\partial x_f}{\partial p_i}\right|}}. \tag{4.47}$$

The amplitude of the Van Vleck propagator is thus related to the derivative of the final position with respect to the initial momentum. To make the transition from the one-dimensional situation considered so far to multiple dimensions, we have to replace the second derivative of the classical action, $\partial^2 S_{cl}(x_i, x_f; t_f - t_i)/\partial x_i \partial x_f$, in Eq. (4.44) by the respective Jacobian determinant, $\det(\partial^2 S_{cl}(\mathbf{x}_i, \mathbf{x}_f; t_f - t_i)/\partial \mathbf{x}_i \partial \mathbf{x}_f)$. This expression involves the inverse of the submatrix $\mathbf{M}_{21}$ of the *monodromy matrix* $\mathbf{M}$:

$$\begin{pmatrix} \delta\mathbf{p} \\ \delta\mathbf{x} \end{pmatrix} = \mathbf{M} \begin{pmatrix} \delta\mathbf{p}_i \\ \delta\mathbf{x}_i \end{pmatrix}. \tag{4.48}$$

The matrix $\mathbf{M}$ relates initial deviations from a reference trajectory to those found at some arbitrary later time. It governs the extent to which perturbations grow or decrease as a set of trajectories evolves in time. The submatrix $\mathbf{M}_{21}$ can be written symbolically as $\frac{\partial \mathbf{x}}{\partial \mathbf{p}_i} = (\partial^2 S_{cl}/\partial \mathbf{x}_i \partial \mathbf{x}_f)^{-1}$, the multidimensional extension of $\frac{\partial x}{\partial p_i}$ in Eq. (4.45a). Since, however, a multiplicity of different trajectories, each characterized by a different initial momentum, may lead to the same final position, $\det \mathbf{M}_{21}$ is not guaranteed to remain finite. This concern may be substantiated by considering an extreme example: imagine a set of trajectories on a three-dimensional sphere, all of them leaving from the south pole of this sphere and fanning out into different directions along great circles. All of these paths differ with respect to their initial momenta, and all of them coincide in at least one further point on the sphere if they are allowed to evolve for a sufficiently long period of time. This is the north pole, or the antipodal place of the initial location.

As pointed out by Gutzwiller [106], these loci of coincidence, or conjugate points, give rise to additional phase factors in the semiclassical propagator formula. In a development analogous to the procedure described by Eq. (4.28) and the corresponding commentary, we can factorize the contribution of the second variation of S_{cl} to the propagator, $\int Dx \exp\left(\frac{i}{\hbar}\delta^2 S_{cl}\right)$, into a product of integrals of the form

$$\int d\eta \exp\left(\frac{i}{\hbar}\lambda\frac{\eta^2}{2}\right) = \sqrt{\frac{2\pi i\hbar}{|\lambda|}} \exp\left[\frac{i}{4}\pi(\text{sign}(\lambda) - 1)\right]. \quad (4.49)$$

The symbol λ stands here for an eigenvalue of the matrix that defines the quadratic form $\delta^2 S_{\text{cl}}$ (see p. 133). The term on the right-hand side contains a phase factor $\exp(-\frac{i\pi}{2})$ if λ happens to be negative. An interesting relation between the number of negative eigenvalues λ and that of conjugate points along the trajectory of reference has been established by Morse [106, 107]: the second variation of S_{cl}, considered a quadratic form in the displacements $\mathbf{y}(\mathbf{t})$ of all possible paths around a given trajectory from $\mathbf{x}_i$ to $\mathbf{x}_f$ in the time t, has as many negative eigenvalues as there are conjugate points along the trajectory. Each conjugate point encountered along the classical trajectory subtracts thus an amount of $\frac{\pi}{2}$ from the phase of the semiclassical propagator. This manipulation may be viewed as a measure to guarantee the continuity of the square root contained in Eq. (4.44) along any path in the complex plane. Combining these statements with the observation that a given final position may be reached from a well-defined initial position in a fixed time interval on more than one classical path, we write a generalized form of the Van Vleck propagator:

$$K_{\text{V}}(x_f, x_i; t_f - t_i) = \sum_{\textit{classical paths}} \left(\frac{1}{2\pi i\hbar}\right)^{\frac{N}{2}}$$

$$\times \frac{1}{\sqrt{\det(\mathbf{M}_{21})}} \exp\left(\frac{i}{\hbar}S_{\text{cl}} - i\nu\frac{\pi}{2}\right). \quad (4.50)$$

The number N and the *Maslov index* ν denote here the spatial dimension of the considered problem and the number of conjugate points encountered on each path included in the sum, respectively. This expression was introduced by Gutzwiller [106] and is referred to as the *Van Vleck–Gutzwiller propagator*. While its main constituents, a segment of the monodromy matrix and the classical action evaluated along multiple trajectories, are firmly rooted in classical mechanics, the summation over alternative paths invokes the superposition principle and thus imports an element of quantum mechanics into the formalism.

We conclude our discussion of the semiclassical propagator with a look at the Green's function $G(x_f, x_i; E)$, an object which is defined in the energy instead of the time domain and thus relevant for problems that are not explicitly time dependent. Transition to E as an independent variable is accomplished by computing the Fourier integral of Eq. (4.15):

$$G(x_f, x_i; E) = \int_0^\infty dt \exp\left(\frac{i}{\hbar}Et\right) \langle x_f, t_f | x_i, 0\rangle$$
$$= \int_0^\infty dt \int Dx \exp\left(\frac{i}{\hbar}(S[x(t); t_f] + Et)\right), \qquad (4.51)$$

setting $t_i = 0$. Comparison with Eqs. (4.16) and (4.19) demonstrates that in going from $K(x_f, x_i; t_f - t_i)$ to $G(x_f, x_i; E)$, we trade S for the transformed action $\widetilde{S}$, defined as $\widetilde{S}[x(t); E] = S[x(t); t_f - t_i] + E(t_f - t_i)$. The change of variables involved in the transition from S to $\widetilde{S}$ may be described as the result of a Legendre transformation [103]. The propagator G is useful if the path integral treatment involves classical trajectories of constant energy rather than time of traversal.

In Chapter 5, we will reencounter path integrals in the context of thermodynamics interpreted as dynamics in imaginary time. As it will turn out, path integrals provide an efficient and convenient formulation of rate theory, associated with the description of molecular reactions.

4.2 The WKB Approximation

The semiclassical regime may be considered the grey zone where the quantum and the classical sphere merge into each other. For quantitative characterization of this realm, one may ask for the conditions that turn the TDSE into its classical analog, the Hamilton–Jacobi equation of classical mechanics. Intuitively, one will expect this transition to proceed as the action associated with the case at hand becomes large by the standard of $\hbar$. According to this notion, one will change from the quantum mechanical to the classical equation of motion upon taking the limit $\hbar \to 0$. In the following, this transition will be inspected in greater detail.

4.2.1 *The WKB wave function*

Starting with the TDSE in one spatial dimension,

$$i\hbar\frac{\partial\psi(x,t)}{\partial t} = \left[-\frac{\hbar^2}{2m}\frac{\partial^2}{\partial x^2} + V(x)\right]\psi(x,t), \qquad (4.52)$$

we use a polar approach to write the wave function, namely

$$\psi(x,t) = R(x,t)\exp\left(\frac{i}{\hbar}S(x,t)\right). \tag{4.53}$$

The two determining functions, the amplitude R and the phase S that carries the unit of action, are chosen real. As we insert Eq. (4.53) into Eq. (4.52) and separate the real and the imaginary part, we obtain the two equations

$$-\frac{\partial S(x,t)}{\partial t} = \frac{1}{2m}\left(\frac{\partial S(x,t)}{\partial x}\right)^2 + V(x) - \frac{\hbar^2}{2m}\frac{\frac{\partial^2\rho(x,t)}{\partial x^2}}{\rho(x,t)} \tag{4.54a}$$

and

$$\frac{\partial\rho(x,t)}{\partial t} + \frac{\partial}{\partial x}\left(\frac{\rho(x,t)}{m}\frac{\partial S(x,t)}{\partial x}\right) = 0, \tag{4.54b}$$

with the assignment $\rho(x,t) = R^2(x,t)$. The first of these two identities expresses energy conservation, while the second is a statement of continuity. We are now in a position to use the semiclassical argument and declare that $\hbar^2 \approx 0$ in Eq. (4.54a). Neglecting the term that contains $\hbar^2$, we eliminate the *quantum potential* $V_Q \equiv \frac{\hbar^2}{2m}\frac{\partial^2\rho(x,t)}{\partial x^2}/\rho(x,t)$ and are left with

$$-\frac{\partial S(x,t)}{\partial t} = \frac{1}{2m}\left(\frac{\partial S(x,t)}{\partial x}\right)^2 + V(x). \tag{4.55}$$

This is the classical Hamilton–Jacobi equation [108]. In the classical limit, the phase S acquires the meaning of the classical action, justifying our choice of nomenclature.[4] This is readily proved by employing the relation $S(x,t) = \widetilde{S}(x,E) - Et$, connecting the classical action in the time and the energy domain. We obtain from Eq. (4.55)

$$E = \frac{1}{2m}\left(\frac{\partial\widetilde{S}(x,E)}{\partial x}\right)^2 + V(x). \tag{4.56}$$

Obviously, the function $\widetilde{\psi}(x,E) = \exp[\frac{i}{\hbar}\widetilde{S}(x,E)]$ reduces the stationary Schrödinger equation

$$E\widetilde{\psi}(x,E) = \left[\frac{\hbar^2}{2m}\frac{\partial^2}{\partial x^2} + V(x)\right]\widetilde{\psi}(x,E) \tag{4.57}$$

to the Hamilton–Jacobi equation, as accomplished by the choice $\psi(x,t) = \exp[\frac{i}{\hbar}S(x,t)]$ in the time-dependent case. A wave packet solution of

[4]It is noteworthy that Eq. (4.55) is recovered from Eq. (4.52) by setting the amplitude R equal to unity.

Eq. (4.52) may be constructed from a continuum of $\widetilde{\psi}$ functions at different energies using the transform

$$\psi(x,t) = \int_0^\infty dE\, g(E) \exp\left[\frac{i}{\hbar}(\widetilde{S}(x,E) - Et)\right], \tag{4.58}$$

with $g(E)$ as the respective energy density distribution. We now recall the semiclassical argument that $\hbar$ is small as compared to the action typical for the case of interest. The phase factor in (4.58) will therefore oscillate strongly and wash out the integral except for places of stationary phase, characterized by

$$t = \frac{\partial \widetilde{S}(x,E)}{\partial E}. \tag{4.59}$$

This, however, is the classical prescription for determining the location x of a particle at a time t.

To derive an explicit form of the wave function valid under the semiclassical condition of small $\hbar$, we solve Eq. (4.56), describing a system with well-defined energy. Hence, $\psi(x,t) = \psi(x)\exp\left(-\frac{i}{\hbar}Et\right)$, and the amplitude factor R, or $\sqrt{\rho}$, is time independent. The continuity equation (4.54b) then yields

$$\rho = \frac{\text{const}}{\frac{\partial \widetilde{S}}{\partial x}}. \tag{4.60}$$

Relation (4.56) leads to

$$\widetilde{S}(x,x',E) = \pm \int_x^{x'} dx'' \sqrt{2m(E - V(x''))}. \tag{4.61}$$

From (4.60) and (4.61), we obtain

$$\rho = \frac{const}{W(x)}, \tag{4.62}$$

where $W(x) \equiv \sqrt{2m(E - V(x))}$ has been used. We obtain the so-called *WKB wave function* ψ_{WKB} ([109]–[111]) by superposing the two alternatives given by Eq. (4.61) according to

$$\psi_{\mathrm{WKB}}(x) = \frac{1}{\sqrt{W(x)}} \left\{ A \exp\left[\frac{i}{\hbar} \int_x^{x'} dx'' W(x'') \right] \right.$$

$$\left. + B \exp\left[-\frac{i}{\hbar} \int_x^{x'} dx'' W(x'') \right] \right\}. \qquad (4.63)$$

The integration range is in the classically allowed region, i.e. an interval $[x, x']$ where $W(x)$ adopts real values. This domain may be bounded by two turning points x_1 and x_2, as illustrated in Figure 4.2.

We may identify x' in Eqs. (4.61) and (4.63) with x_2 in Figure 4.2. Repeating the arguments that led to the wave function (4.63), we derive the WKB wave function on the right of the right turning point, extending it into classically forbidden territory. Specifically,

$$\psi_{\mathrm{WKB}}(x) = \frac{C}{\sqrt{|W(x)|}} \exp\left[-\frac{\int_{x_2}^{x} |W(x'')| dx''}{\hbar} \right]. \qquad (4.64)$$

As expected, the wave function decays exponentially when the system enters the regime with $E < V(x)$. We point out that in most texts that include the WKB approximation, a wave function of the form $\exp\left[\frac{i}{\hbar} S(x)\right]$, with complex-valued action S, is adopted, and $\psi_{\mathrm{WKB}}(x)$ is derived from an

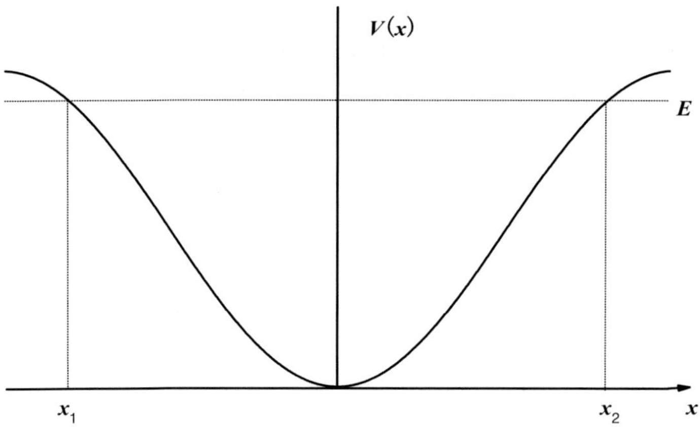

Figure 4.2: Example of a potential energy function $V(x)$. The classical motion of a particle with total energy E proceeds between the turning points x_1 and x_2 where $V(x) = E$.

expansion of $S(x)$ in powers of $\hbar$:

$$S(x) = S_0(x) + \hbar S_1(x) + \hbar^2 S_2(x) + \cdots . \tag{4.65}$$

Inserting this ansatz into Eq. (4.57) and neglecting in this series all orders beyond the linear yields the WKB wave function as given by Eqs. (4.63) and (4.64).

The oscillatory and the exponential solution (Eqs. (4.63) and (4.64), respectively) are divided by a turning point where $P = 0$. The vanishing of the classical momentum at the upper integration limit in Eq. (4.63) and the lower in Eq. (4.64), presents an obvious problem, as the prefactor $\frac{1}{\sqrt{W(x)}}$ is singular at this demarcation point. This situation calls for a procedure of connecting both partial solutions smoothly at $x = x_2$, i.e. the function (4.63) from the left and the function (4.64) from the right of x_2. This task may be confronted by a closer inspection of the stationary Schrödinger equation in the vicinity of the turning point. In a sufficiently small environment of x_2, the potential function is described by the linear approximation

$$V(x) \approx V(x_2) + \left.\frac{dV(x)}{dx}\right|_{x_2} (x - x_2). \tag{4.66}$$

Since $V(x_2) = E$, the stationary Schrödinger equation for the truncated potential (4.66) reads

$$\left[-\frac{\hbar^2}{2m}\frac{d^2}{dx^2} + \left.\frac{dV(x)}{dx}\right|_{x_2} (x - x_2) \right] \psi_{\mathrm{WKB}}(x) = 0, \tag{4.67}$$

which is valid in the vicinity of x_2. The identity (4.67) is the Airy differential equation in disguise, as may be verified by the substitution

$$w = \left(\frac{2m}{\hbar^2} \left.\frac{dV(x)}{dx}\right|_{x_2} \right)^{\frac{1}{3}} (x - x_2), \tag{4.68}$$

transforming Eq. (4.67) into

$$\left[\frac{d^2}{dw^2} - w \right] \psi_{\mathrm{WKB}} = 0. \tag{4.69}$$

This equation is solved by the Airy function (e.g. [112])

$$\mathrm{Ai}(w) = -\frac{1}{\pi} \int_0^{\infty} dv \cos\left(\frac{1}{3} v^3 + wv \right). \tag{4.70}$$

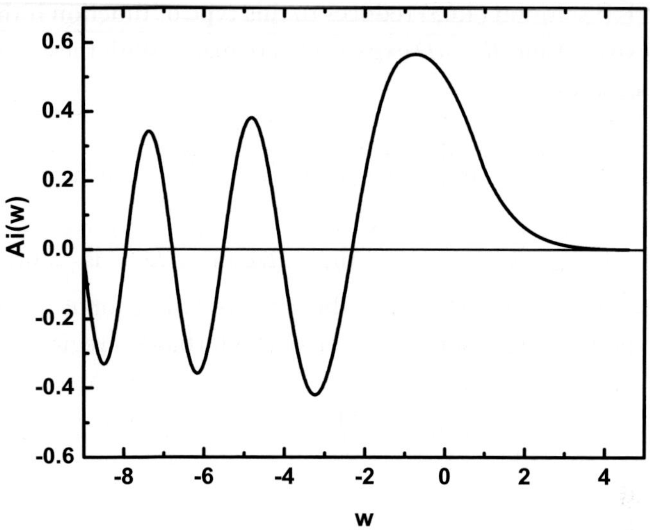

Figure 4.3: The Airy function Ai(w) around the point $w = 0$.

We exclude here the alternative solution of Eq. (4.69), Bi(w), which grows exponentially as w gets large. The Airy function, as represented in Figure 4.3, bears similarity to the wave function of a one-dimensional harmonic oscillator close to a turning point. On the right of x_2, it fades out exponentially and is periodic on the left side. The asymptotic behavior of the Airy function on the two sides of the turning point is given by the formulae

$$\mathrm{Ai}(w) \sim \frac{1}{2}\frac{1}{\sqrt{\pi} w^{\frac{1}{4}}} \exp\left(-\frac{2}{3}w^{\frac{3}{2}}\right), \qquad \text{if } w \gg 0, \tag{4.71a}$$

$$\mathrm{Ai}(w) \sim \frac{1}{\sqrt{\pi}(-w)^{\frac{1}{4}}} \sin\left(\frac{2}{3}(-w)^{\frac{3}{2}} + \frac{\pi}{4}\right), \quad \text{if } w \ll 0. \tag{4.71b}$$

Resubstituting for w (see Eq. (4.68)) in the latter equation which describes the Airy function in the classically allowed region, we find

$$\mathrm{Ai}(w) = \left[\frac{\hbar\left(2m \left.\frac{dV(x)}{dx}\right|_{x_2} \Big/ \hbar^2\right)^{\frac{1}{3}}}{\pi\,(2m(E-V))^{\frac{1}{2}}}\right]^{\frac{1}{2}} \sin\left(\frac{1}{\hbar}\int_x^{x_2} dx''\sqrt{2m(E-V)} + \frac{\pi}{4}\right). \tag{4.72}$$

The WKB solution (4.63) reduces to this type of function if the choices $A = -iD\exp(i\frac{\pi}{4})$ and $B = iD\exp(-i\frac{\pi}{4})$ are made. With these coefficients, $\psi_{\mathrm{WKB}}(x)$ becomes

$$\psi_{\mathrm{WKB}}(x) = \frac{2D}{\sqrt{W(x)}}\sin\left[\frac{1}{\hbar}\int_x^{x_2} dx'' W(x'') + \frac{\pi}{4}\right], \quad \text{if } x < x_2. \tag{4.73}$$

From this derivation, the factor D is proportional to $\sqrt{\frac{\hbar}{\pi}(2m\frac{dV(x)}{dx}|_{x_2}/\hbar^2)^{\frac{1}{3}}}$. Likewise, the Airy function for $w \gg 0$ (4.71a) turns into the WKB solution in the classically forbidden region $x > x_2$:

$$\psi_{\mathrm{WKB}}(x) = \frac{D}{\sqrt{|W(x)|}}\exp\left[-\frac{1}{\hbar}\int_{x_2}^{x} |W(x'')|dx''\right], \quad \text{if } x > x_2, \tag{4.74}$$

such that the coefficients C and D are identical.

Exercise 4.5. Show that the WKB wave functions (4.73) and (4.74) indeed have the forms (4.71b) and (4.71a), respectively, by explicit substitution for the variable w into the latter two equations.

Comparing Eqs. (4.73) and (4.74) yields a simple relation that connects the solution in the classically allowed and forbidden zones. The coefficient of the Airy function in the former case ($x < x_2$) is twice its coefficient in the latter ($x > x_2$), resulting in the *connection rule*

$$2D \leftrightarrow D \tag{4.75}$$

across the WKB function singularity at x_2. Strictly analogous observations are made with respect to the WKB solution in the vicinity of the left turning point, x_1, where ψ_{WKB} is proportional to $\frac{2}{\sqrt{W(x)}}\sin[\frac{1}{\hbar}\int_{x_1}^{x} W(x'')dx'' + \pi/4]$ $(x > x_1)$ within the classically allowed region and to $\frac{1}{\sqrt{|W(x)|}}\exp[-\frac{1}{\hbar}\int_x^{x_1} |W(x'')|dx'']$ $(x < x_1)$ beyond that region. The remaining unknown coefficient D is found from matching the two WKB wave function segments in the allowed region (see later, Eq. (4.81)). As shown in the following subsection, the matching condition implies quantization rules for bound systems described by the WKB approach.

Expanding the result (4.73) according to

$$\psi_{\mathrm{WKB}}(x) \sim \frac{\exp\left(i\frac{\pi}{4}\right)}{\sqrt{W(x)}}\left[\exp\left(\frac{i}{\hbar}\int_x^{x_2} W(x'')dx''\right)\right.$$
$$\left. - \exp\left(-i\frac{\pi}{2}\right)\exp\left(-\frac{i}{\hbar}\int_x^{x_2} W(x'')dx''\right)\right], \quad \text{if } x < x_2, \tag{4.76}$$

demonstrates that the wave function undergoes a phase jump by $\frac{\pi}{2}$ as it is reflected at the turning point. This phenomenon may be compared to the phase shift experienced by a light wave upon reflection from a medium with higher optical density.

To examine in further detail the conditions under which the WKB approximation is valid, we recall that we disregarded the quantum potential V_{Q} in (4.54a), thus reducing the real part of the TDSE to a Hamilton–Jacobi equation. This is justified if the quantum potential is everywhere small compared with the variation of the potential over an interval equal to the local de Broglie wave length $\lambda = \frac{h}{P}$ [103], defining a characteristic spatial dimension for the considered problem:

$$V_{\mathrm{Q}} \ll \lambda \left|\frac{dV}{dx}\right|. \tag{4.77}$$

Utilizing relation (4.62), the quantum potential is readily approximated by

$$V_{\mathrm{Q}} \approx \frac{\hbar^2}{2m}\frac{\left(\frac{dV}{dx}\right)^2}{(E - V(x))^2}, \tag{4.78}$$

where the variation of $V(x)$ has been assumed to be sufficiently small to justify the neglect of its second derivative. We express relation (4.78) through λ and arrive at

$$V_{\mathrm{Q}} \approx \sqrt{\frac{\hbar^2}{2m}}\frac{\left(\frac{dV}{dx}\right)^2}{(E - V(x))^{\frac{3}{2}}}\lambda. \tag{4.79}$$

It follows that

$$\sqrt{\frac{\hbar^2}{2m}}\frac{\frac{dV}{dx}}{(E - V(x))^{\frac{3}{2}}} \ll 1 \tag{4.80}$$

has to be satisfied for Eq. (4.77) to hold. This condition, however, is equivalent to $|\frac{\partial \lambda}{\partial x}| \ll 1$, implying that the WKB theory is appropriate only in a domain where the local de Broglie wave length does not change too rapidly. An identical requirement is that the fractional change of the de Broglie wave length, $\frac{\delta\lambda}{\lambda}$, is small in this region. This criterion is in accord with the intuitive notion that in the classical regime, the de Broglie wave length is extremely small as compared with the typical distances involved in the problem, and therefore the potential causes a negligible local change of the wave length. On an elementary level, this is exemplified by the harmonic oscillator in the limit of large quantum numbers n. The scale at which the potential varies is here given by the oscillation amplitude which, at large n, exceeds by far the de Broglie wave length, as the latter is on the order of the distance between adjacent nodes of a highly oscillatory wave function. Obviously, the constraint (4.80) is violated at the turning points, since the local de Broglie wave length is singular at positions where the total and the potential energy are equal.

Qualitatively similar arguments are used in classical physics to motivate the transition from wave optics to geometric, or ray optics. If the change of the parameters relevant to the optical problem over one wave length of light is negligible, the latter applies. Replacing in this *eikonal approximation* the light wave by a de Broglie wave, one may generate the conditions for the validity of the WKB approach. In Section 8.4, we will describe an extension of this notion to the quantum dynamics of electrons and nuclei, involving situations where the electronic wave functions and their gradients undergo sufficiently small variations over a typical nuclear de Broglie wave length.

4.2.2 *The Bohr–Sommerfeld quantization rules for bound WKB states*

We show in the following that closing the remaining gap in our construction of the WKB wave function, i.e. joining the two oscillatory solution segments for $x > x_1$ and $x < x_2$ (see Figure 4.2), provides a natural way of deriving quantization rules in the phase space spanned by (x, p). Specifically, we require that

$$\sin\left[\frac{1}{\hbar}\int_{x_1}^{x} W(x'')dx'' + \frac{\pi}{4}\right] = D\sin\left[\frac{1}{\hbar}\int_{x}^{x_2} W(x'')dx'' + \frac{\pi}{4}\right]. \tag{4.81}$$

We simplify this equation by setting

$$\alpha \equiv \frac{1}{\hbar}\int_{x_1}^{x_2} W(x'')dx'', \quad \beta \equiv \frac{1}{\hbar}\int_{x}^{x_2} W(x'')dx'' + \frac{\pi}{4}, \tag{4.82}$$

and recast Eq. (4.81) into

$$\sin\left(\alpha + \frac{\pi}{2} - \beta\right) = D\sin\beta \tag{4.83}$$

which becomes

$$\sin\left(\alpha + \frac{\pi}{2}\right)\cos(\beta) - \cos\left(\alpha + \frac{\pi}{2}\right)\sin(\beta) = D\sin(\beta). \tag{4.84}$$

From this identity, one concludes that $D = -\cos(\alpha + \frac{\pi}{2})$ and $\alpha + \frac{\pi}{2} = (n+1)\pi$, with $n = 0, 1, 2, \ldots$, where the choice of $n+1$ $(n = 0, 1, \ldots)$ instead of n $(n = 1, 2, \ldots)$ is conventional. It follows

$$\alpha = \frac{1}{\hbar}\int_{x_1}^{x_2} W(x'')dx'' = \left(n + \frac{1}{2}\right)\pi. \tag{4.85}$$

This relation may be extended to involve an integral over a *closed loop in phase space*, traversed by a particle that moves from the left to the right turning point and returns with reversed momentum from x_2 to x_1:

$$\int_{x_1}^{x_2} W(x'')dx'' + \int_{x_2}^{x_1} -W(x'')dx'' = \oint W(x'')dx'' = \left(n + \frac{1}{2}\right)h. \tag{4.86}$$

The latter integral describes the area encircled by the closed loop of the orbiting particle. From

$$\frac{1}{h}\oint W(x'')dx'' = \oint \frac{dx''}{\lambda} = \left(n + \frac{1}{2}\right), \tag{4.87}$$

it is evident that $(n + \frac{1}{2})$ counts the number of wave lengths contained in the orbit perimeter. The quantization condition (4.87) is distinguished from the Bohr–Sommerfeld rule $\frac{1}{h}\oint W(x'')dx'' = n$ by a number $\frac{1}{2}$ added on the right-hand side. This modification reflects the fact that the WKB wave function is not constrained to disappear at the turning points, while the Bohr–Sommerfeld prescription is based on wave function nodes as boundary conditions.

4.3 The Wigner Function: A Quantum Mechanical Phase Space Distribution

Quantum mechanics operates with state vectors. These may be represented in position or momentum space to yield wave functions $\psi(x)$ or $\widetilde{\psi}(p)$ as well as the related density distributions $\rho(x)$ or $\widetilde{\rho}(p)$. Knowledge of ψ or ρ leads to statements about the probabilities of finding the considered system either at a place x or with a momentum p. Classical statistics involves probability functions of the form $P(x, p)$, involving probabilities defined in *phase space*. The position-momentum uncertainty relation prevents us from importing this concept into the dominion of quantum theory. A function that assigns a definitive value of both position and momentum to a system does not make quantum mechanical sense. Nothing, however, forbids us to introduce a quantum mechanical distribution that depends on both position and momentum, as long as we do not associate this mathematical entity with simultaneously sharp values of both observables. Such a phase space representation of quantum mechanics was suggested by Wigner [113].

In what follows, we will construct the *Wigner function*, or *Wigner transform*, and attempt to understand the specific balance that it keeps between the classical and the quantum world. Within this effort, we will explore the quasi-classical nature of the Wigner function and, at the same time, emphasize the way in which it deviates from a true probability distribution. Most prominently, as it will turn out, the quantum mechanical phase space distribution admits negative values, while probabilities are required to be positive definite. We will underscore the importance of the Wigner function as a versatile tool of semiclassical physics and delineate the domain of its most relevant applications.

4.3.1 *Defining properties of the Wigner function*

For a quantum system characterized by the wave function $\psi(x)$, we may write the transition amplitude related to the position eigenstates $|x'\rangle$ and $|x''\rangle$ as $\langle x''|\psi\rangle\langle\psi|x'\rangle$. This description presumes that the system in question is realized as a pure state. To generalize, we replace the pure state with a statistical mixture, expressing the amplitude as $\langle x''|\hat{\rho}|x'\rangle$ with $\hat{\rho}$ as density operator. The case of a pure state is recovered by setting $\hat{\rho}$ equal to $|\psi\rangle\langle\psi|$.

We continue exploring the Wigner function by relabeling the positions x' and x'' in terms of the midpoint $x = \frac{x'+x''}{2}$ as well as the distance $\widetilde{x} = x'' - x'$ traversed by the system as the transition takes place. Thus,

$$x' = x - \frac{1}{2}\widetilde{x}, \quad x'' = x + \frac{1}{2}\widetilde{x}, \tag{4.88}$$

turning the density operator matrix element $\rho(x', x'') = \langle x''|\hat{\rho}|x'\rangle$ into

$$\widetilde{\rho}(x, \widetilde{x}) = \left\langle x + \frac{1}{2}\widetilde{x}|\hat{\rho}|x - \frac{1}{2}\widetilde{x}\right\rangle, \tag{4.89}$$

and thus a function of the midpoint x and the distance $\widetilde{x}$. We might eliminate the dependence on the latter by performing a Fourier transformation with respect to the variable $\widetilde{x}$, arriving at [114]

$$\rho_{\mathrm{W}}(x, p) = \frac{1}{2\pi\hbar}\int_{-\infty}^{\infty} d\widetilde{x} \exp\left(-\frac{i}{\hbar}p\widetilde{x}\right)\widetilde{\rho}(x, \widetilde{x}). \tag{4.90}$$

This is the Wigner function. As noted before, it represents a quantum mechanical distribution defined in phase space, i.e. on the classical variables x and p. The prefactor of the distribution (4.90) is chosen such that the normalization condition

$$\int_{-\infty}^{\infty} dx \int_{-\infty}^{\infty} dp \rho_{\mathrm{W}}(x, p) = 1 \tag{4.91}$$

is met.

We understand the Wigner function as a proper generalization of both position and momentum space probability distributions. Its Janus-headed nature, overseeing simultaneously the quantum mechanically disjoint regimes of position and momentum space, manifests itself in a variety of characteristic properties. First of all, the reduced Wigner functions obtained by integration over the variables x or p must yield the quantum mechanical probabilities of finding the system at a certain place x or with a certain momentum p, respectively. It is easily shown that this is indeed the case.

To start with the derivation of the position space distribution, integration over p yields:

$$\int_{-\infty}^{\infty} dp\,\rho_{\mathrm{W}}(x,p) = \int_{-\infty}^{\infty} d\widetilde{x}\,\widetilde{\rho}(x,\widetilde{x})\frac{1}{2\pi\hbar}\int_{-\infty}^{\infty} dp\,\exp\left(-\frac{i}{\hbar}p\widetilde{x}\right). \tag{4.92}$$

Since the factor $\frac{1}{2\pi\hbar}\int_{-\infty}^{\infty} dp\,\exp\,(-\frac{i}{\hbar}p\widetilde{x})$ is identical with a delta function in the variable $\widetilde{x}$, expression (4.92) condenses to

$$\int_{-\infty}^{\infty} dp\,\rho_{\mathrm{W}}(x,p) = \int_{-\infty}^{\infty} d\widetilde{x}\left\langle x+\frac{1}{2}\widetilde{x}|\rho|x-\frac{1}{2}\widetilde{x}\right\rangle\delta(\widetilde{x}) = \langle x|\rho|x\rangle = P(x) \tag{4.93}$$

with $P(x)$ as the probability of finding the system at x. Likewise, reducing $\rho_{\mathrm{W}}(x,p)$ with respect to x, we are left with the analogous probability $P(p)$. This is established by considering

$$\int_{-\infty}^{\infty} dx\,\rho_{\mathrm{W}}(x,p) = \frac{1}{2\pi\hbar}\int_{-\infty}^{\infty} dx\int_{-\infty}^{\infty} d\widetilde{x}\,\exp\left(-\frac{i}{\hbar}p\widetilde{x}\right)\widetilde{\rho}(x,\widetilde{x}). \tag{4.94}$$

Upon changing integration variables, we obtain

$$\int_{-\infty}^{\infty} dx\,\rho_{\mathrm{W}}(x,p) = \frac{1}{2\pi\hbar}\int_{-\infty}^{\infty} dx'\int_{-\infty}^{\infty} dx''\,\exp\left[-\frac{i}{\hbar}p(x''-x')\right]\rho(x',x''). \tag{4.95}$$

Recalling that $\sqrt{\frac{1}{2\pi\hbar}}\exp\,(-\frac{i}{\hbar}px) = \langle p|x\rangle$, and using the assumed completeness of the states $|x'\rangle$ and $|x''\rangle$, we find

$$\int_{-\infty}^{\infty} dx\,\rho_{\mathrm{W}}(x,p) = \langle p|\hat{\rho}|p\rangle = P(p), \tag{4.96}$$

demonstrating that the projection of the distribution $\rho_{\mathrm{W}}(x,p)$ into momentum space results in the usual probability distribution as a function of momentum.

Exercise 4.6. Show that

$$\mathrm{Tr}(\hat{\rho}_1\hat{\rho}_2) = 2\pi\hbar\int_{-\infty}^{\infty} dx\int_{-\infty}^{\infty} dp\,\rho_{\mathrm{W},\rho_1}(x,p)\rho_{\mathrm{W},\rho_2}(x,p). \tag{4.97}$$

Proceed by inserting the explicit form of the Wigner function (4.90) into the right-hand side of the equation. Substitute the delta function and change the integration variables to x' and x''. Finally, use the definition of the trace: $\mathrm{Tr}(\hat{A}) = \int_{-\infty}^{\infty} dx \langle x|\hat{A}|x\rangle$.

Statements (4.93) and (4.96) establish the Wigner function as a legitimate extension of the familiar quantum mechanical probabilities defined in position or momentum space. However, as hinted earlier, any attempt to understand the Wigner function consistently in terms of probability is doomed to fail since $\rho_W(x,p)$ can adopt negative values. This distinctive feature of the Wigner function which provides a signature of its non-classical character is a consequence of the relation to be proven in Exercise 4.6. If the density operators $\hat{\rho}_1$ and $\hat{\rho}_2$ are selected in such a way that the trace of their product vanishes, i.e. $Tr(\hat{\rho}_1\hat{\rho}_2) = 0$, the right-hand side of Eq. (4.97) vanishes too, resulting in

$$\int_{-\infty}^{\infty} dx \int_{-\infty}^{\infty} dp\, \rho_{W,\rho_1}(x,p)\rho_{W,\rho_2}(x,p) = 0. \tag{4.98}$$

This condition is satisfied if ρ_1 and ρ_2 are associated with orthogonal states. The simplest respective example is realized if the systems involved are described by pure states Ψ_1 and Ψ_2, with $\hat{\rho}_1 = |\Psi_1\rangle\langle\Psi_1|$, $\hat{\rho}_2 = |\Psi_2\rangle\langle\Psi_2|$ and $\langle\Psi_1|\Psi_2\rangle = 0$. This, as well as more complex cases governed by the equation $\mathrm{Tr}(\hat{\rho}_1\hat{\rho}_2) = 0$, imply that integrating the product of the two Wigner functions over all of phase space yields zero. One infers that phase space domains must exist where one of the factors $\rho_{W,\rho_1}(x,p)$ or $\rho_{W,\rho_2}(x,p)$ is negative. This conclusion, needless to say, does not hold if $\mathrm{Tr}(\hat{\rho}_1\hat{\rho}_2) > 0$.

4.3.2 *Time dependence of the Wigner function*

Focusing on the time behavior of the Wigner function, we invoke the *Liouville–von Neumann equation*:

$$\frac{\partial\hat{\rho}}{\partial t} = -\frac{i}{\hbar}[\hat{H},\hat{\rho}]. \tag{4.99}$$

This implies, with $\hat{H} = \hat{T} + V$

$$\frac{\partial}{\partial t}\left\langle x + \frac{1}{2}\widetilde{x}|\hat{\rho}|x - \frac{1}{2}\widetilde{x}\right\rangle = -\frac{i}{\hbar}\left\langle x + \frac{1}{2}\widetilde{x}\right| [\hat{H},\hat{\rho}] \left|x - \frac{1}{2}\widetilde{x}\right\rangle. \tag{4.100}$$

Carrying out the Fourier transformation of both sides with respect to $\widetilde{x}$ as in Eq. (4.90), we derive a statement about the time variation of the Wigner function, namely

$$\frac{\partial \rho_{\mathrm{w}}}{\partial t} = -\frac{i}{\hbar}\int_{-\infty}^{\infty} d\widetilde{x}\frac{1}{2\pi\hbar}\exp\left(-\frac{i}{\hbar}p\widetilde{x}\right)$$

$$\times\left\langle x+\frac{1}{2}\widetilde{x}\middle|[\hat{T},\hat{\rho}]+[V,\hat{\rho}]\middle|x-\frac{1}{2}\widetilde{x}\right\rangle. \tag{4.101}$$

The right-hand side is partitioned into two commutators, involving the potential and kinetic energy operators. We outline briefly how to evaluate the former and leave the parallel calculation for the latter to the reader. Applying the potential V to both the bra and the ket of the matrix element in Eq. (4.101), we turn it into $[V(x+\frac{1}{2}\widetilde{x})-V(x-\frac{1}{2}\widetilde{x})]\langle x+\frac{1}{2}\widetilde{x}|\hat{\rho}|x-\frac{1}{2}\widetilde{x}\rangle$. The value of the potential V at $x\pm\frac{1}{2}\widetilde{x}$ is found from the Taylor expansion

$$V\left(x\pm\frac{1}{2}\widetilde{x}\right)=\sum_{m=0}^{\infty}\frac{1}{m!}\frac{d^mV(x)}{dx^m}\left(\pm\frac{1}{2}\widetilde{x}\right)^m. \tag{4.102}$$

For the difference $V(x+\frac{1}{2}\widetilde{x})-V(x-\frac{1}{2}\widetilde{x})$ we then obtain the expression

$$V\left(x+\frac{1}{2}\widetilde{x}\right)-V\left(x-\frac{1}{2}\widetilde{x}\right)$$

$$=i\hbar\sum_{m=0}^{\infty}\frac{\left(i\frac{\hbar}{2}\right)^{2m}}{(2m+1)!}\frac{d^{2m+1}V(x)}{dx^{2m+1}}\left(-i\frac{\widetilde{x}}{\hbar}\right)^{2m+1}. \tag{4.103}$$

With these arguments, the potential energy portion of the right-hand side of (4.101) becomes

$$-\frac{i}{\hbar}\int_{-\infty}^{\infty}d\widetilde{x}\frac{1}{2\pi\hbar}\exp\left(-\frac{i}{\hbar}p\widetilde{x}\right)\left\langle x+\frac{1}{2}\widetilde{x}\middle|[V,\hat{\rho}]\middle|x-\frac{1}{2}\widetilde{x}\right\rangle$$

$$=i\hbar\sum_{m=0}^{\infty}\frac{\left(i\frac{\hbar}{2}\right)^{2m}}{(2m+1)!}\frac{d^{2m+1}V(x)}{dx^{2m+1}}\frac{\partial^{2m+1}\rho_{\mathrm{w}}(x,p;t)}{\partial p^{2m+1}}, \tag{4.104}$$

where we have used the definition (4.90) of the Wigner function and represented the polynomial $(-\frac{i}{\hbar}\widetilde{x})^{2m+1}$ by the differential operator $\frac{\partial^{2m+1}}{\partial p^{2m+1}}$ acting on $\exp(-\frac{i}{\hbar}p\widetilde{x})$.

Exercise 4.7. Show that $-\frac{i}{\hbar}\int_{-\infty}^{\infty}d\widetilde{x}\frac{1}{2\pi\hbar}\exp(-\frac{i}{\hbar}p\widetilde{x})\langle x+\frac{1}{2}\widetilde{x}|[\hat{T},\hat{\rho}]|x-\frac{1}{2}\widetilde{x}\rangle=-\frac{p}{M}\frac{\partial\rho_{\mathrm{w}}(x,p;t)}{\partial x}$, where $\hat{T}=\frac{\hat{p}^2}{2M}$, and $\hat{p}$ is the momentum operator in the phase space variable x. Proceed in a manner analogous to the derivation of the corresponding potential energy commutator term, as outlined in the text.

We combine Eqs. (4.101) and (4.104) with the result to be proven in Exercise 4.7 to write down the quantum Liouville equation:

$$\left(\frac{\partial}{\partial t} + \frac{p}{M}\frac{\partial}{\partial x} - \frac{dV(x)}{dx}\frac{\partial}{\partial p}\right)\rho_{\rm W}(x,p;t)$$

$$= \sum_{m=1}^{\infty}(-1)^m \frac{\left(\frac{\hbar}{2}\right)^{2m}}{(2m+1)!}\frac{d^{2m+1}V(x)}{dx^{2m+1}}\frac{\partial^{2m+1}\rho_{\rm W}(x,p;t)}{\partial p^{2m+1}}. \qquad (4.105)$$

We arranged this relation such that all terms containing powers of $\hbar$ are isolated on the right-hand side. This ordering makes it easy to identify the classical limit. As $\hbar$ shrinks to zero, the right-hand side vanishes, assuming that the derivatives of $\rho_{\rm W}$ with respect to p are non-singular. We recall the definition of the classical Liouville equation, i.e.

$$\frac{d\rho}{dt} = \{\rho, H\}_{x,p} + \frac{\partial\rho}{\partial t} = 0, \qquad (4.106)$$

where $\{\rho, H\}_{x,p}$ stands for the Poisson bracket $\{A_1, A_2\}_{x,p} = \frac{\partial A_1}{\partial x}\frac{\partial A_2}{\partial p} - \frac{\partial A_1}{\partial p}\frac{\partial A_2}{\partial x}$ involving the variables x and p, the classical phase space density $\rho = \rho(x,p;t)$, and the Hamilton function H. As is readily demonstrated, the limit $\hbar \to 0$ of relation (4.105) yields

$$\frac{\partial}{\partial t}\rho_{\rm W}(x,p;t) + \frac{p}{M}\frac{\partial}{\partial x}\rho_{\rm W}(x,p;t) - \frac{dV(x)}{dx}\frac{\partial}{\partial p}\rho_{\rm W}(x,p;t)$$

$$= \{\rho_{\rm W}, H\}_{x,p} + \frac{\partial\rho_{\rm W}}{\partial t} = \frac{d\rho_{\rm W}}{dt} = 0. \qquad (4.107)$$

Closer inspection of Eq. (4.105) shows that physical situations exist where the classical equation of motion (4.106) is recovered without any reliance on the condition $\hbar \to 0$. This is related to the fact that the lowest potential derivative occurring on the right-hand side is of third order. Thus, the harmonic potential $V(x) = kx^2, k = \text{constant}$, provides an example of the validity of Eq. (4.107) even in the quantum realm. While the Wigner functions of harmonic problems in classical and quantum physics are governed by the same evolution law, it must be remembered that not all sets of initial conditions acceptable for the classical case are compatible with their quantum counterparts.

From Eq. (4.105), the Wigner function is the extension of the classical phase space density into the quantum domain, reducing correctly to Liouville mechanics as $\hbar \to 0$. The correspondence between the Wigner function and the density operator is paralleled by the relation between a phase space function $A_{\mathrm{W}}(x,p)$ and an operator $\hat{A}$ if one utilizes the following rule for translating $\hat{A}$ into $A_{\mathrm{W}}(x,p)$:

$$A_{\mathrm{W}}(x,p) = \int_{-\infty}^{\infty} d\tilde{x} \exp\left(-\frac{i}{\hbar}p\tilde{x}\right)\left\langle x + \frac{1}{2}\tilde{x}\right| \hat{A} \left| x - \frac{1}{2}\tilde{x}\right\rangle. \tag{4.108}$$

This is the *Weyl–Wigner correspondence* [114]. We employ this rule to establish a relation between the Wigner function and the quantum mechanical expectation value of the operator $\hat{A}$. This goal is attained by use of the definition of the expectation value $\langle \hat{A} \rangle$,

$$\langle \hat{A} \rangle = \mathrm{Tr}(\hat{A}\hat{\rho}). \tag{4.109}$$

We recast the right-hand side of this identity with the help of the trace product formula that was proved in Exercise 4.6, replacing $\hat{\rho}_1$ with $\hat{A}$ and $\hat{\rho}_2$ with $\hat{\rho}$. This results in

$$\langle \hat{A} \rangle = \mathrm{Tr}(\hat{A}\hat{\rho}) = \int_{-\infty}^{\infty} dx \int_{-\infty}^{\infty} dp\, A(x,p)_{\mathrm{W}}\rho_{\mathrm{W}}(x,p). \tag{4.110}$$

We emphasize that formula (4.108) comprises a well-defined strategy for translating a quantum mechanical operator into a c number. Different choices can be made in the case of non-vanishing commutators between observables. The simplest example of this ambiguity is given by the operator product $\hat{x}\hat{p}$ which may be represented by the c number xp or, considering the commutation relation $[\hat{x},\hat{p}] = i\hbar$, likewise as $px + i\hbar$. Still other realizations are conceivable [114]. The correspondence (4.108), however, defines a unique operator mapping procedure. Applying it to the case of $\hat{A} = \hat{x}\hat{p}$, we arrive at the representation $A_{\mathrm{W}}(x,p) = xp + \frac{i}{2}\hbar$, as the reader may verify by working out Exercise 4.8. The Weyl–Wigner correspondence yields symmetric ordering, in accordance with the operator arrangement $\hat{A} = \frac{1}{2}\{\hat{x},\hat{p}\} + \frac{1}{2}[\hat{x},\hat{p}]$.

Exercise 4.8. Use the Weyl–Wigner correspondence (4.108) to show that the phase space representation of $\hat{A} = \hat{x}\hat{p}$ is $A_{\mathrm{W}}(x,p) = xp + \frac{i}{2}\hbar$.

4.3.3 *The Moyal formalism*

To conclude our presentation of the Wigner function, we mention the *Moyal approach* that relates the time dependence of ρ_w to the energy spectrum of the physical system considered. Thus we assume that the problem

$$\hat{H}\psi = E\psi \tag{4.111}$$

has been solved such that the energy eigenvalues and eigenfunctions for both the discrete and the continuous part of the spectrum are known. If $\hat{H}$ is time independent, the evolution of the density operator is governed by the formal solution of the Liouville–von Neumann equation (4.99) according to

$$\hat{\rho}(t) = \exp\left(-\frac{i}{\hbar}\hat{H}t\right)\hat{\rho}(0)\exp\left(\frac{i}{\hbar}\hat{H}t\right). \tag{4.112}$$

We substitute this expression for $\hat{\rho}$ in Eq. (4.89) and subsequently use the definition of ρ_w, Eq. (4.90), to generate a formula for ρ_w as a function of time:

$$\rho_w(x,p;t) = \frac{1}{2\pi\hbar}\int_{-\infty}^{\infty} d\tilde{x}\exp\left(-\frac{i}{\hbar}p\tilde{x}\right)\left\langle x+\frac{1}{2}\tilde{x}\right|\exp\left(-\frac{i}{\hbar}Ht\right)\hat{\rho}(0)$$
$$\times\exp\left(\frac{i}{\hbar}\hat{H}t\right)\left|x-\frac{1}{2}\tilde{x}\right\rangle. \tag{4.113}$$

In the following step we insert the identity

$$\int_\Sigma dE\psi_E^*\psi_E = \int_\Sigma dE|E\rangle\langle E| = 1 \tag{4.114}$$

twice into the scalar product contained in Eq. (4.113). Here ψ_E denotes the eigenfunction of $\hat{H}$ with eigenvalue E and the symbol $\int_\Sigma$ signifies summation over the discrete and integration over the continuous part of the spectrum. As this twofold projection is executed, Eq. (4.113) becomes

$$\rho_w(x,p;t) = \int_\Sigma dE' \int_\Sigma dE''$$
$$\times\exp\left(-\frac{i}{\hbar}(E''-E')t\right)\langle E''|\rho(0)|E'\rangle\rho_{w,E'',E'}(x,p) \tag{4.115}$$

and

$$\rho_{w,E'',E'}(x,p) = \frac{1}{2\pi\hbar}\int_{-\infty}^{\infty} d\tilde{x}\exp\left(-\frac{i}{\hbar}p\tilde{x}\right)\left\langle x+\frac{1}{2}\tilde{x}|E''\right\rangle\left\langle E'|x-\frac{1}{2}\tilde{x}\right\rangle, \tag{4.116}$$

which defines the *Moyal function*. In the diagonal element $\rho_{w,E,E}(x,p)$ we recognize the Wigner function for the pure state ψ_E. As detailed, for instance, in [114], partial differential equations may be derived for the Moyal function from the commutation and anticommutation relations between the operators $\hat{H}$ and $|E''\rangle\langle E'|$. This finding is of major relevance for the use of the Wigner function in physical applications since it allows us to obtain $\rho_{w,E'',E'}(x,p)$ without having previously solved the stationary Schrödinger equation (4.111).

4.4 Coherent States

The notion of the *coherent state* as well as that of the related *squeezed state* have proven to be extremely useful for exploring the intermediate regime between quantum and classical physics. The term *coherent* points back to the place of origin of the coherent state concept, the field of quantum optics [115, 116] where it was found to provide an adequate and convenient tool for the description of a well-stabilized laser oscillator operating in a single mode. Since the latter is not a pure photon number state, the Fock space occupation number formalism [117] is ill-adjusted to capture its characteristics. In particular, one finds that the mean-square fluctuation of the electric field tends to infinity as the occupation number does (as detailed for instance in [118]) which makes the particle number basis an inappropriate means to represent large coherent photon ensembles, as realized in a laser. A salient feature of coherent states, in contrast, is that they are associated with a minimal uncertainty product for momentum and position, and, by the same token, with a constant mean-square fluctuation which equals the zero-point fluctuation of the vacuum. In terms of semiclassical physics, their constant minimal uncertainty is the hallmark of coherent states, assigning them a place at the border between the unsharp quantum and the "certain" classical sphere.

Coherent states cover a far wider realm than quantum optics. They have been found instrumental for understanding various aspects of quantum electrodynamics in general, as well as nuclear, atomic, and solid state physics. Their nature as quantum states on the way to the classical regime makes them interesting for the study of semiclassical problems. In the context of quantum dynamics, it is of primary interest that both electronic and nuclear coherent states can be found. Expressing the equations of motion in terms of these states gives rise to a nonlinear, quasi-classical approximation to the linear, quantum mechanical TDSE, as will be outlined in Chapter 9. In the following, we define coherent states and introduce their properties as far as they are of major relevance to the purpose of this text.

4.4.1 *Coherent and particle number states*

Coherent states are conveniently specified by use of occupation or particle number states. We understand the latter as eigenfunctions of the regular Fock space operator and employ the one-dimensional harmonic oscillator (see, for instance, [58]) as reference system. The particle annihilation and creation operators are denoted $\hat{a}$ and $\hat{a}^\dagger$, respectively, while the occupation states $|n\rangle$ are labeled by their particle numbers.

One may define an eigenstate $|z\rangle$ of $\hat{a}$ by

$$\hat{a}|z\rangle = z|z\rangle, \tag{4.117}$$

where $|z\rangle$ is assumed to be normalized. We ask for an explicit representation of $|z\rangle$ in terms of the complex eigenvalue z as well as the Fock space vacuum $|0\rangle$. Expanding $|z\rangle$ in the basis $\{|n\rangle\}$, we find

$$|z\rangle = \sum_n |n\rangle\langle n|z\rangle. \tag{4.118}$$

It follows immediately that

$$\hat{a}|z\rangle = \sum_{n=0}^{\infty}\langle n+1|z\rangle\sqrt{n+1}|n\rangle, \tag{4.119}$$

where $\hat{a}|n\rangle = \sqrt{n}|n-1\rangle$ has been used. Further, we have from Eqs. (4.117) and (4.118)

$$\hat{a}|z\rangle = z|z\rangle = \sum_{n=0}^{\infty}\langle n|z\rangle z|n\rangle. \tag{4.120}$$

Comparing (4.119) and (4.120), we find

$$\sum_{n=0}^{\infty}\langle n+1|z\rangle\sqrt{n+1}|n\rangle = \sum_{n=0}^{\infty}\langle n|z\rangle z|n\rangle. \tag{4.121}$$

Using the orthonormality of the particle number states $|n\rangle$, we conclude from this identity that

$$\langle n+1|z\rangle = \frac{\langle n|z\rangle}{\sqrt{n+1}} z. \quad (4.122)$$

This is a recursion relation which allows us to write all scalar products $\langle n|z\rangle$ as functions of the vacuum projection $v_0 \equiv \langle 0|z\rangle$. Inserting the results into the expansion (4.118), we obtain

$$|z\rangle = v_0 \sum_{n=0} \frac{z^n}{\sqrt{n!}} |n\rangle. \quad (4.123)$$

In order to express v_0 in terms of z we employ the normalization of the states $|z\rangle$. This implies

$$\langle z|z\rangle = |v_0|^2 \sum_{m,n} \left[\frac{(z^*)^m z^n}{\sqrt{m!n!}} \right] \langle m|n\rangle = |v_0|^2 \exp\left(|z|^2\right)^n = 1 \quad (4.124)$$

and therefore $v_0 = \exp\left(-\frac{1}{2}|z|^2\right)$. In summary, the expansion of the coherent state $|z\rangle$ in particle number states reads

$$|z\rangle = \exp\left(-\frac{1}{2}|z|^2\right) \sum_{n=0} \frac{z^n}{\sqrt{n!}} |n\rangle = \exp\left(-\frac{1}{2}|z|^2 + z\hat{a}^\dagger\right) |0\rangle, \quad (4.125)$$

where the step-up relation $|n\rangle = \frac{(\hat{a}^\dagger)^n}{\sqrt{n!}} |0\rangle$ has been utilized to arrive at the term on the right-hand side. A more symmetric representation of $|z\rangle$ is accomplished by realizing that

$$\exp\left(-\frac{1}{2}|z|^2 + z\hat{a}^\dagger\right)|0\rangle = \exp\left(-\frac{1}{2}|z|^2\right) \exp\left(z\hat{a}^\dagger\right) \exp\left(-z^*\hat{a}\right)|0\rangle, \quad (4.126)$$

which follows directly from the annihilation operator property $\hat{a}^n\, |0\rangle = 0$ if $n > 0$. We recall the commutation relation $[\hat{a}^\dagger, \hat{a}] = -1$. Since the commutator $[\hat{A}, \hat{B}]$ of $\hat{A} \equiv \hat{a}^\dagger$ and $\hat{B} \equiv \hat{a}$ commutes with both of these operators, the Baker–Campbell–Hausdorff formula for $\hat{A}$ and $\hat{B}$ (for a full statement and proof of this identity, see [119]) reduces to

$$\exp\left(\hat{A} + \hat{B}\right) = \exp\left(\hat{A}\right) \exp\left(\hat{B}\right) \exp\left(-\frac{1}{2}[\hat{A}, \hat{B}]\right). \quad (4.127)$$

Setting $\hat{A} \equiv z\hat{a}^\dagger$ and $\hat{B} \equiv -z^*\hat{a}$, we find that $|z\rangle$ can be written as

$$|z\rangle = \exp(z\hat{a}^\dagger - z^*\hat{a})|0\rangle. \tag{4.128}$$

Exercise 4.9. Using the definition of the coherent state $|z\rangle$, prove the following statements:

(a) $\langle N\rangle_c \equiv \langle z|\hat{N}|z\rangle = |z|^2$, where $\hat{N} \equiv \hat{a}^\dagger\hat{a}$ is the number operator and $\langle N\rangle_c$ is the average number of particles contained in the state $|z\rangle$.
(b) $\langle N^2\rangle_c \equiv \langle z|\hat{N}^2|z\rangle = |z|^4 + |z|^2$, with the definitions of $\hat{N}^2$ and $\langle N^2\rangle_c$ analogous to those given in (a).
(c) Conclude from Eq. (4.125) that $\langle z| = \langle 0|\exp(-\frac{1}{2}|z|^2 + z^*a)$. Then show that $\langle z|z'\rangle = \exp[-\frac{1}{2}(|z|^2 + |z'|^2)]\exp(z^*z')$ and $|\langle z|z'\rangle|^2 = \exp(-|z-z'|^2)$, with z' as the eigenvalue of the coherent state $|z'\rangle$, $\hat{a}|z'\rangle = z'|z'\rangle$.

As is plausible from their definition, coherent states are non-orthogonal. From Exercise 4.9(c), the expression for the scalar product of two arbitrary states $|z\rangle$ and $|z'\rangle$ with eigenvalues z and z' is

$$\langle z|z'\rangle = \exp\left[-\frac{1}{2}(|z|^2 + |z'|^2)\right]\exp(z^*z'). \tag{4.129}$$

However, a closure relation for coherent states is established with elementary arguments, as we show in the following. Consider the integral

$$\frac{1}{\pi}\int |z\rangle\langle z|d^2z = \frac{1}{\pi}\sum_{m,n}\frac{|m\rangle\langle n|}{\sqrt{m!}\sqrt{n!}}\int \exp(-|z|^2)z^m z^{*n}d^2z, \tag{4.130}$$

where d^2z is to be understood as the infinitesimal element of the complex plane, $d\Re z d\Im z$. In polar coordinates, this translates into

$$d^2z = rdrd\theta, \quad \text{with } z = r\exp(i\theta). \tag{4.131}$$

We perform the integration on the right-hand side of Eq. (4.130) to find

$$\int \exp(-|z|^2)z^m z^{*n}d^2z = 2\pi\int_0^\infty dr\exp(-r^2)r^{m+n+1}\delta_{mn} = \pi n!. \tag{4.132}$$

Remembering that $m = n$, as dictated by the integral over the angular coordinate, we find by inserting Eq. (4.132) into Eq. (4.130):

$$\frac{1}{\pi}\int |z\rangle\langle z|d^2z = \sum_n |n\rangle\langle n| = 1. \tag{4.133}$$

This result implies that a general state $|g\rangle$ may be expanded into a continuous coherent state basis according to

$$|g\rangle = \frac{1}{\pi}\int |z\rangle\langle z|g\rangle d^2z. \tag{4.134}$$

4.4.2 *Coherent states as minimal uncertainty solutions*

We discuss now coherent states in terms of the uncertainty concept. For comparison, we recall the position-momentum uncertainty relation valid for any particle number state $|n\rangle$ of Fock space [114], namely

$$(\Delta x \Delta p)_n = \frac{\hbar}{2}(2n+1), \tag{4.135}$$

where the products of the variances Δx and Δp are given by the following combinations of scalar products:

$$\Delta x^2 = \langle n|\hat{x}^2|n\rangle - \langle n|\hat{x}|n\rangle^2, \tag{4.136a}$$

$$\Delta p^2 = \langle n|\hat{p}^2|n\rangle - \langle n|\hat{p}|n\rangle^2. \tag{4.136b}$$

Further, the relation between the operators $\hat{x}$ and $\hat{p}$ and the ladder operators $\hat{a}$ and $\hat{a}^\dagger$ is

$$\hat{x} = \sqrt{\frac{\hbar}{2m\omega}}(\hat{a}^\dagger + \hat{a}), \tag{4.137a}$$

$$\hat{p} = i\sqrt{\frac{m\hbar\omega}{2}}(\hat{a}^\dagger - \hat{a}), \tag{4.137b}$$

with ω as the harmonic oscillator frequency. Combining Eqs. (4.136a)–(4.137b) with the rules for applying $\hat{a}$ and $\hat{a}^\dagger$ to the occupation states, one establishes Eq. (4.135) and thus proves that the position-momentum uncertainty of Fock space states increases linearly with their occupation number. The analogous formulae for coherent states are derived with the help of the relation

$$\langle z|(\hat{a}^\dagger)^i(\hat{a})^j|z\rangle = z^{*i}z^j, \tag{4.138}$$

which follows directly from the defining property of coherent states, Eq. (4.117). Using this statement in conjunction with Eqs. (4.137a) and

(4.137b), one derives the coherent state analogs of the Fock space scalar products on the right-hand sides of Eqs. (4.136a) and (4.136b), namely

$$\langle z|\hat{x}|z\rangle = \sqrt{\frac{\hbar}{2m\omega}}(z^* + z), \tag{4.139a}$$

$$\langle z|\hat{p}|z\rangle = i\sqrt{\frac{m\hbar\omega}{2}}(z^* - z), \tag{4.139b}$$

$$\langle z|\hat{x}^2|z\rangle = \frac{\hbar}{2m\omega}(z^{*2} + 2|z|^2 + z^2 + 1), \tag{4.139c}$$

$$\langle z|\hat{p}^2|z\rangle = \frac{m\hbar\omega}{2}(z^{*2} - 2|z|^2 + z^2 - 1). \tag{4.139d}$$

Taking these four identities together, we see that the coherent state equivalent of Eq. (4.135) is

$$(\Delta x \Delta p)_c = \frac{\hbar}{2}. \tag{4.140}$$

Consequently, the position-momentum uncertainty product is a constant for any arbitrary coherent state.

4.4.3 *The nuclear coherent state*

In the following, we address the explicit construction of coherent states by use of prescription (4.128). As an example, we choose the phase space of nuclear position and momentum coordinates X, P, with associated operators $\hat{X}, \hat{P}$. Again, we refer to a single degree of freedom. Inverting Eqs. (4.137a) and (4.137b):

$$\hat{a}^\dagger = \sqrt{\frac{2m\omega}{\hbar}}\hat{X} - i\sqrt{\frac{2}{m\hbar\omega}}\hat{P}, \tag{4.141a}$$

$$\hat{a} = \sqrt{\frac{2m\omega}{\hbar}}\hat{X} + i\sqrt{\frac{2}{m\hbar\omega}}\hat{P}, \tag{4.141b}$$

and defining

$$z^* = \sqrt{\frac{2m\omega}{\hbar}}X' + i\sqrt{\frac{2}{m\hbar\omega}}P', \tag{4.142a}$$

$$z = \sqrt{\frac{2m\omega}{\hbar}}X' - i\sqrt{\frac{2}{m\hbar\omega}}P', \tag{4.142b}$$

we map the operator $\exp(z\hat{a}^\dagger - z * \hat{a})$ into $\exp[\frac{i}{\hbar}(\hat{P}'\hat{X} - \hat{X}'\hat{P})]$. The latter generates coherent states in the nuclear coordinate space when acting on the vacuum, i.e. the zero-particle wave function $\chi_0(X)$ as given by

$$\chi_0(X) = \left(\frac{m\omega}{\pi\hbar}\right)^{\frac{1}{4}} \exp\left(-\frac{m\omega}{2\hbar}X^2\right). \tag{4.143}$$

The general coherent state in this phase space is produced by applying $\exp[\frac{i}{\hbar}(\hat{P}'\hat{X} - \hat{X}'\hat{P})]$ to χ_0(X). We find [120]

$$\begin{aligned}\chi(X) &= \exp\left[\frac{i}{\hbar}(\hat{P}'\hat{X} - \hat{X}'\hat{P})\right]\chi_0(X)\\ &= \left(\frac{m\omega}{\pi\hbar}\right)^{\frac{1}{4}} \exp\left(-\frac{i}{\hbar}\left(\frac{X'}{2} - X\right)P'\right)\exp\left(-\frac{m\omega}{2\hbar}(X - X')^2\right)\end{aligned} \tag{4.144}$$

where again use has been made of the identity (4.127). We arrive at a translated Gaussian wave packet with momentum P' and center position at the place X'. The convenient features of Gaussian functions make the nuclear coherent state a popular representation of the nuclear system wherever the semiclassical formalism is applicable. We will supplement the discussion of the nuclear coherent state by introducing its electronic counterpart in Chapter 9 where we will explore the dynamic properties of coherent states.

Chapter 5

Open Systems: Elements of Rate Theory

In the previous chapters we have considered isolated quantum systems, such as a molecule or a small group of molecules in the gas phase. In these cases, the impact of the environment on the studied system was not taken into account. The system itself is typically characterized by a low number of degrees of freedom whose evolution may be described by the TDSE, or, if a density matrix formalism is employed, by the Liouville–von Neumann equation (4.99). A solute in a solvent, in contrast, must be regarded as an open system, being in contact with a reservoir whose degrees of freedom are, in general, far more numerous than its own. For many problems, this reservoir is macroscopic, making any attempt at a consistent quantum mechanical treatment futile. Clearly, isolated systems are the exception rather than the rule, and the neglect of the environment oversimplifies the actual physical situation in most cases. Including the medium, on the other hand, calls for a new methodology that addresses the dynamics of the two strongly asymmetric system parts as well as their interaction.

The coupling between the system and its reservoir may involve the exchange of particles or of energy. In what follows, we will make admission only for the latter channel, taking into account the flow of energy between the two subsystems. This phenomenon may manifest itself in the form of relaxation or excitation and deexcitation of the solute while no noticeable influence is exerted on the reservoir. If, for instance, an initially excited solute is in contact with a solvent at a well-defined temperature T, it will relax into a state of thermal equilibrium with the solvent that acts as a *heat bath*. The energy deposited into the solvent is distributed over an enormous number of degrees of freedom, or dissipated, involving an irreversible change.

In the context of electron dynamics, environmental effects are of major significance for a large variety of redox processes, involving electron transfer between two centers within a solvent. These centers might be two atoms or molecules, or two different sites within one molecule or polymer. Electron transfer has been found to be at the heart of numerous important reactions in inorganic and organic chemistry, in biology and physiology. This theme is expounded in some detail in Chapter 20. The present chapter will adopt a more general viewpoint and describe models of elementary reactions, namely one-step processes involving the transition from a reactant to a product system, laying the fundament for the theoretical treatment of electron transfer.

As a general trait of reactions, electron transfer processes are determined by an energy barrier, separating the products from the reactants. The need to overcome this hurdle makes them generally rare, or slow by the standard of environmental processes such as molecular vibrations. Direct dynamical modeling of these events is therefore often impracticable, and a statistical scheme is favored. The typical time scale of activated processes implies that a semiclassical approach based on trajectories in the sense of classical mechanics is rarely feasible. Instead, these systems are efficiently captured by use of stochastic variables, time-dependent functions $q(t)$ that are governed by probabilistic instead of deterministic laws. Statistical reaction models make it possible to deduce *rate constants* for transmission processes between equilibrium states. These constants are in principle experimentally accessible and thus allow for the contact between model and measurement.

The first section of this chapter will give a brief outline of how to obtain rate constants in the framework of classical transition state theory, which has been proven to be an efficient tool for analyzing activated processes. When applied to electron transfer, however, the classical approach, neglecting the quantal nature of electronic degrees of freedom, yields necessarily an incomplete picture. Specifically, it does not include the effect of tunneling and thus disregards the dominant transfer mechanism in the low-temperature regime. Much effort has been invested into the extension of classical activated rate theory into the quantum regime (see, for instance, [121]), as discussed in Section 5.2. Essential ingredients of this theory are a quantum representation of the reactive flux as well as the partition function localized at the position of the energy barrier top. We will show in Section 5.3 that evaluating partition functions reduces to a problem of *dynamics in imaginary time*. In particular, the imaginary-time counterpart of the real-time quantum propagator (4.16) can be associated with the statistical density matrix and realized by an appropriately adjusted path integral technique. The partition function is obtained from the density matrix by imposing a periodicity condition on the paths and summing over their initial points which coincide with their end points. For the practical application of this formalism, it turns out to be of great importance that the partition function of a quantum particle can be mapped onto that of a classical polymer.

For the problem at hand, namely the derivation of rate constants within quantum transition state theory, a method has to be devised that allows us to describe the partition function fixed to the barrier top. For this purpose, the *centroid* of an imaginary-time trajectory turns out to be an extremely useful notion, as pointed out in Section 5.4. Subsequently, we address the interaction between reactants and products in terms of the respective electronic states and obtain an approximate rate constant from a golden rule approach to the diabatic coupling between these states. As a statistical approach that transcends the golden rule and allows for explicit inclusion of the environment, *Redfield theory* is discussed in the final section of this chapter.

5.1 Classical Rate Theory

To introduce the basic tenets of classical rate theory, we assume a bistable system, involving two local minima separated by an energy barrier, as sketched in Figure 5.1. Admission is made for energy, but not particle exchange with a surrounding medium. This configuration may be realized by a solute that undergoes a change from a reactant (R) into a product (P) species while being embedded in some medium.

The transition from the reactant to the product regime, or vice versa, requires fluctuations in a reaction coordinate $q(t)$ that connects the reactant and the product state. Classically, for a complete transfer between these regimes, the system has to acquire a certain amount of activation energy from the environment (E_{bar}^{+} in Figure 5.1) to cross the barrier and to lose again energy (E_{bar}^{-} in the figure) in order to become trapped in the other minimum. Acknowledging its complex coupling to the environment, one treats the reaction coordinate connecting the equilibrium positions R and P (see Figure 5.1)

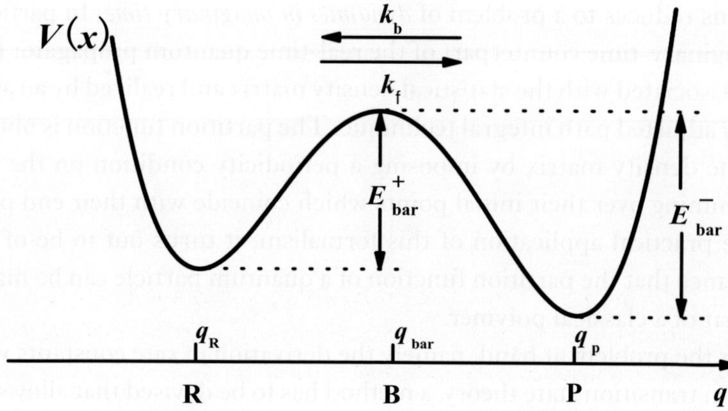

Figure 5.1: One-dimensional representation of a bistable system with reactant (left) and a product (right) minimum. Activated transport processes, characterized by rate constants $k_{\rm f}$ and $k_{\rm b}$, allow for transitions between the two regions of local stability.

not as a deterministic but as a stochastic variable. By monitoring the time behavior of this variable, we do not gain insight into the evolution of the system in the sense of classical or quantum dynamics, but we obtain statistical information that leads to characteristic rate constants.

Let the functions $N_{\rm R}(t)$ and $N_{\rm P}(t)$ label reactant and product state populations respectively. These stand for non-equilibrium quantities while $\langle N_{\rm R}\rangle$ and $\langle N_{\rm P}\rangle$ refer to the corresponding equilibrium averages. The equilibrium constant K is therefore

$$K = \frac{\langle N_{\rm P}\rangle}{\langle N_{\rm R}\rangle} = \frac{x_{\rm P}}{x_{\rm R}} \tag{5.1}$$

with $x_{\rm R}$ and $x_{\rm P}$ as the reactant and product equilibrium mole fractions, respectively. We refer to a gas phase reaction or a process with a solute concentration low enough to justify an elementary linear approach, namely

$$\frac{d\langle N_{\rm R}(t)\rangle}{dt} = -k_{\rm f}\langle N_{\rm R}(t)\rangle + k_{\rm b}\langle N_{\rm P}(t)\rangle, \tag{5.2}$$

to be complemented by the corresponding equation for the change of $\langle N_{\rm P}(t)\rangle$ with time. Here we have introduced the forward (backward) rate constants $k_{\rm f}$ ($k_{\rm b}$). The solution of this so-called *Master equation* is easily found to be

$$\langle N_{\rm R}(t)\rangle = \langle N_{\rm R}\rangle + \left[\langle N_{\rm R}(0)\rangle - \langle N_{\rm R}\rangle\right]\exp(-\gamma t), \tag{5.3a}$$

$$\text{where} \quad \gamma = k_{\rm f} + k_{\rm b}, \tag{5.3b}$$

or, with $\Delta\langle N_R(t)\rangle = \langle N_R(t)\rangle - \langle N_R\rangle$,

$$\frac{\Delta\langle N_R(t)\rangle}{\Delta\langle N_R(0)\rangle} = \exp(-\gamma t). \tag{5.4}$$

From Eqs. (5.1–5.3):

$$K = \frac{k_f}{k_b}. \tag{5.5}$$

An explicit formula for k may be derived employing the language of time correlation functions. We obtain this expression by defining step functions $h_R(q)$ and $h_P(q)$ in the reaction coordinate through

$$h_P(q) = \Theta(q - q_{bar}), \tag{5.6a}$$

$$\text{and} \quad h_R(q) = 1 - h_P(q). \tag{5.6b}$$

The regimes of the reactants and the products are separated by a *dividing surface* that intersects the reaction coordinate at $q = q_{bar}$, specifying the transition state. We set $q_{bar} = 0$. Obviously, the quantities h_R and h_P are normalized reactant and product state populations N_R and N_P, respectively. The reactant equilibrium mole fraction x_P can be written as the average of $h_R(q)$, namely

$$x_P = \langle h_P\rangle, \tag{5.7}$$

and correspondingly for x_R. The stochastic "trajectory" of a particle in a bistable potential as shown in Figure 5.1 can be described by the functional $h_P[q(t)]$ which adopts the value zero in the reactant ($q < q_{bar}$) and one in the product region ($q > q_{bar}$). The *random walk* described by the trajectory reflects statistically fluctuating processes of energy exchange with the environment whose degrees of freedom are too numerous for deterministic or quantum mechanical treatment. The time average of the resulting random sequence is given by x_P (see Figure 5.2).

In this context, the symbol $\langle\ldots\rangle$ may be understood as denoting classical averaging over the canonical phase space density. The average $\langle h_P\rangle$ is thus obtained from integrating over the two phase space variables involved in the process, namely the reaction coordinate q and its conjugate momentum p where the domain of q is restricted to $q > q_{bar}$, and where $\exp[-\beta\{p^2/2m + V(q)\}]$ serves as weight function (e.g. [122]). With reference to Figure 5.2, we

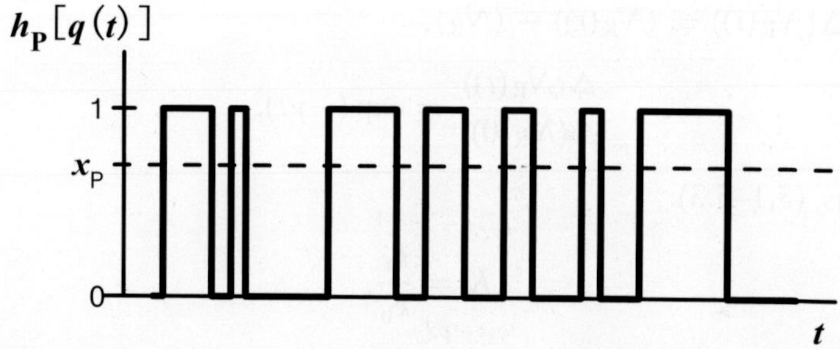

Figure 5.2: Time behavior of the product functional $h_P[q(t)]$. The time average $x_P = \langle h_P \rangle$ is indicated.

may alternatively interpret $\langle h_P \rangle$ as the fraction of time spent by the system in the product state [123]. This reasoning assumes that the studied system is *ergodic* [124], and thus ensemble averages of system properties are the same as time averages. This assumption is justified if the system samples all possible states in the limit of very long times.

Next we consider the instantaneous fluctuation $\delta h_P[q(t)] = h_P[q(t)] - \langle h_P \rangle$. The properties of the step function imply that $\langle h_P^2 \rangle = x_P$. Since $x_R + x_P = 1$, the squared variance of h_P turns out to be

$$\langle \delta h_P^2 \rangle = \langle h_P^2 \rangle - \langle h_P \rangle^2 = x_R x_P. \tag{5.8}$$

We introduce the time correlation $C(T)$ between the spontaneous fluctuation of some variable A at an initial time $t = 0$ and a later time $t = T$ as

$$C(T) \equiv \langle \delta A(0) \delta A(T) \rangle = \langle A(0) A(T) \rangle - \langle A \rangle^2. \tag{5.9}$$

For times sufficiently close to $t = 0, C(T)$ reduces to $\langle \delta A(0)^2 \rangle$. As T approaches infinity, however, the earlier and the later fluctuation become uncorrelated, and therefore

$$\lim_{T \to \infty} C(T) = \lim_{T \to \infty} \langle \delta A(0) \rangle \langle \delta A(T) \rangle = 0. \tag{5.10}$$

Sufficiently close to equilibrium, the equilibrium time correlation function of a fluctuation $\delta A(t)$ and the non-equilibrium average $\Delta \langle A(t) \rangle$ decay according to the same dynamic law, which is the content of *Onsager's regression hypothesis* [125]. Applying this statement to $h_P[q(t)]$, and combining it with Eq. (5.4), one concludes [126] that

$$\frac{\Delta\langle N_{\mathrm{R}}(t)\rangle}{\Delta\langle N_{\mathrm{R}}(0)\rangle} = \frac{\langle \delta h_{\mathrm{P}}[q(t)]|_{t=0}\delta h_{\mathrm{P}}[q(t)]\rangle}{\langle \delta h_{\mathrm{P}}^2\rangle} = \exp(-\gamma t). \tag{5.11}$$

We have exploited here the constancy of $\langle \delta h_{\mathrm{P}}^2\rangle$, established by Eq. (5.8) and implying that $\langle \delta h_{\mathrm{P}}^2\rangle = \langle \delta h_{\mathrm{P}}^2[q(0)]\rangle$. Relation (5.11), however, is not universally applicable. It is not adequate for the time scale of fast transient processes due to the coupling of the reactant and product systems to their environment. The time τ_{E} characteristic of these interactions is expected to be comparable to the period of oscillation in the potential energy wells, as shown in Figure 5.1 [126]. It is therefore clearly separated from the time needed for a passage through the barrier region, the bottleneck of the transition from the reactant to the product species. We thus specify the time range of validity for (5.11) as $t \gg \tau_{\mathrm{E}}$. In this regime, the time derivative of the left-hand side of (5.11) defines the *reactive flux*:

$$\begin{aligned}\frac{d}{dt}\frac{\langle \delta h_{\mathrm{P}}[q(t)]|_{t=0}\delta h_{\mathrm{P}}[q(t)]\rangle}{\langle \delta h_{\mathrm{P}}^2\rangle} &= \frac{\langle \delta h_{\mathrm{P}}[q(t)]|_{t=0}\dot{h}_{\mathrm{P}}[q(t)]\rangle}{\langle \delta h_{\mathrm{P}}^2\rangle}\\ &= -\gamma_{\mathrm{cl}}\exp(-\gamma_{\mathrm{cl}}t)\end{aligned} \tag{5.12}$$

with γ_{cl} as the classical full rate constant (see Eq. (5.3b)), and $\dot{h}_{\mathrm{P}}(=\delta\dot{h}_{\mathrm{P}})$ as the time derivative of h_{P}. In the so-called *plateau regime* [129], it holds that $\tau_{\mathrm{E}} \ll t \ll \tau_{\mathrm{R}} \equiv 1/\gamma_{\mathrm{cl}}$. For some time t_{pl} in this regime, $\exp(-\gamma t_{pl}) \approx 1$. If this condition is met, the reactive flux remains almost constant, and the full rate γ_{cl} is given to a good approximation by

$$\frac{\langle \dot{h}_{\mathrm{P}}[q(t)]\big|_{t=0} h_{\mathrm{P}}[q(t)]|_{t=t_{pl}}\rangle}{\langle \delta h_{\mathrm{P}}^2\rangle} = \frac{\langle\,(\delta[q]\dot{q}(t))\,|_{t=0}\, h_{\mathrm{P}}[q(t)]_{t=t_{pl}}\rangle}{\langle \delta h_{\mathrm{P}}^2\rangle} \approx \gamma_{\mathrm{cl}}. \tag{5.13}$$

The symbol $\delta[]$ in the second term refers to the delta function as the derivative of the step function h_{P}, defined by Eq. (5.6). The term $\langle(\delta[q]\dot{q}(t))|_{t=0}h_{\mathrm{P}}[q(t_{\mathrm{pl}})]\rangle$ can be cast into the form of a conditional average [123], namely as the product of the average of $\dot{q}(t)|_{t=0}h_{\mathrm{P}}[q(t_{\mathrm{pl}})]$ constrained to $q = q_{\mathrm{bar}} = 0$ and the probability $P(q_{\mathrm{bar}})$ for finding the system at the transition point q_{bar}. In summary,

$$\gamma_{\mathrm{cl}} \approx \frac{1}{(x_{\mathrm{R}}x_{\mathrm{P}})}P(q_{\mathrm{bar}})\langle\,\dot{q}(t)|_{t=0}\, h_{\mathrm{P}}[q(t_{\mathrm{pl}})]\rangle \tag{5.14}$$

where use has been made of Eq. (5.8). Expanding the coordinate q around $q_{\text{bar}} = 0$, we observe that $h_{\text{P}}[q(\Delta t)] = 1$ if $\dot{q}(t)|_{t=0} > 0$ and $h_{\text{P}}[q(\Delta t)] = 0$ if $\dot{q}(t)|_{t=0} < 0$ (see Figure 5.2) if the time interval Δt is sufficiently small. Therefore, we have

$$\begin{aligned} &\frac{1}{x_{\text{R}}x_{\text{P}}} P(q_{\text{bar}}) \lim_{t\to 0^+} \langle \dot{q}(t)|_{t=0} h_{\text{P}}(q(t))\rangle \\ &= \frac{1}{x_{\text{R}}x_{\text{P}}} P(q_{\text{bar}}) \langle \dot{q}(t)\Theta[\dot{q}(t)]\rangle|_{t=0} \\ &= \frac{1}{x_{\text{R}}x_{\text{P}}} P(q_{\text{bar}}) \frac{1}{2} \langle |\dot{q}(t)|\rangle|_{t=0} \equiv \gamma_{\text{cl}}^{\text{TST}}. \end{aligned} \tag{5.15}$$

The expression $\gamma_{\text{cl}}^{\text{TST}}$ is the classical full rate constant in the framework of the *transition state approximation* [126, 127]. For the classical forward rate constants $k_{\text{f,cl}}$ and $k_{\text{f,cl}}^{\text{TST}}$, statements analogous to Eqs. (5.14) and (5.15) can be made, namely

$$k_{\text{f,cl}} \approx \frac{1}{x_{\text{R}}} P(q_{\text{bar}}) \langle \dot{q}(t)|_{t=0} h_{\text{P}}[q(t_{pl})]\rangle, \tag{5.16a}$$

$$k_{\text{f,cl}}^{\text{TST}} = \frac{1}{x_{\text{R}}} P(q_{\text{bar}}) \frac{1}{2} \langle |\dot{q}(t)|\rangle|_{t=0}, \tag{5.16b}$$

as we conclude from Eq. (5.3b) and $k_{\text{f}}/k_{\text{b}} = x_{\text{P}}/x_{\text{R}}$. To avoid proliferation of indices, from now on we will identify k with k_{f} unless stated otherwise. Computing the rate constant in the framework of the transition state approximation does not require any knowledge about the plateau regime for the studied reaction. This approximation rests on two crucial assumptions: (1) The entire system is in a state of thermal equilibrium, encompassing all degrees of freedom. (2) No recrossing of the dividing surface. Any reactive trajectory traverses the transition state only once [126].

It has been shown that the transition state approximation (5.16b) always overestimates the rigorous rate theory (5.16a) [124]:

$$k_{\text{cl}}^{\text{TST}} > k_{\text{cl}}. \tag{5.17}$$

In the following step, we substitute for x_{R} and $P(q_{\text{bar}})$ by use of the classical partition function Z_{cl} for the complete system and $Z_{\text{cl,R}}$ for the system in the reactant state. Specifically, we identify x_{R} with $Z_{\text{cl,R}}/Z_{\text{cl}}$ and $P(q_{\text{bar}})$

with Z_{cl}^{bar}/Z_{cl}. The symbol Z_{cl}^{bar} stands for the classical partition function constrained to the barrier top position. With these substitutions,

$$k_{cl}^{TST} = \frac{Z_{cl}^{bar}}{2Z_{cl,R}} \langle|\dot{q}(t)|\rangle|_{t=0}. \tag{5.18}$$

The quantity Z_{cl}^{bar} may be understood as the reduced classical phase space density at the position of the dividing surface along the reaction coordinate q. The factor $\langle|\dot{q}(t)|\rangle|_{t=0}$ represents a flux, determining the rate at which the system leaves the transition state in the product direction [128]. It refers to the classical average over the magnitude of the velocity along the reaction coordinate. In both versions of k_{cl}^{TST}, Eq. (5.16b) as well as Eq. (5.18), the statistical approach to the transition problem, as the preferable alternative to a dynamical treatment, is made explicit.

5.2 Quantum Transition State Theory

In what follows, we discuss the quantum rate expression that corresponds to the classical formula (5.16a). We consider a canonical ensemble characterized by a Hamiltonian $\hat{H}$ and a respective set of eigenfunctions $\{|n\rangle\}$. In the quantum case, the product step functions h_P become population operators, and the classical averages are replaced by the trace with respect to the eigenfunctions of the Hamiltonian $\hat{H}$. With these assignments, the equilibrium mole fraction x_R is defined as

$$x_R = \frac{\mathrm{Tr}\{\hat{h}_R \exp(-\beta\hat{H})\}}{Z} = \frac{Z_R}{Z} \tag{5.19}$$

with $\beta = (k_B T)^{-1}$ and Z_R as the partition function of the reactant state while Z is the overall partition function of the considered ensemble: $Z = Tr\exp(-\beta\hat{H})$. It then holds that

$$k \approx \frac{i}{\beta\hbar x_R}\langle[\hat{h}_P(t_{pl}), \hat{h}_P(0)]\rangle_{av}. \tag{5.20}$$

The subscript av emphasizes that in the present context, the angular brackets denote averaging in the sense of $\langle\hat{A}\rangle = \mathrm{Tr}\{\exp(-\beta\hat{H})\hat{A}\}/Z$ and should not be associated with a scalar product. In order to demonstrate that the expression on the right-hand side of Eq. (5.20) indeed reduces in the classical limit $\hbar \to 0$ to Eq. (5.12), we expand it in the following manner:

$$
\begin{aligned}
&\frac{iZ}{\beta\hbar Z_{\mathrm{R}}}\langle[\hat{h}_{\mathrm{P}}(t_{pl}), \hat{h}_{\mathrm{P}}(0)]\rangle_{av} \\
&= \frac{i}{\beta\hbar Z_{\mathrm{R}}}\left[\sum_{m,n} \exp(-\beta E_m)\langle m|\hat{h}_{\mathrm{P}}(t_{\mathrm{pl}})|n\rangle\langle n|\hat{h}_{\mathrm{P}}(0)|m\rangle\right. \\
&\quad \left. - \sum_{m,n} \exp(-\beta E_n)\langle n|\hat{h}_{\mathrm{P}}(0)|m\rangle\langle m|\hat{h}_{\mathrm{P}}(t_{\mathrm{pl}})|n\rangle\right] \\
&= \frac{i}{\beta\hbar Z_{\mathrm{R}}}\left[\sum_{m,n} (\exp(-\beta E_m) - \exp(-\beta E_n))\langle n|\hat{h}_{\mathrm{P}}(0)|m\rangle\langle m|\hat{h}_{\mathrm{P}}(t_{\mathrm{pl}})|n\rangle\right] \\
&= \frac{i}{\beta\hbar Z_{\mathrm{R}}}\left[\sum_{m,n} \exp(-\beta E_n)\frac{\exp[\beta(E_n - E_m)] - 1}{E_n - E_m}\right. \\
&\quad \left. \times\langle n|[\hat{H}, \hat{h}_{\mathrm{P}}(0)]|m\rangle\langle m|\hat{h}_{\mathrm{P}}(t_{pl})|n\rangle\right].
\end{aligned} \tag{5.21}
$$

In the first step, we have made use of the cyclic interchangeability of operators under the trace, and in the last, we exploited the eigenvalue equations $\hat{H}|n\rangle = E_n|n\rangle$ and $\hat{H}|m\rangle = E_m|m\rangle$. At this stage, the quantum rate constant may be written in the form of an integral, namely as

$$
\begin{aligned}
&\frac{i}{\beta\hbar Z_{\mathrm{R}}}\int_0^\beta d\lambda\left[\sum_{m,n} \exp(\lambda E_n)\exp(-\lambda E_m)\langle n|[\hat{H}, \hat{h}_{\mathrm{P}}(0)]|m\rangle\right. \\
&\quad \left. \times\langle m|\hat{h}_{\mathrm{P}}(t_{pl})|n\rangle\exp(-\beta E_n)\right]
\end{aligned}
$$

where λ is a real parameter of dimension energy^{-1}. The factors $\exp(\lambda E_{\mathrm{N}})$ and $\exp(-\lambda E_m)$ are incorporated into the scalar product and thus turned into the corresponding evolution operators, leading to

$$
\begin{aligned}
&\frac{1}{\beta Z_{\mathrm{R}}}\int_0^\beta d\lambda\sum_n \langle n|\exp(\lambda\hat{H})\frac{i}{\hbar}[\hat{H}, \hat{h}_{\mathrm{P}}(0)] \\
&\quad \times\exp(-\lambda\hat{H})\hat{h}_{\mathrm{P}}(t_{\mathrm{pl}})|n\rangle\exp(-\beta E_n).
\end{aligned}
$$

In the final move, we change the integration variable from the reciprocal energy λ to the time τ, defined by the relation $\lambda = \tau/\hbar$. Therefore

$$k = \frac{1}{\hbar\beta Z_R} \int_0^{\hbar\beta} d\tau \sum_n \left\langle n \right| \exp\left(\frac{i}{\hbar}(-i\tau)\hat{H}\right) \frac{i}{\hbar}[H, \hat{h}_P(0)]$$

$$\times \exp\left(-\frac{i}{\hbar}(-i\tau)\hat{H}\right) \hat{h}_P(t_{pl}) \left| n \right\rangle \exp(-\beta E_n)$$

$$= \frac{Z}{\hbar\beta Z_R} \int_0^{\hbar\beta} d\tau \langle \dot{\hat{h}}_P(-i\tau)\hat{h}_P(t_{pl})\rangle_{av}. \tag{5.22}$$

The term $\exp(\frac{i}{\hbar}\hat{H}(-i\tau))\frac{i}{\hbar}[\hat{H}, \hat{h}_P(0)]\exp(-\frac{i}{\hbar}(-i\tau))$ has been interpreted as the time derivative of the operator $\hat{h}_P$ evaluated at the imaginary time $t = -i\tau$. This final form of the quantum transition rate has the advantage that the classical limit $\hbar \to 0$ is easily inferred from it. We find

$$\lim_{\hbar\to 0} \frac{Z}{\hbar\beta Z_R} \int_0^{\hbar\beta} d\tau \left\langle \frac{d\hat{h}_P}{dt}(-i\tau)\hat{h}_P(t_{pl}) \right\rangle_{av}$$

$$= \frac{Z}{Z_R} \left\langle \frac{d\hat{h}_P}{dt}|_{t=0} \hat{h}_P(t_{pl}) \right\rangle_{av} \approx k, \tag{5.23}$$

which turns into the classical rate (5.16a), as the trace operation implied in Eq. (5.23) is replaced by classical averaging, and with $Z_R/Z = x_R$.

5.2.1 *The quantum transition state approximation*

What is the quantum equivalent of the classical transition state approximation? As suggested by relation (5.20), the limiting procedure that yields k_{cl}^{TST} from the classical rate constant k_{cl} has no direct analog in quantum mechanics. Thus, as we let t_{pl} go towards 0^+ we obtain zero, which is a meaningless result. A suitable reformulation of the quantum rate formula (5.22) is needed to clarify how to proceed from the general formalism to the transition state approximation. The identity (5.18) will serve as a guideline for this search since it provides a classical counterpart of the desired quantum approach.

Carrying out the integral in the final expression of (5.22) explicitly leads to [129]

$$k \approx \frac{Z}{i\hbar\beta Z_{\mathrm{R}}} \langle [\hat{h}_{\mathrm{P}}(0) - \hat{h}_{\mathrm{P}}(i\hbar\beta)]\hat{h}_{\mathrm{P}}(t_{\mathrm{pl}})\rangle_{\mathrm{av}}. \tag{5.24}$$

This in turn is written as the imaginary part of the trace of an operator chain, namely

$$k \approx \frac{2}{\hbar\beta Z_{\mathrm{R}}} \Im \left\{ \mathrm{Tr} \left\{ \hat{h}_{\mathrm{R}} \exp(-\beta\hat{H})\hat{h}_{\mathrm{P}} \exp\left(\frac{i}{\hbar}\hat{H}t_{\mathrm{pl}}\right) \hat{h}_{\mathrm{P}} \exp\left(-\frac{i}{\hbar}\hat{H}t_{\mathrm{pl}}\right)\right\}\right\}. \tag{5.25}$$

Exercise 5.1. Establish Eq. (5.25) by expressing Eq. (5.24) in terms of matrix elements. Show in particular that
(a) $\mathrm{Tr}\{\exp(-\beta\hat{H})\hat{h}_{\mathrm{P}}\hat{h}_{\mathrm{P}}(t_{\mathrm{pl}})\} = \int dq \int dq' \hat{h}_{\mathrm{P}}(q)\langle q'| \exp(-\beta\hat{H})|q\rangle\langle q|\hat{h}_{\mathrm{P}}(t_{\mathrm{pl}})|q'\rangle$, and derive an equivalent term for $Tr\{\hat{h}_{\mathrm{P}} \exp(-\beta\hat{H})\hat{h}_{\mathrm{P}}(t_{\mathrm{pl}})\}$.
(b) $\int dq \int dq' (\hat{h}_{\mathrm{P}}(q) - \hat{h}_{\mathrm{P}}(q'))\langle q'| \exp(-\beta\hat{H})q\rangle\langle q|\hat{h}_{\mathrm{P}}(t_{\mathrm{pl}})|q'\rangle = \int dq \int dq' (\hat{h}_{\mathrm{R}}(q')\hat{h}_{\mathrm{P}}(q) - \hat{h}_{\mathrm{R}}(q)\hat{h}_{\mathrm{P}}(q'))\langle q'| \exp(-\beta\hat{H})|q\rangle\langle q|\hat{h}_{\mathrm{P}}(t_{\mathrm{pl}})|q'\rangle$, by using the relation between $\hat{h}_{\mathrm{P}}$ and $\hat{h}_{\mathrm{R}}$.

We compare this finding with the classical rate according to transition state theory, $k_{\mathrm{cl}}^{\mathrm{TST}}$. It is straightforward to cast the exact quantum rate (5.25) into a form strictly analogous to Eq. (5.18) by writing it as

$$k^{QTST} = \frac{1}{2} \langle|\dot{q}(t)|\rangle|_{t=0} \frac{Z_c^{\mathrm{bar}}}{Z_{\mathrm{R}}}, \tag{5.26}$$

with the reactive flux expectation value $\langle|\dot{q}(t)|\rangle|_{t=0}$ given by

$$\langle|\dot{q}|\rangle|_{t=0} = \frac{4}{\hbar\beta Z_c^{\mathrm{bar}}} \Im \left\{ Tr \left[\hat{h}_{\mathrm{R}} \exp(-\beta\hat{H})\hat{h}_{\mathrm{P}} \times \exp\left(\frac{i}{\hbar}\hat{H}t_{\mathrm{pl}}\right) \hat{h}_{\mathrm{P}} \exp\left(-\frac{i}{\hbar}\hat{H}t_{\mathrm{pl}}\right)\right]\right\}. \tag{5.27}$$

Here we have introduced the quantity Z_c^{bar} as a proper generalization of the classical partition function at the barrier top. A transition state approximation to the exact rate (5.26) and thus a quantum analog to Eq. (5.18) may be attained as one replaces the time correlation function (5.27) by a temporally local expression. In the simplest realization of this proposal, one might insert for $\langle|\dot{q}(t)|\rangle|_{t=0}$ the average speed $\langle|v|\rangle_{\mathrm{FP}}$ of a free particle along the reaction coordinate q [130],

$$\langle|v|\rangle_{\mathrm{FP}} = \sqrt{\frac{2}{\pi m\beta}}, \tag{5.28}$$

to arrive at

$$k^{QTST} = \sqrt{\frac{1}{2\pi m\beta}}\frac{Z_c^{\mathrm{bar}}}{Z_{\mathrm{R}}}. \tag{5.29}$$

This relation is easily cast into the Arrhenius form by use of the identity $F = -\frac{1}{\beta}\ln Z$ that connects the free energy with the partition function. For this purpose, we reformulate Eq. (5.29) according to

$$k^{\mathrm{QTST}} = \frac{1}{2\pi\hbar\beta}\sqrt{\frac{2\pi\beta}{m}}\hbar\frac{Z_c^{\mathrm{bar}}}{Z_{\mathrm{R}}}. \tag{5.30a}$$

or

$$k^{\mathrm{QTST}} = \frac{1}{2\pi\hbar\beta Z_{\mathrm{R}}}\exp\left(-\beta F_c^{\mathrm{bar}}\right). \tag{5.30b}$$

Here we have defined

$$F_c^{\mathrm{bar}} \equiv -\frac{1}{\beta}\ln\left[Z_c^{\mathrm{bar}}\sqrt{\frac{2\pi\hbar^2\beta}{m}}\right] \tag{5.31}$$

as the free energy that corresponds to the quantum statistical partition function Z_c^{bar}. The *centroid free energy* and the *centroid partition function* or *reduced centroid density* at the dividing surface along the reaction coordinate q have been proposed as the desired quantum extensions of the related classical concepts [121]. These entities assume a pivotal role for specifying the quantum transition rate. In Section 5.4 we attempt a brief characterization of the reduced centroid density and in particular its realization through path integrals. In the course of this discussion, it will become plausible how to construct a suitably constrained partition function Z_c^{bar} and thus to obtain the quantum rate k^{QTST} in parallel to the classical result (5.18). These arguments require some familiarity with the *Euclidean path integral*, which the following section is intended to provide.

5.3 The Euclidean Path Integral

We begin by writing the probability $P(q)$ of locating a system in thermal equilibrium at the position q as the trace [100]

$$P(q) = \frac{1}{Z} \sum_i \psi_i^*(q)\psi_i(q) \exp(-\beta E_i), \tag{5.32}$$

assuming a canonical ensemble, and adopting a complete set of quantum states $\{\psi_i(q)\}$. Likewise, the expectation value of some observable $\hat{C}$ is found from

$$\langle \hat{C} \rangle = \frac{1}{Z} \sum_i \int dq\, \psi_i^*(q)\hat{C}\psi_i(q) \exp(-\beta E_i). \tag{5.33}$$

In Eqs. (5.32) and (5.33) the limitation to a discrete set of wave functions and a single configuration space variable are for simplicity of nomenclature. We may write $\langle \hat{C} \rangle$ in a more compact form with the help of the statistical matrix element $\rho(q', q)$, stipulating that

$$\langle \hat{C} \rangle = \frac{1}{Z} \mathrm{Tr}\{\hat{C}\rho(q, q)\}, \tag{5.34}$$

where ρ is defined by

$$\rho(q', q) = \sum_i \psi_i(q')\psi_i^*(q) \exp(-\beta E_i), \tag{5.35}$$

and it is understood that in Eq. (5.34) $\hat{C}$ acts on the wave function ψ_i after which q' is set equal to q, followed by an integration over the coordinate q. Thus, for a complete description of a system in thermal equilibrium, the statistical matrix has to be known. Once this is the case, the partition function is readily derived from

$$P(q) = \frac{1}{Z} \rho(q, q). \tag{5.36}$$

Using the fact that the integration of $P(q)$ over all values of q yields one, we find

$$Z = \int dq \rho(q, q) = \mathrm{Tr}(\rho) \tag{5.37}$$

or, adopting the Dirac notation and setting $|\psi_i\rangle = |E_i\rangle$,

$$Z = \mathrm{Tr}\{\exp(-\beta\hat{H})\} = \sum_i \langle E_i|\exp(-\beta\hat{H})|E_i\rangle. \tag{5.38}$$

The formulae (5.35), (5.37), and (5.38) make reference to a complete set of quantum states that must be known for evaluation of the statistical matrix and the partition function. This methodological impasse, however, can be avoided by use of the path integral formalism, as introduced in Section 4.1. In the following, we will discuss the representation of both Z and ρ in terms of path integrals. Specifically, we propose a relation between the right-hand side of Eq. (5.38) and the quantum propagator Eq. (4.16) connecting the states $|q_a, t_a\rangle$ and $|q_\mathrm{b}, t_\mathrm{b}\rangle$. Both $|q_a, t_a\rangle$ and $|q_\mathrm{b}, t_\mathrm{b}\rangle$ are understood as eigenstates of the Heisenberg picture operator $\hat{q}(t)$. The functions $|q_a\rangle$, $|q_\mathrm{b}\rangle$ satisfy the continuum normalization constraint $\langle q_a|q_\mathrm{b}\rangle = \delta(q_a - q_\mathrm{b})$. The propagator reads

$$\langle q_\mathrm{b}, t_\mathrm{b}|q_\mathrm{a}, t_\mathrm{a}\rangle = \left\langle q_\mathrm{b}\right|\exp\left(-\frac{i}{\hbar}\hat{H}(t_\mathrm{b} - t_\mathrm{a})\right)\left| q_a\right\rangle. \tag{5.39}$$

We exploit the formal analogy between the time evolution operator $\exp(-\frac{i}{\hbar}\hat{H}t)$ and the statistical operator $\exp(-\beta\hat{H})$, and extend the time into the imaginary domain, declaring $t = -i\tau$, where the real variable τ has been introduced in the context of Eq. (5.22). In particular, we stipulate

$$t_\mathrm{b} - t_\mathrm{a} = -i\beta\hbar. \tag{5.40}$$

Upon inserting Eq. (5.40) into Eq. (5.39) and equating the final and the initial coordinate, $q_\mathrm{b} = q_\mathrm{a}$, we arrive at

$$\rho_\mathrm{aa} = \langle q_\mathrm{a}|\exp(-\beta\hat{H})|q_\mathrm{a}\rangle. \tag{5.41}$$

From the assumed completeness of both sets, $\{|q\rangle\}$ and $\{|E_n\rangle\}$ [131] one concludes that

$$Z = \int dq_\mathrm{a}\rho_\mathrm{aa} = \int dq_\mathrm{a}\langle q_\mathrm{a}|\exp(-\beta\hat{H})|q_\mathrm{a}\rangle = \sum_i \langle E_i|\exp(-\beta\hat{H})|E_i\rangle. \tag{5.42}$$

In a final step, we insert the explicit path integral form of the amplitude Z_{aa} (for comparison with the real-time case, see Eq. (4.16)) into the term on the right side of Eq. (5.41), and arrive at

$$Z = \int dq_a \int_{q_a}^{q_a} Dq \exp\left\{-\frac{1}{\hbar}\int_0^{\beta\hbar} d\tau \left[\frac{m}{2}\left(\frac{dq}{d\tau}\right)^2 + V(q)\right]\right\}. \qquad (5.43)$$

Note that the metric changes from (4.17) to $Dq \equiv \lim_{n\to\infty} [\frac{m}{2\pi\hbar\delta\tau}]^{\frac{n}{2}} dq_1 dq_2 \ldots dq_{n-1}$ as one goes from the real-time integral to the Euclidean form. The upper and lower integration limits are identical, indicating that the path integral is taken along closed trajectories. The time derivative of the coordinate q is evaluated with respect to $\tau = it$. From Eq. (5.43) and with reference to Eqs. (5.35) and (5.37), we are now able to write down a path integral representation of the statistical matrix $\rho(q', q)$. It holds that

$$\rho(q', q) = \int_q^{q'} Dq \exp\left\{-\frac{1}{\hbar}\int_0^{\beta\hbar} d\tau \left[\frac{m}{2}\left(\frac{dq}{d\tau}\right)^2 + V(q)\right]\right\}. \qquad (5.44)$$

The integral $\frac{1}{\hbar}\int_0^{\beta\hbar} d\tau \frac{m}{2}(\frac{dq}{d\tau})^2 + V(q)$ in Eq. (5.44) is termed the *Euclidean action integral*. We denote this expression by the symbol S_E and refer to the variable τ as *Euclidean time*. The Lagrangian that defines the real time path integral (see Eq. (4.19)) turns into a Hamilton function as one transforms to Euclidean time. The reader may carry out this step explicitly, as outlined in Exercise 5.2. From Eq. (5.43), the trace structure of Z is recovered for paths that start and terminate at the coordinate q_a with subsequent integration over q_a. Mathematically, the Euclidean time integral (5.44) is of greater simplicity than its real-time counterpart, the propagator K described in Section 4.1. It assigns small weights to paths for which the exponent in Eq. (5.44) is large. Only magnitudes matter in this straightforward weighting procedure, while the "histories" that enter the complex-valued propagator integral have to be added coherently, allowing for the mutual cancellation of close-lying paths.

A high-temperature limit of the partition function (5.43) can be directly inferred from this relation. This limit is attained for T high enough and thus β sufficiently small to guarantee that the closed loops $q(\tau)$ never stray far from their initial positions. For motions proceeding in a given physical time interval, it holds that objects traversing large distances have high kinetic energies. This analogy clarifies that the high-temperature case involves high values of $\frac{m}{2}(\frac{dq}{d\tau})^2$ for extended loops $q(\tau)$, resulting in a negligible contribution of the respective path to the integral (5.43). In this case, the approximation $V(q) \approx V(q_a)$ is justified, and Eq. (5.43) becomes

$$Z = \int dq_a \exp[-\beta V(q_a)] \int_{q_a}^{q_a} Dq \exp\left\{-\frac{1}{\hbar}\int_0^{\beta\hbar} d\tau \, \frac{m}{2}\left(\frac{dq}{d\tau}\right)^2\right\}. \tag{5.45}$$

The closed loop path integral over the kinetic energy term is easily evaluated by realizing that [132]

$$\int_{q_a}^{q_b} Dq \, \exp\left\{-\frac{1}{\hbar}\int_0^{\beta\hbar} d\tau \, \frac{m}{2}\left(\frac{dq}{d\tau}\right)^2\right\}$$
$$= \sqrt{\frac{m}{2\pi\beta\hbar^2}} \exp\left[-\frac{m}{2\beta\hbar^2}(q_b - q_a)^2\right], \tag{5.46}$$

in analogy to the treatment of the free particle motion in real time, as discussed in Section 4.1 (see Eqs. (4.21a)–(4.21c)). Since we want to obtain the partition function, we let the two end points q_a and q_b coincide and find that the kinetic energy path integral reduces to $\sqrt{m/2\pi\beta\hbar^2}$. Therefore, in the high-temperature case, Z turns out to be

$$Z = \sqrt{\frac{m}{2\pi\beta\hbar^2}} \int dq_a \exp[-\beta V(q_a)]. \tag{5.47}$$

This expression may be characterized as the partition function in the limit of classical mechanics [100]. While for real-time processes this limit is determined by the trajectory that results from Lagrangian dynamics, the classical path in the Euclidean time domain collapses to a point.

Exercise 5.2. Show that the real-time path integral for a single coordinate

$$\langle q_b, t_b | q_a, t_a \rangle = \int_{q_a}^{q_b} Dq \, \exp\left[\frac{i}{\hbar}\int_{t_a}^{t_b} dt \, L(q, dq/dt)\right]$$

(see Eq. (4.16)) turns into the Euclidean path integral

$$\langle q_b, \beta\hbar | q_a, 0 \rangle = \int_{q_a}^{q_b} Dq \, \exp\left\{-\frac{1}{\hbar}\int_0^{\beta\hbar} d\tau \, \frac{m}{2}\left(\frac{dq}{d\tau}\right)^2 + V(q)\right\},$$

upon the transformation (*Wick rotation*) $t = -i\tau$, $t_b - t_a = -i\beta\hbar$.

5.3.1 *Classical polymer isomorphism*

Of great importance for evaluating Euclidean path integrals is a structural analogy between the statistics of a quantum ensemble and a classical polymer. This technique is based on a discretization of the partition function path integral (5.43) and leads to an implementation of the path integral formalism discussed earlier in terms of a classical simulation of a molecular fluid. Since the theoretical groundwork has been laid for this mapping from quantum to classical statistics ([133]– [136]), ample use has been made of it in a diversity of applications ([137]– [139]). The basic ideas that underlie this technique will be sketched in the following.

Restricting the discussion once more to a single spatial dimension, we consider the canonical partition function (5.38) and express Z as a trace over the coordinate q:

$$Z = \mathrm{Tr}\{\exp(-\beta \hat{H})\} = \int dq \, \langle q| \exp(-\beta \hat{H})|q\rangle. \tag{5.48}$$

Further we introduce a discrete representation of Z by subdividing the interval $[0, \beta]$ into P segments and utilizing the *Trotter theorem* [101, 140]. This theorem gives a prescription for factorizing the representation $\exp(t(A+B))$ of two bounded linear operators $\hat{A}$ and $\hat{B}$. It holds that

$$\exp[t(\hat{A}+\hat{B})] = \lim_{P\to\infty} \left[\exp\left(\frac{t}{P}\hat{A}\right)\exp\left(\frac{t}{P}\hat{B}\right)\right]^P, \tag{5.49}$$

with P as a natural number. For three operators $\hat{A}$, $\hat{B}$, and $\hat{C}$, this formula is generalized in the following way:

$$\exp[t(\hat{A}+\hat{B}+\hat{C})] = \lim_{P\to\infty} \left[\exp\left(\frac{t}{P}\hat{A}\right)\exp\left(\frac{t}{P}\hat{B}\right)\exp\left(\frac{t}{P}\hat{C}\right)\right]^P. \tag{5.50}$$

We use the latter relation to perform a symmetrized factorization of the Euclidean propagator $\exp(-\beta\hat{H})$. For this purpose, we set $\hat{A} \equiv V/2$, $\hat{B} \equiv \hat{T}$, and $\hat{C} \equiv \frac{V}{2}$, to arrive at

$$\exp(-\beta H) = \lim_{P\to\infty} \left[\exp\left(-\frac{\beta V}{2P}\right)\exp\left(-\frac{\beta \hat{T}}{P}\right)\exp\left(-\frac{\beta V}{2P}\right)\right]^P. \tag{5.51}$$

We replace $\exp(-\beta\hat{H})$ in Eq. (5.48) by the right-hand side of this identity and insert unity in the form of $\int dq|q\rangle\langle q|$ between each of the P factors of the operator product. This yields

$$Z = \lim_{P\to\infty} \int dq_1 \dots dq_P \prod_{j=1}^{P} \left\langle q_j \left| \exp\left(-\frac{\beta V}{2P}\right) \exp\left(-\frac{\beta \hat{T}}{P}\right) \right.\right.$$
$$\left.\left. \times \exp\left(-\frac{\beta V}{2P}\right) \right| q_{j+1} \right\rangle \Bigg|_{q_{P+1}=q_1}, \tag{5.52}$$

which realizes the periodicity constraint imposed on the coordinate representation (5.37) of Z. As V is assumed to be a configurational energy and thus to be diagonal in q, each of the factors on the right-hand side of Eq. (5.52) adopts the form [132]

$$\left\langle q_j \left| \exp\left(-\frac{\beta V}{2P}\right) \exp\left(-\frac{\beta \hat{T}}{P}\right) \exp\left(-\frac{\beta V}{2P}\right) \right| q_{j+1} \right\rangle$$
$$= \sqrt{\frac{mP}{2\pi\beta\hbar^2}} \exp\left[-\frac{\beta}{2P}[V(q_j) + V(q_{j+1})]\right]$$
$$\times \exp\left[-\frac{mP}{2\pi\beta\hbar^2}(q_j - q_{j+1})^2\right]. \tag{5.53}$$

Summarizing, we find

$$Z = \lim_{P\to\infty} \left(\frac{mP}{2\pi\beta\hbar^2}\right)^{\frac{P}{2}} \int dq_1 \dots dq_P \, \exp[-\beta U_{\mathrm{eff}}(q_1, \dots, q_P)]. \tag{5.54}$$

We point out that this discretized partition function has the same structure as the classical, or high-temperature limit Eq. (5.47). All coordinate-dependent terms are comprised in the effective potential U_{eff} which is defined as

$$U_{\mathrm{eff}}(q_1, \dots, q_P) \equiv \sum_{j=1}^{P} \left[\frac{1}{2} m\omega_P^2 (q_j - q_{j+1})^2 + \frac{V(q_j)}{P}\right], \tag{5.55}$$

with $\omega_P = \frac{\sqrt{P}}{\beta\hbar}$. Amazingly, this is the potential for a ring polymer whose particles are connected by harmonic springs, and thus the quantum partition function has been mapped into that of a cyclic classical system. This is illustrated in Figure 5.3 for the case of a single quantum particle.

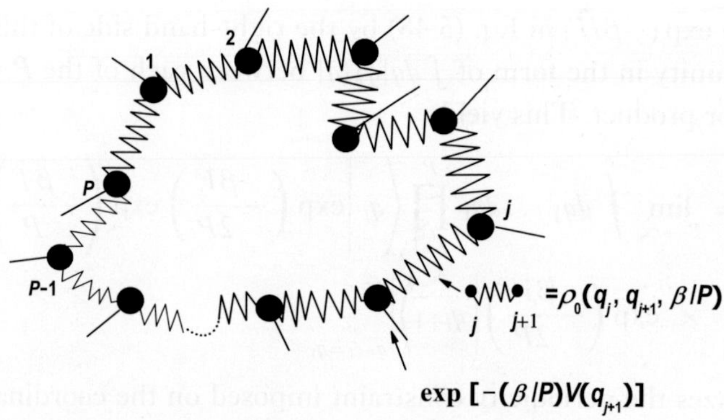

Figure 5.3: Representation of a classical ring polymer isomorphic to the Euclidean path integral of a quantum particle. Each polymer bead corresponds to a time slice of the discretized integral while the wavy lines connecting the beads stand for the harmonic interactions that emerge from the kinetic energy contribution to each integral segment according to Eqs. (5.51)–(5.55). By $\rho_0(q_j, q_{j+1}, \beta/P)$, we understand the matrix element of the statistical operator (see Eq. (5.44)) with vanishing potential V.

For the sake of simplicity, the current discussion is restricted to only one dimension, and so no reference is made to a specific physical system. For application to an atom, q_j would have to be replaced by a three-dimensional vector $\mathbf{X}_j$, for a molecule, by an array whose dimension equals the number of included degrees of freedom. It is easy to see how this approach can be generalized to capture the case of many interacting particles. In this situation, the coordinate q is labeled by a particle index i in addition to the polymer loop index j. If two quantum systems of equal masses are considered, the effective potential $U_{\rm eff}$ is composed of two contributions, namely $\sum_{j=1}^{P} \frac{1}{P} V(q_{2j} - q_{1j})$ and $\frac{1}{2} m\omega_{\rm P}^2 \sum_{j=1}^{P} (q_{1j} - q_{1j+1})^2 + (q_{2j} - q_{2j+1})^2$. The first of these two sums involves terms characterized by one index j only, since the described interaction is restricted to one particular point in Euclidean time. Figure 5.4 demonstrates the equivalent problem of two interacting polymer rings.

The explicit form of this effective potential in combination with Figure 5.4 clarifies that the quantum-classical isomorphism involves a classical polymer with extremely uncommon properties (compare the related discussion in [141]). Within each polymer, the equilibrium distance between adjacent particles is zero. Further, each particle in one loop interacts with one and only one particle in the other loop, such that the two polymers do not become entangled in the course of their evolution.

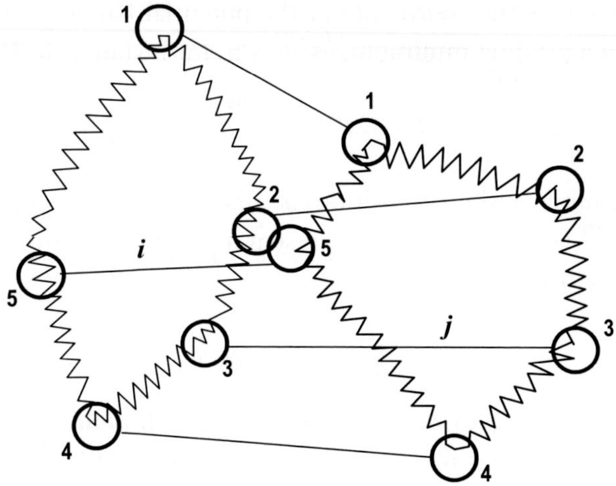

Figure 5.4: Two ring polymers isomorphic to two interacting quantum particles. The beads and the spring lines have the same meaning as in Figure 5.3. The straight lines connecting the two rings denote the two-particle interactions.

5.4 Centroid Dynamics

In Section 5.3 we argued that in the classical or high-temperature limit of the partition function, it suffices to evaluate $V(q)$ at the initial point q_a of the Euclidean time path which coincides with its end point. Formula (5.45) gives the corresponding approximation for Z. For intermediate and low temperatures, one has to go beyond this approach and thus to integrate the potential energy along the trajectory. To this end, it seems natural to expand the function $V(q)$ around the point q_a. However, it turns out to be more advantageous to choose another point of reference on the Euclidean path, namely its centroid, or mean position $\widetilde{q}_c$, as defined by the functional [100]

$$\widetilde{q}_c[q(\tau)] \equiv \widetilde{q}_c = \frac{1}{\beta\hbar}\int_0^{\beta\hbar} d\tau q(\tau). \qquad (5.56)$$

Before motivating the proposed expansion of $V(q)$ around $\widetilde{q}_c$ instead of q_a, we point out that the centroid concept allows for an alternative procedure of summing over the paths. The respective method is based on subdividing all possible paths into groups that share the same centroid. As this is done, one may define a partial density $\widetilde{\rho}(q_a, q_b, q_c)$ with the centroid variable constrained to some position q_c [121]. In the context of transition state theory,

q_c can be chosen as the coordinate of the potential barrier that separates a reactant from a product minimum, as shown, for instance, in Figure 5.1:

$$\widetilde{\rho}(q_a, q_b, q_c) = \int_{q_a}^{q_b} Dq\, \delta(q_c - \widetilde{q}_c) \exp\left\{-\frac{1}{\hbar}S_E[q]\right\}. \tag{5.57}$$

As we let the initial and final positions q_a and q_b coincide and integrate over the centroid variable, we obtain a representation of the partition function in terms of the centroid density:

$$Z = \int dq_c\, \widetilde{\rho}(q_a, q_a, q_c). \tag{5.58}$$

As mentioned earlier, to evaluate the partition function (5.43) in cases where the high-temperature approximation does not hold, one may consider representing the potential $V(q)$ as a power series, where the centroid is preferable to the initial position as the reference point of expansion. To demonstrate this, we expand $V(q)$ about q_c and find, up to second order,

$$Z = \int dq_c \exp[-\beta V(q_c)] \times \int_{q_a}^{q_a} Dq \exp\left\{-\frac{1}{\hbar}\int_0^{\beta\hbar} d\tau \left[\frac{1}{2}m\left(\frac{dq}{d\tau}\right)^2 + (q(\tau) - q_c)^2 \left.\frac{d^2V}{dq^2}\right|_{q=q_c}\right]\right\}. \tag{5.59}$$

The linear order cancels as a direct consequence of the centroid definition (5.56), which is the major advantage of the centroid summation procedure. One further implements the centroid constraint by multiplying the right-hand side of Eq. (5.59) with the delta function

$$\delta\left(\int_0^{\beta\hbar} d\tau\,(q(\tau) - q_c)\right) = \int_{-\infty}^{\infty} d\tau \frac{dk}{2\pi} \exp\left[ik\int_0^{\beta\hbar}(q(\tau) - q_c)\right]. \tag{5.60}$$

The resulting expression is found to be [100]

$$Z = \text{const}\int dq_c \exp\left\{-\beta\left[V(q_c) + \frac{\beta\hbar^2 \left.\frac{d^2V}{dq^2}\right|_{q=q_c}}{24m}\right]\right\}. \tag{5.61}$$

The correction term that appears in the argument of the exponential function contains β^2 which may be neglected in the high-temperature regime.

By comparing Eq. (5.61) with the high-temperature limit Eq. (5.47) we thus find for the constant factor the value $\sqrt{\frac{m}{2\pi\beta\hbar^2}}$. The centroid expansion of $V(q)$ up to second order conserves the form of the classical partition function. A quantum correction, however, is added to $V(q_c)$.

Relation (5.61) emphasizes the special status of the harmonic potential: in this case, $\left.\frac{d^2V}{dq^2}\right|_{q=q_c}$ is independent of the integration coordinate, and Z factors into a classical and a quantum contribution, $Z = Z_{cl}Z_q$.

It should be noted that the centroid may be understood as the zero-frequency Fourier component of the path $q(\tau)$ that satisfies the periodic boundary condition $q(0) = q(\beta\hbar)$. We subject $q(\tau)$ to a Fourier expansion:

$$q(\tau) = q_0' + \sum_{\substack{n=-\infty \\ n\neq 0}}^{\infty} q_n' \exp(i\nu_n\tau) \tag{5.62}$$

and
$$\nu_n = \frac{2\pi n}{\beta\hbar},$$

with complex-valued amplitudes q_n' which obey $q_{-n}' = q_n'^*$, implying that q_0' is real-valued. Further, we apply the prescription (5.56) for calculating the centroid of $q(\tau)$ in this representation and conclude that $\widetilde{q}_c = q_0'$, since the time average taken over any term with $n \neq 0$ vanishes.

We will employ the path integral formalism summarized by the identities (5.57) and (5.58) to specify the quantum extension Z_c^{bar} of the classical partition function at the barrier top Z_{cl}^{bar} (see Eqs. (5.18) and (5.29)). This requires defining a particular potential form in the barrier region. The most popular choice is the parabolic shape [142] which is frequently a good approximation in the thermally activated regime:

$$V(q) = V_0 - \frac{1}{2}m\omega_b^2(q - q_{bar})^2. \tag{5.63}$$

Here m stands for the mass of the particle that crosses the barrier and ω_b for the magnitude of the frequency associated with the barrier coordinate. The dividing surface intersects the reaction coordinate at the top of the parabolic barrier. To calculate Z_c^{bar} as a path integral, we let the centroid $\widetilde{q}_c$ coincide with the barrier top position q_{bar}. The expression (5.63) can then be viewed as an expansion of a general barrier potential up to second order with respect to the centroid position, a case that is discussed in detail by Feynman and

Hibbs [100]. Formulae (5.57) and (5.58) may be applied once the Euclidean action $S_{\mathrm{E,pbar}}$ is known, where pbar is shorthand for *parabolic barrier*.

The parabolic potential in conjunction with the path expansion (5.62) leads to [143]

$$S_{\mathrm{E,pbar}} = \beta\hbar\left[V_0 - \frac{1}{2}m\omega_b^2 q_0'^2 + m\sum_{n=1}^{\infty}(\nu_n^2 - \omega_b^2)|q_n'|^2\right]. \tag{5.64}$$

Evidently, $S_{\mathrm{E,pbar}}$ depends directly on the amplitudes $q_0', q_1', q_2', \ldots$, and evaluating $\exp\{-\frac{1}{\hbar}S_{\mathrm{E,pbar}}[q']\}$ results in a product of Gaussians in the parameters $|q_n'|$ with $n = 1, 2, 3, \ldots$. The real and imaginary parts of the amplitudes define a natural set of new integration variables for the evaluation of the barrier top partition function [100]. We transform the Euclidean path integral for the partition function constrained to the barrier top accordingly and obtain

$$Z_c^{\mathrm{bar}} \equiv Z^{\mathrm{pbar}} = \mathcal{J}\int\prod_{-\infty}^{\infty} d\Re(q_n')d\Im(q_n')\delta(q_{\mathrm{bar}} - q_0')\exp\left\{-\frac{1}{\hbar}S_{\mathrm{E,pbar}}\right\} \tag{5.65}$$

where $\mathcal{J}$ is the Jacobian associated with the change of integration variables. Evaluating the integral in Eq. (5.65) yields

$$Z^{\mathrm{pbar}} = \lim_{N\to\infty}\mathcal{J}(\pi/\beta m)^N\prod_{n=1}^{N}\nu_n^{-2}\prod_{n'=1}^{N} \times\frac{1}{1-\omega_b^2/\nu_n'^2}\exp\left[-\beta\left(V_0 - \frac{1}{2}m\omega_b^2 q_{\mathrm{bar}}^2\right)\right]. \tag{5.66}$$

The Jacobian is difficult to calculate directly. However, in a move similar to the one that yielded the prefactor in (5.61), we readily arrive at a final result for Z^{pbar} by realizing that none of the factors on the left of the product $\prod_{n'=1}^{N}$ in Eq. (5.66) depends on the frequency ω_b. We thus let ω_b tend towards zero and equate the result to the partition function Z_0 for the free particle case (see Exercise 5.3). From this calculation, the Jacobian turns out to be a prefactor $\sqrt{\frac{m}{2\pi\hbar^2\beta}}$.

Exercise 5.3. Calculate the partition function for special cases relevant for the path integral formulation of the transition rate.

(a) Evaluate the partition function Z_0 for a free particle by applying the path integration methods outlined in Section 4.1. Use the density matrix (5.57) and $S_{\mathrm{E}} = \frac{1}{\hbar}\int_0^{\beta\hbar} d\tau(\frac{m}{2}(dq/d\tau)^2 + V_0)$.
(b) The classical limit $Z_{\mathrm{cl}}^{\mathrm{bar}}$ of the partition function at the barrier top.

Assume that the potential (5.63) remains constant along the paths $q(\tau)$. With this premise, use the density matrix (5.57) to obtain the partition function as given by Eq. (5.58).

Note that the solutions of 5.3(a) and 5.3(b) coincide.

Employing further the infinite product representation of the sine function,

$$\sin(x) = x\prod_{k=1}^{\infty} 1 - \frac{x^2}{k^2\pi^2}, \tag{5.67}$$

we conclude that

$$\prod_{n'=1}^{\infty} \frac{1}{1-\omega_b^2/\nu_{n'}^2} = \frac{\omega_b\beta\hbar/2}{\sin(\omega_b\beta\hbar/2)}. \tag{5.68}$$

Using the definition of the frequencies the parabolic potential (5.64) with the centroid constrained to the barrier top reads

$$Z^{\mathrm{pbar}} = \frac{u/2}{\sin(u/2)}\sqrt{\frac{m}{2\pi\hbar^2\beta}}\exp(-\beta V_0), \tag{5.69}$$

where $u \equiv \omega_b\beta\hbar$. The last two factors describe the classical limit $Z_{\mathrm{cl}}^{\mathrm{bar}}$ of the barrier partition function (see Exercise 5.3(b)). As implied by our discussion of the classical case in Section 5.3 (see Eq. (5.45)), this limit is attained for T high enough and thus β small enough to ensure that the paths $q(\tau)$ remain close to the barrier position. This behavior is clearly reflected by expression (5.69) since Z^{pbar} merges into the classical approximation as $T \to \infty$ and thus $\frac{u/2}{\sin(u/2)} \to 1$. The parabolic barrier partition function takes the compact form

$$Z^{\mathrm{pbar}} = \frac{u/2}{\sin(u/2)} Z_{\mathrm{cl}}^{\mathrm{bar}}. \tag{5.70}$$

This result may be used to derive the transition rate for the case of a parabolic barrier from quantum transition state theory by inserting Z^{pbar}

into Eq. (5.29) or, if the Arrhenius form of the rate equation is used, into Eq. (5.30b) and Eq. (5.31). It holds that

$$k_{\text{pbar}}^{QTST} = \frac{1}{2\pi\hbar\beta Z_{\text{R}}} \exp\left(-\beta F^{\text{pbar}}\right), \tag{5.71a}$$

$$F^{\text{pbar}} = -\frac{1}{\beta} \ln\left[Z^{\text{pbar}} \sqrt{\frac{2\pi\hbar^2\beta}{m}} \right]. \tag{5.71b}$$

We point out that the temporally non-local rate expression may be recovered from the transition state approximation k_{pbar}^{QTST} by applying a correction factor κ to the right-hand side of Eq. (5.71a). Inserting the Hamiltonian for the parabolic case into Eq. (5.25) and defining the corresponding rate constant k_{pbar}, one verifies immediately that

$$k_{\text{pbar}} = \kappa k_{\text{pbar}}^{\text{QTST}} \tag{5.72a}$$

with

$$\kappa = 4\pi\Im\left\{ \text{Tr}\left\{ \hat{h}_{\text{R}} \exp\left[-\beta(\hat{H} - F^{\text{pbar}}\hat{I})\right]\hat{h}_{\text{P}} \right.\right.$$
$$\left.\left. \times \exp\left(\frac{i}{\hbar}\hat{H}t_{\text{pl}}\right)\hat{h}_{\text{P}} \exp\left(-\frac{i}{\hbar}\hat{H}t_{\text{pl}}\right) \right\}\right\}. \tag{5.72b}$$

We emphasize that the validity of the parabolic barrier formalism is limited to activated processes. For sufficiently low temperatures, tunneling through the barriers becomes the predominant transition mechanism. The equation $\beta = 2\pi/\hbar\omega_b$ [142] draws a demarcation line between the regimes of thermal activation and tunneling. Rate calculations for temperatures lower than $\hbar\omega_b/(2\pi k_{\text{B}})$ have been successfully performed by use of the periodic orbit [144], or instanton [145] approach which emerges from a stationary phase approximation to a semiclassical rate expression [142]. In this model, the Euclidean action is expanded around the *bounce trajectory*, which is the path of a classical particle under the influence of the inverted potential. A nonadiabatic generalization of instanton theory has been proposed that allows us to describe transfer processes in a landscape of multiple potential energy surfaces [146].

5.5 The Path Integral Form of the Golden Rule Rate Constant

In the preceding sections of this chapter, we have not made any explicit mention of the electronic system. If, however, we assume reactants and products to be characterized by well-defined electronic states $|R\rangle$ and $|P\rangle$, respectively, we can write down a general diabatic two-state Hamiltonian for the overall system as a sum of projection operators, namely

$$\hat{H} = \hat{H}_{\mathrm{R}}|R\rangle\langle R| + \hat{H}_{\mathrm{P}}|P\rangle\langle P| + V|R\rangle\langle P| + V|P\rangle\langle R|$$
$$= \hat{H}_0 + V(|R\rangle\langle P| + |P\rangle\langle R|). \tag{5.73}$$

Here, $\hat{H}_{\mathrm{R}}$ and $\hat{H}_{\mathrm{P}}$ stand for the Hamiltonians of the reactant and the product system, respectively, and V for the electronic coupling. As usual, the condition of zero coupling, $V = 0$, defines the reference Hamiltonian $\hat{H}_0$. We adopt the completeness relation

$$|R\rangle\langle R| + |P\rangle\langle P| = \hat{I}. \tag{5.74}$$

The most elementary and possibly also the most prominent realization of this scheme will be discussed in Chapter 20, namely that of electron transfer between a donor (D) and an acceptor (A) system. In this case, the state $|R\rangle(|P\rangle)$ may be labeled $|D^-A\rangle(|DA^-\rangle)$, as the reaction consists in a transition of an electron from an initial to a final localized state. Weak electronic coupling V, corresponding to nonadiabatic interaction between the two states, justifies the use of a *golden rule approximation* to the rate constant. To deduce the respective rate formulae, we start with the general expression (5.20). Direct expansion of the propagators implicitly contained in $\hat{h}_{\mathrm{P}}(t_{\mathrm{pl}})$ yields [129, 132]

$$k = \frac{2\pi}{Z_{\mathrm{R}}\hbar}\sum_{i,f}|\langle i, R|V|f, P\rangle|^2 \exp(-\beta E_{i,\mathrm{R}})\delta(E_{i,\mathrm{R}} - E_{f,\mathrm{P}}) + O(V^4). \tag{5.75}$$

The energies $E_{i,\mathrm{R}}$ and $E_{f,\mathrm{P}}$ refer to rovibrational levels of the reactant and the product system, respectively, which may be determined by the rovibrational spectrum of the considered species. The delta function in this rate expression guarantees energy conservation in the transition, eliminating any process for which the initial and the final energy do not match.

To make the fundamental result (5.75) plausible, we recall Eq. (3.26) which, to lowest order in the perturbing operator $\hat{H}_{\mathrm{I}'}$, implies

$$\exp\left(\frac{i}{\hbar}\hat{H}_0 t\right)\exp\left(-\frac{i}{\hbar}\hat{H}t\right)\psi(0) = \left[1 - \frac{i}{\hbar}\int_0^t dt'\exp\left(\frac{i}{\hbar}\hat{H}_0 t'\right)\right.$$
$$\left.\times \hat{H}'(t')\exp\left(-\frac{i}{\hbar}\hat{H}_0 t'\right)\right]\psi(0) + \cdots, \tag{5.76}$$

where the Schrödinger picture is employed for the wave functions and operators instead of the interaction picture used in Eq. (3.26). The corresponding expansion of the propagator $\exp(-\frac{i}{\hbar}\hat{H}t)$ is

$$\exp\left(-\frac{i}{\hbar}\hat{H}t\right) = \exp\left(-\frac{i}{\hbar}\hat{H}_0 t\right) - \frac{i}{\hbar}\int_0^t dt'\exp\left(\frac{i}{\hbar}\hat{H}_0(t'-t)\right)$$
$$\times \hat{H}'(t')\exp\left(-\frac{i}{\hbar}\hat{H}_0 t'\right) + \cdots. \tag{5.77}$$

In the present context, the perturbation is given by the coupling V between the diabatic states $|R\rangle$ and $|P\rangle$ characteristic of the reactant and the product region. The two states are eigenfunctions of the regional Hamiltonians $\hat{H}_{\mathrm{R}}$ and $\hat{H}_{\mathrm{P}}$ and are assumed to be highly localized in different regions of space.

We express the population operator $\hat{h}_{\mathrm{P}}(t_{\mathrm{pl}})$ in Eq. (5.20) by use of the propagator (5.77) as

$$\hat{h}_{\mathrm{P}}(t_{\mathrm{pl}}) = \exp\left(\frac{i}{\hbar}\hat{H}t_{\mathrm{pl}}\right)\hat{h}_{\mathrm{P}}(0)\exp\left(-\frac{i}{\hbar}\hat{H}t_{\mathrm{pl}}\right) \tag{5.78}$$

and substitute for the propagator the right-hand side of Eq. (5.77) with $\hat{H}' = V(|R\rangle\langle P| + |P\rangle\langle R|)$. An expansion analogous to that for $\exp(-\frac{i}{\hbar}Ht)$ is performed for the statistical operator $\exp(-\beta\hat{H})$. Truncating this series after the term linear in V and inserting it into the rate formula (5.20) while taking into account Eqs. (5.76)–(5.78) yields up to quadratic order in V:

$$k = \frac{2}{Z_{\mathrm{R}}\hbar^2}\Re\left\{\frac{1}{\beta\hbar}\int_0^{\beta\hbar} d\tau \int_0^{t_{\mathrm{pl}}} dt\,\mathrm{Tr}\left\{\exp(-\beta\hat{H}_{\mathrm{R}})(\exp(\tau\hat{H}_{\mathrm{R}})V\right.\right.$$
$$\left.\left.\times\exp(-\tau\hat{H}_{\mathrm{P}}))\left(\exp\left(\frac{i}{\hbar}\hat{H}_{\mathrm{P}}t\right)V\exp\left(-\frac{i}{\hbar}\hat{H}_{\mathrm{R}}t\right)\right)\right\}\right\}. \tag{5.79}$$

Expressing the trace in terms of the eigenfunctions $|iR\rangle$ and $|fP\rangle$ where the indices i and f label the nuclear and environmental components of the diabatic reactant and product states, respectively, we find that Eq. (5.79) simplifies to[1]

$$k = \frac{2\pi}{Z_R \hbar} V^2 \sum_{i,f} \exp(-\beta E_{i,R}) \delta(E_{i,R} - E_{f,P}). \tag{5.80}$$

This is the golden rule approximation to the rate constant. The difference between this statement and formula (5.75) lies in the assumption of a matrix element $\langle i, R|V|f, P\rangle = V$ that does not depend on the states $\{|i\rangle, |f\rangle\}$. This implements the *Condon approximation.*

Alternative versions of the golden rule formula that are more appropriate for numerical evaluation than Eq. (5.80) are frequently encountered in the literature. Rephrasing the delta function in Eq. (5.80) as a time integral over a trace, we cast the rate constant into the following form:

$$k = \frac{2\pi}{\hbar} V^2 \int_{-\infty}^{+\infty} dt\; C(t), \tag{5.81a}$$

and $$C(t) \equiv \frac{1}{Z_R} \mathrm{Tr} \left\{ \exp\left[-\left(\beta + \frac{i}{\hbar} t\right) \hat{H}_R \right] \exp\left[\frac{i}{\hbar} \hat{H}_P t \right] \right\}. \tag{5.81b}$$

In view of the generally highly oscillatory nature of the integrand which, after all, describes the fluctuations of the environmental degrees of freedom, the integral can be evaluated by the method of steepest descents. For this purpose, we let the integration contour pass through a point t^* where the first derivative of $C(t)$ vanishes. We expand $C(t)$ around this stationary phase point as

$$\exp[\ln C(t)] = \exp\left[\ln C(t^*) + \frac{(t - t^*)^2 \left.\frac{d^2 C(t)}{dt^2}\right|_{t=t^*}}{2C(t^*)} + \cdots \right] \tag{5.82}$$

---

[1]The operators $\hat{H}_R(\hat{H}_P)$ are replaced by the corresponding energy eigenvalues $E_{m,R}(E_{n,P})$. For sufficiently long times t_{pl}, the integral $2\Re\{\int_0^{t_{pl}} dt\; \exp(\frac{i}{\hbar} E_{i,R} t) \exp(-\frac{i}{\hbar} E_{f,P} t)\}$ can be approximated by $2\pi \frac{1}{2\pi} \int_{-\infty}^{+\infty} dt\; \exp(\frac{i}{\hbar} E_{i,R} t) \exp(-\frac{i}{\hbar} E_{f,P} t) = 2\pi\hbar\delta(E_{i,R} - E_{f,P})$. The integration with respect to τ contributes a factor $\frac{\hbar\{\exp[\beta(E_{i,R} - E_{f,P})] - 1\}}{E_{i,R} - E_{f,P}}$ which is reduced to $\hbar\beta$ due to the presence of the delta function in the eigenenergies $E_{i,R}$ and $E_{f,P}$. Taken together, these arguments lead to the result (5.80).

and employ Gaussian integration of the second term to obtain

$$k = \frac{V^2}{\hbar^2}\left[2\pi\left|\frac{C(t^*)}{\left.\frac{d^2C(t)}{dt^2}\right|_{t=t^*}}\right|\right]^{\frac{1}{2}} C(t^*). \tag{5.83}$$

Numerical results can be obtained from this expression if the Hamiltonian of the transfer system is known. In the context of electron-exchange processes, to be discussed in Chapter 20, we will interpret $C(t^*)$ as an Arrhenius exponential, implying that its prefactor in Eq. (5.84) is to be understood as a frequency.

The theoretical work on electron transfer systems has relied largely on the golden rule approach, and our presentation of this theme in Chapter 20 will be based for the most part on relations (5.80)–(5.83). To the extent that electron transfer can be understood as a slow process, as is usually the case when a transition between reactant and product systems in their ground states is considered, the premise of weak electronic coupling is in general justified, and the golden rule can be used for evaluating the transition rate. The transfer reaction, however, may also occur from an excited initial state, which can give rise to a fast transition. If the electronic degrees of freedom change distinctly faster than those of the nuclear and environmental system, the interaction is in the adiabatic regime and thus beyond the validity domain of the golden rule approximation. The following section focuses on a procedure to obtain rate constants that is free from the premises of the golden rule method.

5.6 Beyond the Golden Rule: Reduced Density Matrix Theory

In all of the models considered earlier, the heat bath or reservoir that interacts with the quantum system provided the rationale for a statistical procedure to compute the rate constant for a reaction of interest, but it did not enter the calculations explicitly. Depending on the size and the nature of the environment, however, a direct Hamiltonian approach may be favored. In this case, one may partition the overall Hamiltonian according to

$$\hat{H} = \hat{H}_{\mathrm{S}} + \hat{H}_{\mathrm{E}} + \hat{H}_{\mathrm{SE}}, \tag{5.84}$$

distinguishing the contribution of the system $\hat{H}_{\mathrm{S}}$ from that of the environment $\hat{H}_{\mathrm{E}}$ and the coupling $\hat{H}_{\mathrm{SE}}$ between these two components.

Further, we understand the sum $\hat{H}_S + \hat{H}_E$ as reference Hamiltonian $\hat{H}_0$. The dynamics of this composite is usually not tractable by solving the TDSE or the Liouville–von Neumann equation (4.99) for all variables of the system and its environment. We will address this challenge by making two procedural decisions. Specifically, we will (1) eliminate the environmental degrees of freedom from the set of dynamic parameters by employing a *reduced density operator* formalism, and (2) work in the interaction picture (see Section 3.2) to guarantee that the time propagator of the system involves only the reference Hamiltonian.

To implement this methodology, we ask for an equation that describes the evolution of a reduced density in the interaction picture. A general time-dependent density operator $\hat{\rho}^I$, where the superscript I indicates the interaction picture, obeys the Liouville–von Neumann equation

$$i\hbar\frac{\partial\hat{\rho}^{\mathrm{I}}}{\partial t} = [\hat{H}^I_{\mathrm{SE}}, \hat{\rho}^{\mathrm{I}}]. \tag{5.85}$$

Exercise 5.4. Demonstrate the validity of Eq. (5.85), using the interaction picture (see Eq. (3.15)) for $\hat{H}$ and $\hat{\rho}$, according to

$$\hat{H}^I_{\mathrm{SE}}(t) = \exp\left(\frac{i}{\hbar}\hat{H}_0 t\right)\hat{H}_{\mathrm{SE}}\exp\left(-\frac{i}{\hbar}\hat{H}_0 t\right) \tag{5.86}$$

and

$$\hat{\rho}^{\mathrm{I}}(t) = \exp\left(\frac{i}{\hbar}\hat{H}_0 t\right)\hat{\rho}(t)\exp\left(-\frac{i}{\hbar}\hat{H}_0 t\right). \tag{5.87}$$

Formally the solution of Eq. (5.85) can be written as

$$\hat{\rho}^{\mathrm{I}}(t) = \hat{\rho}^{\mathrm{I}}(0) - \frac{i}{\hbar}\int_0^t dt_1[\hat{H}^{\mathrm{I}}_{\mathrm{SE}}(t_1), \hat{\rho}^{\mathrm{I}}(t_1)]. \tag{5.88}$$

Repeating the argument that led to Eq. (3.23), we expand the integral on the right-hand side to represent $\hat{\rho}^{\mathrm{I}}$ by the series

$$\hat{\rho}^{\mathrm{I}}(t) = \hat{\rho}^{\mathrm{I}}(0) - \frac{i}{\hbar}\int_0^t dt_1[\hat{H}^{\mathrm{I}}_{\mathrm{SE}}(t_1), \hat{\rho}^{\mathrm{I}}(0)]$$

$$-\frac{1}{\hbar^2}\int_0^t dt_1\int_0^{t_1} dt_2[\hat{H}_{\mathrm{SE}}(t_1), [\hat{H}^{\mathrm{I}}_{\mathrm{SE}}(t_2), \hat{\rho}^{\mathrm{I}}(0)]] + \cdots. \tag{5.89}$$

For further simplification, let us assume that $\hat{H}_{\mathrm{SE}}$ is a linear function of the system and the environment coordinates, and that these coordinates do not correlate with $\hat{\rho}^{\mathrm{I}}(0)$. Specifically, we write the coupling operator as a sum over operator products according to

$$\hat{H}_{\mathrm{SE}} = \sum_q \hat{K}_q \hat{\Lambda}_q. \tag{5.90}$$

Here $\hat{K}_q$ and $\hat{\Lambda}_q$ stand for operators in the system and the environment coordinates. An example is given by dipolar interaction of a molecule with an external electric field, as described by the coupling Hamiltonian

$$\hat{H}_{\mathrm{SE}} = -\hat{\boldsymbol{\mu}} \cdot \mathbf{E}, \tag{5.91}$$

with $\hat{\boldsymbol{\mu}}$ as the molecular dipole moment operator and $\mathbf{E}$ as the electric field vector. As implied by the bilinear property of $\hat{H}_{\mathrm{SE}}$, averaging over the bath variables by taking the trace Tr_{E} with respect to the environmental degrees of freedom will eliminate all terms in the expansion (5.89) that contain odd powers of $\hat{H}_{\mathrm{SE}}$. Truncating the sum (5.89) at second order, we therefore derive the following equation of motion:

$$\frac{\partial \hat{\tilde{\rho}}^{\mathrm{I}}(t)}{\partial t} = -\frac{1}{\hbar^2} \int_0^t dt_2 \, \mathrm{Tr}_{\mathrm{E}}\{[\hat{H}_{\mathrm{SE}}(t), [\hat{H}_{\mathrm{SE}}(t_2), \hat{\rho}^{\mathrm{I}}(0)]]\}, \tag{5.92}$$

where the symbol $\hat{\tilde{\rho}}^{\mathrm{I}}$ was introduced to denote the *reduced density operator* that results from averaging over the coordinates of the environment: $\hat{\tilde{\rho}}^{\mathrm{I}} \equiv \mathrm{Tr}_{\mathrm{E}}\{\hat{\rho}^{\mathrm{I}}\}$.[2] The time dependence of the coupling operator is defined in accordance with Eq. (3.15) and thus given by $\exp(\frac{i}{\hbar}\hat{H}_0 t)\hat{H}_{\mathrm{SE}} \exp(-\frac{i}{\hbar}\hat{H}_0 t)$. We neglect the back action from the system on the environment and constrain the latter to maintain equilibrium during the considered process. Inserting the explicit form of the coupling operator (5.90), we rewrite this equation in the following way:

$$\frac{\partial \hat{\tilde{\rho}}^{\mathrm{I}}(t)}{\partial t} = -\frac{1}{\hbar^2} \sum_{mn} \int_0^t dt_2 \, \langle \hat{\Lambda}_m(t) \hat{\Lambda}_n(t_2) \rangle_{\mathrm{E}} [\hat{K}_m(t), [\hat{K}_n(t_2), \hat{\tilde{\rho}}^{\mathrm{I}}(0)]], \tag{5.93}$$

where the bracket $\langle , \rangle_{\mathrm{E}}$ indicates an ensemble average with respect to the environment. We interpret this average as a correlation function C_{mn}:

$$\langle \hat{\Lambda}_m(t) \hat{\Lambda}_n(t_2) \rangle_{\mathrm{E}} \equiv C_{mn}(t - t_2). \tag{5.94}$$

[2] To avert the proliferation of commutator brackets, one may introduce the *Liouville operator* $\hat{L}_{\mathrm{SE}}$ by making the stipulation $\hat{L}_{\mathrm{SE}} f \equiv [\hat{H}_{\mathrm{SE}}, f]$. In terms of the Liouville operator Eq. (5.92) reads $\frac{\partial \hat{\tilde{\rho}}^{\mathrm{I}}_{\mathrm{E}}(t)}{\partial t} = -\frac{1}{\hbar^2} \int_0^t dt_2 \, \mathrm{Tr}_{\mathrm{E}}\{\hat{L}_{\mathrm{SE}}(t) \hat{L}_{\mathrm{SE}}(t_2) \hat{\rho}^{\mathrm{I}}(0)\}$.

Note that the above stipulation of an environment at equilibrium provides a sufficiently strong rationale to justify this assignment [147]. Tackling Exercise 5.5, the reader might trace the steps that establish Eq. (5.94).

Exercise 5.5. Identify the expression

$$\langle \hat{\Lambda}_m(t)\hat{\Lambda}_n(t_2)\rangle_{\mathrm{E}}$$

in Eq. (5.94) with

$$\mathrm{Tr}\{\hat{\Lambda}_m(t)\hat{\Lambda}_n(t_2)\hat{\rho}_{\mathrm{E}}\}.$$

Assume that $[\hat{H}_{\mathrm{E}}, \hat{\rho}_{\mathrm{E}}] = 0$, where $\hat{H}_{\mathrm{E}}$ and $\hat{\rho}_{\mathrm{E}}$ are the Hamiltonian and the statistical operator of the environment, respectively. Show

$$\mathrm{Tr}\{\hat{\Lambda}_m(t)\hat{\Lambda}_n(t_2)\hat{\rho}_{\mathrm{E}}\} = \mathrm{Tr}\{\hat{\Lambda}_m(t-t_2)\hat{\Lambda}_n\hat{\rho}_{\mathrm{E}}\} = C_{mn}(t-t_2).$$

Go on to prove: $C_{mn}(t-t_2) = C_{mn}(t_2-t)$.

Upon changing to $\tau \equiv t - t_2$ as integration variable and increasing the upper integration limit to infinity, we turn Eq. (5.93) into

$$\frac{\partial \hat{\tilde{\rho}}^{\mathrm{I}}(t)}{\partial t} = -\frac{1}{\hbar^2}\sum_{mn}\int_0^{\infty} d\tau\, C_{mn}(\tau)[\hat{K}_m(t), [\hat{K}_n(t-\tau), \hat{\tilde{\rho}}^{\mathrm{I}}(0)]]. \tag{5.95}$$

A differential equation of first order in time for $\hat{\tilde{\rho}}^{\mathrm{I}}$ is obtained from this relation if one assumes that the correlation $C_{mn}(\tau)$ decays during a characteristic time that is too short to allow the reduced density operator to change substantially. This assumption about the behavior of the environment justifies the extension of the integration limit to infinity. In component notation, the equation of motion becomes now

$$\frac{\partial \hat{\tilde{\rho}}^{\mathrm{I}}_{\alpha\alpha'}}{\partial t} = -\frac{1}{\hbar^2}\sum_{mn}\int_0^{\infty} d\tau\, C_{mn}(\tau)\langle\alpha|[\hat{K}_m(t), [\hat{K}_n(t-\tau), \hat{\tilde{\rho}}^{\mathrm{I}}(t)]]|\alpha'\rangle. \tag{5.96}$$

Inspecting the structure of the right-hand side, one realizes that it contains scalar products of the type

$$\begin{aligned}&\langle\alpha|\hat{K}_m(t)\hat{K}_n(t-\tau)\hat{\tilde{\rho}}^{\mathrm{I}}(t)|\alpha'\rangle \\ &\quad = \sum_{\beta\beta'}\langle\alpha|\hat{K}_m(t)|\beta\rangle\langle\beta|\hat{K}_n(t-\tau)|\beta'\rangle\langle\beta'|\hat{\tilde{\rho}}^{\mathrm{I}}(t)|\alpha'\rangle.\end{aligned} \tag{5.97}$$

Using Eq. (3.15), we make the time dependence of the system operators $\hat{K}_m, \hat{K}_n$ explicit and write

$$\langle\alpha|\hat{K}_m(t)\hat{K}_n(t-\tau)\tilde{\hat{\rho}}^{\mathrm{I}}(t)|\alpha'\rangle = \sum_{\beta\beta'} \exp\left[\frac{i}{\hbar}(E_\alpha - E_\beta)t\right]$$

$$\times \exp\left[\frac{i}{\hbar}(E_\beta - E_{\beta'})(t-\tau)\right]$$

$$\times K_{m,\alpha\beta}K_{n,\beta\beta'}\tilde{\rho}^{\mathrm{I}}(t)_{\beta'\alpha'}. \qquad (5.98)$$

The energies E_α, E_β are eigenvalues of the system Hamiltonian $\hat{H}_{\mathrm{S}}$. Note that by assumption the environment remains in equilibrium during the evolution of the system, such that it does not make any contribution to the energy differences in formula (5.98). We now define

$$\frac{1}{\hbar}\int_0^\infty d\tau C_{mn}(\tau)\exp\left(\frac{i}{\hbar}\omega\tau\right) \equiv \tilde{\mathcal{J}}(\omega). \qquad (5.99)$$

The integral on the left-hand side can be easily evaluated if an *isotropic medium* with an exponentially decaying correlation function is realized. In this case all components $\hat{\Lambda}_n$ are found to tend towards the same equilibrium average, exhibiting the same time behavior. It then holds that

$$C_{mn}(\tau) = \delta_{mn}\langle\hat{\Lambda}^2\rangle \exp(-\tau/\tau_{\mathrm{c}}), \qquad (5.100)$$

with τ_{c} as the correlation time of the medium. One concludes:

$$\tilde{\mathcal{J}}(\omega) = \frac{\langle\hat{\Lambda}^2\rangle}{\hbar}\frac{\tau_{\mathrm{c}}}{1 - i\omega\tau_{\mathrm{c}}}. \qquad (5.101)$$

Adopting this model, and making the transition from $\tilde{\hat{\rho}}^{\mathrm{I}}(t)$ to $\tilde{\hat{\rho}}(t)$, i.e. the reduced density operator in the Heisenberg picture, we cast the equation of motion into the following form [148]:

$$\frac{\partial\tilde{\rho}_{\alpha\alpha'}}{\partial t} = -\frac{i}{\hbar}(E_\alpha - E_{\alpha'})\tilde{\rho}_{\alpha\alpha'}(t) - \frac{1}{\hbar}\sum_{\beta,\beta'}\left[\delta_{\alpha'\beta'}\sum_\gamma \tilde{\mathcal{J}}(E_\beta - E_\gamma)\mathbf{K}_{\gamma\beta}\cdot\mathbf{K}_{\alpha\gamma}\right.$$

$$- [\tilde{\mathcal{J}}(E_{\alpha'} - E_{\beta'}) + \tilde{\mathcal{J}}(E_\beta - E_\alpha)]\mathbf{K}_{\beta'\alpha'}\cdot\mathbf{K}_{\alpha\beta}$$

$$\left. + \delta_{\alpha\beta}\sum_\gamma \tilde{\mathcal{J}}(E_\gamma - E_{\beta'})\mathbf{K}_{\beta'\gamma}\cdot\mathbf{K}_{\gamma\alpha'}\right]\tilde{\rho}_{\beta\beta'}(t). \qquad (5.102)$$

The dot product joining matrix elements of the vector operator $\hat{\mathbf{K}}$ implies summation over the index *n* according to $\sum_n K_{n,\beta'\alpha'} K_{n,\alpha\beta}$. Relation (5.100) is known as the *Redfield equation* [149]. A more compact formulation of this equation is achieved as one defines the *Redfield tensor* $R_{\mu\nu\kappa\lambda}$ by

$$R_{\mu\nu\kappa\lambda} = \Gamma^{+}_{\lambda\nu\mu\kappa} + \Gamma^{-}_{\lambda\nu\mu\kappa} - \delta_{\nu\lambda} \sum_{\alpha} \Gamma^{+}_{\mu\alpha\alpha\kappa} - \delta_{\mu\kappa} \sum_{\alpha} \Gamma^{-}_{\lambda\alpha\alpha\nu} \tag{5.103}$$

with

$$\Gamma^{+}_{\lambda\nu\mu\kappa} = \frac{1}{\hbar^2} \int_0^{\infty} dt \langle\langle\lambda|\hat{H}_{\mathrm{SE}}(t)|\nu\rangle\langle\mu|\hat{H}_{\mathrm{SE}}|\kappa\rangle\rangle_{\mathrm{E}} \exp\left(-i\omega_{\mu\kappa} t\right) \tag{5.104}$$

and

$$\Gamma^{-}_{\lambda\nu\mu\kappa} = \frac{1}{\hbar^2} \int_0^{\infty} dt\, \langle\langle\lambda|\hat{H}_{\mathrm{SE}}|\nu\rangle\langle\mu|\hat{H}_{\mathrm{SE}}(t)|\kappa\rangle\rangle_{\mathrm{E}} \exp\left(-i\omega_{\lambda\nu} t\right). \tag{5.105}$$

The frequency $\omega_{\lambda\nu}$ stands here for $(E_\lambda - E_\nu)/\hbar$. Use of the **R** tensor reduces the complexity of the Redfield equations considerably, as they simplify to

$$\frac{\partial \tilde{\rho}_{\alpha\alpha'}}{\partial t} = -i\omega_{\alpha\alpha'} \tilde{\rho}_{\alpha\alpha'}(t) - \sum_{\beta,\beta'} R_{\alpha\alpha'\beta\beta'} \tilde{\rho}_{\beta\beta'}(t). \tag{5.106}$$

We point out that symmetry with respect to time inversion, as guaranteed by the TDSE and the Liouville–von Neumann equation, is broken by Eq. (5.106). This feature is a consequence of the averaging over the environmental degrees of freedom (see Eq. (5.92)) on which the Redfield equation relies. Averaging destroys phase relations that a full quantum treatment of the system and its environment would preserve and thus is the cause of the irreversibility of the equation [150].

5.6.1 *A two-state problem*

To place the Redfield approximation into the context of rate theory, we consider the interaction Hamiltonian (5.91) which may be expressed as

$$\hat{H}_{\mathrm{SE}} = -\boldsymbol{\mu} \cdot \mathbf{E}\{|0\rangle\langle 1| + |1\rangle\langle 0|\}. \tag{5.107}$$

The unperturbed Hamiltonian $\hat{H}_0$ is given in the adiabatic representation and defined by both the ground state ($\langle 0|$) and the first excited state ($\langle 1|$)

energies E_0 and E_1, respectively:

$$\hat{H}_0 = E_0|0\rangle\langle 0| + E_1|1\rangle\langle 1|. \tag{5.108}$$

Writing $\hat{H}_{\mathrm{SE}}$ as a matrix defined in the space spanned by the electronic ground (first excited) state, according to

$$\hat{H}_{\mathrm{SE}} = -\boldsymbol{\mu} \cdot \mathbf{E}\boldsymbol{\sigma}_x = -\boldsymbol{\mu} \cdot \mathbf{E}\begin{pmatrix} 0 & 1 \\ 1 & 0 \end{pmatrix}, \tag{5.109}$$

we may now identify the system operator $\hat{\mathbf{K}}$ with $\boldsymbol{\sigma}$. Further we confine the Redfield equation (5.102) to the case $\alpha = \alpha' = 0$ and obtain a compact statement about the variation of the *ground state population* $\tilde{\rho}_{00}$ as a function of time, namely

$$\frac{\partial \tilde{\rho}_{00}}{\partial t} = -\frac{1}{\hbar}[\tilde{\mathcal{J}}(\Delta E) + \tilde{\mathcal{J}}(-\Delta E)][\tilde{\rho}_{00}(t) - \tilde{\rho}_{11}(t)], \tag{5.110}$$

and $\Delta E \equiv E_1 - E_0$.

Exercise 5.6. Adopting the operators (5.108) and (5.109) as system and interaction Hamiltonians, respectively, derive Eq. (5.110) from the Redfield equation (5.102). In particular, find expressions for the products $\mathbf{K}_{ij} \cdot \mathbf{K}_{kl}$.

Evidently, the factor $\frac{1}{\hbar}[\tilde{\mathcal{J}}(\Delta E) + \tilde{\mathcal{J}}(-\Delta E)]$ is to be considered as a rate constant k_{01} that dictates the time scale for the changes of the two populations. Making the equilibrium and isotropic medium assumptions, as made explicit by Eqs. (5.100) and (5.101), k_{01} turns out to be

$$k_{01} = \frac{\langle \hat{\Lambda}^2 \rangle}{\hbar} \frac{\tau_c}{1 + \omega^2 \tau_c^2}. \tag{5.111}$$

From inspecting the Redfield tensor (5.103) one finds that $k_{01} = -R_{0000}$. The evolution of the first excited state population $\tilde{\rho}_{11}$ is derived from $\tilde{\rho}_{00} + \tilde{\rho}_{11} = 1$ and thus $\frac{\partial \tilde{\rho}_{11}}{\partial t} = -\frac{\partial \tilde{\rho}_{00}}{\partial t}$. In an analogous manner we obtain the differential equations that govern the evolution of the non-diagonal elements of $\tilde{\boldsymbol{\rho}}$, or the *coherences*. Specifically, we have

$$\frac{\partial \tilde{\rho}_{01}}{\partial t} = \frac{i}{\hbar}\Delta E \tilde{\rho}_{01}(t) - \frac{2}{\hbar}[\tilde{\mathcal{J}}(-\Delta E)\tilde{\rho}_{01}(t) + \tilde{\mathcal{J}}(\Delta E)\tilde{\rho}_{10}(t)], \tag{5.112}$$

and a corresponding relation for $\tilde{\rho}_{10}$.

A comparative assessment of the Redfield theory and the golden rule approach to rate processes leads to the conclusion that the former contains more information than the latter. The golden rule yields the probability for a certain final state to evolve in a given time from a subset of initial states if both are part of a complete set. Redfield theory, in contrast, draws a more detailed picture of the considered process, as is evident from Eqs. (5.110)–(5.112). Since it derives from reduced density matrix theory, it allows us to monitor not only the evolution of state populations, but also that of the coherences involving pairs of states. The latter feature is further discussed in Section 14.1. It is of major relevance to the study of decoherence, and in particular of dephasing mechanisms as they determine, for instance, the relaxation behavior of spin systems.

We reiterate that the earlier presentation of reduced density matrix theory applied to open quantum systems rests on several simplifying assumptions. In particular, the derivation relied on a bilinear interaction Hamiltonian $\hat{H}_{\mathrm{SE}}$ to describe the coupling between the quantum system and its environment, and the coupling strength was assumed to be weak enough to justify truncation of the expansion of $\hat{\rho}^{\mathrm{I}}$ at second order in $\hat{H}_{\mathrm{SE}}$. Further, an equilibrium constraint was imposed on the environment, and the premise of a rapidly decaying correlation between environmental variables was used. In particular, this decay was assumed to be fast as compared with the time scale typical for temporal changes of the reduced matrix operator $\tilde{\hat{\rho}}$. The validity of each one of these conditions has to be examined whenever the Redfield formalism is to be employed. In Section 20.7 we will discuss the use of Redfield theory to address problems related to ultrafast electron transfer in condensed media.

Part II: Methods

Chapter 6

Time-Independent Theory of Molecular Collisions I: Multichannel Scattering

The time-independent approach to molecular collisions operates ideally with a basis rich enough to capture the interaction between the potentially large numbers of electronic and nuclear degrees of freedom involved. The broadest theoretical fundament for this approach is given by the energy representation of multichannel scattering between molecules. In this chapter, we extend the results presented in Chapter 3 on single-channel scattering to situations for which multiple channels have to be included. For molecular interactions, three types of channels are distinguished, related to rotational, vibrational, and electronic internal states of the molecules. The typical spacing of energy levels are 10^{-2}, 10^{-1}, and 1 eV, respectively, for the excitations associated with these three pathways. While the explicit treatment of electronic degrees of freedom, the focus of this text, is reserved for the following chapter, in the present one we adopt a more formal point of view, developing the main tenets of multichannel scattering without assigning physical meaning to the channels. We will concentrate in Sections 6.1–6.5 on the case of inelastic scattering, which includes the transfer of energy between the scattering partners, but no particle or charge transfer. In Sections 6.6 and 6.7, however, we will introduce the fundamentals of rearrangement collisions, as they are essential for the understanding of reactive scattering, involving interactions which change the identity of the colliding systems.

6.1 The Multichannel Problem

In a first approach to the multichannel scattering problem, we distinguish a scattering, or *external*, coordinate $\boldsymbol{\rho}$ from a set of coordinates $\mathbf{Q}$ that refer summarily to the internal degrees of freedom characteristic of the entities involved in the scattering process. The coordinate $\boldsymbol{\rho}$ stands for the relative position of the scattering partners which may involve the distance between the centers of mass of both the projectile and the target in a situation of inelastic scattering between two molecules. The wave functions in the internal coordinates, $\{\phi_i(\mathbf{Q})\}$, might be introduced as products of the respective nuclear and electronic molecular states. Other assignments are possible. In fact, the distinction between external and internal coordinates is analogous to that between nuclear and electronic degrees of freedom made in Section 2.1. In both cases, we have the choice between a diabatic and an adiabatic approach, and the form of the stationary Schrödinger equation for the case of multiple channels depends on which frame of description is selected. In the context of inelastic scattering, the set $\{\phi_i(\mathbf{Q})\}$ represents the diabatic alternative. A corresponding sequence of adiabatic basis functions is denoted by $\zeta_i(\mathbf{Q}, \boldsymbol{\rho})$. As the interaction between the scattering partners is assumed to vanish at infinite distance between them, the two bases obey the asymptotic condition

$$\phi_i(\mathbf{Q}) = \lim_{\rho \to \infty} \zeta_i(\mathbf{Q}, \boldsymbol{\rho}). \tag{6.1}$$

As soon as admission is made for rearrangement of the scattering partners as a result of their interaction, $\boldsymbol{\rho}$ can be specified in various plausible ways, as will be described in further detail in Section 6.6.

The excitations associated with the internal degrees of freedom are associated with a multiplicity of channels that might be *open* or *closed* in the scattering situation at hand, depending on the total energy available and on the symmetries of the problem. The multichannel Hamiltonian is written as

$$\hat{H} = \hat{T}(\boldsymbol{\rho}) + \hat{H}_{\text{int}}(\mathbf{Q}) + V(\mathbf{Q}, \boldsymbol{\rho}), \tag{6.2}$$

where the kinetic energy operator $\hat{T}$ refers to the relative motion of the scattering partners, $\hat{H}_{\text{int}}$ incorporates the internal degrees of freedom, and V the interaction between the $\boldsymbol{\rho}$ and $\mathbf{Q}$-dependent subsystems. In accordance with Eq. (6.1), we require that V reduces to zero as $\rho \to \infty$.

Turning to the multichannel equations, or *coupled-channel equations*, we first investigate the diabatic case, adopting a series $\{\phi_i(\mathbf{Q})\}$ as channel basis functions. Let us further assume that these form a complete basis set given by the eigenfunctions of $\hat{H}_{\text{int}}$:

$$\hat{H}_{\text{int}}\phi_i(\mathbf{Q}) = \varepsilon_i^0\phi_i(\mathbf{Q}). \tag{6.3}$$

In parallel to the elastic case, we express the scattering state Ψ in a product basis consisting of $\phi_i(\mathbf{Q})$ as well as functions $\Phi_i(\boldsymbol{\rho})$:

$$\Psi = \sum_i \phi_i(\mathbf{Q})\Phi_i(\boldsymbol{\rho}). \tag{6.4}$$

For the sake of simplicity, we assume a spherically symmetric situation, letting the potential V in Eq. (6.2) depend only on the magnitude, not on the orientation of ρ. We thus treat the orbital angular momentum l, associated with the relative motion of the scattering partners, and the total rotational angular momentum j, resulting from internal nuclear degrees of freedom, both as constants of motion. This is not possible in the general case where both components couple to the total nuclear angular momentum. Since l is a conserved quantum number, we can represent the *scattering functions* Φ_m as sums over partial waves:

$$\Phi_m(\boldsymbol{\rho}) = \frac{1}{\rho}\sum_{l=0}^{\infty} A_l\psi_{ml}(\rho)P_l(\cos\theta). \tag{6.5}$$

Inserting Eqs. (6.4) and (6.5) into the stationary Schrödinger equation

$$\hat{H}\Psi = E\Psi, \tag{6.6}$$

and using the assumed completeness of the functions $\phi_i(\mathbf{Q})$ in Eq. (6.4) as well as that of the Legendre polynomials in expansion (6.5), we find

$$\left[\mathbf{I}\frac{d^2}{d\rho^2} + \mathbf{k}^2 - \mathbf{I}\frac{l(l+1)}{\rho^2} - \frac{2\mu}{\hbar^2}\mathbf{V}\right]\boldsymbol{\psi}_l(\rho) = 0. \tag{6.7}$$

In this equation, $\mathbf{I}$ represents the unity matrix, $\mathbf{V}$ the potential energy matrix with elements

$$V_{mn} = \langle\phi_m|V|\phi_n\rangle, \tag{6.8}$$

$\mathbf{k}^2$ a diagonal matrix with elements

$$k_m^2 = \frac{2\mu}{\hbar^2}(E - \varepsilon_m^0), \tag{6.9}$$

and $\boldsymbol{\psi}_l$ a vector that contains the functions ψ_{ml} as components. Further, Eq. (6.7) contains the translational reduced mass μ associated with the scattering coordinate. We emphasize that every value of the orbital angular momentum quantum number l generates a new set of coupled differential equations. Comparison of Eq. (6.7) with Eq. (3.78) clarifies that the former can be viewed as a set of elastic scattering equations coupled by the non-diagonal elements of the matrix **V**.

If the channel energy ε_m^0 exceeds the total energy E, the latter is insufficient to populate the respective channel. In that case, the asymptotic wave function of the system emerging from the collision will not include any admixture of the state described by ϕ_m. The respective channel is closed. The condition $E > \varepsilon_m^0$, on the other hand, distinguishes open channels. If this inequality is satisfied, k_m is real. The state ϕ_m is then physically accessible and may contribute the final wave function of the scattered system.

Although they are asymptotically unpopulated, closed channels may be of considerable impact on the scattering process since they are included in the coupling matrix (6.8). In particular, a bound level whose energy lies slightly above that of the scattering system may give rise to Feshbach resonance behavior (see Section 3.5). A detailed quantitative discussion of the interaction between an open and a closed channel, where the latter is realized by a bound state, is presented, for instance, in [92].

In the following step we derive the adiabatic multichannel equations. For this purpose, we substitute the expansion

$$\Psi = \sum_i \zeta_i(\mathbf{Q}, \boldsymbol{\rho})\Phi_i(\boldsymbol{\rho}) \tag{6.10}$$

into the stationary Schrödinger equation (6.6), where the basis set $\{\zeta_i(\mathbf{Q}, \boldsymbol{\rho})\}$ is determined by the relation

$$(\hat{H}_{\text{int}}(\mathbf{Q}) + V(\mathbf{Q}, \rho))\zeta_i(\mathbf{Q}, \rho) = \varepsilon_m(\rho)\zeta_i(\mathbf{Q}, \rho). \tag{6.11}$$

Again, the assumption of sphericality reduces the vector $\boldsymbol{\rho}$ as parameter in Eq. (6.11) to its magnitude ρ. In contrast to the diabatic eigenvalue ε_m^0, ε_m is a function of the external variable and as such, it is understood as an adiabatic potential. The counterpart of expression (6.8), i.e. the matrix $\langle\zeta_m|V|\zeta_n\rangle$, is then diagonal. Making the transition from the diabatic to the adiabatic system of equations is tantamount to shifting the coupling terms from the side of the potential energy to that of the kinetic energy, as described in

Section 2.1. Specifically, the coupling matrices are given by

$$A_{mn} = \left\langle \zeta_m \left| \frac{\partial}{\partial \rho} \right| \zeta_n \right\rangle \tag{6.12a}$$

and

$$B_{mn} = \left\langle \zeta_m \left| \frac{\partial^2}{\partial \rho^2} \right| \zeta_n \right\rangle, \tag{6.12b}$$

with **B** as the kinetic energy matrix associated with the relative motion of the two systems. Repeating the steps that led to Eq. (6.7) yields the adiabatic multichannel equation

$$\left[\mathbf{I}\frac{\partial^2}{\partial \rho^2} + \mathbf{k}^2 - \mathbf{I}\frac{l(l+1)}{\rho^2} - \left(2\mathbf{A}\frac{\partial}{\partial \rho} + \mathbf{B}\right)\right] \boldsymbol{\psi}_l(\rho) = \mathbf{0}, \tag{6.13}$$

where the diagonal elements of the matrix $\mathbf{k}^2$ are now defined as

$$k_m^2 = \frac{2\mu}{\hbar^2}(E - \varepsilon_m). \tag{6.14}$$

Exercise 6.1. Write down the diabatic (Eq. (6.7)) as well as the adiabatic (Eq. (6.13)) multichannel equations for general, i.e. non-spherical, potentials.

Whether the diabatic or the adiabatic approach should be used in a given scattering situation has to be decided by judicious choice rather than by application of a rigorous criterion. Both frames are equivalent if the respective sets of basis functions, $\{\phi_i(\mathbf{Q})\}$ and $\{\zeta_i(\mathbf{Q}, \boldsymbol{\rho})\}$, are complete. Numerical solutions of scattering problems, however, necessitate the use of a truncated basis. The resulting system of N multichannel equations is frequently referred to as *close-coupling equations*. The need to operate with a finite number of equations can make it difficult to favor one mode of description over the other. While the mathematical structure of the system (6.7) is simpler than that of (6.13), the adiabatic coupling terms are often smaller than the diabatic ones since the eigenfunctions $|\zeta_n\rangle$ tend to vary slowly with the external variables. Thus, the set $\{\zeta_i(\mathbf{Q}, \boldsymbol{\rho})\}$ may be chosen to be less extensive than the set $\{\phi_i(\mathbf{Q})\}$. For the presentation of the formal multichannel theory, however, the less complex diabatic equations appear preferable to the adiabatic ones. Therefore, the following will be based on the diabatic model, unless mentioned otherwise.

6.2 The Lippmann–Schwinger Equation for Inelastic Scattering

This section focuses on the fundamental tenets of multichannel theory which will be built up by using its single-channel analog, which was outlined in Chapter 3, as a guideline. In particular, we will generalize the concept of the Green's function introduced there and recast the multichannel formalism by employing the Lippmann–Schwinger equation:

$$\psi^+ = \varphi + \hat{G}_0^+ V \psi^+, \tag{6.15}$$

with φ as the solution of the homogeneous problem which is obtained by setting the potential V in Eq. (6.7) equal to zero. In the following development, we will refer to a general Green's operator $\hat{G}_0$ instead of $\hat{G}_0^+$. The latter can be recovered by specifying a set of outgoing boundary conditions. We convert the Green's operator into a matrix by representing it in a basis given by the channel basis functions $\phi_i(\mathbf{Q})$ (see Eqs. (6.3) and (6.4)):

$$\hat{G}_0 = \sum_{ik} |\phi_i\rangle G_{0,ik} \langle \phi_k|. \tag{6.16}$$

We recall from the analysis of the single-channel problem in Chapter 3 that $\hat{G}_0$ is determined by the operator identity

$$(E - H_0)\hat{G}_0 = \hat{I}, \tag{6.17}$$

where the zeroth-order Hamiltonian $\hat{H}_0$ involves no potentials besides the centripetal, corresponding to a spherically symmetric problem. We thus ask for a radial Green's function $G_0 = G_0(\rho, \rho')$. Expanding G_0 into the channel basis functions, we obtain the matrix equation

$$(E\mathbf{I} - \mathbf{H}_0)\mathbf{G}_0 = \mathbf{I}. \tag{6.18}$$

Making the radial dependence of $\mathbf{G}_0$ explicit, we arrive at the analog of Eq. (6.7) on the level of multichannel Green's function theory,

$$\left[\mathbf{I}\frac{\partial^2}{\partial\rho^2} + \mathbf{k}^2 - \mathbf{I}\frac{l(l+1)}{\rho^2}\right]\mathbf{G}_0^l(\rho,\rho') = \frac{2\mu}{\hbar^2}\delta(\rho-\rho')\mathbf{I}, \tag{6.19}$$

which is based on a partial wave expansion for the elements of $\mathbf{G}_0$, more specifically

$$G_{0,ik}(\rho,\rho') = \frac{1}{\rho}\sum_{l=0}^{\infty} c_l G_{0,ik}^l(\rho,\rho') P_l(\cos\theta). \tag{6.20}$$

The meaning of $\mathbf{k}^2$ is the same as in Eq. (6.9). A formal solution of Eq. (6.19) can be readily given in terms of the regular Riccati–Bessel functions $j_{lm}(k_m\rho)$ as well as the irregular Riccati–Neumann functions $n_{lm}(k_m\rho)$. If we arrange these functions into arrays $\mathbf{j}(\rho)$ and $\mathbf{n}(\rho)$, $\mathbf{G}_0^l(\rho, \rho')$ is described by

$$\mathbf{G}_0^l(\rho, \rho') = \begin{cases} \mathbf{j}(\rho)\mathbf{M}_j(\rho') & (\rho < \rho') \\ \mathbf{n}(\rho)\mathbf{M}_n(\rho') & (\rho > \rho'), \end{cases} \tag{6.21}$$

where $\mathbf{M}_j$ and $\mathbf{M}_n$ stand for coefficient matrices. From Eq. (6.19), $\mathbf{j}(\rho)$ as well as $\mathbf{n}(\rho)$ may be understood as diagonal matrices with elements that asymptotically turn into

$$j_{jj}(\rho) = j_l(k_j\rho) \sim \frac{\sin\left(k_j\rho - \frac{l\pi}{2}\right)}{k_j^{\frac{1}{2}}} \tag{6.22a}$$

and

$$n_{jj}(\rho) = n_l(k_jR) \sim \frac{\cos\left(k_j\rho - \frac{l\pi}{2}\right)}{k_j^{\frac{1}{2}}}, \tag{6.22b}$$

where the factor $k_i^{-\frac{1}{2}}$ has been included for later convenience and the $\sim$ sign refers to the behavior of the respective function in the asymptotic regime $\rho \to \infty$.

To derive the matrices $\mathbf{M}_j$ and $\mathbf{M}_n$, we use the continuity of the Green's function as well as the discontinuity of its first derivative at the place $\rho = \rho'$:

$$\mathbf{j}(\rho')\mathbf{M}_j(\rho') = \mathbf{n}(\rho')\mathbf{M}_n(\rho') \tag{6.23a}$$

and

$$\mathbf{n}'(\rho')\mathbf{M}_n(\rho') - \mathbf{j}'(\rho')\mathbf{M}_j(\rho') = \frac{2\mu}{\hbar^2}. \tag{6.23b}$$

In the latter statement, we employ the compact notation $\mathbf{j}'(\rho')$ and $\mathbf{n}'(\rho')$ to denote the derivatives of the Riccati–Bessel and Riccati–Neumann functions with respect to ρ, evaluated at the place $\rho = \rho'$. The statement (6.23b) generalizes the jump condition that has to be satisfied by the first derivative of a single-channel Green's function $G(\rho, \rho')$ [90] to a multichannel situation. A generalized Wronskian for the two classes of solutions, regular and irregular, is obtained from

$$\mathbf{W}(\rho) = \mathbf{n}^{\mathrm{T}}(\rho)\mathbf{j}'(\rho) - \mathbf{n}'^{\mathrm{T}}(\rho)\mathbf{j}(\rho). \tag{6.24}$$

As this Wronskian matrix is assumed to be invertible, the coefficient matrices $\mathbf{M}_j(\rho')$ and $\mathbf{M}_n(\rho')$ are found from the prescription:

$$\mathbf{M}_j(\rho') = -\frac{2\mu}{\hbar^2}\mathbf{W}^{-1}(\rho')\mathbf{n}^{\mathrm{T}}(\rho'), \tag{6.25a}$$

$$\mathbf{M}_n(\rho') = -\frac{2\mu}{\hbar^2}\mathbf{W}^{-1}(\rho')\mathbf{j}^{\mathrm{T}}(\rho'). \tag{6.25b}$$

The Green's function in the multichannel case is therefore expressed as

$$\mathbf{G}_0^l(\rho, \rho') = \begin{cases} -\dfrac{2\mu}{\hbar^2}\mathbf{j}(\rho)\mathbf{W}^{-1}(\rho')\mathbf{n}^{\mathrm{T}}(\rho') & (\rho < \rho'), \\ -\dfrac{2\mu}{\hbar^2}\mathbf{n}(\rho)\mathbf{W}^{-1}(\rho')\mathbf{j}^{\mathrm{T}}(\rho') & (\rho > \rho'). \end{cases} \tag{6.26}$$

Here we point out that the function $\mathbf{n}(\rho)$, with the asymptotic behavior defined by Eq. (6.22), is not the only choice for the irregular solution. Thus, it is readily seen that replacing $\mathbf{n}(\rho)$ by $\mathbf{h}^+(\rho) \equiv \mathbf{n}(\rho) + i\mathbf{j}(\rho)$ leaves the Wronskian matrix unaltered. More specifically, we distinguish between the Riccati–Hankel functions $\mathbf{h}^+(\rho)$ and $\mathbf{h}^-(\rho)$ by setting

$$h_{ii}^+(\rho) = h^+(k_i\rho) = n(k_i\rho) + ij_i(k_i\rho) \sim \frac{\exp\left[+i\left(k_iR - \frac{l\pi}{2}\right)\right]}{(k_i)^{\frac{1}{2}}} \tag{6.27a}$$

and

$$h_{ii}^-(\rho) = h^-(k_i\rho) = n(k_i\rho) - ij(k_i\rho) \sim \frac{\exp\left[-i(k_i\rho - \frac{l\pi}{2})\right]}{(k_i)^{\frac{1}{2}}}. \tag{6.27b}$$

The normalization (6.22) guarantees that the Wronskian $\mathbf{W}$ tends towards the unit matrix as $\rho \to \infty$, in accordance with Eq. (3.96), and regardless of whether $\mathbf{n}$ or $\mathbf{h}^+$ is chosen to represent the irregular alternative. Using $\mathbf{h}^+$ instead of $\mathbf{n}$ in Eq. (6.26), and thus considering $\mathbf{G}_0^{l+}(\rho, \rho')$ instead of

$\mathbf{G}_0^l(\rho, \rho')$, the Lippmann–Schwinger equation in coordinate representation

$$\boldsymbol{\psi}^+(\rho) = \mathbf{j}(\rho) + \int_0^\infty d\rho' \mathbf{G}_0^+(\rho, \rho')\mathbf{V}(\rho')\boldsymbol{\psi}^+(\rho') \tag{6.28}$$

can be written explicitly for the asymptotic regime as

$$\boldsymbol{\psi}^+(\rho) \sim \mathbf{j}(\rho) - \frac{2\mu}{\hbar^2}\int_0^\infty d\rho' \mathbf{h}^+(\rho)\mathbf{j}^{\mathrm{T}}(\rho')\mathbf{V}(\rho')\boldsymbol{\psi}^+(\rho'). \tag{6.29}$$

The scattering function may as well be expressed in terms of the transition matrix, which reads

$$\mathbf{T} = \frac{2\mu}{\hbar^2}\int_0^\infty d\rho \mathbf{j}^{\mathrm{T}}(\rho)\mathbf{V}(\rho)\boldsymbol{\psi}^+(\rho) = \frac{2\mu}{\hbar^2}\langle \mathbf{j}|\mathbf{V}(\rho)|\boldsymbol{\psi}^+\rangle. \tag{6.30}$$

This is analogous to Eq. (3.102). It should be noted, however, that the factor $\frac{1}{k}$ present there does not occur in Eq. (6.30) since it is absorbed by the wave functions according to the definitions (6.22). Comparison with Eq. (6.29) yields

$$\boldsymbol{\psi}^+(\rho) \sim \mathbf{j}(\rho) - \mathbf{h}^+(\rho)\mathbf{T} = \frac{1}{2i}[-\mathbf{h}^-(\rho) + \mathbf{h}^+(\rho)\mathbf{S}], \tag{6.31}$$

where the last line follows from the relation between the transition and the scattering matrix, $\mathbf{S} = \mathbf{I} - 2i\mathbf{T}$. It is worth noting that as $\mathbf{j}, \mathbf{h}^{\pm}, \mathbf{S}$, and $\mathbf{T}$ are matrices, so is $\boldsymbol{\psi}^+$. The element ψ_{ij} describes the scattering solution for the channel with label j if i is the entrance channel index. All these matrices depend on the value of the orbital angular momentum quantum number. For the sake of simplicity, however, we suppress the index l in $\mathbf{S}_l, \mathbf{T}_l$, and $\boldsymbol{\psi}_l^+$. Imposing a real outgoing boundary condition on the system instead of the complex one of Eq. (6.29) which is based on selecting $\mathbf{h}^+(\rho)$ as outgoing radial function, we draw an analogy to Eq. (6.31), asking for a real-valued wave function array $\boldsymbol{\psi}(\rho)$ that satisfies

$$\boldsymbol{\psi} \sim \mathbf{j}(\rho) + \mathbf{n}(\rho)\mathbf{K}, \tag{6.32}$$

with an appropriately defined matrix $\mathbf{K}$ as the real counterpart of the transition matrix $\mathbf{T}$. We infer the explicit form of the K-matrix from the single-channel case by use of formula (3.109) to arrive at

$$\mathbf{K} = -\frac{\mathbf{T}}{\mathbf{I} - i\mathbf{T}}. \tag{6.33}$$

The K- and the S-matrix are thus related by

$$\mathbf{K} = i\frac{\mathbf{I}-\mathbf{S}}{\mathbf{I}+\mathbf{S}}. \tag{6.34}$$

The asymptotic scattering solution (6.32) is found immediately from the K-matrix. In the following section, we will highlight the *Born approximation* as a scheme that leads to an approximate solution of the basic multichannel equation (6.28).

6.3 The Born Approximation

Rewriting Eq. (6.28), we denote the jth diagonal element of $\boldsymbol{\psi}^+$ by ψ_j^+. As described in Chapter 3 for the case of a single channel, the potential dependent expression on the right-hand side of (3.91) may be viewed as an inhomogeneous term, giving rise to a solution in integral form, namely

$$\psi_j^+(\rho) = \delta_{ji}\psi_i^{+(0)}(\rho) + \frac{2\mu}{\hbar^2}\sum_m \int_0^\infty d\rho' G_{0j}^+(\rho,\rho')V_{jm}(\rho')\psi_m^+(\rho')d\rho', \tag{6.35}$$

where outgoing solutions of the homogeneous problem, corresponding to $\mathbf{V}=\mathbf{0}$, are assembled into the diagonal matrix $\boldsymbol{\psi}^{+(0)}$. If the incident channel component $\psi_i^+(\rho)$ clearly outweighs the contributions due to the other channels ($\psi_i^+(\rho) \gg \psi_j^+(\rho), j \neq i$), and is not too different from the plane wave expression $\psi_i^{+(0)}(\rho)(\psi_i^+(\rho) \approx \psi_i^{+(0)}(\rho))$, the Born approximation applies:

$$\psi_i^+(\rho) \approx \psi_i^{+(0)}(\rho) + \frac{2\mu}{\hbar^2}\int_0^\infty d\rho' G_{0i}^+(\rho,\rho')V_{ii}(\rho')\psi_i^{+(0)}(\rho') \tag{6.36a}$$

and

$$\psi_j^+(\rho) \approx \frac{2\mu}{\hbar^2}\int_0^\infty d\rho' G_{0j}^+(\rho,\rho')V_{ji}(\rho')\psi_i^{+(0)}(\rho'), \quad j \neq i. \tag{6.36b}$$

Extending formula (3.99), we write the explicit form of the outgoing Green's function as:

$$G_{0j}^+(\rho,\rho') = -\frac{2\mu}{\hbar^2 k_j}\psi_i^{(0)}(\rho)\psi_j^{+(0)}(\rho'), \quad \rho < \rho', \tag{6.37a}$$

$$= -\frac{2\mu}{\hbar^2 k_j}\psi_i^{(0)}(\rho')\psi_j^{+(0)}(\rho) \quad \rho > \rho', \tag{6.37b}$$

where we specify

$$\psi_j^{(0)}(\rho) \sim \sin\left(k_j R - \frac{l\pi}{2}\right), \tag{6.38a}$$

$$\tilde{\psi}_j^{(0)}(\rho) \sim \cos\left(k_j R - \frac{l\pi}{2}\right), \tag{6.38b}$$

and

$$\psi_j^{+(0)}(\rho) \sim \tilde{\psi}_j^{(0)}(\rho) + i\psi_j^{(0)}(\rho), \tag{6.38c}$$

which result from Eq. (6.7) for vanishing potential **V**, and obtain the solutions for large radii ρ:

$$\psi_i^{+}(\rho) \sim \sin\left(k_i\rho - \frac{l\pi}{2}\right) - \frac{2\mu}{k_i\hbar^2}\exp\left(ik_i\rho - i\frac{l\pi}{2}\right) \times \int_0^\infty d\rho'\psi_i^{+(0)}(\rho')V_{ii}(\rho')\psi_i^{+(0)}(\rho'), \tag{6.39a}$$

and

$$\psi_j^{+}(\rho) \sim -\frac{2\mu}{k_j\hbar^2}\exp\left(ik_j\rho - i\frac{l\pi}{2}\right) \times \int_0^\infty d\rho'\psi_j^{+(0)}(\rho')V_{ji}(\rho')\psi_i^{+(0)}(\rho'), \quad j \neq i. \tag{6.39b}$$

We point out that Eq. (6.39a) is identical with the single-channel Born approximation involving a potential $V = V_{ii}$, while the functions $\psi_j^{+(0)}$ may be obtained from the system of equations

$$\left[\frac{\partial^2}{\partial\rho^2} + k_i^2 - \frac{l(l+1)}{\rho^2}\right]\psi_i^{+(0)}(\rho) = 0, \tag{6.40a}$$

$$\left[\frac{\partial^2}{\partial\rho^2} + k_j^2 - \frac{l(l+1)}{\rho^2}\right]\psi_j^{+(0)}(\rho) = \frac{2\mu}{\hbar^2}V_{ji}\psi_i^{+(0)}(\rho), \quad j \neq i. \tag{6.40b}$$

The latter relation results from the assumption $\psi_i^{+}(\rho) \gg \psi_j^{+}(\rho), j \neq i$, which justifies the Born approximation. From the integral expressions (6.39a) and (6.39b), one easily derives a formula for the transition operator matrix elements by comparison with Eq. (6.30), namely:

$$T_{ij} = (k_i k_j)^{-\frac{1}{2}} \frac{2\mu}{\hbar^2} \int_0^\infty d\rho' \psi_j^{+(0)}(\rho') V_{ji}(\rho') \psi_i^{+(0)}(\rho'). \tag{6.41}$$

It should be noted that the approach (6.41) does not guarantee the unitarity of the S-matrix. Algebraic regularization procedures have been proposed to remedy this shortcoming [151].

6.3.1 *The distorted-wave Born approximation (DWBA)*

The Born approximation employs the asymptotic wave functions $\psi_j^{(0)} (j = 0, 1, \ldots)$ as solutions of zeroth order. However, more elaborate choices can be made for these solutions, the most prominent among which involves the use of *distorted waves*. The basic idea is to replace the sequence $\{\psi_j^{(0)}\}$ by another set of functions $\{\breve{\psi}_j^{(0)}\}$ which are defined as the full solutions for the respective channel j in separation from other channels. Thus, generating the distorted waves implies solving the scattering problem for all channels while neglecting their mutual interaction. Formally, the approach (6.7) remains unchanged. However, we partition the scattering equation in the following way:

$$\left[\frac{\partial^2}{\partial\rho^2} + k_j^2 - \frac{l(l+1)}{\rho^2} - \frac{2\mu}{\hbar^2} V_{jj}\right] \psi_j^+(\rho) = \sum_{k \neq j} \frac{2\mu}{\hbar^2} V_{jk} \psi_k^+(\rho). \tag{6.42}$$

Then we define the distorted waves by the system of uncoupled differential equations that results from equating the right-hand side to zero:

$$\left[\frac{\partial^2}{\partial\rho^2} + k_j^2 - \frac{l(l+1)}{\rho^2} - \frac{2\mu}{\hbar^2} V_{jj}\right] \breve{\psi}_j^{+(0)}(\rho) = 0. \tag{6.43}$$

Hence the effect of the diagonal potential energy matrix elements is now included in the zeroth-order approximation to the scattering solution. These solutions are obtained within the distorted-wave Born approximation, in close analogy to Eqs.(6.39a) and (6.39b), as

$$\psi_i^+(\rho) \sim \breve{\psi}_i^{+(0)}(\rho), \tag{6.44a}$$

$$\psi_j^+(\rho) \sim -\frac{2\mu}{k_j\hbar^2} \exp(ik_j\rho - i\delta_l) \int_0^\infty d\rho' \breve{\psi}_j^{+(0)}(\rho') V_{ji}(\rho') \breve{\psi}_i^{+(0)}(\rho'), \quad j \neq i. \tag{6.44b}$$

By comparing with Eq. (6.41), we find the transition matrix element to be

$$T_{ij} = (k_i k_j)^{-\frac{1}{2}} \frac{2\mu}{\hbar^2} \int_0^\infty d\rho' \breve{\psi}_j^{+(0)}(\rho') V_{ji}(\rho') \breve{\psi}_i^{+(0)}(\rho'). \tag{6.45}$$

The importance of the distorted-wave Born approximation lies chiefly in the subdivision of the potential into a distorting and a coupling part such that the two respective problems can be solved in isolation from each other. This approach is of particular relevance for the treatment of molecular dynamics on coupled potential energy surfaces. Such a situation is frequently analyzed by computing, in an initial step, the distorted waves $\breve{\psi}_j^{+(0)}$ which result from Eq. (6.43) for each surface V_{jj}, neglecting the interdependence of the surfaces. Subsequently, the coupling effect is included in the framework of an approximation scheme which may be the Born formalism or another procedure. In Section 7.4, we will encounter an example of scattering in a landscape of multiple potential energy surfaces, where a distorted-wave approach is combined with a variational assessment of the coupling effect. We point out that interchannel coupling may be, to a certain extent, included in the definition of the distorted waves. Thus, the coupling could be partitioned into a nuclear, i.e. vibrational, rotational, and orbital part and an electronic contribution, with the intent to incorporate into the functions $\breve{\psi}_j^{(0)}$ the effect of the nuclear, but not of the electronic coupling (see for instance [152]).

6.4 Microreversibility

The S operator allows us to assign a certain probability to a process that starts in a channel i and ends in a channel j. Intuitively, one expects to find the same probability for the time-reversed process with j as the initial and i as the final channel. In what follows, we will subject this expectation to systematic scrutiny and establish the feature of *microreversibility*. This basic property of multichannel scattering processes also provides a valuable criterion for the validity of approximate schemes in quantum dynamics, such as semiclassical models.

We recall that for any arbitrary channel k, the scattering function Φ_k (see (6.4)) reduces for sufficiently large values of the scattering coordinate ρ to a superposition of an incoming and outgoing plane wave, according to

$$\Phi_k \sim A_k h_k^- - B_k h_k^+, \tag{6.46}$$

with h^-, h^+ as Riccati–Hankel functions. Each channel index k refers to a well-defined set of quantum numbers, including that of the orbital angular momentum and possibly the electronic spin of the system with their associated magnetic quantum numbers m_l and m_s, respectively. It further comprises a linear channel momentum $\hbar\mathbf{k}_k$. Analysis of rearrangement collisions such as $A + B \rightarrow C + D$ may additionally require the use of a channel coordinate $\boldsymbol{\rho}_k$. Solving the scattering problem by applying the formalism outlined in the preceding sections, one may assign a solution to any entrance channel i, such that ([153], see (6.4))

$$\Psi_i^+ = \sum_j \phi_j(\mathbf{Q})\Phi_{ij}^+(\boldsymbol{\rho}). \tag{6.47}$$

For a particular channel j that is populated during the scattering process, the function Φ_{ij}^+ depends asymptotically on the scattering coordinate like

$$\Phi_{ij}^+(\boldsymbol{\rho}) \sim -\frac{B_j}{A_i} h_j^+(\rho) = -S_{ij} h_j^+(\rho). \tag{6.48}$$

Here we make use of the relation $B_j = S_{ij}A_i$ which is recognized as a generalization of the elastic single-channel case (3.104), where the scattering matrix element appears as the amplitude of the outgoing wave. The linearity of the Schrödinger equation allows us to write a general solution of the multichannel problem as a linear combination of the partial solutions (6.47), namely

$$\Psi^+ = \sum_i A_i \Psi_i^+. \tag{6.49}$$

The incoming channel amplitudes A_i are related to the outgoing amplitude B_j through the identity

$$B_j = \sum_i S_{ij} A_i. \tag{6.50}$$

For the special case of elastic scattering, i.e. no transition between the channels during the interaction, relation (6.50) reduces to $B_j = S_{jj}A_j$.

We will now place the subject of multichannel scattering into the context of time reversal. Assuming that a solution (6.46) of the scattering problem at hand has been found, can we generate its equivalent for the time-reversed problem? The asymptotic incoming wave obeys the TDSE for $t \to -\infty$, the outgoing wave for $t \to +\infty$. As a function of time, the scattering solution Ψ satisfies

$$i\hbar \frac{\partial \Psi(t)}{\partial t} = \hat{H}\Psi(t), \tag{6.51}$$

where no boundary conditions have been specified. Upon time reversal ($t \to -t$), this turns into

$$i\hbar \frac{\partial \Psi(-t)}{\partial t} = \hat{H}\Psi(-t), \tag{6.52}$$

which holds under the premise of a time-independent Hamiltonian. As this equation is complex-conjugated, one arrives at

$$i\hbar \frac{\partial \Psi^*(t)}{\partial t} = \hat{H}^*\Psi^*(-t) \tag{6.53}$$

and thus concludes that the time-reversed and complex-conjugated wave function $\Psi^*(-t)$ solves a Schrödinger equation governed by the complex-conjugated Hamiltonian $\hat{H}^*$. This equation reduces to Eq. (6.51) if $\hat{H} = \hat{H}^*$, which is the case if $\hat{H}$ does not contain spin operators. Setting $\Psi^{\mathrm{r}}(t) \equiv \Psi^*(-t)$, we introduce Wigner's *time-reversal operator* $\hat{W}$ by

$$\Psi^{\mathrm{r}}(t) \equiv \hat{W}\Psi(-t). \tag{6.54}$$

It is easy to see that for a real Hamiltonian, the $\hat{W}$ operation reduces to complex conjugation and thus $\hat{W}$ to the conjugation operator $\hat{C}$. The eigenvalues and expectation values of the linear and the angular momentum as well as spin have to reverse their sign if computed for the time-reversed system. An example is given by the linear momentum operator $\hat{p}$ and its expectation value $\langle \hat{p} \rangle = \langle \Psi(t)|\hat{p}|\Psi(t)\rangle$. For the time-reversed solution $\Psi^{\mathrm{r}}(t)$, $\langle \hat{p} \rangle$ switches its sign:

$$\langle \Psi^{\mathrm{r}}(t)|\hat{p}|\Psi^{\mathrm{r}}(t)\rangle = -\langle \hat{p} \rangle. \tag{6.55}$$

In particular, the magnetic quantum numbers of the system described by $\Psi^{\mathrm{r}}(t)$ reverse their signs. Assuming that Ψ is an eigenfunction of L_z as well as σ_z with eigenvalues m_l and m_s, respectively, it holds for the time-reversed wave function that

$$L_z \Psi^{\mathrm{r}} = -m_l \Psi^{\mathrm{r}}, \tag{6.56a}$$

and

$$\sigma_z \Psi^{\mathrm{r}} = -m_s \Psi^{\mathrm{r}}. \tag{6.56b}$$

Applying the rules (6.54) and (6.56), we generate the time-reversed form of the asymptotic wave function (6.46) by complex conjugation and replacement of the channel index k with k' such that

$$\Phi_k \sim A_k h_k^- - B_k h_k^+ \tag{6.57a}$$

turns into

$$\Phi_{k'}^{\mathrm{r}} \sim -B_{k'}^* h_{k'}^- + A_{k'}^* h_{k'}^+. \tag{6.57b}$$

The label k' stands for the same set of quantum numbers as k, excepting all magnetic quantum numbers. These have to be reversed, in keeping with Eqs. (6.56a) and (6.56b). As a matter of convention, the transition from Eq. (6.57a) to Eq. (6.57b) may involve multiplying the wave function by a phase factor. Following, for instance, the definition of Condon and Shortley [154], one attaches a coefficient $(-1)^M$ to the right-hand side of (6.57b), where M is the total magnetic quantum number of the system. Since, in discussing microreversibility, we are ultimately concerned with the absolute values of the incoming and outgoing components, we will not consider the phase factor in further detail. Applying the time reversal rules $A_k \to -B_{k'}^*$ and $-B_k \to A_{k'}^*$ to Eq. (6.50) between the incoming and outgoing wave amplitudes, we obtain

$$A_{j'}^* = \sum_i S_{ij} B_{i'}^*. \tag{6.58}$$

Upon using Eq. (6.50) once more, the relation

$$A_{j'}^* = \sum_{i,k'} S_{ij} S_{k'i'}^* A_{k'}^* \tag{6.59}$$

is derived, implying

$$\sum_\alpha S_{\gamma'\alpha'} S_{\alpha\beta}^* = \delta_{\beta'\gamma'}. \tag{6.60}$$

In matrix form, Eq. (6.60) is written as

$$\mathbf{S}'^{\mathrm{T}} \mathbf{S}^* = \mathbf{I}, \tag{6.61}$$

where $\mathbf{S}'^{\mathrm{T}}$ is the transpose of the matrix with elements $S_{i'k'}$. From the unitarity of the S matrix (see Section 3.3.3) one finds

$$\mathbf{S}'^{\mathrm{T}} = \mathbf{S}. \tag{6.62}$$

Or, in terms of the matrix elements:

$$S_{\beta'\alpha'} = S_{\alpha\beta}. \tag{6.63}$$

This identity encodes the *reciprocity theorem* which states that the transition processes $i \to j$ for a certain system and $j' \to i'$ for the same system under the condition of time reversal have equal probabilities. For the special case that all magnetic quantum numbers are zero, this theorem implies

$$S_{\beta\alpha} = S_{\alpha\beta} \tag{6.64}$$

and therefore adds to the properties of the S-matrix, such as unitarity, that of symmetry. Relations (6.63) and (6.64) encapsulate the *principle of microreversibility* which, from the derivation given earlier, has turned out to be a consequence of the linearity of the Schrödinger equation and the unitarity of the S-matrix. In terms of transition probabilities P_{ij} and P_{ji}, this principle may be expressed as

$$P_{j'i'} = |S_{j'i'}|^2 = |S_{ij}|^2 = P_{ij}. \tag{6.65}$$

Or, in the case of S-matrix symmetry:

$$P_{ji} = \left|S_{ji}\right|^2 = \left|S_{ij}\right|^2 = P_{ij}. \tag{6.66}$$

The latter relation applies not only when the studied process does not involve magnetic quantum numbers but also when the corresponding spins are not measured. In this case Eq. (6.66) contains a summation over both the initial and the final magnetic quantum numbers.

6.5 R-matrix and Log Derivative Propagation

We have considered the differential coupled channel as well as as the integral equation formalism to generate the scattering solution ψ, as given by the relations (6.7) and (6.29), respectively. These approaches have in common that the whole system of coupled equations has to be solved for any energy E of interest for the considered scattering problem. Alternatively, more economic methods are based on the propagation of the so-called R-matrix or the log derivative matrix that, once obtained, make it possible to derive the solutions at arbitrary energy without recalculation of the matrix. Both schemes employ recursion relations that are derived from the multichannel scattering equations. The R-matrix method, however, propagates the potential characteristic of the problem, while the log derivative procedure advances the solution. The basic methodology underlying these methods may be clarified by inspecting the one-dimensional Hamiltonian in the scattering coordinate

$$\hat{H} = -\frac{\hbar^2}{2\mu}\frac{\partial^2}{\partial\rho^2} + V(\rho). \tag{6.67}$$

We generate a real-valued complete basis $\{v_i\}$ by solving the eigenvalue problem

$$\hat{H}v_i = E_i v_i \tag{6.68}$$

on the interval $[0, L]$, where the $\{v_i\}$ may obey a *Neumann boundary condition*:

$$\frac{\partial}{\partial\rho}v_i(\rho)|_{\rho=L} = 0. \tag{6.69}$$

Next, the solution of $\hat{H}\Phi = E\Phi$ in $[0, L]$ is expanded as

$$\Phi = \sum_{i=1} a_i v_i. \tag{6.70}$$

Combining Eqs. (6.67), (6.68), and (6.70), we conclude that

$$\langle v_i|H|\Phi\rangle = -\frac{\hbar^2}{2\mu}\left\langle v_i \left| \frac{\partial^2}{\partial\rho^2} \right| \Phi \right\rangle + \langle v_i|V|\Phi\rangle. \tag{6.71}$$

Subjecting the first term on the right-hand side to twofold integration by parts, one finds

$$\begin{aligned}\langle v_i|H\Phi|\rangle &= \langle \hat{H}v_i|\Phi\rangle - \frac{\hbar^2}{2\mu}\left[v_i(\rho)\frac{\partial}{\partial\rho}\Phi(\rho) - \left(\frac{\partial}{\partial\rho}v_i(\rho)\right)\Phi(\rho)\right]\Bigg|_{\rho=L} \\ &= E_i\langle v_i|\Phi\rangle - \frac{\hbar^2}{2\mu}\left[v_i(\rho)\frac{\partial}{\partial\rho}\Phi(\rho)\right]_{\rho=L}. \end{aligned} \tag{6.72}$$

From this equation, we realize that the expansion coefficients a_i in (6.70) can be expressed in terms of the values adopted by the eigenfunction v_i as well as the derivative of the solution at the boundary L, namely

$$\langle v_i|\Phi\rangle = \frac{1}{E_i - E}\frac{\hbar^2}{2\mu}v_i(\rho)\frac{\partial}{\partial\rho}\Phi(\rho)|_{\rho=L}. \tag{6.73}$$

Finally, the solution in the interval $[0, L]$ is written as

$$|\Phi\rangle = \frac{\hbar^2}{2\mu}\sum_i \frac{|v_i\rangle}{E_i - E}v_i(\rho)\frac{\partial}{\partial\rho}\Phi(\rho)|_{\rho=L} \tag{6.74}$$

The ratio between the solution and its derivative at the boundary L of the interval is referred to as the R-matrix. In the present one-dimensional case, it reduces to a scalar. Its explicit form is

$$R = \frac{\hbar^2}{2\mu}\sum_i \frac{|v_i(L)|^2}{E_i - E}. \tag{6.75}$$

In order to extract scattering information from the R-matrix, one chooses L in the asymptotic region where, assuming K-matrix boundary conditions (see Eq. (3.108)), we have

$$R(L) = \frac{\sin(kL) + K\cos(kL)}{k\cos(kL) - k\,K\sin(kL)}. \tag{6.76}$$

This, in turn, yields the K-matrix in terms of the R-matrix:

$$K = \frac{R(L)k\cos(kL) - \sin(kL)}{\cos(kL) - R(L)\,k\,\sin(kL)}. \tag{6.77}$$

Prescription (6.77) implies that the K-matrix elements (and therefore those of the S- and T-matrices as well) can be obtained by evaluating $R(L)$. This requires solving the eigenvalue problem (6.68) and thus deriving the set $\{v_i\}$ along with the energy-shifted eigenvalues $E_i - E$, as they enter into (6.75). Once the $\{v_i\}$ and the E_i are obtained, they do not need to be recalculated for a different choice of E. From these considerations, it is desirable to extend the one-dimensional R-matrix formalism, as sketched before, to the multidimensional, and, in particular, to the multichannel case. Further, there is obvious interest in a procedure that allows us to advance the R-matrix into the asymptotic region where it exhibits a direct relation to the K-matrix. This need is addressed by the R-matrix and log derivative methods which also provide a remedy for the problem of exponentially divergent closed channels, a shortcoming of the close-coupling technique which aims at generating the scattering solution itself. In the following subsection, we will highlight salient features of the log derivative techniques. Details on R-matrix propagation may be found in Section 6.5*.

6.5.1 *The log derivative method*

As pointed out before, the occurrence of exponentially increasing solutions, related to closed channels, poses a major obstacle to any direct treatment of the close-coupling equations. The log derivative method alleviates this problem by replacing propagation of the wave function with that of its logarithmic derivative which remains finite for any exponentially growing solution of the form $\exp(k_m\rho)$. In its original form as put forward in [155], the *log derivative method* proceeds along the following lines.

To begin, we express the close coupling equations once more in the compact form

$$\frac{\partial^2 \mathbf{\Phi}(\rho)}{\partial \rho^2} = \mathbf{C}(\rho)\mathbf{\Phi}(\rho), \tag{6.78}$$

with a symmetric coupling matrix $\mathbf{C}$. As, for instance, in Eq. (6.28), the array $\mathbf{\Phi}$ is here understood as an $N \times N$ matrix, corresponding to the inclusion of N channels in the expansion (6.4). The columns of $\mathbf{\Phi}$ thus represent different solutions according to N different initial conditions resulting from the possible choices of $k_m^2, m = 1, \ldots, N$. We realize that the logarithmic derivative of $\mathbf{\Phi}$,

$$Y_{ij}(\rho) = \frac{\partial \Phi_{ij}(\rho)}{\partial \rho} \Big/ \Phi_{ij}(\rho) \tag{6.79}$$

satisfies the Riccati equation

$$\frac{\partial \mathbf{Y}(\rho)}{\partial \rho} = \mathbf{C}(\rho) - \mathbf{Y}^2(\rho). \tag{6.80}$$

It may seem tempting to use the latter relation for advancing $\mathbf{Y}(\rho)$ directly. However, the singularities of $\mathbf{Y}(\rho)$, arising at places where $\det \mathbf{\Phi}(\rho) = 0$, endanger the success of this procedure. This difficulty is circumvented as one defines a propagator scheme in close analogy to that utilized in the R-matrix method, breaking down the ρ coordinate into segments $[\rho_i, \rho_{i+1}]$

where $\rho_i = \rho_0 + ih, i = 1, 2, \ldots$. For any interval thus defined we determine block matrices $\mathbf{y}_i, i = 1, \ldots, 4$, [156, 157] such that

$$\begin{pmatrix} \dfrac{\partial \mathbf{\Phi}(\rho_i)}{\partial \rho} \\ \dfrac{\partial \mathbf{\Phi}(\rho_{i+1})}{\partial \rho} \end{pmatrix} = \begin{pmatrix} \mathbf{y}_1(\rho_i, \rho_{i+1}) & \mathbf{y}_2(\rho_i, \rho_{i+1}) \\ \mathbf{y}_3(\rho_i, \rho_{i+1}) & \mathbf{y}_4(\rho_i, \rho_{i+1}) \end{pmatrix} \begin{pmatrix} -\mathbf{\Phi}(\rho_i) \\ \mathbf{\Phi}(\rho_{i+1}) \end{pmatrix}. \tag{6.81}$$

Evidently, in the limiting case $h \to 0$, the matrices $\mathbf{y}_i$ have to reduce to the simple form $\frac{1}{h}\mathbf{I}$. Relation (6.81) implies a recursion formula for the logarithmic derivative of $\mathbf{\Phi}(\rho)$ in the interval $[\rho_i, \rho_{i+1}]$, namely

$$\begin{aligned} \mathbf{Y}(\rho_{i+1}) &= \mathbf{y}_4(\rho_i, \rho_{i+1}) - \mathbf{y}_3(\rho_i, \rho_{i+1}) \\ &\quad \times [(\mathbf{Y}(\rho_{i+1}) - \mathbf{y}_1(\rho_i, \rho_{i+1})]^{-1} \mathbf{y}_2(\rho_i, \rho_{i+1}). \end{aligned} \tag{6.82}$$

Exercise 6.2. Demonstrate that the logarithmic derivative (6.79) solves the Riccati equation (6.80) and that Eq. (6.82) follows from Eq. (6.81).

Implementing Eq. (6.82) requires the block matrix propagators $\mathbf{y}_i, i = 1-4$, to be known explicitly. Following Manolopoulos [157], we divide this task into two steps. Specifically, we partition the coupling matrix $\mathbf{C}(\rho)$ into a reference term $\mathbf{C}_{ref}$ and a residual interaction term $\mathbf{D}$ such that

$$\mathbf{C}(\rho) = \mathbf{C}_{\mathrm{ref}}(\rho) + \mathbf{D}(\rho). \tag{6.83}$$

The first step involves considering the homogeneous equation

$$\frac{\partial^2 \widetilde{\mathbf{\Phi}}(\rho)}{\partial \rho^2} = \mathbf{C}_{\mathrm{ref}}(\rho)\widetilde{\mathbf{\Phi}}(\rho) \tag{6.84}$$

on each one of the intervals. The reference term may be chosen as the constant matrix $\delta_{ij}p_i^2$ with $p_i^2 = W_{ii}(c)$ where c denotes the sector midpoint. This choice leads to simple expressions for the elements of a propagation matrix $\widetilde{\mathbf{y}}$ defined in analogy with Eq. (6.81):

$$\widetilde{y}_1(\rho_i, \rho_{i+1})_{ij} = \widetilde{y}_4(\rho_i, \rho_{i+1})_{ij} = \delta_{ij} \begin{cases} |p_j| \coth |p_j|h & \text{if } p_j^2 > 0 \\ |p_j| \cot |p_j|h & \text{if } p_j^2 \leq 0 \end{cases} \tag{6.85a}$$

and

$$\widetilde{y}_2(\rho_i, \rho_{i+1})_{ij} = \widetilde{y}_3(\rho_i, \rho_{i+1})_{ij} = \delta_{ij} \begin{cases} |p_j| \operatorname{csch} |p_j|h & \text{if } p_j^2 > 0. \\ |p_j| \csc |p_j|h & \text{if } p_j^2 \leq 0. \end{cases} \tag{6.85b}$$

It is easily shown that the elements of $\tilde{\mathbf{y}}_i$ ($i = 1, \ldots, 4$) thus defined indeed tend towards $\frac{1}{h}$ for sufficiently small arguments $|p_i|h$. The effect of the residual interaction matrix $\mathbf{D}$ is taken into account by rewriting the basic differential Eq. (6.78) as

$$\frac{\partial^2 \mathbf{\Phi}(\rho)}{\partial \rho^2} = (\mathbf{C}_{\text{ref}}(\rho) + \mathbf{D}(\rho))\mathbf{\Phi}(\rho), \tag{6.86}$$

and interpreting Eq. (6.84) as the homogeneous part. Relation (6.86) is recast into the form of an integral equation, where the integral is approximated by a quadrature sum, as outlined in [158] for the example of a collinear collision of a particle with a harmonic oscillator. The quadrature contributions from the three characteristic sector points a, b, and c are:

$$\mathbf{Q}(a) = \frac{h}{3}\mathbf{D}(a), \tag{6.87a}$$

$$\mathbf{Q}(c) = \frac{1}{2}\left[\mathbf{I} - \frac{h^2}{6}\mathbf{D}(c)\right]^{-1}\frac{4h}{3}\mathbf{D}(c), \tag{6.87b}$$

$$\mathbf{Q}(b) = \frac{h}{3}\mathbf{D}(b), \tag{6.87c}$$

with a as the initial and b as the end point of the interval. Combining the statements (6.85a) and (6.85b) with Eqs. (6.87a) and (6.87b), one obtains approximate expressions for the matrices $\mathbf{y}_i, i = 1, \ldots, 4$ in both half sectors of each segment, i.e. in $[a, c]$ as well as $[c, b]$. These are

$$\mathbf{y}_1(\rho_i, \rho_{i+1}) = \widetilde{y}_1(\rho_i, \rho_{i+1}) + \mathbf{Q}(\rho_i), \tag{6.88a}$$

$$\mathbf{y}_2(\rho_i, \rho_{i+1}) = \widetilde{y}_2(\rho_i, \rho_{i+1}), \tag{6.88b}$$

$$\mathbf{y}_3(\rho_i, \rho_{i+1}) = \widetilde{y}_3(\rho_i, \rho_{i+1}), \tag{6.88c}$$

$$\mathbf{y}_4(\rho_i, \rho_{i+1}) = \widetilde{y}_4(\rho_i, \rho_{i+1}) + \mathbf{Q}(\rho_{i+1}), \tag{6.88d}$$

where (ρ_i, ρ_{i+1}) refers to (a, c) or (c, b). We point out that a purely *wave function following* method results for the specific choice of $\mathbf{C}_{\text{ref}} = \mathbf{0}$, as originally suggested by Johnson [155]. The presently described version of the formalism, in contrast, is associated with reference potentials that vary from sector to sector and thus mixes elements of the wave function and the *potential following* prototype.

The explicit form of the homogeneous contributions $\widetilde{\mathbf{y}}_i, i = 1 - 4$, to the overall propagator matrices implies a limitation of this numerical scheme. The matrices $\widetilde{\mathbf{y}}_i$ become singular as $|p_i|h$ equals a multiple of π in the classically allowed regime, i.e. for $p_i^2 \leq 0$. Therefore, the step size h has to be chosen small enough in each interval to exclude this case. This upper bound condition may make the R-matrix method, which is free of a comparable constraint, preferable to the log derivative procedure as presented here if the problem at hand involves a long-range potential. The latter technique, however, has been shown to involve a smaller number of algebraic operations than the former [157], and may, for most applications, be viewed as more economic in terms of computer time requirements than R-matrix propagation.

6.6 Reactive Scattering I: The Differential Equation Approach

So far, we have treated elastic and inelastic scattering processes under the tacit assumption that they leave the reactant configurations unchanged. Chemical reactions, however, may be viewed as *rearrangement collisions*, which alter the identities of the interaction partners. One may think of electron exchange in the scattering of a proton by a hydrogen atom as the most elementary example of such a reactive process:

$$\mathrm{H} + \mathrm{H}^+ \rightarrow \mathrm{H}^+ + \mathrm{H}. \tag{6.89a}$$

A general rearrangement collision between two species A and B is denoted by

$$\mathrm{A} + \mathrm{B} \rightarrow \mathrm{C} + \mathrm{D}. \tag{6.89b}$$

Here the transition from the initial to the final system gives rise to new species with new masses, which require a change of coordinates for adequate description. The scheme (6.89b) implies that no unique partition of the Hamiltonian is valid any longer. Instead, we have to operate with two alternative partitions, namely

$$\hat{H} = \hat{H}_{0,\mathrm{AB}} + V'_{\mathrm{AB}} = \hat{H}_{0,\mathrm{CD}} + V'_{\mathrm{CD}}, \tag{6.89c}$$

The symbols $\hat{H}_{0,\mathrm{AB}}, \hat{H}_{0,\mathrm{CD}}$ are understood as the Hamiltonians for the free systems in asymptotic conditions for both possible arrangements, and $V'_{\mathrm{AB}}, V'_{\mathrm{CD}}$ are the respective interaction potentials. The form of the reactant Hamiltonian thus differs from that of the product Hamiltonian. As a consequence of this ambiguity, the adapted coordinates, describing the system before and after the interaction, deviate from each other. In the following, we will discuss various choices of coordinates by the example of the simplest type of collisions that allow for atomic rearrangement, namely those involving three centers.

Irrespective of the grouping of the three atoms, their interaction may be described in a space-fixed or a more economic body-fixed frame of reference. Both representations will be derived and characterized for the triatomic problem. The outline given in this and the next section will lay the ground for the extended discussion of this problem in the following chapter where electronic degrees of freedom will be explicitly included.

6.6.1 *Jacobi coordinates*

The definition of appropriate coordinates to be used in the case of collisions involving the complexes with three or more atoms is not a trivial task. A natural choice of coordinates for three-center problems, including bound units as well as scattering systems as realized, for instance, in atom-diatom scattering, is given by the *Jacobi variables*. These are illustrated in Figure 6.1. As they require grouping the three constituents into a diatomic and a monatomic subsystem, three basic choices can be made in selecting a

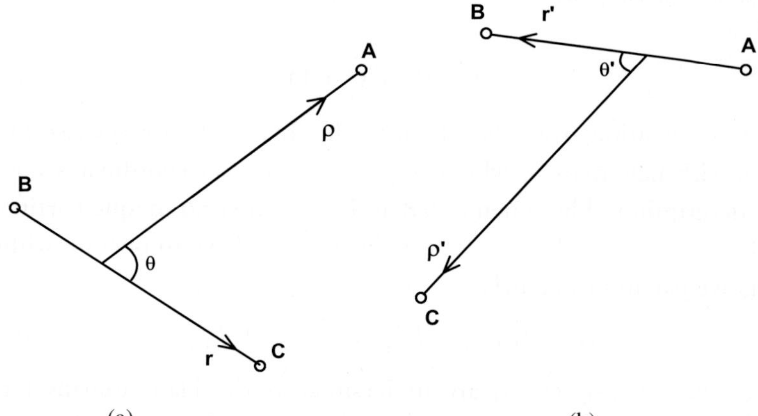

Figure 6.1: Jacobi coordinates for the initial (a) and the final state (b) of a rearrangement collision involving three centers.

set of Jacobi variables. If a reaction of the type

$$\mathrm{A} + \mathrm{BC} \rightarrow \mathrm{AB} + \mathrm{C} \tag{6.90}$$

is to be described, the configuration of the products suggests the use of different Jacobi coordinates than that of the reactants. In this case, it is desirable to determine a transformation that translates one type of coordinates into the other.

The vector $\boldsymbol{\rho}$ connects the locations of the single atom and the diatom center of mass according to

$$\boldsymbol{\rho} = \mathbf{X}_{\mathrm{A}} - \mathbf{X}_{\mathrm{BC}} = \mathbf{X}_{\mathrm{A}} - \frac{m_B \mathbf{X}_{\mathrm{B}} + m_{\mathrm{C}} \mathbf{X}_{\mathrm{C}}}{m_{\mathrm{B}} + m_{\mathrm{C}}} \tag{6.91a}$$

and

$$\mathbf{r} = \mathbf{X}_{\mathrm{C}} - \mathbf{X}_{\mathrm{B}}, \tag{6.91b}$$

while the angle between both vectors is labeled Θ:

$$\Theta = \cos^{-1}\left(\frac{\boldsymbol{\rho} \cdot \mathbf{r}}{\rho r}\right). \tag{6.91c}$$

Adapted coordinates for the product side are defined in an analogous manner. The triatomic Hamiltonian is separable in the variables (6.91a) and (6.91b). Expressing the potential energy surface in a space spanned by $\boldsymbol{\rho}$ and $\mathbf{r}$ we find

$$\hat{H} = \hat{T} + V = -\frac{\hbar^2}{2\mu_\rho}\Delta_\rho - \frac{\hbar^2}{2\mu_r}\Delta_{\mathrm{r}} + V(\boldsymbol{\rho} \cdot \mathbf{r}), \tag{6.92}$$

where the prescriptions to calculate the reduced masses μ_ρ and μ_{r} are

$$\mu_{\mathrm{r}} = \frac{m_{\mathrm{B}} m_{\mathrm{C}}}{m_{\mathrm{B}} + m_{\mathrm{C}}}, \tag{6.93a}$$

and

$$\mu_\rho = \frac{m_{\mathrm{A}}(m_{\mathrm{B}} + m_{\mathrm{C}})}{M}, \tag{6.93b}$$

with

$$M = m_{\mathrm{A}} + m_{\mathrm{B}} + m_{\mathrm{C}}. \tag{6.93c}$$

We label the angular momentum of the diatomic subunit $\mathbf{j}$ and associate an angular momentum $\mathbf{L}$ with the coordinate $\boldsymbol{\rho}$, which refers to the motion of the atom about the center of mass of the diatom. In isotropic space, the overall nuclear orbital angular momentum $\mathbf{J} = \mathbf{j} + \mathbf{L}$ is a constant of motion. In terms of $R, r, \mathbf{j}$, and $\mathbf{L}$, the Hamiltonian becomes [7]

$$\hat{H} = -\frac{\hbar^2}{2\mu_\rho \rho}\frac{\partial^2}{\partial \rho^2}\rho + \frac{\hbar^2}{2\mu_\rho \rho^2}\hat{L}^2 - \frac{\hbar^2}{2\mu_r r}\frac{\partial^2}{\partial r^2} r + \frac{\hbar^2}{2\mu_r r^2}\hat{j}^2 + V(\boldsymbol{\rho}, \mathbf{r}). \tag{6.94}$$

We point out that the scattering problem may be formulated in terms of the alternative product Jacobi coordinates sketched in Figure 6.1(b). The variables $(\boldsymbol{\rho}'\mathbf{r}')$ adapted to the latter arrangement channel are obtained from $(\boldsymbol{\rho}, \mathbf{r})$ by the transformation [92]

$$\begin{pmatrix} \mathbf{r}' \\ \boldsymbol{\rho}' \end{pmatrix} = \begin{pmatrix} -\dfrac{m_{\mathrm{A}}}{m_{\mathrm{A}} + m_{\mathrm{B}}} & -\dfrac{m_{\mathrm{B}} M}{(m_{\mathrm{A}} + m_{\mathrm{B}})(m_{\mathrm{B}} + m_{\mathrm{C}})} \\ -1 & \dfrac{m_{\mathrm{C}}}{m_{\mathrm{B}} + m_{\mathrm{C}}} \end{pmatrix} \begin{pmatrix} \mathbf{r} \\ \boldsymbol{\rho} \end{pmatrix}. \tag{6.95}$$

A basic symmetry between the three possible atom–diatom combinations emerges as one introduces mass-scaled Jacobi coordinates. For this purpose, we add to the list (6.93) of mass-related parameters the three-body reduced mass

$$\mu = \mu_\rho \mu_r = \left(\frac{m_{\mathrm{A}} m_{\mathrm{B}} m_{\mathrm{C}}}{M}\right)^{\frac{1}{2}} \tag{6.96}$$

and define

$$\widetilde{\boldsymbol{\rho}} = \sqrt{\frac{\mu_\rho}{\mu}}\boldsymbol{\rho}, \tag{6.97a}$$

as well as

$$\widetilde{\mathbf{r}} = \sqrt{\frac{\mu_r}{\mu}}\mathbf{r}. \tag{6.97b}$$

In mass-scaled coordinates, the operator of the kinetic energy, as contained in Eq. (6.92), assumes the compact form

$$\hat{T} = -\frac{\hbar^2}{2\mu}(\Delta_{\widetilde{\boldsymbol{\rho}}} + \Delta_{\widetilde{\mathbf{r}}}). \tag{6.98}$$

In spite of the simplified appearance of the kinetic energy operator, the rearrangement problem and specifically the transition from the *reactant set* $(\boldsymbol{\rho}, \mathbf{r})$ to the *product set* $(\boldsymbol{\rho}', \mathbf{r}')$ remains challenging. At the boundary between the domains of these two coordinate groups, the wave function is required to be continuous. An alternative treatment of this problem consists in asking for a unified system of scattering coordinates that covers all possible configurations, thus effectively reducing a reactive to a non-reactive scattering situation. These are *Delves* or *hyperspherical* coordinates, which will be discussed in the following subsection.

6.6.2 *Hyperspherical coordinates*

Considering reactive collisions between three atoms, we have to make allowance for shifting alliances in the course of the interaction, associated with deviating reactant and product configurations. We focus now on the question of a single scattering coordinate that does justice to both the initial and the final arrangement. For this purpose, we refer to the transformation (6.95) that maps the Jacobi vectors characteristic of two different configurations into each other. This transformation adopts a more symmetric appearance when the coordinates (6.97) are used, namely

$$\begin{pmatrix} \tilde{\mathbf{r}}' \\ \tilde{\boldsymbol{\rho}}' \end{pmatrix} = \begin{pmatrix} \cos\alpha_{\mathrm{AC}} & \sin\alpha_{\mathrm{AC}} \\ \sin\alpha_{\mathrm{AC}} & -\cos\alpha_{\mathrm{AC}} \end{pmatrix} \begin{pmatrix} \tilde{\mathbf{r}} \\ \tilde{\boldsymbol{\rho}} \end{pmatrix}, \tag{6.99}$$

where

$$\cos\alpha_{\mathrm{AC}} = -\left(\frac{m_{\mathrm{A}}m_{\mathrm{C}}}{(m_{\mathrm{A}}+m_{\mathrm{B}})(m_{\mathrm{B}}+m_{\mathrm{C}})}\right)^{\frac{1}{2}}, \tag{6.100a}$$

$$\sin\alpha_{\mathrm{AC}} = -\left(\frac{m_{\mathrm{B}}M}{(m_{\mathrm{A}}+m_{\mathrm{B}})(m_{\mathrm{B}}+m_{\mathrm{C}})}\right)^{\frac{1}{2}}. \tag{6.100b}$$

The nomenclature chosen for the transformation angle, α_{AC}, makes reference to the reactant and the product partitioning as shown in Figures 6.1(a) and 6.1(b) which are connected by the mapping (6.99). In the initial state, A is the monatomic species, in the final state C. As a significant implication of Eq. (6.99), the use of mass-scaled Jacobi coordinates leads to a quantity that remains invariant with respect to the rearrangement channel. Thus, it holds that [159, 160]

$$\rho_H^2 \equiv \tilde{\rho}^2 + \tilde{r}^2 = \tilde{\rho}'^2 + \tilde{r}'^2. \tag{6.101}$$

This relation defines the *hyperspherical radius* or *hyperradius*. The important feature of this parameter is that it represents a unified scattering coordinate which does not depend on the reaction channel. The variables that correspond to the $AC + B$ subdivision of the three atoms (not shown in Figure 6.1) combine in an analogous manner to yield the invariant ρ_H^2. The corresponding *hyperangle* is given by

$$\theta_H = \tan^{-1}\left(\frac{\tilde{r}_\gamma}{\tilde{\rho}_\gamma}\right). \tag{6.102}$$

While ρ_H is a universal parameter, θ_H is defined with reference to a specific channel γ. This implies that θ_H, referring to Figure 6.1(a), deviates from the respective quantity θ'_H that describes the alternative frame shown in Figure 6.1b.

Exercise 6.3. Show that Eqs. (6.100a) and (6.100b) are valid. Prove that the matrix in Eq. (6.99) is the equivalent of the transformation (6.95) for mass-scaled coordinates. Verify Eq. (6.101).

The variables ρ_H and θ_H are known as *Delves* or *hyperspherical coordinates*. Expressing the Hamiltonian (6.92) in terms of this coordinate pair, one obtains [159, 160]

$$\hat{H}_{\mathrm{Del}} = -\frac{\hbar^2}{2\mu\rho_H^5}\frac{\partial}{\partial\rho_H}\rho_H^5\frac{\partial}{\partial\rho_H} + \frac{1}{2\mu\rho_H^2}\hat{\Lambda}^2 + V(\rho_H, \theta_H, \Theta), \tag{6.103}$$

where $\hat{\Lambda}^2$ stands for the square of *Smith's grand angular momentum operator* [161, 162]:

$$\hat{\Lambda}^2 = -\frac{\hbar^2}{\sin^2 2\theta_H}\frac{\partial}{\partial\theta_H}\sin^2 2\theta_H + \frac{(\hat{\mathbf{J}} - \hat{\mathbf{j}}_\gamma)^2}{\cos^2\theta_H} + \frac{\hat{\mathbf{j}}_\gamma^2}{\sin^2\theta_H}, \tag{6.104}$$

with $\hat{\mathbf{J}}$ as the total angular momentum of the system and $\hat{\mathbf{j}}_\gamma$ as the angular momentum of the diatom in the reference configuration γ. The Hamiltonian (6.103) involves the distinction between a channel-independent radial and a channel-specific angular part. The hyperradius ρ_H represents a universal scattering coordinate that is defined in complete analogy to the radial coordinate ρ of inelastic scattering theory.

Besides the hyperangle θ_H, four additional angles are needed to specify the selected channel completely. One of these is the angle Θ between the vectors $\boldsymbol{\rho}_\gamma$ and $\mathbf{r}_\gamma$ (see Figure 6.1(a) and Eq. (6.91c)), given by

$$\Theta = \cos^{-1}\left(\frac{\boldsymbol{\rho}_\gamma \cdot \mathbf{r}_\gamma}{\rho_\gamma r_\gamma}\right), \tag{6.105}$$

which is identical in the mass-scaled and the unscaled Jacobi coordinate systems. Any one of the different asymptotic arrangements is accessible as the angles $\theta_{\rm H}$ and Θ, in the course of the triatomic interaction, sweep through their domains $0 \le \theta_{\rm H} \le \frac{\pi}{2}$ and $0 \le \Theta \le \pi$. Further, the three Euler angles are used to define the spatial orientation of the triatomic arrangement as a whole. The asymptotic configurations on both the reactant and the product side are characterized by the condition $\rho_{\rm H} = \infty$ and distinguished by appropriate choices of the angular variables. We point out that other assignments of the internal angles $\theta_{\rm H}$ and Θ have been suggested [163] and are utilized by some authors.

Expansion schemes equivalent to the adiabatic representation have been employed to solve the rearrangement problem in terms of hypercoordinates [164]. Thus, one introduces an adiabatic Hamiltonian $\hat{H}^{\rm ad}$ as the sum over the angle-dependent terms of the overall Hamiltonian (6.103) and specifies adiabatic eigenfunctions $\Phi_{\gamma I}(\rho_{\rm H}; \theta_{\rm H}, \Theta)$ where the subscripts γ, I refer to the channel index and the magnitude of the total angular momentum ($\mathbf{J}$) projection on the body-fixed z-axis, respectively. The solution is expanded into a basis provided by these functions.

The hyperspherical coordinate approach to the reactive scattering problem is far from simple. This is not just to be ascribed to the complexity of the hyperspherical Schrödinger equation based on the Hamiltonian (6.103), but also to the difficulty of generating the appropriate adiabatic basis functions $\Phi_{\gamma I}$ which are non-local bound-state wave functions that encompass both the reactant and the product configurations. Hence, viable alternatives to this differential equation approach are of substantial interest. In Section 6.7, we discuss an integral equation formalism based on the Green's operator [7].

6.7 Space-Fixed and Body-Fixed Frames of Reference

Jacobi coordinates as discussed so far are compatible with various frames for the description of triatomic interaction. The most fundamental distinction concerns the difference between the space-fixed and the body-fixed representation, where the former corresponds to the usual assumption of a coordinate system with rigid axes which define the components of the atomic position vectors, and the latter makes allowance for a variable frame of reference that rotates with the considered molecule. In the following, we will introduce both models, emphasizing the advantages of the body-fixed alternative which will be shown to provide a natural basis for the inclusion of electronic degrees of freedom into the analysis of the three-center problem.

For the treatment of the general multiparticle system, we distinguish between three basic reference frames, namely the laboratory, the space-fixed, and the body-fixed systems. The first of these choices deals with the problem in its full $3N$ dimensionality, where N stands for the number of particles. The second excludes the three center-of-mass coordinates, reducing the number of dimensions to $3N - 3$. The same holds for the body-fixed frame which, in addition, introduces a separation between vibrational and rotational motion. Comparing the space-fixed and the body-fixed frames, we will refer, for simplicity, to the three-center problem, as illustrated by Figure 6.1.

6.7.1 *Space-fixed representation*

Without specifying any particular arrangement, we may assign to the coordinates $\tilde{\boldsymbol{\rho}}$ and $\tilde{\mathbf{r}}$ the angle pairs $(\theta_\rho, \varphi_\rho)$ and (θ_r, φ_r), respectively. The angular momenta associated with these two position vectors, related to the translational motion of the single atom, $\mathbf{L}$, and to the vibrational motion of the diatom, $\mathbf{j}$, are then given by

$$\hat{L}^2 = -\hbar^2 \left(\frac{1}{\sin\theta_\rho} \frac{\partial}{\partial\theta_\rho} \sin\theta_\rho \frac{\partial}{\partial\theta_\rho} + \frac{1}{\sin\theta_\rho^2} \frac{\partial^2}{\partial\varphi_\rho^2} \right) \tag{6.106}$$

and

$$\hat{j}^2 = -\hbar^2 \left(\frac{1}{\sin\theta_r} \frac{\partial}{\partial\theta_r} \sin\theta_r \frac{\partial}{\partial\theta_r} + \frac{1}{\sin\theta_r^2} \frac{\partial^2}{\partial\varphi_r^2} \right). \tag{6.107}$$

The angle Θ that connects the translational and the vibrational coordinate vectors is expressed in terms of the four angles as follows:

$$\Theta = \cos^{-1} (\cos\theta_\rho \cos\theta_r + \sin\theta_\rho \sin\theta_r \cos(\varphi_\rho - \varphi_r)). \tag{6.108}$$

The angles $(\theta_\rho, \varphi_\rho, \theta_r, \varphi_r)$ and, correspondingly, the angular momenta $\hat{\mathbf{L}}$ and $\hat{\mathbf{j}}$ refer to a coordinate system fixed in space. The total angular momentum is

$$\hat{\mathbf{J}} = \hat{\mathbf{L}} + \hat{\mathbf{j}}. \tag{6.109}$$

Eigenfunctions of $\hat{\mathcal{J}}^2$ and $\hat{\mathcal{J}}_z$ are readily obtained by combining those of $(\hat{L}^2, \hat{L}_z)$ and $(\hat{j}^2, \hat{j}_z)$ according to the well-known formula

$$\tilde{\Gamma}_{Lj}^{\mathcal{J}M_\mathcal{J}}(\theta_\rho, \varphi_\rho, \theta_r, \varphi_r) = \sum_{m_L, m_j} c(jm_j, Lm_L|\mathcal{J}M)\, Y_{jm_j}(\theta_r, \varphi_r)\, Y_{Lm_L}(\theta_\rho, \varphi_\rho), \tag{6.110}$$

with the factors $c(Lm_L, jm_j|\mathcal{J}M_\mathcal{J})$ as Clebsch–Gordan coefficients whose symmetries constrain the quantum number m_j to the value $M_\mathcal{J} - m_L$. The indices $\mathcal{J}$ and $M_\mathcal{J}$ are good quantum numbers and are thus used to designate the eigenfunctions of the operators $\hat{H}, \hat{\mathcal{J}}^2$, and $\hat{\mathcal{J}}_z$ [165]:

$$\Phi^{\mathcal{J}M_\mathcal{J}}(\widetilde{\boldsymbol{\rho}}, \widetilde{\mathbf{r}}) = \Phi^{\mathcal{J}M_\mathcal{J}}(\widetilde{\rho}, \widetilde{r}, \omega_\rho, \omega_r) = \sum_{L,j} G_{Lj}^{\mathcal{J}M_\mathcal{J}}(\widetilde{\rho}, \widetilde{r})\tilde{\Gamma}_{Lj}^{\mathcal{J}M_\mathcal{J}}(\omega_\rho, \omega_r), \tag{6.111}$$

where ω_ρ and ω_r label the pairs of angular coordinates $(\theta_\rho, \varphi_\rho)$ and (θ_r, φ_r), respectively. The radius-dependent function $G_{Lj}^{\mathcal{J}M_\mathcal{J}}(\widetilde{\rho}, \widetilde{r})$ is to be derived from the stationary Schrödinger equation, as detailed later. The general solution Φ is composed of the eigenfunctions $\Phi^{\mathcal{J}M_\mathcal{J}}$ according to

$$\Phi(\widetilde{\boldsymbol{\rho}}, \widetilde{\mathbf{r}}) = \sum_{\mathcal{J}=0} \sum_{M_\mathcal{J}=-\mathcal{J}}^{\mathcal{J}} d^{\mathcal{J}M_\mathcal{J}}\, \Phi^{\mathcal{J}M_\mathcal{J}}(\widetilde{\boldsymbol{\rho}}, \widetilde{\mathbf{r}}) \tag{6.112}$$

with expansion coefficients $d^{\mathcal{J}M_\mathcal{J}}$. Inserting this function into the stationary Schrödinger equation with the Hamiltonian

$$\hat{H} = \hat{T} + V(\widetilde{\rho}, \widetilde{r}, \Theta), \tag{6.113}$$

involving the kinetic energy operator given by Eq. (6.98), we arrive at the radial equation

$$\left(-\frac{\hbar^2}{2\mu}\left[\frac{1}{\widetilde{\rho}}\frac{\partial^2}{\partial\widetilde{\rho}^2}\widetilde{\rho} + \frac{1}{\widetilde{r}}\frac{\partial^2}{\partial\widetilde{r}^2}\widetilde{r} - \frac{j(j+1)}{\widetilde{r}^2} - \frac{L(L+1)}{\widetilde{\rho}^2}\right] - E\right) G_{Lj}^{\mathcal{J}M_\mathcal{J}}(\widetilde{\rho}, \widetilde{r})$$
$$= -\sum_{L',j'} \langle Lj|V(\widetilde{\rho}, \widetilde{r})|L'j'\rangle G_{L'j'}^{\mathcal{J}M_\mathcal{J}}(\widetilde{\rho}, \widetilde{r}). \tag{6.114}$$

The scalar product on the right-hand side is defined as

$$\langle Lj|V(\widetilde{\boldsymbol{\rho}},\widetilde{\mathbf{r}})|L'j'\rangle = \int d\omega_\rho d\omega_r \widetilde{\Gamma}_{Lj}^{\mathcal{J}M_\mathcal{J}}(\omega_\rho,\omega_r)V(\widetilde{\rho},\widetilde{r})\widetilde{\Gamma}_{L'j'}^{\mathcal{J}M_\mathcal{J}}(\omega_\rho,\omega_r), \tag{6.115}$$

involving the angular elements $d\omega = d(\cos\theta)d\varphi$. We recognize in Eq. (6.114) a generalization of the diabatic equation system (6.7) from one scattering coordinate ρ to two such coordinates $\widetilde{\rho},\widetilde{r}$. In the context of atom-molecule scattering involving a well-defined scattering potential V, the role of the channel functions $|\phi_i\rangle$ is played by the rotational states $|Lj\rangle$ which span the basis for the expansion of the coupling matrix. Imposing the respective boundary conditions on the solution of Eq. (6.114), one derives from it the elements of the T- or the K-matrix [92] that yield the probabilities for scattering from channel (L,j) to channel (L',j'). The expressions resulting from this approach for the corresponding differential cross sections are quite complex [166]. The formalism, however, can be simplified if the body-fixed system is adopted as a frame of reference. A brief outline of this representation is given in the following.

6.7.2 *Body-fixed representation*

In keeping with the conventional choice of the quantization axis in the body-fixed system (X', Y', Z'), we let the Z'-axis coincide with the vector $\boldsymbol{\rho}$, connecting the atom to the center of mass of the diatom, as shown in Figure 6.2.

While the eigenfunctions of the operator set $(\hat{\mathcal{J}}^2, \hat{\mathcal{J}}_Z, \hat{j}^2, \hat{L}^2)$ are the basis functions in the space-fixed frame, the corresponding set for the body-fixed frame is $(\hat{\mathcal{J}}^2, \hat{\mathcal{J}}_Z, \hat{j}^2, \widetilde{\hat{\mathcal{J}}}_{Z'})$. In both representations, the total angular momentum $\mathcal{J}$ and its projection on the quantization Z-axis, $\mathcal{J}_Z$, are good quantum numbers. As the Z'-axis is oriented along $\boldsymbol{\rho}$, the $\hat{\mathcal{J}}_{Z'}$ component is identical with $\hat{j}_{Z'}$ since the projection of $\mathbf{L}$ on $\boldsymbol{\rho}$ vanishes. Consequently, m_L reduces to zero if the body-fixed representation is adopted, reducing considerably the quantum number bureaucracy involved in the space-fixed description [92]. The angular momentum eigenfunction $\Gamma_{Lj'}^{\mathcal{J}M_\mathcal{J}}(\omega_\rho,\omega_r)$ in the body-fixed frame is related to its space-fixed counterpart — see Eq. (6.110) — by rotation

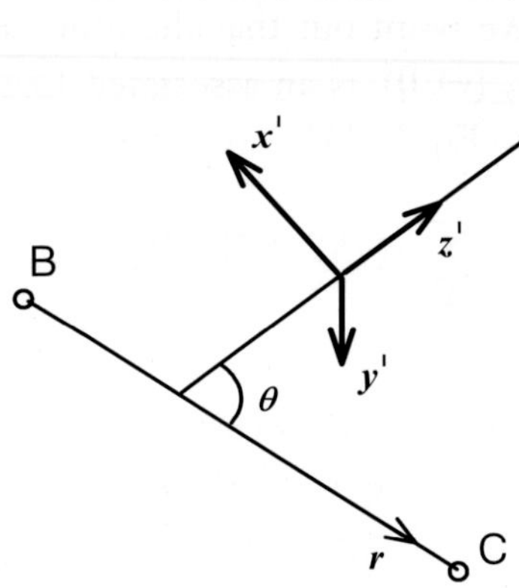

Figure 6.2: The $(X'Y'Z')$ axes of the body-fixed system.

through the Euler angles φ, θ, and χ:

$$\Gamma_{Lj}^{\mathcal{J}M_{\mathcal{J}}}(\omega_\rho, \omega_r) = \sum_K D_{KM_{\mathcal{J}}}^{\mathcal{J}}(\varphi, \theta, \chi)\tilde{\Gamma}_{Lj}^{\mathcal{J}K}(\omega_\rho, \omega_r), \tag{6.116}$$

where $D_{KM_{\mathcal{J}}}^{\mathcal{J}}$ is a rotation matrix element and the label K stands for the quantum number associated with $\mathcal{J}_{Z'}$. The components $\tilde{\Gamma}_{Lj}^{\mathcal{J}K}$ are to be understood as unrotated wave functions in the body-fixed frame. With the geometry chosen as shown in Figure 6.2, the Euler angles θ and φ are identical with θ_ρ and φ_ρ. Since $\boldsymbol{\rho}$ is oriented along the Z'-axis, it holds that $\theta_\rho = \varphi_\rho = 0$ for the unrotated functions. The angle Θ (see Eq. (6.108)) may be understood as the polar angle associated with the diatom vector $\mathbf{r}$, thus replacing the space-fixed quantity θ_r; likewise, the Euler angle χ adopts the role of the azimuthal angle φ_r in the space-fixed representation. With $\omega_\rho = (0,0)$ and $\omega_r = (\Theta, 0)$, we find from Eq. (6.110) the following expression for the unrotated functions:

$$\begin{aligned}\tilde{\Gamma}_{Lj}^{\mathcal{J}K}(\omega_\rho, \omega_r) &= \tilde{\Gamma}_{Lj}^{\mathcal{J}K}(0,0;\Theta,0)\\ &= \sum_{m_j} c(jm_jLK - m_j|\mathcal{J}K)\, Y_{jm_j}(\Theta,0)\, Y_{LK-m_j}(0,0)\\ &= \sqrt{\frac{2L+1}{4\pi}}\, c(jKL0|\mathcal{J}K)\, Y_{jK}(\Theta,0).\end{aligned} \tag{6.117}$$

The factor $Y_{LK-m_j}(0,0)$ which vanishes unless $K = m_j$ makes the sum collapse to a single term, greatly simplifying the structure of the body-fixed unrotated wave function. We point out that the other spherical harmonic function in Eq. (6.117), $Y_{jm_j}(\Theta, 0)$, is an associated Legendre polynomial. Substituting Eq. (6.117) into Eq. (6.116), we obtain

$$\Gamma_{Lj}^{\mathcal{J}M_\mathcal{J}}(\omega_\rho, \omega_r) = \sqrt{\frac{2L+1}{4\pi}} \sum_K c(jKL0|\mathcal{J}K) D_{KM_\mathcal{J}}^{\mathcal{J}}(\varphi, \theta, \chi) Y_{jK}(\Theta, 0). \tag{6.118}$$

This identity is recast into the compact form [7]

$$\Gamma_{Lj}^{\mathcal{J}M_\mathcal{J}} = \sum_K C_{LK} Y_{jK}^{\mathcal{J}M_\mathcal{J}}, \tag{6.119}$$

where we have introduced coefficients C_{LK} defined by

$$C_{LK} = \sqrt{\frac{2L+1}{2\mathcal{J}+1}} c(jKL0|\mathcal{J}K), \tag{6.120}$$

while the functions $Y_{jK}^{\mathcal{J}M_\mathcal{J}}$ are composites of the rotation matrix elements and the normalized associated Legendre polynomials, $P_{l|m|}(\cos\Theta) = \sqrt{2\pi}\, Y_{l|m|}(\Theta, 0)$, and may be written in the following way:

$$Y_{jK}^{\mathcal{J}M_\mathcal{J}} = \sqrt{\frac{2\mathcal{J}+1}{8\pi^2}} D_{KM_\mathcal{J}}^{\mathcal{J}} \sqrt{2\pi}\, Y_{jK}(\Theta, 0). \tag{6.121}$$

Finally, inversion of (6.118) leads to

$$Y_{jK}^{\mathcal{J}M_\mathcal{J}} = \sum_L C_{LK} \Gamma_{Lj}^{\mathcal{J}M_\mathcal{J}}. \tag{6.122}$$

This construction provides us with angular momentum functions that depend on the good quantum numbers of the body-fixed representation. The counterpart of (6.111) for this representation may then be written as

$$\Psi^{\mathcal{J}M_\mathcal{J}}(\boldsymbol{\rho}, \mathbf{r}) = \sum_{j,K} F_{jK}^{\mathcal{J}M_\mathcal{J}}(\rho, r) Y_{jK}^{\mathcal{J}M_\mathcal{J}}(\varphi, \theta, \chi, \Theta), \tag{6.123}$$

where, for clarity of notation, $\widetilde{\rho}$ and $\widetilde{r}$ have been changed into ρ and r, but mass-scaled radii are implied. The radial solution $F_{jK}^{\mathcal{J}M_\mathcal{J}}$ is yet to be determined (see Eq. (6.133)). The summation over K is restrained to the range $-\mathcal{J}$ to $+\mathcal{J}$, that over j to $j \geq |K|$, as one concludes from the appearance of the spherical harmonic factor Y_{jK} in Eq. (6.121). For completeness, we point out that parity-adapted basis functions may be generated by substituting for $Y_{jK}^{\mathcal{J}M_\mathcal{J}}$ linear combinations of $Y_{jK}^{\mathcal{J}M_\mathcal{J}}$ and $Y_{j-K}^{\mathcal{J}M_\mathcal{J}}$.

Since in the body-fixed frame, L loses the status of a good quantum number, the respective Hamiltonian does not contain the operator $\hat{L}^2$ explicitly any longer. It is to be expressed through the operators $\hat{\mathbf{J}}$ and $\hat{\mathbf{j}}$, acknowledging that the new basis states $Y_{jK}^{JM_J}$ are eigenfunctions of both $(\tilde{\hat{J}}^2, \tilde{\hat{J}}_Z)$ and $(\hat{j}^2, \hat{J}_{Z'})$. To distinguish the two operator groups, here and in the following we let the tilde symbol refer to an operator in the space-fixed representation. By exploiting the representation invariance of angular momentum operator squares, we find the body-fixed equivalent of the $\hat{L}^2$-dependent part in the space-fixed Hamiltonian. Thus,

$$\hat{L}^2 = \tilde{\hat{L}}^2 = (\hat{\mathbf{J}} - \hat{\mathbf{j}})^2 = \hat{J}^2 + \hat{j}^2 - \hat{J}_- \hat{j}_+ - \hat{J}_+ \hat{j}_- - 2\hat{J}_{Z'} \hat{j}_{Z'}, \tag{6.124}$$

which involves the angular momentum raising and lowering operators $\hat{J}_+ (\hat{j}_+)$ and $\hat{J}_- (\hat{j}_-)$, respectively. These are related to the X' and Y' components of the operators in the following way:

$$\hat{J}_\pm = \hat{J}_{X'} \pm i\hat{J}_{Y'}; \quad \hat{j}_\pm = \hat{j}_{X'} \pm i\hat{j}_{Y'}. \tag{6.125}$$

We thus expand the Hamiltonian (6.113) as

$$\begin{aligned}\hat{H} = &-\frac{\hbar^2}{2\mu}\left[\frac{1}{\rho}\frac{\partial^2}{\partial\rho^2}\rho + \frac{1}{r}\frac{\partial^2}{\partial r^2}r\right] + \frac{1}{2\mu}\left(\frac{1}{\rho^2} + \frac{1}{r^2}\right)\hat{j}^2 \\ &+ \frac{1}{2\mu\rho^2}(\hat{J}^2 - \hat{J}_- \hat{j}_+ - \hat{J}_+ \hat{j}_- - 2\hat{J}_{Z'} \hat{j}_{Z'}) + V(\rho, r, \Theta),\end{aligned} \tag{6.126}$$

with $\hat{J}_{Z'} = \hat{j}_{Z'} = -i\hbar\frac{\delta}{\delta\chi}$. The body-fixed representations of $\hat{J}^2, \hat{j}^2$, and $\hat{J}_\pm (\hat{j}_\pm)$ in terms of angular differential operators may be found, for instance, in [165]. To understand the action of the ladder operators on the basis functions Eq. (6.123), it is essential to realize that the body-fixed components $\hat{J}_{X'}, \hat{J}_{Y'}$, and $\hat{J}_{Z'}$ obey reversed commutation relations [167], namely

$$[\hat{J}_{P'}, \hat{J}_{Q'}] = \hat{J}_{P'} \hat{J}_{Q'} - \hat{J}_{Q'} \hat{J}_{P'} = -i\hbar \sum_{R'} \varepsilon_{P'Q'R'} \hat{J}_{R'}, \tag{6.127}$$

where each of the labels (P', Q', R') runs through the set of coordinates (X', Y', Z'), and $\varepsilon_{P'Q'R'}$ stands for the Levi-Civita tensor. The commutation behavior of the components of $\hat{\mathbf{J}}$ are in opposition to that of $\hat{j}_{X'}, \hat{j}_{Y'}$, and $\hat{j}_{Z'}$

which commute in accord with regular angular momentum algebra [167]:

$$[\hat{j}_{P'}, \hat{j}_{Q'}] = \hat{j}_{P'}\hat{j}_{Q'} - \hat{j}_{Q'}\hat{j}_{P'} = i\hbar \sum_{R'} \varepsilon_{P'Q'R'}\hat{j}_{R'}. \tag{6.128}$$

The inversion observed for the total angular momentum components has the consequence that the index K is lowered (raised) by the ladder operator $\hat{\mathcal{J}}_{+}$ ($\hat{\mathcal{J}}_{-}$). Specifically, one finds

$$\hat{\mathcal{J}}_{\pm} D^{\mathcal{J}}_{KM_{\mathcal{J}}} = \sqrt{\mathcal{J}(\mathcal{J}+1) - K(K \mp 1)} D^{\mathcal{J}}_{K\mp 1 M_{\mathcal{J}}} \equiv \lambda^{\mp}_{\mathcal{J}K} D^{\mathcal{J}}_{K\mp 1 M_{\mathcal{J}}}. \tag{6.129}$$

To derive a differential equation for the radial part of Eq. (6.123), we use the orthonormality of the $Y^{\mathcal{J}M_{\mathcal{J}}}_{jK}$ functions. Substituting Eq. (6.123) into the stationary Schrödinger equation based on the Hamiltonian (6.126), and contracting from the left by multiplication with $Y^{\mathcal{J}M_{\mathcal{J}}}_{j'K'}$, which amounts to integration over the Euler angles as well as Θ, one obtains [165]

$$\begin{aligned}(\hat{t}^{\mathcal{J}j}_{KK} - E)F^{\mathcal{J}M_{\mathcal{J}}}_{jK}(\rho, r) &+ \hat{t}^{\mathcal{J}j}_{K,K+1}F^{\mathcal{J}M_{\mathcal{J}}}_{jK+1}(\rho, r) + \hat{t}^{\mathcal{J}j}_{K,K-1}F^{\mathcal{J}M_{\mathcal{J}}}_{jK-1}(\rho, r) \\ &+ \sum_{j'>K} V^{K}_{jj'} F^{\mathcal{J}M_{\mathcal{J}}}_{j'K}(\rho, r) = 0. \end{aligned} \tag{6.130}$$

Here it holds that

$$\begin{aligned}\hat{t}^{\mathcal{J}j}_{KK} \equiv& -\frac{\hbar^2}{2\mu}\left[\frac{1}{\rho}\frac{\delta^2}{\delta\rho^2}\rho + \frac{1}{r}\frac{\delta^2}{\delta r^2}r\right] \\ &+ \frac{\hbar^2}{2\mu}\left[\frac{j(j+1)}{r^2} + \frac{1}{\rho^2}(\mathcal{J}(\mathcal{J}+1) + j(j+1) - 2K^2)\right], \end{aligned} \tag{6.131a}$$

as well as

$$\hat{t}^{\mathcal{J}j}_{KK\pm 1} \equiv -\frac{\hbar^2}{2\mu\rho^2}(\lambda^{\pm}_{\mathcal{J}K}\lambda^{\pm}_{jK})^{\frac{1}{2}}. \tag{6.131b}$$

The matrix elements $V^{K}_{jj'}$ are

$$\begin{aligned}V^{K}_{jj'} &= \langle jK|V(\rho, r, \Theta)|j'K\rangle \\ &= 2\pi \int d\cos(\Theta)\, Y_{jK}(\Theta, 0) V(\rho, r, \Theta) Y_{j'K}(\Theta, 0), \end{aligned} \tag{6.132}$$

while the terms $\lambda^{\pm}_{\mathcal{J}K}$, $\lambda^{\pm}_{jK}$ are defined implicitly by Eq. (6.129). Scattering probabilities are derived from this approach by solving Eq. (6.130) for the translational degree of freedom to obtain radial functions that depend on the variable ρ. For this purpose, we represent the radial solutions $F_{jK}^{\mathcal{J}M_\mathcal{J}}(\rho, r)$ in the following manner:

$$F_{jK}^{\mathcal{J}M_\mathcal{J}}(\rho, r) = \frac{1}{\rho r} \sum_v u_{vjK}^{\mathcal{J}M_\mathcal{J}}(\rho)\phi_{vj}(r). \tag{6.133}$$

In this expression, the scattering wave functions $u_{v_jK}^{\mathcal{J}M_\mathcal{J}}$ play the role of the coefficients for expanding $F_{jK}^{\mathcal{J}M_\mathcal{J}}$ into the set of diatomic vibration basis states ϕ_{vj} which satisfy

$$\left[-\frac{\hbar^2}{2\mu}\frac{\partial^2}{\partial r^2} + \frac{\hbar^2}{2\mu r^2} j(j+1) + v(r) - \varepsilon_{vj}\right]\phi_{vj}(r) = 0, \tag{6.134}$$

if $v(r)$ is the asymptotic diatomic potential energy function, corresponding to $\rho \to \infty$, and ε_{vj} are the vibrational eigenvalues. Inserting Eq. (6.133) into Eq. (6.130) and applying closure with respect to the basis ϕ_{vj} establishes a relation between $u_{vjK}^{\mathcal{J}M_\mathcal{J}}$, $u_{vjK-1}^{\mathcal{J}M_\mathcal{J}}$, and $u_{vjK+1}^{\mathcal{J}M_\mathcal{J}}$ (see Exercise 6.4) involving the K-matrix elements K_{if} for the considered reactive process.

Exercise 6.4. *Asymptotic equation system for $u_{vjK}^{\mathcal{J}M_\mathcal{J}}(\rho)$.*

Show by substituting the series given by Eq. (6.133) for $F_{jK}^{\mathcal{J}M_\mathcal{J}}$ into Eq. (6.130) the following: for $V = 0$, it holds that

$$\left[\frac{\partial^2}{\partial \rho^2} - \frac{1}{\rho^2}(\mathcal{J}(\mathcal{J}+1) + j(j+1) - 2K^2) + k_{vj}^2\right] u_{v_jK}^{\mathcal{J}M_\mathcal{J}}(r)$$

$$-\frac{1}{\rho^2}\left(\hat{x}_{KK+1}^{\mathcal{J}j} u_{v_jK+1}^{\mathcal{J}M_\mathcal{J}}(\rho) + \hat{x}_{KK-1}^{\mathcal{J}j} u_{jK-1}^{\mathcal{J}M_\mathcal{J}}(\rho)\right) = 0,$$

with $k_{vj}^2 = \frac{2\mu}{\hbar^2}(E - \varepsilon_{vj})$ and $\hat{x}_{KK\pm 1}^{\mathcal{J}j} \equiv \frac{2\mu}{\hbar^2}\rho^2 \hat{t}_{KK\pm 1}^{\mathcal{J}j}$.

The resulting system of coupled equations looks less intimidating in the regime where ρ is large enough that $\frac{1}{\rho^2} \to 0$. The corresponding translational solution may be written as

$$\lim_{\rho\to\infty} u_{vjK}^{\mathcal{J}M_\mathcal{J}}(\rho) = \frac{1}{\sqrt{k_{vj}}}\left[\sin\left(k_{vj}\rho - \frac{(\mathcal{J}+j)\pi}{2}\right)\delta_{vv_i}\delta_{jj_i}\delta_{KK_i}\right.$$

$$\left. + K_{if}\cos\left(k_{vj}\rho - \frac{(\mathcal{J}+j)\pi}{2}\right)\right], \tag{6.135}$$

involving the K-matrix element K_{if}. The indices i and f refer here to an initial and a final state characterized by quantum numbers v_i, j_i, K_i and

v, j, K, respectively. Translating the K-matrix into the T- or the S-matrix (by following the rules (6.33) and (6.34)), and expressing the latter in the helical or $|\mathcal{J}K\rangle$ representation, we derive the scattering amplitudes for the transition from an initial $|\mathcal{J}K_i\rangle$ to the final $|\mathcal{J}K\rangle$ state. Quite compact formulae result for the differential and the total cross section describing this process [91].

6.8 Reactive Scattering II: The Integral Equation Approach

As an integral alternative to the differential equation approach outlined in the previous subsections, we extend the Lippmann–Schwinger scheme to rearrangement situations. Continuing to refer to the simplest possible example of this class of problems, namely atom–diatom interactions, we label the three possible reactant and product configurations by the index γ. The Hamiltonian of the system may thus be partitioned in the following way:

$$\hat{H} = \hat{H}_\gamma + V_\gamma, \tag{6.136}$$

with $\hat{H}_\gamma$ as the asymptotic Hamiltonian for the rearrangement channel γ and V_γ as the atom–diatom interaction potential for this channel. Using as scattering coordinate a vector $\boldsymbol{\rho}_\gamma$ that connects the atom position with the diatom center of mass, we associate $\hat{H}_\gamma$ with the limit of $\hat{H}$ for large values of ρ_γ: $\lim_{\rho_\gamma \to \infty} \hat{H} = \hat{H}_\gamma$. We focus on the initial configuration, letting the index i denote a set of quantum numbers characteristic of the entrance channel, including the momentum of the two subsystems relative to each other. The Lippmann–Schwinger equation then reads, according to Eq. (3.51),

$$\Psi^+(\gamma, i) = \varphi(\gamma, i) + \hat{G}_\gamma^+ V_\gamma \Psi^+(\gamma, i), \tag{6.137}$$

with $\varphi(\gamma, i)$ as the asymptotic wave function, and with the Green's operator given by the resolvent

$$\hat{G}_\gamma^+ = \frac{1}{E - \hat{H}_\gamma + i\varepsilon}. \tag{6.138}$$

We may reformulate the statement (6.138) in terms of the full Green's operator

$$\hat{G}^+ = \frac{1}{E - \hat{H} + i\varepsilon}, \tag{6.139}$$

and write

$$\hat{G}^+ = \hat{G}^+_\gamma + \hat{G}^+_\gamma V_\gamma \hat{G}^+. \tag{6.140}$$

From the identity

$$(E - \hat{H} + i\varepsilon) = (E - \hat{H}_\gamma + i\varepsilon - V_\gamma), \tag{6.141}$$

we infer that Eq. (6.140) is equivalent to

$$\hat{G}^+ = \hat{G}^+_\gamma + \hat{G}^+ V_\gamma \hat{G}^+_\gamma. \tag{6.142}$$

Exercise 6.5. Show the equivalence of Eqs. (6.140) and (6.142).

This operator equation translates into the following relation between the wave functions Ψ^+ and φ [165]:

$$\Psi^+(\gamma, i) = \varphi(\gamma, i) + \hat{G}^+ V_\gamma \varphi(\gamma, i). \tag{6.143}$$

Since Eq. (6.141) is valid for any other partition β as well, the Green's operator might also be written as $\hat{G}^+ = \hat{G}^+_\beta + \hat{G}^+_\beta V_\beta \hat{G}^+$. Inserting this form of $\hat{G}^+$ into Eq. (6.143), we arrive at

$$\Psi^+(\gamma, i) = \varphi(\gamma, i) + \hat{G}^+_\beta (V_\gamma + V_\beta \hat{G}^+ V_\gamma) \varphi(\gamma, i). \tag{6.144}$$

This line makes a statement about the transition from an arrangement γ to a rivaling arrangement β. The term $V_\gamma + V_\beta \hat{G}^+ V_\gamma$ which operates on the asymptotic wave function for channel γ has a simple interpretation, as is evident from the relation

$$\hat{T}|\varphi(\gamma, i)\rangle = V_\gamma |\Psi^+(\gamma, i)\rangle \tag{6.145}$$

combined with the Lippmann–Schwinger equation given by Eq. (6.137). Substituting for $V_\gamma |\Psi^+(\gamma, i)\rangle$ in the latter, we find that Eq. (6.144) involves a generalized, i.e. a reactive transition operator that mixes the partitions γ and β:

$$\hat{T}_{\beta\gamma} \equiv V_\gamma + V_\beta \hat{G}^+ V_\gamma. \tag{6.146}$$

The operator $\hat{T}_{\beta\gamma}$ is equivalent to $\hat{T}'_{\beta\gamma} \equiv V_\beta + V_\beta \hat{G}^+ V_\gamma$ in the sense that both give rise to the same transition matrix elements, i.e. $\langle\varphi(\beta, j)|\hat{T}_{\beta\gamma}|\varphi(\gamma, i)\rangle = \langle\varphi(\beta, j)|\hat{T}'_{\beta\gamma}|\varphi(\gamma, i)\rangle$. For demonstration of this identity, it is sufficient to realize that $\langle\varphi(\beta, j)|V_\gamma - V_\beta|\varphi(\gamma, i)\rangle = \langle\varphi(\beta, j)|\hat{H}_\gamma + V_\gamma - \hat{H}_\beta - V_\beta|\varphi(\gamma, i)\rangle$. Since the latter scalar product involves the same Hamiltonian $\hat{H}$

in two alternative representations, it vanishes. This establishes the equality of the transition matrix elements that contain the transition operator in *prior form* ($\hat{T}_{\beta\gamma}$) and in *post form* ($\hat{T}'_{\beta\gamma}$).

Although expressions (6.137)–(6.146) are formally correct, their practical usefulness is limited. The Lippmann–Schwinger formalism as encapsulated by these relations is actually incompatible with the multichannel approach defined by Eq. (6.4). The root of this difficulty lies in the fact that Eq. (6.4) and correspondingly the matrix representation of the Green's operator (6.16) are restricted to a discrete spectrum, associated with a channel index i. As soon as we make admission for rearrangement of the interaction partners, realized, for instance, by the triatomic situation illustrated in Figure 1(a) and (b), this approach is insufficient and has to be extended by including a continuous spectrum. This is due to the dissociation of the diatomic species involved in the $A + BC \rightarrow AB + C$ reconfiguration and analogous triatomic processes. A particular complication is immediately obvious from the simple reaction scheme: any adequate description of the considered transition has not only to include the continuum of the BC system, but also that of the newly formed diatom AB. This double continuum problem gives rise to numerical instabilities [7].

In spite of these reservations, the reactive scattering boundary conditions may be correctly stated within the framework of the Lippmann–Schwinger approach. To achieve this goal, only a small adjustment of the inelastic scattering boundary conditions, as formulated in Eqs. (6.29) and (6.31), is necessary. More specifically, it has to be recognized that the three possible partitions of the triatomic problem in general correspond to three different reduced masses μ_λ, $\lambda = 1, 2, 3$. These masses are incorporated into the definition of the asymptotic wave functions by stipulating that

$$j_{ii,\lambda}(\rho_\lambda) \sim \sin\left(k_i\rho_\lambda - \frac{l\pi}{2}\right) \Big/ v_i^{\frac{1}{2}}, \tag{6.147}$$

where $v_i = \frac{\hbar k_i}{\mu_\lambda}$, with $\lambda = 1, 2, 3$. Analogous formulae hold for n_{ii} and $h_{ii}^{\pm}$. A mass normalization is used for the Wronskian, which was chosen in Eq. (6.24) to tend towards unity in the asymptotic regime. Here, in contrast, we have

$$\lim_{\rho_\lambda \rightarrow \infty} W(\mathbf{j}_\lambda, \mathbf{h}_\lambda^+) = \frac{\mu_\lambda}{\hbar}, \quad \lambda = 1, 2, 3. \tag{6.148}$$

With these modifications, the inelastic Green's function scheme of Eqs. (6.28) and (6.29) can be used for reactive scattering situations, where the asymptotic Green's function depends on a scattering coordinate ρ_λ characteristic of the rearrangement channel λ:

$$\mathbf{G}_\lambda^+(\rho_\lambda, \rho'_\lambda) \sim -\frac{2}{\hbar}\mathbf{h}_\lambda^+(\rho_\lambda)\mathbf{j}_\lambda^{\mathrm{T}}(\rho'_\lambda) \quad \text{as } \rho_\lambda \to \infty. \tag{6.149}$$

The reactive counterpart for the inelastic scattering solution given by Eq. (6.31) is then

$$\boldsymbol{\Psi}^+(\gamma, \rho_\gamma) \sim \mathbf{j}(\rho_\gamma, \gamma)\delta_{\gamma\beta} - \mathbf{h}^+(\rho_\beta, \beta)\mathbf{T}_{\beta\gamma} \quad \text{as } \rho_\gamma \to \infty, \tag{6.150}$$

with a matrix $T_{\beta\gamma}$ derived from the reactive transition operator defined by Eq. (6.146). This matrix may be written as

$$\mathbf{T}_{\beta\gamma} = \frac{2}{\hbar}\int_0^\infty d\rho'_\beta \mathbf{j}^{\mathrm{T}}(\rho'_\beta)\mathbf{V}_\beta(\rho'_\beta)\boldsymbol{\Psi}^+(\gamma, \rho'_\beta). \tag{6.151}$$

The corresponding S-matrices are generated by use of $\mathbf{S}_{\beta\gamma} = \boldsymbol{\delta}_{\beta\gamma} - 2i\mathbf{T}_{\beta\gamma}$, from which the cross-section for rearrangement scattering from channel γ into channel β is obtained.

A comprehensive integral formalism of rearrangement collisions in strict analogy to inelastic scattering can be devised by adopting an algebraic approach, based on a description of the scattering wave function in terms of differing reactant states. Specifically, we denote by Ψ_γ a superposition of channel wave functions according to

$$\Psi_\gamma = \sum_\alpha \psi_{\alpha\gamma}, \tag{6.152}$$

where the index γ stands for the initial arrangement, and α runs through all accessible final arrangements. The channel wave functions are nonorthogonal:

$$\langle \psi_{\alpha\gamma}, \psi_{\beta\gamma} \rangle \neq 0, \quad \beta \neq \alpha. \tag{6.153}$$

A parallel may be drawn between the components $\psi_{\alpha\gamma}$ and the nonorthogonal atomic orbitals of *ab initio* theory (see Section 1.1.1) which provide a basis for the delocalized molecular orbitals. In a similar sense, the scattering solution is described as a linear combination of channel wave functions.

With some basic arguments, we can deduce a Lippmann–Schwinger equation for the matrix $\boldsymbol{\psi}$ with elements $\psi_{\alpha\gamma}$. Inserting the ansatz (6.152) into the stationary Schrödinger equation leads to

$$(E - \hat{H}) \sum_{\gamma} \psi_{\alpha\gamma} = 0. \tag{6.154}$$

Using $\hat{H} = \hat{H}_{\beta} + V_{\beta}$, we recast this identity into the following form:

$$(E - \hat{H}_{\beta})\psi_{\beta\gamma} = V_{\beta}\psi_{\beta\gamma} + \sum_{\alpha \neq \beta} (\hat{H} - E)\psi_{\alpha\gamma} \equiv \sum_{\alpha} \hat{U}_{\beta\alpha}\psi_{\alpha\gamma}. \tag{6.155}$$

The matrix operator, as defined by Eq. (6.155), is given by

$$\hat{U}_{\beta\alpha} = V_{\beta}\delta_{\beta\alpha} + (\hat{H} - E)(\hat{I} - \delta_{\beta\alpha}). \tag{6.156}$$

The symmetry of the Kronecker delta implies immediately that $\mathbf{U}$ is Hermitian. From Eq. (6.155) we conclude that the outgoing channel wave functions obey the integral equation

$$\psi^{+}_{\beta\gamma} = \varphi_{\beta}\delta_{\beta\gamma} + \sum_{\alpha} \hat{G}^{+}_{\beta}\hat{U}_{\beta\alpha}\psi^{+}_{\gamma\alpha}, \tag{6.157}$$

where we have used the resolvent

$$\hat{G}^{+}_{\beta} = \frac{1}{E - \hat{H}_{\beta} + i\varepsilon}. \tag{6.158}$$

In matrix form, Eq. (6.157) may be written as

$$\boldsymbol{\psi}^{+} = \boldsymbol{\varphi} + \mathbf{G}^{+}_{\beta}\mathbf{U}\boldsymbol{\psi}^{+}, \tag{6.159}$$

with $\mathbf{G}^{+}_{\beta}$ as a diagonal matrix whose elements are defined by Eq. (6.158). The matrix $\boldsymbol{\varphi}$ is the likewise diagonal array of asymptotic solutions. The multiple arrangement problem is thus reduced to the standard form of the Lippmann–Schwinger equation for inelastic scattering, and the matrix identity (6.159) is an obvious extension of the single-arrangement approach (3.51). The potential energy matrix $\mathbf{U}$ is understood as an agent that facilitates (symmetric) coupling of the various channels present in the problem. Generalizing Eq. (6.145), one finds

$$\mathbf{T}\boldsymbol{\varphi} = \mathbf{U}\boldsymbol{\psi}^{+} \tag{6.160}$$

as the defining equation for the reactive $\hat{T}$ operator in matrix form. It is easily demonstrated that this relation yields the expected transition operator matrix elements in the sense of Eqs. (6.145) and (6.146). Using the explicit form of $\hat{U}$ as given by Eq. (6.156), we write

$$\langle\varphi_\beta|\hat{T}_{\beta\gamma}|\varphi_\gamma\rangle = \sum_\alpha \langle\varphi_\beta|\hat{U}_{\beta\alpha}|\psi^+_{\alpha\gamma}\rangle = \langle V_\beta\varphi_\beta|\sum_\alpha \psi^+_{\alpha\gamma}\rangle = \langle\varphi_\beta|V_\beta\Psi^+_\gamma\rangle. \tag{6.161}$$

Exercise 6.6. Show the validity of Eq. (6.161). Discuss the action of $\hat{U}_{\beta\gamma}$ on φ_β.

With Eqs. (6.160) and (6.161), a reactive T-matrix is generated which is formally equivalent to the inelastic T-matrix. The problem of rearrangement collisions is thus reduced to that of inelastic scattering processes.

We assembled in this chapter some basic concepts and formulae that define multichannel scattering with emphasis on inelastic as well as reactive processes, but did not make explicit mention of the electronic system. The following chapter will make amends for this omission by highlighting the electronic degrees of freedom within the framework of time-independent molecular collision theory, as laid down in the preceding pages.

Chapter 7

Time-Independent Theory of Molecular Collisions II: The Electronic Problem

A single potential energy surface may be sufficient to describe reactive scattering processes, as considered schematically in the previous chapter. In many cases, however, these reactions involve dramatic changes of the electronic system which requires extension to multiple surfaces. Thus, elementary rearrangement collisions of the type

$$AB + B \rightarrow A + B_2, \tag{7.1}$$

as analyzed in Sections 6.6 and 6.7, may rival alternative pathways, involving electron transfer between the two product species:

$$AB + B \rightarrow A^+ + B_2^-. \tag{7.2}$$

Here admission has to be made for two qualitatively differing electronic configurations in the product stage of the reaction, necessitating inclusion of at least two potential energy surfaces to simulate the process. Early experimental research has given proof for the emergence of ionized products from reactions between atoms and diatomic molecules, involving alkali atoms and halogen molecules [168, 169]. The two electronic channels are sketched in Figure 7.1. If the system is characterized by a diabatic crossing, the product state may be covalent or ionic. Evolution of the system along the adiabatic curve of lowest energy is guaranteed only in the case of sufficiently large diabatic coupling between the electronic states (see Section 2.5.1). The energy difference between the reaction branches $A + B_2$ and $A^+ + B_2^-$ is asymptotically equal to the difference between the ionization energy of A and the electron affinity of B_2.

In what follows, we continue our presentation of time-independent molecular collision theory by integrating the electronic system into the

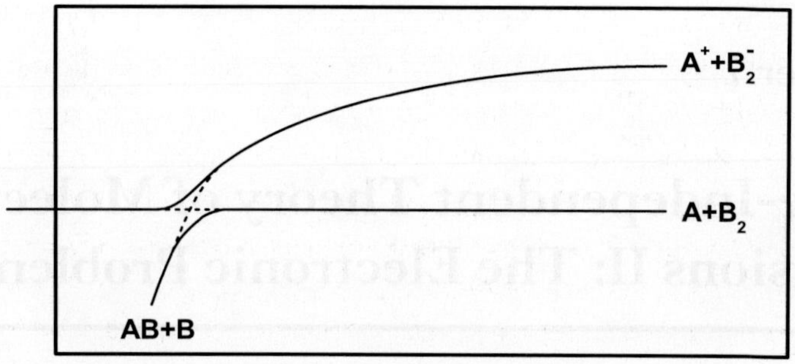

Figure 7.1: Alternative electronic pathways for a rearrangement reaction between the species AB and B, resulting in A and B_2. The asymptotic product state is ionic or covalent, depending on whether the diabatic or the adiabatic route is taken.

formalism introduced in the previous chapter. Section 7.1 discusses the generalization of the differential equation approach outlined in Sections 6.5 and 6.6. An alternative methodology for arriving at information about collision systems is described in Section 7.3 where we focus on variational strategies. To illustrate these contrasting procedures, we highlight in Sections 7.2 and 7.4 recent studies on the systems F-H_2 and Na-H_2, respectively, which can be considered as touchstone cases for the treatment of electronic degrees of freedom in quantum dynamics. In particular, the latter system allows for the analysis of effects related to the interaction between a bound and a dissociative potential energy surface.

These approaches are implementations of rigorous quantum theory. They are exact in the (unattainable) limit of complete basis sets. A different viewpoint is the subject of Section 7.5. Here we introduce the influential Landau–Zener–Stückelberg theory of electronic transitions in the presence of a crossing between two diabatic potential energy curves. This model is here presented from the viewpoint of the WKB approximation which exemplifies the use of semiclassical arguments in the framework of time-independent collision theory.

7.1 Inclusion of the Electronic System

In this section we discuss the incorporation of electronic degrees of freedom into the differential equation approach to molecular scattering. As a general observation, the body-fixed model as introduced in Section 6.7.2 lends itself quite naturally to the description of the electronic system in molecules. Considering the simplest possible case, we refer first to a diatom with the body-fixed axis Z' given by the internuclear vector $\mathbf{r}$ which provides an evident choice for the quantization axis of the electronic motion in the adiabatic representation. Further, we modify the angular momentum formalism (see Eq. (6.109) and the related commentary) to include the electronic angular momentum into the kinetic part of the Hamiltonian, and cast the coupling between the electrons and the nuclei into a physically plausible form. In particular, we use the decomposition of the total angular momentum into a nuclear and an electronic part:

$$\hat{\mathbf{J}} = \hat{\mathbf{j}} + \hat{\mathbf{P}}, \tag{7.3}$$

where the vector operator $\hat{\mathbf{P}}$ is understood as the body-fixed representation of the total electronic angular momentum of the molecule, composed of an orbital and a spin contribution. As before, $\hat{\mathbf{j}}$ stands for the angular momentum associated with the diatom centers. We thus obtain the following expression for the nuclear kinetic energy operator:

$$\hat{T}_{\mathrm{N}} = -\frac{\hbar^2}{2\mu r^2}\left[\frac{\partial}{\partial r}r^2\frac{\partial}{\partial r} - (\hat{\jmath}_{X'} - \hat{P}_{X'})^2 - (\hat{\jmath}_{Y'} - \hat{P}_{Y'})^2\right]. \tag{7.4}$$

It should be noted that $\hat{T}_{\mathrm{N}}$ does not contain the Z' components of the angular momentum operators, since $\hat{\jmath}_{Z'}$ equals $\hat{P}_{Z'}$ in the diatomic case. Eq. (7.4) clearly exhibits the source of both radial and angular nonadiabatic effects in diatomic molecules. Expanding the angular part, we obtain

$$(\hat{\jmath}_{X'} - \hat{P}_{X'})^2 + (\hat{\jmath}_{Y'} - \hat{P}_{Y'})^2 = \hat{\jmath}^2 + \hat{P}^2 - \hat{\jmath}_{-}\hat{P}_{+} - \hat{\jmath}_{+}\hat{P}_{-} - 2\hat{\jmath}_{Z'}\hat{P}_{Z'} \tag{7.5}$$

which correlates ladder operators built from the total and the electronic angular momentum of the molecule. In the following step, we will generalize the theory of electronic-nuclear coupling in a rotating coordinate frame to the triatomic problem in Jacobi coordinates, as described in Sections 6.6 and 6.7.

7.1.1 *The triatomic case*

As the body-fixed frame is the natural choice for the electronic coordinates, they depend on the Euler angles φ, θ, and χ according to $\mathbf{x}'_i = \mathbf{D}(\varphi, \theta, \chi)\mathbf{x}_i$, where $\mathbf{x}_i$ denotes the space-fixed Cartesian position vector of the ith electron. This implies that the nuclear angular momentum operators discussed in Section 6.7.2 act on both the nuclear and the electronic coordinates. For instance, the operator $\hat{\mathcal{J}}_{Z'} = -i\hbar\partial/\partial\chi$ may be associated with a generalized body-fixed system derivative that involves the rotated electronic coordinates [170]:

$$\begin{aligned}\frac{\partial}{\partial\chi}_{\text{tot}} &= \frac{\partial}{\partial\chi} + \sum_i \left[\frac{\partial x'_i}{\partial\chi}\frac{\partial}{\partial x'_i} + \frac{\partial y'_i}{\partial\chi}\frac{\partial}{\partial y'_i} + \frac{\partial z'_i}{\partial\chi}\frac{\partial}{\partial z'_i}\right] \\ &= \frac{\partial}{\partial\chi} + \frac{i}{\hbar}\hat{l}_z.\end{aligned} \tag{7.6}$$

The corresponding expressions for the remaining two Euler angles are

$$\frac{\partial}{\partial\theta}_{\text{tot}} = \frac{\partial}{\partial\theta} - \frac{1}{2\hbar}(\exp(i\chi)\hat{l}_+ + \exp(-i\chi)\hat{l}_-). \tag{7.7}$$

$$\frac{\partial}{\partial\varphi}_{\text{tot}} = \frac{\partial}{\partial\varphi} - \frac{i}{\hbar}\cos\theta\hat{l}_z + \frac{i}{2\hbar}\sin\theta(\exp(i\chi)\hat{l}_+ - \exp(-i\chi)\hat{l}_-) \tag{7.8}$$

where $\hat{\mathbf{l}}$ denotes the electron orbital angular momentum in Cartesian representation, replacing the vector operator $\hat{\mathbf{P}}$ of Eq. (7.3) which refers to the diatomic case. It is defined as

$$\hat{\mathbf{l}} = -i\hbar\sum_i \mathbf{x}'_i \times \nabla_{\mathbf{x}'_i}. \tag{7.9}$$

The orbital operator $\hat{\mathbf{l}}$ may be augmented by the electron spin operator $\hat{\mathbf{s}}$ to yield the operator of the total electronic angular momentum. In the present section, we will confine the electronic angular momentum to its orbital part, while the spin will be included in the discussion of a triatomic reaction in the following section.

Exercise 7.1. Establish Eq. (7.6) using the definition of the electronic angular momentum (7.9) and the mapping from the space- to the body-fixed system.

$$\mathbf{x}'_i = \begin{pmatrix}\cos\chi & -\sin\chi & 0\\ \sin\chi & \cos\chi & 0\\ 0 & 0 & 1\end{pmatrix}\begin{pmatrix}\cos\theta & 0 & -\sin\theta\\ 0 & 1 & 0\\ \sin\theta & 0 & \cos\theta\end{pmatrix}\begin{pmatrix}\cos\varphi & -\sin\varphi & 0\\ \sin\varphi & \cos\varphi & 0\\ 0 & 0 & 1\end{pmatrix}\mathbf{x}_i. \tag{7.10}$$

By passing from $(\partial/\partial u)$ to $(\partial/\partial u)_{\mathrm{tot}'}$, with $u = \varphi, \theta, \chi$, we arrive at a generalized, or total, angular momentum $\hat{\mathcal{J}}_{\mathrm{tot}}$ that encompasses both the nuclear and the electronic contributions. The Z' component of this new operator is given by $\hat{\mathcal{J}}_{\mathrm{tot},Z'} = -i\hbar(\partial/\partial\chi)_{\mathrm{tot}}$. As we substitute for the differential operators $(\partial/\partial u)$ according to the transformation rules (7.6)–(7.8), we find a simple decomposition of the vector $\hat{\mathbf{J}}_{\mathrm{tot}}$ into a nuclear and an electronic part, namely

$$\hat{\mathbf{J}}_{\mathrm{tot}} = \hat{\mathbf{J}} + \hat{\mathbf{l}}. \tag{7.11}$$

This result is the natural extension of Eq. (6.109) for the case that the studied scattering problem involves electronic degrees of freedom. Likewise, we introduce a total internal angular momentum $\hat{\mathbf{j}}_{\mathrm{tot}}$ in addition to the overall nuclear angular momentum. The transformations (7.6)–(7.8) preserve the relation $\widetilde{\hat{\mathcal{J}}}_{Z',\mathrm{tot}} = \hat{j}_{Z',\mathrm{tot}}$ between the overall and the internal nuclear angular momentum in the body-fixed system. Specifically, both components result as the direct generalization of the nuclear expression indicated in Section 6.7.2:

$$\hat{\mathcal{J}}_{Z'} = \hat{j}_{Z'} = -i\hbar\frac{\partial}{\partial\chi}. \tag{7.12}$$

Making use of the transformation formula (7.6), this becomes

$$\hat{\mathcal{J}}_{Z',\mathrm{tot}} = \hat{j}_{Z',\mathrm{tot}} = -i\hbar\frac{\partial}{\partial\chi}_{\mathrm{tot}} = -i\hbar\frac{\partial}{\partial\chi} + \hat{l}_{Z'}. \tag{7.13}$$

Further, replacing $\hat{\mathbf{J}}$ by $\hat{\mathbf{J}}_{\mathrm{tot}}$ and $\hat{\mathbf{j}}$ by $\hat{\mathbf{j}}_{\mathrm{tot}}$, we obtain the generalized operator of the nuclear kinetic energy directly from Eq. (6.126):

$$\begin{aligned}\hat{T}_{\mathrm{N}} = &-\frac{\hbar^2}{2\mu}\left[\frac{1}{\rho}\frac{\partial^2}{\partial\rho^2}\rho + \frac{1}{r}\frac{\partial^2}{\partial r^2}r\right] + \frac{1}{2\mu}\left(\frac{1}{\rho^2} + \frac{1}{r^2}\right)\hat{j}^2_{\mathrm{tot}} \\ &+ \frac{1}{2\mu\rho^2}(\hat{\mathcal{J}}^2_{\mathrm{tot}} - \hat{\mathcal{J}}_{tot,-}\hat{j}_{tot,+} - \hat{\mathcal{J}}_{tot,+}\hat{j}_{tot,-} - 2\hat{\mathcal{J}}^2_{Z',tot}).\end{aligned} \tag{7.14}$$

The kinetic energy operator for the electrons maintains its regular form:

$$\hat{T}_{\mathrm{el}} = -\frac{\hbar^2}{2m_e}\left[\frac{\partial^2}{\partial x'^2} + \frac{\partial^2}{\partial y'^2} + \frac{\partial^2}{\partial z'^2}\right]. \tag{7.15}$$

Using Eqs. (7.14) and (7.15), one may derive the explicit form of the stationary Schrödinger equation, in both the adiabatic and the diabatic representation, taking into account the electronic subsystem. This task is undertaken in the next two subsections. For ease of notation, we drop the subscript tot and understand $\mathcal{J}$ by default as the total angular momentum operator of the system considered, while $\mathcal{J}_{\mathrm{N}}$ will be used as a symbol for the total nuclear angular momentum in the following. We also remove the primes from the electronic coordinates, stipulating that they refer to the body-fixed system, unless stated otherwise. The overall wave function in the body-fixed frame, denoted $\widetilde{\Psi}$, depends on both the nuclear Jacobi coordinates and the electronic coordinates, summarily indicated as $\mathbf{x}$. It solves the equation

$$(\hat{T}_{\mathrm{N}} + \hat{H}_{\mathrm{el}})\widetilde{\Psi}^{\mathcal{J}M_{\mathcal{J}}}(\mathbf{x}, \rho, r, \Theta) = E\widetilde{\Psi}^{\mathcal{J}M_{\mathcal{J}}}(\mathbf{x}, \rho, r, \Theta) \tag{7.16}$$

with

$$\hat{H}_{\mathrm{el}} = \hat{T}_{\mathrm{el}} + V(\mathbf{x}, \rho, r, \Theta) \tag{7.17}$$

where V comprises all electrostatic interaction terms of the problem. In accordance with Eq. (6.116), we use the transformation behavior of the wave function with respect to rotation involving the three Euler angles to establish a connection between the wave function in the space-fixed and that in the body-fixed frame:

$$\widetilde{\Psi}^{\mathcal{J}M_{\mathcal{J}}} = \sum_K D^{\mathcal{J}}_{M_{\mathcal{J}}K}(\varphi, \theta, \chi)\Psi'^{\mathcal{J}K}. \tag{7.18}$$

We substitute

$$\Psi'^{\mathcal{J}K} = \frac{1}{r\rho}\Psi^{\mathcal{J}K} \tag{7.19}$$

into the stationary Schrödinger equation (7.16) and insert Eq. (7.14) for $\hat{T}_{\mathrm{N}}$. Taking advantage of relation (7.18), the orthogonality of the elements $D^{\mathcal{J}}_{KM_{\mathcal{J}}}$, as well as the independence of $\hat{H}_{\mathrm{el}}$ of the Euler angles, we deduce the following equation for $\Psi^{\mathcal{J}K}$ [170]:

$$\hat{t}_{K,K}\Psi^{\mathcal{J}K} + \hat{t}_{K,K+1}\Psi^{\mathcal{J}K+1} + \hat{t}_{K,K-1}\Psi^{\mathcal{J}K-1} + \frac{2\mu}{\hbar^2}(\hat{H}_{\mathrm{el}} - E)\Psi^{\mathcal{J}K} = 0 \tag{7.20}$$

with

$$\hat{t}_{K,K} = -\left(\frac{\partial^2}{\partial\rho^2} + \frac{\partial^2}{\partial r^2}\right) + \frac{1}{\rho^2}(\mathcal{J}(\mathcal{J}+1) - 2M_{\mathcal{J}}^2)$$
$$-\left(\frac{1}{\rho^2} + \frac{1}{r^2}\right)\left(\frac{\partial^2}{\partial\Theta^2} + \cot\Theta\frac{\partial}{\partial\Theta} - \frac{M_{\mathcal{J}}^2}{\sin\Theta^2}\right) \tag{7.21}$$

$$\hat{t}_{K,K\pm1} = -\frac{1}{\rho^2}\lambda^{\pm}_{\mathcal{J}K}\left[(K \pm 1)\cot\Theta \pm \frac{\partial}{\partial\Theta}\right]. \tag{7.22}$$

While the structure of Eq. (7.20) is analogous to that of relation (6.131) which was derived for nuclear degrees of freedom alone, the former deviates from the latter as the terms involving the potential $V(R, r, \Theta)$ have been replaced by $\hat{H}_{\mathrm{el}}$ in Eq. (7.20). Further, we recall that the quantum numbers $\mathcal{J}$, K and $M_{\mathcal{J}}$ change their meaning as one goes from Eq. (6.131) to Eq. (7.20), now referring to the sum of nuclear and electronic components. In what follows, we will discuss how to eliminate the electronic degrees of freedom to arrive at a system of coupled differential equations for the nuclear wave functions. This procedure will be outlined first adopting the adiabatic representation (Section 7.1.1) which may be considered as a natural basis of operation for addressing the electronic problem since the majority of molecular electronic structure calculations make use of the adiabatic approximation. However, as pointed out in the context of the general multichannel formalism (see for instance the comparison between Eq. (6.7) and Eq. (6.13)), a considerable technical simplification can be achieved by transforming into the diabatic frame, as will be detailed in Section 7.1.3.

7.1.2 *The adiabatic case*

We introduce a sequence of adiabatic electronic states $\zeta_n(\mathbf{x}, \mathbf{X}) = \zeta_n(\mathbf{x}, \rho, r, \Theta)$, assuming that they can be enumerated by the collective quantum number n which contains, among other specifications of the electronic system, the total electronic orbital angular momentum, l, and its projection onto the body-fixed quantization axis, m_l. The adiabatic representation of the overall wave function is then written as

$$\Psi^{\mathcal{J}K}(\mathbf{x}, \rho, r, \Theta) = \sum_{n'} \zeta_{n'}(\mathbf{x}, \rho, r, \Theta)\Psi_{n'}^{\mathcal{J}K}(\rho, r, \Theta). \tag{7.23}$$

In keeping with the basic tenets of adiabatic theory, we stipulate that the functions Ψ_n form an orthogonal set, satisfying the eigenvalue equation

$$\hat{H}_{\mathrm{el}}\zeta_n(\mathbf{x}, \rho, r, \Theta) = V_n^{\mathrm{A}}(\rho, r, \Theta)\zeta_n(\mathbf{x}, \rho, r, \Theta) \tag{7.24}$$

where the energy eigenvalue V_n^{A} defines the nth adiabatic potential energy surface characteristic of the problem. Further, the ζ_n are constrained to be eigenvectors of the operators $\hat{l}^2$ and $\hat{l}_z$:

$$\hat{l}^2\zeta_n(\mathbf{x}, \rho, r, \Theta) = l(l+1)\hbar^2\zeta_n(\mathbf{x}, \rho, r, \Theta), \tag{7.25}$$

$$\hat{l}_z\zeta_n(\mathbf{x}, \rho, r, \Theta) = m_l\hbar\zeta_n(\mathbf{x}, \rho, r, \Theta). \tag{7.26}$$

We insert the expansion (7.23) into Eq. (7.20), take on both sides the scalar product with ζ_n^* from the left, and arrive at

$$\hat{t}_{K,K}\Psi_n^{JK} + \hat{t}_{K,K+1}\Psi_n^{JK+1} + \hat{t}_{K,K-1}\Psi_n^{JK-1} + \frac{2\mu}{\hbar^2}(V_n^{\mathrm{A}}(\rho, r, \Theta) - E)\Psi_n^{JK}$$
$$+ \sum_{n'}\left[2\langle\zeta_n|\nabla|\zeta_{n'}\rangle \cdot \nabla + \langle\zeta_n|\nabla^2|\zeta_{n'}\rangle + \left\langle\zeta_n\left|\frac{\partial}{\partial\Theta}\right|\zeta_{n'}\right\rangle\hat{\bigcirc}\right]\Psi_n^{JK} = 0. \tag{7.27}$$

The scalar products $\langle|\rangle$ are here defined with respect to the electronic coordinates, and the operator $\hat{\bigcirc}$ is specified by [170]

$$\hat{\bigcirc}\Psi_n^{JK} \equiv \frac{1}{\rho^2}\left[\lambda_{JK}^{+}\Psi_n^{JK+1} - \lambda_{JK}^{-}\Psi_n^{JK-1} + \frac{\rho^2 + r^2}{r^2}\cot\Theta\Psi_n^{JK}\right] \tag{7.28}$$

with the factors $\lambda_{JK}^{+,-}$ as given by Eq. (6.129). The definition of the gradient ∇ is

$$\nabla \equiv \left(\frac{\partial}{\partial r}, \frac{\partial}{\partial\rho}, \frac{\sqrt{\rho^2 + r^2}}{\rho r}\frac{\partial}{\partial\Theta}\right). \tag{7.29}$$

Cast into matrix form, (7.27) becomes

$$\mathbf{I}[\hat{t}_{K,K}^{JK}\mathbf{\Psi}^{JK} + \hat{t}_{K,K+1}\mathbf{\Psi}^{JK+1} + \hat{t}_{K,K-1}\mathbf{\Psi}^{JK-1}]$$
$$+ \frac{2\mu}{\hbar^2}(\mathbf{V}^A(R, r, \Theta) - E\mathbf{I})\mathbf{\Psi}^{JK} + 2\mathbf{A}\nabla\mathbf{\Psi}^{JK}$$
$$+ \mathbf{B}\mathbf{\Psi}^{JK} + \mathbf{C}\hat{\bigcirc}\mathbf{\Psi}^{JK} = 0 \tag{7.30}$$

where the adiabatic potential energy matrix $\mathbf{V}^A$ is diagonal and the matrices **A**, **B**, and **C** incorporate the nonadiabatic effects characteristic of the problem. The matrix elements are given by

$$A_{nn'} \equiv \langle\zeta_n|\nabla|\zeta_{n'}\rangle, \quad B_{nn'} \equiv \langle\zeta_n|\nabla^2|\zeta_{n'}\rangle, \quad C_{nn'} \equiv \left\langle\zeta_n\left|\frac{\partial}{\partial\Theta}\right|\zeta_{n'}\right\rangle. \tag{7.31}$$

From the definitions (7.31), **A** is a three-dimensional vector (or vector potential, see Section 2.4) whose components are matrices. For instance, we specify the elements of the first component $\mathbf{A}_r$ as $< \zeta_{\mathrm{N}}|\partial/\partial r|\zeta_{n'} >$, etc. It is pointed out that asymptotically, both J_{N} and l are good quantum numbers, provided the electronic state of the system is well defined for large positive and negative times. This implies a multiplicity of $N_J = 2min(J_{\mathrm{N}}, l)+1$ possible values for J. For each one of these choices, the equation system (7.30) is to be solved.

7.1.3 *The diabatic case*

A substantial simplification of the formalism described in the previous subsection is accomplished by going from the adiabatic to the diabatic representation, thus shifting the nonadiabatic coupling from the nuclear kinetic energy to the potential energy terms of the problem. The goal of this operation is eliminating the expressions in Eq. (7.30) that contain the derivative matrices **A**, **B**, and **C**, which often make the equation numerically intractable. For this purpose, we will make use of the transformation from the adiabatic to the diabatic frame (see Section 2.4). Defining diabatic nuclear wave functions χ^{JK}, we recall the orthogonal matrix **W** that communicates between both bases:

$$\mathbf{\Psi}^{JK} = \mathbf{W}\boldsymbol{\chi}^{JK}. \tag{7.32}$$

As this expression for $\mathbf{\Psi}^{JK}$ is substituted into Eq. (7.30), we derive after some remodeling the respective differential equation for the functions χ^{JK} [170]

$$\begin{aligned}
&\mathbf{W}[\hat{t}_{K,K}\boldsymbol{\chi}^{JK} + \hat{t}_{K,K+1}\boldsymbol{\chi}^{J,K+1} + \hat{t}_{K,K-1}\boldsymbol{\chi}^{J,K-1}]\\
&\quad + \frac{2\mu}{\hbar^2}(\mathbf{V}^{\mathrm{A}}(\rho, r, \Theta) - E\mathbf{I})\mathbf{W}\boldsymbol{\chi}^{JK} + 2(\nabla\mathbf{W} + \mathbf{A}\mathbf{W})\cdot\nabla\boldsymbol{\chi}^{JK}\\
&\quad + \left[\nabla^2\mathbf{W} + 2\mathbf{A}\cdot\nabla\mathbf{W} + \mathbf{B}\mathbf{W} + \left(\mathbf{C}\mathbf{W} + \frac{\partial\mathbf{W}}{\partial\Theta}\right)\hat{\mathrm{O}}\right]\boldsymbol{\chi}^{JK} = \mathbf{0}.
\end{aligned} \tag{7.33}$$

We gauge the transformation **W** in such a way that the dependence of (7.33) on the derivatives of the $\nabla\boldsymbol{\chi}^{JK}$ disappears. This succeeds by imposing the constraint

$$\nabla\mathbf{W} + \mathbf{A}\mathbf{W} = \mathbf{0}. \tag{7.34}$$

In accord with the definition of the gradient given by Eq. (7.29), this identity is to be understood as shorthand notation, condensing the three relations

$$\frac{\partial\mathbf{W}}{\partial\rho} + \mathbf{A}_\rho\mathbf{W} = \mathbf{0}, \tag{7.35}$$

$$\frac{\partial\mathbf{W}}{\partial r} + \mathbf{A}_r\mathbf{W} = \mathbf{0}, \tag{7.36}$$

$$\frac{\sqrt{\rho^2 + r^2}}{\rho r}\left[\frac{\partial\mathbf{W}}{\partial\Theta} + \mathbf{A}_\Theta\mathbf{W}\right] = \mathbf{0} \tag{7.37}$$

into one. The symbol $\mathbf{A}_\Theta$ is to be understood as the matrix **C** defined in Eq. (7.31). Implementing Eqs. (7.34–7.37), and making reference to the explicit form of the operators $\hat{t}_{K,K}$ and $\hat{t}_{K,K\pm1}$, as defined by Eqs. (7.21) and (7.22), we arrive at the final equation for the diabatic nuclear wave functions χ^{JK},

$$\mathbf{I}[\hat{t}_{K,K}\chi^{JK} + \hat{t}_{K,K+1}\chi^{JK+1} + \hat{t}_{K,K-1}\chi^{JK-1}] + \frac{2\mu}{\hbar^2}(\mathbf{V}^{\mathrm{D}} - E\mathbf{I})\chi^{JK} = 0, \tag{7.38}$$

which is much less complex than the adiabatic version (7.30). The non-diagonal diabatic potential energy matrix $\mathbf{V}^{\mathrm{D}}$ is constructed by

$$\mathbf{V}^D = \mathbf{W}^\dagger \mathbf{V}^{\mathrm{A}} \mathbf{W}. \tag{7.39}$$

The reader is referred to [170] for technical details related to the derivation of the crucial diabatic equation (7.38). However, the reader is encouraged to work through the reduced case of collinear interaction, fixing the angle Θ at 180°, as suggested in Exercise 7.2.

Exercise 7.2. Discuss the collinear case, restraining the gradient (7.29) to $\nabla \equiv (\partial/\partial r, \partial/\partial \rho)$. Derive the counterpart of (7.33) for this situation. Show by explicit calculation that the combination $\nabla^2\mathbf{W} + 2\mathbf{A}\cdot\nabla\mathbf{W} + \mathbf{B}\mathbf{W}$ reduces to zero as the condition $\nabla\mathbf{W} + \mathbf{A}\mathbf{W} = \mathbf{0}$ is imposed on **W**. From this information, obtain the analog of (7.38) for collinear scattering.

In this special situation the constraint (7.34) involves only the first two equations of the sequence (7.35)–(7.37) as the Jacobi coordinates reduce to r and ρ. Differentiating (7.35) with respect to r and (7.36) with respect to ρ, and subtracting the resulting equations from each other, we deduce the identity

$$\frac{\partial \mathbf{A}_\rho}{\partial r} - \frac{\partial \mathbf{A}_r}{\partial \rho} = [\mathbf{A}_\rho, \mathbf{A}_r]. \tag{7.40}$$

This non-abelian commutator relation is readily extended to the case of three-dimensional collisions. If ρ, r, and Θ are admitted as variables, it reads

$$\frac{\partial \mathbf{A}_u}{\partial v} - \frac{\partial \mathbf{A}_u}{\partial v} = [\mathbf{A}_u, \mathbf{A}_v] \tag{7.41}$$

where $u, v = \rho, r, \Theta$. Equations (7.40) and (7.41) represent necessary conditions for the existence of a unique and analytic solution of Eq. (7.34) for the problems of collinear and three-dimensional collisions, respectively [170]. These relations have to be obeyed by the derivative coupling matrices $\mathbf{A}$ if an adiabatic–diabatic transformation $\mathbf{W}$ is to be established.

The condition (7.34) can be used to generate $\mathbf{W}$. Confining ourselves for simplicity to the collinear situation, we may assume that both $\mathbf{A}_\rho$ and $\mathbf{A}_r$ are given for any combination of ρ and r. Further, the desired transformation matrix may be known at a position of reference (ρ_i, r_i) such that $\mathbf{W}_i \equiv \mathbf{W}(\rho_i, r_i)$. The transformation at some arbitrary position (ρ_f, r_f) is then found from

$$\mathbf{W}(\rho, r) = \mathbf{W}(\rho_i, r_i) + \int_{\rho}^{\rho_i} \mathbf{A}_\rho(\rho, r)\mathbf{W}(\rho, r)d\rho + \int_{r}^{r_i} \mathbf{A}_r(\rho_i, r)\mathbf{W}(\rho_i, r)dr. \tag{7.42}$$

Here a specific path in the (ρ, r) plane has been chosen, connecting the initial with the final location. The constraint (7.34), however, ensures that any other continuous path could be employed as well, since it implies that the integral $\int_{\mathbf{c}_i}^{\mathbf{c}_f} \mathbf{A}_c \mathbf{W} d\mathbf{c}$ does not depend on the selection of $\mathbf{c}$, where $\mathbf{A}_c$ stands for the projection of the vector matrix on a given trajectory $\mathbf{c}$ in the underlying coordinate space. Therefore we might view the commutator relations (7.40) and (7.41) as necessary conditions for the path independence of this integral.

Various authors (e.g. [171–173]) have employed the differential approach to the problem of coupled nuclear and electronic degrees of freedom in molecular scattering by adopting Delves coordinates (see Section 6.6.2). To illustrate the treatment of concrete physical systems in the frame of this methodology we give a brief outline of an *ab initio* study on a fluorine atom interacting with a hydrogen molecule in the following subsection.

7.2 Case Study: The $F + H_2$ Reaction

A plethora of experimental as well as theoretical ([174]–[179]) and references therein) tools has been used to investigate the reaction

$$F + H_2 \rightarrow HF(v', j') + H, \tag{7.43}$$

resulting in highly detailed information on state-to-state scattering processes that occur in this rearrangement collision and make it a touchstone for models of reactive molecular scattering. This exchange process is distinctly exothermic, involving an energy release of about 1.4 eV. At the same time, the reaction is fast, with an Arrhenius activation barrier of only 0.04 eV. Accordingly, even at low and moderate scattering energies, a sizeable number of rovibrational channels of the product HF are open and thus will be populated in the course of the collision.

For adequate description of this process, more than one potential energy surface needs to be included. This requirement is rooted in the degeneracy of the $F(^2P)$ atomic configuration which is split into a $^2P_{3/2}$ ground state and a $^2P_{1/2}$ excited state as the H_2 molecule approaches the atom. With respect to the reaction $F + H_2 \rightarrow FH + H$, three electronic wave functions are distinguished. Two of them emerge from an initial $F(^2P_{3/2}) + H_2(j = 0)$ state. They diverge into products of composition $HF(^3\Pi) + H$ and $HF(^1\Sigma) + H$, corresponding to electronic states with symmetries $1^2A'$ and $1^2A''$, respectively. A further state of $2^2A'$ symmetry evolves from the initial condition $F(^2P_{1/2}) + H_2(j = 0)$. The correlations between the reactants and products are schematically shown in Figure 7.2.

To treat the problem at hand, one has to step beyond the frame set by the Hamiltonian that combines from the operators given by Eqs. (7.14)

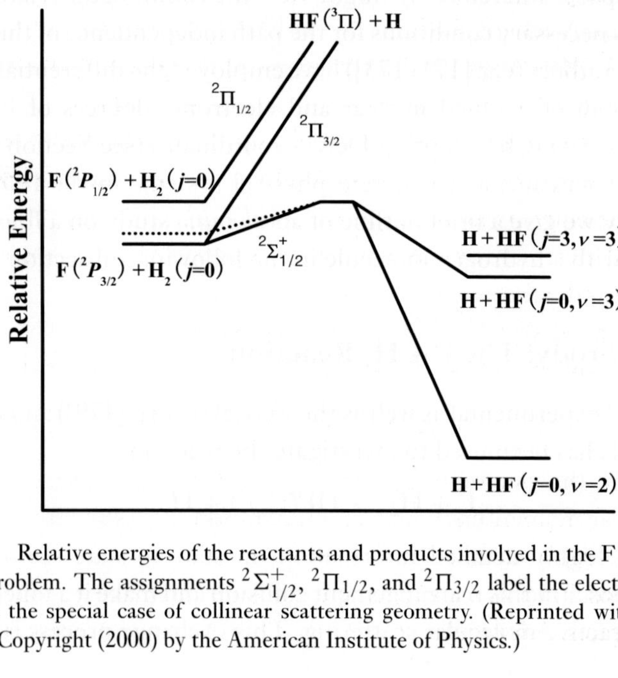

Figure 7.2: Relative energies of the reactants and products involved in the $F + H_2$ reactive scattering problem. The assignments $^2\Sigma^+_{1/2}$, $^2\Pi_{1/2}$, and $^2\Pi_{3/2}$ label the electronic product functions in the special case of collinear scattering geometry. (Reprinted with permission from [173]. Copyright (2000) by the American Institute of Physics.)

and (7.15). So far, our discussion has not included the electronic spin s which is a defining element of the various initial and final configurations of the $F + H_2 \rightarrow HF(v',j') + H$ process. Its effect is quantitatively taken into account by extending the electronic angular momentum operator from $\hat{l}$ to $\hat{j}_{el} = \hat{l} + \hat{s}$, and by a spin-orbit coupling term $\hat{H}_{SO}$ (see Section 1.1.2), augmenting the Hamiltonian (7.16) which now reads

$$\hat{H}(\mathbf{x}, \rho'_H, \theta_H, \Theta) = \hat{T}_{Del}(\rho'_H, \theta_H, \Theta) + \hat{H}_{el}(\mathbf{x}, \rho'_H, \theta_H, \Theta) + \hat{H}_{SO}(\mathbf{x}, \rho'_H \theta_H, \Theta) \tag{7.44}$$

where $\hat{H}_{Del}$ stands for the nuclear kinetic energy in Delves coordinates, as introduced by Eqs. (6.103) and (6.104). The electronic coordinates are comprised by the position vector $\mathbf{x}$. The kinetic part of $\hat{H}$ depends on the reduced mass defined in Eq. (6.96) which in the present problem is given by

$$\mu = \sqrt{\frac{m_F m_H^2}{m_F + 2m_H}} \tag{7.45}$$

with m_F and m_H as the masses of a fluorine and a hydrogen atom, respectively.

Since the body-fixed frame is employed, the axis of quantization is defined by the Jacobi vector $\boldsymbol{\rho}$. This leads to basis functions of the form $|JM_J Kvjkm_l m_s\rangle$ where, in keeping with the assignments introduced in Section 6.7.1, M_J stands for the projection of the total angular momentum quantum number J on the z-axis of the space-fixed system while K refers to the corresponding projection on the $\boldsymbol{\rho}$ vector. The quantum numbers j and k denote the angular momentum of the diatomic subsystem and its projection along the body-fixed z-axis (Z'), respectively. The index v characterizes a vibrational state, and m_l, m_s are related to the electronic orbital angular momentum and spin quantum numbers l and s, respectively, as described in the body-fixed system. Thus K may be written as the sum of the three projection quantum numbers with respect to the body-fixed axis:

$$K = k + m_l + m_s. \tag{7.46}$$

Alternatively, the electronic system may be described by the overall electronic angular momentum quantum number j_{el} that results from spin-orbit coupling and replaces the sum $m_l + m_s$ by m_{el}.

The basis functions comprise products of vibrational, rotational, and electronic states. Adopting a diabatic frame, the latter are easily enumerated for the initial $F + H_2$ arrangement. More specifically, the three alternative orientations of the ^2p hole, described by the atomic states p_x, p_y, and p_z, may be combined with each of the two electronic spin projections, resulting in six states. On the side of the FH + H products, only two electronic functions contribute, if we restrict ourselves to the Σ^+ ground state of the FH molecule and exclude the energetically well-separated excitations of Π character, leaving an electronic doublet that corresponds to the two spin orientations of the free H atom.

The expression for the nuclear orbital angular momentum, replacing $(\hat{\mathcal{J}} - \hat{j})$ in Eq. (6.124), becomes

$$\hat{\mathbf{L}} = \hat{\mathbf{J}} - \hat{\mathbf{l}} - \hat{\mathbf{s}} - \hat{\mathbf{j}}. \tag{7.47}$$

To derive reaction probabilities for the $F + H_2$ system, one diagonalizes the Hamiltonian (7.44), excepting the part of $H_{\text{Del}}(\rho'_{\text{H}}, \theta_{\text{H}}, \Theta)$ that depends exclusively on ρ'_{H} (see Eq. (6.103)), thus arriving at a set of orthogonal *surface functions* [162] coupled by the purely radial contribution to $\hat{H}$. Technical details of the construction of the relevant matrix elements, i.e. those of the nuclear angular contribution to $\hat{H}$, of $\hat{H}_{\text{el}}$ and $\hat{H}_{\text{SO}}$ between functions $\lfloor \mathcal{J} M_{\mathcal{J}} K v j k m_l m_s \rangle$ and $\lfloor \mathcal{J} M_{\mathcal{J}} K' v' j' k' m_{l'} m_{s'} \rangle$ for a single as well as for two different arrangements, are outlined in [173]. The latter task is efficiently performed by expressing the basis functions of the product in terms of the reactant state basis functions. This involves redefinition of the Jacobi vectors $\mathbf{r}$ and $\boldsymbol{\rho}$ which is accomplished by turning the reactant wave function by an angle ω around a body-frame axis perpendicular to the triatomic plane, conventionally taken as the Y'-axis (see Figure 6.2). In this case, the transition between the two states is associated with a change from the set of Euler angles Ω to a new set Ω':

$$D^{\mathcal{J}}_{M_{\mathcal{J}}K'}{}^{*}(\Omega') = \sum_{K} D^{\mathcal{J}}_{M_{\mathcal{J}}K}{}^{*}(\Omega) d^{\mathcal{J}}_{KK'}(\omega) \tag{7.48}$$

where $D^{\mathcal{J}}_{M_{\mathcal{J}}K'}{}^{*}(\Omega')$ designates a Wigner rotation matrix element. Similarly, the electronic spin functions transform according to

$$|sm'_s\rangle = \sum_{m_s} |sm_s\rangle d^{s}_{m_s m'_s}(\omega). \tag{7.49}$$

The functions $D^{\mathcal{J}}_{M_{\mathcal{J}}K}{}^{*}(\Omega)$ $(D^{\mathcal{J}}_{M_{\mathcal{J}}K'}{}^{*}(\Omega))$ and $|sm_s\rangle$ $(|sm_{s'}\rangle)$ are contained as factors in the states $\lfloor \mathcal{J} M_{\mathcal{J}} K v j k m_l m_s \rangle$ $(\lfloor \mathcal{J} M_{\mathcal{J}} K' v' j' k' m_{l'} m_{s'} \rangle)$. Thus the prescriptions (7.48) and (7.49), as well as analogous relations for the remaining angular quantum numbers, allow us to express the product functions in terms of reactant functions.

The matrix elements $\langle \mathcal{J}M_{\mathcal{J}}K'v'j'k'm_{l'}m_{s'}|\hat{H}_{\text{el}}|\mathcal{J}M_{\mathcal{J}}Kvj\ km_lm_s\rangle$ of the electronic Hamiltonian and the spin-orbit elements $\langle \mathcal{J}M_{\mathcal{J}}K'v'j'\ k'm_{l'}m_{s'}|\hat{H}_{\text{SO}}|\mathcal{J}M_{\mathcal{J}}Kvjkm_lm_s\rangle$ comprise integrals over the electronic coordinates, namely $\langle m_{l'}m_{s'}|\hat{H}_{\text{el}}|m_lm_s\rangle = \delta_{m_{s'}m_s}V^{\text{D}}_{m_{l'}m_l}(\rho'_{\text{H}},\theta_{\text{H}},\Theta)$ in the former and $\langle m_{l'}m_{s'}|\hat{H}_{\text{SO}}|m_lm_s\rangle = SO_{m_{s'}m_{l'}m_sm_l}(\rho'_{\text{H}},\theta_{\text{H}},\Theta)$ in the latter case. For the reactant configuration, the term $V^{\text{D}}_{m_{l'}m_l}(\rho'_{\text{H}},\theta_{\text{H}},\Theta)$ corresponds to four independent elements. By the symmetry of the problem, the wave functions with an $F(^2P)$ hole in the p_y orbital and those with holes in the p_x and p_z orbitals do not couple. This reduces the number of potential energy surfaces to four, related to the p_x, p_y and p_z alternatives incremented by the non-diagonal term that describes the interaction between p_x and p_z. The elements of the diabatic $\mathbf{V}^{\text{D}}$ matrix are then $V^{\text{D}}_{00} = V_{zz}$, $V^{\text{D}}_{11} = V^{\text{D}}_{-1-1} = \frac{1}{2}\,(V_{xx} + V_{yy})$, $V^{\text{D}}_{01} = V^{\text{D}}_{10} = -V^{\text{D}}_{0-1} = -V^{\text{D}}_{-10} = -\frac{1}{\sqrt{2}}V_{xz}$, $V^{\text{D}}_{1-1} = V^{\text{D}}_{-11} = \frac{1}{2}\,(V_{yy}+V_{xx})$. Similar considerations lead to the construction of the spin-orbit matrix **SO**. The four potential energy surfaces that govern the reaction were generated using a variety of quantum chemical techniques [177, 180, 181], in particular the multireference configuration interaction method (MRCI, see Section 1.6), as applied by Stark and Werner [177]. Since we expand the product states in a reactant basis, all electronic interactions relevant to the $\text{F} + \text{H}_2 \rightarrow \text{FH} + \text{H}$ problem are included in the matrices $\mathbf{V}^{\text{D}}$ and **SO**.

We add for completeness that further symmetries are exploited to reduce the number of basis functions. Using the observation that the projection quantum numbers involved in the problem switch their sign upon space reflection within the body-fixed frame, one introduces a linear combination of definite parity by

$$|\mathcal{J},M_{\mathcal{J}},K,v,j,k,m_l,m_s,p\rangle = \frac{1}{\sqrt{2}}(|\mathcal{J},M_{\mathcal{J}},K,v,j,k,m_l,m_s\rangle + p|\mathcal{J},M_{\mathcal{J}},-K,v,j,-k,-m_l,-m_s\rangle) \quad (7.50)$$

with $p = \pm 1$. One also invokes the symmetry of the situation with respect to the exchange of the two H atoms, which multiplies the wave function by a phase factor $(-1)^j$. This partitions the H_2 component into para-H_2 with even j and ortho-H_2 with odd j. Expanding the overall wave function, one may thus eliminate all even j or all odd j terms. This wave function modification requires some adjustment for the product rearrangement which does not include an intact H_2 molecule any longer.

Coupled-channel equations are written down employing the diabatic basis $|\mathcal{J}M_{\mathcal{J}}Kvjkm_lm_sp\rangle$, adapted to include well-defined values of both diatomic and triatomic parity. They are solved by propagating the overall scattering wave function within a preset interval $[\rho'_{\mathrm{H_{min}}}, \rho'_{\mathrm{H_{max}}}]$, using the log derivative formalism as characterized in Section 6.5.1. This operation is continued until the ultimate boundary $\rho'_{\mathrm{H_{max}}}$ is reached, which is chosen large enough to be representative of the asymptotic region. The final wave function emerging from this procedure is rewritten in terms of spin-orbit-coupled electronic states by use of the relation

$$|\mathcal{J}M_{\mathcal{J}}Kvjkj_{\mathrm{el}}m_{\mathrm{el}}p\rangle = \sum_{m_lm_s} C(lm_lsm_s|j_{\mathrm{el}}, m_{\mathrm{el}})|\mathcal{J}M_{\mathcal{J}}Kvjkm_lm_sp\rangle \quad (7.51)$$

where the factors C signify Clebsch–Gordan coefficients. From the asymptotic wave function in this representation, one extracts the scattering matrix elements for the process. The reaction probability as a function of the total energy is then calculated from

$$P^{\mathcal{J}}_{j_{\mathrm{el}},jv}(E_{\mathrm{tot}}) = \frac{1}{(2j+1)(2j_{\mathrm{el}}+1)} \sum_{m_{\mathrm{el}}kj'k'v'p} |S^{\mathcal{J}p}(j_{\mathrm{el}}m_{\mathrm{el}}jkv \to j'k'v', E_{\mathrm{tot}})|^2. \quad (7.52)$$

This formula refers to the integral probability for the occurrence of a $\mathrm{F} + \mathrm{H}_2 \to \mathrm{FH} + \mathrm{H}$ reaction, involving a summation over all accessible rotational and vibrational levels of the product configuration, as well as the two choices of triatomic parity. The quantum numbers $\mathcal{J}$ and j_{el} are constants of motion, the value of j determines if H_2 is realized as para- or ortho-hydrogen. We average over the orientations of the diatomic and the total electronic angular momentum. More differential results, comprising state-to-state scattering, are obtained as one eliminates some of the summation indices on the right-hand side of (7.52), for instance, the final state vibrational quantum number v'. Reaction probabilities computed by use of formula (7.52) are shown in Figure 7.3 for $v = j = 0$ in the energy window $6\,\mathrm{kcal/mol} \le E_{\mathrm{tot}} \le 10.5\,\mathrm{kcal/mol}$. Here the spin-orbit coupled ground and first excited state of the F atom are distinguished by the labels $j_{\mathrm{el}} = \frac{3}{2}$ and $\frac{1}{2}$,

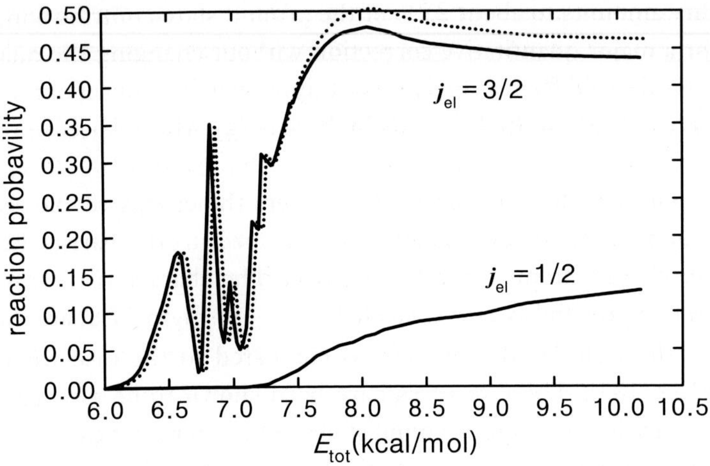

Figure 7.3: Probabilities for the reaction $F + H_2 \to FH + H$, according to formula (7.52). Parameters refer to the total electronic angular momentum $j_{el} = \frac{3}{2}$ and $\frac{1}{2}$, correlating with the ground and the first excited state of the F atom, respectively, and also to the quantum number $\mathcal{J}$ of the total angular momentum. The solid (dotted) curve indicates the choice $\mathcal{J} = \frac{1}{2}(\frac{3}{2})$, where the two curves for $j_{el} = \frac{1}{2}$ nearly coincide. The quantum numbers v and j are set equal to zero. (Reprinted with permission from [173]. Copyright (2000) by the American Institute of Physics.)

respectively. As a further parameter, the total angular momentum $\mathcal{J}$ is used, adopting the values $\frac{1}{2}$ (solid lines) and $\frac{3}{2}$ (dotted lines).

The inclusion of the electronic degrees of freedom into the present calculations, as opposed to the restriction to one potential energy surface only, turns out to have a major impact on the size of the reaction probability. This effect, however, is due to the simple selectivity of reaction pathways that can be inferred from Figure 7.2. If the system is initially prepared in the F atom ground state, only two of the four alternative electronic states ($\Sigma_{\pm\frac{1}{2}}$, $\Pi_{\pm\frac{3}{2}}$ in collinear geometry, see Figure 7.2) will contribute to the reaction since the barrier for the Π states is dramatically higher than for the Σ states. It is instructive to compare the findings based on the multisurface approach with those emerging from a calculation by Castillo *et al.* [182] where a single reactive surface was employed and thus the electronic angular momentum was ignored. This approach preserves most of the essential features of the ground state reactivity as captured by the curve labeled $j_{el} = \frac{3}{2}$ in Figure 7.3. The nonadiabatic admixture of the state with $j_{el} = \frac{1}{2}$ to the overall reaction probability amounts to about 25% of the ground state contribution, which makes for a major quantitative correction without changing the qualitative conclusions derived from a single surface approach. The onset of the $j_{el} = \frac{1}{2}$ channel is markedly shifted towards higher energy with respect to the $j_{el} = \frac{3}{2}$ case.

Incorporating the spin-orbit effect lowers the energy of the reactants in the F atom ground state channel with respect to the barrier by about one-third of the F atom spin-orbit splitting. The characteristic features of the respective probability curve, labeled $j_{el} = \frac{3}{2}$ in Figure 7.3, are therefore shifted to the right by this amount as compared to the case where $\hat{H}_{SO}$ is neglected [183]. These features have been shown to be associated with resonances arising from quasi-bound states. The latter can be traced back to minima in the coordinates related to the H-F stretch and the FH-H van der Waals stretch as well as to the hindered rotation of FH [182].

Schatz *et al.* analyzed a system similar to that described in this case study in essentially the same manner, i.e. by use of the differential equation method [171, 184]. In this case, the coupled-channel equations for the exchange reaction

$$Cl(^2P) + HCl \rightarrow ClH + Cl(^2P) \tag{7.53}$$

were solved by implementing a formalism that is essentially identical with that presented in Section 7.1.2 but augmented by the addition of spin-orbit terms. Again, the reactivities of the electronic states related to $Cl(^2P_{1/2})$ and $Cl(^2P_{3/2})$ were compared, and the latter was found to outweigh the former by a large margin.

The differential equation approach to molecular scattering has proved to be capable of reproducing experimental branching ratios for state-to-state processes, involving the population of the lowest-lying vibrational product states in the reaction $F + H_2 \rightarrow FH + H$, as well as the angular distributions of the respective cross sections [174, 182]. Despite its success in the treatment of triatomic scattering problems, the hyperspherical approach, as exemplified in this section, is of limited use due to its considerable complexity which is largely rooted in the complicated form of the Delves Hamiltonian. Also, generating the surface functions which are required to span both the reactant and the product minima is an extremely time-consuming endeavor. These drawbacks motivate interest in more efficient solution strategies. A substantially different methodology is available with the variational technique which will be highlighted in the next section.

7.3 Variational Procedures

In order to define algebraic schemes that lead to numerical results for inelastic or reactive scattering systems, frequent use is made of variational techniques. In this section, we first give an overall idea of the variational principle applied to scattering situations and then review the rearrangement problem in the light of this principle. In Section 7.4 this discussion will be extended to the challenge of interactions that involve electronic transitions and thus require inclusion of multiple potential energy surfaces.

Several variational methods have been proposed in the context of atomic and molecular scattering theory. These differing schemes require, as a common feature, a certain quantity that carries the essential information about the studied process to be stationary with respect to infinitesimal variations in the scattering function. The quantity of reference, understood as a functional, may be the transition (T), the scattering (S) or the reactance (K) matrix. We thus search for solutions $\widetilde{\Psi}$ ($\widetilde{\Psi}^+$, $\widetilde{\Psi}^-$), such that changing these wave functions by a small linear deviation $\delta\Psi^{(+,-)}$ leaves the variation functional invariant to linear order. In symbolic notation,

$$\frac{\partial F[\Psi]}{\partial \Psi} = 0, \quad \text{if } \Psi = \widetilde{\Psi}, \tag{7.54}$$

with F as the functional of choice and corresponding relations for the states Ψ^+ and Ψ^-. The trial wave functions depend on a set of parameters which may be given by the expansion coefficients $\{a_k\}$ in a linear combination of orthogonal states $\{\Phi_k\}$. Optimal values of these parameters will realize the condition (7.54), determining the state $\widetilde{\Psi}$.

As implied by its very definition, the variational principle is to be used judiciously. It is designed to yield the *best solution*, which may not always be good. The primary task in performing a variational calculation is therefore an adequate assessment of the physical situation investigated and, based on this, the construction of trial functions that are suitable estimates of the true solution.

We illustrate the variational method in scattering theory by the example of the transition matrix as functional of reference [185]. The matrix element

T_{if} may be written as (see Exercise 3.6)

$$T_{if} = \langle \varphi_f | V | \Psi_i^+ \rangle = \langle \Psi_f^- | V | \varphi_i \rangle \tag{7.55}$$

such that

$$2T_{if} = \langle \varphi_f | V \Psi_i^+ \rangle + \langle \Psi_f^- | V \varphi_i \rangle. \tag{7.56}$$

We vary T_{if} with respect to the solutions Ψ_i^+ or Ψ_f^- and demonstrate that the variation δT_{if} in response to small deviations from these states vanishes, where the latter are realized by adding $\delta\Psi_i^+$ ($\delta\Psi_f^-$) to Ψ_i^+ (Ψ_f^-). This is seen as follows: using the Lippmann–Schwinger equation in the form given by Eq. (3.51),

$$\Psi_{i,f}^{+(-)} = \varphi_{i,f} + \hat{G}_0^{+(-)} V \Psi_{i,f}^{+(-)}, \tag{7.57}$$

to replace the asymptotic solution ψ_i in Eq. (7.55), we arrive at

$$T_{if} = \langle \Psi_f^- | V | \Psi_i^+ \rangle - \langle \Psi_f^- | V \hat{G}_0^+ V | \Psi_i^+ \rangle. \tag{7.58}$$

Combining Eqs. (7.56) and (7.58) we conclude that

$$T_{if} = \langle \varphi_f | V | \Psi_i^+ \rangle + \langle \Psi_f^- | V | \varphi_i \rangle - \langle \Psi_f^- | V | \Psi_i^+ \rangle + \langle \Psi_f^- | V G_0^+ V | \Psi_i^+ \rangle. \tag{7.59}$$

The first variation of T_{if} upon small alterations of Ψ_i^+ and Ψ_f^- is accordingly

$$\begin{aligned} \delta T_{if} = {} & \langle \delta\Psi_f^- | V | (\varphi_i - \Psi_i^+ + \hat{G}_0^+ V \Psi_i^+) \rangle \\ & + \langle \varphi_f - \Psi_f^- + \hat{G}_0^- V \Psi_f^- | V | \delta\Psi_i^+ \rangle. \end{aligned} \tag{7.60}$$

Invoking once more the Lippmann–Schwinger equation (7.57), we find that δT_{if} is indeed zero.

Instead of summing the two representations of the T-matrix element given in Eq. (7.55) to make a statement about δT_{if}, we might employ their product and write T_{if} as

$$T_{if} = \frac{\langle \varphi_f | V | \Psi_i^+ \rangle \langle \Psi_f^- | V | \varphi_i \rangle}{\langle \Psi_f^- | V | \Psi_i^+ \rangle - \langle \Psi_f^- | V \hat{G}_0^+ V | \Psi_i^+ \rangle}. \tag{7.61}$$

Since it may be established that the right-hand side is invariant upon a small change in Ψ_i^+ and Ψ_f^-, (7.61) lays the basis for a fractional version of the T-matrix variational principle. The particular virtue of this formula rests on its independence of the scattering state normalization. Variational approaches involving the S- or the K- instead of the T-matrix are obtained in the same manner as Eqs. (7.59) and (7.61) (see Exercise 7.3).

Exercise 7.3. Show that (7.61) is stationary with respect to small variations of Ψ_i^+ and Ψ_f^-. Derive expressions analogous to (7.59) and (7.61) for the K instead of the T-matrix.

The condition $\delta T_{if} = 0$ (or $\delta S_{if} = \delta K_{if} = 0$) can be utilized to derive the scattering solutions as the arguments that make the considered functional stationary. We add that variational expressions such as (7.59) and (7.61) are preferable to more compact formulae for the same quantity, as given by Eq. (7.55). This is related to the fact that, in general, only approximate solutions of a scattering problem are available, such that the result of an actual computation may be properly described as $\Psi_i^+ + \delta\Psi_i^+$ ($\Psi_f^- + \delta\Psi_f^-$) if Ψ_i^+ (Ψ_f^-) stands for the exact scattering state. But the error of the transition matrix, or analogous entities, as induced by a small imprecision of the wave functions, vanishes to first order if the matrix element is defined as in Eqs. (7.59) and (7.61). For the alternative Eq. (7.55), in contrast, the leading order of the error is the first.

In the following, we discuss the algebraic form of the variational equation. As an essential observation in this context, the scattering solution may be expanded by writing

$$\Psi_i^+ = \sum_k a_{ki} v_k \tag{7.62}$$

where the functions $\{v_k\}$ are appropriately chosen basis states. The salient point about this linear combination is that the v_k can be selected as square integrable functions although the scattering solution contains periodic contributions. That restriction to an $\mathcal{L}^2$ basis is indeed sufficient for the purpose at hand can be inferred from the structure of Eqs. (7.59) and (7.61) where each of the scalar products involves at least one potential V multiplying the wave function. Since we stipulate that V tends towards zero asymptotically, it suffices to expand the solutions within an appropriately chosen finite domain.

Defining two matrices **B** and **C** by their elements

$$B_{ki} = \langle v_k | V | \psi_i \rangle \tag{7.63}$$

and

$$C_{mn} = \langle v_m | (V - V\hat{G}_0^+ V) | v_n \rangle, \tag{7.64}$$

we rewrite Eq. (7.59) as a matrix equation:

$$\mathbf{T} = \mathbf{a}^{\mathrm{T}}\mathbf{B} + \mathbf{B}^{\mathrm{T}}\mathbf{a} - \mathbf{a}^{\mathrm{T}}\mathbf{C}\mathbf{a} \tag{7.65}$$

where $\mathbf{a}$ denotes the array of coefficients a_{ki} in the expansion (7.62). A straightforward way of enforcing the stationary character of $\mathbf{T}$ upon variation of the scattering function is to set the first derivative of $\mathbf{T}$ with respect to the coefficient matrix $\mathbf{a}^T$ equal to zero. This constraint yields

$$\mathbf{a} = \mathbf{C}^{-1}\mathbf{B} \tag{7.66}$$

and leads to the following expression for the transition matrix:

$$\mathbf{T} = \mathbf{B}^T\mathbf{C}^{-1}\mathbf{B}. \tag{7.67}$$

We underscore that the importance of formula (7.67) does not lie in its algebraic but in its variational character. Obviously, generating an algebraic expression for the T-matrix does not present any difficulty [7] and can be achieved immediately by use of the relation $T_{if} = \langle \psi_f | V | \Psi_i^+ \rangle$ in conjunction with the Lippmann–Schwinger equation. More specifically, we rely on the argument used before and expand $V\Psi_i^+$ in terms of an $\mathcal{L}^2$ basis $\{u_k\}$, which is legitimate as the potential vanishes asymptotically. Thus,

$$V\Psi_i^+ = \sum_k a_{ki} u_k. \tag{7.68}$$

We multiply the Lippmann–Schwinger equation (7.57) for Ψ_i^+ from the left by V and cast the result in matrix form, arriving at

$$\mathbf{O}\mathbf{a} = \mathbf{B} + \mathbf{F}\mathbf{a} \tag{7.69}$$

with

$$B_{ji} = \langle u_j | V | \psi_i \rangle, \tag{7.70}$$

and

$$F_{mn} = \langle u_m | V G_0^+ | u_n \rangle \tag{7.71}$$

and $\mathbf{O}$ as the overlap matrix defined by $O_{jk} = \langle u_j, u_k \rangle$ which is not necessarily diagonal as the basis vectors $\{u_k\}$ may be chosen as non-orthogonal. As in Eq. (7.65), $\mathbf{a}$ stands for the array of the expansion coefficients. With the assignment

$$\mathbf{M} \equiv \mathbf{O} - \mathbf{F} \tag{7.72}$$

we find

$$\mathbf{a} = \mathbf{M}^{-1}\mathbf{B} \tag{7.73}$$

and, consequently,

$$\mathbf{T} = \mathbf{D}^{\dagger}\mathbf{M}^{-1}\mathbf{B} \tag{7.74}$$

or

$$T_{fi} = \sum_{k} \langle \varphi_f | u_k \rangle a_{ki} \tag{7.75}$$

where the matrix **D** comprises the projections of the asymptotic solutions on the basis vectors $\{u_k\}$:

$$D_{kf} \equiv \langle u_k | \varphi_f \rangle . \tag{7.76}$$

Comparing the results (7.67), (7.74), and (7.75), we note that, by construction, the error in the wave function is of second order in the former while it is of first order in the latter, making Eq. (7.67) preferable over Eqs. (7.74) and (7.75) in numerical applications.

7.3.1 *The Kohn variational principle*

In numerous cases, the reactive scattering of molecular species has been described by use of the *Kohn variational principle* [186] which, in its originally proposed form, relies on the K-matrix. More specifically, a variational expression for a K-matrix element is given by [187]

$$K = K_{\mathrm{t}} - \frac{2}{\hbar} \langle \Psi_{\mathrm{t}} | \hat{H} - E | \Psi_{\mathrm{t}} \rangle \tag{7.77}$$

with K_{t} as a trial matrix element, obtained from the asymptotic limit of the trial wave function Ψ_{t}. Accordingly,

$$\Psi_{\mathrm{t}} \sim v_0 + K_{\mathrm{t}} v_1 \tag{7.78}$$

if the asymptotic solution is constrained to be real-valued (see Eq. (3.108)). For simplicity, we address here a single-channel problem, involving s-wave scattering, to illustrate the basic notions underlying the variational approach in scattering theory.

The variation of the *Kohn integral* $I = \langle \Psi_{\mathrm{t}} | \hat{H} - E | \Psi_{\mathrm{t}} \rangle$ is easily evaluated in the assumed case of s-wave scattering. Introducing a trial wave function

$\Psi_t = \Psi + \delta\Psi$, where Ψ solves the stationary Schrödinger equation, we expand I to first order in $\delta\Psi$ and find

$$\delta I = \langle\Psi|\hat{H} - E|\delta\Psi\rangle + \langle\delta\Psi|\hat{H} - E|\Psi\rangle = \langle\Psi|\hat{H} - E|\delta\Psi\rangle \tag{7.79}$$

since the terms $\langle\Psi|\hat{H} - E|\Psi\rangle$ as well as $\langle\delta\Psi|\hat{H} - E|\Psi\rangle$ vanish due to the premise $\hat{H}\Psi = E\Psi$.[1] Inserting into the remaining scalar product $\langle\Psi|H - E|\delta\Psi\rangle$ the explicit form of the Hamiltonian, $-\frac{\hbar^2}{2m}\frac{\partial^2}{\partial\rho^2} + V(\rho)$, and performing partial integration twice, we arrive at

$$\delta I = -\frac{\hbar^2}{2m}\left(\Psi^*\frac{\partial\delta\Psi}{\partial\rho} - \frac{\partial\Psi^*}{\partial\rho}\delta\Psi\right)\Bigg|_0^\infty . \tag{7.80}$$

The regularity constraint $\Psi(\rho = 0) = \delta\Psi(\rho = 0) = 0$ reduces the lower limit in this expression to zero, leaving

$$\delta I = -\frac{\hbar^2}{2m}\left(\Psi^*\frac{\partial\delta\Psi}{\partial R} - \frac{\partial\Psi^*}{\partial\rho}\delta\Psi\right)\Bigg|_{\rho=\infty} . \tag{7.81}$$

Since we choose real functions for both v_0 and v_1 the asymptotic form of Ψ is written according to Eqs. (3.108) and (3.111) as

$$\Psi(\rho) \sim A[j_0(k\rho) + \tan(\delta)n_0(k\rho)] = A[j_0(k\rho) + Kn_0(k\rho)] \tag{7.82}$$

which corresponds to

$$\delta\Psi(\rho) \sim A\delta K\, n_0(k\rho). \tag{7.83}$$

If we further adopt $A = w^{-1/2}$, with $w = \hbar k/m$, as normalization factor and insert Eqs. (7.82) and (7.83) into Eq. (7.81), we derive the variational condition (see Exercise 7.4)

$$\delta\left(\frac{2}{\hbar}I - K\right) = 0. \tag{7.84}$$

Since $\delta K = K_t - K$, relation (7.84) implies immediately the variational expression for the K-matrix element as given by Eq. (7.77).

Exercise 7.4. Use relation (7.81) to establish variational conditions analogous to (7.81) for (a) the transition matrix element T, (b) the scattering matrix element S. For case (a), employ $\lim_{\rho\to\infty}\Psi = \sin(k\rho) + T\exp(ik\rho)$ as the

[1]Note that no assumption has been made with respect to the range of the basis set functions. Some of these functions might be chosen as periodic, such that the Hamiltonian is not guaranteed to be Hermitian.

asymptotic form of the wave function. For case (b), choose $\lim_{\rho\to\infty} \Psi = \exp(-ik\rho) + S\exp(ik\rho)$.

7.3.2 *Kohn anomalies*

In the following step, we supplement the basis set $\{v_n\}$ with short-range functions ($n > 1$) besides real-valued periodic ($n = 0, 1$) functions. This generalization leaves the asymptotic condition (7.81) unaffected. The extended basis may provide an expansion of the trial function Ψ_t:

$$\Psi_t = v_0 + \sum_k a_k v_k \tag{7.85}$$

which, by suitable choice of the basis functions, vanishes at the origin. Simple realizations of v_0 and v_1 are given by $\varpi^{-\frac{1}{2}} \sin(k\rho)$ and with $\varpi^{-\frac{1}{2}} f(\rho)\cos(k\rho)$, respectively, where $f(\rho)$ enforces a regularization constraint, subjected to the requirement that $f(0) = 0$ and that f is asymptotically constant with $\lim(\rho \to \infty) f(\rho) = 1$. Identifying the coefficient a_1 with the trial K-matrix element K_t, as dictated by the asymptotic structure of Ψ_t (see Eq. (7.78)), one concludes that

$$K = -\frac{2}{\hbar}\left[\langle v_0|\hat{H} - E|v_0\rangle + \sum_{i>0} a_i\langle v_i|\hat{H} - E|v_0\rangle + \sum_{i,j>0} a_i a_j\langle v_i|\hat{H} - E|v_j\rangle\right] \tag{7.86}$$

where we have used the relation

$$\langle v_0|\hat{H} - E|v_1\rangle - \frac{\hbar}{2} = \langle v_1|\hat{H} - E|v_0\rangle \tag{7.87}$$

as obtained by twofold partial integration. In the following step, we implement the variational condition $\partial K/\partial a_k = 0$, and arrive at linear equations for the coefficients which are solved by matrix inversion. Inserting these solutions into (7.86), we find [187]

$$K = -\frac{2}{\hbar}\left[M_{00} - \sum_{m,n=1} M_{m0}\widetilde{M}^{-1}_{mn} M_{n0}\right] \tag{7.88}$$

with

$$M_{ij} \equiv \langle v_i|\hat{H} - E|v_j\rangle, \quad i,j = 0, 1, 2, \ldots; \quad \text{and} \quad \widetilde{M}_{mn} = \langle v_m|\hat{H} - E|v_n\rangle, \tag{7.89}$$

where $m, n > 0$. By definition of the basis set $\{v_n\}$, the matrix $\widetilde{M}_{mn}$ $(m, n > 0)$ is real and symmetric. Consequently, the inverse of this matrix exhibits singularities at certain selected energy values, rendering a numerical procedure based on (7.88) and (7.89) unstable.

The divergences in expression (7.88) are known as Kohn anomalies [153]. A remedy against these anomalies is available with an extension of the Kohn variational method from the K- to the S-matrix [188, 189]. A prelude to the S-matrix version of the Kohn principle is provided by Exercise 7.4. In analogy to (7.77), the S-matrix element may be written in variational form as [190]

$$S = S_t - \frac{i}{\hbar}\int_0^\infty d\rho\, \Psi_t(\hat{H} - E)\Psi_t \tag{7.90}$$

where S_t represents a trial S-matrix element and Ψ_t again a trial wave function which, however, is chosen as complex-valued, according to S-matrix boundary conditions. Once more, we expand the trial wave function into a basis:

$$\Psi_t = -u_0 + \sum_k c_k u_k \tag{7.91}$$

where, as before, the sequence $\{u_k\}$ with $k > 1$ consists of real-valued short-range functions. The functions with $k = 0, 1$ are periodic,

$$u_0 = f(\rho)\exp(-ik\rho), u_1 = f(\rho)\exp(ik\rho), \tag{7.92}$$

but regularized by a factor that reduces them to zero at $\rho = 0$ and tends towards one in the asymptotic regime. Repeating the steps made in the context of the K-matrix-based variational principle, we arrive at

$$S = \frac{i}{\hbar}\left[\mathrm{N}_{00} - \sum_{m,n=1} \mathrm{N}_{m0}(\widetilde{\mathrm{N}}^{-1})_{mn}\mathrm{N}_{n0}\right] \tag{7.93}$$

with

$$N_{ij} = \int_0^\infty d\rho\, u_i(H - EI)u_j, \quad i, j = 0, 1, 2, \ldots,$$

$$\widetilde{N}_{mn} = \int_0^\infty d\rho\, u_m(H - EI)u_n, \quad m, n > 0. \tag{7.94}$$

The salient feature of this formula is that the matrix $\widetilde{\mathrm{N}}$ is not Hermitian (i.e. $\widetilde{N}_{ij}$ differs from $\widetilde{N}^*_{ji}$). It is, instead, complex symmetric [191]. We make this property explicit in Eq. (7.94) by writing the matrix elements as integrals. The complex symmetry feature ensures that the determinant of $\tilde{\mathbf{N}}$ becomes zero at values of E with non-vanishing imaginary parts, as far as the continuous spectrum is considered [191]. In the S-matrix version of the Kohn variational principle, the divergences plaguing the original K-matrix formulation of this principle are thus eliminated, making the former procedure preferential to the latter in numerical applications. As pointed out by Robicheaux [187], the use of complex-valued short-range basis functions $\{v_k\}$ with $k > 1$ in conjunction with the K-matrix formalism equally frees the Kohn variational principle from singularities. We add that the majority of the matrix elements entering the variational prescriptions (7.88) and (7.93) involve $\mathcal{L}^2$ basis functions only. If two functions with indices m and n are of finite range, $\langle u_m|\hat{H}|u_n\rangle$ ($\langle v_m|\hat{H}|v_n\rangle$) is energy-independent, and thus the matrix element does not need to be recalculated as the energy E is varied.

Exercise 7.5. Establish the complex symmetric property of the matrix $\tilde{\mathbf{N}}$ as defined by (7.94). Show also that the matrices $\mathbf{M}$ (7.89) and $\mathbf{N}$ (7.94) are not Hermitian [7]. Discuss the modified case where one of the two extended functions v_0, v_1 (u_0, u_1) is reduced to a finite range.

The following section deals with the application of a variational approach to a nonadiabatic situation.

7.4 Case Study: Quenching of the Sodium Atom 3*p* State by Interaction with Hydrogen Molecules

The deexcitation process

$$\mathrm{Na}(^2P) + \mathrm{H}_2(X^1\Sigma_g^+, v, j) \rightarrow \mathrm{Na}(^2S) + \mathrm{H}_2(X^1\Sigma_g^+, v', j') \tag{7.95}$$

provides a prototypical example of the transfer of energy from electronic to rovibrational degrees of freedom. The relative simplicity of this reaction in conjunction with its experimental accessibility ([192–195]) makes it a preferred subject of study for the purpose of benchmarking quantum chemical theories of dynamic molecular interaction. In the present section, we will highlight the treatment of this process in the framework of time-independent variational theory according to the comprehensive study by

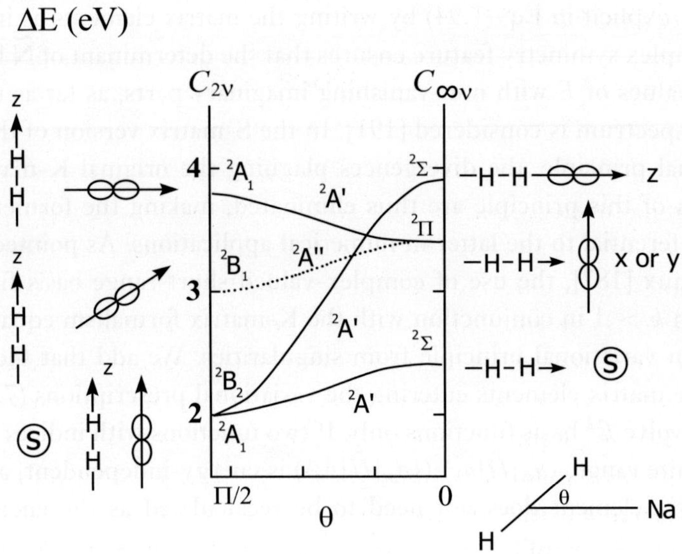

Figure 7.4: Four diabatic potential energy curves resulting from the interaction between H_2 with the 3*s* or the three 3*p* atomic orbitals of an Na projectile, as a function of the bending angle, leading from C_{2v} to $C_{\infty v}$ symmetry. The ground and the first excited state are degenerate in C_{2v} symmetry, while degeneracy between the two remaining excited states is found at $C_{\infty v}$ symmetry, corresponding to a collinear arrangement of the three atoms. An empirical model was parametrized with values extracted from *ab initio* computations [196, 197] to generate these curves. The energies are in eV. (Reprinted with permission from [198]. Copyright (1997) by the American Institute of Physics.)

Tawa *et al.* [152]. In Chapter 11 we will present a complementary exploration of the same reaction from the viewpoint of time-dependent nonadiabatic quantum dynamics.

The reactant side of (7.95) involves the first excited electronic state of the triatomic system combined of Na and H_2, the product side of the ground state. To describe the possible geometric configurations of the three atoms, we use Jacobi coordinates, as sketched in Figure 7.4.

Adiabatic potential energy surfaces for the system $Na + H_2$ are obtained from the eight states of lowest energy as generated within the *diatomics in molecules* (DIM) procedure [199], and adopting the valence-bond model of electronic structure theory (see Section 1.5.1). The two lowest surfaces are retained, corresponding to the ansatz

$$\Psi = \sum_{i=1,2} \xi_i^{\mathrm{A}}(\mathbf{x}, \widetilde{\mathbf{R}})\chi_i(\widetilde{\mathbf{R}}) \tag{7.96}$$

for the overall wave function of the system, which is expressed as a sum over products of electronic and nuclear states, labeled ξ_i and χ_i, respectively, and includes the electronic ground and the first excited state. Here, $\widetilde{\mathbf{R}}$ is a short-hand notation for the set of mass-scaled Jacobi coordinates $\widetilde{\rho}$, $\widetilde{r}$ and the angle Θ included by $\widetilde{\boldsymbol{\rho}}$ and $\widetilde{\mathbf{r}}$. The study of Tawa *et al.* [152] treats the Na-H_2 electronic system within a two-state model. Finer details of the electronic system are not taken into account. Thus, the spin-orbit splitting between the $^2P_{1/2}$ and $^2P_{3/2}$ states of Na is neglected, and the electronic angular momentum not included explicitly, i.e. the total angular momentum $\mathcal{J}$ contains only nuclear components.

Within the simplified frame thus defined, the dynamical computations are based on a diabatic model as obtained by means of an adiabatic-to-diabatic transformation [200, 201]. The matrix $\mathbf{W}$ (see Eq. (2.55)) maps the two adiabatic electronic states $\{\xi_i^{\mathrm{A}}\}$ into their diabatic counterparts $\{\xi_i^{\mathrm{D}}\}$ $(i = 1, 2)$ which are used to generate the two-by-two potential energy matrix $\mathbf{V}^{\mathrm{D}}$. We point out that isolating a subgroup of states for treatment within the frame of a diabatic model, as is done here with the two Na + H_2 states of lowest energy, requires explicit justification. A rationale for this procedure was supplied by a study of Vibok *et al.* [202] who performed a comprehensive analysis of the conical intersections associated with the four states indicated in Figure 7.4 and concluded that the diabatic coupling between the two lower and the two higher states is sufficiently weak to allow neglecting the latter group when diabatizing the former. The reader can find a wider discussion of this issue in [203].

The nuclear coordinates that are needed to derive the transformation angle θ according to Eq. (2.57) are $\widetilde{\rho}$, $\widetilde{r}$ and the angle Θ. The two coupled equations for the nuclear wave functions on the two surfaces are then given by

$$(\hat{T} + V_{11}^{\mathrm{D}}(\widetilde{\mathbf{R}}) - E)\chi_1(\widetilde{\mathbf{R}}) = -V_{12}^{\mathrm{D}}(\widetilde{\mathbf{R}})\chi_2(\widetilde{\mathbf{R}}), \tag{7.97}$$

$$(\hat{T} + V_{22}^{\mathrm{D}}(\widetilde{\mathbf{R}}) - E)\chi_2(\widetilde{\mathbf{R}}) = -V_{12}^{\mathrm{D}}(\widetilde{\mathbf{R}})\chi_1(\widetilde{\mathbf{R}}). \tag{7.98}$$

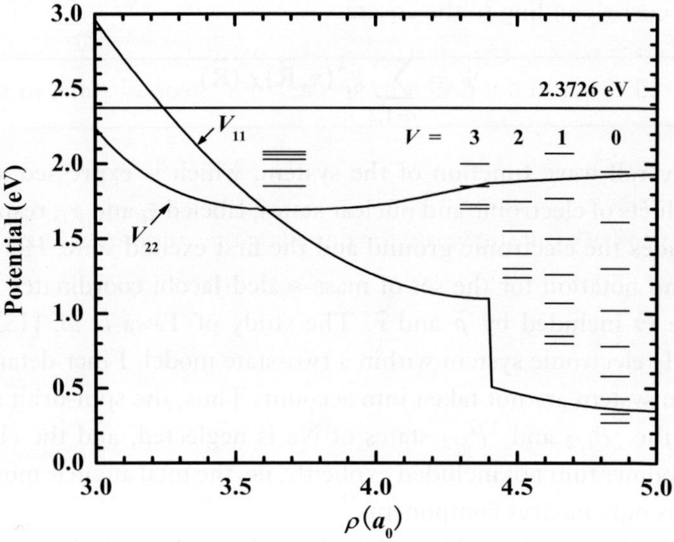

Figure 7.5: Diabatic potential energy curves V_{11}^{D} and V_{22}^{D} for the interaction between Na and H_2 are plotted versus the distance of approach, where C_{2v} symmetry is imposed on the system. The H_2 bond length r has been optimized at each distance ρ to attain the minimal energy V_{22}^{D}. Both V_{11}^{D} and V_{22}^{D} are calculated at the resulting optimum distance r. The discontinuity of V_{11}^{D} and the discontinuous derivative of V_{22}^{D} at $\rho = 4.33\ a_0$ are both consequences of this procedure. The horizontal line at $E = 2.3726$ eV signifies the dissociation energy of the complex $\mathrm{Na}(^2P) + H_2$. The levels in the intersection region denote resonance (exciplex) energies, those on the right refer to ground state rovibrational energies. (Reprinted with permission from [204]. Copyright (1993) by the American Chemical Society.)

The symbol $\hat{T}$ refers to the kinetic energy operator in the coordinate set $\tilde{\boldsymbol{\rho}}, \tilde{\mathbf{r}}$, as given by Eq. (6.98). In terms of the diabatic electronic functions $\xi_1^{\mathrm{D}}, \xi_2^{\mathrm{D}}$, the potential operator may be written as

$$\hat{V}^{\mathrm{D}} = \sum_{k,k'=1,2} |\xi_k^{\mathrm{D}}\rangle V_{kk'} \langle \xi_{k'}^{\mathrm{D}}|. \tag{7.99}$$

The potential energy matrix is diagonal in C_{2v} geometry. Figure 7.5 gives an impression of the two components V_{11}^{D} and V_{22}^{D} as a function of the distance of approach ρ as C_{2v} symmetry is imposed on the scattering system. A point of intersection is found close to the minimum of V_{22}^{D}. Since $V_{12}^{\mathrm{D}} = 0$, the image presents both the diabatic and the adiabatic curves. If we denote the latter by V_1^{A} and V_2^{A}, it holds that $V_1^{\mathrm{A}} = V_{11}^{\mathrm{D}}$ and $V_2^{\mathrm{A}} = V_{22}^{\mathrm{D}}$ to the right, but $V_1^{\mathrm{A}} = V_{22}^{\mathrm{D}}$ and $V_2^{\mathrm{A}} = V_{11}^{\mathrm{D}}$ to the left of the intersection. As indicated in the figure, the funnel region supports a set of metastable states. These are attributed to the formation of an exciplex structure, a complex consisting of an excited entity and a different species in its ground state, in this case realized by $Na(^2P)$ and $H_2(X^1\Sigma_g^+)$.

For simplicity, we will drop in the following the index D and let the symbol V refer to the diabatic potential. This potential is separated into an internal and an asymptotic part determined by the vibrational motion of the H_2 molecule in infinite distance from the Na atom:

$$V = V^{\text{vib}} + V^{\text{int}}. \tag{7.100}$$

With this partition, the Hamiltonian reduces asymptotically to

$$\lim_{\rho\to\infty} \hat{H} = -\frac{\hbar^2}{2\mu}\Delta_{\widetilde{r}} + V^{\text{vib}}, \tag{7.101}$$

since V^{int} vanishes as $\widetilde{\rho} \to \infty$. This internal potential is further subdivided into a distortion and a coupling part, labeled V^{Di} and V^C, respectively. The distortion potential governs all scattering processes that are unrelated to the mutual interaction between the two electronic channels. This interaction is thus obtained as the remainder of the internal potential, after V^{Di} has been subtracted:

$$V^C = V^{\text{Int}} - V^{\text{Di}}. \tag{7.102}$$

The advantage of this partition is that it allows us to choose distorted waves $\breve{\psi}^{(0)}$ (see Section 6.3.1) as states of reference, replacing the asymptotic wave functions. The wave functions $\breve{\psi}_i^{(0)}$ are defined as eigenstates of the distorted-wave Hamiltonian $\hat{H}^{\text{Di}} = \hat{T} + V^{\text{vib}} + V^{\text{Di}}$. With this approach, the variational treatment is limited to the coupling V^{C} which, in many cases, is a small perturbation. This in turn makes it possible to operate with a basis of reduced size when implementing the variational procedure.

Correspondingly, the scattering solution satisfies the Lippmann–Schwinger equation

$$\Psi^{\pm} = \breve{\psi}^{(0)\pm} + \hat{G}^{\text{Di}\pm} U^{\text{C}} \Psi^{\pm} \tag{7.103}$$

with $U^{\text{C}} \equiv -\frac{2\mu}{\hbar^2} V^{\text{C}}$. An outgoing wave is chosen, in keeping with S-matrix boundary conditions. The Green's operator in Eq. (7.103) contains the distortion Hamiltonian according to

$$\hat{G}^{\mathrm{Di}\mp} = \lim_{\varepsilon\to 0^+} \frac{\hbar^2}{2\mu} \frac{1}{E - \hat{H}^{\mathrm{Di}} \pm i\varepsilon}. \tag{7.104}$$

Tawa *et al.* [152] approximate the outgoing scattered wave $\Psi^+_{\mathrm{OW}} = \hat{G}^{\mathrm{Di}+} U^{\mathrm{C}} \Psi^+$ by use of the *outgoing wave variation principle* (OWVP) [205], the S-matrix version of a variational principle first proposed by Schlessinger in the context of the T-matrix [206]. Exercise 7.6 provides a basic introduction to this principle in its original form.

Exercise 7.6. Consider a scattering solution Ψ_k of the stationary Schrödinger equation $(\hat{H} - E)\Psi_k = 0$, as well as the solution φ_k of the equation $(\hat{H}_0 - E)\varphi_k = 0$, involving a reference Hamiltonian $\hat{H}_0$ realized by the asymptotic Hamiltonian of the problem. Use the decomposition $\Psi_k = \varphi_k + \psi_k$ to show that the transition operator matrix element[2] $T_{kk'} = \langle \varphi_k | \hat{T} | \varphi_{k'} \rangle$ may be written as

$$T_{k'k} = \langle \varphi_k | V | \varphi_{k'} \rangle + \langle \varphi_k | V | \psi_{k'} \rangle + \langle \psi_k | V | \varphi_{k'} \rangle - \langle \psi_k | E - \hat{H} | \psi_{k'} \rangle \tag{7.105}$$

where $\hat{H} = \hat{H}_0 + \hat{V}$. Show also that this expression is stationary as ψ_k is varied.

The S-matrix for the considered scattering of H_2 by Na may be separated in analogy to the internal potential partition described by Eq. (7.102) [152]:

$$S_{k,n,k_0,n_0} = \delta_{k,k_0}\, {}^0S_{k,n,k_0,n_0} + {}^{\mathrm{C}}S_{k,n,k_0,n_0}. \tag{7.106}$$

The index k labels the diabatic electronic states, as in Eq. (7.99), while n is a collective index summarizing all nuclear quantum numbers that are not constants of motion. These comprise, in the body-fixed frame, the vibrational and rotational quantum numbers v and j, respectively, as well as the projection K on the body-fixed axis, defined by $\widetilde{\boldsymbol{\rho}}$. The indices k_0 and n_0 refer to the entrance channel values of k and n. The identity (7.106) subdivides the S-matrix element into the contributions ${}^0S$ and ${}^{\mathrm{C}}S$, accounting for

[2]The symbol $\hat{T}$ in this matrix element must not be confused with that for the kinetic energy.

the distortion and the coupling effect, respectively. Since distortion does not give rise to any electronic transition, the ${}^{0}S$ matrix is diagonal in the indices k and k_0.

The second term on the right-hand side of Eq. (7.106) stands for the scalar product

$$ {}^{\mathrm{C}}S_{k,n,k_0,n_0} = \langle \breve{\psi}^{(0)-k,n} | U^{\mathrm{C}} | \Psi^{+k_0,n_0} \rangle, \tag{7.107} $$

linking a regular incoming distorted wave to an outgoing scattering solution with the entrance channel quantum numbers k_0 and n_0. This formula, however, is not variational. Tawa *et al.* [152] base their computation on an alternative expression for ${}^{\mathrm{C}}S_{k,n,k_0,n_0}$ which is stationary with respect to the variation of the *outgoing wave* $\Psi_{\mathrm{OW}}^{\pm k,n} \equiv \Psi^{\pm k,n} - \breve{\psi}^{(0)\pm k,n}$, namely

$$ \begin{aligned} {}^{\mathrm{C}}S_{knk_0n_0} &= \langle \breve{\psi}^{(0)-k,n} | U^{\mathrm{C}} | \breve{\psi}^{(0)+k_0,n_0} \rangle + \langle \breve{\psi}^{(0)-k,n} | U^{\mathrm{C}} | \Psi_{\mathrm{OW}}^{+k_0,n_0} \rangle \\ &\quad + \langle \Psi_{\mathrm{OW}}^{-k,n} | U^{\mathrm{C}} | \breve{\psi}^{(0)+k_0,n_0} \rangle - \langle \Psi_{\mathrm{OW}}^{-k,n} | \frac{2\mu}{\hbar^2} (\hat{H} - E) | \Psi_{\mathrm{OW}}^{+k_0,n_0} \rangle. \end{aligned} \tag{7.108} $$

This form of the S-matrix element due to coupling embodies the outgoing wave variation principle [207]. Other variational schemes may be considered. Thus, from Eq. (7.103), it follows that $\Psi_{\mathrm{OW}}^{+} = \hat{G}^{\mathrm{Di}+} U^{\mathrm{C}} \Psi^{+}$. It is possible to recast the variational expression (7.108) into a form that employs the *amplitude densities* $\zeta \equiv U^{\mathrm{C}}\Psi$ which are governed by the modified Lippmann–Schwinger equation

$$ \zeta^{\pm} = U^{\mathrm{C}} \breve{\psi}^{(0)\pm} + U^{\mathrm{C}} \hat{G}^{\mathrm{Di}\pm} \zeta^{\pm}. \tag{7.109} $$

As the multiplying potential is of finite range, the advantage of this reformulation is once more that it permits choosing $\zeta^{\pm}$ as $\mathcal{L}^2$ functions. It presents no difficulty to formulate Eq. (7.108) in terms of the amplitude densities, replacing, for instance, the scalar product $\langle \Psi_{\mathrm{OW}}^{-k,n} | (\hat{H} - E) \Psi_{\mathrm{OW}}^{+k_0,n_0} \rangle$ with $\langle \zeta^{-k,n} | \hat{G}^{\mathrm{Di}+} (\hat{H} - E) \hat{G}^{\mathrm{Di}+} | \zeta^{+k_0,n_0} \rangle$. The resulting expression is correct to second order in the amplitude density error.

7.4.1 *Basis sets*

Depending on whether ζ or Ψ_{OW} is varied, one will use different approaches to arrive at an algebraic form of the variational procedure. In the first case, a possible expansion is [207]

$$\zeta^+(\widetilde{\rho}, \mathrm{v}) = \sum_m A_m \varphi_m(\widetilde{\rho}, \mathrm{v}) \tag{7.110}$$

where we have suppressed the quantum numbers n and k, and separated the scattering coordinate $\widetilde{\rho}$ from the collective argument v which includes all electronic, vibrational, rotational, and orbital variables. The product states $\varphi_m(\widetilde{\rho}, \mathrm{v})$ are defined by

$$\varphi_m(\widetilde{\rho}, \mathrm{v}) = \frac{1}{\widetilde{\rho}} t_m(\widetilde{\rho})\phi_m(\mathrm{v}) \tag{7.111}$$

with t_m as the translational part and ϕ_m as a channel eigenfunction. The basis states φ_m are understood as $\mathcal{L}^2$-integrable.

Non-$\mathcal{L}^2$ basis functions are generated by mapping the amplitude density onto the outgoing wave and expanding according to

$$\begin{aligned}\Psi^+_{\mathrm{OW}}(\widetilde{\rho}, \mathrm{v}) &= \hat{G}^{\mathrm{Di}+}\zeta^+(\widetilde{\rho}, \mathrm{v}) \\ &= \sum_m A_m \hat{G}^{\mathrm{Di}+}\varphi_m(\widetilde{\rho}, \mathrm{v}) = \sum_m A_m g'^{+}_{m}(\widetilde{\rho}, \mathrm{v}).\end{aligned} \tag{7.112}$$

Here g'_m denotes *a half-integrated Green's function* which obeys the relation

$$\frac{2\mu}{\hbar^2}(\hat{H}^{\mathrm{Di}} - E)g'^{+}_{m} = \varphi_m \tag{7.113}$$

and is, for open channels, constrained to satisfy complex boundary conditions. By definition, these basis functions are energy-dependent. The term *half-integrated Green's function* refers to the expansion of $\hat{G}^{\mathrm{Di}+}$ in terms of $\widetilde{\rho}$ and v dependent states, namely [208]

$$\langle\widetilde{\rho}, \mathrm{v}|\hat{G}^{\mathrm{Di}+}|\widetilde{\rho}', \mathrm{v}'\rangle = \frac{1}{\widetilde{\rho}\widetilde{\rho}'}\sum_{nn'}\Delta_{nn'}\phi^*_n(\mathrm{v})\phi_{n'}(\mathrm{v}')g^+_{nn'}(\widetilde{\rho}, \widetilde{\rho}'). \tag{7.114}$$

The Δ symbol is here to be understood as a Kronecker delta that vanishes only if n and n' are channel indices that involve different electronic states. Details of this convention can be found in [152, 208]. The Green's function $g^+_{nn'}(\widetilde{\rho}, \widetilde{\rho}')$ is composed of radial solutions for the distorted channel equations, disregarding the electronic coupling potential. Integrating over the $\widetilde{\rho}'$ dependent functions used to represent the channel states φ_m and the propagator, we construct the half-integrated Green's function appearing in Eqs. (7.112) and (7.113) by use of the prescription

$$g_m^{\prime +}(\widetilde{\rho}, \mathbf{v}) = \frac{1}{\widetilde{\rho}} \sum_n \Delta_{nm} \left[\int d\rho' g_{nm}^{+}(\widetilde{\rho}, \widetilde{\rho}') t_m(\widetilde{\rho}) \right] \phi_n(\mathbf{v}). \tag{7.115}$$

Both the finite-range basis set $\{\varphi_m\}$ and its extended counterpart $\{g_m^{\prime +}\}$ may be exploited to derive a set of equations for the expansion coefficients A_m. As a result of the assessment presented in [205], the second set achieves better convergence than the first. A gain in flexibility is accomplished by employing a well-balanced hybrid basis that mixes both prototypes and may be further supplemented by arbitrary additional basis states, as deemed suitable for the problem at hand [152]. The outgoing wave is then expanded into a basis $\{\Gamma_m\}$

$$\Psi_{\mathrm{OW}}^{+k_0,n_0}(\mathbf{x}, \widetilde{\mathbf{R}}) = \sum_m A_m^{k_0,n_0} \Gamma_m(\mathbf{x}, \widetilde{\mathbf{R}}) \tag{7.116}$$

that contains both finite and extended elements.

7.4.2 *Algebraic realization of the outgoing wave variational principle*

The ansatz (7.116) is used for an algebraic formulation of the coupling contribution to the S-matrix element (7.108) in terms of the expansion coefficient vector **A** with elements $A_m^{k_0,n_0}$, as well as the arrays **B** and **C**, defined as

$$B_m^{k_0,n_0} = \langle \Gamma_m | U^{\mathrm{C}} | \breve{\psi}^{(0)+k_0,n_0} \rangle, \tag{7.117}$$

$$C_{mm'} = \left\langle \Gamma_m \left| \frac{2\mu}{\hbar^2} (\hat{H} - E) \right| \Gamma_{m'} \right\rangle, \tag{7.118}$$

and

$${}^{\mathrm{C}}S^0_{knk_0n_0} = \langle \breve{\psi}^{(0)-k,n} | U^{\mathrm{C}} | \breve{\psi}^{(0)+k_0,n_0} \rangle. \tag{7.119}$$

Therefore,

$${}^{\mathrm{C}}S_{knk_0n_0} = {}^{\mathrm{C}}S^0_{knk_0n_0} + \mathbf{B}^{\mathrm{T}}\mathbf{A} + \mathbf{A}^{\mathrm{T}}\mathbf{B} - \mathbf{A}^{\mathrm{T}}\mathbf{C}\mathbf{A}. \tag{7.120}$$

Using the stationarity condition

$$\frac{\delta\, {}^{\mathrm{C}}S_{knk_0n_0}}{\delta A_m^{k',n'}} = 0 \qquad (7.121)$$

with $k' = k_0$ or k, and $n' = n_0$ or n, one obtains the coefficient vector as

$$\mathbf{A} = \mathbf{C}^{-1}\mathbf{B} \qquad (7.122)$$

which, when inserted into (7.108), leads to the variational expression for ${}^{\mathrm{C}}S_{knk_0n_0}$ in the compact form

$${}^{\mathrm{C}}S_{knk_0n_0} = {}^{\mathrm{C}}S^{0}_{knk_0n_0} + \mathbf{B}^{\mathrm{T}}\mathbf{C}\mathbf{B}. \qquad (7.123)$$

7.4.3 *Exciplex funnel dynamics*

The formalism described in the preceding subsections has been used to compute state-to-state quenching probabilities involving the decay of $\mathrm{Na}(^2P)$ into $\mathrm{Na}(^2S)$ [152]. The respective calculations were performed at the relatively high total energy of 2.426 eV, and the values of the total angular momentum $\mathcal{J}$ were restricted to zero and one. From Figure 7.5 it can be seen that this energy is about 0.75 eV above that of the lowest-lying conical intersection of the system. It is also clear from that figure that the selected total energy makes a large multiplicity of final rovibrational states accessible to the decaying complex. For various initial conditions, related to the vibrational and rotational quantum numbers of the system on the excited potential energy surface, the final ground state rovibrational distributions were recorded. Further, the results of the OWVP approach were compared with those of the R-matrix propagation method (see Section 6.5), and the predictions of both procedures were found to be in excellent agreement.

Detailed consideration has been given to the nonadiabatic passage through the lowest-lying conical intersection of the Na + H_2 system and in particular to transitions between H_2 rovibrational levels upon the approach of Na, imposing C_{2v} symmetry on the whole system. These processes are characteristic of the conical intersection type realized in this problem, involving a bound and a dissociative electronic state. The intersection funnel thus supports metastable states, related to exciplex formation, as shown in Figure 7.5 [209]. These states may give rise to Feshbach resonances (see Section 3.5) in which the system is temporarily trapped prior to dissociation involving the transfer of energy between electronic and nuclear degrees of

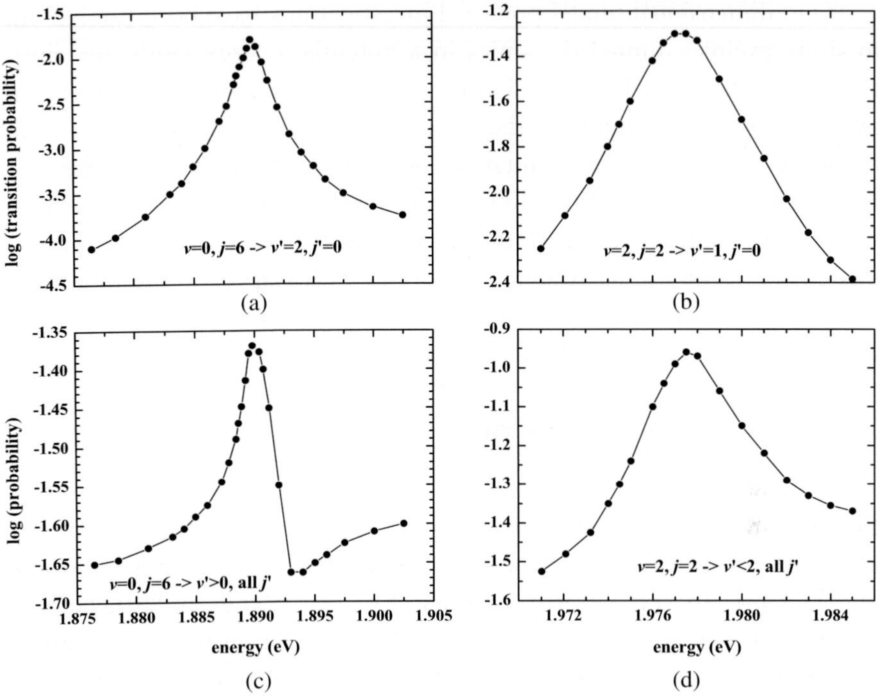

Figure 7.6: Logarithmic state-to-state transition probabilities for the collision of Na with H_2 versus the total energy. The resonance structures are due to Na(^2P)-H_2 exciplex formation. The quantum number pairs (v, j) and (v', j') refer to the initial and the final state of H_2, respectively. The reaction is initiated on the ground state potential energy surface, i.e. with Na in its ^2S state. (Reprinted with permission from [204]. Copyright (1993) by the American Chemical Society.)

freedom. The efficiency of this process is strongly enhanced in the vicinity of the exciplex levels as compared to the case of non-resonant scattering. Figure 7.6 shows the typical resonance behavior associated with these funnel states. Various examples of transitions between initial and final states labeled by the H_2 vibrational and rotational quantum numbers $v(v')$ and $j(j')$, respectively, are included.

The calculated life times of the metastable states are in the range of 0.1–1.0 ps. This exceeds by far the vibrational periods typical for the problem. It is, however, shorter than the radiative life time of the respective levels, such that fluorescence is unlikely to occur as a deexcitation mechanism for these exciplex states.

One-dimensional model potential investigations have been carried out to study exciplex funnel dynamics in a potential energy landscape similar to that shown in Figure 7.5 [210], with emphasis on the competition between direct reflection from the upper diabatic curve and trapping due to funnel resonances. Partial qualitative and even quantitative agreement was found between the full three-dimensional model and the approach involving reduced dimensionality. Both calculations, for instance, led to approximately the same funnel resonance energies with respect to the minimum of the excited adiabatic surface, motivating the use of low-dimensional test computations to assess salient features of exciplex funnel dynamics.

7.5 The Landau–Zener–Stückelberg Model of Nonadiabatic Transitions

We conclude our presentation of the time-independent treatment of molecular scattering involving electronic degrees of freedom by giving an outline of the historically very influential theory of Landau, Zener, and Stückelberg (LZS) ([14–16]). This early attempt to treat the problem of two potential curves crossing at one point analytically in the frame of a semiclassical approach has been formulated in the diabatic representation under rather restrictive assumptions about the potential curve shapes as well as the dynamic situation. In more detail, the conditions for the validity of this model are (1) the diabatic potentials are linear in the vicinity of their crossing point R_0, (2) the coupling potential V_{12} is non-zero and constant close to R_0, and (3) the nuclear velocity v is constant.

For illustration, we discuss the case of a certain diatomic system where these conditions apply. Landau, Zener, and Stückelberg independently pioneered the quantitative treatment of a two-state problem, specifying the diabatic potential matrix in the vicinity of the crossing point and the relevant nuclear velocity in the following way:

$$\begin{aligned} V_{ii}^{\mathrm{D}}(R) = -F_i R_1, \quad R_1 = R - R_0, \;\; i = 1, 2, \\ F_i > 0, \\ V_{12}^{\mathrm{D}} = V_{21}^{\mathrm{D}} = \mathrm{const} \\ v = \mathrm{const}. \end{aligned} \tag{7.124}$$

For clarity, reference is made here to a one-dimensional case. By definition, the two diabatic potentials V_{11} and V_{22} cross at an internuclear distance

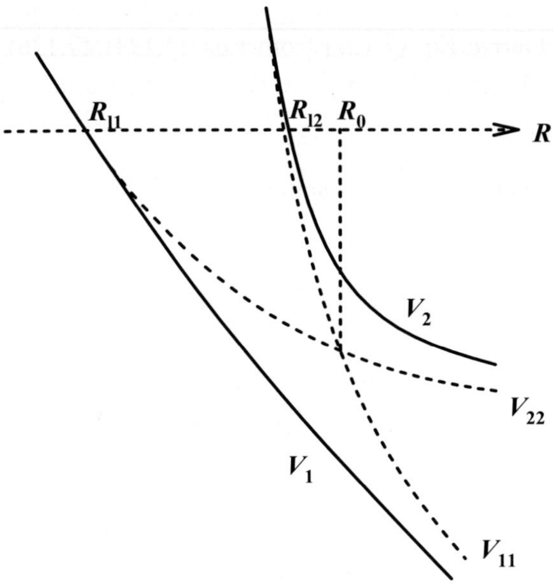

Figure 7.7: Diabatic curves crossing in the two-state model system described in the text. Adiabatic and diabatic potential energy curves, represented by solid and dashed lines, respectively, are shown for a diatomic molecule.

R_0, as shown in Figure 7.7. Since $V_{12} \neq 0$ at R_0, this case involves an avoided crossing of the two corresponding adiabatic potentials.

We refer to the diabatic states corresponding to the two considered electronic levels by the vector $\boldsymbol{\phi} = (\phi_1, \phi_2)$. We further introduce a pair of adiabatic wave functions denoted by $\boldsymbol{\psi} = (\psi_1, \psi_2)$ and connect the two sets of basis functions by a transformation matrix $\mathbf{W}$ as in Eq. (2.57):

$$\boldsymbol{\psi} = \mathbf{W}\boldsymbol{\phi} \tag{7.125}$$

where

$$\mathbf{W} = \begin{pmatrix} \cos[\theta(R)] & \sin[\theta(R)] \\ -\sin[\theta(R)] & \cos[\theta(R)] \end{pmatrix}. \tag{7.126}$$

In Exercise 7.7 it is to be verified that the angle $\theta(R)$ may be written in terms of the potential parameters in (7.124) as

$$\tan[2\theta(R)] = \frac{2V_{12}^{\mathrm{D}}}{V_{22}^{\mathrm{D}}(R) - V_{11}^{\mathrm{D}}(R)}. \tag{7.127}$$

Exercise 7.7. Derive Eq. (7.127) from Eqs. (7.124)–(7.126). Use the fact that the matrix $\mathbf{W}$ diagonalizes $\mathbf{V}^{\mathrm{D}}$, i.e. the diabatic coupling matrix as given by (7.124).

As a consequence of what is stated in Eq. (7.127) and to be proved in Exercise 7.7, one finds

$$\cot[2\theta(R)] = \frac{F_2 - F_1}{2V_{12}}(R - R_0), \tag{7.128}$$

with $V_{12} = V_{12}^{\mathrm{D}}$. Likewise, it is readily established that the derivative of $\mathbf{W}$ with respect to R can be represented in the following form:

$$\frac{\partial}{\partial R}\mathbf{W} = \mathbf{W}\boldsymbol{\gamma} \tag{7.129}$$

and

$$\boldsymbol{\gamma} = \frac{\partial\theta(R)}{\partial R}\begin{pmatrix} 0 & 1 \\ -1 & 0 \end{pmatrix}. \tag{7.130}$$

This substitution simplifies the task of setting up the multichannel equations for the two-state problem considered here. They read, in vectorial notation,

$$\mathbf{I}\frac{\partial^2}{\partial R^2}\boldsymbol{\psi} + 2\boldsymbol{\gamma}\frac{\partial}{\partial R}\boldsymbol{\psi} + \mathbf{I}\left(\frac{\partial}{\partial R}\theta(R)\right)^2\boldsymbol{\psi} + \frac{\partial\boldsymbol{\gamma}}{\partial R}\boldsymbol{\psi}$$
$$+\left[\mathbf{k}^2 - \left(\frac{2\mu}{\hbar^2}\right)\mathbf{V}^{\mathrm{A}} - \mathbf{I}\frac{l(l+1)}{R^2}\right]\boldsymbol{\psi} = \mathbf{0}. \tag{7.131}$$

The non-vanishing elements of the adiabatic potential matrix $\mathbf{V}^{\mathrm{A}}$ are given by $V_{1/2}^{A} = \frac{1}{2}[V_{11}^{\mathrm{D}}(R) - / + V_{22}^{\mathrm{D}}(R)] - V_{12}(1 + \cot^2[2\theta(R)])^{\frac{1}{2}}$. The matrix $\mathbf{k}^2$ is defined as in Eq. (6.9). Only those terms of Eq. (7.131) which contain the antisymmetric matrix $\boldsymbol{\gamma}$ can give rise to coupling between the two components of $\boldsymbol{\psi}$. Evidently, the coupling strength is determined by the parameter $\frac{\partial\theta(R)}{\partial R}$ as well as its derivative $\frac{\partial^2\theta(R)}{\partial R^2}$. In this context, we point out that $\frac{\partial\theta(R)}{\partial R}$ has the profile of a Lorentzian curve with the center at $R = R_0$:

$$\frac{\partial\theta(R)}{\partial R} = \frac{F_2 - F_1}{4V_{12}(1 + \cot^2[2\theta(R)])}. \tag{7.132}$$

For given slopes F_1, F_2, the amplitude of this coupling function increases as V_{12}^{-1}, which is plausible since the separation of the adiabatic potential curves V_{11}^{A} and V_{22}^{A} at $R = R_0$ grows linearly with V_{12}, and so does the chance for a purely adiabatic process.

Let us assume that distorted-wave solutions at first order have been determined for both electronic channels, with the respective approximate wave functions labeled $\breve{\psi}_1^{(0)}$ and $\breve{\psi}_2^{(0)}$. Further, these solutions may be approximated by use of the WKB formalism as summarized in Section 4.2. We employ the respective WKB wave functions, where it should be noted that, with reference to Figure 7.7, the classically allowed region is on the right of the turning points of the two adiabatic potential curves. By adapting formula (4.71) to the case at hand we find [91]

$$\sqrt{\frac{1}{k_i}}\breve{\psi}_i^{(0)}(R) \sim \frac{1}{\sqrt{k_i(R)}}\sin\left[\int_{R_{li}}^{R} dR'' k_i(R'') + \frac{\pi}{4}\right], \quad \text{if } R > R_{li}, \; i = 1, 2, \tag{7.133}$$

where R_{li} stands for the two left-hand turning points and

$$k_i(R) = \left(k^2 - \frac{2\mu}{\hbar^2}V_{ii}^{\mathrm{ad}} - \frac{l(l+1)}{R^2}\right)^{\frac{1}{2}}. \tag{7.134}$$

Using the transition matrix element (6.45) as given by the distorted-wave Born approximation [151] we write

$$T_{12} = \int_0^\infty dR\breve{\psi}_1^{(0)}\left(2\gamma_{12}\frac{\partial}{\partial R} + \frac{\partial}{\partial R}\gamma_{12}\right)\breve{\psi}_2^{(0)}, \tag{7.135}$$

extracting the coupling operator from Eq. (7.131). Upon inserting the non-diagonal element of $\boldsymbol{\gamma}$ as well as some rearrangement, we turn Eq. (7.135) into the antisymmetrized expression

$$T_{12} = \int_0^\infty dR\left(\breve{\psi}_1^{(0)}\frac{\partial}{\partial R}\breve{\psi}_2^{(0)} - \breve{\psi}_2^{(0)}\frac{\partial}{\partial R}\breve{\psi}_1^{(0)}\right)\frac{\partial\theta(R)}{\partial R}. \tag{7.136}$$

Considering the criterion for the validity of the WKB wave functions (7.133) the conditions for the use of these approximations to evaluate the integral (7.136) are easily specified. In view of the Lorentz shape of the function $\frac{\partial\theta(R)}{\partial R}$

which is peaked at the crossing point of the diabatic curves, the main contribution to T_{12} is expected to come from the region around R_0. Therefore, employing the WKB approach is justified if the turning points R_{1i}, $i = 1, 2$, are far enough to the left of R_0 for the functional form (7.133) to hold at a sufficient level of accuracy. Let us assume that the latter requirement is met. It will then be legitimate to use Eq. (7.133) for calculating T_{12} according to Eq. (7.136). We observe that

$$\breve{\psi}_1^{(0)}\frac{\partial}{\partial R}\breve{\psi}_2^{(0)} - \breve{\psi}_2^{(0)}\frac{\partial}{\partial R}\breve{\psi}_1^{(0)} = \frac{k_2(R)\sin(I_1)\cos(I_2) - k_1(R)\cos(I_1)\sin(I_2)}{\sqrt{k_1(R)k_2(R)}}, \tag{7.137}$$

where the substitutions

$$I_i = \int_{R_{li}}^{R} dR'' k_i(R'') + \frac{\pi}{4}, \quad i = 1, 2, \tag{7.138}$$

have been made.

A further simplification of the integrand of T_{12} is achieved in the special case that $\frac{2\mu}{\hbar^2}V_{12}/\widetilde{k}(R) \ll 1/a_0$ is satisfied in the vicinity of the crossing point R_0, with $\widetilde{k}$ as the average of k_1 and k_2 and $\frac{1}{a_0}$ as the unit of the wave number. In more detail, we consider

$$k_1(R) - k_2(R) = \frac{\frac{2\mu}{\hbar^2}V_{12}(1 + \cot^2[2\theta(R)])^{\frac{1}{2}}}{\widetilde{k}(R)} \tag{7.139}$$

as inferred from Eq. (7.134). Since the integral is dominated by the contribution near the crossing point, and $\cot[2\theta(R_0)] = 0$ (see Eq. (7.128)), the assumption that close to R_0 the diabatic coupling element is small as compared to the average wave number implies that we can regard $k_1(R)$ and $k_2(R)$ as approximately equal. Setting $k_1(R) \approx k_2(R)$, we reduce Eq. (7.137) to

$$\breve{\psi}_1^{(0)}\frac{\partial}{\partial R}\breve{\psi}_2^{(0)} - \breve{\psi}_2^{(0)}\frac{\partial}{\partial R}\breve{\psi}_1^{(0)} \approx \sin\left(\delta^0 + \int_{R_0}^{R} dR''[k_1(R'') - k_2(R'')]\right) \tag{7.140}$$

with

$$\delta_0 = \int_{R_{l1}}^{R_0} dR'' k_1(R'') - \int_{R_{l2}}^{R_0} dR'' k_2(R''). \tag{7.141}$$

Using Eq. (7.139), and replacing the function $\cot[2\theta(R)]$ with the variable α, we evaluate the integral over the wave vector difference analytically to find[3]

$$\int_{R_0}^{R} dR''[k_1(R'') - k_2(R'')] = \Gamma[\alpha(1+\alpha^2)^{\frac{1}{2}} + \sinh^{-1}(\alpha)] \tag{7.142}$$

and

$$\Gamma \equiv \frac{2V_{12}^2}{\hbar v_0(F_2 - F_1)}, \tag{7.143}$$

where $v_0 = \frac{\hbar\tilde{k}}{\mu}$ denotes the nuclear velocity at the crossing point. The magnitude of the factor Γ is the *Massey parameter* [8], also known as the *adiabaticity factor*. As a consequence of Eqs. (7.133)–(7.143), the integral T_{12} is transformed into

$$T_{12} = \frac{\sin\delta_0}{2}\int_{-\infty}^{+\infty} d\alpha \frac{\cos(\Gamma[\alpha(1+\alpha^2)^{\frac{1}{2}} + \sinh^{-1}\alpha])}{1+\alpha^2} \tag{7.144}$$

where we have substituted α for R as integration variable and extended the lower limit to $-\infty$. This can be done since, by construction, the domain $R < 0$ does not contribute to the integral. Finally, Eq. (7.144) turns into

$$T_{12} = \left(\frac{\pi}{2} + G\right)\sin\delta_0 \exp\left(-\frac{\pi\Gamma}{2}\right) \tag{7.145}$$

with

$$G \equiv \int_1^{\infty} d\alpha \frac{\sin(\cosh^{-1}\alpha - \alpha(\alpha^2-1)^{\frac{1}{2}})}{\alpha^2 - 1}. \tag{7.146}$$

The quantity δ_0 is termed the *Stückelberg phase*. The overall probability for a transition between the two adiabatic states is obtained from

$$P_{12} = |S_{12}|^2 = 4|T_{12}|^2 = 4\left(\frac{\pi}{2} + G\right)^2 \sin^2\delta_0 \exp(-\pi\Gamma). \tag{7.147}$$

[3]The skeptical reader may want to examine this result using the integral $\int dx\sqrt{1+(ax)^2} = \frac{ax\sqrt{(ax)^2+1}+\sinh^{-1}ax}{2a}$, where a is a constant.

In the domain $\Gamma \geq 3$, G reduces to $-\frac{\pi}{6}$. Adopting this limit and averaging over $\sin^2\delta_0$ yields the compact expression

$$P_{12} \approx 2\frac{\pi^2}{9}\exp(-\pi\Gamma). \tag{7.148}$$

This is close to the transition probability $P_{\mathrm{LZ}} = \exp(-\pi\Gamma)$ originally proposed by Landau and Zener [14, 15] for a single traversal of the crossing region. In most cases, the interacting systems will pass through this region twice, on their way to and from the distance of closest approach. Only one of these two traversals of the crossing zone should give rise to a transition, if a change from one to the other state is to be the overall result of the interaction. In terms of the expression given by Landau and Zener, the respective probability is $2\,P_{\mathrm{LZ}}(1-P_{\mathrm{LZ}})$. If P_{LZ} is small, this is approximately equal to formula (7.148). We recall that the unitarity of the S-matrix is not guaranteed in the framework of the distorted-wave Born approximation (see Section 6.2.1). The error arising from this limitation is small if P_{12} is small. This condition favors higher over lower values of V_{12}, corresponding to well-separated adiabatic potential curves. Physical situations that represent a compromise between this accuracy requirement and the constraint $\frac{2\mu}{\hbar^2}V_{12}/\widetilde{k}(R) \ll 1/a_0$ will be best suited for application of the result (7.147).

In general, the validity of the LZS model is governed by the Massey parameter Γ which measures the degree of adiabaticity of the considered process and has to be sufficiently large for Eq. (7.147) to apply. This means that small values of $\widetilde{k}$ are preferred over large ones. The latter condition is intuitively obvious since low relative velocities will favor adiabatic over nonadiabatic processes and thus suppress transitions between electronic states.

The Massey parameter comes in various guises. Thus, the ratio $|V_{12}/(F_2 - F_1)|$ may be equated to a *Landau–Zener length* L_{LZ}, understood as the distance from the crossing point where the difference between the adiabatic potentials is of the same magnitude as the diabatic coupling element. In terms of this length, the Massey parameter reads

$$\Gamma = 2\frac{|V_{12}|}{\hbar v_0}L_{\mathrm{LZ}}. \tag{7.149}$$

Since L_{LZ}/v_0 can be interpreted as a time t_{p} characteristic of the passage of the nuclei through the crossing zone and $\hbar/2|V_{12}|$ approximates the electron transition time t_{el} from one adiabatic state to the other, which diminishes with the difference between the respective adiabatic potentials, the Massey parameter can lastly be cast into the form of the time ratio t/t_{el}.

A complete solution of the LZS curve-crossing problem that eliminates the restrictions of validity mentioned in the previous paragraph and an extensive generalization of the LZS theory were put forward by Zhu and Nakamura [211, 212]. A variety of semiclassical procedures, most notably among them the trajectory surface hopping method, to be addressed in Chapter 11, have been augmented by incorporating this accurate solution [212].

With slight modifications, the LZS formalism may be used to evaluate the probability of electron transfer between two atomic or molecular species approaching each other (see e.g. [213]), as schematically shown in Figure 7.8 for the case of a diatomic molecule which exhibits ionic bonding at the ground state equilibrium bond length. As the internuclear distance grows sufficiently large, however, the bonding character of the system is found to be covalent rather than ionic. Introducing the labels A and B for the two centers, one may identify a critical distance R_0 where the ionic solution $\mathrm{A}^+ + \mathrm{B}^-$ turns into the covalent bonding alternative $\mathrm{A} + \mathrm{B}$. As seen in the

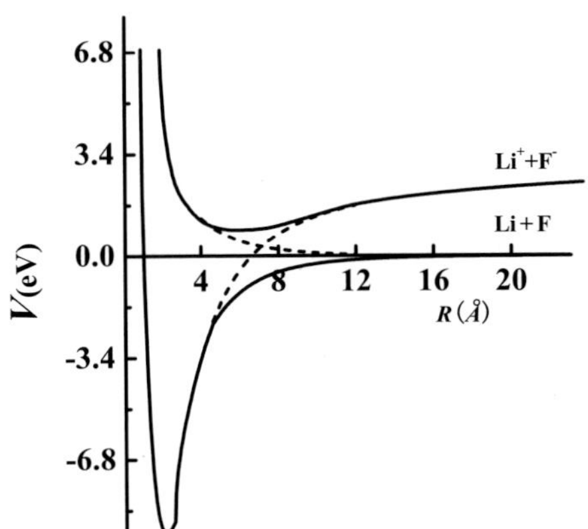

Figure 7.8: Crossing of diabatic potential energy curves involving electron transfer within a diatomic molecule. The adiabatic ground state solution is ionic in the neighborhood of the equilibrium bond length and covalent at large internuclear distance $R \gg R_{\text{eq}}$. These assignments are reversed in the first adiabatic excited state. The crossing point of the diabatic curves marks the critical distance for the transition between bonding types in the two adiabatic states.

figure, the bonding features of the ground and the first excited adiabatic state behave in a complementary fashion, since the latter is covalent close to the ground state equilibrium distance and asymptotically ionic.

Within the frame of the adiabatic approximation, the two atoms approach each other so slowly that the ground state is adopted at any internuclear distance, and the ionic solution is realized for $R < R_0$, the covalent one for $R > R_0$. For the diabatic approximation to apply, the same process would have to happen so fast that the initial bonding configuration is maintained for all internuclear distances. Therefore, the chance that the dissociating system moves from the adiabatic ground state to the adiabatic excited state is identical with that for its persistence in the diabatic ionic state. The problem of deriving probabilities for this process has often been treated within the LZS model, and we will make reference to this approximation when discussing electron transfer in Chapter 20, specifically in the commentary surrounding Eqs. (20.30)–(20.33).

Chapter 8

The Time-Dependent Self-Consistent Field Theory

As we go from the previous to the present chapter, the role of the fundamental variable passes from energy to time, and the focus shifts from the stationary to the time-dependent Schrödinger equation. Dynamic problems may be addressed with a higher degree of uniformity and consistency by adopting the TDSE rather than its time-independent analog. More specifically, the solution of the TDSE is a square-integrable wave packet, making it possible to treat discrete and continuum solutions on the same footing, and eliminating the separation between square-integrable bound and non-normalizable scattering states that complicates the time-independent theory. The problems associated with the latter distinction were exemplified in Section 6.8 where the role of the dissociation continuum in rearrangement collisions was mentioned. As will be demonstrated here, the time-dependent approach to electron dynamics leads in general to equations of lesser technical intricacy and is efficient in avoiding the byzantine formalism based on the stationary scheme, as presented in Section 7.1. Moreover, a wave packet evolving in time may be viewed as a direct quantum generalization of the classical trajectory concept, and thus provides an intuitively appealing solution that naturally opens the gate to semiclassical extensions of the quantum treatment.

This chapter begins our outline of the time-dependent methodology by giving an account of nonadiabatic molecular dynamics based on the time-dependent Hartree–Fock (TDHF) theory. To begin, we recall the derivation of the Hartree–Fock (HF) equations from a variational principle, as presented in Section 1.3. Arguing along the same lines, we will obtain a TDHF equation from a time-dependent variational principle. As in the stationary case, this derivation will be guided by the idea that we cannot solve the full

TDSE for molecular systems, but may be able to approximate the solution in a suitably defined Hilbert space segment.

Section 8.2 views the TDHF theory from a quantum chemical angle and discusses the solutions of the TDHF equations in the context of the HF solutions, laying the ground for the two following sections which explore the TDHF formalism as a tool for quantum dynamics. In particular, we characterize the nonadiabatic Ehrenfest approach by comparison with the adiabatic Born–Oppenheimer model which involves local solutions of the stationary Schrödinger equation at every step of the nuclear propagation process. The Ehrenfest approach is further contrasted with the Car–Parrinello method which employs electron dynamics to simulate the adiabatic evolution of a molecular system with much higher efficiency than the Born–Oppenheimer procedure, as it restricts the use of electronic structure methods to the initial state of the system's evolution. We conclude our survey of electron dynamics in the TDHF frame with an outline of molecular quantum dynamics in the *eikonal representation* which yields a hierarchy of approaches to the problem of the coupling between electrons and nuclei. These range from the *eikonal approximation* which yields Ehrenfest dynamics to the explicit inclusion of electronic momenta and kinetic energies.

All models discussed in this chapter rest on the time-dependent variational principle which is the theme of the following section.

8.1 Time-Dependent Variational Principles

Attempting to define a suitable extension of the HF formalism for the time-dependent case, we build on the variational approach that leads to the canonical HF equations (1.59). Since the derivation of these equations involves a static variational principle based on the stationary Schrödinger equation, we consequently ask now for a *time-dependent variational principle* that yields optimal solutions to the TDSE in a given Hilbert subspace. For this purpose, we borrow from classical mechanics and apply once more Hamilton's principle that was invoked in Section 4.1.2 to characterize the classical trajectory connecting two given locations. We adopt

$$S[\Psi^*, \Psi] = \int_{t_1}^{t_2} dt\, L[\Psi^*, \Psi] \tag{8.1}$$

as the action functional that depends on the Lagrangian L for which the formal expression

$$L \equiv \int d\mathbf{X}\, \Psi^* \left(\hat{H} - i\hbar \frac{\partial}{\partial t} \right) \Psi \tag{8.2}$$

is chosen. To motivate this choice, we note that the first-order variation of S leads to the solution of the TDSE if no restrictions are imposed on Ψ. The variation ∂S is determined to be

$$\delta S = \int_{t_1}^{t_2} dt \int d\mathbf{X} \left[\delta\Psi^* \left(\hat{H} - i\hbar \frac{\partial}{\partial t} \right) \Psi + c.c. \right], \tag{8.3}$$

and setting it equal to zero, thus requiring the action to be stationary, gives the desired variational condition on Ψ and thus the Euler–Lagrange equations for the considered problem. The variations in Ψ and Ψ^* are here treated as independent, since Euler–Lagrange equations may be obtained from either of them. If the unrestrained Hilbert space is admitted as the space of the wave function, Ψ will coincide with the exact solution of the TDSE. As this limit is unattainable in practical applications, the condition $\delta S = 0$ will yield a set of equations characteristic of the considered subspace.

Exercise 8.1. Derive the expression (8.3) for the first variation of the action $S[\Psi^*, \Psi]$ given by (8.1). Note that the variation in Ψ and Ψ^* vanishes at the end points of the time interval $[t_1, t_2]$.

So far, no normalization constraint has been forced on the solution Ψ. This condition, however, can be included in a simple and elegant fashion by reflection on the *gauge invariance* of the action integral [214]. More specifically, the value of S does not change as Ψ is replaced by $\Psi_\alpha \equiv \exp[-\frac{i}{\hbar} \int_{t_0}^{t} dt' \alpha(t')]\, \Psi$ if simultaneously the Hamiltonian is augmented by a term $\alpha(t)$:

$$S_\alpha \equiv \int_{t_1}^{t_2} dt \int d\mathbf{X} \left[\Psi_\alpha^* \left(\hat{H} + \alpha(t) - i\hbar \frac{\partial}{\partial t} \right) \Psi_\alpha \right] = S. \tag{8.4}$$

The first variation of this action functional may be understood as derived from the modified Lagrangian $\tilde{L} = L + \alpha(t)(\langle \Psi_\alpha | \Psi_\alpha \rangle - 1)$. This implies that the real-valued function $\alpha(t)$ can be interpreted as a time-dependent Lagrangian multiplier that imposes normalization on the wave function Ψ_α.

Formally, $\alpha(t)$ represents a potential that modifies the Hamiltonian. It varies in time, but is homogeneous in space. We base the following development on the observation that, for a general wave function Ψ, the variational constraint $\delta S = 0$ is trivially satisfied if the first variation of the integrand $\langle\Psi|i\hbar\frac{\partial}{\partial t} - \hat{H}|\Psi|\rangle$ vanishes, yielding

$$\left(\hat{H} + \alpha(t) - i\hbar\frac{\partial}{\partial t}\right)\Psi_\alpha = 0. \quad (8.5)$$

By use of the normalization constraint imposed on Ψ_α, the function $\alpha(t)$ turns out to be [215, 216]

$$\alpha(t) = \left\langle \Psi_\alpha|\hat{H}|\Psi_\alpha\rangle - i\hbar\langle\Psi_\alpha|\frac{\partial\Psi_\alpha}{\partial t}\right\rangle. \quad (8.6)$$

Inserting the right-hand term into Eq. (8.5), the latter turns into

$$\langle\delta\Psi_\alpha|\hat{P}_\perp\left(\hat{H} + \alpha(t) - i\hbar\frac{\partial}{\partial t}\right)|\Psi_\alpha\rangle + c.c. = 0, \quad (8.7)$$

where we have introduced the operator $\hat{P}_\perp$, defined as

$$\hat{P}_\perp = \hat{I} - |\Psi_\alpha\rangle\langle\Psi_\alpha|. \quad (8.8)$$

Clearly, $\hat{P}_\perp$ projects an arbitrary wave function on the orthogonal complement of $|\Psi_\alpha\rangle$. Whenever the variations $\delta\Psi_\alpha$ and $\delta\Psi_\alpha^*$ are linearly independent, both terms in Eq. (8.7) reduce separately to zero, leading to the equation

$$\hat{P}_\perp\left(\hat{H} + \alpha(t) - i\hbar\frac{\partial}{\partial t}\right)|\Psi_\alpha\rangle = 0. \quad (8.9)$$

Reverting from Ψ_α to Ψ, we have

$$\hat{P}_\perp\left(\hat{H} - i\hbar\frac{\partial}{\partial t}\right)|\Psi\rangle = 0. \quad (8.10)$$

We emphasize that this condition involves only the perpendicular component of $\frac{\partial}{\partial t}|\Psi$, while no statement is made about the parallel component. The indeterminacy of the latter reflects the freedom of choosing an arbitrary real-valued gauge function $\alpha(t)$. Since the norm of Ψ is fixed, the undefined parallel component corresponds to an indeterminate phase angle [216]. If the variation is completely unrestrained and thus no normalization condition imposed, ($\alpha(t) = 0$), the *Frenkel variational principle* [217] results:

$$\langle \delta\Psi | \hat{H} - i\hbar \frac{\partial}{\partial t} | \Psi \rangle = 0. \tag{8.11}$$

We conclude this section by noting that various alternatives to the time-dependent variational principles introduced here have been considered in the literature. McLachlan and Ball [218] do not employ the action (8.1) as variational functional but define the expression

$$I = \int d\mathbf{X} |\hat{H}\Psi - \Lambda|^2. \tag{8.12}$$

In contrast to the action, this is local in time. The variational equations are obtained from the constraint

$$\delta I = \delta \int d\mathbf{X} |\hat{H}\Psi - \Lambda|^2 = 0, \tag{8.13}$$

where Λ is implicitly defined by the relation

$$\Psi(t + \tau) = \Psi(t) - \frac{i}{\hbar}\Lambda\tau. \tag{8.14}$$

The latter two statements encapsulate the essence of the principle which optimizes the time derivative of a wave function assumed to be known at a time t. It thus generates the best possible value of Ψ at a slightly later time $t + \tau$ by use of the function Λ that emerges from the condition (8.13), rendering the integral stationary.

Exercise 8.2. Use the variational principle (8.13) to derive the representation of the Schrödinger equation in a finite basis $\{\phi_i\}$. Proceed by describing both Ψ and Λ as linear combinations of the basis vectors ϕ_i. If $\Lambda = \sum_{i=1}^{N} b_i\phi_i$, write down the first variation of $\hat{H}\Psi - \Lambda$ with respect to the coefficients $\{b_i\}$.

8.1.1 *Time-dependent perturbations*

In what follows, we will demonstrate how the principle (8.11) can be employed to obtain an equation of motion for the considered quantum system. Prior to any reflection on molecules, we will perform this derivation *in abstracto*, making allowance for the dependence of Ψ on a parameter vector **c**, such that

$$\Psi \equiv \Psi(c_1, c_2, \ldots; x_1, \ldots, x_N; X_1, \ldots, X_M) \tag{8.15}$$

without assigning any physical meaning to the components $c_1, c_2, \ldots$ which are assumed to depend on time. The set $\{c_1, c_2, \ldots\}$ here defines the Hilbert space segment from which the wave function is selected. Later, we will understand these parameters as coefficients multiplying the substitutional determinants that emerge from the self-consistent field treatment of a molecular system. The two blocks of variables $\{x_1, \ldots, x_N\}$ and $\{X_1, \ldots, X_M\}$ will then be identified with the electronic and nuclear coordinates of this system, as was done in the preceding chapters. We further partition the Hamiltonian into a static term $\hat{H}_0$ and a perturbing contribution $\hat{H}_1$ that contains the time dependence of the studied problem:

$$\hat{H}(t) = \hat{H}_0 + \hat{H}_1(t). \tag{8.16}$$

We stipulate that the specific choice $\{c_1, c_2, \ldots\} = \{c_{01}, c_{02}, \ldots\}$ determines the ground state Ψ_0 of the unperturbed problem:

$$\hat{H}_0\Psi_0 = E_0\Psi_0. \tag{8.17}$$

and

$$\Psi_0 \equiv \Psi(c_{01}, c_{02}, \ldots; x_1, \ldots, x_N; X_1, \ldots, X_M). \tag{8.18}$$

To first order, the effect of the interaction denoted by $\hat{H}_1$ is accommodated by a small deviation **d** from the reference vector $\mathbf{c}_0$:

$$\mathbf{c}(t) = \mathbf{c}_0 + \mathbf{d}(t), \quad c_i(t) = c_{0i} + d_i(t). \tag{8.19}$$

It is now possible to expand both Ψ and $\delta\Psi$ up to linear order with respect to the parameters $c_1, c_2, \ldots$ around the vector $\mathbf{c}_0$:

$$\Psi = \Psi_0 + \sum_j \left(\frac{\partial\Psi}{\partial c_j}\right)\bigg|_{\mathbf{c}=\mathbf{c}_0} d_j(t), \tag{8.20}$$

$$\delta\Psi = \sum_j \frac{\partial\Psi}{\partial c_j}\bigg|_{\mathbf{c}=\mathbf{c}_0} \delta c_j = \sum_j \left[\frac{\partial\Psi}{\partial c_j}\bigg|_{\mathbf{c}=\mathbf{c}_0} + \sum_i \frac{\partial^2\Psi}{\partial c_i \partial c_j}\bigg|_{\mathbf{c}=\mathbf{c}_0} d_j(t)\right] \delta c_j. \tag{8.21}$$

Introducing the vector $\partial/\partial\mathbf{c}$ of the first and the matrix $\partial^2/\partial\mathbf{c}\partial\mathbf{c}'$ of the second derivatives with respect to the parameters $c_1, c_2, \ldots$, we express these identities in a more compact form, namely

$$\Psi = \Psi_0 + \left(\frac{\partial\Psi}{\partial\mathbf{c}}\right)\bigg|_{\mathbf{c}=\mathbf{c}_0} \cdot \mathbf{d}. \tag{8.22}$$

$$\delta\Psi = \left(\frac{\partial\Psi}{\partial\mathbf{c}}\bigg|_{\mathbf{c}=\mathbf{c}_0} + \frac{\partial^2\Psi}{\partial\mathbf{c}\partial\mathbf{c}'}\mathbf{d}\right) \cdot \delta\mathbf{c}. \tag{8.23}$$

At this point, equations of motion may be derived from the time-dependent variational principle. Employing relation (8.10) in the form $\langle\delta\Psi|\hat{P}'_\perp(\hat{H} - i\hbar\frac{\partial}{\partial t})\Psi\rangle$, we make the normalization constraint on the wave function through the modified projection operator $\hat{P'}_\perp = \langle\Psi|\Psi\rangle^{-1}(\hat{I} - |\Psi\rangle\ \langle\Psi|\Psi\rangle^{-1}\langle\Psi|)$. The reader verifies easily that

$$\left\langle\frac{\partial\Psi}{\partial c_j}|\hat{P}'_\perp\hat{H}|\Psi\right\rangle = \left(\left\langle\frac{\partial\Psi}{\partial c_j}|\hat{H}|\Psi\right\rangle - \left\langle\frac{\partial\Psi}{\partial c_j}|\Psi\right\rangle\langle\hat{H}\rangle\right)/\langle\Psi|\Psi\rangle, \tag{8.24}$$

with $\langle\hat{H}\rangle$ as the energy expectation value, and thus $\langle\hat{H}\rangle = \langle\Psi|\hat{H}|\Psi\rangle/\langle\Psi|\Psi\rangle$. The analogous formula for the ket derivative as well as the continuation of Eq. (8.24) to the second derivative, as contained in formula (8.23), lead to the statement

$$\delta\mathbf{c}^* \cdot \left(\frac{\partial\hat{H}}{\partial\mathbf{c}} + \mathbf{M}_a\mathbf{d} + \mathbf{M}_b\mathbf{d}^* - i\hbar\mathbf{W}\frac{\partial\mathbf{d}}{\partial t}\right) = 0. \tag{8.25}$$

The vector $\frac{\partial\hat{H}}{\partial\mathbf{c}}$ and the matrices $\mathbf{M}_a$, $\mathbf{M}_b$, and $\mathbf{W}$ are defined as follows:

$$\frac{\partial\hat{H}}{\partial c_j} \equiv \left(\left\langle\frac{\partial\Psi}{\partial c_j}|\hat{H}|\Psi\right\rangle - \left\langle\frac{\partial\Psi}{\partial c_j}|\Psi\right\rangle\langle\hat{H}\rangle\right)/\langle\Psi|\Psi\rangle|_{\mathbf{c}=\mathbf{c}_0}, \tag{8.26}$$

$$M_{a,ij} \equiv \left(\left\langle\frac{\partial\Psi}{\partial c_i}|(\hat{H} - \langle\hat{H}\rangle)|\frac{\partial\Psi}{\partial c_j}\right\rangle - \frac{\partial\hat{H}}{\partial c_i}\left\langle\Psi|\frac{\partial\Psi}{\partial c_j}\right\rangle\right.$$

$$\left. - \left\langle\frac{\partial\Psi}{\partial c_i}|\Psi\right\rangle\frac{\partial\hat{H}}{\partial c_j}\right)/\langle\Psi|\Psi\rangle|_{\mathbf{c}=\mathbf{c}_0}, \tag{8.27}$$

$$M_{b,ij} \equiv \left(\left\langle \frac{\partial^2 \Psi}{\partial c_i \partial c_j} | (\hat{H} - \langle \hat{H} \rangle) | \Psi \right\rangle - \frac{\partial \hat{H}}{\partial c_i} \left\langle \frac{\partial \Psi}{\partial c_j} | \Psi \right\rangle \right.$$

$$\left. - \left\langle \frac{\partial \Psi}{\partial c_i} | \Psi \right\rangle \frac{\partial \hat{H}}{\partial c_j} \right) / \langle \Psi | \Psi \rangle |_{\mathbf{c}=\mathbf{c}_0}, \tag{8.28}$$

$$W_{ij} = \left\langle \frac{\partial \Psi}{\partial c_i} | \hat{P}'_{\perp} | \frac{\partial \Psi}{\partial c_j} \right\rangle \bigg|_{\mathbf{c}=\mathbf{c}_0}. \tag{8.29}$$

In the final step, we exploit the assumption that the variations $\delta\mathbf{c}$ are arbitrary and write down the equation of motion:

$$\frac{\partial \hat{H}}{\partial \mathbf{c}} + \mathbf{M}_{\mathrm{a}}\mathbf{d} + \mathbf{M}_{\mathrm{b}}\mathbf{d}^* = i\hbar \mathbf{W} \frac{\partial \mathbf{d}}{\partial t}. \tag{8.30}$$

We point out that the matrix $\mathbf{M}_{\mathrm{a}}$ is Hermitian while $\mathbf{M}_{\mathrm{b}}$ is symmetric. Until now, we have not given any consideration to the perturbing interaction described by the operator $\hat{H}_1(t)$ in Eq. (8.16). If this contribution is factorized into a time-dependent real-valued strength parameter $\gamma(t)$ and a constant Hermitian operator $\hat{h}$, such that

$$\hat{H}_1(t) = \gamma(t)\hat{h}, \tag{8.31}$$

the arrays that determine the dynamical behavior of the system as given by Eq. (8.30) can be expanded in powers of $\gamma(t)$. In this way, a hierarchy of coupled equations is derived that allow inspecting the impact of the perturbation to any order deemed relevant in the context of the problem at hand. Specifically, to first order,

$$\mathbf{d}(t) = \mathbf{d}_0(t) + \gamma(t)\mathbf{d}'(t), \tag{8.32}$$

$$\frac{\partial \hat{H}}{\partial \mathbf{c}} = \frac{\partial \hat{H}_0}{\partial \mathbf{c}} + \gamma(t) \frac{\partial \hat{H}_1}{\partial \mathbf{c}}. \tag{8.33}$$

Analogous relations apply for $\mathbf{M}_{\mathrm{a}}$ and $\mathbf{M}_{\mathrm{b}}$, introducing the correction terms $\mathbf{M}'_a$ and $\mathbf{M}'_b$. By inserting these expansions into Eq. (8.30), truncating at linear order in $\gamma(t)$, and collecting zero- and first-order terms, one finds:

$$\mathbf{M}_{a0}\mathbf{d}_0 + \mathbf{M}_{b0}\mathbf{d}_0^* = i\hbar \mathbf{W} \frac{\partial \mathbf{d}_0}{\partial t} \tag{8.34}$$

$$\frac{\partial \hat{H}_1}{\partial \mathbf{c}} + \mathbf{M}'_{\mathrm{a}}\mathbf{d}_0 + \mathbf{M}_{a0}\mathbf{d}' + \mathbf{M}'_b\mathbf{d}_0^* + \mathbf{M}_{b0}\mathbf{d}'^* = i\hbar \mathbf{W} \frac{\partial \mathbf{d}'}{\partial t}. \tag{8.35}$$

Equation (8.34), involving the zero-order effect, can be solved without recourse to (8.35), while the quantities of zero order have to be known in order to calculate the solutions $\mathbf{d}'$ and $\mathbf{d}'^*$ of the second equation.

8.1.2 *Free and forced oscillations*

An elementary but important application of the scheme described by (8.34) and (8.35) is the response of a molecular system to a small periodic perturbation as might be realized by exposing it to a weak electric field with harmonic time dependence. The displacement $\mathbf{d}_0$ can then be assumed to vary with time as $\exp(-i\omega t)$, or, adopting a more general ansatz,

$$\mathbf{d}_0 = \boldsymbol{\Xi} \exp(-i\omega t) + \mathbf{Z}^* \exp(i\omega t) \tag{8.36}$$

with $\boldsymbol{\Xi}$ and $\mathbf{Z}$ as suitably defined time-independent vector amplitudes. We substitute for $\mathbf{d}_0$ in (8.34), separate terms according to their factors $\exp(i\omega t)$ or $\exp(-i\omega t)$, and obtain the matrix equation

$$\begin{pmatrix} \mathbf{M}_{a0} & \mathbf{M}_{b0} \\ \mathbf{M}^*_{b0} & \mathbf{M}^*_{a0} \end{pmatrix} \begin{pmatrix} \boldsymbol{\Xi} \\ \mathbf{Z} \end{pmatrix} = \omega \begin{pmatrix} \mathbf{W} & 0 \\ 0 & -\mathbf{W} \end{pmatrix} \begin{pmatrix} \boldsymbol{\Xi} \\ \mathbf{Z} \end{pmatrix}. \tag{8.37}$$

It has been shown that this system of equations is solved by vectors $(\boldsymbol{\Xi}_{\omega_i}, \mathbf{Z}_{\omega_i})$ that correspond to a discrete set of angular frequencies ω_1, ω_2 [215]. From the structure of Eq. (8.37), it is obvious that any solution $(\boldsymbol{\Xi}_{\omega_i}, \mathbf{Z}_{\omega_i})$ for ω_i can be mapped into another solution $(\mathbf{Z}_{\omega_i^*}, \boldsymbol{\Xi}_{\omega_i^*}) = (\boldsymbol{\Xi}_{-\omega_i}, \mathbf{Z}_{-\omega_i})$ for $-\omega_i$. These results describe *free oscillations* since our analysis is based on the zero-order identity (8.34) that does not include any perturbing interaction. As a convenient strategy of extending the treatment to the case of non-zero $\hat{H}_1(t)$, the function $\gamma(t)$ is expressed through its Fourier transform $\Gamma(\omega)$, leading to the following integral form of the perturbing operator:

$$\hat{H}_1(t) = \frac{1}{2\pi} \int_{-\infty}^{\infty} d\omega \frac{1}{2} \Gamma(\omega) [\hat{h}_\omega \exp(-i\omega t) + \hat{h}_{-\omega} \exp(i\omega t)]. \tag{8.38}$$

To guarantee the Hermitian character of H_1, we impose the constraint $\hat{h}_{-\omega} = \hat{h}^\dagger_{-\omega}$ on the Fourier transform $\hat{h}_\omega$ of the factor $\hat{h}$ and set $\Gamma(-\omega) = \Gamma(\omega)$. Note that the premise of a real-valued $\gamma(t)$ implies that $\Gamma^*(\omega) = \Gamma(-\omega)$, which, in conjunction with the conditions defined earlier for $\hat{h}_{-\omega}$ and $\Gamma(-\omega)$, readily establishes the term on the right-hand side of Eq. (8.38) as Hermitian.

One might solve Eq. (8.35) which, in contrast to (8.34), includes the perturbation operator $\hat{H}_1(t)$ by considering in an initial approach a single component $\frac{1}{2}(\hat{h}_{\omega} \exp(-i\omega t) + h_{-\omega} \exp(i\omega t))$. An example of the physical realization of this operator is provided by the dipole approximation to the interaction of a molecule with a laser ray in the monochromatic limit, as further discussed in Chapters 18 and 19. The system is assumed to react to this harmonic perturbation with an oscillatory time behavior that parallels the evolution of $\mathbf{d}_0$ as given by (8.36). In particular, the component $\mathbf{d}'_{\omega}$ of $\mathbf{d}'$ is cast into the form

$$\mathbf{d}'_{\omega} = \boldsymbol{\Xi}' \exp(-i\omega t) + \mathbf{Z}'^{*} \exp(i\omega t) \tag{8.39}$$

which describes a *forced oscillation* as opposed to the free oscillation given by Eq. (8.36). After the amplitudes $\boldsymbol{\Xi}'$ and $\mathbf{Z}'$ are obtained, the overall displacement vector $\mathbf{d}'$ is constructed by evaluating the inverse Fourier transform of $\mathbf{d}'_{\omega}$. If the perturbation is due to a periodically changing electric field, the response of the molecule will naturally consist in forced oscillations, in accordance with the approach (8.39).

8.2 The Time-Dependent Hartree–Fock Theory: Application to Molecules

As a virtue of the variational principle, its usefulness does not hinge on the availability of a complete set of basis vectors but rather on the judicious selection of a Hilbert space segment that is of relevance to the problem under study. Its success thus depends critically on physically adequate choices of the trial functions involved in the problem. "A variational procedure is a blind and dumb procedure that always provides an answer, but its accuracy depends crucially on the choice of the trial function," as Kramer and Saraceno observe in their treatise on the geometry of the time-dependent variational principle in quantum mechanics [219]. So far, we have not addressed the question of the trial functions optimally suited to represent dynamical molecular systems. It appears natural, however, to select a trial function Ψ_{t} in such a way that it can be understood in the familiar terms of stationary ground and excited states. A plausible form of Ψ_{t} for a system of N electrons is therefore

$$\Psi_{\mathrm{t}} = \Psi_0 + \sum_{m,s} c_m^s(t)\Psi_m^s + \sum_{m \neq n, s \neq t} c_m^s(t)c_n^t(t)\Psi_{mn}^{st} + \cdots,$$

$$m, n \leq N, \quad s, t > N, \tag{8.40}$$

with Ψ_0 as the HF ground state determinant involving N occupied states, Ψ^s_m as the substitutional determinant that arises from interchanging the molecular orbitals ψ_m and ψ_s, and with Ψ^{st}_{mn} as well as the higher contributions defined accordingly. The formal parameters introduced in (8.19) acquire here the interpretation of expansion coefficients for the substitutional states $\Psi^{st\ldots}_{mn\ldots}$.

Equations of motion for the molecular system described by Ψ_t can be found as the ansatz (8.40) is combined with Eqs. (8.34) and (8.35). This is achieved by evaluating the arrays $\frac{\partial H}{\partial \mathbf{c}}$, $\mathbf{M}_a$, $\mathbf{M}_b$, and $\mathbf{W}$ in the basis $\{\Psi_0, \Psi^s_m, \Psi^{st}_{mn}, \ldots\}$ at $\mathbf{c} = \mathbf{c}_0$, identifying the parameters c_b with the expansion coefficients c^s_m. This task can be readily performed for the zero-order equation (8.37). The rules for calculating Hamilton operator expectation values for Slater determinants, as outlined in Chapter 1 (see especially Section 1.4.3) yield

$$M^{st}_{a0,mn} = \langle \Psi^s_m | \hat{H} - \langle \hat{H} \rangle_{\mathbf{c}=\mathbf{0}} | \Psi^t_n \rangle = \delta_{sm,tn}(\varepsilon_s - \varepsilon_m) + \langle sn || mt \rangle, \tag{8.41}$$

$$M^{st}_{b0,mn} = \langle \Psi^{st}_{mn} | \hat{H} | \Psi_0 \rangle = \langle st || mn \rangle. \tag{8.42}$$

Further, $\mathbf{W}$ on the right-hand side of Eq. (8.37) turns into the unit matrix.

Exercise 8.3. Prove relations (8.41) and (8.42). Show also that $(\frac{\partial \hat{H}}{\partial \mathbf{c}})_{ms} \equiv \langle \Psi^s_m | \hat{H} | \Psi_0 \rangle = \langle s | \hat{F} | m \rangle$, where $\hat{F}$ stands for the Fock operator, as introduced in Section 1.3. Remember that the Fock operator only contains the occupied orbitals of the HF ground state determinant.

As the relevant matrix elements are evaluated in the basis given by HF ground and excited state determinants, the TDHF equations for the free oscillation case (8.37) can be formulated. We find

$$\sum_{tn} [\delta_{sm,tn}(\varepsilon_s - \varepsilon_m) + \langle sn || mt \rangle] \Xi_{tn} + \sum_{tn} \langle st || mn \rangle Z_{tn} = \omega \Xi_{sm}, \tag{8.43}$$

$$\sum_{tn} [\delta_{sm,tn}(\varepsilon_s - \varepsilon_m) + \langle mt || sn \rangle] Z_{tn} + \sum_{tn} \langle mn || st \rangle \, \Xi_{tn} = -\omega \, Z_{sm}. \tag{8.44}$$

The TDHF equations for the case of forced oscillations, as described by (8.35), are derived with strictly analogous arguments. Both realizations of the TDHF formalism are linear since the underlying equations of motion (8.34) and (8.35) were obtained under the premise of a first-order deviation from the parameter set $\{c_{01}, c_{02}, \ldots\}$ that specify the stationary HF ground state.

An alternative version of the TDHF scheme is derived by applying the Frenkel principle (8.11) directly to a single-determinant wave function [215], composed of molecular orbitals. In contrast to the previous approach, we do not prescribe here expansion in the basis $\{\Psi_0, \Psi_m^s, \Psi_{mn}^{st}, \ldots\}$, but make allowance for explicit time dependence of all molecular orbital states, $\psi_i = \psi_i(t)$. With the trial function thus defined, use of the Frenkel procedure requires deriving $\langle\delta\Psi|\hat{H}|\Psi\rangle$ and $\langle\delta\Psi|\frac{\partial\Psi}{\partial t}\rangle$. The results are as follows:

$$\langle\delta\Psi|\hat{H}|\Psi\rangle = \sum_i \langle\delta\psi_i|\hat{F}|\psi_i\rangle, \tag{8.45}$$

$$\left\langle\delta\Psi\left|\frac{\partial\Psi}{\partial t}\right.\right\rangle = \sum_i\sum_{j\neq i}\left\langle\delta\psi_i\left|\frac{\partial\psi_i}{\partial t}\right.\right\rangle + \left\langle\delta\psi_i|\psi_i\right\rangle\left\langle\psi_j|\frac{\partial\psi_j}{\partial t}\right\rangle. \tag{8.46}$$

These expressions lead to the variational equation

$$\left[\hat{F} - i\hbar\frac{\partial}{\partial t} - i\hbar\sum_{j\neq i}\left\langle\psi_j\left|\frac{\partial\psi_j}{\partial t}\right.\right\rangle\right]|\psi_i\rangle - \sum_j|\psi_j\rangle\alpha_{ji} = 0, \tag{8.47}$$

where an orthonormality constraint on the molecular orbitals, in parallel to Eqs. (1.43)–(1.46), has been included, with the factors α_{ji} as Lagrange multipliers. The array $\boldsymbol{\alpha}$ is found by bracketing Eq. (8.47) with $\langle\psi_k|$. The structure of the matrix elements α_{ji} is different for the cases $k = i$ and $k \neq i$. Specifically,

$$\alpha_{ki} = \langle\psi_k|\hat{F}|\psi_i\rangle - i\hbar\left\langle\psi_k\left|\frac{\partial\psi_i}{\partial t}\right.\right\rangle \quad \text{if } k \neq i, \tag{8.48}$$

$$\alpha_{ii} = \langle\psi_i|\hat{F}\psi_i\rangle - i\hbar\left\langle\psi_i\left|\frac{\partial\psi_i}{\partial t}\right.\right\rangle - i\hbar\sum_{j\neq i}\left\langle\psi_j\left|\frac{\partial\psi_j}{\partial t}\right.\right\rangle \quad \text{if } k = i. \tag{8.49}$$

Relations (8.47), (8.48), and (8.49) are summarized into a simple identity, namely

$$\left(\hat{F} - i\hbar\frac{\partial}{\partial t}\right)|\psi_i\rangle = \sum_j|\psi_j\rangle\varepsilon_{ji}. \tag{8.50}$$

For convenience, we have introduced the matrix $\boldsymbol{\varepsilon}$, given by

$$\varepsilon_{ji} = \langle\psi_j|\hat{F}|\psi_i\rangle - i\hbar\left\langle\psi_j\left|\frac{\partial\psi_i}{\partial t}\right.\right\rangle. \tag{8.51}$$

With this substitution, a direct parallel is drawn between time-dependent and stationary self-consistent field theory. This becomes obvious as we compare the TDHF equation (8.50) with the HF equation (1.52). Both are formally identical excepting the occurrence of a time derivative term in Eq. (8.50).

It is easy to see the connection between the two trial functions explored so far, an expansion into configurations as given by Eq. (8.40) and a determinant of the form $\det[\psi_1(t), \psi_2(t), \ldots, \psi_m(t), \ldots, \psi_N(t)]$. To make their relation explicit, we stipulate that each time-dependent orbital ψ_m is represented by a superposition of the stationary orbitals $\{\psi_{0,1}, \psi_{0,2}, \ldots\}$, such that

$$\psi_k(t) = b_k(t)\psi_{0,k} + \sum_s b_{ks}(t)\psi_{0,s}, \quad k \leq N, \ s > N. \tag{8.52}$$

The set $\{\psi_{0,s}\}$ refers to virtual orbitals that are unoccupied at the time t but may become populated in the course of the system's evolution. We insert the $\psi_k(t)$ into $\Psi(t) = \det[\psi_1(t)\psi_2(t)...\psi(t)...\psi_N(t)]$ and expand Ψ into the stationary ground state as well as the excited determinants to find

$$\Psi(t) = \prod_{k=1}^{N} b_k(t) \det[\psi_{0,1}, \psi_{0,2}, \ldots, \psi_{0,N}] + \sum_{m,s} c_m^s(t) \det[\psi_{0,1}, \psi_{0,2}, \ldots, \psi_{0,m-1}, \psi_{0,s}, \psi_{0,m+1}, \ldots, \psi_{0,N}] \ldots \quad m \leq N, \ s > N, \tag{8.53}$$

where the sum extends over all possible single substitutions. We recognize the product of the diagonal elements b_{kk} as the amplitude of the stationary ground state determinant adopted by the system at the instant t. The coefficients of single substitutions emerging from this state as one replaces an occupied orbital ψ_i by a virtual orbital ψ_s are identified with the corresponding amplitudes in Eq. (8.40), and it holds that

$$c_m^s(t) = b(t) \ldots b_{m-1} b_{ms} b_{m+1} \ldots b_{\mathrm{N}} \tag{8.54}$$

which offers an interpretation of the expansion coefficients $\{c_m^s\}$ in terms of individual molecular orbitals. We emphasize that, from the foregoing discussion, the wave function (8.53) is the expanded form of one single determinant. The time-dependent counterpart of the stationary HF model is likewise solved by a determinantal wave function, more specifically by a wave function that remains a single determinant at any time of the system's evolution.

The sum on the right-hand side of Eq. (8.53) can in principle be continued to include excitations of double, triple, and higher order. The first-order TDHF approach, as obtained by truncating after the single substitutions, as in Eq. (8.53), is also known in theoretical physics under the title of *random phase approximation*. This name derives from a Fourier transformation of the electron repulsion term in the many-electron Hamiltonian performed by Bohm and Pines [220], translating the interaction between electrons into the language of momenta and momentum transfer. Neglecting terms with non-zero phase difference in the **k** vector, they retained the momentum transfer as an ensemble average. This treatment yields the average of the interelectronic repulsion and thus the result generated by self-consistent theory utilizing a single determinant.

From the definition of $\boldsymbol{\varepsilon}$ in Eq. (8.50) as well as the Hermitian property of $\hat{F}$, we conclude that $\boldsymbol{\varepsilon}$ is Hermitian too. To eliminate this matrix, we multiply Eq. (8.50) from the right by $\langle\psi_i|$ and subsequently sum over the index i. This yields

$$\sum_i \left(\hat{F} - i\hbar\frac{\partial}{\partial t}\right) |\psi_j\rangle\langle\psi_i| = \sum_{j,i} |\psi_j\rangle\varepsilon_{ji}\langle\psi_i|. \tag{8.55}$$

Writing down the adjoint of Eq. (8.55) and subtracting the result from Eq. (8.55), we eliminate the Lagrange multipliers and arrive at

$$\hat{F}\hat{\rho} - \hat{\rho}\hat{F} = i\hbar\frac{\partial\hat{\rho}}{\partial t}, \tag{8.56}$$

involving the density operator

$$\hat{\rho} = \sum_i |\psi_i\rangle\langle\psi_i|. \tag{8.57}$$

This result is the TDHF form of the Liouville–von Neumann equation, the density operator equivalent of the wave-function-based approach (8.50). For a computational realization of either equation, we approximate the molecular orbitals by expansion into a finite basis $\{\phi_i\}$, according to

$$\psi_m = \sum_i a_{im}\phi_i \tag{8.58}$$

or

$$\Psi = \boldsymbol{\phi}^{\mathrm{T}}\mathbf{a}. \tag{8.59}$$

This leads to the matrix form of the TDHF Liouville–von Neumann equation which reads

$$\mathbf{F}(t)\mathbf{P}(t) - \mathbf{P}(t)\mathbf{F}(t) = i\hbar\frac{\partial \mathbf{P}}{\partial t} \tag{8.60}$$

where $\mathbf{P}$ is the charge density matrix defined by Eq. (1.27), with elements $P_{pq} = \sum_i a_{pi}a_{qi}^*$. In the remainder of this chapter, we will explore realizations of the TDHF formalism in both versions discussed earlier, involving the propagation of the electronic wave function, as in Eq. (8.50), or of the electronic density matrix, as in Eq. (8.56).

8.3 Wave-Function-Based *Ab Initio* Molecular Dynamics

Any ambition at simulating chemical reactions, or, more generally, molecular scattering situations based on the formalism introduced earlier is faced with the task of including the nuclear subsystem and extending the TDHF scheme into an *ab initio molecular dynamics* approach. On the following pages, we point at proposals to implement the simultaneous evolution of electrons and nuclei in a framework defined by the TDHF scheme. We will base this discussion on Eq. (8.50) and thus concentrate on a methodology that involves molecular orbital coefficients as dynamic variables, while the presentation of procedures exploiting the density matrix version of the TDHF equations is relegated to Section 8.4.

Most important, the notion of *Ehrenfest dynamics* will be introduced, following the statement and interpretation of the *Hellmann–Feynman theorem* and the definition of the forces that determine the course of the moving nuclei. The concept of Ehrenfest dynamics will be developed by comparison with two adiabatic methods, namely Born–Oppenheimer (Section 8.3.4) and Car–Parrinello (Section 8.3.5) dynamics.

8.3.1 *Direct molecular dynamics in the time-dependent Hartree–Fock framework*

To exemplify the use of the TDHF framework for modeling the time behavior of molecular systems, we follow the development proposed by Field [221] for the case of a closed-shell system with $2M$ electrons. Let the electronic state be given by a single determinant composed of twofold occupied molecular orbitals of the form

$$\psi_m = \sum_{i=1}^{N} c_{mi}\phi_i. \tag{8.61}$$

In this expression, the ϕ_i stand for N basis functions which may be chosen as atomic orbitals, i.e. centered on the atomic nuclei of the considered molecule. Realizing the approach (8.58), we thus distinguish M occupied from $N - M$ virtual molecular orbitals. With these assignments, the electronic wave function is written as

$$\Psi = \det\left[\psi_1\alpha(1), \psi_1\beta(2), \psi_2\alpha(3), \ldots, \psi_M\beta(2M)\right]. \tag{8.62}$$

While Eq. (8.62), in contrast to Eq. (8.40), describes the electronic system by a single determinant, this determinant is not constrained to the ground state of the evolving molecule, nor any other eigenstate of the electronic Hamiltonian. The molecular orbital coefficients c_{mi} are defined as time-dependent, complex-valued parameters. These coefficients and the atomic coordinates of the molecule are understood as *independent sets of dynamic variables* that govern the temporal variation of the system [222]. The orthonormality requirement imposed on $\{\psi_1, \psi_2, \ldots, \psi_N\}$ demands that they satisfy the relation

$$\sum_{\mu\nu} c_{i\mu}^{\dagger} S_{\mu\nu} c_{\nu j} = \sum_{\mu\nu} c_{\mu i}^{*} S_{\mu\nu} c_{\nu j} = \delta_{ij} \tag{8.63}$$

with $S_{\mu\nu}$ as the overlap integral between the basis functions, defining the scalar product between the molecular orbital functions. For simplicity, one may adopt the *neglect of diatomic differential overlap* (NDDO) approximation that assumes vanishing overlap between two basis functions ϕ_μ and ϕ_ν centered on different nuclei. This reduces $S_{\mu\nu}$ to the identity matrix, and Eq. (8.63) becomes conveniently

$$\sum_{\mu} c_{\mu i}^{*} c_{\mu j} = \delta_{ij}. \tag{8.64}$$

Retaining the more cumbersome original constraint (8.63) does not, in principle, present any serious problem. Employing the Frenkel formula (8.11) in conjunction with the expansion (8.61) and the constraint (8.64), we repeat the steps that led to the TDHF equation (8.50) and find

$$\mathbf{F}\mathbf{c}_i - \sum_{j=1}^{M} \varepsilon_{ji}\mathbf{c}_j = i\hbar \left\{ \frac{\partial \mathbf{c}_i}{\partial t} + \left(\mathbf{c}_i^\dagger \frac{\partial \mathbf{c}_i}{\partial t} \right) \mathbf{c}_i + 2 \sum_{j=1, j\neq i}^{M} \left(\mathbf{c}_j^\dagger \frac{\partial \mathbf{c}_j}{\partial t} \right) \mathbf{c}_i \right\} \tag{8.65}$$

for the chosen closed-shell configuration. The indices i and j refer here to occupied orbitals. This relation introduces the matrix of Lagrange multipliers ε_{ji} which ensures that the constraints (8.64) are satisfied. Contracting Eq. (8.65) from the left-hand side with a molecular orbital vector $\mathbf{c}_j$ yields

$$\varepsilon_{ii} = \mathbf{c}_i^\dagger \mathbf{F}\mathbf{c}_i - 2i\hbar \sum_{j=1}^{M} \mathbf{c}_j^\dagger \frac{\partial \mathbf{c}_j}{\partial t}, \quad \text{if } j = i, \tag{8.66}$$

$$\varepsilon_{ji} = \mathbf{c}_j^\dagger \mathbf{F}\mathbf{c}_i - 2i\hbar \left(\mathbf{c}_j^\dagger \frac{\partial \mathbf{c}_i}{\partial t} \right), \quad \text{if } j \neq i. \tag{8.67}$$

Reinserting ε_{ji} into Eq. (8.65), we obtain a set of differential equations of first order in time for the molecular coefficients, namely

$$\mathbf{F}\mathbf{c}_i - \sum_{j=1}^{M} F_{ji}\mathbf{c}_j = i\hbar \left\{ \mathbf{I} - \sum_{j=1}^{M} (\mathbf{c}_j \mathbf{c}_j^\dagger) \right\} \frac{\partial \mathbf{c}_i}{\partial t} \tag{8.68}$$

where $\mathbf{I}$ denotes the identity matrix. In terms of the basis functions $\{\psi_i\}$ as introduced in Eq. (8.61), the Fock matrix element F_{ji} is expressed by Eq. (1.88) as

$$F_{ji} = h_{ji} + \sum_{m,n,k} c_{mk} c_{nk}^* \{2(ji|nm) - (jm|ni)\}. \tag{8.69}$$

Exercise 8.4. Derive the closed-shell TDHF equation (8.65) from the Frenkel variational principle (8.11) by expanding the time-dependent molecular orbitals as in Eq. (8.61) and evaluating the scalar products (8.45) and (8.46).

8.3.2 *Classical trajectories within TDHF dynamics*

The nuclear counterpart of the electronic dynamics described so far might be implemented by use of a quantum mechanical model, involving nuclear wave packets, or by a classical approach, adopting nuclear trajectories. Most current versions of TDHF dynamics realize the latter alternative. Correspondingly, we will assume a classical description of the nuclei in this section, while Chapter 15 will deal in detail with quantum mechanical solutions of the nuclear problem in the context of wave packet propagation. Plausibly, the classical approximation for the nuclei is legitimate wherever the trajectory concept is well-adjusted to the considered physical situation. This holds in general for scattering between atoms moving at sufficiently high kinetic energy. As far as molecular motion is concerned, the assumption of classical nuclei needs to be justified for each individual application. If the physics of the studied system is strongly influenced by multichannel reaction dynamics or tunneling, the case for classical trajectories will be difficult to make. If, on the other hand, these distinctly quantum mechanical features are absent from the investigated molecular interaction, the classical picture often captures adequately the essence of the process (see e.g. [223]). Adopting a quantum mechanical perspective, the choice of trajectories over wave packets will appear passable if the width of the latter is small enough to regard the nuclei as localized. The width ΔX of the wave packet may, for example, be close to that of a one-dimensional oscillator in the ground state, such that $(\Delta X)^2 = \frac{\hbar}{\mu\omega}$, with μ as the effective nuclear mass and ω as the oscillation frequency. For sufficiently large μ, the wave packet can be assumed as localized, and the trajectory model as valid. If this has been established, employing the classical equations of motion for the nuclear dynamics is justified. These are:

$$\frac{d\mathbf{X}_k}{dt} = \frac{\mathbf{P}_{X_k}}{M_k}, \tag{8.70}$$

$$\frac{d\mathbf{P}_{X_k}}{dt} = -\frac{\partial E}{\partial \mathbf{X}_k}, \tag{8.71}$$

where the index k labels the nuclei. As will be shown in the following, the numerical realization of the two equations of motion does not require global information about the potential energy surfaces of the ground or excited electronic states involved in the studied problem, as the energy gradients are evaluated locally, or *on the fly*. In accordance with Eq. (8.69), the total energy E in the assumed closed-shell case is written as

$$E = 2\sum_{jik} c_{jk}^* c_{ik} h_{ji} + \sum_{jimnkl} c_{jk}^* c_{ik} c_{ml}^* c_{nl} \{2(ji|nm) - (jm|ni)\} + E_{\mathrm{N}} \tag{8.72}$$

with

$$E_{\rm N} = \frac{1}{2}\sum_k \frac{P_{X_k}^2}{M_k} + V_{\rm N}. \qquad (8.73)$$

The nuclei therefore feel a force given by

$$-\frac{\partial E}{\partial \mathbf{X}_k} = -2\sum_{jik} c_{jk}^* c_{ik} \frac{\partial h_{ji}}{\partial \mathbf{X}_k} - \sum_{jimnkl} c_{jk}^* c_{ik} c_{ml}^* c_{nl} \frac{\partial}{\partial \mathbf{X}_k}\{2(ji|nm) - (jm|ni)\} - \frac{\partial E_{\rm N}}{\partial \mathbf{X}_k}. \qquad (8.74)$$

The nuclear coordinates and the molecular orbital coefficients are here understood as independent variables. The time evolution specified by the differential equations (8.65), (8.70, and 8.71) thus proceeds in a space spanned by the nuclear positions and momenta on one hand, and by the electronic molecular orbital coefficients on the other.

Recognizing the generally encountered highly oscillatory features of the molecular orbital coefficients c_{ij}, one may separate a periodic component p_i from the c_{ij} and write these variables in a factorized form:

$$\mathbf{c}_i = p_i \mathbf{a}_i, \qquad (8.75)$$

defining the phase factors p_i as

$$p_i = \exp\left[-\frac{i}{\hbar}\int_0^t \gamma_i(t')dt'\right] \qquad (8.76)$$

where the γ_i are chosen as real functions and t is the propagation time that occurs in the evolution equations (8.68), (8.70), and (8.71) as an independent variable. Since a factor p_i is applied to each one of the molecular orbitals contained in the closed-shell wave function Ψ (see Eq. (8.62)), one can write the overall phase factor associated with Ψ as

$$p_\Psi = \exp\left[-2\frac{i}{\hbar}\int_0^t dt' \sum_{i=1}^{M} \gamma_i(t')\right]. \qquad (8.77)$$

We conclude from Eqs. (8.75) and (8.76) that the phase factor (8.77) can be derived from the total energy of the system. In more detail,

$$\begin{aligned} E = \left\langle \Psi | i\hbar \frac{\partial \Psi}{\partial t}\right\rangle &= 2i\hbar \sum_{i=1}^{M} \mathbf{c}_i^\dagger \frac{\partial \mathbf{c}_i}{\partial t} \\ &= 2i\hbar \sum_{i=1}^{M} \mathbf{a}_i^\dagger \frac{\partial \mathbf{a}_i}{\partial t} + 2\sum_{i=1}^{M} \gamma_i \end{aligned} \qquad (8.78)$$

at any time t. Since the set of molecular orbitals $\mathbf{c}_i$, $i = 1, \ldots, N$, is assumed to be complete, so is the set $\mathbf{a}_i$, $i = 1, \ldots, N$. We can therefore describe the time derivative of any $\mathbf{a}_i$ by a linear combination of these vectors:

$$\frac{\partial \mathbf{a}_i}{\partial t} = \sum_{k=1}^{N} \theta_{ki} \mathbf{a}_k. \tag{8.79}$$

Contracting Eq. (8.68) with a virtual molecular orbital, we arrive at a simple relation that describes θ_{ni} in terms of Fock matrix elements:

$$\theta_{ni} = -\frac{i}{\hbar} F_{ni}, \quad n > M, \quad i \leq M, \tag{8.80}$$

where the Fock matrix element is evaluated in the basis $\{\mathbf{a}_k\}$. From the time derivative of the total energy, as given by Eq. (8.78), one concludes that the θ_{ij} have to obey an anti-Hermitian constraint in the subspace of occupied orbitals:

$$\theta_{ji} + \theta_{ij}^* = 0. \tag{8.81}$$

With Eq. (8.79), the expression for the total energy may be reformulated to read

$$E = 2i\hbar \sum_{i=1}^{M} \theta_{ii} + 2 \sum_{i=1}^{M} \gamma_i. \tag{8.82}$$

To simplify this relation even further, one may exploit molecular orbital redundancy, associated with the freedom of choosing the $M \times M$ parameters that define a unitary transformation among the occupied orbitals (see Eq. (1.53)) to reduce all elements θ_{ij} with $i, j \leq M$ to zero. This particular representation leads to

$$E = 2 \sum_{i=1}^{M} \gamma_i \tag{8.83}$$

and casts the dynamic equations for the electronic degrees of freedom into a very compact shape:

$$\frac{\partial \mathbf{a}_i}{\partial t} = -\frac{i}{\hbar} \sum_{k=M+1}^{N} F_{ni} \mathbf{a}_n. \tag{8.84}$$

The phase of the electronic wave function becomes in this frame

$$p_\Psi = \exp\left[-\frac{i}{\hbar}\int_0^t dt' E(t')\right], \tag{8.85}$$

which reduces to the dynamic phase factor $\exp(-\frac{i}{\hbar}Et)$.

Exercise 8.5. Relation (8.81) implies that the matrix $\boldsymbol{\theta}$ with elements θ_{ij}, $i,j \leq M$, is anti-Hermitian. Prove this statement by considering the expressions (8.78) and (8.82) for the total energy and the fact that the total time derivative of E vanishes.

Propagating the molecular orbital coefficients according to Eq. (8.84) meets with the problem that first-order integration algorithms, employing finite time steps, will in general not conserve the orthonormality of the molecular orbital vectors. One may, however, combine the relations (8.80), (8.81), and (8.84) to construct an approximate propagation scheme that keeps the molecular orbitals, represented by $\mathbf{a}_i$, orthonormal in the sense of Eq. (8.63) [221]. Advancing these vectors linearly from t to $t + \delta t$, one may use the direct procedure

$$\begin{aligned}\mathbf{a}_i(t+\delta t) &\approx \mathbf{a}_i(t) + \left.\frac{\partial \mathbf{a}_i(t')}{\partial t'}\right|_{t=t'} \delta t \\ &= \mathbf{a}_i(t) + \sum_{n=M+1}^{N} \theta_{ni}\mathbf{a}_n(t).\end{aligned} \tag{8.86}$$

We can write this relation in a more condensed form by defining $\mathbf{A}(t)$ as an $N \times N$ matrix composed of the M occupied orbitals $\mathbf{a}_i(t)$ and the $N - M$ unoccupied orbitals $\mathbf{a}_n(t)$. In addition, we introduce the corresponding matrix $\boldsymbol{\Theta}$ that contains the θ_{kl} as elements. Since the defining relation (8.79) does not restrain our choice of the virtual subspace elements θ_{mn} $(M < m, n \leq N)$ in any way, molecular orbital redundancy [224] implies that these can be set equal to zero, just as the elements θ_{ij} $(i,j \leq M)$ that pertain to the occupied subspace, reducing $\boldsymbol{\Theta}$ to the two remaining blocks that mix both subspaces. We reformulate Eq. (8.86) accordingly:

$$\mathbf{A}(t+\delta t) \approx \mathbf{A}(t)(\mathbf{I} + \delta t\boldsymbol{\Theta}). \tag{8.87}$$

For sufficiently small δt, this can be replaced by

$$\mathbf{A}(t+\delta t) \approx \mathbf{A}(t)\exp(\delta t\boldsymbol{\Theta}) \equiv \mathbf{A}(t)\mathbf{U}. \tag{8.88}$$

It can easily be seen that $\mathbf{U}$ is a unitary matrix and thus preserves the orthonormality of $\mathbf{A}(t)$. This follows immediately from the anti-Hermitian character of $\boldsymbol{\Theta}$ via

$$\mathbf{U}^{\dagger} = \exp(\boldsymbol{\Theta}^{\dagger}\delta t) = \exp(-\boldsymbol{\Theta}\delta t). \tag{8.89}$$

The task of describing the time evolution of the TDHF wave function is in principle solved by the relations (8.88) and (8.89). The challenge associated with the implementation of TDHF-based dynamics, as developed by Field [221], is the construction of the matrix $\mathbf{U}$ at each time step of the simulation which, by Eq. (8.80), amounts to evaluating the Fock matrix elements between occupied and virtual molecular orbitals. The dynamics encoded in these relations is free from the constraints imposed by the Born–Oppenheimer approximation on molecular motion.

Implementing this scheme successfully requires a correct representation of the forces that propel the nuclei. The nuclear force term adopted in the present context, given by expression (8.74), may be understood as originating from an *efficient potential energy surface* defined by applying the TDHF formalism to the electronic system. This is a realization of *Ehrenfest dynamics* which will be introduced in Section 8.3.4. In preparation for this account, we discuss the *Hellmann–Feynman force concept* which provides the theoretical basis for extending Newton's force law into quantum theory.

8.3.3 *The Hellmann–Feynman theorem*

To obtain the force exerted by the electrons on the nuclei, one may use the Newtonian prescription, representing this force as the negative gradient of total electronic energy, according to

$$F_i = -\frac{dE_{\text{el}}}{dX_i} = -\frac{d}{dX_i}\langle\Psi_0|\hat{H}_{\text{el}}|\Psi_0\rangle, \tag{8.90}$$

where the electrons are described by a wave function Ψ_0. As the derivative in this equation is carried out, the dependence of the wave function on the nuclear degrees of freedom must be taken into account. Thus

$$F_i = -\left(\left\langle \frac{\partial \Psi_0}{\partial X_i} | \hat{H}_{\mathrm{el}} | \Psi_0 \right\rangle + \left\langle \Psi_0 \left| \frac{\partial \hat{H}_{\mathrm{el}}}{\partial X_i} \right| \Psi_0 \right\rangle + \left\langle \Psi_0 | \hat{H}_{\mathrm{el}} | \frac{\partial \Psi_0}{\partial X_i} \right\rangle\right). \quad (8.91)$$

If Ψ_0 is an exact eigenfunction of $\hat{H}_{\mathrm{el}}$, this expression reduces to its second term, as the reader confirms easily, thus verifying the *Hellmann–Feynman theorem* [225, 226].

Exercise 8.6. *Establishing the Hellmann–Feynman theorem.* (a) Prove the Hellmann–Feynman theorem by showing that Eq. (8.91) reduces to $F_i = -\langle \Psi_0 | \frac{\partial \hat{H}_{\mathrm{el}}}{\partial X_i} | \Psi_0 \rangle$ if Ψ_0 is an eigenfunction of $\hat{H}_{\mathrm{el}}$. (b) Derive the Hellmann–Feynman theorem from Ehrenfest's principle (e.g. [58]) applied to force expectation values:

$$\frac{d\langle \mathbf{P}(t) \rangle}{dt} = \frac{i}{\hbar} \langle \Psi_0 | [\hat{H}_{\mathrm{el}}(\mathbf{X}(t)), \hat{\mathbf{P}}] | \Psi_0 \rangle, \quad (8.92)$$

with $\hat{\mathbf{P}}$ as momentum operator.

In numerical calculations, the force terms in (8.91) that involve wave function gradients will never be zero since the eigenfunction of $\hat{H}_{\mathrm{el}}$ is only accessible within a certain approximation. This situation implies the existence of two qualitatively different corrections to the Hellmann–Feynman limit of Exercise 8.6 [227, 228]. The first of these effects is due to the necessity of representing the one-particle states ψ_k in terms of a finite basis $\{\phi_n\}, n \leq N$, according to $\psi_k = \sum_{n=1}^{N} a_{kn}\phi_n$, and thus to the unavoidable basis set incompleteness. The second effect is the additional lack of self-consistency. Within Hartree–Fock theory, each of these two factors implies a mere approximate validity of the equation $\hat{F}\psi_k = \varepsilon_k \psi_k$, with $\hat{F}$ as the Fock operator, for any choice of $\{\psi_k\}$.

The partial derivative $-\frac{\partial E}{\partial X_i}$ is identical with the *Hellmann–Feynman force*, as the force term derived in Exercise 8.6 is called. In practical calculations, however, the implicit dependence of E on the nuclear coordinates cannot be neglected. We will assume in the following that this effect arises from a parametric dependence of the basis functions $\{\phi_n\}$ on $\{X_i\}$. Expanding the total derivative of the energy accordingly, and using the chain rule for functional derivatives (compare with Eq. (1.113)), we obtain

$$\frac{dE}{dX_i} = \frac{\partial E}{\partial X_i} + 2\Re\left\{\sum_m \int d\mathbf{x}\left[\frac{\delta E}{\delta\psi_m}\frac{\partial\psi_m}{\partial X_i}\right]\right\}. \tag{8.93}$$

To guarantee normalization of the molecular orbitals, we employ here the HF energy functional (1.44), assuming a basis where the Lagrangian multiplier matrix $\boldsymbol{\varepsilon}$ is diagonal. It follows by Eq. (1.52) that

$$\frac{dE}{dX_i} = \frac{\partial E}{\partial X_i} + 2\Re\left\{\sum_m \int d\mathbf{x}\left[\frac{\partial\psi_m}{\partial X_i}(\hat{F} - \varepsilon_m)\psi_m\right]\right\}. \tag{8.94}$$

The equation confirms the conclusion drawn in Exercise 8.6: if we have generated a solution of $\hat{F}\psi_m = \varepsilon_m\psi_m$, the right-hand side of Eq. (8.93) reduces to the first term, or the Hellmann–Feynman force. In practical applications, however, one has to operate with incomplete basis sets, implying that the eigenvalue equation for ψ_m is only approximately satisfied, and giving rise to a potentially significant additional force component beyond the Hellmann–Feynman limit.

Even if we adopt the utopian case that a complete basis set is available, the two-electron contribution to $\hat{F}$ may not attain full self-consistency. Comparing an ideal fully self-consistent single-particle operator $\hat{F}^{\mathrm{SCF}}$ with its real, not completely converged counterpart, $\hat{F}^{\mathrm{NSF}}$, we obtain by simultaneous addition and subtraction of the latter in the integral of Eq. (8.93)

$$\frac{dE}{dX_i} = \frac{\partial E}{\partial X_i} + 2\Re\left\{\sum_m \int d\mathbf{x}\left[\frac{\partial\psi_m}{\partial X_i}(\hat{F}^{\mathrm{NSF}} - \varepsilon_m)\psi_m\right]\right.$$
$$\left. + \int d\mathbf{x}\frac{\partial\psi_m}{\partial X_i}(\hat{F}^{\mathrm{SCF}} - \hat{F}^{\mathrm{NSF}})\psi_m\right\}, \tag{8.95}$$

where the operator difference $\hat{F}^{\mathrm{SCF}} - \hat{F}^{\mathrm{NSF}}$ reduces to the deviation between the completely and the incompletely self-consistent potential terms. By Eq. (8.95), the ith component of the overall force felt by the nuclei can be partitioned into three terms, according to

$$F_i = F_i^{\mathrm{HF}} + F_i^{\mathrm{IBS}} + F_i^{\mathrm{NSF}}, \tag{8.96}$$

with the three ingredients F_i^{HF}, F_i^{IBS}, and F_i^{NSF} standing for the Hellmann–Feynman force and the forces due to basis set incompleteness and lack of self-consistency, respectively, where the latter two are attributed to methodological imperfections. The second term was first quantitatively described by Pulay in the context of molecular physics [230]. With few and obvious modifications, the arguments outlined here for the case of HF theory may be transferred to other electronic structure methods, such as density functional theory (see Chapter 16).

8.3.4 *Ehrenfest dynamics*

The quantum-classical model reviewed in the first two segments of this section may be seen as prototypical for the *Ehrenfest representation* which describes the electronic and nuclear subsystems as averages. As a common element shared by Ehrenfest procedures, they are *effective path methods*, as an average trajectory reacts to an average electronic configuration, or equivalently, the nuclei evolve on a mean potential energy surface to be understood as the sum of properly weighted individual surfaces.

To pinpoint the characteristics of the Ehrenfest representation, we will operate with the minimal model that permits developing the relevant arguments. Let the electronic wave function of some system consisting of electrons and nuclei be written as a sum over orthonormal electronic states $\{|\Psi_k\rangle\}$ weighted with time-dependent coefficients $a_i(t)$:

$$\Psi(t) = \sum_k a_k(t)\Psi_k. \tag{8.97}$$

This wave function may evolve subject to the TDSE:

$$i\hbar\frac{\partial\Psi(t)}{\partial t} = \hat{H}_{\text{el}}[\mathbf{X}(t)]\Psi(t), \tag{8.98}$$

where the parametric dependence of the electronic Hamiltonian $\hat{H}_{\text{el}}$ on the nuclear coordinates $\mathbf{X}(t)$ is made explicit. Inserting the expansion (8.97) into Eq. (8.98) yields

$$\frac{\partial a_k(t)}{\partial t} = -\frac{i}{\hbar}\sum_l \hat{H}_{\text{el},kl}[\mathbf{X}(t)]a_l(t) \tag{8.99}$$

with $H_{\text{el},kl}(\mathbf{X}(t)) = \langle\Psi_k|\hat{H}_{\text{el}}(\mathbf{X}(t))|\Psi_l\rangle$ as electronic Hamiltonian matrix. The matrix elements $H_{\text{el},kl}(\mathbf{X}(t))$ thus depend on the trajectory traced out by the nuclei, understood as classical particles. As a classical object, the trajectory is formally described by the Hamilton equations:

$$\frac{d\mathbf{X}(t)}{dt} = \frac{\partial H(t)}{\partial \mathbf{P}}, \quad \frac{d\mathbf{P}}{dt} = -\frac{\partial H(t)}{\partial \mathbf{X}}, \tag{8.100}$$

where we have introduced a Hamilton function, defined by

$$H(t) = T(P) + \langle \Psi(t)|\hat{H}_{\text{el}}(\mathbf{X}(t))|\Psi(t)\rangle \tag{8.101}$$

and $T(\mathbf{P})$ denotes the kinetic energy function of the nuclei.

So far we have not settled for a specific representation of the electronic system. The mathematical expressions for the forces felt by the nuclei are particularly simple if the diabatic picture is adopted. The adiabatic representation, however, is just as compatible with the Ehrenfest method. Deferring any decision in favor of either frame, we invoke first Ehrenfest's principle for the definition of the force acting on the nuclei:

$$\begin{aligned}\langle \mathbf{F}\rangle &= \frac{d\langle \mathbf{P}(t)\rangle}{dt} = \frac{i}{\hbar}\langle \Psi(\mathbf{X},t)|[\hat{H}_{\text{el}}(\mathbf{X}(t)), \hat{\mathbf{P}}]\Psi(\mathbf{X},t)\rangle \\ &= -\langle \Psi(\mathbf{X},t)|\frac{\partial \hat{H}_{\text{el}}}{\partial \mathbf{X}}(\mathbf{X}(t))|\Psi(\mathbf{X},t)\rangle. \end{aligned}\tag{8.102}$$

We now adopt the adiabatic picture and assume the vectors $|\Psi_k\rangle = |\Psi_k(\mathbf{X})\rangle$ to be eigenfunctions of $\hat{H}_{\text{el}}[\mathbf{X}(t)]$ with eigenvalues E_k. The nuclear gradient of the scalar product $\langle \Psi_k|\hat{H}_{\text{el}}(\mathbf{X}(t))|\Psi_l\rangle$ is now readily evaluated. For ease of notation, we drop the argument $\mathbf{X}$ from the labels for the adiabatic states and find

$$\frac{\partial}{\partial \mathbf{X}}\langle \Psi_k|\hat{H}_{\text{el}}(\mathbf{X}(t))|\Psi_l\rangle = \frac{\partial E_k(\mathbf{X})}{\partial \mathbf{X}}\delta_{lk}. \tag{8.103}$$

On the other hand,

$$\begin{aligned}\frac{\partial}{\partial \mathbf{X}}\langle \Psi_k|\hat{H}_{\text{el}}[\mathbf{X}(t)]|\Psi_l\rangle &= \left\langle \frac{\partial \Psi_k}{\partial \mathbf{X}}|\hat{H}_{\text{el}}[\mathbf{X}(t)]|\Psi_l\right\rangle + \left\langle \Psi_k \left| \frac{\partial \hat{H}_{\text{el}}[\mathbf{X}(t)]}{\partial \mathbf{X}} \right| \Psi_l\right\rangle \\ &\quad + \left\langle \Psi_k|\hat{H}_{\text{el}}[\mathbf{X}(t)] \left| \frac{\partial \Psi_l}{\partial \mathbf{X}}\right.\right\rangle \\ &= \left\langle \Psi_k \left| \frac{\partial \hat{H}_{\text{el}}[\mathbf{X}(t)]}{\partial \mathbf{X}} \right| \Psi_l\right\rangle + (E_k - E_l)\mathbf{A}_{kl}. \end{aligned}\tag{8.104}$$

The nonadiabatic coupling vectors $\mathbf{A}_{lk}$ have the same meaning as in Eq. (2.10), namely

$$\mathbf{A}_{kl} = \left\langle \Psi_k \left| \frac{\partial}{\partial \mathbf{X}} \right| \Psi_l \right\rangle. \tag{8.105}$$

Equating expressions (8.103) and (8.104) leads to

$$\left\langle \Psi_k \left| \frac{\partial \hat{H}_{\mathrm{el}}[\mathbf{X}(t)]}{\partial \mathbf{X}} \right| \Psi_l \right\rangle = \frac{\partial E_k}{\partial \mathbf{X}} \delta_{lk} - (E_k - E_l)\mathbf{A}_{kl}. \tag{8.106}$$

Using Eq. (8.102), one obtains the Ehrenfest force in the adiabatic representation:

$$\langle \mathbf{F} \rangle = -\sum_k |a_k|^2 \frac{\partial E_k}{\partial \mathbf{X}} + \sum_{k,l} a_k^* a_l (E_k - E_l)\mathbf{A}_{kl}. \tag{8.107}$$

Exercise 8.7. Show that the differential equation for the time behavior of the coefficients a_k in Eq. (8.97) is

$$i\hbar \frac{da_k(t)}{dt} = a_k(t) E_k - i\hbar \sum_j a_j(t) \frac{d\mathbf{X}}{dt} \cdot \mathbf{A}_{kj}. \tag{8.108}$$

The right-hand side of Eq. (8.107) is the sum of a force average taken over the adiabatic states, weighted with the respective populations, and a contribution from transitions between these states, involving changes of the populations, to the overall force.

The Ehrenfest force expression differs strongly from that for *Born–Oppenheimer dynamics* where the nuclear propagation is confined to the potential surface of lowest energy, corresponding to the minimization constraint:

$$\langle \mathbf{F} \rangle = -\frac{\partial \min_{\{\Psi_0\}} E}{\partial \mathbf{X}}. \tag{8.109}$$

The Ehrenfest force (8.107) extends the Hellmann–Feynman force (8.91) from a single state to multiple electronic states. While the nuclei move on a potential energy surface that results as a composite of all contributing states, they are once more confined to motion on a single surface which, in the case of the Ehrenfest model, is representative of the potential energy landscape for the problem under study. The limits of this methodology will be scrutinized in Section 10.1. We point out that no self-consistency correction (see (8.95)) applies to the Ehrenfest force, since the latter results from an instantaneously realized wave function which may be obtained from the propagation scheme (8.108) without being subject to any minimization procedure.

Some recent implementations of the Ehrenfest scheme employ a multiconfigurational approach where the wave function (8.97) is generalized to include time-dependent basis states (see e.g. [229], and references therein).

8.3.5 *Car–Parrinello dynamics*

The defining difference between the Ehrenfest and the Born–Oppenheimer methodology, lies in the inclusion of electron dynamics by the former, but not by the latter scheme. The Born–Oppenheimer procedure is thus clearly the method of choice if molecular evolution is confined to the electronic ground state of the studied system. From a practical point of view, the time step required for integrating the relevant force equation (8.109) is dictated by the nuclear motion. As the nuclei move much more slowly than the electrons, this parameter may be chosen as much larger than the electronic step (for instance, δt in Eq. (8.87)). On the other hand, the minimization constraint imposed on Born–Oppenheimer (BO) dynamics implies that the solution to the stationary Schrödinger equation has to be approximated at each step of the nuclear trajectory. If the propagation is restricted to the lowest potential energy surface, however, recourse to electronic structure calculation beyond the determination of the initial state can be avoided by a theory that allows us to advance both the nuclear and the electronic degrees of freedom with equal time steps while subjecting the electronic system to a ground state constraint.

Such a theory is available with the *Car–Parrinello* (CP) approach [18]. The most basic decision that takes CP theory beyond the level of Born–Oppenheimer (BO) dynamics is a definition of the electronic energy not only as a function of the nuclear degrees of freedom, but also as a *functional of the single-particle orbitals* $\{\psi_i\}$ associated with the model chosen to describe the electronic subsystem. Thus

$$E_{\mathrm{el}} = E_{\mathrm{el}}\{\psi_1, \psi_2, \ldots, \psi_N; \mathbf{X}\}. \tag{8.110}$$

The nuclear coordinates are time dependent, and so are the orbitals $\{\psi_i\} = \{\psi_i(\mathbf{X}, t)\}$. Each of these orbitals adds an infinite continuous set to the domain of variables determining E_{el}. This is to be understood in the sense that the value assigned by any function ψ_i to each argument $(\mathbf{X}, t)$ defines a variable. While, as in Born–Oppenheimer theory, the expectation value of the electronic energy specifies a potential energy surface for the motion of the nuclei, the corresponding nuclear kinetic energy T_{N} must be augmented by an electronic analog of the form

$$T_{\text{el}} = \sum_j \mu \left\langle \frac{\partial \psi_j}{\partial t} \middle| \frac{\partial \psi_j}{\partial t} \right\rangle. \tag{8.111}$$

These premises lead to the following Lagrangian:

$$L_{\text{CP}} = T_{\text{N}} + T_{\text{el}} - E + \sum_{j,k} \Lambda_{jk}(\langle \psi_j | \psi_k \rangle - \delta_{jk}), \tag{8.112}$$

where E is the sum of E_{el} and the nuclear potential V_{N}. The electronic ground state Ψ_0 is composed of the orbitals ψ_j whose orthonormality is enforced by the last term in Eq. (8.112), introducing the matrix Λ_{jk} of Lagrange multipliers. The Euler–Lagrange equations engendered by L_{CP} in the two groups of variables read

$$M_i \frac{d^2 \mathbf{X}_i}{dt^2} = -\frac{\partial E}{\partial \mathbf{X}_i} \tag{8.113}$$

and

$$\mu \frac{d^2 \psi_j(\mathbf{X}, t)}{dt^2} = -\frac{\delta E}{\delta \psi_j^*(\mathbf{X}, t)} + \sum_k \Lambda_{jk} \psi_k(\mathbf{X}, t). \tag{8.114}$$

We point out that the nuclear equation (8.113) has to be augmented by a "nuclear constraint force" if the molecular orbitals are chosen to depend on the coordinates $\mathbf{X}_i$. This force is constructed in analogy to the second term on the right-hand side of the electronic equation (8.114), i.e. as the derivative of the orthogonality constraint with respect to $\mathbf{X}_i$. The explicit form of the first term on the right-hand side of Eq. (8.114) depends on whatever electronic structure theory is adopted. Relations (1.40), (1.50a, b), and (1.52) imply that in the case of HF theory, this functional derivative is given by the Fock operator $\hat{F}$ according to[1]

$$\frac{\delta E}{\delta \psi_j^*(\mathbf{X}, t)} = \hat{F}(t)\psi_j(\mathbf{X}, t). \tag{8.116}$$

The parameter μ imports *fictitious electron dynamics* into the theory, which makes it possible for the electronic subsystem to adjust to the varying nuclear configurations, circumventing the necessity of forcing a minimization constraint on the electronic energy. Just as with the nuclear kinetic energy, the kinetic energy associated with the electronic degrees of freedom can be interpreted as a measure of a characteristic temperature. If this nomenclature is adopted, molecular dynamics may be confined to the ground state Born–Oppenheimer surface by keeping the electronic temperature sufficiently low.

At this point, a more rigorous comparison between the CP scheme and the other direct dynamics procedures discussed so far, i.e. Ehrenfest and Born–Oppenheimer dynamics, can be made. The Ehrenfest formalism modifies the equation for the nuclear motion, (8.113), by substituting a multistate average, as in Eq. (8.107), and replaces the equation for the electronic degrees of freedom, (8.114), by the TDSE for the time-dependent component states (see Exercise 8.7).

In the BO procedure, the term $-\frac{\partial \min_{\{\psi_0\}} E}{\partial \mathbf{X}_i}$ determines the force felt by the nuclei, indicating the need to perform a full electronic minimization at each nuclear step, while CP dynamics demands this often time-consuming effort only once, at the beginning of the propagation. The force expression (8.116) thus shares with the Ehrenfest force the advantage of being free from corrections for lack of self-consistency. As the BO approach operates without electron dynamics, the left-hand side of Eq. (8.114) vanishes, corresponding to $\mu = 0$, leaving, after a unitary transformation of the type (1.53) has been performed, the stationary Schrödinger equation.

The chief difference between the BO and the CP scheme lies in the absence of an explicit minimization constraint imposed on the electronic energy and the presence of a fictitious mass μ in the latter. The crucial

[1] The CP method was first implemented in the framework of Kohn–Sham density functional theory (DFT, see Chapter 16), and most of its present-day realizations employ this approach. If Kohn–Sham (KS) DFT is the underlying theory, the derivative becomes

$$\frac{\delta E}{\delta \psi_j^*(\mathbf{X}, t)} = \hat{V}^{KS}\psi_j(\mathbf{X}, t). \tag{8.115}$$

question of how to control adiabaticity in CP simulations has been examined in great detail ([231]–[233]). A technical criterion for minimal energy transfer between the nuclear and the electronic subsystems makes reference to the *vibrational density of states* characteristic of both [11]. If the lowest orbital frequency, as obtained by inspecting $\widetilde{T}_{\mathrm{el}}(\omega)$, the Fourier transform of $T_{\mathrm{el}}(t)$, is by a sufficient margin larger than the highest nuclear vibration frequency, the coupling between the two subsystems is small enough to establish adiabaticity at a good level of approximation.

Evidently, if adiabatic evolution is to be simulated, the parameter μ should not be chosen too large. Strategically, the drawback of a small μ is that the time step at which the equations of motion are integrated must also be chosen small, usually lower than the limit of 1–10 fs which is customary in BO dynamics.

The CP method is frequently used to perform *simulated annealing* studies. This application makes use of the correlation between the nuclear motion and the nuclear temperature. The hot, i.e. very fast, nuclei of a finite structure probe many alternative geometries within a short time. Cooling the system down at a sufficiently small rate will eventually trap the nuclei in a state of low energy which may be the global as opposed to a local minimum, although this is not guaranteed. The respective capabilities of the CP procedure have been demonstrated [234], for instance, by subjecting the small alkali atom cluster Na_4 to simulated annealing in spin singlet and triplet conditions, where the restricted Hartree–Fock (RHF) and the restricted open-shell Hartree–Fock (ROHF) procedures were used, respectively, as underlying electronic structure methods. If the formalism is employed for the purpose of global geometry optimization, the details of the nuclear trajectories are unimportant, and a higher fictitious mass may be chosen, making it possible to use larger simulation time steps.

Added flexibility in utilizing this procedure was achieved by including the generalized valence bond (GVB) formalism to represent the electronic system [235], and by implementing the *reference system propagator algorithm* (RESPA) [236] for simultaneous propagation of both electronic and nuclear degrees of freedom at separated time scales [237]. Electronic state-specific molecular dynamics has been studied, as the approach has been used for both ground and excited state propagation while no allowance was made for inter-state transitions. A later extension of the CP methodology to nonadiabatic situations ([238], see Chapter 16) operates with the originally proposed density functional theory (DFT) as opposed to the HF approach for the description of the electronic system, utilizing a surface hopping scheme (see Chapter 11) to couple different potential energy surfaces.

8.4 Time-Dependent Hartree–Fock Dynamics in the Eikonal Approximation

In this section, we will continue and extend the discussion given earlier on electron dynamics in the framework of TDHF theory. The material to be introduced here combines the TDHF approach with the *eikonal representation* of the molecular wave function. We will demonstrate that the resulting model establishes a hierarchy of approximations to the coupling between electronic and nuclear degrees of freedom, the lowest of which, the *eikonal approximation*, has been used successfully to address complex electron rearrangement processes in dynamic problems. In this context, it will be shown how to incorporate the impact of the nuclear momenta on the evolution of the electrons into the theory, a theme that has been omitted in the various realizations of TDHF-based dynamics as presented in the previous sections. While more fundamental solutions of this problem will be introduced in the following chapters, the transformation into a traveling atomic orbital basis (Section 8.4.3) addresses this issue in a practicable and consistent manner. We will further focus on TDHF time evolution equations of the Liouville–von Neumann type, involving density matrix elements as dynamic variables and thus providing a fundamental alternative to the time propagation of molecular orbitals, as was outlined in Sections 8.3.1 and 8.3.2. Also, we will emphasize procedures to approximate asymptotic state-to-state transition probabilities in the electronic system and derive expressions for the corresponding differential cross sections.

In the initial step, we define the relevant Hamilton operators and electronic as well as nuclear basis functions by use of the stationary Schrödinger equation. For this purpose, we introduce a molecular wave function $\Psi = \Psi(\mathbf{x}, \mathbf{X})$ which satisfies

$$\hat{H}\Psi(\mathbf{x}, \mathbf{X}) = E\Psi(\mathbf{x}, \mathbf{X}) \tag{8.117}$$

where $\mathbf{x}$ and $\mathbf{X}$ stand for the electronic and nuclear coordinate sets, respectively. For clarity, we refer here to the minimal case of a system with two centers, A and B, and we work in a center-of-mass frame with $\mathbf{X}$ as relative coordinate. Introducing μ_{AB} as the reduced mass, we write the Hamiltonian as [20]

$$\hat{H} = -\frac{\hbar^2}{2\mu_{\mathrm{AB}}}\frac{\partial^2}{\partial \mathbf{X}^2} + \hat{H}_X\left(\frac{\hbar}{i}\frac{\partial}{\partial \mathbf{x}}, \mathbf{x}\right). \tag{8.118}$$

The second term in this sum describes the Hamiltonian for fixed nuclear positions and as a function of both the electronic positions and momenta. If one considers scattering processes, as we will in several examples for applications of the eikonal approximation, it is convenient to partition $\hat{H}_X$ into two terms, analogous to Eq. (6.136):

$$\hat{H}_X = \hat{H}^0_\gamma + \hat{H}'_\gamma. \tag{8.119}$$

The subscript γ stands here for a certain scattering channel, related to a well-defined asymptotic arrangement of electrons. The first term, H^0_γ, denotes the corresponding asymptotic Hamiltonian, while the second term refers to the interaction between the scattering system components. In the simplest case, which will be considered in the following, these are two atoms.

The total energy for a system with two scattering centers A and B that assumes asymptotically the electronic state i is given by the sum

$$E = E_i + \frac{\hbar^2 k_i^2}{2\mu_{\mathrm{AB}}} \tag{8.120}$$

with $\mathbf{P}_i = \hbar\mathbf{k}_i$ as relative momentum. The free relative motion is described by the plane wave nuclear state

$$\chi^0_{k_i} = \frac{1}{(2\pi\hbar)^{3/2}}\exp(i\mathbf{k}_i \cdot \mathbf{X}). \tag{8.121}$$

We include the configuration label i and the wave vector $\mathbf{k}_i$ into a set of quantum numbers $\gamma = (i, \mathbf{k}_i)$ and write the asymptotic electronic states as Φ^0_γ, such that

$$H^0_\gamma \Phi^0_\gamma = E_i \Phi^0_\gamma. \tag{8.122}$$

Using this notation, the outgoing molecular wave is expressed in terms of alternative channels $\beta = (j, \mathbf{k}_j)$ as

$$\Psi^+_\gamma(\mathbf{x}, \mathbf{X}) = \sum_\beta \Phi^0_\beta(\mathbf{x}, \mathbf{X})\chi^+_{\beta\gamma}(\mathbf{X}) \tag{8.123}$$

and

$$\chi^{+}_{\beta\gamma}(\mathbf{X}) \propto \frac{1}{(2\pi\hbar)^{3/2}} \exp(-i\mathbf{k}_i \cdot \mathbf{X})\delta_{ji} + \frac{f^{+}_{ji}(\Omega)}{X} \exp(i\mathbf{k}_i \cdot \mathbf{X}). \quad (8.124)$$

Once the asymptotic solutions are known, the differential cross section for a process accompanied by a rearrangement from an initial electronic configuration i to a final configuration j is written in terms of the corresponding scattering amplitude $f^{+}_{ji}(\Omega)$ as

$$\frac{d\sigma_{ji}}{d\Omega} = \frac{k_j}{k_i}|f^{+}_{ji}(\Omega)|^2. \quad (8.125)$$

Operating with these premises, we will introduce in the following an approximate treatment of molecular quantum dynamics that, while relying on a semiclassical trajectory model, describes the co-evolution of both the electronic and nuclear degrees of freedom and allows us to include the coupling between both subsystems at varying levels of accuracy.

8.4.1 *The eikonal approximation*

Firstly, we introduce the *eikonal representation* of the molecular wave function, writing Ψ in the polar form introduced before in the context of the WKB approximation (see Eq. (4.53)):

$$\Psi(\mathbf{x}, \mathbf{X}) = \tilde{\psi}(\mathbf{x}, \mathbf{X}) \exp\left[\frac{i}{\hbar} S(\mathbf{X})\right]. \quad (8.126)$$

The function Ψ is complex, the action S and the amplitude ψ are real-valued. Inserting Eq. (8.126) into the stationary Schrödinger equation, we arrive at

$$\left[\frac{1}{2\mu_{\mathrm{AB}}}\left(\frac{\hbar}{i}\frac{\partial}{\partial \mathbf{X}} + \frac{\partial S}{\partial \mathbf{X}}\right)^2 + \hat{H}_X - E\right]|\tilde{\psi}\rangle = 0. \quad (8.127)$$

Exercise 8.8. Derive the form of the stationary Schrödinger equation (8.127) from the ansatz (8.126).

Applying $\langle\tilde{\psi}|$ from the left, where the scalar product $\langle|\rangle$ involves integration with respect to the electronic coordinates, yields the Hamilton–Jacobi equation (compare with Eq. (4.56))

$$\frac{1}{2\mu_{AB}}\left(\frac{\partial S}{\partial \mathbf{X}}\right)^2 + V\left(\frac{\partial S}{\partial \mathbf{X}}, \mathbf{X}\right) = E \quad (8.128)$$

which may be read as a differential equation for the action function S. The potential V is partitioned into two terms:

$$V = V_0 + V_{\text{qu}}, \tag{8.129}$$

where the leading contribution V_0 is the *Ehrenfest potential*

$$V_0 = \frac{\langle\tilde{\psi}|\hat{H}_X|\tilde{\psi}\rangle}{\langle\tilde{\psi}|\tilde{\psi}\rangle}. \tag{8.130}$$

The quantum potential V_{qu} (see Eq. (4.54a)) is represented as a sum of two components:

$$V_{\text{qu}} = V_1 + V_2, \tag{8.131}$$

with

$$V_1 = -\frac{i\hbar}{2\mu_{\text{AB}}}\frac{\partial S}{\partial \mathbf{X}} \cdot \left[\left\langle \tilde{\psi} \middle| \frac{\partial \tilde{\psi}}{\partial \mathbf{X}} \right\rangle - \left\langle \frac{\partial \tilde{\psi}}{\partial \mathbf{X}} \middle| \tilde{\psi} \right\rangle\right] \Big/ \langle\tilde{\psi}|\tilde{\psi}\rangle, \tag{8.132}$$

$$V_2 = -\frac{\hbar^2}{2\mu_{\text{AB}}}\frac{1}{2}\left[\left\langle \tilde{\psi} \middle| \frac{\partial^2 \tilde{\psi}}{\partial \mathbf{X}^2} \right\rangle + \left\langle \frac{\partial^2 \tilde{\psi}}{\partial \mathbf{X}^2} \middle| \tilde{\psi} \right\rangle\right] \Big/ \langle\tilde{\psi}|\tilde{\psi}\rangle. \tag{8.133}$$

We point out that the term $-i\hbar\langle\tilde{\psi}|\frac{\partial\tilde{\psi}}{\partial\mathbf{X}}\rangle/\langle\tilde{\psi}|\tilde{\psi}\rangle$ in Eq. (8.132) can be identified with the momentum imparted by the nuclei to the electronic system. This feature is further discussed in the context of electron nuclear dynamic theory, and specifically in the commentary on Eq. (9.38).

If V is truncated after V_0 the energy balance (8.128) simplifies to

$$\frac{1}{2\mu_{\text{AB}}}\left[\frac{\partial S}{\partial \mathbf{X}}\right]^2 + \frac{\langle\tilde{\psi}|\hat{H}_X\tilde{\psi}\rangle}{\langle\tilde{\psi}|\tilde{\psi}\rangle} = E. \tag{8.134}$$

Reduction of V to V_0 is equivalent to the Ehrenfest approximation based on the electronic wave function $\tilde{\psi}$. If $\tilde{\psi}$ is a well-defined electronic state of the molecular system, Eq. (8.134) expresses the Born–Oppenheimer approximation. Correspondingly, $\partial S/\partial\mathbf{X}$ describes a nuclear momentum, as was argued before (see Section 4.2.1):

$$\mathbf{P} = \frac{\partial S}{\partial \mathbf{X}}. \tag{8.135}$$

The potential component V_1 couples electronic and nuclear momenta, while V_2 is determined by the electronic momenta alone. We consider now the limiting case of short nuclear de Broglie wave lengths λ, as discussed when introducing the WKB approach in Section 4.2. The meaning of *short* in the present context is that the wave functions as well as their gradients change little over a nuclear wave length. This condition defines the *eikonal approximation* which is evidently valid for sufficiently high nuclear momenta only. Two inequalities provide the criterion for the short wave length case [20]:

$$\frac{|\langle\tilde{\psi}|\frac{\partial}{\partial\mathbf{X}}|\tilde{\psi}\rangle|}{\langle\tilde{\psi}|\tilde{\psi}\rangle} \ll \lambda^{-1}, \tag{8.136}$$

$$\frac{|\langle\tilde{\psi}|\frac{\partial^2}{\partial\mathbf{X}^2}|\tilde{\psi}\rangle|}{|\langle\tilde{\psi}|\frac{\partial}{\partial\mathbf{X}}|\tilde{\psi}\rangle|} \ll \lambda^{-1}. \tag{8.137}$$

On the basis of these relations, and using $P = \frac{h}{\lambda}$ in conjunction with Eq. (8.135), we conclude that both V_1 and V_2 in Eq. (8.131) are substantially smaller than the nuclear kinetic term $\frac{1}{2\mu_{AB}}[\frac{\partial S}{\partial\mathbf{X}}]^2 = \frac{1}{2\mu_{AB}}\mathbf{P}^2$. The eikonal approximation consists in disregarding both these terms in the energy (8.128), reducing V to V_0. To develop the equations of motion from this approach, we insert the resulting expression for the total energy E (8.128) into (8.127). The inequalities (8.136) and (8.137) justify the neglect of the term proportional to $\partial^2\tilde{\psi}/\partial\mathbf{X}^2$ in comparison to $\partial^2 S/\partial\mathbf{X}^2$. The remaining terms may be summarized as follows:

$$\left[\frac{\mathbf{P}}{\mu_{\mathrm{AB}}}\cdot\frac{\hbar}{i}\frac{\partial}{\partial\mathbf{X}} + \hat{H}_X - \hat{W}(\mathbf{X})\right]\tilde{\psi}(\mathbf{x},\mathbf{X}) = 0 \tag{8.138}$$

with

$$\hat{W}(\mathbf{X}) = V_0(\mathbf{X}) + \frac{i\hbar}{2\mu_{\mathrm{AB}}}\frac{\partial}{\partial\mathbf{X}}\cdot\mathbf{P}. \tag{8.139}$$

This potential contains an imaginary part that describes the divergence of streamlines. A formal solution of Eq. (8.138) can readily be written down if the trajectory momentum

$$\mathbf{P} = \mu_{\mathrm{AB}}\frac{\partial\mathbf{X}}{\partial t} \tag{8.140}$$

is introduced. Substituting for $\mathbf{P}$ in Eq. (8.138), we make the transition into a time-dependent frame and find

$$\left[\frac{\hbar}{i}\frac{\partial}{\partial t}\bigg|_X + \hat{H}_X - \hat{W}(\mathbf{X})\right]\tilde{\psi}(\mathbf{x},\mathbf{X},t) = 0. \tag{8.141}$$

The reader may verify explicitly that the following ansatz for $\tilde{\psi}$ expressed as a function of the trajectory $\mathbf{X}(t)$:

$$\tilde{\psi}(\mathbf{x}, \mathbf{X}(t), t) = \left(\frac{1}{2\pi\hbar}\right)^{3/2} \exp\left[\frac{i}{\hbar}\int_{t_i}^{t} dt' \hat{W}(\mathbf{X}(t'))\right] D(\mathbf{x}, t), \qquad (8.142)$$

satisfies Eq. (8.141) if the function $D(\mathbf{x}, t)$ which carries the information about the electronic subsystem is constrained by the TDSE according to

$$\left[\frac{\hbar}{i}\frac{\partial}{\partial t}\bigg|_X + \hat{H}_X\right] D(\mathbf{x}, t) = 0. \qquad (8.143)$$

The factorization achieved by the solution (8.142) separates the electronic and the nuclear coordinates of the problem. In this expression, the exponential factor of $\tilde{\psi}$ is a product of two contributions, corresponding to the real and the imaginary part of $\hat{W}(\mathbf{X})$ (see Eq. (8.139)). Specifically, $\Im\{W(\mathbf{X})\}$ generates a real exponential function that decreases with increasing divergence of the streamlines and reduces the amplitude of the outgoing wave. In the following, we will focus on the second time-dependent factor in the wave function (8.142) and address the electronic problem posed by Eq. (8.143) in the framework of TDHF theory.

8.4.2 *TDHF approach to the electronic problem within the eikonal approximation*

Let us assume an initial electronic state given by the quantum numbers $\gamma = (k_i, i)$, where i characterizes a particular electronic configuration. In this case, we ask for a specific solution $D_\gamma(t)$ of Eq. (8.143) that obeys the initial condition

$$D_\gamma(t) = \Phi_\gamma^0 \exp\left[-\frac{i}{\hbar}E_i^0(t - t_i)\right] \quad \text{if } t \to t_i. \qquad (8.144)$$

Equation (8.143) is to be solved along the nuclear trajectory $\mathbf{X}(t)$ which can be generated as one augments the classical trajectory momentum Eq. (8.140) by the following two relations:

$$\frac{d\mathbf{P}}{dt} = -\frac{\partial V_\gamma}{\partial \mathbf{X}}, \qquad (8.145)$$

$$V_\gamma \equiv \frac{\langle \Phi_\gamma | \hat{H}_X | \Phi_\gamma \rangle}{\langle \Phi_\gamma | \Phi_\gamma \rangle}. \qquad (8.146)$$

Thus, treating the dynamic problem in the framework of the eikonal approximation amounts to solving Eqs. (8.143), (8.145) and (8.146) self-consistently.

In the following, we will address the description of the electronic system. Since the differential equation for the electronic part of the overall wave function has been cast into the form of the TDSE (8.143), an approximate solution can be derived on the basis of TDHF theory [20, 239]. Adopting this approach, we assume the form of a time-dependent determinant composed of molecular spin orbitals ζ_i for $D_\gamma(\mathbf{x}, t)$:

$$D_\gamma(\mathbf{x}, t) = (N!)^{-1/2}\hat{A}\left[\prod_{i=1}^{N}\zeta_i(\mathbf{x}_i, t)\right], \tag{8.147}$$

$$\zeta_i(\mathbf{x}, t) = \psi_i(\mathbf{x}, t)\sigma(\omega_i), \tag{8.148}$$

with $\hat{A}$ as antisymmetrization operator, σ as electron spin function, and ω_i as spin coordinate of electron i. The molecular orbitals are subjected to orthonormality constraints

$$\langle\psi_i, \psi_j\rangle = \delta_{ij}. \tag{8.149}$$

Use of the time-dependent variational principle leads to the TDHF equation (see Eq. (8.50))

$$\hat{F}\psi_i - \sum_j \psi_j \varepsilon_{ji} = i\hbar\frac{\partial\psi_i}{\partial t} \tag{8.150}$$

with the Hermitian matrix ϵ_{ji} given by Eq. (8.51). Making reference to an unrestricted case, as in Section 1.4.2, we allow $\hat{F}$ to be spin-dependent, with $\hat{F}^\alpha(\hat{F}^\beta)$ acting on alpha (beta) spin orbitals. The generalized Fock operator $\hat{F}^\sigma$ ($\sigma = \alpha, \beta$) may be written in the form

$$\hat{F}^\sigma = \hat{h} + \hat{G}^\sigma[\hat{\rho}^\alpha, \hat{\rho}^\beta], \tag{8.151}$$

and thus in terms of the one-electron Hamiltonian $\hat{h}$ and the HF self-consistent two-electron interaction operator $\hat{G}$, whose matrix form is explicitly given by Eq. (1.88). This term is here formulated as a function of both the spin α and the spin β one-particle density operator, defined as

$$\hat{\rho}^\sigma(t) = \sum_{\text{occupied } i} |\psi_i^\sigma\rangle\langle\psi_i^\sigma| \quad \text{with } \sigma = \alpha \quad \text{or} \quad \beta. \tag{8.152}$$

From a computational point of view, advancing $\rho^\sigma(t)$ in time appears preferable to calculating $\psi_i^\sigma(t)$ by solving Eq. (8.150). The time evolution of $\rho^\sigma(t)$ is governed by the TDHF Liouville–von Neumann equation (8.56):

$$\hat{F}^\sigma \hat{\rho}^\sigma - \hat{\rho}^\sigma \hat{F}^\sigma = i\hbar \frac{\partial \rho^\sigma}{\partial t}. \tag{8.153}$$

The state-to-state cross section (8.125) can be cast into the language of density matrices by projecting the electronic determinant that has evolved from an initial configuration i, $D_i(t_f)$, on the asymptotic state D^0 with label j, recognizing that the square magnitude of this projection yields the probability for an electronic transition from i to j:

$$\begin{aligned}\frac{d\sigma_{ji}}{d\Omega} &= \frac{d\sigma_i}{d\Omega}|\langle D_j^0(t_f), D_i(t_f)\rangle|^2 \\ &= \frac{d\sigma_i}{d\Omega}|\det[\langle \psi_j^0(t_f)|\psi_i(t_f)\rangle]|^2 \\ &= \frac{d\sigma_i}{d\Omega}\det[\langle \psi_j^0(t_f)\hat{\rho}_i(t_f)\psi_j^0(t_f)\rangle].\end{aligned} \tag{8.154}$$

In the second step, advantage is taken of the fact that the overlap integral of two determinantal states composed of molecular orbitals is the determinant of the molecular orbital overlap integrals.

8.4.3 *The Liouville–von Neumann equation in a traveling orbital basis*

In what follows, we address the finite basis installation of the Liouville–von Neumann equation according to the scheme (8.60) and comment particularly on the choice of a special basis for numerical evaluation of the electronic densities, taking into account the complicating condition that the molecular orbitals are composed of atomic orbitals ϕ_j that generally contain an explicit dependence on time:

$$\psi_i^\sigma(\mathbf{x}, t) = \sum_j c_{ij}^\sigma(t)\phi_j(\mathbf{x}, t). \tag{8.155}$$

Expressing ρ^σ in the basis of the atomic orbitals, one finds

$$\hat{\rho}^\sigma = \sum_{p,q} |\phi_p\rangle P_{pq}^\sigma \langle \phi_q| \tag{8.156}$$

where P_{pq}^σ refers to the elements of the one-particle charge density matrix $\mathbf{P}^\sigma$. Representing also the Fock operator in the atomic basis, the TDHF

Eq. (8.153) becomes

$$i\hbar\frac{\partial \mathbf{P}^\sigma}{\partial t} = \mathbf{S}^{-1}(\mathbf{F}^\sigma - i\hbar\mathbf{\Omega})\mathbf{P}^\sigma - \mathbf{P}^\sigma(\mathbf{F}^\sigma - i\hbar\mathbf{\Omega})\mathbf{S}^{-1}. \tag{8.157}$$

The structure of the (non-orthogonal) atomic basis $\{\phi_i\}$ gives rise to the appearance of the matrices **S** and **Ω** in the TDHF equation. With **S**, we have introduced the overlap matrix between atomic basis functions $S_{ij} = \langle\phi_i|\phi_j\rangle$ with **Ω** the matrix of nonadiabatic orbital coupling $\Omega_{ij} = \langle\phi_i|\frac{\partial\phi_j}{\partial t}\rangle$.

In contrast to our presentation of wave function propagation based on the TDHF equations in Sections 8.3.1 and 8.3.2, we do not make here any assumption of vanishing overlap between atomic basis functions centered on different atoms. In the asymptotic region where the scattering partners cease to interact, the Ω_{ij} will not necessarily reduce to zero since the atomic basis functions $\{\phi_j\}$, being attached to nuclear centers, keep their dependence on time. This can lead to spurious couplings between functions $|\phi_j\rangle$ that are assigned to the same center. This artifact, however, may be eliminated by a special choice of the atomic basis functions. Defining, as usual, the molecular orbitals $\{\psi_i\}$ in terms of atomic orbitals (AOs), we can ensure the correct inclusion of both the nuclear coordinates and momenta by transformation of static atomic orbitals $\{\varphi_i\}$ into traveling atomic orbitals (TAOs). The atomic basis states $\{\phi_i\}$ are then products of the static AOs and the corresponding *electron translation factors* [240], labeled *ETs*:

$$\phi_i(\mathbf{x}, t) = \varphi_i(\mathbf{x})ET_n(\mathbf{x}, t) \tag{8.158}$$

and

$$ET_n(\mathbf{x}, t) \equiv \exp\left(-\frac{i}{\hbar}m\left[\mathbf{v}_n(t)\mathbf{x} - \frac{1}{2}\int_{t_i}^{t} dt' v_n^2(t')\right]\right). \tag{8.159}$$

The index n refers to the nth nucleus, and m is the electron mass. The overall decomposition of the molecular orbital into atomic orbitals is therefore

$$\psi_i^\sigma(\mathbf{x}, t) = \sum_\nu c_{i\nu}^\sigma(t)\varphi_\nu(\mathbf{x})ET_n(\mathbf{x}, t). \tag{8.160}$$

Thus, time enters the molecular orbitals explicitly through the coefficients $c_{i\nu}^\sigma$ which may be used as dynamic parameters (as outlined in Section 8.3.1) and through the electron translation factors. It is understood here that the atomic orbital φ_ν is attached to the nth nucleus. Factorizing the atomic basis functions according to Eq. (8.160), one may now repeat the procedure that led to Eq. (8.157). The result is similar with an important modification:

$$i\hbar\frac{\partial \mathbf{P}^{\sigma}}{\partial t} = \mathbf{S}^{-1}\mathbf{F}_{\mathrm{T}}^{\sigma}\mathbf{P}^{\sigma} - \mathbf{P}^{\sigma}(\mathbf{F}_{\mathrm{T}}^{\sigma})^{\dagger}\mathbf{S}^{-1} \tag{8.161}$$

where we define

$$\mathbf{F}_{\mathrm{T}}^{\sigma} \equiv \mathbf{H}_{\mathrm{T}} + \mathbf{G}^{\sigma} \tag{8.162}$$

and

$$\mathbf{H}_{\mathrm{T}} \equiv \mathbf{T}_{\mathrm{T}} + \sum_{i} \mathbf{V}_{i} + \mathbf{L}_{\mathrm{T}}. \tag{8.163}$$

The kinetic energy matrix, written in terms of the traveling atomic basis functions, reads

$$(T_{\mathrm{T}})_{\mu\nu} = \langle \phi_{\mu}|\hat{T}_{\mathrm{T}}|\phi_{\nu}\rangle = \int d\mathbf{x}[\phi_{\mu}(\mathbf{x},t)]^{*}[ET_{n}(\mathbf{x},t)]^{*}\,\hat{T}ET_{n}(\mathbf{x},t)\phi_{\nu}(\mathbf{x},t). \tag{8.164}$$

The matrices $\mathbf{V}_i$ stand for the one-particle interactions between the electrons and the nuclei. Correspondingly, the summation runs over all nuclear centers involved, comprising only projectile and target in the diatomic scattering situation considered here. The additional term $\mathbf{L}_{\mathrm{T}}$:

$$(L_{\mathrm{T}})_{\mu\nu} = -m\mathbf{a}_{n}\cdot \int d\mathbf{x}[\phi_{\mu}(\mathbf{x},t)]^{*}[ET_{n}(\mathbf{x},t)]^{*}\mathbf{x}ET_{n}(\mathbf{x},t)\phi_{\nu}(\mathbf{x},t), \tag{8.165}$$

where

$$\mathbf{a}_{n} = \frac{d\mathbf{v}_{n}}{dt}, \tag{8.166}$$

is generated by the time derivative of the electron translation factors. Obviously, it reduces to zero if the nuclear acceleration $\mathbf{a}_n$ vanishes. This condition, however, is satisfied in the asymptotic regions of the problem considered. Thus, the representation (8.158) of the AO achieves the goal of restricting the nonadiabatic coupling to the zone of interaction between the atomic centers involved. Expression (8.165) may be interpreted as the work associated with the dragging force exerted by the nucleus with index n on the electron [20].

Applying the substitution

$$\mathbf{Y}^{\sigma} \equiv \mathbf{S}^{-1}\mathbf{F}_{\mathrm{T}}^{\sigma} \tag{8.167}$$

in Eq. (8.161), one arrives at a simplified differential equation for the charge density matrix, namely

$$\frac{\partial \mathbf{P}^{\sigma}}{\partial t} = \frac{1}{i\hbar}[\mathbf{Y}^{\sigma}\mathbf{P}^{\sigma} - \mathbf{P}^{\sigma}\mathbf{Y}^{\sigma\dagger}]. \tag{8.168}$$

The formal solution of Eq. (8.168) may be written in terms of the time evolution operator

$$\hat{U}(t, t_i) = \exp\left(\frac{i}{\hbar}\int_{t}^{t_i} dt' \hat{Y}^{\sigma}(t')\right) \tag{8.169}$$

as

$$\mathbf{P}^{\sigma}(t) = \mathbf{U}(t, t_0)\mathbf{P}^{\sigma}(t_0)[\mathbf{U}(t, t_0)]^{\dagger}. \tag{8.170}$$

The charge density matrix $\mathbf{P}^{\sigma}(t_0)$ at the initial time is found from a solution of the static HF equations at $t = t_i$. For $t > t_i$ a numerical approximation scheme has to be used to construct the matrix $\mathbf{P}^{\sigma}(t)$, since $\mathbf{W}^{\sigma}(t)$ depends on the changing nuclear positions as well as momenta, and thus, the operator $\hat{U}(t, t_0)$ is not known *a priori*. The solution procedure involves a time step size Δt that is typically chosen long by comparison with the electronic time scale but short with respect to the nuclear motion. This makes it possible to treat the electronic relaxation as the primary dynamic effect and include the change of nuclear positions in a subsequent correction step which is the basic strategy of the *relax and drive* method, as detailed in [241].

The eikonal approximation in conjunction with TDHF theory, or the *Eik-TDHF* approach [242], is particularly well adjusted to treating molecular problems that arise in the course of scattering situations. Considering the reliance of the model on the classical trajectory concept, its success primarily in the physical rather than the chemical domain appears natural, as an adequate representation of molecular reaction dynamics often calls for quantum mechanical treatment of nuclear degrees of freedom. Within a given scattering situation, a trajectory picture tends to be better adjusted to the high energy than to the low energy range. In the following, we will refer to the simplest case of atomic scattering, namely the interaction of an H atom with a proton projectile, as addressed by Eik-TDHF calculation, and underscore the capacity of the model to represent the interplay of atomic and molecular characteristics which are typical for ion–atom collisions at low energies [243].

8.5 Case Study: Light Emission in Slow Proton–Hydrogen Collisions

In the final section of this chapter, we present an application of the Eik-TDHF model, intended to show how molecular signatures of a prototypical atomic scattering process can be extracted by TDHF analysis of the respective process. Salient features of the diatomic interaction are obtained by generating spectral information as the light emission caused by the collision is simulated.

The H atom target, assumed to be initially stationary, is prepared in its ground state. It is located at the origin of a Cartesian coordinate system, as displayed in Figure 8.1. The initial projectile velocity is oriented along the $+z$-axis. Further, the direction of projectile deflection is given by the solid angle $\Omega = (\Theta, \Phi)$ with Θ as the polar angle and Φ as the time-independent orientation angle of the scattering plane, characterized by the coordinate pair (x, z). Using the premises of the Eik-TDHF model, we will describe the light emitted during the H^+-H interaction, asking specifically for the

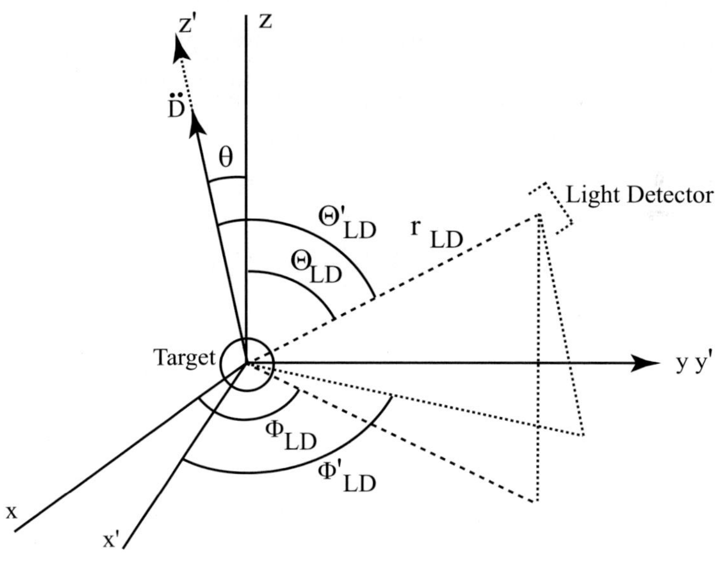

Figure 8.1: The coordinate systems used for the Eik-TDHF treatment of the $H^+ \rightarrow H$ collision discussed in the text (see [243]). (Reprinted with permission from [243]. Copyright (1997) by the American Institute of Physics.)

radiation scattered into the direction of a light detector placed under a solid angle $\Omega_{\rm LD} = (\Theta_{\rm LD}, \Phi_{\rm LD})$ and at a distance $r_{\rm LD}$ from the origin. Besides the fixed laboratory frame (x, y, z), Figure 8.1 shows a rotating frame (x', y', z'), where y' and y coincide and the direction of the z'-axis is defined as that of the second time derivative vector of the electric dipole moment $\frac{\partial^2 \mathbf{D}}{\partial t^2}(t_{\rm r})$. This choice is motivated by the functional forms of the electric and magnetic fields (see Eqs. (8.175) and (8.176)) associated with the dipolar emission considered here which involve the second dipole derivative. The parameter $t_{\rm r}$ refers to the emission time of the radiation.

With these assignments, one may write the differential cross section for the light power emission as a product of the cross section for the ion–atom collision, involving the Ehrenfest potential (8.146) if the eikonal approximation is employed, and a term that accounts for the energy E emitted per unit time into the direction of the light detector while the H^+ projectile is deflected into the solid angle $d\Omega$:

$$\frac{d^2(dE/dt)}{d\Omega d\Omega_{\rm LD}} = \frac{d\sigma}{d\Omega}\left(\frac{dI}{d\Omega_{\rm LD}}\right)_{b,\Phi}. \tag{8.171}$$

A classical treatment may be utilized to approximate the quantity I which denotes the radiated power. Adopting SI units, we express the power emitted into the solid angle $\Omega_{\rm LD}$ by use of the Poynting vector, or vector of the electromagnetic flux:

$$\mathbf{S}(\mathbf{r}_{\rm LD}, t) = \frac{1}{\mu_0}\mathbf{E} \times \mathbf{B} \tag{8.172}$$

with μ_0 as the magnetic permeability of free space. In keeping with the decomposition

$$\mathbf{S} = \mathbf{S}_\perp + \mathbf{S}_z, \tag{8.173}$$

we distinguish two components of the radiated power, namely I_z and $I_\perp$, where the index $k = \perp, z$ indicates the flux components perpendicular and parallel to the direction of the incoming projectile. Accordingly,

$$\left.\frac{dI_k}{d\Omega_{\rm LD}}\right|_{b,\Phi} = \mathbf{S}_k \cdot \mathbf{n}_{\rm LD} r_{\rm LD}^2, \tag{8.174}$$

with b as impact parameter. The vector **S** combines from the electric and magnetic fields at the location of the light detector:

$$\mathbf{E}(\mathbf{r}_{\mathrm{LD}}, t) = \frac{\mu_0}{4\pi r_{\mathrm{LD}}}\left[\left(\mathbf{n}_{\mathrm{LD}} \cdot \frac{\partial^2 \mathbf{D}}{\partial t^2}(t_r)\right)\mathbf{n}_{\mathrm{LD}} - \frac{\partial^2 \mathbf{D}}{\partial t^2}(t_r)\right], \tag{8.175}$$

$$\mathbf{B}(\mathbf{r}_{\mathrm{LD}}, t) = \frac{\mu_0}{4\pi r_{\mathrm{LD}} c}\mathbf{n}_{\mathrm{LD}} \times \frac{\partial^2 \mathbf{D}}{\partial t^2(t_{\mathrm{r}})}. \tag{8.176}$$

With c as the speed of light, $\mathbf{n}_{\mathrm{LD}} = \mathbf{r}_{\mathrm{LD}}/r_{\mathrm{LD}}$ and $t_{\mathrm{r}} = t - r_{\mathrm{LD}}/c$. The total energy radiated into the solid angle Ω_{LD} is obtained by integration of the power factor in Eq. (8.174) with respect to time. We set

$$\left.\frac{dI_k}{d\Omega_{\mathrm{LD}}}\right|_{b,\Phi} \equiv |A_k(t; b, \Phi)|^2. \tag{8.177}$$

For a given impact parameter and scattering plane orientation, the total energy emitted into the angular element is then expressed as

$$\left.\frac{dE_k}{d\Omega_{\mathrm{LD}}}\right|_{b,\Phi} = \int_{-\infty}^{\infty} dt |A_k(t; b, \Phi)|^2 \tag{8.178}$$

or, alternatively, as

$$\left.\frac{dE_k}{d\Omega_{\mathrm{LD}}}\right|_{b,\Phi} = \int_{-\infty}^{\infty} d\omega |\tilde{A}_k(\omega; b, \Phi)|^2 \tag{8.179}$$

where we denote the Fourier transform of A_k by $\tilde{A}_k$. Evaluating the electromagnetic flux and consequently the power radiated per solid angle is greatly simplified by the use of rotating frame quantities [243]. The energy radiated per solid angle and frequency can then be found from

$$\left.\frac{d^2E_k}{d\omega d\Omega_{\mathrm{LD}}}\right|_{b,\Phi} = 2|\tilde{A}_k(\omega; b; \Phi)|^2. \tag{8.180}$$

Exercise 8.9. Verify Eq. (8.179) using the relation $\tilde{A}_k(\omega)^* = \tilde{A}_k(-\omega)$ which holds for real $A_k(t)$.

Finally, summing over the two components $k = \perp, z$ and integrating over both the impact parameter and the scattering plane orientation angle Φ, one may deduce the differential cross section for energy emission in the frequency domain and find

$$\frac{d^2E}{d\omega d\Omega_{\mathrm{LD}}} = \sum_k 2\int_0^\infty db\, b \int_0^{2\pi} d\Phi |\tilde{A}_k(\omega; b, \Phi)|^2. \tag{8.181}$$

The electromagnetic flux and thus the functions $\tilde{A}_k(\omega; b, \Phi)$, or, equivalently, $A_k(t; b, \Phi)$, depend on the dipolar moment $\mathbf{D}$ through the identities (8.172)–(8.177). Partitioning the dipolar moment into nuclear and electronic contributions, we write

$$\mathbf{D} = \mathbf{D}_{\mathrm{nuc}} + \mathbf{D}_{\mathrm{el}}, \tag{8.182}$$

where the nuclear term is simply given by the projectile and target coordinates as $\mathbf{D} = Z_{\mathrm{T}}\mathbf{X}_{\mathrm{T}}(t) + Z_{\mathrm{P}}\mathbf{X}_{\mathrm{P}}(t)$ with nuclear charges $Z_{\mathrm{T}} = Z_{\mathrm{P}} = 1$ for the present example of proton-hydrogen scattering. The forces felt by the nuclei are obtained from the Newtonian equation (8.145). The electronic term involves the time-dependent electronic populations which result from the Liouville–von Neumann equation (8.168). Identifying the electronic dipolar moment with its expectation value, we get

$$\mathbf{D}_{\mathrm{el}}(t) = Tr[\hat{\rho}\mathbf{D}_{\mathrm{el}}] = -e\sum_{i,j} P_{ij}(t)\langle\phi_i|\mathbf{x}|\phi_j\rangle \tag{8.183}$$

where e is the elementary charge, and the density operator is represented in the basis of traveling atomic orbitals, as in Eq. (8.156).

Figure 8.2 shows the frequency dependence of the energy emission cross section (8.181) for the examples of three different light detector angles Θ_{LD} and for a total energy of 250 eV. The spectrum is dominated by a narrow peak that is of atomic origin, corresponding to the $2p \to 1s$ transition of the H atom. The structures surrounding this peak, however, are signatures of light emission from an intermediate H_2^+ complex. Thus, to the left of the atomic line, one can identify a signal stemming from the $3d\sigma_{\mathrm{g}} \to 2p\sigma_{\mathrm{u}}$ transition of H_2^+, to the right appear peaks related to the $2s\sigma_{\mathrm{g}} \to 2p\sigma_{\mathrm{u}}$ and $3d\pi_{\mathrm{g}} \to 2p\sigma_{\mathrm{u}}$ transitions. Further molecular lines are obliterated by the prevailing atomic emission.

The Eik-TDHF model draws a detailed picture of the deexcitation and, by inference, excitation processes observable in the scattering of protons by hydrogen targets. It should be noted, however, that the formalism outlined before is not in keeping with energy conservation, since it does not diminish

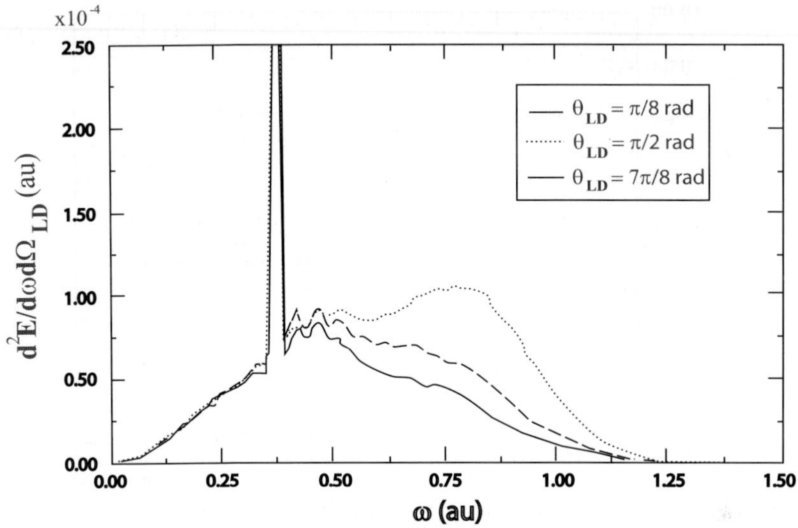

Figure 8.2: Differential cross section for energy emission per unit frequency and solid angle as a function of frequency for $H^+ \rightarrow H$ scattering at $E = 250\,\text{eV}$ [243]. The dominant peak corresponds to the $2p \rightarrow 1s$ transition of the H atom. (Reprinted with permission from [243]. Copyright (1997) by the American Institute of Physics.)

the total energy E of the $H^+ \rightarrow H$ scattering system by the energy portion ΔE lost due to electromagnetic radiation. This limitation restricts the validity of the method to the regime $\Delta E \ll E$. We will readdress this issue in Chapter 9 from the viewpoint of the electron nuclear dynamics theory. Eik-TDHF computations on ion–atom scattering systems have been found to compare favorably with experimental data as well as with competing theoretical approaches in numerous cases. This is documented by Figure 8.3 which shows the integral cross section for charge transfer resulting from the collision between He_2^+ projectiles and H target atoms as a function of the projectile energy.

It can be seen from the figure that the Eik-TDHF method is in excellent agreement with the experimental findings over the whole energy scale $2.5\,\text{keV} \leq E \leq 9.5\,\text{keV}$, exceeding the precision of the available close-coupled-channels calculations. The examples for Eik-TDHF results as shown in Figures 8.2 and 8.3 both involve only one active electron. The theory, however, has been generalized successfully to include cases of slow

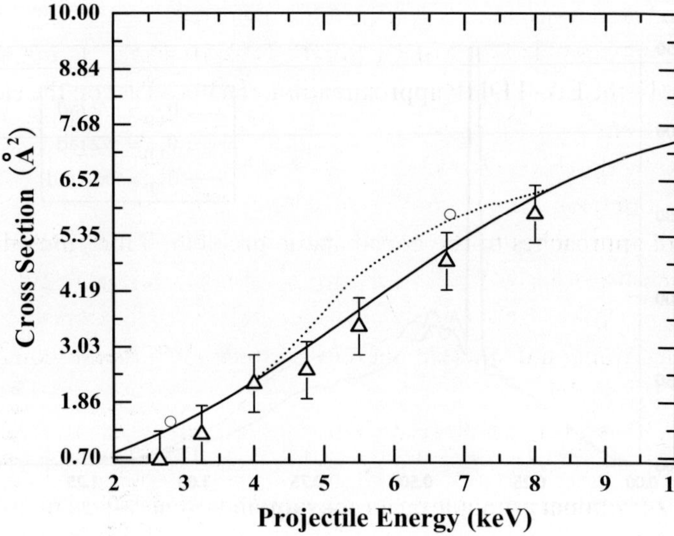

Figure 8.3: Integral charge transfer cross section for the scattering system $He_2^+ \to H$ versus the projectile energy [239]. The solid line refers to Eik-TDHF computations [244], crosses and triangles to experimental data [245, 246]. The dotted line as well as the circles denote results of close-coupled channel calculations ([247]–[250]). (Reprinted with permission from [239]. Copyright (1999) by the American Chemical Society.)

ion–atom collisions with several active electrons, such as the scattering of He^+ projectiles by deuterium targets [242].

The Eik-TDHF theory, presented earlier with emphasis on its simplest class of molecular applications, namely the interaction of an ion and an atom, is readily extendable to more complex systems, involving intermolecular processes. It is noteworthy that the eikonal approximation has also been used successfully in conjunction with potential energy surfaces, for instance in the description of electronically diabatic photodissociation of CH_3I [251].

The dynamic TDHF approaches that have been discussed in this chapter all operate with an average force applied by the evolving electronic system on the nuclei. Self-consistency is accomplished by advancing the nuclear positions and momenta simultaneously with and in reaction to the electronic variables. The coupling between electronic and nuclear degrees of freedom proceeds through the total TDHF energy. More specifically, one could characterize the influence exerted by the electrons on the nuclei as arising from a force associated with an average potential energy surface. Conversely, in all models described in Section 8.3, the electrons feel the nuclei only through their instantaneous positions which enter the Fock operator (see, for instance, Eq. (8.69)). In the Eik-TDHF approximation, the nuclei act on the electrons through both their positions and momenta, where the latter dependence is engendered by the *ad hoc* introduction of electron translation factors.

It is an interesting aspect of the eikonal representation that it admits a hierarchy of approaches to the nonadiabatic problem. Thus, including the full quantum force derived from the sum of all potential contributions V_0, V_1, and V_2 in Eqs. (8.130)–(8.133), introduces the nonadiabatic interaction between electronic and nuclear degrees of freedom into the equation of motion for the nuclei.

We will show in the following chapter how all nonadiabatic terms can be recovered in a consistent and organic fashion by application of the time-dependent variational principle on an appropriately generalized phase space, eliminating the systematic need for electron translation factors. This is the achievement of the electron nuclear dynamics (END) theory.

Chapter 9

Evolution of Coherent Molecular States: Electron Nuclear Dynamics Theory

Both the goal and the methodology of the electron nuclear dynamics (END) theory, the focus of this chapter, are closely related to the TDHF-based models reviewed in the preceding chapter. Like the approaches surveyed in Sections 8.3 and 8.4, END is rooted in the Ehrenfest representation of quantum dynamics, combining a classical treatment of the nuclei with a fully quantum mechanical description of the electrons. Instantaneous electronic configurations define an averaged potential energy surface that dictates the forces felt by the nuclei. Conversely, energy flow from the nuclear to the electronic degrees of freedom may be modeled by a variety of devices, such as electron translation factors (see Eq. (8.159)), providing "nuclear boosts" to atomic orbitals, or extensions of the basic Ehrenfest formalism by inclusion of the quantum terms V_1 and V_2 (Eqs. (8.132) and (8.133)) as higher-order corrections to the averaged potential V_0 (see Eq. (8.130)).

Developed by Deumens and Öhrn [2], END theory reduces the asymmetry between the electrons and the nuclei as typical for TDHF-based methodology, and the associated *ad hoc* character of the coupling between the two sets of variables by letting both electronic and nuclear dynamics grow out of a common root, namely the theory of coherent states, as introduced in Section 4.4. Holding a delicate balance between the classical and the quantum world, END treats both subsystems in a unified theoretical frame. Once this frame is identified, well-defined approximations can be made to recover the classical trajectory picture that underlies the models discussed in Chapter 8.

Like the realizations of TDHF dynamics reviewed earlier, END transcends the limits of the Born–Oppenheimer approximation and does not contain any reference to potential energy surfaces. Although the END formalism has been extended to include multiconfiguration schemes for representation of the electrons [252] and nuclear wave functions [253], we will focus on the theory in its currently implemented form. This involves a one-determinantal description of the electronic system in conjunction with a classical trajectory approach for the nuclei. These traits define the *minimal END theory* [254]. They are shared by the variants of TDHF theory discussed in the preceding chapter and thus facilitate direct comparison between the two approaches. For the most significant steps made by END beyond the limits of TDHF molecular dynamics, we single out two features:

(1) END dynamics takes place in a generalized phase space spanned by four sets of nuclear and electronic parameters. This phase space is not constrained to have the flat — or canonical — metric adopted by TDHF theory. The curved metric of the generalized phase space is shown to contain coupling terms between electronic and nuclear degrees of freedom that are naturally contained in the END equations of motion, but have to be artificially imported into theories that adopt flat phase spaces.
(2) END theory makes use of the *Thouless representation* to describe the electronic system as a determinantal coherent state for N electrons. This choice turns out to provide the non-redundant electronic parameters that are ideally suited to capture the dynamic evolution of the electrons. The description of these processes in terms of molecular orbital coefficients, as outlined in the previous chapter, in contrast, is ambiguous since these coefficients are only determined up to a transformation of the occupied and the unoccupied orbitals among themselves. In this situation, uniqueness has to be ensured by imposing additional constraints.

In the following, we will develop the fundamental tenets of END theory, starting with the construction of the END electronic wave function. If not stated otherwise, the term *END theory* will refer to the minimal END theory [254]. The first two sections are meant to familiarize the reader with the specific viewpoint of the END methodology, invoking coherent states for both the nuclear and the electronic system, and leading to the END equations. In the remainder of this chapter we discuss special versions of END, that arise from simplifying conditions, among them the reduction to TDHF (Section 9.3), the *a posteriori* inclusion of nuclear quantum effects (Section 9.4), as well as the application of the theory to motion in colliding (Section 9.4) and bound (Section 9.5) systems.

9.1 The Thouless Representation

We start by defining the electronic state $|z\rangle$ as a single determinant composed of spin orbitals $\zeta_h(\mathbf{x}_n)$:

$$|z\rangle = \det[\zeta_h(\mathbf{x}_n)]. \tag{9.1}$$

$$\zeta_h = \phi_h + \sum_{p=M+1}^{N} \phi_p z_{ph}. \tag{9.2}$$

The structure of the dynamic orbitals ζ_k is determined by a subdivision of N atomic spin orbitals ϕ_k into an *occupied* range of dimension M and an $(N-M)$ dimensional *unoccupied or excitation* range. Consequently, we distinguish an occupied atomic orbital ϕ_h, where the index h has been chosen for *hole*, from an excited atomic orbital ϕ_p, the index p referring to *particle*. The so-called *Thouless parameters* z_{ph} thus acquire the meaning of excitation amplitudes. The functions ϕ_k are non-orthogonal. Adopting the representation in (9.1) and (9.2) for the electronic system, and using the Thouless parameters z_{ph} as dynamic variables, one removes the ambiguity associated with the use of time-dependent molecular orbitals to describe the electronic degrees of freedom and their temporal evolution. Every spin orbital that contributes to the electronic determinant is characterized by a well-defined atomic orbital of reference, ϕ_h, that remains occupied as the system advances in time. As the dynamic orbitals are uniquely determined, so are the Thouless parameters.

The notion of the dynamic orbital given by (9.1) and (9.2), involving a constant state of reference that is assumed to be occupied throughout the propagation of the system, leads to an obvious difficulty if a parameter z_{ph}

adopts a value that is large as compared to one during any stage of the dynamic process. If the time-dependent variables are allowed to become exceedingly large, the stability of the numerical procedure is endangered. In this case, however, a switch of reference states is possible. If the norm of a Thouless parameter reaches a certain preset threshold value, the respective orbital is selected as the new reference state and the remaining parameters are scaled accordingly.

We will now ask how to construct the Thouless parameters, starting from a conventional single electronic determinant composed of occupied molecular orbitals. In this context, it is of advantage to employ the formalism of particle-hole creation and annihilation operators. Thus, we write a single electronic determinant as

$$|\Psi_0\rangle = |\hat{b}_1^\dagger \cdots \hat{b}_N^\dagger\rangle = \prod_{h=1}^{M} \hat{b}_h^\dagger |\mathrm{vac}\rangle. \tag{9.3}$$

The creation operators $\hat{b}_i^\dagger$ and the corresponding annihilation operators $\hat{b}_j$ obey Fermi anticommutation relations, i.e.

$$[\hat{b}_i, \hat{b}_j^\dagger]_+ = \delta_{ij}. \tag{9.4}$$

The reference determinant provided by the wave function $|\Psi_0\rangle$ implies a natural distinction between occupied and unoccupied orbitals. To introduce electronic excitations, as they may occur in the course of a dynamic process, we define creation operators $\hat{b}_h^{\mathrm{O}}\dagger$, $(h = 1, \cdots, M)$ that operate in the occupied range, and their analogs $\hat{b}_p^{\mathrm{U}}\dagger$ $(p = M+1, \ldots, N)$, for the $(N - M)$ dimensional unoccupied space. The symbols $\hat{b}_h^{\mathrm{O}}$ and $\hat{b}_p^{\mathrm{U}}$ are assigned to the respective annihilation operators. A general unitary transformation $\hat{U}$ that maps a spin orbital basis $\{\zeta_i(\mathbf{x})\}$ into a basis $\{\zeta_i'(\mathbf{x})\}$ can likewise be applied to the creation operators, generating a new operator set $(\hat{\mathbf{c}}^{\mathrm{O}+}, \hat{\mathbf{c}}^{\mathrm{U}+})$ by

$$(\hat{\mathbf{c}}^{\mathrm{O}\dagger}, \hat{\mathbf{c}}^{\mathrm{U}\dagger}) = (\hat{\mathbf{b}}^{\mathrm{O}\dagger}, \hat{\mathbf{b}}^{\mathrm{U}\dagger}) \begin{pmatrix} \mathbf{U}^{\mathrm{OO}} & \mathbf{U}^{\mathrm{OU}} \\ \mathbf{U}^{\mathrm{UO}} & \mathbf{U}^{\mathrm{UU}} \end{pmatrix}. \tag{9.5}$$

The $M \times M$ submatrix $\mathbf{U}^{\mathrm{OO}}$ acts entirely within the occupied, the $(N - M) \times (N - M)$ submatrix within the unoccupied region, and both regions are mixed by the segments $\mathbf{U}^{\mathrm{UO}}$ and $\mathbf{U}^{\mathrm{OU}}$. Using this nomenclature, a general

M-dimensional electronic determinant is expressed as [255]

$$
\begin{aligned}
|\Psi\rangle &= \prod_{h=1}^{M} c_h^{\mathrm{O}\dagger}|\mathrm{vac}\rangle \\
&= \prod_{h=1}^{M}\left[\sum_{l=1}^{M}\hat{b}_l^{\mathrm{O}\dagger}U_{lh}^{\mathrm{OO}} + \sum_{p=M+1}^{N}\hat{b}_p^{\mathrm{U}\dagger}U_{ph}^{\mathrm{UO}}\right]|\mathrm{vac}\rangle \\
&= \prod_{h=1}^{M}\left[\sum_{l=1}^{M}\left(\hat{b}_l^{\mathrm{O}\dagger} + \sum_{p=M+1}^{N}\sum_{k=1}^{M}\hat{b}_p^{\mathrm{U}\dagger}U_{pk}^{\mathrm{UO}}(U_{kl}^{\mathrm{OO}})^{-1}\right)U_{lh}^{\mathrm{OO}}\right]|\mathrm{vac}\rangle \\
&= \alpha\prod_{h=1}^{M}\left[\hat{b}_h^{\mathrm{O}\dagger} + \sum_{p=M+1}^{N}\sum_{k=1}^{M}\hat{b}_p^{\mathrm{U}\dagger}U_{pk}^{\mathrm{UO}}(U_{kh}^{\mathrm{OO}})^{-1}\right]|\mathrm{vac}\rangle \\
&= \alpha\prod_{h=1}^{M}\left[1 + \sum_{p=M+1}^{N}\sum_{k=1}^{M}\hat{b}_p^{\mathrm{U}\dagger}U_{pk}^{\mathrm{UO}}(U_{kh}^{\mathrm{OO}})^{-1}\hat{b}_h^{\mathrm{O}}\right]\prod_{l=1}^{M}\hat{b}_l^{\mathrm{O}\dagger}|\mathrm{vac}\rangle \\
&= \alpha\prod_{h=1}^{M}\left[1 + \sum_{p=M+1}^{N}\sum_{k=1}^{M}\hat{b}_p^{\mathrm{U}\dagger}U_{pk}^{\mathrm{UO}}(U_{kh}^{\mathrm{OO}})^{-1}\hat{b}_h^{\mathrm{O}}\right]|\Psi_0\rangle,
\end{aligned}
\tag{9.6}
$$

introducing the reference state (9.3) in the last line. Proceeding from the third to the fourth line of the above derivation, we have exploited the invariance of a wave function determinant under a linear transformation among its occupied spin orbitals (U_{lh}^{OO}), except for a constant factor α to be determined by the normalization requirement for $|\Psi\rangle$. The reader confirms easily that the expression for $|\Psi\rangle$ derived in (9.6) contains all possible excitations that can arise from the M dimensional reference determinant $|\Psi_0\rangle$. Further, the structure of this formula implies that the excitation amplitudes for a transition leading from an occupied orbital with index h to an unoccupied orbital labeled p are given by

$$
z_{ph} = \sum_{k=1}^{M} U_{pk}^{\mathrm{UO}}(U_{kh}^{\mathrm{OO}})^{-1}. \tag{9.7}
$$

Exercise 9.1. Demonstrate the validity of the last step in the chain of equations (9.6) by use of the anticommutation relations (9.4).

This relation defines the Thouless parameters as complex-valued excitation amplitudes of a ground state $|\Psi_0\rangle$. Substituting Eq. (9.7) in the last line of Eq. (9.6) and choosing the symbol $|z\rangle$ for the non-normalized electronic state, we have:

$$|z\rangle = \prod_{h=1}^{M}\left[1 + \sum_{p=M+1}^{N} \hat{b}_p^{\mathrm{U}\dagger} z_{ph} \hat{b}_h^{O}\right]|\Psi_0\rangle. \tag{9.8}$$

This demonstrates that the determinant representing this state is indeed composed of the dynamic orbitals ζ_h specified in Eq. (9.2). We note that the operator product is nilpotent, i.e. the application of $(\hat{b}_p^{\mathrm{U}\dagger}\hat{b}_h^{O})^K$ on $|\Psi_0\rangle$ will yield zero for $K > 1$. Therefore, Eq. (9.8) may be written as

$$\begin{aligned} |z\rangle &= \prod_{h=1}^{M}\prod_{p=M+1}^{N} [1 + \hat{b}_p^{\mathrm{U}\dagger} z_{ph} \hat{b}_h^{O}]|\Psi_0\rangle \\ &= \prod_{h=1}^{M}\prod_{p=M+1}^{N} \exp(z_{ph}\hat{b}_p^{\mathrm{U}\dagger}\hat{b}_h^{O})|\Psi_0\rangle \\ &= \exp\left(\sum_{h=1}^{M}\sum_{p=M+1}^{N} z_{ph}\hat{b}_p^{\mathrm{U}\dagger}\hat{b}_h^{O}\right)|\Psi_0\rangle. \end{aligned} \tag{9.9}$$

Exercise 9.2. Trace the steps that lead from (9.8) to the final expression of (9.9). Use the nilpotent property of $(\hat{b}_p^{\mathrm{U}\dagger}\hat{b}_h^{O})$ as well as the Campbell–Baker–Hausdorff formula [256] in conjunction with the commutation relation $[\hat{b}_p^{\mathrm{U}\dagger}\hat{b}_h^{\mathrm{O}}, \hat{b}_{p'}^{\mathrm{U}\dagger}\hat{b}_{h'}^{\mathrm{O}}] = 0$ for $h, h' \leq M$ and $M + 1 \leq p, p' \leq N$ (why does this relation hold?).

One may understand the reference determinant $|\Psi_0\rangle$ as a lowest weight state for the irreducible representation $[1^M \quad 0^{(N-M)}]$ of the unitary group $U(N)$ of single-particle transformations. The generators of this group are the products $\hat{b}_j^\dagger \hat{b}_i$. Correspondingly, the general group element g of $U(N)$ is

$$g = \exp\left(\sum_{i,j} g_{ij}\hat{b}_j^\dagger \hat{b}_i\right). \tag{9.10}$$

The operator acting on the reference state to produce the non-normalized determinantal state $|z\rangle$, according to Eq. (9.9), can thus be interpreted as a representation of an element of $U(N)$ with $g_{ji} = z_{ji}$ if $M+1 \leq j \leq N$ and $1 \leq i \leq M$. This restriction imposed on the indices j, i selects those operator products $\hat{b}_j^\dagger \hat{b}_i$ that act as ladder operators on the state of lowest weight. We point out that $|z\rangle$ is *identical with the determinantal coherent state for M electrons*. One may verify this by following a general prescription given by Perelomov [257] to construct coherent states associated with a Lie group G and a unitary representation $\hat{T}$ of this group in Hilbert space.

The Thouless representation of a single determinantal electronic state is found from $|z\rangle$ as given by Eq. (9.9). As emphasized before, however, this state is not normalized. To evaluate the normalization factor of $|z\rangle$, we calculate the overlap integral of two electronic states $|z(\mathbf{X})\rangle$ and $|z'(\mathbf{X}')\rangle$, where the dependence of the state $z(z')$ on the nuclear coordinate $\mathbf{X}(\mathbf{X}')$ is made explicit and $\mathbf{X}$, $\mathbf{X}'$ stand for average nuclear positions.

$$\begin{aligned} S_{\text{el}}(z'^*, \mathbf{X}'; z, \mathbf{X}) &\equiv \langle z'(\mathbf{X}')|z(\mathbf{X})\rangle = \left\langle \text{vac}\left|\prod_{g=1}^{M} \hat{c}_g^{\text{O}}(z') \prod_{h=1}^{M} \hat{c}_h^{\text{O}\dagger}(z)\right|\text{vac}\right\rangle \\ &= \det\left[\left\langle \phi_g + \sum_{q=M+1}^{N} \phi_q z'_{qg} \middle| \phi_h + \sum_{p=M+1}^{N} \phi_{\text{p}} z_{ph}\right\rangle\right] \\ &= \det\left[\Delta_{gh} + \sum_{q=M+1}^{N} \Delta_{qh} z'^*_{qg} + \sum_{p=M+1}^{N} \Delta_{gp} z_{ph}\right. \\ &\qquad \left. + \sum_{q=M+1}^{N} \sum_{p=M+1}^{N} \Delta_{qp} z'^*_{qg} z_{ph}\right] \end{aligned} \tag{9.11}$$

with $1 \leq g, h \leq M$. In Eq. (9.11), we have introduced the matrix describing the overlap between the basis functions $\{\phi_l\}$:

$$\Delta_{ij} = \langle \phi_i(\mathbf{X}')|\phi_j(\mathbf{X})\rangle.$$

Different nuclear coordinates have to be admitted in the bra and the ket part of the scalar product, since the set $\{\phi_i\}$ may be selected as an atomic basis, where each basis function is attached to a definite nuclear center. It should be noted that in principle, the members of the set $\{\phi_i\}$, $1 \leq i \leq N$, can be chosen as an orthonormal basis fixed in space, assuming $\mathbf{X}' = \mathbf{X}$. In view of the ranges of the indices g, h, q, and p in Eq. (9.11), this choice simplifies the expression for the scalar product $\langle z'|z\rangle$ considerably, yielding

$$S_{\text{el}}(z'^*, z) = \langle z'|z\rangle = \det\left[\delta_{gh} + \sum_{p=M+1}^{N} z'^*_{pg} z_{ph}\right]. \tag{9.12}$$

The result (9.11) can be recast using the simpler notation

$$\begin{aligned} S_{\text{el}}(z'^*, \mathbf{X}'; z, \mathbf{X}) &= \langle z'(\mathbf{X}')|z(\mathbf{X})\rangle \\ &= \det[\Delta^{\text{OO}} + \mathbf{z}'^{\dagger}\Delta^{\text{UO}} + \Delta^{\text{OU}}\mathbf{z} + \mathbf{z}'^{\dagger}\Delta^{\text{UU}}\mathbf{z}] \end{aligned} \tag{9.13}$$

where the symbols O (for occupied) and U (for unoccupied) are used in the same sense as in the unitary transformation of (9.5), i.e. to specify the overlap between two occupied (OO) and two unoccupied (UU) basis functions, as well as the mixed cases OU and UO. The factor $\mathbf{z}$ is to be understood as an $((N - M) \times M)$-dimensional array of Thouless coefficients. Applying these conventions to the orthonormal alternative, we arrive at the compact formula

$$S_{\text{el}}(z'^*, z) = \langle z'|z\rangle = \det[\mathbf{I}^{\text{OO}} + \mathbf{z}'^{\dagger}\mathbf{z}] \tag{9.14}$$

with $\mathbf{I}^{\text{OO}}$ denoting the identity matrix in the occupied space with dimension M. The normalization factors $|\langle z'(\mathbf{X}')|z(\mathbf{X})\rangle|_{z'=z,\mathbf{X}'=\mathbf{X}}$ and $|\langle z'|z\rangle|_{z'=z}$ for the non-orthogonal and the orthonormal case, respectively, are found immediately from Eqs. (9.13) and (9.14). In the following, we will choose the $\{\phi_i\}$, with $1 \leq i \leq N$, as a non-orthogonal basis set, consistent with a basis of atomic orbitals centered at the nuclear positions.

9.2 The END Equations

END theory describes both the electrons and the nuclei as coherent states. The *nuclear coherent state*, as outlined in Section 4.4.3, assumes distinguishable nuclei, labeled by an index k. It is given by a product of traveling Gaussian functions, centered at average positions $\tilde{\mathbf{X}}_k$ and involving nuclear momenta $\mathbf{P}_k$:

$$|\chi\rangle = \langle \mathbf{X}|\tilde{\mathbf{X}}, \mathbf{P}(t)\rangle = \prod_k \exp\left[-\frac{1}{2}a(\mathbf{X}_k - \tilde{\mathbf{X}}_k)^2 + \frac{i}{\hbar}\mathbf{P}_k \cdot (\mathbf{X}_k - \tilde{\mathbf{X}}_k)\right]. \tag{9.15}$$

This approach may be regarded as a semiclassical treatment of the nuclei and is reminiscent of the *frozen Gaussian approximation* proposed by Heller [17].[1] Processes involving the splitting of the nuclear wave packet or nuclear tunneling are not accessible within a theory that utilizes the Gaussian product (9.15). However, this definition of $|\chi\rangle$ can be extended to a properly symmetrized, fully quantum mechanical nuclear wave function [258]. On the other hand, the transition to a completely classical description of the nuclei is made by taking the limits $\hbar \to 0$, $a \to \infty$, which reduces the Gaussians to delta functions in $(\mathbf{X}_k - \tilde{\mathbf{X}}_k)$ and thus defines the extreme of an infinitely narrow wave packet.

Having introduced the coherent electronic determinant as a function of the Thouless parameters z_{ph} as well as the coherent nuclear state that is determined by average positions $\tilde{\mathbf{X}}_k$ and momenta $\mathbf{P}_k$, one defines a generalized phase space spanned by two sets of mutually conjugate evolution variables, namely $\{z_{ph}, z^*_{ph}\}$ and $\{\tilde{\mathbf{X}}_k, \mathbf{P}_k\}$. Addition of the former pair of variables may be viewed as a direct extension of classical phase space, and END as an approximation of linear quantum dynamics, involving the evolution of a wave function in Hilbert spaces, by means of nonlinear phase space dynamics.

Application of the time-dependent variational principle (TDVP, see Section 8.1) will yield equations that entangle these basic dynamic quantities and thus couple electronic and nuclear degrees of freedom. The equations of motion that are at the heart of END theory will be derived and interpreted in the next step.

9.2.1 *Derivation of the END equations*

Once more, we employ the Lagrangian Eq. (8.2) associated with the TDSE

$$L(\Psi^*, \Psi) = \frac{\left\langle \Psi | i\hbar\frac{\partial}{\partial t} - \hat{H} | \Psi \right\rangle}{\langle \Psi | \Psi \rangle}, \tag{9.16}$$

[1] For further information and commentary on the frozen Gaussian approach, refer to Section 12.8.

involving the molecular wave function Ψ without imposing normalization. Subjecting the Lagrangian to the extremal condition

$$\int_{t_1}^{t_2} dt\, \delta L = 0, \tag{9.17}$$

we vary $L(\Psi^*, \Psi)$ with respect to both the bra and the ket vector, stipulating

$$\delta|\Psi\rangle = \delta\langle\Psi| = 0 \quad \text{for } t = t_1, t_2. \tag{9.18}$$

To accentuate the symmetry between the bra and the ket variation, we reformulate Eq. (9.16) by use of partial integration as

$$L(\Psi^*, \Psi) = \frac{\frac{i\hbar}{2}\left\langle\Psi\middle|\frac{\partial}{\partial t}\Psi\right\rangle - \frac{i\hbar}{2}\left\langle\frac{\partial}{\partial t}\Psi\middle|\Psi\right\rangle - \langle\Psi|\hat{H}|\Psi\rangle}{\langle\Psi|\Psi\rangle}, \tag{9.19}$$

or shorter:

$$L(\Psi^*, \Psi) = \frac{[\langle\Psi|\frac{i\hbar}{2}\left(\frac{\partial}{\partial t} - \frac{\overleftarrow{\partial}}{\partial t}\right) - \hat{H}|\Psi\rangle]}{\langle\Psi, \Psi\rangle}. \tag{9.20}$$

The symbol $\frac{\overleftarrow{\partial}}{\partial t}$ signifies action of the time derivative operator on the bra instead of the ket vector.

The molecular wave function $|\Psi\rangle$ is defined as the product of the electronic and nuclear contributions, that is, the coherent states (9.8) and (9.15):

$$|\Psi\rangle = |z(z_{ph}, \mathbf{X}_k, \mathbf{P}_k)\rangle|\chi(\mathbf{X}_k, \tilde{\mathbf{X}}_k, \mathbf{P}_k)\rangle \tag{9.21}$$

which in the classical limit of infinitely narrow nuclear wave packets reduces to $|z(z_{ph}, \tilde{\mathbf{X}}_k, \mathbf{P}_k)\rangle$. In the following, we will refer to this limit and thus, for ease of notation, replace $\tilde{\mathbf{X}}_k$ with $\mathbf{X}_k$ which now denotes the average position of the kth nucleus. The reader will note that the electronic determinant in Eq. (9.21) contains a dependence on the nuclear momenta. While the functional relationship between $|z\rangle$ and $\mathbf{X}_k$ is dictated by our choice of atomic basis functions anchored in certain nuclear positions, the influence of $\mathbf{P}_k$ on the electronic state can be included by selecting a traveling atomic basis $\{\phi_i(\mathbf{R}_k, \mathbf{P}_k)\}$ where a particular basis function attached to nucleus k moves with a velocity given by $\mathbf{P}_k/M$ if M is the relevant nuclear mass. Current implementations of the END theory adopt traveling atomic basis functions with Gaussian profiles. The general structure of the atomic basis function is

$$\phi(\mathbf{x}, \mathbf{X}, \mathbf{P}) = (x - X_x)^l (y - X_y)^m (z - X_z)^n \times \exp\left[-a(\mathbf{x} - \mathbf{X})^2 - \frac{i}{\hbar}\mathbf{P}(\mathbf{x} - \mathbf{X})\right], \tag{9.22}$$

with $l, m, n = 0, 1, \ldots$. Applying the time-dependent variational principle, we attempt to identify the conditions under which the quantum mechanical action

$$S = \int_{t_1}^{t_2} dt\, L(\Psi^*, \Psi) \tag{9.23}$$

is stationary in the generalized END phase space. For the sake of brevity, we introduce the operator

$$\hat{\Theta} = \frac{i\hbar}{2}\left(\frac{\partial}{\partial t} - \frac{\overleftarrow{\partial}}{\partial t}\right) - \hat{H} \tag{9.24}$$

and obtain for the variation of the action S

$$\delta S = \delta \int_{t_1}^{t_2} dt \frac{\langle z|\hat{\Theta}|z\rangle}{\langle z|z\rangle} = \int_{t_1}^{t_2} dt \left[\left\langle \delta z | i\hbar \frac{\partial}{\partial t} z\right\rangle - \langle \delta z|\hat{H} z\rangle - \frac{\langle z|\hat{\Theta}|z\rangle}{\langle z|z\rangle}\langle \delta z|z\rangle + c.c.\right] \langle z|z\rangle^{-1} \tag{9.25}$$

where again use has been made of partial integration. The stationarity condition $\delta S = 0$ yields the TDSE if completely general variations are considered. In this case, an arbitrary state $|\zeta\rangle$ is constrained to satisfy the equation

$$\left(i\hbar \frac{\partial}{\partial t} - \hat{H}\right)|\zeta\rangle = \frac{\langle \zeta|\hat{\Theta}|\zeta\rangle}{\langle \zeta|\zeta\rangle}|\zeta\rangle. \tag{9.26}$$

This is reconciled with the TDSE by imposing the constraint

$$\frac{\langle \zeta|\hat{\Theta}|\zeta\rangle}{\langle \zeta|\zeta\rangle} = 0 \tag{9.27}$$

or

$$\langle \zeta|\hat{\Theta}|\zeta\rangle = 0. \tag{9.28}$$

The validity of this equation can be guaranteed by applying a phase factor $\exp(i\gamma)$ such that

$$\langle \zeta | \exp(-i\gamma)\hat{\Theta} \exp(i\gamma) | \zeta \rangle = 0. \tag{9.29}$$

This factor is the END analog of the gauge function $\exp[-\frac{i}{\hbar}\int_{t_0}^{t} dt' \ \alpha(t')]$ introduced in Section 8.1 when discussing the general time-dependent variational principle. Upon inserting the explicit form of the operator $\hat{\Theta}$ and carrying out the time derivatives, we find

$$\left\langle \zeta \left| \exp(-i\gamma)\frac{1}{2}[i(i\dot{\gamma}) - i(-i\dot{\gamma})] \exp(i\gamma) \right| \zeta \right\rangle + \langle \zeta | \hat{\Theta} | \zeta \rangle$$

$$= -\langle \zeta | \zeta \rangle \dot{\gamma} + \langle \zeta | \hat{\Theta} | \zeta \rangle = 0, \tag{9.30}$$

with $\dot{\gamma} = \frac{\partial \gamma}{\partial t}$. The time derivative of the desired phase is therefore

$$\dot{\gamma} = \frac{\langle \zeta | \hat{\Theta} \zeta \rangle}{\langle \zeta | \zeta \rangle}. \tag{9.31}$$

It follows that the molecular wave function $\Psi(\mathbf{x}, \mathbf{X}, \mathbf{P}, t) = \exp(i\gamma)|\zeta\rangle$ obeys the TDSE:

$$\left[i\hbar \frac{\partial}{\partial t} - \hat{H} \right] \Psi(\mathbf{x}, \mathbf{X}, \mathbf{P}, t) = 0. \tag{9.32}$$

In a second step, we restrict the variation to the phase space as defined earlier, taking into account the reaction of $|\zeta\rangle$ to changes of the Thouless parameters $\{z_{ph}\}$ and their complex conjugates as well as nuclear positions and momenta. To facilitate the derivation of a differential equations system for these groups of dynamic variables, we cast the Lagrange function (9.20) into a form that makes its dependence on these groups of parameters explicit:

$$\begin{aligned} L = \sum_{i,j} \left[P_{ij} + \frac{i\hbar}{2} \left(\frac{\partial \ln S_{\text{el}}}{\partial X_{ij}} - \frac{\partial \ln S_{\text{el}}}{\partial X'_{ij}} \right) \right] \dot{X}_{ij} \\ + \frac{i\hbar}{2} \left(\frac{\partial \ln S_{\text{el}}}{\partial P_{ij}} - \frac{\partial \ln S_{\text{el}}}{\partial P'_{ij}} \right) \dot{P}_{ij} \\ + \frac{i\hbar}{2} \sum_{p,h} \left(\frac{\partial \ln S_{\text{el}}}{\partial z_{ph}} \dot{z}_{ph} - \frac{\partial \ln S_{\text{el}}}{\partial z^*_{ph}} \dot{z}^*{}_{ph} \right) - E \end{aligned} \tag{9.33}$$

with $S_{\text{el}} = \langle z(z'_{ph}, \mathbf{X}'_k, \mathbf{P}'_k) | z(z_{ph}, \mathbf{X}_k, \mathbf{P}_k) \rangle$ and $j = x, y, z$.

Exercise 9.3. Derive the Lagrangian (9.33) from Eqs. (9.19) and (9.21).

Varying the action (9.23) locally, i.e. with respect to the END parameters $\{\mathbf{z}, \mathbf{z}^*, \mathbf{X}_k, \mathbf{P}_k\}$, leads to Euler–Lagrange equations in these parameters:

$$\frac{d}{dt}\frac{\partial L}{\partial \dot{q}} = \frac{\partial L}{\partial q}, \qquad q = (z_{ph}, z_{ph}^*, X_k, P_k). \tag{9.34}$$

We focus in the following on the special case $q = \mathbf{X}_k$, the average position of the nucleus with label k. Inserting Eq. (9.33) into Eq. (9.34) yields after a straightforward calculation

$$\begin{aligned}\dot{\mathbf{P}}_k + 2\sum_l \left[\Im\left(\frac{\partial^2 \ln S_{\text{el}}}{\partial X_k' \partial X_l}\right)\dot{X}_l + \Im\left(\frac{\partial^2 \ln S_{\text{el}}}{\partial X_k' \partial P_l}\right)\dot{P}_l\right]&\\ + 2\sum_{p,h}\Im\left(\frac{\partial^2 \ln S_{\text{el}}}{\partial X_k' \partial z_{ph}}\right)\dot{z}_{ph} = -\frac{\partial E}{\partial X_k}.&\end{aligned} \tag{9.35}$$

By use of vectorial notation, and introducing the tensors

$$(C_{UV})_{ik;jl} = -2\hbar\, \Im\left(\frac{\partial^2 \ln S_{\text{el}}}{\partial U_{ik}' \partial V_{jl}}\right)\Big|_{\mathbf{X}'=\mathbf{X},\mathbf{P}'=\mathbf{P}}, \tag{9.36}$$

$$(C_U)_{ph} = \hbar \frac{\partial^2 \ln S_{\text{el}}}{\partial z_{ph}^* \partial U_{ik}}\Big|_{\mathbf{X}'=\mathbf{X},\mathbf{P}'=\mathbf{P}}, \tag{9.37}$$

and $U_{ik}, V_{ik} = X_{i,k}$ or $P_{i,k}$ with $i = 1, 2, 3$, Eq. (9.35) simplifies to

$$\dot{\mathbf{P}}_k - \sum_l [\mathbf{C}_{\mathbf{X}_k\mathbf{X}_l}\dot{\mathbf{X}}_l + \mathbf{C}_{\mathbf{X}_k\mathbf{P}_l}\dot{\mathbf{P}}_l] - 2\Im\, \mathrm{Tr}(\mathbf{C}_{\mathbf{X}_k}^T \dot{\mathbf{z}}) = -\frac{\partial E}{\partial \mathbf{X}_k}. \tag{9.38}$$

It is instructive to compare this identity with its TDHF counterparts as described in the previous chapter (Eq. (8.71) or (8.145)). In these relations, the nuclear dynamics is entirely governed by the total energy gradient, and therefore determined by the Newtonian equations of classical mechanics. From Eq. (9.38), in contrast, it is obvious that END includes a diversity of additional terms that connect the temporal rate of change of the nuclear momenta to the electronic system as represented by the scalar product S_{el} as well as the Thouless parameters z_{ph}. To achieve a similarly rich description of the coupling between electronic and nuclear degrees of freedom on the level of the eikonal representation, one would have to include the quantum potential terms V_1 and V_2 in Eqs. (8.131)–(8.135) and thus transcend the limits of the eikonal approximation. Before inquiring into the physical meaning of the matrices C_{UV} and C_U, we derive the Euler–Lagrange equations for the remaining three groups of dynamic variables and arrange the resulting four equations in a matrix scheme

$$\begin{pmatrix} i\mathbf{C} & 0 & i\mathbf{C}_\mathbf{X} & i\mathbf{C}_\mathbf{P} \\ 0 & -i\mathbf{C}^* & -i\mathbf{C}_\mathbf{X}^* & -i\mathbf{C}_\mathbf{P} \\ i\mathbf{C}_X^\dagger & -i\mathbf{C}_\mathbf{X}^\mathrm{T} & \mathbf{C}_\mathbf{XX} & -\mathbf{I}+\mathbf{C}_\mathbf{XP} \\ i\mathbf{C}_\mathbf{P}^\dagger & -i\mathbf{C}_\mathbf{P}^\mathrm{T} & \mathbf{I}+\mathbf{C}_\mathbf{PX} & \mathbf{C}_\mathbf{PP} \end{pmatrix} \begin{pmatrix} \dot{\mathbf{z}} \\ \dot{\mathbf{z}}^* \\ \dot{\mathbf{X}} \\ \dot{\mathbf{P}} \end{pmatrix} = \begin{pmatrix} \partial E/\partial \mathbf{z}^* \\ \partial E/\partial \mathbf{z} \\ \partial E/\partial \mathbf{X} \\ \partial E/\partial \mathbf{P} \end{pmatrix}. \tag{9.39}$$

We will refer to this set of coupled differential equations as the *END equations*. Besides the matrices already introduced in the preceding subsection, Eq. (9.39) contains the Hermitian matrix **C**, defined as

$$C_{pq} = \hbar \frac{\partial^2 \ln S_\mathrm{el}}{\partial z_p^* \partial z_q}. \tag{9.40}$$

What is the role of the **C** matrices in Eq. (9.39) that make the END equations of motion strongly differ from Hamiltonian dynamics?

9.2.2 *Interpretation of the END equations*

Focusing on the propagation matrix in Eq. (9.39), one may realize that the END equations contain certain characteristic traits of a classical Hamiltonian system. Setting all the **C** matrices in the END equations equal to zero, irrespective of their indices, one recovers the Hamilton equations of classical mechanics with the total energy E as the Hamilton function. Exploring the opposite extreme, one may reduce the phase space $\{\mathbf{z}, \mathbf{z}^*, \mathbf{X}, \mathbf{P}\}$ to the Thouless parameters. The END equation for $\mathbf{z}$, $\mathbf{z}^*$ alone read

$$\begin{pmatrix} i\mathbf{C} & 0 \\ 0 & -i\mathbf{C}^* \end{pmatrix} \begin{pmatrix} \dot{\mathbf{z}} \\ \dot{\mathbf{z}}^* \end{pmatrix} = \begin{pmatrix} \partial E/\partial \mathbf{z}^* \\ \partial E/\partial \mathbf{z} \end{pmatrix}. \tag{9.41}$$

Inverting this equation system, we introduce a generalized Poisson bracket $\{\}_\mathrm{g}$ for two arbitrary functions f and g that depend on $\mathbf{z}$ and $\mathbf{z}^*$:

$$\{f, g\}_\mathrm{g} = \begin{pmatrix} \dfrac{\partial f}{\partial \mathbf{z}^\mathrm{T}} & \dfrac{\partial f}{\partial \mathbf{z}^\dagger} \end{pmatrix} \begin{pmatrix} -i\mathbf{C}^{-1} & 0 \\ 0 & i\mathbf{C}^{*-1} \end{pmatrix} \begin{pmatrix} \partial g/\partial \mathbf{z}^* \\ \partial g/\partial \mathbf{z} \end{pmatrix}. \tag{9.42}$$

By comparison with Eq. (9.41), it is immediately seen that the choice of $f = z_{ph}$ (i.e. any one of the Thouless parameters) and $g = E$ leads to the Hamilton equations

$$\begin{aligned} \dot{\mathbf{z}} &= \{\mathbf{z}, E\}_{\mathrm{g}}, \\ \dot{\mathbf{z}}^* &= \{\mathbf{z}^*, E\}_{\mathrm{g}}. \end{aligned} \tag{9.43}$$

These identities are aided by the additional relations

$$\{\mathbf{z}, \mathbf{z}\}_{\mathrm{g}} = \mathbf{0}, \quad \{\mathbf{z}^*, \mathbf{z}^*\}_{\mathrm{g}} = \mathbf{0}, \quad \{\mathbf{z}, \mathbf{z}^*\}_{\mathrm{g}} = -i\mathbf{C}^{-1}. \tag{9.44}$$

Here $\{\mathbf{z}, \mathbf{z}\}_{\mathrm{g}}$ is a shorthand notation for the matrix defined by the elements $\{z_{ph}, z_{ph}\}_{\mathrm{g}}$, and correspondingly for the two remaining brackets. The relations (9.44) demonstrate that the parameter sets $\mathbf{z}$ and $\mathbf{z}^*$ are indeed mutually conjugate variables in the sense of classical mechanics.

To probe the meaning of the $\mathbf{C}$ tensor in greater depth, we express the generalized Poisson bracket (9.42) in real coordinates. Thus,

$$\mathbf{z} = \frac{\mathbf{q} + i\mathbf{p}}{\sqrt{2}}, \tag{9.45}$$

$$\mathbf{z}^* = \frac{\mathbf{q} - i\mathbf{p}}{\sqrt{2}}, \tag{9.46}$$

where $\mathbf{q}$ and $\mathbf{p}$ denote real variables. From the Hermitian nature of the matrix $\mathbf{C}$ we conclude that its inverse may be split into a symmetric real and an antisymmetric imaginary part according to

$$\mathbf{C}^{-1} = \Re\{\mathbf{C}^{-1}\} + i\Im\{\mathbf{C}^{-1}\} = \mathbf{B} + i\mathbf{A} \tag{9.47}$$

and recover the generalized Poisson bracket in the $(\mathbf{q}, \mathbf{p})$ space as

$$\{f, g\}_{\mathrm{g}} = \begin{pmatrix} \dfrac{\partial f}{\partial \mathbf{q}^{\mathrm{T}}} & \dfrac{\partial f}{\partial \mathbf{q}^{\mathrm{T}}} \end{pmatrix} \begin{pmatrix} \mathbf{A} & \mathbf{B} \\ -\mathbf{B} & \mathbf{A} \end{pmatrix} \begin{pmatrix} \dfrac{\partial g}{\partial \mathbf{q}} \\ \dfrac{\partial g}{\partial \mathbf{p}} \end{pmatrix}. \tag{9.48}$$

Exercise 9.4. (a) Employ the definition of the generalized Poisson bracket (9.42) to verify the relations (9.44). (b) Show that Eq. (9.42) turns into Eq. (9.48) upon the substitutions (9.45) and (9.46) in conjunction with (9.47).

For the special case of $\mathbf{A} = \mathbf{0}, \mathbf{B} = \mathbf{I}$, one obtains the symplectic form of Hamiltonian mechanics. The connection between the END equations and TDHF dynamics, as, for instance, elaborated by Field ([221], see Section 8.3.1), can be made by comparing Eq. (9.39) with Eqs. (8.68), (8.70), and (8.71). From this assessment, one realizes that the latter set of equations does not contain any counterpart of the matrices $\mathbf{C}_{\mathrm{U}}$ and $\mathbf{C}_{\mathrm{UV}}$ with $\mathbf{U}, \mathbf{V} = \mathbf{X}, \mathbf{P}$. The TDVP applied to the coherent basis in the generalized END phase space leads naturally to coupling terms between the electronic and the nuclear degrees of freedom which go beyond the frame set by TDHF dynamics. The latter adds classical nuclear motion to the quantum mechanical model of the electronic system. The END methodology, in contrast, emerges from a unified approach for the electrons and the nuclei, as both subsystems are represented as coherent states. The coupling between these subsystems that remains as the classical limit is taken thus echoes the originally uniform quantum description of both particle types.

For a direct physical understanding of the coupling terms that govern the END equations, we consider the matrix $\mathbf{C}_X$, as given by Eq. (9.37). Thus,

$$\begin{aligned}\frac{\partial^2 \ln S_{\mathrm{el}}}{\partial \mathbf{z}^{*\prime} \partial \mathbf{X}_k} &= \frac{\partial}{\partial \mathbf{z}^{*\prime}} \frac{\partial}{\partial \mathbf{X}_k} \ln\langle \mathbf{z}', \mathbf{X}', \mathbf{P} | \mathbf{z}, \mathbf{X}, \mathbf{P}\rangle|_{\mathbf{X}=\mathbf{X}'} \\ &= \frac{\partial}{\partial \mathbf{z}^{*\prime}} \left\langle \mathbf{z}', \mathbf{X}', \mathbf{P} \left| \frac{\partial}{\partial \mathbf{X}_k} \right| \mathbf{z}, \mathbf{X}, \mathbf{P} \right\rangle / \langle \mathbf{z}', \mathbf{X}', \mathbf{P} | \mathbf{z}, \mathbf{X}, \mathbf{P}\rangle|_{\mathbf{X}=\mathbf{X}'}.\end{aligned} \tag{9.49}$$

Since the electronic basis is of the form (9.22), i.e. the electrons are described as attached to nuclear centers, the dependence of $\langle \mathbf{z}, \mathbf{X}, \mathbf{P}|$ on the electronic coordinate $\mathbf{x}$ is given by the difference $\mathbf{x} - \mathbf{X}_k$ for a particular center k. It therefore holds that

$$\begin{aligned}i\hbar \sum_{k=1}^{N_{\mathrm{nuc}}} \frac{\partial}{\partial \mathbf{X}_k} \ln\langle \mathbf{z}, \mathbf{X}', \mathbf{P} | \mathbf{z}, \mathbf{X}, \mathbf{P}\rangle|_{\mathbf{X}=\mathbf{X}'} &= \sum_{m=1}^{M} -i\hbar \frac{\langle \mathbf{z}, \mathbf{X}', \mathbf{P}| \frac{\partial}{\partial \mathbf{x}_m} |\mathbf{z}, \mathbf{X}, \mathbf{P}\rangle}{\langle \mathbf{z}, \mathbf{X}', \mathbf{P} | \mathbf{z}, \mathbf{X}, \mathbf{P}\rangle|_{\mathbf{X}=\mathbf{X}'}} \\ &= \langle \mathbf{p} \rangle.\end{aligned} \tag{9.50}$$

The last sum of Eq. (9.50) is easily recognized as the expectation value of the electronic momentum. Thus, the third of the END equations (9.39) relates the overall rate of change of the electronic momentum in the space

$\{z_{ph}, z^*_{ph}, \mathbf{X}_k, \mathbf{P}_k\}$, $i\mathbf{C}^\dagger_{\mathrm{X}}\dot{\mathbf{z}} - i\mathbf{C}^{\mathrm{T}}_{\mathrm{X}}\dot{\mathbf{z}}^* + \mathbf{C}_{\mathrm{XX}}\dot{\mathbf{X}} + \mathbf{C}_{\mathrm{XP}}\dot{\mathbf{P}}$, directly to the nuclear force, $-\frac{\partial E}{\partial \mathbf{X}}$. Through this ingredient, the nuclear force equation of END includes the impact of electronic motion on the change of nuclear momenta, a facet of dynamic theory that is absent from TDHF. The more detailed description of the electronic back action on the nuclear degrees of freedom may be understood as the most essential step of END beyond the level of TDHF dynamics.

Exercise 9.5. Use the definition of the **C** matrices given in Eqs. (9.36) and (9.37) to show that the four END equations (9.39) express Newton's law in the END phase space, relating generalized forces to the rates of change of generalized nuclear and electronic momenta.

In Section 9.2*, the matrices that determine the END equations (9.39) are interpreted in terms of their metric, or more generally, geometric significance. A direct relation of the END **C** matrices to the metric tensor of the END phase space is demonstrated. This connection makes it possible to understand the generalized trajectory $(\mathbf{z}(t), \mathbf{z}^*(t), \mathbf{X}(t), \mathbf{P}(t))$ that solves the END equations as solution of a line minimization problem. Specifically, the trajectory is shown to describe an extremal curve in the non-Euclidean space spanned by the END parameters.

9.3 Two Special Cases: The Boosted Self-Consistent Field and the Linearized END Equations

The two cases highlighted in this section are obtained by imposing certain limiting conditions on the END equations (9.39). While these cases are not of major systematic relevance to the END formalism, they are crucial for the preparation of initial states that involve moving nuclei, i.e. for simulating atomic and molecular scattering situations. Further, both cases shed light on characteristic traits of END theory, and are thus instrumental for a deeper understanding of END.

9.3.1 *The boosted electronic system*

We will address now the problem of the initial conditions for the electrons, associated with the mixed quantum-classical description of the dynamic system adopted by END. While the point-like nuclei of the minimal END theory trace out classical trajectories, the electrons are treated on the level of unrestricted self-consistent field (SCF) theory. In the initial state of the considered process, an SCF solution has to be obtained for the electronic system. This solution, however, refers to a static nuclear framework and is thus in conflict with the dynamic condition of non-zero initial nuclear momenta. The traditional reaction to this difficulty, as outlined in the preceding chapter, is the use of *ad hoc* electron translation factors (ETFs). For example, an atomic wave function ϕ may be adjusted to the condition of initial motion with a certain velocity v by multiplication with an exponential factor such that (compare with Eq. (8.159))

$$\phi'(\mathbf{x}) = \exp\left[-\frac{i}{\hbar}\left(M\mathbf{v}\cdot\mathbf{x} - \left(\frac{Mv^2}{2} - \varepsilon\right)t\right)\right]\phi(\mathbf{x}-\mathbf{X}) \qquad (9.51)$$

where the ETF $\exp\left[\frac{i}{\hbar}M\mathbf{v}\cdot\mathbf{x} - (\frac{Mv^2}{2} - \varepsilon)t\right]$ involves the SCF energy ε and the mass M of the atom. The application of ETFs provides a solution to the dynamic problem but leads to highly complicated equations of motion (see, for instance, [259]). END theory, in contrast, does not require the use of ETFs. The description of the electrons in a basis of traveling orbitals is optional, not mandatory. Working with such a basis is thus not essential for the adequacy of the END formalism, although, as a numerical assessment [2] has shown, it may improve the convergence properties of the END method.

We will demonstrate in the following that the electron nuclear coupling terms contained in the END equations (9.39) solve the problem of the initial conditions for the electronic parameters, providing a viable alternative to the customary ETF treatment. With reference to the geometric meaning of the END equations, one might also state that END theory incorporates the correct initial conditions for the electronic system by including the metric of the generalized phase space.

In more detail, if a static instead of a traveling atomic basis is employed, the momentum $\mathbf{P}$ in Eq. (9.22) vanishes, and so do the metric matrices $\mathbf{C_P}$, $\mathbf{C_{XP}}$, and $\mathbf{C_{PP}}$ in Eq. (9.39). While this move involves the description of the electronic system it does not imply that the nuclear momenta are reduced to zero. The first END equation then becomes

$$i\mathbf{C}\dot{\mathbf{z}} + i\mathbf{C_X}\dot{\mathbf{X}} = \frac{\partial E}{\partial \mathbf{z}^*}. \qquad (9.52)$$

Initially, the Thouless coefficients satisfy $\dot{z}_{ph} = 0$, and the time derivatives of the averaged nuclear coordinates are constrained by $\dot{\mathbf{X}}_k = \mathbf{V}_k$, with the velocities $\mathbf{V}_k$ given by the initial nuclear momenta. Thus

$$i \sum_k \mathbf{C}_{\mathbf{X}_k} \mathbf{V}_k = \frac{\partial E}{\partial \mathbf{z}^*}. \tag{9.53}$$

Expressing both the left- and the right-hand side of this identity through Thouless coefficients [2], one rewrites Eq. (9.53) as

$$i \begin{pmatrix} -\mathbf{z} & \mathbf{I}^{UU} \end{pmatrix} \sum_k \mathbf{V}_k \cdot \frac{\partial}{\partial \mathbf{X}_k}^{|\rangle} \Delta \begin{pmatrix} \mathbf{I}^{OO} \\ \mathbf{z} \end{pmatrix} = \begin{pmatrix} -\mathbf{z} & \mathbf{I}^{UU} \end{pmatrix} \mathbf{F} \begin{pmatrix} \mathbf{I}^{OO} \\ \mathbf{z} \end{pmatrix}. \tag{9.54}$$

To make the connection between Eq. (9.53) and Eq. (9.54), it is important to realize that both relations are matrix identities. Although a quasi-vectorial notation has been chosen to designate the force term on the right-hand side of Eq. (9.53), this term stands for a matrix with elements $\partial E/\partial z^*_{ph}$. Likewise, the corresponding element on the left-hand side involves differentiation with respect to z^*_{ph}. The symbol $\frac{\partial}{\partial \mathbf{X}_k}^{|\rangle} \Delta$ denotes the derivative of the ket part of the Δ elements, where the matrix Δ has the same meaning as in Eq. (9.11), the overlap between the atomic basis functions. The right-hand side of Eq. (9.54) contains the Fock matrix $\mathbf{F}$. This equation has been derived assuming an orthonormal electronic basis, but could be as well, with greater mathematical effort, formulated for the case of non-orthogonal basis functions [2]. The identity (9.54) can be condensed into

$$\begin{pmatrix} -\mathbf{z} & \mathbf{I}^{UU} \end{pmatrix} \mathbf{F}' \begin{pmatrix} \mathbf{I}^{OO} \\ \mathbf{z} \end{pmatrix} = \mathbf{0} \tag{9.55}$$

with

$$\mathbf{F}' = \mathbf{F} - \sum_k \mathbf{V}_k \cdot \frac{\partial}{\partial \mathbf{X}_k}^{|\rangle} \Delta. \tag{9.56}$$

The latter two relations comprise the *boosted self-consistent field equations*. Determining the initial values for the Thouless parameters z_{ph} by use of these equations, one prepares an initial solution for the electronic system that follows the moving nuclei. This goal is achieved within END theory without any use of ETFs which, nevertheless, can be utilized if desired. We add that the derivation of the boosted SCF equations from the general END formalism (9.39) is analogous to that of the static SCF from the TDHF equations.

9.3.2 *The linear version of the electronic END equations*

END theory goes beyond the TDHF formalism in its explicit coupling of electronic and nuclear degrees of freedom. The END equations (9.39) thus separate into the TDHF equations and those of Hamiltonian mechanics upon setting the matrices $\mathbf{C}_{\mathrm{U}}$ and $\mathbf{C}_{\mathrm{UV}}(\mathrm{U}, \mathrm{V} = \mathbf{X}, \mathbf{P})$ equal to zero (see Section 9.2.2). The Thouless coefficient block is then obtained from Eq. (9.42), specifically by

$$i\mathbf{C}\dot{\mathbf{z}} = \frac{\partial E}{\partial \mathbf{z}^*} \tag{9.57}$$

and its complex conjugate. As outlined in [2], the explicit form (9.40) of the metric matrix may be substituted for $\mathbf{C}$, and the dependence of the energy derivative on the Fock matrix may be made explicit, to result in the following relation:

$$i\hbar\,\dot{\mathbf{z}} = \begin{pmatrix} -\mathbf{z} & \mathbf{I}^{\mathrm{UU}} \end{pmatrix} \mathbf{F} \begin{pmatrix} \mathbf{I}^{\mathrm{OO}} \\ \mathbf{z} \end{pmatrix}. \tag{9.58}$$

Outwardly, this equation is not too reminiscent of TDHF theory. To adapt the appearance of this equation to the TDHF scheme as, for instance, implemented by the linearized form (8.43) and (8.44), we map the array $\mathbf{z}$ into a vector $\boldsymbol{\xi}$ and linearize Eq. (9.58) in this vector. The result is [255]

$$i\hbar\,\dot{\boldsymbol{\xi}} = \mathbf{f} + \mathbf{A}(\boldsymbol{\xi} - \boldsymbol{\xi}_0) + \mathbf{B}(\boldsymbol{\xi}^* - \boldsymbol{\xi}_0^*) \tag{9.59}$$

with

$$\mathbf{f} = \frac{\partial E}{\partial \boldsymbol{\xi}^*}, \quad \mathbf{A} = \frac{\partial^2 E}{\partial \boldsymbol{\xi}^* \partial \boldsymbol{\xi}}, \quad \mathbf{B} = \frac{\partial^2 E}{\partial \boldsymbol{\xi}^* \partial \boldsymbol{\xi}^*}, \tag{9.60}$$

where all three derivatives are to be evaluated at $\boldsymbol{\xi} = \boldsymbol{\xi}_0$, $\boldsymbol{\xi}^* = \boldsymbol{\xi}_0^*$. Expression (9.59), along with its complex conjugate, turns out to be equivalent to the linear *free-oscillation* TDHF equations (8.43) and (8.44) when normal mode amplitudes are introduced as new coordinates, substituting for the vectors $\boldsymbol{\xi}, \boldsymbol{\xi}^*$. While the applications of Eq. (9.59) to molecular interactions are limited, considering this extreme case aids the physical interpretation of the Thouless parameters. They are understood as linearized TDHF, or *random phase approximation* amplitudes ([260], compare with the related discussion in Section 8.2).

9.4 Inclusion of Nuclear Quantum Effects

Since END theory in its present realization treats the nuclei classically, it is best suited to describe situations in which the concept of a well-defined trajectory is a good approximation to physical reality. This is the case for problems involving atomic or molecular scattering processes at sufficiently high kinetic energies. In fact, for collision systems, END has been shown to produce results that are accurate by experimental standards ([261]–[263]). The reliance on average nuclear trajectories, however, limits the applicability of the theory to experimental situations in three chief respects:

(1) END does not incorporate any interaction between trajectories, related to nuclear interference effects, as they are built into the very foundations of, for instance, the *initial value representation* (IVR) model, to be surveyed in Chapter 12.
(2) The rovibrational state of the nuclei cannot be obtained directly from END analysis.
(3) END does not address any cases of nuclear tunneling or splitting of the nuclear wave packet, which may become essential for the proper description of reactions where several channels compete.

While the restrictions indicated under (3) are systematic and removable only in the framework of the *complete END* theory that operates with nuclear wave functions, the quantum effects (1) and (2) can be included *a posteriori* within the framework of certain approximations by adequate processing of the END results. In this way, the coarseness of the average trajectory construct can be partially overcome. In the following, we will review methods to implement the phenomena mentioned under (1) and (2), involving procedures of accounting for trajectory interference and extracting details on the rovibrational nuclear state from END simulation. In both cases, we will illustrate the methodology with examples of actual END computations, and assess their quality by comparison with measurements. For the sake of simplicity, we will assume in the following development an atomic projectile. However, extensions to molecular projectiles are readily available.

9.4.1 *Trajectory interference*

A classical definition of the total cross section for a given process may be written with the help of the *opacity function* $P(E, b)$ which indicates the probability of finding the final asymptotic END wave function in some state characterized by a specific set of quantum numbers. Evidently, $P(E, b)$ can be found from the projection of the final END state on the state of interest. The cross section is then calculated from

$$\sigma(E) = 2\pi \int_0^{b_{\max}} db\, P(E, b)b \tag{9.61}$$

where b stands for the impact parameter of the considered scattering process. The expression for the respective classical differential scattering cross section is

$$d\sigma(E, \theta) = \frac{P(E, b)b}{\sin\theta |\frac{d\theta}{db}|} \tag{9.62}$$

with θ as the scattering angle. The relation between b and θ is not necessarily unique. In general, several impact parameter values b_i can be assigned to the same θ. If so, Eq. (9.62) is replaced by the sum [254]

$$d\sigma(E, \theta) = \sum_i \frac{P(E, b_i)b_i}{\sin\theta |\frac{d\theta}{db}|_{b=b_i}}. \tag{9.63}$$

This formula admits singularities, as there are physically well-defined scattering situations that correspond to a vanishing denominator in Eq. (9.63). These include forward peak scattering or *glory scattering*, associated with a finite impact parameter where the deflection of the projectile is zero. The differential cross section also diverges at extremal points of the map $b \to \theta$, occurring at *rainbow angles* defined by $\frac{d\theta}{db} = 0$. As is well known, the deficiency of the classical expression (9.63) can be remedied by recourse to a semiclassical treatment.

To apply an appropriate correction to Eq. (9.63) and, by the same token, to improve the purely classical description of the nuclei in END theory, we consider a scattering configuration of an atomic projectile interacting with a stationary target in the laboratory frame. The projectile is deflected into the solid angle element $d\Omega$, defined by a scattering (polar) angle θ and an azimuthal angle φ. The relevant Hamiltonian may then be partitioned in the following way [264]:

$$\hat{H} = -\frac{\hbar^2}{2M_1}\Delta + \hat{H}_1 + \hat{H}_2 + V, \tag{9.64}$$

where the index 1 refers to the projectile and 2 to the target. More specifically, $\hat{H}_1$ is the Hamiltonian for the electronic structure of the projectile, while $\hat{H}_2$ includes both the electronic and the nuclear structure of the target. The potential V denotes the interaction between both. With E as the total energy of the system, the stationary Schrödinger equation reads

$$\left(-\frac{\hbar^2}{2M_1}\Delta + \hat{H}_1 + \hat{H}_2 + \hat{V}\right)|\Psi\rangle = E|\Psi\rangle. \tag{9.65}$$

In the following step, we propose a trial function that separates the electronic from the nuclear component of the solution. Instead of a multichannel approach, as outlined in Chapter 6, we make use of a general electronic END wave function $|z'\rangle$ in conjunction with a nuclear contribution $|\chi\rangle$ and adopt a simple factorization to represent the overall solution. Explicitly,

$$|\Psi\rangle = |z'\rangle|\chi\rangle. \tag{9.66}$$

The complex multichannel wave function of the electronic subsystem, as discussed in Chapter 7, is here replaced by a single, *averaged*, time-dependent function. We may now substitute the separation (9.66) into the stationary Schrödinger equation (9.65) and subsequently project (9.65) on some particular state $|\Psi'_{\mathrm{s}}\rangle$ to obtain

$$(\Delta + k_{\mathrm{s}}^2)|\chi\rangle = U_{\mathrm{s}}|\chi\rangle \tag{9.67}$$

where we have introduced the symbol U_{s} for an *effective screened interaction potential* defined as

$$U_{\mathrm{s}} = \frac{2M_1}{\hbar^2}\frac{\langle z'_{\mathrm{s}}|V|z'\rangle}{\langle z'_{\mathrm{s}}|z'\rangle} \tag{9.68}$$

as well as the projectile wave number during the collision, namely

$$k_{\mathrm{s}}^2 = \frac{2M_1}{\hbar^2}(E - E_1 - E_2) \tag{9.69}$$

with

$$E_1 = \frac{\langle z_s'|\hat{H}_1|z'\rangle}{\langle z_s'|z'\rangle}$$

and

$$E_2 = \frac{\langle z_s'|\hat{H}_2|z'\rangle}{\langle z_s'|z'\rangle}.$$

The nuclear wave function $|\chi\rangle$ can formally be written in terms of asymptotic outgoing wave functions, using the scattering amplitude notation as in Eq. (3.115). Thus,

$$|\chi\rangle_{R\to\infty} \propto \exp(i\mathbf{k}_0 \cdot \mathbf{R})|\chi\rangle\delta_{0s} - \frac{\exp(i\mathbf{k}_s \cdot \mathbf{R})}{R} f(\mathbf{k}_s, \mathbf{k}_0). \tag{9.70}$$

The vector $\mathbf{k}_0$ points into the direction of incidence, while the orientation of the vector $\mathbf{k}_s$ is determined by the pair of angles θ and φ. The prefactor of the scattering amplitude is proportional to the outgoing Green's function G for the operator $\Delta + k_s^2$ since

$$G(\rho) = -\frac{\exp(ik_s\rho)}{4\pi\rho}. \tag{9.71}$$

With the help of this function, a recursive solution for the scattering amplitude can be formulated, namely

$$\begin{aligned} f(\mathbf{k}_s, \mathbf{k}_0) = -\frac{1}{4\pi}\sum_{n=1}^{\infty}\int\cdots\int d\mathbf{R}_1\cdots d\mathbf{R}_n\ \exp(-i\mathbf{k}_s\cdot\mathbf{R}_n)U_s(\mathbf{R}_n)\\ \times G(\mathbf{R}_n - \mathbf{R}_{n-1})U_s(\mathbf{R}_{n-1})G(\mathbf{R}_{n-1} - \mathbf{R}_{n-2})\cdots U_s(\mathbf{R}_2)\\ \times G(\mathbf{R}_2 - \mathbf{R}_1)U_s(\mathbf{R}_1)\exp(i\mathbf{k}_0\cdot\mathbf{R}_1). \end{aligned} \tag{9.72}$$

By truncating this sum after the first term, one obtains the first Born approximation. An approach of wider validity, however, consists in retaining all terms of Eq. (9.72) while approximating each one of them by use of the stationary phase method. We follow here a procedure described by Schiff [265] who demonstrated that, for small angles θ, Eq. (9.72) reduces to the following compact expression:

$$f(k_s, k_0) = ik\int_0^{\infty} db\, b\mathcal{J}_0(qb)\left\{1 - \exp\left[-\frac{i}{2k}\int_{-\infty}^{+\infty} U_s(b, Z)dZ\right]\right\} \tag{9.73}$$

where Z refers to the polar axis which defines the direction of incidence, $\mathcal{J}_0$ is the Bessel function of order zero, and $q = |\mathbf{k}_s - \mathbf{k}_0|$ measures the momentum transfer during the collision.[2] As θ is the angle between $\mathbf{k}_s$ and $\mathbf{k}_0$, the scattering amplitude may be written as $f(\theta)$. Further, we simplify Eq. (9.73) by introducing the impact parameter dependent phase

$$\zeta(b) = \frac{1}{4k}\int_{-\infty}^{+\infty} U_s(b,x)dx \tag{9.74}$$

to obtain

$$f(\theta) = ik\int_0^{\infty} \mathcal{J}_0(qb)\{1 - \exp[-2i\zeta(b)]\}bdb. \tag{9.75}$$

As shown by Mason *et al.* [266], $\zeta(b)$ is the negative of the phase shift $\delta(b)$ which, due to its reference to particle trajectories, is to be interpreted as a *semiclassical phase shift*, corresponding to the *semiclassical scattering amplitude* given by Eq. (9.75). We point out a simple relation between $\delta(b)$ and the deflection function Θ, as implicitly defined by the relation

$$\theta = |\Theta(b)|. \tag{9.76}$$

It holds that

$$\Theta(b) = \frac{2}{k}\frac{d\delta(b)}{db}. \tag{9.77}$$

In view of relations (9.75)–(9.77), the primary task in determining the scattering amplitude from END simulation consists in evaluating the deflection function $\Theta(b)$ which makes it necessary to process a sufficiently large number of trajectories at different impact parameters. In addition, for comparison with experimental data, an average over different orientations of both the projectile and the target has to be formed. From the scattering amplitude (9.75), the semiclassical differential cross section is given by

$$\frac{d\sigma}{d\Omega} = \frac{k_s}{k_0}|f(\theta)|^2. \tag{9.78}$$

This equation, in conjunction with the semiclassical scattering amplitude (9.75), lays the ground for the treatment of collision processes involving

[2] In the small angle approximation utilized here, the Z component of the momentum transfer vector $\mathbf{q}$ is related to the scattering angle and the initial projectile wave number k by $q_Z \approx \frac{1}{2}k\theta^2$ if the direction of incidence coincides with that of the Z-axis.

atoms and molecules by END theory. For more detailed analysis of the scattering situation, which may involve a variety of rearrangement reactions between the constituents, electronic excitation or charge transfer, the special process of interest can be assessed by multiplying the right-hand side of Eq. (9.78) with the probability for its occurrence during the interaction. For this purpose, one projects the evolved asymptotic END wave function, $|\Psi_f\rangle$, on the state under consideration, as given by a function $|\Psi_{\text{ref}}\rangle$, evaluated in the same basis as the END solution. The square magnitude $|\langle\Psi_f|\Psi_{\text{ref}}\rangle|^2$ then yields the transition probability for the studied channel.

9.4.2 *Case study: H impact on molecular and atomic targets by END theory*

As an elementary application of the formalism outlined earlier, we discuss an END study on the interaction of atomic H projectiles with an H_2 target [267], assumed to be initially in its ${}^1\Sigma_g^+$ ground state, at various projectile energies, ranging from 10 eV/amu to 25 keV/amu. The scattering geometry is schematically represented in Figure 9.1.

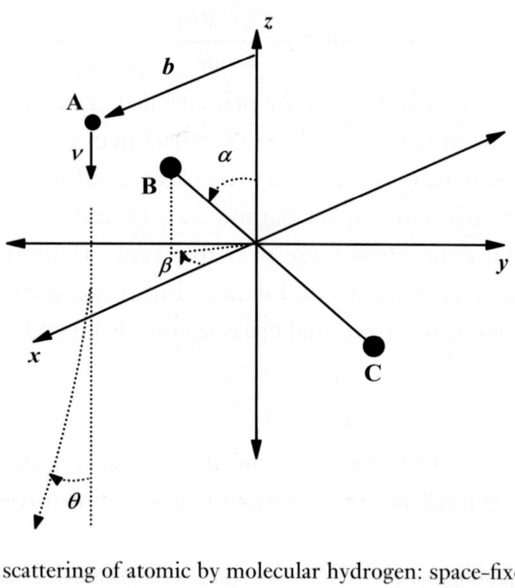

Figure 9.1: The scattering of atomic by molecular hydrogen: space-fixed molecular coordinate frame. The angles α and β specify the orientation of the target molecule. (Reprinted with permission from [267]. Copyright (2002) by the American Physical Society.)

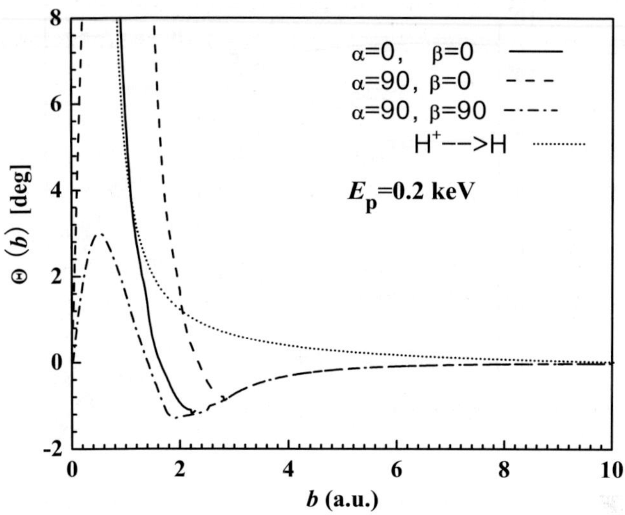

Figure 9.2: Deflection functions $\Theta(b)$ versus the impact parameter at an initial projectile energy $E_p = 0.2\,\text{keV}$ for three orientations of an H_2 target: ($\alpha = 0^o$, $\beta = 0^o$), ($\alpha = 90^o$, $\beta = 0^o$) and ($\alpha = 90^o, \beta = 90^o$) interacting with H^+ projectiles. The corresponding function for $H^+ \rightarrow H$ is included for comparison. (Reprinted with permission from [267]. Copyright (2002) by the American Physical Society.)

For angular averaging, three basic target orientations were considered, corresponding to ($\alpha = 0^o$, $\beta = 0^o$), ($\alpha = 90^o$, $\beta = 0^o$), and ($\alpha = 90^o$, $\beta = 90^o$). The respective deflection functions, as obtained from END trajectories, are presented in Figure 9.2.

For all three $H^+ \rightarrow H_2$ collision geometries distinguished in Figure 9.2, pronounced glory and rainbow scattering effects were observed. These reflect the changing nature of the interaction between the projectile ion and the target molecule as a function of the distance between both. For small impact parameters, the H^+ ions penetrate deeply enough into the electronic cloud of H_2 to experience a repulsive force. For sufficiently large values of b, however, this effective repulsion turns into attraction, where the trajectory of vanishing deflection provides a demarcation line between these two regimes. As the impact parameter increases further, the case of maximum attraction between the interaction partners is encountered, and thus rainbow angle scattering. The deflection function for $H^+ \rightarrow H$, in contrast, is less complex, resulting from consistently repulsive interaction between the two centers that gradually diminishes with distance.

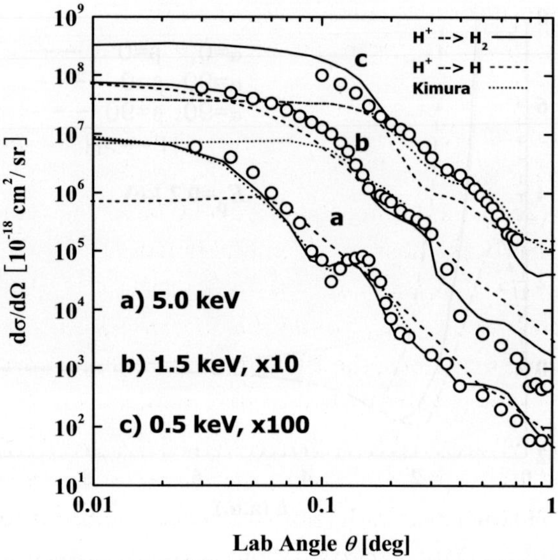

Figure 9.3: Direct differential cross sections for the reaction $H^+ \rightarrow H_2$ versus the scattering angle. The END results are represented by the solid line, the measured values [268] by open circles. Also included are theoretical findings based on the close-coupling approach employed by Kimura *et al.* [269], indicated by the dotted lines. The long dashed lines refer to the scattering of H^+ by an atomic H target as found from END theory. (Reprinted with permission from [267]. Copyright (2002) by the American Physical Society.)

The differential cross section, as computed according to Eq. (9.78), versus the scattering angle θ, evaluated in the laboratory frame, is shown in Figure 9.3 for three initial projectile energies. The calculated values are averages over the three considered target orientations.

Comparison is made with close-coupling calculations by Kimura *et al.* [269] as well as measurement. The agreement between END results and experiment is seen to be particularly good for small scattering angles, implying adequate treatment of the large impact parameter region. In the close collision regime, characterized by large θ, the accuracy of the END cross sections deteriorates. This trend has been ascribed to two unrelated methodological limitations:

(1) The averaging over a relatively coarse grid of target orientations. This shortcoming of the END application presented here obviously affects the description of close collisions more seriously than that of distant ones.

(2) The pronounced occurrence of inelastic scattering events at small impact parameters, involving electronic excitation and exchange processes.

Both of these characteristic challenges to practical implementation of END theory can, in principle, be alleviated: problem (1) by refining the grid, problem (2) by extending the electronic basis set, which in the case presented here has been chosen as a combination of five *s* and two *p* functions on each atomic center with additional diffuse *s* and *p* orbitals. The treatment of ionization effects, however, requires the inclusion of electronic continuum functions. In some situations, the effect of these functions may be simulated by adding diffuse orbitals with high angular momentum.

Comparing the close and the far region of projectile-target interaction, one further notices that the molecular scattering results approach the atomic case of $H^+ \rightarrow H$ scattering in the limit of close encounters between the two partners, as is intuitively plausible. Also, the similarity of atomic and molecular interaction increases as the initial projectile energy does. The oscillatory features in the zone of small θ are attributed to trajectory interference effects, which are reproduced by the Schiff approximation. It is noteworthy that, by the standard of experiment, the performance of END theory is comparable to that of the close-coupling approach, but in the regime of small scattering angles superior to the latter.

Among the electronic phenomena associated with atomic or molecular scattering, electron exchange effects are of particular interest as they have been extensively and systematically explored in a wide variety of experiments. Charge transfer processes between projectile and target determine the effective charge of the projectile and thus are, in general, of sensitive influence on the projectile trajectory. By the same token, the understanding of electron exchange is essential for in-depth knowledge of the behavior of projectiles traversing media, be those gases, liquids, or solids. The basic parameters characteristic for the passage of particles through matter, such as the range and the energy loss of ions or the straggling of ion tracks, are critically dependent on the average projectile charge. Conversely, charge exchange probabilities are of high relevance to any assessment of the influence exerted by moving particles on their host media.

Within the END formalism, the projectile charge as a function of a particular impact parameter b and kinetic energy $E_{\rm p}$ is available through population analysis. More specifically, the Fock operator of the evolving system can be constructed from the dynamic orbitals (9.2) at every instant of the system's evolution. Subsequently, the one-particle density can be subjected to conventional Mulliken population analysis ([270]–[272], see Chapter 1). The overall electronic population of the projectile, $\langle n_{\text{p}} \rangle$, averaged over the target orientations, is used to obtain an approximate value of the projectile charge q_{p}. For protons, this approximation yields simply

$$q_{\text{p}} = 1 - \langle n_{\text{p}} \rangle. \tag{9.79}$$

Further, the probability for the capture of a target electron by a proton projectile $P_{if}(b, E_{\text{p}}) = P_{10}(b, E_{\text{p}})$ where the indices i and f stand for the initial and the final projectile charge, respectively, is

$$P_{10}(b, E_{\text{p}}) = \langle n_{\text{p}} \rangle. \tag{9.80}$$

An analogous definition applies to the probability for the loss of an electron from the projectile shell to that of the target. Weighting the probability of interest by the differential cross section (9.62) and integrating over the solid angle Ω, one arrives at a compact expression for the total cross section of the considered process:

$$\sigma_{if}(E_{\text{p}}) = \int db\, d\phi\, bP_{if}(b, E_{\text{p}}). \tag{9.81}$$

Applications of the END theory to atom–atom and atom–molecule charge exchange scattering have resulted in qualitative and, in some cases, quantitative agreement with experimental data [267, 273].

From measured electron capture and loss cross sections, one may infer experimental charge fractions present in beams traversing media. According to the charge state approach, for hydrogen beams these fractions are obtained from the relations ([274]– [276]).

$$f(H^{+}) = \frac{\sigma_{01}}{\sigma_{01} + \sigma_{10}}, \tag{9.82}$$

$$f(H^{-}) = \frac{\sigma_{10}}{\sigma_{01} + \sigma_{10}}. \tag{9.83}$$

From END computation of the relevant charge transfer cross sections according to Eq. (9.81) one obtains the calculated equilibrium charge state fractions [273, 277]. These are compared with available experimental data

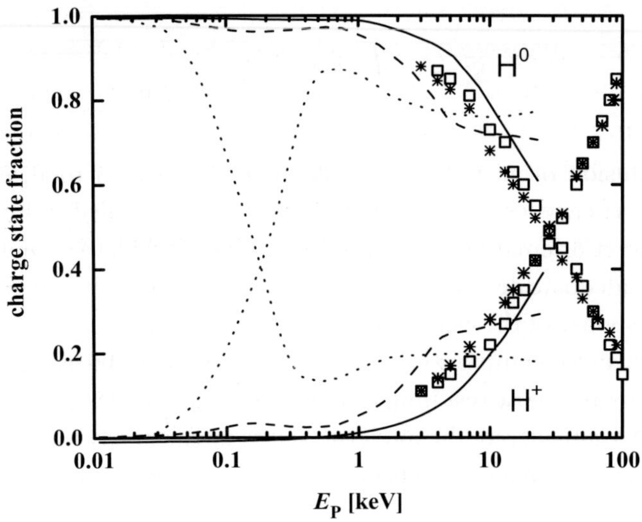

Figure 9.4: Experimental data for equilibrium charge state fractions found in a hydrogen beam interacting with nitrogen (open squares) and oxygen (stars) versus the projectile energy [275], in comparison with results from END modeling. The solid line refers to END simulation for the case of atomic nitrogen, the dashed and the dotted line to atomic oxygen and fluorine, respectively. (Reprinted with permission from [273]. Copyright (2000) by the American Physical Society.)

for hydrogen beam interaction with nitrogen and oxygen targets, as shown in Figure 9.4. Although the simulation is based on an atom–atom scattering model while the experiment involves the target species in molecular form, the overall agreement between calculated and measured data is good. For the nitrogen target case, it is excellent throughout, while in the case of oxygen, salient trends of the experimental values are reproduced for $E_{\rm p} \leq 5\,{\rm keV}$.

We point out that the equilibrium charge state fraction is related to another quantity characteristic of the dynamic interaction between the projectiles and their host media. This is the stopping cross section S, or energy loss per unit length, which has been experimentally recorded for a large multiplicity of projectile-host combinations over a wide range of $E_{\rm p}$ values (for references, see, for instance, [278]). Assuming again a hydrogen beam, we write the stopping cross section averaged over the projectile charge state as

$$S_x = f(H^+)S_x^+ + f(H^0)S_x^0 \tag{9.84}$$

with

$$S_x^i = -\int db\, d\phi\, b\langle \Delta E_x(b)\rangle$$

and $i = +, 0$.

By inclusion of an index x, we distinguish between total, electronic, and nuclear energy loss ΔE. As before, the average implied in Eq. (9.84) is over the target orientations. Any understanding of the electronic stopping requires an adequate description of electronic excitations and, in general, the dynamic response of the electronic subsystem, as included in the END formalism. Extensive comparison has been made between measured stopping cross sections and those resulting from END calculation [262,263,277,279]. In most cases considered, the model accounts satisfactorily for the orders of magnitude and the major trends of the observed effects, in many instances also for quantitative details.

9.4.3 *Rovibrational analysis of the nuclear system*

As END theory operates with classical nuclei, no direct information about the evolution of nuclear quantum states is available from END treatment. An approximate *a posteriori* resolution of END trajectories into vibrational or rotational states that exploits the properties of coherent states [261], however, has been shown to be successful. More specifically, and focusing on the vibration problem only, one may associate each vibrational degree of freedom of the considered system with an evolving canonical coherent state (see Section 4.4.1) of the form

$$|\alpha\rangle = \exp\left(-\frac{|\alpha|^2}{2}\right)\sum_v \frac{\alpha^v}{\sqrt{v!}}|v\rangle. \tag{9.85}$$

The quantity α is here a continuous complex time-dependent parameter while $|v\rangle$ refers to an eigenfunction of the harmonic oscillator with frequency ω. Therefore, the function $|\alpha\rangle$ consists of a sum over harmonic oscillator states, each of which is occupied by a population

$$P(v) = \exp\left(-|\alpha|^2\right)\frac{|\alpha|^{2v}}{v!}. \tag{9.86}$$

In expression (9.86), one recognizes a Poisson distribution. Evaluating the energy expectation value of the harmonic oscillator Hamiltonian $\hat{H}_{\mathrm{HO}}$ with respect to the wave function (9.85), we find

$$E_{\mathrm{CS}} \equiv \langle\alpha|\hat{H}_{\mathrm{HO}}|\alpha\rangle = \hbar\omega\left(|\alpha|^2 + \frac{1}{2}\right). \tag{9.87}$$

Exercise 9.6. Use prescription (9.85) for constructing the vibrational coherent state $|\alpha\rangle$, and prove the statement (9.87).

The square magnitude of α thus assumes the interpretation of an averaged vibrational quantum number. For classical nuclei, Eq. (9.87) has to be modified to yield

$$E_{\mathrm{CS}} = \hbar\omega|\alpha|^2 \tag{9.88}$$

since they do not possess any zero-point energy. Combining Eqs. (9.85), (9.86), and (9.88), we conclude that

$$P(v) = \exp\left(-\frac{E_{\mathrm{vib}}}{\hbar\omega}\right)\frac{(E_{\mathrm{vib}}/(\hbar\omega))^v}{v!}, \tag{9.89}$$

with $E_{\mathrm{vib}} = E_{\mathrm{CS}}$. This formula describes the occupation probability of an oscillator state $|v\rangle$ as a function of the respective frequency and the energy of the investigated vibrational mode E_{vib}. The assumption of a harmonic oscillator profile, underlying Eq. (9.89), is in many cases an oversimplification, but often adequate in situations where predominantly the lowest vibrational states are populated. For problems involving strong coupling among vibrational degrees of freedom or vibrational and rotational degrees of freedom, it will not be possible to define E_{vib} in a meaningful way. In an application of coherent state analysis of END results related to scattering of H^+ projectiles by H_2 molecules [261] at a collision energy of 30 eV in the laboratory frame, the latter complication does not exist. In this case, one has to deal with only one vibrational coordinate, associated with the internal motion of H_2 (or H_2^+), and the contribution of rotational motion has been found small. The respective computations were carried out in response to a vibrationally resolved experimental observation of $H^+ \rightarrow H_2$ scattering [280] and two theoretical attempts to account for the measured effects, using the *trajectory surface hopping* (TSH, see Chapter 11) [280] model and the *infinite-order sudden approximation* (IOSA) [281].

Figure 9.5 demonstrates that END theory supplemented by *a posteriori* coherent state analysis is capable of accounting for the major trends of the vibrationally resolved differential cross section for $H^+ \rightarrow H_2$ as obtained by experiment. It should be noted that finer trends of the measured data are also reproduced, apparently to a higher extent than by the competing methods, namely the TSH and the IOSA procedure. The figures, however, do not allow for an absolute quantitative assessment of the END results since the measured overall differential cross section has been adjusted to fit the END computation at the rainbow angle of 5°, as shown in the upper diagram of Figure 9.5. Attempts to apply vibrational analysis to the charge transfer channel of the considered $H^+ \rightarrow H_2$ scattering process have failed, as the direct channel dominates and thus largely determines the nuclear trajectories. For the direct, or non-transfer pathway, good agreement with experiment has been found for the six lowest frequencies of H_2.

As recently demonstrated [282], a realistic rovibrational analysis of END results is feasible by application of an advanced numerical scheme based on a fitting method originally proposed by French engineer and mathematician Gaspard-Marie Riche de Prony (1755–1839) [283]. The Prony procedure was devised to match each point within a set of data to a sum over exponentials. The strategy underlying this technique consists in identifying the time-dependent factors in this sum with the roots of a polynomial that can be derived from the data set. Blass *et al.* [282] generalized this model for the situation of a multiplicity of vibrational degrees of freedom coupled by a rotational mode. The mass-weighted coordinates $\mathbf{R}_k(t_n)$, as constructed from END simulation results, where the index k labels the nuclei, are decomposed in the following fashion:

$$\mathbf{R}_k(t_n) = \mathbf{O}(t_n - 1)\left[\mathbf{E}_k(t_n) + \sum_{j=1}^{p} \mathbf{T}_{k,j} c_j \exp\left(2\pi i \Omega_j (t_n - 1)\Delta t + \phi_j\right)\right]. \tag{9.90}$$

In this formula, p stands for the number of vibrational modes of the investigated molecule, $\mathbf{O}(t_n)$ for a rotation matrix, with $\mathbf{O}(0) = \mathbf{I}$, and $\mathbf{E}_k(n)$ for the equilibrium position of the nucleus k. The displacement vector of the nucleus k in the jth normal mode is denoted by $\mathbf{T}_{k,j}$, with c_j as the weight factor for this mode. The symbols Ω_j and ϕ_j refer to the respective frequencies and phases.

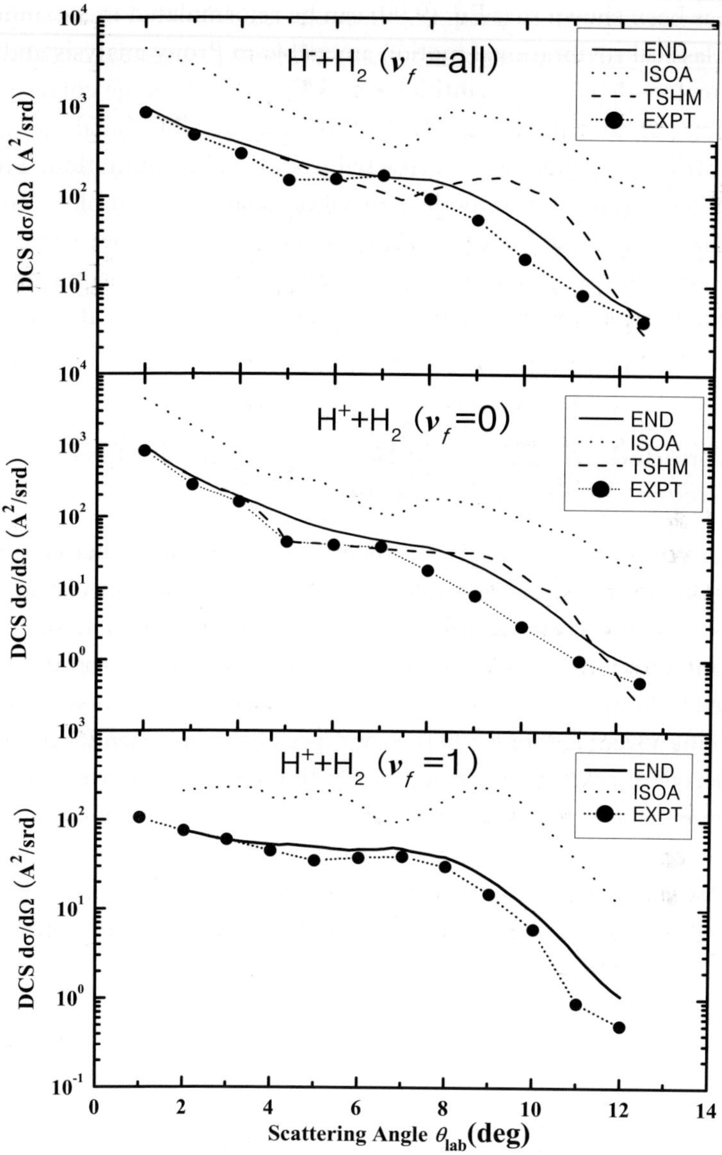

Figure 9.5: Vibrationally resolved differential cross sections $\frac{d\sigma}{d\Omega}$ for direct scattering of protons by H_2 versus the scattering angle θ_{lab}. The results of vibrational analysis of END trajectories in terms of coherent states are displayed for $v_f = 0, 1$. Shown in the upper diagram is the overall differential cross section from END calculation. At the rainbow angle of 5°, the experimental data have been matched to the END results. Comparisons with alternative theoretical procedures are included. (Reprinted with permission from [261]. Copyright (1995) by the American Institute of Physics.)

It has been shown that Eq. (9.90) can be reformulated in a manner that makes classical rovibrational motion amenable to Prony analysis and yields the normal mode related quantities c_j and $\mathbf{T}_{k,j}$ as well as the corresponding frequencies Ω_j and the phases ϕ_j. In this way, the vibrational energies for each contributing mode can be extracted from END computation. The subsequent resolution of any mode into vibrational states employs coherent state analysis based on Eq. (9.85). Successful applications of these concepts include the interpretation of time-of-flight energy loss spectra for vibrational excitations of water molecules colliding with protons at 46 eV in the center-of-mass system [282].

9.5 Nonadiabatic Effects in Bound Systems by END Theory: The Pseudorotation of H_3^+

The trajectory approach of END theory makes it appear better suited to address scattering situations than dynamical problems involving bound systems. For the latter, a fully quantum mechanical model, such as the multiconfiguration time-dependent Hartree (MCTDH) procedure, to be surveyed in Chapter 15, seems to be the more appropriate choice. However, the specific advantages of END theory, especially its independence from any potential energy surface construction as well as its consistent treatment of the interaction between electronic and nuclear degrees of freedom, also come to bear when applied to bound systems. In particular, qualitative insight can be expected from END analysis of bound cases that involve sizeable coupling between electronic and nuclear degrees of freedom and, in response to this, a substantial amount of electronic excitation.

In the following, we summarize some findings that have emerged from END analysis of a specific motional phenomenon, namely the pseudorotation of the light triatomic cations H_3^+ and Li_3^+. As these molecules are closed-shell systems, there is less systematic interest in them than in their open-shell counterparts H_3 and Li_3 which exhibit a Jahn–Teller effect related to orbital ground state degeneracy in D_{3h} geometry, giving rise to the accumulation of an electronic geometric phase upon pseudorotation of the molecule, as described in Section 2.5.3. Regardless of its relevance with respect to the geometric phase phenomenon the pseudorotation of triatomic molecules has proven to be a worthwhile topic of dynamic studies, since it involves an extraordinary range of interactions of nuclear degrees of freedom, both among themselves and with electronic degrees of freedom.

As outlined in greater detail earlier (see Section 2.5.3), pseudorotation consists of the coordinated, phase-correlated periodic motion of the atoms within a molecule or cluster. Figure 2.5 shows the model used to set up the initial condition for the simulation of pseudorotating H_3^+: each one of the three nuclei moves in a circle centered on one corner of an equilateral triangle, where initially the three circles have equal radii. The individual rotations are shifted with respect to each other by an angle of $2\pi/3$ in cyclic order. Ideally, the pseudorotation of a triatomic molecule proceeds entirely in a space defined by two orthogonal vibration modes, namely the coordinates Q_θ, conserving C_{2v} symmetry, and Q_ε, breaking this symmetry, as displayed in Figure 9.6 along with the other translational, vibrational, and rotational modes that collectively determine the molecule's dynamic evolution.

As a function of the initial nuclear kinetic energy ΔE_i imparted to the system, one can distinguish a hierarchy of dynamic regimes, where the spatial angular momentum $L \equiv \langle \hat{L} \rangle$ serves as a guideline. Prior to including the electronic contribution to L, we focus on the nuclear component of this quantity which is composed of three elements:

$$L_{\text{nuc}} = L_{\text{pseu}} + L_{\text{rot}} + L_{\text{disp}} \tag{9.91}$$

where

$$L_{\text{pseu}} = Q_\varepsilon P_\theta - Q_\theta P_\varepsilon \tag{9.92}$$

$$L_{\text{rot}} = A_2 P_{A_1} - A_1 P_{A_2} \tag{9.93}$$

$$L_{\text{disp}} = X P_Y - Y P_X. \tag{9.94}$$

Here, L_{pseu} is the purely pseudorotational admixture to the overall nuclear angular momentum, while L_{rot} refers to the internal spatial rotation of the molecule. The last term in Eq. (9.91) denotes the angular momentum related to the translational coordinates X and Y, i.e. to the displacement of the rigid molecule. The direction of L_{nuc} is perpendicular to the plane of the three nuclei. For initial kinetic energies $\Delta E_i \ll 2.0\,\text{eV}$, L_{nuc} is of chiefly pseudorotational character [285]. This is documented in Figure 9.7 which displays L_{nuc} versus time along with the angular momentum contributions listed in (9.92)–(9.94). Obviously, the process is dominated by L_{pseu} .

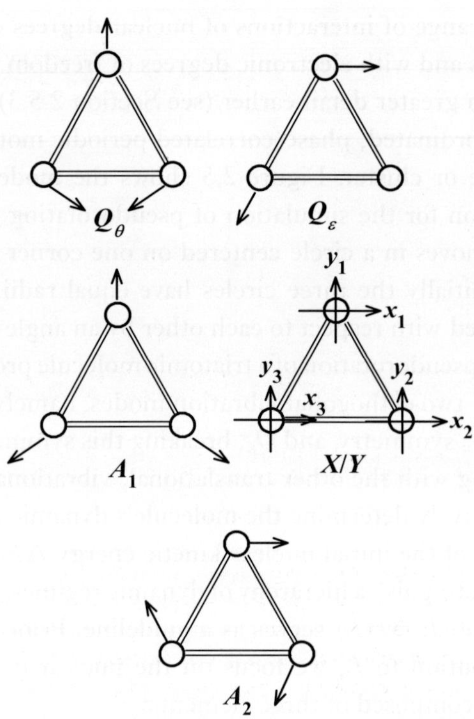

Figure 9.6: The six normal modes employed in the description of the pseudorotation of a triatomic molecule: the vibrational coordinates Q_θ and Q_ε, the totally symmetric coordinate A_1, the rotational and translational coordinates A_2 and X/Y, respectively. (Reprinted with permission from [284]. Copyright (2002) by the American Physical Society.)

This, however, does not preclude a contribution from $L_{\rm rot}$. In other words, the molecule acquires rotational motion in the course of its pseudorotation. The origin of this effect can be found in both the anisotropy and the anharmonicity of the H_3^+ ground state potential energy surface $V(Q_\theta, Q_\varepsilon)$ [285], with the remaining normal modes displayed in Figure 9.6 kept constant. These deviations from sphericality induce a pseudorotational torque that changes $L_{\rm pseu}$. Since, in the considered motional regime, the total nuclear angular momentum $L_{\rm nuc}$ is conserved to a good approximation, the system responds with an equal and opposite change of $L_{\rm rot}$, while $L_{\rm disp}$ remains zero as no center-of-mass motion of the molecule is involved.

A different dynamic regime is entered as one increases the initial kinetic energy of H_3^+ beyond a value of 2 eV, as demonstrated in Figure 9.8. The pseudorotation of the molecule is now superposed by a very pronounced

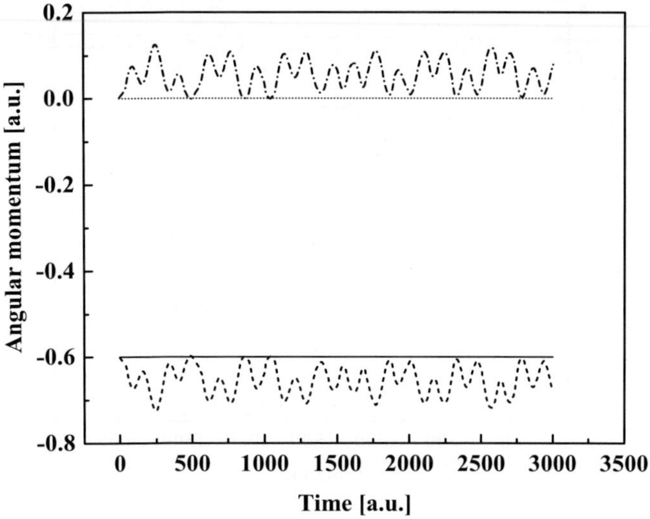

Figure 9.7: Time dependence of the nuclear angular momentum of pseudorotating H_3^+ at an initial kinetic energy $\Delta E_i = 0.21$ eV. Solid line: total nuclear angular momentum $L_{\rm nuc}$; dashed line: pseudorotational angular momentum, $L_{\rm pseu}$; dashed-dotted line: rotational angular momentum, $L_{\rm rot}$; dotted line: nuclear angular momentum related to the displacement of the molecule, $L_{\rm disp}$. (Reprinted with permission from [284]. Copyright (2002) by the American Physical Society.)

spatial rotation, and the interplay between the components $L_{\rm pseu}$ and $L_{\rm rot}$ is much more complex than that shown in Figure 9.7 for lower initial kinetic energy. Although $L_{\rm nuc}$ is subjected to small changes with the elapse of time, it still can be considered as a constant of motion, taking the large variations undergone by $L_{\rm pseu}$ and $L_{\rm rot}$ individually as a standard. This situation changes, however, when ΔE_i is enhanced further, as demonstrated in Figure 9.9 which shows $L_{\rm nuc}$ as a function of time for pseudorotating H_3^+ at $\Delta E_i = 4.47$ eV.

As the system evolves, $L_{\rm nuc}$ deviates dramatically from its initial value, signaling the emergence of strong coupling between the nuclear and electronic degrees of freedom. This implies that the total angular momentum, defined as

$$L_{\rm tot} = L_{\rm nuc} + \langle L_{el,z} \rangle, \tag{9.95}$$

is the constant of motion in this new dynamic regime, as vindicated by Figure 9.9, where $L_{\rm nuc}$ is seen to fluctuate rapidly around an unchanging $L_{\rm tot}$.

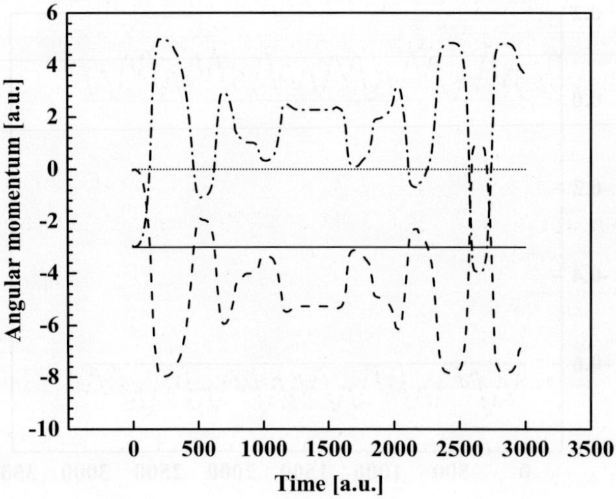

Figure 9.8: Time dependence of the nuclear angular momentum of pseudorotating H_3^+ at $\Delta E_i = 2.34$ eV. Dashed line: rotational angular momentum, $L_{\rm rot}$; dashed-dotted line: pseudorotational angular momentum, $L_{\rm pseu}$. The remaining assignments are as in Figure 9.7. (Reprinted with permission from [284]. Copyright (2002) by the American Physical Society.)

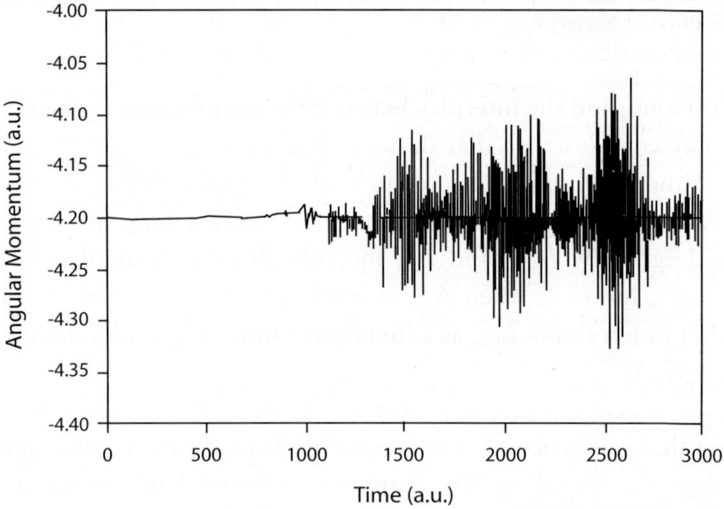

Figure 9.9: Comparison between the total nuclear angular momentum $L_{\rm nuc}$ and the total angular momentum $L_{\rm tot}$ versus time at $\Delta E_i = 4.47$ eV. The former is subjected to large fluctuations while the latter is a constant of motion. (Reprinted with permission from [284]. Copyright (2002) by the American Physical Society.)

The expectation value $\langle L_{\mathrm{el},z}\rangle$ of the z component (defined as perpendicular to the plane of the nuclei) of the electronic orbital angular momentum $\langle L_{\mathrm{el}}\rangle$ is computed according to

$$\langle L_{\mathrm{el},z}\rangle = \mathrm{Tr}(\mathbf{L}_z\boldsymbol{\rho}) \tag{9.96}$$

with $\boldsymbol{\rho}$ as the electronic density matrix. Generalizing the discussion of the linear momentum given earlier (see Eq. (9.50)), we cast this expression into the following form:

$$\langle L_{\mathrm{el},i}\rangle = -i\hbar\ Tr\left(\left[\sum_{k,j,m} \varepsilon_{ijm}(X_{k,j} + (x_j - X_{k,j}))\frac{\partial}{\partial X_{k,m}}{}^{|\rangle}\Delta\right]\rho\right). \tag{9.97}$$

In this formula, ε_{ijm} is the Levi-Civita tensor, and the symbol $\frac{\partial}{\partial X_{k,m}}{}^{|\rangle}\Delta$ has the same meaning as in Eq. (9.54).

As the foregoing discussion shows, the various dynamic stages of pseudorotation can be characterized in terms of the total nuclear angular momentum, L_{nuc}. This quantity is mostly pseudorotational in nature at low ΔE_i where kinetic effects are sufficiently small. At intermediate energy, it combines both pseudorotation and spatial rotation. If the energy is increased further, it ceases to be a constant of motion, which is attributed to the onset of sizeable dynamic coupling between the nuclear and the electronic system.

To make END theory more generally applicable to processes in the chemical regime, the two most restraining conditions of its minimal implementation must be overcome. The reliance of the minimal model on classical nuclei is to be replaced by a quantum treatment of the nuclear degrees of freedom. Further, the representation of the electronic system must be extended beyond the self-consistent field level, such that correlation effects are included. These are the essential elements of the *complete electron nuclear dynamics* (CEND) program [286]. A numerical realization of this generalized END scheme that employs complete active space multiconfigurational (CASMC) electronic wave functions in conjunction with a hierarchy of nuclear orbitals has been devised. Other versions of END theory have been proposed, such as the approach of Morales who combined the END formalism with a density functional theory description of the electrons ([287], see Chapter 16).

While present efforts at generalizing END theory aim at turning it into a fully quantum mechanical model, and thus at emancipating it from the reliance of the minimal END theory on the classical trajectory concept, some limitations of END are rooted in its most basic fundament, the Ehrenfest representation of quantum theory. In the following chapter, we will present a critical examination of the Ehrenfest approach to quantum molecular dynamics, to be followed by a discussion of methodological alternatives in Chapters 10–15.

Chapter 10

The Classical Electron Analog

Two types of Ehrenfest dynamics have been discussed so far, namely the TDHF approach reviewed in Sections 8.3 and 8.4, and END theory, as summarized in the preceding chapter. While the TDHF-based formalism conjoins classical nuclear and quantum mechanical electronic degrees of freedom, END theory offers, in its various stages of implementation, different representations of the nuclear system which lead up to the limit of nuclear coherent states that evolve simultaneously with the electronic system. The full quantum mechanical treatment of both the electrons and the nuclei is, at this juncture, prohibitively expensive in terms of computational effort except for the very smallest systems. Thus, quantum-classical theories are indispensable tools of quantum molecular dynamics, and they are challenged to establish a viable compromise between feasibility and rigor. While often successful in the physical and not rarely in the chemical regime, combinations of classical and quantum dynamics treat the electronic and the nuclear degrees of freedom inconsistently and therefore do not account correctly for the coupling between electrons and nuclei.

The first section of this chapter will examine the premises and implications of the Ehrenfest approach that was specified in Section 8.3.3 as an *effective path method* in a more precise terminology, and clarify its major shortcomings. The essential idea pursued in this chapter was introduced as a proposal to remedy these deficiencies.

At the root of the problems encountered by the Ehrenfest model is a disparity in the treatment of electronic and nuclear degrees of freedom. This gap can be closed by devising a uniformly quantum mechanical model, or a *uniformly classical model*. The latter alternative is developed by the *classical electron analog* (CEA) model [21] which is outlined in Section 10.2. In the remaining section of this chapter, we comment on the capacity of this model to account for electron dynamics in the vicinity of basic Jahn–Teller systems (Section 10.3).

10.1 Critique of the Ehrenfest Representation

Utilizing the Hellmann–Feynman theorem in conjunction with the adiabatic representation, we arrived at formula (8.107) for the force acting on the nuclei if an average potential energy surface is assumed. The most severe limitation of this approximation consists in its reliance on one averaged or "best" nuclear trajectory, advancing in an effective potential energy surface that may be generated from the physically relevant multiple surfaces by an appropriate weighting procedure. The Ehrenfest methodology, therefore, is expected to be most reliable in cases of nuclear kinetic energies that are high by the standard of the energy differences between the contributing potential energy surfaces. This makes the Ehrenfest approach more successful in the physical than the chemical regime. Processes that involve different reaction channels or multiple transitions between electronic states are more appropriately described by multiple trajectories than a single averaged nuclear path. Such a refined version of Ehrenfest dynamics will be discussed in Chapter 11, and further in Chapters 13 and 14.

The basic Ehrenfest model leads to wrong conclusions wherever a process allows for strongly differing motional alternatives. For instance, the interaction of a particle with the surface of a solid might result in particle adsorption or in its reflection by the surface [288]. A realistic treatment of this situation would involve a superposition of these two possible solutions, while a single averaged trajectory would bear a resemblance to neither of them.

A further shortcoming of the mean-field approach encapsulated by Eqs. (8.97)–(8.101) is that it violates the principle of microreversibility (see Section 6.4), as can be seen from the gedanken experiment [238] sketched in Figure 10.1. Shown are two curves on weakly coupled adiabatic potential energy curves along with their populations. The left half of the figure refers to a forward path, involving initial preparation of the system in state $|\Psi_1\rangle$, while the right half presents the corresponding backward path where the state $|\Psi_2\rangle$ is occupied initially. As the system evolves, the two states mix, giving rise to different averaged potential energy surfaces, labeled E_{eff}. For the forward path, E_{eff} stays close to the lower potential energy curve E_1. For the backward path, however, E_{eff} defines a curve that is intermediate between

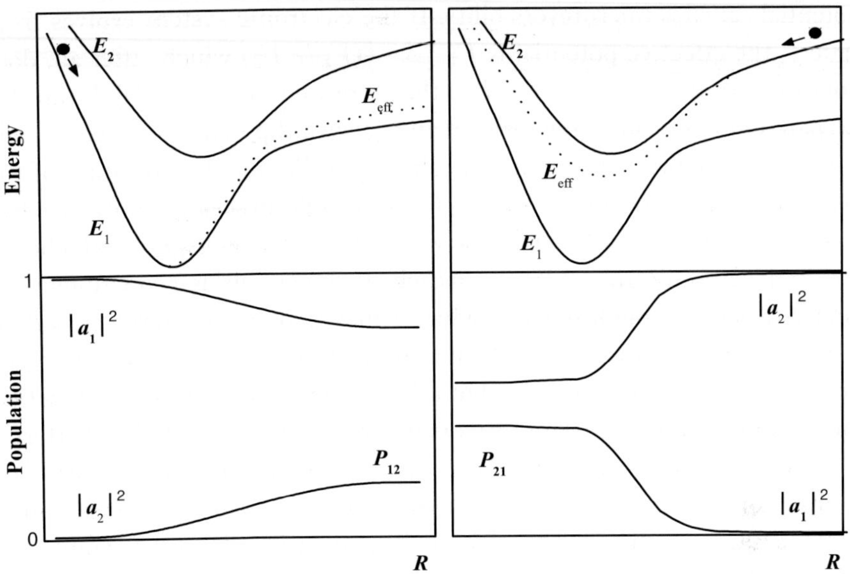

Figure 10.1: Adiabatic potential energy curves (upper panel) and associated populations (lower panel) for forward (left) and backward (right) reaction paths versus a reaction coordinate demonstrating qualitatively that Ehrenfest dynamics is in conflict with the principle of microreversibility. (Reprinted with permission from [238]. Copyright (2002) by the American Chemical Society.)

E_1 and E_2. These features are reflected by the evolutions of the populations $|a_1|^2$ and $|a_2|^2$. The forward path leads to only a small asymptotic value, $|a_2^\infty|^2$, which equals the transition probability P_{12}. The backwards path, in contrast, yields a very substantial value, $|a_1^\infty|^2$, as the inverse probability P_{21}. Clearly, both differ sizeably from each other, demonstrating that the mean-field model underlying Ehrenfest dynamics is not constrained to obey the principle of microreversibility which requires $P_{21} = P_{12}$.

It is, however, possible to generalize the effective potential such that microreversibility is recovered. This succeeds, in principle, by averaging over the forward and the backward reaction. For two electronic states the respective average potential may be written as [289]

$$\bar{V} = \frac{1}{2}(V_1 + V_2) \tag{10.1}$$

where V_1 results from an initial electronic state $|\Phi_1^0\rangle$, and V_2, correspondingly, from $|\Phi_2^0\rangle$. It can be easily seen that this modification of the quantum potential satisfies microreversibility. If the electronic system evolves from state 1, the effective potential is $\bar{V}_{1\rightarrow} = \frac{1}{2}(V_1 + V_2)$ which, after all, does not differ from $\bar{V}_{2\rightarrow} = \frac{1}{2}(V_2 + V_1)$, the corresponding potential if state 2 is prepared initially. Since the forward and backward potentials are equal, one will arrive at equal state-to-state probabilities P_{12} for the evolution of the electronic system from 1 to 2 and P_{21} for the reverse process. If more than two electronic states are considered, the weight factors, which were both chosen as $\frac{1}{2}$ in the example, may differ from each other, possibly as a consequence of degeneracies. Although this procedure achieves formally the reconciliation of the Ehrenfest scheme with the principle of microscopic reversibility, its practical use requires simulating both the forward and the backward reaction for any case investigated, thus increasing greatly the computational effort.

Numerous proposals have been put forward for a more fundamental solution of the problem by introducing a quantum mechanical description of the nuclei into the dynamical formalism at various levels of complexity. Chapters 13–15 of this text will deal in some detail with these attempts. This chapter, however, will highlight a strikingly original attempt at reconciling nuclear and electronic dynamics, suggested in 1979 by Meyer and Miller [21]. This approach is based on the realization that not only the nuclear, but also the electronic degrees of freedom might be subjected to a Hamiltonian formalism in the spirit of classical mechanics. Implementing this methodology accomplishes a treatment of both types of variables on the same footing and eliminates the inconsistencies in the description of the coupling between them that compromise any mixed quantum-classical scheme.

Asked to consider a classical analog for the electronic motion, of course, we are not invited to think of electrons as classical particles. The following section will show, however, that not only a formally consistent but also a useful and practicable model of nonadiabatic dynamics can be obtained from a classical electron analog approach in which the evolution of the electronic subsystem is governed by Hamilton's equations.

10.2 The Classical Electron Analog

In this section we will demonstrate that a description of the electron system that is equivalent to formulae (8.98)–(8.100) can be developed on the ground of classical mechanics. A *classical electron analog* (CEA) theory with well-defined rules of translation between quantum mechanical and classical elements emerges. As we will see, the computational task posed by this model is essentially reduced to determining trajectories in the sense of classical mechanics. This methodological feature is of great practical relevance, as it solves the problem of including large numbers of nuclear degrees of freedom. The computational effort required by a classical approach increases linearly with the number of modes while this increase is exponential if the full quantum mechanical formalism is used [290].

What is the value of the CEA beyond its advantageous scaling with the number of included nuclear modes? The theory may be considered as important for a deeper understanding of the relation between classical and quantum mechanics, but, not surpassing the level of the Ehrenfest model, it seems to do little to advance the cause of nonadiabatic molecular dynamics. We will, however, argue that the CEA in conjunction with the classical S-matrix formalism is capable of overcoming the main deficiencies of Ehrenfest dynamics as sketched in the preceding section. Our presentation of these ideas will largely follow their original presentation by Meyer and Miller [21].

To begin, one expresses M complex coefficients $a_k(t)$, as given by Eq. (8.97), through $2N$ real variables $q_k(t)$ and $n_k(t)$ using the polar decomposition:

$$a_k(t) = \sqrt{n_k(t)}\exp[-iq_k(t)]. \tag{10.2}$$

Expanding the electronic wave function $\Phi(t)$ as in Eq. (8.97), we write the expectation value $\langle\Phi(t)|\hat{O}|\Phi(t)\rangle$ of an arbitrary quantum mechanical operator $\hat{O}$ defined in M-dimensional Hilbert space as

$$O(\mathbf{n},\mathbf{q}) = \sum_{k,k'=1}^{M} \sqrt{n_k n_{k'}}\exp[i(q_k - q_{k'})]O(k,k'). \tag{10.3}$$

Applying this expression to the electronic Hamiltonian, and interpreting $q_k(t)$ and $n_k(t)$ as independent classical action-angle variables,[1] one derives

[1] The variables n_k are assumed to be dimensionless in Eqs. (10.2) and (10.3), as well as all relations later that contain these variables. Dimensional consistence is recovered upon $n_k \to n_k/\hbar$. For simplicity, we will omit this dimensional correction throughout this chapter.

the electronic Hamilton function $H_{\mathrm{el}}(\mathbf{n}, \mathbf{q}; t)$ in the diabatic representation:

$$H_{\mathrm{el}}(\mathbf{n}, \mathbf{q}; t) = \sum_{k,k'=1}^{M} \sqrt{n_k n_{k'}} \exp\left(i(q_k - q_{k'})\right) H_{\mathrm{el},kk'}[\mathbf{X}(t)]. \tag{10.4}$$

With these assignments, the time evolution of $q_k(t)$ and $n_k(t)$ is governed by Hamilton's equations:

$$\frac{dq_k(t)}{dt} = \frac{1}{\hbar}\frac{\partial H_{\mathrm{el}}(\mathbf{n}, \mathbf{q}; t)}{\partial n_k}, \quad \frac{dn_k(t)}{dt} = -\frac{1}{\hbar}\frac{\partial H_{\mathrm{el}}(\mathbf{n}, \mathbf{q}; t)}{\partial q_k}. \tag{10.5}$$

Deriving Eq. (10.2) with respect to time, and inserting the time derivatives of $q_k(t)$ and $n_k(t)$ as given by Hamilton's equations, the reader will easily generate the relation (8.99) for the expansion coefficients $\{a_k(t)\}$ (see Exercise 10.1). Therefore, the time propagation procedure (8.99), based on the TDSE, and Eq. (10.5), based on the classical electron analog, turn out to be formally equivalent.

Exercise 10.1. Demonstrate the equivalence of the Hamilton equations (10.5) and the TDSE (8.99). For this purpose, consider the time derivative of Eq. (10.2) and use the explicit form of the CEA Hamilton function (10.4).

10.2.1 *The CEA equations of motion*

In order to make the connection between the CEA and the Ehrenfest method, we adopt a diabatic frame. The full Hamilton function is then obtained by adding a nuclear kinetic energy term to the electronic Hamilton function (10.4) which allows us to complement the two sets of Hamilton equations for the electrons by the corresponding two sets for the nuclei. We write the Hamilton function extended by the nuclear contribution as

$$H(\mathbf{P}, \mathbf{X}, \mathbf{n}, \mathbf{q}; t) = T(\mathbf{P}) + \sum_{k,k'=1}^{M} \sqrt{n_k n_{k'}} \exp\left[i(q_k - q_{k'})\right] H_{\mathrm{el},kk'}[\mathbf{X}(t)]. \tag{10.6}$$

The time evolution of the full system is then determined by the derivatives of the Hamilton function with respect to the four groups of independent variables, i.e. **P**, **X**, **n**, and **q**:

$$\frac{dX_k(t)}{dt} = \frac{\partial H}{\partial P_k}, \tag{10.7}$$

$$\frac{dP_k(t)}{dt} = -\frac{\partial H}{\partial X_k}, \tag{10.8}$$

$$\frac{dq_k(t)}{dt} = \sum_{k'=1}^{M} \frac{1}{\hbar}\sqrt{\frac{n_{k'}}{n_k}}\, \Re\{\exp\left[i(q_k - q_{k'})\right]\} H_{el,kk'}(\mathbf{X}(t)), \tag{10.9}$$

$$\frac{dn_k(t)}{dt} = -\sum_{k'=1}^{M} 2\frac{1}{\hbar}\sqrt{n_k n_{k'}}\, \Im\{\exp\left[i(q_k - q_{k'})\right]\} H_{el,kk'}(\mathbf{X}(t)). \tag{10.10}$$

This system of equations is strictly equivalent to the Ehrenfest equations of motion given by (8.99) and (8.100). At this juncture, a new perspective on the formalism of Ehrenfest dynamics is obtained, but no additional physical insight seems to be gained. It will be shown, however, how the crucial problems plaguing the Ehrenfest method can in principle be remedied by adopting and extending the classical analog introduced.

10.2.2 *Adiabatic representation of the Hamilton function*

Prior to discussing further the relation between the CEA and the Ehrenfest representation we point out that the classical analog formalism described in this section does not depend on the specific representation chosen. It is possible to change within this formalism from the diabatic to the adiabatic frame and so to derive an adiabatic Hamilton function. In general terms, the classical counterpart of a quantum mechanical basis set transformation, such as that from a set $|\Phi_k\rangle$ to the set $|\Psi_K\rangle$, is a canonical transformation that maps the variables $(\mathbf{n}, \mathbf{q})$ onto the new variables $(\mathbf{N}, \mathbf{Q})$. Both groups of classical variables are related according to

$$\sum_k \exp\left(-iq_k\right)\sqrt{n_k}\, |\Phi_k\rangle = \sum_K \exp\left(-iQ_K\right)\sqrt{N_K}\, |\Psi_K\rangle. \tag{10.11}$$

We use this translation rule to obtain the explicit form of the adiabatic Hamiltonian on the ground of classical electron theory. Focusing first on the electronic system, we write the respective Hamiltonian as a function of the variables **N** and **Q**, and specify the functions $|\Psi_K\rangle$ as a set of adiabatic basis states:

$$
\begin{aligned}
H_{\rm el}(\mathbf{N},\mathbf{Q}) &= \sum_{k,k'} \sqrt{n_k n_{k'}} \exp\left[i(q_k - q_{k'})\right] H_{{\rm el},kk'} \\
&= \sum_{k,k'} \sum_{K,K'} \sqrt{N_K N_{K'}} \exp\left[i(Q_K - Q_{K'})\right] \\
&\quad \times \langle \Psi_K | \Phi_k \rangle H_{el,kk'} \langle \Phi_{k'} | \Psi_{K'} \rangle \\
&= \sum_{K,K'} \sqrt{N_K N_{K'}} \exp\left[i(Q_K - Q_{K'})\right] H_{el,KK'} \\
&= \sum_K N_K E_K. \qquad (10.12)
\end{aligned}
$$

The last step is based on the fact that, by assumption, the basis $|\Psi_K\rangle$ diagonalizes $\mathbf{H}_{\rm el}$. One realizes from Eq. (10.12) that $H_{\rm el}$ is cyclic in $\mathbf{Q}$, so that $H_{\rm el} = H_{\rm el}(\mathbf{N})$. This does not, however, hold for the full Hamilton function in the adiabatic representation which reads

$$
\begin{aligned}
H(\mathbf{P},\mathbf{X},\mathbf{N},\mathbf{Q}) &= \sum_i \frac{P_i^2}{2M_i} \\
&\quad + \sum_j \left[\frac{1}{M_j} \mathbf{P}_j \cdot \mathbf{F}_j(\mathbf{N},\mathbf{Q},\mathbf{X}) + \frac{1}{2M_j} |\mathbf{F}_j(\mathbf{N},\mathbf{Q},\mathbf{X})|^2 \right] \\
&\quad + \sum_K E_K(\mathbf{X}) N_K \qquad (10.13)
\end{aligned}
$$

with $\mathbf{F}_j(\mathbf{N},\mathbf{Q},\mathbf{X}) = -i\hbar \sum_{K,K'} \sqrt{N_K N_{K'}} \exp\left[i(Q_K - Q_{K'})\right] \mathbf{A}_{KK',j}$, where $\mathbf{A}_j$ is the coupling matrix (2.10), providing a classical electron formulation of the nonadiabatic coupling.

It is instructive to examine the action variables n_k which, by their relation to the expansion coefficients a_k in Eq. (10.2), may be interpreted in terms of probability. From the foregoing discussion, it is plausible that the counterpart of the completeness relation

$$
\sum_k |a_k|^2 = 1 \qquad (10.14)
$$

for the classical analog is

$$
\sum_k n_k = 1. \qquad (10.15)
$$

In the special case of the adiabatic Hamilton function $H_{\mathrm{el}}(\mathbf{N})$, Eq. (10.15) becomes

$$\sum_K N_K = 1. \tag{10.16}$$

Since $H_{\mathrm{el}}(\mathbf{N})$ does not depend on the angle coordinates Q_K, the latter relation leads to an interesting conclusion about the validity of the quasi-classical approach in adiabatic situations. The parameters N_K have the status of *adiabatic invariants*. We apply the semiclassical action-angle quantization rule ([58], see Section 4.2.2) on the action variables N_K,

$$\oint \tilde{P}_k dQ_k = 2\pi\hbar N_k, \tag{10.17}$$

with $\tilde{P}_k$ as the kth generalized momentum, and thus constrain them to adopt only integral and non-negative values. The normalization condition (10.16) then implies that all N_K are zero except for one. The semiclassical electronic energies resulting from Eq. (10.12) are thus identical with the quantum mechanical eigenvalues of the adiabatic electronic Hamiltonian.

Exercise 10.2. By use of the CEA equation of motion (10.10), show that

$$\frac{d}{dt}\sum_k n_k(t) = 0. \tag{10.18}$$

From Eqs. (10.14) and (10.15), this relation expresses the conservation of probability within the CEA.

10.2.3 *The classical analog of the electronic two-state problem*

The invariance of the sum over the action variables, made explicit by the constraint (10.15) and implied by Exercise 10.2 in terms of the conservation of probability, can be used to introduce adapted action-angle coordinates $\tilde{\mathbf{q}}, \tilde{\mathbf{n}}$ by declaring

$$\tilde{n}_1 = \sum_{k=1}^{M} n_k, \quad \tilde{q}_1 = q_1, \tag{10.19}$$

and

$$\tilde{n}_k = n_k, \quad \tilde{q}_k = q_k - q_1, \quad \text{if } 2 \le k \le M. \tag{10.20}$$

The transition from the original set of variables $\{q_k, n_k\}$ to the new set $\{\tilde{q}_k, \tilde{n}_k\}$ is a canonical transformation, facilitated by the generating function $q_1\tilde{n}_1 + \sum_{k=2}^{M}(q_k - q_1)\tilde{n}_k$ (see, for instance, [291]).

The advantage of this coordinate transformation is most easily illustrated by the example of two electronic states with Hamilton function $H_{\text{el}}(\mathbf{n}, \mathbf{q})$. In the diabatic representation, H_{el} takes the form

$$H_{\text{el}}(n_1, n_2, q_1, q_2) = n_1 H_{\text{el},11} + n_2 H_{\text{el},22} + 2\sqrt{n_1 n_2}\cos(q_2 - q_1) H_{\text{el},12} \tag{10.21}$$

where $H_{\text{el},kk'}$, $k, k' \le 2$, is real-valued. Applying the rules (10.19) and (10.20), one obtains

$$n_1 = 1 - \tilde{n}_2, \tag{10.22}$$

$$n_2 = \tilde{n}_2, \tag{10.23}$$

and

$$q_2 - q_1 = \tilde{q}_2. \tag{10.24}$$

Upon making these three substitutions in Eq. (10.21), H_{el} becomes solely dependent on $(\tilde{q}_2, \tilde{n}_2) \equiv (q, n)$. The form of the Hamilton function in these new coordinates is

$$H_{\text{el}}(n, q) = (n - 1)H_{\text{el},11} + nH_{\text{el},22} + 2\sqrt{n(n-1)}\cos(q) H_{\text{el},12}. \tag{10.25}$$

With this expression, a very simple description of the electronic two-state problem is accomplished, involving only one degree of freedom.

So far, our presentation of the classical analog has not gone beyond the confines of the Ehrenfest method, and merely introduced a new parametrization for its basic equations. It remains to be shown in the following that the switch from M quantum mechanical to $2M$ classical variables is actually capable of overcoming the main restrictions of the Ehrenfest model. More specifically, the CEA procedure allows us to associate one particular electronic transition with one particular nuclear path, replacing the notion of the trajectory that results from an average over the time-dependent electronic states with an average over trajectories adequately correlated with well-defined electronic processes.

This can be achieved by recourse to *classical S-matrix theory* [292] which employs classical equations of motion but generates transition probabilities according to quantum mechanical superposition, involving the addition of probability amplitudes. The classical limit, expressed in terms of n_k and q_k, for the probability of a specific transition is

$$P_{k_2 \leftarrow k_1} = \left[(2\pi)^M \det \left\{ \frac{\partial \mathbf{n}(t_2)}{\partial \mathbf{q}(t_1)} \right\} \right]^{-1}, \tag{10.26}$$

with the index k ranging from 1 to F. This relation will be motivated in the context of the semiclassical propagator, as discussed in Section 12.1. The transition is characterized by boundary conditions imposed on n_k:

$$n_k(t_1) = \delta_{k,k_1}, \tag{10.27}$$

$$n_k(t_2) = \delta_{k,k_2}. \tag{10.28}$$

These conditions do not seem to involve the other group of dynamic electronic variables, q_k. It is, however, crucial for our discussion that $q_k(t_1)$ can be chosen as to guarantee that the boundary condition (10.28) for the final state is satisfied. Identifying the initial values of q_k for all possible configurations (10.27) and (10.28) is therefore the key to a consistent inclusion of individual electronic transitions into CEA theory.

The example of an electronic two-state problem determined by a Landau–Zener–Stückelberg Hamiltonian, as presented in the following section, will clarify how this methodology may be implemented. Equations (10.7)–(10.10) in conjunction with boundary conditions (10.27, 10.28) make it possible to compute a specific nuclear trajectory corresponding to a specific electronic process.

At this juncture, a serious objection against the strategy put forward in the preceding paragraph needs to be addressed. Inspecting the structure of the equation system (10.7)–(10.10) with special attention to the action variables, one realizes that the value of any action variable in the final state, $n_k(t_2)$, does not depend on the initial angle $q_k(t_1)$, which makes it impossible to satisfy Eqs. (10.27) and (10.28) in the general case. This argument is readily understood by referring to Eq. (10.27) which implies that all equations in the subset (10.10), determining the evolution of the action variables, vanish unless $k = k_1$. From the form of the right-hand side of Eq. (10.10), one further realizes that even $n_{k_1}(t)$ cannot depend on the respective angle variable, $q_{k_1}(t)$. This dilemma seems to thwart the program outlined, aiming at a one-to-one assignment of nuclear trajectories and electronic transitions.

The minimal change that overcomes this problem consists in applying a *Langer modification* to the diabatic electronic Hamilton function (10.4), changing it into

$$H_{\rm el}(\mathbf{n},\mathbf{q};\mathbf{X}(t)) = \sum_{k,k'=1}^{N} \sqrt{\left(n_k + \frac{1}{2}\right)\left(n_{k'} + \frac{1}{2}\right)} \times \exp\left[i(q_k - Q_{K'})\right] H_{{\rm el},kk'}[\mathbf{X}(t)]. \tag{10.29}$$

For a theoretical justification of this intervention, the reader is once more referred to the discussion of the semiclassical propagator in Chapter 12, and in particular to Section 12.5. As is obvious from Eq. (10.10), the action variables initially set to zero will, with the Hamilton function modified according to Eq. (10.29), in general adopt finite values with the elapse of time and depend on the choice made for $q_k(t_1)$.

Although the prescription (10.29) appears to cure the ailment that plagues the CEA dynamics, as diagnosed earlier, it is still not viable. A remaining weakness is exposed when we switch to the adiabatic Hamilton function which now reads

$$H_{\rm el}(\mathbf{N},\mathbf{Q}) = \sum_{K} \left(N_K + \frac{1}{2}\right) E_K(\mathbf{X}). \tag{10.30}$$

With this form of $H_{\rm el}(\mathbf{N},\mathbf{Q})$, it is not possible to recover the adiabatic electronic energy eigenvalues, i.e. one does not arrive at $H_{\rm el}(\mathbf{N},\mathbf{Q}) = E_{K_1}(\mathbf{X})$ given that $N_{K_1} = \delta_{K,K_1}$. However, subtracting the quantity $\frac{1}{2}\sum_K E_K(\mathbf{X})$ from (10.30) yields at once the correct eigenvalues. To generalize this result, we use the invariance of the trace with respect to the basis set transformation $|\Phi_k\rangle \to |\Psi_K\rangle$:

$$\frac{1}{2}\sum_{K} E_K(R) = \frac{1}{2}{\rm Tr}\{\mathbf{H}_{\rm el}(\mathbf{X})\} = \frac{1}{2}\sum_{k} H_{{\rm el},kk}(\mathbf{X}). \tag{10.31}$$

Subtracting the right-hand term in (10.31) from the Hamilton function (10.29) leads to

$$H_{\rm el}(\mathbf{n},\mathbf{q};\mathbf{X}(t)) = \sum_{k=1}^{M} n_k H_{{\rm el}_{kk}}(\mathbf{X}(t)) + \sum_{k,k'=1,k\neq k'}^{M} \sqrt{\left(n_k + \frac{1}{2}\right)\left(n_{k'} + \frac{1}{2}\right)} \times \exp\left[i(q_k - Q_{K'})\right] H_{{\rm el}_{kk'}}(\mathbf{X}(t)). \tag{10.32}$$

With this correction, the Langer modification is merely applied to the non-diagonal elements of the electronic Hamilton function. This solution appears *ad hoc*, as a makeshift device borne out of the necessity to reconcile the dynamic equation system (10.7)–(10.10) with the constraints (10.27) and (10.28) rather than as a firmly grounded result, deduced from the basic principles of CEA. When discussing in Section 12.5 the mapping approach of semiclassical quantum dynamics, we will return to the formalism presented here and embed the two versions of the electronic Hamilton function, (10.29) and (10.32), into a wider context, replacing the action variable n_k with $n_k + \gamma$. The choice $\gamma = \frac{1}{2}$ will be justified from the perspective of a deeper layer of dynamic theory. In Section 10.2*, the reader finds an elementary example that provides a test of the classical analog model by comparison with a quantum mechanical result.

10.3 CEA Theory Applied to a Conical Intersection Problem

In order to assess the validity of any procedure designed to capture nonadiabatic phenomena, physical situations involving comparable electronic and nuclear time scales have to be addressed. This condition is easily realized close to places of degeneracy or near-degeneracy in the potential energy landscape. In this section, we will discuss a simulation that probes the CEA formalism by applying it to such a touchstone configuration. Specifically, we will follow the discussion presented by Zwanziger *et al.* [293], related to a Jahn–Teller $E \times e$ system treated in a CEA framework.

As in the previous section, the result of this model will be compared with the respective quantum mechanical solutions. The latter are readily available, since the intensely studied model $E \times e$ Jahn–Teller Hamiltonian in linear approximation is employed (see Section 2.5.2). In particular, the well-known energy levels of this system can be compared with those obtained by combining the CEA with a semiclassical quantization procedure. Further, the example of this elementary Jahn–Teller system will provide insight into the physical significance of the angle variables $\{q_k\}$ which have been formally introduced in Section 10.2 as phases arising from the polar decomposition of the expansion coefficients $\{a_k\}$ for the electronic wave function.

From Section 2.5.2 we recall the adiabatic Hamiltonian for the $E \times e$ Jahn–Teller problem up to the first order in the nuclear coordinates:

$$\hat{H}^{\mathrm{A}} = \frac{\hbar}{2}\left[\left(\hat{P}_\rho^2 + \hat{P}_\alpha^2 + \frac{1}{4\rho^2} + \rho^2\right)\boldsymbol{\sigma}_0 - P_\alpha \rho^2 \boldsymbol{\sigma}_y\right] + V_E \rho \boldsymbol{\sigma}_z, \quad (10.33)$$

with assignments as in Chapter 2, i.e. ρ and α stand for the radial and the angular vibronic coordinate, respectively, both defined as dimensionless variables, and the constant V_E for the nonadiabatic coupling. The symbols $\hat{P}_\rho$, $\hat{P}_\alpha$ denote the corresponding conjugate momenta. The harmonic frequency ω of the vibrational mode has been set equal to one. $\hat{H}^{\mathrm{A}}$ operates in a space defined by the upper and the lower adiabatic electronic states, which are, using the notation of Section 2.5.2, labeled as

$$|\psi_+\rangle = \frac{1}{\sqrt{2}}\left[\exp\left(-i\frac{\alpha}{2}\right)|E_1\rangle + \exp\left(i\frac{\alpha}{2}\right)|E_2\rangle\right],$$

$$V_+^{\mathrm{APES}} = \frac{\hbar}{2}\rho^2 + V_E\rho, \quad (10.34)$$

$$|\psi_-\rangle = \frac{i}{\sqrt{2}}\left[-\exp\left(-i\frac{\alpha}{2}\right)|E_1\rangle + \exp\left(i\frac{\alpha}{2}\right)|E_2\rangle\right],$$

$$V_-^{\mathrm{APES}} = \frac{\hbar}{2}\rho^2 - V_E\rho, \quad (10.35)$$

where $|E_{1,2}\rangle$ refer to the electronic states in the diabatic representation. Expression (10.33) contains the coupling between nuclear and electronic degrees of freedom in the diagonal as well as non-diagonal elements. Neglect of the nonadiabaticity in $\hat{H}^{\mathrm{A}}$ leads to the stationary Schrödinger equation (2.116) for the vibronic solution that corresponds to the lower potential energy surface, V_-^{APES}.

The translation rules of the classical analog model, as given by Eq. (10.6) in combination with the Langer modification Eq. (10.32) for non-diagonal elements, can now readily be applied to Eq. (10.33). It should be noted that the alternative prescription outlined in Section 10.2.1 which involves constructing the classical analog of the diabatic Hamiltonian, followed by the transition from the diabatic to the adiabatic representation, does not lead to the correct classical analog of $\hat{H}^{\mathrm{A}}$. It turns out that these two procedures result in somewhat differing expressions for the nuclear–electronic coupling. If a problem demands a change between different representations, it is therefore advisable to operate within the realm of quantum mechanics as long as possible, and to apply the CEA translation rules only as the last step in a chain of transformations. With this caveat, one finds the classical analog of Eq. (10.33) to be

$$H(\rho, P_\rho, q, n) = \frac{\hbar}{2}P_\rho^2 - \frac{\hbar}{\rho^2}P_\alpha\left[\left(n + \frac{1}{2}\right)\left(\frac{3}{2} - n\right)\right]^{\frac{1}{2}} \sin(q)$$
$$+ \frac{\hbar}{2\rho^2}\left(P_\alpha^2 + \frac{1}{4}\right) + \frac{\hbar}{2}\rho^2 + V_E\rho(2n-1). \quad (10.36)$$

Since the angular coordinate of the nuclear system, α, is cyclic in $H(\rho, P_\rho, q, n)$, the total vibronic or pseudorotational (see Section 9.5) angular momentum P_φ is a constant of motion. As required by the geometric features of the Jahn–Teller $E \times e$ problem (see Section 2.5.2), the eigenvalue j of P_α is set to a constant half-odd integer in the simulation based on Eq. (10.33).

Arriving at energy levels that may be compared with those found from the exact quantum mechanical solution requires the use of a semiclassical quantization method on $H(\rho, P_\rho, q, n)$. Zwanziger *et al.* employ the *adiabatic switching method* (e.g. [294]) for this purpose. This method, which is based on the adiabatic invariance of the actions characteristic of the investigated system, is in general not applicable to non-separable problems but can still be assumed to yield realistic results if the physical situation is not dominated by chaotic regions in phase space or by strong resonances of low order [293]. In the present case, the Hamilton function (10.36) is partitioned into a separable (H_0) and a non-separable (H_1) segment according to

$$H(\mathbf{J}, \mathbf{\Phi}; t) = H_0(\mathbf{J}) + \lambda(t)H_1(\mathbf{J}, \mathbf{\Phi}), \quad (10.37)$$

where we have introduced the collective notation $\{\mathbf{J}, \mathbf{\Phi}\}$ for the manifold of action-angle variables. The function $\lambda(t)$ increases from zero to one as a trajectory is run during a time interval T chosen sufficiently long. At $t = T$, the perturbation H_1 is switched on completely and the energy of the trajectory is expected to approximate the corresponding semiclassical eigenvalue of the full Hamiltonian. The separable part of Eq. (10.36) is chosen as the zeroth-order Hamilton function H_0, namely

$$H_0 = \frac{\hbar}{2}P_\rho^2 + \frac{\hbar}{2\rho^2}\left(P_\varphi^2 + \frac{1}{4}\right) + \frac{\hbar}{2}\rho^2 + V_E\rho(2n-1), \quad (10.38)$$

while the remainder defines the perturbation:

$$H_1 = -\frac{\hbar}{\rho^2} P_\alpha \left[\left(n + \frac{1}{2} \right) \left(\frac{3}{2} - n \right) \right]^{\frac{1}{2}} \sin(q). \tag{10.39}$$

Implementing this methodology, one has to take into account that the conservation of actions that underlies the procedure cannot be guaranteed for the non-separable problem at hand. Since a single trajectory, depending on its initial conditions, might encounter resonances that do not leave the initial actions invariant, it is mandatory to run an ensemble of trajectories with initial conditions sampling the zeroth-order $\{\mathbf{J}, \mathbf{\Phi}\}$ manifold that is governed by $\hat{H}_0(\mathbf{J})$. The latter is specified by quantizing the three actions characteristic of the problem studied here. The initial values for two of them follow immediately from the physical situation: as the motion starts on the lower one of the two adiabatic potential energy surfaces, the electronic action n is set initially to zero. Further, the action associated with the pseudorotational angular momentum is defined by the constraint $j = \frac{1}{2}(2N_j + 1)$ with $N_j = 0, 1, 2, \ldots$. This leaves the radial action for the nuclear dynamics. The latter component is determined by the quantization condition for the corresponding phase integral as

$$\oint P_\rho d\rho = 2\pi \left(N_\rho + \frac{1}{2} \right), \quad N_\rho = 0, 1, 2, \ldots. \tag{10.40}$$

This equation is used to define the energy eigenvalues of the zeroth order Hamiltonian. With j as a constant of motion and n chosen initially as zero, one may vary the energy until the condition (10.40) is met for a particular index N_ρ.

Table 10.1 summarizes some of the results of this procedure as reported in [293]. For six vibronic angular momentum eigenvalues, the lowest energy levels are included. More specifically, the table compares the quantum mechanical energy eigenvalue for the fully coupled system with that for the lower of the two adiabatic surfaces alone, and also with the result of semiclassical quantization, as discussed before. The agreement between quantum mechanics and the semiclassical quantization based on the classical analog is generally excellent. The gaps in Table 10.1, however, are related to states sufficiently close in energy to both adiabatic surfaces. In this case, the adiabatic switching method may fail due to strongly pronounced chaotic behavior of the respective trajectories.

Table 10.1: Quantum mechanical and semiclassical energy eigenvalues of the Hamiltonian (10.36) with $V_E = 3.0$. For each value of the pseudorotational angular momentum j, the lowest eigenvalues are shown [293]. The second column refers to the quantum mechanical result for the coupled system, the third to the lower potential energy surface alone, the fourth to the eigenvalues emerging from semiclassical quantization, as described in the text.

j	E_Q	E_L	E_{SE}
$\frac{1}{2}$	−3.9829	−3.9826	−3.9673
	−2.9723	−2.9698	−2.9510
	−1.9484	−1.9357	
	−0.8980	−0.8584	
$\frac{3}{2}$	−3.8544	−3.8534	−3.8429
	−2.7917	−2.7881	−2.7795
	−1.6946	−1.6845	
	−0.5547	−0.5326	
$\frac{5}{2}$	−3.6231	−3.6215	−3.6153
	−2.5001	−2.4958	−2.4928
	−1.3371	−1.3279	−1.331
	−0.1319	−0.1148	
$\frac{7}{2}$	−3.3140	−3.3118	−3.3086
	−2.1357	−2.1309	−2.1309
	−0.9182	−0.9095	−0.9102
$\frac{9}{2}$	−2.9448	−2.9422	−2.9410
	−1.7183	−1.7134	−1.7150
	−0.4557	−0.4474	−0.4526
$\frac{11}{2}$	−2.5282	−2.5253	−2.5253
	−1.2602	−1.2551	−1.2577
	0.0408	0.0486	0.0431

By definition of the quantities compared, the entries of columns 1 and 2 deviate increasingly from each other as the energy level increases. This trend accentuates the enhanced importance of nonadiabatic coupling effects for higher energies. It is interesting to observe that for the third levels of the series with $j = \frac{5}{2}, \frac{7}{2}, \frac{9}{2}, \frac{11}{2}$, the agreement between the second and the fourth column is better than that between the second and the third. This feature indicates that the classical analog is capable of accounting reasonably well for the nonadiabatic admixtures which are characteristic of the respective states, both with regard to the qualitative trend of the nonadiabatic effect and to quantitative accuracy.

The presently discussed example also clarifies how nonadiabatic behavior manifests itself within the CEA model. For a closer understanding of this crucial issue, one may compute frequency distributions of nuclear as well as electronic degrees of freedom and compare them for two states, namely one that is predominantly adiabatic, and another one with a pronounced nonadiabatic component. This procedure has been applied to the ground state of the $j = \frac{1}{2}$ series (see Figure 10.2) and the third level of the $j = \frac{5}{2}$ series (see Figure 10.3). For both states, the Fourier transform of ρ, the radial nuclear coordinate, as well as $\sin(q)$, governing the dependence of the Hamiltonian (10.36) on the electronic angle variables, have been displayed. Commenting first on the $j = \frac{1}{2}$ ground state, Figure 10.2 shows that the frequency distribution for ρ is sharply peaked at a value of about 1.0 while the corresponding picture for $\sin(q)$ exhibits a wider distribution centered around the much higher value of 18.0. This marked difference of the nuclear and the electronic degrees of freedom is the fingerprint of an adiabatic situation:

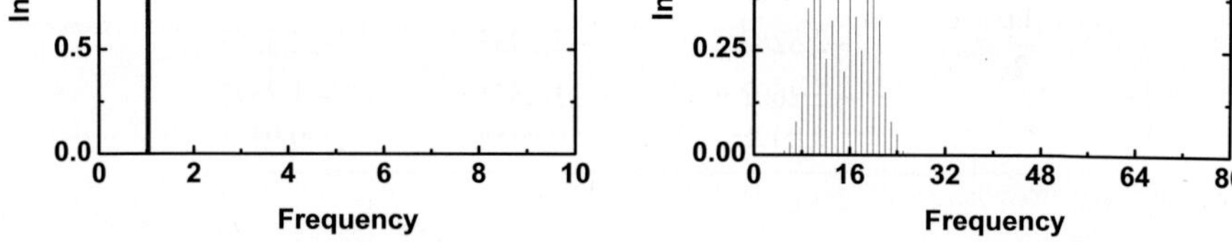

(a) (b)

Figure 10.2: Fourier transforms of ρ and $\sin(q)$ for the ground state of the j = 1/2 series. (a) The transform of ρ, (b) the transform of $\sin(q)$. The insert shows a model for $\sin(q)$, as explained in the text [293]. No overlap between the nuclear and the electronic distributions is visible, as is typical for predominantly adiabatic behavior. (Reprinted with permission from [293]. Copyright (1986) by the American Institute of Physics.)

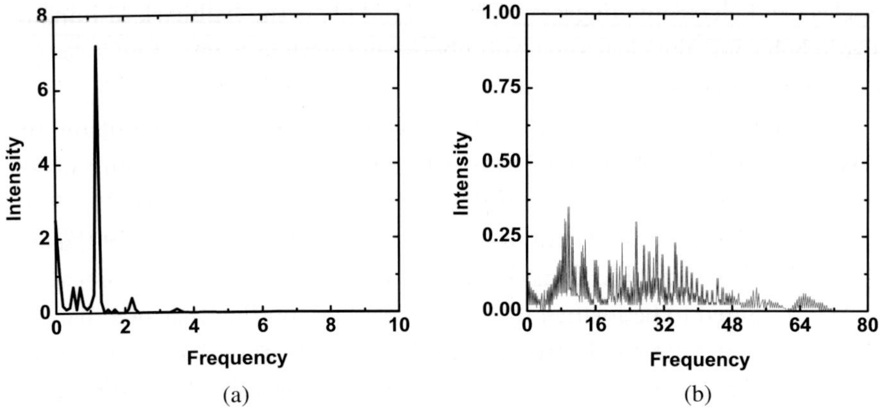

Figure 10.3: Fourier transforms of ρ, shown in (a), and sin(q), shown in (b), for the third level of the $j = \frac{5}{2}$ series. In this case, overlap is found between the nuclear and the electronic distribution, revealing pronounced nonadiabatic features of the investigated state. (Reprinted with permission from [293]. Copyright (1986) by the American Institute of Physics.)

nuclear and electronic motions proceed at clearly separated time scales. It should be noted that Figure 10.2(b) allows for a simple interpretation in terms of electron dynamics. The central frequency corresponds to the vertical energy difference of size $2\rho V_E$ between the two Born–Huang surfaces as measured at the minimum of the lower surface, occurring at $\rho_{\mathrm{MIN}} = V_E/\hbar$ (compare with Eq. (2.78)), which is chosen as 3.0 in the presented example. Likewise, the sidebands of the transform of $\sin(q)$ can be related to electronic transitions between the surfaces, either at $\rho < \rho_{\mathrm{MIN}}$, leading to lower, or at $\rho > \rho_{\mathrm{MIN}}$, leading to higher frequencies (compare also with Figure 2.2). The insert in Figure 10.2(b) is a model distribution based on the simple formula $\sin[n(t)] = \sin\{[F + a\sin(ft)]t\}$ where F stands for the average electronic frequency and a, f are constants chosen as $a = 0.02$ and $f = 1.01$. This function obviously captures the essential properties of the transform derived from the classical analog.

The results obtained for the third level of the $j = \frac{5}{2}$ series, subjected to the same analysis, are shown in Figures 10.3(a) and (b). The contrast between Figures 10.2(b) and 10.3(b) is evident, as the transform of $\sin(q)$ appears much more irregular in the latter than in the former image, and further, it spreads over a much wider frequency interval. It is clear from 10.3(a) that there is now a non-negligible overlap between frequencies assigned to nuclear and electronic degrees of freedom. This is the hallmark of nonadiabatic behavior: the clear separation between the time scales of nuclear and electronic processes has been lost.

The discrepancies remaining between the full quantal treatment and the classical analog can be ascribed mostly to the quantization procedure which may be improved by adopting a uniform semiclassical approach ([295–299]).

Meyer treated a more complex conical intersection problem [290] using the classical analog. He considered the ultrafast non-radiative decay of $C_2H_4^+$, assumed to be formed by photoionization, on the basis of a diabatic representation for two electronic states coupled to three or ten vibration modes. Comparison was made with the results of earlier work [300] where wave packet propagation was used to treat the same system (see Chapter 15). The relevant coupling constants as well as nuclear vibration frequencies were taken from this source. Adopting a histogram procedure, Meyer evaluated the electronic population in the diabatic state of higher energy, inferring essential information about the details of the non-radiative transition under study from the time dependence of this quantity. Several hundred trajectories were run with statistically distributed initial values of the electronic angle variable as well as the nuclear variables and the constraint that initially only the second diabatic state be occupied. The occupation number of this state, averaged over the number of trajectories, was recorded as a function of time. The time dependence predicted by the CEA approach turned out to be well compatible with that based on the quantum mechanical treatment.

We emphasize that these investigations subject the classical analog theory to a more rigorous assessment than most scattering computations could, since a bound system is characterized by longer interaction times than a scattering system, corresponding to longer accumulation periods for the errors induced by the approximations made in the model. Further, these applications are not located in the "favorable territory" of the classical analog, as they do not involve nuclear kinetic energies that are large as compared to the average splitting between the relevant potential energy surfaces. In spite of these challenges and impediments, the CEA leads to results that are qualitatively adequate throughout and of reasonable quantitative accuracy in most of the test cases discussed in this section.

Chapter 11

Hopping and Spawning

The most serious limitation of the Ehrenfest method as described in the preceding chapters consists in its reliance on an average potential energy surface on which the "best" trajectory evolves, since this average may not always be physically meaningful. Propagating the trajectory on one given well-defined potential energy surface, on the other hand, cannot do justice to a physical situation that involves electronic transitions. Including the latter seems to call for a more comprehensive scheme that treats both the electronic and the nuclear subsystem with quantum mechanical rigor, as prototypically realized by the propagation of nuclear wave packets in a landscape of coupled potential energy surfaces (see Chapter 15). There are, however, strong reasons for continued exploration of classical trajectories in the context of nonadiabatic problems which usually call for extensive computational effort, giving the quasi-classical approach a distinct advantage over the quantum treatment. While the computational demands of the latter grow exponentially with the number of non-separable degrees of freedom, the former exhibits linear scaling. On the other hand, classical trajectories in a quantum setting have to be employed judiciously, as they fail to account for essential quantum phenomena such as tunneling or interference. These shortcomings necessitate a careful, case-specific use of the quasi-classical methodology. A classical description of the nuclei may, for instance, be legitimate where only an average over the initial conditions of the nuclear system is known and thus its phase information is lost.

How can one incorporate transitions between discrete electronic levels, as a hallmark of quantum physics, into the trajectory-based theory? A conceptually simple realization of this notion was devised in 1967 by Bjerre and Nikitin [22] with the *trajectory surface hopping* (TSH) method and subsequently adopted by Tully and Preston [301]. In this approach, the nuclear trajectory is constrained to move on one potential energy surface at any given moment of the system's evolution, but switches between different surfaces are possible. This method, along with its later variations and refinements, has developed into a popular choice for molecular dynamics simulations of nonadiabatic processes. It has the advantages of a quasi-classical, trajectory-based model in terms of implementation and interpretation while allowing for quantum discontinuities as the system evolves in a landscape of multiple potential energy surfaces. In this chapter, we will present the salient features of the surface hopping method and then proceed to approaches spawned by this method and intended to lead, in a well-defined asymptotic limit, to the exact solution of the quantum problem.

11.1 The Trajectory Surface Hopping Method

In Tully and Preston's original proposal [301], electronic transitions were confined to *a priori* selected places, chiefly the locations of avoided crossings between two surfaces. A subsequent generalization, however, extended the procedure to allow for surface hopping at any point of the potential energy landscape [288]. In the following summary of the model, we will mostly refer to this more flexible version. Before presenting some characteristic applications of the surface hopping method, we will highlight the main arguments on which it rests.

As in Eq. (2.14), we partition the overall Hamiltonian of the electronic-nuclear system into a nuclear contribution that depends solely on classical parameters and an electronic term that contains the positions of the classical nuclei as parameters:

$$\hat{H} = \hat{T}_{\mathrm{N}} + V_{\mathrm{N}}[\mathbf{R}(t)] + \hat{H}_{\mathrm{el}}[\mathbf{r}, \mathbf{R}(t)] \tag{11.1}$$

with $\hat{T}_{\mathrm{N}}$ and V_{N} as nuclear kinetic and potential energy, respectively, $\mathbf{R}(t)$ as vector of the internal nuclear coordinates, or, alternatively, the nuclear positions encoded as a string of Cartesian coordinates, and $\mathbf{r}$ as the electronic coordinate vector.

The adiabatic representation has been shown to provide a natural basis of operation for the surface hopping method in its original form [301]. This is related to the most characteristic feature of this method, namely the inclusion of instantaneous switches between surfaces, where the strength of the local coupling between the surfaces provides the switching criterion. In this respect, the adiabatic frame is usually more suitable than the diabatic one, since the adiabatic coupling term tends to be more sharply localized than the diabatic coupling. We operate with an electronic basis $\{\psi_i\}$ that depends parametrically on the nuclear coordinates $\mathbf{R}(t)$, such that the solution $\Psi(\mathbf{r}, \mathbf{R}, t)$ is given by the superposition

$$\Psi(\mathbf{r}, \mathbf{R}, t) = \sum_k a_k(t)\psi_k(\mathbf{r}, \mathbf{R}) \exp\left[-\frac{i}{\hbar}\int_{t_0}^{t} dt' E_k(\mathbf{R}(t'))\right]. \tag{11.2}$$

As the adiabatic representation is chosen, the stationary Schrödinger equation assigns to each basis state ψ_k a potential energy surface V_k^{A}, by Eq. (2.17):

$$\hat{H}_{\mathrm{el}}\psi_k(\mathbf{r}, \mathbf{R}) = V_k^{\mathrm{A}}(\mathbf{R})\psi_k(\mathbf{r}, \mathbf{R}). \tag{11.3}$$

Inserting Eq. (11.2) into the TDSE, we obtain upon integrating over $\mathbf{r}$ and introducing the nonadiabatic coupling vector $\mathbf{A}_{kj}$

$$\frac{da_k}{dt} = \sum_j a_j \mathbf{A}_{kj} \cdot \frac{d\mathbf{R}}{dt} \exp\left[-\frac{i}{\hbar}\int_{t_0}^{t} \Delta V_{k,j}^{\mathrm{APES}} dt'\right]. \tag{11.4}$$

The difference $\Delta V_{k,j}^{\mathrm{APES}} \equiv V_k^{\mathrm{APES}} - V_j^{\mathrm{APES}}$ is identical with $\Delta V_{k,j}^{\mathrm{A}} \equiv V_k^{\mathrm{A}} - V_j^{\mathrm{A}}$ since V_k^{APES} and V_j^{A} differ only by the nuclear potential energy (see Eqs. (2.17) and (2.18)). Once the initial state of the system is specified we can use Eq. (11.4) to calculate the probability of finding the system in another state at a given time for any trajectory $\mathbf{R}(t)$. Let us assume an avoided crossing at the point $\mathbf{R}_0$ and a constant vector of nuclear velocity $\frac{d\mathbf{R}}{dt}$ as well as an energy gap $\Delta V_{k,i}^{\mathrm{APES}}$ in the vicinity of this position. Solving Eq. (11.4) to linear order in time around $\mathbf{R}_0$ yields then

$$\mathbf{a} = (\mathbf{I} + i\mathbf{M})\mathbf{a}(t = t_0) \tag{11.5}$$

and

$$M_{ij} = \hbar \mathbf{A}_{ij} \cdot \frac{d\mathbf{R}}{dt} \frac{\exp\left[-\frac{i}{\hbar}\int_{t_0}^{t} dt'\, \Delta V_{i,j}^{\mathrm{APES}}\right]}{\Delta V_{i,j}^{\mathrm{APES}}}, \tag{11.6}$$

The reciprocal value of the prefactor on the right-hand side of Eq. (11.6):

$$\xi_{ij} = \frac{\Delta V_{i,j}^{\mathrm{APES}}}{\hbar\, \mathbf{A}_{ij} \cdot \frac{d\mathbf{R}}{dt}}, \tag{11.7}$$

is essentially the Massey parameter (7.143) in adiabatic disguise [302] (see Exercise 11.1).

Exercise 11.1. Show that for an avoided crossing problem of the Landau–Zener type, involving two electronic states and a single coupling coordinate, it holds that

$$|\xi_{12}| = 4|\Gamma| \tag{11.8}$$

where ξ_{12} is defined by Eq. (11.7) and $|\Gamma|$ is the Massey parameter (7.143).

The quantities ξ_{ij} or Γ provide adiabaticity criteria in the sense that large values of these parameters indicate adiabatic motion. In this case, the trajectory is that of a classical particle subjected to a certain potential energy function. Wherever these quantities are found to be small, however, transitions between potential energy surfaces might occur. This feature is exploited by the TSH technique in its initial formulation which makes allowance for changes of the trajectory from one to another surface in regions of sufficiently strong nonadiabatic coupling. As the system passes through the seam (see Section 2.6) of the problem or, more generally, through a regime of sizeable coupling between two adiabatic surfaces, hopping from one to the other becomes allowed by a suitably defined criterion. In the realization by Tully and Preston, comparison of the transition probability with some random number between zero and one decides if the hop takes place or not.

The cases studied by Tully and Preston for explorative purposes, $H^+ + H_2$ and $H^+ + D_2$ [301,303], provide prototypes for systems with well-localized regions of nonadiabatic coupling.

The diabatic potential energy curves of H_2 and H_2^+ as constituents of a linear H_3^+ system cross at an interatomic distance of $R_{12} = 1.3$ Å, as shown in Figure 11.1. With the third center infinitely far removed, the diabatic coupling term V_{12} vanishes. As the third center is fixed at a large but finite distance from the diatoms, however, both configurations acquire the same symmetry, $^1\Sigma^+$, if spin singlet character is imposed on the molecule. The coupling is now different from zero, and it grows in magnitude as the atomic and the diatomic fragments approach each other, giving rise to an avoided crossing with an increasing energy gap and thus favoring increasingly adiabatic dynamics. Continuing in the same adiabatic potential curve while passing through the site of the diabatic intersection is associated with charge transfer according to $H_2 + H^+ \leftrightarrow H_2^+ + H$. Using the diatomics-in-molecules (DIM, [304, 305]) method, Preston and Tully [303]

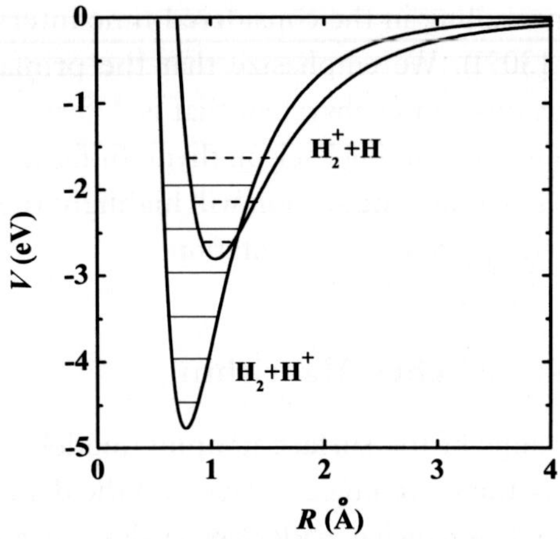

Figure 11.1: Asymptotic H_3^+ potential energy curves, with H and H^+ in infinite distance from H_2^+ and H_2, respectively. The zero of energy is set at the sum of the energies of the three separate atoms: $E(H^+ + H + H) = 0$.

calculated the nonadiabatic coupling vector $\mathbf{A}_{12}$ and found its magnitude sharply peaked in a narrow regime around $R_{12} = 1.3$ Å for excitation energies up to about 10 eV. From this analysis, nonadiabatic transitions affect the dynamics of collinear H_3^+ whenever the bond length of a diatomic fragment is in the vicinity of the diabatic crossing and the remaining atom is sufficiently far from the diatom. Transition probabilities computed on the basis of Eq. (11.4) were found to agree very well with the prediction of the Landau–Zener–Stückelberg formula (7.148).

The rest of this chapter is devoted to several variations and offshoots of the basic TSH methodology. In all of these variants, three quantum subsystems can be distinguished that are advanced simultaneously as the TSH simulation proceeds: the *primary*, the *reference*, and the *auxiliary wave function* [306]. The first of these is the mixed-state solution from which transition probabilities are derived, the second describes the momentarily occupied electronic state which may or may not be depopulated as the hopping criterion is applied. Lastly, the auxiliary wave function determines the classical motion of the nuclear subsystems, e.g. by providing the Hellmann–Feynman force that propels the nuclei. In Tully's original realization of the hopping method, as reviewed in this section, the reference and the auxiliary wave functions coincide, since at each stage of the propagation, some adiabatic electronic state is occupied, and this state also dictates the nuclear dynamics prevailing in the considered time interval. Other choices can be made (e.g. [307]). We emphasize that the primary wave function describes a fictitious quantum subsystem that is determined by the nuclear degrees of freedom without acting back on them, fulfilling a mere bookkeeping task [308,309]. The following section will highlight the hopping strategy in its most frequently applied implementation.

11.2 The Fewest Switches Algorithm

The numerous versions of the surface hopping model differ essentially in two respects, namely the switching criterion and the definition of the seam, understood here, in a somewhat wider sense than in Section 2.6, as the region of maximum transition probability. In cases where precise information is available on the regime of pronounced nonadiabatic interaction, this knowledge can be used to confine the occurrence of hops between surfaces to well-specified zones of the configuration space. This may greatly simplify the application of the surface hopping procedure. In general, however, one cannot assume the nonadiabatic coupling to be sharply localized, and permitting surface hopping at any stage of the systems evolution is preferable to restricting it to selected regions.

In this spirit, Tully generalized the original TSH formalism to the *fewest switches algorithm* [288] which has developed into the widely applied standard form of the surface hopping methodology. Again, a basis of electronic functions $\psi_i(\mathbf{r},\mathbf{R})$ is introduced that depend on time through the nuclear coordinates $\mathbf{R}(t)$. As the method does not make explicit reference any longer to the adiabatic representation, we eliminate from Eq. (11.2) the dynamic phase factor which contains the adiabatic energies E_k and expand the primary wave function instead according to[1]

$$\Psi(\mathbf{r},\mathbf{R},t) = \sum_i a_i(t)\psi_i(\mathbf{r},\mathbf{R}). \tag{11.9}$$

The corresponding electronic Hamiltonian matrix is given by

$$H_{\mathrm{el},ij}(\mathbf{R}(t)) = \langle\psi_i(\mathbf{r},\mathbf{R})|\hat{H}_{\mathrm{el}}[\mathbf{R}(t)]|\psi_j(\mathbf{r},\mathbf{R})\rangle. \tag{11.10}$$

[1] Since, however, the adiabatic representation leads to simpler nuclear propagation equations and minimizes the number of transitions at low energy, it may be considered superior to the diabatic representation [310] and will be adopted as the default frame for the TSH method in the following development.

Integrating the TDSE now yields

$$i\hbar \frac{da_k}{dt} = \sum_j a_j \left[H_{\text{el},kj} - i\hbar \frac{d\mathbf{R}}{dt} \cdot \mathbf{A}_{kj} \right]. \tag{11.11}$$

This equation may be reformulated in terms of the electronic state populations and coherences, understood as the diagonal and non-diagonal elements, respectively, of the density matrix ρ,

$$\rho_{kj} = a_k a_j^*, \tag{11.12}$$

turning Eq. (11.11) into

$$i\hbar \frac{d\rho_{kj}}{dt} = \sum_c \left[\rho_{lj} \left(H_{el,kl} - i\hbar \frac{d\mathbf{R}}{dt} \cdot \mathbf{A}_{kl} \right) - \rho_{kl} \left(H_{el,lj} - i\hbar \frac{d\mathbf{R}}{dt} \cdot \mathbf{A}_{lj} \right) \right]. \tag{11.13}$$

Relation (11.13) implies that the temporal change of the population ρ_{kk} of an electronic state can be expressed as a sum over matrix elements b_{kl} such that

$$\frac{d\rho_{kk}}{dt} = \sum_{l \neq k} b_{kl} \tag{11.14}$$

with

$$b_{kl} = \frac{2}{\hbar} \Im\{\rho_{kl}^* H_{\text{el},kl}\} - 2\Re \left\{ \rho_{kl}^* \frac{d\mathbf{R}}{dt} \mathbf{A}_{kl} \right\}. \tag{11.15}$$

From Eq. (11.14), b_{kl} carries information about the rate of population change from a state ψ_k to another state ψ_l. Therefore, the ratio

$$g_{kj} = \Delta t \frac{b_{jk}}{\rho_{kk}} \tag{11.16}$$

adopts the meaning of a switching probability for the reference state ψ_k. Here the small time interval Δt may be the integration step chosen to propagate ρ_{kj} in Eq. (11.5), or a few integration steps. In any case, Δt must be short enough to guarantee that the population change during this interval is only slight. The surface hopping method can now be described as a sequence of steps based on Eqs. (11.10)–(11.16):

(1) Preparation of the initial state. Classical positions and momenta are assigned to all nuclei involved. The electronic system is characterized through initial values of the density matrix ρ_{jl}. At the beginning of its

evolution, the system usually occupies one well-defined potential energy surface labeled k, implying $\rho_{jl}(t = 0) = \delta_{jk}\delta_{lk}$ which will be assumed in the following. More complex choices for the initial state of the electronic system, associated with mixtures of electronic states, are compatible with the formalism.

(2) The nuclei are propagated in time on the initially occupied potential energy surface V_{kk} utilizing classical equations of motion. Likewise, the electronic system is advanced, integrating Eq. (11.4) over a sufficiently small time interval.

(3) The switching probabilities for electronic transitions from the reference state $|k\rangle$ to alternative states $|j\rangle$ are evaluated according to Eq. (11.16). Simultaneously, a uniform random number is generated such that

$$0 > \zeta > 1, \tag{11.17}$$

providing a criterion for the occurrence of electronic transitions. In the simplified scenario of a two-state problem, the transition from the initial state $|1\rangle$ to the state $|2\rangle$ is invoked if $g_{21} > \zeta$. Otherwise, the system will continue to evolve in $|1\rangle$. Extending the argument to the case of N states, we define the expressions

$$g_k^{(m)} = \sum_j^m g_{kj}, \quad m \leq N, \tag{11.18}$$

and stipulate that a switch from $|k\rangle$ to $|m\rangle$ occurs if

$$g_k^{(m-1)} < \zeta < g_k^{(m)}. \tag{11.19}$$

The reader may consider some special cases to develop an intuitive understanding of the way in which rule (11.19) operates. If, for instance, three states are involved, and the system has a high probability of changing from the initial state $|1\rangle$ to state $|2\rangle$, while the transition $|1\rangle \rightarrow |3\rangle$ proceeds with a much smaller likelihood, the widths of corresponding intervals for the random number, namely $[0, g_1^{(2)}]$ and $[g_1^{(2)}, g_1^{(3)}]$, reflect this order of probabilities, i.e. the second interval is much smaller than the first. We add that other realizations of the threshold term (11.16) have been suggested (see, for instance, [311]).

(4) The procedure restarts from step 2. If no switch has been invoked when executing step 3, the nuclei are propagated on the previous potential

energy surface with label k, otherwise on the newly occupied one with label m. This prescription raises the question of energy conservation for the case that $H_{\text{el},kk}[\mathbf{R}(t)] \neq H_{\text{el},mm}[\mathbf{R}(t)]$ at the point of transition $\mathbf{R}(t)$. The surface hopping method responds to this problem by an adjustment of the nuclear velocities in the direction of the nonadiabatic coupling vector $\mathbf{A}_{km}(\mathbf{R})$. After implementing this adjustment, steps 2–4 are repeated until the computation of the trajectory has been completed.

(5) The whole procedure is reiterated for a new trajectory. Termination is attained when a statistically meaningful sample of trajectories has been generated in this fashion. Nuclear populations of the contributing potential energy surfaces are obtained by averaging over the resulting trajectory swarm.

The steps 1–5 specify a self-consistent set of rules for selecting classical nuclear trajectories in a landscape of potential energy surfaces. It should be noted that, for an infinitely large ensemble of trajectories, the fraction propagated on the potential energy surface with label k is identical with the population ρ_{kk} as generated by Eq. (11.4) [288]. Step 3 comprises the *fewest switches* requirement. This feature is manifest by the proportionality of the switching threshold (11.16) to the time step Δt. Thus, decreasing Δt decreases the switching probability at any step, but due to the enhanced number of steps in a given propagation period, the net switching probability remains unaffected. In contrast, a switching criterion that ignores the step size will lead to an increase of the hopping frequency upon reduction of Δt [288]. Constraining the instances of transitions to a minimum is motivated by the recognition that an exorbitant number of changes between the respective surfaces would defy the purpose of the method, since this would amount to averaging over the surfaces and thus effectively reduce the surface hopping model to a version of the Ehrenfest method.

The major strength of the TSH scheme lies in its shift from a single optimized trajectory evolving on an averaged potential energy surface to a series of correlated trajectories, distributing themselves in the course of the interaction over the full multiplicity of the potential energy surfaces considered. Artificial effects stemming from the averaging procedure are therefore eliminated. Having left the interaction zone where transitions are likely, the system can be analyzed in the asymptotic region of its evolution in terms of well-defined final electronic states.

The most sensitive premise of the surface hopping model lies in its assumed sudden switching mechanism that has to be balanced by appropriate *ad hoc* corrections of the nuclear velocities. If a uniform scaling factor κ is to be applied to the nuclear momenta in the direction of the nonadiabatic coupling vector $\mathbf{A}_{ij}(\mathbf{R})$ [312], the energy conservation constraint leads to a quadratic equation for this factor [313]. If this equation yields two real solutions for κ, the one with the smaller magnitude may be selected. A more serious difficulty, however, is encountered in situations where no real solution for κ exists. Such a case arises whenever the hop from a lower to a higher potential energy surface leads into classically forbidden territory, i.e. the available kinetic nuclear energy is smaller than the energy difference $H_{\text{el},k'k'} - H_{\text{el},kk}$ at the point of transition. If this is the case, energy conservation cannot be satisfied, and no hopping process is carried out. Tully's original suggestion to forego any manipulation of the nuclear momenta in these situations has turned out to be more robust and resistant [313] than alternative proposals [314].

Several authors [313, 315, 316] diagnosed a lack of consistency in the fewest-switches hopping method in practical applications which are necessarily restricted to a finite number of trajectories. The fraction of trajectories occupying a certain potential energy surface at an arbitrary stage of the simulation was found to deviate from the population probability of the surface (the respective diagonal element of the density matrix) at that stage. This shortcoming has been related to the problem of classically forbidden hops and, more fundamentally, to the lack of interaction between trajectories, as inherent in elementary realizations of the TSH scheme [23].

The fewest switches algorithm, as encoded in the five-step sequence indicated earlier, may be characterized in terms of the three quantum subsystems underlying the TSH model, namely the primary, the reference, and the auxiliary wave function (see Section 11.1). The primary quantum subsystem, defined by the superposition (11.3), determines the switching probabilities used in step 3 of the fewest switches algorithm. The primary subsystem undergoes an entirely coherent evolution, governed by Eq. (11.13), while decoherence is introduced by the final averaging over the resulting set of trajectories. As in the original TSH method by Tully and Preston (see

Section 11.1), the reference wave function, as given by the momentarily occupied basis state ψ_k, is identical with the auxiliary wave function, specifying the classical environment in the TSH simulation [317]. This choice results in adiabatic dynamics punctuated by jumps between reference states. In the context of the semiclassical propagator concept, we will show that the TSH approach can be recovered from a more general nonadiabatic model of quantum dynamics which involves a temporally delocalized force, implying divergence between the reference and the auxiliary system (see Section 12.2).

Microreversibility (see Section 6.4) can be satisfied rigorously if the locations for hops between surfaces are well defined, which is among the premises of the original TSH scheme described in Section 11.1 [310]. The fewest switches algorithm does not restrain the hopping locations to pre-determined places in the potential energy surface landscape. Therefore, the identity of the classical paths in forward and reverse directions, as a prerequisite for microreversibility to hold in mixed quantum-classical approaches, is not guaranteed in this case. One observes, however, that TSH in its various realizations deviates from this principle much less than Ehrenfest dynamics [310].

It has been shown that the fewest switches algorithm obeys the principle of detailed balance [309], in contrast to Ehrenfest dynamics. The ratio of the equilibrium populations of two quantum states, as obtained by fewest switches simulation, are thus simply given by Boltzmann statistics.

As a further extension of the trajectory surface hopping scheme, inclusion of non-vertical hopping between surfaces has been suggested [318]. Besides position-conserving jumps, momentum-conserving and thus horizontal jumps may be admitted. Introducing the classically forbidden non-vertical hops allows for an approximate treatment of quantal single-surface phenomena, and chiefly of tunneling. We will revisit this proposal in the context of the spawning technique in Section 11.3.1.

11.2.1 *Three test cases*

Assessing the fewest switches TSH method as summarized earlier, Tully discusses three one-dimensional two-state model problems for which exact quantal solutions are available [288]. These problems are designed to address prototypical nonadiabatic situations that may arise from dynamic

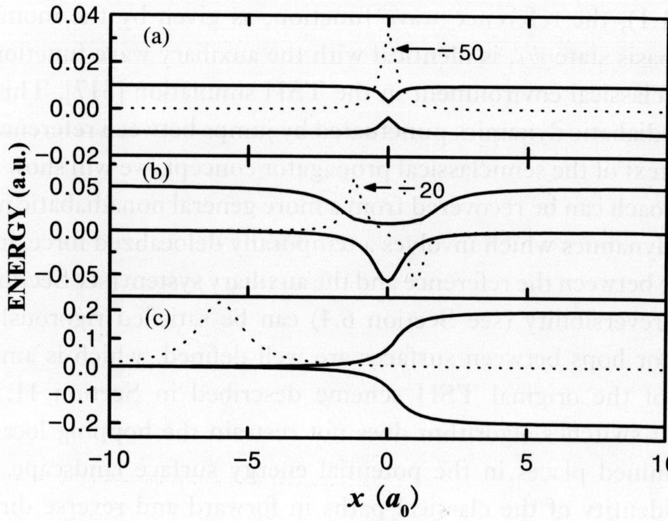

Figure 11.2: Three test situations for the fewest switches TSH algorithm. (a) Simple avoided crossing, (b) dual avoided crossing, (c) extended coupling with reflection. The solid lines represent adiabatic potential energy curves, the dashed lines the nonadiabatic coupling strength A_{12}. (Reprinted with permission from [288]. Copyright (1990) by the American Institute of Physics.)

interactions between molecules. Specifically, they involve (1) a simple avoided crossing, (2) a dual avoided crossing, (3) a region of extended coupling with reflection. This "Tully canon" [319] has served subsequently as a touchstone for various models of nonadiabatic dynamics ([311], [319]–[322]). Adiabatic potential energy curves for these three basic test cases are shown in Figure 11.2(a–c) along with the nonadiabatic coupling strength A_{12}.

The quantal solutions were generated by wave packet propagation using the fast Fourier transform technique (FFT), see Section 15.1.2. The initial wave packet was modeled as a Gaussian, namely

$$\Psi(X,0) = \exp(ikX)\exp\left[-\left(\frac{X}{\sigma}\right)^2\right], \tag{11.20}$$

where σ was chosen about 20 times larger than the inverse of the momentum k. In the following, we comment on each of the three models with emphasis on the comparison between the surface hopping simulation and the quantal results.

(1) *Simple avoided crossing*

For this fundamental case, the diabatic matrix elements are written as

$$H_{\text{el},11}(X) = a[1 - \exp(-bX)], \quad \text{if } X > 0;$$
$$H_{\text{el},11}(X) = -a[1 - \exp(bX)], \quad \text{if } X < 0; \tag{11.21}$$
$$H_{\text{el},22}(X) = -H_{\text{el},11}(X); \tag{11.22}$$
$$H_{\text{el},12}(X) = H_{\text{el},21}(X) = c\exp(-DX^2); \tag{11.23}$$

with appropriate choices for the constants a, b, c, and d. Initially, the system was set to occupy the lower one of the two adiabatic states in the asymptotic region with large negative X. The evolution of the system was recorded until the asymptotic region for positive X was reached. The agreement between the surface hopping approach and the fully quantum mechanical calculation is very satisfactory for the process shown in Figure 11.3, namely reflection in the lower adiabatic state. Displayed is the reflection probability versus the wave number k of the propagated system. Not surprisingly, the coincidence between the results from both approaches is nearly perfect in the realm of high kinetic energies. Moreover, the threshold behavior of the reflection function is essentially reproduced by the surface hopping method. Also, finer details of the interaction are well accounted for. Thus, trajectories trapped for a certain time in the minimum of the upper curve may be transmitted

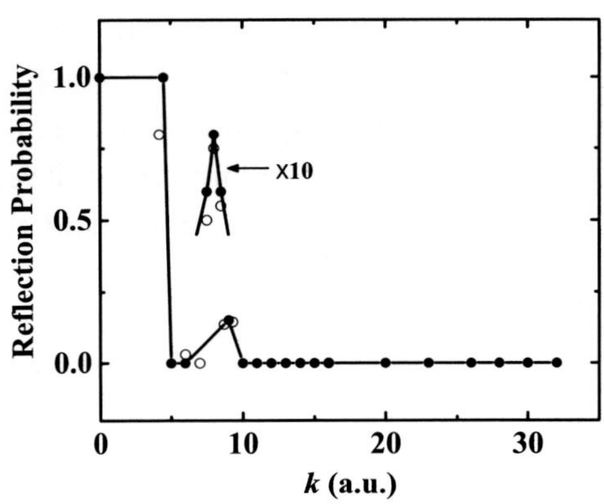

Figure 11.3: Test case 1: Comparison between the fewest switches TSH model (filled circles, joined by solid lines) and wave packet propagation computations (open circles) for the simple avoided crossing displayed in Figure 11.2(a). The reflection probability in the lower state is shown versus the wave number. (Reprinted with permission from [288]. Copyright (1990) by the American Institute of Physics.)

or reflected. The latter possibility produces the clearly discernible peak in Figure 11.3. Obviously, any model that relies on a single optimized trajectory is incapable of capturing this particularity. Similar observations are made for transmission processes in the lower and the upper state.

(2) *Dual avoided crossing*

The diabatic matrix elements for the second, more challenging test case are chosen as

$$H_{\mathrm{el},11}(X) = 0 \tag{11.24}$$

$$H_{\mathrm{el},22}(X) = -a\exp(-bX^2) + E_0 \tag{11.25}$$

$$H_{\mathrm{el},12}(X) = H_{\mathrm{el},21}(X) = c\exp(-dX^2) \tag{11.26}$$

with a, b, c, d, and E_0 as constants. The diabatic potentials $H_{\mathrm{el},11}(X)$ and $H_{\mathrm{el},22}(X)$ cross at two values of X which translate into two positions of avoided crossing in the adiabatic scenario of Figure 2(b). This test case is substantially more demanding than the first, involving a more genuinely quantum mechanical situation. Thus, *Stückelberg oscillations* are to be expected, quantum phenomena that originate from the interference of excitation probabilities. In a TSH simulation, they may be observed when nonadiabatic transitions occur at different times along a trajectory. However, this quantum coherence effect is described by the surface hopping formalism at sufficiently high kinetic energy in full agreement with the quantal calculation, as shown in Figure 11.4 by the example of transmission in the lower state. If the kinetic energy is so low that the excited state is not asymptotically accessible, all possible transitions will lead into the well region of the upper surface where the superposition of the incoming and the reflected component gives rise to strong quantum interference effects. While the TSH includes the oscillation phenomenon *per se*, the results predicted for this low energy regime ($E \leq 0.05$ a.u.) are not in phase with the oscillations obtained from quantum theory.

(3) *Extended coupling with reflection*

For the third test case, the diabatic matrix elements are defined as:

$$H_{\mathrm{el},11}(X) = a \tag{11.27}$$

$$H_{\mathrm{el},22}(X) = -a \tag{11.28}$$

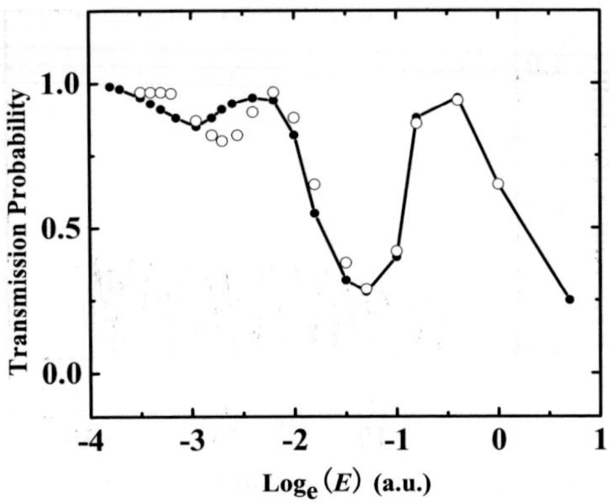

Figure 11.4: Test case 2: The transmission probability in the lower state versus energy in logarithmic representation. The meaning of the symbols is the same as in Figure 11.3. (Reprinted with permission from [288]. Copyright (1990) by the American Institute of Physics.)

$$H_{\mathrm{el},12}(X) = H_{\mathrm{el},21}(X) = b\exp(cX), \qquad \text{if } X < 0; \qquad (11.29)$$

$$H_{\mathrm{el},12}(X) = H_{\mathrm{el},21}(X) = b[2 - \exp(-cX)], \qquad \text{if } X > 0. \qquad (11.30)$$

The constant a is chosen as small in comparison with b and c (6×10^{-4} a.u. versus 0.1 and 0.9 a.u., respectively). This choice determines the characteristic shapes of the adiabatic curves shown in Figure 11.2(c) which are mirror-symmetric with respect to the X-axis. The two branches nearly coincide for $X \ll 0$ and open up in the neighborhood of $X = 0$ to maintain a sizeable near-constant distance from each other for $X \gg 0$. This profile implies a twofold challenge for quantum dynamical theory. First, strong nonadiabatic coupling between the two states stretches over an extended zone instead of being localized at well-defined places, which was among the premises of the surface hopping method in its original form [301]. Second, for sufficiently low kinetic energy, all trajectories on the excited potential curve will be reflected while all those on the ground state curve will be transmitted. Obviously, this behavior cannot be reproduced by any formalism that relies on a single averaged classical path. The surface hopping treatment, in contrast, arrives at the same probabilities of transmission in the lower state as quantum theory. The reflection probabilities on either of the two potential energy curves as displayed in Figure 11.5 for the excited state, however, exhibit rapid oscillations that are not present in the quantal results. These

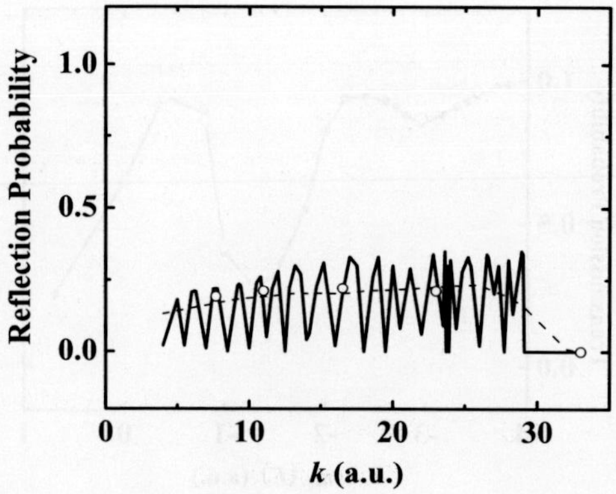

Figure 11.5: Test case 3: The reflection probability in the upper state versus the wave number. The meaning of the symbols is the same as in Figure 11.3. The dashed curve refers to an average of the TSH results over the initial momenta. (Reprinted with permission from [288]. Copyright (1990) by the American Institute of Physics.)

are Stückelberg oscillations which emerge through interference between the forward and the backward traversal of the coupling region. Surprisingly, the quasi-classical theory strongly exaggerates this typical quantum effect. Averaging the TSH results over a Gaussian distribution of initial momenta with a width σ as employed in the wave packet (11.20) eliminates the oscillations and yields full agreement with the quantum approach.

11.2.2 *Complex-valued trajectories*

As a more formal and more rigidly justified version of the TSH methodology we mention the model of Miller and George [323]. This approach is based on an analytical continuation of the potential energy surfaces and thus on complex-valued trajectories. In this fashion, as Miller and George were able to show, the discontinuities of the TSH scheme can be avoided. The trajectories do not hop between real-valued potential energy surfaces, but they move from one to the other smoothly through crossing points in the multidimensional complex plane. As in the TSH procedures summarized before, the nuclear coordinates are propagated by the Hamilton equations, but admission is made for a complex-valued time variable. The semiclassical path integral formalism is used to generate transition probabilities. Qualitatively, this probability can be related to the separation between the complex-valued branch point and the real axis [324]. If this distance is large, so is the action accumulated by the system in reaching the branch point, which reduces the likelihood for the transition. As the hallmark of the Miller–George theory, nonadiabatic dynamics results entirely from the structure of the analytically continued potential energy surfaces, without reliance on the coupling vector $\mathbf{A}_{ij}$ as propagation device. Although conceptually elegant, the Miller–George theory is not easy to apply, which is mostly due to the requirement of a carefully chosen path in complex time. Successful implementations of the scheme for $H^+ + H_2$ are presented in [325, 326]. Admitting complex-valued coordinates and momenta has recently been shown to be a promising way of extending quantum hydrodynamics to nonadiabatic situations (see Section 13.3).

The discussion about the theoretical fundaments of the TSH model and in particular the quest for a consistent treatment of classical nuclear and the quantum mechanical electronic system goes on (e.g. [23]). The following two sections deal with an early proposal to incorporate nuclear coherence effects into the TSH methodology. At its various levels of realization, the resulting formalism bridges the gap between the mixed quantum-classical setting of the original TSH scheme and the fully quantal regime.

11.3 Spawning

While the TSH method, as summarized in the two previous sections, provides a means to eliminate unphysical effects related to the concept of an average nuclear trajectory, it retains a purely classical description of the nuclear system. How could one augment the surface hopping procedure as a quantum-classical hybrid in the minimal, most economic fashion to incorporate nuclear wave packet dynamics? In response to this question, Levine, Martinez, and coworkers ([327]–[329]) made admission for an approximate quantal treatment of the nuclei, representing them by the nuclear coherent state (see Sections 4.4 and 9.2) and thus by use of the local harmonic approximation to the "true" nuclear wave function. The resulting construct may be viewed as the TSH equivalent of the END nuclear wave function (see Chapter 9). On one hand, the nuclei are represented by wave packets. On the other hand, these states preserve a classical flavor since they become sharply localized in the limit of vanishing Gaussian width and, further, the time-dependent average values of the position and momentum variables are constrained to obey classical equations. Although in accordance with quantum mechanics, this approach contains the trajectory model of classical dynamics as a limit and is thus of sufficient flexibility to mediate between a strictly quantum and a semiclassical treatment of a given dynamic problem. The local harmonic approximation therefore appears as an appropriate choice for the nuclear wave function if one tries to design an "optimally classical" simulation procedure for dynamical processes while preserving essential quantal characteristics of the nuclear system, such as Stückelberg oscillations in the vicinity of potential energy surface intersections or avoided crossings.

The *spawning method*, in contrast to surface hopping, involves a sequence of approximations that, in a well-defined limit, leads to the exact quantum solution of the problem at hand. In what follows, we will outline the basic tenets of the spawning concept, which may be characterized as a synthesis of the TSH model and the local harmonic approximation for the nuclear degrees of freedom.

Let the overall molecular wave function be given by the expansion

$$\Psi(\mathbf{R}, \mathbf{r}, t) = \sum_i C_i(t)\psi_i(\mathbf{r}, \mathbf{R})\chi^i(\mathbf{R}, t). \tag{11.31}$$

As usual, the electronic basis functions $\psi_i(\mathbf{r}, \mathbf{R})$ are assumed to be orthonormal. No orthogonality constraint is imposed on the $\chi^i(\mathbf{R}, t)$ which represent the corresponding set of normalized nuclear wave functions, chosen as Gaussian wave packets. The square magnitudes of the complex coefficients C_i are interpreted as populations of the electronic states ψ_i, while the non-diagonal products $C_i^* C_j$ describe the coherences of the reduced density matrix $\boldsymbol{\rho}$ associated with the studied problem. We insert Eq. (11.31) into the TDSE and write the result in matrix form, arriving at

$$i\hbar\frac{\partial}{\partial t}\begin{pmatrix} C_1\,\chi^1 & \\ \cdots & \cdots \\ C_N\,\chi^N & \end{pmatrix} = \begin{pmatrix} H_{1,1} & H_{1,2} & \cdots \\ H_{2,1} & H_{2,2} & \cdots \\ \cdots & \cdots & \cdots \end{pmatrix}\begin{pmatrix} C_1\,\chi^1 & \\ \cdots & \cdots \\ C_N\,\chi^N & \end{pmatrix}, \tag{11.32}$$

where $\mathbf{H}$ is represented in the basis $\{\psi_i\}$. A preliminary step towards the spawning procedure is entirely classical: the energetically available phase space is sampled by running a sufficient number of trajectories on the potential energy surfaces of interest for the studied problem. At this stage, no consideration is given to any coupling between the electronic states. In a second step, non-spreading traveling Gaussian wave packets are assigned to the trajectories. The nuclear wave function χ^i, corresponding to the ith electronic state, is defined as a linear combination of all contributing nuclear wave functions [321]:

$$\chi^i(\mathbf{R};t) = \sum_j b^i_j(t)\chi^i_j(\mathbf{R}, \bar{\mathbf{R}}^i_j(t), \bar{\mathbf{P}}^i_j(t), \bar{\gamma}^i_j(t), \alpha^i_j), \tag{11.33}$$

with j as the index for nuclear wave functions. The bar denotes a mean of the respective quantity, position or momentum, as given by the corresponding classical trajectory. The parameter γ^i_j represents the phase of the nuclear basis state χ^i_j in the absence of any electronic interstate coupling. Conveniently, one uses a Cartesian system with $3N$ coordinates for N nuclei. The explicit form of the traveling Gaussians χ^i_j is then:

$$\chi^i_j(\mathbf{R}, \bar{\mathbf{R}}^i_j(t), \bar{\gamma}^i_j(t), \alpha^i_j) = \left(\frac{2\alpha^i_{\rho_j}}{\pi}\right)^{\frac{1}{4}} \exp(i\bar{\gamma}^i_j)$$
$$\times \prod_{\rho=1}^{3N} \exp[-\alpha^i_{\rho_j}(R_{\rho_j} - \bar{R}^i_{\rho_j})^2 + i\bar{P}^i_{\rho_j}(R_{\rho_j} - \bar{R}^i_{\rho_j})], \tag{11.34}$$

where the index ρ ($\rho = 1, \ldots, 3N$) refers to the nuclear coordinates. Each Gaussian state is thus associated with a classical path defined by the average values $\bar{R}^i_j(t)$ and $\bar{P}^i_j(t)$ which are in turn derived from the potential energy surface of the ith electronic state. As indicated in Eq. (11.33), the width parameter $\alpha^i_{\rho j}$ is not allowed to vary with time. This feature draws a parallel between the presently introduced formalism and the "frozen Gaussian" approximation by Heller which we will further explore in Section 12.8. As in the latter scheme, the local harmonic approximation constrains the time-dependent variables according to

$$\frac{\partial \bar{R}^i_{\rho_j}(t)}{\partial t} = \frac{\bar{P}^i_{\rho_j}(t)}{M_\rho}, \tag{11.35}$$

$$\frac{\partial \bar{P}^i_{\rho_j}(t)}{\partial t} = -\left[\frac{\partial V_{ii}(\mathbf{R})}{\partial R_{\rho_j}}\right]_{\bar{R}^i_{\rho_j}(t)}, \tag{11.36}$$

$$\frac{\partial \bar{\gamma}^i_j(t)}{\partial t} = \sum_{\rho=1}^{3N} \frac{1}{2M_\rho}[(\bar{P}^i_{\rho_j}(t))^2 - 2\alpha^i_{\rho j}] - V_{ii}[\bar{\mathbf{R}}^i_j(t)], \tag{11.37}$$

where V_{ij} denotes the potential energy matrix elements in the basis of electronic states $\{\psi_i\}$, with reference to Eq. (11.31). That the center of the Gaussian wave packet evolves along the classical trajectory is a rigorous result for harmonic potentials, as implied by the Ehrenfest theorem (e.g. [58]). By virtue of the relations (11.35)–(11.37), the method reduces to the frozen Gaussian propagation procedure [330] if restricted to a single potential energy surface. It may therefore be considered as a direct generalization of the latter approach. We note that the coefficients $b^i_j(t)$ in Eq. (11.33) and the phase factors $\exp(i\bar{\gamma}^i_j)$ are both complex-valued variables, depending on the same pair of indices. For this reason, the phase factors may be viewed as redundant, and thus they can be fixed arbitrarily. While Eq. (11.37) achieves this goal by means of the local harmonic approximation, alternative schemes rooted in classical dynamics may be employed as well to remove the redundancy.

From Eqs. (11.31) and (11.33), one realizes that the time-dependent factor that determines the probability of finding the nuclear system in the basis state χ^i_j is given by $D^i_j(t) \equiv C_i(t)b^i_j(t)$. Substituting for χ_i in Eqs. (11.31) and (11.32) the explicit form Eq. (11.33), we obtain the differential equation that governs the time behavior of D^i_j:

$$\frac{dD^i_j}{dt} = \frac{dC_i(t)b^i_j(t)}{dt}$$

$$= -\frac{i}{\hbar}\sum_{k,l}(S_{ii}{}^{-1})_{j,k}\left\{\left(H_{ii} - i\hbar\frac{\vec{\partial}}{\partial t}S_{ii}\right)_{k,l} D^i_l + \sum_{i'\neq i'}(H_{ii'})_{k,l}D^{i'}_j\right\}. \tag{11.38}$$

Here $(H_{ii'})_{k,l}$ stands for $\langle\chi^i_k|H_{ii'}|\chi^{i'}_l\rangle$, and $(S_{ii})_{j,k}$ for $\langle\chi^i_j|\chi^i_k\rangle$. The symbol $\frac{\vec{\partial}}{\partial t}$ denotes the right-acting derivative, and thus $(\frac{\vec{\partial}}{\partial t})_{k,l}S_{ii} = \langle\chi^i_k|\frac{\partial\chi^i_l}{\partial t}\rangle$. The expressions on the right-hand side of Eq. (11.38) are arranged to separate the intrastate coupling, characterized by index pairs ii, from the interstate coupling terms, labeled ii'. Obviously, only the former group is of relevance as long as the propagation proceeds effectively on a single potential energy surface. The initial conditions for the expansion coefficients D^i_j have to be chosen in accordance with the initial molecular wave function $\Psi(t = 0)$. Specifically, one declares a particular electronic state ψ_i to be occupied at $t = 0$, such that $|C_{i'}(t = 0)|^2 = \delta_{ii'}$. Further, the initial position and momentum values $\bar{\mathbf{R}}^i_j(t)$ and $\bar{\mathbf{P}}^i_j(t)$ on which the nuclear basis states χ^i_j are centered may be obtained from a Monte Carlo sampling procedure. In preparation for this step, the Wigner transform of the initial molecular wave function (see Section 4.3) has been employed [321]. This defines the mean positions and momenta of each Gaussian function. The Gaussian widths $\alpha^i_{\rho j}$, in contrast, enter as arbitrary parameters, and the initial values of the nuclear phases $\bar{\gamma}^i_j$ are set to zero which, due to their redundancy with the trajectory amplitudes, does not limit the generality of the procedure. Once the values of the nuclear basis functions χ^i_j at $t = 0$ have been determined, the corresponding probability amplitudes b^i_j follow from

$$b^i_j(0) = \sum_k (S^{-1}_{ii})_{j,k} \langle \chi^i_j(t = 0) | \Psi(t = 0) \rangle. \tag{11.39}$$

Since the χ^i_j are not constrained to be orthogonal, they might develop a high degree of linear dependence in the course of the propagation, making it difficult to evaluate the inverted overlap matrix S^{-1}_{ii} in Eq. (11.38). This situation, however, can in general be remedied by use of the singular value decomposition technique [331].

Exercise 11.2. From Eqs. (11.31)–(11.33), derive the differential equation (11.38) for the expansion coefficients D^i_j.

The central idea of the multiple spawning method consists in leaving the number of the nuclear trajectories and basis functions variable during the simulation of the dynamic process, and thus operating with an *adaptive basis*. The major advantage of this methodology is that at any given moment, a flexible basis is adjusted to the size of the relevant configuration space segment while a fixed basis must be designed to access all of this space at any time.

As mentioned before, the multiple spawning procedure naturally begins by selecting a potential energy surface with index i. As the electronic states

$\psi_{i'}$ with $i' \neq i$ are unoccupied at this juncture, the trajectories assigned to them are unused. This changes, however, as the system encounters a region of pronounced nonadiabatic interaction. As these places are reached, additional trajectories associated with an index $i' \neq i$ may be *spawned*, i.e. generated on the previously empty potential energy surface. The spawning technique thus builds on the hopping methodologies and extends the latter as it operates with classical trajectories *dressed with frozen Gaussians*. Once created, these functions are transported along the trajectories of the newly populated potential energy surface. From there, back-spawning into the original electronic state is possible, as well as spawning into a third electronic state. In this manner, the size of the nuclear basis that actively participates in the propagation can be kept minimal and at each stage of the computation decidedly smaller than the total number of available basis functions.

We point out that the mechanism of population transfer between electronic states in the spawning model differs from that operative in the TSH scheme where it is realized by hops between surfaces. Spawning, in contrast, refers only to the creation of basis states without implying that these states are populated. It is the evolution of the time-dependent coefficients D_j^i, as described by Eq. (11.38), that decides if, and to what extent, the spawned wave functions become occupied.

When should spawning events take place? The question of a spawning criterion is of central importance for the application of the method. In analogy to the TSH method, the local strength of the effective interstate coupling must decide if spawning occurs at any given stage of the simulation. The appearance of the coupling term $\Lambda_{ii'}^{\text{eff}}$ varies with the frame chosen to represent the molecular wave function, diabatic or adiabatic. More specifically, the following terms have been suggested as measures for the nonadiabatic interaction strength [332]:

$$\Lambda_{ii'}^{\text{eff}}(\mathbf{R}) = \left| \frac{V_{ii'}^{\text{D}}}{V_{ii}^{\text{D}} - V_{i'i'}^{\text{D}}} \right| \quad \text{(diabatic case)}, \tag{11.40}$$

$$\Lambda_{ii'}^{\text{eff}}(\mathbf{R}) = \left| \mathbf{A}_{ii'} \cdot \frac{d\mathbf{R}}{dt} \right| \quad \text{(adiabatic case)}, \tag{11.41}$$

with $\mathbf{A}_{ii'}$ as the vector of nonadiabatic coupling. At some place $\mathbf{R}$, the necessary spawning condition is that $\Lambda_{ii'}^{\text{eff}}(\mathbf{R})$ exceeds a certain threshold value which is assigned prior to the simulation. If the electronic state with index i' is unpopulated at $\mathbf{R}$, the sufficient spawning condition is met [198]. The time during which $\Lambda_{ii'}^{\mathrm{eff}}(\mathbf{R})$ is above the spawning threshold defines the time it takes a certain trajectory – and, by the same token, its associated basis function – to traverse a region of nonadiabaticity. This time is subdivided into N_{S} intervals. Since one basis function is spawned within each of these intervals, N_{S} gives the number of spawned basis functions in a nonadiabatic event. Convergence with respect to N_{S} ensures that the method has a well-defined limit. The procedure has to converge not only with respect to the number of spawned Gaussians but also with respect to the spawning threshold. The latter criterion is to be understood as the condition that further lowering of the threshold is of negligible impact on the results of the calculation at hand.

11.3.1 *Applications to model problems*

There is no unique spawning methodology. However, guidelines for implementing this procedure have been formulated and followed in actual applications of the method (e.g. [321]). In particular, rules for the optimum choice of a spawned function $\chi_{j'}^{i'}$, given a parent function χ_j^i, in terms of the phase space parameters $\bar{\mathbf{P}}_{\rho_j}^i$ and $\bar{\mathbf{R}}_{\rho_j}^i$, are required. Among these is the constraint that the coupling between the parent and the spawned function should be maximum in the phase space region where nonadiabatic transitions are likely to take place. The conservation of classical energy with respect to the phase space variables around which the parent and the spawned functions are centered, i.e. the positions and momenta of the nuclear trajectories, is to be added to this requirement. The resulting set of *spawning conditions*, to complement the nonadiabaticity criterion, is

$$\frac{\partial|\langle\chi_j^i|H_{ii'}|\chi_{j'}^{i'}\rangle|}{\partial\bar{\mathbf{P}}_{j'}^i} = 0, \tag{11.42}$$

$$\frac{\partial|\langle\chi_j^i|H_{ii'}|\chi_{j'}^{i'}\rangle|}{\partial\bar{\mathbf{R}}_{j'}^i} = 0, \tag{11.43}$$

$$\sum_\rho \frac{(\bar{P}_{\rho_j}^i)^2}{2M_\rho} + V_{ii}(\bar{\mathbf{R}}_j^i) = \sum_\rho \frac{(\bar{P}_{\rho_j}^{i'})^2}{2M_\rho} + V_{i'i'}(\bar{\mathbf{R}}_j^{i'}). \tag{11.44}$$

Applying conditions (11.42)–(11.44) in conjunction with solving the propagation equation (11.38) encompasses the essence of the *full multiple spawning* (FMS) method. A graphic illustration of the spawning events central to this method is given in Figure 11.6 for the basic case of a collinear $A + BC \rightarrow AB + C$ reaction [333]. The figure presents the dynamic evolution of this system in four consecutive stages.

Diabatic potential energy surfaces are displayed as functions of the Jacobi coordinates r and ρ (see the legend of Figure 11.6) for the initially occupied product channel $A + BC$ as well as the initially unpopulated $AB + C$ reactant state. As time passes, spawning events create nuclear basis functions on the diabatic surface that corresponds to the $AB + C$ configuration, involving population transfer between the two coupled states. As suggested by the second pair of panels, the initial overlap between the spawned and the parent wave functions is high. Subsequently, the different nature of the two interacting electronic states gives rise to very different dynamic patterns on the two potential energy surfaces, corresponding to the two competing disintegration modes of the three-atom complex. Comparison of the resulting populations in both channels yields the branching ratio for the considered reaction. This quantity has been obtained by the multiple spawning procedure for a two-dimensional model problem [321] and found accurate by the standard of fully quantal calculations.

We point out that the set of conditions (11.42)–(11.44) does not encode an unambiguous prescription for spawning but allows in general for a variety of solutions. In the following, we will discuss different possible spawning mechanisms by considering the first challenge of the "Tully canon" (see Figure 11.2(a)), namely the avoided crossing between two one-dimensional diabatic potential curves shown in Figure 11.7(b). The solid arrows refer to two different trajectories entering the interaction region with energies E_1 and E_2. The phase space representations of these trajectories shown in Figures 11.7(a) and 11.7(c) demonstrate that the set of possible spawning events is energy-dependent. For E_1 (Figure 11.7(a)) which is above the adiabatic barrier of the lower state, these events may be position- or momentum-conserving, or may realize a combination of these basic types. This is understood by inspection of the occupied and the virtual trajectory, which share the same energy but move under the influence of different diabatic potentials. These two trajectories are found to be spatially separated as one considers

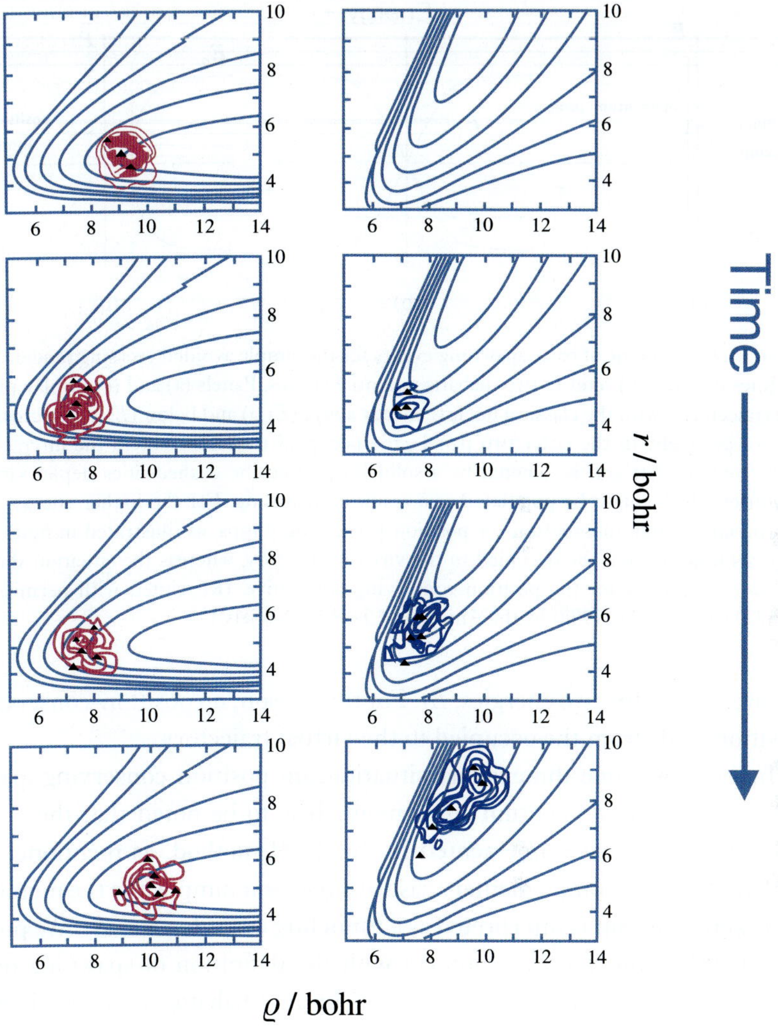

Figure 11.6: Treatment of a collinear reaction of the form A + BC → AB + C by the multiple spawning technique. The four pairs of panels show four stages of this collision in temporal succession. Diabatic potential energy surfaces are indicated for two states, where the images on the left refer to the A + BC, and those on the right to the AB + C branch. Jacobi coordinates r and ρ are used, denoting the B–C distance and the distance between the location of A and the BC center of mass, respectively. Representations of the nuclear wave functions are superposed on the equipotential lines. Each one of the black triangles stands for the center of one Gaussian basis function. While the diabatic state on the right initially does not carry any nuclear basis functions, spawning events create these functions with the elapse of time. (Reprinted with permission from [333]. Copyright (1990) by the American Chemical Society.)

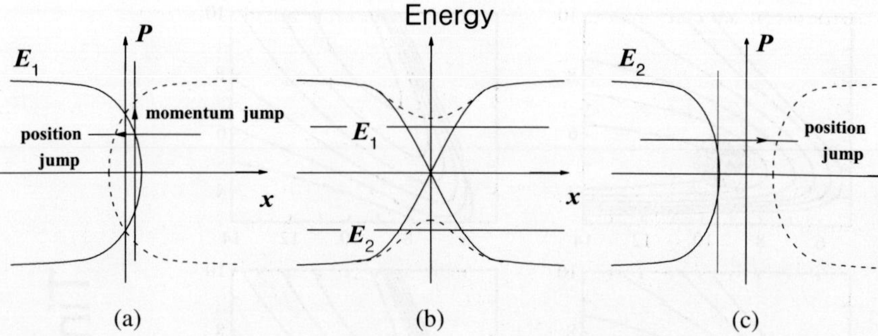

Figure 11.7: Scheme of basic spawning events for the simple avoided crossing model. The solid lines in panel (b) refer to two diabatic potential curves. Panels (a) and (c) display phase space trajectories with the classical nuclear energy above (E_1,a) and below (E_2,c) the crossing point, respectively. In the latter two panels, the occupied trajectory enters the interaction region from the left and is denoted by a solid line, while the dashed lines depict virtual trajectories which might be populated as a result of spawning. For the higher energy, this can proceed through momentum- or position-preserving jumps, as illustrated in figure (a) by arrows leading from the occupied to the virtual trajectory, whereas the situation shown in (c) does not allow for the position-conserving alternative. (Reprinted with permission from [321]. Copyright (1998) by the American Chemical Society.)

the lower one of the two energies, E_2. For this reason, no position-conserving transition leads from the occupied to the virtual trajectory.

In the chosen one-dimensional situation, any position-conserving spawn leads to a unique momentum adjustment. It is to be noted that the latter is identical with that implemented in the TSH method, as mentioned in Section 11.2. The complementary case of a position jump while the momentum is kept constant is interpreted as a tunneling event for $E = E_2$. We point out that this transition can conform with the definition of spawning only if the diabatic representation is utilized. In an adiabatic frame, evidently, tunneling processes associated with trajectories below the adiabatic barrier cannot give rise to the population of unoccupied potential energy surfaces but redistribute the population of the surface of reference. However, if the propagation procedure includes intrastate coupling, as contained in the matrix elements $(H_{ii})_{k,l}$ in Eq. (11.38), these processes can still be accounted for.

Intrastate interference effects, on the other hand, may be considerably reduced by averaging over the initial conditions that refer to a selected asymptotically occupied electronic state. This reasoning motivates an approximation to the equation of motion (11.38), consisting in the neglect of non-diagonal intrastate terms. This yields the simplified relation

$$\frac{dD_j^i}{dt} = -\frac{i}{\hbar}\left[\left(H_{ii} - i\hbar\frac{\vec{\partial}}{\partial t}S_{ii}\right)_{j,j} D_j^i\right] - \frac{i}{\hbar}\sum_{i'\neq i}\sum_k (H_{ii'})_{j,k}D_k^{i'}. \quad (11.45)$$

The nuclear basis functions assigned to an electronic state with label i evolve now independently of each other. This circumvents the need for inversion of the overlap matrix S which may be numerically cumbersome. The reduced form (11.45) represents the *multiple independent spawning* (MIS) approximation [334] to the FMS approach (11.38) which implements the full time evolution equations.

11.4 Case Study: The Dynamics of Na*-Quenching by Collision with Hydrogen Molecules

As an example of the application of the multiple spawning procedure, we describe here the FMS treatment of the electronic quenching in the collision of excited sodium atoms by H_2 molecules [198], reexamining the deexcitation process $Na(^2P) + H_2(X^1\Sigma_g^+, v, j) \rightarrow Na(^2S) + H_2(X^1\Sigma_g^+, v', j')$ that has been discussed earlier in the framework of a time-independent variational approach (see Section 7.4).

The left-hand side of the reaction formula involves the first excited electronic state of the triatomic system combined of Na and H_2, the product side refers to the ground state. Again, we use Jacobi coordinates to describe the possible geometric configurations of the three atoms. The ground state of the Na-H_2 system is repulsive throughout, whereas the first excited state is bonding, as seen from Figure 11.8. In C_{2v} symmetry conditions, a conical intersection emerges, connecting the ground state with the first excited state potential energy surface.

The electronic ground and the first excited states along the seam realize different irreducible representations of the C_{2v} group, with labels $^2A_1$ and $^2B_2$, respectively. In the former state, the Na valence electron occupies the 3s orbital; in the latter, it is located in the 3p orbital that aligns with the H_2 axis. The diabatic coupling between these states vanishes on the seam which, by Eq. (2.63), is preserved as one changes from the diabatic to the adiabatic representation. This observation is important for understanding the Na* quenching dynamics through collision with an H_2 molecule. By Eqs. (11.40)

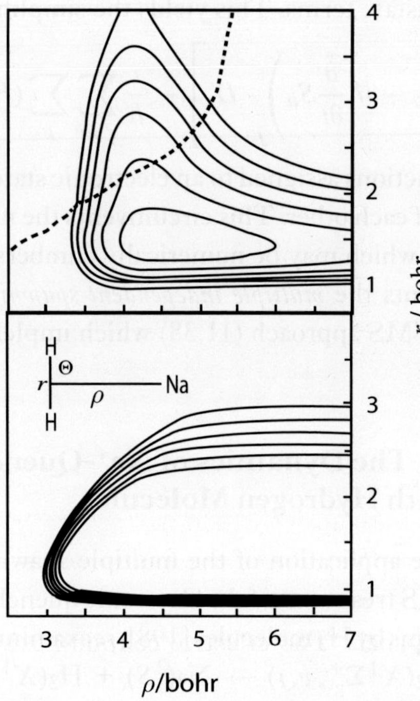

Figure 11.8: The two lowest diabatic potential energy surfaces for Na-H_2 in C_{2v} symmetry, represented as functions of the Jacobi coordinates r and ρ. The contour lines displayed in both panels are regularly spaced, spanning the interval from 1.75 to 3.00 eV. While the ground state (lower panel) is repulsive, the excited state (upper panel) exhibits a shallow minimum. The dashed line shown in the upper panel refers to the crossing seam between the ground and the first excited electronic state in C_{2v} conditions. (Reprinted with permission from [198]. Copyright (1997) by the American Chemical Society.)

and (11.41), the coupling between the two states is expected to be strong in the vicinity of the intersection seam, as the energy difference between the ground and the first excited state is small in this region. However, at the seam location, no transition takes place, since the diabatic coupling element for the two states reduces to zero at these positions.

The dynamic simulation reported in [198] was performed in a landscape defined by the four diabatic potential energy surfaces indicated in Figure 7.4. These were obtained using empirical functional forms parametrized by values from *ab initio* computations. The calculation involved nine Cartesian nuclear coordinates and a relative kinetic energy of 0.039 eV which is in the thermal region. The H_2 molecule was prepared as a coherent rovibrational state at the start of the simulation, separated sufficiently far from the Na* projectile. The excited $3p$ state of Na* was oriented along a chosen axis and decomposed into the three orthogonal space-fixed $3p$ orbitals. With these premises, FMS propagation was performed for a large range of impact parameters.

Figure 11.9 combines some characteristic observations. In particular, the lowest panel monitors the gradual buildup of the electronic ground state population. This process is correlated in the top panel with the time behavior of the H_2 bond length expectation value as well as that of the bending angle Θ. In the middle, the effective nonadiabatic coupling Λ_{eff} between the lowest excited state and the ground state is shown according to formula (11.40) for the diabatic case.

From the time dependence of the quantities assembled in Figure 11.9, one immediately concludes that the population transfer from $\text{Na}(^2P) + H_2$ to $\text{Na}(^2S) + H_2$ is a slow process by the standard of an average H_2 vibrational period. Peculiar features of the quenching dynamics can be extracted from the upper two panels. The effective coupling exhibits an oscillatory time dependence. In particular, the coupling vanishes whenever the angle $\langle\Theta\rangle$ crosses the $\frac{\pi}{2}$-line. This demonstrates by direct dynamic simulation that nonadiabatic transition requires a deviation of the considered system from C_{2v} symmetry. The periodic alternation of the H_2 bond length is another factor that determines the population transfer, since it affects the energy difference between the interacting states. In general, bond length elongation of H_2 moves the system towards the conical intersection region and thus decreases the energy gap between the states. The interplay of the two geometric parameters $\langle r\rangle$ and $\langle\Theta\rangle$ results in multiple traversals of the conical intersection neighborhood, leading to the step-like profile of the ground state population growth shown in the lowest panel of Figure 11.9. The simulation thus provides a detailed model of the Na*-quenching mechanism. Rather than a single event, a complex sequence of interactions between the participating electronic states, involving fluctuations of the nonadiabatic coupling between them over an extended period, leads eventually to the decay of the $\text{Na}^* + H_2$ exciplex.

In the presently described application, the FMS method is used to analyze in time-resolved detail the dynamic interaction of the two strongly differing electronic states shown in Figure 11.8. In this manner, various basic

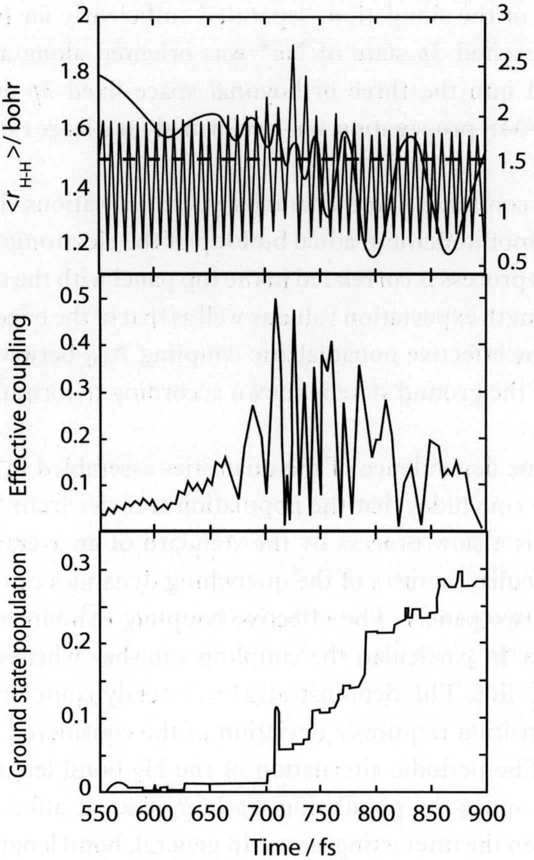

Figure 11.9: Quenching dynamics of the lowest electronic excited state of Na* by collision with an H_2 molecule at an impact parameter of $6a_0$. Further details of the depicted FMS simulation are given in the text. Upper panel: The expectation values of the H_2 bond length r (thin, strongly oscillating line) and the bending angle Θ (thick, slowly oscillating line) as a function of time. Middle panel: The effective coupling between the two electronic states. Lower panel: The ground state population versus time. The dashed horizontal line in the upper panel refers to $\Theta = \frac{\pi}{2}$, corresponding to the C_{2v} limit of the Na-H_2 system. (Reprinted with permission from [198]. Copyright (1997) by the American Chemical Society.)

wave-function-related properties of a quantum dynamical system may be obtained, such as the time behavior of ground or excited state delocalization or vibrational and rotational excitation.

Various dynamic effects as described in the foregoing paragraphs will be found to weaken or even vanish if the results are averaged over many possible initial geometries, defined by the polar and azimuthal angles of the H_2 molecule orientation. This averaging procedure, on the other hand, allows for comparison between experimental [335] and calculated quenching cross sections for the Na^*-H_2 reaction. Both are seen to be in the order of 20-25 Å^2 [198], making the prediction of the spawning model compatible with the experimental finding.

Simulations employing the spawning method have also been reported for various organic reactions. Thus, the *cis-trans* isomerization and fragmentation dynamics of ethylene following photoexcitation have been investigated besides electronic absorption and resonance Raman spectra [336, 337]. As a further phenomenon related to a photoinduced *cis-trans* isomerization process, the non-radiative relaxation of the all-*trans* retinal protonated Schiff base through electronic funnels has been subjected to spawning analysis [338]. In addition, the time behavior of the electrocyclic ring-opening reaction of cyclobutene to 1,3-butadiene has been modeled by spawning computations describing the interplay of the two electronic states that dominate this process [333, 339]. Recently, a three-state intersection was discovered in malonaldehyde by use of a spawning technique ([340]–[342]), and the conditions for accessing such a point were discussed. This analysis confirms the expectation that relaxation through three-state intersections is likely to be preceded by passage through a two-state intersection.

11.5 Comparison with Other Methods

The spawning procedure and the various trajectory hopping schemes realize very similar methodological principles. In both approaches, the notion of a nonadiabatic coupling threshold that decides if possible discontinuous changes of some of the propagated quantities take place or not plays a pivotal role. However, the spawning formalism allows for a wave function representation of the nuclear system by traveling Gaussians and thus introduces a semiclassical description of the nuclear degrees of freedom which are treated as entirely classical parameters by the hopping approach as presented in the first two sections of this chapter. Genuine quantum phenomena involving the nuclear subsystem, such as intrastate tunneling [332], are thus within the reach of the spawning method.

The spawning model has been demonstrated to brave the third and most demanding test case of the "Tully canon", involving an extended region of

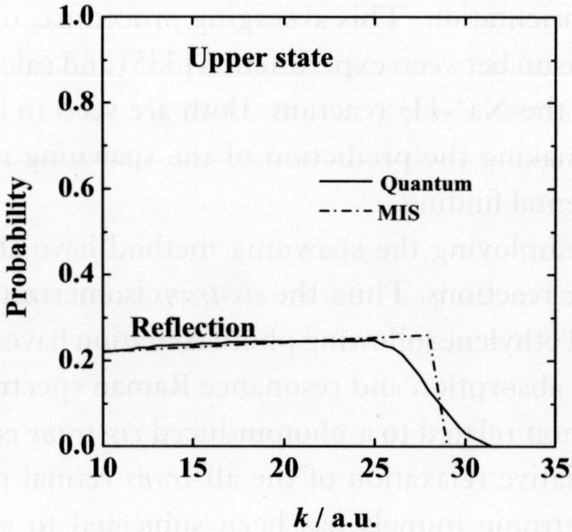

Figure 11.10: Upper state reflection probability for the situation depicted in Figure 11.2c (extended coupling with reflection versus wave number). Shown are the results of the fully quantal computation as compared with those of the multiple independent spawning (MIS) model. (Reprinted with permission from [320]. Copyright (1995) by the American Chemical Society.)

diabatic coupling with reflection (see Figure 11.2(c)), with better success than the fewest switches TSH formalism [320]. The restriction of the latter to a single trajectory, along which the phase evolution of the system is recorded, leads to an exaggeration of quantum interference effects, as seen in Figure 11.5. Introducing nuclear basis functions and applying the spawning technique remedies this deficiency. The results of the multiple independent spawning model for the configuration shown in Figure 11.2(c) are free of artificial quantal oscillations. This is demonstrated in Figure 11.10.

A detailed comparison between hopping and spawning for triatomic model systems [343] confirms the expected similarity of the results generated by both methods. In these studies, distributions of hopping and spawning events were recorded as functions of characteristic geometric and energetic parameters of the systems, and in general, no pronounced deviation between the two procedures was found. As a systematic difference, however, it is noted that the density distributions yielded by the FMS method tend to be more narrowly centered around small energy gaps than those obtained by the TSH scheme. This trend may be related to the different nonadiabatic coupling parameters utilized by the two methods. Thus, the explicit dependence of $\Lambda_{ii'}^{\mathrm{eff}}$ on $V_{ii}^{\mathrm{D}} - V_{i'i'}^{\mathrm{D}}$ in the diabatic representation (see Eq. (11.40)) favors places of small energy gaps as spawning sites, while no such dependence is built into the fewest switches hopping algorithm (see Section 11.2).

As a distinguishing feature, the spawning procedure admits a flexible nuclear basis set size. This gives it an advantage over comparable hopping methods. The number of basis functions may be reduced in regions of the potential energy landscape where single-surface propagation is adequate, and it may be appropriately extended in zones of complex nonadiabatic interaction. In these respects, spawning techniques are able to meet the demands of both precision and practicability.

Further, the spawning methodology involves a hierarchy of approaches. By systematic augmentation of the nuclear basis set and lowering of the spawning threshold, it is, in principle, possible to arrive at the solution of the TDSE, although for a physical system, this limit is hard to attain with current computational means. Still, the spawning method contains a prescription for the assessment and systematic improvement of its results. The quality of predictions obtained by the hopping procedure, in contrast, can only be gauged by comparison with experiment or another method of higher accuracy.

places of small energy gaps as spawning sites, while no such dependence is built into the fewest switches hopping algorithm (see Section 11.2).

As a distinguishing feature, the spawning procedure admits a flexible nuclear basis set size. This gives it an advantage over comparable hopping methods. The number of basis functions may be reduced in regions of the potential energy landscape where single surface propagation is adequate, and it may be appropriately extended in zones of complex nonadiabatic interaction. In these respects, spawning techniques are able to meet the demands of both precision and practicability.

Further, the spawning methodology involves a hierarchy of approaches. By systematic augmentation of the nuclear basis set and lowering of the spawning threshold, it is, in principle, possible to arrive at the solution of the TDSE, although for a physical system, this limit is hard to attain with current computational means. Still, the spawning method contains a prescription for the assessment and systematic improvement of its results. The quality of predictions obtained by the hopping procedure, in contrast, can only be gauged by comparison with experiment or another method of higher accuracy.

Chapter 12

Semiclassical Propagator Techniques

The present chapter might be viewed as a synthesis of the two preceding ones, as it lays a foundation for both the classical electron analog and the hopping methodology. Both techniques emerge from the *semiclassical propagator concept* which is reviewed in this chapter.

The classical or mixed quantum-classical approximations to nonadiabatic quantum dynamics are capable of achieving remarkable qualitative and, in some cases, even quantitative agreement with full quantal calculations. For a more complete understanding of these methods, as well as for their improvement, it is desirable to clarify their connection with quantum theory. A systematic search for a dynamic scheme that grows out of a rigorous quantum approach by a sequence of well-motivated and well-defined semiclassical approximation steps will naturally include the propagator, or Feynman representation of quantum mechanics, as outlined in Section 4.1. In the following sections we will refer to recent attempts at making this methodology productive for the purposes of nonadiabatic molecular dynamics.

As mentioned earlier, this chapter aims specifically at providing a fundament for the semiclassical ideas and methods treated in the two preceding chapters. In particular, the surface hopping procedure will receive a more rigorous motivation than given at the beginning of the previous chapter, where it was rationalized in a pragmatic fashion. Also, we will extend the program underlying the classical electron analog, building quantum mechanical tools from classical elements, by taking it to the level of propagator theory, and we will give a rigorous justification for the CEA method.

The central character of this chapter is the Van Vleck–Gutzwiller propagator (4.50). As described in Section 4.1, this agent, while emerging from the consistently quantum mechanical path integral formalism, is entirely composed of classical entities, namely the monodromy matrix and the action functional, with the Maslov index reflecting instances of conjugate points on the respective classical path. Section 12.1 tackles the question of how to include both electronic and nuclear degrees of freedom in the frame of the Van Vleck–Gutzwiller model, where trajectory surface hopping is obtained as a special case of the resulting semiclassical propagator methodology (Section 12.2). The initial value representation (IVR), as introduced in Section 12.3, rids this approach of the need for nonlinear root search and therefore removes one of the major stumbling blocks that hamper its practical implementation. In the subsequent two sections, we discuss various ways of mapping quantum mechanical onto classical observables in order to arrive at a classical representation of electronic degrees of freedom within propagator theory, providing a rationale for the CEA. An application of these ideas to a conical intersection problem is highlighted in Section 12.6. We conclude this chapter by focusing on some techniques to remedy the practical difficulties marring the realization of semiclassical propagator strategies (Section 12.7) and on the concept of cellular dynamics that deals with some of these difficulties by capitalizing on the properties of Gaussian functions (Section 12.8).

12.1 The Path Integral Approach to Molecular Dynamics

In a pioneering move, Pechukas [344, 345] formulated a semiclassical model of molecular quantum dynamics in the language of Feynman path integrals. Accordingly, the wave function of the molecule is written in the usual propagator form as

$$\Psi(\mathbf{x}_t, \mathbf{X}_t, t) = \int d\mathbf{x}_0 d\mathbf{X}_0\, K(\mathbf{x}_t, \mathbf{X}_t, t|\mathbf{x}_0, \mathbf{X}_0, t_0)\Psi(\mathbf{x}_0, \mathbf{X}_0, t_0) \tag{12.1}$$

where $\mathbf{x}_0$ ($\mathbf{X}_0$) refer to the electronic (nuclear) coordinates in the state realized at the initial time t_0 and at some later time t, and K denotes the system propagator in coordinate representation,

$$\begin{aligned} K(\mathbf{x}_t, \mathbf{X}_t, t|\mathbf{x}_0, \mathbf{X}_0, t_0) &\equiv \left\langle \mathbf{x}_t, \mathbf{X}_t \left| \exp\left(-\frac{i}{\hbar}\hat{H}t'\right) \right| \mathbf{x}_0, \mathbf{X}_0 \right\rangle \\ &= \int_{\mathbf{X}_0}^{\mathbf{X}_t} D\mathbf{X} \int_{\mathbf{x}_0}^{\mathbf{x}_t} D\mathbf{x} \exp\left(\frac{i}{\hbar} S[\mathbf{x}(t'), \mathbf{X}(t'), t']\right) \end{aligned} \tag{12.2}$$

with $t' \equiv t - t_0$. While the spatial variables associated with the quantum propagator discussed in Section 4.1 were left without concrete physical interpretation, we now make admission for two disparate sets of coordinates, corresponding to the electronic and the nuclear subsystem.

Introducing an instantaneous nuclear basis $\chi_\nu(\mathbf{X}, t)$ as well as its electronic counterpart $\psi^\mu(\mathbf{x}, t)$, we express the molecular wave function in diabatic representation as

$$\Psi(\mathbf{x}, \mathbf{X}, t) = \sum_\nu \chi_u(\mathbf{X}, t)\psi^u(\mathbf{x}, t). \tag{12.3}$$

Also, a *reduced propagator* $K_{\beta\alpha}$ that advances the nuclear system from $(\mathbf{X}_0, t_0)$ to $(\mathbf{X}_t, t)$ [344, 346] is defined by

$$\chi_\beta(\mathbf{X}_t, t) = \sum_\alpha \int d\mathbf{X}_0 K_{\beta\alpha}(\mathbf{X}_\beta, t|\mathbf{X}_0, t_0)\chi_\alpha(\mathbf{X}_0, t_0). \tag{12.4}$$

Using this relation, one may compute the probability $P_\beta(t)$ for finding the system in the electronic basis state β at some time t as

$$P_\beta(t) = \langle\chi_\beta(t)|\chi_\beta(t)\rangle = \int d\mathbf{X}\, |\chi_\beta(\mathbf{X}, t)|^2 \tag{12.5}$$

where the normalization condition requires that $\sum_\beta P_\beta(t) = 1$. The reduced propagator $K_{\beta\alpha}$ is given explicitly by

$$K_{\beta\alpha}(\mathbf{X}_\beta, t|\mathbf{X}_0, t_0) = \int_{\mathbf{X}_0}^{\mathbf{X}_t} D\mathbf{X} \exp\left(\frac{i}{\hbar}S_{\text{cl}}[\mathbf{X}(t)]\right) T_{\beta\alpha}[\mathbf{X}(t)], \tag{12.6}$$

as demonstrated in [344, 345]. Formula (12.6) involves the classical action integral S_{cl} for a free particle (compare with Eq. (4.21a)),

$$S_{\text{cl}}[\mathbf{X}(t)] = \int_{t_1}^{t_2} dt \frac{M}{2}\left(\frac{\partial\mathbf{X}}{\partial t}\right)^2, \tag{12.7}$$

as well as the electronic transition amplitude $T_{\beta\alpha}$,

$$T_{\beta\alpha}[\mathbf{X}(t)] = \left\langle \psi^\beta(t) \left| \exp\left(-\frac{i}{\hbar}\int_{t_0}^{t} dt\, \hat{H}_{\text{el}}\left[\mathbf{x}, \mathbf{X}(t)\right]\right) \right| \psi^\alpha(t_0) \right\rangle. \tag{12.8}$$

The scalar product (12.8) describes the overlap between an instantaneous electronic basis state $\psi^\beta(t)$ with a mixed state $\psi^\alpha(t, t_0)$ that is generated by advancing an original basis state $\psi^\alpha(t_0)$ through the time interval $t - t_0$.

In this process, use is made of a time-dependent electronic Hamiltonian $\hat{H}_{\rm el}[\mathbf{x}, \mathbf{X}(t)]$.

So far, the development has been quantum mechanically exact. At this point, however, the notion of a definitive classical trajectory may be invoked by using a stationary phase approximation, in analogy to our treatment of the propagator (4.26) in Section 4.1.2. Only paths that contribute maximally are retained in the expression (12.6) for the reduced propagator, i.e. those that are singled out by the stationarity condition. This yields a semiclassical approximation to the reduced propagator.

To implement this procedure, we derive a phase expression from the integrand of (12.6). For this purpose, we cast the product $\exp\left(\frac{i}{\hbar}S_{\rm cl}[\mathbf{X}(t)]\right)$ $T_{\beta\alpha}[\mathbf{X}(t)]$ into polar form and find

$$\exp\left(\frac{i}{\hbar}S_{\rm cl}[\mathbf{X}(t)]\right) T_{\beta\alpha}[\mathbf{X}(t)] = |T_{\beta\alpha}[\mathbf{X}(t)]| \exp\left(\frac{i}{\hbar}\Phi[\mathbf{X}(t)]\right)$$

with

$$\Phi[\mathbf{X}(t)] = S_{\rm cl}[\mathbf{X}(t)] + \hbar\Im(\ln T_{\beta\alpha}[\mathbf{X}(t)]). \tag{12.9}$$

The classical paths satisfy the stationary phase condition

$$\delta\Phi[\mathbf{X}(t)] = 0 \tag{12.10}$$

which leads to a Newtonian equation involving the electronic states $\psi^\alpha(\psi^\beta)$ that are advanced along the trajectory $\mathbf{X}(t)$ (see Exercise 12.1(b)).

$$\begin{aligned} M\frac{\partial^2\mathbf{X}(t')}{\partial t'^2} &= \mathbf{F}_{\beta\alpha}(t') \\ &= -\Re\left\{\frac{\langle\psi^\beta(t',t)|\frac{\partial\hat{H}_{\rm el}}{\partial\mathbf{X}}(t')|\psi^\alpha(t',t_0)\rangle}{\langle\psi^\beta(t',t)|\psi^\alpha(t',t_0)\rangle}\right\}. \end{aligned} \tag{12.11}$$

The force $\mathbf{F}_{\beta\alpha}$ is associated with an effective potential $V_{\rm eff}$, given by

$$V_{\rm eff}(t') = \Re\left\{\frac{\langle\psi^\beta(t',t)|\hat{H}_{\rm el}(t')|\psi^\alpha(t',t_0)\rangle}{\langle\psi^\beta(t',t)|\psi^\alpha(t',t_0)\rangle}\right\}. \tag{12.12}$$

Exercise 12.1. *Newton's equations from a semiclassical path integral approach.* (a) Derive the explicit form of the phase $\Phi[\mathbf{X}(t)]$ given by Eq. (12.9). (b) Use the stationary phase condition (12.10) in conjunction with the definition of the transition amplitude $T_{\beta\alpha}[\mathbf{X}(t)]$ (12.8) to obtain the Newtonian equation (12.11).

The wave functions in Eqs. (12.11) and (12.12) are to be considered solutions of the TDSE governed by the Hamiltonian $\hat{H}_{\text{el}}(\mathbf{x}, \mathbf{X}(t'))$. More specifically, the mixed state $\psi^{\alpha}(t', t_0)$ results from propagation of the initial electronic basis state $\psi^{\alpha}(t_0)$ over the time interval $t' = t - t_0$, and in an analogous way, $\psi^{\beta}(t', t)$ emerges from the instantaneous basis state $\psi^{\beta}(t)$ through propagation backwards in time. Formula (12.11) describes a temporally non-local, nonadiabatic equation of motion that conserves both energy and angular momentum. It provides a prescription for self-consistent evaluation of both the classical trajectory of the system and the wave function. The force acting along this trajectory depends on $\psi^{\alpha}(\psi^{\beta})$, which are determined by a Hamiltonian that, in turn, is defined by the trajectory. Equation (12.11) may therefore be solved by an iteration strategy, starting with a guess of the trajectory, $\mathbf{X}_{\text{trial}}(t')$, which leads to a first approximation to the Hamiltonian, $\hat{H}_{\text{el}}(\mathbf{x}, \mathbf{X}_{\text{trial}}(t))$, and thus to first approximate solutions $\psi^{\alpha}_{\text{trial}}(\psi^{\beta}_{\text{trial}})$, completing the initial iteration cycle.

Once all trajectories compatible with the stationary phase constraint have been obtained from Eq. (12.11), the reduced propagator may be approximated by adopting the Van Vleck–Gutzwiller scheme discussed in Section 4.1.2:

$$K_{\text{V},\beta\alpha}(\mathbf{X}, t|\mathbf{X}_0, t_0) = \sum_{\text{P}} \left(\frac{M}{2\pi i\hbar}\right)^{\frac{F}{2}} \left|\det\left(\frac{\partial \mathbf{X}'_t}{\partial \dot{\mathbf{X}}'_0}\right)_{\text{P}}\right|^{-\frac{1}{2}}$$

$$\times \exp\left[\frac{i}{\hbar}(S_{\text{cl}}[\mathbf{X}_{\text{P}}(t')]) - \nu_{\text{P}}\frac{\pi}{2}\right] T_{\beta\alpha}[\mathbf{X}(t')] \qquad (12.13)$$

where $\dot{\mathbf{X}}_0$ denotes the time derivative of $\mathbf{X}_0$ and F denotes the number of degrees of freedom involved in the problem. It is characteristic of the transition from the quantum to the semiclassical description that the path integral occurring in Eq. (12.6) is in Eq. (12.13) replaced by a sum that extends over all stationary phase trajectories. For each of these trajectories, the Jacobian $\det\left(\frac{\partial \mathbf{X}'_t}{\partial \dot{\mathbf{X}}'_0}\right)$, involving derivatives of the final nuclear positions with respect to the initial nuclear velocities, has to be computed. In analogy to Eq. (4.50), the index ν_{P} indicates how many times this determinant vanishes in the time interval $[t_0, t]$ along the trajectory $\mathbf{X}_{\text{P}}(t')$.

In replacing the quantum mechanically exact propagator (12.2) by its semiclassical counterpart (12.13), one sacrifices unitarity, implying that the probabilities assigned to alternative terminal states with labels β are not guaranteed to sum up to one as the semiclassical approximation (12.13) is implemented. This is the price to pay for reducing the continuum of interfering paths that determines the exact propagator to a set of discrete trajectories that enter with maximum weight. This limitation of the formalism has to be taken into account in practical propagation schemes realizing the semiclassical approach, as discussed in the following section.

12.2 Semiclassical Propagation and Surface Hopping

Given an electronic state ψ_α at the initial time $t = 0$, the nuclear wave function component χ_β can be semiclassically approximated at any time t according to

$$\chi_\beta(\mathbf{X}_t, t) = \int d\mathbf{X}_0 K_{\mathrm{V},\beta\alpha}(\mathbf{X}_t, t|\mathbf{X}_0, 0)\chi_\alpha(\mathbf{X}_0, 0). \tag{12.14}$$

Advancing the wave function by use of Eq. (12.14) is not an easy task. In principle, construction of the semiclassical propagator requires identifying all stationary paths that connect a pair $(\mathbf{X}_0, \mathbf{X}_t)$ of initial and final nuclear coordinates. This has to be done for a sufficiently large set of $\mathbf{X}_0$ as well as $\mathbf{X}_t$ values. The determination of all paths that connect these two positions may proceed through a root search based on relation (12.11). As $\mathbf{X}_0$ is defined, one may scan the initial velocities $\dot{\mathbf{X}}_0$ and select all paths that converge to the desired final position $\mathbf{X}_t$:

$$\mathbf{X}_t = \mathbf{X}_t(\mathbf{X}_0, \dot{\mathbf{X}}_0). \tag{12.15}$$

Numerical realizations of this path integral approach have been proposed in the framework of various mixed quantum-classical schemes [346, 347]. Specifically, Coker and Xiao [346] recognized that the formalism introduced by Pechukas, in particular its pivotal concepts of the temporally delocalized force (12.11) and the effective potential (12.12), can be related to the basic tenets of the surface hopping method as outlined in the preceding chapter. This appears surprising since the TSH operates with the notion of a set of trajectories, each of which feels a well-defined instantaneous force at any given moment. It can be shown, however, that the surface hopping model may be interpreted as a *temporally localized approximation* to the stationary phase path integral theory. Following the arguments put forward by Coker and Xiao, we will now outline the connection between both approaches and emphasize that the formalism proposed by Pechukas supplies a theoretical fundament for the surface hopping model, accounting in a systematic way for what appears *ad hoc* in this scheme, dictated by the need to reconcile it with energy and momentum conservation.

We start by expressing the forward and backward propagated electronic wave functions, $\psi^{\alpha}(t', t_0)$ and $\psi^{\beta}(t', t)$, that appear in Eqs. (12.11) and (12.12), in terms of an instantaneous orthonormal basis $\{\psi^{i}(t)\}$, where the adiabatic representation is adopted:

$$|\psi^{\alpha}(t', t_0)\rangle = \sum_{n} a_n^{\alpha}(t', t_0)|\psi^{n}(t')\rangle, \tag{12.16}$$

$$\langle\psi^{\beta}(t', t)| = \sum_{m} b_m^{\beta *}(t', t)\langle\psi^{m}(t')|. \tag{12.17}$$

In an effort to reduce Eq. (12.11) to a local force, we consider the evolution of a trajectory swarm member during the small time interval Δ, such that $t_0 \leq t' \leq t = t_0 + \Delta$. The *Pechukas force* (12.11) is then determined by the three time parameters t_0, t, and Δ according to

$$\mathbf{F}_{\beta\alpha}(t') = -\Re\left\{\frac{\sum_{m,n} b_m^{\beta *}(t', t_0+\Delta)a_n^{\alpha}(t', t_0)\langle\psi_m(t')|\frac{\partial H_{\text{el}}(t')}{\partial \mathbf{X}}|\psi_n(t')\rangle}{\sum_m b_m^{\beta *}(t', t_0+\Delta)a_m^{\alpha}(t', t_0)}\right\}. \tag{12.18}$$

Taking into account the conditions adopted by the surface hopping model, Eq. (12.18) may be simplified considerably. As it is generally used, this model constrains the quantal system to be in a unique adiabatic state at any time of the evolution. Thus, during the interval Δ, the system may undergo a transition from $|\psi^{\alpha}(t_0)\rangle$ to $|\psi^{\beta}(t)\rangle$, corresponding to a change of the nuclear trajectory from $\mathbf{X}(t_0)$ to $\mathbf{X}'(t)$. This is tantamount to setting

$$a_{\alpha}^{\alpha}(t_0, t_0) = 1, \qquad a_n^{\alpha}(t_0, t_0) = 0, \quad \text{if } n \neq \alpha, \tag{12.19}$$

$$b_{\beta}^{\beta}(t, t) = 1, \qquad b_m^{\beta}(t, t) = 0, \quad \text{if } m \neq \beta. \tag{12.20}$$

With these substitutions, $\mathbf{F}_{\beta\alpha}(t')$ is evaluated at the initial time t_0 as

$$\mathbf{F}_{\beta\alpha}(t_0) = -\Re\left\{\frac{\sum_n b_n^{\beta*}(t_0, t_0+\Delta)\langle\psi_n(t_0)|\frac{\partial H_{\text{el}}(t_0)}{\partial \mathbf{X}}|\psi_\alpha(t_0)\rangle}{b_\alpha^{\beta*}(t_0, t_0+\Delta)}\right\}. \tag{12.21}$$

In the spirit of time-dependent perturbation theory, we assume Δ to be small enough for the approximation

$$b_m^\beta(t,t) = b_m^\beta(t_0+\Delta, t_0+\Delta) \approx b_m^\beta(t_0, t_0+\Delta) \tag{12.22}$$

to hold, implying

$$b_\beta^\beta(t_0, t_0+\Delta) \approx 1, \quad b_m^\beta(t_0, t_0+\Delta) \approx 0, \quad \text{if } m \neq \beta. \tag{12.23}$$

Therefore, $\mathbf{F}_{\beta\alpha}(t_0)$ is governed by the term resulting for $m = \beta$. Equating the Pechukas force with this single, dominant term, one obtains

$$\mathbf{F}_{\beta\alpha}(t_0) \approx -\Re\left\{\frac{b_\beta^{\beta*}(t_0, t_0+\Delta)\langle\psi_\beta(t_0)|\frac{\partial H_{\text{el}}(t_0)}{\partial \mathbf{X}}|\psi_\alpha(t_0)\rangle}{b_\alpha^{\beta*}(t_0, t_0+\Delta)}\right\}. \tag{12.24}$$

The derivative of the electronic Hamiltonian is eliminated by use of relation (2.61):

$$\left\langle \phi_i \left| \frac{\partial \hat{H}_{\text{el}}}{\partial \mathbf{X}} \right| \phi_j \right\rangle = (E_j - E_i)\left\langle \phi_i \left| \frac{\partial}{\partial \mathbf{X}} \right| \phi_j \right\rangle, \quad \text{if } i \neq j,$$

leading to

$$\mathbf{F}_{\beta\alpha}(t_0) \approx -\Re\left\{\frac{b_\beta^{\beta*}(t_0, t_0+\Delta)(E_\beta - E_\alpha)\langle\psi_\beta(t_0)|\frac{\partial}{\partial \mathbf{X}}|\psi_\alpha(t_0)\rangle}{b_\alpha^{\beta*}(t_0, t_0+\Delta)}\right\},$$
$$\text{if } \alpha \neq \beta. \tag{12.25}$$

In the case that no transition takes place during the time interval, Eq. (12.24) reduces to

$$\mathbf{F}_{\alpha\alpha}(t_0) \approx -\Re\left\{\left\langle \psi_\alpha(t_0)\left|\frac{\partial \hat{H}_{\text{el}}}{\partial \mathbf{X}}(t_0)\right|\psi_\alpha(t_0)\right\rangle\right\}$$

$$= -\left\langle \psi_\alpha(t_0)\left|\frac{\partial \hat{H}_{\text{el}}}{\partial \mathbf{X}}(t_0)\right|\psi_\alpha(t_0)\right\rangle$$

$$= -\frac{\partial E_\alpha}{\partial \mathbf{X}}. \tag{12.26}$$

Exercise 12.2. Demonstrate the validity of Eq. (12.26) by use of $\frac{\partial}{\partial \mathbf{X}}\langle\psi_\beta|\hat{H}_{\text{el}}|\psi_\alpha\rangle = \frac{\partial E_\alpha}{\partial \mathbf{X}}\delta_{\beta\alpha}$.

Obviously, Eq. (12.26) recovers Born–Oppenheimer dynamics on a single potential energy surface, in keeping with the principles of the original surface hopping approach (see Section 11.1).

For the more interesting problem of a transition between the electronic states with indices α and β, Eq. (12.25) may be further developed by evaluating the coefficients $b_\alpha^{\beta *}(t_0, t_0 + \Delta)$. Since they are obtained by back-propagation from a well-defined final condition as given by (12.20), they can be determined by applying of the TDSE to the wave function $|\psi^\beta(t', t)\rangle$ (see Eq. (12.17)). This implies

$$\frac{db_k^\beta}{dt'} = \sum_m \left(-i\frac{E_{km}}{\hbar} - \left\langle \psi_k \left| \frac{\partial \psi_m}{\partial t'}\right.\right\rangle\right) b_m^\beta = \sum_m c_{km} b_m^\beta, \tag{12.27}$$

and $E_{km} = \langle\psi^k|\hat{H}_{\text{el}}|\psi^m\rangle$. By assumption, the coefficient b_β^β outweighs the remaining coefficients strongly enough to justify the approximation

$$\frac{db_k^\beta}{dt'} = c_{k\beta} b_\beta^\beta. \tag{12.28}$$

The time dependence of b_k^β is readily found if the matrix elements $c_{k\beta}$ can be considered as time independent in the interval Δ. In this case Eq. (12.28) yields

$$b_\beta^\beta(t', t_0 + \Delta) = \exp[c_{\beta\beta}(t' - (t_0 + \Delta))], \tag{12.29}$$

$$b_k^\beta(t', t_0 + \Delta) = \frac{c_{k\beta}}{c_{\beta\beta}}\{\exp[c_{\beta\beta}(t' - (t_0 + \Delta))] - 1\}, \quad \beta \neq k. \tag{12.30}$$

From the definition of the $c_{k\beta}$ coefficients, as implied by Eq. (12.27), one concludes that $c_{\beta\beta} = -\frac{i}{\hbar}E_\beta$ and $c_{k\beta} = -\langle\psi_k|\frac{\partial\psi_\beta}{\partial t'}\rangle$, with $k \neq \beta$. With these substitutions, the ratio $b_\beta^{\beta*}/b_\alpha^{\beta*}$ in Eq. (12.25) is written as

$$\frac{b_\beta^{\beta*}(t_0, t_0 + \Delta)}{b_\alpha^{\beta*}(t_0, t_0 + \Delta)} = \frac{iE_\beta/\hbar}{\{1 - \exp[\frac{i}{\hbar}E_\beta\Delta]\}\langle\psi_\beta|\frac{\partial\psi_\alpha}{\partial t'}\rangle}. \tag{12.31}$$

Making use of the nonadiabatic coupling vector $\mathbf{A}_{\beta\alpha}$, we rewrite the scalar product $\langle\psi_\beta|\frac{\partial\psi_\alpha}{\partial t'}\rangle$ according to

$$\left\langle\psi_\beta\left|\frac{\partial}{\partial t'}\right|\psi_\alpha\right\rangle = \left\langle\psi_\beta|\frac{\partial\psi_\alpha}{\partial\mathbf{X}}\right\rangle \cdot \frac{d\mathbf{X}}{dt'} = \mathbf{A}_{\beta\alpha}\cdot\mathbf{v}. \tag{12.32}$$

Inserting Eq. (12.31) into Eq. (12.25), we complete the time localization of the Pechukas force by letting Δ tend toward zero. This results in

$$\mathbf{F}_{\beta\alpha}(t_0) \approx -\left\{\frac{E_\beta - E_\alpha}{\mathbf{A}_{\beta\alpha}^\wedge \cdot \mathbf{v}\Delta}\right\}\mathbf{A}_{\beta\alpha}^\wedge, \quad \alpha \neq \beta, \tag{12.33}$$

where the nonadiabatic coupling vector has been normalized: $\mathbf{A}_{\beta\alpha}^\wedge \equiv \frac{\mathbf{A}_{\beta\alpha}}{|\mathbf{A}_{\beta\alpha}|}$. This finding is remarkable in several respects. First, the time-localized Pechukas force acts in the direction of the nonadiabatic coupling vector. This provides a theoretical basis for the momentum scaling procedure introduced by Tully [288] on the level of an *ad hoc* manipulation for restoring the momentum balance after a transition between potential energy surfaces (see Section 11.2). In the same vein, it can be shown that the action of the Pechukas force reconciles the trajectory surface hopping procedure with the energy conservation law. Thus, we find that the energy difference associated with the transition between electronic states with labels α and β is compensated for by an equal and opposite change of the nuclear kinetic energy.

In quantitative detail, the work done during the time interval Δ by the instantaneous Pechukas force (12.33) is approximated as

$$\int_{t_0}^{t_0+\Delta} dt\, \mathbf{F}_{\beta\alpha} \cdot \mathbf{v} \approx \mathbf{F}_{\beta\alpha} \cdot \mathbf{v}\Delta = -(E_\beta - E_\alpha). \tag{12.34}$$

Also,

$$\int_{t_0}^{t_0+\Delta} dt \mathbf{F}_{\beta\alpha} \cdot \mathbf{v} = \frac{1}{2} M v^2(t_0 + \Delta) - \frac{1}{2} M v^2(t_0). \tag{12.35}$$

Combining the two preceding formulae, it is seen that the kinetic nuclear energy adjustment following any surface hop, as proposed by Tully, emerges naturally within the stationary phase path integral formalism as the work done by the time-localized Pechukas force. After the transition, the kinetic nuclear energy turns out to be changed by the electronic energy difference:

$$\frac{1}{2} M v^2(t_0 + \Delta) = \frac{1}{2} M v^2(t_0) - (E_\beta - E_\alpha). \tag{12.36}$$

From the foregoing, the limit of small Δ yields a force expression that is local in time and therefore in accord with the methodological premise of instantaneous hops between potential energy surfaces, as adopted by the fewest switches algorithm or earlier realizations of the TSH idea. Time localization of the nonadiabatic force (12.18) therefore recovers the model of propagation on some well-defined adiabatic surface at any given time, interrupted by hopping events. As momentum and energy scaling prescriptions in TSH simulations emerge as natural consequences of the resulting approximate force (12.33), designing an extended TSH algorithm that incorporates this force appears desirable. A sketch of such an algorithm is found among the supplementary sections in Section 12.2*.

12.3 The Initial Value Representation

From the foregoing, it may be plausible that the path integral approach proposed by Pechukas, while offering fundamental insight into the semiclassical formalism, is hardly a convenient way of treating real quantum dynamical systems. Both the time non-locality of Eqs. (12.11) and (12.12) and the nonlinear root search implied by Eq. (12.14) limit the effectiveness of the method severely and prohibit its implementation for problems involving molecular interactions. In what follows we will focus on an alternative propagator method that combines semiclassical accuracy with practical feasibility, the *initial value representation*. Conceived in the early seventies [292, 348, 349], this model has found strong renewed interest in the wake of classical molecular dynamics simulations of complex systems and processes.

To introduce the seminal idea of the semiclassical IVR (SC-IVR), we begin with the transition amplitude, or matrix element of the quantum propagator (4.6) in the time interval $[0, t]$:

$$K_{n_1,n_2}(t) = \int d\mathbf{X}_0 \int d\mathbf{X}_t \Psi^*_{n_2}(\mathbf{X}_t)\Psi_{n_1}(\mathbf{X}_0)\left\langle \mathbf{X}_t \left| \exp\left(-\frac{i}{\hbar}\hat{H}t\right)\right| \mathbf{X}_0 \right\rangle. \tag{12.37}$$

This expression describes the probability amplitude for a transition from a state n_1 at the time $t = 0$ to a state n_2 at the time t, with the respective wave functions Ψ_{n_1} and Ψ_{n_2}. Once more, we invoke the Van Vleck semiclassical propagator (see Eq. (4.50)) in the coordinate representation, while not yet making explicit reference to molecular parameters.

$$K_{\mathrm{V},n_1n_2}(t) = \sum_{\text{roots}} \int d\mathbf{X}_0 \int d\mathbf{X}_t \Psi^*_{n_2}(\mathbf{X}_t)\Psi_{n_1}(\mathbf{X}_0) \times \left[(2\pi i\hbar)^F det\left(\frac{\partial \mathbf{X}_t}{\partial \mathbf{P}_0}\right)\right]^{-\frac{1}{2}} \exp\left(\frac{i}{\hbar}\mathrm{S}_{\mathrm{cl}}(\mathbf{X}_t, \mathbf{X}_0)\right) \tag{12.38}$$

where $S_{\mathrm{cl}}(\mathbf{X}_t, \mathbf{X}_0)$ stands for the classical action along the trajectory that leads from $\mathbf{X}_0$ to $\mathbf{X}_t$ in a time interval $[0, t]$, according to

$$S_{\mathrm{cl}}(\mathbf{X}_t, \mathbf{X}_0) = \int_0^t dt' \, \mathbf{P}(t') \cdot \dot{\mathbf{X}}(t') - H(\mathbf{P}(t'), \mathbf{X}(t')). \tag{12.39}$$

Here and in the following development, the Maslov index is, for ease of notation, not explicitly indicated, but its inclusion in the propagator phase is implied. The summation over the roots takes into account that, for a given $\mathbf{X}_0$, various choices of $\mathbf{P}_0$ may lead to the same $\mathbf{X}_t$.

This ambiguity and the inconvenience of root searching *in toto* is circumvented by adopting the initial value representation (IVR) [24,292]. This formalism avoids the sum over the roots along with the integral over the end point coordinates $\mathbf{X}_t$ by the following coordinate transformation:

$$\sum_{\text{roots}} \int d\mathbf{X}_t = \int d\mathbf{P}_0 \det\left(\frac{\partial \mathbf{X}_t(\mathbf{X}_0, \mathbf{P}_0)}{\partial \mathbf{P}_0}\right). \tag{12.40}$$

The sum over the trajectories is absorbed in the integral over the initial momenta, making the nonlinear root search obsolete. Relation (12.40) introduces $\mathbf{P}_0$ as an independent variable while in the original formulation (12.38), $\mathbf{P}_0$ is bound by the condition $\mathbf{X}_t = \mathbf{X}_t(\mathbf{P}_0, \mathbf{X}_0)$. Applying this change of variables to (12.38), we find

$$K_{\mathrm{V},n_1n_2}(t) = \int d\mathbf{X}_0 \int d\mathbf{P}_0\, \Psi^*_{n_2}(\mathbf{X}_t)\Psi_{n1}(\mathbf{X}_0)\left[\frac{1}{(2\pi i\hbar)^F}\det\left(\frac{\partial \mathbf{X}_t}{\partial \mathbf{P}_0}\right)\right]^{\frac{1}{2}}$$
$$\times \exp\left(\frac{i}{\hbar}S_{\mathrm{cl}}(\mathbf{X}_0, \mathbf{P}_0)\right). \tag{12.41}$$

The integration now must be carried out over the phase space of the initial coordinates and momenta. As the pair $(\mathbf{X}_0, \mathbf{P}_0)$ uniquely determines a classical trajectory, this manipulation eliminates any multivalued terms from the integrand, in contrast to formulation (12.38) where the potential multiplicity of $\mathbf{P}_0$ values necessitates the summing over the roots. Going from Eq. (12.38) to Eq. (12.41), one notices that the singularities associated with the vanishing of the Jacobian determinant $\det\left(\frac{\partial \mathbf{X}_t}{\partial \mathbf{P}_0}\right)$ are turned into zeroes. This change aids the stability of numerical simulations employing the semiclassical propagator.

Acknowledging that

$$K_{n_1,n_2}(t) = \left\langle \Psi_{n_2} \left| \exp\left(-\frac{i}{\hbar}\hat{H}t\right) \right| \Psi_{n_1} \right\rangle, \tag{12.42}$$

one may write the IVR propagator explicitly as [24]

$$\exp\left(-\frac{i}{\hbar}\hat{H}t\right) = \int d\mathbf{X}_0 \int d\mathbf{P}_0 \left[\frac{1}{(2\pi i\hbar)^F}\, det\left(\frac{\partial \mathbf{X}_t(\mathbf{X}_0, \mathbf{P}_0)}{\partial \mathbf{P}_0}\right)\right]^{\frac{1}{2}}$$
$$\times \exp\left(\frac{i}{\hbar}S_t(\mathbf{X}_0, \mathbf{P}_0)\right)|\mathbf{X}_t\rangle\langle \mathbf{X}_0|, \tag{12.43}$$

which defines the IVR approach in the Cartesian coordinate representation. Likewise, the corresponding momentum representation is readily written as

$$\exp\left(-\frac{i}{\hbar}\hat{H}t\right) = \int d\mathbf{X}_0 \int d\mathbf{P}_0 \left[\frac{1}{(2\pi i\hbar)^F}\, det\left(\frac{\partial \mathbf{X}_t(\mathbf{X}_0,\mathbf{P}_0)}{\partial \mathbf{P}_0}\right)\right]^{\frac{1}{2}}$$

$$\times \exp\left(\frac{i}{\hbar}\tilde{S}_t(\mathbf{X}_0,\mathbf{P}_0)\right) |\mathbf{P}_t\rangle\langle\mathbf{P}_0| \qquad (12.44)$$

with $\tilde{S}_t$ defined as

$$\tilde{S}_t \equiv -\int_0^t dt'\mathbf{X}(t')\dot{\mathbf{P}}(t') + H(\mathbf{X}(t'),\mathbf{P}(t')). \qquad (12.45)$$

Besides Eqs. (12.43) and (12.44), a third representation has found frequent use in applications of the IVR formalism. Here, mixed position-momentum states are employed, such as the Glauber coherent state $|\mathbf{P}_0,\mathbf{X}_0\rangle$ (see Section 4.4) whose projection into coordinate space reads

$$\langle\mathbf{X}|\mathbf{P}_0,\mathbf{X}_0\rangle = \left(\frac{\gamma}{\pi}\right)^{\frac{F}{4}} \exp\left(-\frac{\gamma}{2}|\mathbf{X}-\mathbf{X}_0|^2\right)\exp\left(\frac{i}{\hbar}\mathbf{P}_0\cdot(\mathbf{X}-\mathbf{X}_0)\right). \qquad (12.46)$$

These hybrid wave functions are simultaneously localized in position and momentum space. The parameter γ determines the nature of these states, which tend towards a position eigenfunction as $\gamma \to \infty$ and a momentum eigenfunction as $\gamma \to 0$. Herman and Kluk [350] demonstrated that the extension of Eqs. (12.43) and (12.44) for these mixed states is

$$\exp\left(-\frac{i}{\hbar}\hat{H}t\right) = (2\pi\hbar)^{-F}\int d\mathbf{X}_0\int d\mathbf{P}_0\, C_t(\mathbf{X}_0,\mathbf{P}_0)$$

$$\times \exp\left(\frac{i}{\hbar}\tilde{S}_t(\mathbf{X}_0,\mathbf{P}_0)\right) |\mathbf{P}_t,\mathbf{X}_t\rangle\langle\mathbf{P}_0,\mathbf{X}_0| \qquad (12.47)$$

where the prefactor $C_t(\mathbf{X}_0,\mathbf{P}_0)$, given by

$$C_t(\mathbf{X}_0,\mathbf{P}_0) \equiv \left(\det\left[\frac{1}{2}\left(\frac{\partial\mathbf{X}_t}{\partial\mathbf{X}_0} + \frac{\partial\mathbf{P}_t}{\partial\mathbf{P}_0} + \left(\frac{\hbar\gamma}{i}\right)\frac{\partial\mathbf{X}_t}{\partial\mathbf{P}_0} + \left(\frac{i}{\hbar\gamma}\right)\frac{\partial\mathbf{P}_t}{\partial\mathbf{X}_0}\right)\right]\right)^{\frac{1}{2}}, \qquad (12.48)$$

contains all elements of the monodromy matrix introduced by Eq. (4.48). So far, our description of the IVR model has not made explicit reference to the molecular problem. If we identify the variables **X** and **P** with Cartesian nuclear coordinates and momenta, the formalism has to be extended to include electronic degrees of freedom. Obviously, using an approximation in the spirit of the classical electron analog to this end, as outlined in Chapter 10, would introduce a hiatus into the model, an inconsistency between the representations of the nuclear and the electronic degrees of freedom: a wave function as prescribed by Eqs. (12.43), (12.44), and (12.47) on the nuclear versus a set of action-angle variables with imposed semiclassical quantization conditions on the electronic side. The two subsystems would thus be treated at different levels of dynamic theory. In contrast, a uniform semiclassical model for the propagator is desired that establishes symmetry between the nuclear and electronic degrees of freedom. A dynamically consistent implementation of the semiclassical propagator formalism requires incorporating an electronic wave function such that the initial and the final state of the system may be characterized by both a nuclear and an electronic contribution according to

$$\Psi_{n,i}(\mathbf{x}, \mathbf{X}) = \chi_n(\mathbf{X})\psi_i(\mathbf{x}). \tag{12.49}$$

Sun and Miller [319] used heuristic arguments to demonstrate that inclusion of an electronic wave function into the IVR model does not require solving the full quantum mechanical problem for the electrons. Instead, an analog of this wave function can be found on semiclassical grounds. Stock and Thoss [351, 352] gave a more rigorous derivation of the same result, using a mapping formalism based on Schwinger's theory of angular momentum [353]. In the following, we will present a condensed account of the latter approach.

12.4 The Mapping Approach to Electronic Degrees of Freedom

In this section we will discuss a semiclassical scheme to imitate electronic degrees of freedom, thus placing them on the same footing as the nuclear position and momentum variables. This methodological idea is informed by the classical electron analog theory surveyed in Chapter 10. Once more, the objective is to represent the electronic subsystem with sets of continuous coordinates that can be subjected to classical treatment. Here, however, the pivotal move consists in designing a map between two quantum mechanical systems, involving a transition from the system of interest to an ensemble of N bosonic oscillators. As result of this procedure, electronic states are obtained in the Hilbert space of the N-boson problem which are used to construct the propagator matrix element (12.37). At the same time, a natural way emerges to assign a classical counterpart to all operators of relevance for the situation at hand, such that Hamiltonian dynamics in the sense of the classical electron analog can be performed. In particular, the physical Hamiltonian of the system is, as in Eq. (10.6), mapped onto a classical function that depends on continuous parameters. The essential contents of the mapping idea are best illustrated by demonstrating the equivalence of a system with arbitrary spin and an oscillator pair, as will be done in the following paragraphs.

12.4.1 *The Schwinger mapping formalism*

To begin, we define the spin operators $\hat{S}_1$, $\hat{S}_2$, and $\hat{S}_3$ in a Hilbert space of spin states $|sm\rangle$, where $-s \leq m \leq s$ and $s = \frac{1}{2}, 1, \frac{3}{2} \ldots$. These states are eigenfunctions of $\hat{S}_3$:

$$\hat{S}_3 = |sm\rangle = m|sm\rangle. \tag{12.50}$$

The spin operators satisfy the commutation relations

$$[\hat{S}_j, \hat{S}_k] = i\hbar\delta_{jkl}\hat{S}_l \tag{12.51}$$

and combine to form ladder operators according to

$$\hat{S}_\pm = \hat{S}_1 \pm i\hat{S}_2. \tag{12.52}$$

The essence of Schwinger's theory of angular momentum consists in a one-to-one correspondence between the spin operators $\hat{S}_i$, $i = 1, 2, 3$, and well-defined combinations of the creation and annihilation operators $\hat{a}_j^\dagger$ and $\hat{a}_j$, where $j = 1, 2$, for the bosonic oscillator. These operators obey the commutation relation

$$[\hat{a}_i, \hat{a}_j^\dagger] = \delta_{ij}. \tag{12.53}$$

Their products $\hat{a}_i^\dagger \hat{a}_i$ $(i = 1, 2)$ operate in a basis $|n_1 n_2\rangle$, such that

$$\hat{a}_1^\dagger \hat{a}_1 |n_1 n_2\rangle = n_1 |n_1 n_2\rangle, \tag{12.54}$$

$$\hat{a}_2^\dagger \hat{a}_2 |n_1 n_2\rangle = n_2 |n_1 n_2\rangle. \tag{12.55}$$

The mapping relations establish the equivalence between a discrete degree of freedom associated with the spin s and a system of two bosonic oscillators. More specifically, the rules for translation between the respective two types of operators are

$$\hat{S}_1 \to \frac{1}{2}(\hat{a}_2^\dagger \hat{a}_1 + \hat{a}_1^\dagger \hat{a}_2), \tag{12.56}$$

$$\hat{S}_2 \to \frac{1}{2i}(\hat{a}_2^\dagger \hat{a}_1 - \hat{a}_1^\dagger \hat{a}_2), \tag{12.57}$$

$$\hat{S}_3 \to \frac{1}{2}(\hat{a}_2^\dagger \hat{a}_2 - \hat{a}_1^\dagger \hat{a}_1), \tag{12.58}$$

and therefore

$$\hat{S}_+ \to \hat{a}_2^\dagger \hat{a}_1, \tag{12.59}$$

$$\hat{S}_- \to \hat{a}_1^\dagger \hat{a}_2. \tag{12.60}$$

It can be readily verified that the bosonic counterparts of $\hat{S}_i$ ($i = 1, 2, 3$, as listed in Eqs. (12.56) and (12.58)), indeed satisfy the spin operator commutation relations (12.51) (see Exercise 12.3).

Exercise 12.3. Show that the commutation relations (12.51) are valid for the image operators that arise from the Schwinger mapping procedure, as defined by Eqs. (12.56)–(12.58).

On the basis of the mapping relations, one can construct the image of the $(2s + 1)$-dimensional spin Hilbert space to arrive at

$$|sm\rangle \to \frac{(\hat{a}_1^\dagger)^{s-m}(\hat{a}_2^\dagger)^{s+m}}{\sqrt{(s+m)!(s-m)!}}|0_1 0_2\rangle = |(s-m)_1(s+m)_2\rangle. \tag{12.61}$$

The space of state functions for a spin s is mapped into the subspace of two oscillators in an excited state characterized by a constant sum of $2s$ excitation quanta. Inspecting the mapping rules (12.56)–(12.58), as well as (12.59) and (12.60), it is evident that any composite spin operator $\hat{A}(\hat{S}_1, \hat{S}_2, \hat{S}_3)$, when translated by use of the Schwinger mapping rules, will preserve the subspace spanned by the states $|(s-m)_1, (s+m)_2\rangle$. Under the action of such an operator, no state within this subspace can transcend its limits. With respect to the semiclassical propagator, the most relevant conclusion is that Schwinger's mapping procedure preserves the matrix elements of the operator $\hat{A}$ in the sense of the following identity:

$$\langle sm|\hat{A}|sm'\rangle = \langle (s-m)_1, (s+m)_2|\tilde{A}|(s-m')_1, (s+m')_2\rangle. \tag{12.62}$$

Thus, if the Hamiltonian $\hat{H}$ acts on the basis functions $|sm\rangle$, one may express the propagator of the system in the basis $|(s-m)_1, (s+m)_2\rangle$ using

$$\left\langle sm \left| \exp\left(-\frac{i}{\hbar}\hat{H}t\right) \right| sm' \right\rangle = \left\langle (s-m)_1, (s+m)_2 \left| \exp\left(-\frac{i}{\hbar}\tilde{H}t\right) \right| \right. \\ \left. \times\, (s-m')_1, (s+m')_2 \right\rangle \tag{12.63}$$

where $\tilde{H}$ is the image of $\hat{H}$.

Other mapping procedures have been proposed that achieve the same goal as Schwinger's theory of angular momentum. Possibly best known among these is the *Holstein–Primakoff transformation* [354] that is based on a similar methodological idea as Schwinger's theory of angular momentum. Holstein and Primakoff use the image of the identity in the spin Hilbert space which translates into the operator $(\hat{a}_1^\dagger \hat{a}_1 + \hat{a}_2^\dagger \hat{a}_2)$ in the two-oscillator Hilbert space. The latter represents an invariant of this space, since, from Eqs. (12.56), (12.57), and (12.61), its eigenvalue is $2s$. This constant is used to eliminate one oscillator degree of freedom and thus to construct a new set of mapping rules that lead from the spin Hilbert space to the single-oscillator Hilbert space. Alternative realizations of this variant of the mapping technique have been conceived (see, for instance, [355]).

12.4.2 *Extension to general N-level systems*

As pointed out in the previous section, Schwinger's theory of angular momentum might be utilized to represent an N-level system by two oscillators. Likewise, the Holstein–Primakoff or an equivalent transformation accomplishes a similar mapping on a single oscillator. For $N > 2$, however, these strategies involve Hamiltonians that contain higher powers of the spin operators $\hat{S}_i$, $i = 1, 2, 3$, yielding nonlinear equations of motion which are difficult to handle by semiclassical theory.

An alternative representation advocated by Stock and Thoss [351, 352] circumvents this problem by extending the mapping procedures mentioned so far to the case of N oscillators assigned to an N-level system. To clarify this concept, we consider a general N-level system as given by the Hamiltonian

$$\hat{H} = \sum_{n,m} |\Psi_n\rangle H_{nm} \langle \Psi_m| \quad m, n \leq N, \tag{12.64}$$

involving N basis states $\{\Psi_i\}$. Generalizing Schwinger's theory of angular momentum, the new approach makes allowance for N creation/annihilation operators $\hat{a}_i^\dagger/\hat{a}_i, i \leq N$, instead of one or two, as described before. This is achieved by the mapping rules

$$|\Psi_n\rangle\langle\Psi_m| \rightarrow \hat{a}_n^\dagger \hat{a}_m, \tag{12.65}$$

$$|\Psi_n\rangle \rightarrow |0_1 \ldots 1_n \ldots 0_N\rangle, \tag{12.66}$$

i.e. $|\Psi_n\rangle$ is a state composed of N independent oscillators with $N-1$ modes i $(i \neq n)$ in their ground states and a single quantum of excitation in mode n. Thus, in the coordinate representation, these functions are written as

$$\Psi_n(\mathbf{x}) = \Psi_1(x_n) \prod_{k=1, k\neq n}^{N} \Psi_0(x_k) \tag{12.67}$$

with Ψ_0 and Ψ_1 as the one-dimensional harmonic oscillator eigenfunctions for the ground state and the first excited state. The creation and annihilation operators $a_n^\dagger$, a_m satisfy the bosonic commutation relations (12.53). The rules (12.65) and (12.66) define a boson representation of the Hamiltonian (12.64), namely

$$\tilde{H} = \sum_{n,m} \hat{a}_n^\dagger H_{nm} \hat{a}_m. \tag{12.68}$$

In exact analogy to the parallel identity valid in Schwinger's theory of angular momentum, one finds for the propagator

$$\left\langle \Psi_n \left| \exp\left(-\frac{i}{\hbar}\hat{H}t\right) \right| \Psi_m \right\rangle$$

$$= \left\langle 0_1 \ldots 1_n \ldots 0_N \left| \exp\left(-\frac{i}{\hbar}\tilde{H}t\right) \right| 0_1 \ldots 1_m \ldots 0_N \right\rangle. \tag{12.69}$$

With reference to formula (12.63), it is plausible that for the special case $N = 2$, the mapping procedure of Stock and Thoss reduces to Schwinger's theory of angular momentum. The two-level problem corresponds to a spin

$\frac{1}{2}$ with $m = -\frac{1}{2}, +\frac{1}{2}$. Both approaches lead to the oscillator image states $\langle 0_1 1_2|$ and $\langle 1_1 0_2|$.

The image Hamiltonian (12.68) can also be shown to emerge in a natural way from a second quantization scheme. For this purpose, we represent the time-dependent wave function of the N-level system as

$$\Psi(t) = \sum c_n(t)|\Psi_n\rangle. \tag{12.70}$$

The TDSE applied to $\Psi(t)$ reads

$$i\hbar\frac{\partial c_n}{\partial t} = \sum_m H_{mn}c_m. \tag{12.71}$$

Upon substituting

$$c_n = \frac{1}{\sqrt{2}}(x_n + ip_n) \tag{12.72}$$

one readily proves that the expectation value of the Hamiltonian $\hat{H}$ can be expressed as

$$H = \langle\Psi(t)|\hat{H}|\Psi(t)\rangle = \sum_{n,m} c_n^* c_m H_{nm} = \frac{1}{2}\sum_{n,m}(x_n x_m + p_n p_m)H_{nm}. \tag{12.73}$$

Exercise 12.4. Derive Eq. (12.73) from Eqs. (12.70) and (12.72).

Using H as a classical Hamilton function and x_n, p_m as classical continuous variables, one finds complete equivalence between the TDSE (12.71) and Hamilton's equations for $\partial x_n/\partial t$ and $\partial p_m/\partial t$. The reader may compare these arguments with the parallel development outlined in Section 10.2, when introducing the fundamentals of the classical electron analog (see Eq. (10.5) and the related commentary).

Second quantization may be applied to expression (12.73) by identifying the coefficients c_n^* and c_m with the well-known expansions of the harmonic oscillator creation and annihilation operators

$$\hat{a}_n^\dagger = \frac{1}{\sqrt{2}}(\hat{x}_n - i\hat{p}_n), \tag{12.74}$$

$$\hat{a}_n = \frac{1}{\sqrt{2}}(\hat{x}_n + i\hat{p}_n), \tag{12.75}$$

in terms of the position and momentum operators $\hat{x}$ and $\hat{p}$.

12.5 The Mapping Technique Applied to Nonadiabatic Dynamics

So far, our account of mapping procedures has made no explicit reference to the nonadiabatic problem of molecular physics. At this juncture, however, it is tempting to interpret $\hat{x}_n$ and $\hat{p}_n$ as position and momentum operators, respectively, for the nth electronic degree of freedom. Adopting a diabatic representation, we further make the following choice for the Hamiltonian matrix elements:

$$H_{nm} = (T(\tilde{\mathbf{P}}) + V_0(\tilde{\mathbf{X}}))\delta_{nm} + H_{\mathrm{el},nm}(\tilde{\mathbf{X}}). \tag{12.76}$$

For simplicity of notation, here and in the following, dimensionless coordinates and momenta are employed for the nuclear/electronic subsystem, denoted by $\tilde{\mathbf{X}}, \tilde{\mathbf{P}}/\tilde{\mathbf{x}}, \tilde{\mathbf{p}}$. The element $H_{\mathrm{el},nm}$ stands for the diabatic potential energy matrix and V_0 for a state-independent additional potential, the nuclear kinetic energy being given by $T(\tilde{\mathbf{P}})$. The term V_0 might, for instance, be given by the harmonic approximation to the potential energy in the electronic ground state, resulting in a state-independent part of the Hamilton function (12.76) of the form

$$T(\tilde{\mathbf{P}}) + V_0(\tilde{\mathbf{X}}) = \sum_{i=1}^{F} \frac{1}{2}\hbar\omega((\tilde{\mathbf{P}}_i)^2 + (\tilde{\mathbf{X}}_i)^2). \tag{12.77}$$

Inserting Eqs. (12.74)–(12.76), into Eq. (12.68) and using $[\tilde{x}_n, \tilde{p}_m] = i\delta_{nm}$, one derives

$$H = H_0(\tilde{\mathbf{X}}, \tilde{\mathbf{P}}) + \frac{1}{2}\sum_{n,m} H_{\mathrm{el},nm}(\tilde{\mathbf{X}})(\tilde{x}_n\tilde{x}_m + \tilde{p}_n\tilde{p}_m) \tag{12.78}$$

and

$$H_0(\tilde{\mathbf{X}}, \tilde{\mathbf{P}}) = T(\tilde{\mathbf{P}}) + V_0(\tilde{\mathbf{X}}) - \frac{1}{2}\sum_{n} H_{\mathrm{el},nn}(\tilde{\mathbf{X}}). \tag{12.79}$$

Exercise 12.5. Use the decomposition given in (12.74) and (12.75) to obtain a representation of the Hamiltonian (12.68) in terms of electronic position and momentum operators. Show that the transition from operators to classical variables yields the Hamilton function (12.78).

This result invites again comparison with the corresponding expression obtained within the classical electron analog. It is noteworthy that the Langer modification that was introduced into the electronic Hamiltonian (10.4) to retrieve the correct energy eigenvalues appears in Eqs. (12.78) and (12.79) as a natural consequence of the operator mapping approach. Thus, the term $\frac{1}{2}\sum_n H_{\mathrm{el},nn}(\tilde{\mathbf{X}})$, describing the zero-point energy associated with the electronic degrees of freedom, emerges from the commutation relations $[\hat{a}_n, \hat{a}_m^\dagger] = \delta_{nm}$ which are not present in the CEA in its original formulation outlined in Chapter 10. Without any *ad hoc* manipulation, the CEA Hamiltonian (10.32) is recovered by substituting $\sqrt{n_m + \frac{1}{2}}\exp(iq_m)$ for the operator $\hat{a}_m$ in Eq. (12.74) and inserting the resulting expressions for $\tilde{x}_n$ and $\tilde{p}_n$ into Eq. (12.78) [356] (see Exercise 12.7).

The scalar product identity (12.69) makes the construction of the propagator for the nonadiabatic molecular problem formally straightforward. Using this identity, one may write

$$
\begin{aligned}
K_{1,2}(t) &= \left\langle \tilde{\mathbf{X}}' \left| \left\langle \psi_n \left| \exp\left(-\frac{i}{\hbar}\hat{H}t\right) \right| \psi_m \right\rangle \right| \tilde{\mathbf{X}} \right\rangle \\
&= \left\langle \chi_1(\tilde{\mathbf{X}}') \left| \langle 0_1 \dots 1_n \dots 0_N \left| \exp\left(-\frac{i}{\hbar}\tilde{H}t\right) \right| 0_1 \dots 1_m \dots 0_N \rangle \right| \chi_2(\tilde{\mathbf{X}}) \right\rangle.
\end{aligned}
\tag{12.80}
$$

This expression can be evaluated in the framework of the IVR approximation to the Van Vleck propagator. Describing in addition the electronic system by use of the classical coordinates $\tilde{\mathbf{x}}$ and $\tilde{\mathbf{p}}$, one obtains the semiclassical initial value representation (SC-IVR). This is done by extending the SC-IVR propagator (12.41) to the case of two sets of coordinates and momenta, nuclear and electronic ones:

$$
\begin{aligned}
K_{\mathrm{V},1,2}(t) = &\int d\tilde{\mathbf{x}}_0 d\tilde{\mathbf{X}}_0 \int d\tilde{\mathbf{p}}_0 d\tilde{\mathbf{P}}_0 \left[det \left(\frac{\partial(\tilde{\mathbf{x}}_t, \tilde{\mathbf{X}}_t)}{\partial(\tilde{\mathbf{p}}_0, \tilde{\mathbf{P}}_0)} \right) \left(\frac{1}{2\pi i} \right)^{F+N} \right]^{\frac{1}{2}} \\
&\times \chi_2^*(\tilde{\mathbf{X}}_t) \psi_{n_2}^*(\tilde{\mathbf{x}}_t) \psi_{n_1}(\tilde{\mathbf{x}}_0) \chi_1(\tilde{\mathbf{X}}_0) \\
&\times \exp\left[\frac{i}{\hbar} S_{\mathrm{cl},t}(\tilde{\mathbf{x}}_0, \tilde{\mathbf{p}}_0, \tilde{\mathbf{X}}_0, \tilde{\mathbf{P}}_0) \right]
\end{aligned}
\tag{12.81}
$$

where F and N refer to the nuclear and electronic degrees of freedom, respectively, and the expression contains the generalization of the Jacobian in Eq. (12.41) for the molecular case of both nuclear and electronic sets of coordinates and momenta. From Eq. (12.81), the nuclear wave function χ is advanced by projecting out the electronic wave functions. Thus, the nuclear contribution to a target state characterized by an electronic wave function ψ_{n2} is obtained via [319]

$$\begin{aligned}\chi_{n2\leftarrow n1}(\tilde{\mathbf{X}},t) &= \int d\tilde{\mathbf{x}}_0 d\tilde{\mathbf{X}}_0 \int d\tilde{\mathbf{p}}_0 d\tilde{\mathbf{P}}_0 \left[det\left(\frac{\partial(\tilde{\mathbf{x}}_t,\tilde{\mathbf{X}}_t)}{\partial(\tilde{\mathbf{p}}_0,\tilde{\mathbf{P}}_0)}\right)\left(\frac{1}{2\pi i}\right)^{F+N}\right]^{\frac{1}{2}} \\ &\quad\times \delta(\tilde{\mathbf{X}}-\tilde{\mathbf{X}}_t)\psi^*_{n2}(\tilde{\mathbf{x}}_t)\psi_{n1}(\tilde{\mathbf{x}}_0)\chi_1(\tilde{\mathbf{X}}_0) \\ &\quad\times \exp\left[\frac{i}{\hbar}S_{\mathrm{cl},t}(\tilde{\mathbf{x}}_0,\tilde{\mathbf{p}}_0,\tilde{\mathbf{X}}_0,\tilde{\mathbf{P}}_0)\right]. \end{aligned} \tag{12.82}$$

Remedies for the typical numerical problems encountered when evaluating the integrals (12.81) and (12.82), which are chiefly associated with the oscillating term $\exp(\frac{i}{\hbar}S_t)$, will be addressed in Section 12.7.

The formulae (12.79) and (12.82) encapsulate a dynamically consistent model of nonadiabatic processes. Both nuclear and electronic degrees of freedom are treated on the same footing, as indicated by the symmetric arrangement of F and N. For implementation of this formalism, one interprets Eq. (12.78) as a Hamilton function that generates classical trajectories for the nuclei and electrons involved in the studied problem and subsequently uses Eqs. (12.79) and (12.82) to add quantum effects to this classical description. For each trajectory, the classical action as well as the monodromy matrix $\mathbf{M}$ is generated as a function of time by solving the differential equations [357]

$$\frac{d}{dt}S_{\mathrm{cl},t} = \boldsymbol{\pi}_t \cdot \dot{\mathbf{q}}_t - H, \tag{12.83}$$

$$\frac{d}{dt}\mathbf{M}_t = \begin{pmatrix} \dfrac{\partial^2 H}{\partial\mathbf{q}\partial\boldsymbol{\pi}} & \dfrac{\partial^2 H}{\partial\mathbf{q}\partial\mathbf{q}} \\ \dfrac{\partial^2 H}{\partial\boldsymbol{\pi}\partial\boldsymbol{\pi}} & \dfrac{\partial^2 H}{\partial\boldsymbol{\pi}\partial\mathbf{q}} \end{pmatrix}\mathbf{M}, \tag{12.84}$$

where, for ease of notation, we have consolidated the nuclear and electronic coordinates into the arrays $\mathbf{q}$ and $\boldsymbol{\pi}$, where $\mathbf{q} \equiv (\mathbf{X},\mathbf{x})$, $\boldsymbol{\pi} \equiv (\mathbf{P},\mathbf{p})$. If this nomenclature is adopted, the four submatrices of $\mathbf{M}$ are written $\mathbf{M}_{11} = \partial\boldsymbol{\pi}_t/\partial\boldsymbol{\pi}_0$, $\mathbf{M}_{12} = \partial\boldsymbol{\pi}_t/\partial\mathbf{q}_0$, $\mathbf{M}_{21} = \partial\mathbf{q}_t/\partial\boldsymbol{\pi}_0$, and $\mathbf{M}_{22} = \partial\mathbf{q}_t/\partial\mathbf{q}_0$.

Exercise 12.6. Derive the differential matrix equation (12.84) and relate the submatrices of $\mathbf{M}$ to the elements of the Herman–Kluk factor $C_t(\mathbf{X}_0, \mathbf{P}_0)$ (12.48).

A comparison between the SC-IVR approach and the path integral formalism of Pechukas, as summarized at the beginning of this chapter, may shed light on both models. Pechukas arrives at an exact expression for the propagator that consists essentially of a nuclear path integral over the electronic transition amplitude $T_{\beta\alpha}$, defined as a functional of the nuclear path (see Eq. (12.8)). This propagator is simplified by a stationary phase procedure that allows us to approximate the integral by a sum over the classical trajectories of maximum weight. The SC-IVR representation that employs the mapping Hamiltonian (12.78), in contrast, yields, by virtue of the scalar product identity (12.69), an exact result for the electronic transition amplitude. The Pechukas model, involving the separate treatment of nuclear and electronic degrees of freedom, may be recovered as an approximation to SC-IVR theory which achieves symmetry between both sets of dynamic variables.

12.5.1 *The SC-IVR approach applied to nonadiabatic model cases*

In order to assess the SC-IVR representation as sketched in the foregoing paragraphs with respect to its ability to capture prototypical nonadiabatic situations, we will discuss applications of this formalism to the test cases proposed by Tully (see Section 11.2.1), following the account given by Sun and Miller [319]. Each of these three challenges to nonadiabatic theory involves the interaction of one nuclear mode with two electronic states defined in a two-parameter space. The Hamilton function (12.78) therefore reads

$$H = H(\tilde{x}_1, \tilde{p}_1, \tilde{x}_2, \tilde{p}_2, \tilde{X}, \tilde{P}) = \frac{\tilde{P}^2}{2M} + \frac{1}{2}H_{\text{el},11}(\tilde{p}_1^2 + \tilde{x}_1^2 - 1)$$
$$+\frac{1}{2}H_{\text{el},22}(\tilde{p}_2^2 + \tilde{x}_2^2 - 1) + H_{\text{el},12}(\tilde{p}_1\tilde{p}_2 + \tilde{x}_1\tilde{x}_2) \qquad (12.85)$$

with the symmetric matrix $H_{\mathrm{el},ij}(\tilde{x})(i,j = 1,2)$ varying from problem to problem. With appropriate substitutions for $\tilde{x}_n$ and $\tilde{p}_n$, this Hamiltonian is identical with the classical electron analog Hamiltonian (10.32) (see Exercise 12.7). Constructing the propagator, one instantly finds from Eq. (12.67) the relevant harmonic oscillator states, representing the electronic subsystem, as

$$\psi_1(\tilde{x}_1, \tilde{x}_2) = \sqrt{\frac{2}{\pi}}\tilde{x}_1 \exp\left[-\frac{1}{2}(\tilde{x}_1^2 + \tilde{x}_2^2)\right], \tag{12.86}$$

$$\psi_2(\tilde{x}_1, \tilde{x}_2) = \sqrt{\frac{2}{\pi}}\tilde{x}_2 \exp\left[-\frac{1}{2}(\tilde{x}_1^2 + \tilde{x}_2^2)\right]. \tag{12.87}$$

Exercise 12.7. Identify the coefficients c_n introduced in Eq. (12.70) and described as functions of x_n and p_n by Eq. (12.72) with the coeffcients a_k of Eq. (10.2). Generalize the Hamiltonian (12.85) to include M electronic degrees of freedom and prove that the result is identical with the classical electron analog Hamiltonian (10.32). Include the Langer modification when expressing $\tilde{x}_n$ and $\tilde{p}_n$ in terms of q_n and n_n.

The initial nuclear wave function is chosen as a Gaussian wave packet, i.e. as a nuclear coherent state:

$$\chi_1(\tilde{X}) = \left(\frac{f}{\pi}\right)^{\frac{1}{4}} \exp\left[-\frac{f}{2}(\tilde{X} - \tilde{X}_0)^2 + i\tilde{P}_0\tilde{X}\right] \tag{12.88}$$

where $\tilde{X}_0$ is located, with reference to Figure 11.2, in the asymptotic region $\tilde{X} \to -\infty$, and f is a constant. With these assignments, the wave function can in principle be advanced in time by use of Eq. (12.82), and thus by computing the evolution of the nuclear wave functions correlated with the electronic states. The quantities relevant for comparison both with exact quantum theory and with competing nonadiabatic approaches, namely the transmission and reflection probabilities associated with each of the two potential energy surfaces, are finally calculated from

$$P_k^{\mathrm{trans}} = \lim_{t\to\infty} \int_0^{\infty} d\tilde{X}|\chi_k(\tilde{X}, t)|^2, \tag{12.89}$$

$$P_k^{\mathrm{refl}} = \lim_{t\to\infty} \int_{-\infty}^{0} d\tilde{X}|\chi_k(\tilde{X}, t)|^2, \tag{12.90}$$

and $k = 1, 2$. Most recent applications of SC-IVR theory, however, do not utilize the propagator in the Van Vleck–Gutzwiller approximation but the coherent state (or Herman–Kluk) version which involves the propagator as given by Eq. (12.47) [357]. The corresponding requirement of phase space wave functions is already satisfied for the nuclear subsystem by the choice (12.88). Phase space representations of the electronic states are generated by subjecting the functions (12.86) and (12.87) to a coherent state (or *Husimi*) transform according to

$$\psi^g(\tilde{\mathbf{x}}, \tilde{\mathbf{p}}) = \langle g|\psi\rangle = \int d\tilde{\mathbf{x}}' \psi(\tilde{\mathbf{x}}') \left(\frac{\gamma}{\pi}\right)^{\frac{F}{4}} \exp\left[-\frac{\gamma}{2}(\tilde{\mathbf{x}} - \tilde{\mathbf{x}}')^2 + i\tilde{\mathbf{p}}_0(\tilde{\mathbf{x}} - \tilde{\mathbf{x}}')\right]. \tag{12.91}$$

Combining Eqs. (12.88) and (12.91) with Eq. (12.48) and Eq. (12.82), one can write the nuclear contribution to the state n_2 as a function of time:

$$\begin{aligned} \chi_{n_2 \leftarrow n_1}(\tilde{X}_t, t) = \int & d\tilde{X} d\tilde{P} d\tilde{x}_1 d\tilde{p}_1 d\tilde{x}_2 d\tilde{p}_2 \, C_t \chi_1^g(\tilde{X}, \tilde{P}) \psi_{n1}^g(\tilde{x}_1, \tilde{p}_1, \tilde{x}_2, \tilde{p}_2) \\ & \times g(\tilde{X}_t, \tilde{P}_t; \tilde{X}_1) \psi_{n2}^{g*}(\tilde{x}_{1t}, \tilde{p}_{1t}, \tilde{x}_{2t}, \tilde{p}_{2t}) \\ & \times \exp\left[iS_{\mathrm{cl},t}(\tilde{X}, \tilde{P}, \tilde{x}_1, \tilde{p}_1, \tilde{x}_2, \tilde{p}_2)\right]. \end{aligned} \tag{12.92}$$

The factor $g(\tilde{X}_t, \tilde{P}_t; \tilde{X})$ which may be read as the Husimi transform of the delta function contained in the state (12.82) projects the advanced nuclear wave function on the coordinate $\tilde{X}_1$:

$$g(\tilde{X}_t, \tilde{P}_t; \tilde{X}_1) = \langle \tilde{X}_1, g\rangle = \left(\frac{\gamma}{\pi}\right)^{\frac{1}{4}} \exp\left[-\frac{\gamma}{2}(\tilde{X}_1 - \tilde{X}_t)^2 + i\tilde{P}_t(\tilde{X}_1 - \tilde{X}_t)\right]. \tag{12.93}$$

Figure 12.1 shows the SC-IVR results, evaluated according to Eqs. (12.89) and (12.90) for the first test case, involving a simple avoided crossing. Similar conclusions are drawn as from the parallel treatment of this situation by the surface hopping model (Section 11.2.1). The comparison between the exact quantum calculation yields excellent agreement for sufficiently high nuclear momenta but noticeable deviations in the low energy region. This dwindling of precision with the decrease of energy reflects the diminishing accuracy of the SC-IVR approximation as the simulation time becomes long, which is obviously the case for low nuclear momenta.

Essentially the same observations are made for the two remaining test cases. For case 3 (Figure 11.2(c)), this is documented by the results shown

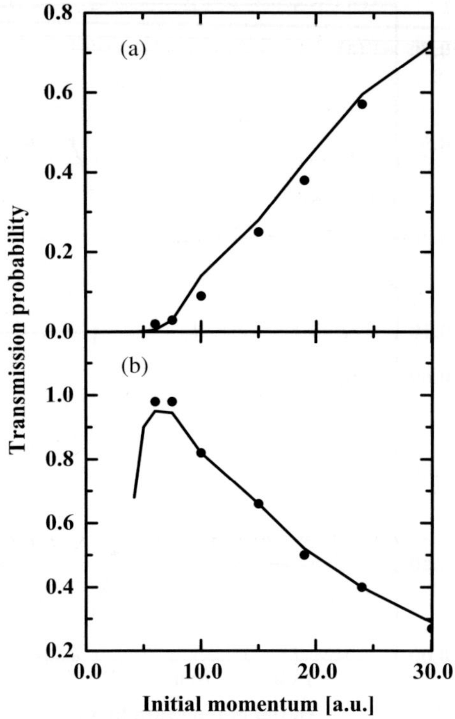

Figure 12.1: Results from the SC-IVR treatment of test case 1: Single avoided crossing (Figure 11.2(a)), in comparison with results of quantum mechanical calculations, represented by the filled circles. In the upper and the lower panel, transmission probabilities are shown as a function of the initial nuclear momentum for the upper and the lower adiabatic state, respectively. The solid lines refer to the quantum mechanical solution, the filled circles to the SC-IVR results. (Reprinted with permission from [319]. Copyright (1997) by the American Institute of Physics.)

in Figure 12.2 which displays probabilities for transmission and reflection due to the two diabatic states, involving an extended coupling region with reflection. Since the states are very close in energy, their respective transmission and reflection probabilities deviate only minimally from each other. Thus, each of the results displayed in Figure 12.2 refers to both states. The threshold behavior of the quantal reflection curve, as is inherent in the physical situation (see Figure 11.2(c)), is well reproduced by the SC-IVR approximation. Also, the latter does not give rise to the unphysical oscillation phenomena which are generated by the surface hopping model discussed in Chapter 11 (see Figure 11.5).

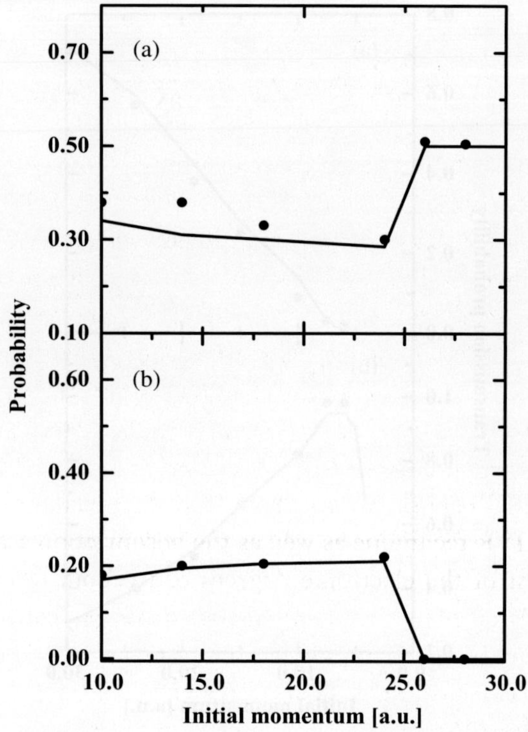

Figure 12.2: Results for test case 3: Extended coupling with reflection (see Figure 11.2(c)) based on SC-IVR approximation and quantal treatment. The upper and lower panels refer to transmission and reflection probabilities, respectively, of both diabatic states. The meaning of the lines and the dots is as in Figure 12.1. (Reprinted with permission from [319]. Copyright (1997) by the American Institute of Physics.)

12.5.2 *Comparison with the Ehrenfest model*

Concluding our introduction to the use of the mapping technique to model nonadiabatic dynamics, we consider the force experienced by the nuclei, as derived from the Hamiltonian (12.85):

$$\begin{aligned} M\frac{\partial^2\tilde{\mathbf{X}}}{\partial t^2} &= -\frac{1}{2}(\tilde{p}_1^2+\tilde{x}_1^2-1)\frac{\partial H_{\mathrm{el},11}}{\partial\tilde{\mathbf{X}}} \\ &\quad -\frac{1}{2}(\tilde{p}_2^2+\tilde{x}_2^2-1)\frac{\partial H_{\mathrm{el},22}}{\partial\tilde{\mathbf{X}}} - (\tilde{p}_1\tilde{p}_2+\tilde{x}_1\tilde{x}_2)\frac{\partial H_{\mathrm{el},12}}{\partial\tilde{\mathbf{X}}}. \qquad (12.94) \end{aligned}$$

In view of the relation between the electronic position as well as momentum coordinates and the coefficients c_k, made explicit in relation (12.72), this becomes

$$M\frac{\partial^2 \tilde{\mathbf{X}}}{\partial t^2} = -\sum_{k,k'=1}^{2} c_k^* \frac{\partial H_{\text{el},kk'}}{\partial \tilde{\mathbf{X}}} c_{k'}. \tag{12.95}$$

This expression is readily identified as an average force whose components are weighted according to the coefficients of the contributing electronic states. Thus, the SC-IVR mapping approach yields an Ehrenfest force expression (compare with Eq. (8.104)), a realization that may, at face value, disappoint hopes that this approach represents a substantial step beyond the pure Ehrenfest methodology as realized, for instance, by the TDHF and END procedures discussed in Chapters 8 and 9, respectively. This impression, however, is deceiving, since the average arises here from nuclear motions *proceeding on distinct potential energy surfaces*, not on an artificial averaged potential energy surface. For deeper understanding of this feature of the SC-IVR technique as well as the bosonization strategy followed in the treatment of the electronic degrees of freedom (their mapping on harmonic oscillator states), one may invoke a comparison with the WKB scheme [356]. For bounded motion in a one-dimensional potential, the classical action variable emerges as quantized (see Section 4.2.2). In the case of the mapping Hamiltonian, however, the action variable corresponds to the electronic degree of freedom n_k (see Exercise 12.7) which is understood as a measure of the population of the electronic state with label k. Further, the dynamics of the problem conserves the single quantum of excitation that is present in the two oscillators considered, such that (n_1, n_2) will be realized by (1, 0) or (0, 1), corresponding to the electronic ground state or the first excited state, respectively. As a result, the nuclei propagate exclusively on the physical potential energy surfaces, and Ehrenfest trajectories that advance on an average potential energy surface can be shown to vanish as a result of destructive interference. These findings have been confirmed by a simulation employing the *forward-backward version* of the SC-IVR formalism (see Section 12.7.3) [358]. Among other tests, a scenario involving two electronic states was investigated for the case of a simple avoided crossing, as shown in Figure 11.2(a). Analysis of the distribution of final momenta yields, for sufficiently high initial energy, two clearly separated maxima, corresponding to propagation on the two distinct potential energy surfaces, whereas Ehrenfest dynamics gives rise to a single peak.

12.6 Case Study: The S_1–S_2 Transition in Pyrazine: SC-IVR Treatment of a Conical Intersection Problem

While the test cases described in Section 11.2.1 set standards that any method of nonadiabatic quantum dynamics is expected to satisfy, they involve highly idealized paradigmatic situations and thus convey nothing about the practicability of the examined method. Ultimately, any useful procedure has to prove viable not only when compared with theories of higher accuracy but also with experimental data. In this section, we describe an application of the SC-IVR formalism in conjunction with a bosonization technique to a physical system, namely the pyrazine molecule (shown in Figure 12.3) which exhibits nonadiabatic dynamics following photoexcitation [359]. The ${}^1B_{3u}(n\pi^*)$ and ${}^1B_{2u}(\pi\pi^*)$ excited states in pyrazine, termed S_1 and S_2, exhibit a conical intersection [360].

The existence of only one coupling mode (labeled $\tilde{X}_{10a}$) besides five totally symmetric tuning modes ($\tilde{X}_1, \tilde{X}_2, \tilde{X}_{6a}, \tilde{X}_{8a}, \tilde{X}_{9a}$) makes this system a touchstone for the treatment of conical intersections in models of nonadiabatic dynamics. Figure 12.4 gives illustrations of the pyrazine S_1-S_2 conical intersection as represented in the reduced space spanned by the vibronic coordinates $\tilde{X}_{10a}$ and $\tilde{X}_{6a}$. Over the last two decades, the problem of the electronic–vibronic coupling between the first two excited states in pyrazine has been considered at various increasingly complex levels of theory which greatly differ in their response to the challenge posed by the large vibrational space of this molecule, comprising 24 coordinates. The most active among these modes have been included selectively by some authors ([362]– [367]). Alternative approaches incorporate the four most important modes, while the remaining 20 are simulated as a bath that is weakly coupled to the system

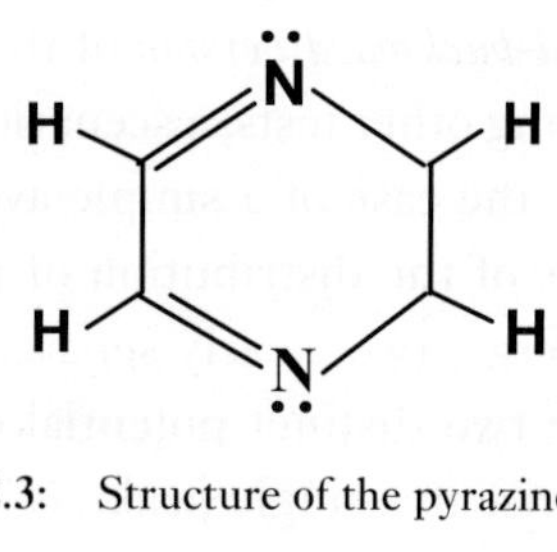

Figure 12.3: Structure of the pyrazine molecule.

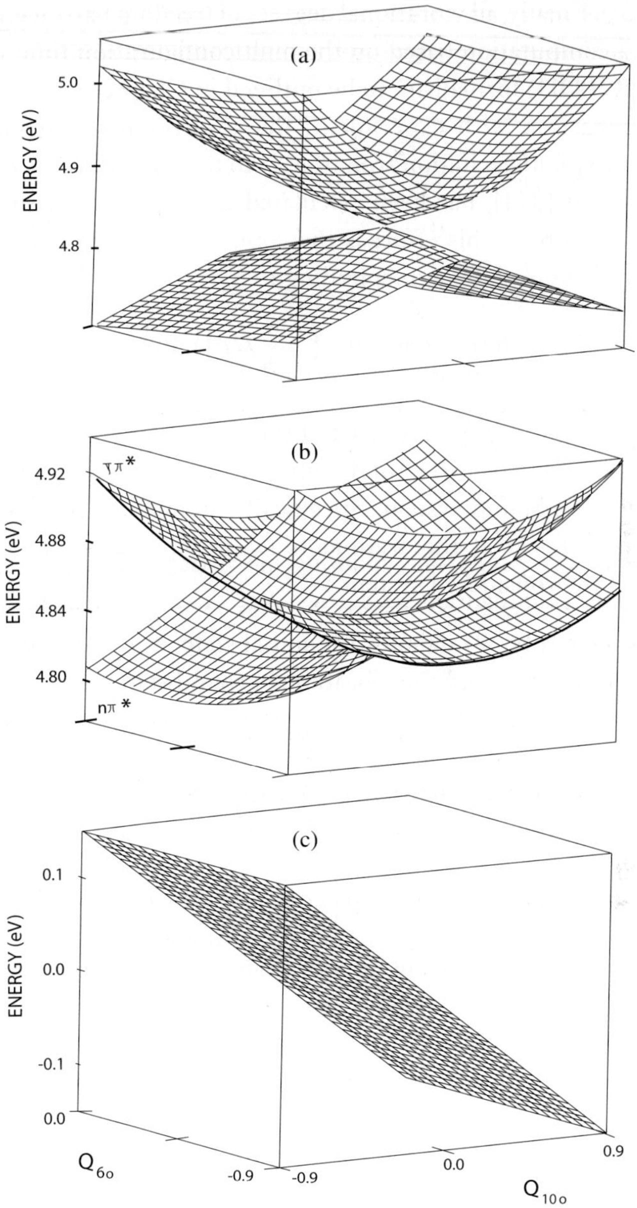

Figure 12.4: Various perspectives of the S_1–S_2 conical intersection in pyrazine in a subspace defined by the vibration coordinates $\tilde{X}_{6a}$ and $\tilde{X}_{10a}$: (a) the adiabatic surfaces, (b) the diagonal elements $H_{11}(n\pi^*)$ and $H_{22}(\pi\pi^*)$ of the diabatic Hamiltonian, (c) the non-diagonal diabatic matrix element H_{12}. (Reprinted with permission from [361]. Copyright (1994) by the American Institute of Physics.)

([368]–[373]). Finally, all vibrational degrees of freedom have been taken into account in a computation based on the multiconfiguration time-dependent Hartree (MCTDH) formalism, to be outlined in Chapter 15.

For comparison with experiment, the absorption spectrum intensity $I(\omega)$ related to photoexcitation from the S_1 to the S_2 surface, as determined by measurement [374], has been calculated as a function of the radiation energy. At the core of this effort is the computation of the *autocorrelation function* of the target state corresponding to S_2:

$$\mathcal{J}(t) \equiv \left\langle \Psi_2 \left| \exp\left(-\frac{i}{\hbar}\hat{H}t\right) \right| \Psi_2 \right\rangle. \tag{12.96}$$

Exercise 12.8. Use the discrete part of the expansion (3.1) to demonstrate that the Fourier transform $\tilde{\mathcal{J}}(\omega)$ of the autocorrelation function (12.96) yields the spectrum of a bound system. Specifically, show that the maxima of the function $\tilde{\mathcal{J}}(\omega)$ appear at the arguments $\omega_i = E_i$, with E_i as the energy eigenvalue of the considered system.

More specifically, Ψ_2 denotes the initial state, a product of the second excited electronic state and the vibrational ground state:

$$|\Psi_2\rangle = |\psi_2\rangle |v_0\rangle. \tag{12.97}$$

With these assignments, one may obtain the absorption spectrum from a relation derived originally by Heller [375]:[1]

$$I(\omega) \propto \omega \Re \left\{ \int_0^\infty dt \, \exp\left[i\left(\omega + \frac{\varepsilon_2}{\hbar}\right) t \right] \mathcal{J}(t) \right\}, \tag{12.98}$$

where ε_2 stands for the energy of the vibrational ground state of the molecule. The formulae (12.93)–(12.98) reduce the essential computational task to time propagation of the second excited state according to (12.96). The mapping approach (12.69) yields for this situation:

$$\left\langle \psi_2 \right| \left\langle v_0 \left| \exp\left(-\frac{i}{\hbar}\hat{H}t\right) \right| v_0 \right\rangle \left| \psi_2 \right\rangle$$
$$= \langle 0_1 \ldots 1_n \ldots 0_N | \langle v_0 | \exp\left(-\frac{i}{\hbar}\tilde{H}t\right) |v_0\rangle |0_1 \ldots 1_n \ldots 0_N\rangle. \tag{12.99}$$

---

[1] See [376] for the definition of the absorption cross section in the framework of time-independent theory.

Since the case at hand involves two electronic states, $N = 2$. The harmonic oscillator eigenfunctions $\langle 1|\langle 0|$ and $\langle 0|\langle 1|$ represent the lower and the higher excited state of pyrazine, respectively. One might now write the auto-correlation function in analogy to the propagator (12.81). Employing the Herman–Kluk representation, one finds

$$\mathcal{J}(t) = \int d\tilde{\mathbf{x}}_0 d\tilde{\mathbf{p}}_0 \int d\tilde{\mathbf{X}}_0 d\tilde{\mathbf{P}}_0 \left(\frac{1}{2\pi}\right)^{F+N} \langle 0|\tilde{x}_{1t}\tilde{p}_{1t}\rangle\langle 1|\tilde{x}_{2t}\tilde{p}_{2t}\rangle\langle v_0|\tilde{\mathbf{X}}_t\tilde{\mathbf{P}}_t\rangle$$
$$\times C_t \exp(iS_t[\tilde{\mathbf{x}}_0, \tilde{\mathbf{p}}_0, \tilde{\mathbf{X}}_0, \tilde{\mathbf{P}}_0])\langle\tilde{\mathbf{X}}_0\tilde{\mathbf{P}}_0|v_0\rangle\langle\tilde{x}_{10}\tilde{p}_{10}|0\rangle\langle\tilde{x}_{20}\tilde{p}_{20}|1\rangle. \tag{12.100}$$

As in the case of propagator calculations, the presence of an oscillating factor in the integrand makes evaluating this multidimensional integral a formidable task. As demonstrated in [359], this challenge can be overcome by an adaptation of a smoothing method due to Walton and Manolopoulos [377]. We will further comment on this and related auxiliary techniques in Section 12.7.

In the following, we want to summarize the hierarchy of SC-IVR approaches that have been used to describe the $S_1 \rightarrow S_2$ transition in pyrazine, where both the fully quantal MCTDH simulation (see Section 15.3 for details) and the experimental absorption spectrum will serve as criteria for the performance of the SC-IVR method. For the most basic implementation of this method [359], Thoss *et al.* discuss a truncated model that reduces the full space of vibration coordinates to the non-totally symmetric coupling mode ($\tilde{X}_{10a}$) and the subclass G of the three most strongly coupled symmetric modes ($G = \{\tilde{X}_1, \tilde{X}_{6a}, \tilde{X}_{9a}\}$), as introduced in [361]. The Hamiltonian (12.78) is then determined by the diabatic matrix:

$$\mathbf{H}_{\text{el}} = \begin{pmatrix} E_1 + \sum_{j\varepsilon G} \kappa_j^{(1)}\tilde{X}_j & \lambda\,\tilde{X}_{10a} \\ \lambda\,\tilde{X}_{10a} & E_2 + \sum_{j\varepsilon G}\kappa_j^{(2)}\tilde{X}_j \end{pmatrix}. \tag{12.101}$$

The variables $\tilde{X}_j$, $\tilde{X}_{10a}$ refer here to properly symmetrized vibronic coordinates. The coupling parameters λ and $\kappa_j (j = 1, 6a, 9a)$ are obtained from *ab initio* calculations [361] based on the multireference configuration interaction method (MRCI, see Section 1.6).

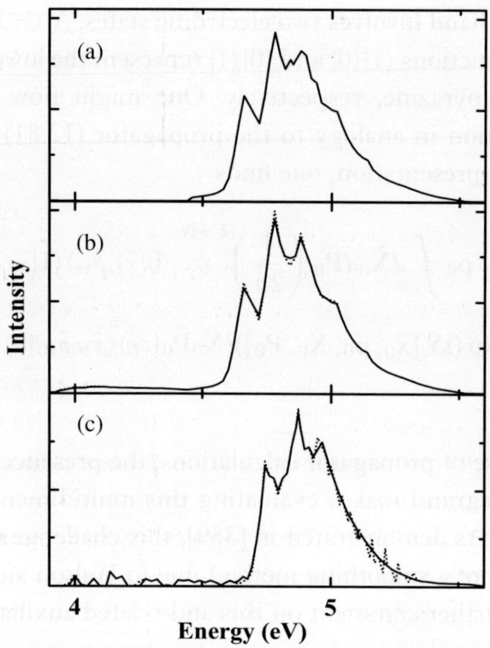

Figure 12.5: Pyrazine absorption spectrum in the vicinity of the S_1-S_2 conical intersection. Comparison is made between experimental data (panel (a), [374]) and the result of the four-mode approach described in the text (b). Further, panel (c) refers to a computation based on a 24 mode model. The solid lines in (b) and (c) refer to MCTDH [378], the dotted lines to SC-IVR computation. Dephasing factors $\exp(-\frac{t}{T_2})$ have been applied with $T_2 = 30$ fs (b) and $T_2 = 150$ fs (c) to smooth the calculated distributions. (Reprinted with permission from [359]. Copyright (2000) by the American Institute of Physics.)

Evaluating the autocorrelation function (12.96) and subsequently applying Eq. (12.98) yields the absorption spectrum that may be compared with the respective experimental finding. The main discrepancy between model and measurement lies in the finer structural details of the distribution which turn out to be exaggerated by the theoretical treatment. However, inclusion of an empirical dephasing factor by substituting $\mathcal{J}(t)\exp(-\frac{t}{T_2})$ for $\mathcal{J}(t)$ in (12.98), with $T_2 = 30$ fs, leads to acceptable agreement [359].

The consequence to be drawn from this finding is that the dissipative dynamics of the $S_1 - S_2$ interaction in pyrazine is insufficiently described by including only four modes. An improved approximation may be considered that accounts qualitatively for the remaining 20 normal modes by implementing a system-bath model, as suggested in [368] and [369]. This approach employs the Hamiltonian

$$\mathbf{H}_{\mathrm{S-B}} = \mathbf{H} + \sum_{j=1}^{20} \frac{\omega_j}{2}(\tilde{P}_j^2 + \tilde{X}_j^2)\mathbf{I} + \begin{pmatrix} \sum_{j=1}^{20} \kappa_j^{(1)} \tilde{X}_j & 0 \\ 0 & \sum_{j=1}^{20} \kappa_j^{(2)} \tilde{X}_j \end{pmatrix}. \tag{12.102}$$

The symbol **H** stands here for the four-mode model Hamiltonian, as used in the minimal realization discussed earlier. Formula (12.102) includes the bath modes on a phenomenological level. Thus, 18 of the 20 modes that comprise the bath are non-totally symmetric and cannot contribute to the linear coupling assumed in (12.102). Still this expression makes it possible to study the overall impact of dissipation on the shape of the absorption spectrum as a function of the number of contributing modes. The constants $\kappa_j^{1(2)} (j = 1, 20)$ are chosen randomly, subject to the constraint of weak coupling between the system and the bath. Coupling to the bath leads to a broadened distribution similar to what is shown in Figure 12.5(b).

The model (12.101) has further been generalized to include all 24 modes of the pyrazine problem. Acknowledging that the non-totally symmetric vibrational coordinates give rise to quadratic terms in the Hamiltonian, the diabatic matrix elements were truncated at second order. The result of the full 24 mode approach is seen in Figure 12.5(c). A phenomenological broadening of $T_2 = 150$ fs is included, accommodating the finite experimental resolution as well as rotational motion. For each of the three stages of implementation considered — the four-mode approach, the system-bath model, and the extension to all 24 modes — the semiclassical computation compares well with the quantal results.

12.7 Numerical Procedures for Semiclassical Propagation Methods

Although the SC-IVR method is a distinct improvement over procedures that necessitate root search, its usefulness is limited by the complexity of the integrals to be evaluated, as exemplified by Eqs. (12.81) and (12.100). To find application in reaction dynamics, the SC-IVR approximation must be capable of dealing successfully with expressions considerably more complex than the latter examples. Computing transition rates, for instance, involves the time correlation function (see Eq. (5.81b))

$$C_{AB}(t) = \mathrm{Tr}\left\{\hat{A}\exp\left(\frac{i}{\hbar}\hat{H}t\right)\hat{B}\exp\left(-\frac{i}{\hbar}\hat{H}t\right)\right\}. \tag{12.103}$$

In this formula, $\hat{A}$ might be given by the canonical density

$$\hat{A} = \frac{\exp(-\beta\hat{H})}{\mathrm{Tr}\exp(-\beta\hat{H})}, \tag{12.104}$$

while $\hat{B}$ is determined by the considered interaction. To translate the time correlation function into the SC-IVR formalism, one may use the time propagator representation (12.47) twice, arriving at [379]

$$\begin{aligned} C_{AB}(t) = {} & \left(\frac{1}{2\pi\hbar}\right)^{2F} \int dX_0 \int dP_0 \int dX'_0 \int dP'_0 \; C_t(X_0, P_0) C_t(X'_0, P'_0) \\ & \times \exp\left[\frac{i}{\hbar}(S'_t(X_0, P_0) - S_t(X'_0, P'_0))\right] \\ & \times \langle P_0, X_0|\hat{A}|P'_0, X'_0\rangle\langle P'_t, X'_t|\hat{B}|P_t, X_t\rangle. \end{aligned} \tag{12.105}$$

Applying SC-IVR theory in the context of transition processes thus meets with the substantial difficulty of performing the double phase space average over an oscillating integrand. Besides various customary numerical tools utilized in IVR theory, this section will highlight the so-called *forward-backward* (FB-IVR) variant of the IVR approximation that has been introduced to reduce the dimensionality of IVR integrals as exemplified by Eq. (12.105), and has been shown to be capable of overcoming some of the chief numerical obstacles marring the SC-IVR method.

By the semiclassical nature of the SC-IVR approximation, i.e. its reliance on classical trajectories, it encounters difficulties if the classical equivalent of the studied system is chaotic. As has been shown, for instance by considering the two-dimensional quartic oscillator potential [380] or the two-dimensional Hénon–Heiles bound state problem [381], long-lived chaotic trajectories give rise to highly singular contributions to the SC-IVR propagator in the Herman–Kluk representation. As this tendency is related to the irregular behavior of the Herman–Kluk prefactor $C_t(X_0, P_0)$ in the chaotic case, and, more specifically, an exponential increase of the underlying monodromy matrix elements with time [380], the suggestion has been made to eliminate pathological cases from the trajectory bundle by imposing a cutoff function on $|C_t(X_0, P_0)|^2$, such that a given trajectory is discarded upon crossing the numerical threshold thus defined. This pragmatic procedure, which is based on the assumption that the contributions from unstable trajectories will cancel out, has met with considerable success when applied to some model systems [380]. However, the cutoff function is chosen arbitrarily, and therefore the method does not provide criteria that would allow us to draw the demarcation line between retained and eliminated trajectories on grounds of physical arguments. In this section, we introduce several alternative ways to address the problem of trajectory instability in the limit of long propagation times. This problem is interwoven with the two other difficulties to be faced by any numerical realization of the SC-IVR formalism, as mentioned earlier. From the discussion given so far, the main tasks are associated with

(1) the occurence of chaotic trajectories,
(2) the high dimensionality of the SC-IVR integrals, and
(3) the presence of strongly oscillatory integrands.

In general, implementation of SC-IVR theory will have to stand the test of these three challenges. In what follows, we present some computational tools that have been proven valuable in these pursuits. Specifically, we will emphasize *Monte Carlo integration* as a means to alleviate the high-dimensionality problem, and the *stationary Monte Carlo* as well as the *Filinov filtering* procedure, involving strategies to reduce the strong time dependence of IVR-SCF integrands. The *forward-backward* version of IVR theory is effective in addressing both types of problems.

12.7.1 *Monte Carlo integration*

The main distinction of the Monte Carlo method as compared to conventional integration schemes, such as the trapezoidal or Simpson algorithms, is the use of random numbers. For example, a one-dimensional integral

$$I = \int_{\mathrm{a}}^{b} dx\, f(x) \tag{12.106}$$

is approximated as

$$I \approx \frac{b-a}{N} \sum_{n=1}^{N} f(x_i) = (b-a)\langle f \rangle \qquad (12.107)$$

where the arguments x_i are chosen at random within the interval $[a, b]$ and $\langle \rangle$ denotes the arithmetic mean over N sample points. A measure of the uncertainty of this procedure is given by the variance σ_f which indicates the mean of the squared deviation between f and its average value in $[a, b]$:

$$\sigma_f^2 = \frac{1}{N} \sum_{n=1}^{N} f(x_i)^2 - \left(\frac{1}{N} \sum_{n=1}^{N} f(x_i) \right)^2. \qquad (12.108)$$

The Monte Carlo integration error is then found from

$$\sigma_I^2 = \frac{(b-a)^2}{N} \sigma_f^2 \qquad (12.109)$$

and, therefore, the error decreases as $\sigma_I \approx 1/\sqrt{N}$. By the standard of traditional integration algorithms, this scaling with the number of points may not be viewed as too favorable. For instance, the error of the iterated trapezoidal method scales with $1/N^2$, that of the iterated Simpson method with $1/N^4$. However, the Monte Carlo algorithm has the advantage that σ_I does not depend on the dimensionality of the problem, which makes it, in most cases, the evaluation method of choice for multidimensional integrals.

The statistical nature of the Monte Carlo strategy makes it possible to reduce the task of evaluating integrals over an arbitrarily shaped m-dimensional domain to integrating over a regular reference volume, such as a hypercube in m dimensions which contains the domain. This scheme amounts to approximating the integral over some multidimensional region V by a sum over N values adopted by the integrand in the simpler reference volume V_{ref}, according to

$$\int_V d^m x\, f(\mathbf{x}) = V_{\mathrm{ref}} \frac{1}{N} \sum_{i=1}^{N} f(\mathbf{x}_i) \delta_{\mathbf{x}_i, V} \qquad (12.110)$$

with $\delta_{x_i,V} = 1$ if x_i lies in V, and $\delta_{x_i,V} = 0$ otherwise [382]. The fundamental formula for numerical integration by use of the Monte Carlo technique reads

$$\int dV f = \lim_{N \to \infty} V \left(\langle f \rangle_N \pm \sqrt{\frac{\langle f^2 \rangle_N - \langle f \rangle_N^2}{N}} \right). \qquad (12.111)$$

To hold the dimension independence of the Monte Carlo error against conventional numerical integration techniques that may be characterized by a step size h and an error $O(h^k)$, we observe that the latter involve $N = (\frac{L}{h})^m$ integration points if the integral is m-dimensional and carried out over a hypercube of side length L. It follows that the error of such a method scales as $N^{-\frac{k}{m}}$ and thus Monte Carlo integration is superior to a traditional algorithm of order k if $m > 2k$.

The stationary phase Monte Carlo method

Several recent proposals aim at overcoming, or mellowing, the computational problems posed by the oscillatory nature of the integrands occurring in semiclassical propagator theory. Various rivaling approaches may be credited with substantial progress in making the unruly integrals more tractable, while, at this juncture, none of them distinguishes itself as the most efficient one for all applications. Here we will give an account of a prototypical Monte Carlo-based method that, designed to address the integrability problem, was also shown to alleviate the instability problem, and thus to be of substantial use for treating divergent trajectories in the limit of long propagation times.

The idea of the *stationary phase Monte Carlo* method is introduced most easily by considering a one-dimensional case, namely the integral

$$I(t) \equiv \int_{-\infty}^{\infty} dx\, \rho(x) \exp\left[it\Phi(x)\right]. \tag{12.112}$$

In the next step, one reformulates I by introducing a *filter function* $D(x)$ such that [383]

$$I(t) = \int_{-\infty}^{\infty} dx\, \rho(x) D(x) \exp\left[it\Phi(x)\right] \tag{12.113}$$

where $D(x)$ is defined as

$$D(x) \equiv \int_{-\infty}^{\infty} dy\, P(y) \frac{\rho(x-y)}{\rho(x)} \exp\left[it(\Phi(x-y) - \Phi(x))\right]. \tag{12.114}$$

In this expression, the function $P(y)$ stands for an arbitrary normalized probability distribution. With this assignment, the right-hand sides of Eqs. (12.112) and (12.113) are identical, provided the variable x extends over an interval that is infinite or in which the function $\Phi(x)$ is periodic.

Exercise 12.9. Using the definition (12.114), show that the integrals (12.112) and (12.113) are equal.

In general, Eq. (12.114) is clearly not more amenable to evaluation than Eq. (12.112). We may, however, specify $P(y)$ as the narrow Gaussian function

$$P(y) = \frac{1}{\varepsilon\sqrt{2\pi}} \exp\left(-\frac{y^2}{2\varepsilon^2}\right)$$

which tends towards the delta function $\delta(y)$ in the limit of vanishing ε. Thus, for sufficiently small ε, we may set $\rho(x-y) \approx \rho(x)$ and also expand the function $\Phi(x)$ up to first order, thereby reducing the integral (12.114) to

$$D(x) \approx \frac{1}{\varepsilon\sqrt{2\pi}} \int_{-\infty}^{\infty} dy \exp\left(-\frac{y^2}{2\varepsilon^2}\right) \exp\left(-it\Phi'(x)y\right)$$

with $\Phi'(x) = \frac{d\Phi(x)}{dx}$. Applying the formula

$$\int_{-\infty}^{\infty} \exp\left(-u\xi^2\right) \exp\left(v\xi\right) d\xi = \sqrt{\frac{\pi}{u}} \exp\left(\frac{v^2}{4u}\right),$$

we arrive at

$$D(x) \approx \exp\left[-\frac{1}{2}(\varepsilon t \Phi'(x))^2\right]. \tag{12.115}$$

Evidently, $D(x)$ is sensitive to the phase in the integrand of Eq. (12.112), acting as a filter that discriminates against regions where $\Phi(x)$ varies strongly, and becoming transparent where $\Phi(x)$ is approximately constant. Therefore, replacing the distribution $\rho(x)$ with $\rho(x)D(x)$ serves the desired purpose of smoothing the integrand by deemphasizing the oscillatory nature of $\exp(it\Phi(x))$. Although these arguments have been developed with reference to a one-dimensional situation, they carry over readily to the multidimensional case.

12.7.2 *Filinov filtering*

In the following, we will give a brief outline of an expedient smoothing, or filtering technique, originally introduced by Filinov [384], which has turned out to be conveniently adaptable to the challenge of SC-IVR integrals.

As in the previous section, the method applies to a general F-dimensional integral of the type

$$I = \int_{-\infty}^{\infty} d\mathbf{z}\, R(\mathbf{z}) \exp\left[i\Phi(\mathbf{z})\right] \tag{12.116}$$

where $R(\mathbf{z})$ denotes a slowly varying function, and Φ a complex function,

$$\Phi(\mathbf{z}) \equiv \phi(\mathbf{z}) + i\vartheta(\mathbf{z}), \tag{12.117}$$

with real-valued ϕ and ϑ. The Filinov approach aims at eliminating from I the high-frequency contributions due to $\exp[i\phi(\mathbf{z})]$ which are of minor influence on the value of I but greatly complicate evaluating the integral by Monte Carlo methods. The basic idea is to incorporate into Eq. (12.116) the following representation of unity [385]:

$$1 = \sqrt{\frac{\det(\boldsymbol{\alpha})}{\pi^F}} \int_{-\infty}^{\infty} d\mathbf{z}_0 \exp\left[-(\mathbf{z}-\mathbf{z}_0)^{\mathrm{T}}\boldsymbol{\alpha}(\mathbf{z}-\mathbf{z}_0)\right]. \tag{12.118}$$

This identity introduces a $F \times F$ Gaussian width parameter matrix $\boldsymbol{\alpha}$ which is to be adjusted to suppress the oscillations of the integrand in expression (12.117) with maximum efficiency. One further replaces $\Phi(\mathbf{z})$ with its expansion about $\mathbf{z}_0$ up to second order in $(\mathbf{z}-\mathbf{z}_0)$,

$$\Phi(\mathbf{z}) \approx \Phi(\mathbf{z}_0) + \boldsymbol{\Phi}'(\mathbf{z}_0)^{\mathrm{T}}(\mathbf{z}-\mathbf{z}_0) + \frac{1}{2}(\mathbf{z}-\mathbf{z}_0)^{\mathrm{T}}\boldsymbol{\Phi}''(\mathbf{z}_0)(\mathbf{z}-\mathbf{z}_0),$$

with

$$\boldsymbol{\Phi}'(\mathbf{z}_0) = \frac{\partial\Phi(\mathbf{z}_0)}{\partial\mathbf{z}_0}$$

as the gradient and

$$\boldsymbol{\Phi}''(\mathbf{z}_0) = \frac{\partial^2\Phi(\mathbf{z}_0)}{\partial\mathbf{z}_0\partial\mathbf{z}_0}$$

as the Hessian matrix of Φ, to approximate I by

$$I \approx \int_{-\infty}^{\infty} d\mathbf{z}_0\, R(\mathbf{z}_0)\sqrt{\frac{\det(2\boldsymbol{\alpha})}{\det(2\boldsymbol{\alpha} - i\boldsymbol{\Phi}''(\mathbf{z}_0))}}$$

$$\times \exp\left[i\Phi(\mathbf{z}_0) - \frac{1}{2}\boldsymbol{\Phi}'(\mathbf{z}_0)^{\mathrm{T}}[2\boldsymbol{\alpha} - i\boldsymbol{\Phi}''(\mathbf{z}_0)]^{-1}\boldsymbol{\Phi}'(\mathbf{z}_0)\right] \tag{12.119}$$

which holds under the condition $R(\mathbf{z}) \approx R(\mathbf{z}_0)$. This term clarifies the role of the matrix $\boldsymbol{\alpha}$ as an agent that determines the accuracy of the approximation to the original integral (12.116) and, at the same time, plays the crucial role of a filter that reduces the high-frequency oscillations. Thus, if $|\boldsymbol{\alpha}| \to \infty$, Eq. (12.116) is recovered. On the other hand, the smaller the values adopted by $\boldsymbol{\alpha}$, the more effective it is in suppressing the undesired oscillatory features. This can be seen by inspecting the phase of the integrand, substituting the decomposition (12.117) for $\Phi(\mathrm{z}_0)$ and neglecting the Hessian of the real part of $\Phi(\mathrm{z}_0)$, i.e. $\boldsymbol{\phi}''(\mathrm{z}_0)$. The result is a form of Eq. (12.119) in which the periodic part of the exponential function becomes explicit:

$$I \approx \int_{-\infty}^{\infty} d\mathrm{z}_0\, R(\mathrm{z}_0)\sqrt{\frac{\det(2\boldsymbol{\alpha})}{\det(2\boldsymbol{\alpha} - \boldsymbol{\vartheta}''(\mathrm{z}_0))}}\, \exp[i\{\phi(\mathrm{z}_0) - \boldsymbol{\vartheta}'(\mathrm{z}_0)^{\mathrm{T}} \times [2\boldsymbol{\alpha} + \boldsymbol{\vartheta}''(\mathrm{z}_0)]^{-1}\boldsymbol{\phi}'(\mathrm{z}_0)\}] \times \exp\left[-\vartheta(\mathrm{z}_0) - \frac{1}{2}\boldsymbol{\phi}'(\mathrm{z}_0)^{\mathrm{T}} \times [2\boldsymbol{\alpha} + \boldsymbol{\vartheta}''(\mathrm{z}_0)]^{-1}\boldsymbol{\phi}'(\mathrm{z}_0) + \frac{1}{2}\boldsymbol{\vartheta}'(\mathrm{z}_0)^{\mathrm{T}}[2\boldsymbol{\alpha} + \boldsymbol{\vartheta}''(\mathrm{z}_0)]^{-1}\boldsymbol{\vartheta}'(\mathrm{z}_0)\right]. \tag{12.120}$$

To minimize oscillations, one has to ask for the conditions that make the phase of the integral (12.120) stationary. The requirement is that

$$\frac{d}{d\mathrm{z}_0}\{\phi(\mathrm{z}_0) - \boldsymbol{\vartheta}'(\mathrm{z}_0)^{\mathrm{T}}[2\boldsymbol{\alpha} + \boldsymbol{\vartheta}''(\mathrm{z}_0)]^{-1}\boldsymbol{\phi}'(\mathrm{z}_0)\} = \mathbf{0} \tag{12.121}$$

or, using the assumption $\boldsymbol{\phi}''(\mathrm{z}_0) \approx \mathbf{0}$ and $\boldsymbol{\vartheta}'''(\mathrm{z}_0) \approx \mathbf{0}$,

$$\boldsymbol{\phi}'(\mathrm{z}_0) - \boldsymbol{\vartheta}''(\mathrm{z}_0)^{\mathrm{T}}[2\boldsymbol{\alpha} + \boldsymbol{\vartheta}''(\mathrm{z}_0)]^{-1}\boldsymbol{\phi}'(\mathrm{z}_0) = \mathbf{0}. \tag{12.122}$$

This equation implies $\boldsymbol{\alpha} = \mathbf{0}$. The limit of infinite $|\boldsymbol{\alpha}|$, restoring the original integral (12.116), retains its oscillatory character, while the opposite extreme of vanishing $\boldsymbol{\alpha}$ minimizes the phase variation but is, by the dependence of Eq. (12.120) on $\boldsymbol{\alpha}$, clearly not a viable choice. Thus, the optimal selection of $\boldsymbol{\alpha}$ will establish a compromise between filtering efficiency and approximation accuracy.[2]

[2] In [385], a generalized transformation is discussed that adds flexibility to the Filinov filtering method as sketched in the preceding paragraphs. For this purpose, the decomposition of unity (12.118) is modified to include a term linear in $(\mathrm{z} - \mathrm{z}_0)$ according to

$$\exp[-(\mathrm{z} - \mathrm{z}_0)^{\mathrm{T}}\boldsymbol{\alpha}(\mathrm{z} - \mathrm{z}_0)] \to \exp[-(\mathrm{z} - \mathrm{z}_0)^{\mathrm{T}}\boldsymbol{\alpha}(\mathrm{z} - \mathrm{z}_0) - \boldsymbol{\beta}^{\mathrm{T}}(\mathrm{z} - \mathrm{z}_0)] \tag{12.123}$$

12.7.3 *The forward-backward initial value representation*

At the beginning of this section, we pointed out that the ubiquitous time correlation function $C_{AB}(t)$ can be calculated in the framework of SC-IVR theory only at the cost of very sizeable computational expense, since its SC-IVR representation (12.105) involves a double phase space average, complicated further by the oscillatory character of the integrand. This problem has been addressed by a linearized approach to the SC-IVR model (LSC-IVR) [386] which is based on the assumption that the only substantial contribution to the time correlation function stems from trajectories infinitesimally close to each other. This approximation operates with Dirac position eigenstates rather than with coherent states and involves a transformation from the coordinates $\mathbf{X}_0$ and $\mathbf{X}'$ to their midpoint $\mathbf{X}_m = \frac{1}{2}(\mathbf{X}+\mathbf{X}')$ and their difference $\Delta\mathbf{X} = \mathbf{X} - \mathbf{X}'$. The approach yields the time correlation function as an integral over a product of Wigner functions ([356], [386], [387], see Section 4.3.1):

$$C_{AB}(t) = \left(\frac{1}{2\pi\hbar}\right)^F \int d\mathbf{X}_0 \int d\mathbf{P}_0\, A_\mathrm{w}(\mathbf{X}_0, \mathbf{P}_0) B_\mathrm{w}(\mathbf{X}_t, \mathbf{P}_t), \qquad (12.125)$$

where $\mathbf{X}_t = \mathbf{X}_t(\mathbf{X}_0, \mathbf{P}_0)$. This expression reduces the double phase space integral contained in Eq. (12.105) to a single one. Results based on this model have been found to compare very well with the corresponding quantal findings, e.g. for reaction rate constants [386] and electron transfer probabilities in the solid phase, as realized by a two-state system coupled to a harmonic bath [388]. The ability of the LSC-IVR theory to account for quantum coherence effects, however, is strongly compromised by the assumed close proximity of the trajectories $(\mathbf{X},\mathbf{P})$ and $(\mathbf{X}', \mathbf{P}')$ [356].

Making this replacement, and invoking, with the same assumptions as before, once more the stationarity condition for the phase factor of the integral, one derives a relation between the matrix $\boldsymbol{\alpha}$ and the complex-valued constant vector $\boldsymbol{\beta}$, namely

$$\boldsymbol{\beta} = 2\boldsymbol{\alpha}\vartheta''(\mathrm{z}_0)^{-1}\boldsymbol{\phi}'(\mathrm{z}_0). \qquad (12.124)$$

Adjusting Eq. (12.118) according to Eq. (12.123) shows that the original integral (12.116) is recovered for infinite $\boldsymbol{\alpha}$ and vanishing $\boldsymbol{\beta}$. However, within the limits of relation (12.124), a sufficiently large $\boldsymbol{\alpha}$ will, in general, not be compatible with a sufficiently small $\boldsymbol{\beta}$, which implies another trade off between precision and feasibility. We note that in practical applications of the generalized Filinov transformation, it may turn out to be favorable to assign to $|\boldsymbol{\beta}|$ a smaller value than obtained from relation (12.124) [385].

As demonstrated in [24] and [379], an elementary reinterpretation of the time correlation function (12.103), along with a reformulation of the propagator used to express $C_{AB}(t)$ semiclassically, can reduce the numerical effort substantially. In the following, we will, by way of a simplifying example, discuss the basic idea of the *forward-backward IVR* (FB-IVR) representation, which was originally introduced in the context of anharmonic influence functionals within a Feynman path integral approach [389, 390].

The example to be considered involves a particular choice for the operator $\hat{B}$ in $\mathrm{Tr}[\hat{A}\exp(\frac{i}{\hbar}\hat{H}t)\hat{B}\exp(-\frac{i}{\hbar}\hat{H}t)]$, namely

$$\hat{B} = \exp\left[\frac{i}{\hbar}\phi(\mathbf{X})\right], \tag{12.126}$$

with $\phi(\mathbf{X})$ as some local function, defined in the respective coordinate space. The time correlation function then becomes

$$C_{AB}(t) = \mathrm{Tr}[\hat{A}\hat{U}] \tag{12.127}$$

where $\hat{U}$ stands for a unitary operator product,

$$\hat{U} = \exp\left(\frac{i}{\hbar}\hat{H}t\right)\exp\left[\frac{i}{\hbar}\phi(\mathbf{X})\right]\exp\left(-\frac{i}{\hbar}\hat{H}t\right). \tag{12.128}$$

This operator takes the system successively from an initial time 0 forward to the time t and backward to the time 0. The initial and final coordinates that enter the IVR representation of this operator refer exclusively to the time 0, whereas their values at the time t are not involved, in contrast to expression (12.105). The double phase space average of Eq. (12.105) is thus reduced to a single phase space integration over the coordinates and momenta at the initial time:

$$\hat{U} = \left(\frac{1}{2\pi\hbar}\right)^F \int d\mathbf{X}_0 \int d\mathbf{P}_0\, C_0(\mathbf{X}_0, \mathbf{P}_0) \times \exp\left(\frac{i}{\hbar}S_0[\mathbf{X}_0, \mathbf{P}_0]\right) |\mathbf{P}_0', \mathbf{X}_0'\rangle\langle\mathbf{P}_0, \mathbf{X}_0|, \tag{12.129}$$

where we have adopted the Herman–Kluk version of the IVR propagator (12.47). The monodromy matrix elements of the prefactor are calculated recognizing that $\mathbf{X}_0' = \mathbf{X}_0'(\mathbf{X}_0, \mathbf{P}_0)$ and $\mathbf{P}_0' = \mathbf{P}_0'(\mathbf{X}_0, \mathbf{P}_0)$. In Eq. (12.129), the action S_0 is given by

$$S_0(\mathbf{X}_0, \mathbf{P}_0) = \int_0^t dt'[\mathbf{P} \cdot \dot{\mathbf{X}} - H(\mathbf{X}, \mathbf{P})] + \phi(\mathbf{X})$$
$$+ \int_t^0 dt'[\mathbf{P} \cdot \dot{\mathbf{X}} - H(\mathbf{X}, \mathbf{P})], \qquad (12.130)$$

reflecting the three-step structure of the propagation process, i.e. a forward motion from 0 to t followed by an instantaneous change at the time t and subsequent backward motion. The corresponding classical trajectories are determined by the time-dependent Hamilton function

$$H(t') = \begin{cases} H - \delta(t - t')\phi(\mathbf{X}) & \text{for } 0 \to t \\ H & \text{for } t \to 0. \end{cases} \qquad (12.131)$$

Hamilton's equation implies that a system advanced according to Eq. (12.131) will undergo a discontinuous momentum change at the time t according to

$$\mathbf{P}_t \to \mathbf{P}_t + \frac{\partial \phi(\mathbf{X})}{\partial \mathbf{X}}. \qquad (12.132)$$

Accordingly, the backward path will not simply retrace the forward path unless $\phi(\mathbf{X}) = 0$, in which case the action (12.130) vanishes. Incorporating the operator $\hat{A}$ in keeping with Eq. (12.127), we can now write the FB-IVR representation of the time correlation function as

$$C_{AB}(t) = \left(\frac{1}{2\pi\hbar}\right)^F \int d\mathbf{X}_0 \int d\mathbf{P}_0 \, C_0(\mathbf{X}_0, \mathbf{P}_0)$$
$$\times \exp\left(\frac{i}{\hbar} S_0(\mathbf{X}_0, \mathbf{P}_0)\right) \langle \mathbf{X}_0, \mathbf{P}_0 | \hat{A} | \mathbf{P}_0', \mathbf{X}_0' \rangle. \qquad (12.133)$$

Although this description of the FB-IVR strategy refers to a special operator type, namely $\exp[\frac{i}{\hbar}\phi(\mathbf{X})]$, and thus to the simplest conceivable case, it emphasizes the key element for the use of FB-IVR theory, namely the definition of a unitary operator product analogous to (12.128). The operator $\hat{B}$ of the time correlation function $C_{AB}(t)$ therefore must be expressed through a phase factor that can be absorbed by the action $S_0(\mathbf{X}_0, \mathbf{P}_0)$ in the semiclassical formula for $C_{AB}(t)$. The essence of the FB-IVR approximation lies not so much in an amendment to the underlying SC-IVR theory but in a resourceful use of this theory for the computation of $C_{AB}(t)$. Thus, the time-translated operator $\hat{B}(t) = \exp(\frac{i}{\hbar}\hat{H}t)\hat{B}\exp(-\frac{i}{\hbar}\hat{H}t)$ (see Eq. (12.103)) is interpreted as a multistep time propagation operator, as exemplified by the operator product (12.128), to which regular SC-IVR theory is applied.

In contrast to the LSC-IVR approximation, the FB-IVR model has been shown to capture quantum coherence effects to a high degree of precision [24]. From the foregoing discussion, the root of this success is the momentum jump that mediates between the forward and the backward process, making them potentially different enough for destructive interference to occur. On an elementary level, this notion has been examined by a simulation of a double-slit experiment [391]. Since the LSC-IVR constrains the contributing trajectories to infinitesimally close distances from each other, it excludes events in which the forward and the backward path lead through different slits. The resulting angular probability distribution $P(\theta)$ for a particle to be detected behind the two-slit arrangement essentially reproduces the classical prediction, i.e. $P(\theta)$ is dominated by two peaks, corresponding to transmission through the left and through the right slit. This is shown in Figure 12.6 along with the angular distributions obtained from the exact quantal and the FB-IVR computation. Obviously, the latter accounts well for the interference effects associated with the quantum mechanical two-slit problem. This success is attributed to the momentum jump that follows the forward motion through one of the slits and, in some cases, gives rise to a return through the second one.

The FB-IVR formalism has been shown to agree well with exact quantum mechanical computation when applied to a variety of model systems [358]. There are also numerous examples of the use of this method to treat physical molecular systems ([392]–[395]), including a full-dimensional computation of hydrogen atom transfer in hydroxyphenyl oxazole [393] and a simulation of a resonance Raman spectrum of I_2 in clusters of up to ten xenon atoms [395].

12.8 Cellular Dynamics

A particularly convenient choice for the shape of a semiclassical wave packet is that of a Gaussian function. The nuclear coherent state, as introduced in Section 4.4.3, gives an example of a frequently used function of the

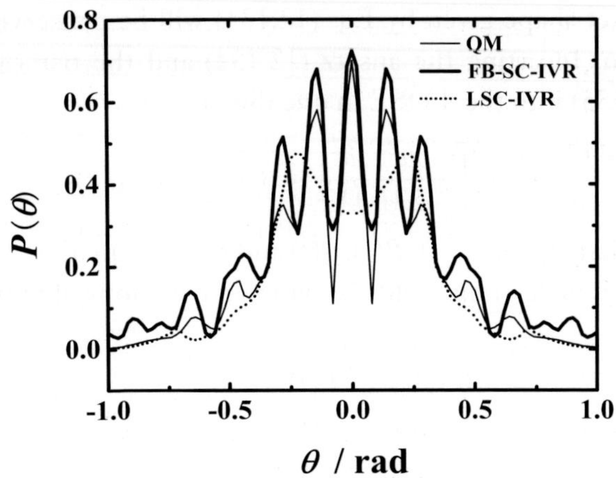

Figure 12.6: Probability of detecting a particle transmitted by a double-slit arrangement as a function of the deflection angle. Shown are the results of the FB-IVR (thick solid line) and the LSC-IVR (dashed line) approximations as compared to the quantal computation (thin solid line). (Reprinted with permission from [24]. Copyright (2001) by the American Chemical Society.)

Gaussian type in quantum dynamics. In the *semiclassical Gaussian wave packet propagation* developed by Heller [17], Gaussians are advanced along classical trajectories that "guide" them by determining both their phases and their widths. In the very simplest one-dimensional application of this model, a single wave packet of the form

$$\Phi(X,t) = \exp\left[-\alpha(t)(X - X_c(t))^2 + iP(t)(X - X_c(t)) + i\beta(t)\right] \tag{12.134}$$

is assumed. Here, $X_c(t)$ stands for the center of the Gaussian wave packet, and $\alpha(t)$ as well as $\beta(t)$ are time-dependent, complex-valued functions. Heller's approach is based on the fact that a Gaussian wave function evolving under the influence of a quadratic potential will retain its Gaussian shape. If, therefore, the potential that governs the time propagation can be approximated by expanding it up to second order about the center of the wave packet, i.e.

$$V(X) \approx V(X_c) + \left.\frac{dV(X)}{dX}\right|_{X=X_c} (X - X_c(t)) + \frac{1}{2}\left.\frac{d^2V(X)}{dX^2}\right|_{X=X_c} (X - X_c)^2, \tag{12.135}$$

the wave packet shape given by Eq. (12.134) will be conserved during the time evolution. Inserting the ansatz (12.134) and the truncated potential function (12.135) into the TDSE in one dimension

$$i\hbar\frac{\partial\Phi(X,t)}{\partial t} = \left[-\frac{\hbar^2}{2M}\frac{\partial^2}{\partial X^2} + V(X)\right]\Phi(X,t), \tag{12.136}$$

one derives relations for $X_c(t)$, $P(t)$, $\alpha(t)$ and $\beta(t)$ by equating the coefficients of equal powers in X. This yields Newtonian equations of motion for $X_c(t)$ and $P(t)$:

$$\frac{\partial X_c(t)}{\partial t} = \frac{P(t)}{M}, \tag{12.137}$$

$$\frac{\partial P(t)}{\partial t} = -\frac{dV(x)}{dX}|_{X=X_c}, \tag{12.138}$$

as well as differential equations for the functions $\alpha(t)$ and $\beta(t)$ that determine the width and the phase of the Gaussian wave packet, respectively. Thus, in semiclassical Gaussian wave packet propagation, the center of the Gaussian traces out a classical trajectory which in turn defines the shape parameters of the wave packet.

Considering the approximations on which it is based, this model meets with several limitations. As it is bound to a single real-valued trajectory, it cannot account for classically forbidden processes, and further, the assumed quadratic potential is insufficient in cases where nonlinear dynamics is prevalent. These problems can be partially alleviated by replacing the single Gaussian function with a sum over Gaussians (e.g. [396]) which mitigates the difficulties associated with nonlinear evolution without eliminating them. If the propagation of Gaussian functions is to be adequate, the dynamics must be, to a good approximation, linear in a phase space cell, i.e. a region with $\Delta X \Delta P = h$, about the center of each Gaussian. Formally, one may break up this cell into smaller subunits and let a Gaussian originate from each of them, thus creating arbitrarily small zones in which the dynamics always can be linearized. This is the basic idea of the *cellular dynamics approach* of Heller [397].

The principle is demonstrated in Figure 12.7. While the initial phase space cell of size h undergoes substantial distortion as it advances in time, the cell may be decomposed into subunits sufficiently small for approximately linear evolution.

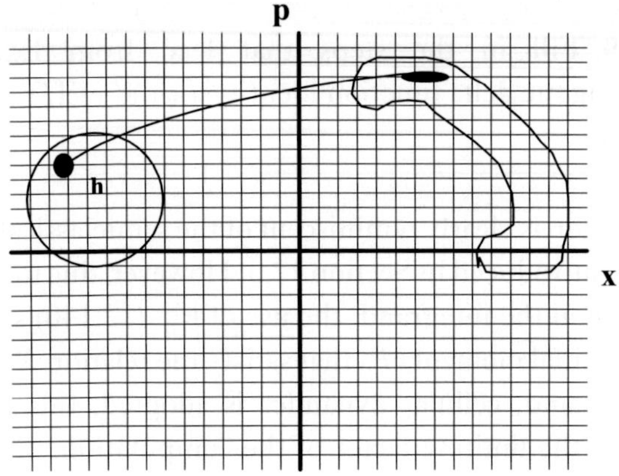

Figure 12.7: The basic strategy of cellular dynamics: breaking down a finite phase space volume into a set of subdivisions chosen sufficiently small to display approximately linear behavior with the elapse of time. (Reprinted with permission from [397]. Copyright (1991) by the American Institute of Physics.)

The linearization thus achieved makes it possible to implement a propagation formalism that shares some features with the IVR procedure. It entails, however, an interesting solution to the root search problem that differs significantly from the avenue taken by the IVR approach. The essence of this method lies in using the locally linear dynamics of the phase space cells to perform the root search analytically. In the following, we will sketch the tenets of this scheme.

The elementary semiclassical Van Vleck–Gutzwiller propagator in one-dimensional coordinate space (see Eq. (4.50)):

$$K_{\mathrm{V}}(X, X_0; t) = \left(\frac{1}{2\pi i\hbar}\right)^{\frac{1}{2}} \sum \left|\frac{\partial P_0}{\partial X}|_{X=X_0}\right|^{\frac{1}{2}} \exp\left[\frac{i}{\hbar}S(X, X_0) - i\nu\frac{\pi}{2}\right], \tag{12.139}$$

may be expressed as an integral over the initial momentum P_0, transforming Eq. (12.139) into

$$K_{\mathrm{V}}(X, X_0; t) = \left(\frac{1}{2\pi i\hbar}\right)^{\frac{1}{2}} \int dP_0 \left|\frac{\partial X_t}{\partial P_0}|_{X=X_0}\right|^{\frac{1}{2}} \delta(X - X_t(X_0, P_0))$$
$$\times \exp\left[\frac{i}{\hbar}S(X_0, P_0) - i\nu\frac{\pi}{2}\right]. \tag{12.140}$$

Exercise 12.10. Fill in the steps that lead from Eq. (12.139) to Eq. (12.140), using the delta function identity $\int dy\, g(y)\delta[f(y)-a] = g(y)/|\frac{df}{dy}|_{y=y_0, f(y_0)=a}$.

This description is clearly reminiscent of the semiclassical propagator in the IVR form, as introduced in Section 12.3. However, due to the presence of the delta function in the integrand, the use of Eq. (12.140) still necessitates searching for all initial momenta P_0 that satisfy the relation $X = X_t(X_0, P_0)$. If, on the other hand, cellular dynamics is adopted, this root search no longer represents the limiting factor of the computation. This can be seen by writing the semiclassical wave function as obtained by propagation with $K_{\rm V}(X, X_0; t)$:

$$\Psi^{\rm SC}(X,t) = \int dX_0 K_{\rm V}(X, X_0; t)\Psi^{\rm SC}(X_0, 0)$$

$$= \left(\frac{1}{2\pi i\hbar}\right)^{\frac{1}{2}} \int dX_0 \int dP_0 \left|\frac{\partial X_t}{\partial P_0}|_{X=X_0}\right|^{1/2} \delta(X - X_t(X_0, P_0))$$

$$\times \exp\left(\frac{i}{\hbar}S(X_0, P_0) - i\nu\frac{\pi}{2}\right)\Psi^{\rm SC}(X_0, 0). \qquad (12.141)$$

At this step, the dynamics is linearized by introduction of "cells". This proceeds in a move very similar to the Filinov filtering procedure presented in Section 12.7.2. While the latter scheme operates with an integral over Gaussians, as given by Eq. (12.118), cellular dynamics sums over Gaussians. Thus one inserts into the semiclassical propagator the expression

$$\eta \sum_n \exp\left[-\beta(y - n\alpha)^2\right] \qquad (12.142)$$

and chooses the parameters β and α, determining the width and the center location of the Gaussians, respectively, as well as the normalization factor η such that the sum is approximately unity. Multiplying the integrand of Eq. (12.141) by the two sums that are obtained from substituting X_0 and P_0 for y in Eq. (12.142), one arrives at a cellularized form of the semiclassical propagator, transforming (12.141) into

$$\Psi^{\rm SC}(X,t) = \eta\eta'\left(\frac{1}{2\pi i\hbar}\right)^{1/2}\sum_n\sum_m\int dX_0\int dP_0\left|\frac{\partial X_t}{\partial P_0}\right|_{X=X_0}^{1/2}$$
$$\times\,\delta(X - X_t(X_0,P_0))\exp(-\alpha(X_0 - X_n)^2)$$
$$\times\exp(-\beta(P_0 - P_m)^2)$$
$$\times\exp\left(\frac{i}{\hbar}S[X_0,P_0] - i\nu\frac{\pi}{2}\right)\Psi^{\rm SC}(X_0,0). \qquad (12.143)$$

The dynamical problem can now be solved for each cell separately, each of them being distinguished by a pair of indices (n,m). Sufficiently large exponential factors α and β will confine the initial variables X_0 and P_0 to the vicinities of the grid points (X_n,P_m). This makes it possible to expand the advanced positions X_t as functions of the initial variables (X_0,P_0) with respect to the grid point (X_n,P_m) according to

$$X_t(X_0,P_0) = X_t(X_n,P_m) + \left.\frac{\partial X_t}{\partial P_0}\right|_{P_0=X_m}(P_0 - P_m)$$
$$+\left.\frac{\partial X_t}{\partial X_0}\right|_{X_0=X_n}(X_0 - X_n). \qquad (12.144)$$

Alternatively, using the properties of the monodromy matrix

$$X_t(X_0,P_0) = X_t(X_n,P_m) + m_{21}(P_0 - P_m) + m_{22}(X_0 - X_n). \qquad (12.145)$$

The requirement $X = X_t(X_0,P_0)$, as built into the propagator $K_V(X,X_0;t)$, implies

$$P_0 = P_m + \frac{1}{m_{21}}(X - X_t(X_n,P_m)) - \frac{m_{22}}{m_{21}}(X_0 - X_n). \qquad (12.146)$$

For each cell (n,m), this equation replaces the cumbersome numerical root search by swift analytical root evaluation. To implement Eq. (12.146), one must know the monodromy matrix elements. These are derived for any given trajectory by solving the system (12.84) of auxiliary equations, namely

$$\frac{d}{dt}\mathbf{M} = \mathbf{K}\mathbf{M}, \tag{12.147}$$

while simultaneously advancing the trajectory. The matrix $\mathbf{K}$ is defined as in (12.84). Since

$$\mathbf{M}(0) = \begin{pmatrix} 1 & 0 \\ 0 & 1 \end{pmatrix}$$

and $\frac{d}{dt}(\det \mathbf{M}) = 0$, it follows that $\det \mathbf{M} = 1$. With this condition, it is easy to express the second derivative of the action, $\det\left(\frac{\partial^2 S}{\partial X_0 \partial P_0}\right)$, which, in the multi-dimensional case, is to be replaced by the Van Vleck–Gutzwiller determinant (compare with Eqs. (4.46) and (4.50)), as a function of the monodromy matrix elements. More explicitly, the linear expansion of $X_t(X_0, P_0)$ around (X_n, P_m) corresponds to an expansion of $S(X_0, P_0)$ up to second order. The second derivatives occurring in the latter are conveniently written as products of monodromy matrix elements. Using the identity

$$m_{21} = \frac{\partial X_t}{\partial P_0}, \tag{12.148}$$

which is implicit in Eq. (12.146) to recast Eq. (12.143), and also inserting the second-order expansion of $S(X_0, P_0)$ (see [377]), one arrives at the ultimate, much simplified structure of the cellularized semiclassical propagator:

$$K_V(X, X_0; t) = \sum_{m,n} |m_{12}|^{-\frac{1}{2}} g_{nm}(X, X_0). \tag{12.149}$$

The functions $g_{nm}(X, X_0)$ stand for Gaussians that are quadratic in both X and X_0, and it has been assumed that initially, the wave function is described by a linear combination of coherent states, namely

$$\Psi^{SC}(X_0, 0) = \sum_n a_n \exp\left[-\alpha_n (X_0 - X_n)^2 + \frac{iP_n(X_0 - X_n)}{\hbar}\right]. \tag{12.150}$$

Since it was conceived, cellular dynamics has been the subject of various studies aiming at an extension of the method or its advantageous combination with alternative methods. Thus, Walton and Manolopoulos [377] considered a cellularized version of the Herman–Kluk propagator in the context of a simulation of Franck–Condon spectra. In this work, the autocorrelation function was computed for the two-dimensional Hénon–Heiles bound state problem. The authors found stability even for highly chaotic test cases and, moreover, were able to converge the autocorrelation function by use of relatively modest numbers of trajectories. As suggested by this success, their approach is efficient as a means to avoid the "runaway trajectory" problem of the uncellularized Herman–Kluk procedure and, at the same time, the residual sensitivity to conjugate points where $m_{12} = 0$.

problem. The authors found stability even for highly chaotic test cases and, moreover, were able to converge the autocorrelation function by use of relatively modest numbers of trajectories. As suggested by this success, their approach is efficient as a means to avoid the "runaway trajectory" problem of the uncellularized Herman–Kluk procedure and, at the same time, the residual sensitivity to conjugate points where $m_{21} = 0$.

Chapter 13

Quantum Hydrodynamics I: Coupled Trajectories in Bohmian Mechanics

The dynamic models surveyed in the previous five chapters may be characterized as *independent trajectory approximations*, involving classical nuclear subsystems that propagate either on an average potential energy surface (Chapters 8 and 9) or in a landscape of multiple surfaces (Chapters 10–12). Naturally, trajectory-based approaches appear ill-suited to account for hallmark quantum features of the nuclei, such as delocalization associated with the splitting of the nuclear wave packet. These features, however, may be incorporated into dynamic theory by "dressing" the classical path with nuclear wave functions, as is the strategy of the spawning approach presented in Chapter 11, or by extending it into an ensemble of simultaneously evolving, mutually interacting paths. Adopting the latter notion, one makes the step from classical to *quantum trajectories* [398], or from mixed quantum-classical dynamics to *quantum hydrodynamics*.

The hydrodynamic formulation of quantum mechanics has grown out of the causal interpretation of quantum theory, as proposed and developed along a line of tradition that reaches from De Broglie [399] to Bohm [260] and Bell [400], followed by later proponents who performed the first numerical simulations of quantum mechanical problems, utilizing the conceptual tools provided by the causal interpretation [401]. According to this perspective on quantum theory, the wave function that solves the TDSE only provides a partial solution to any quantum problem, acting as a *guiding wave* that directs the entirely deterministic motion of the particles whose velocities are derived from the wave function. The basic methodological idea of quantum hydrodynamics is, correspondingly, to assemble a wave packet by tracing the evolution of correlated trajectories.

The following section will provide a very condensed introduction to the basics of the *quantum theory of motion* [25] as far as they are of relevance for the purposes of molecular dynamics. Section 13.2 will show how the Bohm–de Broglie formalism can be exploited as a theoretical frame for models of quantum trajectory propagation. The final two sections consider extensions of quantum hydrodynamics to nonadiabatic situations.

13.1 Elements of the Quantum Theory of Motion

The basic idea underlying *Bohmian mechanics* or the *quantum theory of motion* [25] may be summarized as follows: individual physical systems consist of point particles whose trajectories are determined by waves $\Psi(\mathbf{X}, t)$. These *guiding* or *pilot* waves are solutions of the TDSE. The action of the wave on a particle and the resulting law of motion obeyed by the particle will now be outlined in further detail.

The Bohm–de Broglie theory is formulated in terms of the functions R and S, the modulus and the phase, respectively, of $\Psi(\mathbf{X}, t)$ in polar representation (see Eq. (4.53)):

$$\Psi(\mathbf{X}, t) = R(\mathbf{X}, t) \exp\left[\frac{i}{\hbar} S(\mathbf{X}, t)\right]. \tag{13.1}$$

The variable $\mathbf{X}$ could refer to nuclear degrees of freedom as elsewhere in this text, but at this stage of our presentation of the Bohm–de Broglie theory no specific physical system is implied. Inserting (13.1) into the TDSE and separating the real from the imaginary part yields the two equations (4.54a) and (4.54b) which express the principles of energy conservation and continuity, respectively.

Addressing a particular physical problem requires that a well-defined initial state be specified:

$$\Psi(\mathbf{X}, t)|_{t=0} \equiv \Psi_0(\mathbf{X}).$$

Equivalently,

$$R(\mathbf{X}, t)|_{t=0} \equiv R_0(\mathbf{X}),$$
$$S(\mathbf{X}, t)|_{t=0} \equiv S_0(\mathbf{X})$$

have to be known if the two dependent equations (4.54a) and (4.54b) are to be solved. The polar representation of the wave function (13.1) implies that the phase $S_0(\mathbf{X})$ is defined only if $R_0(\mathbf{X}) \neq 0$. This condition, of course, is not just valid at the start of the motion but holds at any later time as well. From the single-valuedness of $\Psi(\mathbf{X}, t)$ one concludes that $R(\mathbf{X}, t)$ has to be defined unambiguously. The function $S(\mathbf{X}, t)$, however, is naturally multivalued. If $S(\mathbf{X}, t)$ has been identified, all choices $\tilde{S}(\mathbf{X}, t)$ with

$$\tilde{S}(\mathbf{X}, t) = S(\mathbf{X}, t) + 2\pi n\hbar, \tag{13.2}$$

where n is an integer, are equally compatible with Eq. (13.1). Therefore a net phase of

$$\oint_c dS = 2\pi n\hbar \tag{13.3}$$

is accumulated as a closed loop c is traversed. The continuous curve c can be chosen arbitrarily but must not cross itself or contain any point where the wave function vanishes, i.e. it must not coincide with a node of Ψ. Nodal regions are given by the constraint

$$\Re(\Psi) = \Im(\Psi) = 0 \tag{13.4}$$

or, more concisely,

$$R = 0. \tag{13.5}$$

The vanishing of the real and the imaginary part of Ψ in separation from each other specifies a surface while the more stringent constraint $R = 0$ defines the crossing line of these two surfaces, a *nodal line*. From Eq. (13.3) it can be concluded that the case $n \neq 0$ implies the occurrence of a discontinuity in the phase S on any surface bounded by the closed loop c (see Exercise 13.1). Such a phase jump, in turn, can only happen at a place where $R = 0$, and thus on a nodal line.

Exercise 13.1. Use Eq. (13.3) to show that the condition $n \neq 0$ requires the phase S to be discontinuous at some place on any surface bounded by the closed loop c. To demonstrate this, rewrite $\oint_c dS$ as $\oint_c \partial S/\partial\mathbf{X} \cdot d\mathbf{X}$ with $\mathbf{l}$ as a tangent vector to a point on the curve c, and recast the circulation as an integral over a surface bounded by c.

We turn now to a closer inspection of Eq. (4.54a). As pointed out in Section 4.2.1, neglect of the third term on the right-hand side yields the classical Hamilton–Jacobi equation. In what follows, we will view Eq. (4.54a) as a generalized Hamilton–Jacobi equation, maintaining the term proportional to $\hbar^2$ which will be interpreted as a *quantum potential* as in Eqs. (8.128)–(8.133). This quantum potential is defined as

$$Q(\mathbf{X}, t) \equiv -\frac{\hbar^2}{2m}\frac{1}{R}\frac{\partial^2 R}{\partial \mathbf{X}^2}. \tag{13.6}$$

As the Hamilton–Jacobi equation of classical mechanics, Eq. (4.54a) expresses energy conservation with

$$E[\mathbf{X}(t), t] = -\left.\frac{\partial S(\mathbf{X}, t)}{\partial t}\right|_{\mathbf{X}=\mathbf{X}(t)} \tag{13.7}$$

and

$$\frac{1}{2m}\left.\left(\frac{\partial S}{\partial \mathbf{X}}\right)^2\right|_{\mathbf{X}=\mathbf{X}(t)} = \frac{m}{2}\left(\frac{\partial \mathbf{X}}{\partial t}\right)^2. \tag{13.8}$$

A particle momentum is therefore given by

$$\mathbf{P} = \left.\frac{\partial S(\mathbf{X}, t)}{\partial \mathbf{X}}\right|_{\mathbf{X}=\mathbf{X}(t)}, \tag{13.9}$$

completing the definition of phase space coordinates $(\mathbf{X}, \mathbf{P})$. Consequently, a *flow velocity field* can be defined in analogy to that of classical fluid dynamics, namely

$$\mathbf{V}(\mathbf{X}, t) = \frac{1}{m}\frac{\partial S(\mathbf{X}, t)}{\partial \mathbf{X}}. \tag{13.10}$$

Considering this expression, one may read Eq. (13.3) as a statement about the strength of a *quantized hydrodynamic vortex* which is measured by the circulation integral over the velocity field, evaluated along a curve that encloses the vortex. Thus,

$$\oint_c \mathbf{V}(\mathbf{X}) \cdot d\mathbf{l} = \frac{1}{m}\oint_c \Delta S \cdot d\mathbf{l} = \frac{2}{m}\pi n\hbar, \tag{13.11}$$

with $\mathbf{l}$ as a tangent vector to a point on the curve c. By Exercise 13.1, the above circulation integral may be transformed into a surface integral over the curl of the velocity field $\mathbf{V}(\mathbf{X})$, or the *vorticity* [398]. The sign of n in (13.3) acquires the meaning of the rotational sense of the velocity field around the vortex, the *vortex chirality*. The condition for *irrotational flow* is thus $n = 0$. In classical hydrodynamics, the particle flux $\mathbf{j}$ is defined as the product of the particle density and the velocity field. Consequently, a tentative definition of the *current density field* is

$$\mathbf{j}(\mathbf{X}, t) = R^2(\mathbf{X}, t)\mathbf{V}(\mathbf{X}, t). \tag{13.12}$$

As is to be shown in Exercise 13.2, the current density of the guiding wave,

$$\mathbf{j} = \frac{\hbar}{m}\Im\left\{\Psi^* \frac{\partial \Psi}{\partial \mathbf{X}}\right\} \tag{13.13}$$

reduces indeed to the right-hand side of (13.12). From relations (13.10)–(13.13), it is seen to be determined by spatial variations of the phase $S(\mathbf{X}, t)$.

Exercise 13.2. Show that the current density (13.13) of the pilot wave (13.1) is identical with the right-hand side of (13.12), in accordance with classical hydrodynamics.

13.1.1 *Quantum trajectories*

We turn our attention now from the Bohm–de Broglie wave function to the trajectories $\mathbf{X}(t)$ along which the wave function evolves. Combining the two assignments (13.10) and (13.12) with Eq. (4.54b) makes explicit that the real part of the TDSE expresses the continuity principle. Specifically, we may write the left-hand side of Eq. (4.54b) as

$$\frac{\partial R^2}{\partial t} + \frac{\partial}{\partial \mathbf{X}} \cdot (R^2\mathbf{V}) = \frac{\partial R^2}{\partial t} + \frac{\partial}{\partial \mathbf{X}} \cdot \mathbf{j}. \tag{13.14}$$

Although the phase S is multivalued, the gradient $\frac{\partial S(\mathbf{X},t)}{\partial \mathbf{X}}$ is a single-valued function of position, so that the trajectory momentum $\mathbf{P}$ is well defined by Eq. (13.9). It follows that each trajectory has a unique tangent vector at any time. This implies that trajectories can neither cross nor touch at any given instant. But how can this notion be reconciled with the possible occurrence of "pathological situations" where $R = 0$? Wherever this condition is satisfied, the phase S is indefinite, and so is its gradient. It may be shown, however, that the trajectories do not cross any nodal points either. To see this, we revert to relation (4.54b) which may be cast into the form

$$\frac{\partial R^2}{\partial t} + \mathbf{V} \cdot \frac{\partial R^2}{\partial \mathbf{X}} = -R^2\frac{\partial}{\partial \mathbf{X}} \cdot \mathbf{V}. \tag{13.15}$$

We now transform from the fixed-in-space *Eulerian* to the *Lagrangian frame* that moves with the velocity field. If this picture is adopted the time derivative is written as

$$\frac{d}{dt} = \frac{\partial}{\partial t} + \mathbf{V} \cdot \frac{\partial}{\partial \mathbf{X}}. \tag{13.16}$$

In consequence,

$$\frac{d}{dt} \ln R^2(\mathbf{X}, t) = -\frac{\partial}{\partial \mathbf{X}} \cdot \mathbf{V}|_{\mathbf{X}=\mathbf{X}(t)} \tag{13.17}$$

which relates the density R^2 to the divergence of streamlines $\frac{\partial}{\partial \mathbf{X}} \cdot \mathbf{V}$ [25]. The formal solution of this equation may be written as

$$R^2(\mathbf{X}(t), t) = R_0^2(\mathbf{X}_0) \exp\left[-\int_0^t dt' \frac{\partial}{\partial \mathbf{X}} \cdot \mathbf{V}(\mathbf{X}, t') \right]. \tag{13.18}$$

The structure of the density R^2 as described by Eq. (13.18) implies that this function does not vanish at any point along the trajectory $\mathbf{X}(t)$ unless it is zero at the very start of the motion. No trajectory, however, can begin at a location where the wave function assumes a node and thus the particle momentum is completely undefined. Positive divergence of the streamlines ($\frac{\partial}{\partial \mathbf{X}} \cdot \mathbf{V} > 0$) implies a decrease in the density.

Since $\frac{\partial S(\mathbf{X},t)}{\partial \mathbf{X}}|_{\mathbf{X}=\mathbf{X}(t)}$ is related to a particle momentum, Eq. (4.54a) assumes a simple form if the operator $\frac{\partial}{\partial \mathbf{X}}$ is applied. One readily obtains

$$\left\{ \frac{\partial}{\partial t} + \frac{1}{m} \frac{\partial S}{\partial \mathbf{X}} \cdot \frac{\partial}{\partial \mathbf{X}} \right\} \frac{\partial S}{\partial \mathbf{X}} = -\left. \frac{\partial (V + Q)}{\partial \mathbf{X}} \right|_{\mathbf{X}=\mathbf{X}(t)}. \tag{13.19}$$

By use of Eqs. (13.10) and (13.16), this becomes

$$\frac{d}{dt}\left(m \frac{\partial \mathbf{X}}{\partial t} \right) = -\frac{\partial}{\partial \mathbf{X}} (V + Q)|_{\mathbf{X}=\mathbf{X}(t)}. \tag{13.20}$$

This Newtonian equation describes the motion of a particle that moves under the influence of a force $-\frac{\partial}{\partial \mathbf{X}}(V + Q)$ derived from the classical potential V and the quantum potential Q. The relation gives meaning to the term *guiding wave* since it refers to a particle that responds to the quantum force $-\frac{\partial Q}{\partial \mathbf{X}}$ that through Eq. (13.6) depends on the wave modulus R. It follows from Eq. (13.20) that a classically free motion, corresponding to $V = 0$, is, in general, not free in quantum mechanics, since a free solution of the TDSE does not necessarily lead to a vanishing quantum potential.

The wave function Ψ, as given by Eqs. (4.54a) and (4.54b) is determined by both the classical and the quantum potential. Since the quantum potential, in turn, depends on the wave function or, more specifically, its modulus R, it is not obvious that the solution of Eqs. (4.54a) and (4.54b) is a ray $A\Psi$ where Ψ stands for a specific solution of these equations and A is any complex-valued constant. This, however, is to be expected since (4.54a) and (4.54b) are nothing but a reformulation of the TDSE. Closer inspection of the quantum potential clarifies indeed that it depends on the form of the wave function rather than its square magnitude. This is immediately inferred from Eq. (13.6) as

$$Q = -\frac{\hbar^2}{2m}\frac{1}{R}\frac{\partial^2 R}{\partial \mathbf{X}^2} = -\frac{\hbar^2}{2m}\frac{1}{|A|^2 R}\frac{\partial^2 |A|^2 R}{\partial \mathbf{X}^2}. \tag{13.21}$$

Thus one may normalize the wave function without changing the quantum potential. Further, if the wave function tends towards zero in the asymptotic region of the considered problem, the quantum potential may stay finite in this regime and continue to influence the physical process. This is of consequence in particular for the description of particle interference phenomena within the quantum theory of motion where it is shown that the recorded interference patterns are determined by the specific form adopted by the quantum potential in the considered physical situation.

In this context we point out that the quantum potential may be understood as a contribution to the local kinetic energy T_{local} of the guiding wave that is not associated with the wave momentum but rather with the shape of the wave. This is seen by evaluating [25]

$$T_{\text{local}} = \Re\left\{\frac{\Psi^*(-\frac{\hbar^2}{2m}\frac{\partial^2\Psi}{\partial \mathbf{X}^2})}{\Psi^*\Psi}\right\}. \tag{13.22}$$

Adopting the polar representation (13.1) of Ψ, one partitions T_{local} into two components, labeled T_{momentum} and T_{shape} [402]. More specifically,

$$T_{\text{momentum}} = \frac{1}{2m}\left(\frac{\partial S}{\partial \mathbf{X}}\right)^2 = \frac{1}{2}m\mathbf{V}\cdot\mathbf{V} \tag{13.23}$$

and

$$T_{\text{shape}} = -\frac{\hbar^2}{2m}\frac{1}{R}\frac{\partial^2 R}{\partial \mathbf{X}^2}. \tag{13.24}$$

In expression (13.24), we recognize the quantum potential. Therefore it is seen that the latter quantity originates from the kinetic energy component of the guiding wave that is determined by the curvature of the wave amplitude but independent of the phase.

Exercise 13.3. Show that the local kinetic energy can be partitioned according to $T_{\text{local}} = T_{\text{momentum}} + T_{\text{shape}}$, where the three kinetic energy terms are defined by Eqs. (13.22)–(13.24).

13.1.2 *The pilot wave and the guided particle*

According to the Bohm–de Broglie interpretation of quantum theory, an individual physical system involves a pilot wave that guides point particle motion. The wave results as a solution of the TDSE, and the equation of motion for the particle is given by Eq. (13.10). To reconcile this view with the laws of quantum mechanics, a further interpretative step needs to be taken. Evidently, the notion of a point particle moving along a classical trajectory $\mathbf{X}(t)$, as implied by relation (13.10), is alien to quantum theory where a system is represented by its wave function. According to the Copenhagen interpretation of quantum theory, the latter does not make any statement about objectively existing physical entities but about the probabilities for observing these entities under well-defined experimental conditions. The probability concept is easily recovered on the ground of the Bohm–de Broglie theory if one replaces the single particle with an ensemble of particles and postulates that

$$p(\mathbf{X}, t) = R^2(\mathbf{X}, t)d\mathbf{X} \tag{13.25}$$

with $p(\mathbf{X}, t)$ as the probability that a particle in the ensemble is located between $\mathbf{X}$ and $\mathbf{X}+d\mathbf{X}$ [25]. It should be noted that it is sufficient to make the claim (13.25) for the initial time of the evolution. At $t = 0$, it then holds that

$$p(\mathbf{X}, t = 0) \equiv p_0(\mathbf{X}) = R_0^2\, d\mathbf{X}. \tag{13.26}$$

This means that the initial probability density associated with the particle ensemble is identical with the squared wave function modulus at time $t = 0$. The probability density of the particle ensemble at later times is uniquely defined by condition (13.26) combined with the continuity equation. Since both the ensemble probability density and the squared wave function modulus have to satisfy the equation of evolution (13.14) and share by stipulation (13.26) their initial conditions, they are identical. By implication, a particle in the ensemble has zero probability of being located at the place of a wave function node.

The term *ensemble* that has acquired systematic importance for the basic tenets of the Bohm–de Broglie theory is in need of further clarification. This theory involves sets consisting of individual physical systems, each one specified by a pair of a pilot wave and a guided particle. If the individual pilot wave is denoted by $\Psi_i(\mathbf{X}, t) = R_i(\mathbf{X}, t) \exp\left[\frac{i}{\hbar} S_i(\mathbf{X}, t)\right]$, the corresponding particle trajectory satisfies the equation of motion

$$\dot{\mathbf{X}}_i = \mathbf{V}_i = \frac{1}{m} \left. \frac{\partial S_i(\mathbf{X}, t)}{\partial \mathbf{X}} \right|_{\mathbf{X}_i = \mathbf{X}_i(t)} . \tag{13.27}$$

The pilot waves define probability distributions for their associated particles. We can thus have complete knowledge about the pilot wave and incomplete knowledge about particle location. The quantum mechanical equivalent of this case is a well-defined wave function that yields the probability density for finding a particle at the place $\mathbf{X}$. In this case all particles of the ensemble are guided by the same pilot wave. Their spatial distribution reflects our lack of information about the precise particle location.

The quantum theory of motion provides an interpretative and mathematical framework for understanding quantum processes, as well as for their quantitative prediction. Due to its inherent realism, built on the notion of objectively existent particles and waves, and restoring the notion of individual events governed by causality, it seems to turn back the dial advanced by the Copenhagen interpretation of quantum mechanics. The touchstone for the success or failure of the quantum theory of motion is, however, its ability to provide a valid description of paradigmatic quantum phenomena. Among the latter, the effects of particle interference and tunneling through potential barriers are of particular consequence for quantum dynamics. The reader can find an outline of the specific ways in which the quantum theory of motion accommodates these phenomena in Section 13.1*.

13.2 Lagrangian Quantum Hydrodynamics

In contrast to the trajectory-based procedures discussed in Chapters 8–12, the quantum trajectory method (QTM) as devised by Wyatt *et al.*

([403]–[405]) advances all trajectories simultaneously. This *motion in lockstep* of the ensemble members does not imply a many-particle situation. As outlined earlier in the context of our presentation of the tunneling problem, the approach, in its simplest form, involves single-particle propagation. It may, however, be generalized to many particles.

The QTM as a computational scheme aims at deriving the spatio-temporal evolution of a probability fluid, determined by the density $\rho = R^2$ and the current density ($\mathbf{j} = \frac{R^2}{m}\frac{\partial S}{\partial \mathbf{X}}$). Addressing this task is simplified by adopting the Lagrangian as opposed to the Eulerian view. As implied before in the context of formula (13.16), both refer to basic modes of representing hydrodynamic phenomena. In the Eulerian frame, one describes changes at a fixed place in the fluid, while the Lagrangian viewpoint applies to changes occurring as one follows a fluid particle along its path. The QTM proceeding within the Lagrangian frame has been termed *Lagrangian quantum hydrodynamics* [406]. The "particles" to which we have referred so far become in this formalism fluid cells whose quantum trajectories are to be found. In practical applications of the QTM it turns out to be advantageous to cast the wave function into the compact form

$$\Psi(\mathbf{X}, t) = \exp\left[\frac{i}{\hbar}\xi(\mathbf{X}, t)\right] \tag{13.28}$$

with the complex-valued generalized phase

$$\xi(\mathbf{X}, t) = -i\hbar C(\mathbf{X}, t) + S(\mathbf{X}, t). \tag{13.29}$$

Both the amplitude function C and the action function S are real-valued. Comparison of ansatz (13.28) and (13.29) with Eq. (13.1) yields

$$\rho(\mathbf{X}, t) = R^2(\mathbf{X}, t) = \exp\left[2C(\mathbf{X}, t)\right]. \tag{13.30}$$

The reason for the manipulation (13.28) and (13.29) is that C and S often vary more slowly than the real and imaginary parts of Ψ, making the former quantities more easily accessible to numerical treatment than the latter. Utilizing (13.28) instead of (13.1) as the form of the pilot wave and substituting into the TDSE, we arrive at the same real part as before (see Eq. (4.54a)):

$$-\frac{\partial S}{\partial t} = \frac{1}{2m}\left|\frac{\partial S}{\partial \mathbf{X}}\right|^2 + V + Q, \tag{13.31}$$

where the quantum potential Q is now given in terms of the amplitude function C as

$$Q(\mathbf{X},t) = -\frac{\hbar^2}{2m}\left[\frac{\partial^2 C(\mathbf{X},t)}{\partial \mathbf{X}^2} + \left|\frac{\partial C}{\partial \mathbf{X}}\right|^2\right]. \tag{13.32}$$

The imaginary part of the TDSE becomes

$$\frac{\partial C}{\partial t} = -\frac{1}{2}\frac{\partial}{\partial \mathbf{X}}\cdot \mathbf{V} - \mathbf{V}\cdot\frac{\partial C}{\partial \mathbf{X}}. \tag{13.33}$$

Changing from the Eulerian to the Lagrangian frame, we make the substitution (13.16) to simplify Eq. (13.33) which now reads

$$\frac{dC}{dt} = -\frac{1}{2}\frac{\partial}{\partial \mathbf{X}}\cdot \mathbf{V}. \tag{13.34}$$

This equation describes the evolution of the scalar field C in a frame that moves with the fluid. The *convective operator* $\mathbf{V}\cdot\partial/\partial\mathbf{X}$ that enters Eq. (13.33) has been eliminated. Likewise, as done before in the development that results in relation (13.20), we derive the flow acceleration in the Lagrangian picture and obtain

$$\frac{d\mathbf{V}}{dt} = -\frac{1}{m}\frac{\partial V}{\partial \mathbf{X}} + \frac{\hbar^2}{2m^2}\frac{\partial}{\partial \mathbf{X}}\left[\frac{\partial^2 C(\mathbf{X},t)}{\partial \mathbf{X}^2} + \left|\frac{\partial C}{\partial \mathbf{X}}\right|^2\right]. \tag{13.35}$$

The equation pair (13.34) and (13.35) may be used for the numerical implementation of Lagrangian quantum hydrodynamics as presented, for instance, in [406]. As the scalar field C and the vector field $\mathbf{V}$ is propagated, the position of each discretized fluid element or "particle" is updated by integrating $d\mathbf{X} = \mathbf{V}dt$. It is important to note that the trajectories for the N fluid elements considered are correlated by the non-local contribution $\frac{\hbar^2}{2m^2}\frac{\partial}{\partial \mathbf{X}}[\frac{\partial^2 C(\mathbf{X},t)}{\partial \mathbf{X}^2} + |\frac{\partial C}{\partial \mathbf{X}}|^2]$ to Eq. (13.35), the gradient of the quantum potential. In a physical situation where the action of the quantum force is restricted to the initial phase of the process, the quantum fluid turns quickly into a classical fluid as manifested by the uncorrelated motion of the discretized elements.

Each fluid cell i is assigned a weight through

$$W_i = \rho(\mathbf{X}_i,t)\,\delta\tilde{V}_i = \exp\left[2C(\mathbf{X}_i,t)\right]\delta\tilde{V}_i, \tag{13.36}$$

where $\delta \tilde{V}_i$ stands for an infinitesimal volume centered on a position $\mathbf{X}_i$, with i as trajectory index. By the continuity equation (4.54b), (13.34), the weights W_i are conserved. The conservation of the total density, i.e. the quantum hydrodynamical counterpart of wave function normalization, is expressed by

$$\sum_i \rho(\mathbf{X}_i, t) d\mathbf{X}_i = \sum_i W_i = 1. \tag{13.37}$$

Exercise 13.4. Use the continuity equation (4.54b) to demonstrate the conservation of the weight factors W_i as introduced by Eq. (13.36).

In this terminology, the transmission coefficient T of a wave packet through a barrier is simply written as the sum

$$T(t) = \sum_j^{\text{trans}} W_j, \tag{13.38}$$

where the sum extends over all trajectories that have traversed the barrier region [407].

13.2.1 *Assembling the wave function*

In the following, we turn to constructing the wave function by integrating the fluid cell propagation equations (13.34) and (13.35). The former yields immediately the time dependence of the amplitude function, namely

$$C(\mathbf{X}, t) = C_0(\mathbf{X}_0) - \frac{1}{2} \int_0^t dt' \frac{\partial}{\partial \mathbf{X}} \cdot \mathbf{V}\,. \tag{13.39}$$

From Eq. (13.35) one obtains the fluid velocity field as a function of time, and, by $\mathbf{V} = \frac{1}{m} \frac{\partial S}{\partial \mathbf{X}}$ (see Eq. (13.27)), the gradient of the phase function. This leads to an equation analogous to (13.39) for the action S. Like C, the function S can be represented as a propagation time integral. To demonstrate this, we inspect the total differential of S, using Eq. (13.31):

$$\begin{aligned} dS &= \frac{\partial S}{\partial \mathbf{X}} \cdot d\mathbf{X} + \frac{\partial S}{\partial t} dt \\ &= \frac{\partial S}{\partial \mathbf{X}} \cdot d\mathbf{X} - \frac{1}{2m} \left(\frac{\partial S}{\partial \mathbf{X}} \right)^2 - V - Q. \end{aligned} \tag{13.40}$$

Further,

$$\frac{\partial S}{\partial \mathbf{X}} \cdot d\mathbf{X} = \mathbf{P} \cdot d\mathbf{X} = \mathbf{P} \cdot \mathbf{V}dt$$
$$= 2T_{\text{momentum}}. \tag{13.41}$$

Therefore,

$$dS = (T_{\text{momentum}} - V - Q)dt$$
$$\equiv L_{\text{q}}dt, \tag{13.42}$$

where L_{q} denotes a quantum Lagrangian that is related to the classical Lagrangian $L_{\text{c}} = T_{\text{momentum}} - V$ by

$$L_{\text{q}} \equiv L_{\text{c}} - Q. \tag{13.43}$$

The quantum Lagrangian governs the evolution of S by

$$S(\mathbf{X}, t) = S_0(\mathbf{X}_0) + \int_0^t dt' \, L_{\text{q}}. \tag{13.44}$$

Integrating both the amplitude and the action function according to Eqs. (13.39) and (13.44), respectively, for each one of the discretized fluid elements, one may assemble the wave function as given by Lagrangian quantum hydrodynamics. From definition (13.43) combined with the explicit form of the quantum potential (13.32) one concludes that the quantum Lagrangian couples nearby trajectories. The same holds true for the integrand $\frac{\partial}{\partial \mathbf{X}} \cdot \mathbf{V}$ of the amplitude exponential. Thus, the quantum Lagrangian L_{q} and the divergence of the velocity field $\frac{\partial}{\partial \mathbf{X}} \cdot \mathbf{V}$ may be viewed as agents through which the quantum trajectories communicate, although these paths never coincide and may be separated by nodal surfaces that they cannot cross.

At the position $\mathbf{X}_j$ of the jth fluid element the full wave function results at the time t as [402]

$$\Psi(\mathbf{X}_j(t)) = \Psi(\mathbf{X}_j(0)) \exp\left[-\frac{1}{2}\int_0^t dt' \frac{\partial}{\partial \mathbf{X}} \cdot \mathbf{V}|_{\mathbf{X}_j(t')} + \frac{i}{\hbar}\int_0^t dt' \, L_q[\mathbf{X}_j(t')]\right]. \tag{13.45}$$

13.2.2 *Technical challenges for quantum trajectory propagation*

In practice, quantum trajectories are generated by employing the canonical relation $\mathbf{V}(\mathbf{X}) = \frac{1}{m}\frac{\partial S(\mathbf{X})}{\partial \mathbf{X}}$ (Eq. (13.27)) rather than the Newtonian equation (13.35). In this way, one circumvents the need to evaluate the third spatial derivative of the quantum potential as contained in Eq. (13.35). A particular methodological challenge for the computational realization of the QTM formalism lies in the non-uniform character of the mesh defined by the particle locations. Although at $t = 0$, all particles are evenly spaced, their distribution exhibits zones of compression or inflation as time passes (see [404], [405]). Compression develops in regions where $R = \exp(C)$ is close to a local maximum, and conversely, inflation is associated with local minima of R. This behavior of the quantum fluid is related to the tendency of quantum trajectories to avoid nodes, i.e. zeroes of the density function R^2. As the trajectories are repelled from the nodal regions, they cluster in the spaces between these regions. Thus, discretizing R or the other quantities relevant to the propagation over the moving unstructured mesh can lead to a sampling problem. In particular, advancing the scalar fields C and S in time according to Eqs. (13.31), (13.34), and (13.27) presents a difficulty, as it requires evaluating first and second spatial derivatives. Given the unstructured particle topology, this is a highly non-trivial task.

In many cases, this situation may be remedied by use of a *propagating least-squares approximation*, as outlined in [404]. Further computational techniques that respond to the difficulty of computing the derivatives that occur in quantum hydrodynamics have been devised. In particular, techniques of particle mesh regularization [405] or of transformation between the physical, non-uniform particle grid and a uniform grid that allows us to determine derivatives with higher accuracy [406] have been successfully implemented.

The arguably most severe difficulty faced by quantum trajectory propagation is the *node problem*. From the definition of the quantum potential (13.6), it becomes singular at a nodal point $R = 0$. Close to such a point, the quantum potential changes rapidly, and the quantum trajectories tend to become unstable. Among the various remedies proposed to address the node problem is the *counterpropagation wave method* (CPWM) ([408]–[410]). The idea underlying this method is to modify the polar representation of the wave function (13.1) according to

$$\Psi(\mathbf{X}, t) = P(\mathbf{X}, t) \exp\left[\frac{i}{\hbar} S_0(\mathbf{X}, t)\right]. \tag{13.46}$$

The adapted phase S_0 which is constrained to be a smooth function at any position defines a *carrier momentum* $\mathbf{P}_0 = \frac{\partial S_0}{\partial \mathbf{X}}$, and the factor P, in contrast to the amplitude R, is no longer positive or zero, but can adopt negative values. It is decomposed as

$$P(\mathbf{X}, t) = \frac{1}{2} q(\mathbf{X}, t) \left\{ \exp\left[\frac{i}{\hbar} \delta(\mathbf{X}, t) \right] + \exp\left[-\frac{i}{\hbar} \delta(\mathbf{X}, t) \right] \right\}, \tag{13.47}$$

with a suitably defined function q.

The substitution (13.47) introduces two distinct momenta, $(\mathbf{P}_0 \pm \partial\delta/\partial\mathbf{X})$, into the propagation problem, corresponding to two wave function components, Ψ_+ and Ψ_-. While each of these two is free of nodes, their superposition to create the overall wave function Ψ can give rise to node formation. Alternative ways of coping with the node problem can be found in ([411]–[413]).

13.3 Nonadiabatic Lagrangian Quantum Hydrodynamics

Although borrowing vocabulary from classical hydrodynamics, the quantum theory of motion is an exact quantum theory. In the limit of infinitely many fluid cell trajectories of infinitely dense spacing, the precise solution of the TDSE is synthesized. In what follows we will outline the first steps towards the application of Lagrangian quantum hydrodynamics to the evolution of molecules in time and especially emphasize the nonadiabatic problem of two or more coupled potential energy surfaces. This move requires subdividing the coordinate space into nuclear (**X**) and electronic coordinates (**x**). For the sake of simplicity, both the nuclear and the electronic subspace will be assumed to be one-dimensional in the following development. We consider an idealized situation that involves two electronic states and adopt the diabatic representation, stipulating that $\langle \psi_i | \frac{\partial \psi_j}{\partial \mathbf{X}} \rangle = 0$, with $i, j = 1, 2$. The wave function is then given by the expansion

$$\Psi(x, X, t) = \psi_1(x)\chi_1(X, t) + \psi_2(x)\chi_2(X, t). \tag{13.48}$$

We partition the Hamiltonian$\hat{H}$ into the nuclear and the electronic contribution:

$$\hat{H} = \hat{H}_{\mathrm{N}} + \hat{H}_{\mathrm{el}} = -\frac{\hbar^2}{2M} \frac{\partial^2}{\partial X^2} + V_{\mathrm{N}}(X) + \hat{H}_{\mathrm{el}}, \tag{13.49}$$

where M stands for the nuclear mass corresponding to the coordinate X. Contracting the TDSE for $\Psi(X, x, t)$ with the wave function component $\psi_i(x)$ ($i = 1, 2$), we denote the electronic matrix element $\langle \psi_i | \hat{H}_{\text{el}} | \psi_j \rangle$ by V_{ij}. Further, the polar representation is used for the nuclear wave packets: $\chi_i(X,t) = R_i(X,t) \exp[\frac{i}{\hbar} S_i(X,t)]$. Relation (4.54b) is readily generalized, as one generates two equations from the imaginary part of the TDSE for $\Psi(x, X, t)$ [414], namely

$$\frac{d\rho_1}{dt} = -\rho_1 \frac{\partial V_1}{\partial X} - \lambda_{12} \tag{13.50a}$$

and

$$\frac{d\rho_2}{dt} = -\rho_2 \frac{\partial V_2}{\partial X} + \lambda_{21}, \tag{13.50b}$$

where ρ_i has been substituted for R_i^2 and the Lagrangian frame has been used. The terms λ_{ij} stand for $\frac{2}{\hbar} V_{ij} (\rho_i \rho_j)^{1/2} \sin(\Delta)$ with $i,j = 1, 2$, and $\Delta(X,t) \equiv \frac{S_1 - S_2}{\hbar}$. The symbol V_i labels the velocity field associated with S_i, such that $V_i = \frac{1}{M} \frac{\partial S_i}{\partial X}$. We refer to the factor $(\rho_1 \rho_2)^{1/2}$ as the *transition density*. The identities (13.50a) and (13.50b) are interpreted as *continuity equations for compressible flow* with added source or sink terms that create or destroy density, respectively. The only item that distinguishes these relations from their adiabatic counterpart (13.17) is the involvement of the rates λ_{ij} which account for density changes due to the interaction between the two electronic states.

The real part of the TDSE determines the evolution of the action functions S_i (i = 1, 2), yielding a pair of equations analogous to (13.50a) and (13.50b), namely

$$\frac{dS_1}{dt} = \frac{1}{2} M V_1^2 - (V_{11} + Q_1 + Q_{12}), \tag{13.51a}$$

and

$$\frac{dS_2}{dt} = \frac{1}{2} M V_2^2 - (V_{22} + Q_2 + Q_{21}). \tag{13.51b}$$

This is the direct generalization of (4.54a), again augmented by a transition term. The expressions Q_{12} (Q_{21}) represent *transfer quantum potentials* given by

$$Q_{12} = V_{12}\sqrt{\frac{\rho_2}{\rho_1}}\cos\Delta \tag{13.52}$$

with Q_{21} obtained by interchanging the indices on the right-hand side of (13.52).

Exercise 13.5. For the case of two electronic states coupled by the diabatic potential matrix **V**, derive the imaginary and real parts of the TDSE, i.e Eqs. (13.50a) and (13.50b), as well as (13.51a) and (13.51b), respectively.

The time derivatives of the action functions indicate the kinetic energy excess over the potential energy contributions, including the transfer components, for each of the two potential energy surfaces. They can therefore be understood as generalized Lagrangians that characterize situations involving two coupled electronic states.

From the definitions of the coupling terms, it is seen that they satisfy the symmetry relations:

$$\lambda_{ij} = \lambda_{ji} \tag{13.53}$$

and

$$\rho_i Q_{ij} = \rho_j Q_{ji}, \quad i \neq j. \tag{13.54}$$

Taking the gradient with respect to the nuclear coordinate on both sides of Eqs. (13.51a) and (13.51b), we obtain the nonadiabatic version of the Newtonian equation (13.20):

$$\frac{dP_1}{dt} = -\frac{\partial(V_{11} + Q_1 + Q_{12})}{\partial X} \tag{13.55a}$$

and

$$\frac{dP_2}{dt} = -\frac{\partial(V_{22} + Q_2 + Q_{21})}{\partial X}, \tag{13.55b}$$

introducing transition quantum forces of the form $-\frac{\partial Q_{ij}}{\partial X}$, where $i,j = 1,2$ and $i \neq j$.

Relations (13.50a) and (13.50b) to (13.55a) and (13.55b) encapsulate the fundamentals of *nonadiabatic Lagrangian quantum hydrodynamics*. As has been pointed out in the context of elementary Bohm–de Broglie theory (Section 13.2), this formalism is exact in the sense of quantum mechanics.

Thus, Eqs. (13.50a) and (13.50b) together with (13.51a) and (13.51b) are equivalent to the TDSE applied to the same physical situation. The QTM simulation, however, represents necessarily an approximation to the true quantum mechanical solution since a finite number of particles, or fluid cells, is used.

We turn now to an example of the numerical implementation of the nonadiabatic model introduced in the preceding paragraphs. Lopreore and Wyatt [415] discuss the evolution of a Gaussian wave packet on two diabatic potential energy surfaces defined by

$$V_{11}(X) = -\frac{c_{11}}{1 + \exp[-a(X - X_0)]}, \qquad V_{22}(X) = 0, \tag{13.56a}$$

and

$$V_{12}(X) = V_{21}(X) = c_{12} \exp(-\beta_{12}(X - X_0)^2), \tag{13.56b}$$

as non-diagonal terms. The symbols c_{11}, c_{12}, a and β_{12} refer to the constants that determine the potential energy surface. The potentials defined in (13.56a) and (13.56b) are graphically represented in Figure 13.1.

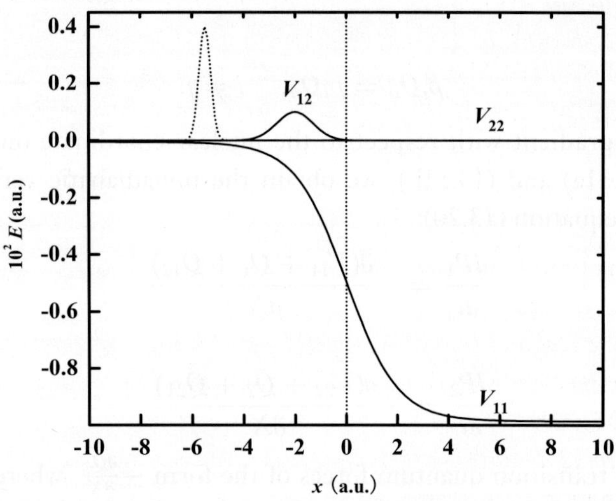

Figure 13.1: The diabatic potentials used in the Lagrangian quantum hydrodynamics model calculation described in [415]. The dotted distribution shows the incoming Gaussian wave packet. (Reprinted with permission from [415]. Copyright (2002) by the American Institute of Physics.)

It is not advisable to constrain the initial wave packet to a single electronic channel. This is due to the structure of the coupling terms Q_{ij} and λ_{ij} $(i \neq j)$ which contain contributions from both potential energy surfaces. For instance, the transfer quantum potential Q_{21} is initially ill-defined if the Gaussian wave packet starts on surface 1, implying $\rho_2(t = 0) = 0$. This difficulty can be circumvented if one propagates simultaneously two coherent superpositions $\Psi^+ \approx \Psi_1 + \Psi_2$ and $\Psi^- \approx \Psi_1 - \Psi_2$, replacing the single wave function Ψ_1. By their definitions, Ψ^+ and Ψ^- initially populate both channels. The special case of single-channel initial conditions can be recovered for any time of the scattering process by constructing a suitable linear combination of Ψ^+ and Ψ^-. Thus, the evolution of a wave packet that starts on surface 1 is obtained by tracing $\Psi^+ + \Psi^-$.

The hydrodynamic simulation was carried out with an ensemble of 61 fluid elements. The probabilities for transmission and reflection were calculated, as for the Eckart barrier problem considered in [403], by evaluating the relevant populations. The trajectories are further distinguished according to whether or not the system has undergone a transition from the lower to the higher potential energy surface. Thus, the populations are computed by integrating the density on both surfaces in the product and in the reactant region ($X > 0$ and $X < 0$, respectively). The probabilities obtained at a constant time for the four different processes, i.e. reflection or transmission with or without electronic transition, are represented as functions of the initial translational energy in Figure 13.2.

Shown is a comparison between the QTM data and the exact quantum mechanical results as obtained from integrating the TDSE. Excellent agreement is found between both approaches throughout the studied range of energies. An equally satisfying picture emerges as one extends the comparison to the real and the imaginary part of the guiding wave [415], lending credence to the capability of Lagrangian quantum hydrodynamics to predict both the phase and the modulus of the wave function with high accuracy. From Figure 13.2, the probability for reflection on the initial surface drops sharply with increasing translational energy, and correspondingly, the transmission probability rises. While these trends are intuitively plausible, one also notices a very weak energy dependence of the transition probabilities $P_{1\to 2,r}$ and $P_{1\to 2,t}$. This feature has been attributed to the relatively small

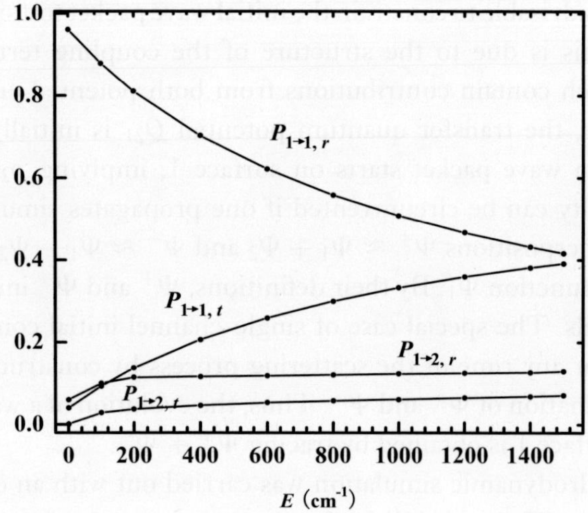

Figure 13.2: Probabilities for transmission or reflection with or without electronic transition as found from nonadiabatic Lagrangian quantum hydrodynamics versus the translational energy for the model system discussed in the text [415]. The symbol $P_{1\to 1,t}$ means the probability for transmission on the initially populated surface (1), $P_{1\to 2,r}$ refers to the probability for reflection combined with transition to the initially unpopulated surface (2). The remaining quantities, $P_{1\to 1,r}$ and $P_{1\to 2,t}$, are defined accordingly. The dots represent the results of the QTM simulation, while the continuous curves have been generated by integrating the TDSE. (Reprinted with permission from [415]. Copyright (2002) by the American Institute of Physics.)

size of the diabatic coupling considered in the simulation ($c_{11} = 2,110\,\text{cm}^{-1}$, $c_{12} = 200\,\text{cm}^{-1}$).

From QTM calculations within Lagrangian quantum hydrodynamics one may derive useful information about the investigated nonadiabatic process by analyzing the time behavior of individual trajectories. This is demonstrated by Figure 13.3 which shows the transition rate λ_{12} for selected trajectories as a function of time.

The respective fluid elements are initially lined up with equal spacings of $\delta = 0.016$ a.u. such that element 1 is located farthest from the interaction zone. Among the trajectories included in Figure 13.3, the one associated with element 50 is therefore the first to reach the coupling region, centered at $X_0 = -2.0$ a.u. This causes the maximum in the transition rate at $t = 11$ fs. The peak belonging to trajectory 30 is attained 17 fs later. It is distinctly higher than the other maxima visible in Figure 13.3. This finding is

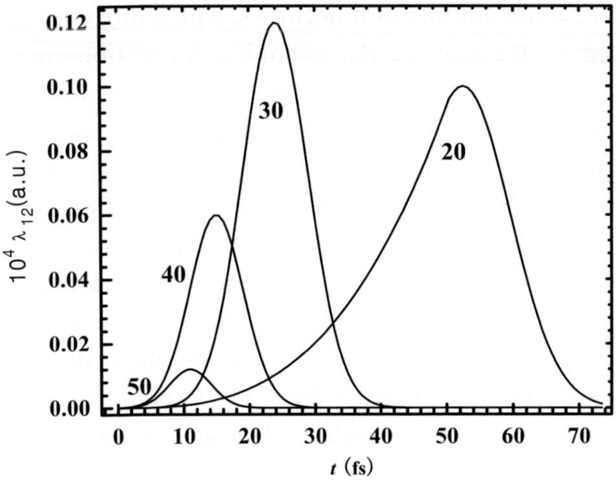

Figure 13.3: Transition rates λ_{12} (in a.u.) from surface 1 to surface 2 versus time for the trajectories 20, 30, 40, and 50 out of an ensemble of 61 trajectories. (Reprinted with permission from [415]. Copyright (2002) by the American Institute of Physics.)

readily explained by the proportionality of λ_{12} to $\rho_1^{1/2}$. Originating from the center of the initial Gaussian wave packet, trajectory 30 carries the highest density ρ_1.

In this manner, the QTM technique allows us to generate differential trajectory-by-trajectory information that provides insight into the details of the scattering or reaction mechanism studied.

Lagrangian quantum hydrodynamics, as outlined in the previous paragraphs, is of particular interest in our survey of nonadiabatic models for molecular interactions as it is quantum mechanically exact, consisting of a mere reformulation of the TDSE. On the level of numerical simulation, this feature of the model implies that swarms of coupled trajectories are propagated instead of independent trajectories as considered by classical or mixed quantum-classical methods for which the surface hopping approach outlined in Chapter 11 provides a typical example. The rearrangement of population among the contributing electronic channels, as described by Lagrangian quantum hydrodynamics, is a smooth process that does not involve any discontinuous jumps between surfaces.

In spite of its higher degree of formal rigor, the practical application of Lagrangian quantum hydrodynamics in the framework of QTM relies necessarily on a finite number of trajectories, which introduces inaccuracies into the procedure. We point out that not only computations on simple model systems have generated very encouraging results. Nonadiabatic quantum hydrodynamics has also braved the challenge posed by more realistic physical systems, such as a simulation of the reaction $O(^3P, ^1D) + H_2 \rightarrow OH + H$ [416] which included four potential energy surfaces and yielded good agreement with a wave-packet propagation approach (see Chapter 15).

We further mention a recent model of nonadiabatic dynamics that operates with quantum trajectories [423, 424] and rests on a continuation of Bohmian mechanics from a real-valued to a complex-valued action [425]. Newton's laws determine complex-valued positions and momenta, and each trajectory advances on a single surface. The interaction between two surfaces is governed by their coupling potential as well as the difference in the complex phase between them. Diabatic [423] and adiabatic [424] versions of the formalism have, with good success, been exposed to the challenges of the Tully canon (Section 11.2.1).

13.4 The Classical Limit of the Quantum Theory of Motion

So far we have emphasized the basic tenets of the quantum theory of motion, developing the formally correct nonadiabatic equations for the case of a highly simplified, idealized physical system. We will now apply the nonadiabatic Bohm–de Broglie theory to a general system without imposing any simplifying constraints. Having derived the explicit form of the respective equations, we will explore the classical limit of these equations to generate a computationally tractable scheme for solving nonadiabatic problems. The resulting formalism will turn out to be an interesting hybrid of classical and quantum mechanical methodology, involving the propagation of delocalized wave packets but governed by the classical Hamilton–Jacobi equation [417]. A multidimensional situation with several nuclear centers will be considered in the frame of the adiabatic representation, thus sacrificing the assumption $\langle \psi_i | \frac{\partial \psi_j}{\partial \mathbf{X}} \rangle = 0$ made in the previous section.

A physical system is given in the adiabatic representation as

$$\Psi(\mathbf{x}, \mathbf{X}, t) = \sum_i \chi_i(\mathbf{X}, t)\psi_i(\mathbf{x}, \mathbf{X}). \qquad (13.57)$$

The functions $\psi_i(\mathbf{x}, \mathbf{X})$ are now defined as adiabatic eigenstates of the electronic subsystem at a specified nuclear coordinate $\mathbf{X}$. We consider the case of real and non-degenerate $\psi_i(\mathbf{x}, \mathbf{X})$. Inserting Eq. (13.57) into the TDSE, multiplying by $\psi_j(x, \mathbf{X})^*$ from the left, and subsequently integrating over the electronic coordinates $\mathbf{x}$ leaves us with

$$i\hbar \frac{\partial \chi_j}{\partial t} = -\sum_a \frac{\hbar^2}{2M_a} \frac{\partial^2 \chi_j}{\partial \mathbf{X}_a^2} + V_j^{\mathrm{APES}}(\mathbf{X})\chi_j - \sum_{a,i} \frac{\hbar^2}{2M_a} \left(B_{ji,a}\chi_i + 2\mathbf{A}_{ji,a} \cdot \frac{\partial \chi_i}{\partial \mathbf{X}_a} \right) \tag{13.58}$$

with

$$\mathbf{A}_{ji,a}(\mathbf{X}) = \left\langle \psi_j \middle| \frac{\partial \psi_i}{\partial \mathbf{X}_a} \right\rangle \tag{13.59}$$

and

$$B_{ji,a}(\mathbf{X}) = \left\langle \psi_j \middle| \frac{\partial^2 \psi_i}{\partial \mathbf{X}_a^2} \right\rangle. \tag{13.60}$$

The index a labels the nuclear centers. We assume that the adiabatic potential energy surfaces $V_j^{\mathrm{APES}}(\mathbf{X})$ are known. At this point we introduce the polar representation for the nuclear wave functions $\chi_j(\mathbf{X}, t)$. Substituting this ansatz into Eq. (13.58), one finds that the TDSE separates into the following two equations:

$$\frac{\partial R_j}{\partial t} = -\sum_a \left[\frac{1}{M_a} \frac{\partial R_j}{\partial \mathbf{X}_a} \cdot \frac{\partial S_j}{\partial \mathbf{X}_a} + \frac{1}{2M_a} R_j \frac{\partial^2 S_j}{\partial \mathbf{X}_a^2} \right] - \sum_{a,i} \left[\frac{R_i}{M_a} \mathbf{A}_{ji,a} \cdot \frac{\partial S_i}{\partial \mathbf{X}_a} - \frac{i\hbar}{M_a} \mathbf{A}_{ji,a} \cdot \frac{\partial R_i}{\partial \mathbf{X}_a} + \frac{i\hbar}{2M_a} R_i B_{ji,a} \right] \exp\left[\frac{i}{\hbar}(S_i - S_j) \right], \tag{13.61}$$

$$\frac{\partial S_j}{\partial t} = -V_j^{\mathrm{APES}}(\mathbf{X}) - \sum_a \left[\frac{1}{2M_a} \left(\frac{\partial S_j}{\partial \mathbf{X}_a} \right)^2 - \frac{\hbar^2}{2M_a} \frac{\partial^2 R_j}{\partial \mathbf{X}_a^2} \frac{1}{R_j} \right]. \tag{13.62}$$

One recognizes in these identities the equation pair (4.54a) and (4.54b) as applied to a wave function in the adiabatic representation. The classical limit of these relations is easily found from Eq. (13.62). Neglecting the quantum potential, we arrive at the classical Hamilton–Jacobi equation

$$\frac{\partial S_j}{\partial t} = -V_j^{\mathrm{APES}}(\mathbf{X}) - \sum_a \frac{1}{2M_a}\left(\frac{\partial S_j}{\partial \mathbf{X}_a}\right)^2. \tag{13.63}$$

In direct generalization of Eq. (13.27), $\frac{\partial S_j}{\partial \mathbf{X}_a}$ is understood as the momentum on surface j, associated with the set $\mathbf{X}_a$ of nuclear coordinates. Combining Eq. (13.63) with Eq. (13.61), we retain a quantum mechanical wave packet description of the nuclear subsystem which is nevertheless subjected to classical dynamics.

In the following, we will move the equation system (13.61) and (13.62) one step further towards the pole of classical mechanics while maintaining the wave packet model. This is achieved by investigating the case of small $\hbar$ ($\hbar \to 0$) in Eq. (13.61). The last two terms in this equation, containing $\hbar$ as a factor, are negligible in the limit of infinitesimal $\hbar$. This leaves

$$\begin{aligned}\frac{\partial R_j}{\partial t} &= -\sum_a \left[\frac{1}{M_a}\frac{\partial R_j}{\partial \mathbf{X}_a}\cdot\frac{\partial S_j}{\partial \mathbf{X}_a} + \frac{1}{2M_a}R_j\frac{\partial^2 S_j}{\partial \mathbf{X}_a^2}\right] \\ &\quad - \sum_{a,i}\left[\frac{1}{M_a}R_i\mathbf{A}_{ji,a}\cdot\frac{\partial S_i}{\partial \mathbf{X}_a}\right]\exp\left[\frac{i}{\hbar}(S_i - S_j)\right].\end{aligned} \tag{13.64}$$

It should be noted that the last term in this identity is responsible for the coupling between the various potential energy surfaces. As we have assumed $\hbar \to 0$, the exponential $\exp[\frac{i}{\hbar}(S_i - S_j)]$ is highly oscillatory unless the two corresponding action functions are sufficiently similar, $S_i \approx S_j$. The equation pair (13.61) and (13.62) is referred to as the *classical limit Schrödinger equation*, the pair (13.63) and (13.64) as the *velocity coupling approximation* [417]. With the latter scheme we have still not entered the territory of mixed quantum-classical approaches, in spite of the two classical assumptions on which it rests, namely the constraints of vanishing quantum potential and $\hbar \to 0$. The borderline to the mixed quantum-classical model is crossed as one adopts a point particle description of the nuclei.

From relation (13.64) we see that the velocity coupling approximation is not norm-conserving. This result is obtained from an inspection of the sum $\sum_i p_{ii}$ with $p_{jj} = \int d\mathbf{X}\, R_j(\mathbf{X})^* R_j(\mathbf{X})$. While norm conservation implies $\sum_j p_{jj} = 1$ and thus $\sum_i \frac{\partial p_{ii}}{\partial t} = 0$, one obtains from Eq. (13.64) after some rearrangement

$$\sum_j \frac{\partial p_{jj}}{\partial t} = -\sum_{a,j,j<i} \int \frac{2}{M_a} \Re \left\{ R_j^* R_i \exp\left[\frac{i}{\hbar}(S_i - S_j)\right] \mathbf{A}_{ji,a} \right\}$$

$$\cdot \left(\frac{\partial S_i}{\partial \mathbf{X}_a} - \frac{\partial S_j}{\partial \mathbf{X}_a} \right) d\mathbf{X}. \tag{13.65}$$

Exercise 13.6. Derive Eq. (13.65) from Eq. (13.64), using the anti-Hermitian nature of $\mathbf{A}_{ji,a}$.

Relation (13.65) clarifies that norm conservation within the velocity coupling approximation is only guaranteed if $\frac{\partial S_i}{\partial \mathbf{X}_a} = \frac{\partial S_j}{\partial \mathbf{X}_a}$. However, one also infers from Eq. (13.65) that norm conservation is restored as one replaces both $\frac{\partial S_i}{\partial \mathbf{X}_a}$ and $\frac{\partial S_j}{\partial \mathbf{X}_a}$ with their average $\frac{1}{2}(\frac{\partial S_i}{\partial \mathbf{X}_a} + \frac{\partial S_j}{\partial \mathbf{X}_a})$. This operation leads to the *averaged velocity coupling approximation* (AVCA) [417]. The latter model was found to be compatible with exact quantum mechanical results when exposed to the test suite of the Tully canon (see Section 11.2.1). In the chemical energy regime, it turned out to be significantly more precise than the surface hopping approach described in Sections 11.1 and 11.2.

The demonstrated advantage of models that rely on the quantum theory of motion over independent trajectory approximations may be traced back to their freedom from any assumption of sharply defined nuclear positions. As these evolve in time, the system is confined to a single potential energy surface at any given moment, with jumps between surfaces to simulate electronic transitions. Quantum hydrodynamics *per se* as well as its classical limits defined by (13.61) and (13.62) or (13.63) and (13.64), respectively, abandon any discontinuous switch of population from one potential energy surface to another in favor of gradually changing amplitudes R_i, allowing for a realistic account of delocalization and for continuous energy exchange between the electrons and the nuclei of the considered system [417].

Generalizing the quantum trajectory formulation of nonadiabatic dynamics, Garashchuk and Rassolov considered semiclassical propagation involving an *approximate quantum potential* ([418]–[422]). In these models, the degree of rigor in the description of the quantum potential becomes the basic parameter that may be adjusted to yield full equivalence to the quantum mechanical solution of the problem at hand or, in the opposite extreme, to provide a linearized quantum force that is exact in the specific case of Gaussian wave packets in locally quadratic potentials. For molecular applications, the regime of intermediate accuracy where leading quantum effects may be included at moderate computational effort is of primary interest. The method has been shown to scale linearly with the number of dimensions [422].

Chapter 14

Quantum Hydrodynamics II: The Semiclassical Liouville–Von Neumann Equation

The quantum hydrodynamical approach, as presented in the previous chapter, yields a prescription for how to construct the solution of the TDSE by advancing ensembles of coupled quantum trajectories in a landscape of multiple potential energy surfaces. In this chapter, we discuss the same methodological principle from a different perspective, emphasizing the *nuclear density* as opposed to the *nuclear wave packet*. Rather than relying on the TDSE and interpreting its solution in classical terms, we will inspect the semiclassical limit of a quantum mechanical equation, namely the quantum Liouville–von Neumann equation (8.56), and deduce a dynamical formalism from this approach. We will show that the resulting scheme is well adjusted to simulate the evolution of the density by tracing quantum trajectories as a function of time. As implied by our use of Liouville theory, these trajectories are defined in a classical phase space.

Density distributions as opposed to wave functions are employed as elements of quantum-statistical models which provide the conceptual framework for analyzing open systems [426]. This feature of Liouville theory will be made explicit in Sections 14.3 and 14.4. The preceding sections focus on the nonadiabatic density matrix formalism in the coherent limit, as introduced by Martens and Fang [427]. Dynamical variables are here the density matrix elements. Liouville theory yields a set of coupled equations that describe the time behavior of the studied system in terms of populations (diagonal density matrix elements) and coherences (non-diagonal elements), supplying direct information about electronic transition processes in the course of molecular evolution. Section 14.2 outlines how both the relevant populations and coherences may be assembled by propagating quantum trajectories defined in a classical phase space as compared to the position space which was considered in the previous chapter. Correspondingly, semiclassical representation of the density is achieved by invoking the Wigner transform.

The remainder of this chapter extends our presentation of semiclassical dynamics based on Liouville theory by including mixed quantum states, corresponding to the transition from the density to the statistical operator. While the relation between pure state and mixed state dynamics is outlined in Section 14.3, the final section of this chapter applies the resulting methodology to the situation of chief interest in this text, namely coupled electronic states. In particular, it will be shown that non-local effects in coordinate space may be described in terms of local hydrodynamic fields. These fields are represented in the form of a *hierarchy of moments* obtained by integration over weighted powers of the phase space momentum [428, 429]. In this framework, the Liouville–von Neumann equation generates equations of motion for the sequence of moments thus obtained. Essential for the practical use of this scheme is the definition of truncation conditions for an infinite system of equations of first order in time, reflecting an infinite hierarchy of moments. We will demonstrate that the assumption of pure quantum states, as distinguished from statistical mixtures, naturally makes the problem tractable by terminating the set of equations, and discuss other truncation conditions.

14.1 The Semiclassical Liouville Formalism for Multistate Problems

From Section 4.3 we recall that quantum dynamic observables may be approximated by distributions defined on the classical phase space following the prescription by Wigner. The Wigner transform $A_{\mathrm{w}}(X, P, t)$ of an operator $\hat{A}(t)$, assuming a one-dimensional situation as will be considered in the following, is (see Eq. (4.108))

$$A_{\mathrm{w}}(X, P, t) = \int_{-\infty}^{\infty} \left\langle X + \frac{\tilde{X}}{2} |\hat{A}(t)| X - \frac{\tilde{X}}{2} \right\rangle \exp\left(-\frac{i}{\hbar} P\tilde{X}\right) d\tilde{X}. \quad (14.1)$$

It is important to note that the Wigner transform of an operator product $\hat{A}\hat{B}$ differs from the product of the Wigner transforms of $\hat{A}$ and $\hat{B}$. Instead, one finds that [430, 431]

$$(\hat{A}\hat{B})_{\mathrm{w}} = A_{\mathrm{w}} \exp\left[\frac{\hbar}{2i}\hat{\Lambda}\right] B_{\mathrm{w}} = B_{\mathrm{w}} \exp\left[-\frac{\hbar}{2i}\hat{\Lambda}\right] A_{\mathrm{w}} \tag{14.2}$$

with $\hat{\Lambda}$ as *Poisson bracket operator*,

$$\hat{\Lambda} = \frac{\partial}{\partial P^{\leftarrow}}\frac{\partial}{\partial X^{\rightarrow}} - \frac{\partial}{\partial X^{\leftarrow}}\frac{\partial}{\partial P^{\rightarrow}}, \tag{14.3}$$

where the arrow superscripts indicate if the derivative is to be applied to the left or the right. From Eq. (14.2) we infer immediately the Wigner representation of the commutator $[\hat{A}, \hat{B}]$:

$$[\hat{A}, \hat{B}]_{\mathrm{w}} = 2iA_{\mathrm{w}}(X, P, t) \sin\left(\frac{\hbar}{2}\hat{\Lambda}\right) B_{\mathrm{w}}(X, P, t), \tag{14.4}$$

and, likewise, of the anticommutator

$$[\hat{A}, \hat{B}]_{+,w} = 2\, A_{\mathrm{w}}(X, P, t) \cos\left(\frac{\hbar}{2}\hat{\Lambda}\right) B_{\mathrm{w}}(X, P, t). \tag{14.5}$$

Exercise 14.1. Use the definition of the Wigner transform as given by Eq. (14.1) in conjunction with the explicit form of the $\hat{\Lambda}$ operator (14.3) to derive the expressions (14.2), (14.4), and (14.5) for the Wigner transforms of the product between two operators as well as that of their commutator and anticommutator.

It follows from Eq. (14.4) that the commutator $[\hat{A}, \hat{B}]_{\mathrm{w}}$ is approximated up to second order by the Poisson bracket of A_{w} and B_{w} according to

$$[\hat{A}, \hat{B}]_{\mathrm{w}} = i\hbar\{A_{\mathrm{w}}, B_{\mathrm{w}}\}_{X,P} + O(\hbar^3). \tag{14.6}$$

We will refer to this representation of the commutator as the *Wigner–Moyal expansion* [432]. The analogous relation for the anticommutator is

$$[\hat{A}, \hat{B}]_{+,\mathrm{w}} = 2A_{\mathrm{w}}(X, P, t)B_{\mathrm{w}}(X, P, t) + O(\hbar^2). \tag{14.7}$$

In this section, we will explore the Liouville formalism beyond the confines of Ehrenfest dynamics, as introduced in Section 8.3.4 in the context of the TDHF. We will thus deviate from the mean trajectory model, as presented in Chapters 8 and 9, and derive equations of motion for the density matrix elements characteristic of problems that involve a multiplicity of electronic surfaces. We will consider the classical, or rather semiclassical, limit of this theory which leads to a numerically tractable scheme of nonadiabatic quantum dynamics. The precise meaning of *semiclassical* in the present context will be clarified in the following.

Adopting scalar density and Hamilton operators $\hat{\rho}$ and $\hat{H}$, respectively, which are adequate for the situation of a single potential energy surface, we write the Liouville–von Neumann equation as

$$i\hbar\frac{\partial}{\partial t}\hat{\rho} = [\hat{H}, \hat{\rho}]. \tag{14.8}$$

In the classical limit, $\hat{\rho}$ and $\hat{H}$ become functions of the phase space variables X and P, and the commutator turns into the Poisson bracket. Applying the Wigner–Moyal expansion to the right-hand side of Eq. (14.8), i.e. upon replacing $[\hat{H}, \hat{\rho}]$ with $i\hbar\{H_{\rm w}, \rho_{\rm w}\}$, we arrive at (compare with Eq. (4.106)):

$$\frac{\partial}{\partial t}\rho_{\rm w} = \{\hat{H}_{\rm w}, \rho_{\rm w}\}, \tag{14.9}$$

where $\rho_{\rm w} = \rho_{\rm w}(X, P, t)$ and $H_{\rm w} = H_{\rm w}(X, P, t)$. Proceeding to the case of two potential energy surfaces, Eq. (14.8) becomes

$$i\hbar\frac{\partial}{\partial t}\hat{\boldsymbol{\rho}} = [\hat{\mathbf{H}}, \hat{\boldsymbol{\rho}}] \tag{14.10}$$

with

$$\hat{\boldsymbol{\rho}} = \begin{pmatrix} \hat{\rho}_{11} & \hat{\rho}_{12} \\ \hat{\rho}_{21} & \hat{\rho}_{22} \end{pmatrix}, \quad \hat{\mathbf{H}} = \begin{pmatrix} \hat{H}_{11} & \hat{H}_{12} \\ \hat{H}_{21} & \hat{H}_{22} \end{pmatrix}. \tag{14.11}$$

We emphasize that all components $\hat{\rho}_{ij}$ and $\hat{H}_{ij}$ of these matrices are operators, satisfying the equation

$$i\hbar\frac{\partial}{\partial t}\hat{\rho}_{ij} = \sum_{k=1}^{2} \hat{H}_{ik}\hat{\rho}_{kj} - \hat{\rho}_{ik}\hat{H}_{kj}. \tag{14.12}$$

With the assignments

$$\hat{H}_{12} = \hat{V}_{\rm c}, \quad \hat{H}_{21} = \hat{V}_{\rm c}^{\dagger}, \tag{14.13}$$

we introduce the coupling potential operator $\hat{V}_{\rm c}$ and secure the Hermitian nature of the Hamiltonian matrix. Martens and Fang [427] suggest a set of substitutions for both the density and Hamiltonian matrix operator elements that make it possible to express the right-hand side of Eq. (14.12) entirely in terms of commutators and anticommutators of Hermitian operators.

If this goal is achieved, the Wigner–Moyal prescription for the transition from quantum mechanical operators to classical functions may be applied (see Eqs. (14.4) and (14.5)). Using the terminology of Martens and Fang, we define

$$\hat{\nu} = \frac{1}{2}(\hat{V}_{\mathrm{c}} + \hat{V}_{\mathrm{c}}^{\dagger}), \qquad \hat{\mu} = \frac{1}{2i}(\hat{V}_{\mathrm{c}} - \hat{V}_{\mathrm{c}}^{\dagger}), \tag{14.14a}$$

$$\hat{H}_0 = \frac{1}{2}(\hat{H}_{11} + \hat{H}_{22}), \quad \hat{H}' = \hat{H}_{11} - \hat{H}_{22}, \tag{14.14b}$$

$$\hat{\rho}_0 = \frac{1}{2}(\hat{\rho}_{11} + \hat{\rho}_{22}), \qquad \hat{\delta} = \hat{\rho}_{11} - \hat{\rho}_{22}, \tag{14.14c}$$

$$\hat{\alpha} = \frac{1}{2}(\hat{\rho}_{12} + \hat{\rho}_{21}), \qquad \hat{\beta} = \frac{1}{2i}(\hat{\rho}_{12} - \hat{\rho}_{21}). \tag{14.14d}$$

With $\hat{\rho}_{21} = \hat{\rho}_{12}^{\dagger}$ and $\hat{V} = \hat{V}^{\dagger}$, one may readily prove the Hermitian property for all operators introduced in Eq. (14.14). Substituting these operators for $\hat{H}_{ij}$ and $\hat{\rho}_{ij}(i,j = 1,2)$ on the right-hand side of Eq. (14.12) accomplishes the desired expansion into commutators and anticommutators. The resulting equations of motion for the three independent components of $\hat{\boldsymbol{\rho}}$ are

$$i\hbar\frac{\partial}{\partial t}\hat{\rho}_{11} = [\hat{H}_{11}, \hat{\rho}_{11}] + [\hat{\nu}, \hat{\alpha}] + [\hat{\mu}, \hat{\beta}] + i[\hat{\mu}, \hat{\alpha}]_{+} - i[\hat{\nu}, \beta]_{+}, \tag{14.15a}$$

$$i\hbar\frac{\partial}{\partial t}\hat{\rho}_{22} = [\hat{H}_{22}, \hat{\rho}_{22}] + [\hat{\nu}, \hat{\alpha}] + [\hat{\mu}, \hat{\beta}] + i[\hat{\nu}, \hat{\beta}]_{+} - i[\hat{\mu}, \hat{\alpha}]_{+}, \tag{14.15b}$$

$$i\hbar\frac{\partial}{\partial t}\hat{\rho}_{12} = [\hat{H}_0, \hat{\rho}_{12}] + [\hat{V}_{\mathrm{c}}, \hat{\rho}_0] + \frac{1}{2}[\hat{H}', \hat{\rho}_{12}]_{+} - \frac{1}{2}[\hat{V}_{\mathrm{c}}, \hat{\delta}]_{+}. \tag{14.15c}$$

The equation for $\hat{\rho}_{21}$ is obtained by Hermitian conjugation of the last of these three equations. The Wigner–Moyal limit is now taken by expanding all contributions on the right-hand side of Eq. (14.15) into powers of $\hbar$ and replacing the operators with functions defined on the phase space $\{X, P\}$. To lowest order, this operation leads to

$$\frac{\partial}{\partial t}\rho_{11} = \hat{L}_{11}\rho_{11} + \{\nu, \alpha\} + \{\mu, \beta\} - \frac{2}{\hbar}(\nu\beta - \mu\alpha), \tag{14.16a}$$

$$\frac{\partial}{\partial t}\rho_{22} = \hat{L}_{22}\rho_{22} + \{\nu, \alpha\} + \{\mu, \beta\} + \frac{2}{\hbar}(\nu\beta - \mu\alpha), \tag{14.16b}$$

$$\frac{\partial}{\partial t}\rho_{12} = \hat{L}_0\rho_{12} + \{V_{\mathrm{c}}, \rho_0\} - \frac{i}{\hbar}H'\rho_{12} + \frac{i}{\hbar}V_{\mathrm{c}}\delta. \tag{14.16c}$$

For simplicity of notation, we have dropped here the subscript w that was used in Eqs. (14.1)–(14.2) and (14.4)–(14.7) to designate phase space functions. The set of equations (14.16) introduces the *Liouville operator* $\hat{L}_\mu$, defined as

$$\hat{L}_\mu f \equiv \{H_\mu, f\}, \quad \mu = 0, 11, 22. \tag{14.17}$$

Using $\rho_{12} = \alpha + i\beta$ and $V_c = \nu + i\mu$, one separates Eq. (14.16c) into two real-valued equations of motion for the functions α and β, namely

$$\frac{\partial}{\partial t}\alpha = \{H_0, \alpha\} + \{\nu, \rho_0\} + \frac{1}{\hbar}H'\beta - \frac{1}{\hbar}\mu\delta, \tag{14.18a}$$

$$\frac{\partial}{\partial t}\beta = \{H_0, \beta\} + \{\mu, \rho_0\} - \frac{1}{\hbar}H'\alpha + \frac{1}{\hbar}\nu\delta. \tag{14.18b}$$

The Wigner–Moyal limit of the Liouville–von Neumann equation thus yields four dynamic relations: Eqs. (14.16a) and (14.16b), governing the time behavior of the diagonal density matrix elements and therefore the *populations* of the two considered surfaces, and Eqs. (14.18a) and (14.18b), yielding the non-diagonal density matrix elements, or *coherences*. The structure shared by all four equations sheds light on the connection of the present formalism with classical mechanics. The partial time derivatives of the dynamic quantities are described by the respective Poisson bracket according to $\frac{\partial f}{\partial t} = \{H, f\}$, but inhomogeneous terms are added to this bracket. These involve non-classical objects, as the functions α, β and μ, ν are related to the coherences and the coupling potential, respectively. It is in this sense that we refer to the emerging dynamic scheme as *semiclassical*. A further step towards classical dynamics is taken by enforcing the Hermitian nature of the coupling potential operator $\hat{V}_c$ so that its classical counterpart $V_c = V_c(X, P)$ is real. In this case, the canon of the four semiclassical dynamic equations (14.16) and (14.18) takes the simplified form

$$\frac{\partial \rho_{11}}{\partial t} = \hat{L}_{11}\rho_{11} + \{V_c, \alpha\} - \frac{2}{\hbar}V_c\beta, \tag{14.19a}$$

$$\frac{\partial \rho_{22}}{\partial t} = \hat{L}_{22}\rho_{22} + \{V_c, \alpha\} + \frac{2}{\hbar}V_c\beta, \tag{14.19b}$$

$$\frac{\partial \alpha}{\partial t} = \hat{L}_0\alpha + \{V_c, \rho_0\} + \frac{1}{\hbar}H'\beta, \tag{14.19c}$$

$$\frac{\partial \beta}{\partial t} = \hat{L}_0\beta - \frac{1}{\hbar}H'\alpha + \frac{1}{\hbar}V_c\delta. \tag{14.19d}$$

Exercise 14.2. Show that Eqs. (14.19a)–(14.19d) are equivalent to:

$$\frac{\partial \rho_{11}}{\partial t} = \hat{L}_{11}\,\rho_{11} + \{V_c, \Re[\rho_{12}]\} - \frac{2V_c}{\hbar}\Im[\rho_{12}], \tag{14.20}$$

$$\frac{\partial \rho_{22}}{\partial t} = \hat{L}_{22}\,\rho_{22} + \{V_c, \Re[\rho_{12}]\} + \frac{2V_c}{\hbar}\Im[\rho_{12}], \tag{14.21}$$

$$\frac{\partial \rho_{12}}{\partial t} = (\hat{L}_0 - i\omega)\rho_{12} + \frac{1}{2}\{V_c, \rho_{11} + \rho_{22}\} + \frac{iV_c}{\hbar}(\rho_{11} - \rho_{22}), \tag{14.22}$$

where $\omega = \omega(X) = \frac{H'(X)}{\hbar}$.

A dramatic additional reduction of the basic equation system is accomplished if the densities $\rho_{ii}, i = 1, 2$ of the two electronic states are replaced by the respective total diabatic populations P_{ii}^{D}. The latter quantities are derived from the former by taking the phase space trace according to

$$P_{ii}^{\mathrm{D}}(t) = \int dXdP\, \rho_{ii}(X, P, t), \quad i = 1, 2. \tag{14.23}$$

Likewise, the coherence-related functions become

$$A^{\mathrm{D}}(t) = \int dXdP\, \alpha(X, P, t), \tag{14.24a}$$

$$B^{\mathrm{D}}(t) = \int dXdP\, \beta(X, P, t). \tag{14.24b}$$

We assume that all the functions involved in this development vanish at infinity, i.e. their phase space integrals are finite. Therefore, the phase space trace of their Poisson bracket is zero. Performing the integration of Eqs. (14.19a)–(14.19d), we see that all brackets vanish, and we are left with a much less complex set of equations [427]:

$$\frac{dP_{11}^{\mathrm{D}}}{dt} = -\frac{2}{\hbar}V_c B^{\mathrm{D}}, \tag{14.25a}$$

$$\frac{dP_{22}^{\mathrm{D}}}{dt} = \frac{2}{\hbar}V_c B^{\mathrm{D}}, \tag{14.25b}$$

$$\frac{dA^{\mathrm{D}}}{dt} = \frac{1}{\hbar}H' B^{\mathrm{D}}, \tag{14.25c}$$

$$\frac{dB^{\mathrm{D}}}{dt} = -\frac{1}{\hbar}H' A^{\mathrm{D}} + \frac{1}{\hbar}V_c(P_{11}^{\mathrm{D}} - P_{22}^{\mathrm{D}}), \tag{14.25d}$$

where a constant coupling potential matrix element V_c has been assumed. Denoting $P_{11}^{D} - P_{22}^{D}$ by D^{D}, we condense these four into three equations, written in matrix form:

$$\frac{d}{dt}\begin{pmatrix} D^{D} \\ A^{D} \\ B^{D} \end{pmatrix} = \begin{pmatrix} 0 & 0 & -4V_c/\hbar \\ 0 & 0 & H'/\hbar \\ V_c/\hbar & -H'/\hbar & 0 \end{pmatrix}\begin{pmatrix} D^{D} \\ A^{D} \\ B^{D} \end{pmatrix} \equiv \mathbf{M}\begin{pmatrix} D^{D} \\ A^{D} \\ B^{D} \end{pmatrix}. \tag{14.26}$$

Going from Eqs. (14.25) to (14.26), we have eliminated the normalization condition $P_{11}^{D} + P_{22}^{D} = 1$ which results from adding Eq. (14.25a) and Eq. (14.25b). The case of constant **M**, defining the *Rabi limit*, is exactly solvable, as further detailed in Section 14.1*.

14.1.1 *Two coupled states: A model problem*

For a numerical assessment of the semiclassical Liouville formalism we discuss the dynamics of a particle with mass m in a problem involving two coupled potential energy curves, $V_1(X)$ and $V_2(X)$ [427]. It is assumed that one of the corresponding states is populated at the beginning of the process ($t = 0$) by excitation of a wave packet from a harmonic ground state using an ultrashort laser pulse. One potential energy curve is repulsive, the other one a bound Morse potential according to

$$V_1(X) = V_{10}\exp[-\alpha(X - X_1)] - C, \tag{14.27a}$$

$$V_2(X) = V_{20}[\exp(-2\beta(X - X_2)) - 2\exp(-\beta(X - X_2))], \tag{14.27b}$$

where $V_{10}, V_{20}, \alpha, \beta$, and C are constants. The particular choice made for V_1 and V_2 in [427] is displayed in Figure 14.1.

The coupling potential is chosen as a Gaussian, adopting its maximum value at the crossing point X_c of the potential curves shown in Figure 14.1:

$$V_{12}(X) = V_0\exp[-c(X - X_c)^2], \tag{14.28}$$

with constant c. Since we assume instantaneous excitation at $t = 0$ from a harmonic oscillator, the initial wave packet is also a Gaussian. The respective density function $\rho_{11}(X, P, t = 0)$ is defined as the Wigner transform of this wavepacket, and thus the initial partial derivatives with respect to the phase space variables can be determined. At the beginning of the motion, the system is localized exclusively on the repulsive potential curve $V_1(X)$.

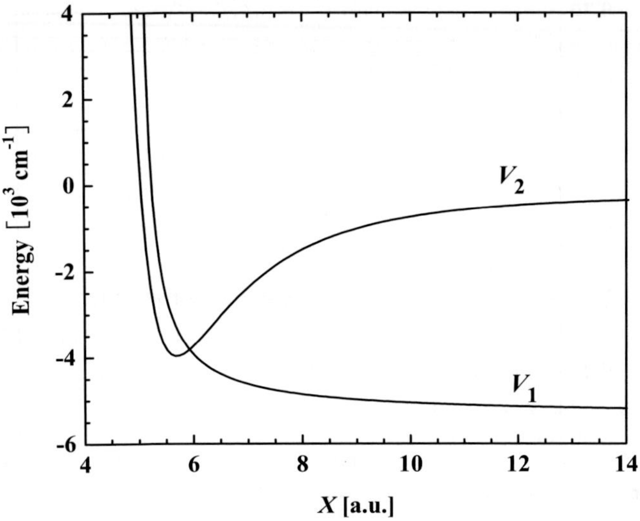

Figure 14.1: The two potential energy curves, corresponding to two excited states, used in a numerical simulation in order to assess the semiclassical Liouville approach. (Reprinted with permission from [427]. Copyright (1997) by the American Institute of Physics.)

With the physical situation thus described, the equation system (14.19) is solved. Martens and Fang [427] employ for this purpose a fast Fourier transform (*FFT*) scheme (see Section 15.1.2), advancing each of the four distributions $\rho_{11}, \rho_{22}, \alpha$, and β on a segmented domain that consists of 256×512 points of the phase space (X, P). Comparison is made with the exact quantum mechanical solution of the TDSE for the problem of the two coupled potential energy curves V_1 and V_2. The latter computation is carried out by generalizing the *FFT* grid procedure of Kosloff and Kosloff [433] to deal with the two-state problem at hand.

Figure 14.2 compares solutions emerging from both approaches, the semiclassical Liouville approximation and exact quantum mechanical theory. More specifically, the population of the bonding state (state 2) is displayed as a function of time. The interpretation of the time behavior shown is as follows: the wave packet, initially occupying V_1, is exposed to a repulsive force at $t = 0$, moving the system towards dissociation. With the elapse of time, it passes through the crossing region, which occurs with the parameters chosen in [427] around $t = 500$ a.u. In this time regime, a dramatic increase of the state 2 population is observed. As the evolution proceeds, this

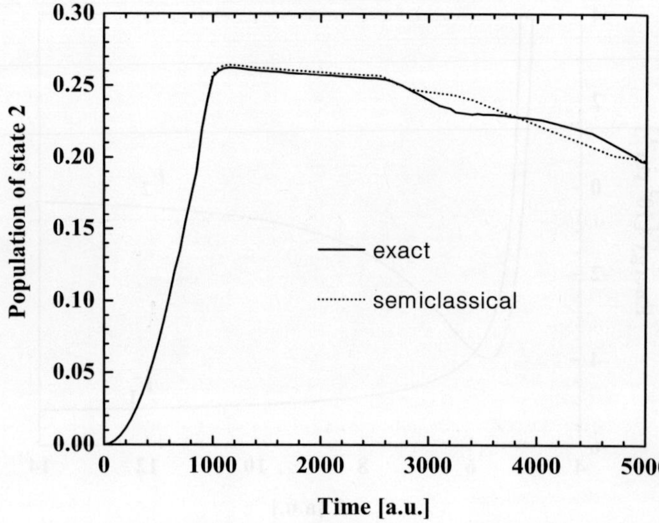

Figure 14.2: Time dependence of the bound state (state 2) population. The semiclassical Liouville approach is compared with the exact quantum mechanical solution. (Reprinted with permission from [427]. Copyright (1997) by the American Institute of Physics.)

population partially decays into the dissociative state 1. This trend is well reproduced by the semiclassical Liouville theory; in fact, it agrees almost quantitatively with the quantum mechanical calculation.

A more detailed comparison between the two methods involves the reduced configuration space densities $\widetilde{P}_{ij}$ $(i,j = 1,2)$. These quantities are defined by integration over the momentum variable of the phase space densities: $\widetilde{P}_{ij}(X,t) = \int dP\ \rho_{ij}(X,P,t)$. Two snapshots of the real part of the reduced coherence, taken at $t = 2,000$ a.u. and $t = 3,000$ a.u., demonstrate that the qualitative agreement between semiclassical Liouville and full quantum theory remains good even for long simulation times (see Figure 14.3). As seen from Figure 14.3, however, quantitative deviations between the two solutions grow with the elapse of time and become quite prominent. We emphasize once more that the semiclassical Liouville scheme surveyed in Section 14.1 and exemplified here uses the density components $\rho_{ij}(X,P,t)$ as primary dynamic variables. In the following section, we will outline how to represent the semiclassical Liouville theory in a more conventional frame of molecular dynamics, involving ensembles of trajectories.

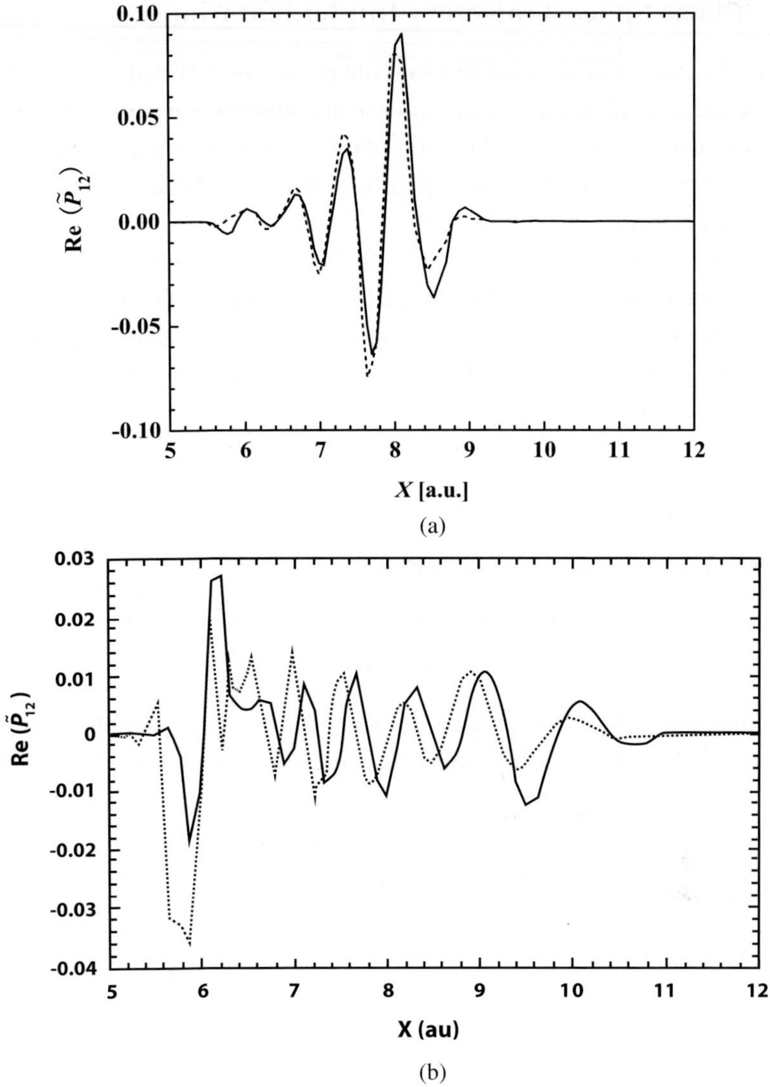

Figure 14.3: The real part of the reduced configuration space coherence, $\Re[\widetilde{P}_{ij}(X,t)]$, as a function of the spatial coordinate at (a) $t = 2{,}000$ a.u., (b) $t = 3{,}000$ a.u. The solid and the dashed lines refer to the exact quantum mechanical and the semiclassical solutions, respectively. (Reprinted with permission from [427]. Copyright (1997) by the American Institute of Physics.)

14.2 Phase Space Trajectory Implementation

The realization of semiclassical Liouville theory as outlined earlier relies on a phase space grid method. The expense of computation time to be invested into sufficiently accurate applications of this scheme, however, defies the purpose of approximating the exact quantum mechanical theory, whose implementation requires comparable effort. To make the semiclassical approach practicable, another numerical technique has to be applied. In the trivial situation of vanishing coupling and difference potentials ($V_c = H' = 0$), the equations (14.16) are truncated after the first Poisson bracket on the right-hand side. In this case, the semiclassical equations take the form

$$\frac{\partial}{\partial t}\rho_\mu = \hat{L}_\mu \rho_\mu \quad (\mu = 11, 22, 12), \tag{14.29}$$

involving the classical Hamiltonian flow of phase space trajectories. If an ensemble of N_μ trajectories is defined, an approximate expression for the corresponding density is given by [434]

$$\widetilde{\rho}_\mu(\Gamma, t) \equiv \frac{1}{N_\mu}\sum_{j=1}^{N_\mu} \delta[\Gamma - \Gamma_j^\mu(t)]. \tag{14.30}$$

This formula reduces the density associated with the ensemble μ to a sum over two-dimensional delta functions centered on $\Gamma_j^\mu(t) = \{X_j^\mu(t), P_j^\mu(t)\}$ with $j = 1, 2, \ldots, N_\mu$, where N_μ stands for the number of trajectories. Using Eq. (14.30), one may trace the density as a function of time by solving Hamilton's equations for a set of initial coordinates and momenta $\{X_j^\mu(t = 0), P_j^\mu(t = 0)\}$ with $j = 1, 2, \ldots, N_\mu$, which are sampled from the density $\widetilde{\rho}_\mu(\Gamma, t = 0)$. For numerical tractability, one may smooth the singularities associated with the sum over delta functions by convoluting $\widetilde{\rho}_\mu(\Gamma, t)$ with two-dimensional Gaussian functions defined on the phase space, obtaining

$$\rho_\mu(\Gamma, t) = \int d^2\Gamma' \, \phi(\Gamma - \Gamma')\widetilde{\rho}_\mu(\Gamma', t), \tag{14.31}$$

where

$$\phi(\Gamma - \Gamma') = \frac{1}{2\pi\sigma_X\sigma_P} \exp\left[-\frac{(X - X')^2}{2\pi\sigma_X^2} - \frac{(P - P')^2}{2\pi\sigma_P^2}\right]. \tag{14.32}$$

One confirms that $\rho_\mu(\Gamma, t)$ obeys the normalization constraint

$$\int d\Gamma^2 \, \rho_\mu(\Gamma, t) = \mathrm{Tr}\rho_\mu = 1, \tag{14.33}$$

according to the conservation of phase space trajectories for each index μ considered. If, however, we proceed to nonadiabatic dynamics and take into account the coupling between the states 1 and 2, trajectories are allowed to move from one potential surface to the other. Under these conditions, only the total trace of both contributing states is conserved. It holds that

$$\mathrm{Tr}\rho = \mathrm{Tr}\rho_{11} + \mathrm{Tr}\rho_{22} = 1. \tag{14.34}$$

In this case, the classical ansatz (14.29) is obviously insufficient to capture the dynamic process, and the corresponding density function (14.30) has to be generalized. In [435] such an extension is achieved by introducing arrays of dynamical variables $\{a_j^\mu(t)\}$ with $j = 1, 2, \ldots, N_\mu$, that permit us to assign a time-dependent weighting coefficient to each trajectory of each group μ. The new density function $\widetilde{\rho}_\mu$ is then written as

$$\widetilde{\rho}_\mu(\Gamma, t) = \sum_{j=1}^{N_\mu} a_j^\mu(t)\delta(\Gamma - \Gamma_j^\mu(t)). \tag{14.35}$$

The structure of the density function (14.35), or of its smoothed equivalent, defined in analogy to (14.31), implies that the task of density propagation is subdivided into two steps: an update of the coefficients $a_j^\mu(t)$, replacing them with $a_j^\mu(t + \Delta t)$, followed by Hamiltonian dynamics to advance the phase space coordinates by proceeding from $\Gamma_j^\mu(t)$ to $\Gamma_j^\mu(t + \Delta t)$. For this purpose, we write the semiclassical Liouville–von Neumann equations as given by Eqs. (14.20)–(14.22) (see Exercise 14.2) in integral form and apply the short time limit to arrive at the propagation equations to first order in Δt [434]:

$$\rho_{11}(\Gamma, t + \Delta t) = \exp(\Delta t \hat{L}_{11})[\rho_{11}(\Gamma, t) + b_{11}(\Gamma, t)\Delta t + O(\Delta t^2)], \tag{14.36a}$$

$$\rho_{22}(\Gamma, t + \Delta t) = \exp(\Delta t \hat{L}_{22})[\rho_{22}(\Gamma, t) + b_{22}(\Gamma, t)\Delta t + O(\Delta t^2)], \tag{14.36b}$$

$$\begin{aligned}\rho_{12}(\Gamma, t + \Delta t) &= \exp(\Delta t \hat{L}_0)\exp(-i\omega(\Gamma)\Delta t) \\ &\quad \times[\rho_{12}(\Gamma, t) + b_{12}(\Gamma, t)\Delta t + O(\Delta t^2)],\end{aligned} \tag{14.36c}$$

with

$$b_{11}(\Gamma, t) = \{V_c(\Gamma), \Re[\rho_{12}(\Gamma, t)]\} - \frac{2V_c}{\hbar}\Im[\rho_{12}(\Gamma, t)], \tag{14.37a}$$

$$b_{22}(\Gamma, t) = \{V_c(\Gamma), \Re[\rho_{12}(\Gamma, t)]\} + \frac{2V_c}{\hbar}\Im[\rho_{12}(\Gamma, t)], \tag{14.37b}$$

$$b_{12}(\Gamma, t) = \frac{1}{2}\{V_c(\Gamma), \rho_{11}(\Gamma, t) + \rho_{22}(\Gamma, t)\}$$
$$+ \frac{i}{\hbar} V_c(\Gamma)[\rho_{11}(\Gamma, t) - \rho_{22}(\Gamma, t)]. \tag{14.37c}$$

In these identities, as in Eqs. (14.16a)–(14.16c) and (14.20)–(14.22), classical and quantum dynamics appear clearly partitioned. While $\exp(\Delta t \hat{L}_\nu)$, $\nu = 11, 22, 0$, plays the role of a classical propagator, the inhomogeneous terms b_μ, $\mu = 11, 22, 12$, take into account quantum effects related to coherence and coupling between potential energy surfaces. The set of weighting coefficients $\{a_j^\mu(t)\}$ with $j = 1, 2, \ldots, N_\mu$, is introduced to accommodate the influence of these latter agents. Focusing on the example $\mu = 11$, we obtain the advanced weighting coefficients $\{a_j^\mu(t + \Delta t)\}$, $j = 1, 2, \ldots, N_\mu$, by inspecting (14.36a) and (14.37a). The new set of coefficients is found from

$$\sum_{j=1}^{N_{11}} a_j^{11}(t + \Delta t)\delta(\Gamma - \Gamma_j^{11}(t))$$
$$= \sum_{j=1}^{N_{11}} a_j^{11}(t)\delta(\Gamma - \Gamma_j^{11}(t)) + \Delta t \sum_{k=1}^{N_0} \Re\{a_k^0(t)\} V_c'[\Gamma_k^0(t)]\delta'[\Gamma - \Gamma_k^0(t)]$$
$$- \frac{2}{\hbar}\Im\{a_k^0(t)\}\, V_c[\Gamma_k^0(t)]\delta[\Gamma - \Gamma_k^0(t)] \tag{14.38}$$

with

$$V_c'[\Gamma] \equiv \frac{dV_c(X)}{dX} \tag{14.39}$$

and

$$\delta'(\Gamma - \Gamma_k^0(t)) \equiv \frac{\partial}{\partial P}\delta(\Gamma - \Gamma_0) \equiv \delta(X - X_0)\frac{d}{dP}\delta(P - P_0). \tag{14.40}$$

In the latter relation, we replace the delta function with the smoothed distribution $\phi(\Gamma - \Gamma_0)$, defined by Eq. (14.32), insert the resulting expression into the right-hand side of Eq. (14.38), and equate this right-hand side with a function $f^{11}(\Gamma)$. Since the goal of these operations is to select an optimum set of parameters $a_j^{11}(t + \Delta t)$, $j = 1, 2, \ldots, N_{11}$, we seek to minimize the phase space norm of the difference between $\sum_{j=1}^{N_{11}} a_j^{11}(t + \Delta t)\phi(\Gamma - \Gamma_j^{11}(t))$

and $f(\Gamma)$:

$$\int d^2\Gamma |f^{11}(\Gamma) - \sum_{j=1}^{N_{11}} \alpha_j \phi(\Gamma - \Gamma_j^{11}(t))|^2 = \min, \tag{14.41}$$

where α_j has been substituted for $a_j^{11}(t + \Delta t)$ to simplify notation. This requirement yields the constraint

$$\frac{\partial}{\partial \alpha_m} \int d^2\Gamma |f^{11}(\Gamma) - \sum_{j=1}^{N_{11}} \alpha_j \phi(\Gamma - \Gamma_j^{11}(t))|^2 = 0. \tag{14.42}$$

Using Eq. (14.38) and the definition of $f(\Gamma)$, one reformulates this condition such that it adopts the convenient form of a matrix-vector equation [434]:

$$\begin{aligned} \mathbf{S}^{(11,11)}\mathbf{a}^{(11)}(t + \Delta t) = \mathbf{S}^{(11,11)}\mathbf{a}^{(11)}(t) &+ \Delta t[\mathbf{D}^{(11,11)}\mathbf{V}'^{(0)}\Re\{\mathbf{a}^{(0)}(t)\} \\ &- \frac{2}{\hbar}\mathbf{S}^{(11,11)}\mathbf{V}^{(0)}\Im\{\mathbf{a}^{(0)}(t)\}], \end{aligned} \tag{14.43}$$

with the definitions

$$S_{mj}^{(\mu\nu)} \equiv \int d^2\Gamma \phi(\Gamma - \Gamma_m^{\mu})\phi(\Gamma - \Gamma_j^{\nu}), \tag{14.44a}$$

$$D_{mj}^{(\mu\nu)} \equiv \int d^2\Gamma \phi(\Gamma - \Gamma_j^{\mu})\frac{\partial}{\partial P}\phi(\Gamma - \Gamma_j^{\nu}), \tag{14.44b}$$

and

$$V_{c,ij}^{(0)} \equiv V_c(\Gamma_i^{(0)})\delta_{ij}, \quad V'^{(0)}_{c,ij} \equiv V'_c(\Gamma_i^{(0)})\delta_{ij}. \tag{14.44c}$$

In Eqs. (14.44a)–(14.44c), the time dependence of Γ_j^{μ} has been suppressed. We point out that the Gaussian approach (14.32) makes it possible to evaluate the matrix elements $S_{mj}^{(\mu\nu)}$ and $D_{mj}^{(\mu\nu)}$ analytically. In analogy to Eq. (14.43), matrix-vector equations for $\mathbf{a}^{(22)}(t + \Delta t)$ and $\mathbf{a}^{(0)}(t + \Delta t)$ are readily derived. Simultaneously solving the emerging three sets of equations updates the weighting coefficients introduced by Eq. (14.35). The evolution equation (14.43) and its counterparts for $\mu = 22, 12$ couple the trajectories within a given ensemble and further entangle the various ensembles with each other, as Eq. (14.43) connects the ensembles with $\mu = 11$ and $\mu = 12$. Therefore, all trajectories involved in the simulation have to be advanced simultaneously, in contrast to semiclassical schemes that employ ensembles of non-interacting trajectories, such as the original surface hopping or initial value representation theories summarized in Chapters 11 and 12, respectively, but in analogy to the quantum hydrodynamic approach outlined in Chapter 13.

It is to be noted that propagating the coefficients a_j^μ leaves the phase space coordinates unaffected. In the subsequent step, however, the trajectories are subjected to classical dynamics, evolving from $\Gamma_j^\mu(t)$ to $\Gamma_j^\mu(t+\Delta t)$ with $\mu = 11, 22, 12$ and $j = 1, \ldots, N_\mu$, under the Hamiltonian of the physical system considered.

In the trajectory-based implementation of the semiclassical Liouville theory, as discussed in this section, the trajectory approach is used as a vehicle to advance the distributions $\rho_{ij}(\Gamma, t), i, j = 1, 2$. The three ensembles model the time behavior of the relevant densities. This methodological viewpoint differs from that of the various semiclassical models which employ trajectories to trace the location of the nuclear centers for the purpose of direct simulation of the nuclear dynamics involved in the studied process. Electronic coherence emerges in semiclassical Liouville theory from the interference of trajectories which in turn approximate the development of the density function, and thus cannot be associated with any individual member of the nuclear ensemble.

The trajectory propagation technique has been employed to solve the test problem involving the two potential energy curves (14.27a) and (14.27b) (see also Figure 14.1) [435]. The parameters characterizing the physical system in terms of the crossing point between the curves, the initial wave packet, etc., were chosen as similar to those of the case discussed earlier (see Figures 14.1–14.3, and the respective commentary). For each of the three ensembles, 900 trajectories were used, and the simulation extended over a time interval of 3,000 a.u. At any time, the population $P_\mu(t)$, $\mu = 11, 22$, of the states 1 or 2 can be evaluated from

$$P_\mu(t) \equiv \mathrm{Tr}\rho_\mu(\Gamma, t) = \sum_{j=1}^{N_\mu} a_j^\mu(t), \quad \mu = 11, 22. \tag{14.45}$$

Information about reduced semiclassical configuration space densities is obtained by computing partial traces in phase space, as introduced in Section 14.1.1. The semiclassical Liouville formalism reproduces these quantities by the standard of the fully quantum mechanical, TDSE-based results in great detail. In particular, we find that the agreement is quantitative at early simulation times. The representation of the distributions

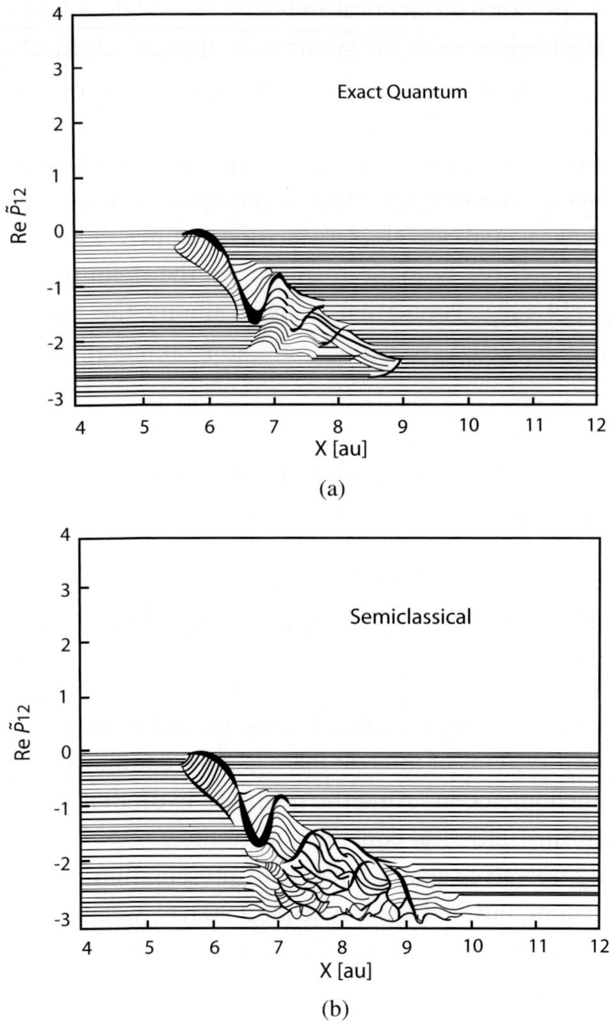

Figure 14.4: Reduced configuration space densities $\Re\{P_{12}(X)\}$ from the exact solution of the TDSE (a) and from the trajectory-based semiclassical Liouville approach (b). Each curve is shifted in time from its predecessor by an interval of $\Delta t = 60$ a.u. (Reprinted with permission from [434]. Copyright (1998) by the American Institute of Physics.)

$\widetilde{P}_{ij}(X,t)(=\int \rho(X,P,t)dP)$ as rendered by the semiclassical scheme can be assessed from the comparison given in Figure 14.4 where 50 $\widetilde{P}_{11}(X)$ curves, each one displaced from the adjacent curve by a time interval of 60 a.u., are plotted in temporal succession, spanning an overall time interval of 3,000 a.u.

While the quantum mechanical and semiclassical distributions are nearly identical at the beginning of the simulation, the semiclassical results accumulate admixtures of noise at large propagation times. However, Figure 14.4(b) demonstrates the capacity of a dynamic approach operating with classical phase space trajectories to simulate the development of a coherence, and thus a genuinely quantum mechanical entity, over an extended time interval in a qualitatively satisfactory manner. This is achieved by trajectory interference, as made manifest by the definition of the arrays (14.44a) and (14.44b). Population transfer among the two states considered has been modeled over the same time interval and with similar accuracy.

14.3 Generalized Quantum Hydrodynamics: Mixed States

The language of density distributions lends itself naturally to making the transition from pure to mixed quantum states. This transition is of relevance as one goes from closed to open systems, thus allowing for dissipative effects that might give rise to decoherence, turning a pure quantum state into a statistical mixture. Discussing mixed states, we will address the challenge of describing open systems of interest for, e.g., non-equilibrium dynamics and dissipation phenomena, by using the quantum hydrodynamic formalism to generate quantum-statistical equations of motion.

Confining ourselves again, for ease of presentation, to a single spatial degree of freedom, but including non-local effects, we introduce the statistical operator element $\rho(X, X')$ as incoherent superposition of projection operators $|i\rangle\langle i|$,

$$\rho(X, X') = \sum_i W_i \langle X|i\rangle\langle i|X'\rangle, \tag{14.46}$$

where the set of states $\{|i\rangle\}$ is assumed to be complete and the coefficients W_i stand for statistical weights, indicating the probability for the occurrence of a particular state $|i\rangle$. Obviously, the diagonal elements of Eq. (14.46), $\rho(X, X' = X)$, yield the local probability density P as a function of the coordinate X. If a pure state $|j\rangle$ is realized, the diagonal element of $\rho(X, X')$ reduces to the product $\langle X|j\rangle\langle j|X\rangle = \psi_j(X)\psi_j^*(X)$. In the context of generalized quantum hydrodynamics, the probability density is understood as the *moment of zeroth order*, labeled as

$$P(X) = \langle\rho\rangle(X). \tag{14.47}$$

In what follows, we will show that non-local effects can be accommodated by a *hierarchy of moments* that are natural extensions of the hydrodynamic quantities introduced in Chapter 13, and whose lowest order is given by $P(X)$. Further, the order of these moments is mirrored by a system of equations of motion that generalize the dynamical relations (4.54a) and (4.54b) which underlie wave-function-based Lagrangian quantum hydrodynamics.

We begin by expressing the Liouville–von Neumann equation (14.8) in the coordinate representation (14.46) and find

$$\begin{aligned} i\hbar\frac{\partial}{\partial t}\rho(X,X') = &-\frac{\hbar^2}{2m}\left[\frac{\partial^2}{\partial X^2} - \frac{\partial^2}{\partial X'^2}\right]\rho(X,X') \\ &+ [V(X) - V(X')]\rho(X,X'). \end{aligned} \tag{14.48}$$

Exercise 14.3. Prove relation (14.48).

The relation may be written in a more compact manner by adopting the superoperators $\hat{P}_+$ and $\hat{P}_-$ [426] defined by

$$\hat{P}_+ \equiv \frac{\hbar}{2i}\left[\frac{\partial}{\partial X} - \frac{\partial}{\partial X'}\right], \tag{14.49a}$$

$$\hat{P}_- \equiv \frac{\hbar}{i}\left[\frac{\partial}{\partial X} + \frac{\partial}{\partial X'}\right]. \tag{14.49b}$$

In terms of $\hat{P}_\pm$ the Liouville–von Neumann equation assumes the form

$$i\hbar\frac{\partial}{\partial t}\rho(X,X') = \frac{1}{m}\hat{P}_+\hat{P}_-\rho(X,X') + [V(X) - V(X')]\rho(X',X). \tag{14.50}$$

The operators $\hat{P}_+$ and $\hat{P}_-$ obey simple commutation relations with the corresponding position superoperators $\hat{X}_+$ and $\hat{X}_-$ where $\hat{X}_+ = \frac{1}{2}(X + X')$ and $\hat{X}_- = X - X'$. More specifically,

$$[\hat{X}_+, \hat{P}_+] = [\hat{X}_-, \hat{P}_-] = 0, \tag{14.51a}$$

$$[\hat{X}_+, \hat{P}_-] = [\hat{X}_-, \hat{P}_+] = i\hbar. \tag{14.51b}$$

Therefore, $\{\hat{X}_\pm, \hat{P}_\mp\}$ are to be viewed as conjugate operator pairs. Successive application of the *moment operator* $\hat{P}_+$ on the density $\rho(X,X')$ generates the quantities of relevance to quantum hydrodynamics. This can be seen as we define the moments engendered by the action of $\hat{P}_+$ as follows [428]:

$$\langle \hat{P}_+^n \rho \rangle(q) = \int_{-\infty}^{\infty} dr\, \delta(r) \left[\hat{P}_+^n \rho \left(q + \frac{r}{2}, q - \frac{r}{2} \right) \right]. \tag{14.52}$$

This relation is based on a switch of coordinates from (X', X) to (q, r) where $q \equiv \frac{1}{2}(X + X')$ and $r \equiv X - X'$. In this representation, $\hat{P}_+$ and $\hat{P}_-$ assume the simple form

$$\hat{P}_+ = \frac{\hbar}{i} \frac{\partial}{\partial r}, \tag{14.53a}$$

$$\hat{P}_- = \frac{\hbar}{i} \frac{\partial}{\partial q}. \tag{14.53b}$$

According to formula (14.52), the nth moment of $\rho\,(X', X)$ is obtained by applying $\hat{P}_+$ n times to $\rho\,(X', X)$ and then reducing the density to its diagonal element $\rho\,(q, q)$. We point out that the moments associated with $\hat{P}_+$ are closely related to the Wigner representation of the density $\rho_{\mathrm{W}}(q, p)$ (see Eq. (4.90)) where

$$\rho_{\mathrm{W}}(q, p) = \frac{1}{2\pi\hbar} \int_{-\infty}^{\infty} dr \rho \left(q + \frac{r}{2}, q - \frac{r}{2} \right) \exp\left(-\frac{i}{\hbar} pr \right). \tag{14.54}$$

Expressing the moments $\langle P_+^n \rho \rangle$ through the Wigner transform, we find

$$\langle \hat{P}_+^n \rho \rangle = \int_{-\infty}^{\infty} dp\, p^n\, \rho_{\mathrm{W}}(q, p). \tag{14.55}$$

The latter identity suggests that the basic hydrodynamic quantities may be formulated in terms of the moments obtained by the action of $\hat{P}_+^n$ on ρ. Just as the local density $\langle \rho \rangle_q$, according to Eq. (14.55), is understood as the moment of zeroth order, we can express the current density by use of a first-order moment, namely as

$$\mathcal{J}(q) = \frac{\langle \hat{P}_+ \rho \rangle_q}{m}. \tag{14.56}$$

The symbol $\langle \hat{P}_+ \rho \rangle_q$ denotes $\langle \hat{P}_+ \rho \rangle(q)$, with reference to Eq. (14.52). The second moment is related to the kinetic energy density K through

$$K(q) = \frac{\langle \hat{P}_+^2 \rho \rangle_q}{2m}. \tag{14.57}$$

The time behavior of the moments in a given physical process is governed by the Liouville–von Neumann equation (14.8) and (14.48). The moment construction rule (14.52) can be readily applied to this equation as given in the form (14.48). This operation leads to *moment dynamics*, i.e. to a system of time-dependent equations for moments of various order. The first three of these equations of motion read [428]

$$\frac{\partial}{\partial t}\langle \rho \rangle_q = -\frac{1}{m}\frac{\partial}{\partial q}\langle \hat{P}_+ \rho \rangle_q, \tag{14.58a}$$

$$\frac{\partial}{\partial t}\langle \hat{P}_+ \rho \rangle_q = -\frac{1}{m}\frac{\partial}{\partial q}\langle \hat{P}_+^2 \rho \rangle_q - \frac{\partial V}{\partial q}\langle \rho \rangle_q, \tag{14.58b}$$

$$\frac{\partial}{\partial t}\langle \hat{P}_+^2 \rho \rangle_q = -\frac{1}{m}\frac{\partial}{\partial q}\langle \hat{P}_+^3 \rho \rangle_q - 2\frac{\partial V}{\partial q}\langle \hat{P}_+ \rho \rangle_q. \tag{14.58c}$$

These relations may be extended to yield the general dynamic equation for moments of arbitrary order n [428]:

$$\begin{aligned}\frac{\partial}{\partial t}\langle \hat{P}_+^n \rho \rangle_q &= -\frac{1}{m}\frac{\partial}{\partial q}\langle \hat{P}_+^{n+1} \rho \rangle_q - n\frac{\partial V}{\partial q}\langle \hat{P}_+^{n-1} \rho \rangle_q \\ &\quad - \sum_{k=1}^{\frac{n-1}{2}} \left(\frac{\hbar}{2i}\right)^{2k} \binom{n}{2k+1} \frac{\partial^{2k+1} V}{\partial q^{2k+1}} \langle \hat{P}_+^{n-(2k+1)} \rho \rangle_q. \end{aligned}\tag{14.58d}$$

The result is an infinite sequence of interdependent equations. Comparing Eqs. (14.58d) and (14.48), one notes that we have traded the non-local solution $\rho(X, X')$ of the Liouville–von Neumann equation for a hierarchy of local quantities, namely the moments $\langle \hat{P}_+^n \rho \rangle_q$, to be extracted from an infinite system of coupled equations. The moments thus defined have turned out to be directly related to local hydrodynamic fields, such as the density, the current, and the kinetic energy density. It will be useful and interesting to determine conditions that reduce the first identities of the system (14.58) to a closed set and thus truncate this system, making it potentially practicable for the computational treatment of physical cases.

It is to be noted that the *quantum-classical moment approach* [436], as summarized by Eq. (14.58d), can be readily extended to include environmental effects acting on the system of interest. This may be accomplished by introducing a *dissipative superoperator* $\hat{L}_{\text{diss}}$ into the Liouville–von Neumann equation, according to

$$\frac{\partial \hat{\rho}}{\partial t} = -\frac{i}{\hbar}[\hat{H}, \hat{\rho}] + \hat{L}_{\text{diss}}. \tag{14.59}$$

A basic example of the use of Eq. (14.59) to include dissipation into the hierarchy of moments and subsequently the equations of motion (14.58d) is discussed in [428] where the Markov approximation [437], related to the separation of time scales between the system and its environment, is assumed to be valid. In Sections 17.1 and 17.2, we will examine Eq. (14.59) as a tool for augmenting the description of a mixed quantum-classical system by a realistic representation of the decoherence effect.

Exercise 14.4. Demonstrate that the expression $\frac{\langle \hat{P}_+\rho\rangle_q}{m}$ represents a one-dimensional current density, as stated by Eq. (14.56). Do this (a) by deriving relation (14.58a), (b) by direct evaluation of $\int_{-\infty}^{\infty} dp\, p\rho_{\mathrm{W}}(q,p)$, according to Eq. (14.55).

14.3.1 *Pure states*

It turns out that the system (14.58) can be closed if the density (14.46) reduces to the pure case, i.e. $\rho = \rho(X, X') = \langle X|j\rangle\langle j|X'\rangle$ for some j. We indicate the explicit form of the first and the second moment as derived from prescription (14.52):

$$\langle \hat{P}_+\rho\rangle_q = p(q)\langle\rho\rangle_q, \tag{14.60a}$$

$$\langle \hat{P}_+^2\rho\rangle_q = p^2(q)\langle\rho\rangle_q - \frac{\hbar^2}{4}\langle\rho\rangle_q\frac{\partial}{\partial q^2}\ln\langle\rho\rangle_q. \tag{14.60b}$$

We emphasize that the second moment is given in terms of the zeroth and the first, a property that carries over to all higher moments [428]. This feature is of consequence for the hydrodynamic equations. Inserting the right-hand sides of Eqs. (14.60a) and (14.60b) into Eqs. (14.58a) and (14.58b), one realizes that the resulting two equations for the pure case describe a closed set: they can be solved without recourse to any moments higher than the first. Specifically, Eqs. (14.58a) and (14.58b) turn into

$$\frac{\partial}{\partial t}\langle\rho\rangle_q = -\frac{1}{m}\frac{\partial}{\partial q}[p(q)\langle\rho\rangle_q], \tag{14.61a}$$

$$\frac{\partial}{\partial t}[p(q)\langle\rho\rangle_q] = -\frac{1}{m}\frac{\partial}{\partial q}[p^2(q)\langle\rho\rangle_q] + \frac{1}{m}\frac{\partial}{\partial q}\left[\frac{\hbar^2}{4}\langle\rho\rangle_q\frac{\partial^2}{\partial q^2}\ln\langle\rho\rangle_q\right] - \frac{\partial V}{\partial q}\langle\rho\rangle_q. \tag{14.61b}$$

We recognize in Eq. (14.61a) the continuity equation. Equation (14.61b) describes the temporal change of the momentum density as resulting from three agents: the momentum flow, the external force, and, given by the term between these two components, the effect of the surrounding medium which exerts pressure on the fluid cell. It should be noted that the latter element of the theory is of quantum origin, as is evident from its factor $\hbar^2$, while all other terms are classical. Within the dynamics of the moments, this term is the sharply defined locus of quantum effects in an otherwise classical formalism. It is therefore the counterpart of the quantum force or the quantum potential of Bohmian mechanics which acts as the sole generator of quantum phenomena in a trajectory-based model. Closer inspection of the term $\frac{1}{m}\frac{\partial}{\partial q}\left[\frac{\hbar^2}{4}\langle\rho\rangle_q \frac{\partial^2}{\partial q^2}\ln\langle\rho\rangle_q\right]$ in (14.61b) reveals that it may be interpreted as a *quantum force density*. More specifically, one verifies that

$$\frac{1}{m}\frac{\partial}{\partial q}\left[\frac{\hbar^2}{4}\langle\rho\rangle_q \frac{\partial^2}{\partial q^2}\ln\langle\rho\rangle_q\right] = -\langle\rho\rangle_q \frac{\partial}{\partial q}Q \tag{14.62}$$

with

$$Q = -\frac{\hbar^2}{2m}\langle\rho\rangle_q^{-\frac{1}{2}} \frac{\partial^2}{\partial q^2}\langle\rho\rangle_q^{\frac{1}{2}}. \tag{14.63}$$

This is the quantum potential (13.6) as introduced in the context of the quantum theory of motion. Likewise, it is possible to recover from Eq. (14.61b) the Newtonian equation of motion (13.20). For this purpose, we turn Eq. (14.61b) into a statement about the momentum field rather than the momentum density. This succeeds as one realizes that

$$\frac{\partial}{\partial t}[p(q)\langle\rho\rangle_q] = \frac{\partial}{\partial t}[p(q)]\langle\rho\rangle_q - \frac{p(q)}{m}\frac{\partial}{\partial q}[p(q)\langle\rho\rangle_q], \tag{14.64}$$

where Eq. (14.61a) has been used. Introducing the total derivative through $\frac{\partial}{\partial t} = \frac{d}{dt} - v\frac{\partial}{\partial q}$, and inserting Eqs. (14.62) and (14.64) into Eq. (14.61b), the latter relation adopts the well-known compact form

$$\frac{d}{dt}p = -\frac{\partial}{\partial q}(V + Q). \tag{14.65}$$

As demonstrated in Section 13.1, the quantum force acting on a particle originates from the component of its kinetic energy that is due to the curvature of its wave function modulus. Similarly, the quantum force density is rooted in the quantum contribution to the kinetic energy density (14.57), as reflected by the decomposition of the second moment given by (14.60b).

14.3.2 *Mixed states*

The derivation of the hydrodynamic equation (14.61b) and the Newtonian equation (14.65) relied on the assumption of a pure case, and thus the development so far did not transcend the wave function viewpoint of Bohmian mechanics laid out in Chapter 13. To widen this perspective, we have to remove the restriction imposed so far on the density which we write now in the generalized form

$$\rho(X, X') = \tilde{r}(X, X') \exp\left[\frac{i}{\hbar}\tilde{s}(X, X')\right], \tag{14.66}$$

with the conditions $\tilde{r}(X', X) = \tilde{r}(X, X')$ and $\tilde{s}(X, X') = -\tilde{s}(X', X)$ to guarantee the Hermitian property of the density operator. Clearly, Eq. (14.66) reduces to the pure case if $\tilde{r}(X, X') = R(X)R(X')$ and $\tilde{s}(X, X') = S(X) - S(X')$ with a wave function Ψ given in the polar representation $\Psi(X) = R(X)\exp[\frac{i}{\hbar}S(X)]$. What is the form of the quantum potential and the quantum force as the pure case is replaced by mixed quantum states? Pursuing this question, we will utilize the generalized density (14.66) to deduce hydrodynamic equations analogous to (14.61a) and (14.61b). Using Eq. (14.66) to derive the first few moments according to the rule (14.52) yields in zeroth order the function $\tilde{r}(X', X)$ with $X = X' = q$ as the density $\langle\rho\rangle_q$. The corresponding moments of first and second order (compare with Eqs. (14.60a) and (14.60b)) are

$$\langle \hat{P}_+ \rho \rangle_q = p(q)\langle\rho\rangle_q, \tag{14.67a}$$

$$\langle \hat{P}_+^2 \rho \rangle_q = p^2(q)\langle\rho\rangle_q + \langle \hat{P}_+^2 \tilde{r}\rangle_q. \tag{14.67b}$$

We note that $\langle \hat{P}_+^2 \tilde{r}\rangle_q$ may be interpreted as a squared, q-dependent momentum variance, σ_q^2. Employing (14.67a) in combination with (14.52), one concludes that the momentum $p(q)$ can be represented as

$$p(q) = \frac{\langle P_+ \rho\rangle_q}{\langle\rho\rangle_q} = \left\langle \frac{\partial}{\partial X}\tilde{s}(X, X')\right\rangle_q . \tag{14.68}$$

For the pure case, this reduces to $p(q) = \frac{\partial}{\partial X}S(X)$ and thus to the equation of motion for a particle trajectory in the Bohm–de Broglie theory. We point out that the second moment is partitioned into a classical and a quantum component which is due to the appearance of the operator $\hat{P}_+$, as found previously for the pure case. In contrast to the pure case, however, the second and higher moments are not exclusively given by the density $\langle\rho\rangle_q$ and the momentum $p(q)$ any longer for the mixed state.

With the moments defined by Eqs. (14.67a) and (14.67b), the second of the two hydrodynamic equations (14.61a) and (14.61b) reads

$$\frac{\partial}{\partial t}[p(q)\langle\rho\rangle_q] = -\frac{1}{m}\frac{\partial}{\partial q}[p^2(q)\langle\rho\rangle_q] - \frac{1}{m}\frac{\partial}{\partial q}\langle\hat{P}_+^2\tilde{r}\rangle_q - \frac{\partial V}{\partial q}\langle\rho\rangle_q, \tag{14.69}$$

while the continuity statement (14.61a) remains unaltered. Comparing Eqs. (14.69) and (14.61b), one observes a characteristic modification of the intermediate term on the right-hand side, the contribution that accounts for the influence of the environment on the fluid element. The nature of this difference is clarified as we perform an operation analogous to the step that leads from Eq. (14.61b) to Eq. (14.65), deriving the Newtonian equation for the mixed state case:

$$\frac{dp}{dt} = -\frac{\partial V}{\partial q} - \frac{1}{m}\langle\rho\rangle_q^{-1}\frac{\partial}{\partial q}\langle\hat{P}_+^2 r\rangle_q. \tag{14.70}$$

By inspecting this identity, we arrive at the generalized quantum force F_{qu} which turns out to be

$$F_{\text{qu}} = -\frac{\partial}{\partial q}Q = -\frac{1}{m}\langle\rho\rangle_q^{-1}\frac{\partial}{\partial q}\langle\sigma_q^2\rangle_q. \tag{14.71}$$

Exercise 14.5. Recover the pure state quantum force (see Eqs. (14.63) and (14.65)) from Eq. (14.71), inserting $R(X)R(X')$ for $\tilde{r}(X', X)$.

Cast into the form (14.71), F_{qu} appears as a direct quantum extension of the classical hydrodynamic force

$$F_{\text{cl}} = -\frac{1}{m}\langle\rho_{\text{cl}}\rangle_q^{-1}\frac{\partial}{\partial q}\langle\sigma_{q,\text{cl}}^2\rangle_q, \tag{14.72}$$

where $\sigma_{q,\text{cl}}$ is the variance of the classical phase space distribution $\rho_{\text{cl}}(q, p)$ [429]. The hierarchy of moments derived from the quantum Liouville approach thus implies a natural transition from quantum to classical hydrodynamics. We note that the wave-function-based theory discussed in Chapter 13 introduces F_{qu} as a pure quantum phenomenon that does not have a classical analog. This might be seen from the quantum potential (13.6) and the respective force expression which vanish as $\hbar \to 0$. Likewise, the pure state version of (14.71), as obtained by combining Eqs. (14.63) and (14.65), reduces to zero in the classical limit.

For using the hydrodynamic formalism in the case of mixed quantum states it is of crucial importance to identify conditions that reduce the hierarchy of the time-dependent coupled moment equations (14.58) to a closed set. We have seen that the chain of recursions is broken if a pure state is assumed. Similarly, imposing a certain mathematical form on the mixed state density (14.66) can finalize the series of moment equations. As shown in [428], a Gaussian approach terminates this series at second order. Thus, we propose

$$\rho(X, X') = \exp\,[b_0 + b_1 X + b_1^* X' + b_2 X^2 + b_2^* X'^2 + b_3 XX'], \qquad (14.73)$$

with b_0, b_3 as real and b_1, b_2 as complex parameters. A pure state representation is recovered if $b_3 = 0$. The ansatz (14.73) implies that $\tilde{r}(X', X) = \exp\,[b_0 + \Re\{b_1\}(X + X') + \Re\{b_2\}(X^2 + X'^2) + b_3 XX']$ and $\tilde{s}(X', X) = \hbar[\Im\{b_1\}(X - X') + \Im\{b_2\}(X^2 - X'^2)]$. We notice that the structure of $\tilde{s}(X', X)$ satisfies $\tilde{s}(X, X') = S(X) - S(X')$ which is a feature shared with the pure case. Upon computation of the moments by applying Eq. (14.52) on the specific choice (14.73) one concludes that $\langle \hat{P}_+^3 \tilde{s} \rangle_q = 0$ which yields a truncation of the kinetic equation hierarchy, such that the first three equations (14.58) constitute a closed subset. To substantiate this statement, we indicate the explicit form of Eq. (14.58c) with the Gaussian density (14.73) substituted for ρ:

$$\frac{\partial}{\partial t}\langle \hat{P}_+^2 \rho \rangle_q = -\frac{1}{m}\frac{\partial}{\partial q}[p^3(q)\langle \rho \rangle_q + 3p(q)\langle \hat{P}_+^2 \tilde{r} \rangle_q] - 2\frac{\partial V}{\partial q}\langle \hat{P}_+ \rho \rangle_q. \qquad (14.74)$$

No moments of higher order than the second are contained in this relation. Further, Hughes *et al.* [436] proposed a termination strategy based on the *maximum entropy method* [438] that has proved instrumental for the purpose of moment closure in classical hydrodynamics [439]. This scheme involves approximating an optimal distribution function $\tilde{\rho}_{\mathrm{w}}$ in terms of Hermite polynomials.

14.4 Coupled Electronic States

We turn now to the application of density-function-based quantum hydrodynamics to nonadiabatic interactions between electronic states. In the frame of the Bohm–de Broglie theory, i.e. the wave-function-based quantum theory of motion, this problem was addressed by considering a superposition such as (13.48) as opposed to a single state. If the physical system can be described by a superposition state $\Psi(x, X, t) = \chi_1(X, t)\psi_1(x) + \chi_2(X, t)\psi_2(x)$, the density-based hydrodynamic theory of nonadiabatic processes will have to reduce to this expression in the pure state limit, as much as the model introduced so far was shown to reduce to $\Psi(x, X, t) = \chi(X, t)\psi(x)$ in that limit. As we are about to include several electronic states into the density-based hydrodynamic formalism it is helpful to consider the density operator in a basis of electronic states with labels m, n:

$$\hat{\rho}(X, X') = \sum_{m,n} \rho_{mn}(X, X')|m\rangle\langle n|. \tag{14.75}$$

The population of a state $\langle n|$ is determined by the diagonal elements $\rho_{nn}(X, X')$ while the non-diagonal elements account for nuclear coherence effects, related to interference between the states $\langle n|$ and $\langle m|$ with $n \neq m$. Evidently, the task at hand is to extend the development that leads from the Liouville–von Neumann equation (14.48) to the coupled system of hydrodynamic equations (14.58d) such that it applies to density matrices $\rho_{mn}(X, X')$ rather than a density function $\rho(X, X')$. If we are concerned with a pure case, i.e. a superposition state, and represent it in polar form as $\Psi(X) = \sum_j c_j R_j(X) \exp[\frac{i}{\hbar} S_j(X)]$, the non-diagonal elements may be written explicitly as $\rho_{mn}(X, X') = c_m c_n^* R_m(X) R_n(X') \exp[\frac{i}{\hbar}(S_m(X) - S_n(X'))]$. For the sake of clarity, we will confine ourselves in the following to two electronic states. Also, a diabatic frame with a real-valued coupling potential $V_{12} = V_{21}$ will be assumed. Building on Eq. (14.48), we employ the Liouville–von Neumann equation to derive an equation of motion for the elements of the density matrix $\rho_{mn}(X, X')$. For ρ_{11} and ρ_{12} one finds [440]

$$\begin{aligned} i\hbar\frac{\partial}{\partial t}\rho_{11}(X, X') = &-\frac{\hbar^2}{2m}\left[\frac{\partial^2}{\partial X^2} - \frac{\partial^2}{\partial X'^2}\right]\rho_{11}(X, X') \\ &+ [V_1(X) - V_1(X')]\rho_{11}(X, X') + V_{12}(X)\rho_{21}(X, X') \\ &- V_{21}(X')\rho_{12}(X, X'), \end{aligned} \tag{14.76a}$$

$$\begin{aligned} i\hbar\frac{\partial}{\partial t}\rho_{12}(X, X') = &-\frac{\hbar^2}{2m}\left[\frac{\partial^2}{\partial X^2} - \frac{\partial^2}{\partial X'^2}\right]\rho_{12}(X, X') \\ &+ [V_1(X) - V_2(X')]\rho_{12}(X, X') + V_{12}(X)\rho_{22}(X, X') \\ &- V_{12}(X')\rho_{11}(X, X'), \end{aligned} \tag{14.76b}$$

and corresponding equations are obtained for the remaining two elements. We will now follow the same program as realized in the first section of this chapter for the density-function-based dynamics, i.e. we will introduce a procedure to localize the matrix elements $\rho_{mn}(X,X')$ in order to define diagonal as well as non-diagonal hydrodynamic field components, such as the local density $R_{mn}(X)$ or the current density $j_{mn}(X)$. This is achieved by the following generalization of Eq. (14.52):

$$\langle \hat{P}_+^N \rho_{nm} \rangle_q = \int_{-\infty}^{\infty} dr\delta(r) \left[\hat{P}_+^N \rho_{nm} \left(q + \frac{r}{2}, q - \frac{r}{2} \right) \right]. \tag{14.77}$$

Again, $\langle \rho_{nm} \rangle_q$ is associated with a probability density, in this context related to the nuclear population of an electronic level n if $n = m$, and to a transition probability density if $n \neq m$. Similar remarks can be made about the current and the kinetic energy density.

Equations of motion for the moments of zeroth order are generated by reformulating Eqs. (14.76a) and (14.76b) in terms of the variables q, r and considering $\rho_{mn}(X,X')$ at the place $X = X' = q$. It follows that

$$\frac{\partial}{\partial t}\langle \rho_{11} \rangle_q = -\frac{1}{m}\frac{\partial}{\partial q}\langle \hat{P}_+ \rho_{11} \rangle_q - \frac{i}{\hbar} V_{12}(q)[\langle \rho_{21} \rangle_q - \langle \rho_{12} \rangle_q], \tag{14.78a}$$

$$\begin{aligned} \frac{\partial}{\partial t}\langle \rho_{12} \rangle_q &= -\frac{1}{m}\frac{\partial}{\partial q}\langle \hat{P}_+ \rho_{12} \rangle_q - \frac{i}{\hbar} V_{12}(q)[\langle \rho_{22} \rangle_q - \langle \rho_{11} \rangle_q] \\ &\quad - \frac{i}{\hbar}(V_1(q) - V_2(q))\langle \rho_{12} \rangle_q, \end{aligned} \tag{14.78b}$$

with similar relations for ρ_{21} and ρ_{22}. The nuclear coherence $\rho_{mn}(X,X')$ explicit in Eq. (14.76) transforms into an infinite series of local fields, as seen before for the case of a single electronic state.

Exercise 14.6. Prove Eqs. (14.78a) and (14.78b). First establish that $\frac{\partial^2}{\partial X^2} - \frac{\partial^2}{\partial X'^2} = \frac{\partial}{\partial q}\frac{\partial}{\partial r}$.

Inspecting Eq. (14.78a), one notices that it deviates from the continuity equation (14.58a) by an additional term that carries the meaning of a source or sink of the population residing in the electronic state 1. It thus depends on both the transition densities and the coupling potential. Writing down the corresponding identity for the population of the second electronic state, i.e. for ρ_{22}, one concludes that a continuity equation holds for the sum of both populations, namely

$$\frac{\partial}{\partial t}[\langle\rho_{11}\rangle_q + \langle\rho_{22}\rangle_q] = -\frac{1}{m}\frac{\partial}{\partial q}[\langle\hat{P}_+\rho_{11}\rangle_q + \langle\hat{P}_+\rho_{22}\rangle_q], \tag{14.79}$$

as a straightforward extension of (14.58a) for the case of two instead of one electronic state. Equation (14.78b) for the evolution of the off-diagonal density ρ_{12} is structured in analogy to Eq. (14.78a) but contains an oscillatory term proportional to the difference between the potentials V_1 and V_2, reflecting the phase coherence between the two electronic states.

The relations that govern the time behavior of the first moment are derived by applying $\hat{P}_+$ to the Liouville–von Neumann equations (14.76a) and (14.76b). The result generalizes the identity (14.58b) obtained for the case of a single electronic state. One arrives at [440]

$$\begin{aligned}\frac{\partial}{\partial t}\langle\hat{P}_+\rho_{11}\rangle_q &= -\frac{1}{m}\frac{\partial}{\partial q}\langle\hat{P}_+^2\rho_{11}\rangle_q - \frac{\partial}{\partial q}V_1(q)\langle\rho_{11}\rangle_q\\ &\quad - \frac{1}{2}\frac{\partial}{\partial q}V_{12}(q)[\langle\rho_{21}\rangle_q + \langle\rho_{12}\rangle_q]\\ &\quad - \frac{i}{\hbar}V_{12}(q)[\langle\hat{P}_+\rho_{21}\rangle_q - \langle\hat{P}_+\rho_{12}\rangle_q],\end{aligned} \tag{14.80a}$$

$$\begin{aligned}\frac{\partial}{\partial t}\langle\hat{P}_+\rho_{12}\rangle_q &= -\frac{1}{m}\frac{\partial}{\partial q}\langle\hat{P}_+^2\rho_{12}\rangle q - \frac{1}{2}\frac{\partial}{\partial q}[V_1(q) + V_2(q)]\langle\rho_{12}\rangle_q\\ &\quad - \frac{1}{2}\frac{\partial}{\partial q}V_{12}(q)[\langle\rho_{11}\rangle_q + \langle\rho_{22}\rangle_q] - \frac{i}{\hbar}[V_1(q) - V_2(q)]\langle\hat{P}_+\rho_{12}\rangle_q\\ &\quad - \frac{i}{\hbar}V_{12}(q)[\langle\hat{P}_+\rho_{22}\rangle_q - \langle\hat{P}_+\rho_{11}\rangle_q].\end{aligned} \tag{14.80b}$$

These identities, describing the evolution of the momentum fields, are of considerably higher complexity than Eq. (14.58b). However, the structure of Eq. (14.58b) is readily recovered from Eq. (14.80a) for the case of vanishing coupling potential $V_{12} = 0$ in conjunction with identical potentials V_1, V_2. A *coupling force* $-\frac{\partial}{\partial q}V_{12}(q)$, weighted by the average of the transition density, emerges as an additional term. Further, the source/sink component of Eq. (14.78) is seen to reappear in Eq. (14.80a) with the order of the transition moments enhanced by one. It is noteworthy that the regular force expression $-\frac{\partial}{\partial q}V_1(q)$ in Eq. (14.80a), corresponding to the moment dynamics in the first electronic state, is replaced by the average over both states, $-\frac{1}{2}\frac{\partial}{\partial q}(V_1(q)+V_2(q))$, as one considers the transition moment dynamics given by Eq. (14.80b).

The generalization of Eq. (14.80) for moments of N*th* order is made explicit in [441]. As a pervading feature of moment dynamics, the right-hand sides of the equations of motion may be subdivided into a *hydrodynamic part* and a *Liouville part*, where the latter encompasses all terms in Eq. (14.78) and (14.80) that contain a prefactor $i/\hbar$. We point out that these expressions are determined by the Liouville–von Neumann equation for the electronic two-level system and thus are strictly analogous to the corresponding components of the Wigner–Moyal equations (14.16).

At this point, as discussed before in the context of Eq. (14.58), the problem of the termination conditions for the moment coupled equations in the nonadiabatic case arises. While we were previously dealing with the challenge of closing an infinite system of moment coupled equations, the task to be confronted now consists in finalizing moment coupled systems of interaction potential coupled systems. In view of the observations made earlier for a single state, we expect closure of these multiply entangled equations if a pure case is realized. Representing the density matrix elements, accordingly, as $\rho_{nm}(X, X') = R_n(X)R_m(X') \exp[\frac{i}{\hbar}(S_n(X) - S_m(X'))]$, we can calculate the first moments for the situation of two electronic states. The structure of the resulting expressions for any one of the two states does not differ from the single-state case (see Eqs. (14.60)). The alternative of closing the system of equations (14.80) by imposing the form of a Gaussian on the individual component densities $\rho_{ij}(X, X')$ is described in [440].

Chapter 15

Wave Packet Propagation Methods

The time-dependent theories surveyed in Chapters 8–14 involve, to varying degrees, classical notions. These pertain to the representation of the nuclei, with the exception of models that use or are based on the *classical electron analog* (see Chapters 10 and 12) where they pervade also the description of the electronic degrees of freedom. Each of these models was illustrated by characteristic applications that demonstrated their ability to capture certain facets of electron dynamics in molecular interactions appropriately and, in many cases, with quantitative accuracy by the standards of fully quantum mechanical calculation or experimental observation. In spite of this success, from a theoretical as well as applied point of view, a dynamically consistent formalism that treats both the electronic and the nuclear degrees of freedom on the basis of quantum mechanics is ultimately preferable. Thus, the accounts given in Chapters 8 and 9 of the TDHF and END methods, respectively, as realizations of Ehrenfest dynamics, underscored the success of these procedures in generating solutions to problems of molecular scattering at sufficiently high energy where the trajectory concept is a natural mode of description. These schemes, however, become inadequate wherever typical quantum mechanical phenomena, related to tunneling, zero-point energy, or reaction path branching with possible interference between alternative processes, come into play. In the context of quantum hydrodynamics, reference was made to nuclear wave packets which might be assembled following the description given in Section 13.2.1. While it is to be conceded that this technique, as well as wave function spawning (see Section 11.3), recovers the solution of the TDSE asymptotically, applications of these methods have to operate with finite sets of nuclear trajectories, and thus maintain a distinctly classical flavor.

Dynamical consistency is guaranteed by any theory that does justice to the quantum mechanical nature of both the electrons and the nuclei. Acknowledging that the solution of the TDSE is a wave packet, one arrives at the most elementary realization of consistent quantum molecular dynamics, namely propagation of nuclear wave packets in a landscape of potential energy surfaces.

The first section of this chapter will focus on direct wave packet propagation schemes, designed to solve the TDSE. At this stage of our account, we will also introduce techniques for the efficient spatial representation, namely the — in many ways complementary — *discrete variable representation* (DVR) and the *fast Fourier transform* (FFT) methods (Section 15.1), followed by an overview of frequently employed numerical wave packet propagation schemes in Section 15.2. Applying grid procedures to solve the TDSE for molecular systems, however, meets its limits of tractability at relatively moderate sizes, corresponding currently to about six dimensions. This raises the question for a dynamic theory that yields approximate solutions of the nuclear wave packet propagation problem. The nineties saw rapid progress in the development and refinement of such a theory, namely the *multiconfigurational time-dependent Hartree* (MCTDH) formalism, as documented by about 250 publications in major journals of physics or physical chemistry by 2013 (for an internet source on MCTDH publications, see [442]). As outlined in Section 15.3, this methodology trades the previously employed multidimensional grid or basis set approach to represent the wave packet for optimized time-dependent basis sets. The flexibility gain in the description of the wave packet time behavior translates into a reduction of computational effort. Correspondingly, the MCTDH method has been shown to be highly efficient in treating comparatively large systems. The final section of this chapter highlights a case study related to the photostability of biological molecules as an example of nuclear wave packet propagation in a prototypical nonadiabatic situation.

15.1 The Grid Representation

Prior to discussing propagation techniques, we will address the problem of casting both the TDSE and its solution into a form that is optimal in terms of efficient numerical treatment. Expanding operators as well as wave packets in an infinite basis that spans the whole Hilbert space of the molecular interaction problem is not feasible. Therefore, one has to reduce the infinite-dimensional Hilbert space to finite size, corresponding to a finite basis set. Alternatively, both operators and wave packets may be defined on a finite grid of points in space. In this case, solving the time-dependent problem amounts to tracing the development of complex amplitudes at a set of grid points as a function of time. In the following, we will first contrast the usual finite (or *spectral*, see Chapter 11 in [443]) basis representation of the Hamiltonian matrix with a grid (or *pseudospectral*) representation and discuss the transformation that connects these two frames. The remaining two subsections will introduce two important implementations of the pseudospectral formalism, the DVR and the FFT schemes.

The starting point is the TDSE which governs the motion of a nuclear wave packet $\chi(X_1, \ldots, X_f, t)$,

$$i\hbar \frac{\partial}{\partial t}\chi(X_1, \ldots, X_f, t) = \hat{H}\chi(X_1, \ldots, X_f, t). \tag{15.1}$$

Employing the expansion

$$\chi(X_1, \ldots, X_f, t) = \sum_n a_n(t)\phi_n(X_1, \ldots, X_f), \tag{15.2}$$

into the orthonormal basis $\{\phi_n\}$, we arrive at

$$i\hbar \frac{\partial}{\partial t} a_n(t) = \sum_m H_{nm} a_m(t). \tag{15.3}$$

Equation (15.3) is the TDSE in a finite basis representation. We ask now for efficient strategies to evaluate the matrix H_{nm} and explore a procedure that replaces the integral involved in H_{nm} by a sum over spatial grid points. For the sake of simplicity, we will assume at first a one-dimensional wave function χ in spatial representation $\chi(X) = \langle X | \chi \rangle$ and later generalize to the multidimensional case. As will be related in Section 15.3, the one-dimensional case is of interest not just for demonstrative purposes but also for practical application, as the basis vectors used in the MCTDH approach are frequently chosen as products of one-dimensional wave packets described by a DVR scheme.

Using a finite orthonormal basis in space representation, $\langle X|\phi_n\rangle$ with $n = 1, \ldots, N$, to decompose $\langle X|\chi\rangle$, we write

$$\langle X|\chi\rangle = \sum_n a_n \langle X|\phi_n\rangle \tag{15.4}$$

and

$$a_n = \langle \phi_n|\chi\rangle = \int dX\, \langle \phi_n|X\rangle\langle X|\chi\rangle. \tag{15.5}$$

Given this latter identity, the expansion coefficients may be approximated on a grid obtained by discretizing the X coordinate and replacing integration with quadrature. Thus

$$a'_n = \sum_{j=1}^{M} \omega_j \langle \phi_n|X_j\rangle\langle X_j|\chi\rangle, \tag{15.6}$$

with X_j as grid points and the factors ω_j as quadrature weights. Both sets of quantities are chosen such that the functions $\langle X|\phi_n\rangle$ continue to satisfy an orthonormality constraint, namely

$$\sum_{j=1}^{M} \omega_j \langle \phi_n|X_j\rangle\langle X_j|\phi_m\rangle = \delta_{nm}. \tag{15.7}$$

In the limit of sufficiently fine subdivision of the X coordinate, a'_n and a_n will be equal. Assuming that this limit has been attained, in the following, we will make the transition between a finite basis representation of $\langle X|\chi\rangle$ to a grid representation [382, 444, 445]. Combining relations (15.4) and (15.6), we conclude that

$$\begin{aligned}\langle X|\chi\rangle &= \sum_{n=1}^{N} a'_n \langle X|\phi_n\rangle \\ &= \sum_{n=1}^{N}\sum_{j=1}^{M} \omega_j \langle \phi_n|X_j\rangle\langle X_j|\chi\rangle\langle X|\phi_n\rangle \\ &= \sum_{j=1}^{M} \chi'_j f_j(X).\end{aligned} \tag{15.8}$$

The final expression in this chain of equations implicitly introduces

$$\chi'_j = \sqrt{\omega_j}\langle X_j|\chi\rangle$$
$$\text{and} \qquad f_j(X) = \sqrt{\omega_j}\sum_{n=1}^{N}\langle\phi_n|X_j\rangle\langle X|\phi_n\rangle. \tag{15.9}$$

The set $\{f_1(X), \ldots, f_j(X), \ldots\}$ is interpreted as a new basis that defines a grid representation. Evidently, one basis vector is assigned to every grid point X_j. At the same time, the vectors $\{f_j\}$ are functions that depend continuously on the variable X. The transformation from the original or *variational basis* $\langle X|\phi_n\rangle$ to the discrete basis $f_j(X)$ is readily derived from Eq. (15.9):

$$f_j(X) = \sum_{n=1}^{N}\langle X|\phi_n\rangle T^{\dagger}_{nj}, \tag{15.10}$$

where $T^{\dagger}_{nj} = \sqrt{\omega_j}\langle\phi_n|X_j\rangle$.

The matrix $T^{\dagger}_{nj}$ communicates between the variational basis and the discrete basis. In the next step, we will explore the salient properties of this matrix. From the orthonormality requirement (15.7), one infers immediately that

$$(T^{\dagger}T)_{im} = \delta_{im}. \tag{15.11}$$

It should be noted that the dimensions of the two finite bases, $\{\langle X|\phi_1\rangle, \ldots, \langle X|\phi_N\rangle\}$ and $\{f_1(X), \ldots, f_j(X), \ldots\}$ can be chosen as identical for simplification of algebraic manipulations involving $\mathbf{T}$, which consequently becomes a square matrix. We conclude that the product $\mathbf{TT}^{\dagger}$ is idempotent since

$$(\mathbf{TT}^{\dagger})^2 = \mathbf{T}(\mathbf{T}^{\dagger}\mathbf{T})\mathbf{T}^{\dagger} = \mathbf{TIT}^{\dagger} = \mathbf{TT}^{\dagger}. \tag{15.12}$$

Since the eigenvalues of idempotent matrices are either one or zero, and since

$$\mathrm{Tr}(\mathbf{T}^{\dagger}\mathbf{T}) = \mathrm{Tr}(\mathbf{TT}^{\dagger}) = N, \tag{15.13}$$

the product matrix $\mathbf{TT}^{\dagger}$ must have nothing but ones in diagonal and zeroes in off-diagonal places. Therefore

$$(\mathbf{TT}^{\dagger})_{kl} = \delta_{kl}. \tag{15.14}$$

This establishes the unitary property for $\mathbf{T}$. In view of Eq. (15.10), it follows further that the discrete or *grid basis* $\{f_1(X), \ldots, f_N(X)\}$ is orthonormal, just as the variational basis $\{\langle X|\phi_1\rangle, \ldots, \langle X|\phi_N\rangle\}$.

Exercise 15.1. Establish the orthonormality property for the discrete basis $\{f_1(X), \ldots, f_N(X)\}$.

We will now explore the difference between the variational basis (VB) and the grid basis (GB) representation with respect to the Hamiltonian, and show that the use of a coordinate grid offers an alternative and in many instances more efficient way of evaluating the matrix elements H_{nm} in Eq. (15.3). More specifically, the potential energy, given in the variational basis as

$$\langle \phi_n|V|\phi_m\rangle = \int dX\, \langle \phi_n|X\rangle V(X)\langle X|\phi_m\rangle, \tag{15.15}$$

may be approximated by

$$\begin{aligned}\langle \phi_n|V|\phi_m\rangle &\approx \sum_{i=1}^{N} \omega_i \langle \phi_n|X_i\rangle V(X_i)\langle X_i|\phi_m\rangle \\ &= \sum_{i,j=1}^{N} T_{ni}^{\dagger} V_{ij} T_{jm},\end{aligned} \tag{15.16}$$

where we have used

$$V_{ij} = V(X_j)\delta_{ij}. \tag{15.17}$$

Denoting the Hamiltonian that acts on the variational basis by $\mathbf{H}^{VB}$, we make the transition to the grid basis observing that

$$\begin{aligned}\mathbf{H}^{\mathrm{GB}} &= \mathbf{T}\mathbf{H}^{\mathrm{VB}}\mathbf{T}^{\dagger} = \mathbf{T}\mathbf{T}_{\mathrm{kin}}^{\mathrm{VB}}\mathbf{T}^{\dagger} + \mathbf{T}\mathbf{V}^{\mathrm{VB}}\mathbf{T}^{\dagger} \\ &= \mathbf{T}\mathbf{T}_{\mathrm{kin}}^{\mathrm{VB}}\mathbf{T}^{\dagger} + \mathbf{T}\mathbf{T}^{\dagger}\mathbf{V}\mathbf{T}\mathbf{T}^{\dagger} = \mathbf{T}\mathbf{T}_{\mathrm{kin}}^{\mathrm{VB}}\mathbf{T}^{\dagger} + \mathbf{V}.\end{aligned} \tag{15.18}$$

In the last step, we have used the unitary property of the transformation matrix $\mathbf{T}$. From Eq. (15.18), the grid representation reduces the task of computing the potential energy to evaluating the function V at the grid points. An analogous statement can be made for any multiplicative operator, as, for example, the position operator. It is pointed out that some authors use the label DVR — discrete value representation — for what we denote by GB in the present context, since we prefer to reserve DVR for the particular choice of basis functions to be introduced in the following subsection.

15.1.1 *The discrete variable representation (DVR)*

The two realizations of the grid basis representation discussed in this survey, DVR and FFT, are distinguished by different criteria for the selection of the grid points. DVR chooses these points from the diagonalization of the position operator as given in the VB representation. Thus

$$\mathbf{X}^{\mathrm{DVR}} = \mathbf{T}\mathbf{X}^{\mathrm{VB}}\mathbf{T}^{\dagger}. \tag{15.19}$$

As the potential energy in the grid basis (see Eqs. (15.16) and (15.18)), $\mathbf{X}^{DVR}$ is diagonal, consisting of position eigenvalues in the grid basis. The coordinates specified by $\mathbf{X}^{\mathrm{DVR}}$ are used as DVR grid points. These points depend on the type of basis functions chosen, and so do the corresponding quadrature weights which define the DVR basis $\{f_1(X), \ldots, f_j(X), \ldots\}$. These DVR points are tabulated in [446] for a variety of orthogonal polynomials $P_n(X)$ in selected ranges of the variable X. All of these functions satisfy an orthogonality relation of the form

$$\int dX\, \omega(X) P_n(X) P_m(X) = \gamma_n^2 \delta_{nm}, \tag{15.20}$$

where the weight function $\omega(X)$ and the real parameter γ_n^2 are determined by the chosen polynomial type: Chebyshev, Legendre, Laguerre, etc. The Gaussian quadrature points $X_k, k = 1, \ldots, N$, and the related weights $\omega(X_k)$ are selected such that the relation

$$\int dX\, \omega(X) F(X) = \sum_{k=1}^{N} \omega(X_k) F(X_k) \tag{15.21}$$

is an exact identity for a polynomial $F(X)$ of order $2N - 1$ or less [446]. This implies the validity of the discrete counterpart of Eq. (15.20), namely

$$\sum_{k=1}^{N} \omega(X_k) P_n(X_k) P_m(X_k) = \gamma_n^2 \delta_{nm} \tag{15.22}$$

if $n, m < N$. The DVR of an operator $\hat{O}$ can therefore be found directly once the orthogonal polynomials are defined and the grid points X_k as well as their associated weights $\omega(X_k)$ are determined. Thus, one has to compute

$$O_{ii'} = \sum_{n=1}^{N} \sqrt{\omega(X_i)\omega(X_{i'})} P_n(X_i)(\hat{O}P_n(X))|_{X=X_{i'}}. \quad (15.23)$$

Exercise 15.2. Prove relation (15.23).

As an example, we calculate the DVR of the one-dimensional kinetic energy operator adopting an orthogonal polynomial basis consisting of *particle-in-a-box* eigenfunctions as defined in an interval $[a, b]$:

$$P_n(X) = \left(\frac{2}{b-a}\right)^{\frac{1}{2}} \sin\left(\frac{n\pi}{b-a}(X-a)\right), \quad n = 1, N-1. \quad (15.24)$$

These functions are classified as Chebyshev polynomials of the second kind which are characterized by even spacing of the quadrature points X_n,

$$X_n = a + n\frac{b-a}{N} = a + n\Delta X, \quad n = 1, N-1, \quad (15.25)$$

as well as $\omega(X_n) = \frac{b-a}{N}$ [382] and $\gamma_n^2 = 1$ for all n. The boundary conditions $P_n(a) = P_n(b) = 0$ are clearly satisfied for the orthogonal polynomials (15.24). Applying relation (15.23), we can now determine the DVR form of the kinetic energy matrix, employing the one-dimensional operator $\hat{T}_{\mathrm{kin}} = -\frac{\hbar^2}{2m}\frac{d^2}{dX^2}$ [447]:

$$T_{Kin,ii'} = -\frac{\hbar^2}{2m}\frac{b-a}{N}\sum_{n=1}^{N-1} P_n(X_i)\left.\frac{d^2P_n(X)}{dX^2}\right|_{X=X_{i'}}. \quad (15.26)$$

Colbert and Miller [447] derive an analytic expression for the sum resulting from Eq. (15.26) upon inserting the basis functions (15.24). It is found that

$$T_{\mathrm{kin},ii'} = \frac{\hbar^2}{2m}\frac{(-1)^{i-i'}}{(b-a)^2}\frac{\pi^2}{2}\left[\frac{1}{\sin^2\left(\frac{\pi(i-i')}{2N}\right)} - \frac{1}{\sin^2\left(\frac{\pi(i+i')}{2N}\right)}\right], \quad \text{if } i \neq i',$$

$$T_{\mathrm{kin},ii} = \frac{\hbar^2}{2m}\frac{1}{(b-a)^2}\frac{\pi^2}{2}\left[\frac{2N^2+1}{3} - \frac{1}{\sin^2\left(\frac{\pi i}{N}\right)}\right]. \quad (15.27)$$

This example demonstrates that the DVR approach can lead to analytic kinetic energy matrix elements of extremely compact form if an appropriate choice of the orthogonal polynomials is made, while the corresponding potential energy matrix is diagonal by construction.

The described one-dimensional example is easily generalized to several Cartesian coordinates. Thus, for a three-dimensional problem that involves evenly spaced grids extending along the X-, Y-, and Z-axes with respective step sizes ΔX, ΔY, and ΔZ, the Hamiltonian matrix is given by [382]

$$\begin{aligned} H_{iX,i'X,iY,i'Y,iZ,i'Z} &= T_{\mathrm{kin},iX,i'X}\delta_{iY,i'Y}\delta_{iZ,i'Z} + T_{\mathrm{kin},iY,i'Y}\delta_{iX,i'X}\delta_{iZ,i'Z} \\ &\quad + T_{\mathrm{kin},iZ,i'Z}\delta_{iX,i'X}\delta_{iY,i'Y} + V(X_{iX}, Y_{iY}, Z_{iZ})\delta_{iX,i'X}\delta_{iY,i'Y}\delta_{iZ,i'Z}. \end{aligned} \tag{15.28}$$

It has to be mentioned, however, that the direct product ansatz (15.28) is of limited applicability. Thus, it fails in cases where the dimensions are associated with quantum numbers that exhibit some dependence on each other, such as the orbital angular momentum quantum number l and the azimuthal quantum number m in problems with spherical symmetry.

15.1.2 *The fast Fourier transform (FFT)*

It is among the salient features of the DVR that, in principle, any set of basis functions can be employed to represent the basis states $\langle X|\phi_n\rangle$ and thus to construct the DVR basis $\{f_1(X), \ldots, f_j(X), \ldots\}$. Also, the condition (15.19) implies that the spacing of the grid points is not necessarily regular. None of these two characteristics of DVR holds for the FFT method ([433], see [448] for further reading). Limiting our discussion again to the one-dimensional case, FFT operates in an interval $[X_i, X_f]$ which is evenly subdivided into grid points with coordinates $X_k = X_i + k\Delta X = X_i + k(X_f - X_i)/N$, $k = 0, \ldots, N$. Further, the basis $\langle X|\phi_n\rangle$ is given by plane waves:

$$\langle X|\phi_n\rangle = |P_n\rangle = \frac{1}{\sqrt{N}}\exp\left(-\frac{i}{\hbar}P_nX\right). \tag{15.29}$$

The values of the momentum P_n are confined to the interval $[P_i, P_i + 2\pi\hbar/\Delta X]$ and, as in the case of the X coordinate, evenly spaced. Having made the choice (15.29) for $\langle X|\phi_n\rangle$, we can now proceed to construct a grid basis by applying the discrete Fourier transform:

$$f_m(X) = \sqrt{\omega_m}\sum_{n=1}^{N}\langle\phi_n|X_m\rangle\langle X|\phi_n\rangle$$
$$= \frac{1}{\sqrt{N}}\sum_{n=1}^{N}\exp\left(\frac{i}{\hbar}P_nX_m\right)|P_n\rangle, \tag{15.30}$$

where the same weight is given to each grid point $m(\omega_m = 1)$. To emphasize the symmetry between the plane wave basis and the spatial grid basis, we may set $f_m(X) = |X_m\rangle$ and write

$$|X_m\rangle = \frac{1}{\sqrt{N}}\sum_{n=1}^{N}\exp\left(\frac{i}{\hbar}P_nX_m\right)|P_n\rangle \tag{15.31a}$$

and

$$|P_n\rangle = \frac{1}{\sqrt{N}}\sum_{m=1}^{N}\exp\left(-\frac{i}{\hbar}P_nX_m\right)|X_m\rangle. \tag{15.31b}$$

As both bases are related by the discrete Fourier transform, the transformation matrix $\mathbf{T}$ assumes a particularly simple form. It is given by

$$T_{mn} = \langle X_m|P_n\rangle = \frac{1}{\sqrt{N}}\exp\left(-\frac{i}{\hbar}P_nX_m\right). \tag{15.32}$$

The use of the momentum basis, as implied in the above derivation, makes the evaluation of the Hamiltonian $\mathbf{H}^{\mathrm{GB}}$, defined in Eq. (15.18), a particularly straightforward task. This is due to the fact that the kinetic energy operator, $\hat{T}_{\mathrm{kin}}$, is diagonal in the momentum representation.

It should be noted that the wave function described in the grid basis is periodic outside the interval $[X_i, X_f]$. In momentum space, it repeats correspondingly with the parameter $2\pi\hbar/\Delta X$. Park and Light [449] discuss an extension of the FFT procedure to the case of non-evenly spaced grids.

The FFT combines the crucial advantage of the grid representation, eliminating the need for any evaluation of the potential energy in a basis, with a particularly attractive scaling feature. Thus, the transformation between the space and the momentum representation (15.32) can be shown to involve $O(N \times \log_2 N)$ instead of N^2 operations which would be required for direct matrix multiplication.

15.2 Numerical Wave Packet Propagation Techniques

A plethora of methods is available for solving coupled ordinary differential equations that are of first order in time, and the TDSE falls under this category. Thus, one might employ an explicit integration scheme that makes it possible to generate the $(n+1)$st evolution step directly from values obtained for the nth step. The corresponding propagation equations are of the form $y_{n+1} = y_n + \Delta t \left(\frac{\partial y}{\partial t}\right)_n$. Implicit — or *backwards* — methods, in contrast, are based on integration schemes that involve both the solution and first time derivative at the $(n+1)$st step, as realized by the relation of the type $y_{n+1} = y_n + \Delta t \left(\frac{\partial y}{\partial t}\right)_{n+1}$. The *Adams–Bashforth* procedure gives an example of the former, the *Adams–Moulton* method of the latter category [450]. Both schemes are usually realized as multistep algorithms where the following value results from a number of predecessor values. Implicit methods are known to have better stability properties than explicit ones while the strength of explicit methods lies in their ease of application. The virtues of both strategies are combined in *predictor-corrector* procedures [331]. Here, an explicit algorithm is employed for an initial estimate of the solution at the $(n+1)$st step. Subsequently, the information thus obtained is used in the application of an implicit scheme that is expected to yield a result of improved stability and accuracy.

Equation (15.3), however, and the TDSE in general, may be solved with particularly high efficiency by use of an integration scheme that is adjusted to the special structure of this equation. In the following, we will introduce some of these techniques.

15.2.1 *The Crank–Nicolson scheme*

In the case of low-dimensional problems, the time dependence of the wave packet may be directly derived from the propagator $\hat{U}(t, t_0)$:

$$|\chi(t)\rangle = \hat{U}(t, t_0)|\chi(t_0)\rangle$$
$$\text{where} \qquad \hat{U}(t, t_0) = \exp\left(-\frac{i}{\hbar}\hat{H}(t - t_0)\right). \tag{15.33}$$

A conservative system has been assumed and the dependence of the wave packet χ on spatial coordinates has been suppressed. Further, if a basis $\{|n\rangle\}$, consisting of eigenvectors of $\hat{H}$, can be identified, $|\chi(t_0)\rangle$ may be expanded according to

$$|\chi(t_0)\rangle = \sum_n |n\rangle\langle n|\chi(t_0)\rangle. \tag{15.34}$$

Relations (15.33)–(15.34) lead to

$$\begin{aligned} |\chi(t)\rangle &= \sum_n \exp\left(-\frac{i}{\hbar}\hat{H}(t-t_0)\right)|n\rangle\langle n|\chi(t_0)\rangle \\ &= \sum_n \exp\left(-\frac{i}{\hbar}E_n(t-t_0)\right)|n\rangle\langle n|\chi(t_0)\rangle. \end{aligned} \tag{15.35}$$

Since the utility of Eq. (15.35) in actual computations hinges on the availability of an eigenstate basis $|n\rangle$ this scheme is of formal rather than practical relevance for wave packet propagation. As, in general, a complete basis $|n\rangle$ cannot be generated, one will have to work with an approximation to Eq. (15.34), and the precision of the procedure will deteriorate as errors add up with the elapse of time. If, in contrast, the elementary propagation step involves only a small time interval, one may employ an approximate but direct formalism by expanding the propagator $\hat{U}(t, t_0)$:

$$\begin{aligned} |\chi(t_0+\Delta t)\rangle &= \hat{U}(t_0+\Delta t, t_0)|\chi(t_0)\rangle \\ &= \sum_{j=0}^{\infty}\frac{1}{j!}\left(-\frac{i}{\hbar}\hat{H}\Delta t\right)^j|\chi(t_0)\rangle \\ &\approx \left(1-\frac{i}{\hbar}\hat{H}\Delta t\right)|\chi(t_0)\rangle. \end{aligned} \tag{15.36}$$

This approach, however, is not feasible, since the resulting operator is non-unitary and therefore does not conserve the norm of the wave packet. This situation, luckily, does not pose a serious problem, as rationalized by the following argument.

The propagator equation

$$|\chi(t_0+\Delta t)\rangle = \exp\left(-\frac{i}{\hbar}\hat{H}\Delta t\right)|\chi(t_0)\rangle$$

may be reformulated to read

$$\exp\left(\frac{i}{2\hbar}\hat{H}\Delta t\right)|\chi(t_0+\Delta t)\rangle = \exp\left(-\frac{i}{2\hbar}\hat{H}\Delta t\right)|\chi(t_0)\rangle. \tag{15.37}$$

Expanding the propagation operators on the left as well as the right of (15.37) up to first order, we arrive at

$$\left(1+\frac{i}{\hbar}\hat{H}\frac{\Delta t}{2}\right)|\chi(t_0+\Delta t)\rangle \approx \left(1-\frac{i}{\hbar}\hat{H}\frac{\Delta t}{2}\right)|\chi(t_0)\rangle$$

$$\text{and}\quad |\chi(t_0+\Delta t)\rangle \approx \left(1+\frac{i}{\hbar}\hat{H}\frac{\Delta t}{2}\right)^{-1}\left(1-\frac{i}{\hbar}\hat{H}\frac{\Delta t}{2}\right)|\chi(t_0)\rangle. \qquad (15.38)$$

It is easily seen that the approximate propagator $[1+\frac{i}{\hbar}\hat{H}\frac{\Delta t}{2}]^{-1}[1-\frac{i}{\hbar}\hat{H}\frac{\Delta t}{2}]$ is indeed unitary. The propagation scheme (15.38) encapsulates the *Crank–Nicolson method*. It may be viewed as a combination of a forward and a backward approach which turns out to be of second order in time, i.e. the discretization error decreases with Δt^2 [331].

15.2.2 *Split operator propagation*

For a small time step Δt, the symmetrized representation of the propagator has been shown to be correct up to third order in Δt [451]. Thus, partitioning the Hamiltonian as $\hat{H} = \hat{H}_0 + V$, one may write

$$\exp\left(-\frac{i}{\hbar}\hat{H}\Delta t\right) = \exp\left(-\frac{i}{2\hbar}V\Delta t\right)\exp\left(-\frac{i}{\hbar}\hat{H}_0\Delta t\right)\exp\left(-\frac{i}{2\hbar}V\Delta t\right) + O(\Delta t^3). \qquad (15.39)$$

The operator $\hat{H}_0$ might be chosen as $\hat{T}_{\text{kin}}$ or as some other reference Hamiltonian. We will assume in the following that $\hat{H}_0$ is a partial Hamiltonian whose eigenstates $|\phi_n\rangle$ and eigenvalues $E_{0,n}$ are known. The central factor of the operator product (15.39) can then be evaluated in the basis $|\phi_n\rangle$, while a grid method is employed to represent both surrounding operators, resulting in

$$\exp\left(-\frac{i}{\hbar}\hat{H}\Delta t\right) = \left[\sum_k |f_k\rangle \exp\left(-\frac{i}{2\hbar}V(x_k)\Delta t\right)\langle f_k|\right] \times \left[\sum_n |\phi_n\rangle \exp\left(-\frac{i}{\hbar}E_n\Delta t\right)\langle\phi_n|\right] \times \left[\sum_m |f_m\rangle \exp\left(-\frac{i}{2\hbar}V(x_m)\Delta t\right)\langle f_m|\right] + O(\Delta t^3). \qquad (15.40)$$

Cast into this shape, both contributing exponential operators, $\exp(-\frac{i}{\hbar}\hat{H}_0\Delta t)$ and $\exp(-\frac{i}{2\hbar}V\Delta t)$, are diagonal matrices. The propagator acts on a wave function given in the grid basis, i.e. coordinate representation. After the diagonal potential energy operator has been applied, the wave function is transformed from the DVR into the basis of eigenstates $|\phi_n\rangle$ where the diagonal operator associated with $\hat{H}_0$ acts on it. In the subsequent step, the wave function is retransformed into the coordinate basis, and the potential energy contribution $\exp(-\frac{i}{2\hbar}V\Delta t)$ operates on it a second time.

Propagation over a finite time interval is then achieved by use of a product of operator products, namely

$$\exp\left(-\frac{i}{\hbar}\hat{H}t\right) = \prod_{k=1}^{K}\exp\left(-\frac{i}{2\hbar}V\Delta t_k\right)\exp\left(-\frac{i}{\hbar}\hat{H}_0\Delta t_k\right)\exp\left(-\frac{i}{2\hbar}V\Delta t_k\right)$$
$$+O(\Delta t^3) \quad \text{and} \quad \Delta t_k = \Delta t, K = t/\Delta t. \tag{15.41}$$

This scheme is unitary. Since, therefore, the norm of the wave packet advanced in time is not changed, applying Eq. (15.41) will result in a phase alteration only. Correspondingly, the strength of this propagation technique is stability in the limit of long times. As it is, however, of only second-order precision, it necessitates the use of very small time steps.

15.2.3 *Propagator expansion techniques*

An obvious way of approximating the propagator $\exp\left(-\frac{i}{\hbar}\hat{H}\Delta t\right)$ is by expanding and truncating the exponential function. Thus,

$$\exp\left(-\frac{i}{\hbar}\hat{H}\Delta t\right) \approx \sum_{n=0}^{N}\frac{(-i\Delta t)^n}{n!}\hat{H}^n. \tag{15.42}$$

For a given N, this sum is accurate up to a term of order $O(\Delta t^{N+1})$, providing a simple relation between the number of terms included in the series and the precision of the approach. The truncated expansion (15.42), however, is not unitary. A unitary propagation method based on polynomial expansion can be constructed with the help of the *Lanczos recursion scheme* [449]. In what follows, we summarize the ideas underlying this methodology and continue by giving a condensed account of polynomial expansion by the example of the Chebyshev technique.

The Lanczos scheme

This method involves the operation of the Hamiltonian in the *Krylov space* whose basis vectors are generated by repeated application of $\hat{H}$ on a selected state, commonly chosen as the initial wave function $|\chi(t_0)\rangle$ which is to be advanced through the time interval Δt. The basis of the Krylov space thus has the form $\{|\chi(t_0)\rangle, \hat{H}|\chi(t_0)\rangle, \ldots, \hat{H}^N|\chi(t_0)\rangle\}$, reflecting the order of terms of the series expansion (15.42). It is assumed that these vectors are linearly independent, which in general is the case. The Lanczos recursion procedure [449] makes it possible to express the Hamiltonian matrix in a particularly simple, namely tridiagonal, form. Specifically, one introduces a first basis vector $\boldsymbol{\Xi}_1$ by setting

$$\boldsymbol{\Xi}_1 = \left(\frac{|\chi(t_0)\rangle}{|\chi(t_0)|}, 0, \ldots, 0\right). \tag{15.43}$$

Application of $\mathbf{H}$ on $\boldsymbol{\Xi}_1$ yields

$$\mathbf{H}\boldsymbol{\Xi}_1 = \alpha_1\boldsymbol{\Xi}_1 + \beta_1\boldsymbol{\Xi}_2. \tag{15.44}$$

The factors α_1 and β_1 are the first members of a series of coefficients α_k, β_k defined as $\alpha_k \equiv \langle\boldsymbol{\Xi}_k|\mathbf{H}|\boldsymbol{\Xi}_k\rangle, \beta_{k-1} \equiv \langle\boldsymbol{\Xi}_{k-1}\mathbf{H}|\boldsymbol{\Xi}_k\rangle$, where $\alpha_j, \beta_i = 0$ if $i,j < 1$, and $\boldsymbol{\Xi}_1$ and $\boldsymbol{\Xi}_2$ are orthonormal.

Generalizing Eq. (15.44), we write

$$\mathbf{H}\boldsymbol{\Xi}_k = \beta_{k-1}\boldsymbol{\Xi}_{k-1} + \alpha_k\boldsymbol{\Xi}_k + \beta_k\boldsymbol{\Xi}_{k+1}. \tag{15.45}$$

For clarity, we assume here that the basis functions $\{\boldsymbol{\Xi}_k\}$ are not merely linearly independent but also orthonormal, such that $\langle\boldsymbol{\Xi}_k|\boldsymbol{\Xi}_l\rangle = \delta_{kl}$. By repeated use of Eq. (15.45), an $N+1$-dimensional basis $\{\boldsymbol{\Xi}_1, \boldsymbol{\Xi}_2, \ldots, \boldsymbol{\Xi}_{N+1}\}$ is derived that spans the Krylov space. One concludes that in this basis $\mathbf{H}$ is represented in the tridiagonal form:

$$\mathbf{H}_{N+1} = \begin{pmatrix} \alpha_1 & \beta_1 & 0 & \cdots & 0 & 0 \\ \beta_1 & \alpha_2 & \beta_2 & \cdots & 0 & 0 \\ 0 & \beta_2 & \alpha_3 & \cdots & 0 & 0 \\ \vdots & \vdots & \vdots & \vdots & \vdots & \vdots \\ 0 & 0 & 0 & \cdots & \alpha_N & \beta_N \\ 0 & 0 & 0 & \cdots & \beta_N & \alpha_{N+1} \end{pmatrix}. \tag{15.46}$$

Subsequently, one diagonalizes the matrix $\mathbf{H}_{N+1}$ by solving the eigenvalue equation

$$\mathbf{H}_{N+1}\boldsymbol{\psi}_i = E_{N+1,i}\boldsymbol{\psi}_i. \tag{15.47}$$

One then reverts to the propagator (15.42) that is straightforwardly evaluated by representing $\mathbf{H}_{N+1}$ in the basis of its eigenstates $\boldsymbol{\psi}_i$:

$$\exp\left(-\frac{i}{\hbar}\mathbf{H}_{N+1}\Delta t\right)\boldsymbol{\Xi}_1 = \sum_{i=1}^{N+1}\boldsymbol{\psi}_i \exp\left(-\frac{i}{\hbar}E_{N+1,i}\Delta t\right)\langle\boldsymbol{\psi}_i, \boldsymbol{\Xi}_1\rangle + O(\Delta t^{N+1}). \tag{15.48}$$

With this relation, a unitary propagation scheme is recovered. Obviously, the accuracy afforded by the Lanczos iteration procedure is determined by the dimension of the model space and can thus be adjusted to the desired precision.

Chebyshev polynomial expansion

Employing polynomials of the Chebyshev type makes it possible to reduce the full propagation to one single polynomial expansion. Representing some arbitrary function $f(z)$ in a basis of Chebyshev polynomials $T_n(z)$, we may write

$$f(z) = \frac{b_0}{2} + \sum_{n=1}^{\infty} b_n T_n(z), \tag{15.49}$$

where the coefficients b_n are given by

$$b_n = \frac{2}{\pi}\int_{-1}^{+1} dz\, f(z)\frac{T_n(z)}{\sqrt{1-z^2}}. \tag{15.50}$$

As one applies this formalism to the propagator problem, the variable z acquires the meaning of the Hamilton operator $\hat{H}$, and we set $f(\hat{H}) = \exp(-\frac{i}{\hbar}\hat{H}t)$. The nature of the Chebyshev expansion, however, restricts the domain of the argument z in Eq. (15.50) to the interval $[-1, +1]$. For the expansion of the propagator this implies that the full range of the eigenvalues of $\hat{H}$ must be mapped on this interval. This is achieved by introducing a normalized operator as

$$\hat{H}' = \frac{2(\hat{H} - E_{\min}\hat{I})}{E_{\max} - E_{\min}} - \hat{I}. \tag{15.51}$$

Exercise 15.3. Show that the eigenvalues of $\hat{H}'$ are bounded by $[-1, +1]$.

To use the Chebyshev formalism for wave packet propagation, we have to express the propagator as a function of $\hat{H}'$. Setting $\Delta E = E_{\max} - E_{\min}$, $E' = (E_{\max} + E_{\min})/2$, and recognizing that

$$\hat{H}' = \frac{\hat{H} - E_{\min}\hat{I} - \frac{\Delta E}{2}\hat{I}}{\Delta E/2},$$

we write

$$\begin{aligned}
\exp\left(-\frac{i}{\hbar}\hat{H}t\right) &= \exp\left(-\frac{\Delta E}{2}\frac{i}{\hbar}\frac{\hat{H} - \frac{\Delta E}{2}\hat{I} - E_{\min}\hat{I}}{\Delta E/2}t\right) \\
&\quad \times \exp\left(-\frac{i}{\hbar}\left(\frac{\Delta E}{2} + E_{\min}\right)\hat{I}t\right) \\
&= \exp\left(-\frac{i}{\hbar}\frac{\Delta E}{2}\hat{H}'t\right)\exp\left(-\frac{i}{\hbar}E't\right).
\end{aligned} \tag{15.52}$$

In Eq. (15.52), the modified propagator $\exp\left(-\frac{i}{\hbar}\frac{\Delta E}{2}\hat{H}'t\right)$ is multiplied by a phase factor. In this form, the propagator can be inserted into Eq. (15.50) for direct evaluation of the Chebyshev coefficients. The result is [452]:

$$b_n = (-i)^n(2 - \delta_{n0})\exp\left(-\frac{i}{\hbar}E't\right)\mathcal{J}_n(\alpha), \quad \alpha = \frac{\Delta E\, t}{2\hbar}, \tag{15.53}$$

with $\mathcal{J}_n$ as the nth order Bessel function of the first kind. For $n > \alpha$, the b_n tend to zero exponentially. This behavior establishes a basis for an estimate of the minimal number of terms to be taken into account in the series (15.49) for a given propagation time t.

From the preceding discussion, it follows that wave packet propagation according to the Chebyshev scheme can be cast into a compact form:

$$\begin{aligned}\chi(t) &= \exp\left(-\frac{i}{\hbar}\hat{H}t\right)\chi(t=0)\\ &= \exp\left(-\frac{i}{\hbar}E't\right)\sum_{n=0}^{N}(-i)^n(2-\delta_{n0})\mathcal{J}_n(\alpha)T_n\left(\frac{\hat{H}-E'\hat{I}}{\Delta E/2}\right)\chi(t=0).\end{aligned} \tag{15.54}$$

The Chebyshev procedure has been proven to be stable, efficient, and precise [453]. While the requirement of prior knowledge of the quantities E' and ΔE, and thus the values that bound the spectrum of $\hat{H}$, may be understood as a limitation of the method, experience has shown that estimates of $E_{\rm min}$ and $E_{\rm max}$ can already lead to Chebyshev propagation results of very acceptable accuracy [453].

15.3 The Multiconfiguration Time-Dependent Hartree (MCTDH) Method

The preceding sections of this chapter dealt with numerically exact procedures of solving the TDSE by means of wave packet propagation. In all of these methods, basis sets or multidimensional grids are used to describe the wave function, which leads to exponential scaling with the number of degrees of freedom in either case. This limitation restricts the practical application of these schemes in the context of molecular quantum dynamics with current computational tools to systems of a few atoms only. Semiclassical methods that are capable of operating at significantly reduced computational cost have been outlined in Chapters 8–12. To extend the scope of wave packet propagation to larger systems, one has to approximate not only the solution but also the problem and identify models that improve upon the efficiency of the direct approaches to solving the TDSE, as discussed earlier, without sacrificing too much of their accuracy. An obvious candidate for such an approach is a mean-field theory, as represented by the time-dependent Hartree (TDH) or time-dependent self-consistent field [454, 455] model. Whereas the mean-field formalism introduced in Section 8.2 was applied to the electronic system, here we employ it for the treatment of the nuclear degrees of freedom.

In this section, the TDH theory will serve as a point of reference rather than an example of a successful procedure of quantum dynamics. Since the nuclear wave packet is given by the TDH model as a single product of one-dimensional functions, the correlation between the nuclear degrees of freedom is described inadequately. An obvious generalization of the basic TDH method is accomplished with a multiconfiguration scheme, replacing the single product with multiple products [456]. Among these procedures, the MCTDH [5, 457] theory deserves particular emphasis, as the high efficiency and accuracy of this approximation has been demonstrated by numerous applications in areas of molecular quantum dynamics as diverse as photodissociation ([458]– [460]) and photoabsorption [378, 461, 462], predissociation [463, 464], reactive scattering ([465]– [468]), and molecule-surface interaction ([469]– [472]). It should be noted that this method does not overcome the problem of exponential scaling which resurges if all possible products of one-particle functions are taken into account. In practice, however, the numerical effort required by the MCTDH formalism increases less strongly with dimensionality than that involved in direct wave packet propagation schemes. In particular, the use of optimized time-dependent expansion functions makes the MCTDH procedure more economic than standard methods, as will be further detailed in Section 15.3.2. In the following, we will summarize the basic tenets of the time-dependent Hartree method and its extension to many configurations.

15.3.1 *The time-dependent Hartree (TDH) approach*

The TDH theory utilizes the simplest form of the nuclear wave packet: a product of time-dependent single-particle functions [5]

$$\chi(X_1,\ldots,X_N,t) = A(t)\chi_1(X_1,t)\ldots\chi_N(X_N,t). \tag{15.55}$$

We may associate the nuclear coordinates $X_i, i = 1,\ldots,N$, with the normal modes of the investigated system. The term single-particle function is to be understood as a *façon de parler*, underscoring the one-dimensional character of the χ_i which are not to be associated with physically real particles. The single-particle wave functions are not unique. A set $\{\tilde{\chi}_i\}$ may be chosen instead of $\{\chi_i\}$ if both are related by a unitary transformation. The function $A(t)$ might be used to absorb the normalization or phase factors contained in the single-particle functions χ_i. For instance, the χ_i may be subjected to the constraint

$$\left\langle \chi_1 \middle| \frac{\partial \chi_1}{\partial t} \right\rangle = \dots = \left\langle \chi_N \middle| \frac{\partial \chi_N}{\partial t} \right\rangle = 0. \tag{15.56}$$

Straightforwardly, the conditions (15.56) impose norm conservation on the single particle wave functions. If initially $\|\chi_i\| = 1$, $i = 1, \dots, N$, this normalization will be preserved throughout the evolution of the system. As, in general, the one-dimensional functions χ_i as well as the factor $A(t)$ are complex-valued, the Frenkel variational principle (8.11) may be invoked to derive time-dependent coupled differential equations for the functions χ_i. These equations are generated successively by varying the product (15.55) with respect to each of its factors and employing the identity

$$\left\langle \delta\chi \middle| \left(i\hbar \frac{\partial}{\partial t} - \hat{H} \right) \chi \right\rangle = 0$$

or

$$\left\langle \delta\chi \middle| i\hbar \frac{dA}{dt} \chi_1 \dots \chi_N + i\hbar A \frac{\partial \chi_1}{\partial t} \dots \chi_N \right.$$
$$\left. + \dots + i\hbar A \chi_1 \dots \frac{\partial \chi_N}{\partial t} - \hat{H} A \chi_1 \dots \chi_N \right\rangle = 0. \tag{15.57}$$

Since variation of χ with respect to A yields $\chi_1 \dots \chi_N$, we conclude that

$$i\hbar \frac{dA}{dt} = A \langle \chi_1 \dots \chi_N | \hat{H} | \chi_1 \dots \chi_N \rangle = A \langle \hat{H} \rangle, \tag{15.58}$$

using the constraints (15.56). Similarly, the variation with respect to χ_i, $i = 1, \dots, N$ yields

$$\begin{aligned} i\hbar \frac{\partial \chi_i}{\partial t} &= (\langle \chi_1 \dots \chi_{i-1} \chi_{i+1} \dots \chi_N | \hat{H} | \chi_1 \dots \chi_{i-1} \chi_{i+1} \dots \chi_N \rangle - \langle \hat{H} \rangle \hat{I}) \chi_i, \\ &= (\hat{H}^{(i)} - \langle \hat{H} \rangle \hat{I}) \chi_i. \end{aligned} \tag{15.59}$$

The expressions $\hat{H}^{(i)}$ in Eq. (15.59) are interpreted as mean-field operators. Equations (15.59) may be formulated in an even more compact fashion by making use of the projection operators $\hat{P}_i = |\chi_i\rangle\langle\chi_i|$, namely

$$i\hbar \frac{\partial \chi_i}{\partial t} = (\hat{I} - \hat{P}_i) \hat{H}^{(i)} \chi_i. \tag{15.60}$$

Combining relations (15.58) and (15.59), we observe that the product χ obeys a time-dependent equation of the form

$$i\hbar \frac{\partial \chi}{\partial t} = \hat{H}_{\text{eff}} \chi, \tag{15.61}$$

with $\hat{H}_{\text{eff}} = \sum_{i=1}^{N} \hat{H}^{(i)} - (N-1)\langle \hat{H} \rangle \hat{I}$.

If the Hamiltonian is separable, the error associated with the replacement of the true with the mean-field Hamiltonian $\hat{H}_{\text{eff}}$ vanishes. In the physically relevant case where this condition is not satisfied, the error is small if the widths of the non-separable contributions to the Hamiltonian are nearly constant over the spatial extension of the single-particle functions χ_i [5]. This implies an advantage of the time-dependent over the time-independent Hartree method, as the former operates with wave packets which are frequently well localized, while the latter involves energy eigenstates which tend to be more widely spread out in space.

15.3.2 *The multiconfiguration time-dependent Hartree (MCTDH) approach*

In terms of the correlation between the nuclear degrees of freedom the elementary TDH approximation is greatly improved by including multiple single-particle products into the wave function χ. In summarizing this approach, we will refer to the MCTDH formalism as introduced by Meyer *et al.* [473] and follow the account on this theory given by Beck *et al.* [5]. The product wave function χ as defined in Eq. (15.55) is readily generalized to

$$\chi(X_1 \dots X_N, t) = \sum_{j_1=1}^{n_1} \cdots \sum_{j_N=1}^{n_N} A_{j_1 \dots j_N}(t) \prod_{\kappa=1}^{N} \chi_{j_\kappa}^{(\kappa)}(X_\kappa, t). \tag{15.62}$$

The index j_κ ($j_\kappa = 1, \dots, n_\kappa$) assigns a single-particle function $\chi_{j_\kappa}^{(\kappa)}$ to the mode κ. The upper limit n_κ, correspondingly, determines the maximum number of functions associated with this mode. For representation of the functions $\chi_{j_\kappa}^{(\kappa)}$, a grid scheme is usually employed, where both the FFT and the DVR formalism, as introduced in Section 15.2, have been used. Each product appearing in the sum (15.62), as well as the respective amplitude $A_{j_1 \dots j_N}(t)$, is defined by a set of N indices j_κ ($j_\kappa = 1, \dots, n_\kappa$). The limiting case of $n_1 = \dots = n_N = 1$ is easily identified as the TDH approximation. As in the ansatz (15.55), redundant choices of the single-particle functions have to be eliminated from the wave function (15.62) by introducing suitable constraints. More specifically, the initial values of the $\chi_{j_\kappa}^{(\kappa)}$ are chosen such that

$$\langle \chi_{i_\kappa}^{(\kappa)}(t=0)|\chi_{j_\kappa}^{(\kappa)}(t=0)\rangle = \delta_{i_\kappa, j_\kappa}. \tag{15.63}$$

Generalizing the conditions (15.56), we introduce *Hermitian constraint operators* $\hat{g}^{(\kappa)}$ which act on the mode with label κ and stipulate

$$\left\langle \chi_{i_\kappa}^{(\kappa)}(t) \middle| \frac{\partial \chi_{j_\kappa}^{(\kappa)}(t)}{\partial t} \right\rangle = -\frac{i}{\hbar}\langle \chi_{i_\kappa}^{(\kappa)}(t)|\hat{g}^{(\kappa)}|\chi_{j_\kappa}^{(\kappa)}(t)\rangle. \tag{15.64}$$

The latter identity accentuates the fact that, due to the anti-Hermitian character of $\frac{\partial}{\partial t}$, the scalar products $\langle \chi_{i_\kappa}^{(\kappa)}(t)|\frac{\partial \chi_{i_\kappa}^{(\kappa)}(t)}{\partial t}\rangle$ can adopt arbitrary imaginary values. Equations (15.63) and (15.64) guarantee that the orthonormality of the functions $\chi_{i_\kappa}^{(\kappa)}$ is preserved during the evolution of the system. For further development of the theory, it is useful to introduce the abbreviations

$$A_{\mathcal{J}} \equiv A_{j_1 \ldots j_N} \quad \text{and} \quad \Phi_{\mathcal{J}} \equiv \prod_{\kappa=1}^{N} \chi_{j_\kappa}^{(\kappa)}, \tag{15.65}$$

simplifying the MCTDH wave function (15.62) to

$$\chi = \sum_{\mathcal{J}} A_{\mathcal{J}} \Phi_{\mathcal{J}}. \tag{15.66}$$

An alternative formulation of this wave function is obtained in terms of the *single-hole functions* $\Phi_m^{(\kappa)}$. These functions are built in analogy to χ according to Eqs. (15.62) and (15.66); however, the product of single particle-functions leaves out those that depend on the coordinate X_κ. Thus,

$$\begin{aligned}\Phi_m^{(\kappa)} &= \sum_{j_1=1}^{n_1} \cdots \sum_{j_{\kappa-1}=1}^{n_{\kappa-1}} \sum_{j_{\kappa+1}=1}^{n_{\kappa+1}} \cdots \sum_{j_N=1}^{n_N} A_{j_1 \ldots j_{\kappa-1}, m, j_{\kappa+1}, \ldots j_N}(t) \\ &\quad \times \prod_{\kappa=1}^{N} \chi_{j_1}^{(1)} \cdots \chi_{j_{\kappa-1}}^{(\kappa-1)} \chi_{j_{\kappa+1}}^{(\kappa+1)} \cdots \chi_{j_N}^{(N)} \\ &= \sum_{\mathcal{J}^\kappa} A_{\mathcal{J}_m^\kappa} \chi_{j_1}^{(1)} \cdots \chi_{j_{\kappa-1}}^{(\kappa-1)} \chi_{j_{\kappa+1}}^{(\kappa+1)} \cdots \chi_{j_N}^{(N)}. \end{aligned} \tag{15.67}$$

The nomenclature $\mathcal{J}_m^\kappa$ denotes a multi-index with the constant label m in the κth entry, and the symbol $\sum_{\mathcal{J}^\kappa}$ means summation over all degrees of freedom with the exception of the κth. Using the single-hole functions, we express χ in the compact form

$$\chi = \sum_{j=1}^{n_\kappa} \chi_j^{(\kappa)} \Phi_j^{(\kappa)}. \tag{15.68}$$

Also, the mean-field operator expectation values $\langle \hat{H} \rangle_{i,j}^{(\kappa)}$ for the κth degree of freedom assume a simple shape when cast into the language of $\Phi_m^{(\kappa)}$:

$$\langle \hat{H} \rangle_{i,j}^{(\kappa)} = \langle \Phi_i^{(\kappa)} | \hat{H} | \Phi_j^{(\kappa)} \rangle. \tag{15.69}$$

Due to the multiconfiguration character of the theory outlined here, Eq. (15.69) represents a matrix whose elements are operators. In general, $\langle \hat{H} \rangle_{i,j}^{(\kappa)}$ will exhibit some dependence on the mode with label κ. The integration involved in the evaluation of the scalar product (15.69) extends over all nuclear coordinates X_i, omitting X_κ.

The scalar product of two single-hole functions $\Phi_k^{(\kappa)}$ and $\Phi_l^{(\kappa)}$ defines a density matrix element, namely

$$\begin{aligned} \rho_{kl}^{(\kappa)} &= \langle \Phi_k^{(\kappa)} | \Phi_l^{(\kappa)} \rangle \\ &= \sum_{j_1=1}^{n_1} \cdots \sum_{j_{\kappa-1}=1}^{n_{\kappa-1}} \sum_{j_{\kappa+1}=1}^{n_{\kappa+1}} \cdots \sum_{j_N=1}^{n_N} A^*_{j_1 \ldots j_{\kappa-1}, k, j_{\kappa+1}, \ldots j_N} A_{j_1 \ldots j_{\kappa-1}, l, j_{\kappa+1}, \ldots j_N} \\ &= \sum_{\mathcal{J}^\kappa} A^*_{\mathcal{J}_k^\kappa} A_{\mathcal{J}_l^\kappa}. \end{aligned} \tag{15.70}$$

With the definition of the product functions $\Phi_{\mathcal{J}}$, the single-hole functions $\Phi_j^{(\kappa)}$, and the density matrix elements $\rho_{kl}^{(\kappa)}$, the basic vocabulary for expressing the dynamic relations that govern the propagation of the Hartree product (15.62) is in place. This task will be addressed in the following subsection.

15.3.3 *The MCTDH equations*

We assemble now the ingredients for the use of the Frenkel variational principle to derive a set of MCTDH equations of motion. Most importantly, expressions have to be found for the time derivative of χ as well as the variation of χ with respect to both the amplitudes $A_{\mathcal{J}}$ and the single-particle functions $\chi_j^{(\kappa)}$. The latter task is greatly simplified due to the availability of the two equivalent representations of χ, (15.66) and (15.68). Using Eq. (15.66), we find

$$\frac{\delta \chi}{\delta A_{\mathcal{J}}} = \Phi_{\mathcal{J}}. \tag{15.71}$$

From Eq. (15.68), on the other hand, we have

$$\frac{\delta \chi}{\delta \chi_j^{(\kappa)}} = \Phi_j^{(\kappa)}. \tag{15.72}$$

Employing both Eq. (15.66) and Eq. (15.68), we find that the time derivative of the MCTDH wave function can be written as

$$\frac{\partial \chi}{\partial t} = \sum_{\kappa=1}^{N} \sum_{j=1}^{n_\kappa} \frac{\partial \chi_j^{(\kappa)}}{\partial t} \Phi_j^{(\kappa)} + \sum_{\mathcal{J}} \frac{dA_{\mathcal{J}}}{dt} \Phi_{\mathcal{J}}. \tag{15.73}$$

With the help of Eq. (15.71), the variation with respect to the amplitudes $A_{\mathcal{J}}$ can be performed. It yields

$$\langle \Phi_{\mathcal{J}} | \hat{H} | \chi \rangle - i\hbar \left\langle \Phi_{\mathcal{J}} \middle| \frac{\partial \chi}{\partial t} \right\rangle = 0, \tag{15.74}$$

or, by inserting Eq. (15.73),

$$i\hbar \frac{dA_{\mathcal{J}}}{dt} = \langle \Phi_{\mathcal{J}} | \hat{H} | \chi \rangle - \sum_{\kappa=1}^{N} \sum_{m=1}^{n_\kappa} g_{j_\kappa, m}^{(\kappa)} A_{\mathcal{J}_m^\kappa}, \tag{15.75}$$

where $g_{ij}^{(\kappa)} = \langle \chi_i^{(\kappa)} | \hat{g}^{(\kappa)} | \chi_j^{(\kappa)} \rangle$ (see Eq. (15.64)). From Eq. (15.73), variation with respect to the single-particle functions leads to the set of equations

$$\langle \Phi_j^{(\kappa)} | \hat{H} | \chi \rangle = i\hbar \left\langle \Phi_j^{(\kappa)} \middle| \sum_{\kappa'=1}^{N} \sum_{m=1}^{n_{\kappa'}} \frac{\partial \chi_m^{(\kappa')}}{\partial t} \Phi_m^{(\kappa')} \right\rangle + i\hbar \sum_{\mathcal{J}} \langle \Phi_j^{(\kappa)} | \Phi_{\mathcal{J}} \rangle \frac{dA_{\mathcal{J}}}{dt}. \tag{15.76}$$

All scalar products in this relation contain the single-hole functions $\Phi_j^{(\kappa)}$ in their bra parts. Therefore, the integration is to be performed over all modes except X_κ.

Exercise 15.4. Employ the Frenkel variational principle to derive relation (15.74).

Expanding identity (15.76) further by use of Eq. (15.75) as well as the definition of the density matrix (15.70) we arrive at

$$i\hbar \sum_{m=1}^{n_\kappa} \rho_{jm}^{(\kappa)} \frac{\partial \chi_m^{(\kappa)}}{\partial t} = \langle \Phi_j^{(\kappa)} | \hat{H} | \chi \rangle - \sum_{\mathcal{J}} \langle \Phi_j^{(\kappa)} | \Phi_{\mathcal{J}} \rangle \langle \Phi_{\mathcal{J}} | \hat{H} | \chi \rangle$$

$$+ \sum_{n,m=1}^{n_\kappa} \rho_{jn}^{(\kappa)} g_{mn}^{(\kappa)} \chi_m^{(\kappa)}. \tag{15.77}$$

Exercise 15.5. Obtain Eq. (15.77) from Eq. (15.76). Proceed by splitting up the first term on the right-hand side of Eq. (15.76) into a contribution with $\kappa = \kappa'$ and one with $\kappa \neq \kappa'$.

Some final rearrangement of relations (15.75) and (15.77) is needed to generate the MCTDH equations. For this purpose, we introduce the projector $\hat{P}^{(\kappa)}$ on the space defined by the single-particle functions for the κth mode, namely

$$\hat{P}^{(\kappa)} = \sum_{i=1}^{n_\kappa} |\chi_i^{(\kappa)}\rangle \langle \chi_i^{(\kappa)}|. \tag{15.78}$$

From the definitions given earlier for $\Phi_j^{(\kappa)}$, $\Phi_{\mathcal{J}}$, and $\hat{P}^{(\kappa)}$, one concludes that

$$\sum_{\mathcal{J}} \langle \Phi_j^{(\kappa)} | \Phi_{\mathcal{J}} \rangle \langle \Phi_{\mathcal{J}} | = \hat{P}^{(\kappa)} \langle \Phi_j^{(\kappa)} |. \tag{15.79}$$

Further,

$$\langle \Phi_j^{(\kappa)} | \hat{H} | \chi \rangle = \sum_{k=1}^{n_\kappa} \langle \hat{H} \rangle_{jk}^{(\kappa)} \chi_k^{(\kappa)}. \tag{15.80}$$

Combining Eq. (15.66) with Eq. (15.75), we find the first group of MCTDH equations, a set of coupled first-order differential equations in the amplitudes $A_{\mathcal{J}}$:

$$i\hbar \frac{dA_{\mathcal{J}}}{dt} = \sum_L \langle \Phi_{\mathcal{J}} | \hat{H} | \Phi_L \rangle A_L - \sum_{\kappa=1}^{N} \sum_{m=1}^{n_\kappa} g_{j_\kappa, m}^{(\kappa)} A_{\mathcal{J}_m^\kappa}. \tag{15.81}$$

To express the second set, which propagates the single-particle functions, we adopt the convention

$$\boldsymbol{\chi}^{(\kappa)} = (\chi_1^{(\kappa)}, \ldots, \chi_n^{(\kappa)})^{\mathrm{T}},$$

and write

$$i\hbar\frac{\partial\boldsymbol{\chi}^{(\kappa)}}{\partial t} = g^{(\kappa)}\mathbf{I}_{n_\kappa}\boldsymbol{\chi}^{(\kappa)} + (\hat{I} - \hat{P}^{(\kappa)})[(\boldsymbol{\rho}^{(\kappa)})^{-1}\langle\hat{\mathbf{H}}\rangle^{(\kappa)} - \hat{g}^{(\kappa)}\mathbf{I}_{n_\kappa}]\boldsymbol{\chi}^{(\kappa)}, \tag{15.82}$$

where $\mathbf{I}_{n_\kappa}$ stands for the $n_\kappa \times n_\kappa$ unit matrix, and $\langle\hat{\mathbf{H}}\rangle$ for the mean-field operator matrix. The product $\hat{g}^{(\kappa)}\mathbf{I}_{n_\kappa}$ may be interpreted as a *constraint matrix* $\hat{\mathbf{g}}^{\kappa}$.

Since the MCTDH equations are derived from the Frenkel variational principle, they conserve the total energy of the system as well as the norm of the wave function (see Section 8.1).

As pointed out before, the Hermitian constraint matrix $\mathbf{g}^{(\kappa)}$ may be selected arbitrarily, as it does not affect the evolution of the wave function. One can take advantage of this freedom of choice to cast the MCTDH equations into an ultimate form that makes the numerical integration of the equations as efficient as possible.

Evidently, the limiting case of vanishing $\hat{g}^{(\kappa)}$ is very tempting as it generates the much simplified equations of motion

$$i\hbar\frac{dA_{\mathcal{J}}}{dt} = \sum_L \langle\Phi_{\mathcal{J}}|\hat{H}|\Phi_L\rangle A_L \tag{15.83}$$

and

$$i\hbar\frac{\partial\boldsymbol{\chi}^{(\kappa)}}{\partial t} = (\hat{I} - \hat{P}^{(\kappa)})[(\boldsymbol{\rho}^{(\kappa)})^{-1}\langle\hat{\mathbf{H}}\rangle^{(\kappa)}]\boldsymbol{\chi}^{(\kappa)}. \tag{15.84}$$

It is interesting that the condition $g_{ij}^{(\kappa)} = 0$ leads to the same equation for the coefficient derivative $\frac{dA_{\mathcal{J}}}{dt}$ as one obtains for a time-independent basis $\Phi_{\mathcal{J}}$. The extreme case of a complete single-particle basis, corresponding to $\hat{P}^{(\kappa)} = \hat{I}$, leads to $\frac{\partial\boldsymbol{\chi}^{(\kappa)}}{\partial t} = 0$.

Surprisingly, in terms of computational effort, the choice $g^{(\kappa)} = 0$ is not the most favorable one [5]. Alternatively, one may use a partition of the Hamiltonian into a sum over single-particle contributions and a residual part:

$$\hat{H} = \sum_{\kappa=1}^{N} \hat{h}^{(\kappa)} + \hat{H}_{\mathrm{R}}. \tag{15.85}$$

Examples of the one-dimensional components $h^{(\kappa)}$ are given by the kinetic energy terms for the modes κ. The operator $\hat{H}_{\mathrm{R}}$, in contrast, contains the correlations between these modes. Stipulating

$$g^{(\kappa)} = h^{(\kappa)}, \tag{15.86}$$

one finds by inspecting Eq. (15.82) that the separable parts $h^{(\kappa)}$ of the Hamiltonian cancel, leaving

$$i\hbar \frac{dA_J}{dt} = \sum_L \langle \Phi_J | \hat{H}_\mathrm{R} | \Phi_L \rangle A_L. \tag{15.87}$$

The analogous equations of motion for the single-particle functions read

$$i\hbar \frac{\partial \boldsymbol{\chi}^{(\kappa)}}{\partial t} = [\hat{h}^{(\kappa)} \mathbf{I}_{n_\kappa} + (\hat{I} - P^{(\kappa)})][(\boldsymbol{\rho}^{(\kappa)})^{-1} \langle \hat{\mathbf{H}}_\mathrm{R} \rangle^{(\kappa)}] \boldsymbol{\chi}^{(\kappa)}. \tag{15.88}$$

Equation (15.87) is preferable to Eq. (15.83) since only the residual Hamiltonian is employed for propagation. Relation (15.88), on the other hand, is no improvement upon (15.84) which becomes obvious as the latter is reformulated to make explicit both the separable and the residual part of the Hamiltonian.

Formally, the extension of the wave packet propagation formalism outlined earlier to nonadiabatic situations is straightforward. If the evolution of the considered system involves N_α electronic states, the overall wave function may be expressed using one additional degree of freedom, α, to label the electronic level. It may then be written as a sum over α [461] as

$$|\chi\rangle = \sum_{\alpha=1}^{N_\mathrm{s}} |\chi_\alpha\rangle |\alpha\rangle, \tag{15.89}$$

with α as electronic state. To derive a compact equation of motion for this multiset picture, we introduce $\boldsymbol{\chi}$ as the vector of the wave functions $|\chi_\alpha(X_1, X_2, \ldots, X_N, t)\rangle$ which correlate with well-defined electronic states. Correspondingly, the Hamiltonian has to be generalized to act on these states:

$$\hat{\mathbf{H}} = \sum_{\alpha,\beta=1}^{N_\alpha} |\alpha\rangle \hat{H}_{\alpha\beta} \langle \beta|. \tag{15.90}$$

As a consequence, the Frenkel principle assumes the form of a matrix equation:

$$\left\langle \delta\boldsymbol{\chi} \left| \left(i\hbar \mathbf{I} \frac{\partial}{\partial t} - \hat{\mathbf{H}} \right) \right| \boldsymbol{\chi} \right\rangle = \mathbf{0}.$$

The variational optimization of both the amplitudes and the single-particle functions has to be performed now for each electronic state $|\alpha\rangle$ while it was restricted to one specific state of reference in the MCTDH equations (15.87) and (15.88).

The nonadiabatic generalization of the MCTDH equations has been used to investigate the dynamics of various polyatomic species, such as pyrazine [474, 378], allene [475], butatriene [476], benzene [477], and furan [478]. Reference to the latter investigation was made in Section 12.6, and MCTDH results related to the radiationless S_2–S_1 transition in pyrazine were shown in Figure 12.5.

The methodological predicament of the MCTDH procedure applied to physical systems is the need to generate sufficiently precise potential energy surfaces which is a particularly time-consuming task when coupling between these surfaces is important for the studied problem. This impediment is reduced as one substitutes Gaussian-shaped wave functions (see e.g. Eq. (12.134)) for the nuclear single-particle functions $\chi_{j_\kappa}^{(\kappa)}$ ([479]–[481]), each of which is characterized by a set of time-dependent parameters, namely average positions and momenta as well as global phase functions. When this modification is adopted, propagation equations in these variables take the place of Eq. (15.88). The practical advantage of this approach is that it allows for on-the-fly determination of the potential energy surfaces since only surface segments in the vicinity of the mean Gaussian position need to be known at any stage of the propagation.

15.4 Case Study: Photostability of Biologically Relevant Molecules

We illustrate the use of wave packet dynamics in a landscape of multiple electronic surfaces by recent applications to biological species, with particular emphasis on the biogenetic importance of photophysical processes. In the following subsection, we will link the photostability of biological molecules to the presence and accessibility of conical intersections in the respective system and discuss some exemplary species in terms of ultrafast internal conversion of photonic energy. In Section 15.4.2, we will focus on the dynamical aspect of the photostability problem by highlighting a recent wave packet propagation study on pyrrole.

15.4.1 *Ultrafast deexcitation by passage through conical intersections in nucleic acid bases and base pairs*

Among the most elementary chemical conditions for the existence of life is the stability of biochemical molecules. The emergence and persistence of living organisms require that their molecular constituents be sufficiently resistant with respect to external perturbations. In particular, the fact that most living entities are frequently and for long periods exposed to sunlight makes the photostability of biological molecules an essential evolutionary factor. This evident statement, however, appears to be in conflict with the observation that most organic molecules are strongly affected by irradiation with ultraviolet (UV) light. Since this spectral component is energetic enough to break covalent bonds, the exposed species may undergo a variety of chemical changes, such as fragmentation or isomerization in response to UV exposure. It is all the more remarkable that the evolution of early aquatic organisms preceded the formation of the ozone layer which today acts as a shield against the most detrimental segment of the UV range. One concludes that mechanisms preserving the identity of biological molecules under the impact of UV light played a crucial role in facilitating the evolutionary processes that led to life.

The key to understanding these mechanisms lies in the electronic structure and the primary bonding features of the most basic building bricks of biochemical organisms. These are the ingredients of deoxyribonucleic acid (DNA), most prominently the nucleic acid bases adenine, cytosine, guanine, and thymine (see Figure 15.1) whose sequence in the DNA molecule encode the complete genetic information of a living entity.

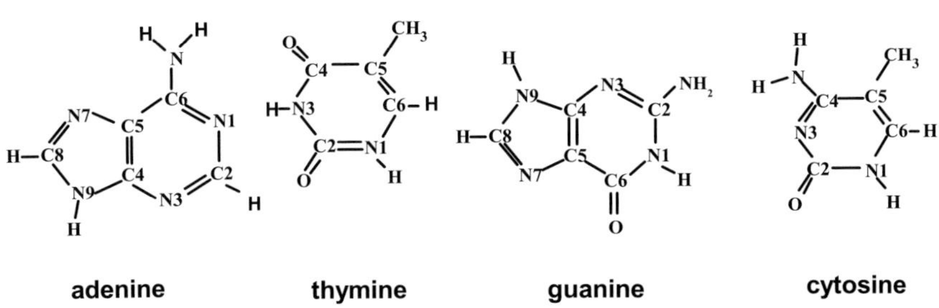

Figure 15.1: Molecular structure of the nucleic acid bases adenine, thymine, guanine, and cytosine.

Similar stability constraints as imposed on the four nucleic acid bases must be in place for the 20 amino acids of which all proteins consist. In spite of their photostability, these species interact with UV radiation, exhibiting strong absorption bands in the regime of 200–300 nm. Thus, the impact of photons in the UV range on these units induces transitions to excited electronic states. The crucial protective mechanism consists in *the subsequent fast decay into the electronic ground state by radiationless deexcitation*. The net result of the interaction of photostable organic species with energetic photons is therefore not a chemical change, but the internal conversion of radiation into heat, or photonic into vibrational energy, dissipated over multiple degrees of freedom.

Ultrafast non-radiative deexcitation proceeds through conical intersections, provided they are accessible by barrier-free relaxation pathways immediately following the transition to the excited potential energy surface. For nucleic acid bases, two major photochemical reaction paths have been described that satisfy this criterion. These are (1) the torsion of C-N bonds in the six-membered rings (see Figure 15.1), and (2) the loss of hydrogen atoms from azine or amino groups of the molecules. The first process is exemplified in Figure 15.2 which combines the S_0 ground state of 9H-adenine, the most abundant adenine tautomer, with its three lowest-lying S_1 excited states. The respective potential energy profiles are shown as functions of reaction coordinates that involve predominantly the torsion of the C2-N3 and the C6-N1 bonds (compare with Figure 15.1). Recent computational work on the radiationless deexcitation of 9H-adenine has extended the geometric parameter space and identified a broad range of qualitatively different S_0-S_1 conical intersections, incorporating all low-lying excited states indicated in Figure 15.2 [482].

Vertical population of the ^1L_a state is a strongly allowed optical transition, in contrast to that of the weakly allowed and nearly forbidden states ^1L_b (a $^1\pi\pi^*$ state) and $^1n\pi^*$.[1] The ^1L_a-S_0 crossing, as shown in Figure 15.2(a), provides a highly efficient non-radiative deexcitation channel, associated with a life time of about 200 fs in aqueous solution [483], [484]. Another

[1]The symbols $^1\pi\pi^*$ and $^1n\pi^*$ stand for excitations involving promotion of an electron from a π-bonding molecular orbital to a π-antibonding orbital and from a non-bonding to a π-antibonding orbital, respectively.

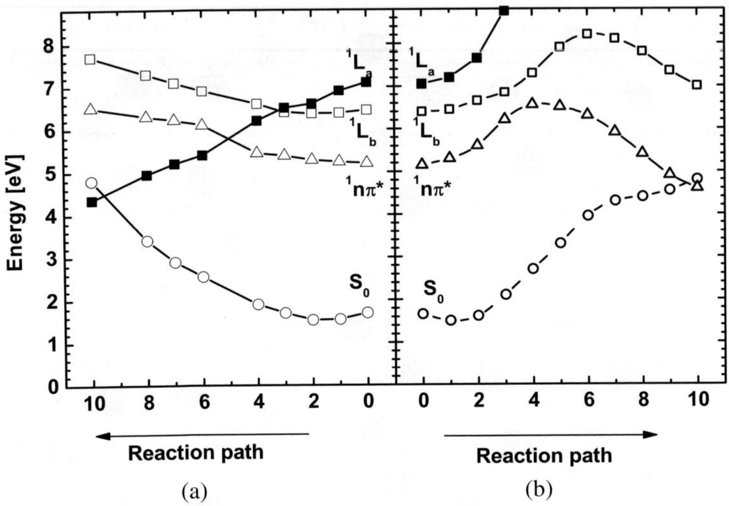

Figure 15.2: Diabatic potential energy curves for the electronic ground state and three low-lying isomeric states of 9H-adenine (a $^1n\pi^*$ as well as two $^1\pi\pi^*$ states, labeled ^1L_a and ^1L_b) by complete active space self-consistent field (CASSCF, see Section 1.6) calculation. The zero of the reaction paths is defined by the equilibrium geometry of the $^1n\pi^*$ state. The reaction coordinates describe torsions of the (a) C2-N3 and the (b) C6-N1 bond. (Reprinted with permission from [485]. Copyright (2004) by the American Chemical Society.)

pathway has been identified that leads to a conical intersection between the S_0 and the $^1n\pi^*$ state [485]. In both cases, the presence of crossing points restricts the excitation life times to the subpicosecond regime.

This feature separates 9H-adenine which occurs in the adenine-thymine base pair [486] from the minor tautomer 7H-adenine. From UV irradiation measurements in aqueous solution, the latter decays into the ground state about 40 times more slowly than the former.

In view of these arguments, the existence and accessibility of conical intersections that connect the electronic ground state with low-lying excited states of an organic species appear as major selection criteria for the evolution of biologically relevant molecules. It is of particular interest to investigate the hydrogen-bonded base pairs adenine-thymine (AT) and guanine-cytosine (GC), the material carriers of the genetic code, from this point of view. The three most stable conformations of the GC pair are displayed in Figure 15.3, where panel (a) represents the famous Watson–Crick geometry, followed by the two most stable GT isomers in (b) and (c). From

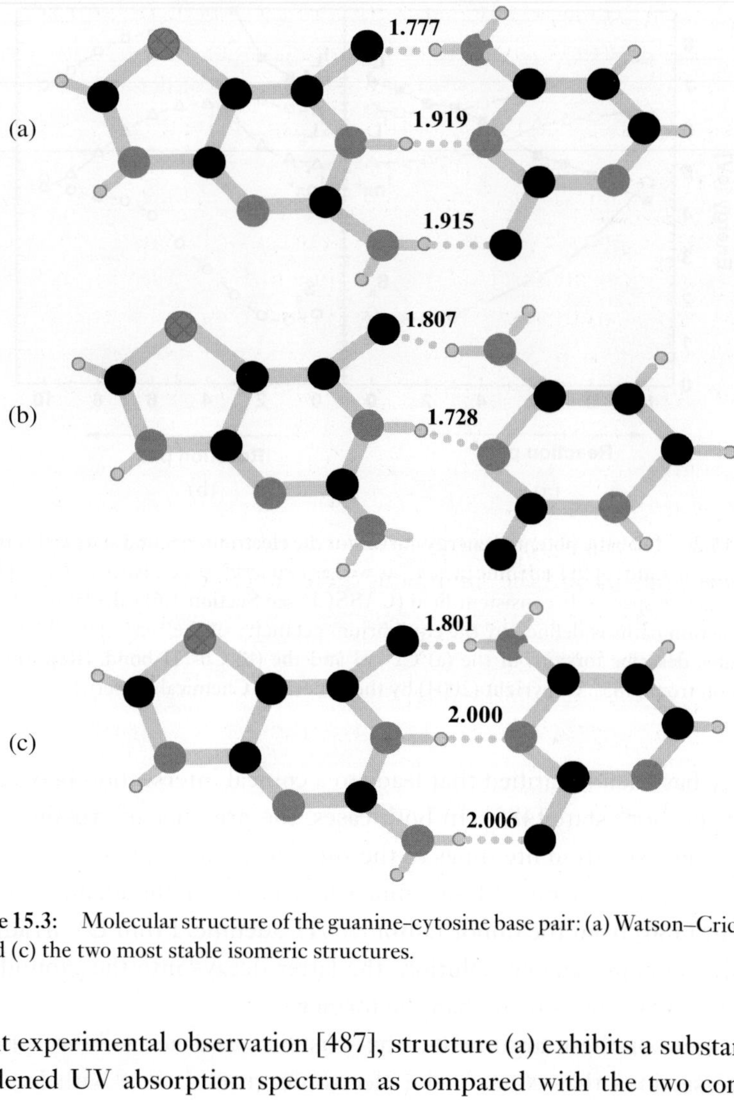

Figure 15.3: Molecular structure of the guanine-cytosine base pair: (a) Watson–Crick form; (b) and (c) the two most stable isomeric structures.

recent experimental observation [487], structure (a) exhibits a substantially broadened UV absorption spectrum as compared with the two competitors. This finding in conjunction with the extremely low intensity of the resonance-enhanced multiphoton ionization signal is a spectral fingerprint of a very short-lived electronic excitation. The latter is identified as the optically absorbing $^1\pi\pi^*$ state of the Watson–Crick form (a) that decays into the electronic S_0 ground state. Being populated by a vertical transition from structure (a) upon photon impact, this state stabilizes through proton transfer from guanine to cytosine.

A suitable reaction coordinate for the GC relaxation following UV excitation is the distance between the intermediate bonding H atom and the adjacent N atom of guanine (see Figure 15.3(a)). The lowest optically accessible excited states of all three conformers displayed in Figure 15.3 are local $^1\pi\pi^*$ states of guanine or cytosine. In the case of structure (a), the lowest excited state interacts with a proton transfer state, giving rise to a crossing between the respective potential energy functions close to the energy minimum of the GC excitation [488], and leading to a barrierless proton transfer reaction. From careful inspection of the S_0 and S_1 potential energy surfaces involved, one infers the presence of a conical intersection between the proton transfer and the ground state [489], providing an ultrafast non-radiative decay channel. As pointed out by Sobolewski *et al.* [488], this configuration is a distinguishing feature of the Watson–Crick form and is not realized by the alternative isomers (b) and (c). Here, the crossings between the potential energy functions of the primary local excitations and the proton transfer state occur far beyond the minima of these functions, resulting in sizeable barriers to proton transfer and, in further consequence, blocking access to the S_0-S_1 conical intersection. Thus, only the Watson–Crick structure is conducive to an efficient ultrafast internal conversion mechanism. Photophysical arguments hold the key to understanding the tautomeric selectivity found in DNA base pairs.

15.4.2 *Dynamics at the $^1\pi\sigma^*$-S_0 conical intersection of pyrrole*

Once the potential energy surfaces that play an active role in the radiationless deexcitation processes mentioned before are known with sufficient accuracy, a full representation of internal conversion dynamics may be accomplished by wave packet propagation. This has been demonstrated by Lan *et al.* [490] and Vallet *et al.* [491] for the cases of phenol and pyrrole, respectively. In what follows, we will focus on the latter study. Pyrrole (C_4H_4NH) is a heterocyclic aromatic species that occurs in more complex macrocycles, including the porphyrins of heme and the bacteriochlorins of chlorophyll.

Adsorption by pyrrole in the near-UV range populates mainly the four lowest excited singlet states which are of $^1A_2(\pi\sigma^*)$, $^1B_2(\pi\pi^*)$, $^1B_1(\pi\sigma^*)$, and $^1A_2(\pi\pi^*)$ character. The two $\pi\pi^*$ states account for a strong absorption band

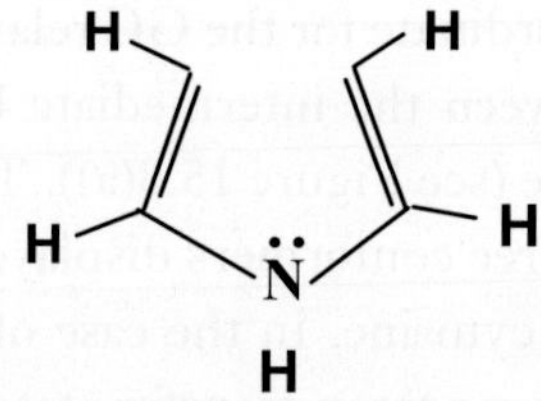

Figure 15.4: Model of pyrrole, C_4H_4NH.

at about 6 eV while those of the $\pi\sigma^*$ type are populated with much smaller probabilities. Being higher in energy, however, the $\pi\pi^*$ decay into the $\pi\sigma^*$ states with short conversion times on the order of 20 fs [492, 493]. Potential energy profiles for the four lowest excited states as well as the ground state are shown in Figure 15.5 as a function of the N-H stretch (compare with the scheme of the pyrrole molecule, Figure 15.4). Evidently, the S_0 as well as the $^1A_1$ and $^1B_2$ curves display deep potential wells, confining the H atom. The two $\pi\sigma^*$ configurations, in contrast, involve a Rydberg 3*s* orbital which makes the N-H interaction strongly antibonding. As the N-H distance is sufficiently elongated, however, the σ^* orbital collapses into a hydrogen 1*s* valence orbital [492]. This change in the chemical nature of the N-H bond is reflected by the emergence of a shallow minimum in the $^1A_2$ and $^1B_1$ potential energy functions.

From Figure 15.5, extension of the N-H bond in both $\pi\sigma^*$ states leads to crossing points between the $^1A_2$ ($^1B_1$) diabatic potential energy curves with that of the ground state. Photoexcitation into either one of the two excited states will therefore result in a competition of two processes, namely radiationless decay by passage through the $^1A_2$ ($^1B_1$)-S_0 funnel and dissociation by hydrogen elimination on the S_0 surface. Traversal of the conical intersection may be followed by dissociation on the $^1A_2$ ($^1B_1$) surface. Vallet *et al.* [491] simulated the photoinduced dynamics of pyrrole by performing wave packet propagation on coupled potential energy surfaces. This work was carried out under two major methodological restrictions. First, the $^1A_2$-S_0^1 and the $^1B_1$-S_0 problems were addressed in separation from each other while a more comprehensive analysis would incorporate the $^1A_2$-$^1B_1$ coupling. Second, the dimensionality of the active nuclear coordinate space was limited to two, namely the N-H stretch and a suitably chosen angular degree of freedom. A Jacobi representation of these variables is shown in Figure 15.6.

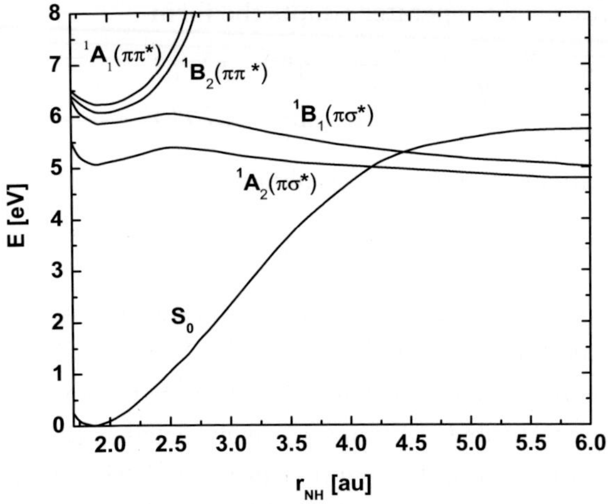

Figure 15.5: Potential energy functions of the pyrrole S_0 ground states as well as the four lowest excited electronic singlet states versus the N-H stretch coordinate, obtained by complete active space self-consistent field/multireference configuration interaction (CASSCF/MRCI) computation. (Reprinted with permission from [491]. Copyright (2004) by the American Institute of Physics.)

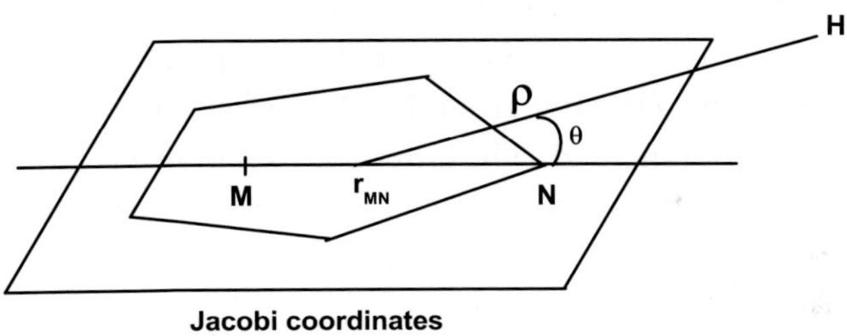

Figure 15.6: Definition of Jacobi coordinates for the pyrrole $^1B_1$-S_0 system [491]. The symbol M stands for the center of mass of all pyrrole constituents excepting N and the H atom attached to it. The distance ρ is given by the separation between the active H atom and the center of mass of the subunit M and the N atom. Θ denotes the angle between r_{MN} and ρ.

In the following, we refer to the $^1B_1$-S_0 alternative. Here the Jacobi coordinates ρ and Θ provide a tuning and a coupling mode, respectively. We point out that the angle Θ describes out-of-plane motion since the coupling matrix element V_{12} vanishes for planar geometry [491]. The diabatic Hamiltonian is written as

$$\hat{H} = \hat{T}_{\mathrm{N}} \begin{pmatrix} 1 & 0 \\ 0 & 1 \end{pmatrix} + \begin{pmatrix} V_{11} & V_{12} \\ V_{21} & V_{22} \end{pmatrix}, \tag{15.91}$$

where the kinetic energy operator adopts the form

$$\hat{T}_{\mathrm{N}} = -\frac{\hbar^2}{2\mu_\rho}\frac{\partial^2}{\partial\rho^2} - \frac{\hbar^2}{2I}\frac{\partial^2}{\partial\Theta^2}. \tag{15.92}$$

The inverse moment of inertia I^{-1} is obtained from

$$I^{-1} = (\mu_\rho \rho^2)^{-1} + (\mu_r r_{\mathrm{MN}}^2)^{-1}, \tag{15.93}$$

and the reduced masses μ_r, μ_ρ are defined in accordance with Eqs. (6.93a) and (6.93b), respectively.

Analytical expressions are adopted for the diabatic potentials V_{11}, V_{22} as well as the coupling term V_{12}. Thus, V_{11} is modeled as the sum of a Morse potential in the distance ρ and a harmonic potential in Θ. A similar functional form is chosen for V_{22}. All parameters that determine the elements of the diabatic potential matrix are extracted by comparing representation (2.63) of the adiabatic ground state potential with a fit to the respective CASSCF computation results. The coupling element V_{12} is assumed to be proportional to Θ. The position of the conical intersection in the reduced vibronic parameter space $\{\rho, \Theta\}$ is thus constrained by the condition $\Theta = 0^0$.

In preparation for wave packet dynamics computation, the vibrational eigenfunctions of the adiabatic ground state surface were evaluated by use of the pseudospectral method. The latter consists in propagating the autocorrelation function (compare with Eq. (12.96)) over a sufficiently long time interval and generating the pseudospectral intensity $I(E)$, defined as

$$I(E) = \int_0^\infty dt \langle \chi(0)|\chi(t)\rangle \exp\left(\frac{iEt}{\hbar}\right). \tag{15.94}$$

The maxima of this distribution occur at the energy eigenvalues of the bound states. For the initial wave packet $\chi(0)$, a product of Gaussians in the coordinates ρ and Θ was chosen.

Once the vibrational states supported by the surface of lowest energy are known, mode-specific photoinduced dynamics can be investigated. This is done by placing selected vibrational eigenfunctions vertically into the $^1B_1$ state and following their time evolution on the two coupled diabatic surfaces. For this purpose, Vallet *et al.* adopted a two-dimensional grid representation of the time-dependent wave packet. Within this approach, the propagation was performed by use of the split-operator scheme

(Section 15.2.2). The FFT procedure (Section 15.1.2) was employed to transform the kinetic energy segment of the propagator.

The development of the system may be characterized by monitoring the populations of the two contributing surfaces as a function of time. These quantities are constructed by applying the projection operators $|\psi_i\rangle\langle\psi_i|$, $i = 0, 1$, to the molecular wave function $\Psi(t)$, describing the evolution of the system in a space reduced to two vibronic parameters and two coupled potential energy surfaces. Here ψ_0 refers to the electronic ground state (or *lower*) and ψ_1 to the $^1B_1$ state (or *upper*) component of the wave packet:

$$P_i^{\mathrm{D}}(t) = \langle\Psi(t)|\psi_i^{\mathrm{D}}\rangle\langle\psi_i^{\mathrm{D}}|\Psi(t)\rangle, \tag{15.95a}$$

$$P_i^{\mathrm{A}}(t) = \langle\Psi(t)|\psi_i^{\mathrm{A}}\rangle\langle\psi_i^{\mathrm{A}}|\Psi(t)\rangle, \quad i = 0, 1. \tag{15.95b}$$

The superscript A(D) denotes, as before, the adiabatic (diabatic) representation. Of great interest is the dissociated population, involving the molecules that have undergone H elimination. This fraction is given by the time-integrated current density at an asymptotic distance ρ^{diss}, chosen sufficiently far beyond the S_0-$^1B_1$ intersection. Explicitly,

$$P^{\mathrm{diss}}(t) = \frac{\hbar}{\mu_\rho}\int_0^t dt' \,\Im\left\{\left\langle \Psi(\rho,\Theta,t') \middle| \frac{\partial\Psi(\rho,\Theta,t')}{\partial\rho}\right\rangle\Big|_{\rho=\rho^{\mathrm{diss}}}\right\}, \tag{15.96}$$

where the scalar product involves integration over Θ. Both the diabatic and the adiabatic population probability were recorded over a time interval of 300 fs, as shown in Figure 15.7. The initial wave packet is labeled by a set of two indices, (n_ρ, n_Θ), corresponding to the number of quanta in the N-H stretch and in the angular vibration mode. By Figure 15.7, the evolution of the adiabatic populations is readily understood as a superposition of two components, distinguished by different time behaviors. The high energy contribution to the wave packet overcomes the barrier in the $^1B_1$ state (see Figure 15.5) and decays within about 10 fs. The more gentle slope of the population functions at later times reflects the tunneling motion of the wave packet component trapped in the $^1B_1$ well. The coexistence of direct and hindered dynamics can be traced through all three configurations depicted in Figure 15.7. They differ by the quantum numbers of the initial vibrational function, with the number of quanta in the angular coordinate varying from zero to two, while none are in the radial coordinate. As n_Θ is increased, the crossing between the P_0^{D} and P_1^{D} population curves is reached more quickly.

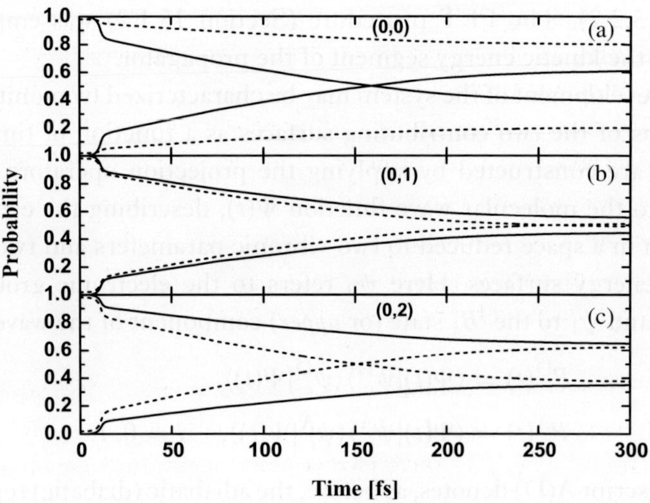

Figure 15.7: Populations for the states S_0 and $^1B_1$ as a function of time. Full lines refer to the adiabatic, dashed lines to the diabatic case. The initial wave packets are of (0, 0) (a), (0, 1) (b) and (0, 2) (c) character, involving excitation of the angular, but not the radial degree of freedom. (Reprinted with permission from [491]. Copyright (2004) by the American Institute of Physics.)

This tendency is plausible since for higher n_Θ, the system explores a larger configuration space, corresponding to large values of the coupling mode Θ. For $\Theta \neq 0$, one encounters a regime of high diabatic coupling, giving rise to an avoided crossing with an energy splitting proportional to Θ at the places where the diabatic S_0 and $^1B_1$ curves intersect. This rationalizes why the point of coincidence between the adiabatic populations P_0^{A} and P_1^{A} is increasingly delayed, as may be seen from Figure 15.7 as one goes from panel (a) to (c). Upon increase of n_Θ, the overall character of the motion becomes more adiabatic. As demonstrated by further simulations [491], the fast decaying component is strongly emphasized on comparison with the hindered fraction if allowance is made for $n_\rho > 0$ in the initial wave packet.

The arguments that explain the change of the population profiles with n_Θ account also for the variation of the branching ratio of dissociation on the lower ($^1B_1$) and the higher (S_0) surface. While dissociation in the $^1B_1$ state prevails in the case $n_\Theta = 0$, for $n_\Theta > 1$, the majority of dissociation events occur in the S_0 state, with a saturation limit of 25%/75% reached at $n_\Theta = 3$.

The simulation of Vallet *et al.* does not yield any bound population in the electronic ground state as a result of the passage through the conical intersection. One could ascribe the absence of this component to the reduced dimensionality assumed in this study, since inclusion of further degrees of freedom may, in principle, lead to energy loss from the N-H stretching mode and thus to an undissociated final system configuration. With sufficient energy available in the stretching mode, however, the fast decay component discussed in the context of Figure 15.7 becomes strongly dominant, and the process occurs substantially faster than vibrational energy redistribution.

The S_0-$^1A_2$ system exhibits a similar but characteristically different behavior. In contrast to the $^1B_1$ case, the $^1A_2$ barrier is somewhat lower than the dissociation limit. The wave packet component with energies between these two bounding values will pass over the barrier but undergo reflection by the attractive upper adiabatic potential and thus cross the region of the conical intersection repeatedly. For this reason, dissociation occurs preferentially from the lower adiabatic surface, making the branching ratio of dissociation events from the lower and from the higher surface consistently larger than one.

Besides the N-H stretching mode, two other nonadiabatic deactivation pathways of pyrrole have been characterized [494]. One of them is associated with N-C cleavage as the ring adopts a puckered structure following $\pi\pi^*$ excitation [495, 496]. The most recently identified mechanism involves a planar N-C ring opening which leads to a conical intersection between the corresponding $\pi\sigma^*_{\mathrm{NC}}$ state and the ground state [497]. Classical *on-the-fly* methodology was paired with multireference configuration interaction [497] and multiconfiguration self-consistent field (MCSCF) [494] computation to model the conical intersection dynamics that strongly determines both mechanisms. Both intersections have been shown to be connected by the same seam [497]. A recent study by Faraji *et al.* [498] employed the MCTDH method to model the evolution of the five lowest singlet states of pyrrole following vertical excitation to the $^1B_2$ state. This study included ten of the twenty-four vibrational modes of the pyrrole molecule.

Chapter 16

Density Functional Dynamics

For complex molecules, the traditional description of the electronic system in terms of wave functions faces insurmountable obstacles even when confronted with the means of modern computation, as it poses a $3N$-dimensional problem, with N as the number of electrons involved. The single-electron density, in contrast, is defined in three-dimensional space, and thus seems to provide an intuitively as well as mathematically much more accessible tool for characterizing the molecular electrons. Electronic structure theory as surveyed in Chapter 1 with reference to stationary problems and extended to the time-dependent case in Chapters 8–10, supplies a set of prescriptions for obtaining the wave function from which the density may be derived.

Conceptually, density functional theory (DFT) reverses this order by establishing the ground state density ρ_0 as the fundamental system quantity. All other properties, including potentials and wave functions for ground and excited states can, in principle, be expressed as functionals of ρ_0. The possibility of such a theory was demonstrated in the historical paper by Hohenberg and Kohn in 1964 [27] that marked the foundation event of DFT. A viable implementation of the formalism in the frame of the Hartree–Fock analogous *Kohn–Sham theory* was proposed one year later [499]. Devised originally in the context of solid state physics, DFT has been applied during the past decades with ever-growing success to finite units [500], to atoms, molecules, clusters, and nanostructures.

Although the *Kohn–Sham equations* operate with the familiar molecular orbital terminology and involve single determinantal wave functions, the latter are not to be interpreted as many-body states but rather as a vehicle for deriving the physically meaningful ground state density. This limitation has consequences for DFT-based dynamics. The generalization of the Hohenberg–Kohn theorem for time-dependent cases was achieved in 1984

[501], but the resulting *time-dependent density functional theory* (TDDFT) differs from *ab initio*-based approaches since it does not readily allow us to interpret a dynamic molecular process in terms of individual many-particle states evolving in time. While TDDFT encompasses time-dependent interactions with external agents, electronic transitions, and nonadiabatic situations as much as comparable *ab initio* schemes, new procedures are needed to interrogate the computational solutions for physically relevant dynamic information.

In the first two sections of this chapter, we will summarize elementary notions of DFT, to be followed by an introduction to TDDFT in Section 16.3. We will focus on two increasingly popular versions of the time-dependent theory: its realization in the linear response domain and its use as an instrument for propagation in real time that is no longer restricted by the constraint of small perturbations.

16.1 Fundamentals of Density Functional Theory

In presenting the basic tenets of DFT in a condensed overview, we will accentuate its parallels to and deviations from *ab initio* theory. Once more, we employ the Hamilton operator $\hat{H}_{\text{el}}$, as introduced in Eq. (1.3), to describe a system of N electrons. Writing the operator in a more abstract form as

$$\hat{H}_{\text{el}} = -\frac{\hbar^2}{2m}\sum_{i=1}^{N}\frac{\partial^2}{\partial \mathbf{x}_i^2} + \sum_{i=1}^{N} V_{\text{ext}}(\mathbf{x}_i) + V_{\text{int}}(\mathbf{x}_1, \mathbf{x}_2, \ldots, \mathbf{x}_N), \qquad (16.1)$$

we distinguish an *external potential* $V_{\text{ext}} = \sum_{i=1}^{N} V_{\text{ext}}(\mathbf{x}_i)$ from a contribution V_{int} due to the mutual repulsion between any two electrons. The external potential accounts for the attraction between the electrons and nuclei but could contain also any other interaction expressible as a simple partition into additive one-particle terms, such as the influence of an electrostatic field. The ground state expectation values of any operator satisfying this condition can be written as functionals of the ground state electron density, as is exemplified in Exercise 16.1 by the case of V_{ext}. This feature, in fact, holds for all ground state variables of a given system and thus also includes the two-particle terms that describe the interelectronic contribution to $\hat{H}_{\text{el}}$. DFT exploits this result using the density as a much less redundant, conceptually more simple, and computationally more tractable descriptor of a system than its wave function.

Exercise 16.1. Show that the ground state expectation value of V_{ext} can be represented as a functional of the ground state electron density ρ, namely as $V_{\text{ext}}[\rho] = \int d\mathbf{x}\, \rho(\mathbf{x}) V_{\text{ext}}(\mathbf{x})$.

Implementations of the idea to replace the wave function as the solution of electronic structure problems by the density go back to the early approaches of Thomas and Fermi [502, 503]. Here, as in later realizations of this concept, the task of identifying the density is reduced to minimizing a total energy functional that can be decomposed into a kinetic energy, an external potential, and an interelectronic repulsion component, according to

$$E[\rho(\mathbf{x})] = T[\rho(\mathbf{x})] + V_{\text{ext}}[\rho(\mathbf{x})] + V_{\text{int}}[\rho(\mathbf{x})], \tag{16.2}$$

where the density is subjected to the normalization constraint $\int d\mathbf{x} \rho(\mathbf{x}) = N$. While the Thomas–Fermi method neglected salient elements of *ab initio* procedures, for instance the exchange interaction, and was merely a qualitative approximation to these, incremental modifications of the density functional methodology improved its accuracy and made it applicable to a wide range of physical systems [504, 505]. The *Hohenberg–Kohn theorems* [27] established the density approach as electron structure theory in its own right, providing an alternative to the wave-function-based formalism surveyed in Chapter 1.

The first of the two theorems states that the position space ground state density of an N-electron system uniquely determines the wave function of this system. This is demonstrated by a concise proof using the *reductio ad absurdum* strategy. More specifically, one assumes the ground state density of the system to be known. As shown by Hohenberg and Kohn, the further assumption that this density is compatible with two different Hamiltonians then leads to a contradiction. Since two Hamiltonians for any ensemble of N electrons can only differ with respect to the external potential, the proof consists in deriving a unique assignment of a given ground state density ρ_0 to an external potential, such that the difference between two choices for V_{ext} that are both consistent with ρ_0 can only be a constant. A constant difference can be admitted, since it will only affect the phase of the wave function, leaving the density invariant. The latter is therefore in principle sufficient to deduce the Hamiltonian. In conclusion, the density defines all ground state properties of the system unanimously, including the energy.

The second theorem extends the variational principle that was used in Chapter 1 to obtain the Hartree–Fock equations (1.59) from wave functions to densities. As much as the ground state wave function Ψ_0, the ground state density ρ_0 minimizes the total energy of the studied system. It therefore holds that

$$E[\rho_0] \leq E[\rho], \quad \rho \neq \rho_0, \tag{16.3}$$

which implies the possibility of computing ρ_0 by use of a variational procedure. The second theorem of Hohenberg and Kohn lays the ground for the practical use of DFT in electronic structure calculations. The pivotal step consists in translating the abstract condition

$$\delta \left\{ E[\rho(\mathbf{x})] - \lambda \left(\int d\mathbf{x}\rho(\mathbf{x}) - N \right) \right\} = 0, \tag{16.4}$$

where the normalization constraint has been included with λ as Lagrange multiplier, into a set of equations that yield the ground state density of a physical system. The proposal put forth by Kohn and Sham [499] acquired a high level of prominence as it yielded a practicable approximation to solving this problem.

As the variation in Eq. (16.4) is to be carried out with respect to ρ, the equation may be reformulated in terms of a *functional derivative*, as introduced in the appendix of Chapter 1, with ρ as the independent quantity:

$$\frac{\delta}{\delta\rho(\mathbf{x})} \left\{ E[\rho(\mathbf{x})] - \lambda \left(\int d\mathbf{x}\, \rho(\mathbf{x}) - N \right) \right\} = 0, \tag{16.5}$$

For the unknown energy functional $E[\rho]$, Kohn and Sham take the following approach:

$$E[\rho(\mathbf{x})] = T[\rho(\mathbf{x})] + \int d\mathbf{x}\, V_{\text{ext}}(\mathbf{x})\rho(\mathbf{x}) + \int d\mathbf{x}\, V_{\text{coul}}(\mathbf{x})\rho(\mathbf{x}) + \tilde{E}_{\text{xc}}[\rho(\mathbf{x})]. \tag{16.6}$$

The expression on the right-hand side contains, besides functionals involving the external potential and the Coulomb potential V_{coul} that governs the mutual repulsion between electrons, two additional functionals: that of the kinetic energy and an *exchange-correlation functional* $\tilde{E}_{\text{xc}}(\mathbf{x})$ that summarily contains the quantum and many-body effects influencing the total energy of the system. Varying $E[\rho(\mathbf{x})]$ up to first, or linear, order in the density, we find by combining Eqs. (16.4) and (16.6) that

$$\frac{\delta T[\rho]}{\delta\rho} + V_{\text{ext}}(\mathbf{x}) + V_{\text{coul}}(\mathbf{x}) + \frac{\delta \tilde{E}_{\text{xc}}[\rho]}{\delta\rho} = \lambda. \tag{16.7}$$

The crucial move made by Kohn and Sham in addressing the challenge posed by this equation was mapping the real system of N interacting electrons on an auxiliary model system of N non-interacting electrons governed by an effective single-particle potential V^{KS}, where both systems have the same ground state density and total energy. Further on we will add some remarks on the existence of a non-interacting companion of a given electronic system, such that both share their ground state density and energy.

Before commenting on the actual construction of the fictitious system, we point out that this device makes it possible to represent the solution of the stationary Schrödinger equation for the problem at hand as in Hartree–Fock theory (Section 1.3), i.e. as a *single determinant of molecular orbitals*:

$$\Psi = \frac{1}{N!}\det[\Phi_1, \Phi_2, \ldots, \Phi_N], \tag{16.8}$$

where the spatial orbitals Φ_i are eigenstates of a single-particle Hamiltonian $\hat{h}$, according to

$$\hat{h}\Phi_i = E_i\Phi_i, \; i = 1, 2, \ldots, N, \tag{16.9}$$

and

$$\hat{h} = -\frac{\hbar^2}{2m}\frac{\partial^2}{\partial\mathbf{x}^2} + V^{\text{KS}}(\mathbf{x}). \tag{16.10}$$

It is important to note that the single-particle states Φ_i, in contrast to their role in *ab initio* theory, have no meaning beyond their auxiliary function of furnishing the ground state density. As in Eq. (1.26), the electronic density results from the spatial orbitals as

$$\rho = \sum_{i=1}^{N} |\Phi_i|^2. \tag{16.11}$$

With these premises, the kinetic energy functional is easily determined to be

$$T_{\text{s}}[\rho] = \sum_{i=1}^{N} \left\langle \Phi_i \left| -\frac{\hbar^2}{2m}\frac{\partial^2}{\partial\mathbf{x}_i^2} \right| \Phi_i \right\rangle, \tag{16.12}$$

where the subscript s stands for *single particle*. This functional differs from its counterpart in Eq. (16.7) where interacting electrons are adopted. An expression analogous to Eq. (16.7), however, is obtained by introducing the single-particle potential

$$V^{\mathrm{KS}}(\mathbf{x}) = \frac{\delta T[\rho]}{\delta\rho} - \frac{\delta T_{\mathrm{s}}[\rho]}{\delta\rho} + V_{\mathrm{ext}}(\mathbf{x}) + V_{\mathrm{coul}}(\mathbf{x}) + \frac{\delta \tilde{E}_{\mathrm{xc}}[\rho]}{\delta\rho}, \tag{16.13}$$

or shorter:

$$V^{\mathrm{KS}}(\mathbf{x}) = V_{\mathrm{ext}}(\mathbf{x}) + V_{\mathrm{coul}}(\mathbf{x}) + V_{\mathrm{xc}}(\mathbf{x}), \tag{16.14}$$

based on the definition

$$V_{\mathrm{xc}}(\mathbf{x}) \equiv \frac{\delta T[\rho]}{\delta\rho} - \frac{\delta T_{\mathrm{s}}[\rho]}{\delta\rho} + \frac{\delta \tilde{E}_{\mathrm{xc}}[\rho]}{\delta\rho} \tag{16.15}$$

of the *exchange-correlation potential*.

This completes the assembly of the single-particle potential. The Kohn–Sham orbitals could be determined by solving Eq. (16.9) if the potential V_{xc} were known. Incremental improvements of DFT have focused on this element of the theory which has been explicitly determined only for the very simplest conceivable system, namely the free electron gas [506]. Nevertheless, approximations to V_{xc} that allow us to represent various relevant quantities with good or even excellent accuracy have been implemented. We note that Kohn–Sham theory rests on the premise that for any arbitrary system and thus any ground state density, an effective single-particle potential exists, or, in more technical terms, that ρ must be *V-representable*. However, exceptions to this rule have been found [507, 508]. Fortunately, the relevant criterion has turned out to be *N-representability* which refers to a weaker condition than V-representability (e.g. [500]). If a density is N-representable it may be derived from an antisymmetric wave function. This requirement leads to a set of constraints that are satisfied by any physical density.

Before commenting on realizations of V_{xc}, we add an observation regarding Eq. (16.7). As can be seen from this equation, the Lagrange multiplier λ may be interpreted as a *chemical potential*, indicating the change of the total electronic energy as the number of electrons changes. Thus,

$$\lambda = \frac{\delta E}{\delta N} = \frac{\delta E[\rho(\mathbf{x})]}{\delta \rho(\mathbf{x})}. \tag{16.16}$$

Evidently, the term on the outermost right of Eq. (16.16) is constrained to be a constant.

16.1.1 *Exchange-correlation potentials*

The main challenge for any application of DFT to physical systems is identifying adequate expressions for the exchange-correlation potential V_{xc}. The most elementary approximation to this quantity is based on the homogeneous electron gas which allows for exact analytical descriptions of the exchange and correlation energy as a function of the density in the high and low density regimes. This defines the *local* solution to the problem of representing V_{xc}, as originally proposed by Kohn and Sham [499]. Later authors refined this approach by suggesting various procedures of interpolating between the known cases of density extremes ([509]–[512]). The *local density approximation* (LDA) treats the exchange-correlation effect as position-independent. In the framework of LDA, the energy E_{xc} becomes (see, for instance, [500])

$$E_{\text{xc}}^{LDA}[\rho] = \int d\mathbf{x}\; \rho(\mathbf{x})\varepsilon_{\text{xc}}[\rho], \tag{16.17}$$

with $\varepsilon_{\text{xc}}[\rho]$ as exchange-correlation energy per electron. Variation of this expression with respect to the density yields the LDA exchange-correlation potential:

$$V_{\text{xc}}^{LDA}(\mathbf{x}) = \varepsilon_{\text{xc}}[\rho(\mathbf{x})] + \rho(\mathbf{x})\frac{\delta \varepsilon_{\text{xc}}[\rho]}{\delta \rho}. \tag{16.18}$$

By construction, the LDA technique should be useful in those cases where the density varies slowly in space. Experience, however, has shown it to be a surprisingly powerful method even where this prerequisite is not satisfied, as is the regular situation for atoms, molecules, and crystals. This success may be rationalized by the observation that many quantities of major interest, such as the total energy of a system, involve integrals over ε_{xc} which are relatively insensitive to the local features of the exchange-correlation effect.

The next higher step on the ladder of DFT implementations is reached as allowance is made for spatial variation of ε_{xc} with the density. The exchange-correlation energy per electron now depends not only on ρ itself, but also on its first or higher derivatives with respect to position. Various realizations of this so-called *generalized gradient approximation* (GGA) have been proposed (e.g. [513]– [515]).

Making reference to spatial orbitals only (see Eqs. (16.8–16.12)), we have operated so far under the tacit assumption of a closed-shell system. If this condition is dropped, the formalism has to be generalized in close analogy to the extension that leads from the restricted to the unrestricted Hartree–Fock model, as discussed in Section 1.4. The density function is then partitioned according to

$$\rho(\mathbf{x}) = \rho^{\alpha}(\mathbf{x}) + \rho^{\beta}(\mathbf{x}) \tag{16.19}$$

to distinguish between the subsets of electrons with alpha and beta spins. The functional of the electronic energy is modified by going from $E[\rho(\mathbf{x})]$ to $E[\rho^{\alpha}(\mathbf{x}), \rho^{\beta}(\mathbf{x})]$ or, alternatively, to $E[\rho(\mathbf{x}), m(\mathbf{x})]$. The quantity $m(\mathbf{x})$ is here defined as *magnetic* or *spin-polarized density*. Procedures that operate with extensions of local exchange-correlation potentials to open-shell situations are labeled *local spin density* (LSD) methods.

We add two remarks on special types of functionals used in DFT. In the context of Hartree–Fock theory, we observed that the diagonal terms of the Coulomb and the exchange integral matrices **J** and **K**, respectively, are identical (see Eqs. (1.41) and (1.42)):

$$K_{ii} = J_{ii}. \tag{16.20}$$

As the Hartree–Fock equations (1.59) depend on the difference between **J** and **K**, they are uncompromised by any spurious interaction of an electron with itself. There is no such cancellation in Kohn–Sham theory since the Coulomb and the exchange-correlation term that appear in formula (16.14) for the single particle potential V^{KS} are of different origin. Specifically, V^{KS} combines the exact Coulomb with an approximate exchange-correlation component, resulting in the appearance of an unphysical *self-interaction effect*. Remedying this introduces a further potential term into the Kohn–Sham equations (16.9). The explicit form of this contribution, labeled $V_{\mathrm{SIC},i}$, where SIC stands for *self-interaction correction*, changes with the theory adopted to approximate the exchange-correlation potential [510]. This addition imports an orbital dependence into the single-particle Hamiltonian, as indicated by the index i, and thus detracts from the formal simplicity of the original approach.

Mixtures between the DFT and Hartree–Fock schemes are widely used (e.g. [516]). These employ *hybrid functionals* which combine a single-particle correlation potential with an exchange potential that incorporates both a DFT and an *ab initio* component. These functionals interpolate between the exact exchange expression (1.50b), which is composed of Kohn–Sham orbitals, and one or various DFT representations of the exchange effect. The frequently applied *Becke three-parameter* approximation, for instance, implements both Hartree–Fock and non-local DFT (specifically GGA, see the earlier discussion) corrections to the LDA (or LSD) approach. In the popular B3LYP potential, two parameters balance the admixtures due to *ab initio* and GGA to the local density exchange, and a third parameter modifies the local density correlation by the correlation functional of Lee, Yang, and Parr [517]. Pragmatic as hybrid functionals are, they are also practical, adding a high degree of flexibility to the DFT methodology, and alleviating the self-interaction problem.

16.2 Excited Electronic States in DFT

It is not evident how to extend DFT methods from the ground state to excited states, in contrast to *ab initio* approaches where this extension is, while computationally demanding, conceptually unambiguous. The Hohenberg–Kohn theorem cannot be directly transferred to the case of electronic excitations. In principle, the ground state energy can be expressed exactly as a functional of the ground state density, but the excited state energy cannot likewise be represented as an exact functional of the excited state density [518, 519]. Among the chief consequences of the first Hohenberg–Kohn theorem is, however, the proposition that the ground state density of a stationary system is sufficient for a *complete description* of the system, including the energies as well as the many-body wave functions of its excitations.

As mere proof of existence, the first Hohenberg–Kohn theorem does not give any guidance for the construction of the respective functionals. Yet, several avenues to make excited state energies accessible by DFT have been explored. Thus, the variational principle as rationalized by the second Hohenberg–Kohn theorem might be applied to states of a well-defined symmetry class, resulting in the lowest energy for each one of these classes [520].

The *restricted open-shell Kohn–Sham* procedure (ROKS) [521] may serve as an illustration of this methodology. In analogy to the restricted Hartree–Fock formalism introduced in Section 1.4, excited determinants can be built from the Kohn–Sham ground state determinant (16.8) by promoting electrons into virtual Kohn–Sham energy levels while assigning the same sets of spatial orbitals to both the alpha and beta spin subsystems. In this way, trial functions for electronic excitations might be defined by superposing selected excited determinants, subject to the constraint that their linear combination is an eigenstate of the $\hat{S}^2$ operator. Kohn–Sham analogous equations, involving orbital-dependent exchange-correlation potentials, have been derived to generate excited state energies and density distributions [238, 521].

An alternative way of using DFT to identify excitation energies consists in exploring the functional $E[\rho]$ for local extrema [522], as the absolute minimum obtained from the variational principle (16.5) specifies the ground state energy and density of the respective system, while the other extrema yield the respective information about excited states.

Of relevance for the problem of excitation energies within DFT is the *adiabatic connection* concept which is also at the core of the hybrid functional methodology mentioned in the previous section. It rests on the notion of a Hamiltonian parametrized by an index α that defines a continuum between the extreme choices of the fully interacting system, characterized by the condition $\alpha = 1$, and the non-interacting alternative that is realized when $\alpha = 0$, according to

$$\hat{H}_\alpha = \hat{T}_s + \alpha V_{xc} + V_\alpha + V_{coul} + V_{ext}, \tag{16.21}$$

where $0 \leq \alpha \leq 1$, and V_α stands for a sum of single-particle operators,

$$V_\alpha = \sum_{i=1}^{N} V_{\alpha,i}. \tag{16.22}$$

Taking α from one of its limits to the other achieves therefore a mapping between the two contrasting system types and merges continuously their eigenfunctions and eigenvalues into each other. Moreover, the systems assigned to all allowed choices of α are constrained to have the same density which can be accomplished by suitable adjustment of the operators $\hat{V}_{\alpha,i}$. In this manner, an adiabatic connection is established between a many-body wave function and a corresponding Kohn–Sham state. Goerling discusses a *generalized adiabatic connection* [523, 524] and specifies its application to excited states within the Kohn–Sham formalism. A scheme is proposed that assigns the states of one given physical system, involving interacting electrons, to the corresponding states of an auxiliary Kohn–Sham system. This leads to a prescription how to obtain the energy of the many-body system from the respective Kohn–Sham solution that is not limited to the lowest energy state.

Alternatively, *ensemble DFT* may be used to derive excited state energies ([525]–[529]). Consider two orthonormal states, Ψ and Ψ', and define the energy functional

$$E^{\omega}[\Psi, \Psi'] = (1 - \omega)\langle\Psi|\hat{H}|\Psi\rangle + \omega\langle\Psi'|\hat{H}|\Psi'\rangle, \tag{16.23}$$

with $0 \leq \omega \leq \frac{1}{2}$. As implied by the variational principle, the *ensemble energy* $E^{\omega}[\Psi, \Psi']$ attains its minimum for $\Psi = \Psi_0$ and $\Psi' = \Psi_1$ if Ψ_0 and Ψ_1 are the ground and the first excited state of the considered system, respectively. Similarly, the *ensemble density*

$$\rho^{\omega}[\mathbf{x}] = (1 - \omega)\rho_{\Psi_0} + \omega\rho_{\Psi_1} \tag{16.24}$$

may be defined within the framework of DFT. As both the DFT ensemble density and the DFT ensemble energy, referring to a non-interacting system, reproduce the respective quantities for the interacting system, the Kohn–Sham formalism may then be employed to compute E^{ω}. From the known ground state energy, the first excited energy can be inferred. A recent version of the method, operating with orbital functionals, arrived at accurate values for excitation energies of several light atom species [529].

A qualitatively different way of obtaining excited state energies within DFT is monitoring the response of the studied system to an appropriately small external perturbation. This is an application of *time-dependent density functional theory* whose principles we sketch in the following section.

16.3 Time-Dependent Density Functional Theory

As in the parallel case of *ab initio* theory, DFT may be extended to include time-dependent processes. Generalizing the stationary theory, we make allowance for a parametric dependence of the nuclear degrees of freedom on time, as well as a possible time dependence of the external potential. The Schrödinger equation to be solved thus assumes the form

$$i\hbar\frac{\partial\Psi(\mathbf{x},t)}{\partial t} = \hat{H}(\mathbf{x},t)\Psi(\mathbf{x},t) \tag{16.25}$$

where we suppress the nuclear coordinates. The corresponding single-particle electron density is written as

$$\rho(\mathbf{x},t) = N\int \mathbf{x}_2\ldots d\mathbf{x}_N|\Psi(\mathbf{x},\mathbf{x}_2\ldots\mathbf{x}_N,t)|^2. \tag{16.26}$$

In the dynamic case, the place of the first Hohenberg–Kohn theorem is taken by the *Runge–Gross theorem* [501] which assigns a unique external potential $V_{\mathrm{ext}}(\mathbf{x},t)$ to a given electronic density $\rho(\mathbf{x},t)$ for any many-body system that develops in time from a well-defined initial wave function $\Psi_{t=0}$. In analogy to Hohenberg–Kohn theory, one asserts that two densities that emerge from the same state $\Psi_{t=0}$ are different if the two corresponding potentials differ from each other by more than a function $f(t)$. While this statement appears like a straight parallel to the static situation, the involvement of a definite initial wave function in the time-dependent formalism must be borne in mind as a subtlety that is absent from the first Hohenberg–Kohn theorem. The Runge–Gross theorem is compatible with the case of two different potentials giving rise to identical densities if these originate from different states. One may plug this loophole by the decree that the system evolves from the ground state, with recourse to the first Hohenberg–Kohn theorem.

Whereas the static Kohn–Sham equations are derived from the energy minimization condition (16.4), this does not hold for their dynamic analogs, as energy conservation is not guaranteed in this case. As discussed in Chapter 8 in the context of TDHF theory, these equations are obtained from a time-dependent variational principle, imposing a stationarity constraint on the quantum mechanical action which is, in view of the Runge–Gross theorem, interpreted as a functional of the density. Thus, the variational condition for dynamic many-body problems is

$$\frac{\delta A[\rho]}{\delta \rho(\mathbf{x}, t)} = 0 \tag{16.27}$$

with

$$A[\rho] = \int_{t_i}^{t_f} dt \langle \Psi[\rho](t) | i\hbar \frac{\partial}{\partial t} - \hat{H}(t) | \Psi[\rho](t) \rangle, \tag{16.28}$$

where, for simplicity of notation, we do not include the spatial coordinates on which $\hat{H}$ and Ψ depend. As the total energy in the case of static DFT, the action (16.28) is a unique functional of the density for TDDFT. Solving the static DFT problem means obtaining the exact density by searching for the extremal points of the total energy, while the solution of the time-dependent problem yields the exact density as a function of time as derived from the stationary points of the action. Relation (16.27) holds under the boundary conditions

$$\delta|\Psi[\rho](t_1)\rangle = \delta|\Psi[\rho](t_2)\rangle = 0. \tag{16.29}$$

The time-dependent exchange-correlation potential is defined as the functional derivative of the corresponding component of the action $A[\rho]$, namely

$$V_{\rm xc}[\rho](\mathbf{x}, t) = \frac{\delta A_{\rm xc}[\rho]}{\delta \rho(\mathbf{x}, t)}. \tag{16.30}$$

The functional relation between the density and both $V_{\rm xc}$ and $A_{\rm xc}$ is unknown. In general, $V_{\rm xc}$ is non-local with respect to both spatial and temporal degrees of freedom. A simplified but computationally efficient explicit form of $A_{\rm xc}$ is given by the *adiabatic local density approximation* to this quantity:

$$A_{\rm xc}^{\rm ALDA}[\rho] = \int_{t_i}^{t_f} dt' \int d\mathbf{x}' \rho(\mathbf{x}', t') \varepsilon_{\rm xc}^{\rm LDA}[\rho(\mathbf{x}', t')]. \tag{16.31}$$

Formally, $V_{\rm xc}$ may be expressed through an *exchange-correlation kernel* $f_{\rm xc}[\rho](\mathbf{x}, t, \mathbf{x}', t')$, which exhibits non-locality in both space and time, according to

$$V_{\rm xc}[\rho](\mathbf{x}, t) = \int dt' d\mathbf{x}' f_{\rm xc}[\rho](\mathbf{x}, t, \mathbf{x}', t') \rho(\mathbf{x}', t'). \tag{16.32}$$

Interpreting this identity in terms of causality, one requires that $V_{\rm xc}(\mathbf{x}, t)$ be determined by any value the density may adopt at a time $t' < t$, i.e. in the "past", but it must vanish if $t < t'$. This implies for the functional derivative of V_{xc} with respect to ρ that

$$\frac{\delta V_{xc}[\rho](\mathbf{x}, t)}{\delta\rho(\mathbf{x}', t')} = 0, \quad \text{if } t < t'. \tag{16.33}$$

On the other hand, we find by Eq. (16.30) that

$$\frac{\delta V_{xc}(\mathbf{x}, t)}{\delta\rho(\mathbf{x}', t')} = \frac{\delta^2 A_{xc}[\rho]}{\delta\rho(\mathbf{x}, t)\delta\rho(\mathbf{x}', t')}. \tag{16.34}$$

The symmetry of the right-hand side with respect to the interchange of the space–time arguments violates causality. A related problem involves the boundary conditions $\delta|\Psi[\rho](t_1)\rangle = \delta|\Psi[\rho](t_2)\rangle = 0$ required for deriving the TDSE by varying the action (16.28). The occurrence of the operator $\frac{\partial}{\partial t}$ implies that $\delta|\Psi[\rho](t_2)\rangle$ is completely determined once $\delta|\Psi[\rho](t_1)\rangle$ is given, making it generally impossible to satisfy both boundary conditions simultaneously. These two interrelated problems, both concerning the time structure of the action principle underlying TDDFT, have drawn much commentary ([530]–[532]). Recently, it was shown that both difficulties may be resolved by replacing the stationarity constraint $\delta A[\rho] = 0$ with the more complex expression $\delta A[\rho] = i\langle\Psi[\rho](t_2)|\delta\Psi[\rho](t_2)\rangle$, and by retaining only the boundary condition for the initial time t_1, while allowing for a non-zero variation at $t = t_2$ [532].

In the context of TDDFT, the term *nonadiabaticity* refers usually to the absence of memory effects in the exchange-correlation kernel (e.g. [533]) that is implicitly defined by Eq. (16.32). We point out that the adiabatic approximation circumvents the causality dilemma since it is local in time and thus does not make admission for memory effects. The kernel here is local in both space and time:

$$f_{xc}[\rho]^{gs}(\mathbf{x}, t, \mathbf{x}', t') = \delta(t - t')\frac{\delta V_{xc}^{gs}[\tilde{\rho}](\mathbf{x})}{\delta\tilde{\rho}(\mathbf{x}')}|_{\tilde{\rho}=\rho(\mathbf{x},t)}, \tag{16.35}$$

where V_{xc}^{gs} is the ground state exchange-correlation potential. In most practical calculations, the time non-locality inherent in V_{xc} is ignored, and the relevant potentials are evaluated by use of the instantaneous density.

16.3.1 *TDDFT in the linear response domain*

A small time-dependent perturbation of the form δV_{ext} is expected to produce a small variation $\delta\rho$ of the ground state density. This process is tractable within linear response theory if cause and effect are related in the following way:

$$\delta\rho = \int dt' \int d\mathbf{x}' \; \tilde{\chi}(\mathbf{x}, t; \mathbf{x}', t')\delta V_{\text{ext}}, \tag{16.36}$$

with $\tilde{\chi}$ as *density–density response function.*[1] TDDFT within the *linear response domain* accounts, in principle, for a wide variety of phenomena in finite systems besides electronic excitations, such as ionization thresholds, transport through molecules, or optical absorption spectra [534]. The linear response (LR) version of TDDFT operates with a Kohn–Sham response function, $\tilde{\chi}_{\text{KS}}$, which is related to the density of non-interacting particles such that

$$\delta\rho = \int dt' \int d\mathbf{x}' \; \tilde{\chi}_{\text{KS}}(\mathbf{x}, t; \mathbf{x}', t')\delta V^{\text{KS}}, \tag{16.37}$$

with V^{KS} as the Kohn–Sham single-particle potential [535] and $\delta\rho$ as $\rho(\mathbf{x}, t) - \rho_0$, where ρ_0 stands for the unperturbed ground state density of the considered system. The uniqueness of the external potential assigned to a certain density, on one hand, and the availability of a unique Kohn–Sham potential for this same density, on the other make, it possible to invert (16.36) and to write the exact density–density response function by use of the chain rule for functional derivatives in the following way:

$$\tilde{\chi}(\mathbf{x}, t; \mathbf{x}', t') = \int d\mathbf{x}_1 dt_1 \frac{\delta\rho(\mathbf{x}, t)}{\delta V^{\text{KS}}(\mathbf{x}_1, t_1)} \frac{\delta V^{\text{KS}}(\mathbf{x}_1, t_1)}{\delta V_{\text{ext}}(\mathbf{x}', t')} |_{V_{\text{ext}}[\rho_0]}. \tag{16.38}$$

This expression can be used as the basis for deriving a relation between the exact response function $\tilde{\chi}$ and its approximation in the framework of Kohn–Sham theory. By substitution for the functional derivatives under the

[1]When we label the density–density response function $\tilde{\chi}$, instead of using the customary notation χ, it is to avoid any confusion with the symbol for the nuclear wave function, $\chi(\mathbf{X}, t)$, as adopted throughout this text.

integral, a Dyson-type equation for $\tilde{\chi}$ may be derived, namely

$$\tilde{\chi}(\mathbf{x},t;\mathbf{x}',t') = \tilde{\chi}_{\mathrm{KS}}(\mathbf{x},t;\mathbf{x}',t') + \int d\mathbf{x}_1 dt_1 \int d\mathbf{x}_2 dt_2\, \tilde{\chi}_{\mathrm{KS}}(\mathbf{x},t;\mathbf{x}_1,t_1) \times \left\{ \frac{e^2}{4\pi\varepsilon_0|\mathbf{x}_2-\mathbf{x}_1|}\delta(t_2-t_1) + f_{\mathrm{xc}}[\rho_0](\mathbf{x}_1,t_1;\mathbf{x}_2,t_2) \right\} \times \tilde{\chi}(\mathbf{x}_2,t_2;\mathbf{x}',t'). \tag{16.39}$$

The exchange-correlation part is here expressed in terms of the respective kernel. Thus, $\tilde{\chi}$ can be approximated through the iterative procedure defined by Eq. (16.39) once $\tilde{\chi}_{\mathrm{KS}}$ and the kernel f_{xc} are known.

Exercise 16.2. Derive Eq. (16.39) from Eq. (16.38). Use the relations

$$\chi_{\mathrm{KS}}(\mathbf{x},t;\mathbf{x}',t') = \frac{\delta\rho[V^{\mathrm{KS}}](\mathbf{x},t)}{\delta V^{\mathrm{KS}}(\mathbf{x}',t')} \tag{16.40}$$

and

$$f_{\mathrm{xc}}(\mathbf{x},t;\mathbf{x}',t') = \frac{\delta V_{\mathrm{xc}}[\rho](\mathbf{x},t)}{\delta\rho(\mathbf{x}',t')} \tag{16.41}$$

in conjunction with $V^{\mathrm{KS}}(\mathbf{x},t) = V_{\mathrm{ext}}(\mathbf{x},t) + V_{\mathrm{coul}}(\mathbf{x},t) + V_{\mathrm{xc}}(\mathbf{x},t)$, generalizing Eq. (16.14).

Changing to the Fourier transform $\tilde{\chi}(\mathbf{x},\mathbf{x}',\omega)$ of $\tilde{\chi}(\mathbf{x},t;\mathbf{x}',t')$, and casting the result into the *Lehmann representation* (see, for instance, [536]) makes the excitations of the system explicit:

$$\tilde{\chi}(\mathbf{x},\mathbf{x}',\omega) = \lim_{\eta\to 0^+} \sum_i \left[\frac{\langle\Psi_0|\hat{\rho}(\mathbf{x})|\Psi_i\rangle\langle\Psi_i|\hat{\rho}(\mathbf{x}')|\Psi_0\rangle}{\hbar\omega-(E_i-E_0)+i\eta} - \frac{\langle\Psi_0|\hat{\rho}(\mathbf{x}')|\Psi_i\rangle\langle\Psi_i|\hat{\rho}(\mathbf{x})|\Psi_0\rangle}{\hbar\omega+(E_i-E_0)+i\eta} \right]. \tag{16.42}$$

The functions $\{\Psi_i\}$ form here a complete set of many-body states, with $\{E_i\}$ as the set of their eigenenergies. Evidently, the excitation energies coincide with the *poles of the response function*. Similarly, the *Kohn–Sham response function* becomes singular wherever $\hbar\omega$ equals the difference between two

Kohn–Sham energy levels. The corresponding response function reads

$$\tilde{\chi}_{KS}(\mathbf{x},\mathbf{x}',\omega) = \lim_{\eta\to 0^+} \sum_{k,m} (n_m - n_k) \frac{\Phi_k(\mathbf{x})\Phi_k^*(\mathbf{x}')\Phi_m(\mathbf{x}')\Phi_m^*(\mathbf{x})}{\hbar\omega - (E_m - E_k) + i\eta}, \tag{16.43}$$

with n_m, n_k as occupation numbers of the ground state Kohn–Sham orbitals Φ_m and Φ_k with energy eigenvalues E_m and E_k, respectively. From Eqs. (16.39) and (16.43), the exact response function for the physical system could be found if the exact exchange-correlation kernel were known and a full self-consistent solution of Eq. (16.39) could be generated. The *random phase approximation* to the response function is obtained by setting f_{xc} equal to zero. This choice generally does not account well for short-range correlations [537]. Various approximations to f_{xc} have been proposed, among them the frequency-independent *PGG exchange-correlation kernel*, where PGG stands for Petersilka, Gossmann, and Gross (see [535]). It has been found, however, that the accuracy of the Kohn–Sham orbitals, and thus the ground state exchange correlation potential, has a greater impact on the quality of linear response results than that of f_{xc} [538]. Thus, in finite systems, employing an orbital-dependent functional that contains the exact exchange contribution (EXX) has turned out to be much more successful than using functionals with approximate exchange, such as LDA [539, 540]. This observation is attributed to the unrealistic exponential decay of the LDA potential in the asymptotic region, while the EXX potential shows the correct $1/r$ behavior, implying a much better definition of the unoccupied states and therefore the excitation energies.

16.3.2 *Time-dependent current density functional theory*

The adiabatic approximation to f_{xc} is adequate for the case of a homogeneous electron gas at low frequency but encounters problems beyond this regime of near-validity. Thus, the adiabatic model fails to capture systems with highly non-local exchange and correlation effects, such as quasi one-dimensional polymer chains [541].

A generalized Kohn–Sham theory that replaces density as basic variable with the *current density* [542, 543] has been shown to alleviate this ailment. Without sacrificing the intrinsic non-locality of the physical system, the approach preserves the essential advantage of the adiabatic scheme as it operates with functionals of quantities that are local in space and time. On the most elementary level, this proposal can be motivated by observing that the continuity equation

$$\frac{\partial \rho(\mathbf{x},t)}{\partial t} = -\frac{\partial}{\partial \mathbf{x}} \cdot \mathbf{j}(\mathbf{x},t) \tag{16.44}$$

relates ρ to $\mathbf{j}$, the density to the current density. Expressing both ρ and $\mathbf{j}$ through their Fourier transforms, this equation becomes

$$\rho(\mathbf{x},\omega) = \frac{\frac{\partial}{\partial \mathbf{x}} \cdot \mathbf{j}(\mathbf{x},\omega)}{i\omega}. \tag{16.45}$$

A solution for the longitudinal component of the current density, i.e. the component that may be written as the derivative of a scalar function, is readily found [544]:

$$\mathbf{j}_{\mathrm{L}}(\mathbf{x},t) = -\frac{i\omega}{4\pi} \int d\mathbf{x}' \rho(\mathbf{x}',\omega) \frac{\partial}{\partial \mathbf{x}} \frac{1}{|\mathbf{x}-\mathbf{x}'|}. \tag{16.46}$$

Likewise, any time-dependent scalar potential can be associated with a longitudinal vector potential $\mathbf{A}$ such that

$$\mathbf{A}(\mathbf{x},t) = -\frac{c}{e} \int_{t_0}^{t} dt' \frac{\partial}{\partial \mathbf{x}} V(\mathbf{x},t'). \tag{16.47}$$

Vignale and Kohn [542] succeeded in reformulating TDDFT linear response theory in terms of a perturbing vector potential and a responding current density, mediated by a current–current response function. The great advantage of this transition from TDDFT to *time-dependent current density functional theory* (TDCDFT) is that the onerous spatial and temporal non-localities are entirely contained in relations of the type (16.46) and (16.47), connecting the density with the current density and the scalar with the vector potential, respectively, and thus do not explicitly appear in the relevant equations. Likewise, the Kohn–Sham equations which have to be solved beyond the linear regime were recast into vector potential terminology [543].

Stepping from TDDFT to TDCDFT widens the range of TDDFT applications considerably, to include, for instance, interactions with magnetic fields and electromagnetic waves. A hydrodynamic interpretation was given to the exchange correlation part of the TDCDFT vector potential whose time derivative is related to the divergence of a viscoelastic stress tensor [543], linking the TDCDFT formalism to the continuum mechanics of the electron liquid.

16.4 Direct Molecular Dynamics Based on DFT

The prospects and limitations of TDDFT as a device for real-time propagation of a quantum system are currently the subject of lively experimentation. As mentioned before, no exact DFT potentials exist for excited states, and further, Kohn–Sham determinants are not to be identified with physical many-body wave functions. This situation makes it difficult to transfer many familiar notions of relevance to electron dynamics, related to the coupling between electronic states, excitation to higher potential energy surfaces, or interaction between electronic and nuclear degrees of freedom, from *ab initio* theory to DFT. As established by an ever-growing body of data, however, TDDFT dynamics is often successful in the ancestral territory of *ab initio* approaches. Whether this success is fortuitous or systematic might be decided in each individual case once a more thorough understanding of the fundamentals of TDDFT is reached.

In the following, we will comment on the time-dependent Kohn–Sham equation as an instrument for propagating a quantum system in real time. As in the case of TDHF (see Chapter 8), the TDDFT Kohn–Sham equations are generated by replacing the full Hamiltonian in the TDSE with the relevant approximation, resulting in

$$i\hbar \frac{\partial \Phi_i(\mathbf{x}, t)}{\partial t} = (\hat{T} + V^{\mathrm{KS}}(\mathbf{x}, t))\Phi_i(\mathbf{x}, t), \tag{16.48}$$

where only electronic degrees of freedom are included and time locality is assumed, with a single-particle potential of the form

$$\begin{aligned} V^{\mathrm{KS}}[\rho, \Phi_{t=0}](\mathbf{x}, t) &= V_{\mathrm{ext}}[\rho, \Psi_{t=0}](\mathbf{x}, t) + V_{\mathrm{coul}}[\rho](\mathbf{x}, t) \\ &\quad + V_{\mathrm{xc}}[\rho, \Psi_{t=0}, \Phi_{t=0}](\mathbf{x}, t), \end{aligned} \tag{16.49}$$

in parallel to Eq. (16.14). Further, the time-dependent extension of Kohn–Sham theory employs the properly modified form of Eq. (16.11), namely

$$\rho(\mathbf{x}, t) = \sum_{i=1}^{N} |\Phi_i(\mathbf{x}, t)|^2. \tag{16.50}$$

As implied by Eq. (16.49), the exchange-correlation functional V_{xc} depends not only on the density and the initial Kohn–Sham determinant $\Phi_{t=0}$ but, as established by the Runge–Gross theorem, also on the initially realized many-body state $\Psi_{t=0}$. With Eq. (16.48), the solution of the TDDFT problem is reduced to propagating the time-dependent Kohn–Sham orbitals, implying that the N initially occupied orbitals are advanced in time. This proceeds ideally within a complete basis which may be chosen as the set $\{\Phi_i(\mathbf{x}, t = t_0)\}$ of all occupied and virtual Kohn–Sham orbitals at the beginning of the time evolution. Since this basis is static, the time dependence is carried by the expansion coefficients:

$$\Phi_i(\mathbf{x}, t) = \sum_{k=1}^{\infty} a_{ik}(t)\Phi_r(\mathbf{x}, t = t_0). \tag{16.51}$$

In practice, the sum must be truncated at some finite index $k_{\max} = K$. At any time, the instantaneous density ρ is defined by the single-particle potential which, being a density functional, can only be evaluated if ρ is known. This mutual dependence sets up a self-consistency loop: the initial guess of the density defines the initial single-particle potential V^{KS}, which then is used to compute the first approximation to the density, etc. The process converges if two consecutive choices of V^{KS} are indistinguishable within a preset margin of accuracy.

If the adiabatic representation is chosen, the time-dependent Kohn–Sham orbitals acquire a dependence on the nuclear coordinates, according to $\Phi_i = \Phi_i(\mathbf{x}, \mathbf{X}, t = t_0)$. The equation of motion for the coefficients a_{ik} then reads

$$i\hbar\frac{\partial a_{ik}(t)}{\partial t} = \sum_{m}^{K} a_{im}\left(\varepsilon_m(\mathbf{X})\delta_{km} - \mathbf{A}_{km}\cdot\frac{\partial \mathbf{X}(t)}{\partial t}\right), \tag{16.52}$$

with $\varepsilon_m(\mathbf{X})$ as Kohn–Sham eigenenergy at the place $\mathbf{X}$, and $\mathbf{A}_{km}$ as the nonadiabatic coupling vector of the Kohn–Sham orbitals with indices k and m.

In the following, we will give an example of a TDDFT application of great relevance for optical spectroscopy, the calculation of optical absorption spectra.

16.4.1 *Calculating molecular photoabsorption spectra*

Time-dependent Kohn–Sham theory has been applied to generate photoabsorption spectra of finite species [536, 545]. More specifically, the photoabsorption cross section as function of the photon frequency is related to the imaginary part of the *dynamic polarizability* $\boldsymbol{\alpha}$ by the formula

$$\sigma(\omega) = \frac{4\pi\omega}{3c}\Im\left\{\sum_q \alpha_q(\omega)\right\}, \quad \text{and} \quad q = x, y, z. \tag{16.53}$$

The dynamic polarizability depends on the change of the system's density in response to an external perturbation δV_{ext}. The latter may be chosen as

$$\delta V_{\text{ext}}(\mathbf{x}, t) = -\mathbf{C} \cdot \mathbf{x}\, \delta(t), \tag{16.54}$$

with a perturbing potential proportional to the position $\mathbf{x}$, and $\mathbf{C}$ as a constant amplitude vector. Utilizing a perturbation of this type has the advantage that all frequencies of the studied system are excited with equal weight [536]. Further, it gives rise to initial Kohn–Sham orbitals of the form

$$\Phi_i(\mathbf{x}, t = 0) = \exp(i\mathbf{C} \cdot \mathbf{x})\tilde{\Phi}_i(\mathbf{x}), \tag{16.55}$$

with $\{\tilde{\Phi}_i\}$ as the Kohn–Sham orbitals of the unperturbed system. The real-time representation of the dynamic polarizability $\boldsymbol{\alpha}$, corresponding to this choice of perturbation, is given by

$$\alpha_q(t) = -\frac{1}{C_q}\int d\mathbf{x}\, x_q \delta\rho(\mathbf{x}, t). \tag{16.56}$$

The density response $\delta\rho$ is here understood as the time-dependent deviation from the density adopted at $t = 0$, $\rho(\mathbf{x}, t) - \rho(\mathbf{x}, 0)$. Kohn–Sham TDDFT propagation of the orbitals may then provide the density at some later time, and, by Eq. (16.56), the respective polarizability. For comparison with experiment, one finally subjects $\boldsymbol{\alpha}$ to Fourier transformation and so arrives, by Eq. (16.53), at the cross section. In the linear response domain, $\alpha(\omega)$ is also available through the density–density response function, as defined in Section 16.3.1. This presumes a sufficiently small perturbation amplitude $\mathbf{C}$. Explicitly, the polarizability may be found from the response function $\tilde{\chi}$ through the identity

$$\alpha_q(\omega) = -\int d\mathbf{x}\int d\mathbf{x}' x_q \tilde{\chi}(\mathbf{x}, \mathbf{x}', \omega)x'_q. \tag{16.57}$$

We continue by addressing the question of how to extract information about molecular bonding and its variations due to the studied interaction from the results of TDDFT computation.

16.4.2 *Molecular bonding properties analyzed by the electron localization function*

Since TDDFT does not involve representations of physical many-particle states it has to rely on density-based methods for interpretation of its results. A powerful instrument for visualizing molecular bonding properties and their change in the course of the dynamic process suitable for DFT computation is given with the *electron localization function* (ELF). Molecular orbitals are only defined up to a unitary transformation which leaves both the electronic density and the total energy of the system unchanged. The description of the electronic system given by a molecular orbital might vary between the extremes of complete delocalization to maximal localization. In the absence of any criterion that would single out one correct representation, it must be concluded that there is no systematic way to analyze chemical bonding by inspection of molecular orbitals. Information about all molecular properties that involve electron localization, such as single or multiple bonds or lone pairs, can therefore not be inferred directly from the orbitals, but may be extracted from the density function. Becke and Edgecombe proposed a technique of processing the density to obtain a measure of electron localization [546]. Specifically, they discussed the probability of finding an electron with spin σ in some volume around a point of reference $\mathbf{x}$ occupied with an electron of the same spin. This quantity varies with the extent of electron localization at $\mathbf{x}$: the more localized the reference electron, the lower the probability of finding another electron with the same spin close to it. The basic concept to be derived from the density is here a *conditional pair probability*, $P^{\sigma\sigma}_{\rm cond}(\mathbf{x}, \mathbf{x}')$. The leading term of the Taylor expansion of the spherical average of $P^{\sigma\sigma}_{\rm cond}$ around $\mathbf{x}$ was shown to be determined by the expression D_σ, defined as

$$D_\sigma \equiv \tau_\sigma - \frac{1}{4}\left(\frac{\partial \rho_\sigma}{\partial \mathbf{x}'}\right)^2 / \rho_\sigma - \frac{j_\sigma^2}{\rho_\sigma}, \tag{16.58}$$

to be evaluated at $\mathbf{x}' = \mathbf{x}$. The first term on the right-hand side is a kinetic energy density, given by

$$\tau_\sigma = \sum_{i=1}^{N_\sigma} \left| \frac{\partial \Phi_i}{\partial \mathbf{x}'} \right|^2, \tag{16.59}$$

where N_σ is the number of electrons with spin σ. The last term involves the absolute value of the current produced by these electrons, j_σ. This contribution is absent if only real-valued orbitals are admitted but becomes important in dynamic situations [547, 548]. The function $D_\sigma(\mathbf{x})$ is positive definite, but not bounded from above. Further, it does not increase with the degree of electron localization but delocalization. Properly inverting and normalizing D_σ yields the ELF, namely

$$\mathrm{ELF} = \left(1 + \left(\frac{D_\sigma}{D_\sigma^0}\right)^2\right)^{-1}. \tag{16.60}$$

The quantity D_σ^0 refers here to a uniform electron gas whose density equals $\rho_\sigma(\mathbf{x})$. In contrast to D_σ, ELF is confined to the interval [0, 1], where the value 1 indicates the case of perfect localization, and $\frac{1}{2}$ the electron-gas case. Graphical representation of the ELF makes it possible to visualize architectural features of a given system that are accessible neither from molecular orbitals nor from the density alone, such as the separation between core and valence electrons or the atomic shell structure [549, 550]. The extension of the ELF concept to the time-dependent case makes it possible to monitor the formation, modulation, or breaking of a bond in the course of a chemical reaction, or to generate a time-resolved model of excitation processes in terms of spatial properties of the contributing electrons [548]. For this purpose, real-time TDDFT propagation has been combined with ELF analysis, substituting time-dependent densities and currents for the respective quantities in Eq. (16.58).

16.4.3 *Combining TDDFT with standard methods of nonadiabatic dynamics*

Generalizing the TDDFT formalism to cases of sizeable coupling between electrons and nuclei is a matter of intense current research activity. Among the traditional methods of nonadiabatic dynamics that have been translated into the language of TDDFT are trajectory surface hopping (Chapter 11) and electron nuclear dynamics (Chapter 9).

The ROKS formalism mentioned in Section 16.2, involving the description of both the electronic ground state and one or several excitations in terms of Kohn–Sham determinants, has been used to derive the nonadiabatic coupling between the evolving states, in close analogy to *ab initio* approaches. If the nuclear degrees of freedom are treated classically, propagation of the nuclei along an averaged potential energy surface, in accordance with Ehrenfest dynamics, or along individual surfaces may be realized.

The latter alternative was realized by incorporating the TSH scheme into a DFT framework [551, 552]. This adaptation of the TSH method was shown to succeed by implementing a relatively small number of plausible alterations to its original form. Thus the Kohn–Sham Hamiltonian was employed in conjunction with a basis $\{\tilde{\Psi}_0, \tilde{\Psi}_o^v\}$ of single-determinantal states built from the Kohn–Sham ground state determinant $\tilde{\Psi}_0 = \det[\Phi_1, \Phi_2, \ldots, \Phi_N]$ by occupying virtual orbitals. In more detail, the Kohn–Sham wave function is then given by [552]

$$\Psi(t) = C_0(t)\tilde{\Psi}_0 + \sum_{j=1} C_j(t) \sum_{o,v} c^j_{o,v} \tilde{\Psi}_o^v, \tag{16.61}$$

where the set $\{\tilde{\Psi}_o^v\}$ refers to single excitations of $\tilde{\Psi}_0$,

$$\tilde{\Psi}_o^v = \hat{a}_v^\dagger \hat{a}_o \tilde{\Psi}_0. \tag{16.62}$$

Tully's fewest switches algorithm, as described in Section 11.2, was utilized, implying that the system advances at any given time on a well-defined potential energy surface i defined by $\langle \Psi_i^{\mathrm{a}} | \hat{H} | \Psi_i^a \rangle$, until executing a hop and continuing its course on an alternative surface. This approach was shown to yield the correct time scales for various processes of distinct nonadiabatic character, such as the non-radiative relaxation of the GFP chromophore and the photoinduced electron injection from the alizarin chromophore into a TiO_2 surface [552]. Also, the method confirmed the experimentally secured *Gomer–Noyes mechanism* for the photochemical ring opening of oxirane [553].

The use of the fewest switches TSH formalism in these and other applications of TDDFT to molecular dynamics (e.g. [554, 555]) is predicated on the availability of nonadiabatic coupling terms within TDDFT. As far as interactions between the electronic ground state and excited states are concerned, these can be calculated within linear response theory (see Section 16.3.1). Recent studies on this topic are found in [556]–[560].

As representative of numerous approaches to realize TDDFT within the framework of Ehrenfest dynamics [561, 562], we mention the effort of Tsereteli *et al.* [287] who proposed a DFT version of END theory, adjusting the time-dependent variational principle (9.17) by admitting Kohn–Sham determinants which were parametrized by use of the Thouless representation.

Chapter 17

Decoherence

At various places in this monograph, the question of how to treat the environment of molecular processes involving electronic multistate dynamics has been addressed. Thus, Section 12.6 deals with the pyrazine absorption spectrum in the vicinity of the $S_1 - S_2$ conical intersection as a function of the number of vibrational modes included in the respective SC-IVR computation. In fact, the majority of the dynamic cases considered in the preceding nine chapters, associated with mixed quantum-classical realizations of time-dependent dynamics, may be interpreted in terms of the interaction between an open quantal system (the electrons) and a classical environment (the nuclei). In this context, one has to be careful to understand the notion of an open system properly. The contrasting concept of a closed system clearly does not imply the absence of any interference due to external agents, such as electric or magnetic fields that act from outside on the system and may be accommodated by a perturbing term in the respective Hamiltonian. Rather, the distinguishing feature that sets apart a closed from an open system is the lack of any *back action* on the environment. This constraint on the interaction between the considered quantum object and its surroundings obviously renders the description of quantum reality incomplete. Dynamically, this idealization is manifested by the constraint of unitary evolution of the system for unlimited times.

Physical quantum systems, however, display the phenomenon of *decoherence*. With the elapse of time, interaction with their environment destroys the coherences between the quantum states and leaves the system asymptotically in a definite state which has been characterized as a vector of a preferred basis or *pointer basis* [563]. This term alludes to the essential role that decoherence assumes in the process of quantum measurement. In this context, decoherence accounts for the transition from a coherent manifold of quantum states to a statistical mixture of possible experimental results, represented by different pointer positions of a classical detector. The environment acts here as a monitor which is comparable to a classical observer. While the pointer basis states remain unaffected by the interaction with the environment, their coherences vanish in the course of this process. The survival of these preferred states has been associated with a quasi-Darwinian selection mechanism, and their stability is a recent focus in the continuing debate on locating the demarcation line between the domains of the quantum and the classical [564]. In the mixed quantum-classical theories discussed so far in this text, the nuclear subsystem may be interpreted as a *classical environment* which, if represented properly, causes the quantum mechanical electronic subsystem to decohere.

A consistent quantum approach that encompasses both the system and its environment will include the effect of decoherence (see e.g. [147]). A truncated quantum-classical scheme that operates with classical nuclei to represent the environment, on the other hand, has to introduce decoherence as an artifice imposed on the quantum equations of motion. Any practicable way of implementing the latter methodology is of great relevance in view of the very stringent size limitations on systems that can be described in a fully quantum mechanical fashion at this juncture. The consistent quantal treatment of objects large enough that their environment could be qualified as macroscopic, such as complex biological molecules, is not feasible with present-day means. This situation makes viable models that build decoherence into the quantum formalism highly desirable.

Various current efforts aim at the definition of a theoretical framework for dealing with decoherence as a basic phenomenon whose effects pervade all of nature, the only system excepted from it being the universe itself. The present chapter emphasizes possible ways of introducing decoherence into the equations that govern nonadiabatic molecular dynamics.

In Section 17.1, we will summarize arguments that lead to a phenomenological extension of the Liouville–von Neumann equation by a dissipative term accounting for decoherence. A model that makes explicit the physical mechanism of decoherence in nonadiabatic molecular dynamics by treating the nuclear degrees of freedom in a semiclassical manner is presented in Section 17.1, yielding an estimate for the time scales characteristic of these processes (Section 17.2). These concepts are applied to a physical system, namely the hydrated electron, in Section 17.3. As will be shown in Section 17.4, various trajectory surface hopping algorithms may be viewed as attempts to incorporate decoherence into the propagation of electronic and nuclear degrees of freedom. A method is outlined that represents a compromise between the entirely coherent Ehrenfest and the surface hopping dynamics and thus imports decoherence into the former procedure. We conclude this chapter with a survey of the *decay-of-mixing* formalism (Section 17.5) that achieves decoherent molecular evolution by use of a suitably augmented Ehrenfest approach.

17.1 The Dissipative Liouville–von Neumann Equation

The mixed quantum-classical simulations described in the foregoing chapters lend themselves naturally to dynamical models of molecular evolution in terms of "histories" as defined by the path integral formalism. Following the prescription given by Feynman (Section 4.1), alternative nuclear trajectories are endowed with amplitudes and phases and thus made capable of interference. Phase coherence between different paths corresponds to a state of quantum superposition. Nonadiabatic coupling in molecular systems may be used to illustrate this observation. Let us consider the case of two electronic states, as described by *ab initio* dynamics. Initially, the system may evolve on only one of these surfaces. Passage through a region of sizeable nonadiabatic coupling, however, will populate the second one as well, giving rise to a superposition of both contributing electronic states. If these two states continue to evolve in the same potential energy landscape as defined by the nuclear degrees of freedom, the coherence between them is preserved. However, each of the two states imparts a different force on the nuclear environment, leading to deviating dynamics in the two cases. If a mixed quantum-classical scenario is assumed, which concerns us here, the increasing divergence between the trajectories associated with the electronic states may be viewed as the physical root of the gradual decoherence between these states.

According to the surface hopping model, as described in Chapter 11, a classical trajectory in a landscape of multiple adiabatic potential energy surfaces is pieced together from a variety of path segments, each involving the propagation of the classical system on one specific surface for a specified time interval. Transferring this construction principle into the domain of path integral theory, one arrives at a rich diversity of differing histories. As is intuitively evident, phase coherence is only maintained between histories that are not too dissimilar. A strong divergence between two classical paths will be reflected by a substantially deviating behavior of the respective phase factors, reducing their coherences to zero. This realization begs for a viable definition of the notions *similar* and *dissimilar paths*. We will address this issue in the following. Subsequently, we will deal with the question of how to quantify these notions.

Following the analysis given by Bittner and Rossky [565] which refers to the foregoing work of Ghirardi, Pearle, and Rimini [566], we distinguish *fine-grained* from *coarse-grained* histories. The former are given by individual paths $Q(\mathrm{t})$ connecting an initial state labeled Q_i and a final state Q_f. Each of these may be decomposed into a temporal sequence of quantum numbers $\{\alpha_k\}$, in addition to a variable X which, in mixed quantum-classical molecular dynamics, is related to the classical configuration embedding the quantum subsystem. Thus, in accordance with the notation chosen in this text, X may denote a nuclear coordinate. For simplicity, we consider the case of a one-dimensional trajectory. The $\{\alpha_k\}$ might then be realized by labels of adiabatic electronic states occupied in consecutive time intervals with index k.

Fine-grained histories are characterized by amplitudes, not probabilities. As a general feature of quantum mechanics, generating probabilities requires an act of amplitude reduction, or summation over amplitudes. We refer to this procedure as coarse graining. The latter achieves a partitioning of the whole space of paths into disjoint sets that bundle coherent fine-grained histories, while two different coarse-grained histories are incoherent. Each coarse-grained history is assigned a probability P_n. By virtue of normalization, the probability of some arbitrary path $Q(t)$ to be contained in one of the coarse-grained sets is one:

$$\sum_n P_n[Q(t)] = 1. \tag{17.1}$$

To guarantee the coherence of all paths that are contained in the same coarse-grained histories, one has to limit the maximum admissible deviation between any two of these paths. This is accomplished by defining a succession of windows Δ_α or Δ_β that restrict the allowed variation between two paths belonging to the same coarse-grained set at a given time of the propagation. This idea may be illustrated again by the problem of molecular dynamics involving two adiabatic surfaces, where α and β are understood as surface indices. In this case, the windows Δ_α or Δ_β are interpreted as preset energy ranges associated with the states α and β, respectively. As displayed in Figure 17.1, a given path may be specified by a series of windows traversed consecutively as the system evolves. A coarse-grained history is constituted by all fine-grained histories passing through the same sequence of windows. By varying the window width Δ one may impose more stringent or more liberal standards on the coarse graining procedure. However, the window size has to be chosen small enough to ensure that the decoherence of the coarse-grained histories is complete in the asymptotic limit of the considered process.

The example of electron diffraction by a double-slit apparatus may illustrate how the definition of windows determines the level of coarse graining imposed on a quantum system. If each slit should act as a window that bundles fine-grained histories, the experimental conditions must allow us to specify which one of the two slits has been traversed by the particle. In this case, the alternatives *left* and *right* are incoherent since they define different coarse-grained histories. If, however, the window comprises both slits, the left and the right path of the particle are capable of interference, and both are part of the same coarse-grained bundle.

As suggested by the scheme shown in Figure 17.1, any coarse-grained history is associated with a trajectory surrounded by a "tube" [566] which contains a coherent ensemble of fine-grained histories. To build decoherence into the dynamical equations describing the evolution of the physical system, one installs a selection mechanism whose mathematical form is given by a *reduction map* on each one of these trajectories, defined by the operation

$$\hat{\rho}(t_i) \rightarrow \hat{P}_i^\alpha \hat{\rho}(t_i) \hat{P}_i^\alpha \equiv \hat{R}^\alpha[\rho(t_i)]. \qquad (17.2)$$

This projection eliminates any path which is outside the coarse-grained set determined by the sequence of windows $\{\Delta_\alpha\}$. The projection operator $\hat{P}^\alpha$ may be implemented by a discrete subdivision of the set of all paths into

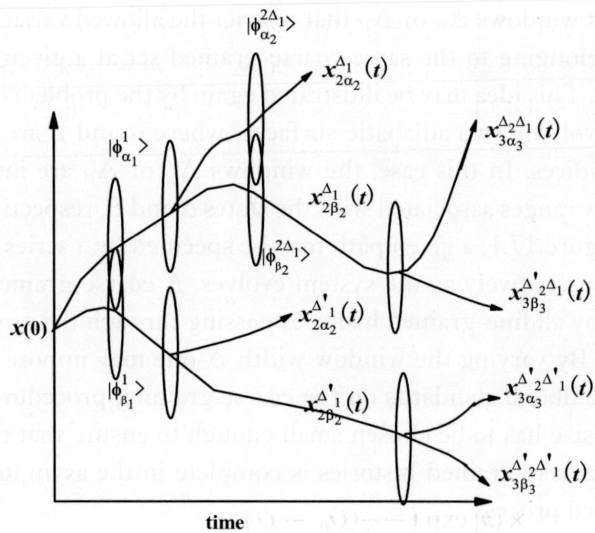

Figure 17.1: Schematic representation of the coarse graining procedure. Histories may be classified according to the sequence of windows traversed as time passes, in the sense that two paths that traverse the same sequence of windows belong to the same coarse-grained history. These windows are displayed by the ovals in the image. Bifurcations, associated with discrete transitions between quantum states α and β, are simulated as discontinuous stochastic events, and each path may be characterized by the sequence of branches followed at the bifurcation points, labeled $\Delta_n\Delta_{n-1}\ldots$. The overlap between two ovals may be viewed as an illustration of the overlap of the wave functions describing the two alternative states. The classical paths diverge in the course of the simulation, resulting in the reduction of this overlap, or in the decoherence of the system as a whole. (Reprinted with permission from [565]. Copyright (1995) by the American Institute of Physics.)

those that belong to $\{\Delta_\alpha\}$ and those that do not:

$$\hat{P}_i^\alpha(Q) = \begin{cases} 1 & \text{if} \quad Q\,\varepsilon\{\Delta_\alpha\} \\ 0 & \text{otherwise.} \end{cases} \tag{17.3}$$

The exact demarcation line drawn by the choice (17.3) between the realms of included and excluded paths is physically unmotivated. A continuous alternative, blurring the sharply edged frame of the window according to Eq. (17.3), is provided by the Gaussian operator

$$\hat{P}_i^{\mathrm{G}}(Q) = \left(\frac{\alpha}{\pi}\right)^{\frac{1}{4}} \exp\left(-\frac{\alpha}{2}(\hat{Q} - Q)^2\right) \tag{17.4}$$

where $\hat{Q}$ stands for a bounded operator in Hilbert space with a discrete spectrum $\{Q_i\}$, and the quantity $\frac{1}{\sqrt{\alpha}}$ takes the place of the discontinuous range Δ_α employed in Eq. (17.3), providing a measure of the window size. Using Eq. (17.4) to perform the mapping operation, we arrive at

$$\hat{R}^{\mathrm{G}}[\hat{\rho}] = \sqrt{\frac{\alpha}{\pi}} \int dQ \, \exp\left(-\frac{\alpha}{2}(\hat{Q} - Q)^2\right) \hat{\rho} \exp\left(-\frac{\alpha}{2}(\hat{Q} - Q)^2\right). \quad (17.5)$$

Inserting the complete basis $\sum_m |m\rangle\langle m|$ of eigenvectors of $\hat{Q}$ twice into this expression one finds

$$\begin{aligned}
\hat{R}^{\mathrm{G}}[\hat{\rho}] &= \sqrt{\frac{\alpha}{\pi}} \sum_{m,n} \int dQ \, \exp\left(-\frac{\alpha}{2}(Q_m - Q)^2\right) |m\rangle\langle m|\hat{\rho}|n\rangle \\
&\quad \times \langle n| \exp\left(-\frac{\alpha}{2}(Q_m - Q)^2\right) \\
&= \sqrt{\frac{\alpha}{\pi}} \sum_{m,n} \int dQ \, \exp\left(-\frac{\alpha}{2}(Q_m - Q)^2\right) |m\rangle \hat{\rho}_{mn} \\
&\quad \times \langle n| \exp\left(-\frac{\alpha}{2}(Q_n - Q)^2\right) \\
&= \exp\left(-\frac{\alpha}{4}(Q_m - Q_n)^2\right) \sum_{m,n} |m\rangle \hat{\rho}_{mn} \langle n| \qquad (17.6)
\end{aligned}$$

where, in the last step, we have utilized the integral formula for the overlap of two Gaussian functions with different centers [567]. We may thus write the projected density operator $\hat{R}^{\mathrm{G}}[\hat{\rho}]$ in the form of a matrix with elements

$$R^{\mathrm{G}}[\hat{\rho}]_{mn} = \exp\left(-\frac{\alpha}{4}(Q_m - Q_n)^2\right) \rho_{mn}. \quad (17.7)$$

This result demonstrates how the reduction map (17.5) diminishes the coherences of the evolving system. Obviously, the populations ρ_{nn} are unaffected by this projection. Formula (17.7) clarifies explicitly the role of the parameter α as determining the extent of decoherence and, by the same token, the level of coarse graining. It ranges between the extreme cases of $\alpha = 0$ and $\alpha = \infty$, corresponding to the alternatives of a coherent quantum system and a statistical mixture, respectively. Below, we will endow this construct with physical content by applying it to nonadiabatic dynamics.

The action of the map $\hat{R}$ can be understood in terms of a (partial) *collapse of the wave function*, restricting quantum interference to the paths contained in the coarse-grained histories. This feature draws a parallel between the effect of the presently discussed reduction map and the act of measurement. Figuratively speaking, the quantum subsystem is periodically monitored by its classical environment, or *bath*, resulting in coherence loss and thus a certain amount of dissipation. In this model, coarse graining may be imprinted on the system at any given moment. In the simplest statistical treatment, these events are random and uncorrelated. In this case, their probability may be described by a Poisson distribution with time scale τ_{D} [565]:

$$P_{cg} = 1 - \exp\left(-\frac{t}{\tau_{\mathrm{D}}}\right) \tag{17.8}$$

where P_{cg} stands for the likelihood of the occurrence of such a coarse-graining event. For sufficiently short times Δt, the reduction to an ensemble described by $\hat{R}^{\mathrm{G}}[\hat{\rho}]$ happens with the probability $\Delta t/\tau_{\mathrm{D}}$. Consequently, the completely coherent system is assigned a probability $1-\Delta t/\tau_{\mathrm{D}}$. In summary, we arrive at the statement

$$\hat{\rho}(t+\Delta t) = \left(1 - \frac{\Delta t}{\tau_{\mathrm{D}}}\right)(\hat{\rho}(t) + \Delta\hat{\rho}_{\mathrm{c}}) + \frac{\Delta t}{\tau_{\mathrm{D}}}\hat{R}^{\mathrm{G}}[\hat{\rho}]. \tag{17.9}$$

Here $\Delta\hat{\rho}_{\mathrm{c}}$ stands for the change undergone by the density matrix under the constraint of entirely coherent evolution during the time interval $[t, t+\Delta t]$. To first order in Δt, this increment is $-\frac{i}{\hbar}[\hat{H},\hat{\rho}]\Delta t$. Therefore,

$$\hat{\rho}(t+\Delta t) = \left(1 - \frac{\Delta t}{\tau_{\mathrm{D}}}\right)\left(\hat{\rho}(t) - \frac{i}{\hbar}[\hat{H},\hat{\rho}]\Delta t\right) + \frac{\Delta t}{\tau_{\mathrm{D}}}\hat{R}^{\mathrm{G}}[\hat{\rho}]. \tag{17.10}$$

Shrinking Δt to dt we obtain

$$\frac{d\hat{\rho}}{dt} = -\frac{i}{\hbar}[\hat{H},\hat{\rho}] - \kappa(\hat{\rho} - R^{\mathrm{G}}[\hat{\rho}]) \tag{17.11}$$

with $\kappa = \frac{1}{\tau_{\mathrm{D}}}$. This relation represents a master equation for the quantum density matrix which includes the effect of decoherence. As one neglects the second term on the right-hand side, Eq. (17.11) reduces to the Liouville–von Neumann equation. Inclusion of this term introduces dissipation into the Liouville–von Neumann equation and destroys its unitarity. The unitary limit is attained in the case of an infinitely long decoherence time

($\tau_D \to \infty$) or, with reference to Eq. (17.7), infinite width of the windows that characterize the coarse-graining procedure applied ($\alpha \to 0$), completely dissolving the coarse-grained sets into multiplicities of fine-grained histories. In the general situation of finite τ_D, corresponding to a deviation between $\hat{\rho}$ and $\hat{R}^G[\hat{\rho}]$, however, (17.11) describes dissipative dynamics, reflecting the decoherence phenomenon.

To apply this formalism to mixed quantum-classical molecular dynamics we parametrize the density operator in Eq. (17.11) by the classical trajectory $X(t)$. We employ a basis of adiabatic eigenfunctions ϕ_i of $\hat{H}$ with eigenenergies E_i. This approach turns the Liouville–von Neumann equation

$$\frac{\partial \hat{\rho}(X(t))}{\partial t} = -\frac{i}{\hbar}[\hat{H}(X(t)), \hat{\rho}(X(t))] \tag{17.12}$$

into

$$\begin{aligned}\frac{\partial \rho_{ij}}{\partial t} = -\frac{i}{\hbar}\sum_k \Bigg[& \left(E_i(X(t))\delta_{ik} + i\hbar\frac{\partial X}{\partial t}A_{ik}\right)\rho_{kj} \\ & - \rho_{ik}\left(E_i(X(t))\delta_{ik} + i\hbar\frac{\partial X}{\partial t}A_{kj}\right)\Bigg],\end{aligned} \tag{17.13}$$

where A_{ik} denotes the nonadiabatic coupling element (2.10):

$$A_{ik} = \langle \phi_i(x, X(t)) \left| \frac{\partial}{\partial X} \right| \phi_j(x, X(t)) \rangle. \tag{17.14}$$

Exercise 17.1. Derive Eq. (17.13) from Eq. (17.12), representing the density operator in the adiabatic basis $\{\phi_i\}$, and exploiting the implicit time dependence of the eigenstates $\phi_i \equiv \phi_i(X(t))$.

Using Eq. (17.11), we extend the pure Liouville formalism to make allowance for decoherence:

$$\begin{aligned}\frac{\partial \rho_{ij}}{\partial t} = -\frac{i}{\hbar}\sum_k \Bigg[& \left(E_i(X(t))\delta_{ik} + i\hbar\frac{\partial X}{\partial t}A_{ik}\right)\rho_{kj} \\ & - \rho_{ik}\left(E_i(X(t))\delta_{ik} + i\hbar\frac{\partial X}{\partial t}A_{kj}\right)\Bigg] \\ & - \kappa(\rho_{ij} - R[\rho(X(t))]_{ij})\end{aligned} \tag{17.15}$$

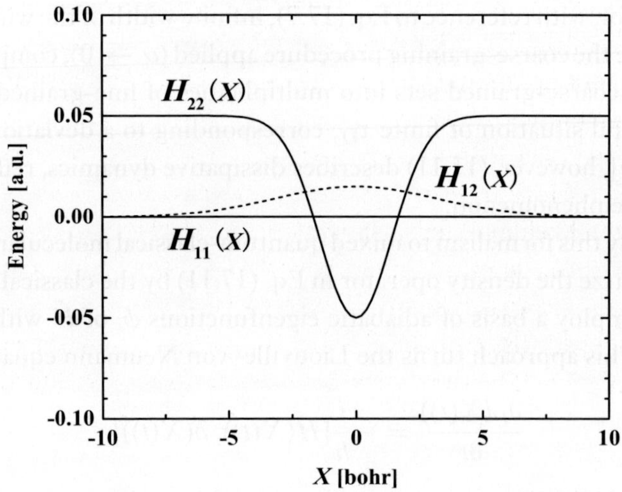

Figure 17.2: Representation of the elements of the diabatic Hamiltonian matrix (17.16). The diabatic coupling H_{12} has been scaled by a factor of 12. (Reprinted with permission from [565]. Copyright (1995) by the American Institute of Physics.)

with R_{ij} as the matrix element of the reduction map $\hat{R}$. For illustration of this approach, we consider an elementary model problem, involving a dual avoided crossing situation, as shown in Figure 17.2.

The Hamiltonian discussed in [565] is given by the diabatic matrix elements

$$H_{11}(X) = 0, \quad H_{22}(X) = -a\exp(-bX^2) + E_0,$$
$$H_{12}(X) = H_{21}(X) = c\exp(-dX^2) \tag{17.16}$$

where the constants have been chosen as $a = 0.1$, $b = 0.28$, $E_0 = 0.28$, $c = 0.05$, and $d = 0.06$, in atomic units. Diagonalizing this Hamiltonian yields the adiabatic energies E_i ($i = 1, 2$) contained in Eq. (17.15). The coarse-graining procedure consists here in a projection on the discrete adiabatic levels. With recourse to Eq. (17.7), and identifying the spectrum $\{Q_i\}$ with the energy sequence $\{E_i\}$, we write the matrix elements of the reduction map as

$$R^{\mathrm{G}}[\rho]_{ij} = \exp\left(-\frac{\alpha}{4}(E_i - E_j)^2\right)\rho_{ij}. \tag{17.17}$$

For a given parameter $\frac{1}{\sqrt{a}}$ which may be understood as an observational inaccuracy, suppression of the coherences will strongly increase with the difference between the energies E_i and E_j. Employing this form of the reduction term, Bittner and Rossky [565] propagated the density matrix elements according to Eq. (17.15). The evolution of the classical subsystem was started on the lowest energy branch, and the trajectories $X(t)$ were generated by application of Tully's fewest switches algorithm ([288], Section 11.2). This approach allows us to compute the transmission probability T_{11} which represents the likelihood for a classical particle to be in the state of lowest energy after traversal of the interaction zone. In the chosen energy range, reflection turns out to be sufficiently small to justify the approximation $T_{12} \approx 1 - T_{11}$ for the probability of finding the system in the excited state after transmission. The simulation has been performed as a function of the initial energy and with the decoherence time as parameter. The result is displayed in Figure 17.3.

In the completely coherent limit, corresponding to $\tau_D \to \infty$, T_{11} is characterized by sizeable oscillations, indicating the presence of marked interference between the two electronic states. Adopting as a standard the time needed by a classical particle to traverse the distance between the two

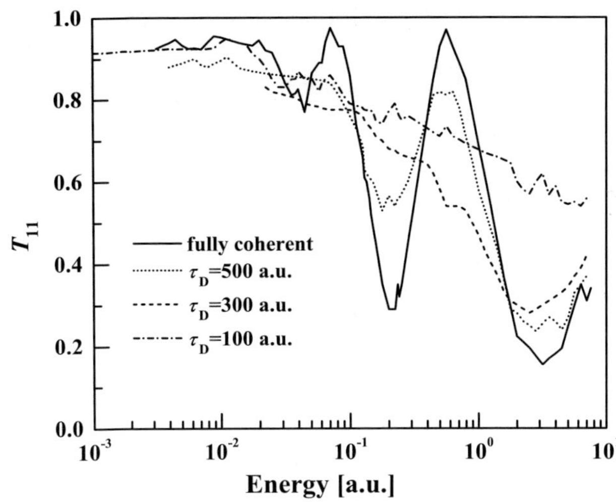

Figure 17.3: The transmission probability T_{11} of a classical particle traversing the two-level quantum system described by Eq. (17.16) versus initial energy. The calculation is based on the decoherence model realized by Eq. (17.15). The decoherence time τ_D is used as parameter. (Reprinted with permission from [565]. Copyright (1995) by the American Institute of Physics.)

potential curve crossing points (see Figure 17.2), which amounts to about 450 a.u. for an intermediate energy of 0.225 a.u., we find the oscillation amplitudes substantially reduced for comparable decoherence times and ultimately vanishing for times that are small by this measure. The considered interval of the quantity τ_{D} ranges from the limiting case of a long-lived quantum superposition through weak coarse graining to the entirely incoherent extreme of strong coarse graining. The effect of decoherence on the asymptotic ground state population following the passage through the interaction zone can be easily inferred from the transmission elements for any initial energy of interest.

From the foregoing, how can we characterize the limit of extremely short decoherence times ($\kappa \to \infty$) in combination with a low level of coarse graining as determined by the parameter $\alpha(\alpha \to 0)$? We revert to Eq. (17.5), identifying the operator $\hat{Q}$ with the Hamiltonian and the variable Q with the energy E. The reduction map is strategically rewritten as

$$\hat{R}^{\text{G}}[\hat{\rho}] = \left(\frac{\alpha}{\pi}\right)^{\frac{1}{2}} \exp\left(-\frac{\alpha}{2}\hat{H}^2\right) \int dE \exp\left(-\frac{\alpha}{2}E^2\right) \exp(\alpha\hat{H}E)\hat{\rho}$$

$$\times \exp(-\alpha\hat{H}E) \exp\left(\frac{\alpha}{2}E^2\right) \exp\left(-\frac{\alpha}{2}\hat{H}^2\right). \qquad (17.18)$$

We use the assumption that α is small by expanding $\exp(\alpha\hat{H}E)$ according to [566]:

$$\exp(\alpha\hat{H}E) = 1 + \alpha\hat{H}E + \frac{1}{2}\alpha^2(\hat{H}E)^2, \qquad (17.19)$$

employing the identities

$$\left(\frac{\alpha}{\pi}\right)^{\frac{1}{2}} \int_{-\infty}^{\infty} dE \exp(-\alpha E^2) = 1, \qquad (17.20)$$

$$\left(\frac{\alpha}{\pi}\right)^{\frac{1}{2}} \int_{-\infty}^{\infty} dEE \exp(-\alpha E^2) = 0, \qquad (17.21)$$

$$\left(\frac{\alpha}{\pi}\right)^{\frac{1}{2}} \int_{-\infty}^{\infty} dEE^2 \exp(-\alpha E^2) = \frac{1}{2\alpha}. \qquad (17.22)$$

In general, the expression $(\frac{\alpha}{\pi})^{\frac{1}{2}} \int_{-\infty}^{\infty} dE\ E^N \exp(-\alpha E^2)$ vanishes if N is odd and behaves as $O(\alpha^{\frac{N}{2}})$ if N is even. Relations (17.19)–(17.22) make it straightforward to verify that $\hat{R}^{\mathrm{G}}$ reduces in the limit of small α to

$$\hat{R}^{\mathrm{G}}[\hat{\rho}] = \exp\left(-\alpha\frac{\hat{H}^2}{2}\right)\left[\hat{\rho} + \frac{1}{2}\alpha\left(\hat{H}\hat{\rho}\hat{H} + \frac{1}{2}[\hat{H}^2, \hat{\rho}]_+\right) + O(\alpha^2)\right] \times \exp\left(-\alpha\frac{\hat{H}^2}{2}\right). \tag{17.23}$$

Expanding $\exp(-\alpha\frac{\hat{H}^2}{2})$ in analogy to Eq. (17.19) and truncating after terms linear in α, one obtains

$$\hat{R}^{\mathrm{G}}[\rho] = \hat{\rho} + \frac{1}{2}\alpha\left(\hat{H}\hat{\rho}\hat{H} - \frac{1}{2}[\hat{H}^2, \hat{\rho}]_+\right) + O(\alpha^2). \tag{17.24}$$

Inserting this expression into the equation of motion (17.11), we find

$$\frac{d\hat{\rho}}{dt} = -\frac{i}{\hbar}[\hat{H}, \hat{\rho}] + \frac{1}{2}\alpha\kappa\left(\hat{H}\hat{\rho}\hat{H} - \frac{1}{2}[\hat{H}^2, \hat{\rho}]_+\right) \tag{17.25}$$

up to linear order in α.

Exercise 17.2. Show that Eq. (17.25) can be cast into the compact form $\frac{d\hat{\rho}}{dt} = -\frac{i}{\hbar}[\hat{H}, \hat{\rho}] - \frac{1}{4}\alpha\kappa[\hat{H}, [\hat{H}, \hat{\rho}]]$.

The product of the two characteristic constants governing the decoherence process, α and κ, determines the *infinite frequency limit* as given by Eq. (17.25). Here the environment continuously monitors the quantum system but does so with very low observation accuracy. The result (17.25) can also be derived by adopting a statistical model for a bath that is coupled linearly to the quantum subsystem in the instantaneous force f [568]. The domain of validity for the approximation (17.25) turns out to be a model that treats the bath as a predefined set of Brownian oscillators and prescribes linear interaction between the quantum and the classical elements of the system. The more general formalism (17.11) removes these two limiting assumptions and allows us to assess the response of the bath to the dynamics of the quantum component. The approach, however, is instantaneous in the bath fluctuations and thus local in time. Extending the picture beyond the Markovian limit by admitting temporally correlated histories complicates the theory considerably.

17.2 Evaluating Decoherence Times in a Semiclassical Framework

Implementing the dissipative Liouville–von Neumann equation (17.15) allows us in principle to simulate both the emergence and the decay of quantum superpositions in a molecular ensemble under conditions of nonadiabaticity. Once a reduction map has been defined, the decoherence time is the crucial piece of information that must be available for propagation of the density matrix in a given landscape of adiabatic potential energy surfaces. It is plausible that the time behavior of the transition rates between adiabatic states, as a measure of the coupling between them, yields a quantitative estimate of this basic constant. In this section, we will outline an elementary approximation for these rates by use of Fermi's golden rule. This approach will clarify how to arrive at the time scale of the decoherence process, and simultaneously shed light on the decoherence mechanism from a less abstract perspective than adopted in the previous discussion.

Naturally, the use of perturbation theory at first order is restricted to cases of weak interaction. The following discussion rests on this assumption. We maintain the earlier subdivision into an electronic quantum system and a nuclear bath described by classical coordinates and introduce an adiabatic electronic basis $\{\phi_i\}$ such that

$$\hat{H}_{\mathrm{el}}(\mathbf{x}, \mathbf{X})|\phi_i\rangle = E_i|\phi_i\rangle, \tag{17.26}$$

where the total Hamiltonian is composed of the electronic contribution and the nuclear kinetic energy operator according to $\hat{H} = \hat{H}_{\mathrm{el}}(\mathbf{x}, \mathbf{X}) + \hat{T}(\mathbf{X})$. We introduce operators $\hat{T}_{ij}$ that act on the nuclear wave functions χ_K by stipulating [569]

$$\hat{T}_{ij} = \langle\phi_i|\hat{T}|\phi_j\rangle = -\sum_n \frac{\hbar^2}{2M_n}\left\langle\phi_i\left|\frac{\partial^2}{\partial\mathbf{X}_n^2}\right|\phi_j\right\rangle. \tag{17.27}$$

The golden rule expression for the rate of nonadiabatic transitions from an initial state to another state, labeled i and j, respectively, is

$$k_{ij} = \frac{2\pi}{\hbar}\sum_F |\langle\phi_i\chi_I|\hat{T}|\phi_j\chi_F\rangle|^2\delta(E_{iI} - E_{jF}) \tag{17.28}$$

with E_{iI}, E_{jF} as the initial and final energies, respectively, of the system as a whole. Introducing a thermal average, denoted by the subscript T, over the electronic and nuclear initial states $|\phi_j\rangle|\chi_F\rangle$, respectively, and inserting (17.27), this becomes ([570], compare with Eq. (5.75))

$$\bar{k}_{ij} = \frac{2\pi}{\hbar}\sum_I \frac{\exp(-\beta E_{iI})}{Z_i}\sum_F\sum_n \frac{\hbar^2}{2M_n} \times \left|\left\langle \phi_i\chi_I \left| \frac{\partial^2}{\partial \mathbf{X}_n^2}\right| \phi_j\chi_F\right\rangle\right|^2 \delta(E_{iI} - E_{jF})$$
$$\equiv \frac{2\pi}{\hbar}\left\langle \sum_F\sum_n \frac{\hbar^2}{2M_n}\left|\left\langle \phi_i\chi_I \left| \frac{\partial^2}{\partial \mathbf{X}_n^2}\right| \phi_j\chi_F\right\rangle\right|^2 \delta(E_{iI} - E_{jF})\right\rangle_T \tag{17.29}$$

where Z_i denotes the partition function for the nuclear states in the initial electronic state $|i\rangle$, and $\bar{k}_{ij}$ denotes the thermally averaged rate constant. We restrict the electronic basis to the two vectors $|1\rangle$ and $|2\rangle$ and recast the respective rate $\bar{k}_{12}$ by employing the Fourier representation of the delta function $\delta(E_{iI} - E_{jF}) = 1/2\pi\hbar \int dt \ \exp[\frac{i}{\hbar}(E_{2F} - E_{1I})t]$, having summed over the final nuclear quantum numbers F to obtain

$$\bar{k}_{12} = \frac{1}{\hbar^2}\int_{-\infty}^{+\infty} dt \langle \chi_I(\mathbf{X})|\hat{T}_{12}\exp\left(\frac{i}{\hbar}\hat{H}_2 t\right)\hat{T}_{21}\exp\left(-\frac{i}{\hbar}\hat{H}_1 t\right)|\chi_I(\mathbf{X})\rangle_T. \tag{17.30}$$

By $\hat{H}_1$ and $\hat{H}_2$ we understand the nuclear Hamiltonians related to the first and the second potential energy surface, respectively:

$$\hat{H}_k = \hat{T} + \langle \phi_k|\hat{H}_{\rm el}|\phi_k\rangle, \quad k = 1, 2, \ldots \tag{17.31}$$

The two Hamiltonians refer to a common energy origin. The scalar product $\langle \chi_I(\mathbf{X})|\hat{T}_{12}|\chi_F(\mathbf{X})\rangle$ is approximated by neglecting any second derivatives of the electronic wave functions. Note that the latter form an orthogonal set, as implied by the assumption of adiabaticity. We arrive at

$$\langle \chi_I(\mathbf{X})|\hat{T}_{12}|\chi_F(\mathbf{X})\rangle \approx \hbar^2 \sum_n \frac{\mathbf{F}_n}{M_n}\langle \chi_I|\hat{\mathbf{P}}_n|\chi_F\rangle \tag{17.32}$$

where we identify $\hat{\mathbf{P}}_n$ with nuclear momentum operators and define

$$\mathbf{F}_n \equiv \frac{1}{\hbar}\langle \phi_1|\mathbf{P}_n|\phi_2\rangle. \tag{17.33}$$

In terms of the force matrix element $\langle\phi_1|\frac{\partial\hat{H}_{\rm el}}{\partial\mathbf{X}_n}|\phi_2\rangle$ involving the two adiabatic states, $\mathbf{F}_n$ becomes by Eq. (2.61)

$$\mathbf{F}_n = i\frac{\left\langle\phi_1\left|\frac{\partial\hat{H}_{\rm el}}{\partial\mathbf{X}_n}\right|\phi_2\right\rangle}{E_1 - E_2}. \tag{17.34}$$

The initial and final nuclear wave functions may be factored into the contributions χ_{Kn} of the individual nuclear modes, with $K = I, F$, and $n = 1, 2, 3, \ldots$ [571],

$$\chi_I = \chi_{I1}\chi_{I2}\chi_{I3}\cdots, \quad \chi_F = \chi_{F1}\chi_{F2}\chi_{F3}\cdots, \tag{17.35}$$

to generate the following expression for the momentum matrix element in Eq. (17.32)

$$\langle\chi_I|\hat{\mathbf{P}}_n|\chi_F\rangle = \frac{\langle\chi_{In}|\hat{\mathbf{P}}_n|\chi_{Fn}\rangle}{\langle\chi_{In}|\chi_{Fn}\rangle}\prod_m\langle\chi_{Im}|\chi_{Fm}\rangle. \tag{17.36}$$

So far, our derivation has not made use of any semiclassical arguments. In the following step, however, we parametrize the $\mathbf{F}$ matrix elements with a classical trajectory: $\mathbf{F}_n = \mathbf{F}_n(\mathbf{X}(t))$. Further, we replace $\langle\chi_I|\mathbf{P}_n|\chi_F\rangle/M_n$ with a classical velocity vector $\mathbf{v}_n$. With these substitutions, Eq. (17.30) turns into

$$\begin{aligned}
\bar{k}_{12} &\approx \int_{-\infty}^{+\infty} dt\left\langle\sum_{n,n'}\left\langle\chi_I\left|\exp\left(\frac{i}{\hbar}\hat{H}_2t\right)\exp\left(\frac{-i}{\hbar}\hat{H}_2t\right)\mathbf{F}_n(\mathbf{X}(0))\cdot\mathbf{v}_n(0)\right.\right.\right.\\
&\quad\left.\left.\left.\times\exp\left(\frac{i}{\hbar}\hat{H}_2t\right)\mathbf{F}_{n'}(\mathbf{X}(0))\cdot\mathbf{v}_{n'}(0)\exp\left(-\frac{i}{\hbar}\hat{H}_1t\right)\right|\chi_I\right\rangle\right\rangle_T\\
&= \int_{-\infty}^{+\infty} dt\left\langle\sum_{n,n'}\left\langle\chi_I(\mathbf{X})\right|\exp\left(\frac{i}{\hbar}\hat{H}_2t\right)\mathbf{F}_n(\mathbf{X}^{(2)}(t))\cdot\mathbf{v}_n^{(2)}(t)\right.\\
&\quad\left.\times\mathbf{F}_{n'}^{(1)}(\mathbf{X}(0))\cdot\mathbf{v}_{n'}^{(1)}(0)\exp\left(-\frac{i}{\hbar}\hat{H}_1t\right)\chi_I\right\rangle\Big\rangle_T\\
&= \int_{-\infty}^{+\infty} dt\left\langle\sum_{n,n'}\mathbf{F}_n(\mathbf{X}^{(2)}(t))\cdot\mathbf{v}_n^{(2)}(t)\mathbf{F}_{n'}^{(1)}(\mathbf{X}(0))\cdot\mathbf{v}_{n'}^{(1)}(0)\mathcal{J}(t)\right\rangle_T.
\end{aligned} \tag{17.37}$$

The factor $\mathcal{J}(t)$ is given by

$$\mathcal{J}(t) \equiv \left\langle \chi_I \exp\left(\frac{i}{\hbar}\hat{H}_2 t\right) \exp\left(-\frac{i}{\hbar}\hat{H}_1 t\right) |\chi_I\right\rangle. \tag{17.38}$$

We recognize in this term the overlap between the nuclear wave functions propagated on the two different surfaces through a time t, where $\mathbf{X}^{(i)}$ ($\mathbf{v}^{(i)}$) stands for the trajectory (the velocity vector) transported on the potential energy surface with label i. Integration with respect to the nuclear coordinates is implied. We conclude from Eq. (17.37) that the nonadiabatic transition rate is determined by the time-dependent correlation between the semiclassical scalar product $\mathbf{F} \cdot \mathbf{v}$ computed on potential energy surface 1 at $t = 0$ and propagated through a time t on surface 2. This correlation is expected to decay as time proceeds. It is further plausible that the rate increases with the projection of $\mathbf{v}$, the nuclear velocity vector, on the vector of nonadiabatic coupling $\mathbf{F}$, since the deviation from adiabatic behavior becomes more pronounced as this nuclear velocity component does. Relation (17.37) makes explicit that not only the strength of the nonadiabatic coupling decides the transition rate between the two surfaces but also the time during which the surfaces remain correlated. The formalism embodied by Eq. (17.37) allows us to distinguish between a semiclassical and a modulating quantum mechanism of decoherence, where the latter is described by the overlap integral (17.38) which provides an estimate for the rate of correlation loss due to the deviation between the nuclear wave functions evolving on the two different electronic surfaces. We point out that the computation of the integral (17.38) requires knowledge only of the initial nuclear wave function χ_I, i.e. at $t = 0$ when $\mathcal{J}(t)$ reduces to one.

For practical implementation of the golden rule approach to decoherence, the frozen Gaussian approximation of Heller [572] has been utilized [570]. This procedure appears justified for processes associated with time scales so short that the effect of wave packet spreading can be disregarded. In the case study presented in Section 17.4, involving the solvated electron problem, typical decoherence times are in the femtosecond range, rationalizing the use of the frozen Gaussian formalism.

We proceed to show that this approximation leads to a simple factorization of the correlation function $\mathcal{J}(t)$. Specifically, the frozen Gaussian approach proposes a nuclear wave function in the form of the product

$$\chi_I^k(\mathbf{X}(t)) = \prod_n G_n^{(k)}(\mathbf{X},\mathbf{P};t)\exp\left[\frac{i}{\hbar}S^{(k)}(\mathbf{X}^{(k)}(t))\right], \tag{17.39}$$

referring to a nuclear subsystem that evolves on an electronic surface with label k. A Gaussian factor, given by (see Eq. (9.22))

$$G_n^{(k)}(\mathbf{X},\mathbf{P};t) = \left(\frac{\alpha_n}{\pi}\right)^{\frac{3}{4}}\exp\left[-\left(\frac{\alpha_n}{2}\right)(\mathbf{X}_n - \mathbf{X}_n^{(k)}(t))^2 + \frac{i}{\hbar}\mathbf{P}_n^{(k)}(t)\cdot(\mathbf{X}_n - \mathbf{X}_n^{(k)}(t))\right], \tag{17.40}$$

is assigned to each nuclear center n. The phase factors depend on the action integral which we write as

$$S^{(k)}(\mathbf{X}^{(k)}(t)) = \int_0^t dt' \sum_n \frac{P_n^{(k)}(t')^2}{2M_n} - \left\langle \prod_n G_n^{(k)}|\hat{H}_k|\prod_n G_n^{(k)}\right\rangle. \tag{17.41}$$

In the subtracted term we recognize an expectation value of the total nuclear energy, and the integration in the scalar product is thus with respect to the nuclear coordinates. Summarizing Eqs. (17.38) to (17.41), the frozen Gaussian approximation yields an expression for $\mathcal{J}(t)$ that is naturally factorized into an overlap and a phase part [571]:

$$\mathcal{J}(t) \approx \left\langle \prod_m G_m^{(2)}(\mathbf{X}_m^{(2)},\mathbf{P}_m^{(2)};t)|\prod_n G_n^{(1)}(\mathbf{X}_m^{(1)},\mathbf{P}_m^{(1)};t)\right\rangle \times \exp\left[\frac{i}{\hbar}(S^{(2)}(\mathbf{X}^{(2)}(t)) - S^{(1)}(\mathbf{X}^{(1)}(t)))\right] = \mathcal{J}_{\text{overlap}}(t)\mathcal{J}_{\text{phase}}(t). \tag{17.42}$$

Using relation (17.41) between the action and the Lagrangian function, we realize that for sufficiently short evolution times, the exponential factor in Eq. (17.42) may be replaced by [571]

$$\exp\left[\frac{i}{\hbar}(\Delta E_{12}(t) - \Delta K_{12}(t))t\right] \tag{17.43}$$

where ΔK_{12} is the difference in nuclear kinetic energy evaluated along the trajectory, and ΔE_{12} the corresponding potential energy difference.

We further separate $\mathcal{J}(t)$ into an electronic part that involves ΔE_{12} and a purely nuclear part $D(t)$, according to

$$\mathcal{J}(t) = \exp\left[\frac{i}{\hbar}\Delta E_{12}(t)t\right] D(t), \tag{17.44}$$

and define

$$D(t) \equiv \exp\left[-\frac{i}{\hbar}\Delta K_{12}(t)t\right] \mathcal{J}_{\text{overlap}}(t) \tag{17.45}$$

as the *decoherence function*. It may be viewed as the time domain counterpart of the Franck–Condon factor which refers to the frequency domain [573]. The separation (17.44) amounts to a distinction between dephasing, given by the first factor on the right-hand side, and decoherence, encapsulated by the function $D(t)$. We emphasize the conceptual difference between these two sources of the considered non-radiative transition. The first component can be derived within the framework of unitary, entirely coherent quantum evolution, as described by the Liouville–von Neumann equation. The second component, however, involves a back action of the electronic degrees of freedom on their nuclear environment, as reflected by the time-dependent discrepancy between the nuclear kinetic energies on the two surfaces as well as the variation of the nuclear overlap. The decoherence function is thus entirely composed of nuclear quantities which, in the semiclassical model considered here, are accessible by molecular dynamics simulations in the given landscape of potential energy surfaces. The divergence of the bath trajectories with the elapse of time affects both ingredients of $D(t)$, reducing the wave function overlap and enhancing the discrepancy between the kinetic nuclear energies on surfaces 1 and 2, and thus determines the overall coherence loss in the studied ensemble. By observing the time dependence of the function D, one derives an estimate of the decoherence time which subsequently can be used to propagate the density matrix according to the extended Liouville–von Neumann scheme (17.15).

The decoherence time may be obtained by a number of consecutive approximations. In the first step, one makes reference to (17.37), assuming the ensemble average of $D(t)$ to be separable from that of the remaining rate expression:

$$\bar{k}_{12} \approx \int_{-\infty}^{+\infty} dt \Big\langle \sum_{n,n'} \mathbf{F}_n(\mathbf{X}^{(2)}(t)) \cdot \mathbf{v}_n^{(2)}(t) \mathbf{F}_{n'}(\mathbf{X}^{(1)}(0)) \cdot \mathbf{v}_{n'}^{(1)}(0)$$

$$\times \exp\left[\frac{i}{\hbar}\Delta E_{12}(t)t\right]\Big\rangle_T \langle D(t)\rangle_T. \quad (17.46)$$

This product involves a separation between the two conceptually distinct sources of coherence loss in a quantum-classical ensemble. The first factor represents the transition rate according to the Liouville–von Neumann equation. It contains the effect of *pure dephasing* [573], resulting from fluctuations of the electronic energy eigenvalues. The other cause for dissipation, namely that arising from the divergent evolution of nuclear degrees of freedom on the two potential energy surfaces, is encoded in the second factor.

An alternative treatment of the nonadiabatic transition was presented by Staib and Borgis [574] who also rely on Fermi's golden rule but do not give explicit consideration to the nuclear wave functions. The result is identical with the right-hand side of Eq. (17.46) without the factor $\langle D(t)\rangle_T$. This approach, in other words, includes electronic relaxation but no loss of coherence on account of the electronic back action on the nuclear or environmental degrees of freedom.

17.2.1 *Ensemble average of the decoherence function*

The task of deriving the ensemble average of $D(t)$ is greatly simplified by casting it into the form of an exponential function:

$$D(t) = \exp\left[-\frac{i}{\hbar}\Delta K_{12}(t)t + \ln\left[\mathcal{J}_{\text{overlap}}(t)\right]\right]. \quad (17.47)$$

In the following step, we make use of the *cumulant expansion* in order to find the average of $D(t)$. Thus,

$$\langle \exp(S)\rangle = \exp\left[\sum_{n=1} \frac{1}{n!}\langle S^n\rangle_{\text{c}}\right] \quad (17.48)$$

where $\langle S^n\rangle_{\text{c}}$ is the nth *cumulant of the quantity* S. The first cumulant of S is the average $\langle S\rangle$, the second is the squared variance $\langle S^2\rangle - \langle S\rangle^2$, etc.

Assuming that $\langle D(t)\rangle_T$ is close to the first term of the cumulant expansion, we write

$$\langle D(t)\rangle_T \approx \exp\left[-\frac{i}{\hbar}\langle \Delta K_{12}(t)\rangle_T t + \langle \ln[\mathcal{J}_{\mathrm{overlap}}(t)]\rangle_T\right]. \tag{17.49}$$

For evaluating the two thermal averages on the right-hand side of this equation, we employ the following series representations of the trajectories $\mathbf{X}_{kn}(t)$ and the corresponding momenta $\mathbf{P}_{kn}(t)$ with $k = 1, 2$ [571]:

$$\mathbf{X}_{kn}(t) = \mathbf{X}_{kn}(0) + \frac{\mathbf{P}_{kn}(0)}{M_n}t + \frac{1}{2M_n}\mathbf{F}'_{kn}(0)t^2 + O(t^3), \tag{17.50}$$

$$\mathbf{P}_{kn}(t) = \mathbf{P}_{kn}(0) + \mathbf{F}'_{kn}(0)t + \frac{1}{2M_n}\frac{\partial \mathbf{F}'_{kn}(0)}{\partial t}t^2 + O(t^3). \tag{17.51}$$

The vector $\mathbf{F}'_{kn}$ stands here for the Hellmann–Feynman force that determines the course of the nth nucleus on the kth potential energy surface. Assuming identical initial conditions for both the nuclear positions and momenta on both potential energy surfaces, and acknowledging the fact that the ensemble average $\langle \mathbf{P}_{kn}(0)\cdot \mathbf{F}'_{kn}(0)\rangle$ vanishes, we conclude (see Exercise 17.3):

$$\langle \Delta K_{12}(t)\rangle_T = 0 \tag{17.52}$$

up to second order.

Exercise 17.3. Verify that Eq. (17.52) is valid up to second order in time, using $\Delta K_{12}(t) = \sum_n \frac{1}{2M_n}[(P_{1n}(t))^2 - (P_{2n}(t))^2]$ as well as $\mathbf{P}_{1n}(0) = \mathbf{P}_{2n}(0)$.

The Gaussian overlap $\mathcal{J}_{\mathrm{overlap}}(t)$ (see Eqs. (17.40) and (17.42)) is obtained by use of the Gaussian overlap formula for the case of two centers. Inserting Eqs. (17.50) and (17.51) into the resulting expression yields, again up to quadratic order in time:

$$\langle \ln[\mathcal{J}_{\mathrm{overlap}}(t)]\rangle_T \approx -\left\langle \sum_n \left[\frac{1}{4\alpha_n\hbar^2}(\mathbf{F}'_{1n}(0) - \mathbf{F}'_{2n}(0))^2\right]\right\rangle_T t^2. \tag{17.53}$$

Summarizing Eqs. (17.49), (17.52), and (17.53), we find a Gaussian profile for the ensemble averaged decoherence function,

$$\langle D(t)\rangle_T \approx \exp\left[-\left\langle \sum_n \frac{1}{4\alpha_n\hbar^2}(\mathbf{F}'_{1n}(0) - \mathbf{F}'_{2n}(0))^2\right\rangle_T t^2\right], \tag{17.54}$$

and identify the width of this Gaussian with the decoherence time τ_{D}, such that

$$\tau_{\mathrm{D}} = \left[\left\langle \sum_n \frac{1}{4\alpha_n \hbar^2} (\mathbf{F}'_{1n}(0) - \mathbf{F}'_{2n}(0))^2 \right\rangle_T \right]^{-\frac{1}{2}}. \quad (17.55)$$

This relation identifies the physical source of the electronic back action on the nuclei as the initial difference between the Hellmann–Feynman forces experienced by the nuclear subsystem in the two electronic states.

Exercise 17.4. Use the trajectory and momentum expansions (17.50) and (17.51) to derive the result (17.53). Start by transforming the overlap term $\mathcal{J}_{\mathrm{overlap}}$ employing the formula for the two-center integral over the product of two Gaussian functions with equal widths:

$$\int d\mathbf{X}\, G_{1n}^* G_{2n} = \exp\left[-\frac{\alpha_{\mathrm{N}}}{4} (\mathbf{X}_{2n}(t) - \mathbf{X}_{1n}(t))^2 - \frac{1}{4\alpha_{\mathrm{n}}\hbar^2} (\mathbf{P}_{2n}(t) - \mathbf{P}_{1n}(t))^2 + \frac{i}{2\hbar} (\mathbf{X}_{2n}(t) - \mathbf{X}_{1n}(t)) \cdot (\mathbf{P}_{2n}(t) - \mathbf{P}_{1n}(t)) \right] \quad (17.56)$$

(see, for instance, [572]). Neglect the overlap between Gaussians belonging to different nuclear centers.

From the formula for the overlap integral of two Gaussian functions of the form (17.40), made explicit in Exercise 17.4, the calculation of $\mathcal{J}_{\mathrm{overlap}}(t)$ yields a product composed of three factors, involving the deviations between the positions on the two surfaces, the respective momentum differences, as well as a cross term of positions and momenta. While the second factor is dominant on a short time scale, leading to the approximation (17.54), the two remaining contributions exert noticeable influence on the behavior of the overlap function in the longer run. The various components of the problem are discussed in [575] in the context of the quantum decoherence undergone by a solvated electron in methanol. From this assessment, the short time expression (17.54) is seen to be appropriate for decoherence times up to about 5 fs. In the 10 fs range which is typical for the problem of an electron in methanol, it still provides a satisfactory description of the decay process. The following section will focus on water and deuterated water as solvent media for a single electron. In both environments, decoherence times smaller than 5 fs are found, justifying the use of the short time approach.

We observe that the widths α_n of the frozen Gaussians exert a sensitive influence on the decoherence time. Given a certain initial difference between the forces associated with the two potential energy surfaces, broad wave packets, characterized by small α_n, sample larger regions of the differing surfaces and thus decohere more rapidly than narrow wave packets. A physically meaningful choice of the set $\{\alpha_n\}$ is therefore crucial for the adequate description of the quantum-classical ensemble evolving in time. Neria and Nitzan [570] present an analytic derivation of the width parameters as a function of temperature within the frozen Gaussian approximation, resulting in

$$\alpha_n = \frac{M_n \omega_n}{\hbar} \left[\coth\left(\frac{\hbar \omega_n}{2 k_{\mathrm{B}} T} \right) - \frac{2 k_{\mathrm{B}} T}{\hbar \omega_n} \right]^{-1}. \tag{17.57}$$

This finding is obtained in the framework of an elementary model that operates with two groups of displaced identical three-dimensional harmonic oscillator potential functions to represent the two coupled surfaces. Thus, the parameter ω_n denotes the eigenfrequency of the nth nucleus. Ample use is made of the displaced harmonic oscillator scheme in the description of many radiationless processes, such as electron transfer (see Chapter 20) in gas and condensed phases. From Eq. (17.57), the high-temperature limit of α_n is found to be [570, 576]

$$\alpha_n = \frac{6 M_n k_{\mathrm{B}} T}{\hbar^2}. \tag{17.58}$$

Likewise, for the extreme of low temperatures, one arrives at

$$\alpha_n = \frac{M_n \omega_n}{\hbar} + \frac{2 M_n k_{\mathrm{B}} T}{\hbar^2}, \tag{17.59}$$

i.e. the width parameter merges into that of the harmonic oscillator ground state as T approaches zero. Interestingly, the high-temperature case is independent of the frequencies ω_n and thus does not require analyzing all contributing nuclear modes. On the other hand, it should be noted that the frozen Gaussian approximation is exact if $T = 0$.

For assessing the validity of the decoherence model surveyed in this and the preceding subsection, and for demarcating its domain of validity, it has been applied to a variety of physical and model situations (see, for instance, [308, 577]). The most elementary quantum system available in physical chemistry is a single electron. What makes it a preferred object of study is not merely the advantage of its relative simplicity but also the availability of much experimental information that has accumulated over several decades [578, 579]. The following section comprises a brief summary of computational research devoted to decoherence phenomena involving the *hydrated electron*.

17.3 Case Study: The Dynamics of Electron Hydration

By a *solvated electron* we understand an electron under the influence of a potential generated by a surrounding condensed matter system (see Chapter 10 in [580]). The electron is confined by its environment whose configuration, in turn, is affected by the electron's presence. The mutual interaction between the subsystems leads to an *autolocalized state* of the electron. The field of the latter gives rise to a local reconfiguration of the embedding medium, creating substructures with characteristic spectroscopic and dynamic signatures. Evidently, the geometric features of the entire system, consisting of solute and solvent, depend sensitively on the nature of the molecular constituents enclosing the electron. For polar liquids such as water, the solvated electron interacts with the inertial polarization of the surrounding molecules, exerting a long-range influence on the medium. For non-polar liquids, in contrast, short-range repulsive interaction [581], modified by effects due to electronic polarization induced by density fluctuations, tend to be dominant. Confinement of the electron in substructures of the medium, however, is counteracted by its substantial kinetic energy that arises from the electron's low mass and favors spreading of the particle over an extended solvent region. These antagonistic effects result in a compromise between electron localization and fluctuation, where the latter couples to structural fluctuations within the environment.

A large portion of the research on solvated electron systems has focused on the special case of the hydrated, or aqueous electron, labeled e_{aq}^{-} [582, 583], which was first identified in the early sixties [584]. The formation of a solvated electron in water has been monitored in detail by femtochemical investigation ([585]–[588]). In these experiments, electrons were generated by multiphoton ionization of liquid water. The equilibration of these electrons was seen to involve at least one intermediate. Thus, a transient system that turned out to absorb in the near-infrared region was shown to relax to the fully solvated final state on a time scale ranging from 240 fs [587] to about 540 fs [585]. Experimental observation, assisted by nonadiabatic simulation [589, 590], suggests that the precursor of the equilibrium state is produced by a temporary trapping of the electron in its lowest excited state, preceding its transition into the ground state. The latter has been identified with an almost spherical, localized *s*-wave-like distribution, while the first excited state consists in a triplet of localized *p*-like waves whose degeneracy is broken due to the deviation of the medium from spherical symmetry [591]. These two electronic states interact in different ways with their environment. It is therefore plausible that the time scale of electron deexcitation is influenced by the decoherence of the two states as caused by the different responses of the vibrational and rotational modes of the medium surrounding the electron to these states. Combining the information about the ground and the first excited state of the hydrated electron with known data on these environmental degrees of freedom can yield a detailed picture of the coherence loss mechanism involving the deexcitation of an electron trapped in H_2O. To achieve this goal, Prezhdo and Rossky [571] write the averaged decoherence function (17.54) in the following form:

$$\langle D(t)\rangle_T = \exp\left[-\left\langle\sum_n d_n t^2\right\rangle\right], \tag{17.60}$$

where

$$d_n \equiv \left\langle \frac{1}{4\alpha_n \hbar^2}[\mathbf{F}'_{1n}(0) - \mathbf{F}'_{2n}(0)]^2 \right\rangle_T \tag{17.61}$$

and n labels the various normal modes of water. Formula (17.57) may be used to determine the width parameters α_n since the eigenfrequencies ω_n are available from experiment [592]. The terms $\mathbf{F}'_{kn}$, $k = 1, 2$, refer to the Hellmann–Feynman forces evaluated with respect to the normal mode n in the ground or the first excited state of the electron. The different contributions of the vibrational and rotational degrees of freedom of H_2O to decoherence are shown in Figure 17.4. The underlying simulation involved an ensemble of 200 simple point charge (SPC, [593]) water molecules and the use of a pseudopotential [594] to describe the electron–water interaction.

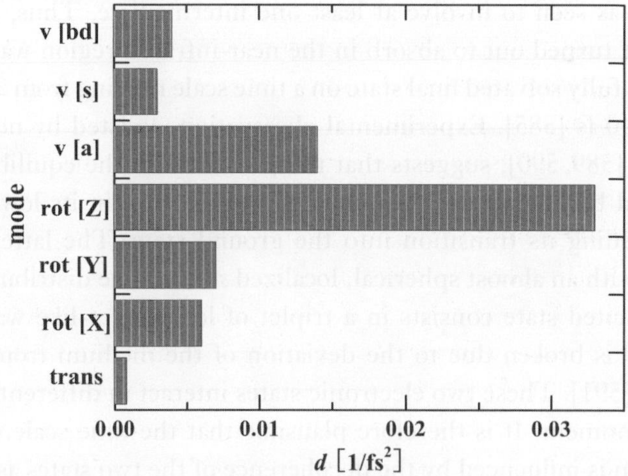

Figure 17.4: Bar representation of the H_2O normal mode contributions to the decoherence process of the hydrated electron undergoing non-radiative transition from the first excited to the ground state. Shown is the coefficient d_n (see Eq. (17.61)) for the following degrees of freedom: bending (bd), symmetric (s), and asymmetric (a) vibrations as well as rotations about the X, Y and Z axes, and translation. The X-axis is chosen as the C_{2v} symmetry axis of H_2O, the Y-axis is perpendicular to the plane of the molecule, and the Z-axis perpendicular to both X and Y. (Reprinted with permission from [571]. Copyright (1997) by the American Institute of Physics.)

As clarified by the diagram, decoherence in the hydrated electron deexcitation process is dominated by the action of two normal modes, namely the asymmetric stretch and a rotation around an axis in the molecular plane and perpendicular to the H_2O symmetry axis. The latter is characterized by the smallest moment of inertia among the three orthogonal rotational degrees of freedom. We note that these two maximally effective motions both change the orientation of the H_2O dipole moment and therefore couple strongly to the field of the electron.

17.3.1 *Isotope effects in hydrated electron relaxation*

The concept of decoherence holds the key for the solution of an experimental riddle related to the solvated electron relaxation in H_2O and D_2O. In photoexcitation measurements ([595]–[597]), both the solvation of the excited state and its transition into the ground state equilibrium was observed. Rearrangement of the solvent particles occurs to accommodate the newly formed p-like charge distribution, decreasing the quantum energy gap. This solvation stage is followed by the radiationless return of the hydrated electron into its ground state as a result of nonadiabatic coupling between the quantum system and its rapidly moving environment. Thus, the overall life time of the excited state is combined of times characteristic of the solvation process of the excited species and of its relaxation into the ground state, namely 250–300 fs and about 1 ps, respectively, as determined by theoretical analysis of experimental spectra [598, 599].

Initial modeling yielded a surprisingly large isotope effect, predicting an increase of about a factor of two of the excited state lifetime in D_2O over that in H_2O [600]. From classical mechanics, one assumes the highest nuclear velocities in D_2O to be smaller by a factor $\sqrt{2}$ than those in H_2O. In accordance with this expectation, Barbara and co-workers reported an excited state lifetime ratio $\tau(D_2O)/\tau(H_2O)$ close to 1.4 [601,602]. In contrast to this observation, nonadiabatic transitions from the first excited into the ground state might be predicted to proceed twice as fast in water than in its deuterated analog, as one considers the velocity dependence of $\bar{k}_{12}$ (see Eqs. (17.37) and (17.46)). This estimate, however, neglects the effect of decoherence.

Inspecting the two quantities that determine the decoherence time according to Eq. (17.55), namely the Gaussian widths α_n and the Hellmann–Feynman forces $\mathbf{F}'_{1n}$, $\mathbf{F}'_{2n}(0)$, we realize that the latter are the same in both media, H_2O and D_2O. The coefficients α_n, however, are mass-dependent (see Eqs. (17.57)–(17.59)). Intuitively, the same effect that gives rise to longer relaxation times in D_2O than in H_2O should cause a quantum subsystem to decohere more slowly in the former than in the latter environment. This is immediately confirmed in the high-temperature limit described by Eq. (17.58), stating a linear increase of α_n with the respective mass parameter. By Eq. (17.55), enhanced values of α_n imply longer decoherence times. This readily suggests a qualitative interpretation of the vanishing isotope effect in the two different liquids: as a consequence of the less rapid molecular motion in D_2O than in H_2O, the nonadiabatic decay proceeds more slowly in the former than in the latter medium. At the same time, however, the slower decoherence in D_2O implies that nonadiabatic coupling in this medium acts coherently through times significantly longer than those available in H_2O. This balance leads to similar transition rates for both systems.

This view was confirmed by simulating the two systems, a solvated electron in H_2O and D_2O [569, 571]. The liquids were represented by a set of 200 classical molecules and a quantum electron confined to a cubic cell at room temperature, imposing periodic boundary conditions. A pseudopotential approach [594] was used to describe the interaction between the electron and the water molecules. At each propagation time step of $\Delta t = 1$ fs, the adiabatic eigenstates were determined employing a plane wave basis. The studied system was equilibrated by running ground state trajectories for a period of 30 ps. From the resulting structure, 20 initial configurations for non-equilibrium excited state trajectories were chosen, where the matching between the experimental laser frequency [595, 596] and the energy gap bridging the ground and the excited state served as a selection criterion.

In order to investigate the solvation and decay process of the hydrated electron in its first excited state, a trajectory surface hopping algorithm, namely the *stationary phase surface hopping* procedure introduced by Webster *et al.* [589, 603, 604] was used as propagation method. Instead of adiabatic Hellmann–Feynman forces, this procedure involves the temporally non-local Pechukas force (12.18) which is not confined to a single potential energy surface but involves a mixture of quantum states. The simulation operates with hopping probabilities computed according to the scheme originally proposed by Tully and Preston [301], and is thus determined by the overlap integral O_{ij} that connects the initial with possible final adiabatic states. With $\hat{U}$ as quantum propagator, this integral is given by

$$O_{ij} = \langle \phi_j(\mathbf{X}(t)) | \hat{U}(t, t_0) | \phi_i(\mathbf{X}(t_0)) \rangle. \tag{17.62}$$

The magnitudes of these matrix elements decide the final adiabatic state to which the system decoheres and which enters the Pechukas force expression. We emphasize that this approach maintains complete quantum coherence during each time step. For the overall probability of a nonadiabatic transition during the N consecutive time intervals of duration Δt we find

$$P_{ij}(\Delta t) = \frac{1}{\Delta t} \left\langle \frac{1}{N} \sum_{n=1}^{N} |O_{ij}(\Delta t_n)|^2 \right\rangle, \tag{17.63}$$

where the average is taken over starting times and trajectories, and Δt_n denotes the nth time step.

The decoherence time is here given by the length of an individual time step Δt. Considering the great impact of decoherence on nonadiabatic processes, as established by the foregoing discussion, this time clearly should not be selected by decree but on the basis of physical arguments. Instead of identifying the decoherence time with the simulation time step, one may describe a fully coherent transition probability by generalizing Eq. (17.63) according to [569]:

$$P_{ij}(\tau_{\mathrm{D}}) = \frac{1}{\tau_{\mathrm{D}}} \left\langle \left| \sum_{n=1}^{N} O_{ij}(\Delta t_n) \right|^2 \right\rangle \tag{17.64}$$

with $\tau_{\mathrm{D}} = N\Delta t$. This expression involves the coherent addition of N transition amplitudes evaluated in N consecutive intervals. Since the matrix elements O_{ij} are recorded during the trajectory surface hopping simulation, transition probabilities and thus excited state life times can be determined *post factum* for arbitrarily chosen τ_{D}. This has been done to elucidate the relaxation dynamics of a solvated electron in its first excited state embedded in H_2O and D_2O, as shown in Figure 17.5, which displays a sketch of the relation between the life time and the decoherence time in both environments. For both media, the filled circles refer to the decoherence times that are "physically correct" by the standard of prescription (17.55).

Evidently, the life times in the two media become similar if adequate decoherence times are employed for their evaluation. Thus, the interplay of nonadiabatic transition and decoherence turns out to reduce the difference between hydrated electron relaxation in the two media. It is to be kept in mind, however, that Eq. (17.55) represents a short-time approximation. The excited state life times, as extracted from Figure 17.5, being the order of several hundred femtoseconds, are considerably lower than the experimental values which are in the picosecond range. We mention that a later simulation [606] that accounted for the significant impact of collective low-frequency unstable vibrational modes on the relaxation behavior of the hydrated electron was able to recover the experimental ratio of excited state life times in D_2O and H_2O, 1.4.

Wong and Rossky [573, 607] have demonstrated that a more realistic description of solvated electron relaxation can be accomplished by implementing the Liouville–von Neumann equation augmented by a dissipative term, as in Eq. (17.15). This is realized in the framework of a

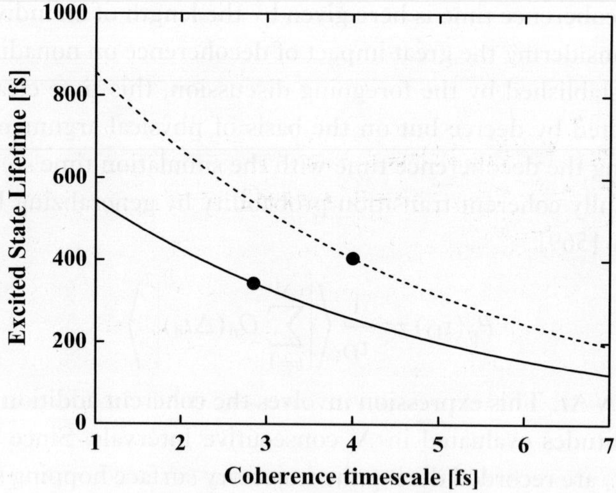

Figure 17.5: The life time of the first excited state of a hydrated electron in H_2O (solid line) and in D_2O (dotted line) versus the decoherence time. The lines are calculated from the relation (17.64). The two filled circles refer to decoherence times derived from Eq. (17.55). (Reprinted with permission from [571]. Copyright (1997) by the American Institute of Physics.)

hybrid scheme that merges the advantages of the Ehrenfest approach with those of the surface hopping procedure. Thus the extended Liouville–von Neumann equation was solved by implementing a mixed mean-field/surface-hopping (MF/SH) treatment [608, 609]. At any time of the simulation, the energy of the total quantum-classical system is given by

$$E_{\text{tot}} = \frac{M}{2}\left(\frac{d\mathbf{X}}{dt}\right)^2 + V_{\text{cl}} + \langle \Psi(\mathbf{x}, \mathbf{X}(t))|\hat{H}_{\text{q}}(\mathbf{x}, \mathbf{X}(t))|\Psi(\mathbf{x}, \mathbf{X}(t))\rangle \quad (17.65)$$

where V_{cl} and $\hat{H}_{\text{q}}$ stand for the classical potential and the quantum Hamiltonian, respectively, and the summation over the classical degrees of freedom has been dropped to simplify the notation. The MF/SH scheme augments the set of tools used to implement trajectory surface hopping, as made explicit in Section 11.2, by an auxiliary wave function Ψ to determine the Ehrenfest force. More specifically, we obtain by deriving Eq. (17.65) with respect to time and by using energy conservation:

$$M\frac{d^2\mathbf{X}}{dt^2} = \mathbf{F}_{\text{cl}} + \mathbf{F}_{\text{q}} \quad (17.66)$$

with

$$\mathbf{F}_{\mathrm{q}} = -\frac{\partial}{\partial \mathbf{X}} \langle \Psi(\mathbf{x}, \mathbf{X}(t)) | \hat{H}_{\mathrm{q}}(\mathbf{x}, \mathbf{X}(t)) | \Psi(\mathbf{x}, \mathbf{X}(t)) \rangle. \tag{17.67}$$

The regular Hellmann–Feynman force results from this expression if Ψ is an eigenfunction of $\hat{H}_{\mathrm{q}}$ which is in general not the case.

In an MF/SH simulation, one propagates both the primary and the auxiliary quantum subsystem where the former decides surface hopping probabilities and the latter provides the mean-field force experienced by the nuclei in the limit of complete coherence. As pointed out in Section 11.2, Tully's trajectory surface hopping approach [288] treats the auxiliary wave function as identical with the reference wave function which describes the currently occupied state of the quantum system. The updated primary solution is used to compute surface hopping probabilities which decide the transition of a trajectory to a specific potential energy surface, according to the utilized hopping criterion. In parallel to these operations, reference trajectories are propagated on all contributing potential energy surfaces. This additional information makes it possible to include an assessment stage, where the validity of the mean-field assumption is scrutinized. More specifically, if the surface hopping step does not result in a change between surfaces, the position and momentum deviations between the average and the reference trajectories are compared with a set of predefined thresholds. Should these thresholds be exceeded, the relevant trajectories are deemed too divergent for mean-field theory to apply. In this case, the auxiliary system is projected on the reference state singled out as currently occupied by the trajectory hopping simulation.

This procedure has been applied to the hydrated electron problem [573, 607] where the effect of decoherence was included in the determination of the primary quantum subsystem according to Eq. (17.15), with various choices for the decoherence factor. In particular, this investigation involved a comparison between a thermal average (see Eq. (17.55)) and an instantaneous decoherence time scale, with a variable value of τ_{D} which was calculated *on the fly* [573, 607]. In both cases, the first excited state relaxation times were found to be on the order of a picosecond, in accordance with the experimental range.

The deexcitation of an equilibrated *p*-electron embedded in a solvent continues to provide an important touchstone for understanding the impact of decoherence on condensed phase nonadiabatic dynamics. A novel form of the golden rule, Eq. (17.29), for the nonadiabatic transition rate between two quantum states was explored by Borgis *et al.* and tested by computing the $p \rightarrow s$ decay undergone by an electron in water and in methanol [610]. A comparative assessment of several mixed quantum-classical algorithms applied to the problem of hydrated electron relaxation dynamics is found in [611].

17.4 Continuous Surface Switching: A Compromise between Mean-Field and Individual Surface Propagation

In Chapter 11 we listed some general arguments for and against adopting various hopping or spawning methodologies in comparison with the Ehrenfest viewpoint which may be characterized as the limit of a *completely coherent treatment of the electronic wave function* in molecular processes. Particularly, the latter approach avoids the ambiguities associated with regions of nonadiabatic coupling as described by hopping or spawning procedures. Thus, Ehrenfest dynamics is not plagued by any dependence on the chosen electronic representation, diabatic versus adiabatic, and does not require any *ad hoc* decisions about the occurrence of nonadiabatic transitions. Its major deficiency as compared with hopping models, however, lies in its reliance on an average potential energy surface or an averaged nuclear trajectory to propagate the nuclear degrees of freedom. On the basis of the foregoing discussion, this problem may be called the *inadequate treatment of decoherence*. As has been shown in Chapter 11, the hopping approach eliminates unphysical mixed states in regions where the nonadiabatic coupling vanishes [29].

As a result of weighing the advantages and drawbacks of the two contrasting strategies, one might desire a synthesis between them that combines their virtues and minimizes their imperfections. Such a method will shift towards the "Ehrenfest limit" in regions of pronounced nonadiabatic interaction and will rely on a treatment based on the hopping scheme for propagation on uncoupled potential energy surfaces. The contours of this idea have been outlined in Section 17.3.1 in the context of the hydrated electron problem.

In the following, we will present the basic tenets of the "eclectic" approach proposed by Volobuev *et al.* [29] and, by the same token, exemplify the construction of a generalized composite from several well-established traditional procedures of nonadiabatic quantum dynamics.

An electronic basis of dimension K is assumed. Since trajectories $\mathbf{X}(t)$ are used, an electronic wave function that depends parametrically on the nuclear path is written as

$$\Phi(\mathbf{x}, \mathbf{X}(t)) = c_1(t)\Psi_1(\mathbf{x}, \mathbf{X}(t)) + \cdots + c_K(t)\Psi_K(\mathbf{x}, \mathbf{X}(t)). \qquad (17.68)$$

The evolution of the time-dependent coefficients is determined by the electronic Hamiltonian $\hat{H}_{\text{el}}$:

$$V_{kk'}(X) = \langle \Psi_k(\mathbf{x}, \mathbf{X}(t))|\hat{H}_{\text{el}}(\mathbf{x}, \mathbf{X}(t))|\Psi_{k'}(\mathbf{x}, \mathbf{X}(t))\rangle \qquad (17.69)$$

and the nonadiabatic coupling vector $\mathbf{A}_{ij}$. In an ideal diabatic model, which will be adopted later, this vector is zero, as the electronic basis functions are independent of the nuclear coordinates $\mathbf{X}(t)$, and the potential energy matrix $V_{kk'}$ has non-vanishing off-diagonal elements. We stipulate that the elements $V_{kk'}$, $k \neq k'$, reduce to zero in the asymptotic regions where consequently the diabatic and adiabatic representations of the dynamic process become identical.

As the starting point for developing the envisaged hybrid formalism, we employ the Ehrenfest method. From Section 8.3.4, the expression for the potential energy in this approximation is given by the expectation value of the electronic Hamiltonian:

$$V(\mathbf{X}) = \langle \Phi(\mathbf{X}(t), \mathbf{x})|\hat{H}_{\text{el}}(\mathbf{x}, \mathbf{X}(t))|\Phi(\mathbf{X}(t), \mathbf{x})\rangle. \qquad (17.70)$$

Inserting Eq. (17.68) into Eq. (17.70), we arrive at

$$V(\mathbf{X}) = \sum_k \sum_{k'} \rho_{kk'} V_{kk'}(\mathbf{X}) \qquad (17.71)$$

where the density matrix elements $\rho_{kk'}$ are subdivided into electronic state populations $\rho_{kk} = |c_k|^2$ and coherences $\rho_{kk'} = c_k^* c_{k'}$, $k \neq k$, as in Section 14.1. The idea put forward by Volobuev *et al.* [29] to unify complementary approaches involving nonadiabatic trajectories consists in replacing the density matrix elements by the generalized expressions $w_{ij}(\rho_{ij})$ [613]. These are designed to recover the original form (17.71) of the Ehrenfest potential energy in situations of pronounced nonadiabatic coupling, while being conducive to trajectory motion on individual potential energy surfaces whenever the coupling is sufficiently small.

To give a more concrete meaning to this proposal, we exemplify Ehrenfest dynamics by the classical analog model outlined in Section 10.2 although we do not impose any double-ended boundary conditions. Specifically, we set $\rho_{kk} = n_k$ and $\rho_{kk'} = \sqrt{n_k n_{k'}} \cos(q_k - q_{k'})$, where the definition of the parameters n_k and q_k, i.e. the action-angle variables that describe the electronic degrees of freedom, is here as in Section 10.2. Further, we rewrite the diabatic Hamilton function Eq. (10.6) of the classical analog model as

$$H = T_{\mathrm{N}} + V'(\mathbf{X}) + V_{\mathrm{c}}(\mathbf{X}), \tag{17.72}$$

with T_{N} as classical nuclear kinetic energy, $V'(\mathbf{X})$ as diabatic potential energy average of the K surfaces

$$V'(\mathbf{X}) = \sum_{k=1}^{K} n_k V_{kk}(\mathbf{X}), \tag{17.73}$$

and the coupling potential

$$V_c(\mathbf{X}) = \sum_{k} \sum_{k' \neq k} \sqrt{n_k n_{k'}} \cos(q_k - q_{k'}) V_{kk'}(\mathbf{X}). \tag{17.74}$$

The root of the unphysical mixed states marring the dynamic theories of the Ehrenfest type lies in the mean potential (17.73). In particular, the violation of the microreversibility principle has been traced back to the limited accuracy of the resulting weighted average over the accessible channels which may deviate strongly from the correct superposition of these alternatives in the quantum mechanical sense (see Section 10.1). We recall that in the classical electron analog model as suggested by Meyer and Miller [614], double-ended boundary conditions are implemented to overcome this obstacle. While formally viable and tested successfully for smaller systems, this solution is hardly practicable in the general case of complex molecular interactions. From an economic point of view, it is more preferable to confront the initial condition problem which provides the motivation initial value representation, as described in Section 12.3.

A key to extending and improving the Ehrenfest approach may be found in introducing generalized weight coefficients w_k for the average potential (17.73). In the model of Meyer and Miller, these weights are fixed at $w_k = n_k$. By relaxing this constraint, can one manipulate the factors w_k in such a way that they allow for mixtures of several states in zones of nonadiabaticity but specify one well-defined state wherever propagation on individual surfaces prevails? The set of candidates for functional forms of w_k is narrowed down by certain restrictions [615], among them the requirement that the limits valid for the action variables of the classical electron analog:

$$\sum_{k=1}^{K} n_k = 1, \quad 0 \le n_k \le 1 \tag{17.75}$$

apply to the weights w_k as well:

$$\sum_{k=1}^{K} w_k = 1, \quad 0 \le w_k \le 1. \tag{17.76}$$

To clarify the main ideas underlying the generalized theory, let us consider the case of two states. Instead of identifying the action variables n_k with the diabatic weights, Volobuev *et al.* define the latter as functions of n_k, namely as [29]

$$w_1 = \frac{Q(\mathbf{X})W[n_1 - n_2; -\eta, Q(\mathbf{X})] + n_1}{Q(\mathbf{X}) + 1}, \tag{17.77}$$

$$w_2 = \frac{Q(\mathbf{X})W[n_2 - n_1; -\eta, Q(\mathbf{X})] + n_2}{Q(\mathbf{X}) + 1}, \tag{17.78}$$

with η as a real parameter. The function W is given by

$$W[n; \eta, Q(\mathbf{X})] = \frac{1}{2} + \frac{1}{\pi} \arctan\left[\frac{n - \eta}{Q(\mathbf{X})}\right]. \tag{17.79}$$

An important rationale for this choice of W is that it merges into the Heaviside step function for large Q:

$$\lim_{Q(\mathbf{X}) \to \infty} W[n; \eta, Q(\mathbf{X})] = \Theta(n - \eta). \tag{17.80}$$

For large values of Q, W will thus tend towards zero or one, depending on the sign of the argument $n_2 - n_1 - \eta$. Under these conditions, one of the two potential energy surfaces will be occupied, and the other left empty, in accordance with the single-surface propagation mode. From Eqs. (17.77) and (17.78) one sees immediately that the opposite extreme of vanishing Q yields the Ehrenfest limit, i.e. w_k reduces to n_k. Thus, the function Q represents a device that allows for a continuous change between the prototypes of single or average surface evolution.

An appropriate definition of Q that accomplishes this task may be illustrated by the case of three atoms. Any element of the diabatic potential energy matrix can be expressed as the sum

$$V_{kk'}(R_1, R_2, R_3) = V^{\text{diat}}_{k,k';1}(R_1) + V^{\text{diat}}_{kk';2}(R_2) + V^{\text{diat}}_{kk';3}(R_3) + \Delta V_{kk'}(R_1, R_2, R_3) \qquad (17.81)$$

where the R_i, $i = 1, 2, 3$, stand for the three internuclear distances present in the problem. The superscript diat denotes a purely diatomic situation, $V^{\text{diat}}_{k,k',i}(R_i)$ is the diabatic matrix element if the ith diatom is at a bond length R_i, and the two remaining internuclear distances are infinite. Consequently, $\Delta V_{kk'}$ represents the residual interaction that involves participation of all three atoms. With these assignments, a plausible choice for the regulating function Q is

$$Q(\mathbf{X}) = \frac{V_0^2}{(\Delta V_{11})^2 + (\Delta V_{12})^2 + (\Delta V_{22})^2} \qquad (17.82)$$

with V_0 as an appropriately chosen constant. The definition of the matrix $\Delta\mathbf{V}$ implies that each of its elements tends towards zero if any of the internuclear distances becomes large. In this asymptotic region Q approaches infinity, and, by relations (17.77)–(17.80), the system will occupy one of the two possible potential energy surfaces. The opposite extreme of small Q, leading to an averaged effective potential, is reached if all three atoms are interacting strongly. By construction, the function Q adjusts the dynamics to match the Ehrenfest model in the region of triatomic coupling, and to favor asymptotically single-surface propagation. The latter feature is accomplished without any reliance on stochastic processes, eliminating the need for hopping criteria and related unphysical discontinuous jumps. Thus, both the Ehrenfest and the hopping approximation have been merged into a *continuous surface switching* (CSS) model [29]. Refinements and extensions of this model are described in Section 17.4*.

17.5 Decay of Mixing

Among the requirements imposed on the continuous surface switching formalism is the vanishing of the electronic coherences in regions that do not exhibit any interstate coupling. Approaches of the CSS type aim at accomplishing this goal with the help of an appropriately modeled function that specifies the dynamic mode in terms of mixed- versus single-surface propagation. This agent makes the unphysical hopping events obsolete but still imports a certain measure of arbitrariness into the theory, since even a stringent set of criteria imposed on the form of the weighting coefficients w_k [615] does not fix them unambiguously, and different results must be expected from different choices for w_k. Further, the self-consistent determination of the parameter η, requiring a sequence of trajectory reruns, can make the procedure computationally cumbersome.

Alternatively, one may revert to the root of the problem that plagues self-consistent potential methods in nonadiabatic quantum dynamics. These approximations tend asymptotically towards quantum superpositions of electronic states (see Section 10.1) when realistically, the system should evolve into a statistical mixture of final states [616], as a consequence of its coupling to environmental degrees of freedom. In Sections 17.1 and 17.2 we discussed proposals that address this problem by incorporating decoherence into the exact quantum mechanical formalism, extending the Liouville–von Neumann equation of motion by a dissipative term. In Section 17.4, we drew the contours of a model that allows us to alternate between two methodological prototypes, namely Ehrenfest and trajectory hopping dynamics, depending on their validity at any stage in the evolution of the considered system. The present section highlights a further facet of the current debate on the appropriate treatment of decoherence in nonadiabatic molecular interactions. Instead of pursuing this goal by balancing two different traditional propagation techniques, as attempted by continuous surface switching approaches, one may build decoherence directly into the self-consistent potential method.

Presenting the fundamentals of this approach, we begin once more with the classical electron analog Hamilton function, written in compact form as (see Eq. (10.6))

$$H_{\mathrm{CEA}} = T_{\mathrm{N}} + \langle \Phi | \hat{H}_{\mathrm{el}}(\mathbf{x}, \mathbf{X}) | \Phi \rangle = T_{\mathrm{N}} + V(\mathbf{X}). \tag{17.83}$$

Parametrizing the expansion coefficients of the electronic wave function in the representation (17.68) with the quasi-classical position and momentum pseudocoordinates x_k and p_k (see Eq. (12.72)),

$$c_k = \frac{1}{\sqrt{2}}(x_k + ip_k), \tag{17.84}$$

we arrive at the following form of the potential $V(\mathbf{X})$:

$$V(\mathbf{X}) = \sum_k n_k V_{kk}(\mathbf{X}) + \sum_k \sum_{k'<k} (x_k x_{k'} + p_k p_{k'}) V_{kk'}(\mathbf{X}). \tag{17.85}$$

Here $|c_k|^2 = \frac{1}{2}(x_k^2 + p_k^2) = n_k$ represents the population of the electronic state with label k. A fully classical description applies to the nuclei, where one may eliminate the center-of-mass momentum and thus equate the overall nuclear momentum to the vibrational and rotational contributions:

$$\mathbf{P} = \mathbf{P}_{\text{vib}} + \mathbf{P}_{\text{rot}}. \tag{17.86}$$

The relative motion between the system components is here included in the local vibrational momentum $\mathbf{P}_{\text{vib}}$. The decomposition Eq. (17.86) amounts to separating the linear momentum into a non-rotational and a rotational component. For the special case of a three-dimensional system, the partition (17.86) is carried out explicitly in [343]. The instantaneous vibrational energy, $T_{\text{vib}} = P_{\text{vib}}^2/2\mu$, depends on the reduced mass μ of the arrangement. In keeping with the classical analog model as described in Chapter 10, the equations of motion for the nuclear degrees of freedom are

$$\frac{d\mathbf{R}}{dt} = \frac{\mathbf{P}}{\mu}, \tag{17.87}$$

$$\frac{d\mathbf{P}}{dt} = \frac{d\mathbf{P}^{\text{C}}}{dt} = -\frac{\partial V^{\text{C}}}{\partial \mathbf{R}} \tag{17.88}$$

where μ is the reduced mass for relative nuclear motion and $\mathbf{R}$ the vector of the relative nuclear coordinates ([615], see Section 2.2). The electronic degrees of freedom evolve according to the TDSE (compare with Eqs. (10.5) and (12.71)):

$$\frac{dx_k}{dt} = \frac{dx_k}{dt}^{\text{C}} = \frac{1}{\hbar}\left(p_k V_{kk} + \sum_{k' \neq k} p_{k'} V_{kk'}\right), \tag{17.89}$$

$$\frac{dp_k}{dt} = -\frac{dp_k}{dt}^{\text{C}} = \frac{1}{\hbar}\left(x_k V_{kk} + \sum_{k' \neq k} x_{k'} V_{kk'}\right). \tag{17.90}$$

The superscript C stands for *coherent*. In the present context, it labels rates of change as defined within the classical analog model. In the following discussion, the decoherent contributions to the rates of change will be introduced.

The *decay-of-mixing algorithm* modifies the relations (17.87)–(17.90) in order to include decoherence into the self-consistent potential formalism. This is achieved as the time derivatives (17.88), (17.89), and (17.90) are augmented by decoherence terms. Specifically [30],

$$\frac{d\mathbf{P}}{dt} = \frac{d\mathbf{P}^{\mathrm{C}}}{dt} + \frac{d\mathbf{P}^{\mathrm{D}}}{dt}, \tag{17.91}$$

$$\frac{dx_k}{dt} = \frac{dx_k^{\mathrm{C}}}{dt} + \frac{dx_k^{\mathrm{D}}}{dt}, \tag{17.92}$$

$$\frac{dp_k}{dt} = \frac{dp_k^{\mathrm{C}}}{dt} + \frac{dp_k^{\mathrm{D}}}{dt}, \tag{17.93}$$

where the superscript D stands for decoherence. We note that the momentum derivative $\frac{d\mathbf{P}^{\mathrm{D}}}{dt}$ is interpreted as a *decoherence force*, to be commented on in further detail later. The requirement that $\frac{d\mathbf{P}^{\mathrm{D}}}{dt}$ exists at all times during the considered molecular process implies that trajectories never undergo discontinuous momentum changes, a feature that distinguishes the decay-of-mixing model from the original trajectory surface hopping scheme, as described in Section 11.1.

Reverting from the Cartesian electronic parameters x_k and p_k to the polar action-angle variables q_k and n_k, we require the system to tend asymptotically towards a particular state with label K, such that $n_k = 0$, if $k \neq K$, and $n_K = 1$ in the limit of sufficiently long times. Besides *dephasing*, this constraint imposes *demixing* on the evolution of the considered system, corresponding to decoherence towards a pure state rather than a coherent mixture. A way to generalize this approach will be indicated in Section 17.5.2.

Note that the separation in Eqs. (17.91)–(17.93) into classical analog and decoherence components immediately leads to a partitioned form of the time derivative of $n_k = \frac{1}{2}(x_k x_k + p_k p_k)$, namely

$$\frac{dn_k}{dt} = \frac{dn_k^{\mathrm{C}}}{dt} + \frac{dn_k^{\mathrm{D}}}{dt}. \tag{17.94}$$

By the same token, the population change of a state with label k due to decoherence may be expressed as

$$\frac{dn_k^{\mathrm{D}}}{dt} = \frac{dn_k}{dt} - \frac{dn_k^{\mathrm{C}}}{dt} = x_k \frac{dx_k^{\mathrm{D}}}{dt} + p_k \frac{dp_k^{\mathrm{D}}}{dt}. \tag{17.95}$$

Assuming first-order kinetics, the decay of populations is described by

$$\frac{dn_k^{\mathrm{D}}}{dt} = -\frac{n_k}{\tau_{kK}}, \quad k \neq K. \tag{17.96}$$

For simplicity, we do not apply a superscript D to the state-specific decoherence times τ_{kK}. Probability conservation requires that

$$\sum_k \frac{dn_k}{dt} = \sum_k \frac{dn_k^{\mathrm{C}}}{dt} + \sum_k \frac{dn_k^{\mathrm{D}}}{dt} = 0. \tag{17.97}$$

Since the classical analog model conserves probability, we conclude that

$$\sum_k \frac{dn_k^{\mathrm{D}}}{dt} = \sum_{k,k\neq K} \frac{dn_k^{\mathrm{D}}}{dt} + \frac{dn_K^{\mathrm{D}}}{dt} = 0 \tag{17.98}$$

such that

$$\frac{dn_K^{\mathrm{D}}}{dt} = -\sum_{k,k\neq K} \frac{dn_k^{\mathrm{D}}}{dt} = \sum_{k,k\neq K} \frac{n_k}{\tau_{kK}}. \tag{17.99}$$

The angle variables q_k remain unaffected by the decoherence process. Constraining the temporal change of $\tan q_k = p_k/x_k$ due to the presence of decoherence terms in Eqs. (17.92) and (17.93) to be zero,

$$\frac{d\tan q_k}{dt}\Big|_{\mathrm{D}} = \frac{x_k \frac{dp_k^{\mathrm{D}}}{dt} - \frac{dx_k^{\mathrm{D}}}{dt} p_k}{x_k^2} = 0, \tag{17.100}$$

we arrive at

$$\frac{dp_k^{\mathrm{D}}}{dt} / p_k = \frac{dx_k^{\mathrm{D}}}{dt} / x_k. \tag{17.101}$$

From Eqs. (17.95) and (17.96) together with Eq. (17.84) we find the explicit form of the decoherence contributions to Eqs. (17.92) and (17.93):

$$\frac{dp_k^{\mathrm{D}}}{dt} = -\frac{p_k}{2\tau_{kK}}, \tag{17.102}$$

$$\frac{dx_k^{\mathrm{D}}}{dt} = -\frac{x_k}{2\tau_{kK}}, \tag{17.103}$$

where $k \neq K$. If $k = K$, it holds that

$$\frac{dp_K^{\mathrm{D}}}{dt} = \frac{p_K}{2n_K} \sum_{k \neq K} \frac{n_k}{\tau_{kK}}, \tag{17.104}$$

$$\frac{dx_K^{\mathrm{D}}}{dt} = \frac{x_K}{2n_K} \sum_{k \neq K} \frac{n_k}{\tau_{kK}}. \tag{17.105}$$

Exercise 17.5. Derive Eqs. (17.102) and (17.103), beginning with $n_k = \frac{1}{2}(x_k^2 + p_k^2)$.

Having outlined its basic architecture, we summarize the decay-of-mixing approach by casting its equations of motion into the compact density matrix form. Using Eq. (17.84) in conjunction with

$$\rho_{ij} = c_i c_j^* \tag{17.106}$$

we observe

$$\Im\{\rho_{ij}\} = \frac{1}{2}(x_j p_i - x_i p_j), \tag{17.107}$$

$$\Re\{\rho_{ij}\} = \frac{1}{2}(x_i p_j + p_i p_j), \quad i \neq j, \tag{17.108}$$

$$n_k = \rho_{kk} = \frac{1}{2}(x_k^2 + p_k^2). \tag{17.109}$$

Combining (17.106) with Eqs. (17.92)–(17.94), we get

$$\frac{d\rho_{ij}}{dt} = \frac{d\rho_{ij}^{\mathrm{C}}}{dt} + \frac{d\rho_{ij}^{\mathrm{D}}}{dt}. \tag{17.110}$$

The first term on the right-hand side is given by

$$i\hbar \frac{d\rho_{kj}^{\mathrm{C}}}{dt} = \sum_l (\rho_{lj} V_{kl} - \rho_{kl} V_{lj}). \tag{17.111}$$

Exercise 17.6. Show that in the adiabatic representation, Eq. (17.111) turns into

$$i\hbar\frac{d\rho_{kj}^{\mathrm{C}}}{dt} = \sum_{l}\left\{\rho_{lj}\left[V_{kl} - i\hbar\frac{d\mathbf{R}}{dt}\cdot\mathbf{A}_{kl}\right] - \rho_{kl}\left[V_{lj} - i\hbar\frac{d\mathbf{R}}{dt}\cdot\mathbf{A}_{lj}\right]\right\} \tag{17.112}$$

with $\mathbf{A}_{kl}$ as the vector of nonadiabatic coupling between the states k and l.

Relations (17.102) and (17.104) yield the corresponding time derivatives for the decoherent components of $\frac{d\rho_{ij}}{dt}$. For the diagonal elements, we obtain

$$\frac{d\rho_{ii}^{\mathrm{D}}}{dt} = -\frac{\rho_{ii}}{\tau_{iK}}, \quad i \neq K, \tag{17.113}$$

$$\frac{d\rho_{ii}^{\mathrm{D}}}{dt} = \sum_{j\neq K}\frac{\rho_{jj}}{\tau_{Kj}}, \quad i = K, \tag{17.114}$$

and arrive at the following results for the non-diagonal elements:

$$\frac{d\rho_{ij}^{\mathrm{D}}}{dt} = -\frac{1}{2}\left(\frac{1}{\tau_{iK}} + \frac{1}{\tau_{jK}}\right)\rho_{ij}, \quad i \neq K, j \neq K, \tag{17.115}$$

$$\frac{d\rho_{ij}^{\mathrm{D}}}{dt} = \frac{1}{2}\left(\frac{1}{\rho_{KK}}\sum_{k\neq K}\frac{\rho_{kk}}{\tau_{Kk}} - \frac{1}{\tau_{jK}}\right)\rho_{ij}, \quad i = K, j \neq K, \tag{17.116}$$

$$\frac{d\rho_{ij}^{\mathrm{D}}}{dt} = -\frac{1}{2}\left(\frac{1}{\rho_{KK}}\sum_{k\neq K}\frac{\rho_{kk}}{\tau_{Kk}} - \frac{1}{\tau_{iK}}\right)\rho_{ij}, \quad i \neq K, j = K. \tag{17.117}$$

The identities (17.110)–(17.117) supported by the relations (17.87) and (17.88), i.e. the laws of classical dynamics that define the nuclear trajectory, constitute the equations of motion in the decay-of-mixing model.

For any use of the decoherence notion in numerical computations, a prescription is needed of how to obtain the physical parameters determining this effect, i.e. the times τ_{kK} as well as, with recourse to Eq. (17.91), the decoherence force $\frac{d\mathbf{P}^{\mathrm{D}}}{dt}$. The latter is derived from arguments based on energy conservation [30], as will be shown in the following. Formally, we equate

$\frac{d\mathbf{P}^{\mathrm{D}}}{dt}$ to $F\mathbf{s}$, the product of a unit vector $\mathbf{s}$ and some scalar F. To specify this factor, we invoke total energy conservation, requiring the time derivative of the Hamilton function to vanish. Thus, we set

$$\frac{d(T_{\mathrm{N}} + V)}{dt} = \frac{\mathbf{P}}{\mu} \cdot \frac{d\mathbf{P}}{dt} + \frac{dV}{dt} = \frac{\mathbf{P}}{\mu} \cdot \frac{d\mathbf{P}^{\mathrm{C}}}{dt} + \frac{dV^{\mathrm{C}}}{dt} + \frac{\mathbf{P}}{\mu} \cdot \frac{d\mathbf{P}^{\mathrm{D}}}{dt} + \frac{dV^{\mathrm{D}}}{dt} = 0 \qquad (17.118)$$

where V is given by Eq. (17.85), and use has been made of the separation $\frac{dV}{dt} = \frac{dV^{\mathrm{C}}}{dt} + \frac{dV^{\mathrm{D}}}{dt}$ which is a consequence of Eqs. (17.92) and (17.93). From the same set of equations, we deduce the form of the decoherence contribution to the time derivative of V to be [615]

$$\frac{dV^{\mathrm{D}}}{dt} = \sum_k \left(x_k \frac{dx_k^{\mathrm{D}}}{dt} + p_k \frac{dp_k^{\mathrm{D}}}{dt} \right) V_{kk} + \sum_k \sum_{k<k'} \left(x_k \frac{dx_{k'}^{\mathrm{D}}}{dt} + x_{k'} \frac{dx_k^{\mathrm{D}}}{dt} + p_k \frac{dp_{k'}^{\mathrm{D}}}{dt} + p_{k'} \frac{dp_k^{\mathrm{D}}}{dt} \right) V_{kk'}. \qquad (17.119)$$

As energy conservation is guaranteed within the confines of the classical analog theory, the sums of the components indexed C and D have to vanish separately in Eq. (17.118). Combining the latter statement with $\frac{d\mathbf{P}^{\mathrm{D}}}{dt} = F\mathbf{s}$, we find

$$F = -\frac{dV^{\mathrm{D}}}{dt} \frac{\mu}{\mathbf{P} \cdot \mathbf{s}}. \qquad (17.120)$$

Thus, the decoherence-related change of the nuclear momentum, or the decoherence force, turns out to be

$$\frac{d\mathbf{P}^{\mathrm{D}}}{dt} = -\mu \frac{dV^{\mathrm{D}}}{dt} \frac{1}{\mathbf{P} \cdot \mathbf{s}} \mathbf{s}. \qquad (17.121)$$

The vector $\mathbf{s}$ denotes a *direction of decoherence*, related to the flux of energy deposited or absorbed during the decoherence process. The force $\frac{d\mathbf{P}^{\mathrm{D}}}{dt}$ is thus understood as the driving agent that causes the redistribution of energy associated with decoherence. Assuming a universal direction of decoherence, indicated by a state-unspecific vector $\mathbf{s}$, however, appears unfounded.

As implied by the expression for $\frac{dV}{dt}^{\mathrm{D}}$ to be derived in Exercise 17.7(b), we may write the decoherent contribution to the time derivative of V as

$$\frac{dV^{\mathrm{D}}}{dt} = \sum_{k \neq K} \frac{dV_k^{\mathrm{D}}}{dt} \tag{17.122}$$

and

$$\frac{dV_k^{\mathrm{D}}}{dt} = \frac{\rho_{kk}}{\tau_{Kk}} V_{KK} + \frac{d\rho_{kk}^{\mathrm{D}}}{dt} V_{kk} + \Re\left\{\frac{d\rho_{Kk}^{\mathrm{D}}}{dt}\right\} V_{Kk} + \sum_{k' \neq K} \Re\left\{\frac{d\rho_{kk'}^{\mathrm{D}}}{dt}\right\} V_{kk'}. \tag{17.123}$$

Equation (17.122) allows us to formulate the expression (17.121) for the decoherent force as a sum over all contributing electronic states with the exception of the *decoherent state* K:

$$\frac{d\mathbf{P}^{\mathrm{D}}}{dt} = -\sum_{k \neq K} \mu \left(\frac{dV_k}{dt}^{\mathrm{D}}\right) \frac{1}{\mathbf{P}\cdot\mathbf{s}_k} \mathbf{s}_k, \tag{17.124}$$

where $\mathbf{s}_k$ refers to a direction of decoherence related to the state with index k. It follows from the restriction $k \neq K$ on the summation in Eq. (17.124) that $\mathbf{s}_k = \mathbf{s}$ if only two electronic states are considered.

Exercise 17.7. (a) Show that

$$\frac{dV^{\mathrm{D}}}{dt} = \sum_{k} \frac{d\rho_{kk}^{\mathrm{D}}}{dt} V_{kk} + \sum_{k} \sum_{k' < k} 2\Re\left\{\frac{d\rho_{kk'}^{\mathrm{D}}}{dt}\right\} V_{kk'} \tag{17.125}$$

starting with Eq. (17.119). (b) Use part (a) of this problem in conjunction with relation (17.113) to prove

$$\frac{dV^{\mathrm{D}}}{dt} = \sum_{k \neq K} \left[\frac{\rho_{kk}}{\tau_{Kk}} V_{KK} + \frac{d\rho_{kk}^{\mathrm{D}}}{dt} V_{kk} + \Re\left(\frac{d\rho_{Kk}^{\mathrm{D}}}{dt}\right) V_{Kk} + \sum_{k' \neq K} \Re\left\{\frac{d\rho_{kk'}^{\mathrm{D}}}{dt}\right\} V_{kk'}\right]. \tag{17.126}$$

Zhu *et al.* [617] suggest an explicit expression for the state-dependent direction of decoherence, $\mathbf{s}_k$. The discussion of this entity is guided by the recognition that the momentum change due to decoherence may not alter the total angular momentum of the dynamic system. On the basis of this notion, a suitably constructed linear combination of $\mathbf{P}_{\text{vib}}$ (see Eq. (17.86)) and the longitudinal portion of the nonadiabatic coupling vector $\mathbf{A}_{Kk}$ that connects an arbitrary state k with the decoherent target state K is proposed. This vector indicates the direction of energy transfer related to decoherence in regions characterized by strong coupling between the states k and K. We point out that, just as $\mathbf{P}_{\text{vib}}$, the longitudinal part of $\mathbf{A}_{Kk}$ inhabits the non-rotating subspace of the nuclear coordinates, such that any linear combination of these two entities will conserve the nuclear angular momentum.

In the decay-of-mixing formalism, the decoherent state takes the role occupied by the instantaneous electronic state in the trajectory surface hopping technique. Sudden jumps in the electronic state, however, are avoided. Instead, the evolution of the system is smoothly guided by the force (17.124) which points towards the decoherent state. This state may change during propagation.

17.5.1 *Decoherence time*

Among the remaining problems is that of determining the decoherence time τ_{Kk}. Various proposals have been made regarding the explicit form of this quantity [30, 617]. While no rigorous derivation of τ_{Kk} has been accomplished, constraints have been specified that any analytic formula describing the decoherence time in the framework of a decay-of-mixing formalism has to obey. First, we note that the assumption of a well-defined decoherent force requires, by Eq. (17.124):

$$\frac{d{V_k}^{\mathrm{D}}}{dt} \sim (\mathbf{P} \cdot \mathbf{s}_k)^n, \quad n > 1, \tag{17.127}$$

ensuring regular behavior of the state-specific decoherent force in the limit of small momenta. From Eqs. (17.122) and (17.123) or Exercise 17.7 in cooperation with the expressions (17.113)–(17.117) for $\frac{d\rho_{ij}^{\mathrm{D}}}{dt}$, one concludes that the constraint (17.124) can only be guaranteed if a parallel statement holds true for the times τ_{kK}, namely

$$\frac{1}{\tau_{kK}} \sim (\mathbf{P}\cdot\mathbf{s}_k)^n, \quad n > 1. \tag{17.128}$$

Further, it is expected that the time scale related to electronic dephasing is the fastest in the dynamic system. This translates into the inequality

$$\tau_{kK} \geq \frac{\hbar}{|V_{kk} - V_{KK}|}, \quad \text{or} \quad \frac{1}{\tau_{kK}} \leq \frac{|V_{kk} - V_{KK}|}{\hbar}. \tag{17.129}$$

A simple choice for τ_{kK} that does justice to both criteria, (17.128) as well as (17.129), is proposed in [617, 618]. It is

$$\tau_{kK} = \frac{\hbar}{|V_{kk} - V_{KK}|}\left(C + \frac{2\mu E_0}{(\mathbf{P}\cdot\mathbf{s}_k)^2}\right) \tag{17.130}$$

where the constant C is dimensionless and E_0 is an energy. Assuming realistic orders of magnitude for τ_{kK}, the right-hand side of Eq. (17.130) does not depend strongly on either of the constants, as demonstrated by various numerical tests [618]. In the term $(\mathbf{P}\cdot\mathbf{s}_k)^2/2\mu$ we recognize the contribution to the instantaneous vibrational energy that results from projecting $\mathbf{P}$ on the decoherence direction $\mathbf{s}_k$. While the prefactor is purely quantal and dictated by the time-energy uncertainty relation, the second term accommodates the effect of the decoherent force due to the classical trajectory. On the basis of formula (17.130) one expects faster reduction of non-diagonal density matrix elements in cases of wide as compared to narrow spacing of the contributing levels, a tendency that has been observed in semiclassical investigations of an oscillator coupled to a bath [619]. The momentum dependence of Eq. (17.130) implies the intuitive trend of an infinite decoherence time as $\mathbf{P}\cdot\mathbf{s}_k \rightarrow 0$, which is realized as the temperature of the environment tends towards zero.

17.5.2 *Determining the decoherent state*

The presented method would be incomplete without a prescription for how to identify the decoherent state. Naturally, at the beginning of the propagation this state coincides with the one initially realized. As the simulation progresses, however, another state may assume this role. Thus, switches of the decoherent state occur along the decay-of-mixing trajectory, while this trajectory itself does not switch, in keeping with the Ehrenfest methodology and in contrast to the trajectory surface hopping paradigm. The latter method, however, may be employed to determine the instantaneous decoherent state [615, 617]. It also assigns to each electronic state a well-defined probability of becoming the decoherent state, thus reconciling the decay-of-mixing formalism with the notion that a system in contact with an environment decays into an incoherent mixture of states rather than a pure state.

To implement this procedure, Tully's fewest switches method [288] has been utilized. With reference to a two-state case, the probability $p_{K\to K'}$ for coherent state switching from label K to label K' in the time interval $[t, t+dt]$ is obtained from[1]

$$p_{K\to K'} = \mathrm{Max}\left\{\left(-\frac{d\rho_{KK}}{dt}\right)dt\frac{1}{\rho_{KK}}, 0\right\}, \quad K, K' = 1, 2. \qquad (17.131)$$

If, by this criterion, population is lost from K to K', the system may start decohering towards the alternative state. We emphasize that the switching criterion Eq. (17.131) does not decide the instantaneous electronic state, as is typical for trajectory surface hopping schemes, but exclusively the decoherent state. The generalization of formula (17.131) to a multistate situation can be found in [618].

A variety of switching techniques have been assessed with respect to their capacity for determining the decoherent state. For details, the reader is referred to a survey article of Zhu *et al.* [617]. The competing approaches have been applied to various model systems comprising two electronic states and examined by comparison with exact quantum mechanical calculations. Specifically, nonadiabatic processes of the form

$$\mathrm{A}^* + \mathrm{BC}(v, j) \to \mathrm{A} + \mathrm{BC}(v', j'), \qquad (17.132)$$

$$\mathrm{A}^* + \mathrm{BC}(v, j) \to \mathrm{B} + \mathrm{AC}(v'', j'') \qquad (17.133)$$

have been considered [29], involving the deexcitation of a model atom A, which may simulate a metal atom, in the course of its interaction with a model diatom BC, and the quenching of A without rearrangement of the scattering partners.

[1] This definition of the coherent state switching probability is referred to as *natural switching*. Replacing the time derivative of the total density with that of its coherent component ρ^{C} specifies the alternative of *self-consistent*, or *coherent switching* [617]. From an assessment of five methods of nonadiabatic molecular dynamics, including a broad variety of test models, the coherent switching method turned out to yield, by the standard of fully quantal computations, the best balance between coherence and decoherence [620].

Three basic situations were studied: (1) systems of the Landau–Zener type, i.e. situations involving two crossing diabatic potential energy curves with non-vanishing diabatic coupling at the crossing point; (2) conical intersections, differing from the model systems included under (1) by diabatic coupling terms set equal to zero [621]; (3) Rosen–Zener–Demkov systems, involving non-crossing diabatic curves which are weakly coupled [622].

Using these prototypes, several variations of the decay-of-mixing formalism, as outlined before, have been tested, among them a linearization of the dynamic equations (17.115)–(17.117) for the non-diagonal density matrix elements. Further, a diversity of switching strategies to define and redefine the decoherent state along the Ehrenfest trajectory has been evaluated. In each case a set of observables was computed, among them the probabilities for the reaction (17.133) and for the non-reactive quenching process (17.132). Average deviation from the fully quantum mechanical standard results on the order of 25% have been obtained. Comparative studies of the atom–diatom collisions indicated earlier have established the decay-of-mixing approach as more accurate than the pure Ehrenfest formalism [615] as well as a diversity of implementations of the trajectory surface hopping scheme [618]. The various flavors of the decay-of-mixing methodology were shown to be of comparable quality, demonstrating a relatively weak dependence of the latter on the details of the computational realization.

Part III: Special Topics

Chapter 18

Ultrafast Optical Spectroscopy

The advent of pulsed lasers with pulse durations in the femtosecond regime has made it possible to trace the evolution of a molecular system by direct observation in the time domain. This success is related to the advances made by nonlinear spectroscopy during the last two decades, and in particular to the development and refinement of pump–probe procedures that yield dynamic information about the considered molecular systems at the time scale of molecular evolution, i.e. typically the femtosecond range. The respective experimental research relies on support from the side of theoretical modeling, as the measured results by themselves usually do not allow for definitive conclusions regarding the number of relevant electronic states or vibrational modes nor the nature of the inspected dynamic process (e.g. [612]). Before discussing some characteristic observations on photonic interactions in nonadiabatic systems along with their theoretical interpretations, we define in Section 18.1 the major spectroscopic notions underlying these studies. The next section (18.2) deals with the interpretation of nonlinear optical spectroscopy experiments in terms of polarization components and introduces perturbative (Section 18.2.1) as well as non-perturbative approaches (Section 18.2.2) to this problem. The latter approximation is exemplified by a theoretical treatment of the *cis-trans*-photoisomerization phenomenon, as presented in Section 18.3. While the schemes discussed in Sections 18.2 and 18.3 employ a wave packet picture and therefore the TDSE as a device to generate the relevant molecular dynamics, models based on density matrices and thus on the Liouville–von Neumann equation will be presented in the two following sections. The respective perturbation theory is outlined in Section 18.4, and its application to the pump–probe spectroscopy of small inorganic species, namely non-stoichiometric alkali-halide clusters, is described in the concluding section of this chapter.

18.1 Linear and Nonlinear Polarization

Optical interactions with matter can be categorized into linear and nonlinear effects, with reference to the dependence of the observed process on the electric radiation field $\boldsymbol{\epsilon}(\mathbf{X}, t)$. In the following, we will describe the coupling between electromagnetic radiation and matter in the framework of the electric dipole approximation, neglecting contributions due to the magnetic field as well as electric terms of quadrupolar and higher order.

The electric radiation field $\boldsymbol{\epsilon}(\mathbf{X}, t)$ associated with a light wave or pulse interacting with a medium tends to displace charges located at the place $\mathbf{X}$ within the medium. This effect may be described classically in terms of a *polarization vector* $\mathbf{P}(\mathbf{X}, t)$, understood as *induced dipole moment* $\boldsymbol{\mu}^{ind}$ *per unit volume*, a representation of a macroscopic oscillating dipole that may act as absorber or emitter of electromagnetic radiation. A convenient integral representation of the polarization vector employs the *susceptibility tensor* $\boldsymbol{\chi}(\mathbf{X}, t)$ that accounts for the response of the medium to the incident field:

$$\mathbf{P}(\mathbf{X}, t) = \varepsilon_0 \int_0^{\infty} dt' \boldsymbol{\chi}(\mathbf{X}, t) \boldsymbol{\varepsilon}(\mathbf{X}, t - t'). \tag{18.1}$$

With respect to the electric field, the polarization may be expanded (e.g. [430]) in the following way:

$$\mathbf{P}(\mathbf{X}, t) = \mathbf{P}^{(1)}(\mathbf{X}, t) + \mathbf{P}^{(2)}(\mathbf{X}, t) + \cdots + \mathbf{P}^{(n)}(\mathbf{X}, t), \tag{18.2}$$

where the component $\mathbf{P}^{(n)}$ depends on the nth power of $\boldsymbol{\varepsilon}$. Thus $\mathbf{P}^{(1)}$ stands for the *linear polarization*. Assembling all terms with $n > 1$, we define the nonlinear polarization $\mathbf{P}^{\mathrm{NL}}$ by partitioning $\mathbf{P}(\mathbf{X}, t)$ according to

$$\mathbf{P}(\mathbf{X}, t) = \mathbf{P}^{(1)}(\mathbf{X}, t) + \mathbf{P}^{\mathrm{NL}}(\mathbf{X}, t). \tag{18.3}$$

The linear polarization governs a multitude of familiar optical phenomena, describing, for instance, the absorption, reflection, and refraction of light. The expansion (18.2) is valid if the nonlinear polarization terms can be understood as corrections to the dominant linear effect. The harmonic time dependence of an elementary monochromatic wave with $\boldsymbol{\varepsilon} \sim \boldsymbol{\varepsilon}(t = 0) \exp(i\omega t)$ implies that the nonlinear polarization involves novel processes that differ fundamentally from the linear ones. Thus, the term $\mathbf{P}^{(2)}$ describes a doubling of the incoming radiation frequency by the oscillating dipoles of the medium, $\mathbf{P}^{(3)}$ a tripling, etc. *Second harmonic generation*, as first reported by Franken *et al.* [623] who observed ultraviolet emission at $\lambda = 3,471$ Å from a quartz crystal traversed by a ruby laser beam at $\lambda = 6,942$ Å, provides a characteristic example of the former phenomenon. This technique represents an elementary case of *three wave mixing*, which more generally refers to the linear combination of the wave vectors $\mathbf{k}_1$, $\mathbf{k}_2$, and frequencies ω_1, ω_2, of two incoming waves, giving rise to difference or sum signals of the form $\mathbf{k}_s = \mathbf{k}_1 - \mathbf{k}_2$, $\omega_s = \omega_1 - \omega_2$, or $\mathbf{k}_s = \mathbf{k}_1 + \mathbf{k}_2$, $\omega_s = \omega_1 + \omega_2$ [430]. If the dipolar approximation is adopted, however, $\mathbf{P}^{(2)}$ vanishes in random isotropic media with inversion symmetry, encompassing glasses, liquids, and gases. The lowest order of nonlinear effects is then three, corresponding to four-wave mixing processes. *Stimulated Raman scattering*, *third harmonic generation*, and *photon echo* are some of the nonlinear optical phenomena that are associated with third-order polarization.

We will start with a condensed overview of the basic ideas underlying the methods that are presently used to explore the adiabatic and nonadiabatic dynamics of molecular systems by nonlinear spectroscopy. The technology of ultrashort laser pulses with durations on the order of about 10 fs made the direct observation of dynamic molecular phenomena accessible to spectroscopic experiment (see e.g. [624]). The time resolution accomplished by these methods matches the time scales of the actual molecular processes. A wealth of applications of ultrafast spectroscopic techniques in a wide diversity of disciplines, ranging from physical chemistry and condensed matter physics to biology and medical application, testify to the success and rapid expansion of these procedures (for reviews see [625]–[629]).

Among them, the most broadly established one is the *pump–probe technique* with its numerous variants. The common principle of all femtosecond pump–probe techniques is the preparation of a non-stationary state by a pump pulse followed by an interrogating probe pulse with a well-defined delay time. This scheme makes it possible to monitor the evolution of excited electronic states in real time. The elementary steps involved in an idealized two-pulse pump–probe experiment comprise three distinct processes: (1) the creation of an excited nonstationary state $|\Psi_{\mathrm{E}}\rangle$ from the ground state $|\Psi_{\mathrm{G}}\rangle$ of the inspected system by application of an ultrashort laser pulse at $t = 0$, (2) the evolution of this intermediate state during a time interval Δt according to $|\Psi_{\mathrm{E}}(t)\rangle = \exp(-\frac{i}{\hbar}\hat{H}_{\mathrm{E}}\Delta t)|\Psi_{\mathrm{E}}\rangle$, (3) the projection of the evolved excited state $|\Psi_{\mathrm{E}}(t)\rangle$ on a specified final state $|\Psi_{\mathrm{F}}\rangle$ by use of the probe pulse [3]. Pump–probe procedures may be classified according to the type of final state employed in the measurement. Thus, the electronic part of $|\Psi_F\rangle$ could be that of the ground state [3, 630], or, alternatively, another electronic state which is higher in energy than $|\Psi_{\mathrm{E}}(\Delta t)\rangle$ (e.g. [631]). Also, the cationic ground state (e.g. [632]) and high-lying Rydberg states (e.g. [633]) have taken the role of the final state in the probe mechanism. The various choices of the final state correspond to a variety of detection strategies. We emphasize that pump–probe spectroscopy usually yields simultaneously time- and frequency-resolved information, as a series of spectra is recorded in temporal succession. In the following, we will discuss the pump–probe signal obtained from probe pulse transmittance as a function of the time delay between the two pulses [634].

18.1.1 *Deriving the pump–probe signal*

The term *pump–probe signal* usually refers to the transmittance of the probe signal, and more specifically, to the difference between this transmittance with and without application of the pump pulse [3]. To introduce the basic concepts needed for the description of this signal, we define a classical time-dependent electric field by

$$\boldsymbol{\varepsilon}(\mathbf{X}, t) = \sum_{i=1,2} \boldsymbol{\varepsilon}_i(\mathbf{X}, t). \tag{18.4}$$

The two terms in this sum denote the pump field ($\boldsymbol{\varepsilon}_1$) and the probe field ($\boldsymbol{\varepsilon}_2$). The functions $\boldsymbol{\varepsilon}_i(\mathbf{X}, t)$, $i = 1, 2$ are further decomposed into

$$\boldsymbol{\varepsilon}_i(\mathbf{X}, t) = \mathbf{E}_i(\mathbf{X}, t) \exp(i\mathbf{k}_i \cdot \mathbf{X}) + \mathbf{E}_i^*(\mathbf{X}, t) \exp(-i\mathbf{k}_i \cdot \mathbf{X}), \tag{18.5}$$

where the factors $\mathbf{E}_i(\mathbf{X}, t)$ contain the pulse envelope functions $E_i(\mathbf{X}, t)$ and the polarization vectors $\mathbf{e}_i$

$$\mathbf{E}_i(\mathbf{X}, t) = \mathbf{e}_i E_i(\mathbf{X}, t) \exp(-i\omega_i t), \tag{18.6}$$

with ω_i as laser frequency. For a theoretical derivation of the pump–probe pulse, a prescription is needed to calculate the electric field after it has passed through the sample of interest. This makes it necessary to investigate the propagation of the probe pulse through a medium with nonlinear polarization by solving the respective Maxwell equation [635]. To facilitate this task, we invoke the *slowly varying envelope approximation* [634, 636], assuming

Gaussian-shaped pulse envelopes whose spatial extension exceeds the pulse wave length significantly. This simplification reduces the differential equation for the propagation of E_2 from second to first order. Adopting a scalar field $E_2(Z, t)$ that is polarized in the X-direction and travels along the Z-axis, the wave equation becomes

$$\left(\frac{\partial}{\partial Z} + \frac{1}{v}\frac{\partial}{\partial t}\right) E_2(Z, t) = \frac{2\pi i \omega_2}{nc} P(Z, t) \tag{18.7}$$

where $n = v/c$ stands for the refractive index of the traversed medium.

The quantity $P(Z, t)$ denotes the macroscopic polarization of the medium. An adequate solution of the problem given by relation Eq. (18.7) still poses a formidable challenge, related to the fact that the macroscopic polarization itself depends on the field E_2. Icsevgi and Lamb [635] discuss the self-consistent treatment of a pulse traversing a nonlinear medium, based on an interplay between the Liouville formalism and the Maxwell equations. An initial guess of the electric field provides an approximation that yields, by use of the Liouville–von Neumann equation as well as statistical summation, a first result for the macroscopic polarization. The latter quantity is used as source term for Maxwell's equations, leading to an improved expression for the electric field. The solutions for E_2 and P are obtained from iterating this cycle.

A dramatic simplification is achieved if the sample can be considered optically thin [634], implying that the incident electric field is not altered as it passes through the medium. In this case, the final field is represented as a sum of the initial field and a contribution due to the polarization induced in the sample. If, in addition, the envelope function does not change appreciably with time, Eq. (18.7) reduces to the elementary relation

$$E_2(L, t) = E_2(0, t) + \frac{2\pi i \omega_2 L}{nc} P(0, t) \tag{18.8}$$

with L as the length traversed by the pulse.

An expression for the pump–probe signal is obtained from the rate of energy dissipated by the probe pulse as it interacts with the medium. As implied by classical electrodynamics, this rate is given by the product $\boldsymbol{\varepsilon}(\mathbf{X}, t) \cdot \frac{\partial}{\partial t}\mathbf{P}(\mathbf{X}, t)$ [636]. Associating a wave vector $\mathbf{k}_2$ with the probe pulse (see Eq. (18.5)), we imagine a photodetector placed at the end of the medium and gathering all photons emitted in the direction $\mathbf{k}_2$. The pump–probe signal is proportional to ΔI, the difference in energy dissipation by the probe pulse with the pump pulse on or off. Integrating with respect to time over the dissipation rate, we express this quantity as

$$\Delta I \sim \int_{-\infty}^{+\infty} dt \frac{\partial}{\partial t} \boldsymbol{\varepsilon}_2(t) \cdot \Delta \mathbf{P}(t) \tag{18.9}$$

where the time derivative has been shifted from the polarization to the field and the dependence of the vectors on the spatial coordinate $\mathbf{X}$ has been suppressed for the sake of simplicity. The symbol $\Delta\mathbf{P}(t)$ stands for the difference between the polarizations of the medium propagating in the $\mathbf{k}_2$ direction as the pump pulse is applied or not [637]:

$$\Delta \mathbf{P}(t) = \mathbf{P}(t)_{\text{pump on}} - \mathbf{P}(t)_{\text{pump off}}. \tag{18.10}$$

The energy ΔI is defined as negative in the absorption case and as positive for emission, corresponding to the loss and the gain of photons, respectively. It should be noted that a dependence on both the delay time Δt between the pump and the probe pulse and the frequency ω_2 are implicitly contained in Eq. (18.9). To make this dependence explicit, we insert the expansion Eq. (18.5) for $\boldsymbol{\varepsilon}_2$ into Eq. (18.9) and proceed in the same way with the polarization field. As the slowly varying envelope approximation is assumed to be valid, reducing the time dependence of the electric field to the periodic factors in Eq. (18.6), the time derivative can be performed. Further, we make the *rotating wave approximation* [638], including only *resonant* terms in the integrand Eq. (18.9) while the contributions with rapidly oscillating time behavior are assumed to be negligible. This eliminates components with factors of the form $\exp[i(\omega_2 + \Delta\omega)t]$, if ω_2 is the probe frequency and $\Delta\omega$ a transition frequency of the studied molecular system, whereas the corresponding $\exp[i(\omega_2 - \Delta\omega)t]$ terms are included.[1] The resulting simplified description makes it possible, using Eqs. (18.3),

[1]The term *rotating wave approximation* refers to the interaction picture. Assuming dipolar interaction of a system with an electromagnetic wave, and applying the transformation (3.11), one may partition the wave into a component that co-rotates and another that counter-rotates with respect to the wave function of the system. The rotating wave approximation consists in keeping only the former and discarding the latter contribution.

(18.5), and Eq. (18.6), to express the signal as a function of ω_2:

$$\Delta I(\omega_2, \Delta t) \sim 2\omega_2 \Im \int_{-\infty}^{+\infty} dt\, \boldsymbol{\varepsilon}_2(t) \cdot \Delta \mathbf{P}^*(t). \qquad (18.11)$$

Pump–probe experiments usually involve both time- and frequency-resolved spectroscopy. To simulate the results of experiments where spectra are recorded for each delay time considered, a dispersed pump–probe signal I_d has to be approximated. This may be done by reformulating Eq. (18.11) in terms of the Fourier transforms $\tilde{\boldsymbol{\varepsilon}}_2$, $\Delta\tilde{\mathbf{P}}$:

$$I_d(\omega, \Delta t) \sim 2\omega\, \Im\{\tilde{\boldsymbol{\varepsilon}}_2(\omega) \cdot \Delta\tilde{\mathbf{P}}\}. \qquad (18.12)$$

We have described the basis for a *heterodyne detection scheme*, related to a signal that involves the product of the probe field and the polarization, $\mathbf{E} \cdot \mathbf{P}^*$. This corresponds to observing the evolution of electronic populations [637]. However, time-resolved information about the studied system may also be extracted by use of a *homodyne scheme*, involving the magnitude of the emitted polarization. This alternative strategy is, as will be clarified in the following discussion of the polarization in terms of quantum mechanics, sensitive to the time evolution of electronic coherence.

18.2 Theory of Nonlinear Polarization in Femtosecond Molecular Spectroscopy

In this section we will derive expressions for the nonlinear polarization which in turn determines the experimentally detected signals according to Eqs. (18.9)–(18.12), including both the perturbative and the non-perturbative treatment of the problem, as proposed by Seidner *et al.* [637]. Starting with the former, we consider the dipolar interaction of a molecule with electromagnetic radiation, adopting the Schrödinger picture, and assuming two diabatic electronic states, ψ_0 and ψ_1, interacting through a coupling potential V. The evolution of the molecular wave function $\Psi(t)$ is governed by the TDSE with a time-dependent Hamiltonian $\hat{H}(t)$, partitioned into a molecular and an electric interaction term:

$$\hat{H}(t) = \hat{H}_M + \hat{H}_{int}(t). \qquad (18.13)$$

Expanding the molecular part $\hat{H}_M$ into the diabatic basis, we find

$$\hat{H}_{\mathrm{M}} = \sum_{k=0,1} |\psi_k\rangle \hat{h}_k \langle\psi_k| + \{|\psi_0\rangle \hat{V} \langle\psi_1| + |\psi_1\rangle \hat{V} \langle\psi_0|\}. \tag{18.14}$$

The operators $\hat{h}_k$ ($k = 1, 2$) stand for the nuclear Hamiltonians of the two diabatic states. Lastly, the interaction with the external electric field is given by the Hamiltonian

$$\hat{H}_{\mathrm{int}} = -\{|\psi_0\rangle \boldsymbol{\mu} \cdot \boldsymbol{\varepsilon}^*(t) \langle\psi_1| + |\psi_1\rangle \boldsymbol{\mu} \cdot \boldsymbol{\varepsilon}(t) \langle\psi_0|\} \tag{18.15}$$

with $\boldsymbol{\mu}$ as the electric transition dipole moment. We point out that $\boldsymbol{\varepsilon}(t)$ may describe the succession of the two signals $\boldsymbol{\varepsilon}_1$ and $\boldsymbol{\varepsilon}_2$ in temporal separation from each other, as will be assumed in the following development. In the general case, however, allowance has to be made for overlapping pulses. To obtain the polarization $\mathbf{P}$, we evaluate

$$\begin{aligned} \mathbf{P}(t) &= \langle\Psi(t)|\hat{\boldsymbol{\mu}}|\Psi(t)\rangle \\ &= 2\Re\{\langle\Psi(t)|\psi_0\rangle \boldsymbol{\mu}_{01} \langle\psi_1|\Psi(t)\rangle\} \\ &\equiv \mathbf{P}_{01}(t) + \mathbf{P}_{10}(t). \end{aligned} \tag{18.16}$$

We now substantiate the remark made in the last paragraph of the previous section, relating the heterodyne detection scheme to monitoring the temporal development of the electronic populations. Introducing the label N_0 (N_1) for the ground state (first excited state) populations, we write

$$\frac{dN_0}{dt} = \langle\dot{\Psi}|\psi_0\rangle\langle\psi_0|\Psi\rangle + \langle\Psi|\psi_0\rangle\langle\psi_0|\dot{\Psi}\rangle. \tag{18.17}$$

Using the TDSE and Eqs. (18.14) and (18.15), the right-hand side of this equation becomes

$$2\Re\{\langle\hat{O}p\Psi|\psi_0\rangle\langle\psi_0|\Psi\rangle\} = 2\Re\{\langle\Psi|\hat{O}p^\dagger\psi_0\rangle\langle\psi_0|\Psi\rangle\}, \tag{18.18}$$

with $\hat{O}p \equiv \frac{1}{i\hbar}\hat{H}_{\mathrm{M}} - \frac{1}{i\hbar}|\psi_0\rangle\boldsymbol{\mu}_{01}\langle\psi_1| \cdot \boldsymbol{\varepsilon}^*$, and thus $\hat{O}p^\dagger = -\frac{1}{i\hbar}\hat{H}_{\mathrm{M}} + \frac{1}{i\hbar}$ $|\psi_1\rangle\boldsymbol{\mu}_{10}\langle\psi_0| \cdot \boldsymbol{\varepsilon}$. This establishes

$$\frac{dN_0}{dt} = \frac{2}{\hbar}\Im\{\langle\Psi|\psi_1\rangle\boldsymbol{\mu}_{10}\langle\psi_0|\Psi\rangle \cdot \boldsymbol{\varepsilon}\} = \frac{2}{\hbar}\Im\{\mathbf{P}_{10} \cdot \boldsymbol{\varepsilon}\}. \tag{18.19}$$

The change of the ground state population is obtained by integrating this expression with respect to time. This yields [443]

$$\Delta N_0 = \frac{2}{\hbar}\Im\left\{\int dt\, \mathbf{P}_{10} \cdot \boldsymbol{\varepsilon}\right\}. \tag{18.20}$$

A parallel argument connecting the product $\mathbf{P}_{10}(t) \cdot \dot{\boldsymbol{\varepsilon}}(t)$ with the energy difference corresponding to the loss of ground state population is to be worked out in Exercise 18.1. As a consequence, the heterodyne observation mode is applicable in the case of population change, as stated before.

Exercise 18.1. Following the lines of the preceding derivation for ΔN_0, prove that this ground state population change is associated with an energy change ΔE described by $-2\Re\{\mathbf{P}_{10}(t) \cdot \dot{\boldsymbol{\varepsilon}}(t)\}$ and conclude that

$$\Delta E(\omega) = -\hbar\omega\Delta N_0(\omega). \tag{18.21}$$

The molecular wave function $|\Psi(t)\rangle$ may be derived by perturbation theory or by a non-perturbative procedure. If the former avenue is taken, each order of approximation can be related to a specific type of radiation–matter interaction, such as stimulated emission or stimulated Raman scattering. The overall polarization is represented as a sum over all the components thus obtained. It is understood as a composite of various processes, each of them individually accessible to spectroscopic detection, and therefore lends itself in a natural way to interpreting experimental observations.

The perturbative approach, however, fails in the limit of strong laser fields, and in those cases, one may take recourse to non-perturbative methods. These yield the molecular wave function $|\Psi(t)\rangle$ and, via Eq. (18.16), the total polarization $\mathbf{P}(t)$. While the different orders of perturbation of the molecule by the laser field assemble $\mathbf{P}(t)$ in a synthetic fashion, an analytic strategy is needed to break up the total polarization resulting from a non-perturbative scheme into physically meaningful segments that can be associated with measured spectra. In the following, we will introduce both techniques and demonstrate their capability to provide information about typical nonadiabatic situations in nonlinear optical spectroscopy.

18.2.1 *The perturbative approach*

Describing $|\Psi(t)\rangle$ by a perturbation series, one may generate the various contributions to the overall polarization. Processes up to third order, corresponding to the nonlinear response to $\mathbf{P}^{(3)}$, will receive particular attention. This means restricting the computational treatment to a weak-field limit that may be justified in the case of low laser intensity. Defining $|\Psi_0\rangle$ as the ground state of the diabatic system and the time-dependent function $|\Psi^{(0)}(t)\rangle = \exp(-\frac{i}{\hbar}\hat{H}_{\mathrm{M}}t)|\Psi_0\rangle$ as the zeroth approximation to the true wave function, we write

$$|\Psi(t)\rangle = \sum_{N=0}^{\infty} |\Psi^{(N)}(t)\rangle. \tag{18.22}$$

Using $-\boldsymbol{\mu}\cdot\boldsymbol{\varepsilon}(t)$ as perturbation, the Nth approximation $|\Psi^{(N)}(t)\rangle$ becomes

$$|\Psi^{(N)}(t)\rangle = -\frac{i}{\hbar}\int_{-\infty}^{t} dt' \exp\left[-\frac{i}{\hbar}\hat{H}_{\mathrm{M}}(t-t')\right]\boldsymbol{\mu}\cdot\boldsymbol{\varepsilon}(t')|\Psi^{(N-1)}(t')\rangle. \tag{18.23}$$

Inserting the series Eq. (18.22) into Eq. (18.16) leads to a hierarchy of polarization terms, beginning with the linear response and followed by the nonlinear orders:

$$\mathbf{P}(t) = \sum_{N=0}^{\infty} \mathbf{P}^{(2N+1)}(t) \tag{18.24}$$

and

$$\mathbf{P}^{(2N+1)}(t) = 2\Re\sum_{j=0}^{N}\langle\Psi^{2(N-j)}(t)|\psi_0\rangle\boldsymbol{\mu}_{01}\langle\psi_1|\Psi^{(2j+1)}(t)\rangle. \tag{18.25}$$

The exclusive occurrence of odd-numbered components in Eq. (18.25) is a direct consequence of the ungerade symmetry property of the dipolar interaction operator in conjunction with the definition of the polarization vector by Eq. (18.16) and the wave function expansion Eq. (18.22). Under these premises, the non-vanishing polarization term of lowest order in Eq. (18.25) is the contribution $\mathbf{P}^{(3)}$. Processes of this category may be realized by several configurations of pump–probe events, as illustrated in Figure 18.1. In spectroscopic experiments, it is possible to distinguish between the various interactions contained in Eq. (18.25) by the direction of the emitted

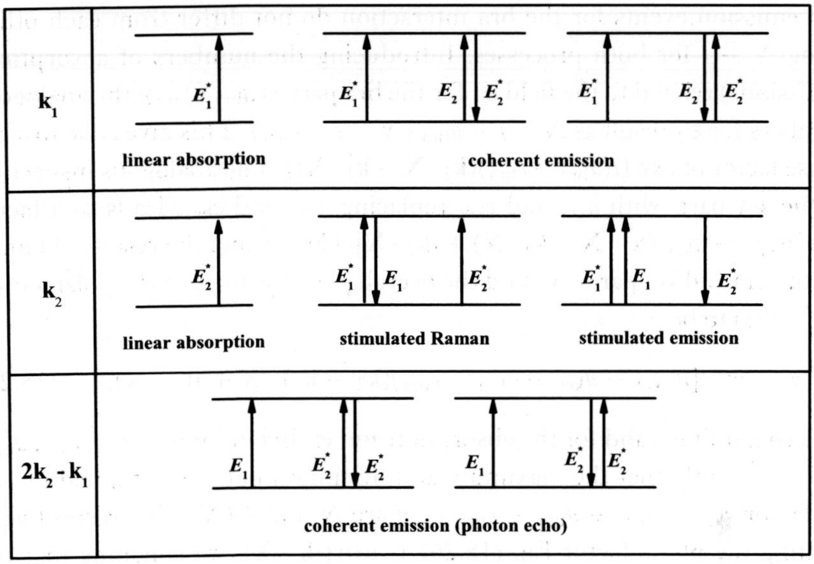

Figure 18.1: Illustration of optical processes contributing to the electric polarization up to third order. Under the simplifying assumptions discussed in the text, namely the rotating wave and non-overlapping pulse approximations, only three wave vectors are possible for the emitted polarization field, namely $\mathbf{k}_1$, $\mathbf{k}_2$, and $2\mathbf{k}_2 - \mathbf{k}_1$, defining directions for the observation of the indicated processes in spectroscopic experiments. (Reprinted with permission from [637]. Copyright (1995) by the American Institute of Physics.)

polarization signal. The wave vector $\mathbf{k}$ that defines the spatial orientation of this radiation is composed of the wave vectors $\mathbf{k}_1$ and $\mathbf{k}_2$ of the pump and probe fields $\boldsymbol{\varepsilon}_1$ and $\boldsymbol{\varepsilon}_2$, respectively. The entirety of the resulting phase factors $\exp[i(n\mathbf{k}_1 + m\mathbf{k}_2)\cdot \mathbf{X}]$ of $\mathbf{P}^{(2N+1)}$ determines the angles under which the polarization signal can be detected. If no further constraints are imposed on the phase factors associated with the $2N + 1$ interaction terms contained in Eq. (18.25), the integers n and m satisfy the relation $|n| + |m| = 2N + 1$ for any N. The multitude of allowed emission directions, however, is greatly reduced if the rotating wave approximation is applicable. Conventionally, one associates with any absorption event included in the ket vector $|\Psi^{(2j+1)}(t)\rangle$ of Eq. (18.25) a factor $\exp(i\mathbf{k}_l \cdot \mathbf{X})$, $l = 1, 2$, and a factor $\exp(-i\mathbf{k}_l \cdot \mathbf{X})$ with any emission. For the bra vector $\langle\Psi^{2(N-j)}(t)|$, the complex conjugates of these factors have to be employed. Utilizing the rotating wave approximation, we subdivide the $2j + 1$ ket interaction terms in Eq. (18.25) into $j+1$ absorptions and j emissions. The corresponding numbers of absorption and emission events for the bra interaction do not differ from each other, being $N - j$ for both processes. Introducing the numbers of absorptions (emissions) related to the field $\boldsymbol{\varepsilon}_1$ for the bra part as a_{bra} (e_{bra}), the analogous numbers for $\boldsymbol{\varepsilon}_2$ result as $N - j - a_{\text{bra}}$ ($N - j - e_{\text{bra}}$). This gives rise to a bra phase factor of $\exp[i(a_{\text{bra}} - e_{\text{bra}})(\mathbf{k}_2 \cdot \mathbf{X} - \mathbf{k}_1 \cdot \mathbf{X})]$. The analogous inspection of the ket part, with a_{ket} and e_{ket} replacing a_{bra} and e_{bra}, leads to a factor $\exp[i(e_{\text{ket}} - a_{\text{ket}})(\mathbf{k}_2 \cdot \mathbf{X} - \mathbf{k}_1 \cdot \mathbf{X}) + i\mathbf{k}_2 \cdot \mathbf{X}]$. Combining the results obtained for the bra and ket parts, we find the overall phase factor φ of the polarization $\mathbf{P}^{(2N+1)}(t)$ to be

$$\varphi = \exp[i(e_{\text{ket}} - a_{\text{ket}} + a_{\text{bra}} - e_{\text{bra}})(\mathbf{k}_2 - \mathbf{k}_1) \cdot \mathbf{X} + i\mathbf{k}_2 \cdot \mathbf{X}]. \qquad (18.26)$$

The constraints valid for the absorption and emission indices a_{ket}, e_{ket}, a_{bra}, and e_{bra}[2] imply that the maximum and minimum numbers attained by the prefactor $e_{\text{ket}} - a_{\text{ket}} + a_{\text{bra}} - e_{\text{bra}} \equiv n$ are N and $-(N + 1)$, respectively. Writing the phase factor Eq. (18.26) as $\exp(i\mathbf{k} \cdot \mathbf{X})$, we conclude that the wave vector varies within limits given by

$$\mathbf{k} = n(\mathbf{k}_2 - \mathbf{k}_1) + \mathbf{k}_2, \quad n = -(N + 1), \ldots, N. \qquad (18.27)$$

In addition to the rotating wave approximation, many pump–probe experiments admit for still simpler treatment, related to the temporal separation of the two pulses. Thus, if the pump and the probe pulse do not overlap in time, each step in the sequence of field–matter interactions included in the series Eq. (18.25) involves only one of the fields $\boldsymbol{\varepsilon}_1$ and $\boldsymbol{\varepsilon}_2$, where it is understood that the pump pulse $\boldsymbol{\varepsilon}_1$ precedes the probe pulse $\boldsymbol{\varepsilon}_2$. This condition introduces further restrictions on the allowed values of the absorption and emission indices. Thus, the rotating wave approximation and its requirement that the ket interactions comprise one more absorption than emission event while the numbers of both electromagnetic processes are equal for the bra interaction, namely $N - j$ (see earlier), imply that the difference between $a_{\text{ket(bra)}}$ and $e_{\text{ket(bra)}}$ is maximally one:

$$0 \le a_{\text{ket}} - e_{\text{ket}} \le 1, \qquad (18.28)$$

$$0 \le a_{\text{bra}} - e_{\text{bra}} \le 1. \qquad (18.29)$$

[2]These are, by the definition of the numbers of absorption (emission) events related to $\boldsymbol{\varepsilon}_2$, $a_{\text{bra}} \le N - j$, $e_{\text{bra}} \le N - j$, $a_{\text{ket}} \le j + 1$, $e_{\text{ket}} \le j$.

The limitations (18.28) and (18.29) confine the prefactor n in Eq. (18.27) to the values -1, 0, $+1$. This singles out three possible wave vectors, namely $\mathbf{k}_1$, $\mathbf{k}_2$, and $2\mathbf{k}_2 - \mathbf{k}_1$. Each of the three directions specified by these vectors is associated with a certain subset of optical interactions. Controlled detection under one of these directions is therefore a way to select processes of interest and, by the same token, to exclude others. Figure 18.1 shows a schematic representation of some optical interactions observable along these three directions. Contributions up to third order are included.

The wave vectors $\mathbf{k}_1$ and $\mathbf{k}_2$ are compatible with linear as well as third-order nonlinear processes, while $2\mathbf{k}_2 - \mathbf{k}_1$ contributes only to the latter. Each arrow stands for an electronic transition induced by the electric fields and their conjugates, each diagram presents, from left to right, a time-ordered succession of events. Adding a further interaction with the field $\boldsymbol{\varepsilon}_2$ at the outermost right of each diagram generates a representation of the dispersed pump–probe signal obtained in frequency-resolved spectroscopy (see Eq. (18.12)). By use of the recipe (18.25), the linear response of the system to the pump or the probe pulse is readily derived from

$$P^{(1)}_{\mathbf{k}_j}(t) = 2\Re\langle\Psi^{(0)}(t)|\hat{\mu}_{01}|\Psi^{(1)}_{\mathbf{k}_j}(t)\rangle, \quad j = 1, 2, \tag{18.30}$$

where we define the scalar operator $\hat{\mu}_{01}$ by use of the scalar product between the dipolar moment operator and the polarization vector of the field $\boldsymbol{\varepsilon}_j$ introduced in Eq. (18.6) as

$$\hat{\mu}_{01} = |\psi_0\rangle \mathbf{e}_j \cdot \boldsymbol{\mu}_{01}\langle\psi_1|. \tag{18.31}$$

Consequently, Eq. (18.30) describes the projection of the electric polarization on the field polarization vector. Applying the perturbative recursion scheme Eq. (18.23) allows us to write $P^{(1)}_{\mathbf{k}_j}$ in fuller detail. We arrive at [3]

$$P^{(1)}_{\mathbf{k}_j}(t) = 2\Re\left\{\frac{i}{\hbar}\exp(i\mathbf{k}_j\cdot\mathbf{X})\int_{-\infty}^{t} dt' E_j(t')\langle\Psi^{(0)}|\hat{\mu}_{01}\exp\left[-\frac{i}{\hbar}\hat{H}_{\mathrm{M}}(t-t')\right]\right.$$
$$\left.\times\,\hat{\mu}_{10}\exp\left[\frac{i}{\hbar}\hat{H}_{\mathrm{M}}(t-t')\right]|\Psi^{(0)}\rangle\right\}. \tag{18.32}$$

If $j = 1$, $\boldsymbol{\varepsilon}_j$ is the pump pulse. In this situation, the detectible signal is, according to Eq. (18.12), proportional to $\boldsymbol{\varepsilon}_1^* \cdot \boldsymbol{\varepsilon}_2$ which will vanish if a phase-averaged observation mode is employed. This, however, is not the case for

$j = 2$, where proportionality to $\boldsymbol{\varepsilon}_2^* \cdot \boldsymbol{\varepsilon}_2 = |\boldsymbol{\varepsilon}_2|^2$ and thus phase independence is found. The pump–probe signal for the linear polarization case includes an implicit dependence on the delay time between $\boldsymbol{\varepsilon}_1$ and $\boldsymbol{\varepsilon}_2$.

We label the wave functions by a linear combination of $\mathbf{k}_1$ and $\mathbf{k}_2$ in order to characterize the electric interactions entailed by $|\Psi^{(N)}\rangle$. Clearly, $|\Psi^{(1)}_{\mathbf{k}_j}(t)\rangle, j = 1, 2$, has to appear in the expression for the linear polarization since a single absorption event is involved. Besides the first-order contribution, processes of third order with polarization signals propagating into the direction $\mathbf{k}_2$ can be identified. As indicated in Figure 18.1, these are stimulated Raman (SR) and emission (SE) scattering. The third-order polarization in the $\mathbf{k}_2$ direction therefore results as

$$P^{(3)}_{\mathbf{k}_2}(t) = 2\Re\{P_{\mathrm{SR}}(t) + P_{\mathrm{SE}}(t)\}. \tag{18.33}$$

While the probe pulse causes an excitation from the ground state for Raman scattering, it induces the transition from an excited level to the ground state if stimulated emission is realized. The latter case can be associated with a transfer of both the ket and the bra wave packet from the ground state to the excited state potential energy surface. After some delay period has passed, during which the wave packets propagate freely on the excited state surface, the probe pulse induces the transition of the bra wave packet to the ground state. This sequence of events is illustrated by Figure 18.2, detailing the steps that are sketched in the respective cartoon of Figure 18.1.

The stimulated emission term of the third-order polarization along $\mathbf{k}_2$ may be written as

$$P^{(3)}_{\mathrm{SE}}(t) = \frac{i}{\hbar}\exp(i\mathrm{k}_2\mathrm{X})\int_{-\infty}^{t} dt'\, E_2(t')\langle\Psi^{(1)}_{\mathbf{k}_1}(t')|\hat{\mu}_{10}\exp\left[\frac{i}{\hbar}\hat{H}_{\mathrm{M}}(t-t')\right]$$
$$\times\,\hat{\mu}_{01}\exp\left[-\frac{i}{\hbar}\hat{H}_{\mathrm{M}}(t-t')\right]|\Psi^{(1)}_{\mathbf{k}_1}(t')\rangle. \tag{18.34}$$

An analogy between this term and expression (18.32) that refers to the much more elementary problem of pump or probe pulse absorption is evident. In fact, the integral appearing in Eq. (18.32) is the same as that in Eq. (18.34) if the indices of the electric dipole operator are exchanged, acknowledging the difference between absorption and emission, and if $\Psi^{(0)}$ is replaced by $\Psi^{(1)}_{\mathbf{k}_1}(t)$, that is, the static molecular ground state function by the excited wave packet evolving in real time under the influence of the pump pulse. Thus,

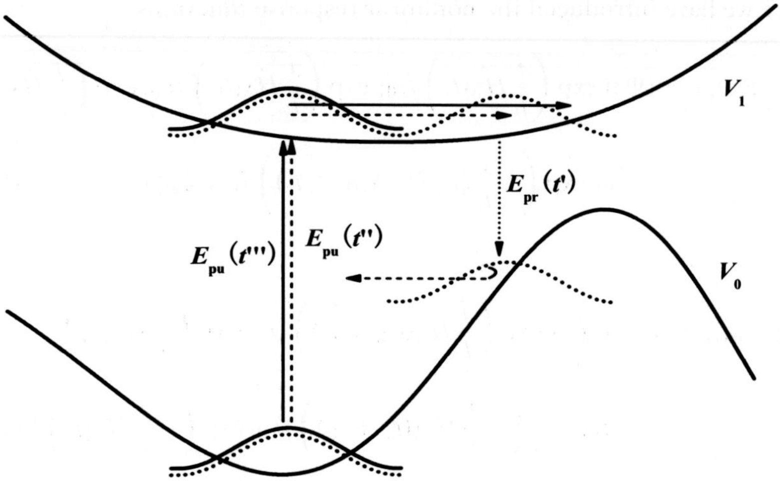

Figure 18.2: Pump–probe detection mechanism for stimulated emission. Vertical arrows represent transitions between the two coupled potential energy surfaces, horizontal arrows free propagation on the respective surface. The solid line refers to the ket wave packet, the dashed line to its bra counterpart. Two pump pulses promote the ket and the bra wave packet from the ground to the excited state, the subsequent probe pulse induces the return of the bra wave packet to the lower surface. (Reprinted with permission from [639]. Copyright 1990 by the American Institute of Physics.)

in the stimulated emission case, the probe pulse acts on the non-stationary excited state wave function. In view of the structural similarity between Eqs. (18.32) and (18.34), the stimulated emission component of third-order polarization may be understood as the linear response of a non-stationary system. Probing this contribution yields information about the time behavior of the excited state $|\Psi^{(1)}_{\mathbf{k}_1}\rangle$. To make the time structure of $P^{(3)}_{\mathrm{SE}}$ as well as its dependence on the electric pulse amplitudes explicit, we expand Eq. (18.34) further and obtain

$$P^{(3)}_{\mathrm{SE}}(t) = \left(\frac{i}{\hbar}\right)^3 \exp(i\mathbf{k}_2 \cdot \mathbf{X}) \int_{-\infty}^{t} dt_3 \int_{-\infty}^{t_3} dt_2 \int_{-\infty}^{t_2} dt_1$$
$$\times E_2(t_3)(E_1^*(t_2)E_1(t_1)R_1(t-t_3, t_3-t_2, t_2-t_1)$$
$$+ E_1(t_2)E_1^*(t_1)R_2(t-t_3, t_3-t_2, t_2-t_1)) \qquad (18.35)$$

where we have introduced the nonlinear response functions

$$R_1(t_a, t_b, t_c) = \langle\Psi_0|\exp\left(\frac{i}{\hbar}\hat{H}_M t_c\right)\hat{\mu}_{01}\exp\left(\frac{i}{\hbar}\hat{H}_M t_b\right)\hat{\mu}_{10}\exp\left(\frac{i}{\hbar}\hat{H}_M t_a\right)$$
$$\times\hat{\mu}_{01}\exp\left(-\frac{i}{\hbar}\hat{H}_M(t_a + t_b + t_c)\right)\hat{\mu}_{10}|\Psi_0\rangle\rangle \tag{18.36}$$

and

$$R_2(t_a, t_b, t_c) = \langle\Psi_0|\hat{\mu}_{01}\exp\left(\frac{i}{\hbar}\hat{H}_M(t_b + t_c)\right)\hat{\mu}_{10}\exp\left(\frac{i}{\hbar}\hat{H}_M t_a\right)$$
$$\times\hat{\mu}_{01}\exp\left(-\frac{i}{\hbar}\hat{H}_M(t_a + t_b)\right)\hat{\mu}_{10}\exp\left(-\frac{i}{\hbar}\hat{H}_M t_c\right)|\Psi_0\rangle\rangle. \tag{18.37}$$

All dynamic information on the stimulated emission process is contained in the functions R_1 and R_2. Determining the polarization is thus reduced to evaluating expressions that depend solely on molecular properties, and subsequent time integration. However, carrying out this program may still be a daunting problem. This task is greatly alleviated if the ultrashort laser pulses are approximated by delta functions. This implies that the pulse envelope function $E_i(\mathbf{X}, t)$ (see Eq. (18.6)) is chosen as proportional to $\delta(t - t_i^*)$, where t_i^* is the instant of application, and the index i labels the three pulses involved in the stimulated emission process. In general, the time order $t_1^* < t_2^* < t_3^*$ is valid. If this simplified model is applied, the time integrals in Eq. (18.35) collapse, and the polarization is obtained from mere calculation of the response functions. A word of caution: delta-shaped pulses, providing a convenient tool for the theoretical elucidation of time-resolved spectroscopy, would not be desirable in the laboratory, even if they could be produced. This is related to the fact that infinitely short pulses would be infinitely broad in the frequency domain and thus induce all dipole-allowed transitions of the system simultaneously.

Exercise 18.2. *Calculating polarization components.* (a) Derive the linear polarization $P_{\mathbf{k}_i}^{(1)}(t)$ as described by Eq. (18.32), the polarization (18.35) for the case of stimulated emission $P_{SE}(t)$, as well as the corresponding expression for stimulated Raman interaction $P_{SR}(t)$, using the prescriptions (18.23)

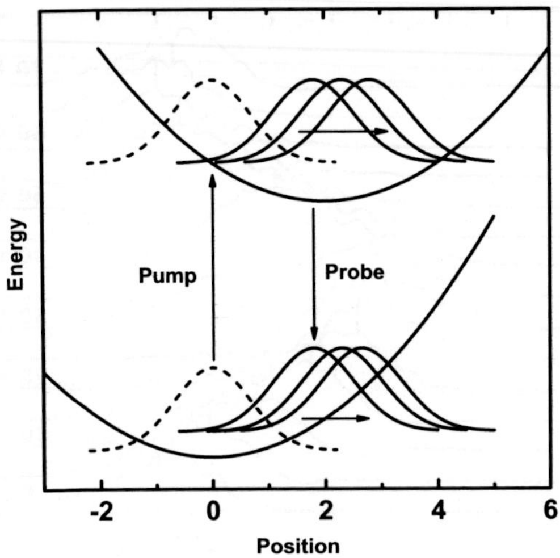

Figure 18.3: Model situation for stimulated emission, assuming harmonic ground state and excited state potential energy surfaces, displaced with respect to each other, but otherwise identical. (Reprinted with permission from [639]. Copyright (1990) by the American Institute of Physics.)

and (18.25). (b) Evaluate $P_{\mathrm{SE}}(t)$ and $P_{\mathrm{SR}}(t)$ by use of Eq. (18.35), assuming delta-shaped pulses.

For harmonic ground and excited state surfaces in conjunction with delta-function-shaped pulses, the nonlinear response functions can be treated analytically. Adopting such a model system, Pollard *et al.* [639] simulated a pump–probe experiment under conditions of stimulated emission dynamics. The harmonic initial state of the molecule is lifted by the pump pulse from the ground to the excited state surface, where it is interrogated by the probe pulse after a certain time of free propagation, as depicted in Figure 18.3.

If the probe pulse is applied without any delay ($\Delta t = 0$), the stimulated emission spectrum does not differ from the ground state absorption spectrum. For $\Delta t > 0$, the oscillatory shift of the emission maximum reflects the motion of the wave packet on the harmonic excited surface, as it traces out a path with alternatingly narrowing and widening vertical energy gaps between the two surfaces (see Figure 18.4).

Along the lines laid out for stimulated emission, stimulated Raman scattering may be treated. In this case, the non-stationary ground state wave

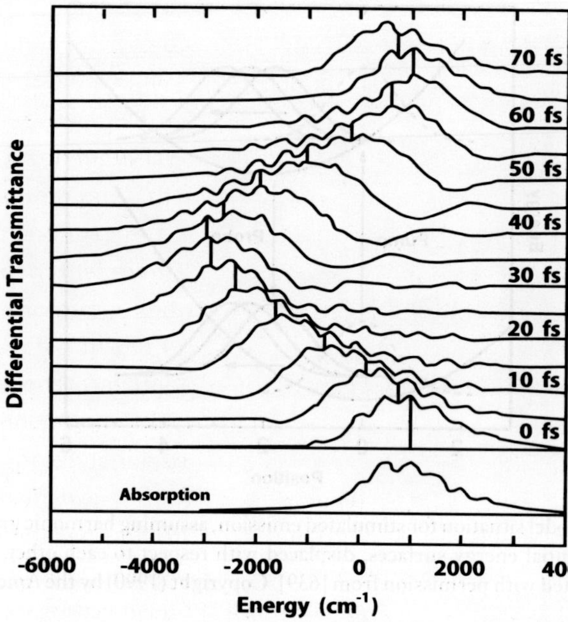

Figure 18.4: Transient differential transmittance for the configuration sketched in Figure 18.3. Stimulated emission is probed with variable delay time, extending over one wave packet oscillation period on the excited potential energy surface. The time delay is indicated as a parameter to the right of each spectrum. The zero of energy is taken as the difference between the minima of the two surfaces. Both the pump and the probe pulse are approximated by delta functions. The absorption spectrum is plotted for comparison. (Reprinted with permission from [639]. Copyright (1990) by the American Institute of Physics.)

function is monitored by the probe pulse, and the corresponding diagram in Figure 18.1 is realized by two terms whose sum yields the stimulated Raman component of the third-order polarization according to

$$P_{\mathrm{SR}}(t) = \langle \Psi^{(0)}(t)|\hat{\mu}_{01}|\Psi^{(3)}_{\mathbf{k}_2}(t)\rangle + \langle \Psi^{(2)}(t)|\hat{\mu}_{01}|\Psi^{(1)}_{\mathbf{k}_2}(t)\rangle. \tag{18.38}$$

The perturbative approach represents the various orders of polarization within a comprehensive mathematical scheme, and also allows for intuitive partition of the polarization terms according to the contributing spectroscopic processes. The application of this method is, on the other hand, limited to weak pulses. To include moderate or strong fields, one desires a more general procedure, as highlighted in the following subsection.

18.2.2 *The non-perturbative approach*

The non-perturbative alternative to the method outlined in the previous subsection consists in solving the TDSE by expanding the molecular wave function $|\Psi(t)\rangle$ into a suitably chosen basis which may be defined as the direct product of electronic and nuclear states [3, 612, 637]. In this basis, $|\Psi(t)\rangle$ reads

$$|\Psi(t)\rangle = \sum_{k\leq N_k, i_1\leq N_1, i_2\leq N_2,\ldots,i_m\leq N_m} C_{k,i_1,i_2,\ldots,i_m} |\psi_k\rangle |\chi_{i_1}\rangle |\chi_{i_2}\rangle \ldots |\chi_{i_m}\rangle, \tag{18.39}$$

where N_k and N_i ($i = 1, m$) stand for the numbers of basis functions associated with the electronic system and the ith nuclear degree of freedom, respectively. A total number of $N_k N_1 \cdots N_m$ equations results from the insertion of Eq. (18.39) into the TDSE. Writing them conveniently in matrix form, we arrive at

$$i\hbar \frac{\partial \mathbf{C}(t)}{\partial t} = \mathbf{H}(t)\mathbf{C}(t) \tag{18.40}$$

with $\mathbf{C}(t)$ as the vector of all coefficients emerging from the projection of $|\Psi(t)\rangle$ onto the direct product basis states, and $\hat{H}(t)$ as the Hamiltonian (18.13) expressed in this basis. With the help of relation (18.16), one infers the overall polarization $\mathbf{P}(t)$ from the molecular wave function. To represent selected polarization signals, however, as described so far and exemplified by the case of stimulated emission in Eqs. (18.34) and (18.35), one has to know the wave function at various orders of its expansion, Eq. (18.22). By definition, these orders are not calculated when the non-perturbative approach is used. Therefore, analyzing experimental data requires the use of a method to extract information about the individual spectroscopic processes from the non-perturbative scheme. This is accomplished by following a prescription given by Seidner *et al.* [637]. One may expand $\mathbf{P}(t)$ as

$$\mathbf{P}(t) = 2\Re\left\{\sum_{n,m} \mathbf{P}_{n,m}(t) \exp\left[i(n\mathbf{k}_1 + m\mathbf{k}_2)\cdot \mathbf{X}\right]\right\}, \tag{18.41}$$

making explicit the dependence of $\mathbf{P}(t)$ on the wave vectors $\mathbf{k}_1$ and $\mathbf{k}_2$ of the pump and the probe pulse, respectively. As non-overlapping pulses up to third order are considered within the rotating wave approximation, we only have to distinguish between signals emitted in three directions, namely

$\mathbf{k}_1, \mathbf{k}_2$, and $2\mathbf{k}_2 - \mathbf{k}_1$, as motivated in the previous subsection. This may be achieved by introducing the variable $\varphi \equiv \mathbf{k}_1 \cdot \mathbf{X}$, i.e. the phase acquired by the polarization as a result of interaction with the pump pulse, and by writing the polarization as the sum

$$\mathbf{P}(\varphi) = \sum_n \mathbf{P}_n \exp(in\varphi) \tag{18.42}$$

with $n = -1, 0, +1$. By Figure 18.1 and expression (18.41), each of the terms in this sum is associated with one of the directions of interest, as $\mathbf{P}_{-1}$, $\mathbf{P}_0$, and $\mathbf{P}_1$ are readily assigned to polarization in the $2\mathbf{k}_2 - \mathbf{k}_1$, $\mathbf{k}_2$ and $\mathbf{k}_1$ directions, respectively. This is evident as the third-order cases selected impose the constraint $n + m = 1$ on the index pair (n, m) in $(n\mathbf{k}_1 + m\mathbf{k}_2)$. Further, formula (18.42) implies that the linear combination $\mathbf{P}(0) + \mathbf{P}(\pi)$ selects the contribution of direction $\mathbf{k}_2$. More specifically, we find[3]

$$\mathbf{P}_{\mathbf{k}_2} = \Re\{\mathbf{P}(0) + \mathbf{P}(\pi)\}. \tag{18.43}$$

Likewise, the difference $\mathbf{P}_{\mathbf{k}_1, 2\mathbf{k}_2 - \mathbf{k}_1} = \Re\{\mathbf{P}(0) - \mathbf{P}(\pi)\}$ filters out the combined components along $\mathbf{k}_1$ and $2\mathbf{k}_2 - \mathbf{k}_1$. Making admission for overlapping pulses and thus enhancing the number of distinct directions of radiation from three to four, one can discriminate between $\mathbf{P}_{\mathbf{k}_1}$ and $\mathbf{P}_{2\mathbf{k}_2 - \mathbf{k}_1}$. The sketched procedure reduces the task of distinguishing between various directions by non-perturbative computation to evaluating $\mathbf{P}(\varphi)$ for certain well-defined arguments φ. Alternatively, the integral pump–probe spectrum may be compared with the calculated diabatic or adiabatic population of the contributing electronic states. In the following section, this procedure is exemplified by the analysis of *cis-trans* isomerization processes.

18.3 Polarization Studies of *cis-trans* Isomerization

In the following, the concepts outlined in the previous section will be exemplified by the photoinduced *cis-trans* isomerization phenomenon. In polyenes and similar organic species, it has been observed that twisting around a carbon–carbon double bond creates a degeneracy or near-degeneracy between the excited electronic state and the ground state [74, 637]. Most

[3] Realize that $\mathbf{P}(0) = \mathbf{P}_{-1} + \mathbf{P}_0 + \mathbf{P}_{+1}$, and that $\mathbf{P}(\pi) = -\mathbf{P}_{-1} + \mathbf{P}_0 - \mathbf{P}_{+1}$. Adding the real parts of both expressions, one constructs, by Eq. (18.41), the polarization component $\mathbf{P}_{\mathbf{k}_2}$.

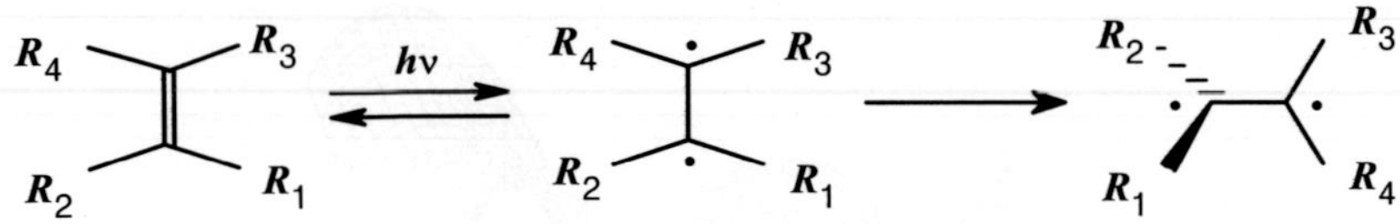

Figure 18.5: Transition from the planar ground state of an alkene to a vertically photoexcited state, followed by relaxation into a three-dimensional geometry on the excited potential energy surface by twisting around the central olefinic bond. The symbols R_i, $i = 1, \ldots, 4$, stand for organic groups.

alkenes in their ground states are strongly stabilized with respect to twisting around their C=C bond (e.g. [640]) which transforms a planar into a three-dimensional isomer. This resistance, however, is found to be broken as the alkene undergoes vertical photoexcitation, associated with the promotion of a bonding π orbital into an antibonding π^* orbital. As a result of this $\pi^* \leftarrow \pi$ transition, the electrostatic repulsion between the p electrons of the two carbon atoms is no longer compensated by bonding, such that a twist around the carbon–carbon axis reduces the total energy of the system, leading to relaxation of the excited molecule from planar to three-dimensional geometry, as sketched in Figure 18.5.

Cis-trans photoisomerization is completed by radiationless deexcitation of the molecule through a *photochemical funnel*, allowing for ultrafast internal conversion into the *trans* structure on the ground state surface, or back into the original *cis* structure. Figure 18.6 illustrates the interplay of these processes in a landscape of two adiabatic potential surfaces spanned by two nuclear degrees of freedom, namely the diabatic coupling coordinate between the two potential surfaces, and the torsion angle.

This mechanism has been found operative in various biological and physiological systems, most spectacularly in the retinal chromophore of the peptide *rhodopsin*, where it was shown to be a decisive step in a chain of processes involved in vision. *Cis-trans* photoisomerization dynamics has been the subject of numerous theoretical studies, aiming both at the understanding of model systems ([641]–[643]), and the interpretation of physical cases ([644]–[647]). In the following, we will present and discuss results communicated in [637] where the quantum dynamical analysis of a model problem is described. It consists of two electronic singlet states S_0 and S_1 which interact with three nuclear degrees of freedom, namely the torsion angle φ, a coupling mode Q_c, and a totally symmetric *tuning mode* Q_t. This approach

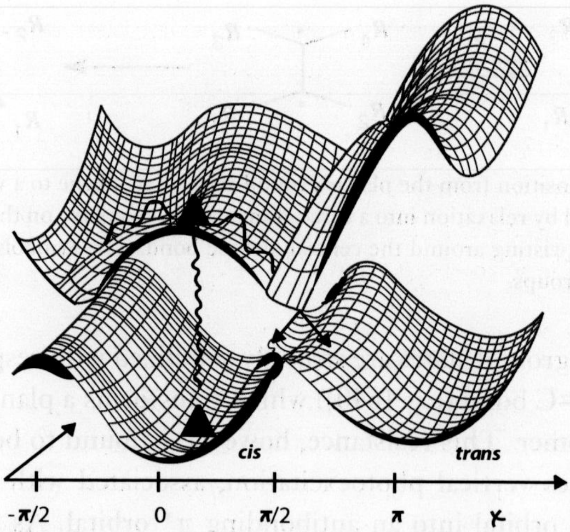

Figure 18.6: Representation of the elementary processes involved in *cis-trans* photoisomerization: A pump pulse induces vertical excitation of a wave packet from an initially occupied adiabatic ground state in *cis* conformation to the upper adiabatic surface. Subsequently, the system deforms along a torsion coordinate ("twisting") and eventually undergoes radiationless decay through a photochemical funnel, stabilizing either in a new ground state geometry, the *trans* conformation, or in the initial *cis* structure. (Reprinted from [3], with permission from Wiley Publishing.)

yields for the diagonal diabatic matrix elements of the Hamiltonian (18.14) the expression

$$\hat{h}_k = \hat{h} + V^{(k)}(\varphi) + \kappa Q_t, \quad k = 0, 1, \tag{18.44}$$

with

$$\hat{h} = -\frac{\hbar^2}{2I}\frac{\partial^2}{\partial\varphi^2} + \frac{\hbar}{2}\sum_{i=\mathrm{c,t}} \omega_i \left(Q_i^2 - \frac{\partial^2}{\partial Q_i^2}\right). \tag{18.45}$$

In (18.45), I stands for the torsional moment of inertia, while the quantities ω_i $(i = \mathrm{c, t})$, are the harmonic vibrational frequencies associated with the coupling and the tuning mode, respectively. It should be noted that both Q_c and Q_t have been chosen as dimensionless coordinates, i.e. $Q_i = \sqrt{m_i\omega/\hbar}R_i$ $(i = \mathrm{c, t})$ (see Section 2.5.2). The factor κ denotes the gradient of the excited state potential energy surface at the ground state equilibrium geometry of the system. The non-diagonal elements are written as

$$V_{01} = V_{10} = \lambda Q_c \tag{18.46}$$

with λ as vibronic coupling constant. The values adopted for the constants I, ω_c, ω_t, and κ are listed in [637]. The torsional potentials $V^{(k)}, k = 0, 1$, for the surfaces S_0 and S_1 are defined as periodic functions, where $V^{(1)}$ exhibits an inverted profile with respect to $V^{(0)}$:

$$V^{(0)} = E_0 + \frac{1}{2}V_0(1 - \cos\varphi), \tag{18.47}$$

$$V^{(1)} = E_1 - \frac{1}{2}V_1(1 - \cos\varphi). \tag{18.48}$$

The parameters E_0 and E_1 (V_0 and V_1) are selected as 0.0 and 2.75 eV (5.0 and 2.5 eV) in the model calculation described here. Both curves are displayed in Figure 18.7.

Within the frame set by these definitions, the *cis-trans* photoisomerization dynamics is investigated using the non-perturbative approach outlined in Section 18.2.2. The expansion (18.39) of the molecular wave function $|\Psi(t)\rangle$ is implemented by adopting a direct product basis composed of

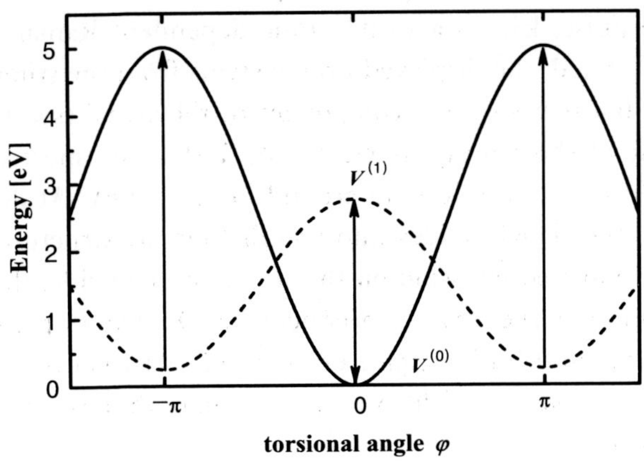

Figure 18.7: Dependence of the diabatic potential energies $V^{(k)}, k = 0, 1$, on the torsional angle φ. The solid line refers to $V^{(0)}$ which defines the ground state in the *cis* region around $\varphi = 0°$, the dashed line to $V^{(1)}$ which in the *trans* region around $|\varphi| = \pi$ becomes lower in energy than $V^{(0)}$, reflecting the essential features of the potential energy landscape for the photoinduced *cis-trans* isomerization problem, as represented in greater detail in Figure 18.6. (Reprinted with permission from [637]. Copyright (1995) by the American Institute of Physics.)

the diabatic electronic wave functions, free rotor states for the torsional coordinate, and harmonic oscillator states to account for the two remaining nuclear degrees of freedom. In summary,

$$|\Psi(t)\rangle = \sum_{k=0,1;i_\varphi \le N_\varphi, i_c \le N_c, i_t \le N_t} C_{k,i_\varphi,i_c,i_t} |\psi_k\rangle |\chi_{i_\varphi}\rangle |\chi_{i_c}\rangle |\chi_{i_t}\rangle. \tag{18.49}$$

Inserting $|\Psi(t)\rangle$ into the TDSE (18.40) yields $2N_\varphi N_c N_t$ coupled equations of motion from which the coefficients $C_{k,i_\varphi,i_c,i_t}(t)$ and, through use of Eq. (18.16), the overall polarization $\mathbf{P}(t)$ are determined.

This approach has been employed to simulate the integral pump–probe signal as a function of both the delay time and the frequency of the probe pulse according to formula (18.9). The probe frequencies $\Delta E/\hbar$ were chosen to cover the interval $\Delta E = 0 - 6\,\text{eV}$, thus scanning the range of the energy differences between the ground state and the excited state potential curve as shown in Figure 18.7. The integral pump–probe signal provides an overall view of the basic processes involved in *cis-trans* photoisomerization dynamics. Figure 18.8 presents this signal for the case of Gaussian laser pulses of 6 fs duration.

The figure shows the pump–probe spectrum, detected in the direction of the probe pulse, $\mathbf{k}_2$. As a weakly time-dependent Raman background has been subtracted, the displayed profile stems from the stimulated emission and the linear absorption component of the signal (see Figure 18.1). In the regime of short delay times, $\Delta t \ll 100\,\text{fs}$, the spectrum is dominated by an absorption maximum around $\hbar\omega = 2.5\,\text{eV}$. At delay times of $\Delta t \approx 100\,\text{fs}$, the signal divides into two distinct substructures, one associated with stimulated emission on the *cis*, or reactant side, the other with linear absorption on the *trans*, or product side. Oscillatory patterns, clearly visible for both contributions, give evidence of quasi-periodic recurrences due to coherent evolution of the wave packets in both the electronic ground and excited state.

Evidently, the pump–probe spectra contain information about the electronic populations of the participating states and their dynamic development. This information can be made more explicit by considering the expectation values $P_k^{\mathrm{D}}(t), k = 0, 1$, of the relevant electron populations. The time-dependent population $P_1^{\mathrm{D}}(t)$ of the diabatic excited state involved in the photoisomerization case outlined earlier, for instance, provides a

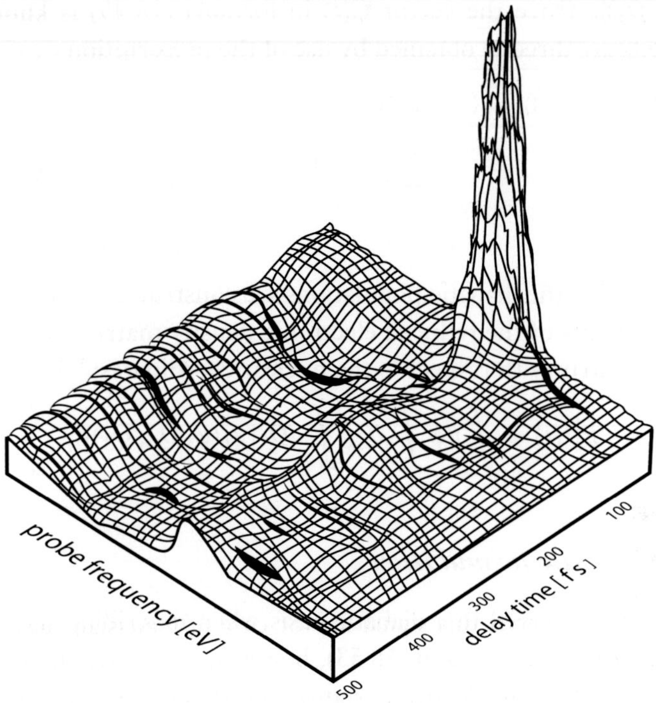

Figure 18.8: Integral pump–probe spectrum for the idealized three-dimensional *cis-trans* photoisomerization process described in the text. Gaussian laser pulses of 6 fs duration are assumed. (Reprinted with permission from [637]. Copyright (1995) by the American Institute of Physics.)

monitor of the interaction between the two electronic states. Thus, if the interstate coupling V_{01} vanishes, the excited state population, as prepared by the pump pulse, will be a constant of motion. The opposite extreme of strong interaction, in contrast, will lead to fast depopulation of the excited electronic level. We define $P_k^{\mathrm{D}}(t)$ through the diabatic projection operator $\hat{P}_k^{\mathrm{D}}$ ([3], also see Eqs. (15.95a,b)):

$$P_k^{\mathrm{D}}(t) = \langle \Psi(t)|\hat{P}_k^{\mathrm{D}}|\Psi(t)\rangle \tag{18.50}$$

and

$$\hat{P}_k^{\mathrm{D}} = |\psi_k\rangle\langle\psi_k|. \tag{18.51}$$

Likewise, the expectation values for the population of the kth electronic surface may be constructed from the respective nuclear wave packets

$|\chi_k(\varphi, Q_c, Q_t)\rangle$. Once the vector $\mathbf{C}(t)$ in formula (18.49) is known, these wave packets are directly obtained by use of the prescription

$$|\Xi_k(\varphi, Q_c, Q_t)\rangle = \langle\psi_k|\Psi(t)\rangle = \sum_{i\leq N, i_c\leq N_c, i_t\leq N_t} C_{k,i_\varphi,i_c,i_t}(t)\chi_{i_\varphi}(\varphi)\chi_{i_c}(Q_c)\chi_{i_t}(Q_t). \tag{18.52}$$

From Eq. (18.50) the diabatic populations are constructed by evaluating the diagonal elements of the reduced electronic density matrix, i.e. the trace of the density matrix with respect to the nuclear coordinates:

$$P_k^D(t) = \int d\varphi \int dQ_c \int dQ_t |\chi_k(\varphi, Q_c, Q_t)|^2. \tag{18.53}$$

18.3.1 *Adiabatic formulation*

So far we have referred to a diabatic basis when describing the electronic system. Equations (18.52) and (18.53), however, can be readily modified to accommodate adiabatic electronic states, as one replaces the diabatic wave function $|\psi_k\rangle, k = 0, 1$, and the corresponding projection operator (18.51) with their adiabatic counterparts (see Eq. (15.95b)). Finally, the adiabatic population in the *cis* or the *trans* region of the two potential energy surfaces are generated by computing the expectation values

$$P^A_{1,\text{cis}}(t) = \langle\Psi(t)|\hat{P}_{\text{cis}}\hat{P}^A_1|\Psi(t)\rangle, \tag{18.54}$$

$$P^A_{0,\text{trans}}(t) = \langle\Psi(t)|\hat{P}_{\text{trans}}\hat{P}^A_0|\Psi(t)\rangle. \tag{18.55}$$

With reference to Figures 18.6 and 18.7, as well as Eqs. (18.47) and (18.48), the projection operators are given by theta functions in the angle as

$$\hat{P}_{\text{trans}} = \Theta(|\varphi| - \pi/2), \tag{18.56}$$

$$\hat{P}_{\text{cis}} = 1 - \hat{P}_{\text{trans}}. \tag{18.57}$$

The entity $P^A_{1,\text{cis}}(t)$, for instance, yields the probability of populating the adiabatic S_1 state in *cis* geometry. It is plausible that the time behavior of this probability should be monitored by the stimulated emission signal in the reactant channel, and thus in the region around the pump

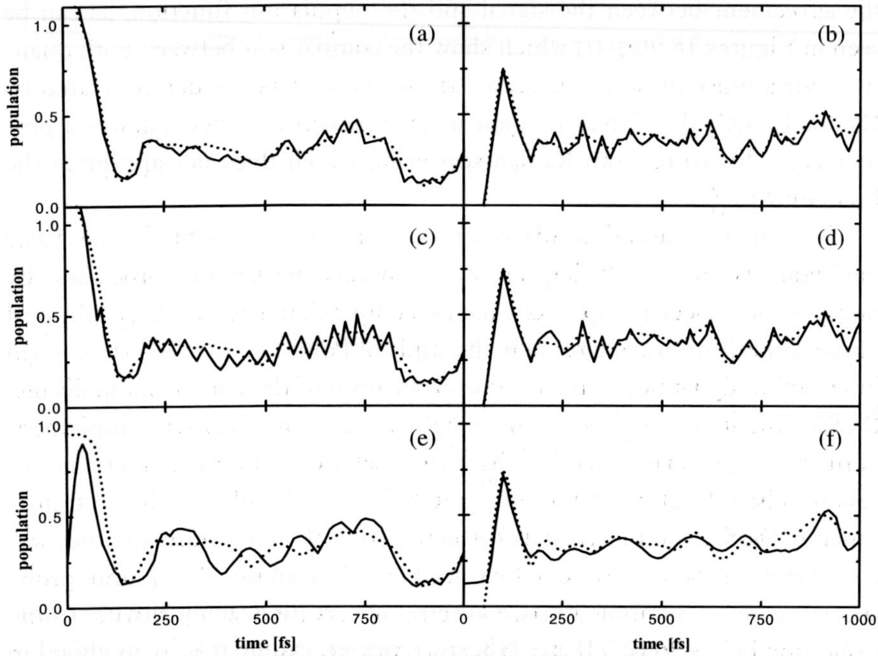

Figure 18.9: The integral pump–probe signal (solid line) is compared with the adiabatic electronic populations $P^{A}_{1,\text{cis}}(t)$ (dotted line, left) and $P^{A}_{0,\text{trans}}(t)$ (dotted line, right). All images refer to well-defined probe frequencies ω_2, namely $\omega_2 = 2.5\,\text{eV}/\hbar$ ($2.75\,\text{eV}/\hbar$) for (a) ((c), (e)) and $\omega_2 = 4.5\,\text{eV}/\hbar$ ($4.75\,\text{eV}/\hbar$) for (b) ((d), (f)). Pulses of 6 fs (40 fs) duration have been applied in the cases (a)–(d) ((e), (f)). (Reprinted with permission from [637]. Copyright (1995) by the American Institute of Physics.)

frequency $2.75\,\text{eV}/\hbar$. Figure 18.9(a) verifies this assumption, comparing $P^{A}_{1,\text{cis}}(t)$ with a cross section through the integral pump–probe spectrum shown in Figure 18.8 at $\omega_2 = 2.5\ \text{eV}/\hbar$. Likewise, the absorption-dominated product channel should give rise to a pump–probe signal whose time dependence reflects that of the population $P^{A}_{0,\text{trans}}(t)$, which is confirmed by Figure 18.9(b). The additional panels of Figure 18.9 continue the comparison between the population probabilities and cuts through the integral pump–probe spectrum at a well-defined probe frequency ω_2. At the resonance frequencies $\omega_2 = 2.75\ \text{eV}/\hbar$ ($\omega_2 = 4.75\ \text{eV}/\hbar$), the reactant signal exhibits a periodic modulation due to interference with the tuning mode which is not seen in the evolution of $P^{A}_{1,\text{cis}}$, as shown in Figures 18.9(c) and (d). Lastly, variation of the pulse duration also diminishes the agreement between the signal and the population function, as can be seen in Figures 18.9(e)–(f) which show the comparison between both quantities for a pulse duration increased from 6 fs to 40 fs. As demonstrated by Figure 18.9(e), the signal resonant in the reactant channel acquires a pronounced admixture from Raman scattering which does not appear in the function $P^{\mathrm{A}}_{1,\mathrm{cis}}(t)$.

The model situation discussed in this section outlines the basic methodology for analyzing *cis-trans* photoisomerization processes by pump–probe spectroscopy. As exemplary for related research on physical photoisomerization systems, note the study of Polli *et al.* [648] on the conical intersection dynamics in the *cis-trans* transition of the rhodopsin molecule. Besides initializing the perception of light, as mentioned earlier, this pigment of the retina performs various other tasks that are vital for the emergence of vision. The relaxation of the molecule following the photoinduced transition into the first excited *cis* state through a conical intersection into the *trans* conformation was experimentally traced by means of ultrafast pump–probe spectroscopy, combining a broad spectral observation window with a time resolution below 20 fs. The latter feature was necessitated by the very short isomerization time of 200 fs. Modeling the observed processes was achieved by adopting a combined quantum and molecular mechanics (QM/MM) approach. In this fashion, measurement supported by numerical simulation gave evidence of the crucial role played by nonadiabatic electron dynamics in the primary photoisomerization event leading to vision.

18.4 The Density Matrix Approach to Simulating Pump–Probe Signals

The semiclassical Liouville–von Neumann theory described in Chapter 14 can be extended in various ways to account for optical interactions and thus to simulate ultrafast dynamical processes as observed in pump–probe femtosecond spectroscopy. This may proceed by generalizing the superoperator formalism introduced in Section 14.1 to include the interaction between the molecule and a radiation field. For clarity, our presentation of this formalism will include only linear optical processes. In contrast to our treatment of wave-function-based optical spectroscopy in the preceding sections of this chapter, however, allowance will be made for three distinct electronic levels, and besides ground and excited states also a final probe state will be considered that may differ from both of them. As in Section 18.1.1, we will refer to cases where the field is weak enough to be adequately treated by use of perturbation theory. More specifically, the Liouville operator $\hat{L}$, defined by $\hat{L}\hat{O} = [\hat{H}, \hat{O}]$, is written as a contribution of zeroth order, $\hat{L}_0$, augmented by a perturbing term $\hat{L}_1$ that involves the time-dependent interaction Hamiltonian $\hat{H}_1$ [649]:

$$\hat{L} = \hat{L}_0 + \hat{L}_1(t). \tag{18.58}$$

Again, the dipole approximation is used for $\hat{H}_1$. With these assumptions, it holds that

$$\mathbf{H}_1 = \begin{pmatrix} 0 & -\boldsymbol{\mu}_{01} \cdot \boldsymbol{\varepsilon} \\ -\boldsymbol{\mu}_{10} \cdot \boldsymbol{\varepsilon} & 0 \end{pmatrix} \tag{18.59}$$

and

$$\boldsymbol{\varepsilon}(\mathbf{X}, t) = \mathbf{E}(t) \exp(-i\omega t) \exp(i\mathbf{k} \cdot \mathbf{X}) + \mathbf{E}(t) \exp(i\omega t) \exp(-i\mathbf{k} \cdot \mathbf{x}). \tag{18.60}$$

Both the pump and the probe pulse are characterized by the same frequency ω and the same wave vector $\mathbf{k}$. We will explicitly refer to short pulses and write the envelope function as a Gaussian centered at some time t_i:

$$E(t) = \frac{E_0}{\sqrt{2\pi}\sigma} \exp\left[-\frac{(t - t_i)^2}{2\sigma^2}\right] \tag{18.61}$$

with σ as pulse duration. Casting the envelope function into this shape has the advantage that the limit of infinitesimally short pulses is well defined, i.e. $E(t)$ tends towards $E_0\delta(t - t_i)$ as σ tends towards zero.

We adopt $\hat{L}_1$ as a first-order perturbation operator and expand the electronic density operator $\hat{\rho}$ into the perturbation series

$$\hat{\rho} = \hat{\rho}^{(0)} + \hat{\rho}^{(1)} + \hat{\rho}^{(2)} \cdots . \tag{18.62}$$

This approach in conjunction with Eq. (18.58) leads to a hierarchy of coupled differential equations for $\hat{\rho}^{(i)}, i = 0, 1, 2, \ldots$. Inserting Eq. (18.58) and Eq. (18.62) into the Liouville–von Neumann equation $i\hbar\frac{\partial\hat{\rho}}{\partial t} = \hat{L}\hat{\rho}$, we find for the zeroth order of the density operator the relation

$$\frac{\partial \hat{\rho}^{(0)}}{\partial t} = -\frac{i}{\hbar}\hat{L}_0\hat{\rho}^{(0)} \tag{18.63}$$

and for some arbitrary order j, with $j = 1, 2,$

$$\frac{\partial \hat{\rho}^{(j)}}{\partial t} = -\frac{i}{\hbar}(\hat{L}_0\hat{\rho}^{(j)} + \hat{L}_1\hat{\rho}^{(j-1)}). \tag{18.64}$$

The formal solution of Eq. (18.63) is

$$\hat{\rho}^{(0)}(t) = \exp\left(-\frac{i}{\hbar}\hat{L}_0 t\right) \equiv \hat{G}^{(0)}(t). \tag{18.65}$$

The operator $\hat{G}^{(0)}(t)$ is analogous to the zeroth-order Hamiltonian propagator $\exp(-\frac{i}{\hbar}\hat{H}_0)$ and therefore referred to as the *zeroth-order Liouvillian propagator*. As for the Hamiltonian case, Eq. (18.64) admits an iterative solution. Assuming $\hat{\rho}^{(j)}(0) = 0$ for $j = 1, 2$, one finds

$$\hat{\rho}^{(j)} = -\frac{i}{\hbar}\int_0^t d\tau\, \hat{G}^{(0)}(t-\tau)\hat{L}_1(\tau)\hat{\rho}^{(j-1)}(\tau). \tag{18.66}$$

Exercise 18.3. Prove relation (18.66) by demonstrating that the partial time derivative of the right-hand expression equals the right-hand side of (18.64).

Applying these concepts to a system with two electronic states, $|0\rangle$ and $|1\rangle$, coupled by the radiation operator (18.59) and describing both the ground state $|0\rangle$ and the excited state $|1\rangle$ within an adiabatic frame, we may cast the full Hamiltonian into the matrix form

$$\mathbf{H} = \begin{pmatrix} H_0 & -\boldsymbol{\mu}_{01}\cdot\boldsymbol{\varepsilon} \\ -\boldsymbol{\mu}_{10}\cdot\boldsymbol{\varepsilon} & H_1 \end{pmatrix}. \tag{18.67}$$

The structure of this Hamiltonian operator matrix is the same as that of the two-state diabatic Hamiltonian (14.11) used earlier to develop the semi-classical Liouville theory, where the radiation field term $\boldsymbol{\mu}_{ij}\cdot\boldsymbol{\varepsilon}$ $(i,j = 0,1)$, here plays the same role as the coupling potential V_c there. In Section 14.1, the respective diabatic Hamiltonian provided the basis for a perturbation treatment within Schrödinger theory while Eq. (18.66) lays the ground for the corresponding Liouvillian perturbation formalism. As in Eq. (14.11), the vibronic density operator matrix may be subdivided into populations (the diagonal entries) and coherences (the non-diagonal entries):

$$\boldsymbol{\rho} = \begin{pmatrix} \rho_{00} & \rho_{01} \\ \rho_{10} & \rho_{11} \end{pmatrix}. \tag{18.68}$$

To establish contact with physical systems and experimentally testable results, we make two steps. So far, we have not specified the density operator matrix in terms of nuclear degrees of freedom. Acknowledging the conceptual clarity and relative computational feasibility of the phase space formulation, as introduced in Section 4.3 and expanded in Chapter 14, we use the Wigner transform of the density, according to Eq. (4.90):

$$\rho_{\mathrm{w}}(\mathbf{X}, \mathbf{P}, t) = \frac{1}{2\pi\hbar} \int_{-\infty}^{\infty} d\tilde{\mathbf{X}} \left\langle \mathbf{X} + \frac{\tilde{\mathbf{X}}}{2} | \hat{\rho}(t) | \mathbf{X} - \frac{\tilde{\mathbf{X}}}{2} \right\rangle \exp\left(-\frac{i}{\hbar} \mathbf{P} \cdot \tilde{\mathbf{X}} \right). \tag{18.69}$$

Developing the Wigner representation density matrix elements relevant to pump–probe spectroscopy, we start with an arbitrary number N of interacting molecular states and subsequently narrow down to the case of interest, i.e. a three-state model. With reference to Eq. (14.2), we expand the Liouville–von Neumann equation in the following manner [650]:

$$i\hbar \frac{\partial \rho_{\mathrm{w},ij}(\mathbf{X}, \mathbf{P}, t)}{\partial t} = H_i(\mathbf{X}, \mathbf{P}) \exp\left[\frac{i\hbar}{2} \hat{\Lambda} \right] \rho_{\mathrm{w},ij}(\mathbf{X}, \mathbf{P}, t) - \rho_{\mathrm{w},ij}(\mathbf{X}, \mathbf{P}, t) \times \exp\left[\frac{i\hbar}{2} \hat{\Lambda} \right] H_j(\mathbf{X}, \mathbf{P}) + [\hat{L}_1 \rho_{\mathrm{w}}(\mathbf{X}, \mathbf{P}, t)]_{ij} \tag{18.70}$$

and

$$[\hat{L}_1 \rho_{\mathrm{w}}(\mathbf{X}, \mathbf{P}, t)]_{ij} = \varepsilon(t) \sum_{k, i \neq k} \mu_{ik} \rho_{\mathrm{w},kj}(\mathbf{X}, \mathbf{P}, t) - \rho_{\mathrm{w},ik}(\mathbf{X}, \mathbf{P}, t) \mu_{kj}, \tag{18.71}$$

where $\varepsilon \mu_{ik}$ is short for $\boldsymbol{\varepsilon} \cdot \boldsymbol{\mu}_{ik}$ and the sum on the right-hand side is to be evaluated under the constraint that the dipole operator matrix $\boldsymbol{\mu}$ does not contain any diagonal elements.

The Condon approximation is made for $\boldsymbol{\mu}$, involving the neglect of any change in the geometric structure of the molecule while it undergoes a transition. As detailed in Section 14.1, taking the classical limit of Eq. (18.70) involves expanding the operator $\exp[i\hbar\frac{\hat{\Lambda}}{2}]$ into powers of $\hbar$ and truncating the resulting series after the lowest non-vanishing order. It is easily verified that this is the zeroth order for $i \neq j$ and the first for $i = j$. Additionally, we utilize the perturbation approach introduced before by adopting the series

$$\rho_{\mathrm{w},ij} = \rho^{(0)}_{\mathrm{w},ij} + \rho^{(1)}_{\mathrm{w},ij} + \rho^{(2)}_{\mathrm{w},ij} + \cdots . \tag{18.72}$$

for the Wigner transform of the vibronic density.

Exercise 18.4. Starting with the multidimensional Liouville–von Neumann equation (14.10), and adopting (18.67) as the Hamiltonian, show that

$$i\hbar\frac{\partial\rho^{(n)}_{ab}}{\partial t} = H_a\rho^{(n)}_{ab} - \rho^{(n)}_{ab}H_b - \varepsilon(t)\sum_c(\mu_{ac}\rho^{(n-1)}_{cb} - \rho^{(n-1)}_{ac}\mu_{cb}). \tag{18.73}$$

Interpreting the electric dipolar interaction term as perturbation operator of first order, we derive from Eq. (18.70) the following equation for the transition density matrix elements $\rho^{(n)}_{\mathrm{w},ij}$ with $i \neq j$:

$$\begin{aligned} i\hbar\frac{\partial\rho^{(n)}_{\mathrm{w},ij}(\mathbf{X},\mathbf{P},t)}{\partial t} &= [H_i(\mathbf{X},\mathbf{P}) - H_j(\mathbf{X},\mathbf{P})]\rho^{(n)}_{\mathrm{w},ij}(\mathbf{X},\mathbf{P},t) \\ &\quad + [\hat{L}_1\rho^{(n-1)}_{\mathrm{w}}(\mathbf{X},\mathbf{P},t)]_{ij}. \end{aligned} \tag{18.74}$$

For the diagonal density matrix elements, or populations, we find

$$\begin{aligned} i\hbar\frac{\partial\rho^{(n)}_{\mathrm{w},ii}(\mathbf{X},\mathbf{P},t)}{\partial t} = \frac{\hbar}{i}\Bigg[&\frac{\partial H_i(\mathbf{X},\mathbf{P})}{\partial\mathbf{P}}\frac{\partial\rho^{(n)}_{\mathrm{w},ii}(\mathbf{X},\mathbf{P},t)}{\partial\mathbf{X}} \\ &- \frac{\partial H_i(\mathbf{X},\mathbf{P})}{\partial\mathbf{X}}\frac{\partial\rho^{(n)}_{\mathrm{w},ii}(\mathbf{X},\mathbf{P},t)}{\partial\mathbf{P}}\Bigg] + [\hat{L}_1\rho^{(n-1)}_{\mathrm{w}}(\mathbf{X},\mathbf{P},t)]_{ii}. \end{aligned} \tag{18.75}$$

One readily converts Eq. (18.74) into an integral equation of the type (18.66), to be solved iteratively, with the function

$G^{(0)}(t-\tau) = \exp[-\frac{i}{\hbar}(H_i(\mathbf{X},\mathbf{P}) - H_j(\mathbf{X},\mathbf{P}))(t-\tau)]$ as propagator:

$$\rho^{(n)}_{\mathrm{w},ij}(\mathbf{X},\mathbf{P},t) = \frac{1}{i\hbar}\int_0^t dt_1 \exp\left[-\frac{i}{\hbar}V_{ij}(\mathbf{X},\mathbf{P})(t-t_1)\right]\varepsilon(t_1)$$
$$\times \sum_k (\mu_{ik}\rho^{(n-1)}_{\mathrm{w},kj}(\mathbf{X},\mathbf{P},t_1) - \rho^{(n-1)}_{\mathrm{w},ik}(\mathbf{X},\mathbf{P},t_1)\mu_{kj}),\quad i \neq j, \tag{18.76}$$

where the substitution $V_{ij}(\mathbf{X},\mathbf{P})$ has been made for the energy gap $H_i(\mathbf{X},\mathbf{P}) - H_j(\mathbf{X},\mathbf{P})$, and the term $(\hat{L}_1\rho^{(n-1)}_{\mathrm{w}}(\mathbf{X},\mathbf{P},t))_{ij}$ has been expanded into its components. Likewise, transforming Eq. (18.75) into an integral equation requires solving the corresponding homogeneous equation that is identical with the unperturbed Liouville–von Neumann equation. Since this relation is classical, the solution is a delta function in the phase space trajectory $\{\mathbf{X}_i(t), \mathbf{P}_i(t)\}$ of a classical particle evolving under the Hamilton function $H_i(\mathbf{X},\mathbf{P})$ with initial conditions $\mathbf{X}_i(t = 0) \equiv \mathbf{X}_{i0}$ and $\mathbf{P}_i(t = 0) \equiv \mathbf{P}_{i0}$ (see also Eq. (14.30)). This defines the zeroth order of the perturbation approach pursued here. Thus,

$$\rho^{(0)}_{\mathrm{w},ij}(\mathbf{X},\mathbf{P},t) = \delta(\mathbf{X} - \mathbf{X}(t,\mathbf{X}_0))\delta(\mathbf{P} - \mathbf{P}(t,\mathbf{P}_0)). \tag{18.77}$$

Setting $\{\mathbf{X},\mathbf{P}\} \equiv \mathbf{\Gamma}$ and $\delta(\mathbf{\Gamma} - \mathbf{\Gamma}(t,\mathbf{\Gamma}_0)) = \delta(\mathbf{X} - \mathbf{X}(t,\mathbf{X}_0))\delta(\mathbf{P} - \mathbf{P}(t,\mathbf{P}_0))$, the integral form of Eq. (18.75) for the diagonal elements of the phase space density matrix results as [650]

$$\rho^{(n)}_{\mathrm{w},ii}(\mathbf{\Gamma},t) = \frac{1}{i\hbar}\int d\mathbf{\Gamma}_{i0}\int_0^t dt_2\delta(\mathbf{\Gamma} - \mathbf{\Gamma}(t-t_2,\mathbf{\Gamma}_{i0}))$$
$$\times\varepsilon(t_2)\sum_{m,m\neq i}(\mu_{im}\rho^{(n-1)}_{\mathrm{w},mi}(\mathbf{\Gamma}_{i0},t_2) - \rho^{(n-1)}_{\mathrm{w},im}(\mathbf{\Gamma}_{i0},t_2)\mu_{mi}) \tag{18.78}$$

with $\mathbf{\Gamma}_{i0}$ standing for $\{\mathbf{X},\mathbf{P}\}$ on surface i at $t = 0$. As the elements $\mu_{im}\rho^{(n-1)}_{\mathrm{w},mi}$ are non-diagonal, they are described by the identity (18.76). By inserting Eq. (18.76) into Eq. (18.78) we obtain an expression for the density matrix elements $\rho^{(n)}_{\mathrm{w},ii}(\mathbf{\Gamma},t)$, related to the population of the molecular state i:

$$\rho^{(n)}_{\mathrm{w},ii}(\mathbf{\Gamma},t) = \frac{2}{\hbar^2}\Re\left\{\int d\mathbf{\Gamma}_{i0}\int_0^t dt_2\int_0^{t_2} dt_1\,\delta(\mathbf{\Gamma}-\mathbf{\Gamma}(t-t_2,\mathbf{\Gamma}_{i0}))\varepsilon(t_2)\varepsilon(t_1)\right.$$

$$\times\sum_{k,m}\mu_{im}(\mu_{mk}\rho^{(n-2)}_{\mathrm{w},ki}(\mathbf{\Gamma}_{i0},t_1)-\rho^{(n-2)}_{\mathrm{w},mk}(\mathbf{\Gamma}_{i0},t_1)\mu_{ki})$$

$$\left.\times\exp\left[-\frac{i}{\hbar}V_{mi}(\mathbf{\Gamma}_{i0})(t_2-t_1)\right]\right\}. \tag{18.79}$$

As we limit the discussion to two states only, labeled 0 for the ground and 1 for the excited state, the possible combinations of the indices m and k are easily enumerated. We ask now for the Wigner function $\rho^{(n)}_{\mathrm{w},11}(\mathbf{\Gamma},t)$, containing the information about the time behavior in the excited state. The non-diagonal nature of the electric dipolar moment matrix demands that $i \neq m$ for all contributing terms, and thus restricts the Wigner function of order $n-2$ to $\rho^{(n-2)}_{\mathrm{w},00}(\mathbf{\Gamma}_{10},t_1)$ and $\rho^{(n-2)}_{\mathrm{w},11}(\mathbf{\Gamma}_{10},t_1)$. Stipulating that the excited state population is initially equal to zero, the only surviving element of $\rho^{(n-2)}_{\mathrm{w},pq}$ is $\rho^{(n-2)}_{\mathrm{w},00}$. Implementing these changes, setting $\mu_{01}=\mu_{01}\equiv\mu$, and applying the rotating wave approximation, one turns Eq. (18.79) into [649]

$$\rho^{(n)}_{\mathrm{w},11}(\mathbf{\Gamma},t) = \frac{2\mu^2}{\hbar^2}\Re\left\{\int d\mathbf{\Gamma}_{10}\int_0^t dt_2\int_0^{t_2} dt_1\delta(\mathbf{\Gamma}-\mathbf{\Gamma}(t-t_2,\mathbf{\Gamma}_{10}))\right.$$

$$\left.\times E(t_1)E(t_2)\rho^{(n-2)}_{\mathrm{w},00}(\mathbf{\Gamma}_{10},t_1)\exp\left[i(\omega-V_{01}(\mathbf{\Gamma}_{01})/\hbar)(t_2-t_1)\right]\right\}, \tag{18.80}$$

with $E_i, i=1,2$, as the electric field envelope functions defined by Eqs. (18.5) and (18.6). Inspection of the product $E(t_1)E(t_2)$ yields

$$E(t_1)E(t_2) = \frac{E_0^2}{2\pi\sigma^2}\exp\left[-(t_1-t_i)^2/2\sigma^2-(t_2-t_i)^2/2\sigma^2\right]$$

$$= \frac{E_0^2}{2\pi\sigma^2}\exp\left[-((t_1-t_2)^2+(t_1+t_2-2t_i)^2)/4\sigma^2\right]. \tag{18.81}$$

From this relation it is obvious that the prevailing contribution to the product stems from the region $t_1 \approx t_2$. Approximating $E(t_1)E(t_2)$ by setting

$t_1 + t_2 = 2t_2$, we find

$$E(t_1)E(t_2) \approx \frac{E_0^2}{2\pi\sigma^2} \exp[-(t_1 - t_2)^2/4\sigma^2] \exp[-(t_2 - t_i)^2/\sigma^2]. \tag{18.82}$$

In the following step, we introduce the correlation time $\tau_c = t_2 - t_1$ as well as the time $\tau_1 = t - t_2$ which provides a measure of the time of propagation on the potential energy surface of the excited state. As we further select $n = 2$, we recognize that the ground state density $\rho_{w,00}^{(n-2)}(\boldsymbol{\Gamma}_{10}, t_1)$ turns into the time-independent zeroth-order expression $\rho_{w,00}^{(0)}(\boldsymbol{\Gamma}_{10})$. Making the respective substitution in (18.80), inserting (18.82), and changing variables in the time integrals, one arrives at

$$\begin{aligned}\rho_{w,11}^{(2)}(\boldsymbol{\Gamma}, t) = \frac{2\mu^2}{\hbar^2}\Re\bigg\{\int d\boldsymbol{\Gamma}_{10}\int_0^t d\tau_1 \int_0^{t-\tau_1} d\tau_c \delta(\boldsymbol{\Gamma} - \boldsymbol{\Gamma}(\tau_1, \boldsymbol{\Gamma}_{10}))\\ \times \frac{1}{\sqrt{4\pi}\sigma} I(t - \tau_1)\exp(-\tau_c^2/4\sigma^2)\\ \times \exp[(i(\omega - V_{01}(\boldsymbol{\Gamma}_{10})/\hbar)\tau_c]\rho_{w,00}^{(0)}(\boldsymbol{\Gamma}_{10})\bigg\}\end{aligned} \tag{18.83}$$

where $I(t) \equiv E_0^2/(\sqrt{\pi}\,\sigma)\exp(-(t - t_i)^2/\sigma^2)$ has been introduced. For sufficiently small pulse duration σ, the upper limit of the integral with respect to τ_c may be extended to infinity. The appearance of the term $\exp(-\tau_c^2/4\sigma^2)$ in the integrand justifies this approximation, as the contributions of the domain $\tau_c > 2\sigma$ to the integral are kept small. This manipulation has the advantage that the integration over τ_c can now be performed analytically, since

$$\begin{aligned}\int_0^\infty d\tau_c \frac{1}{\sqrt{4\pi}\sigma}\exp(-\tau_c^2/4\sigma^2)\exp[(i(\omega - V_{01}(\boldsymbol{\Gamma}_{10})/\hbar)\tau_c)]\\ \approx \exp[-\sigma^2(\omega - V_{01}(\boldsymbol{\Gamma}_{10})/\hbar)^2].\end{aligned} \tag{18.84}$$

The excited state population (18.83) therefore simplifies to [649]

$$\begin{aligned}\rho_{w,11}^{(2)}(\boldsymbol{\Gamma}, t) = \frac{\mu^2}{\hbar^2}\int_0^t d\tau_1 I(t - \tau_1)\int d\boldsymbol{\Gamma}_{10}\delta(\boldsymbol{\Gamma} - \boldsymbol{\Gamma}(\tau_1, \boldsymbol{\Gamma}_{10}))\\ \times \exp[-\sigma^2(\omega - V_{01}(\boldsymbol{\Gamma}_{10})/\hbar)^2]\rho_{w,00}^{(0)}(\boldsymbol{\Gamma}_{10}).\end{aligned} \tag{18.85}$$

This formula may be understood as a prescription for trajectory-based propagation of the density in the Wigner representation. More specifically, an initial ground state distribution has to be defined by an ensemble of appropriately chosen initial values $\mathbf{\Gamma}_{10}$ for classical trajectories. The factor $\exp[-\sigma^2(\omega - V_{01}(\mathbf{\Gamma}_{10})/\hbar)^2]$ acts like a filter for the ensemble, suppressing those members of the trajectory set whose initial values are too far from the resonance condition. It is good economy that this selection is built into the initial step instead of resulting from interference effects in the course of the propagation. Obviously, the latter scheme would be more cost-extensive in terms of computational effort, since the whole of the trajectory swarm would have to be generated before time integration would decide which elements of this swarm contribute to the excited state density at time t and which do not. On the basis of Eq. (18.85), only those trajectories that pass through the "excitation window" set up by the filter factor are subjected to propagation under the action of the classical Hamiltonian H_1 for the excited state potential energy surface. The emerging time-dependent density is folded by the intensity function $I(t - \tau_1)$ which accounts for the finite duration of the pulse.

18.4.1 *The pump–probe signal*

How does one have to modify the derivation of the excited state density to incorporate a third electronic state and thus to simulate the condition of a pump–probe experiment? Achieving this goal requires generalizing Eq. (18.80) by defining two electric fields with envelope functions E_{pu} and E_{pr}, centered at times t_{pu} and t_{pr}:

$$E_{\mathrm{pu}} = \frac{E_0^{pu}}{\sqrt{2\pi}\sigma_{\mathrm{pu}}} \exp\left(-\frac{(t - t_{\mathrm{pu}})^2}{2\sigma_{\mathrm{pu}}^2}\right), \tag{18.86}$$

$$E_{\mathrm{pr}} = \frac{E_0^{pr}}{\sqrt{2\pi}\sigma_{\mathrm{pr}}} \exp\left(-\frac{(t - t_{\mathrm{pr}})^2}{2\sigma_{\mathrm{pr}}^2}\right). \tag{18.87}$$

We assign to these two pulses carrier frequencies ω_{pu} and ω_{pr} and assume that the pump pulse is tuned to cause the transition from the ground state $\langle 0|$ to the excited state $\langle 1|$ while the probe pulse leads from $|1\rangle$ to another excited state $|2\rangle$. Adjusting relation (18.80) to describe this double-step scenario, one derives the time-dependent Wigner distribution $\rho_{w,22}(\mathbf{\Gamma}, t)$ for this case and obtains the pump–probe signal $S(t_{\mathrm{d}})$, with t_{d} as delay time $t_{\mathrm{pr}} - t_{\mathrm{pu}}$, from [649]

$$S(t_d) \sim \lim_{t\to\infty} \int d\mathbf{\Gamma}\, \rho_{\mathrm{w},22}(\mathbf{\Gamma}, t). \tag{18.88}$$

In terms of the initial Wigner distribution, the signal results as

$$\int_0^\infty d\tau_2 I_{\mathrm{pu}}(t-\tau_2-\tau_\beta) I_{\mathrm{pr}}(t-\tau_2) \sim E_{\mathrm{pu},0}^2 E_{\mathrm{pr},0}^2 \sqrt{\frac{\pi\sigma_{\mathrm{pu}}^2\sigma_{\mathrm{pr}}^2}{\sigma_{\mathrm{pu}}^2+\sigma_{\mathrm{pr}}^2}}$$

$$\times \exp\left[-\frac{(\tau_\beta - t_d)^2}{\sigma_{\mathrm{pu}}^2+\sigma_{\mathrm{pr}}^2}\right] \equiv g(\tau_\beta - t_{\mathrm{d}}), \tag{18.89}$$

The steps that lead from Eq. (18.80) to Eq. (18.89), including the approximations made in arriving at the latter expression, are outlined in the appendix of this chapter. The correlation time τ_β and the intensity time correlation function g are defined by Eqs. (18.97) and (18.103), respectively.

Formula (18.89) is readily interpreted in terms of the sequence of events realized in the pump–probe experiment. The initial phase space density $\rho_{\mathrm{w},00}^{(0)}(\mathbf{\Gamma}_{10})$ is subjected to a spectral filter associated with the pump pulse, and a probe filter function is applied to the density transported to and advanced on the excited state surface. Equation (18.105) can be translated into a rule for generating a spectroscopic signal by means of semiclassical dynamics in phase space. Thus, the first step is the selection of an ensemble of trajectories whose initial values represent the ground state density $\rho_{\mathrm{w},00}^{(0)}(\mathbf{\Gamma}_{10})$ multiplied by the weighting function $\exp[-\sigma_{\mathrm{pu}}^2(\omega_{\mathrm{pu}} - V_{10}(\mathbf{\Gamma}_{10})/\hbar)^2]$. Subsequently, the trajectories are propagated over a period τ_β on the excited state potential energy surface by action of the classical Hamiltonian corresponding to $\hat{H}_1$. At this stage, the second window function $\exp[-\sigma_{\mathrm{pr}}^2(\omega_{\mathrm{pr}} - V_{21}(\mathbf{\Gamma}_1(\tau_\beta, \mathbf{\Gamma}_{10}))/\hbar)^2]$ is constructed.

For each time, the product of the initial density and the two filter functions is integrated with respect to the phase space coordinates of the ground state. This procedure can be viewed as averaging the probe window function over the initial density distribution filtered by the pump window function. The ultimate step is a convolution of this time-dependent average with the pump–probe correlation function $g(\tau_\beta - t_{\mathrm{d}})$, blurring the propagation starting time as well as the probing time according to the temporal structure of the pulses.

Several variations of the formalism developed earlier have been considered, differing from each other in the nature of the probe state. The latter may be realized by a second excited state, it may alternatively be produced by ionization of the excited system. This is the principle of the *NeExPo* version of pump–probe spectroscopy (e.g. [651]) where the pump pulse acts on a molecule in the neutral (Ne) ground state and the probe pulse ionizes the excited (Ex) species, leaving it as a positive (Po) ion. Another avenue is taken by the *NeExNe pump–dump* method which employs the ground state in the probing step. For simulating the respective signal, $V_{21}(\mathbf{\Gamma}_1(\tau_\beta;\mathbf{\Gamma}_{10}))$ in Eq. (18.105) is to be replaced by $V_{01}(\mathbf{\Gamma}_1(\tau_\beta;\mathbf{\Gamma}_{10}))$. A succession of isomers with different charges is traversed by the *NeNePo* method which starts with a negative system that is neutralized by photodetachment under the influence of the pump pulse and finally turned into a cation by the ionizing probe pulse. An example of the latter mechanism is provided by a simulation of the spectroscopic signals arising from the sequence $Ag_3^- \rightarrow Ag_3 \rightarrow Ag_3^+$ [650].

18.5 Case Study: Ultrafast Spectroscopy on Non-Stoichiometric Alkali-Halide Clusters

Computations of time-resolved pump–probe signals have been performed for various molecular systems. An obvious challenge to be faced in any such endeavor is the propagation of the trajectory ensemble on the excited potential energy surface, requiring the calculation of excited state gradients with sufficient accuracy. Thus, the relevant excitations of a molecular species suitable for pump–probe simulation must be accessible to computation without forbidding effort. It should further display a variety of phenomena of basic interest in terms of dynamics, such as bond breaking, isomerization, or occurrence of conical intersections. These conditions are largely satisfied by non-stoichiometric alkali-halide clusters of composition Na_nF_{n-1} or Na_nF_{n-2}. In these units, the valence electrons contributed by the alkali atoms participate in strongly polar, almost ionic bonds formed between the sodium and the fluorine constituents. In the case of Na_nF_{n-1}, this leaves one excess alkali electron which determines the absorption spectra of these molecules in the region of low transition energy and thus the low-lying excited states. The single-electron character of these states greatly simplifies their theoretical description. The following two subsections will highlight model computations on both the adiabatic and the nonadiabatic evolution of effective single- (Na_nF_{n-1}) and double- (Na_nF_{n-2}) electron systems induced by pump pulse application.

18.5.1 *Effective single-electron systems of the form Na_nF_{n-1}*

We will first focus on some results obtained for the small system Na_2F [651]. Figure 18.10 illustrates the relevant spectroscopic processes. The pump pulse causes a vertical electronic transition of the molecule to the first excited state at $\Delta E = 1.07$ eV, preserving its C_{2v} equilibrium geometry. On the excited potential energy surface, however, the molecule relaxes into linear shape, since the Na-Na bond is broken. An ionizing probe pulse may interrogate the system on its way to the linear minimum and thus monitor the excited state isomerization of the molecule by use of NeExPo pump–probe spectroscopy.

As the linear geometry is realized, the transition energy between the first excited and the ground state amounts to only 0.3 eV. The excited as well as the ground state at linear structure may be understood as differing

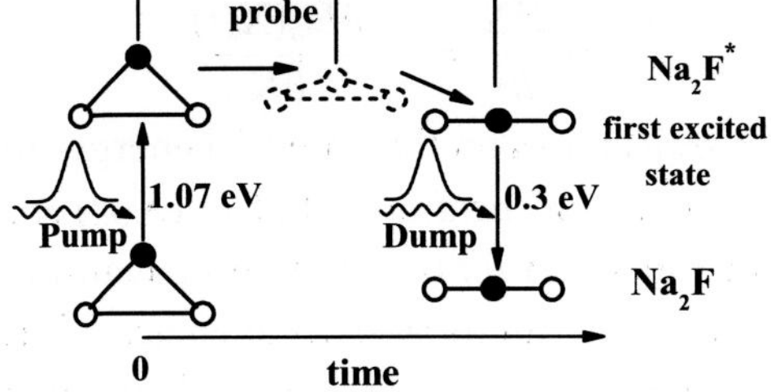

Figure 18.10: Various processes involved in NeExPo pump–probe and NeExNe pump–dump spectroscopy applied to Na_2F. Vertical excitation through the pump pulse is followed by relaxation on the potential energy surface of the first excited state, leading to linear geometry. The pump–probe observation mode interrogates the system by subsequent ionization, the pump–dump channel by vertical deexcitation to the ground state. (Reprinted with permission from [651]. Copyright (2001) by the American Institute of Physics.)

superpositions of the resonant configurations $\dot{\text{Na}}$-F^--Na^+ and Na^+-F^--$\dot{\text{Na}}$. The coupling between these two alternative states is small but not vanishing, giving rise to an avoided crossing for Na_2F at linear geometry.

Relation (18.105) provides the fundament for simulating the relevant pump-probe as well as pump–dump signals. In the simulation performed by Hartmann *et al.* [651], a thermal ensemble was used as input phase space density distribution $\rho^{(0)}_{w,00}(\mathbf{\Gamma}_{10})$, making it possible to study the spectroscopic signals as a function of temperature. Further, the propagation of the trajectory ensemble on the excited state surface proceeded *on the fly*, i.e. by *ab initio* quantum dynamics, thus eliminating the need to compute an extended segment of the potential energy surface as a whole. More specifically, the low-lying excited states were generated as eigenstates of an effective one-excess-electron Hamiltonian. Using this approximation, the dynamics of Na_nF_{n-1} clusters is made tractable by including the remaining $n-1$ valence electrons into the set of core electrons, since they may be considered as nearly localized in Na^+-F^- bonds. Within this frame, the energy E_m of an excited state m is calculated from

$$E_m = E_{\text{SCF}} + \varepsilon_m - \varepsilon_0 \tag{18.90}$$

where E_{SCF} denotes the self-consistent field ground state energy and ε_0, ε_m refer to ground and excited state energy eigenvalues of the single active electron, respectively. The difference $\varepsilon_m - \varepsilon_0$, therefore, approximates the transition energy between the electronic states $|0\rangle$ and $|m\rangle$. This approach in combination with the Newtonian equation of motion

$$M\frac{d^2\mathbf{X}}{dt^2} = -\frac{\partial E_m(\mathbf{X}(t))}{\partial \mathbf{X}} \tag{18.91}$$

was used to advance trajectories on the potential energy surface of state $|m\rangle$. Details on the construction of the gradient of $E_m(\mathbf{X}(t))$ are given in [651].

Applying this methodology for trajectory propagation, the pump–dump signal was generated at two dump laser frequencies, corresponding to the energies $E_{\text{du}} = 0.3$ and 1.0 eV. An ensemble with an initial temperature of 50 K was assumed. The duration of the pump pulse σ_{pu} was set to zero while σ_{du} was chosen as 50 fs. The condition $\sigma_{\text{pu}} = 0$ may be understood as a special case of the *impulsive or sudden approximation* (e.g. [652]) which assumes that the pulse imparts an instantaneous momentum to the studied species. This is warranted in situations where the pump pulse is sufficiently short to justify the neglect of any nuclear dynamics during its application time. As σ_{pu} is set to zero, the first spectral filtering step contained in relation (18.105) is not operative, and therefore, the entire ground state population is assumed to be lifted to the excited state surface instantly upon pump pulse application.

The resulting two signals are shown in Figure 18.11 as a function of the delay time within a time interval of 1.5 ps.

Both curves display oscillating features, reflecting the periodic alternation of the Na_2F structure in the first excited electronic state. The signal with the lower energy, corresponding to the solid line in Figure 18.11, is associated with transitions proceeding at the location of the avoided crossing between the ground and excited state surfaces. As suggested by Figure 18.10, the respective gap size of 0.3 eV matches the dump pulse energy. From our foregoing discussion, we conclude that the maxima of this curve occur at

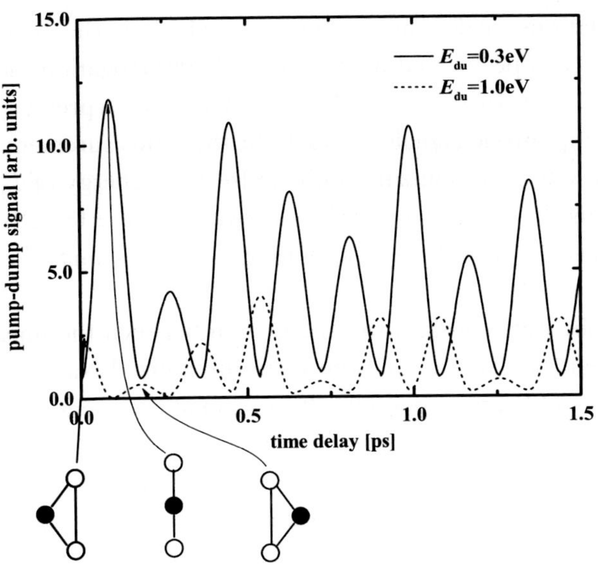

Figure 18.11: The dependence of pump–dump signals for Na_2F clusters at $E_{du} = 0.3$ and 1.0 eV on the dump pulse delay time [651]. The pump pulse is treated as instantaneous, the temperature of the initial ensemble is 50 K. The minima and maxima are related to various geometric shapes adopted by the molecule in the course of its evolution on the potential energy surface of the probed excited state. (Reprinted with permission from [651]. Copyright 2001 by the American Institute of Physics.)

times when the molecule adopts linear geometry. Correspondingly, the maxima of the dashed curve, referring to the dump pulse with $E_{du} = 1.0$ eV, are associated with occurrences of the obtuse ground state equilibrium structure of Na_2F, as the energy of the respective vertical transition, 1.07 eV, is close to E_{du}. From these assignments, it is plausible that the time difference between the first maxima of the dashed ($E_{du} = 1.0$ eV) and solid ($E_{du} = 0.3$ eV) curve provides a time scale for the isomerization of Na_2F from obtuse to linear geometry. This time interval, which is about 90 fs, may also be understood as time required for the breaking of the Na-Na bond and thus contains dynamic information about the electronic process that causes the isomerization effect. A modulation of the periodic bending motion carried out by the Na_2F molecule is clearly visible, particularly in the signal for $E_{du} = 1.0$ eV which exhibits a sequence of three low minima repeating in intervals of approximately 555 fs. This feature has been explained by the presence of an anharmonic admixture to the primary harmonic frequency of the Na_2F bending mode [651].

Computations analogous to those described in the preceding paragraphs have been performed for the NeExPo interrogation mode of Na_2F (see Figure 18.10; [654]), with the objective to interpret experimentally recorded pump–probe signals of Na_2F. In these measurements, the pump pulse is followed by an ionizing probe pulse at an energy of $E_{pr} = 3.06$ eV. For simulating the respective signal, as shown in Figure 18.12(b), an ensemble of 300 classical trajectories was employed at an estimated experimental temperature of 300 K.

The probe pulse bridges the energy gap between the linear geometry on the excited state surface and the vertically ionized Na_2F^+ species. The computation yields a near-periodic alternation of the energy difference separating the excited and the ionized level, an oscillation between the limits 3.06 eV and 3.30 eV (see Figure 18.12(a)) which is the respective difference for Na_2F in the obtuse ground state equilibrium structure. Correspondingly, the NeExPo signal displays a regular succession of maxima at moments when the linear geometry is realized and minima associated with the ground state geometry of Na_2F (see Figure 18.12(b)). In this manner, a stable shape oscillation with a period of 186 fs is monitored. This theoretical prediction is borne out by experiment. As can be seen from Figures 18.12(b) and (c), the matching between the computed and the measured ion signal maxima

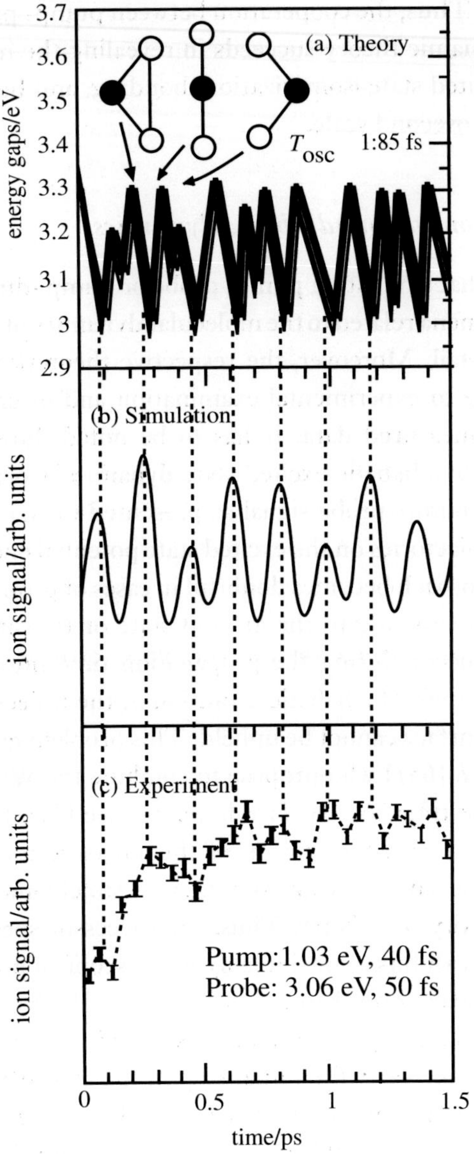

Figure 18.12: Theoretical and experimental results related to NeExPo spectroscopy of Na_2F [654]. The computed energy differences between the first excited and the ionized levels are shown in (a) as a function of time, revealing a Na_2F shape oscillation period of 186 fs. This motion is mirrored by the calculated pump–probe signal shown in (b). As can be seen from (c), it reproduces the salient features of the respective experimental data. (Reprinted with permission from [654]. Copyright (2002) by the American Physical Society.)

is almost perfect. Thus, the cooperation between pump–probe spectroscopy and quantum dynamic theory succeeds in revealing the multifaceted interplay between excited state isomerization, bonding, and bond-breaking processes at the femtosecond scale.

18.5.2 *Extension to nonadiabatic dynamics*

We have argued that simulating pump–probe or pump–dump signals allows us to trace phenomena related to the molecular dynamics of excited electronic states in great detail. Moreover, the respective theoretical predictions are directly accessible to experimental examination and of crucial importance for interpreting measured data. It has to be noted, however, that so far we have referred to adiabatic excited state dynamics. Correspondingly, the derivation of the pump–probe signal as presented earlier involves classical propagation of trajectories on the excited state potential energy surface. The validity of this approach becomes doubtful in cases of pronounced interstate coupling, where recrossing to the ground state or transition to the probe state is likely to occur during the propagation time interval τ_β. In those situations, the premise of adiabatic evolution of the trajectory ensemble on the excited state surface cannot be upheld. This problem has been addressed by Hartmann *et al.* [651] who propose to combine the Wigner distribution approach with the trajectory surface hopping (see Chapter 11) scheme as a viable solution. Such an extension of the adiabatic formalism sketched earlier becomes relevant as one investigates non-stoichiometric alkali halides of higher complexity than Na_2F. Thus, the successor species in the series Na_nF_{n-1}, Na_3F_2, exhibits a considerably wider scope of possible excitation, relaxation, and deexcitation channels than displayed by Na_2F. Schematically, the various routes taken by the system are shown in Figure 18.13.

As in the case of Na_2F, the initial step in the evolution of Na_3F_2 consists in a vertical transition of a thermal ensemble to the first excited state surface, sampling the most stable ground state structure. This is followed by deformation, leading to linear geometry. In contrast to Na_2F, this structural change involves the breaking not only of a metallic Na-Na but also of an ionic Na-F bond. We recall from our discussion of the avoided crossing found for Na-F-Na that the wave functions of linear Na_nF_{n-1} units are superpositions of configurations that differ in the location of the excess electron. For Na_3F_2,

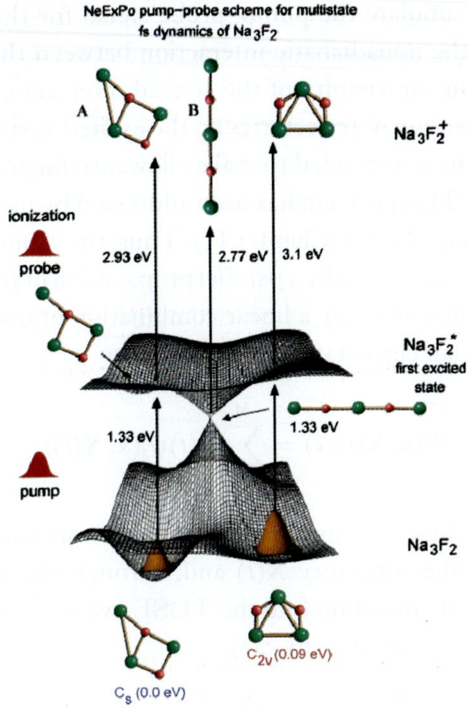

Figure 18.13: Sketch of the processes monitored by NeExPo spectroscopy of Na_3F_2. Upon initial excitation, $Na_3F_2^*$ relaxes into linear geometry which is marked by a conical intersection between the ground state and the first excited state potential energy surfaces. Radiationless passage through the conical intersection may lead the system either to the C_s ground state equilibrium structure or to a C_{2v} isomer. All channels involved in the evolution of the system can be interrogated by NeExPo probe pulses spanning an energy interval from 2.9 eV to 4.8 eV. (Reprinted with permission from [28]. Copyright (2001) by the American Chemical Society.)

the components are $\dot{N}a$-F^--Na^+-F^- -Na^+ and F^--Na^+-F^--Na^+-F^--$\dot{N}a$. Due to the increased length of the alkali-halide chain in comparison to Na_2F, the coupling between the alternative configurations is sufficiently reduced to close the gap between the ground and the first excited state and thus to turn the avoided crossing into a conical intersection. As the system reenters the ground state surface through this funnel, it branches out into two channels, one leading back to the equilibrium geometry, the second to a C_{2v} isomer lying by 0.09 eV above the minimal energy. It is evident from Figure 18.13 that all of these processes can in principle be monitored by an ionizing probe pulse that covers an energy range from 2.9 eV to 4.8 eV.

Any effort to simulate the pump–probe signal for this problem has to take into account the nonadiabatic interaction between the ground and the first excited state in the vicinity of the conical intersection. In this region, the classical propagation of trajectories on the excited surface implied in formula (18.105) has to be extended to make allowance for transitions between the coupled states. This problem has been addressed by use of Tully's fewest switches algorithm ([288], see Chapter 11). Thus, the evolution of the trajectory ensemble is accompanied by a parallel propagation of the electronic wave function which is described by a linear combination of one-excess-electron effective Hamiltonian eigenstates ψ_i, according to

$$\Psi(\mathbf{x}, \mathbf{X}(t); t) = \sum_{i=0}^{M} c_i(t)\psi_i(\mathbf{x}, \mathbf{X}(t)). \tag{18.92}$$

This expansion includes a number M of one-electron states which depend parametrically on the trajectory $\mathbf{X}(t)$ and, through the nuclear variables, indirectly on time. By insertion into the TDSE, we arrive at the equation of motion (11.11) for the coefficients $c_i(t)$:

$$i\hbar\frac{\partial c_i(t)}{\partial t} = \sum_k \left[\varepsilon_k\delta_{ik} - i\hbar\frac{\partial \mathbf{X}}{\partial t}\cdot\left\langle\psi_i \mid \frac{\partial}{\partial \mathbf{X}}\psi_k\right\rangle\right] c_k(t). \tag{18.93}$$

Along the lines of the fewest switches strategy introduced in Section 11.2, a criterion for the transition between two electronic potential energy surfaces with indices i and j is provided by the switching probability Eq. (11.15). The computation of the nonadiabatic coupling term $\langle\psi_i|\frac{\partial}{\partial \mathbf{X}}\psi_j\rangle$ for the single-electron states ψ_k as obtained in the framework of the frozen bond approximation for the non-stoichiometric alkali halides is discussed in [651]. If g_{ij} is larger than a random threshold number, the transition between the states i and j is assumed to take place. Transferring this concept to the case of Na_3F_2, as sketched in Figure 18.13, one has to modify the formula for the pump-probe signal (18.105), replacing $V_{21}(\mathbf{\Gamma}_1(\tau_\beta, \mathbf{\Gamma}_{10}))$ with $V_{+1,x}(\mathbf{\Gamma}_x(\tau_\beta, \mathbf{\Gamma}_{10}))$, where x stands for either 0 or 1, i.e. the ground or the first excited state, and +1 denotes the cationic probe state. An additional gain in accuracy is achieved by averaging the signal, for every given initial condition, over various trajectories generated by a set of N_h random numbers. Using an index ν to enumerate these, one may write the nonadiabatic generalization of Eq. (18.105) as

$$S(t_{\rm d}) \sim \int_0^{\infty} d\tau_\beta\, g(\tau_\beta - t_{\rm d}) \int d\mathbf{\Gamma}_{10} \frac{1}{N_h}$$
$$\times \sum_{\nu} \exp\left[-\frac{\sigma_{\rm pr}^2}{\hbar^2} (\hbar\omega_{\rm pr} - V_{+1,x}(\mathbf{\Gamma}_x^{\nu}(\tau_\beta, \mathbf{\Gamma}_{10})))^2 \right]$$
$$\times \exp\left[-\frac{\sigma_{\rm pu}^2}{\hbar^2} (\hbar\omega_{\rm pu} - V_{10}(\mathbf{\Gamma}_{10}))^2 \right] \rho_{\rm w,00}^{(0)}(\mathbf{\Gamma}_{10}). \qquad (18.94)$$

The implementation of this formalism to simulate the time-resolved spectroscopy of Na_3F_2 was carried out by sampling 100 phase space points. This makes it possible to study the statistical distribution of the transition energies from the first excited state to the cationic state as a function of time. For the canonical 50 K ensemble chosen in [653], and in the initial time interval 0 fs $\leq t_d \leq$ 400 fs, this distribution peaks around 2.9 eV, where the energy gap of maximum abundance gradually moves from $\Delta E = 2.8$ eV to $\Delta E = 3.0$ eV. The corresponding pump–probe signal is shown in Figure 18.14(a) for two excitation energies, 2.8 eV and 3.0 eV.

Each of the two curves exhibits a marked maximum. Inspecting the time development of the phase space distribution, it is possible to assign the maxima unanimously to the two bond-breaking events that characterize the evolution of the system in the excited state. Thus, the first peak, occurring at $t_{\rm d} \sim 90$ fs, is attributed to the rupture of the metallic Na-Na bond, followed by the breaking of the ionic Na-F bond, as indicated by the second peak at $t_d \sim 220$ fs. At $t_{\rm d} \sim 500 - 600$ fs, both signals display a pronounced drop to low intensities. This corresponds to a substantial rise of the signals shown in Figure 18.14(b), associated with probe pulses at $E_{\rm pr} = 4.3$ eV and $E_{\rm pr} =$ 4.8 eV. With reference to the pathways depicted in Figure 18.13, the former pulse probes the ground state equilibrium C_s structure of Na_3F_2 while the latter is compatible with the C_{2v} isomer. The alternation between the signals reflects the time dependence of the intramolecular vibrational energy flow between the structural alternatives after the radiationless deexcitation of the system. By comparison of the excited (Figure 18.14(a)) and the ground state (Figure 18.14(b)) signals, it is possible to infer a time scale for the passage through the conical intersection.

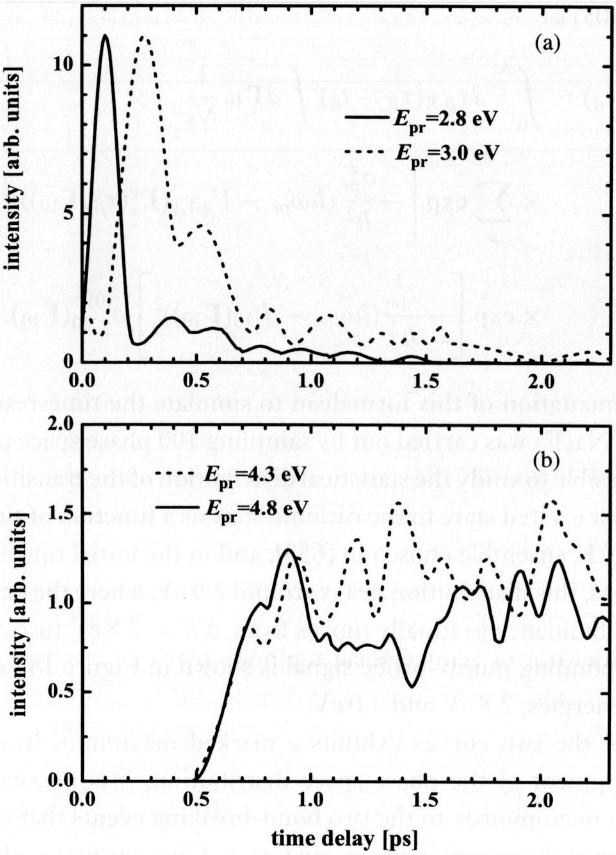

Figure 18.14: The computed Na_3F_2 NeExPo pump–probe signal for four different excitation energies: $E_{pr} = 2.8$ eV (solid line in (a)), 3.0 eV (dashed line in (a)), 4.3 eV (dashed line in (b)), and 4.8 eV (solid line in (b)), assuming a thermal initial ensemble at 50 K. The calculations are based on formula (18.94) that includes nonadiabatic effects. (Reprinted with permission from [653]. Copyright (2001) by the American Institute of Physics.)

This analysis elucidates the complex interplay of widely varied molecular processes, such as the breaking of a metallic and an ionic bond during geometric relaxation of the species in an electronically excited state, the subsequent deexcitation by traversal of a conical intersection and, following this, redistribution of the vibrational energy.

The reader will find a supplementary discussion of modeling pump-probe spectra for non-stoichiometric alkali-halide molecules with two active electrons in Section 18.5*.

In conclusion, this chapter has focused on the methodology of simulating molecular dynamics induced and monitored by nonlinear optical processes. In particular, various pump–probe mechanisms were highlighted. Two types of schemes employed to model the respective spectroscopic signals were introduced, namely perturbative and non-perturbative realizations of the wave-function-centered approach introduced in Section 18.1 that aims at deriving the nonlinear polarization of the process under study, as well as the density matrix alternative based on Liouville theory (Section 17.4). These procedures were illustrated by applications to the spectroscopy of two characteristic effects: *cis-trans* photoisomerization in Section 18.3 and the ultrafast response of small alkali-halogen clusters to excitation and ionization in Section 18.5. Two of the cases discussed, namely *cis-trans* photoisomerization and the excited state evolution of Na_3F_2, involve non-radiative transitions through conical intersections. The corresponding nonadiabatic dynamics can, in principle, be captured by fully quantum mechanical schemes, such as wave packet propagation techniques or the non-perturbative model outlined in Section 18.2.2, but has to be imported into the framework of semiclassical methods that operate with trajectory propagation on potential energy surfaces, exemplified here by Liouville theory rooted in a phase space representation of the nuclear degrees of freedom. The results summarized in Section 18.5 demonstrate that for some physical systems, computational modeling has reached a good level of convergence with measurement. In particular, it accounts for the complex processes of excitation and ionization, bond breaking and making, geometric and electronic relaxation, involved in the interaction of molecules with electromagnetic radiation as studied by femtosecond spectroscopy.

18.6 Appendix: Derivation of the Pump–Probe Signal $S(t_d)$

Extending formula (18.80) for the excited state Wigner distribution to the double-step process involved in pump–probe experiments, one finds

$$\rho_{w,22}(\mathbf{\Gamma}, t) \sim \Re\Bigg(\int d\mathbf{\Gamma}_{10} \int d\mathbf{\Gamma}_{21} \int_0^t dt_4 \int_0^{t_4} dt_3 \int_0^{t_3} dt_2 \int_0^{t_2} dt_1$$

$$\times\, \delta(\mathbf{\Gamma} - \mathbf{\Gamma}_2(t - t_4, \mathbf{\Gamma}_{21})) \exp\left[i(\omega_{pr} - V_{21}(\mathbf{\Gamma}_{21})/\hbar)(t_4 - t_3)\right]$$

$$\times \delta(\mathbf{\Gamma}_{21} - \mathbf{\Gamma}_1(t_3 - t_2, \mathbf{\Gamma}_{10})) \exp[i(\omega_{\text{pu}} - V_{10}(\mathbf{\Gamma}_{21})/\hbar)(t_2 - t_1)]$$
$$\times \rho^{(0)}_{\text{w},00}(\mathbf{\Gamma}_{10}) E_{\text{pu}}(t_1) E_{\text{pu}}(t_2) E_{\text{pr}}(t_3) E_{\text{pr}}(t_4)\Big). \tag{18.95}$$

Although this expression appears quite complex, it clearly depicts the physical sequence of events: initial transition to the first excited state is followed by propagation on the respective potential energy surface. These processes are repeated for the subsequent step, populating the second excited state. The further development proceeds in analogy to that outlined earlier for the case of only one pulse. Inspecting expression (18.95), we distinguish three correlation times, namely

$$\tau_\alpha \equiv t_2 - t_1, \tag{18.96}$$

$$\tau_\beta \equiv t_3 - t_2, \tag{18.97}$$

$$\tau_\gamma \equiv t_4 - t_3, \tag{18.98}$$

with the additional definition $\tau_2 \equiv t - t_4$. The product (18.82) becomes

$$E_{\text{pu}}(t_1) E_{\text{pu}}(t_2) \approx I_{\text{pu}}(t_2) \frac{1}{\sqrt{4\pi}\sigma_{\text{pu}}} \exp[-(t_2 - t_1)^2/4\sigma^2_{\text{pu}}], \tag{18.99}$$

$$E_{\text{pr}}(t_3) E_{\text{pr}}(t_4) \approx I_{\text{pr}}(t_4) \frac{1}{\sqrt{4\pi}\sigma_{\text{pr}}} \exp[-(t_2 - t_{\text{pr}})^2/4\sigma^2_{\text{pr}}], \tag{18.100}$$

where the substitutions $I_{\text{pu}}(t_2) \equiv E^2_{\text{pu},0}/(\sqrt{\pi}\sigma_{\text{pu}}) \exp(-(t_2 - t_{\text{pu}})^2/\sigma^2_{\text{pu}})$ and $I_{\text{pr}}(t_4) \equiv E^2_{\text{pr},0}/(\sqrt{(\pi)}\sigma_{\text{pr}}) \exp(-(t_4 - t_{\text{pr}})^2/\sigma^2_{\text{pr}})$ have been made. Inserting Eqs. (18.99) and (18.100) into Eq. (18.95), changing the integration variables from (t_1, t_2, t_3, t_4) to $(\tau_\alpha, \tau_\beta, \tau_\gamma, \tau_2)$, and integrating over τ_α and τ_γ by use of the strategy that led to Eq. (18.85), one arrives at

$$\rho_{\text{w},22}(\mathbf{\Gamma}, t) \sim \Re\left\{\int_0^{t-\tau_2} d\tau_\beta \left[\int_0^t d\tau_2 I_{\text{pu}}(t - \tau_2 - \tau_\beta) I_{\text{pr}}(t - \tau_2)\right] \int d\mathbf{\Gamma}_{10}\right.$$
$$\times \int d\mathbf{\Gamma}_{21} \delta(\mathbf{\Gamma} - \mathbf{\Gamma}_2(\tau_2, \mathbf{\Gamma}_{21})) \exp[-\sigma^2_{\text{pr}}(\omega_{\text{pr}} - V_{21}(\mathbf{\Gamma}_{21})/\hbar)^2]$$
$$\left.\times \delta(\mathbf{\Gamma}_{21} - \mathbf{\Gamma}_1(\tau_\beta, \mathbf{\Gamma}_{10})) \exp[-\sigma^2_{\text{pu}}(\omega_{\text{pu}} - V_{10}(\mathbf{\Gamma}_{10})/\hbar)^2] \rho^{(0)}_{\text{w},00}(\mathbf{\Gamma}_{10})\right\}. \tag{18.101}$$

In this move, we have exploited the particular dependence of I_{pu} on τ_γ. In greater detail, t_2 is represented as t-τ_2-τ_γ-τ_β, but since the main contribution to $I_{\rm pu}$ stems from the region $t_3 \approx t_4$ and hence $\tau_\gamma = 0$, we set $t_2 \approx t - \tau_2 - \tau_\beta$. The pump–probe signal S is proportional to the asymptotic population of state $|2\rangle$, as obtained by taking the trace over $\rho_{\rm w,22}(\mathbf{\Gamma}, t)$ and simultaneously the limit $t \to \infty$:

$$S(t_{\rm d}) \sim \lim_{t\to\infty} \int d\mathbf{\Gamma}\, \rho_{\rm w,22}(\mathbf{\Gamma}, t). \tag{18.102}$$

The signal depends on the delay time $t_{\rm d} = t_{\rm pr} - t_{\rm pu}$. The time integral over the product of the pump and probe intensities, a convolution of two Gaussian functions, can be written in closed form as the upper limit approaches infinity. Explicitly [649],

$$\int_0^\infty d\tau_2\, I_{\rm pu}(t - \tau_2 - \tau_\beta) I_{\rm pr}(t - \tau_2) \sim E^2_{\rm pu,0} E^2_{\rm pr,0} \sqrt{\frac{\pi\sigma^2_{\rm pu}\sigma^2_{\rm pr}}{\sigma^2_{\rm pu} + \sigma^2_{\rm pr}}}$$

$$\times \exp\left[-\frac{(\tau_\beta - t_{\rm d})^2}{\sigma^2_{\rm pu} + \sigma^2_{\rm pr}}\right] \equiv g(\tau_\beta - t_{\rm d}), \tag{18.103}$$

where g is a correlation function for the pump and probe pulse intensities. Summarizing, we represent the pump–probe signal as

$$S(t_{\rm d}) \sim \Re\left\{\int_0^\infty d\tau_\beta\, g(\tau_\beta - t_{\rm d}) \int d\mathbf{\Gamma}_{21} \int d\mathbf{\Gamma}_{10}\right.$$

$$\times \exp\left[-\sigma^2_{\rm pr}(\omega_{\rm pr} - V_{21}(\mathbf{\Gamma}_{21})/\hbar)^2\right]\delta(\mathbf{\Gamma}_{21} - \mathbf{\Gamma}_1(\tau_\beta, \mathbf{\Gamma}_{10}))$$

$$\left.\times \exp\left[-\sigma^2_{\rm pu}(\omega_{\rm pu} - V_{10}(\mathbf{\Gamma}_{10})/\hbar)^2\right]\rho^{(0)}_{\rm w,00}(\mathbf{\Gamma}_{10})\right\}. \tag{18.104}$$

As the integration over the intermediate coordinate set $\mathbf{\Gamma}_{21}$ is carried out, this becomes

$$S(t_{\rm d}) \sim \Re\left\{\int_0^\infty d\tau_\beta\, g(\tau_\beta - t_{\rm d}) \int d\mathbf{\Gamma}_{10}\right.$$

$$\times \exp\left[-\sigma^2_{\rm pr}(\omega_{\rm pr} - V_{21}(\mathbf{\Gamma}_1(\tau_\beta, \mathbf{\Gamma}_{10}))/\hbar)^2\right]$$

$$\left.\times \exp\left[-\sigma^2_{\rm pu}(\omega_{\rm pu} - V_{10}(\mathbf{\Gamma}_{10})/\hbar)^2\right]\rho^{(0)}_{\rm w,00}(\mathbf{\Gamma}_{10})\right\}. \tag{18.105}$$

With Eq. (18.105), the rather intimidating original formula (18.95) has been reduced to a final relation of remarkably simple structure.

Chapter 19

Optical Control of Electron Multistate Molecular Dynamics

While the preceding chapter dealt with methods of modeling molecular electron dynamics induced by electromagnetic fields in real time, the present chapter makes the step from simulation to manipulation. To what extent, and under what conditions, may the evolution of a molecular system, involving transitions between electronic levels, be steered by applying well-defined principles of *quantum control*? The high practical relevance of any methodology that allows guiding a given system of reactants towards a targeted set of products is obvious. In many situations where a photoinduced molecular process branches out into several reaction pathways, it is of critical interest to select a certain desired outcome and to suppress the alternative results. In this chapter, we will ask for procedures that allow us to maximize a selected reaction route by optimizing the parameters that define the electromagnetic signal, such as the shape and the number of pulses, or the delay time between subsequent pulses.

Specifically, we will highlight two basic control mechanisms in conjunction with exemplary cases of nonadiabatic molecular evolution. The paradigm of reaction control is given by the photodissociation of a triatomic species, ABC. Control involves steering the system towards one of the three final configurations A + BC, B + AC, and C + AB. Section 19.4 focuses on the treatment of such a case in the framework of the coherent control scheme proposed by Brumer and Shapiro, as introduced in Section 19.3. This is preceded by a sketch of the optimal control procedure due to Tannor and Rice (Section 19.2) and a preparatory section on the interaction between a molecule and a pulse of light, supplementing the respective material presented in Chapter 18, and placing it into the more general context of the vector potential. This chapter concludes with a modification of the Tannor–Rice formalism used as a device to route the evolution of the small non-stoichiometric alkali-halogen clusters discussed earlier (Section 19.5) in the context of nonlinear ultrafast spectroscopy.

19.1 Interaction of a Molecule with a Pulse of Light

We consider a molecule exposed to a coherent pulse of light. As in the previous chapter, this interaction between radiation and matter is analyzed by adopting the dipolar approximation. The TDSE is then once more defined by the Hamiltonian (18.13). We consider the classical solution for a free light field traveling in the Z-direction and write the field $\boldsymbol{\varepsilon}(Z,t)$ as

$$\boldsymbol{\varepsilon}(Z,t) = \mathbf{e}\int_0^\infty d\omega \left\{ a(\omega)\exp\left[i\omega\left(\frac{Z}{c} - t\right)\right] + a^*(\omega)\exp\left[-i\omega\left(\frac{Z}{c} - t\right)\right]\right\}. \quad (19.1)$$

As before, the vector $\mathbf{e}$ indicates the direction of the polarization. Since the pulse is assumed to be coherent, we assign the same orientation to all of its modes, labeled by the angular frequency ω. The function $a(\omega)$ is related to the vector potential amplitudes A_k as contained in the expression (see, for instance, [655])

$$\mathbf{A}(Z,t) = \mathbf{e}\int_0^\infty dk\{A_k \exp[-i\omega_k t + ikZ] + A_k{}^* \exp[i\omega_k t - ikZ]\}. \quad (19.2)$$

Including the complex conjugate of the integral over the field modes ensures that the vector potential is a real quantity. Our model operates with a free pulse, or equivalently a field that propagates through a charge-free space. In this case, the field is entirely transversal and thus derived from the vector potential, while no scalar, or longitudinal, components are present. The connection between the field and the vector potential is then

$$\boldsymbol{\varepsilon}(Z,t) = -\frac{1}{c}\frac{\partial \mathbf{A}(Z,t)}{\partial t} \quad (19.3)$$

and we conclude that the function $a(\omega)$ is related to the amplitudes A_k by $a(\omega) = ik/cA_k$. By defining the *retarded time* $\tau \equiv t - Z/c$, we cast the field

(19.1) into the concise form

$$\boldsymbol{\varepsilon}(\tau) = \mathbf{e} \int_{-\infty}^{\infty} d\omega\, a(\omega) \exp(-i\omega\tau). \tag{19.4}$$

The TDSE is solved by expanding $\Psi(t)$ into a basis of eigenfunctions of $\hat{H}_{\mathrm{M}}$:

$$\Psi(t) = \sum_n c_n(t) \Psi_n \exp\left(-\frac{i}{\hbar} E_n t\right) \tag{19.5}$$

and

$$\hat{H}_{\mathrm{M}} \Psi_n = E_n \Psi_n. \tag{19.6}$$

The coefficients $c_n(t)$ are derived by inserting Eq. (19.5) into the TDSE for the solution $\Psi(t)$. This yields

$$\frac{dc_m(t)}{dt} = -\frac{i}{\hbar} \sum_n c_n(t) \exp(-i\omega_{nm} t) \langle \Psi_m | \hat{H}_{\mathrm{int}}(t) | \Psi_n \rangle \tag{19.7}$$

with

$$\omega_{mn} = \frac{(E_m - E_n)}{\hbar}. \tag{19.8}$$

At the beginning of the process, the system may be in a state Ψ_1, such that the initial conditions are

$$c_1(t = -\infty) = 1, \quad c_j(t = -\infty) = 0 \quad \text{if} \quad j \neq 1. \tag{19.9}$$

In cases of sufficiently weak dipole interaction, perturbation theory at first order is applicable. We then conclude from Eqs. (19.7) and (19.9) in combination with the expression (19.4) for the coherent light pulse that the solution for $c_m(t)$ with $m \neq 1$ may, to first order, be written as [656]

$$c_m(t) = \frac{i}{\hbar} \mu_{m1} \int_{-\infty}^{\infty} d\omega\, \tilde{a}(\omega) \int_{-\infty}^{t} dt' \exp\left[i(\omega_{m1} - \omega)t'\right], \tag{19.10}$$

where $\tilde{a}(\omega) \equiv a(\omega) \exp[i\frac{\omega}{c} Z]$ and $\mu_{m1} = \langle \Psi_m | \mathbf{e} \cdot \hat{\boldsymbol{\mu}}_{m1} | \Psi_1 \rangle$. The coefficient $c_1(t)$ is determined by the normalization constraint on $\Psi(t)$. We are now in a position to investigate the behavior of $c_m(t)$ and, by the same token, that of $\Psi(t)$ for asymptotic times. As $t \to \infty$, Eq. (19.10) yields:

$$\lim_{t\to\infty} c_m(t) = 2\pi \frac{i}{\hbar} \tilde{a}(\omega_{m1})\mu_{m1}. \tag{19.11}$$

Further, we express the complex-valued function $a(\omega)$ in polar notation by introducing a phase factor $\alpha(\omega)$ and obtain

$$\lim_{t\to\infty} c_m(t) = 2\pi \frac{i}{\hbar} |\tilde{a}(\omega_{m1})| \, \mu_{m1} \exp\left[i \left(\alpha(\omega_{m1}) + \omega_{m1} \frac{Z}{c} \right) \right]. \tag{19.12}$$

In Eq. (19.12), the effect of the dipolar interaction on the molecule is partitioned into an amplitude $|\tilde{a}(\omega_{m1})|\mu_{m1}$ and a phase $\alpha(\omega_{m1}) + \omega_{m1}Z/c$. For absorption, $E_m > E_1$ and therefore $\omega_{m1} > 0$, while stimulated emission is characterized by $\omega_{m1} < 0$. This implies that the light wave gives rise to a positive phase in the former case and a negative in the latter (see Exercise 19.1).

Exercise 19.1. From the reality requirement imposed on the field $\boldsymbol{\varepsilon}(Z, t)$ (see Eq. (19.4)), conclude that $a(-\omega) = a^*(\omega)$, and therefore $\alpha(-\omega) = -\alpha(\omega)$. Show further that, by Eq. (19.5), absorption of light results in a positive, and stimulated emission of light in a negative asymptotic phase $\alpha(\omega_{m1}) + \omega_{m1}\frac{Z}{c}$.

Numerous applications of laser control methods deal with the manipulation of photodissociation processes. To simulate these processes, one has to extend the description of photoexcited molecular wave functions into the energy continuum, as described more formally in our treatment of molecular dissociation in Section 6.8. In particular, the dipolar moment becomes $\mu_{E\{m\},1} = \langle \Psi_{E\{m\}} | \mathbf{e} \cdot \hat{\boldsymbol{\mu}}_{m1} | \Psi_1 \rangle$, where E denotes a continuum energy, while all internal quantum numbers that may refer to the relevant electronic, vibrational, and rotational states and specify the accessible dissociation channels are summarized by the set of indices $\{m\}$. From Eq. (19.11), we conclude that the continuous part $\Psi_{\text{cont}}(t)$ of the molecular solution $\Psi(t)$ is of the following form:

$$\Psi_{\text{cont}}(t) = 2\pi \frac{i}{\hbar} \sum_{\{n\}} \int dE \, \tilde{a}(\omega_{E1}) \mu_{E\{m\},1} \Psi_{E\{m\}} \exp\left(-\frac{i}{\hbar} Et \right). \tag{19.13}$$

Here we denote by $\Psi_{E\{m\}}$ the solution with energy E and a set of internal quantum numbers $\{m\}$. Based on perturbation theory at first order,

Eq. (19.13) represents a *weak field limit* to the continuum component of the wave function. Laser control of photodissociation aims at exciting selectively a chosen channel characterized by a certain set of quantum numbers $\{m\}$ at a given energy. Various proposals to meet this challenge for photodissociation and general molecular processes involving excitation by electromagnetic radiation are discussed in the following sections.

19.2 The Tannor–Rice Scheme: Optimal Control

The discussion so far has not made any mention of the electronic system. The label m employed in the preceding section was used to denote a set of discrete internal quantum numbers, the electronic state index being one among them. We now make the electronic system explicit, referring to a problem that involves an electronic ground state Ψ_0 and an excited state Ψ_1. The respective diabatic 2×2 Hamiltonian operator matrix (equivalent to what was defined by Eqs. (18.14) and (18.15) in bra-ket notation) contains the coupling between these states by dipolar interaction with a light pulse:

$$\hat{H} = \begin{pmatrix} \hat{H}_0 & -\hat{\boldsymbol{\mu}} \cdot \boldsymbol{\varepsilon}^*(t) \\ -\hat{\boldsymbol{\mu}} \cdot \boldsymbol{\varepsilon}(t) & \hat{H}_1 \end{pmatrix}. \tag{19.14}$$

The dipolar transition moment $\boldsymbol{\mu}_{01} = \langle \Psi_0 | \hat{\boldsymbol{\mu}} | \Psi_1 \rangle = \boldsymbol{\mu}_{10} = \langle \Psi_1 | \hat{\boldsymbol{\mu}} | \Psi_0 \rangle \equiv \boldsymbol{\mu}$ defines here the non-diagonal matrix elements. Setting up the TDSE for (Ψ_0, Ψ_1), and assuming that the system initially occupies the electronic ground state, according to $\Psi_0(t \to -\infty) \equiv \Psi_{g,i}$, we arrive at the following set of formal solutions for the two states (see Eq. (3.20)):

$$\Psi_0 = \exp\left(-\frac{i}{\hbar}\hat{H}_0 t\right)\Psi_{0,i},$$

$$+\frac{i}{\hbar}\int_{-\infty}^{t} dt' \exp\left[-\frac{i}{\hbar}\hat{H}_0\left(t - t'\right)\right] \boldsymbol{\mu} \cdot \boldsymbol{\varepsilon}(t')\Psi_1(t') \tag{19.15}$$

$$\Psi_1 = \frac{i}{\hbar}\int_{-\infty}^{t_2} dt' \exp\left[-\frac{i}{\hbar}\hat{H}_1(t - t')\right] \boldsymbol{\mu} \cdot \boldsymbol{\varepsilon}(t')\Psi_0(t'). \tag{19.16}$$

The TDSE has thus been cast into integral equation form. Inserting Eq. (19.16) into Eq. (19.15), and in accordance with Eq. (3.23), we obtain the second-order expression

$$\Psi_0^{(2)} = -\frac{1}{\hbar^2}\int_{-\infty}^{t} dt_2 \int_{-\infty}^{t_2} dt_1 \exp\left[-\frac{i}{\hbar}\hat{H}_0(t-t_2)\right]\boldsymbol{\mu}\cdot\boldsymbol{\varepsilon}(t_2) \times \exp\left[-\frac{i}{\hbar}\hat{H}_1(t_2-t_1)\right]\boldsymbol{\mu}\cdot\boldsymbol{\varepsilon}(t_1)\Psi_{g,i}\exp\left(-\frac{i}{\hbar}E_0 t_1\right). \tag{19.17}$$

The physical interpretation of this term is evident: the system propagates on the ground state potential energy surface until it undergoes a transition to the excited state at $t = t_1$. During the time interval $t_2 - t_1$ it advances on the excited potential energy surface, returning to the ground state at time t_2.

Expression (19.17) refers to a two-photon transition, associated with a pump and a dump pulse ([657], see Section 18.3), as reflected by the twofold appearance of the coupling term $\boldsymbol{\mu}\cdot\boldsymbol{\varepsilon}$. The basic idea underlying the *Tannor–Rice scheme* is to optimize the yield of a selected product channel by manipulating the dump pulse for a given pump pulse. An elementary example is provided by a photodissociation process involving a triatomic molecule ABC that may follow the alternative routes ABC $\rightarrow$ AB + C or ABC $\rightarrow$ A + BC. The purpose of an optimum control operation might be to direct the system with maximum efficiency along one of these pathways and to suppress the other. A general expression for the transition probability of a molecular system that evolves under the influence of the two pulses, labeled $\boldsymbol{\varepsilon}_1$ and $\boldsymbol{\varepsilon}_2$, from some initial state i, not necessarily associated with the ground state potential energy surface, to a final state Ψ_m is the following [657]:

$$P_{i\to m} \sim \left|\int_{-\infty}^{\infty} dt_2 \int_{-\infty}^{t} dt_1 \left\langle \Psi_m \middle| \boldsymbol{\mu}\cdot\boldsymbol{\varepsilon}_2(t_2) \times \exp\left[-\frac{i}{\hbar}H(t_2-t_1)\right]\boldsymbol{\mu}\cdot\boldsymbol{\varepsilon}_1(t_1)\middle| \Psi_i\right\rangle\right|^2 \tag{19.18}$$

where m stands for the label of the targeted channel. The probability (19.18) may be employed as a variational quantity. More specifically, it may be maximized as a function of $\boldsymbol{\varepsilon}_2(t)$. This optimization proceeds under the normalization constraint

$$\int_{-\infty}^{\infty} dt_2 \, |\boldsymbol{\varepsilon}_2(t_2)|^2 = \bar{I} \tag{19.19}$$

which may be interpreted as a definition of the number of incident photons, or equivalently their total energy. Interpreting $P_{i\to m}$ as a functional $P_{i\to m}[\boldsymbol{\varepsilon}_2(t)]$, and setting the variation $\delta P_{i\to m}[\boldsymbol{\varepsilon}_2(t)]$ equal to zero yields the Lagrange multiplier associated with the constraint (19.19) and results eventually in the solution for the optimum pulse $\boldsymbol{\varepsilon}_2(t)$.

Generalizing these concepts, one may derive the optimum conditions for the realization of a molecular wave function $\boldsymbol{\Psi}(t_f)$ if an initial state $\boldsymbol{\Psi}(t_i) \equiv \boldsymbol{\Psi}_i$ has been defined. We make continued reference to a two-state model and let $\boldsymbol{\Psi}$ denote the two-component vector

$$\boldsymbol{\Psi}(t) = \begin{pmatrix} \Psi_0(t) \\ \Psi_1(t) \end{pmatrix}. \tag{19.20}$$

In its most concise formulation, the variational quantity is provided by the expectation value of the projector $\hat{\mathbf{P}}$ as applied to $\boldsymbol{\Psi}(t_f)$ in the matrix element $\mathcal{J}$:

$$\mathcal{J} \equiv \langle \boldsymbol{\Psi}(t_f) | \hat{\mathbf{P}} | \boldsymbol{\Psi}(t_f) \rangle. \tag{19.21}$$

Here $\hat{\mathbf{P}}$ projects onto the target state $\boldsymbol{\Phi}$, i.e. $\hat{\mathbf{P}} \equiv |\boldsymbol{\Phi}\rangle\langle\boldsymbol{\Phi}|$. The photon number, or the energy constraint (19.19), refers now to the finite time interval $[t_i, t_f]$, according to

$$\int_{t_i}^{t_f} dt \, |\boldsymbol{\varepsilon}(t)|^2 = \bar{I}, \tag{19.22}$$

where we have replaced the dump pulse $\boldsymbol{\varepsilon}_2(t)$ with the overall pulse function, which, in the simplest case, might be composed of two temporal delta functions:

$$\boldsymbol{\varepsilon}(t) = \boldsymbol{\varepsilon}_{\mathrm{p}} \exp(i\omega_{\mathrm{p}} t + i\varphi_{\mathrm{p}})\delta(t - t_{\mathrm{p}}) + \boldsymbol{\varepsilon}_{\mathrm{d}} \exp(i\omega_d t + i\varphi_{\mathrm{d}})\delta(t - t_{\mathrm{d}}). \tag{19.23}$$

The energy constraint is supplemented by a *dynamic constraint* to guarantee that the evolution of $\boldsymbol{\Psi}$ is governed by the TDSE:

$$\left(i\hbar \mathbf{I} \frac{\partial}{\partial t} - \hat{\mathbf{H}}(t) \right) \boldsymbol{\Psi} = 0. \tag{19.24}$$

Incorporating Eqs. (19.22) and (19.24) into Eq. (19.21), the variation functional $\mathcal{J}$ adopts the form [658]

$$\mathcal{J} = \left\langle \mathbf{\Psi}(t_f)|\hat{\mathbf{P}}|\mathbf{\Psi}(t_f)\right\rangle + \lambda \left[\int_{t_i}^{t_f} dt\, |\boldsymbol{\varepsilon}(t)|^2 - \bar{I} \right]$$

$$+ \left[\frac{i}{\hbar} \int_{t_i}^{t_f} dt \left\langle \mathbf{\Phi} \left| i\hbar \boldsymbol{I} \frac{\partial}{\partial t} - \boldsymbol{H}(t) \right| \mathbf{\Psi} \right\rangle + c.c. \right]. \qquad (19.25)$$

The components of the two-vector $\mathbf{\Phi}$, $\Phi_0(t)$ and $\Phi_1(t)$, are understood as two further Lagrange multipliers besides the parameter λ. The factor $i/\hbar$ in front of the dynamic constraint term is conventional, allowing for a direct comparison between the function $\mathbf{\Phi}$, and the projected molecular wave function $\hat{\mathbf{P}}\Psi$. Variation of expression (19.25) is carried out with respect to both the solution and the pulse, i.e. the entities $\mathbf{\Psi}$, $\Re\{\boldsymbol{\varepsilon}\}$, and $\Im\{\boldsymbol{\varepsilon}\}$. Setting $\delta\mathcal{J}$ equal to zero (see Exercise 19.2), we conclude that the components of $\mathbf{\Phi}$ satisfy the TDSE:

$$i\hbar\frac{\partial}{\partial t}\mathbf{\Phi} = \hat{\mathbf{H}}(t)\mathbf{\Phi}. \qquad (19.26)$$

Further, the variational treatment demonstrates that $\mathbf{\Phi}$ obeys the boundary condition

$$\mathbf{\Phi}(t_f) = \hat{\mathbf{P}}\mathbf{\Psi}(t_f), \qquad (19.27)$$

with the implication that $\mathbf{\Phi}(t_f)$ is to be interpreted as the target state. For the shape of the optimized electromagnetic field pulse we obtain

$$\boldsymbol{\varepsilon}(t) = \frac{i}{\lambda}\left[\langle \Psi_0(t)|\boldsymbol{\mu}|\Phi_1(t)\rangle - \langle \Phi_0(t)|\boldsymbol{\mu}|\Psi_1(t)\rangle \right] \equiv \frac{i}{\lambda}\mathbf{O}(\mathrm{t}). \qquad (19.28)$$

Lastly, the Lagrange multiplier λ is determined by the pulse constraint (19.22), which yields

$$\lambda = \pm\sqrt{\frac{1}{\bar{I}}\int_{t_i}^{t_f} dt\, |\mathbf{O}(t)|^2}. \qquad (19.29)$$

It should be noted that the optimized pulse is out of phase by an angle $\frac{\pi}{2}$ with the dipolar overlap matrix element of the molecular wave functions $\Psi_{1(0)}$ and the target functions $\Phi_{0(1)}$.

Exercise 19.2. Write down the variation $\delta\mathcal{J}$ with respect to $\mathbf{\Psi}$, $Re(\boldsymbol{\varepsilon})$, and $Im(\boldsymbol{\varepsilon})$ for $\mathcal{J}$ as given by Eq. (19.25). From the condition $\delta\mathcal{J} = 0$, derive the results (19.26)–(19.29) [658]. In particular, use partial integration of $\langle \frac{\partial}{\partial t}\delta\mathbf{\Psi}|\phi\rangle$ and $\langle\phi|\frac{\partial}{\partial t}\delta\mathbf{\Psi}\rangle$ to arrive at the asymptotic statement (19.27).

To implement this control scheme, an iterative procedure may be employed. This method operates with an initial guess of the electromagnetic pulse as a function of time. Using this estimate, the molecular wave function is propagated through the time interval $[t_i, t_f]$. At $t = t_f$, the projection operator $\hat{\mathbf{P}}$ is applied to $\mathbf{\Psi}$ and thus the state $\mathbf{\Phi}$ is constructed. Subsequently, $\mathbf{\Psi}(t_f)$ and $\mathbf{\Phi}(t_f)$ are both propagated backwards until the initial time t_i is reached. During the latter step, the dipolar overlap element $\mathbf{O}(t)$ (see Eq. (19.28)) is computed from the recorded values of $\mathbf{\Psi}(t)$ and $\mathbf{\Phi}(t)$. By Eqs. (19.28) and (19.29), this result is used to improve upon the original estimate of the pulse. The cycle is repeated until convergence is attained.

This strategy has been applied by Kosloff *et al.* [658] to a model rearrangement process of the form

$$\mathrm{AAB} \rightarrow \begin{cases} \mathrm{AA+B} & \text{(channel 1)} \\ \mathrm{A+AB} & \text{(channel 2)} \end{cases} \tag{19.30}$$

involving the dissociation of a linear triatomic molecule along two alternative pathways. For a landscape of excited and ground state potential energy surfaces that may be adopted to model this reaction, see Figure 19.1. A possible realization of this process is the decay of the HHD isopotomer of H_3:

$$\mathrm{H+HD} \leftarrow \mathrm{HHD} \rightarrow \mathrm{D} + \mathrm{H_2}. \tag{19.31}$$

Figure 19.2 shows the profile of the pulse that emerged from the described iteration scheme as optimally suited to populate channel 1. Figure 19.3 gives some impression of the wave function evolution on the ground state potential energy surface under the influence of this pulse. Contours representing the square magnitude of the wave function superpose the contour lines of the potential energy surface. The corresponding optimization procedure converged at a selectivity ratio of $5 : 2$, favoring channel 1 over channel 2 as exit pathway. This preference is documented by Figure 19.3(b) which presents a snapshot of the evolving system after the pulse has faded out. The image displays a substantially higher population buildup in channel 1 as compared to channel 2.

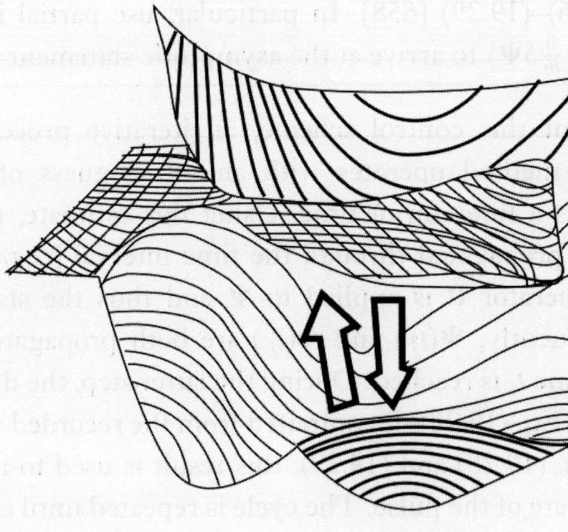

Figure 19.1: Schematic view of the ground state and the first excited state potential energy surface for the collinear disintegration of an AAB model system. The coordinates are given by the distances between A (B) and the centers of mass of AB (AA). (Reprinted from [658] with permission from Elsevier.)

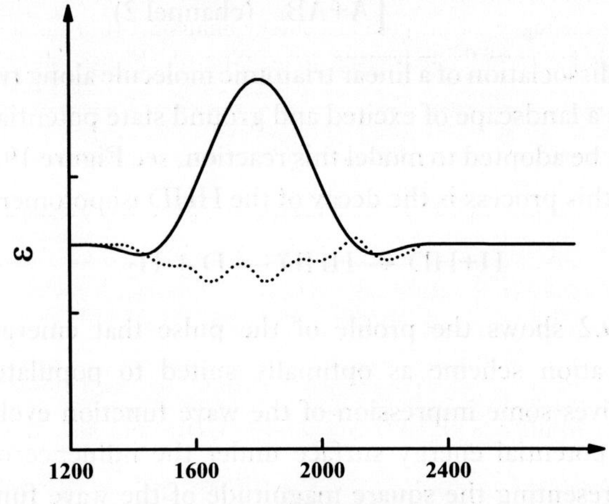

Figure 19.2: The real part (solid line) and the imaginary part (dashed line) of the electromagnetic pulse generated by use of optimum control theory to maximize the yield in the AAB → AA + B channel (channel 1) following the decay of the AAB model system. (Reprinted from [658] with permission from Elsevier.)

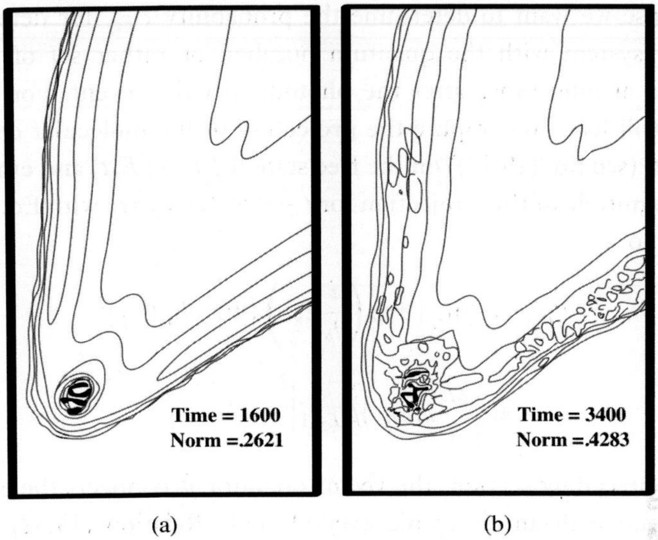

Figure 19.3: Two stages in the evolution of the considered triatom on the ground state surface. Both the square magnitude of the molecular wave function and the potential energy surface are represented by contour lines. The horizontal (vertical) axis refers to the distance between B and AA. (a) At the onset of the pulse (see Figure 19.2) the molecular population is concentrated in the region of the potential energy minimum. (b) After application of the pulse, the molecule has partially undergone dissociation which proceeds preferentially into channel 1. (Reprinted from [658] with permission from Elsevier.)

The Tannor–Rice optimization method has been applied to simulate control procedures for several chemical systems. In this context, the photodissociation of I_2 and HgAr has been considered [659], dynamics studies have been performed in relation to the control of Br_2 photodissociation, and specifically the ratio of the products Br* and Br [660].

19.3 The Brumer–Shapiro Scheme: Coherent Control

The control method devised by Tannor and Rice employs a variational approach to maximize the probability for a selected molecular process. Alternatively, one may use the branching ratio of competing reaction pathways as a starting point for developing a control procedure. Could analysis of the branching ratio yield a prescription for how to tailor the electromagnetic pulse for preferential population of a chosen channel? This question may be considered once more by the example of photodissociation.

Suppose we want to determine the probability $P_{E_{n,1}}$ for detecting the molecular system with the quantum number, or rather set of quantum numbers n at long times after the photodissociation event. For this purpose, we will have to compute the projection of the molecular continuum state $\Psi_E(t)$ (see Eq. (19.13)) on the free state $\Psi^0_{E_n} t(-\frac{i}{\hbar}E_n t)$ and evaluate the square magnitude of this projection for $t \to \infty$ (compare with Eq. (19.11)), according to

$$\begin{aligned} P_{E_{n,1}} &= \lim_{t\to\infty} |\exp\left(\frac{i}{\hbar}E_n t\right)\langle\Psi^0_{E_n}|\Psi_E(t)\rangle|^2 \\ &= \left|\frac{2\pi}{\hbar}\tilde{a}(\omega_{E_1})\mu_{E_{N},1}\right|^2. \end{aligned} \tag{19.32}$$

As in the preceding section, the vector operator $\hat{\mu}$ connects the electronic ground state with an electronic excited state. Relation (19.32) provides the basis for comparing the probabilities of detecting the photodissociation product in the channels n and m at energy E, following excitation by an electromagnetic pulse from an initial state Ψ_1. The respective ratio is

$$\frac{P_{E_{n,1}}}{P_{E_{m,1}}} = \left|\frac{\mu_{E_{n,1}}}{\mu_{E_{m,1}}}\right|^2. \tag{19.33}$$

The finding (19.33) seems to imply a negative answer to the earlier question if a control strategy might be based on the branching ratio, as the latter is evidently independent of the external field. Manipulating this field will have no influence on the relative populations of the exit channels.

However, to conclude from this result that laser control theory cannot operate with the branching ratio would be premature. Scrutinizing the premises which were used to obtain the statement (19.33), one realizes that it was generated under the tacit assumption of a *single initial molecular state*, Ψ_1. As we will see in the following, making admission for a superposition of multiple initial states to interact with the electromagnetic pulse brightens the prospects for laser control based on branching ratio optimization [661].

To substantiate this idea, we recalculate the branching ratio (19.33) under the assumption of a non-stationary linear combination of bound molecular eigenfunctions Ψ_k as initial state. We denote this state with φ_i and describe it as

$$\varphi_i = \sum_k b_k \Psi_{0,k} \exp\left(-\frac{i}{\hbar} E_k t\right). \tag{19.34}$$

We stipulate that the eigenfunctions contained in Eq. (19.34) are nuclear rovibrational states on the electronic ground state potential energy surface, hence the label $\Psi_{0,k}$. By analogy to Eq. (19.13), the continuum contribution to the molecular wave function in the asymptotic time regime, i.e. long after the electromagnetic pulse has faded out, is given by

$$\Psi_E(t) = \frac{2\pi i}{\hbar} \sum_{k=1}^{K} b_k \sum_n \int dE\, \tilde{a}(\omega_{E,k}) \mu_{E_{n,k}} \Psi_{E_n} \exp\left(-\frac{i}{\hbar} Et\right). \tag{19.35}$$

It presents no difficulty to extend the prescription (19.32) for computing the probability to detect the system in a photodissociation channel n to the case of multiple initial states. We obtain

$$P_{E_n,i} = \left(\frac{2\pi}{\hbar}\right)^2 \sum_{k,m} b_k b_m^* \tilde{a}_k(\omega_{E_k}) \tilde{a}_m^*(\omega_{E_m}) \mu_{E_{n,k}} \mu_{E_{n,m}}^*. \tag{19.36}$$

Consequently, the probability of detecting a certain arrangement q of dissociation products, for instance AA + B as opposed to A + AB in the disintegration of the triatom AAB, may be described by summing over the expressions $P_{E_{n,i}}$ where n assumes the interpretation of a variable quantum number that labels all the states belonging to the chosen exit channel. Thus,

$$P_{E_q} = \sum_n P_{E_{n,q}}. \tag{19.37}$$

This yields the branching ratio $R_{qq'} = P_{E_q}/P_{E_{q'}}$ for two such channels, q and q'. We acknowledge that $R_{qq'}$, in contrast to its counterpart Eq. (19.33) for a single initial state, depends on several controllable parameters. In the formula for P_{E_q}, the alternatives $k = m$ and $k \neq m$ carry different information. The first case refers to the probability that excitation from the initial level with index k contributes to populating the selected subspace q. The second case, however, involves the interference of two deviating excitation routes which proceed from the two ground state levels k and m. Both components of Eqs. (19.36) and (19.37) may be sensitively influenced by strategic choices of the pulse profiles, as contained in the coefficients $\tilde{a}_k$, and of the weights of the initial molecular states, as determined by the second set of coefficients, b_k.

We consider now the simplest possible case of control by simultaneous excitation of multiple initial molecular levels, namely *bichromatic control* (see, for instance, [661]). For this purpose, we first ask for the electric field $\boldsymbol{\varepsilon}$ provided by a continuous wave laser that operates at a single frequency, ω_i. In this case, the amplitudes in (19.1) may be described as $a(\omega)\delta(\omega - \omega_i)$. Insertion into Eq. (19.1) yields

$$\boldsymbol{\varepsilon}(Z,t) = 2\mathbf{e}\Re[\tilde{a}(\omega_i)\exp(-i\omega_i t)]. \tag{19.38}$$

If the field is produced by superposition of two continuous waves with frequencies ω_1 and ω_2, the operator of dipolar interaction becomes

$$\hat{H}_{\text{int}} = -2\boldsymbol{\mu}\cdot\mathbf{e}\sum_{i=1}^{2}\Re[\tilde{a}(\omega_i)\exp(-i\omega_i t)]. \tag{19.39}$$

We label the initially occupied molecular energy levels E_1 and E_2 and observe that, for conservation of energy, the final level E has to equal both $E_1 + \hbar\omega_1$ and $E_2 + \hbar\omega_2$. From Eq. (19.37), the probability P_{E_q} for bichromatic excitation is

$$P_{E_q} = \left(\frac{2\pi}{\hbar}\right)^2 [|b_1|^2|\tilde{a}(\omega_1)|^2 M_q(1,1) + |b_2|^2|\tilde{a}(\omega_2)|^2 M_q(2,2) + 2\Re\{b_1 b_2^* \tilde{a}(\omega_1)\tilde{a}(\omega_2)^* M_q(1,2)\}] \tag{19.40}$$

where we have chosen the abbreviation

$$M_q(i,j) = \sum_n \langle\Psi_i|\hat{\boldsymbol{\mu}}\cdot\boldsymbol{\varepsilon}|\Psi_n\rangle\langle\Psi_n|\hat{\boldsymbol{\mu}}\cdot\boldsymbol{\varepsilon}|\Psi_j\rangle, \tag{19.41}$$

and $i, j = 1, 2$. The branching ratio for the subspaces q and q' becomes

$$R_{qq'} = \frac{|x_1|^2|M_q(1,1) + |x_2|^2|M_q(2,2) + 2\Re\{x_1 x_2^* M_q(1,2)\}}{|x_1|^2|M_{q'}(1,1) + |x_2|^2 M_{q'}(2,2) + 2\Re\{x_1 x_2^* M_{q'}(1,2)\}|}, \tag{19.42}$$

where we have substituted x_i for $b_i\tilde{a}(\omega_i)$, $i = 1, 2$. Since the right-hand side of Eq. (19.42) has been derived from the first-order perturbation theory solution (19.40), its range of validity is the weak field regime. The ratio is simplified as one introduces the polar form of $M_q(i,j)$, $|M_q(i,j)| \exp(i\beta_q(i,j))$, and further declares

$$x \equiv \frac{|b_2\tilde{a}(\omega_2)|}{|b_1\tilde{a}(\omega_1)|} \tag{19.43}$$

and

$$\tan\theta_i \equiv \frac{\Im\{b_i\tilde{a}(\omega_i)\}}{\Re\{b_i\tilde{a}(\omega_i)\}} \tag{19.44}$$

to introduce the parameters x and θ_i. With the respective substitutions, the branching ratio assumes a more compact appearance:

$$R_{qq'} = \frac{|M_q(1,1)| + x^2|M_q(2,2)| + 2x\cos[\theta_1 - \theta_2 + \beta_q(1,2)]|M_q(1,2)|}{|M_{q'}(1,1)| + x^2|M_{q'}(2,2)| + 2x\cos[\theta_1 - \theta_2 + \beta_{q'}(1,2)]|M_{q'}(1,2)|}. \tag{19.45}$$

Casting the ratio into this form makes it immediately obvious that the interference between the two alternative excitation routes plays a decisive role in manipulating the reaction of interest. The control scheme exemplified by Eq. (19.45) operates with a superposition of indistinguishable excitation pathways that maintain their phase coherence during the process. It is thus based on the principle of *coherent control*. Shapiro and Brumer [661] employ the quantities $\Delta\theta \equiv \theta_1 - \theta_2$ and $s \equiv x^2/(x^2+1)$ as control parameters. The latter maps all possible laser intensity ratios into the interval $0 \leq s \leq 1$, while the former refers to the relative phases of the two lasers. We point out that the coefficients (b_1, b_2) that characterize the initial superposition of molecular states also determine the parameters x and $\Delta\theta$ and thus are instrumental for coherent bichromatic control. The control parameters may be sensitively influenced by purposeful preparation of the initial state, and specifically by defining the relative weight of the participating molecular levels.

An alternative strategy to steer a chemical reaction along a desired pathway consists in the selective excitation of targeted degrees of freedom. For instance, the product A + AB of AAB photodissociation will be favored over the product AA + B if one succeeds in exciting the specific degree of freedom that corresponds to the bond between A and AB. Obviously, this concept of mode selectivity differs substantially from both optimal and coherent control [662]. This methodology, however, may encounter the challenge that the mode of interest is strongly coupled to other modes. In this case, the energy deposited into the chosen degree of freedom is subject to intramolecular vibrational redistribution (IVR), and the goal of selective excitation cannot be attained.

19.4 Case Study: Coherent Control of ICN Photodissociation

In the weak field limit, compact closed expressions are available for coherent control ratios like the one derived before for the bichromatic case (see Eq. (19.45)). Information about the evolving molecular system enters the ratio through the magnitude and the phase of $M_q(i,j)$, the cumulative state-to-state transition matrix element. The computational task to be faced in determining $R_{qq'}$ is thus reduced to the calculation of $M_q(i,j)$ and $M_{q'}(i,j)$, as long as the weak field regime is not transcended.

As mentioned in the preceding sections, molecular photodissociation is a natural field of application for laser control techniques. In particular, one aims at guiding a disintegration process along a selected pathway in order to maximize the yield of a certain fragmentation product. In the context of our survey on molecular electron dynamics, however, laser control involving electronic excitations is of particular interest. Excellent agreement between experiment and simulation has been attained in cases where dissociation proceeds through one predominant photoexcited electronic channel. This success may be demonstrated by the example of H_2O (for an extensive overview and discussion see [663]). In the majority of cases, however, a clear separation between excited potential energy surfaces that contribute to the molecular decay is not possible, and the dissociation mechanism is determined by several interacting electronic states. An elementary realization of such an electronic interaction is sketched in Figure 19.4. A dipole-allowed transition may lead from the ground state to an excited potential energy surface. Let us imagine that the point of transition is in the vicinity of a diabatic crossing, coupling the primary to a secondary surface, as shown in the figure. In such a nonadiabatic situation, the photoexcited wave packet may split at the crossing into two parts that do not differ with respect to their molecular composition but display different electronic structures. Laser control theory applied to this case is challenged to define conditions that favor one chosen electronic state over others. Such a selection procedure,

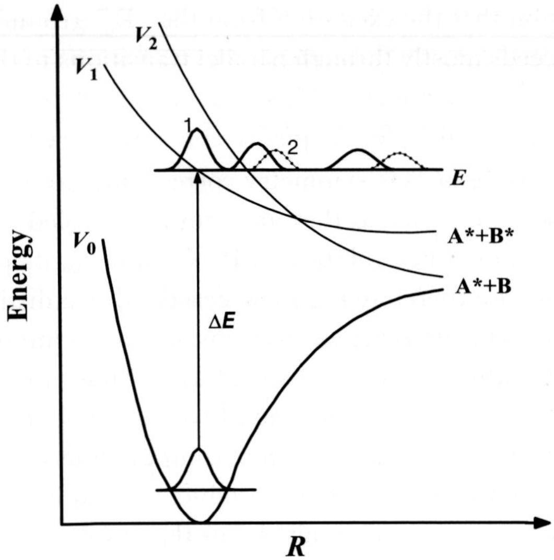

Figure 19.4: Photoexcitation in the vicinity of a diabatic crossing point between two upper potential energy surfaces, schematic view. Two electronic exit channels result from the coupling between the two contributing excited electronic states, giving rise to wave packet bifurcation. A* (B*) stands for an excited A (B) moiety of the complex A+B.

if successful, is of great importance for exploring state-specific reactions, and particularly for the study of chemical processes in well-defined excited electronic states.

In the following, we will focus on computations by Batista and Brumer [664] who used the bichromatic coherent control formalism in the context of the nonadiabatic photodissociation of ICN in the $\tilde{A}$ continuum, a broad absorption band in the 200–300 nm range. Numerous experimental (e.g. [665]–[667]) and theoretical studies (e.g. [668,669]) have been devoted to this process. In particular, it has been shown that photolysis at approximately 266 nm gives rise to two peaks in the translational photofragment spectrum. These are associated with two different photodissociation channels, namely

$$\begin{aligned} \mathrm{ICN} &\rightarrow \mathrm{I}^*(^2P_{\frac{1}{2}}) + \mathrm{CN}(X^2\Sigma^+), \\ \mathrm{ICN} &\rightarrow \mathrm{I}(^2P_{\frac{3}{2}}) + \mathrm{CN}(X^2\Sigma^+). \end{aligned} \tag{19.46}$$

The relative weights of these two channels may be defined by controlling the ratio of ground state and excited iodine atoms, I/I*. The treatment of the ICN photodissociation problem by Batista and Brumer [664] is based on the observation that the excitation from the $^1\Sigma_0^+$ ground state to the $\tilde{A}$ continuum proceeds mostly through parallel transitions to the excited electronic state with $^3\Pi_{0^+}$ symmetry. This pathway correlates with the $I^*(^2P_{\frac{1}{2}})$ dissociation product which, for simplicity of notation, we term $I_{\frac{1}{2}}$. An alternative electronic state of $^1\Pi_1$ symmetry corresponds to the $I(^2P_{\frac{3}{2}})(\equiv I_{\frac{3}{2}})$ channel. Both states interact, as they are strongly coupled to each other by a conical intersection. The evolution of ICN following its photoexcitation into the $\tilde{A}$ continuum therefore has to be described as a distinctly nonadiabatic process, involving two excited electronic states at minimum. Empirical potential energy surfaces [670] corresponding to these two states were utilized in the work of Batista and Brumer. Five potential energy surfaces in the $\tilde{A}$ continuum were derived by *ab initio* computations [671, 672].

Bichromatic control of the relative product yield in ICN photodissociation requires the preparation of an initial state that involves the superposition of two nuclear levels, such as

$$\Psi_0(n,m) = \Phi_0(b_m\chi_0(m) + b_n\chi_0(n)), \tag{19.47}$$

where we have factorized the initial molecular wave function Ψ_0 into the electronic ground state Φ_0 and the vibrational states χ_0 with labels m and n on the ground state surface.

We describe the geometry of ICN by Jacobi coordinates (see Section 6.6.1) and define the variable ρ as the distance between the fragmentation product I and the CN center of mass (see Figure 19.5). The functions $\chi_0(m)$ and $\chi_0(n)$ are associated with the m*th* and the n*th* excitation of the corresponding vibrational degree of freedom.

As the ICN ground state is raised into the energy continuum and the molecule undergoes photodissociation, the CN fragment shows little vibrational excitation, but may be in a rotational state with high angular momentum L. From experimental rotational distributions of CN [673],

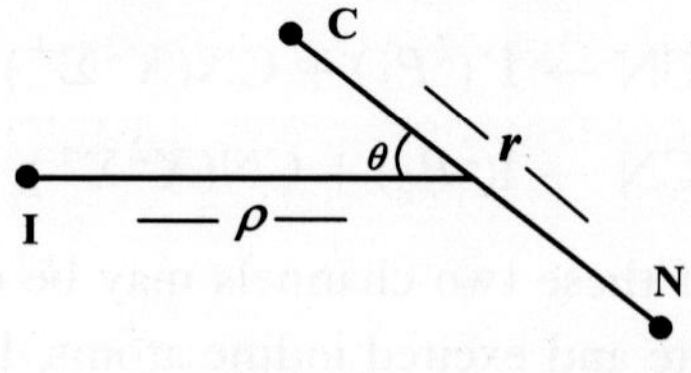

Figure 19.5: Jacobi coordinates chosen to describe ICN.

relatively low angular momentum quantum numbers are found for CN in the $I_{\frac{1}{2}}$ channel while high quantum numbers in the range of $L = 40$–50 are characteristic of the $I_{\frac{3}{2}}$ pathway. The $^3\Pi_{0+}$ potential energy surface exhibits an attractive well at $\rho = 4.92$Å, the conical intersection that couples the $^3\Pi_{0+}$ to the $^1\Pi_1$ state is found at $\rho = 6.42$Å.

The nuclear contributions to the continuum functions Ψ_{E_n} (see Eq. (19.13)) are labeled $\chi_{E,q,L}$ where q adopts the values $\frac{1}{2}$ or $\frac{3}{2}$, the former correlating with the $^3\Pi_{0+}$, the latter with the $^1\Pi_1$ excitation. The primary theoretical task involved in the bichromatic control of the I/I* ratio consists in evaluating the cumulative transition matrix element $M_q(i,j)$. This might be done, as formula (19.41) suggests, by solving the complete time-independent state-to-state scattering problem. More specifically, one has to calculate

$$M_q(m,n) = \sum_{L=0}^{\infty} \langle \chi_0(m) | \hat{\boldsymbol{\mu}} \cdot \boldsymbol{\varepsilon} | \chi_{E,q,L} \rangle \langle \chi_{E,q,L} | \hat{\boldsymbol{\mu}} \cdot \boldsymbol{\varepsilon} | \chi_0(n) \rangle. \tag{19.48}$$

Instead of generating the scattering solutions $\chi_{E,q,L}$, however, one might determine the transition matrix elements on the basis of a time-dependent formalism [664] by propagating the initial wave functions $\chi_0(m)$ and $\chi_0(n)$ and projecting ultimately on asymptotic states χ_{E,q,L_0} instead of scattering states $\chi_{E,q,L}$. Therefore,

$$M_q(m,n) = \lim_{t\to\infty} \sum_{J=0}^{\infty} \langle \chi_t(m) | \boldsymbol{\mu} \cdot \boldsymbol{\varepsilon} | \chi_{E,q,L_0} \rangle \langle \chi_{E,q,L_0} | \boldsymbol{\mu} \cdot \boldsymbol{\varepsilon} | \chi_t(n) \rangle \tag{19.49}$$

with $\chi_t(m)$ and $\chi_t(n)$ as the time-evolved initial vibrational states [674]. By Eq. (19.49), the computation of $M_q(m,n)$ is essentially reduced to determining these states and thus to a problem of nonadiabatic dynamics. Batista and Brumer [664] confront this problem in the framework of the semiclassical initial value representation (SC-IVR) model, combined with the classical electron analog Hamiltonian (see Chapters 12 and 10, respectively). The two contributions to the overall initial wave function $\Psi_0(n,m)$ (see Eq. (19.47)) are treated in separation from each other. The components $\Phi_0\chi_0(i)$, $i = m, n$, are instantaneously excited to the optically active target state in resonance with the excitation energy. Their subsequent evolution is modeled by semiclassical dynamics according to the Herman–Kluk propagation scheme (see Eqs. (12.47) and (12.48)):

$$\Psi_t(\mathbf{Q}) = \int d\mathbf{Q}'\, C_t(\mathbf{Q}, \mathbf{Q}')\Psi_0(\mathbf{Q}') \tag{19.50}$$

with $\mathbf{Q}$, $\mathbf{Q}'$ as collective coordinates that include both the nuclear and the electronic degrees of freedom relevant to the problem, and with C_t as the Herman–Kluk propagator in coordinate representation. In keeping with the classical electron analog methodology, both sets of coordinates are advanced by solving the Hamilton equations. The resulting trajectories are utilized to construct the classical action as a function of time and, concomitantly, the propagator C_t. The Hamilton function H that governs the evolution of the system depends on the Jacobi nuclear coordinates ρ and Θ (see Figure 19.5) and the conjugate momenta K and L, where a rigid rotor approximation is made for the CN subsystem. With $\mathbf{x}$ and $\mathbf{p}$ as electronic degrees of freedom, H reads ([675], compare with Eqs. (12.78) and (12.79))

$$\begin{aligned} H(\rho, K, \Theta, L, \mathbf{x}, \mathbf{p}) &= \frac{K^2}{2\mu_{\mathrm{I-CN}}} + \frac{L^2}{2\mu_{\mathrm{I-CN}}\rho^2} + \frac{L^2}{2\mu_{CN}r^2} \\ &\quad + \frac{1}{2}\sum_{k=1}^{2}\sum_{k'=1}^{2}[p_k p_{k'} + x_k x_{k'}]H_{kk'}(\rho, \Theta) \\ &\quad - \frac{1}{2}\sum_{k=1}^{2} H_{kk}(\rho, \Theta), \end{aligned} \tag{19.51}$$

where $\mu_{\mathrm{I-CN}}$ and μ_{CN} stand for the reduced masses of the I-CN and the CN system, respectively, and the angular momenta of both rotations are of equal magnitude, corresponding to a vanishing total angular momentum of the molecule. At any arbitrary time, the evolving system can be interrogated by bracketing $\Psi_t(\mathbf{Q})$ with the wave function Φ_q of the excited electronic state considered:

$$\Psi_t^q(\rho, \Theta; n) = \int d\mathbf{x} \langle \Phi_q | \mathbf{x} \rangle \langle \mathbf{x}, \rho, \Theta | \Psi_t(n) \rangle, \; q = \frac{1}{2}, \frac{3}{2}. \tag{19.52}$$

The initial wave packet that is advanced according to Eq. (19.50) to yield $\Psi_t(n)$ results from instantaneous photoexcitation of the ground state component with index n, which is $\Phi_0\chi_0(\rho, \Theta; n)$ according to Eq. (19.47).

Electronic states are described here by use of the mapping approach outlined in Section 12.4. Specifically, they are chosen as a product of two one-dimensional harmonic oscillator states which are eigenfunctions of the Hamilton operator counterpart of the Hamilton function (19.51). This allows us to model the nonadiabatic interaction between the two dominant electronic excitations $^3\Pi_{0+}$ and $^1\Pi_1$.

In the following step, the scalar product $\langle \chi_{K,q,L} | \hat{\boldsymbol{\mu}} \cdot \boldsymbol{\varepsilon} | \chi_t(n) \rangle$ is constructed by computing

$$\langle \chi_{K,q,L} | \hat{\boldsymbol{\mu}} \cdot \boldsymbol{\varepsilon} | \chi_t(n) \rangle = A \int_0^\infty d\rho \int_0^\pi d\Theta \; Y_{L0}^*(\Theta) \sin(\Theta) \times \exp(-iK\rho) \hat{\boldsymbol{\mu}} \cdot \boldsymbol{\varepsilon} \; \Psi_t^q(\rho, \Theta; n). \quad (19.53)$$

Here A denotes a normalization constant and Y_{L0} the Lth rotational state of the CN subsystem. For convenience, the second factor that enters the cumulative transition matrix element $M_q(m, n)$ (19.48) has been expressed in terms of the translational momentum K instead of the energy E. The relation between both quantities is

$$K = \left[2\mu_{\text{I-CN}} \left\{ \frac{E - E_0(q)}{\hbar^2} - \frac{L(L+1)}{2\mu_{\text{C-N}} r^2} \right\} \right]^{\frac{1}{2}} \quad (19.54)$$

with $E_0(q)$ as the asymptotic energy of the potential energy surface that corresponds to the electronic dissociation channel q. For assessment of the semiclassical approach to the cumulative transition matrix element, $M_q(m, n)$ was also evaluated by performing full quantum mechanical calculations employing the fast Fourier transform (FFT, see Section 15.1.2) [664]. An extensive comparison between both procedures yielded deviations on the order of 15–20% between the respective numerical values obtained for the modulus and the phase of $M_{\frac{1}{2}}(1, 3)$. The qualitative agreement between the results, however, was seen to be excellent. In spite of this success, the semiclassical treatment reported in [664] required a large number of trajectories on the order of 2,000 for convergence, accentuating the need for more efficient implementations of the semiclassical formalism such that the gain in tractability clearly outweighs the sacrifice of quantitative accuracy. Recent proposals to enhance the feasibility of trajectory-based approaches to problems of quantum dynamics were presented earlier (see Section 12.7).

We turn now to the results of the bichromatic control simulation for the photodissociation of ICN in terms of branching ratios which were generated from an initial superposition of vibrational states with quantum numbers $\nu = 1$ and 3. The parameter x (see Eq. (19.43)) is here defined as $x = |b_1\tilde{a}(\omega_1)|/|b_3\tilde{a}(\omega_3)|$. Figure 19.6 shows the relative yield $\mathrm{I}_{\frac{1}{2}}/(\mathrm{I}_{\frac{1}{2}} + \mathrm{I}_{\frac{3}{2}})$ as a function of both the phase difference $\theta_1 - \theta_2$ and the variable $s \equiv x^2/(x^2+1)$. This quantity is represented by contour plots for six different photoexcitation energies, where the solid lines refer to the semiclassical, the dashed lines to the quantal results.

Upon inspecting these plots, one may distinguish between regimes of high and low control sensitivity in terms of the parameters $\theta_1 - \theta_2$ and s besides the photoexcitation energy E. Thus, the first panel (a) demonstrates that for $E = 4.7\,\mathrm{eV}$ and $s = 0.5$, the relative yield varies between 1% and 40–50% as the phase variable changes from 0° to 180°. For $s > 0.5$, in contrast, the dependence of the relative yield on the phase shift is weak. As E is increased, the displayed profiles undergo substantial alterations. The overall deviation between the semiclassical and the fully quantum mechanical results is within 10–15%. The trends documented in Figure 19.6 emphasize that the photodissociation of ICN represents a highly suitable test case for the extension of laser control to electronic channels in nonadiabatic processes.

Computations analogous to the one outlined have been performed on the related systems [683, 684]

$$\mathrm{I}(^2P_{\frac{3}{2}}) + \mathrm{CH_3} \leftarrow \mathrm{CH_3I} \rightarrow \mathrm{I}^*(^2P_{\frac{1}{2}}) + \mathrm{CH_3} \tag{19.55}$$

and

$$\mathrm{Br}(^2P_{\frac{3}{2}}) + \mathrm{I} \leftarrow \mathrm{IBr} \rightarrow \mathrm{Br}^*(^2P_{\frac{1}{2}}) + \mathrm{I}. \tag{19.56}$$

Application of coherent control strategies to systems of higher dimensionality, as will be highlighted in the following section, is an active field of current research [655, 657].

19.5 Optimal Control in Pump–Probe Spectroscopy

This section focuses on a recent extension of laser control theory to the complex situation of pump–probe spectroscopy. We will once more refer to the nonadiabatic isomerization dynamics of non-stoichiometric alkali-halide

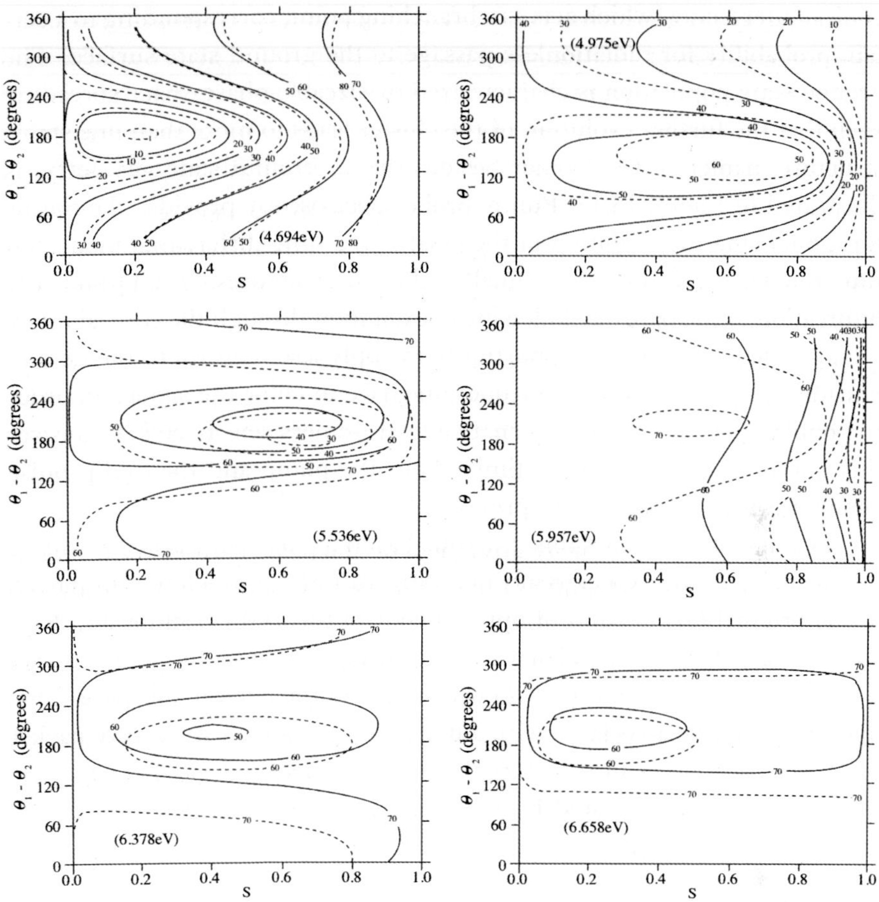

Figure 19.6: Bichromatic coherent control of the photodissociation of ICN at six photon energies where vibrational levels with $\nu = 1$ and $\nu = 3$ of the ICN ground state potential energy surface were used. Contour plots of the relative product yield $\frac{I_{\frac{1}{2}}}{I_{\frac{1}{2}} + I_{\frac{3}{2}}}$ in per cent are shown as a function of both the laser phase difference $\theta_1 - \theta_2$ and the variable $s = \frac{x^2}{x^2+1}$. In each panel, the excitation energy is indicated in parenthesis. (Reprinted with permission from [664]. Copyright (2001) by the American Chemical Society.)

clusters, as introduced in Section 18.5, and in particular to optimal control studies involving the cluster Na_3F_2 [28, 676]. As outlined in the previous chapter, the vertical transition of this system from its ground state to the first excited state is followed by geometric relaxation into the vicinity of a conical intersection which acts as a branching point, corresponding to a certain probability for radiationless passage to the ground state surface. The ground state population is characterized by sizeable phase-space spreading such that the further evolution of the cluster leads to more than one stable structure, namely a C_{2v} isomer besides the C_s ground state geometry, as illustrated in Figure 18.13. Pump–probe spectroscopy provides the means for monitoring the evolution of the cluster in a time-differential mode and thus tracing the routes to its finally adopted structures. In addition, the bifurcation of the pathway following deexcitation through the conical intersection offers an excellent opportunity to apply laser control to the field of nonadiabatic interactions. How does one have to tailor the signals in order to populate one of the two alternative final isomers selectively? To answer this question, a strategy to optimize both the pump and the dump pulse simultaneously will have to be proposed.

In a first step towards this goal, one has to define a variational quantity, i.e. an analog of the expectation value $\langle \boldsymbol{\Psi}(t_f)|\hat{\mathbf{P}}|\boldsymbol{\Psi}(t_f)\rangle$ which was employed in the original formulation of optimal control theory by Tannor and Rice (see Section 19.2). Generalizing this approach, Yan *et al.* [677] introduced a Hermitian target operator $\hat{A}(t_f)$ which acts on the nuclear subspace. This operator could be associated with a dynamical variable, with a wave packet in phase space, or, as in the example presented here, a *projection on a pure state*. The latter is dictated by the *objective*, which may be given in the case of Na_3F_2 photoisomerization by the C_{2v} isomer. The *control yield* is defined as

$$A(t_f) = \mathrm{Tr}[\hat{A}\hat{\rho}(t_f)]. \tag{19.57}$$

This is the quantity that optimal control theory applied to pump–probe spectroscopy seeks to maximize. In the preceding chapter, we have pointed out that semiclassical phase space dynamics based on the Liouville–von Neumann formalism provides a widely applicable and computationally efficient tool for simulating spectroscopic signals that monitor ultrafast molecular processes. Likewise, Liouville–von Neumann theory has been demonstrated to be a convenient venue for formulating and implementing optimization conditions for the control yield $A(t_f)$ ([677]–[679]). This approach not only reduces the essential computational task to propagating phase space trajectories, it also frees the treatment from any assumption of a pure initial state, operating with density matrices defined in Liouville space rather than with Schrödinger wave functions. Following Yan *et al.* [677,679], we specify a *Gaussian target* by considering a target operator whose Wigner phase space representation (see Eq. (4.90)) is a general Gaussian function in the variables X and P:

$$A_w(X,P) = \frac{1}{2\pi\hbar}\int_{-\infty}^{\infty} d\tilde{X}\exp\left(-\frac{i}{\hbar}P\tilde{X}\right)\times\tilde{A}(X,\tilde{X})$$

$$= \frac{1}{2\pi\hbar C}\exp\left(-\frac{1}{2C^2}\left[\Delta_{PP}(X-\bar{X})^2 + \Delta_{XX}(P-\bar{P})^2 - 2\Delta_{XP}(X-\bar{X})(P-\bar{P})\right]\right). \quad (19.58)$$

We have restricted the argument of A_w to a single conjugate pair of space and momentum coordinates, but one may readily extend Eq. (19.58) to the case of multiple variables by employing products of Gaussians. The parameters $\bar{X}, \bar{P}, \Delta_{XX}, \Delta_{PP}$, and $\Delta_{XP}(=\Delta_{PX})$ stand for the centers and variances of the target wave packet, and the phase space orientation of this wave packet with respect to position and momentum. The function $\tilde{A}$ denotes the representation of the target operator in configuration space, $\tilde{A}(X',X'') = \langle X''|A|X'\rangle$ with $X'' = X + \frac{1}{2}\tilde{X}$ and $X' = X - \frac{1}{2}\tilde{X}$ (see Eq. (4.88)). The parameter C is composed of the variances as $C \equiv (\Delta_{XX}\Delta_{PP} - \Delta_{XP}^2)^{\frac{1}{2}}$. The normalization requirement for the target wave packet translates into

$$\mathrm{Tr}\hat{A} = \int dP \int dX\, A_w(X,P) = \int dX'\, \tilde{A}(X',X') = 1. \quad (19.59)$$

If the target is realized as a pure state, it holds that $(\Delta_{XX}\Delta_{PP} - \Delta_{XP}^2)^{\frac{1}{2}} = C = \hbar/2$ (Exercise 19.3). Turning back to the original problem, namely that of the yield optimization for a certain preselected isomer, we realize that in this case a multidimensional target function $A_w(\mathbf{X},\mathbf{P})$ is desired which contains information about the spatial coordinates of the cluster constituents in the chosen final structure.

Exercise 19.3. Pure state limit. Prove the following: if a Gaussian target wave packet Φ is realized, such that $\Phi = (\frac{1}{2\pi\Delta_{XX}})^{\frac{1}{4}}\exp[-\frac{1}{4\Delta_{XX}}(1 - 2i\hbar\Delta_{XP})(X-\bar{X})^2 + \frac{i}{\hbar}\bar{P}(X-\bar{X})]$, it holds that $C = (\Delta_{XX}\Delta_{PP} - \Delta_{XP}^2)^{\frac{1}{2}} = \hbar/2$.

Adopting the pure state limit and integrating over the momentum coordinates yields by (19.58)

$$A_{\mathrm{w}}(X) = \prod_{i=1}^{N} \frac{1}{\sqrt{2\pi}\Delta X_i} \exp\left(-\frac{(X_i - \bar{X}_i)^2}{2\Delta X_i^2}\right) \tag{19.60}$$

where the index i labels the N coordinates of the problem at hand. The parameters $\bar{X}_i$ might be the Cartesian coordinates of the atomic positions in the target geometry, for instance, with reference to the presently discussed example, the C_{2v} isomer of Na_3F_2.

Exercise 19.4. Show that Eq. (19.60) gives the correct expression for the Wigner function of the target operator $\hat{A}$ in configuration space if the target is realized as a pure state. Assume N coordinates and a random momentum distribution.

Mitric *et al.* [676] propose a momentum-constrained target function of the form

$$A_{\mathrm{w}}(X,P) = \prod_{i=1}^{N} \frac{1}{\sqrt{2\pi}\Delta X_i} \exp\left[-\frac{(X_i - \bar{X}_i)^2}{2\Delta X_i^2}\right] \Theta\left(E_{\mathrm{min}} - \sum_{i=1}^{N} \frac{P_i^2}{2m_i}\right). \tag{19.61}$$

The theta function in this expression guarantees that the kinetic energy of the system does not surpass the lowest isomerization threshold E_{min}. In the following we describe the semiclassical simulation of $A(t_f)$ as the final expectation value of $A_{\mathrm{w}}(X,P)$ for an ensemble that evolves under conditions of pump–dump spectroscopy. We assume it to be subjected to a weak laser field given in the rotating wave approximation, defined in Section 18.1, as (see also 19.38):

$$\varepsilon_{\mathrm{p(d)}}(t) = \tilde{a}_{\mathrm{p(d)}}(\omega_{10}) \exp(-i\omega_{10}t) + \tilde{a}^*_{\mathrm{p(d)}}(\omega_{10}) \exp(i\omega_{10}t). \tag{19.62}$$

The frequency ω_{10} corresponds here to the difference between the potential energy minima of the excited and the ground state, the indices p and d stand for *pump* and *dump*. We employ the Liouville–von Neumann formalism in second-order perturbation theory (compare with Eq. (18.95)) in order to make a statement about the target configuration yield at the time t_f:

$$A(t_f) = \int d\boldsymbol{\Gamma}_0 \int_0^{t_f} d\tau \int_0^{\tau} d\tau' \int_0^{\tau'} d\tau'' \int_0^{\tau''} d\tau''' A_{\mathrm{w}}(\boldsymbol{\Gamma}_0'(t_f - \tau; \boldsymbol{\Gamma}_1(\tau' - \tau''; \boldsymbol{\Gamma}_0)))$$
$$\times \exp\left[\frac{i}{\hbar}(\hbar\omega_{10} - \Delta_{10}(\boldsymbol{\Gamma}_1(\tau' - \tau''; \boldsymbol{\Gamma}_0)))(\tau - \tau')\right]$$
$$\times \exp\left[\frac{i}{\hbar}(\hbar\omega_{10} - \Delta_{10}(\boldsymbol{\Gamma}_0))(\tau'' - \tau''')\right]$$
$$\times \rho_{00}(\boldsymbol{\Gamma}_0)\tilde{a}_{\mathrm{p}}(\tau''')\tilde{a}_{\mathrm{p}}^*(\tau'')\tilde{a}_{\mathrm{d}}(\tau')\tilde{a}_{\mathrm{d}}^*(\tau). \tag{19.63}$$

The symbols $\boldsymbol{\Gamma}_0$, $\boldsymbol{\Gamma}_0'$ denote here the manifold of the initial phase space coordinates, $d\boldsymbol{\Gamma}_0 = d\mathbf{X}_0 d\mathbf{P}_0$. In the sense of the initial value representation of semiclassical dynamics (see Section 12.3), the trace (19.57) is carried out by integration over the initial ground state distribution of position and momentum coordinates.

Accordingly, $\rho_{00}(\boldsymbol{\Gamma}_0)$ refers to the initial ground state density in the Wigner representation. The definition of $\boldsymbol{\Gamma}_1$ is analogous to that of $\boldsymbol{\Gamma}_0$, referring to the excited state ensemble, as it evolves under the impact of the pump and dump pulses. As in Eq. (18.95), the propagation of the initial ensemble is determined by the energy gap between the participating electronic states, designated by Δ_{10}.

To maximize $A(t_f)$, one has to vary the right-hand side of Eq. (19.63) with respect to both the pump and the dump pulse. For global optimization, however, two constraints must be taken into account, involving the well-defined energies of the two pulses. The functional may then be written as [28, 676].

$$\mathcal{J}(t_f) = A(t_f) + \lambda_{\mathrm{p}} \int_0^{t_f} dt |\tilde{a}_{\mathrm{p}}(t)|^2 + \lambda_{\mathrm{d}} \int_0^{t_f} dt |\tilde{a}_{\mathrm{d}}(t)|^2. \tag{19.64}$$

We point out that the constraints implemented in Eq. (19.64) modify the Tannor–Rice control method (see Eq. (19.25)) as both pulses enter the variational procedure with their respective Lagrange multipliers. Further, these pulses are represented by their envelope functions $\tilde{a}_{\mathrm{p}}$, $\tilde{a}_{\mathrm{d}}$. The special form of the constraint terms is based on the realization that contributions to the squared pulse magnitude $|\boldsymbol{\varepsilon}_{\mathrm{p(d)}}|^2$ (see Eq. (19.62)) which depend on ω_{10} oscillate strongly and thus may be safely neglected in the time integration [679].

Setting the variation of $\mathcal{J}(t_f)$ with respect to $\tilde{a}_\mathrm{p}(t)$ and $\tilde{a}_\mathrm{d}(t)$ equal to zero yields a pair of coupled integral equations from which the optimal field envelopes can be determined. These equations are

$$\int_0^{t_f} d\tau' M_\mathrm{p}(\tau, \tau'; \tilde{a}_\mathrm{d})\tilde{a}_\mathrm{p}(\tau') = \lambda_\mathrm{p}\tilde{a}_\mathrm{p}(\tau), \tag{19.65}$$

$$\int_0^{t_f} d\tau' M_\mathrm{d}(\tau, \tau'; \tilde{a}_\mathrm{p})\tilde{a}_d(\tau') = \lambda_\mathrm{d}\tilde{a}_\mathrm{d}(\tau). \tag{19.66}$$

The Lagrange multipliers related to the energy constraints on the two pulses are here realized as eigenvalues of integral operators with kernels, or *materials response functions*, $M_\mathrm{p}(\tau, \tau'; \tilde{a}_\mathrm{d})$ and $M_\mathrm{d}(\tau, \tau'; \tilde{a}_\mathrm{p})$. These are defined as

$$\begin{aligned} M_\mathrm{p}(\tau, \tau'; \tilde{a}_\mathrm{D}) = & \int d\mathbf{\Gamma}_0 \int_0^{t_f} d\tau'' \int_0^{\tau''} d\tau''' A_\mathrm{w}(\mathbf{\Gamma}'_0(t_f - \tau''; \mathbf{\Gamma}_e(\tau''' - \tau; \mathbf{\Gamma}_0))) \\ & \times \exp\left[\frac{i}{\hbar}(\hbar\omega_{10} - \Delta_{10}(\mathbf{\Gamma}_1(\tau''' - \tau; \mathbf{\Gamma}_0)))(\tau'' - \tau''')\right] \\ & \times \exp\left[\frac{i}{\hbar}(\hbar\omega_{10} - \Delta_{10}(\mathbf{\Gamma}_0))(\tau - \tau')\right] \rho_{00}(\mathbf{\Gamma}_0)\tilde{a}_\mathrm{d}(\tau''')\tilde{a}^*_\mathrm{d}(\tau''), \end{aligned} \tag{19.67}$$

$$\begin{aligned} M_\mathrm{d}(\tau, \tau'; \tilde{a}_\mathrm{P}) = & \int d\mathbf{\Gamma}_0 \int_0^{\tau'} d\tau'' \int_0^{\tau''} d\tau''' A_\mathrm{w}(\mathbf{\Gamma}'_0(t_f - \tau; \mathbf{\Gamma}_1(\tau' - \tau''; \mathbf{\Gamma}_0))) \\ & \times \exp\left[\frac{i}{\hbar}(\hbar\omega_{10} - \Delta_{10}(\mathbf{\Gamma}_1(\tau' - \tau''; \mathbf{\Gamma}_0)))(\tau - \tau')\right] \\ & \times \exp\left[\frac{i}{\hbar}(\hbar\omega_{10} - \Delta_{10}(\mathbf{\Gamma}_0))(\tau'' - \tau''')\right] \\ & \times \rho_{00}(\mathbf{\Gamma}_0)\tilde{a}_\mathrm{p}(\tau''')\tilde{a}^*_\mathrm{p}(\tau''), \quad \tau \geq \tau'. \end{aligned} \tag{19.68}$$

The iterative solution of the Fredholm-type system of integral equations (19.65) and (19.66) is a formidable task, and they appear too complex to provide a suitable theoretical tool for laser control simulation. If short pulses in the femtosecond time scale are used, however, it is possible to decouple the two equations by assuming delta-function profiles for the envelopes $\tilde{a}_\mathrm{p}$ and $\tilde{a}_\mathrm{d}$. As we specify a characteristic time delay t_d between the two pulses

$\tilde{a}_{\rm p}(t)$ and $\tilde{a}_{\rm d}(t)$ and describe their time behavior with $\delta(t)$ and $\delta(t-t_{\rm d})$, respectively, the integral kernels (19.67), (19.68) simplify to [28, 676, 680]

$$M_{\rm p}^0(\tau,\tau') = \int d\mathbf{\Gamma}_0\, A_{\rm w}(\mathbf{\Gamma}_0'(t_f - t_{\rm d}; \mathbf{\Gamma}_1(t_{\rm d} - \tau; \mathbf{\Gamma}_0)))$$

$$\times \exp\left[\frac{i}{\hbar}(\hbar\omega_{10} - \Delta_{10}(\mathbf{\Gamma}_0))(\tau - \tau')\right]\rho_{00}(\mathbf{\Gamma}_0), \qquad (19.69)$$

$$M_{\rm d}^0(\tau,\tau') = \int d\mathbf{\Gamma}_0\, A_{\rm w}(\mathbf{\Gamma}_0'(t_f - \tau; \mathbf{\Gamma}_e(\tau'; \mathbf{\Gamma}_0)))$$

$$\times \exp\left[\frac{i}{\hbar}(\hbar\omega_{10} - \Delta_{10}(\mathbf{\Gamma}_1(\tau'; \mathbf{\Gamma}_0)))(\tau - \tau')\right]\rho_{00}(\mathbf{\Gamma}_0), \quad \tau \geq \tau'. \qquad (19.70)$$

Replacing $M_{\rm p(d)}(\tau,\tau')$ in Eqs. (19.65) and (19.66) by the approximate kernels $M^0_{\rm p(d)}(\tau,\tau')$, one arrives at two independent equations for the two optimal pulses. This gain in tractability comes at the cost of introducing an additional optimization parameter, the delay time $t_{\rm d}$. In Eq. (19.69), τ is restricted to values smaller than or equal to this delay time. In the first step of the optimization procedure, an estimate of the delay time is generated. As suggested by the structure of the zeroth-order kernels $M^0_{\rm p(d)}(\tau,\tau')$, this can be achieved by advancing the initial ensemble on the excited state surface and back-propagating the final ensemble on the ground state surface to a common time t within the interval $[0, t_f]$ where the overlap between the two resulting distributions is computed. The condition of maximum overlap between the forward and backward propagated ensembles defines the initial guess of the delay time $t_{\rm d}$ as well as the *intermediate target* configuration [28, 676, 680] which may be understood as the excited state trajectory ensemble at the instant of dump pulse application, $\mathbf{\Gamma}_1(t_{\rm d})$.

19.5.1 *Case study: Application to Na_3F_2*

With reference to Na_3F_2, the space coordinates of the intermediate target describe a structure adopted by the molecule in the neighborhood of the conical intersection and, as will be emphasized later, related to the geometry of the transition state between the C_S and the C_{2v} minimum on the ground state surface. The obtained value of $t_{\rm d}$, along with the optimized trajectory ensemble corresponding to this value, is inserted into the kernel (19.69) and used in the following step to approximate the optimal pump pulse by substituting $M_{\mathrm{p}}^{(0)}(\tau, \tau')$ for $M_{\mathrm{p}}(\tau, \tau')$ in Eq. (19.65). This pulse, by construction, will guide the ensemble to the intermediate target. Finally, an analogous calculation, involving relations (19.66) and (19.70), leads to an estimate of the optimal dump pulse that projects the intermediate target to the ground state surface and maximizes the yield of the C_{2v} isomer. At this point, the three steps-determination of t_{d} and the intermediate target as well as optimization of the pump and the dump pulse might be iterated. The pulse profiles generated in the first cycle may now be used to obtain improved materials response functions, defining a starting point for a repetition of the sequence of operations. Practice shows, however, that already the first or even the zeroth approximation can be expected to meet the goal in a satisfactory fashion. Nonadiabatic effects in the propagation of the trajectory ensemble may be taken into account by including the fewest switches TSH algorithm, as described in Section 18.5.2.

This strategy has been used to steer the isomerization of Na_3F_2 from its C_S equilibrium geometry to the C_{2v} isomer on the ground state surface [28, 676, 680]. As mentioned in Section 18.5.1, both isomers are populated by deexcitation through the conical intersection, Optimal control with the objective to maximize the C_{2v} yield will therefore identify a pathway that suppresses the route through the conical intersection. As shown in Figure 19.7, this indeed is found. The intermediate target, as determined by the procedure indicated earlier, bears more resemblance to the structure of the transition state that separates the two minima on the ground state surface than to the conical intersection geometry. This finding emphasizes that the intermediate target is to be interpreted as a step on the path that connects the initial C_S with the final C_{2v} isomer. The intermediate target configuration is generated by averaging over the atomic positions of the whole trajectory ensemble at a delay time $t_{\mathrm{d}} = 250$ fs. For this choice of t_{d}, a residence time at the C_{2v} minimum in excess of 500 fs is reported [28, 676, 680]. This target structure is also attained for longer delay times. In these cases, however, the observed times of residence in the targeted minimum are found to be sensitively shortened, which is plausible in view of the larger amount of kinetic energy gained by the ensemble as it is allowed to approach the conical intersection point more closely before making the

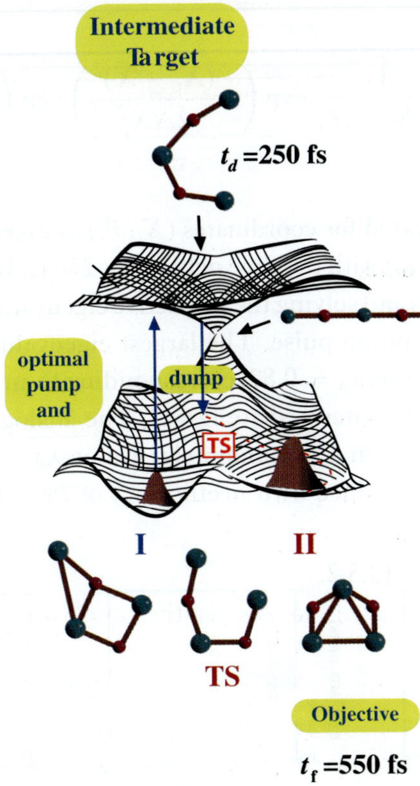

Figure 19.7: Representation of the path taken by Na_3F_2 from its C_S equilibrium structure to the C_{2v} isomer on the ground state surface as induced by optimal pump and dump pulses. An open-ring geometry is characteristic of both the structure of the intermediate target and that of the transition state on the lower surface. The intermediate target is attained at a delay time of 250 fs. (Reprinted with permission from [28]. Copyright (2005) by the American Chemical Society.)

transition to the ground state surface (see Figure 19.7). These results were obtained from a simulation involving about 1,000 spatial coordinates and conjugate momenta. A canonical distribution at a temperature of 50 K was adopted as the initial ensemble [676].

Having identified the intermediate target structure, one proceeds to derive the pump pulse that optimally concentrates phase space density around the spatial coordinates defining this structure. The respective target function in the Wigner representation is chosen as a minimum uncertainty wave packet:

$$A_{\mathrm{w}}(X,P)=\prod_{i=1}^{15}\frac{1}{2\pi\Delta X_i\Delta P_i}\exp\left(-\frac{(X_i-\bar{X})^2}{2\Delta X_i^2}\right)\exp\left(-\frac{(P_i-\bar{P})^2}{2\Delta P_i^2}\right). \tag{19.71}$$

This function is evaluated for coordinates (X_i, P_i) as given by the ensemble $\boldsymbol{\Gamma}'_0(t_f-\tau;\boldsymbol{\Gamma}_1(t_{\mathrm{d}}-\tau;\boldsymbol{\Gamma}_0))$ with the time delay $t_{\mathrm{d}}=250$ fs. Inserting $A_{\mathrm{w}}(X,P)$ into expression (19.69) and solving the respective eigenvalue equation (19.65) generates the desired pump pulse. The largest eigenvalue obtained in this manner for the Na_3F_2 was $\lambda_{\mathrm{p}}=0.82$, corresponding to an ensemble fraction of 82% that reaches the intermediate target. The analogous treatment for the dump pulse resulted in a maximum eigenvalue $\lambda_{\mathrm{d}}=0.78$. The target geometry, therefore, is attained with an efficiency of 78%. From Figure 19.8,

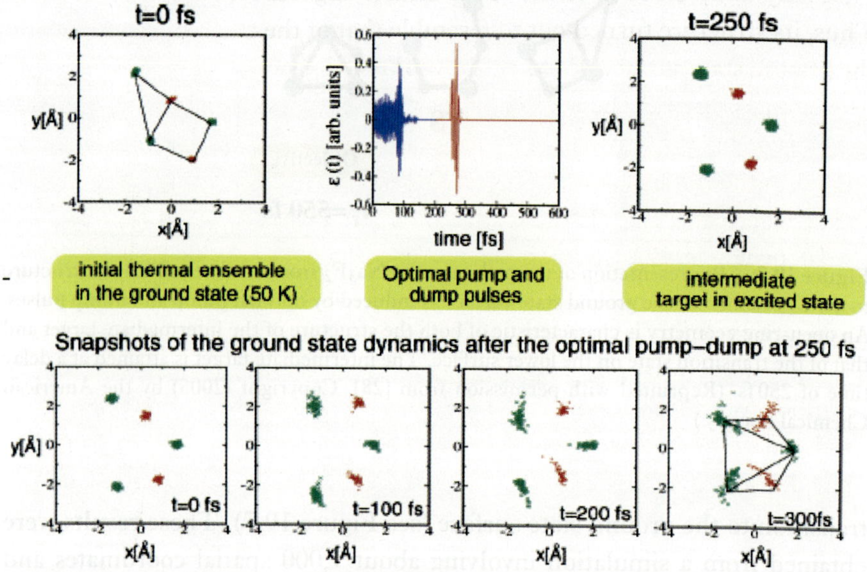

Figure 19.8: Stages of the Na_3F_2 isomerization from the initial C_s (upper left panel) to the final C_{2v} species (lower right panel). The time structures of the optimal pump and the dump pulses are shown in the upper middle panel. The intermediate target is reached at $t=250$ fs (upper right panel). The lower panels trace the evolution of the ground state ensemble after deexcitation, showing the sequence of phase space distributions in two spatial coordinates with a time step of 100 fs. (Reprinted with permission from [28]. Copyright (2005) by the American Chemical Society.)

the optimized dump pulse is very short, not exceeding 20 fs. This feature is explained by the same argument that rationalizes the upper limit on the delay time. A longer dump pulse would make it possible for the ensemble to intrude too far into the region of the conical intersection and to pick up too much kinetic energy in this process to be stabilized in the target configuration.

As can be seen by inspecting the series of phase space distributions in Figure 19.8, the target isomer is reached at approximately 300 fs after the delay time of 250 fs has passed. These results suggest that semiclassical propagation provides a powerful tool for the optimal design of pump–probe experiments intended to guide a molecular species along a chosen isomerization pathway. The pivotal concept of the control simulation method described above is that of the intermediate target which routes the considered system in the direction of the selected final structure. We emphasize that in the presented example of Na_3F_2, the obtained intermediate target could be readily understood in terms of the underlying isomerization mechanism. Thus, its structure turned out to resemble that of the transition complex on the ground state surface. Further, the optimized process was proven to avoid deexcitation through the conical intersection which preferentially populates the most stable ground state isomer. Lastly, the optimal delay time could be related to the requirement of limited kinetic energy of the molecule on the excited surface to ensure the stability of the target configuration.

In principle, the method described here for a small unit could be extended to the optimal laser control of more complex molecules and clusters, depending on the expense associated with generating the respective intermediate target. The ongoing efforts to validate and expand the described formalism appear well motivated in view of the key role played by structure-property relationships typical for the molecular and the nanoscale. Changing the structure of an object in this size region, one may sensitively influence its electronic, energetic, and magnetic characteristics, its reactivity as well as its dissociation behavior. Effective and reliable strategies to alter molecular or cluster geometries in a controlled manner are thus of major interest for basic research and open wide perspectives for technological application.

The treatment of nonadiabatic dynamics in the simulations described here is based on the semiclassical formalism presented in Section 18.5, combining an *on-the-fly* Wigner function approach with a trajectory surface hopping algorithm. This model has recently been extended in two directions, making allowance for a higher degree of flexibility in the quantum chemical description of the molecular system, and removing the constraint of weak laser fields as imposed by the perturbation approach (18.58).

In view of the limited applicability of the *frozen ionic bond* ansatz Eqs. (18.90) and (18.91), it is desirable to reformulate the multistate nonadiabatic dynamics outlined here using a variety of quantum chemical procedures. This generalization involves a change in describing the wave functions, as well as the forces associated with a given potential energy surface and the trajectory surface hopping probabilities. Thus, the time-dependent density functional theory (TDDFT), as addressed in Section 16.3, and its tight-binding realization (*time-dependent density functional theory–tight binding theory*, TDDFT-TB) have been substituted for the original quantum chemical approach, and the modified model has been examined by application to various physical systems [681, 682]. The approximate character of both schemes, where the latter is an approximation to the former, necessitates a careful motivation of their validity in each individual case of application.

Inclusion of laser fields of intermediate strength has been implemented by use of the semiclassical Liouville–von Neumann equations (14.15) [685], with $V_c = -\boldsymbol{\mu} \cdot \boldsymbol{\epsilon}$. From this canon of equations, a hopping probability has been derived that consistently incorporates the electric field $\boldsymbol{\epsilon}$. This is the fundament of the *field-induced surface hopping* (FISH) method which has been successfully tested by comparison with the full quantum dynamical treatment of a two-state model problem. In particular, the Rabi oscillations obtained from the two computations (see Section 14.1), manifesting the interference between the participating electronic states, were found to be in excellent agreement with each other [685, 686]. The capability of this procedure to control the laser driven dynamics in pump–dump experiments on complex molecular systems was exemplified by the photoisomerization of the Schiff base N-methylethaniminium, $[CH_3NH{=}CHCH_3]^+$. Pulse optimization employing a proper target functional of the type (19.64) led to the conditions for maximum yield of the *cis* conformation.

Chapter 20

Electron Transfer in Condensed Media

Electron transfer within molecules or between molecules in solution is of critical importance for a wide range of phenomena, encompassing the electric conductivity of polymers, chemiluminescence, photosynthesis, and many more subjects of interest for both fundamentals and applications of chemistry, biology, physics, and materials science (see, for instance, [580, 687]). These reactions have been found instrumental for energy transport between widely separated sites in biological and physiological systems [33]. Typical electron transfer distances are often on the order of tens of angstroms. The characteristic times for these redox processes are governed by intermediate steps which vary with the transfer mechanism operative in the system of interest. They might involve the formation of a *thermally activated complex*, or tunneling through the barrier that separates two potential wells, making them often slow on the time scale of nuclear motion. In those cases, an adiabatic approach to the description of electron transfer, associated with the nuclear rearrangement from a reactant to a product configuration on a given potential energy surface, is likely to fail, and diabatic or nonadiabatic models are to be utilized.

Electron transfer processes from a donor to an acceptor system are subdivided into two prototypes: *outer* and *inner sphere* mechanisms. In the former case, the transition of an electron from a donor to an acceptor occurs without any changes in their coordination spheres. Both the reduced and the oxidized complex persist as separate chemical entities in the course of the transfer process which implies the notion of an electron traveling from one redox center to the other. This differs from inner sphere mechanisms where the electron is transferred through the formation of an intermediate bridging unit. The interaction between donor and acceptor is generally weaker for outer than for inner sphere reactions. The former, however, tend to be advantaged over the latter in terms of entropy, since they proceed without the participation of a well-ordered bridging structure.

The outer sphere mechanism is typically realized for transition metal complexes [688]. These tend to be stable in a variety of oxidation states which makes them preferred systems for the observation of electron transfer in the laboratory. An example is provided by the heteronuclear reaction

$$[Fe(CN)_6]^{4-} + [Ir(Cl)_6]^{3-} \rightarrow [Fe(CN)_6]^{3-} + [Ir(Cl)_6]^{4-}. \qquad (20.1)$$

Marcus and Sutin [33] proposed a phenomenological approach that describes electron transfer as the transition between two charge-localized states, as realized by the reaction (20.1) for the case of small multiligated transition metal centers. This classical model will be considered in further detail in Section 20.4. In qualitative terms, it refers to a solvated donor-acceptor (DA) pair. The interaction with a polar medium, such as water, may stabilize this redox pair in the state with the exchange electron at the donor (D) site, involving a solvent arrangement that energetically favors this state over the acceptor (A) alternative. A solvent fluctuation, however, can cause both states to become temporally degenerate, creating a situation of resonance where the electron is delocalized between the D and A states. This implies a finite probability for the electron to move from D to A. Since the asymmetry of the charge distribution is lost in this intermediate structure, and therefore the solvation energy is reduced as compared with both the initial (D) and the final (A) configuration, the degenerate complex assumes the character of a transition state between a reactant and a product that differ with respect to the electron location. Due to finite dipole moments in the initial and the final state the transfer process is largely governed by the polarization of the environment. This is illustrated in Figure 20.1 where single energy levels are assigned to the D and A configurations. The vectors $\mathbf{P}_i^{\mathrm{eq}}$ and $\mathbf{P}_f^{\mathrm{eq}}$ stand for the equilibrium polarization values that stabilize the two charge-polarized states of the redox pair, while $\mathbf{P}^*$ denotes a particular polarization at which the donor and the acceptor levels coincide.

The theoretical frame drawn by this model accommodates a bewildering variety of electron transfer processes of relevance for inorganic as well as biological systems. The time range typical for the *ferrous–ferric exchange* in water, involving electron transfer from Fe^{2+} to Fe^{3+} and back is on the order of 0.1 s (e.g. [689]). Marcus theory has also been shown to account for

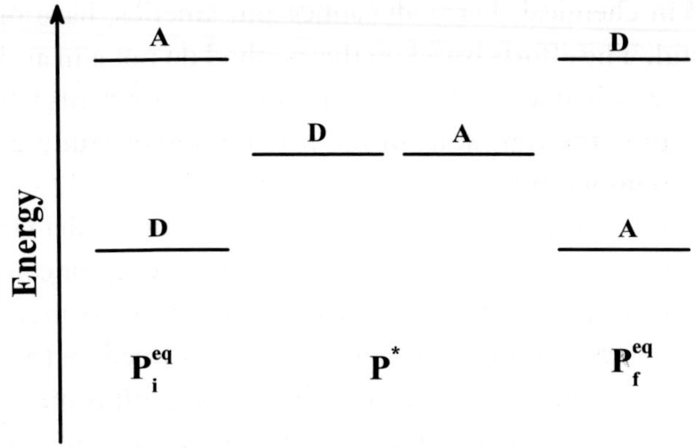

Figure 20.1: Energy levels of donor and acceptor complexes as a function of the solvent polarization. The initial and final equilibrium polarizations are labeled $\mathbf{P}_i^{eq}$ and $\mathbf{P}_f^{eq}$, $\mathbf{P}^*$ refers to the intermediate polarization that leads to degenerate donor and acceptor levels.

essential features of photosynthesis [690,691] where the elementary electron transfer step proceeds within 1–10 ps. In this case, the initial activation energy is supplied by photoexcitation of the biological complex.

The examples indicated have in common that the excess electron remains bound during its displacement between two sites in a certain molecular species or between two different molecules. In other words, the process takes place within the discrete spectrum of the considered complex, and does not involve the excitation of the exchanged electron into the continuum. This is the understanding of electron transfer that will be adopted in the following survey. In particular, we will focus on thermal activation as transfer mechanism, and also, for low temperatures, tunneling through potential barriers characteristic of the energy landscapes of the respective chemical or biological matrices.

With respect to the environment of the transfer process, two sets of variables have to be distinguished: the internal nuclear degrees of freedom of the electron transfer system and those of the environment or the *bath*. The need to include both groups of coordinates as well as their interactions makes redox reactions frequently too complex for treatment by any of the quantum dynamical methods discussed in the previous chapters. While the time-dependent or time-independent standard approaches of quantum dynamics, e.g. wave packet propagation or basis set expansion methods, are often of limited use for solving electron transfer problems, statistical procedures, as employed in chemical thermodynamics and kinetics, have proven to be very successful. The efforts based on this method do not aim at the dynamic details of the investigated system but at thermal rate constants which contain information about the time scale of the reaction under study, as well as its variation with temperature.

In the first two sections of this chapter, we develop model Hamiltonians for the interaction of an excess electron with a donor–acceptor complex. After introducing the purely electronic problem (Section 20.1) we include the coupling between electronic and vibrational degrees of freedom (Section 20.2). The meaning of adiabaticity versus nonadiabaticity with respect to the electron transfer problem will be discussed in Section 20.3, while Sections 20.4 and 20.5 concentrate on the intramolecular electron exchange problem and distinguish between thermally activated transfer and nuclear tunneling, respectively, as chief transfer mechanisms in different regimes of temperature. Emphasis is placed on a diabatic exchange model, involving sufficiently small electronic coupling constants. In many cases, this model is accommodated by the golden rule formalism for the two-state problem, as delineated earlier in skeletal abstraction (see Section 5.4).

Up to this point, our survey focuses on the concepts essential for treating electron transfer theoretically but does not yet involve the action of the solvent which will be considered explicitly in Section 20.6. As the most essential result of the central sections of this chapter, 20.5 and 20.6, we will show that, by proper adjustment of Marcus theory parameters, the same rate formalism applies to an exchange system embedded in a vibrational bath and one surrounded by a polar environment, where the first case is determined by the vibrational Hamiltonian (Section 20.5) and the second by the electrostatic coupling between the solute and the solvent (Section 20.6) as the perturbing agents.[1]

In the interest of readability, various somewhat laborious derivations that nevertheless aid the understanding of the electron transfer rate constant formalism have been assembled into an appendix, Section 20.9.

Following a remark on ultrafast electron transfer (Section 20.7), this chapter closes with the presentation of a case study, involving the task of simulating a physical electron transfer system in a polar environment. Here

[1] A comprehensive survey of the molecular electron transfer problem, including environmental effects, is found in [694].

we will focus on a seminal Monte Carlo approach [692] to the ferrous–ferric exchange problem and subsequently discuss a refined treatment of the quantum effects relevant to this problem by Euclidean path integral analysis [693].

20.1 The Electronic Hamiltonian

The simplest type of electron transfer, involving an excess electron interacting with two symmetric centers, is exemplified by the ferrous–ferric exchange mentioned earlier, where the electron alternates between two $Fe^{3+}(H_2O)_6$ ions. This system realizes a well-known prototype of molecular quantum mechanics. In fact, the respective Hamiltonian $\hat{H}_{\mathrm{el}}$ can be constructed in analogy to that for the H_2^+ molecule. Classically, the electron is confined to the potential trough of one of the two ionic cores, as shown in Figure 20.2.

We adopt the notation ϕ_{D}, ϕ_{A} as the electronic wave function at the donor and the acceptor site, respectively. To cover a wider variety of electron transfer systems, the scenario sketched in Figure 20.2 is to be extended into two directions: making admission for asymmetry between donor and acceptor, and allowing for a set of intermittent bridge sites between the centers D and A. To implement these generalizations, we approximate the scalar product $\langle \phi_i | \phi_j \rangle$ by δ_{ij} where the indices i and j refer to different places in the chain of possible sites of electron attachment. In general, two different

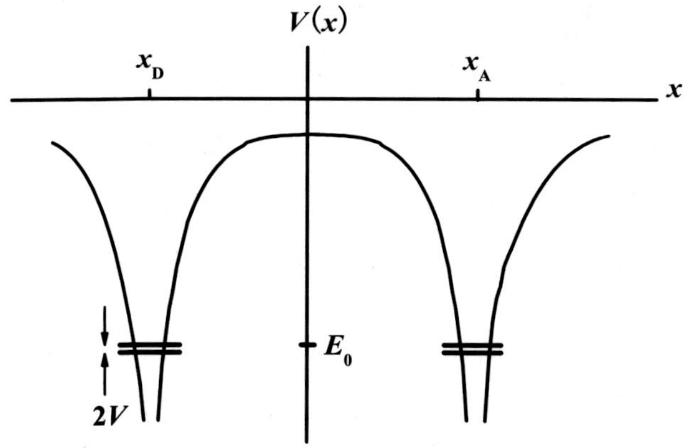

Figure 20.2: Model potential for the excess electron in the case of symmetric transfer between two centers.

states from the set of localized wave functions ϕ_i will exhibit a finite overlap, and the orthonormality condition has to be examined in every individual case considered. Assuming in the following development that this condition is satisfied suggests a convenient formulation of the electronic Hamiltonian $\hat{H}_{\text{el}}$ by use of projection operators $|\phi_i\rangle\langle\phi_j|$, related to sites i and j of the studied complex. Employing this notation, we write the electronic Hamiltonian as

$$\hat{H}_{\text{el}} = \sum_{i,j=\text{D},\text{B}_1,\text{B}_2\ldots\text{B}_N,\text{A}} \langle\phi_i|\hat{H}_{\text{el}}|\phi_j\rangle|\phi_i\rangle\langle\phi_j| \tag{20.2}$$

where we have inserted N bridge sites, $\text{B}_1,\ldots,\text{B}_N$, between the starting point and the end of the electron's journey. For explicit specification of the electronic matrix elements $\langle\phi_i|\hat{H}_{\text{el}}|\phi_j\rangle$ we introduce $N+2$ pseudopotentials that describe the influence of the ionic cores on the transported electron and merge into an overall pseudopotential for the composite consisting of the donor, the acceptor, and the bridging units, as illustrated by Figure 20.3.

Associating with the pseudopotential at the *nth* site in the transfer sequence a function $V_n = V(X_n)$, we use the integrals $\langle\phi_i|V_n|\phi_j\rangle$ to express the electronic matrix element. In this context, we apply the simplifying assumption that three-center integrals can be safely neglected:

$$\langle\phi_i|V_n|\phi_j\rangle \approx 0, \quad \text{if } i \neq j, n;\ j \neq n. \tag{20.3}$$

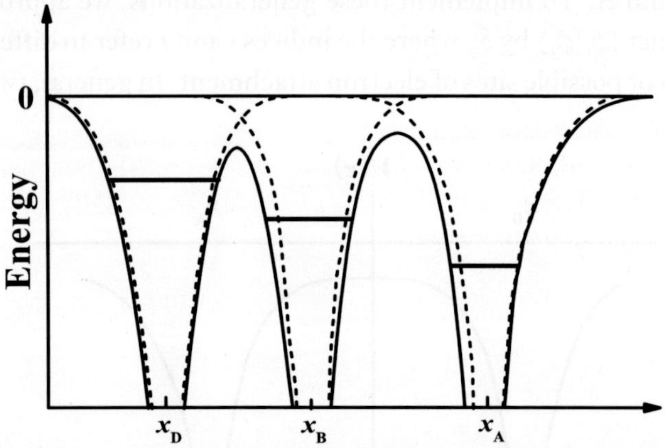

Figure 20.3: Sketch of pseudopotentials (dashed line) centered at consecutive sites of the electron transfer chain, starting at the donor and ending at the acceptor location. The bold line refers to the pseudopotential of the transfer system as a whole. Possible energy levels of the electron are indicated in the individual potential troughs, as it traverses the complex by tunneling through the barriers separating adjacent sites.

Combining this condition with the orthonormality constraint, we write the matrix elements of $\mathbf{H}_{\text{el}}$ as [694]:

$$\langle\phi_i|\hat{H}_{\text{el}}|\phi_j\rangle = \delta_{ij}\left(E_i + \sum_{n\neq j}\langle\phi_i|V_n|\phi_j\rangle\right) + (1-\delta_{ij})\langle\phi_i|\hat{T}_{\text{el}} + V_i + V_j|\phi_j\rangle, \tag{20.4}$$

involving the operator $\hat{T}_{\text{el}}$ of the electronic kinetic energy and the energy level E_i of an excess electron localized at the site i. More specifically, E_i is defined as the lowest energy eigenvalue admitted by the stationary single-particle Schrödinger equation

$$(\hat{T}_{\text{el}} + V_i(\mathbf{x}))\phi_i(\mathbf{x}) = E_i\phi_i(\mathbf{x}). \tag{20.5}$$

We use this relation to recast the off-diagonal portion of Eq. (20.4) into

$$\langle\phi_i|\hat{T}_{\text{el}} + V_i + V_j)|\phi_j\rangle = \frac{1}{2}\langle\phi_i|V_i + V_j|\phi_j\rangle \equiv V_{ij}. \tag{20.6}$$

Exercise 20.1. Use Eq. (20.5) to demonstrate the validity of Eq. (20.6). Show further that $V_{ij} = -\langle\phi_i|\hat{T}_{\text{el}}|\phi_j\rangle$.

On the basis of Eqs. (20.4) and (20.6), and including the diagonal pseudopotential matrix elements in the energies E_i, the full electronic Hamiltonian (20.2) is written in the following form:

$$\hat{H}_{\text{el}} = \sum_{i=\text{D},\text{B}_1,\ldots,\text{B}_N,\text{A}} E_i|\phi_i\rangle\langle\phi_i| + \sum_{i,j=\text{D},\text{B}_1,\ldots,\text{B}_N,\text{A},i\neq j} V_{ij}|\phi_i\rangle\langle\phi_j|. \tag{20.7}$$

This Hamiltonian, involving localized electronic states at well-defined ion sites, is also called the *tight binding* Hamiltonian. It should be noted that we have referred to the electronic ground state level only, as shown in Figure 20.3, while the simple idealized situation illustrated by Figure 20.2 contains the lowest and the first excited energy level, reflecting the effect of tunneling splitting. For gas phase species, narrow spacing between the energies of the ground and the first excited state may exert a distinct impact on the transfer dynamics. In the present discussion, however, our guiding interest is in solvated systems which will be addressed in Section 20.6. Since the first excited state in general involves sizeable charge redistribution, it is often substantially disadvantaged when the system interacts with a polar environment, resulting in a large energy separation from the ground state.

20.2 Electronic–Vibronic Coupling: The Spin-Boson Hamiltonian

The representation of electron transfer systems developed in the preceding section does not explicitly contain the nuclear degrees of freedom. Aiming at inclusion of the vibrational subsystem in order to evaluate its impact on the transfer process, we introduce an electron transfer Hamiltonian $\hat{H}_{\mathrm{ET}}$ by the following extension of the ansatz (20.7):

$$\hat{H}_{\mathrm{ET}} \equiv \hat{H}_{\mathrm{el}}(\mathbf{X}) + \hat{T}_{\mathrm{N}} + V_{\mathrm{N}}(\mathbf{X}). \tag{20.8}$$

The operator $\hat{T}_{\mathrm{N}}$ refers to the kinetic energy associated with the vibrational degrees of freedom present in the system, and $V_{\mathrm{N}}(\mathbf{X})$ denotes the respective electrostatic potential. The right side of Eq. (20.8) is partitioned into a purely nuclear part and a component that involves the electronic states by using Eq. (20.7) in conjunction with the nonadiabaticity matrix $\mathbf{\Lambda}$ (2.12):

$$\begin{aligned}\hat{H}_{ET} = &\sum_{i=\mathrm{D,B_1,\ldots,B_N,A}} [\hat{T}_{\mathrm{N}} + E_i(\mathbf{X}) + V_{\mathrm{N}}(\mathbf{X}) + \Lambda_{ii}(\mathbf{X})]|\phi_i\rangle\langle\phi_i| \\ &+ \sum_{i,j=\mathrm{D,B_1,\ldots,B_N,A},i\neq j} V_{ij}(\mathbf{X})|\phi_i\rangle\langle\phi_j|. \end{aligned} \tag{20.9}$$

Due to the localized, quasi-diabatic nature of the electronic states adopted in this derivation, the off-diagonal contributions to the nonadiabaticity may be assumed to be small and have been neglected. The influence of the kinetic nuclear energy operator on the electronic states has been included in the term $\Lambda_{ii}(\mathbf{X})$, and so $\hat{T}_{\mathrm{N}}$ in (20.9) does not act on the wave functions $|\phi_i\rangle$. We combine the three diagonal potential terms into a potential energy surface related to an excess electron at the site i:

$$U_i(\mathbf{X}) \equiv E_i(\mathbf{X}) + V_{\mathrm{N}}(\mathbf{X}) + \Lambda_{ii}(\mathbf{X}). \tag{20.10}$$

Upon this substitution, $\hat{H}_{\mathrm{ET}}$ assumes the simpler form

$$\begin{aligned}\hat{H}_{\mathrm{ET}} = &\sum_{i=\mathrm{D,B_1,\ldots,B_N,A}} [\hat{T}_{\mathrm{nuc}} + U_i(\mathbf{X})]\,|\phi_i\rangle\langle\phi_i| \\ &+ \sum_{i,j=\mathrm{D,B_1,\ldots,B_N,A},i\neq j} V_{ij}(\mathbf{X})|\phi_i\rangle\langle\phi_j|. \end{aligned} \tag{20.11}$$

As we apply the harmonic approximation to the surfaces U_i and introduce mass-weighted normal coordinates $\tilde{R}_k$ (Section 2.2) by diagonalizing the

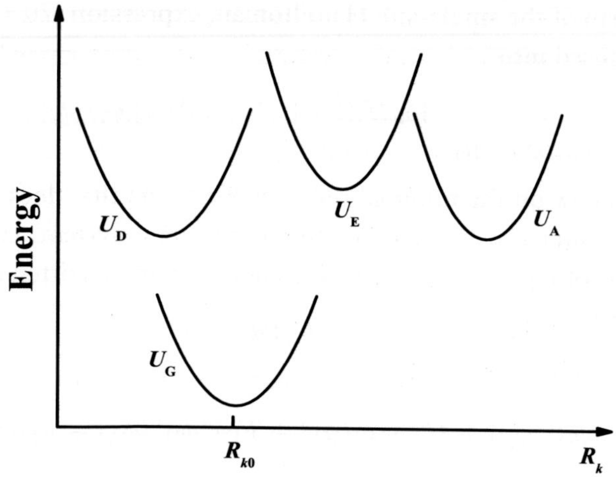

Figure 20.4: Illustration of the potential energy surfaces associated with the sites occupied by the excess electron during the transfer process. The harmonic approximation is applied, the independent variable R_k is the kth normal coordinate while all other coordinates have been fixed at constant values.

Hessian matrix, we obtain

$$U_i(\tilde{\mathbf{R}}) = U_i^{(0)} + \frac{1}{2}\sum_k \omega_{ik}^2(\tilde{R}_k - \tilde{R}_k^{(i)})^2, \quad i = \mathrm{D}, \mathrm{B}_1, \ldots, \mathrm{B}_\mathrm{N}, \mathrm{A} \quad (20.12)$$

where $U_i^{(0)}$ stands for the equilibrium value of the surface U_i which is attained at the nuclear position $\tilde{\mathbf{R}}^{(i)}$. The latter may be defined with respect to a reference potential, which can be chosen as that of the host complex without the excess electron [695].

Figure 20.4 shows various displaced parabolas with different offset values $U_i^{(0)}$ as a one-dimensional representation of the resulting potential energy surfaces due to the donor, the acceptor, and several bridge states. One may think of intramolecular electron transfer as a physical realization of this model. The treatment of intermolecular processes would necessitate introducing two separate sets of vibrational coordinates.

Formula (20.12) provides the potential energy contribution to the local Hamiltonian $\hat{H}_i(\tilde{\mathbf{R}})$ assigned to site i of the transfer chain:

$$\hat{H}_i(\tilde{R}) = U_i^{(0)} + \frac{1}{2}\sum_k \left[-\hbar^2 \frac{\partial^2}{\partial \tilde{R}_k^2} + \omega_{ik}^2(\tilde{R}_k - \tilde{R}_k^{(i)})^2\right],$$
$$i = \mathrm{D}, \mathrm{B}_1, \ldots, \mathrm{B}_\mathrm{N}, \mathrm{A}. \quad (20.13)$$

Using this form of the single-site Hamiltonian, expression (20.11) is conveniently condensed into

$$\hat{H}_{\mathrm{ET}} = \sum_{i,j=D,\mathrm{B}_1,\ldots,\mathrm{B}_\mathrm{N},\mathrm{A}} [\delta_{ij}\hat{H}_j(\tilde{\mathbf{R}}) + (1-\delta_{ij})V_{ij}]|\phi_i\rangle\langle\phi_j|. \tag{20.14}$$

We will focus now on the minimal redox pair, i.e. one that does not contain any bridging units and thus reduces to a two-state system, composed of donor and acceptor only. The respective Hamiltonian is written explicitly as

$$\begin{aligned}\hat{H}_{\mathrm{ET}} &= \sum_{i,j=\mathrm{D,A}} [\delta_{ij}\hat{H}_j(\tilde{\mathbf{R}}) + (1-\delta_{ij})V_{ij}]|\phi_i\rangle\langle\phi_j| \\ &= \hat{H}_\mathrm{D}|\phi_\mathrm{D}\rangle\langle\phi_\mathrm{D}| + \hat{H}_\mathrm{A}|\phi_\mathrm{A}\rangle\langle\phi_\mathrm{A}| + V_{\mathrm{DA}}|\phi_\mathrm{D}\rangle\langle\phi_\mathrm{A}| + V_{\mathrm{DA}}|\phi_\mathrm{A}\rangle\langle\phi_\mathrm{D}|.\end{aligned} \tag{20.15}$$

As a further idealization, we will assume that the donor and acceptor states are associated with the same set ω_k of vibrational frequencies. The potential energy surfaces associated with these states are thus described as N-dimensional parabolas of the same shape albeit with different ground state energies and displaced with respect to each other in the space of the vibrational coordinates. In the following step, we develop the sum (20.15) into the prototypical two-state Hamiltonian that exploits the formal analogy between a two-level system and a spin one-half particle. To this end, we revert to the local operators given by Eq. (20.13) and make use of the creation and annihilation operator formalism by substituting

$$\hat{\tilde{P}}_k = -i\hbar\frac{\partial}{\partial\tilde{R}_k} = -i\sqrt{\frac{\hbar\omega_k}{2}}(\hat{b}_k - \hat{b}_k^\dagger), \tag{20.16}$$

$$\tilde{R}_k = \sqrt{\frac{\hbar}{2\omega_k}}(\hat{b}_k + \hat{b}_k^\dagger), \tag{20.17}$$

where the creation and annihilation operators are defined by

$$\hat{b}_k^\dagger \equiv \sqrt{\frac{\omega_k}{2\hbar}}\tilde{R}_k - i\frac{1}{\sqrt{2\hbar\omega_k}}\hat{\tilde{P}}_k \tag{20.18}$$

and

$$\hat{b}_k \equiv \sqrt{\frac{\omega_k}{2\hbar}}\tilde{R}_k + i\frac{1}{\sqrt{2\hbar\omega_k}}\hat{\tilde{P}}_k, \tag{20.19}$$

respectively. They obey the bosonic commutation relations

$$[\hat{b}_k, \hat{b}^\dagger_{k'}] = \delta_{kk'}. \tag{20.20}$$

Upon replacement of $\tilde{R}_k$ and $-i\hbar\frac{\partial}{\partial\tilde{R}_k}$ in Eq. (20.13), according to the prescriptions (20.16, 20.17), the two-state Hamiltonian (20.15) becomes

$$\hat{H}_{\mathrm{ET}} = U'^{(0)}_{\mathrm{D}}|\phi_{\mathrm{D}}\rangle\langle\phi_{\mathrm{D}}| + U'^{(0)}_{\mathrm{A}}|\phi_{\mathrm{A}}\rangle\langle\phi_{\mathrm{A}}| + \sum_k \hbar\omega_k\left(\hat{b}^\dagger_k\hat{b}_k + \frac{1}{2}\right)$$
$$+ \sum_k \hbar\omega_k(\hat{b}_k + \hat{b}^\dagger_k)(g_{\mathrm{D},k}|\phi_{\mathrm{D}}\rangle\langle\phi_{\mathrm{D}}| + g_{\mathrm{A},k}|\phi_{\mathrm{A}}\rangle\langle\phi_{\mathrm{A}}|)$$
$$+ V_{\mathrm{DA}}|\phi_{\mathrm{D}}\rangle\langle\phi_{\mathrm{A}}| + V_{\mathrm{AD}}|\phi_{\mathrm{A}}\rangle\langle\phi_{\mathrm{D}}|. \tag{20.21}$$

Here we have utilized the completeness relation for the two-state system, namely $|\phi_{\mathrm{D}}\rangle\langle\phi_{\mathrm{D}}| + |\phi_{\mathrm{A}}\rangle\langle\phi_{\mathrm{A}}| = \hat{I}$, and introduced the quantities

$$U'^{(0)}_i \equiv U^{(0)}_i + \frac{1}{2}\sum_k \omega_k^2\tilde{R}^{(i)2}_k, \quad i = \mathrm{D}, \mathrm{A}, \tag{20.22}$$

and

$$g_{i,k} = -\sqrt{\frac{\omega_k}{2\hbar}}\tilde{R}^{(i)}_k, \quad i = \mathrm{D}, \mathrm{A}. \tag{20.23}$$

If the coordinate system is adjusted such that $\tilde{R}^{(\mathrm{D})}_k = -\tilde{R}^{(\mathrm{A})}_k$, the two vibronic-electronic coupling constants g_{D} and g_{A} are found to be equal and opposite, $g_{\mathrm{D},k} = -g_{\mathrm{A},k} \equiv g_k$. Setting further $\overline{U}'^{(0)} \equiv \frac{1}{2}(U'^{(0)}_{\mathrm{A}} + U'^{(0)}_{\mathrm{D}})$, $\varepsilon \equiv U'^{(0)}_{\mathrm{D}} - U'^{(0)}_{\mathrm{A}}$, $\hat{H}_{\mathrm{vib}} = \sum_k \hbar\omega_k(b^\dagger_k b_k + \frac{1}{2})$, and assuming a real-valued coupling element V_{DA}, such that $V_{\mathrm{DA}} = V_{\mathrm{AD}} \equiv V$, we write $\hat{H}_{\mathrm{ET}}$ for a donor-acceptor system in the form

$$\hat{H}_{\mathrm{SB}} = \overline{U}'^{(0)} + \hat{H}_{\mathrm{vib}} + \frac{1}{2}\varepsilon\hat{\sigma}_z + \sum_k \hbar\omega_k g_k(\hat{b}_k + \hat{b}^\dagger_k)\hat{\sigma}_z + V\hat{\sigma}_x. \tag{20.24}$$

The notation $\hat{H}_{\mathrm{SB}}$ refers to the *spin-boson Hamiltonian* in pseudospin nomenclature. This term is related to the boson nature of the harmonic vibrational degrees of freedom to which the electronic states are coupled by Eq. (20.24). The symbols $\hat{\sigma}_x$ and $\hat{\sigma}_z$ stand for two of the three spin-$\frac{1}{2}$ operators which are expressed in terms of ϕ_D and ϕ_A as follows:

$$\hat{\sigma}_x = |\phi_D\rangle\langle\phi_A| + |\phi_A\rangle\langle\phi_D|, \tag{20.25}$$

$$\hat{\sigma}_y = i(|\phi_D\rangle\langle\phi_A| - |\phi_A\rangle\langle\phi_D|), \tag{20.26}$$

$$\hat{\sigma}_z = |\phi_D\rangle\langle\phi_D| - |\phi_A\rangle\langle\phi_A|. \tag{20.27}$$

Alternatively, one may choose the matrix representation of the spin-boson Hamiltonian by substituting the Pauli matrices $\boldsymbol{\sigma}_m$ for the operators σ_m, $m = x, y, z$. The matrix $\mathbf{H}_{SB}$ thus generated acts on a spinor of the space spanned by the basis $(\phi_D, 0), (0, \phi_A)$.

20.3 Adiabatic versus Nonadiabatic Electron Transfer

So far, our discussion of electron transfer has been restricted to a localized or diabatic model, allowing for a direct assignment of the electronic states to the physical components involved in the transfer process: the donor, the acceptor, and various bridge species. Connecting the trough regions of the parabolas shown in Figure 20.5 yields an approximation to the adiabatic ground state potential energy curve of the considered system. Thus, Figure 20.5 contains both the diabatic and the adiabatic curves for a donor-acceptor complex, as analyzed in the previous section. With

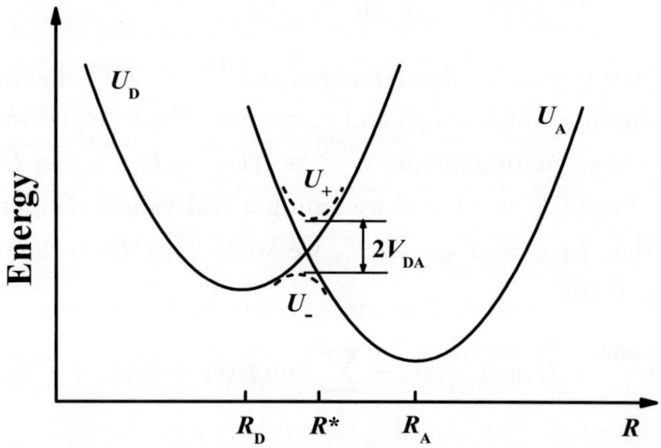

Figure 20.5: Schematic view of donor and acceptor potential energy surfaces in the diabatic (solid line) and the adiabatic (dashed line) representation as a function of a reaction coordinate R. For electron transfer to proceed between the two states, a finite interaction matrix element V is required. This implies that the crossing point of the diabatic curves turns into an avoided crossing of their adiabatic analogs.

reference to the spin-boson Hamiltonian (20.24), the vertical shift between the adiabatic ground and first excited states at the diabatic crossing point amounts to $2V$. If the probability for electron transfer is finite, the coupling matrix element V does not vanish, resulting in an avoided crossing between the adiabatic curves.

From a technical perspective, there is no objection against casting electron transfer as outlined in the previous section into the conceptual framework of an adiabatic approach. However, the diabatic mode of description that subdivides the complex according to the location of the excess electron is a particularly suggestive model for the treatment of electron transfer. Obviously, the size of the interaction element V provides a criterion for the utility of the adiabatic versus the diabatic picture, where the former is naturally preferred in the case of sufficiently large V, allowing for a distinct separation between the electronic ground state and the first excited state in the considered region of the potential energy landscape.

Once the adiabatic representation has been adopted, we may distinguish between the possibilities of electron transfer proceeding on a single potential energy surface or involving more than one surface, corresponding to an adiabatic or a nonadiabatic effect, respectively. The comparison of characteristic electronic and nuclear time scales decides the rubric under which a given electron transfer process falls.

Typical times for the nuclear motion are estimated by the reciprocal value of an average vibrational frequency, $2\pi/\langle\omega_{\text{vib}}\rangle$. In general, however, the various vibrations present in the nuclear subsystem will differ substantially from each other which implies that the average $\langle\omega_{\text{vib}}\rangle$ is of limited physical meaning. It might be replaced by the interval $[2\pi/\omega_{\max}, 2\pi/\omega_{\min}]$ where $\omega_{\min}$ and $\omega_{\max}$ could be understood as the mean values of subsets of highest and lowest frequencies, respectively. A time scale for electron transfer is given by the quantity $\hbar/V$. In the adiabatic regime, this time is sizeably smaller than the characteristic time for nuclear motion. The electronic subsystem adjusts at any moment to the temporal changes of a reaction coordinate R which is composed of nuclear degrees of freedom (see Figure 20.4). If this condition is realized, the electron transfer rate k_{ET} is proportional to an Arrhenius factor according to

$$k_{\text{ET}} \sim \exp\left(-\frac{E_{\text{AD}}}{k_{\text{B}}T}\right) \tag{20.28}$$

with an adiabatic activation energy $E_{AD} = U_G(R^*) - U_G(R_D)$. The nonadiabatic case applies in the time domain whenever the electronic time scale $\hbar/V$ is comparable with or larger than the nuclear time scale $2\pi/\langle\omega_{vib}\rangle$. In this domain, the quantity $\hbar\langle\omega_{vib}\rangle/2\pi$ provides an upper bound for the electronic coupling element V, where appropriate modifications have to be made if averaging over the vibrational frequencies is deemed unsuitable for the treatment of the system under study. While the structure of the nonadiabatic rate constant is not immediately clear from intuition, the nonadiabatic counterpart of relation (20.28) can be found by utilizing the Landau–Zener theory of electronic transitions as a guideline. As outlined in Section 7.5, this model refers to cases of avoided crossings between adiabatic surfaces. Further, we pointed out (see the discussion following formula (7.149)) that the Landau–Zener probability P_{LZ} for the transition from an initial to a final adiabatic state, namely

$$P_{LZ} = \exp(-\pi\Gamma), \qquad \Gamma = \frac{2V^2}{\hbar v^*|F_D - F_A|}, \tag{20.29}$$

can be read also as the probability for the survival of an initial diabatic state which, in the present context, is the donor state. The symbols F_D and F_A denote the slopes of the diabatic curves U_D and U_A at their crossing point, as shown in Figure 20.5, while v^* is the reaction coordinate velocity at this point. For the following argument to apply, we assume that the coupling element is small enough to justify an expansion of P_{LZ} with respect to V, but not so small as to endanger the validity of the Landau–Zener approach.

From Section 7.5 we recall that the Landau–Zener transition probability during one full oscillation period of the nuclear coordinate, assumed classical, is given by $2(1 - P_{LZ})P_{LZ}$. Relying on an elementary model, we might obtain the electron transfer rate from the product of the frequency $\langle\omega_{vib}\rangle/2\pi$ associated with the reaction coordinate, and the Landau–Zener transition probability per period:

$$k_{ET}^{LZ} = \frac{\langle\omega_{vib}\rangle}{2\pi}2(1 - P_{LZ})P_{LZ} = \frac{\langle\omega_{vib}\rangle}{2\pi}2(1 - \exp(-\pi\Gamma))\exp(-\pi\Gamma). \tag{20.30}$$

In the nonadiabatic limit of small V, $\exp(-\pi\Gamma)$ is approximated by $1 - \pi\Gamma$, and it further holds that $\Gamma^2 \approx 0$. The Landau–Zener electron transfer rate for the nonadiabatic case, $k_{\mathrm{NA}}^{\mathrm{LZ}}$ becomes therefore

$$k_{\mathrm{NA}}^{\mathrm{LZ}} = \frac{2\langle\omega_{\mathrm{vib}}\rangle V^2}{\hbar v^* |F_{\mathrm{D}} - F_{\mathrm{A}}|}. \tag{20.31}$$

From Eq. (20.30), we also immediately derive the adiabatic analog of $k_{\mathrm{NA}}^{\mathrm{LZ}}$. Under adiabatic conditions, Γ is large, such that $\exp(-2\pi\Gamma) \approx 0$. This leads to

$$k_{\mathrm{AD}}^{\mathrm{LZ}} = \frac{\langle\omega_{\mathrm{vib}}\rangle}{\pi} \exp(-\pi\Gamma). \tag{20.32}$$

Thus, the dependence of the rate constant on the electronic transition matrix element V turns out to be markedly different in the adiabatic and the non-adiabatic frame.

20.4 Thermally Activated Transfer

In the following we will discuss the transition rate for nonadiabatic processes on the basis of Fermi's golden rule while maintaining the classical description of the vibrational subsystem [696]. The regime of validity for this approach is the *limit of high temperature*, corresponding to the condition $k_{\mathrm{B}}T \gg \hbar\omega_k$, where k encompasses all normal modes of the transfer complex. We will, however, single out a one-dimensional reaction coordinate $\tilde{R}$, as represented in Figure 20.5. As in the preceding development, we consider intramolecular electron transfer. Therefore, the same coordinate is used to describe both the donor and the acceptor component of the species. We carry out a thermal averaging procedure over the vibrational modes of the reactant, i.e. the donor system, by including the Boltzmann–Maxwell distribution function $\tilde{f}(\tilde{R}, \tilde{P}_{\tilde{R}})$. This acknowledges that in any realistic experimental situation the observed system is prepared as a statistical mixture of multiple initial states. The golden rule expression for the rate constant reads accordingly (compare with Eq. (17.29)):

$$\bar{k}_{\mathrm{ET}} = \frac{2\pi}{\hbar} \int d\tilde{R} d\tilde{P}_{\tilde{R}} \tilde{f}(\tilde{R}, \tilde{P}_{\tilde{R}}) V^2 \delta(H_{\mathrm{D}}(\tilde{R}, \tilde{P}_{\tilde{R}}) - H_{\mathrm{A}}(\tilde{R}, \tilde{P}_{\tilde{R}})), \tag{20.33}$$

with H_{D}, H_{A} as Hamilton functions for the donor and the acceptor segment of the unit, respectively. Since

$$H_i(\tilde{R}, \tilde{P}_{\tilde{R}}) = U_i(\tilde{R}) + \frac{1}{2}\tilde{P}_{\tilde{R}}^2, \tag{20.34}$$

$$U_i(\tilde{R}) = U_i^{(0)} + \frac{1}{2}\omega_{\mathrm{vib}}^2(\tilde{R} - \tilde{R}^{(i)})^2, \quad i = \mathrm{D, A}, \tag{20.35}$$

the kinetic energy contributions to H_{D} and H_{A} cancel out in the argument of the delta function in Eq. (20.33). The momentum coordinate is thus integrated out, and the transfer rate reduces to

$$\bar{k}_{\mathrm{ET}} = \frac{2\pi}{\hbar}\int d\tilde{R}\, f(\tilde{R})\, V^2 \delta(U_{\mathrm{D}}(\tilde{R}) - U_{\mathrm{A}}(\tilde{R})) \qquad (20.36)$$

where f is a thermal distribution function in the reaction coordinate alone:

$$f(\tilde{R}) = \sqrt{\frac{\omega_{\mathrm{vib}}^2}{2\pi k_B T}}\exp\left[-\frac{\omega_{\mathrm{vib}}^2}{2k_{\mathrm{B}}T}(\tilde{R} - \tilde{R}^{(\mathrm{D})})^2\right]. \qquad (20.37)$$

As in our previous treatment involving Landau–Zener theory, we assume a constant V, that is, independence of the coupling matrix element of the vibrational variable $\tilde{R}$. From the harmonic approximation (20.35), we find the difference between the two diabatic potential energy curves to be

$$U_{\mathrm{D}}(\tilde{R}) - U_{\mathrm{A}}(\tilde{R}) = \Delta U^{(0)} - \omega_{\mathrm{vib}}^2(\tilde{R}^{(\mathrm{D})} - \tilde{R}^{(\mathrm{A})})\tilde{R} + \frac{1}{2}\omega_{\mathrm{vib}}^2(\tilde{R}^{(D)2} - \tilde{R}^{(A)2}). \qquad (20.38)$$

The quantity $\Delta U^{(0)} \equiv U_{\mathrm{D}}^{(0)} - U_{\mathrm{A}}^{(0)}$, the so-called *driving force* of the considered electron transfer process, measures the "bias" that shifts the energies of the donor and the acceptor well with respect to each other. Upon inserting the right-hand side of Eq. (20.38) into the golden rule integral (20.36) and including the explicit form of the distribution function (20.37), we can easily perform the integration with respect to $\tilde{R}$ and arrive at

$$\bar{k}_{\mathrm{ET}} = \frac{2\pi}{\hbar}\frac{V^2}{\sqrt{2\pi k_{\mathrm{B}} T\omega_{\mathrm{vib}}^2(\tilde{R}^{(\mathrm{D})} - \tilde{R}^{(\mathrm{A})})^2}}\exp\left[-\frac{\omega_{\mathrm{vib}}^2(\tilde{R}^* - \tilde{R}^{(\mathrm{D})})^2}{2k_{\mathrm{B}}T}\right], \qquad (20.39)$$

with $\tilde{R}^*$ as the coordinate of the crossing point between the donor and the acceptor potential curve.

Exercise 20.2. Express the crossing point coordinate $\tilde{R}^*$ as a function of $\tilde{R}^{(\mathrm{D})}$, $\tilde{R}^{(\mathrm{A})}$ and the driving force. Evaluate the integral (20.36) to confirm the validity of the result (20.39).

We notice that the exponential factor exhibits the Arrhenius form $\exp(-\frac{E_A}{k_B T})$ with

$$E_A = \frac{\omega_{vib}^2}{2}(\tilde{R}_0 - \tilde{R}^{(D)})^2 \tag{20.40}$$

as activation energy. Plausibly, the reaction barrier is the potential energy increase encountered when moving from the equilibrium position of the donor trough to the curve crossing point. Rewriting this expression in terms of the driving force, we obtain

$$E_A = \frac{(\Delta U^{(0)} - E_R)^2}{4E_R} \tag{20.41}$$

and

$$E_R \equiv \frac{1}{2}\omega_{vib}^2(\tilde{R}^{(A)} - \tilde{R}^{(D)})^2. \tag{20.42}$$

This relation defines the *reorganization energy* E_R. The physical meaning of this quantity may be clarified with the help of Figure 20.6(a). When the system changes from the donor to the acceptor state at $\tilde{R}^{(D)}$ as initial reaction coordinate value, a sudden rearrangement of the redox pair takes place. Subsequent relaxation or reorganization to the acceptor equilibrium coordinate $\tilde{R}^{(A)}$ releases an amount of energy equal to E_R. As one recasts Eq. (20.39) by use of $\Delta U^{(0)}$ and E_R, the rate constant adopts the more compact *Marcus form* [33],

$$\bar{k}_{ET} = V^2\sqrt{\frac{\pi}{\hbar^2 E_R k_B T}}\exp\left[-\frac{(\Delta U^{(0)} - E_R)^2}{4E_R k_B T}\right]. \tag{20.43}$$

The relation between the driving force and the reorganization energy provides a useful criterion for distinguishing different prototypes of electron transfer. This is illustrated in the three panels of Figure 20.6. In the *normal region*, as shown in Figure 20.6(a), E_R exceeds $\Delta U^{(0)}$. As, however, the acceptor parabola is lowered and thus $\Delta U^{(0)}$ increased, coincidence between $\Delta U^{(0)}$ and E_R can be achieved (Figure 20.6(b)). If this configuration is realized, no activation energy is necessary for the transfer to proceed. Enhancing $\Delta U^{(0)}$ beyond this margin will result in an increase of the activation energy. With this we enter the *inverted region*. A hallmark of this domain, as displayed in Figure 20.6(c), is the sizeable overlap between the donor and acceptor regimes. This suggests tunneling of the nuclear subsystem as

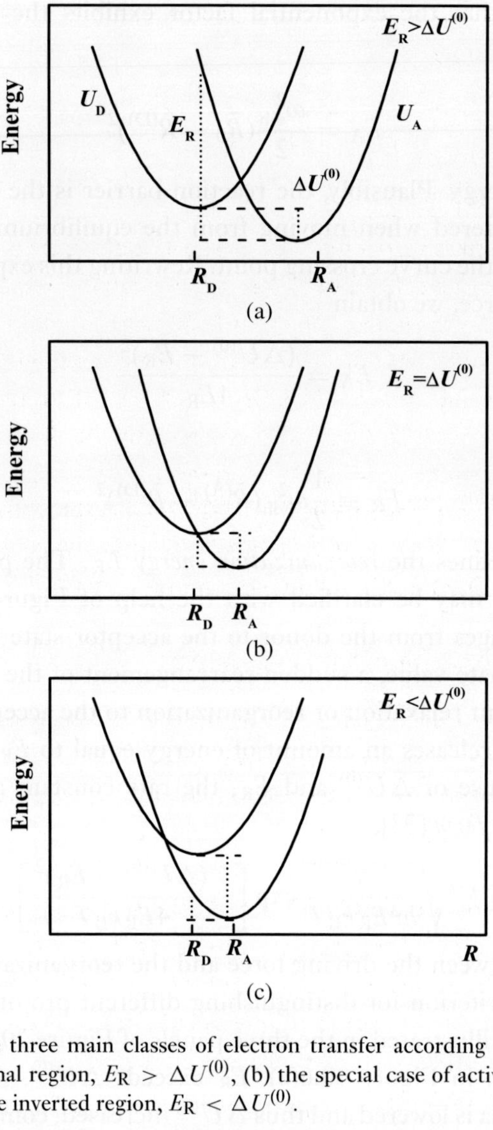

Figure 20.6: The three main classes of electron transfer according to classical Marcus theory: (a) the normal region, $E_R > \Delta U^{(0)}$, (b) the special case of activation-free transfer, $E_R = \Delta U^{(0)}$, (c) the inverted region, $E_R < \Delta U^{(0)}$.

a competitive transfer mechanism, implying that the classical treatment of the nuclear degrees of freedom, as adopted in our discussion so far, is likely to yield an insufficient description of inverted processes.

Exercise 20.3. *Detailed balance between forward and backward reaction.* The Marcus formula (20.43) indicates the rate constant for electron transfer from the donor to the acceptor. If $\bar{k}_{\mathrm{ET}}$ (A $\rightarrow$ D) denotes the rate constant for the reverse reaction, show

$$\bar{k}_{\mathrm{ET}}(D \rightarrow A) = \exp\left(-\Delta U^{(0)}/k_{\mathrm{B}}T\right)\bar{k}_{\mathrm{ET}}(A \rightarrow D)$$

which states the principle of detailed balance for the electron transfer and the corresponding back reaction.

This treatment is restricted to the high-temperature limit $k_{\mathrm{B}}T \gg \hbar\omega_k$, justifying the use of a classical approach for the nuclear subsystem. For low or intermediate temperature, however, nuclear tunneling could become a competitive transfer channel. Moreover, this process is expected to play a prominent role in the inverted Marcus region, as shown by the earlier analysis. In general, we expect both mechanisms, transfer by thermal activation and nuclear tunneling, to coexist. By these arguments, a more comprehensive discussion of electron transfer has to include the quantum structure of the nuclear framework. This extension of the electron transfer theory presented so far will be outlined in the following section.

20.5 Inclusion of Nuclear Tunneling

The electronic matrix element V is understood as a *tunneling factor*, describing the probability of an electronic transition between donor and acceptor states. If the temperature is low enough, however, *nuclear tunneling* through the barriers separating the donor and acceptor parabolas becomes important as well.

The validity of the formalism introduced in the preceding paragraphs is limited to the high-temperature domain. To remove this restriction, we ask for a generalized rate expression that does justice to the quantum nature of the nuclei, and, by the same token, incorporates the vibrational spectrum of the transfer complex. On an elementary level, this is done by including nuclear wave functions $\chi_{\mathrm{D}m}$, $\chi_{\mathrm{A}n}$, related to the donor as well as the acceptor regime, into the golden rule integral (20.36). The extended formula for the electron transfer rate constant reads

$$\bar{k}_{\mathrm{ET}} = \frac{2\pi}{\hbar} \sum_{m=1,M;n=1,N} f(E_{\mathrm{D}m}) V^2 |\langle \chi_{\mathrm{D}m} | \chi_{\mathrm{A}n} \rangle|^2 \times \delta\left(U_{\mathrm{D}}^{(0)} + \sum_k \hbar m_k \omega_k - U_{\mathrm{A}}^{(0)} - \sum_k \hbar n_k \omega_k\right). \quad (20.44)$$

The scalar product in Eq. (20.44) involves integration over the nuclear coordinates. The Condon approximation used in this expression, namely the partitioning of the transition matrix element into an electronic and a nuclear factor, according to $\langle \chi_{\mathrm{D}m}|V|\chi_{\mathrm{A}n}\rangle \approx V\langle \chi_{\mathrm{D}m}|\chi_{\mathrm{A}n}\rangle$, is based on the diabatic character of the transfer process considered here, implying a weak dependence of the electronic wave function on the vibrational degrees of freedom. Since we continue assuming that the frequencies ω_k are independent of the electronic states, the zero-point energies of the donor and the acceptor cancel out in the delta function argument.

We write the rate constant (20.44) in the more compact form

$$\bar{k}_{\mathrm{ET}} = \frac{2\pi}{\hbar} V^2 D\left(\frac{\Delta U^{(0)}}{\hbar}\right). \tag{20.45}$$

With the factor D, we have introduced a density-of-states function weighted by the energy-dependent thermal distribution function f as well as by the Franck–Condon factors, i.e. the square magnitudes of the nuclear overlap terms $\langle \chi_{\mathrm{D}m}|\chi_{\mathrm{A}n}\rangle$. From Eq. (20.44), D is given by

$$D\left(\frac{\Delta U^{(0)}}{\hbar}\right) = \sum_{n,m} f(E_{\mathrm{D}m})|\langle \chi_{\mathrm{D}m}|\chi_{\mathrm{A}n}\rangle|^2 \delta(\Delta U^{(0)} + \sum_k \hbar(m_k - n_k)\omega_k). \tag{20.46}$$

Both for the sake of physical insight and practical application we will cast this expression into a more general form. In particular, we will eliminate from Eq. (20.46) the explicit appearance of nuclear eigenfunctions which makes the formula virtually useless in the case of problems involving the condensed phase or larger molecular systems.

In a first step, the delta function is rewritten as a time integral, resulting in

$$D(\omega) = \frac{1}{2\pi\hbar} \sum_{n,m} \int_{-\infty}^{+\infty} dt\, f(E_{\mathrm{D}m})|\langle \chi_{\mathrm{D}m}|\chi_{\mathrm{A}n}\rangle|^2 \exp\left[i\left(\omega - \frac{E_{\mathrm{A}n} - E_{\mathrm{D}m}}{\hbar}\right)t\right] \tag{20.47}$$

with $\omega = \frac{\Delta U^{(0)}}{\hbar}$ and $E_{\mathrm{D}n} = \sum_k n_k \hbar\omega_k (E_{\mathrm{A}m} = \sum_k m_k \hbar\omega_k)$. Further, recalling that $\hat{H}_i \chi_{in} = E_{in}\chi_{in}$ $(i = \mathrm{D}, \mathrm{A})$, we remodel Eq. (20.47) with the aim to remove the dependence of this formula on the energy eigenvalues of the vibrational spectrum. To this end, we replace the energies E_{An} and E_{Dm} with the respective Hamiltonians $\hat{H}_A$ and $\hat{H}_D$. This succeeds as one realizes that

$$|\langle\chi_{Dm}|\chi_{An}\rangle|^2 \exp\left(\frac{i}{\hbar}(E_{Dm} - E_{An})t\right)$$
$$= \left\langle\chi_{Dm}\right| \exp\left(\frac{i}{\hbar}E_{Dm}t\right) \exp\left(-\frac{i}{\hbar}E_{An}t\right) \left.\chi_{An}\right\rangle \langle\chi_{An}|\chi_{Dm}\rangle$$
$$= \left\langle\chi_{Dm}\right| \exp\left(\frac{i}{\hbar}\hat{H}_D\right) \exp\left(-\frac{i}{\hbar}\hat{H}_A t\right) \left.\chi_{An}\right\rangle \langle\chi_{An}|\chi_{Dm}\rangle. \quad (20.48)$$

In addition, we introduce the equilibrium statistical operator $\hat{\rho}$ related to the donor state, and thus substitute for the thermal distribution function. This operator is defined as

$$\hat{\rho}_D = \frac{\exp(-\beta\hat{H}_D)}{Tr_{vib}\{\exp(-\beta\hat{H}_D)\}}. \quad (20.49)$$

The trace is performed with respect to the vibrational donor states. As Eq. (20.47) is combined with Eq. (20.48) and Eq. (20.49), the density of states turns into

$$D(\omega) = \frac{1}{2\pi\hbar}\sum_m \int dt \exp(i\omega t)\left\langle\chi_{Dm}\left|\hat{\rho}_D \exp\left(\frac{i}{\hbar}\hat{H}_D t\right)\right.\right.$$
$$\times \exp\left(-\frac{i}{\hbar}\hat{H}_A t\right)\left|\chi_{Dm}\right\rangle = \frac{1}{2\pi\hbar}\int dt$$
$$\times \exp(i\omega t)\mathrm{Tr}\left\{\hat{\rho}_D \exp\left(\frac{i}{\hbar}\hat{H}_D t\right) \exp\left(-\frac{i}{\hbar}\hat{H}_A t\right)\right\}, \quad (20.50)$$

where we have used closure with respect to the acceptor states. A generalization of the result (20.50) is obtained by use of relation (3.29) if $\exp(-\frac{i}{\hbar}\hat{H}_D t)$ is identified with the reference propagator $\hat{U}_0(t,0)$ and the perturbation Hamiltonian $\hat{H}'_I$ with the operator difference $\hat{H}_{DA} = \hat{H}_D - \hat{H}_A$. This implies for the propagator $\hat{U}_A \equiv \exp(-\frac{i}{\hbar}\hat{H}_A t)$

$$\hat{U}_A = \hat{U}_D(t,0)\hat{\hat{U}}_I(t,0) \quad (20.51)$$

and, by Eq. (3.27),

$$\hat{\hat{U}}_I(t,0) = \hat{T}\exp\left[-\frac{i}{\hbar}\int_0^t dt'\,\hat{H}_{DA}(t')\right]. \quad (20.52)$$

The density of states (20.50) thus acquires the form

$$D(\omega) = \frac{1}{2\pi\hbar}\int dt\,\exp(i\omega t)\mathrm{Tr}\left\{\hat{\rho}_{\mathrm{D}}\hat{T}\exp\left[-\frac{i}{\hbar}\int_0^t dt'\,\hat{H}_{\mathrm{DA}}(t')\right]\right\}. \tag{20.53}$$

This formula reduces the task of deriving the density of states for more complex physical situations to specifying the counterpart of the driving force $\Delta U^{(0)}$ and the perturbation $\hat{H}_{\mathrm{DA}}$. An example in the context of solvent effects on electron transfer is provided below in Section 20.6.4.

We remark that inserting Eq. (20.50) into Eq. (20.45) yields the rate expression (5.81) for the special case $\omega = 0$ and thus for vanishing driving force if the reactant is identified with the donor and the product with the acceptor species, and if the trace extends over the vibrational subsystem. The nuclear wave functions $\chi_{\mathrm{D}m}$ are represented by products of single-particle oscillator functions:

$$|\chi_{\mathrm{D}m}\rangle = \tilde{N}|\chi_{\mathrm{D}1}\rangle|\chi_{\mathrm{D}2}\rangle\cdots|\chi_{\mathrm{D}k}\rangle\cdots|\chi_{\mathrm{D}K}\rangle, \tag{20.54}$$

with $\tilde{N}$ as normalization factor and K as the number of vibrational degrees of freedom. Using the abbreviated form

$$T(t) \equiv \mathrm{Tr}\left\{\hat{\rho}_{\mathrm{D}}\exp\left(\frac{i}{\hbar}\hat{H}_{\mathrm{D}}t\right)\exp\left(-\frac{i}{\hbar}\hat{H}_{\mathrm{A}}t\right)\right\}, \tag{20.55}$$

we readily find that $T(t)$ separates into a product over the contributions from individual modes, according to

$$T(t) = \prod_{k=1,K} T_k(t). \tag{20.56}$$

After various manipulations that are outlined in Section 20.9.1 of this chapter, one establishes a relation between the single-mode traces and the Bose–Einstein distribution $f_{\mathrm{BE}}(\omega)$:

$$f_{\mathrm{BE}}(\omega) = \frac{1}{\exp\left(\frac{\hbar\omega}{k_{\mathrm{B}}T}\right) - 1}, \tag{20.57}$$

namely a representation of each factor $T_k(t)$ as [694]

$$T_k(t) = \exp\left[W_k(t)\right] \tag{20.58}$$

and

$$W_k(t) \equiv \frac{\Delta g_k^2}{2}[2(1 + f_{\mathrm{BE}}(\omega_k))\{\exp(-i\omega_k t) - 1\} + 2f_{\mathrm{BE}}(\omega_k)\{\exp(i\omega_k t) - 1\}] \tag{20.59}$$

where Δg_k denotes the difference $g_{\mathrm{D},k} - g_{\mathrm{A},k}$ between the electronic–vibronic coupling constants for the donor and the acceptor system as given by Eq. (20.23). With the help of a second auxiliary function, labeled $G_k(T)$ and defined as

$$G_k(t) \equiv \Delta g_k^2[\exp(-i\omega_k t)\{1 + f_{\mathrm{BE}}(\omega_k)\} + \exp(i\omega_k t) f_{\mathrm{BE}}(\omega_k)], \tag{20.60}$$

we write the result (20.59) more concisely as

$$W_k(t) = G_k(t) - G_k(0). \tag{20.61}$$

This relation is combined with the product representation (20.56) and Eq. (20.58) to yield

$$T(t) = \exp(G(t) - G(0)), \quad \text{and} \quad G(t) \equiv \sum_{k=1,K} G_k(t). \tag{20.62}$$

Summarizing the development from Eq. (20.45) to Eq. (20.62), we arrive at the following new formulation of the transfer rate constant:

$$\bar{k}_{\mathrm{ET}} = \frac{V^2}{\hbar^2} \int dt \, \exp\left[\frac{i}{\hbar} \Delta U_{\mathrm{D}}^{(0)} t + G(t) - G(0)\right]. \tag{20.63}$$

So far, an enumerable set of nuclear frequencies was assumed. Generalizing this approach, we may replace this set by a *spectral density function* that covers the cases of both discrete and continuous nuclear frequncies. The following subsection addresses this concept in more detail.

20.5.1 *The continuous limit of nuclear frequencies*

If the electron transfer proceeds in a large molecule or a condensed medium, the number of oscillators coupling to the electronic states becomes large. In this case, summation over the discrete modes implicit in Eq. (20.62) is to be replaced by integration. This is accomplished by introducing the *spectral density* distribution of the donor-acceptor pair.[2] For the presently discussed situation of a finite set of nuclear modes, the spectral density reads

$$\mathcal{J}_{\mathrm{DA}}(\omega) = \sum_{k=1,K} \Delta g_k^2 \delta(\omega - \omega_k). \tag{20.64}$$

We recognize in this expression the density of oscillator states weighted by an electronic–vibronic coupling factor. The function $G(t)$ is now conveniently written in integral form as

$$G(t) = \int_0^{\infty} d\omega [\exp(-i\omega t)\{1 + f_{\mathrm{BE}}(\omega)\} + \exp(i\omega t) f_{\mathrm{BE}}(\omega)] \mathcal{J}_{\mathrm{DA}}(\omega). \tag{20.65}$$

The spectral density may appear as a mere computational device. The product $\omega \mathcal{J}_{\mathrm{DA}}(\omega)$, however, carries essential information about the considered electron transfer process, as it is intimately related to the reorganization energy E_{R} defined by Eq. (20.42) for a single degree of freedom. To see this, we recall the definition Eq. (20.23) of the coupling factors g_k and rewrite the integral $\int_0^{\infty} d\omega\, \omega \mathcal{J}_{\mathrm{DA}}(\omega)$, to arrive at

$$\int_0^{\infty} d\omega\, \omega \mathcal{J}_{\mathrm{DA}}(\omega) = \frac{1}{2\hbar} \sum_k \omega_k^2 (\tilde{R}_k^{(\mathrm{D})} - \tilde{R}_k^{(\mathrm{A})})^2 = \frac{E_{\mathrm{R}}}{\hbar}. \tag{20.66}$$

Evidently, the integrand $\omega \mathcal{J}_{\mathrm{DA}}$ describes the distribution of all components that contribute to the reorganization energy as a function of the frequency. We point out at this place that a suitable change of assignments makes the formalism discussed here applicable to the treatment of optical transition phenomena. In the context of optical interactions, the donor and acceptor states of electron transfer have to be replaced by the ground and the excited state of a molecular species, respectively. Associating again the two contributing electronic states with K-dimensional parabolas of identical shapes, one infers that the rearrangement of the nuclear subsystem following vertical excitation will lead to an emitted line of lower frequency than the absorbed line. If the system returns to its ground state equilibrium, the overall process

[2] The spectral density, customarily written as $\mathcal{J}(\omega) = \frac{\pi}{2} \sum_i C_i \delta(\omega - \omega_i)$ [697], contains complete information about the influence of the environment on a given system, as long as the studied problem involves taking a thermal equilibrium average over the initial states of the environment and summing over the final states. The constant coefficients C_i are related to the coupling strength between the two sets of degrees of freedom.

involves two rearrangement stages, related to nuclear relaxation in the excited state following optical absorption and in the ground state following optical emission. The overall difference S between the absorption and the emission line therefore amounts to

$$2\hbar \int_0^\infty d\omega\, \omega \mathcal{J}_{\mathrm{GE}}(\omega) = S, \tag{20.67}$$

with G, E as labels for the ground and the excited state, respectively. This is the so-called *Stokes shift*, defined as frequency shift between the absorption and the emission line in optical interactions involving the transition from a ground to an excited state and the subsequent return of the system to the ground state equilibrium under release of radiation. By comparison of Eqs. (20.66) and (20.67), this quantity turns out to be the analog of the reorganization energy in optical excitation processes. Specifically, the Stokes shift may be understood as the sum of the ground and excited state reorganization energies.

As the transfer system is coupled to a macroscopic reservoir, a situation that will be studied later in further detail, the structure of the spectral density as given in Eq. (20.64) ceases to be practicable since the vibrational modes are not enumerable any longer. In this case, $\mathcal{J}_{\mathrm{DA}}(\omega)$ may take a variety of explicit forms according to the nature of the problem under consideration. Phenomenological modeling often operates with frequency laws of the type $\mathcal{J}_{\mathrm{DA}}(\omega) \sim \omega^P \exp(-\omega/\omega_{\mathrm{c}})$. A polynomial of degree P is multiplied by a factor that imposes exponential reduction of the spectral density beyond a cut-off frequency ω_c.[3] We will show in the following that the high-temperature limit of the electron transfer rate constant as discussed in Section 20.4 may be obtained from the more general formula (20.63) provided that

$$k_{\mathrm{B}}T \gg \hbar\omega_{\mathrm{c}}. \tag{20.68}$$

Consequently, this inequality may be read as the definition of *high temperature* for electron transfer problems. Subdividing the function $G(t)$ as given by Eq. (20.65) into its real and imaginary parts, we find

$$G(t) = \int_0^\infty d\omega \cos(\omega t)(1 + 2f_{BE})\mathcal{J}_{\mathrm{DA}}(\omega) - i\int_0^\infty d\omega \sin(\omega t)\mathcal{J}_{\mathrm{DA}}(\omega). \tag{20.69}$$

[3]The choice $P = 1$ defines the *Ohmic spectral density*.

Consider short times, satisfying the constraint $\omega_c|t| << \pi/2$. If this condition is met, $G(t)$ is dominated by the real part. We approximate the transfer rate constant (20.63) by

$$\bar{k}_{\mathrm{ET}} \sim \int dt \exp\left(\frac{i}{\hbar}\Delta U^{(0)}t\right)\exp\left(\Re\{G(t) - G(0)\}\right) \qquad (20.70)$$

and

$$\Re\{G(t) - G(0)\} = \int_0^\infty d\omega(\cos(\omega t) - 1)(1 + 2f_{\mathrm{BE}})\mathcal{J}_{\mathrm{DA}}(\omega). \qquad (20.71)$$

In the frequency integrand, the non-positive quantity $(\cos(\omega t) - 1)$ is multiplied by two positive-definite functions, resulting in a factor that falls off exponentially with time in the integrand of the right-hand expression of Eq. (20.70). Relaxing the time constraint but still assuming times short enough to justify the approximations $\sin(\omega t) \approx \omega t$ and $\cos(\omega t) \approx 1 - \frac{1}{2}(\omega t)^2$, we derive

$$G(t) - G(0) \approx -\int_0^\infty d\omega \frac{\omega}{\hbar} t^2 k_{\mathrm{B}}T\,\mathcal{J}_{\mathrm{DA}}(\omega) - i\int_0^\infty d\omega\,\omega t\,\mathcal{J}_{\mathrm{DA}}(\omega) \qquad (20.72)$$

where condition (20.68) has been used to replace $(1 + 2f_{\mathrm{BE}})$ with $\frac{2k_{\mathrm{B}}T}{\hbar\omega}$.[4] On account of Eq. (20.66), this relation may be expressed as

$$G(t) - G(0) \approx -\frac{E_{\mathrm{R}}}{\hbar^2}t^2 k_{\mathrm{B}}T - i\frac{E_{\mathrm{R}}}{\hbar}t. \qquad (20.73)$$

When inserted into the electron transfer rate formula (20.63), this approach yields indeed the high-temperature approximation (20.43) to $\bar{k}_{\mathrm{ET}}$, as is to be verified in the following exercise.

Exercise 20.4. Show that the approximation (20.73) leads to $\bar{k}_{\mathrm{ET}} = V^2\sqrt{\pi/(\hbar^2 E_{\mathrm{R}} k_{\mathrm{B}}T)}\exp[-(\Delta U^{(0)} - E_{\mathrm{R}})^2/(4E_{\mathrm{R}}k_{\mathrm{B}}T)]$, i.e. to the high-temperature rate expression Eq. (20.43), when combined with the identity (20.63).

20.6 Effects of Polar Solvents on Electron Transfer

In the preceding discussion, the electron transfer complex was described as a union of an electronic and a nuclear subsystem, with the latter acting

[4]This is valid if $\omega < \omega_c$. In the range $\omega > \omega_c$, $\mathcal{J}_{\mathrm{DA}}(\omega)$ falls off rapidly.

as a *bath* that consists of coupled vibrational degrees of freedom. Typically, however, an electron transfer reaction proceeds in a polar environment. The approach followed so far is of limited use for analyzing these reactions if the reservoir is a macroscopic solvent, since any attempt to include all potentially relevant coordinates of the surrounding medium into the computation of the rate constant would be futile. Many systems, however, have been successfully treated by simulating the background of the transfer process as a *dielectric continuum* (for instance [698], [699]). In the present section, we will introduce this model and outline some basic tenets of a theory of electron transfer occurring in polar environments.

In non-polar solvents, short-range interaction between the embedded complex and the background molecules prevails. In this case, the contribution of the solvent may be limited to the very first solvation shells, and in some situations, a consistently microscopic approach to capture the overall system combining the electron transfer unit and its medium may be feasible. If the solution is polar, in contrast, electrostatic interactions of long range play an essential role, making any model incomplete that restricts the reservoir to a few molecular layers surrounding the immersed redox pair. Ironically, the most common solvent, water, presents a particularly intricate case as the electron transfer process may have a sensitive impact on hydrogen bonds. Wherever these are involved, possible changes of the solvent bonding and related alterations of the solvent molecules should be considered. If these aspects can be neglected, the polar solution may be treated in the spirit of classical electrostatics.

The first two subsections address the solvent polarization induced by the presence of the solute. The electrostatic agent representing the latter is the dielectric displacement field **D** (Section 20.6.1), and the polarization **P** it causes in the solvent is mediated by the solvent polarizability, understood as the spatio-temporal correlation function (Section 20.6.2). The interaction between both **D** and **P** is captured by a *free energy functional*, to be introduced in Section 20.6.3. The material presented in Sections 20.6.1–20.6.3 sets the stage for deriving a rate constant formula within Marcus theory for electron exchange in a polar environment, as is done in Section 20.6.4. Remarkably, the result is seen to be strictly analogous to that obtained in the previous section for the case of a vibrational instead of a dielectric bath.

20.6.1 *The dielectric displacement field*

Both the reorganization energy and the driving force depend on the nature of the polar environment surrounding the solute. Reversing the direction of inquiry, we will first ask for the influence of the transfer complex on the medium. In this context, the most important among the conceptual tools provided by electrostatics is the *vector of polarization*, **P**. As in Section 18.1, we define this quantity as the volume density of the dipoles present in a dielectric substance, according to $\mathbf{P}(\mathbf{X}) = N_\mathrm{d}\langle\mathbf{d}\rangle(\mathbf{X})$, with N_d as the dipole number density and $\langle\mathbf{d}\rangle(\mathbf{X})$ as the dipole moment averaged over a small volume around the point **X**. The electric field associated with the solute has two main effects on the solvent: it distorts the distribution of the solvent electrons, and it changes the orientation of the permanent solvent dipoles. Correspondingly, the vector **P** is partitioned into two components:

$$\mathbf{P} = \mathbf{P}_\mathrm{e} + \mathbf{P}_\mathrm{o}, \tag{20.74}$$

denoting the *electronic* ($\mathbf{P}_\mathrm{e}$) and the *orientational* ($\mathbf{P}_\mathrm{o}$) polarization. Frequently, a hierarchy of time constants can be established between the processes that determine the polarization components $\mathbf{P}_\mathrm{e}$, $\mathbf{P}_\mathrm{o}$, and the electron transfer. The latter is usually faster than the reorientation of the solvent molecules and slower than the rapid motions of the solvent electrons. If this order of characteristic times is realized, the electronic subsystem and the orientational motion of the solvent can be assumed to be in equilibrium with each other as well as the vibrational motion of the solute.

In the dielectric continuum model the medium is represented by its polarization, understood as a vector field $\mathbf{P}(\mathbf{X})$ that assigns to any place **X** in the solvent a certain value of **P**. The domain of the independent variables of the diabatic Hamiltonians $\hat{H}_i$ is therefore augmented by the *external* coordinate $\mathbf{P}(\mathbf{X})$: $\hat{H}_i = \hat{H}_i(R_1, \ldots R_K, \mathbf{P}(\mathbf{X}))$ with R_k as the set of normal modes.

To specify the long range effects of the solute on the solvent, we may utilize the terminology of electric fields. The agent that governs the impact of the electron transfer system on its polar environment is, rather than the overall electric field **E**, the *dielectric displacement field* **D** caused by the transfer complex. To see this, we consider the Poisson equation for the composite of the solute and the solvent, namely

$$\frac{\partial}{\partial \mathbf{X}} \cdot \mathbf{E}(\mathbf{X}) = 4\pi(\rho_{\mathrm{P}} + \rho_{\mathrm{ext}}), \tag{20.75}$$

where $\mathbf{E}(\mathbf{X})$ is the macroscopic electric field at a place $\mathbf{X}$ in the medium, while ρ_{P} and ρ_{ext} stand for the dipole density and for the density of external charges, respectively. In the scenario described here, ρ_{ext} is related to the charge distribution of the solute. From the structure of the electrostatic potential created by the dipoles alone, one infers

$$\frac{\partial}{\partial \mathbf{X}} \cdot \mathbf{P}(\mathbf{X}) = -\rho_{\mathrm{P}}. \tag{20.76}$$

As the $\mathbf{D}$ field is composed of $\mathbf{E}$ and $\mathbf{P}$ according to $\mathbf{D} = \mathbf{E} + 4\pi\mathbf{P}$, Eq. (20.76) combined with Eq. (20.75) implies the Poisson equation

$$\frac{\partial}{\partial \mathbf{X}} \cdot \mathbf{D}(\mathbf{X}) = 4\pi\rho_{\mathrm{ext}} \tag{20.77}$$

which relates the dielectric displacement field $\mathbf{D}$ to the external charge density ρ_{ext}. Consequently, $\mathbf{D}$ adopts the role of a field related to a source externally added to the medium. In the further development, we will rely on the assumption that the response of the medium to the electric field is linear. This approximation is often warranted, provided the solvent is not dominated by phenomena involving strong dipolar coupling, such as ferroelectricity. If, in addition, the medium is isotropic, the macroscopic electric field and the polarization are related by a simple proportionality:

$$\mathbf{P} = \chi\mathbf{E}, \tag{20.78}$$

with the *linear electric susceptibility* χ, which in the general case is a tensor, reduced to a coefficient. This identity in conjunction with the relation between $\mathbf{D}$, $\mathbf{E}$, and $\mathbf{P}$ results in

$$\mathbf{D} = \varepsilon\mathbf{E}, \quad \text{and} \quad \varepsilon = 1 + 4\pi\chi, \tag{20.79}$$

which introduces the dielectric function ε. For simplicity, we use Gaussian units, such that the dielectric function of the vacuum is unity, $\varepsilon_{\mathrm{vac}} = 1$. As mentioned before, we distinguish between different frequency regimes that are characteristic of the response of the medium to an electric perturbation, motivating the partition (20.74). The quantities that determine the electric interaction between the solute and the solvent are thus described as functions of frequency. For monochromatic fields **E** and **D**, we write (e.g. [700])

$$\mathbf{D}(\mathbf{X},\omega) = \varepsilon(\omega)\mathbf{E}(\mathbf{X},\omega). \tag{20.80}$$

Since the right-hand side is the product of two frequency-dependent quantities, the relation between the electric field and the dielectric displacement field is non-local in time. To see this, we express **D**(**X**, t) as the Fourier transform of **D**(**X**, ω):

$$\begin{aligned}\mathbf{D}(\mathbf{X},t) &= \frac{1}{\sqrt{2\pi}}\int_{-\infty}^{\infty} d\omega\, \mathbf{D}(\mathbf{X},\omega)\exp(-i\omega t)\\ &= \frac{1}{\sqrt{2\pi}}\int_{-\infty}^{\infty} d\omega\, \varepsilon(\omega)\mathbf{E}(\mathbf{X},\omega)\exp(-i\omega t).\end{aligned} \tag{20.81}$$

Proceeding in the analogous way with **E**(**X**, ω) yields

$$\mathbf{D}(\mathbf{X},t) = \frac{1}{2\pi}\int_{-\infty}^{\infty} d\omega\, \varepsilon(\omega)\exp(-i\omega t)\int_{-\infty}^{\infty} dt' \exp(i\omega t')\mathbf{E}(\mathbf{X},t'). \tag{20.82}$$

Assuming that the integrations with respect to ω and t' are interchangeable, we find that Eq. (20.82) adopts the compact form

$$\mathbf{D}(\mathbf{X},t) = \int_{-\infty}^{\infty} dt' \tilde{\varepsilon}(t-t')\mathbf{E}(\mathbf{X},t') \tag{20.83}$$

with

$$\tilde{\varepsilon}(t) \equiv \delta(t) + 4\pi\tilde{\chi}(t) \tag{20.84}$$

and

$$\tilde{\chi}(t) = \frac{1}{8\pi^2}\int_{-\infty}^{\infty} d\omega[\varepsilon(\omega)-1]\exp(-i\omega t). \tag{20.85}$$

Exercise 20.5. Prove Eq. (20.83).

20.6.2 *Polarization and polarizability*

It is readily seen that the susceptibility $\tilde{\chi}(t)$ becomes proportional to $\delta(t)$, or local in time, if ε does not depend on the frequency. Conversely, the time non-locality of Eq. (20.83) is a consequence of the frequency variation of ε.

This feature is associated with the presence of two separate time regimes in the dielectric, reflecting the different behavior of the system components in response to the influence of the perturbing field. While the electronic polarization $\mathbf{P}_e$ adjusts instantaneously to the external influence, the orientational contribution $\mathbf{P}_o$ reacts much more slowly. In order to construct the temporal connection between the dielectric displacement field and the polarization we operate with the counterpart of Eq. (20.80) for the polarization and the dielectric displacement field:

$$\mathbf{P}(\mathbf{X}, \omega) = \alpha(\omega)\mathbf{D}(\mathbf{X}, \omega) \tag{20.86}$$

and

$$\alpha(\omega) \equiv \frac{1}{4\pi}(1 - \varepsilon^{-1}(\omega)), \tag{20.87}$$

and carry out the sequence of steps that led to Eq. (20.83), arriving at

$$\mathbf{P}(\mathbf{X}, t) = \frac{1}{4\pi}\mathbf{D}(\mathbf{X}, t) + \frac{1}{4\pi}\int_{-\infty}^{\infty} dt' \, \tilde{\varepsilon}^{-1}(t - t')\mathbf{D}(\mathbf{X}, t'). \tag{20.88}$$

The response of the polarization to the external field $\mathbf{D}$ is here described as the superposition of an instantaneous and a temporally non-local part. We mention that the causality constraint which restricts $t - t'$ to positive values is satisfied since the integrations involved in Eqs. (20.85) and (20.88), to be performed in the complex plane, give rise to factors of the form $\Theta(t - t')$ (e.g. [700]). The coefficient $\alpha(\omega)$ that determines both the strength and the time structure of the response to the dielectric displacement field is termed *polarizability*. From (20.88), one gathers that the polarizability in the time domain is

$$\tilde{\alpha}(t) = \frac{1}{4\pi}(\delta(t) + \tilde{\varepsilon}^{-1}(t)). \tag{20.89}$$

We interpret the field $\mathbf{P}(\mathbf{X}, t)$ in Eq. (20.88) as the expectation value of a polarization operator $\hat{\mathbf{P}}(\mathbf{X}, t)$. Further, we make admission for spatial besides temporal non-locality of the polarizability function, employing an analog of $\tilde{\alpha}$ that depends simultaneously on time and space coordinates and thus captures the effect of *spatial dispersion*:

$$\mathbf{P}(\mathbf{X}, t) \equiv \langle\hat{\mathbf{P}}(\mathbf{X}, t)\rangle = \int d\mathbf{X}' \int dt' \tilde{\alpha}(\mathbf{X}, t; \mathbf{X}', t')\mathbf{D}(\mathbf{X}', t'). \tag{20.90}$$

This formula is a linearized version of a more complex relation that may be inferred from Eq. (3.29). The polarizability $\tilde{\alpha}(\mathbf{X}, t; \mathbf{X}', t')$ has the status of a *linear response function*. It governs the time behavior of a system returning to equilibrium after a small perturbation has been applied. In the case at hand, involving the action of a solute and the reaction of its dielectric environment, the external field $\mathbf{D}$ causes a deviation from the equilibrium polarization that is, in most cases, sufficiently slight to justify a linear approach.

The explicit form of the polarizability is derived from the nonlinear generalization of (20.90). Using the partitioned propagator (3.29), we write the polarization expectation value as

$$\mathbf{P}(\mathbf{X}, t) = \mathrm{Tr}\{\hat{\rho}_{\mathrm{D}}^{\mathrm{eq}} \hat{U}_I^\dagger(t, t_0) \hat{U}_{\mathrm{solv}}^\dagger(t, t_0) \hat{\mathbf{P}}(\mathbf{X}) \hat{U}_{\mathrm{solv}}(t, t_0) \hat{U}_{\mathrm{I}}(t, t_0)\} \qquad (20.91)$$

with $\hat{U}_{\mathrm{I}}$ as defined in Eq. (3.27), involving a perturbation Hamiltonian $\Delta\hat{H}_{\mathrm{DA}} \equiv \hat{H}_{\mathrm{A}} - \hat{H}_{\mathrm{D}}$ that couples the field generated by the electron transfer complex to the solvent polarization induced by this field, and with $\hat{\rho}_{\mathrm{D}}^{\mathrm{eq}}$ as statistical operator, referring to the donor equilibrium state. The propagator $\hat{U}_{\mathrm{solv}}$ advances the free solvent through the time interval $[t_0, t]$. In the next step, we linearize Eq. (20.91) replacing $\hat{U}_{\mathrm{I}}(t, t_0)$ with $1 - \frac{i}{\hbar}\int_{t_0}^{t} dt' \Delta\hat{H}_{\mathrm{DA}}(t')$, and turn (20.91) into

$$\mathbf{P}(\mathbf{X}, t) = -\frac{i}{\hbar}\int_{t_0}^{t} dt' \, \mathrm{Tr}\{\hat{\rho}_{\mathrm{D}}^{\mathrm{eq}} \Delta\hat{H}_{\mathrm{DA}}(t') \hat{\mathbf{P}}(\mathbf{X}, t) - \hat{\mathbf{P}}(\mathbf{X}, t) \Delta\hat{H}_{\mathrm{DA}}(t')\}. \qquad (20.92)$$

This identity is valid under the premise that the equilibrium expectation value of the polarization vanishes, as is commonly assumed. At this place, we anticipate a result for $\Delta\hat{H}_{\mathrm{DA}}(t')$ which will be motivated in the following subsection (see Eq. (20.114) and the subsequent discussion), namely $\Delta\hat{H}_{\mathrm{DA}} = -\int d\mathbf{X}\, [\mathbf{D}_{\mathrm{A}}(\mathbf{X}, t) - \mathbf{D}_{\mathrm{D}}(\mathbf{X}, t)] \cdot \hat{\mathbf{P}}(\mathbf{X}, t)$. Substituting for $\Delta\hat{H}_{\mathrm{DA}}(t')$ and employing the more concise commutator nomenclature, we obtain

$$\mathbf{P}(\mathbf{X}, t) = -\frac{i}{\hbar}\int d\mathbf{X}' \int_{t_0}^{t} dt' \, \mathrm{Tr}\{\hat{\rho}_{\mathrm{D}}^{\mathrm{eq}} [\hat{\mathbf{P}}(\mathbf{X}, t), \hat{\mathbf{P}}(\mathbf{X}', t')] \Delta\mathbf{D}_{\mathrm{DA}}(\mathbf{X}', t')\}, \qquad (20.93)$$

where $\mathbf{D}_{\mathrm{DA}} \equiv \mathbf{D}_{\mathrm{A}} - \mathbf{D}_{\mathrm{D}}$. Comparing with Eq. (20.90), we conclude that the polarizability may be written as

$$\tilde{\alpha}(\mathbf{X}, t; \mathbf{X}', t') = -\frac{i}{\hbar}\Theta(t - t') \mathrm{Tr}\{\hat{\rho}_{\mathrm{D}}^{\mathrm{eq}} [\hat{\mathbf{P}}(\mathbf{X}, t), \hat{\mathbf{P}}(\mathbf{X}', t')]\}. \qquad (20.94)$$

Here we include a theta function that enforces causality. We add the still more compact formula

$$\tilde{\alpha}(\mathbf{X}, t; \mathbf{X}', t') = -\frac{i}{\hbar}\Theta(t - t')\tilde{C}^{-}_{PP}(\mathbf{X}, t; \mathbf{X}', t') \qquad (20.95)$$

which involves the antisymmetrized correlation function $\tilde{C}^{-}_{PP}(\mathbf{X}, t; \mathbf{X}', t') \equiv \tilde{C}_{PP}(\mathbf{X}, t; \mathbf{X}', t') - \tilde{C}^{*}_{PP}(\mathbf{X}, t; \mathbf{X}', t')$, and $\tilde{C}_{PP}(\mathbf{X}, t; \mathbf{X}', t') \equiv Tr\{\hat{\rho}^{eq}_{\mathrm{D}}\hat{\mathbf{P}}(\mathbf{X}, t)\hat{\mathbf{P}}(\mathbf{X}', t')\} = \langle\hat{\mathbf{P}}(\mathbf{X}, t)\hat{\mathbf{P}}(\mathbf{X}', t')\rangle_{\mathrm{D}}$. Note that $\tilde{C}^{-}_{PP}(\mathbf{X}, t; \mathbf{X}', t')$ is entirely imaginary. Neglect of spatial dispersion and anisotropy leads to the purely time-dependent correlation function $\tilde{C}_{PP}(t - t')$:

$$\tilde{C}_{PP}(\mathbf{X}, t; \mathbf{X}', t') = \delta(\mathbf{X} - \mathbf{X}')\tilde{C}_{PP}(t - t'). \qquad (20.96)$$

By arguments similar to those that established the identity (20.65), one may prove (see Exercise 20.6) the validity of the following representation of $\tilde{C}_{PP}(t)$:

$$\tilde{C}_{PP}(t) = \int_0^{\infty} \frac{d\omega}{2\pi}[\exp(-i\omega t)\{1 + f_{\mathrm{BE}}(\omega)\} + \exp(i\omega t) f_{\mathrm{BE}}(\omega)]C^{-}_{PP}(\omega) \qquad (20.97)$$

which involves the Fourier transform $C^{-}_{PP}(\omega)$ of $\tilde{C}^{-}_{PP}(t - t')$. The corresponding transform $\alpha(\omega)$ of the polarizability is related to $C^{-}_{PP}(\omega)$ in the following way:

$$\alpha(\omega) = \int d\omega' \frac{1}{2\pi\hbar}\frac{C^{-}_{PP}(\omega')}{\omega - \omega' + i\varepsilon}, \qquad (20.98)$$

as is readily derived by use of Eq. (20.95) in conjunction with the theta function transform, namely $-\int \frac{d\omega}{2\pi i}\frac{\exp(-i\omega t)}{\omega + i\varepsilon}$ where ε is an arbitrarily small quantity that by no means should be confused with the dielectric function. By Eq. (20.98), the imaginary part of the polarizability is directly linked to that of the dielectric function, according to

$$\Im\alpha(\omega) = -\frac{C^{-}_{PP}(\omega)}{2\hbar} = -\Im\frac{\varepsilon^{-1}(\omega)}{4\pi} = \Im\frac{\varepsilon(\omega)}{4\pi|\varepsilon(\omega)|^2}, \qquad (20.99)$$

where the last two expressions follow directly from Eq. (20.87).

Exercise 20.6. Retracing the steps that lead to Eq. (20.65), establish relation (20.97).

We are now in the position of writing the polarization function (20.97) in terms of the dielectric constant, obtaining

$$\tilde{C}_{PP}(t) = -\frac{\hbar}{4\pi^2}\int_0^{\infty} d\omega(\exp(-i\omega t)(1 + f_{\rm BE}(\omega)) + \exp(i\omega t)f_{\rm BE}(\omega))\,\Im\frac{\varepsilon(\omega)}{|\varepsilon(\omega)|^2}. \tag{20.100}$$

We add a note on the time dependence of the vector fields **E**, **D**, and **P**. Their variation with time may be characterized as *quasi-static* [701]. This is to say that they are longitudinal or irrotational, deriving from scalar potentials, and thus not giving rise to electromagnetic radiation. We will emphasize this feature of the three fields later when commenting on the generalization of potential energy to free energy surfaces. In the low-frequency domain, **P** contains both the electronic and the orientational contribution, while the latter is filtered out in the high-frequency regime. With $\mathbf{P}_{\rm s}$ as the polarization for small values of $\omega(\omega \to 0)$, and with reference to Eqs. (20.86) and (20.87), we write

$$\mathbf{P}_{\rm s} = \frac{1}{4\pi}\left(1 - \frac{1}{\varepsilon_{\rm s}}\right)\mathbf{D}, \tag{20.101}$$

where the *static dielectric function* $\varepsilon_{\rm s}$ is defined as $\lim_{\omega\to 0}\varepsilon(\omega)$. Likewise, the high-frequency limit $\mathbf{P}_\infty$ of the polarization is given by

$$\mathbf{P}_\infty = \frac{1}{4\pi}\left(1 - \frac{1}{\varepsilon_\infty}\right)\mathbf{D} \equiv \frac{1}{c_{\rm e}}\mathbf{D}, \tag{20.102}$$

with $\varepsilon_\infty = \lim_{\omega\to\infty}\varepsilon(\omega)$ which is known as the *optical dielectric function* and a constant $c_{\rm e}$ that has been introduced for convenience. Since we can identify $\mathbf{P}_\infty$ with the electronic admixture to the polarization, its orientational part in the region of low frequencies results simply as the difference between $\mathbf{P}_{\rm o}$ and $\mathbf{P}_\infty$:

$$\mathbf{P}_{\rm o} = \mathbf{P}_{\rm s} - \mathbf{P}_\infty = \frac{1}{4\pi}\left(\frac{1}{\varepsilon_\infty} - \frac{1}{\varepsilon_{\rm s}}\right)\mathbf{D} = \frac{1}{4\pi}c_{\rm P}\mathbf{D}. \tag{20.103}$$

The constant $c_{\rm P}$ introduced by (20.103) is the *Pekar factor* (e.g. [580]).

Exercise 20.7. *The relation between the polarizability and the correlation function* C_{PP}^-.

(a) Demonstrate the validity of Eq. (20.98).

(b) Acknowledging that $C^{-}_{PP}(\omega)$ is real-valued, and further taking the limit $\varepsilon \to 0$ while using the delta function representation $\delta(\omega - \omega') = \frac{1}{\pi}\frac{\varepsilon}{(\omega-\omega')^2+\varepsilon^2}$, conclude that (see Eq. (20.99))

$$\Im\alpha(\omega) = -\frac{C^{-}_{PP}(\omega)}{2\hbar}. \tag{20.104}$$

20.6.3 *The free energy functional*

Having defined the roles of the electrostatic fields **E**, **D**, and **P** in the problem of a solute immersed in a dielectric solvent, we ask now how to modify the electron transfer formalism outlined in the preceding sections if the transfer complex is immersed in a dielectric substance. Due to the macroscopic environment of the complex, the result for the rate constant, as given by Eq. (20.70), has to be extended to include the interaction between the solute and the solvent. As the energies of the two redox states and the dynamics of the transition between them are no longer determined by an enumerable set of vibrational degrees of freedom but by a continuous dielectric medium, the potential energy U employed in our discussion of electron transfer so far has to be augmented by the *free energy* F of the medium. By the *equilibrium free energy* of the dielectric we understand the work needed to move the solvent into the field **D** of the solute starting at an initial state of infinite separation between these two entities. At constant temperature, this work is reversible.

As mentioned before, the relevant fields **E**, **D**, and **P** are treated as quasi-static or longitudinal. Thus the **D** field as the external contribution to the overall electric field derives from the scalar potential φ_{DA} produced by the solute alone. With $\hat{\rho}_{\mathrm{DA}}$ as the charge density of the redox pair, one finds

$$\varphi_{\mathrm{DA}}(\mathbf{X}) = \int d\mathbf{X}' \frac{\rho_{\mathrm{DA}}(\mathbf{X}')}{|\mathbf{X} - \mathbf{X}'|}, \tag{20.105}$$

such that

$$\mathbf{D}(\mathbf{X}) = -\frac{\partial\varphi_{\mathrm{DA}}(\mathbf{X})}{\partial\mathbf{X}}. \tag{20.106}$$

The second component of the overall electric field is due to the dipole density ρ_{P} and its associated potential

$$\varphi_{\mathrm{P}}(\mathbf{X}) = \int d\mathbf{X}' \frac{\rho_{\mathrm{P}}(\mathbf{X}')}{|\mathbf{X} - \mathbf{X}'|} = -\int d\mathbf{X}' \frac{\frac{\partial}{\partial\mathbf{X}'}\cdot\mathbf{P}(\mathbf{X}')}{|\mathbf{X} - \mathbf{X}'|} \tag{20.107}$$

where Eq. (20.76) is used in the second step. Combining the three preceding equations, we obtain the overall electric field by taking the negative gradient of the potential $\varphi = \varphi_{\mathrm{DA}} + \varphi_{\mathrm{P}}$, according to

$$\mathbf{E}(\mathbf{X}) = -\frac{\partial \varphi(\mathbf{X})}{\partial \mathbf{X}} = -\frac{\partial}{\partial \mathbf{X}} \int d\mathbf{X}' \frac{\rho_{\mathrm{DA}}(\mathbf{X}') - \frac{\partial}{\partial \mathbf{X}'} \cdot \mathbf{P}(\mathbf{X}')}{|\mathbf{X} - \mathbf{X}'|}. \qquad (20.108)$$

Any *free energy functional* suitable for describing the combined system of solute and solvent must be compatible with this electric field expression. The latter statement is meant in a variational sense. Specifically, we ask for a term of the form $F[\hat{\mathbf{P}}]$, i.e. for the free energy understood as a functional of the polarization vector field as independent variable. The constraint (20.108) must result as a consequence of the variational condition $\delta F[\mathbf{P}]/\delta \mathbf{P}(\mathbf{X}, t) = 0$, specifying the polarization field that makes the free energy stationary with respect to $\mathbf{P}$. The appendix of this chapter (Section 20.9.2) provides a rationale for the following choice of an effective free energy functional that depends on the orientational polarization of the medium as well as the solute field:

$$F_{\mathrm{eff}}[\mathbf{P}'_{\mathrm{o}}, \mathbf{D}] = \langle \hat{H}_{\mathrm{DA}} \rangle - \frac{1}{2c_{\mathrm{e}}} \int d\mathbf{X}\, \mathbf{D}^2(\mathbf{X}, t) + \frac{2\pi}{c_{\mathrm{P}}} \int d\mathbf{X}\, \mathbf{P}'^2_{\mathrm{o}}(\mathbf{X}, t)$$
$$- \int d\mathbf{X}\, \mathbf{D}(\mathbf{X}, t) \cdot \mathbf{P}'_{\mathrm{o}}(\mathbf{X}, t). \qquad (20.109)$$

In Eq. (20.109), $\langle \hat{H}_{\mathrm{DA}} \rangle$ stands for the energy expectation value of the solute, and $\mathbf{P}'_{\mathrm{o}}$ denotes the renormalized quantity $\mathbf{P}_{\mathrm{o}}/\varepsilon_{\infty}$.

In what follows, we will relate this result to our discussion of electron transfer in polar environments. This will proceed by substituting *free energy surfaces* $F_i[\mathbf{P}'_{\mathrm{o}}, \mathbf{D}]$, $i = \mathrm{D}, \mathrm{A}$, for the potential energy surfaces U_i, that have been used so far in our outline of electron transfer rate theory. In particular, we identify $\langle \hat{H}_{\mathrm{DA}} \rangle$ with the donor (acceptor) state energy E_{D} (E_{A}). Since the field of the solute changes depending on which of these two alternatives is realized, an index i with $i = \mathrm{D}, \mathrm{A}$ is attached to the $\mathbf{D}$ field. Our approach for the diabatic free energy surfaces therefore reads

$$F_i[\mathbf{P}'_{\mathrm{o}}, \mathbf{D}_i] = E_i - \frac{1}{2c_{\mathrm{e}}} \int d\mathbf{X}\, \mathbf{D}_i^2(\mathbf{X}) + \frac{2\pi}{c_{\mathrm{P}}} \int d\mathbf{X}\, \mathbf{P}'^2_{\mathrm{o}}(\mathbf{X})$$
$$- \int d\mathbf{X}\, \mathbf{D}_i(\mathbf{X}) \cdot \mathbf{P}'_{\mathrm{o}}(\mathbf{X}), \quad i = \mathrm{D}, \mathrm{A}. \qquad (20.110)$$

In the next step, we reduce the functional to a single argument, namely the solute-related field $\mathbf{D}$, by evaluating F_i at the equilibrium positions of $\mathbf{P}'_{oi}$. This is achieved by imposing the constraint $\delta F_i/\delta \mathbf{P}'_o = 0$ on the system, implementing a stationarity condition with respect to the orientational polarization. We obtain immediately

$$\mathbf{P}'^{[eq]}_{o,i} = \frac{c_P}{4\pi}\mathbf{D}_i \tag{20.111}$$

Using this identity to eliminate $\mathbf{P}'_{o,i}$ from Eq. (20.110) leads to the following simple form for the equilibrium free energy surfaces:

$$F_i[\mathbf{D}_i]^{[eq]} = E_i - \frac{1}{8\pi}\left(1 - \frac{1}{\varepsilon_s}\right)\int d\mathbf{X}\, \mathbf{D}_i^2(\mathbf{X})^2. \tag{20.112}$$

By Eq. (20.101), the electrostatic contribution to Eq. (20.112) reproduces the classical equilibrium expression for the energy of a dielectric with polarization $\mathbf{P}_s$, $\frac{1}{2}\int d\mathbf{X}\, \mathbf{D}(\mathbf{X}) \cdot \mathbf{P}_s(\mathbf{X})$.

From Eqs. (20.110)–(20.112), we find the solvent contribution to the reorganization energy E_R^{sol}. Making reference to the graphical representation of E_R in Figure 20.6, and translating from vibrational coordinates to polarization parameters, the calculation of E_R^{sol} amounts to comparing the acceptor free energies evaluated at the donor state equilibrium and at the acceptor equilibrium:

$$E_R^{sol} = F_A[\mathbf{P}'^{[eq]}_{oD}, \mathbf{D}_A] - F_A[\mathbf{P}'^{[eq]}_{oA}, \mathbf{D}_A] = \frac{c_P}{8\pi}\int d\mathbf{X}\, [\mathbf{D}_A(\mathbf{X}) - \mathbf{D}_D(\mathbf{X})]^2. \tag{20.113}$$

The reader may note the structural analogy between the term on the right-hand side and the reorganization energy $\frac{1}{2}\omega_{vib}^2(\tilde{R}^{(A)} - \tilde{R}^{(D)})^2$, as introduced in Eq. (20.42). The latter quantity acquires now the interpretation of a vibrational counterpart of the solvent reorganization energy, and will further be denoted by E_R^{vib}.

Exercise 20.8. Use Eqs. (20.110)–(20.112) to derive the solvent contribution to the rearrangement energy as given by Eq. (20.113).

The total reorganization energy results as the sum of both contributions, $E_R = E_R^{vib} + E_R^{sol}$. The overall driving force $\Delta F^{(0)}$ for an electron transfer reaction, to be compared with the gas phase expression $\Delta U^{(0)}$ in Eq. (20.38) is inferred from Eq. (20.110). We consider the free energy difference

$$F_{\mathrm{A}}[\mathbf{P}'_{\mathrm{o}}, \mathbf{D}_i] - F_{\mathrm{D}}[\mathbf{P}'_{\mathrm{o},i}, \mathbf{D}_i] = -\Delta F^{(0)} - \int d\mathbf{X}[\mathbf{D}_{\mathrm{A}}(\mathbf{X}) - \mathbf{D}_{\mathrm{D}}(\mathbf{X})] \cdot \mathbf{P}'_{\mathrm{o}}(\mathbf{X}) \equiv \Delta F_{\mathrm{DA}}[\mathbf{P}'_{\mathrm{o}}, \mathbf{D}_i] \quad (20.114)$$

where $i = D, A$ and $\Delta F^{(0)}$ is defined by

$$\Delta F^{(0)} = E_{\mathrm{D}} - E_{\mathrm{A}} - \frac{1}{2c_{\mathrm{e}}} \int d\mathbf{X}\, \mathbf{D}^2_{\mathrm{D}}(\mathbf{X}) - \mathbf{D}^2_{\mathrm{A}}(\mathbf{X}). \quad (20.115)$$

When turning in the following section to the rate constant for nonadiabatic electron transfer in polar solvents, $\bar{k}^{\mathrm{sol}}_{\mathrm{ET}}$, we will recognize the free energy difference in Eq. (20.114) as the analog of the difference of the gas phase Hamiltonians $\hat{H}_{\mathrm{D}}$ and $\hat{H}_{\mathrm{A}}$, accommodating a physical situation that includes a dielectric environment beyond the vibrational bath considered in Sections 20.4 and 20.5.

20.6.4 *The electron transfer rate in a polar environment*

All elements are in place for a suitable adjustment of the gas phase formula (20.45) to the case of a solute immersed in a dielectric environment. Evidently the driving force $\Delta U^{(0)}$, based on potential energy surfaces, must be replaced by $\Delta F^{(0)}$, which involves free energy surfaces. Thus

$$\bar{k}^{\mathrm{sol}}_{\mathrm{ET}} = \frac{2\pi}{\hbar} V^2 D\left(\frac{\Delta F^{(0)}}{\hbar}\right). \quad (20.116)$$

To derive an expression for the density of states for electron transfer in a solvent, the scheme (20.53) can be employed. Redefining the donor and acceptor Hamiltonians considered in Section 20.5, $\hat{H}_i$, $i = \mathrm{D}, \mathrm{A}$, we identify their counterparts for a polar environment by the operator analogs of the free energies, $F_i[\hat{\mathbf{P}}'_{\mathrm{o}}, i = \mathrm{D}, \mathrm{A}]$. This yields, by Eq. (20.114):

$$D\left(\frac{\Delta F^{(0)}}{\hbar}\right) = \frac{1}{2\pi\hbar} \int dt \exp\left(\frac{i}{\hbar}\Delta F^{(0)} t\right) \mathrm{Tr}\Big\{\hat{\rho}^{\mathrm{D}} \hat{T} \times \exp\left[\frac{i}{\hbar}\int_0^t dt' \int d\mathbf{X}'\{\mathbf{D}_{\mathrm{A}}(\mathbf{X}') - \mathbf{D}_{\mathrm{D}}(\mathbf{X}')\} \cdot \hat{\mathbf{P}}'_{\mathrm{o}}(\mathbf{X}', t')\right]\Big\}. \quad (20.117)$$

We point out that the perturbation Hamiltonian consists entirely of the polarization-dependent part of the difference (20.114). As in the parallel case of electron transfer in a vibrational bath (see Eq. (20.53)), the driving force term has been absorbed into a prefactor of the trace. Once the trace is known we are able to compute the electron transfer rate according to (20.116). In the following, we concentrate on this final step of our analysis. Recalling that the trace refers to the thermal average of the propagator $\hat{U}_{\rm I}(t, t_0)$ if the solute is in the donor state, we make use of the representation [702]

$$\begin{aligned}\mathrm{Tr}\{\rho^{\rm D}\hat{U}_{\rm I}(t,t_0)\} &= \left\langle \hat{T}\exp\left[-\frac{i}{\hbar}\int_{t_0}^{t} dt'\,\Delta\hat{H}^{\rm D}_{\rm DA}(t')\right]\right\rangle_{\rm D} \\ &= 1 - \frac{i}{\hbar}\int_{t_0}^{t} dt_1\,\langle\Delta\hat{H}^{\rm D}_{\rm DA}(t_1)\rangle_{\rm D} + \left(\frac{i}{\hbar}\right)^2\frac{1}{2}\int_{t_0}^{t} dt_1 \\ &\quad\times\int_{t_0}^{t_1} dt_2\,\langle\Delta\hat{H}^{\rm D}_{\rm DA}(t_1)\Delta\hat{H}^{\rm D}_{\rm DA}(t_2)\rangle_{\rm D} + \cdots, \end{aligned} \tag{20.118}$$

where $\Delta\hat{H}^{i}_{\rm DA} = \Delta F^{(0)} + \Delta F_{\rm DA}[\hat{\mathbf{P}}'_{{\rm o},i}, \mathbf{D}_i]$. In the next step, we equate this expression to a cumulant series (see Eq. (17.48)) which is given by the expansion of the function $\exp[-\sum_{n=1} W_n(t)]$ (see Section 20.9.3, 20.192) and truncate after the term with $n = 2$.

A development sketched in the appendix of this chapter (Section 20.9.3) leads to

$$D\left(\frac{\Delta F^{(0)}}{\hbar}\right) \approx \frac{1}{2\pi\hbar}\int dt\,\exp\left[\frac{i}{\hbar}\Delta F^{(0)}t - W_1(t) - W_2(t)\right], \tag{20.119}$$

with

$$\begin{aligned} W_1(t) &= -\frac{i}{\hbar}\int_{t_0}^{t} dt_1 \int d\mathbf{X}\,[\mathbf{D}_{\rm A}(\mathbf{X}) - \mathbf{D}_{\rm D}(\mathbf{X})]\cdot\langle\hat{\mathbf{P}}'_{\rm D}(\mathbf{X}, t_1)\rangle_{\rm D} \\ &= -\frac{i}{\hbar}\frac{c_{\rm P}}{4\pi}t\int d\mathbf{X}\,\Delta\mathbf{D}_{\rm DA}(\mathbf{X})\cdot\mathbf{D}_{\rm D}(\mathbf{X}), \end{aligned} \tag{20.120}$$

and

$$\begin{aligned} W_2(t) &= \left(\frac{1}{\hbar}\right)^2\int_{t_0}^{t} dt_1\int_{t_0}^{t_1} dt_2\int d\mathbf{X}_1\int d\mathbf{X}_2 \\ &\quad\times\Delta\mathbf{D}_{\rm DA}(\mathbf{X}_1)\tilde{C}_{PP}(\mathbf{X}_1, t_1; \mathbf{X}_2, t_2)\Delta\mathbf{D}_{\rm DA}(\mathbf{X}_2). \end{aligned} \tag{20.121}$$

The structure of $W_2(t)$ invites comparison with the function $G(t)$ as given by Eq. (20.65). To clarify the relation between these two entities, we define the spectral density of the solvent in the following way:

$$\mathcal{J}^{\text{sol}}(\omega) = \frac{2E_{\text{R}}^{\text{sol}}}{c_{\text{P}}\pi\hbar} \frac{1}{\omega^2} \frac{\Im\varepsilon(\omega)}{|\varepsilon(\omega)|^2}. \tag{20.122}$$

This is valid if the elementary *Debye form* of the dielectric function $\varepsilon(\omega)$ is adopted (see Exercise 20.9). To satisfy the solvent analog of Eq. (20.66), any physical choice for $\varepsilon(\omega)$ has to obey the condition

$$\int_0^\infty d\omega\, \hbar\omega \mathcal{J}^{\text{sol}}(\omega) = E_{\text{R}}^{\text{sol}}. \tag{20.123}$$

By solving the problem posed in Exercise 20.9, the reader may confirm that Eq. (20.123) is indeed valid if the Debye form of the dielectric function is used.

Exercise 20.9. Assume a simple dielectric function of the following (Debye) form:

$$\varepsilon(\omega) = \varepsilon_\infty + \frac{\varepsilon_{\text{S}} - \varepsilon_\infty}{1 + \tau_{\text{D}}\omega}.$$

Here τ_{D} is the so-called Debye relaxation time. Note that the static limit $\varepsilon(\omega) = \varepsilon_s$ is reached for $\omega = 0$, while $\lim_{\omega\to\infty} \varepsilon(\omega) = \varepsilon_\infty$. (a) Show:

$$\mathcal{J}^{\text{sol}}(\omega) = \frac{2E_{\text{R}}^{\text{sol}}}{\pi\hbar} \frac{\omega_{long}}{\omega} \frac{1}{\omega^2 + \omega_{long}^2}$$

if $\varepsilon(\omega)$ has the Debye form. The constant ω_{long} (longitudinal frequency) is defined as $\frac{\varepsilon_s}{\varepsilon_\infty}\frac{2\pi}{\tau_{\text{D}}}$. (b) Prove that the spectral density derived under (a) satisfies Eq. (20.123).

Following the arguments presented in Section 20.9.3, and taking advantage of Eqs. (20.122) and (20.123) when carrying out the integral in Eq. (20.201), we arrive at

$$W_2(t) = G(0) - G(t) - \frac{i}{\hbar} E_{\text{R}}^{\text{sol}} t, \tag{20.124}$$

where the function G is defined as in Eq. (20.65), with the solvent spectral density $\mathcal{J}^{\text{sol}}$ taking the place of $\mathcal{J}_{\text{DA}}$. We combine this finding with the result for $W_1(t)$ (see Section 20.9.3, Eq. (20.196)). Recalling further the expressions (20.115) and (20.113) for $\Delta F^{(0)}$ and $E_{\text{R}}^{\text{sol}}$, respectively, we calculate $\bar{k}_{\text{ET}}^{\text{sol}}$.

All ingredients needed to compute finally the exponential $\exp[\frac{i}{\hbar}\Delta F^{(0)}t - W_1(t) - W_2(t)]$ of Eq. (20.119) are assembled. We obtain

$$\frac{i}{\hbar}\Delta F^{(0)}t - W_1(t) - W_2(t) = \frac{i}{\hbar}\Delta F^{(\mathrm{eq})}t + G(t) - G(0). \qquad (20.125)$$

Here $\Delta F^{(eq)}$ refers to the difference between the free energies evaluated at the donor and acceptor equilibrium values of the polarization, as obtained from Eq. (20.112). In the final step the resulting density of states is inserted into the prescription (20.116) for the electron transfer rate in polar solvents to yield

$$\bar{k}_{\mathrm{ET}}^{\mathrm{sol}} = \frac{V^2}{\hbar^2}\int dt \exp\left[\frac{i}{\hbar}\Delta F^{(\mathrm{eq})}t + G(t) - G(0)\right]. \qquad (20.126)$$

The resulting structure of the rate constant formula is the same as that of its gas phase analog $\bar{k}_{\mathrm{ET}}$ (20.63), with the equilibrium free energy difference $\Delta F^{(\mathrm{eq})}$ replacing the vibrational driving force $\Delta U^{(0)}$. We have thus arrived at a method for computing the intramolecular electron transfer rate as a function of three system-specific ingredients of the Marcus model. Besides the free energy difference $\Delta F^{(\mathrm{eq})}$, these are the electronic coupling element V and the solvent reorganization energy $E_{\mathrm{R}}^{\mathrm{sol}}$ which determines the spectral density $\mathcal{J}^{\mathrm{sol}}$.

We underscore that the development of the electron transfer formalism in this and the preceding sections rests on the premise of nonadiabaticity. If the coupling as described by the element V becomes strong enough, the participating states decouple, and electron transfer proceeds on a well-defined potential energy surface.

The premise that the polar medium is appropriately represented by a dielectric continuum is borne out of the need to approximate the solvent by a macroscopic model, as a consistent description at the molecular level is usually too costly in terms of computational effort. A substantial body of theoretical work (e.g. [699], [703]–[705]), testifies to the adequacy of the continuum approach in a wide variety of applications. This success can be largely ascribed to the long-range nature of the electric force. This feature implies that electrostatic interactions with outer solvent layers tend to dominate chemical effects associated with reactions between the solute and the first solvation shell. A more detailed picture of the electron transfer process emerges, however, as the molecular structure of the environment is included at some level. This may be realized by classical molecular dynamics or combinations of classical and quantum procedures (QM/MM; see, for instance, [706]). The final section of this chapter presents electron transfer simulations that transcend the frame of the Marcus model and involve both classical and quantal representations of a molecular environment. Prior to this we focus in the following section on cases where the golden rule approach fails, involving electron transfer problems with strong coupling between reactant and product states. This is the challenge posed by *ultrafast electron transfer* processes.

20.7 Ultrafast Electron Transfer

The analysis of electron transfer presented earlier relies on a hierarchy of separable time regimes, where the time scale of the transfer process is understood to be distinctly slower than the times that govern the solvent relaxation. This distinction motivates treating the interaction between the donor and the acceptor state as a perturbation and thus using a golden rule expression to derive the rate constant for a given process. Electron transfer, however, may proceed from an excited donor level instead of the donor ground state which was assumed in the preceding discussion. If the reaction is initialized by photoexcitation [707, 708] it proceeds typically at a time scale of 10–100 fs [709], and the evolution of the transfer system is no longer governed by the relaxation dynamics of a medium, be it the vibrational subsystem or a polar solvent. This condition is realized in a variety of biological electron transfer complexes, such as photosynthetic reaction centers [710] or metallointercalators in DNA [711]. Besides donor excitation according to the scheme $D - A \rightarrow D^* - A \rightarrow D^+ - A^-$, photoinduced electron transfer may also involve promotion into an excited acceptor state, to be formally described by $D - A \rightarrow D - A^* \rightarrow D^+ - A^-$ [712]. The transfer Hamiltonian Eq. (20.15) has to be extended accordingly [713]:

$$\hat{H}_{\text{ET}} = \sum_{i=\text{D,A}} (\hat{H}_i + \Delta E_i)|\phi_i\rangle\langle\phi_i| + V_{\text{DA}}|\phi_{\text{D}}\rangle\langle\phi_{\text{A}}| + V_{\text{DA}}|\phi_{\text{A}}\rangle\langle\phi_{\text{D}}|, \tag{20.127}$$

with $\Delta E_i, i = \text{D, A}$ as electronic excitation energies. To address the problem of ultrafast electron transfer, reduced density matrix theory has been invoked, and in particular the Redfield formalism summarized in Section 5.5

[709, 713, 714]. This scheme treats the interaction between the solute and its environment at the second order of perturbation theory, but involves no perturbation treatment of the electron transfer element. We exemplify this situation by considering an electron transfer system coupled to a bath of harmonic oscillators, as described by the Hamiltonian

$$\hat{H}_{\mathrm{E}} = \sum_k \frac{\hat{P}_{Ek}^2}{2m_k} + \frac{m_k}{2}\omega_k^2 R_{Ek}^2 \tag{20.128}$$

where the index E distinguishes the environmental coordinates. A bilinear coupling Hamiltonian of the form

$$\hat{H}_{\mathrm{SE}} = R_{\mathrm{s}} \sum_k c_k R_{Ek} \tag{20.129}$$

is assumed, with R_{s} as a reaction coordinate of the electron transfer system. The corresponding spectral density is [122]

$$\mathcal{J}(\omega) = \frac{\pi}{2} \sum_k \frac{c_k^2}{m_k \omega_k} \delta(\omega - \omega_k). \tag{20.130}$$

In practice, one often operates with a continuous spectral density obtained by phenomenological modeling, as mentioned in Section 20.5.1. In the *Markovian case*, defined by frequency-independent damping [122], the spectral density is proportional to the frequency. This choice, however, implies divergence as $\omega \to \infty$. A way to address this difficulty is by allowing for a threshold frequency ω_{c} such that

$$\mathcal{J}(\omega) = \eta\omega \exp\left(-\frac{\omega}{\omega_{\mathrm{c}}}\right), \tag{20.131}$$

which realizes the Ohmic case (see the related discussion in Section 20.5.1). By Eqs. (20.130) and (20.131), the coupling strength parameter η is related to the coefficients c_k that determine the interaction Hamiltonian through [713]

$$\sum_k \frac{c_k^2}{2m_k\omega_k^2} = \int_0^\infty d\omega \frac{\mathcal{J}(\omega)}{\pi\omega} = \frac{\eta\omega_{\mathrm{c}}}{\pi}. \tag{20.132}$$

To derive the equation of motion (5.106) for the reduced density matrix $\tilde{\bar{\rho}}$ one has to evaluate the Redfield tensor (5.103) for the coupling operator given by Eq. (20.129). The result for the components Γ^+, Γ^- (Eqs. (5.104) and (5.105)) is [713]

$$\Gamma^+_{\lambda\nu\mu\kappa} = \frac{1}{\pi\hbar^2} \langle\lambda|R_s|\nu\rangle\langle\mu|R_s|\kappa\rangle \int_0^\infty d\tau\, G(\tau)\exp(-i\omega_{\mu\kappa}\tau) \tag{20.133}$$

and

$$\Gamma^-_{\lambda\nu\mu\kappa} = \frac{1}{\pi\hbar^2} \langle\lambda|R_s|\nu\rangle\langle\mu|R_s|\kappa\rangle \int_0^\infty d\tau\, G(\tau)\exp(-i\omega_{\lambda\nu}\tau), \tag{20.134}$$

where G is the function defined by Eq. (20.69). This quantity was seen earlier to determine the golden rule rate constants for an environment of harmonic oscillators (Eq. (20.70)) as well as a polar environment (Eq. (20.126)). Through the relations that tie rate constants to elements of the Redfield tensor, as discussed in Section 5.5, it turns out to dominate also rate theory in the framework of the Redfield approximation if bilinear coupling is assumed.

If the conditions for the use of Redfield theory, as specified in Section 5.5, are met, Eq. (5.106) may be used to obtain the donor and acceptor populations of interest, as well as the respective coherences. Representing the donor population with a function $P_D(t)$, one can then describe the electron transfer rate as the average life time of the transferred electron in the donor state, according to [694]

$$\bar{k}_{ET} = \left(\int_0^\infty dt\, P_D(t)\right)^{-1}. \tag{20.135}$$

Ultrafast electron transfer proceeding from molecular adsorbates to nanoparticle substrates has been an intense focus of recent research on electron exchange [715]. This interest is mostly motivated by the prospect of highly efficient novel nanodevices, to be used for instance in the areas of molecular electronics [716] or solar power technology [717]. In photovoltaic engineering dye molecules may be utilized as ultrafast electron donors following optical excitation, and the acceptor may be a nanocrystalline semiconductor complex [718]. Among the numerous challenges posed to theory by molecule–nanoparticle exchange systems is not only their size but also the electron transfer path that usually contains bridging units since the donor molecule is anchored in the nanoparticle surface by a connector species.

20.8 Case Study: Aqueous Ferrous–Ferric Exchange

In the final section of our account on electron transfer in condensed media we will address the challenge of simulating nonadiabatic electron transfer in a molecular environment described by a particle instead of a continuum model. We will emphasize the pioneering investigations by Chandler and coworkers [692,693] who in turn made reference to the preceding quantum mechanical approach by Newton *et al.* [719,720] when analyzing the *ferrous–ferric electron exchange*, i.e. the reaction $Fe^{2+} \leftarrow Fe^{3+}$, in water. These explorations were guided by the question of the nature of the transition state, the free energy barrier to overcome as the electron is transferred between spatially localized Fe(3d) levels, the transition mechanism, the impact of the solvent type (H_2O versus D_2O), and in particular the contribution of tunneling as compared to thermal activation to the exchange process. Our summary will focus on a Monte Carlo treatment of these issues (Section 20.8.1, [692]), followed by a path integral approach (Section 20.8.2, [693]), involving centroid dynamics and allowing for a greatly improved exposition of the quantum features inherent in the problem. Lastly, we mention recent attempts at consistent quantum-chemical solutions (Section 20.8.3).

20.8.1 *Monte Carlo modeling*

Making reference to the study of Kuharski *et al.* [692], we will outline a representation of the solvent by means of a statistical procedure, concentrating on the evaluation of the activation free energy that governs the exchange process. As elementary ingredients, the simulation employed a single electron, two Fe^{3+} ions and several hundred water molecules. For the solvent, a simple point charge description (SPC, [593]) was adopted, as each molecule was treated as a rigid classical entity. Semiempirical approaches were implemented to capture the various interactions present in the system as a whole, namely those between the H_2O constituents, between the electron and the solvent as well as the cationic centers, and between the latter and the solvent. Pair interaction potentials were constructed utilizing available data on the solvated ferrous–ferric composite for calibration. For instance, both Fe^{2+} and Fe^{3+} exist as regular octahedral complexes in an aqueous environment (see Figure 20.7), where the distance between the center and the ligand in $Fe^{2+}(H_2O)_6$ exceeds that in $Fe^{3+}(H_2O)_6$ by 0.13 Å. The effective potential

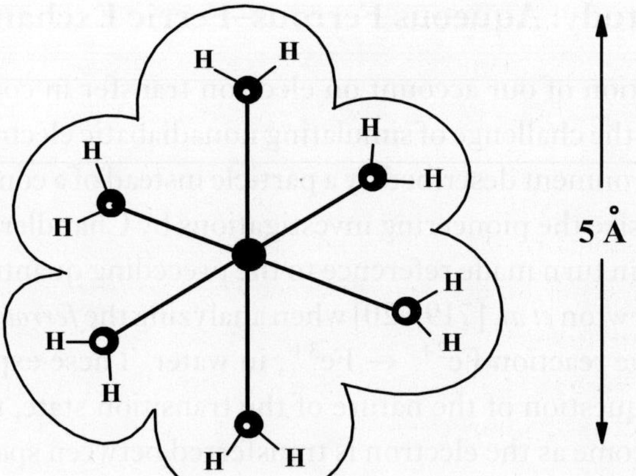

Figure 20.7: Scheme of an octahedral aqueous complex formed by an Fe^{2+} or Fe^{3+} ion. The first solvation shell is shown. The ligand distances are 2.0 Å and 2.1 Å for an Fe^{2+} and Fe^{3+} center, respectively.

used to describe the bonding between the Fe ions and the first solvent shell reproduces this finding.

The one-electron Schrödinger equation was solved to evaluate the tunneling splitting $2V$ (see earlier, Section 20.3, Figure 20.5) as a function of the distance between the iron centers by taking into account the two lowest energy states of the electron in the field of the ions. The regular arrangement of the first solvent shell determines the typical interionic separations. For distances smaller than 4.5 Å between the centers, these structures are disrupted. Separations in the range 5–7 Å are of maximum interest for this study since they are compatible with the constraint imposed by the ligand geometry and yet small enough to allow for electron transfer. In this range, V was seen to be the order of a few $k_B T$, suggesting a potentially sizeable nonadiabatic admixture to the considered exchange process. In a subsequent study [721], the tunneling splitting was reevaluated in the framework of a path integral approach.

A tight binding scheme, involving strongly localized electrons (see the Hamiltonian (20.7)), was employed to compute the free energy difference, or reversible work, required to establish degeneracy between the two redox sites, as the condition for electron transfer. The solute–solvent configurations of relevance for this process were characterized by two elementary parameters, labeled $\mathbf{R}^N$ and z. The first symbol refers to the coordinates of all H_2O molecules in the system, the second to the fraction of electronic charge transferred between the two centers. Specifically, the limiting donor and acceptor states Fe^{2+}+Fe^{3+} and Fe^{3+}+Fe^{2+} correspond to $z = 0$ and $z = 1$, respectively. A general, intermediate charge state is thus written as $Fe^{(2+z)+}$+$Fe^{(3-z)+}$, with $0 \leq z \leq 1$. We will also refer to the iron cores initially and finally occupied by the exchange electron as the left and the right center, respectively. The distance between the two iron ions was held fixed during each individual simulation. With these premises, the total potential energy of a configuration may be described by a function $V_z(\mathbf{R}^N)$, involving a fractional electronic charge $-(1 - z)e$ localized at the left center, and the remaining charge ze at the right.

In the case of resonance between the two redox partners, their potential energies are equal. This situation is described by $z = 0.5$, involving equal amounts of electronic charge on both centers. This degenerate case can be equivalently characterized in terms of the difference $\Delta\Phi$ between the electric potentials at the two iron centers as $\Delta\Phi = 0$. A partition function Z is introduced that involves the canonical distribution of $V_0(\mathbf{R}^N)$ at the equilibrium charge state $z = 0$:

$$Z(\Delta\Phi) \propto \mathrm{Tr}\{\exp[-\beta V_0(\mathbf{R}^N)]\delta(\Delta\Phi(\mathbf{R}^N) - \Delta\Phi)\}. \tag{20.136}$$

The delta function constrains Z to the argument $\Delta\Phi$. The trace is to be performed over the space $\mathbf{R}^N$. The ferrous–ferric exchange is activated by the amount of free energy ΔF that establishes degeneracy between the iron centers. This quantity is readily derived from Eq. (20.136) and given by the expression

$$\Delta F = -\frac{1}{\beta}\ln\left[\frac{Z(0)}{Z(\Delta\Phi_{\mathrm{min}})}\right], \tag{20.137}$$

where $-\frac{1}{\beta}\ln[Z(\Delta\Phi_{\mathrm{min}})]$ is understood as the free energy at the argument $\Delta\Phi$ that minimizes $-\frac{1}{\beta}\ln[Z(\Delta\Phi)]$ while $-\frac{1}{\beta}\ln[Z(0)]$ is that of the transition state, involving degenerate donor and acceptor levels.

To determine the right-hand side of Eq. (20.136), a Monte Carlo sampling scheme may be used. Kuharski *et al.* employed the Metropolis algorithm [722] to generate states with weights $\exp[-\beta V_0(\mathbf{R}^N)]$, as prescribed by Eq. (20.136). This technique involves a random walk in a configuration space where the individual steps are weighted in accordance with the statistics of the equilibrium canonical ensemble. This is done by implementing the following set of rules: At some time t_i of the simulation, the system may be in a configuration $\mathbf{R}^N(t_i) \equiv \mathbf{R}_i^N$. Transition from $\mathbf{R}_i^N$ to $\mathbf{R}_{i+1}^N$ is associated with a change ΔV_0 of the potential energy from $V_0(\mathbf{R}_i^N)$ to $V_0(\mathbf{R}_{i+1}^N)$. It depends on the sign of ΔV_0 if the step from $\mathbf{R}_i^N$ to $\mathbf{R}_{i+1}^N$ is made, or if the system remains in the previous state $\mathbf{R}_i^N$. In greater detail, the move is accepted if the potential energy change is zero or negative:

$$\mathbf{R}^N(t_i + \Delta t) = \mathbf{R}_{i+1}^N \quad \text{if } \Delta V_0 \leq 0 \tag{20.138}$$

with Δt as temporal step size. The procedure for the case of a positive potential energy change is summarized by declaring

$$\mathbf{R}^N(t_i + \Delta t) = \begin{cases} \mathbf{R}_{i+1}^N & \text{if } \exp(-\beta \Delta V_0) \geq y, \\ \mathbf{R}_i^N & \text{if } \exp(-\beta \Delta V_0) < y, \end{cases} \tag{20.139}$$

where it is assumed that $\Delta V_0 > 0$, and the symbol y stands for a random number chosen from the interval [0, 1]. If the Boltzmann weight $\exp(-\beta \Delta V_0)$ is larger or equal to this number, the transition takes place. If it is smaller than y, the change is rejected and the system persists in its former state. Starting from a random initial state and applying this procedure repeatedly will generate a Monte Carlo trajectory as a sequence of configurations that are sampled in keeping with the Boltzmann canonical distribution. From these trajectories, one acquires the likelihood for the occurrence of configurations satisfying $\Delta\Phi(\mathbf{R}^N) = \Delta\Phi$, which is the meaning of the expression $\mathrm{Tr}[\exp(-\beta V_0(\mathbf{R}^N))\delta(\Delta\Phi(\mathbf{R}^N) - \Delta\Phi)]$ in Eq. (20.136).

It should be noted, however, that the free energy difference ΔF in Eq. (20.137) involves the term $Z(0)$ which refers to the transition state characterized by the condition $z = 0.5$. This state is high in potential energy. Consequently, the neighborhood of the configuration with $z = 0.5$, or $\Delta\Phi = 0$, will be visited very infrequently by the Monte Carlo trajectories, and the respective statistical yield will be poor. This concern, however, can be addressed by *non-Boltzmann sampling* [723]. The idea underlying this procedure is a modification of the sampling prescription implied by Eq. (20.136), as one makes admission for fictitious potentials based on electronic configurations far from the equilibrium. In our case, this is achieved by reformulating Eq. (20.136) in the following way:

$$\begin{aligned} Z(\Delta\Phi) \propto \mathrm{Tr}\{&\exp[-\beta V_z(\mathbf{R}^N)]\exp[-\beta V_0(\mathbf{R}^N) \\ &+ \beta V_z(\mathbf{R}^N)]\delta(\Delta\Phi(\mathbf{R}^N) - \Delta\Phi)\}. \end{aligned} \tag{20.140}$$

As $\Delta\Phi(\mathbf{R}^N)$ is understood as the voltage between the two atomic centers, we find, within the limits of the employed tight binding approximation, that

$$V_z(\mathbf{R}^N) = V_0(\mathbf{R}^N) + ze\Delta\Phi(\mathbf{R}^N) \qquad (20.141)$$

and conclude that

$$Z(\Delta\Phi) \propto \exp(\beta ze\Delta\Phi)\langle\delta(\Delta\Phi(\mathbf{R}^N) - \Delta\Phi)\rangle_z. \qquad (20.142)$$

The bracket refers to an ensemble average involving $V_z(\mathbf{R}^N)$ as total potential energy. Operating with Eq. (20.142) removes the reference to the equilibrium charge state $z = 0$ which is present in Eq. (20.136). The sampling method according to Eq. (20.142) may be based on canonical distributions for charge states with general z. In particular, as z approaches 0.5, a bias is created for configurations close to the transition state, greatly improving the statistical yield in this regime. To put the non-Boltzmann sampling scheme into practice, the electron is displaced from the initial configuration with $z = 0$ to the final state with $z = 0.5$, thus spanning the regime from the equilibrium state with maximum polarization to the transition state where the polarization is zero, in well-defined steps while allowance is made for nuclear equilibration at each stage. Another, physically more adequate path with the roles of the nuclear and electronic subsystems interchanged, involving an electron response to the primary change of the water molecule environment, could be chosen to model the charge transfer process. Since, however, the computation aims at determining the reversible work, any reversible path may, in principle, be selected.

Kuharski *et al.* [692] performed the non-Boltzmann technique for four different choices of the parameter z in the interval [0, 0.5]. The four disconnected sampling regimes generated by this procedure were joined smoothly to derive a continuous free energy function [723]. The result is shown in Figure 20.8 for the example of two Fe^{2+} centers kept at a fixed distance of 5.5 Å.

This computation involved a set of 430 water molecules at a temperature of 298 K. Periodic boundary conditions were imposed on the system as a whole. From the figure, it can be seen that the reversible work for the displacement of the transferred electron from one of the iron cores to the midpoint between them amounts to about 20 kcal/mol in the model used. This is the result for the electron transfer activation free energy. Its order

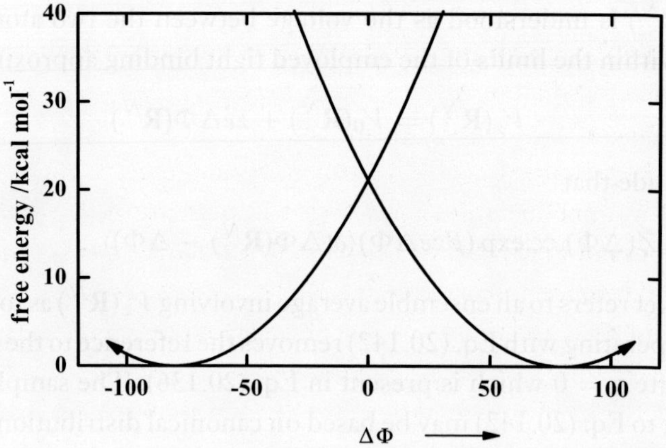

Figure 20.8: Diabatic free energy functions calculated according to Eq. (20.137) for aqueous ferrous–ferric electron transfer. The independent variable is the potential difference $\Delta\Phi$, plotted in units of $\frac{\text{kcal}}{\text{mol}\cdot e}$ (e = elementary charge), as solvent polarization coordinate. (Reprinted with permission from [692]. Copyright (2005) by the American Institute of Physics.)

of magnitude is in accord with an extrapolation from experimental results, around 15 kcal/mol ([724]–[726]). The curves describing ΔF as a function of $\Delta\Phi$ turn out to be parabolic to an excellent approximation [692]. This behavior was confirmed also for a larger distance between the Fe centers. It agrees with the model originally proposed by Marcus [33] which describes the environment of the electron transfer system as dielectric continuum in the framework of linear response theory. The bonds between the ions and the ligands of the first solvation shell are assumed to be harmonic. A quantum mechanical alternative to the calculation of the activation free energy presented here will be discussed in the following subsection along with consequences of this generalization for the exchange rate constant.

20.8.2 *Euclidean path integral simulations*

While an entirely classical description of the electron transfer environment, as considered in the previous subsection, is capable of yielding a realistic estimate of the free energy barrier separating the two terminal charge states of the exchange complex, quantitative modeling of the ferrous–ferric electron transfer rate requires a quantal treatment of the solvent degrees of freedom [693]. Specifically, the tunneling effect which was neglected in the simulation described earlier, turned out to enhance the rate constant very significantly. As tunneling is included, the solvent ensemble can reach the transition state configuration at classically forbidden energies below that of the diabatic crossing point. The free energy of activation is therefore effectively smaller in the quantum mechanical than in the classical case. The quantum calculations on the two iron cores interacting with a single electron yield a tunneling splitting $2V$ that is, at some $k_B T$, small enough to justify the use of Fermi's golden rule to describe the transfer process.

With reference to formula (5.83) for the rate constant, we ask for a stationary point of the time-dependent trace function $C(t)$ in order to determine the exchange rate. As implied by the structure of C (Eq. (5.81b)) and the symmetry of the electron transfer problem addressed here, this point occurs at the imaginary time $t^* = \frac{i}{2}\beta\hbar$ [727]. At this value C becomes $\mathrm{Tr}\{\exp[-\frac{\beta}{2}\hat{H}_{\mathrm{R}}]\exp[-\frac{\beta}{2}\hat{H}_{\mathrm{P}}]\}/Z_{\mathrm{R}}$. In the language of path integrals, the term $C(t^*)$ is expressed as the following ratio:

$$C(t^*) = \frac{\int DR(\tau)\exp(-\frac{1}{\hbar}S^*_{\mathrm{E}}[R(\tau)])}{\int DR(\tau)\exp(-\frac{1}{\hbar}S_{\mathrm{E,R}}[R(\tau)])}, \tag{20.143}$$

with $R(\tau)$ as a Euclidean path traced out in the solvent. By comparison with Eq. (5.81b), the denominator is identified as the partition function of the reactant side in path integral notation. The expression (20.143) may be evaluated as a ratio of the partition functions of interacting polymers, as described in Section 5.3.1. The symbol $S_{E,R}$ denotes the Euclidean action Eq. (5.42) for the system evolving in the diabatic state corresponding to the initial configuration Fe^{2+}+ Fe^{3+}. The respective electron path is

$$\mathbf{r}(\tau) = \mathbf{R}_{\mathrm{R}} \quad \text{if} \quad 0 \le \tau < \beta\hbar, \tag{20.144}$$

i.e. the electron does not leave the initially occupied iron center. The *transition state path* of the electron is, as a consequence of $t^* = \frac{i}{2}\beta\hbar$, equally shared between the two centers according to

$$\begin{aligned} \mathbf{r}(\tau) &= \mathbf{R}_{\mathrm{R}} \quad \text{if} \quad 0 \le \tau < \frac{\beta\hbar}{2}, \\ \mathbf{r}(\tau) &= \mathbf{R}_{\mathrm{P}} \quad \text{if} \quad \frac{\beta\hbar}{2} \le \tau < \beta\hbar. \end{aligned} \tag{20.145}$$

In more general terms, the transition from the reactant to the product will occur at some intermediate value of t which corresponds to the degeneracy between the donor and the acceptor center. From these considerations, it is plausible that $C(t^*)$ is related to the free energy needed to overcome the barrier between reactant and product. To substantiate this, we recall the relation between the partition function and the free energy, $F = -\frac{1}{\beta}\ln Z$, rewriting $C(t^*)$ as

$$C(t^*) = \frac{Z^*}{Z_R} = \exp\left[-\beta(F^* - F_R)\right]. \tag{20.146}$$

where Z^* stands for the transition state partition function in the numerator of Eq. (20.143), and F^* for the respective free energy. In what follows, we set F_R, the free energy of the stable reactant configuration equal to zero, such that F^* becomes the activation free energy. Identity (20.146) clarifies that the ratio $C(t^*)$ in the golden rule rate equation (5.83) plays the role of an Arrhenius factor. In terms of our path integral discussion in Section 5.4, Eq. (20.146) implies that $C(t^*)$ is determined by the expense of free energy required to move the centroid of the electron path from the reactant equilibrium to the transition state. Formally, Eq. (20.143) may be rewritten to express the activation free energy as

$$F^* = -\frac{1}{\beta}\int_0^1 d\lambda \frac{d}{d\lambda}\ln\int DR(\tau)\exp\left(-\frac{1}{\hbar}S_{E,\lambda}[R(\tau)]\right) \tag{20.147}$$

where we have introduced a real parameter λ that interpolates linearly between the actions S_R and S^* which bound the activation process. Correspondingly, S_λ is defined as

$$\begin{aligned} S_{E,\lambda} &\equiv S_{E,R} + \lambda(S_E^* - S_{E,R}) \\ &= S_{E,R} + \lambda\int_0^{\beta\hbar} d\tau\, z(\tau)(V_P[R(\tau)] - V_R[R(\tau)]). \end{aligned} \tag{20.148}$$

The difference $S_E^* - S_{E,R}$ cancels the kinetic energy component of the action. The factor $z(\tau)$ stands for the step function $\Theta(\tau - \frac{\beta}{2}\hbar)$. This Heaviside factor gives a description of the electron transition state path in Euclidean time, as defined by Eq. (20.145). For illustration of the result (20.148), let us inspect the case $\lambda = 1$. The corresponding Euclidean action turns out to be (compare with [727])

$$\begin{aligned} S_{E,1} = &\int_0^{\frac{\beta}{2}\hbar} d\tau \left\{\frac{m}{2}\left(\frac{dR}{d\tau}\right)^2 + V_R[R(\tau)]\right\} \\ &+ \int_{\frac{\beta}{2}\hbar}^{\beta\hbar} d\tau \left\{\frac{m}{2}\left(\frac{dR}{d\tau}\right)^2 + V_P[R(\tau)]\right\}. \end{aligned} \tag{20.149}$$

This is the transition state Euclidean action. If $\tau < \frac{\beta}{2}\hbar$ ($\frac{\beta}{2}\hbar \leq \tau < \beta\hbar$), all solvent paths that enter Eq. (20.147) are evaluated on the reactant (product) potential energy surface. The quantity λ, ranging from 0 to 1, may be viewed as a *charging parameter*: in the interval $\frac{\beta}{2}\hbar \leq \tau < \beta\hbar$, $S_{E,\lambda}$ refers to solvent interaction with a fictitious charge $-(1-\lambda)e$ on the left and $-\lambda e$ on the right Fe center, while the solvent interacts with the full electron charge on the left center during $0 \leq \frac{\beta\hbar}{2}$. An actual electron path is only realized at the end points $\lambda = 0$ and $\lambda = 1$. As in the classical treatment of the solvent summarized earlier, a tight binding approach is used to describe the redox sites, such that the difference $V_P - V_R$ coincides with $-e\Delta\Phi$.

The parametrization Eq. (20.148) makes it possible to compute explicitly the derivative with respect to λ contained in Eq. (20.147) and thus to arrive at a compact closed expression for F^*. We find

$$F^* = \int_0^1 d\lambda \int_0^{\beta\hbar} d\tau \frac{z(\tau)\Delta E_\lambda(\tau)}{\beta\hbar} \tag{20.150}$$

where the symbol $\Delta E_\lambda(\tau)$ denotes the average $\langle V_P[R(\tau)] - V_R[q(t)]\rangle_\lambda$. With this formula, a factorization of the integrand into electron and solvent paths has been achieved.

Exercise 20.10. *Activation free energy.*
(a) Derive Eq. (20.147) from Eq. (20.146), keeping in mind that $\lambda = 0$ ($\lambda = 1$) characterizes the reactant (transition) state. (b) For the extreme case $\lambda = 1$, verify that the Euclidean action $S_{E,\lambda}$ reduces to expression (20.149). (c) Use definition (20.148) to prove Eq. (20.150).

In the following step, we introduce a simple partition of F^* into a zero-frequency and a fluctuating, purely quantum mechanical component. Substituting for $z(\tau)$ and $\Delta E_\lambda(\tau)$ in Eq. (20.150) according to

$$\begin{aligned}\Delta E_\lambda(\tau) &= \Delta E_\lambda^{(0)} + \delta\Delta E_\lambda(\tau),\\ z(\tau) &= z^{(0)} + \delta z(\tau),\end{aligned} \tag{20.151}$$

we arrive at

$$\begin{aligned}F^* &= \frac{1}{\beta\hbar}\int_0^1 d\lambda \int_0^{\beta\hbar} d\tau\{z^{(0)}\Delta E_\lambda^{(0)} + \delta z \cdot \delta\Delta E_\lambda\}\\ &= F^{*(0)} + \delta F^*\end{aligned} \tag{20.152}$$

The relations (20.151) subdivide the electron and solvent paths into a centroid component (see Section 5.4) and deviations from this Euclidean time average. For ΔE_λ, the centroid is defined as

$$\Delta E_\lambda^{(0)} = \frac{1}{\hbar\beta} \int_0^{\hbar\beta} d\tau\, \Delta E_\lambda(\tau) \tag{20.153}$$

and accordingly for z.

Exercise 20.11. Show that F^* can be partioned into a centroid and a fluctuation part, as made explicit by Eq. (20.152). Use Eq. (20.151) and the definition of the centroid, Eq. (5.56).

At this juncture, a connection can be made to the classical solvent model assumed in the earlier analysis of the aqueous ferrous–ferric exchange problem [692], as summarized before. Although the statement $\delta F^* = 0$ is valid for a classical environment, it is in general not justified to equate $F^{*(0)}$ with the activation free energy obtained from the classical treatment of the solvent. This can be done, however, in the case of a quantum solvent that obeys Gaussian statistics and couples linearly to the electronic subsystem, as given by the spin-boson Hamiltonian described earlier in Section 20.2. In particular, we recall from Section 5.4 that for problems involving a harmonic potential the partition function can be separated into a product of a purely classical and a quantum contribution, $Z = Z_{\text{qu}} Z_{\text{cl}}$ (see Eq. (5.69) along with the related discussion), corresponding to $\ln Z = -\frac{1}{\beta}(F_{\text{cl}} + \delta F)$, with δF as quantum correction to the classical free energy. As indicated before, the assumption of a classical medium yields, to a very good approximation, harmonic profiles for the reactant and product free energy surfaces and, further, a realistic prediction of the activation energy. To the extent that the degrees of freedom of the corresponding quantum environment can also be regarded as harmonic, $F^{*(0)}$ is independent of quantum effects and δF^* represents the quantum correction to the activation free energy. This correction can be computed by exploiting the isomorphism between the Euclidean path integral and the partition function of a classical polymer, as outlined in Section 5.3.1. In the simulation reported in [693], each water molecule is described as rigid and thus endowed with six degrees of freedom. Its trajectory in Euclidean time is mapped on a sequence of polymer segments, or *beads*, to create an effective potential U (see Eq. (5.55)) that assigns a statistical weight to any possible polymer configuration. In this way one can generate the potential energy difference $\Delta E_\lambda(t)$ along an average path, or equivalently, an average configuration, for any choice of the charging parameter λ. This information yields the centroids of the products $z^{(0)}\Delta E_\lambda^{(0)}$ and $\delta z\delta\Delta E_\lambda$ by use of Eqs. (20.151) and (20.153), and both $F^{(0)}$ and $\delta F*$ are directly calculated from Eq. (20.152).

This scheme was applied by Bader and coworkers [693] who carried out Monte Carlo computations at room temperature for the aqueous ferrous–ferric exchange. The whole ensemble was equilibrated by molecular dynamics simulations preceding the statistical sampling. The integration over λ was achieved by interpolating both ΔE_λ and $\Delta E_\lambda^{(0)}$ linearly between the limiting points $\lambda = 0$ and $\lambda = 1$. In each case, ten initial configurations were included. As detailed in [693], the whole water environment was subdivided into different regions distinguished by the distances between the water molecules and the electron transfer system. The rigor of the quantum mechanical treatment varied with this parameter. Specifically, the extent of polymer discretization increased with the proximity of a region to the Fe ions. This procedure was used to study the electron exchange in both H_2O and D_2O.

As argued earlier, the term δF^* contains the essential information about the impact of quantum effects on the ferrous–ferric electron exchange. A direct comparison between the classical and the quantal transfer rate may be based on expression (5.83) for k. Since $C(t^*)$ has turned out to be an Arrhenius factor, the term multiplying it is interpreted as frequency factor ν. The ratio Γ between the quantum and the classical rate is seen to be

$$\Gamma = \frac{k}{k_{\text{cl}}} = \frac{\nu}{\nu_{\text{cl}}} \exp(-\beta\delta F^*) \approx \exp(-\beta\delta F^*). \tag{20.154}$$

From the work of Bader *et al.* [693], Γ is surprisingly large. It is found to be 65 +/− 6 for H_2O and 25 +/− 4 for D_2O, reflecting a major influence of tunneling on the electron transfer process. The ratio $\Gamma_{H_2O}/\Gamma_{D_2O}$ results as 2.6 +/− 0.5, in overall satisfactory agreement with experimental assessments of this quantity which vary from about 1.7 [728] to 2.0 [729]. The approximate unity of $\frac{\nu}{\nu_{\text{cl}}}$ is borne out by application of the harmonic model to the rate constants, as justified by the parabolic shape of the free energy curves shown in Figure 20.8. The solvent frequencies entering this model, or rather the spectral density of these frequencies, was obtained from a classical simulation [730] of solvent relaxation immediately following photoexcitation from the reactant equilibrium to the product state at the same configuration. Fourier transforming the relaxation curves yielded the spectral density.

The essential conclusion to be drawn from the path integral analysis of the aqueous ferrous–ferric exchange is that tunneling related to the solvent degrees of freedom increases the electron transfer rate constant dramatically. As far as the precise magnitude of the enhancement factor is concerned, the research summarized in this section provides an estimate rather than a definitive statement. We point out that a semiclassical implementation of the harmonic model which neglects the commutator $[\hat{H}_R, \hat{H}_P]$ led to a massively exaggerated enhancement on the order of 10^4 [731]. A consistent evaluation of k based on the spin-boson Hamiltonian with harmonic oscillator potentials for the solvent degrees of freedom [693], on the other hand, reduces $\Gamma(H_2O)$ to 41 +/− 2 while leaving $\Gamma(D_2O)$ nearly unaffected. In their reexamination of the ferrous–ferric exchange problem in H_2O, Song and Marcus [732] employed the same model but extracted the spectral density from experimental data on the dielectric dispersion of water, i.e. the frequency behavior of ε, the dielectric function. The spectral density for the solvent region beyond the first solvation shell was derived by use of a dielectric continuum approximation in contrast to the microscopic approach proposed by Bader *et al.* [693]. From this procedure, the authors arrived at a more moderate enhancement factor on the order of ten. They were able to show that the simple point charge (SPC) model of water, as utilized in [693], is unable to reproduce the experimental dielectric dispersion in the quantum mechanically relevant region $\beta\hbar\omega \gg 1$, calling for a more complete description of the exchange medium.

20.8.3 *Recent quantum dynamical extensions*

The mismatch between theory and experiment mentioned in the preceding paragraphs, and in particular the overestimation of the tunneling contribution to the ferrous–ferric electron transfer rate, has been blamed largely on a simplified description of the solvent, involving rigid water molecules. An explicit verification of this conjecture, however, comes at the expense of a full inclusion of both the electronic and the nuclear vibrational degrees of freedom into the representation of the solvent molecules, which presents a formidable computational challenge.

The recent increase in computational resources has allowed us to reexamine the aqueous electron transfer problem for a variety of metal complexes in the framework of Car–Parrinello molecular dynamics (CPMD, see Section 8.3.5) ([733]– [738]). This technique achieves a consistently quantum mechanical representation of all solute and solvent degrees of freedom involved in the process. The high computational expense associated with this methodology, however, limits the CPMD procedure with respect to system size and propagation times, such that convergence problems may arise [735]. These issues are alleviated by compromise solutions that mix quantum and classical mechanics (QM/MM, e.g. [706], [739]–[741]). Customarily, the solute is treated here at a high quantum mechanical level while the solvent, or a specified outer domain of the solvent, is described by classical mechanics or a semiempirical quantum method. Zeng *et al.* [742] combined the QM/MM approach with the use of a charging variable (termed λ in the preceding subsection) that defines an order parameter to distinguish between the limiting states of the electron transfer. The latter process is thus viewed as a displacement of the electron charge, understood as continuous, in successive stages from the left to the right metal center.

These studies, involving state-of-the-art simulation tools but still moving along the guidelines set by Marcus theory, have been extended from the ferrous–ferric electron transfer to a variety of other systems. With respect to the metal species, particular attention has been paid to the analogous exchange with ruthenium in the place of iron atoms (see [743, 744] and references therein). The solvent has been varied as well, opening new horizons to the analysis of electron transfer in condensed media. Thus, Rustad *et al.* [745] discuss the dependence of the free energy barrier on the pH value of the solution. For an adequate treatment of this issue, the relative weight of antagonistic factors, such as the stabilization of the Fe^{3+} center by OH^- ions versus the effect of Fe^{3+} hydrolysis, need to be assessed — intricate questions that nevertheless lie within the traditional scope of quantum chemistry.

20.9 Appendix: Formulae Relevant for Electron Transfer Theory within the Marcus Model

In the following we derive some essential results used in our presentation of the electron transfer rate constant formalism in Sections 20.5 and 20.6.

Specifically, Section 20.9.1 supports the discussion of electron transfer assisted by tunneling, as outlined in Section 20.5, while Sections 20.9.2 and 20.9.3 both refer to the problem of electron transfer in a polar environment, adding detail to Sections 20.6.3 and 20.6.4. Throughout this section, we follow largely the expositions given by May and Kühn [746].

20.9.1 *Electron transfer in a vibrational bath: Formal procedures used in the derivation of the rate constant*

We provide here a justification of Eq. (20.62), and thereby derive the form of the rate constant Eq. (20.63), starting with the trace expression Eq. (20.55). In the first step, we define the vibrational donor and acceptor Hamiltonians as

$$\hat{H}_{\rm D} = \hat{D}_{\rm D}^{\dagger}\hat{H}_{\rm vib}\hat{D}_{\rm D} \tag{20.155}$$

and

$$\hat{H}_{\rm A} = \hat{D}_{\rm A}^{\dagger}\hat{H}_{\rm vib}\hat{D}_{\rm A}. \tag{20.156}$$

Here $\hat{H}_{\rm vib}$ denotes the vibrational reference operator

$$\hat{H}_{\rm vib} = \sum_k \hbar\omega_k \hat{b}_k^{\dagger}\hat{b}_k, \tag{20.157}$$

while $\hat{D}_i, \hat{D}_i^{\dagger}, i = {\rm D, A}$ represent shift operators, satisfying

$$\hat{D}_i^{\dagger} = \exp\left[\sum_k g_{i,k}(\hat{b}_k - \hat{b}_k^{\dagger})\right], \tag{20.158}$$

with the factors $g_{i,k}$ as given by Eq. (20.23). The factor $T(t)$ (see Eq. (20.55)) can then be written as

$$T(t) = {\rm Tr}_{\rm vib}\Big\{\hat{D}_{\rm D}\hat{D}_{\rm D}^{\dagger}\hat{\rho}_{\rm vib}\hat{D}_{\rm D}\hat{D}_{\rm D}^{\dagger}\exp\left(\frac{i}{\hbar}\hat{H}_{\rm vib}t\right)\hat{D}_{\rm D}\hat{D}_{\rm A}^{\dagger} \times \exp\left(-\frac{i}{\hbar}\hat{H}_{\rm vib}t\right)\hat{D}_{\rm A}\hat{D}_{\rm D}^{\dagger}\Big\}, \tag{20.159}$$

as $\hat{\rho}_{\rm D}$ is identical with $\hat{D}_{\rm D}^{\dagger}\hat{\rho}_{\rm vib}\hat{D}_{\rm D}$. Using the abbreviation $\hat{D}_{DA} \equiv \hat{D}_{\rm D}\hat{D}_{\rm A}^{\dagger}$, Eq. (20.159) contracts to yield

$$T(t) = Tr_{\rm vib}\Big\{\hat{\rho}_{\rm vib}\exp\left(\frac{i}{\hbar}\hat{H}_{\rm vib}t\right)\hat{D}_{DA}\exp\left(-\frac{i}{\hbar}\hat{H}_{\rm vib}t\right)\hat{D}_{DA}^{\dagger}\Big\} \tag{20.160}$$

or shorter:

$$T(t) = Tr_{\text{vib}}\{\hat{\rho}_{\text{vib}}\hat{D}_{\text{DA}}(t)\hat{D}^{\dagger}_{\text{DA}}(0)\}. \tag{20.161}$$

Since $|\chi\rangle = |N_1\rangle|N_2\rangle\ldots|N_k\rangle\ldots|N_K\rangle$, with $|N_j\rangle$ as a harmonic oscillator eigenfunction associated with the jth vibrational mode, a factorial representation may be used to write the trace as (see Eq. (20.56))

$$Tr(t) = \prod_{k=1,K} T_k(t). \tag{20.162}$$

From the definition of the displacement operator by Eq. (20.158), one concludes:

$$\begin{aligned}\hat{D}_{\text{DA}}(t) &= \hat{D}_{\text{DA}}(g_{\text{D}} - g_{\text{A}}, t) \equiv \hat{D}(\Delta g, t) = \exp\left(\sum_k \Delta g_k[\hat{b}^{\dagger}_k(t) - \hat{b}_k(t)]\right)\\ &= \exp\left(\sum_k \Delta g_k[\exp(i\omega\hat{b}^{\dagger}_k\hat{b}_k t)\hat{b}^{\dagger}_k \exp(-i\omega\hat{b}^{\dagger}_k\hat{b}_k t)\right.\\ &\quad\left. - \exp(i\omega\hat{b}^{\dagger}_k\hat{b}_k t)\hat{b}_k \exp(-i\omega\hat{b}^{\dagger}_k\hat{b}_k t)]\right). \end{aligned} \tag{20.163}$$

For ease of notation, we have replaced here $\omega_{\text{vib},k}$ with ω. We note that the expression $\exp(i\omega\hat{b}^{\dagger}_k\hat{b}_k t)\hat{b}_k \exp(-i\omega\hat{b}^{\dagger}_k\hat{b}_k t)$ reduces to $\hat{b}_k \exp(-i\omega t)$. This follows from the definition of $\hat{b}_k$ as step-down operator for the kth vibrational mode, justifying the representation of $\hat{b}_k$ as $|N\rangle\langle N|\hat{b}_k|N+1\rangle\langle N+1|$ in a basis $\{|N\rangle\}$ of single-particle states for the kth mode, where the subscript k, as used before, has been suppressed. Accordingly, the displacement operator turns into

$$\hat{D}(\Delta g, t) = \exp\left[-\sum_k \Delta g(\hat{b}_k \exp(-i\omega t) - \hat{b}^{\dagger}_k \exp(i\omega t))\right]. \tag{20.164}$$

Any factor in the trace representation Eq. (20.162) may then be written as

$$T_k(t) = \frac{1}{\text{Tr}\left\{\exp\left(-\frac{\hat{H}_{\text{vib}}}{kT}\right)\right\}} \sum_N \left\langle N \left| \exp\left(-N\frac{\hbar\omega}{kT}\right) \hat{D}(\Delta g, t)\hat{D}^{\dagger}(\Delta g, t)\right| N\right\rangle. \tag{20.165}$$

In the following step, we inspect the structure of the scalar products contained in the sum on the right-hand side of Eq. (20.165). Introducing the label $S(N)$, we find

$$S(N) \equiv \langle N|\hat{D}(\Delta g, t)\hat{D}^{\dagger}(\Delta g, 0)|N\rangle$$
$$= \langle N| \exp[-\gamma(t)\hat{b} + \gamma^{*}(t)\hat{b}^{\dagger}] \exp[\gamma(0)\hat{b} - \gamma^{*}(0)\hat{b}^{\dagger}]|N\rangle, \quad (20.166)$$

with $\gamma(t) \equiv \Delta g \exp(-i\omega t)$. For further development, we employ the Baker–Campbell–Hausdorff formula to obtain

$$\exp(\alpha\hat{b} + \beta\hat{b}^{\dagger}) = \exp\left(-\frac{\alpha\beta}{2}\right)\exp(\alpha\hat{b})\exp(\beta\hat{b}^{\dagger})$$
$$= \exp\left(\frac{\alpha\beta}{2}\right)\exp(\beta\hat{b}^{\dagger})\exp(\alpha\hat{b}), \quad (20.167)$$

where α and β are arbitrary complex constants. We now cast $S(N)$ into the form

$$S(N) = \exp\left\{-\frac{1}{2}\{|\gamma(t)|^2 + |\gamma(0)|^2\}\right\}$$
$$\times\langle N| \exp[\gamma^{*}(t)\hat{b}^{\dagger}]\exp[-\gamma(t)\hat{b}]\exp[-\gamma^{*}(0)\hat{b}^{\dagger}]\exp[\gamma(0)\hat{b}]|N\rangle$$
$$= \exp\left\{-\frac{1}{2}\{|\gamma(t)|^2 + |\gamma(0)|^2\}\right\}\langle N| \exp[\gamma^{*}(t)\hat{b}^{\dagger}]$$
$$\times\exp[\gamma(t)\gamma(0)^{*}]\exp[-\gamma^{*}(0)\hat{b}^{\dagger}]\exp[-\gamma(t)\hat{b}]\exp[\gamma(0)\hat{b}]|N\rangle$$
$$= \exp\left\{-\frac{1}{2}\{|\gamma(t)|^2 + |\gamma(0)|^2 - 2\gamma(t)\gamma(0)^{*}\}\right\}\langle N|$$
$$\times\exp[(\gamma^{*}(t) - \gamma^{*}(0))\hat{b}^{\dagger}]\exp[-(\gamma(t) - \gamma(0))\hat{b}]|N\rangle. \quad (20.168)$$

Defining $\Delta\gamma \equiv \gamma(t) - \gamma(0)$, one finds

$$|\gamma(t)|^2 + |\gamma(0)|^2 - 2\gamma(t)\gamma(0)^{*} = |\Delta\gamma(t)|^2 - 2i\Im\{\gamma(t)\gamma(0)^{*}\}. \quad (20.169)$$

$S(N)$ simplifies considerably:

$$S(N) = \exp\left[-\frac{1}{2}|\Delta\gamma(t)|^2 - i\Im\{\gamma(t)^{*}\gamma(0)\}\right]$$
$$\times\langle N|\exp(\Delta\gamma^{*}\hat{b}^{\dagger})\exp(-\Delta\gamma\hat{b})|N\rangle. \quad (20.170)$$

The action of the operator $\exp(-\Delta\gamma(t)\hat{b})$ on a harmonic oscillator eigenfunction $|N\rangle$ is evaluated by Taylor expansion as

$$\exp[-\Delta\gamma\hat{b}]|N\rangle = \sum_{j=0}^{N}\frac{(-\Delta\gamma)^{j}}{j!}\sqrt{\frac{N!}{(N-j)!}}|N-j\rangle. \quad (20.171)$$

This result may be used to build a bridge between the overlap integral $\langle N|\exp(\Delta\gamma^*\hat{b}^\dagger)\exp(-\Delta\gamma\hat{b})|N\rangle$ in Eq. (20.170) and the Laguerre polynomials L_N (see, for instance, [731]):

$$\langle N|\exp(\Delta\gamma^* b^\dagger)\exp(-\Delta\gamma\hat{b})|N\rangle = L_N(y), \tag{20.172}$$

where y has been substituted for $|\Delta\gamma|^2$. The trace factor of Eq. (20.165), expressed as a function of y, becomes

$$T_k(t) = \left[1 - \exp\left(-\frac{\hbar\omega}{kT}\right)\right]\exp\left[-\frac{y}{2} - i\Im(\gamma(t)^*\gamma(0))\right] \times \sum_{j=0}^{\infty}\exp\left(-j\frac{\hbar\omega}{kT}\right)L_j(y). \tag{20.173}$$

The trace of $\exp(-\frac{\hat{H}_{\text{vib}}}{kT})$ has here been expressed as the limit of a geometric series, according to

$$\operatorname{Tr}\left\{\exp\left(-\frac{\hat{H}_{\text{vib}}}{kT}\right)\right\} = \sum_{j=0}\exp\left(-j\frac{\hbar\omega}{kT}\right) = \sum_{j=0}\left[\exp\left(-\frac{\hbar\omega}{kT}\right)\right]^j = \sum_{j=0} q^j = \frac{1}{1-q}, \tag{20.174}$$

with $q \equiv \exp(-\frac{\hbar\omega}{kT})$. We now use the relation between the Laguerre polynomials and their generating function, i.e.

$$\sum_{j=0}^{\infty} q^j L_j(y) = \frac{1}{1-q}\exp\left(-\frac{qy}{1-q}\right), \tag{20.175}$$

assuming $|q| < 1$. This condition is realized for the choice $q = \exp(-\frac{\hbar\omega}{kT})$. Inserting into Eq. (20.173) yields

$$T_k(t) = \exp[W_k(t)] \tag{20.176}$$

where

$$W_k(t) = -\frac{y}{2} - i\Im\{\gamma(t)^*\gamma(0)\} - yf_{\text{BE}}. \tag{20.177}$$

Here the Bose–Einstein distribution function f_{BE} has been substituted for $\frac{1}{\exp(\frac{\hbar\omega}{kT})-1}$. After reinserting $\Delta g\exp(-i\omega t)$ for γ in conjunction with the definition of the variable y, we find by some algebraic rearrangement

$$W_k(t) = \frac{\Delta g_k^2}{2}[2(1 + f_{\rm BE}(\omega_k))(\exp(-i\omega_k t) - 1) + 2f_{\rm BE}(\omega_k)(\exp(i\omega_k t) - 1)], \tag{20.178}$$

where we have reinstated the vibrational mode index k. A compact, convenient formulation of the trace is accomplished by introducing the function G:

$$G_k(t) \equiv \Delta g_k^2[\exp(-i\omega_k t)(1 + f_{\rm BE}(\omega_k)) + \exp(i\omega_k t)f_{\rm BE}(\omega_k)]. \tag{20.179}$$

Obviously,

$$W_k(t) = G_k(t) - G_k(0). \tag{20.180}$$

Finally, with

$$G(t) = \sum_k G_k(t), \tag{20.181}$$

the trace Eq. (20.162) is expressed in a simple and elegant manner as

$$T(t) = \exp[G(t) - G(0)], \tag{20.182}$$

asserting what is stated in Eq. (20.62).

20.9.2 *Derivation of the effective free energy functional Eq.* (20.109)

A free energy functional that depends on the solvent polarization, describing the electrostatic interaction between the solute and its polar environment, is given by [701]

$$F[\mathbf{P}] = \frac{1}{2}\int d\mathbf{X}\,\mathbf{P}(\mathbf{X},t)\int dt'\chi^{-1}(t-t')\mathbf{P}(\mathbf{X},t') + \frac{1}{2}\int d\mathbf{X}\int d\mathbf{X}'\,\frac{\frac{\partial}{\partial\mathbf{X}}\cdot\mathbf{P}(\mathbf{X},t)\,\frac{\partial}{\partial\mathbf{X}'}\cdot\mathbf{P}(\mathbf{X}',t)}{|\mathbf{X}-\mathbf{X}'|} - \int d\mathbf{X}\,\varphi_{\rm DA}(\mathbf{X},t)\frac{\partial}{\partial\mathbf{X}}\cdot\mathbf{P}(\mathbf{X},t). \tag{20.183}$$

This formula does not contain any surface effects, pertaining to the interface between the solute and the solvent. As all quantities combined in Eq. (20.183) are time-dependent, the functional refers to a non-equilibrium situation.

This is appropriate since the electron transfer may proceed at a time scale differing significantly from that typical for the evolution of the polarization present in the dielectric.

We calculate the functional derivative $\frac{\delta F[\mathbf{P}]}{\delta \mathbf{P}(\mathbf{X},t)}$, realizing that the functional F is of the form

$$F[\mathbf{P}] = \int d\mathbf{X} f(\mathbf{P}(\mathbf{X},t), \frac{\partial}{\partial \mathbf{X}} \cdot \mathbf{P}(\mathbf{X},t)). \tag{20.184}$$

A concise summary of the rules that govern the calculus of functional derivatives may, for instance, be found in [747] (see also the appendix of Chapter 1). Applying these rules, one derives the Euler–Lagrange equation

$$\frac{\delta F}{\delta \mathbf{P}(\mathbf{X},t)} = \frac{\partial f}{\partial \mathbf{P}(\mathbf{X},t)} - 2\frac{\partial}{\partial \mathbf{X}} \frac{\partial f}{\partial(\frac{\partial}{\partial \mathbf{X}} \cdot \mathbf{P}(\mathbf{X},t))} = 0. \tag{20.185}$$

This yields the statement (20.108), i.e. the correct expresssion for the overall electric field due to both the solute and the solvent.

Exercise 20.12. *The free energy functional.* (a) Calculate $\frac{\delta F}{\delta \mathbf{P}(\mathbf{X},t)}$ using the explicit form of the free energy functional as given by Eq. (20.183). Note that $\frac{\partial g(\mathbf{x})}{\partial g(\mathbf{y})} = \delta(\mathbf{x}-\mathbf{y})$ and $\frac{\partial g(\mathbf{x})/\partial \mathbf{x}}{\partial g(\mathbf{y})} = \frac{\partial \delta(\mathbf{x}-\mathbf{y})}{\partial \mathbf{x}}$. (b) Show that the condition $\frac{\delta F}{\delta \mathbf{P}(\mathbf{X},t)} = 0$ leads to the electric field (20.108). Use the relation between the quantities $\mathbf{P}$, $\mathbf{E}$, and χ .

With reference to Eq. (20.74), we split in the next step the polarization vector into an electronic and an orientational component. We then approximate the first term on the right-hand side of Eq. (20.183) by a time-localized approach, eliminating the function $\chi(t-t')$ in favor of the low- and high-frequency susceptibility limits $\lim_{\omega\to 0}\chi \equiv \chi_o$ and $\lim_{\omega\to\infty}\chi \equiv \chi_\infty$, respectively. Neglecting further the interference terms between $\mathbf{P}_e$ and $\mathbf{P}_o$ that emerge from the product $\mathbf{P}\cdot\mathbf{P}$ and oscillate rapidly with time (at a rate determined by the difference of the frequencies ω_e and ω_o characteristic of the time behavior of the electronic and the orientational component, respectively), we arrive at [703, 748]

$$\frac{1}{2}\int d\mathbf{X}\left(\frac{\mathbf{P}_e(\mathbf{X},t)^2}{\chi_o} + \frac{\mathbf{P}_o(\mathbf{X},t)^2}{\chi_\infty}\right). \tag{20.186}$$

With the help of formula (20.76), we recognize in the second term of (20.183) the dipole–dipole interaction. We rewrite this contribution to the free energy functional in the following form:

$$\frac{1}{2}\int d\mathbf{X}\int d\mathbf{X}'\cdot\mathbf{P}(\mathbf{X},t)\mathbf{L}(\mathbf{X}-\mathbf{X}')\mathbf{P}(\mathbf{X}',t), \qquad (20.187)$$

where the tensor $\mathbf{L}$ is given by $\mathbf{L}(\mathbf{X},\mathbf{X}')\equiv\frac{\partial}{\partial\mathbf{X}}\frac{\partial}{\partial\mathbf{X}'}\frac{1}{|\mathbf{X}-\mathbf{X}'|}$. As a conclusion from the Helmholtz decomposition theorem (e.g. [90]), this tensor represents the *longitudinal delta function* [749], projecting a vector field on its longitudinal component. Since all fields considered in the present context are assumed to be longitudinal, $L_{ij}(\mathbf{X},\mathbf{X}')$ reduces to $-4\pi\delta_{ij}\delta(\mathbf{X}-\mathbf{X}')$.[5] In summary, the free energy functional reads

$$\begin{aligned} F[\mathbf{P}_{\mathrm{e}},\mathbf{P}_{\mathrm{o}},\mathbf{D}] = \langle\hat{H}_{DA}\rangle &+ \frac{1}{2}\int d\mathbf{X}\left[\frac{\mathbf{P}_{\mathrm{e}}(\mathbf{X},t)^2}{\chi_\infty}+\frac{\mathbf{P}_{\mathrm{o}}(\mathbf{X},t)^2}{\chi_{\mathrm{o}}}\right] \\ &+ 2\pi\int d\mathbf{X}[\mathbf{P}_{\mathrm{e}}(\mathbf{X},t)+\mathbf{P}_{\mathrm{o}}(\mathbf{X},t)]^2 \\ &- \int d\mathbf{X}[\mathbf{P}_{\mathrm{e}}(\mathbf{X},t)+\mathbf{P}_{\mathrm{o}}(\mathbf{X},t)]\cdot\mathbf{D}(\mathbf{X},t), \end{aligned} \qquad (20.188)$$

where Eq. (20.106) has been utilized to include the dielectric displacement field as an independent variable besides the two polarizations $\mathbf{P}_{\mathrm{e}}$ and $\mathbf{P}_{\mathrm{o}}$. Also, we have added the contribution of the redox pair represented by the expectation value of the respective Hamiltonian. For further simplification, we assume that the electronic polarization reacts instantaneously to the field of the solute, corresponding to the premise that the time regimes of all three arguments, $\mathbf{P}_{\mathrm{e}}$, $\mathbf{P}_{\mathrm{o}}$ and $\mathbf{D}$ are well separated from each other. This implies that the system is in a state of equilibrium with respect to $\mathbf{P}_{\mathrm{e}}$, or $\partial F/\partial\mathbf{P}_{\mathrm{e}}=0$. It follows that

$$\mathbf{P}_{\mathrm{e}}=\frac{1}{c_{\mathrm{e}}}(\mathbf{D}-4\pi\mathbf{P}_{\mathrm{o}}) \qquad (20.189)$$

and $c_{\mathrm{e}}=4\pi/(1-\frac{1}{\varepsilon_\infty})$ (see Eq. (20.102)), where use has been made of $\varepsilon_\infty = 1+4\pi\chi_\infty$ (Eq. (20.79). This condition makes it possible to eliminate $\mathbf{P}_{\mathrm{e}}$

[5]For motivation of the additional factor 4π, consider the representation of the delta function $\Delta_{\mathbf{X}}\frac{1}{|\mathbf{X}-\mathbf{X}'|}=-4\pi\delta(\mathbf{X}-\mathbf{X}')$ as scalar counterpart of the tensor $\mathbf{L}$.

from the free energy functional, retaining $\mathbf{P}_o$ and $\mathbf{D}$ as variables. The result is

$$F_{\text{eff}}[\mathbf{P}'_o, \mathbf{D}] = \langle \hat{H}_{\text{DA}} \rangle - \frac{1}{2c_e} \int d\mathbf{X}\, \mathbf{D}^2(\mathbf{X}, t) + \frac{2\pi}{c_P} \int d\mathbf{X}\, \mathbf{P}'^2_o(\mathbf{X}, t) - \int d\mathbf{X}\, \mathbf{D}(\mathbf{X}, t) \cdot \mathbf{P}'_o(\mathbf{X}, t). \tag{20.190}$$

The Pekar factor c_P has been introduced by Eq. (20.103), and $\mathbf{P}'_o$ stands for the renormalized orientational polarization $\mathbf{P}_o/\varepsilon_\infty$.

20.9.3 *The density of states for electron transfer in a solvent: Calculating the trace Eq. (20.118)*

The trace that governs the density of states (20.117) is equated to a cumulant series with functions $W_1(t), W_2(t), \ldots$, according to

$$\left\langle \hat{T} \exp\left[-\frac{i}{\hbar} \int_{t_0}^{t} dt' \Delta \hat{H}^{\text{D}}_{\text{DA}}(t') \right] \right\rangle_{\text{D}} = \exp\left[-\sum_{n=1} W_n(t) \right]. \tag{20.191}$$

Expanding the right-hand side of this identity, we get

$$\begin{aligned} \exp\left[-\sum_{n=1} W_n(t) \right] &= 1 - (W_1(t) + W_2(t) + \cdots) \\ &\quad + \frac{1}{2}(W_1(t) + W_2(t) + \cdots)^2 + \cdots \\ &\approx 1 - W_1(t) - \left(W_2(t) - \frac{1}{2} W_1(t)^2 \right) - \cdots . \end{aligned} \tag{20.192}$$

In what follows, we will neglect in the exponential $\exp[-\sum_{n=1} W_n(t)]$ all W_n terms higher than $n = 2$ and ask for expressions that describe W_1 and W_2. As the exponential is truncated at $n = 2$, the density of states (20.117) is approximated by

$$D\left(\frac{\Delta F^{(0)}}{\hbar} \right) \approx \frac{1}{2\pi\hbar} \int dt \exp\left[\frac{i}{\hbar} \Delta F^{(0)} t - W_1(t) - W_2(t) \right]. \tag{20.193}$$

We will focus now on determining the exponents W_1 and W_2. From comparing (20.118) and (20.192), they are readily found to be

$$W_1(t) \equiv \frac{i}{\hbar}\int_{t_0}^{t} dt_1 \langle \Delta\hat{H}_{\mathrm{DA}}^{\mathrm{D}}(t_1)\rangle_{\mathrm{D}}, \tag{20.194}$$

$$W_2(t) \equiv \frac{1}{2}W_1(t)^2 + \frac{1}{\hbar^2}\int_{t_0}^{t} dt_1 \int_{t_0}^{t_1} dt_2 \langle \Delta\hat{H}_{\mathrm{DA}}^{\mathrm{D}}(t_1)\Delta\hat{H}_{\mathrm{DA}}^{\mathrm{D}}(t_2)\rangle_{\mathrm{D}}. \tag{20.195}$$

From definition (20.194), one finds

$$\begin{aligned} W_1(t) &= -\frac{i}{\hbar}\int_{t_0}^{t} dt_1 \int d\mathbf{X}\, [\mathbf{D}_{\mathrm{A}}(\mathbf{X}) - \mathbf{D}_{\mathrm{D}}(\mathbf{X})] \cdot \langle \hat{\mathbf{P}}_{\mathrm{D}}(\mathbf{X}, t_1)\rangle_{\mathrm{D}} \\ &= -\frac{i}{\hbar}\frac{c_{\mathrm{P}}}{4\pi} t \int d\mathbf{X}\, \Delta\mathbf{D}_{\mathrm{DA}}(\mathbf{X}) \cdot \mathbf{D}_{\mathrm{D}}(\mathbf{X}), \end{aligned} \tag{20.196}$$

where, in the first line, $\langle \hat{\mathbf{P}}_{\mathrm{D}}(\mathbf{X}, t_1)\rangle_{\mathrm{D}}$ has been replaced by the donor state equilibrium polarization according to Eq. (20.111). Here and in the following development we use $\hat{\mathbf{P}}_i$, i = D, A, instead of $\hat{\mathbf{P}}'_{oi}$, for ease of notation. One obtains for W_2,

$$\begin{aligned} W_2(t) = \frac{1}{2}W_1(t)^2 + \frac{1}{\hbar^2}\int_{t_0}^{t} dt_1 \int_{t_0}^{t_1} dt_2 \int d\mathbf{X}_1 \int d\mathbf{X}_2 \Delta\mathbf{D}_{\mathrm{DA}}(\mathbf{X}_1) \\ \times \langle \hat{\mathbf{P}}_{\mathrm{D}}(\mathbf{X}_1, t_1) \cdot \hat{\mathbf{P}}_{\mathrm{D}}(\mathbf{X}_2, t_2)\rangle_{\mathrm{D}} \Delta\mathbf{D}_{\mathrm{DA}}(\mathbf{X}_2). \end{aligned} \tag{20.197}$$

Substituting for $\hat{\mathbf{P}}_{\mathrm{D}}(\mathbf{X}, t)$ with $\mathbf{P}_{\mathrm{D}}^{[\mathrm{eq}]} + \Delta\hat{\mathbf{P}}_{\mathrm{D}}(\mathbf{X}, t)$ we cast W_2 into the form

$$\begin{aligned} W_2(t) = \left(\frac{1}{\hbar}\right)^2 \int_{t_0}^{t} dt_1 \int_{t_0}^{t_1} dt_2 \int d\mathbf{X}_1 \int d\mathbf{X}_2\, \Delta\mathbf{D}_{\mathrm{DA}}(\mathbf{X}_1) \\ \times \tilde{C}_{PP}(\mathbf{X}_1, t_1; \mathbf{X}_2, t_2)\Delta\mathbf{D}_{\mathrm{DA}}(\mathbf{X}_2), \end{aligned} \tag{20.198}$$

where the term $\frac{1}{2}W_1(t)^2$ is cancelled by the contribution of the equilibrium polarization $P_{\mathrm{D}}^{[\mathrm{eq}]}$ to W_2, and the function $\tilde{C}_{PP}$ is defined in analogy to Eq. (20.96), correlating the polarization fluctuations $\Delta\hat{\mathbf{P}}_{\mathrm{D}}$:

$$\begin{aligned} \tilde{C}_{PP}(\mathbf{X}_1, t_1; \mathbf{X}_2, t_2) &= \langle \Delta\hat{\mathbf{P}}_{\mathrm{D}}(\mathbf{X}_1, t_1) \cdot \Delta\hat{\mathbf{P}}_{\mathrm{D}}(\mathbf{X}_2, t_2)\rangle_{\mathrm{D}} \\ &= \delta(\mathbf{X}_1 - \mathbf{X}_2)\tilde{C}_{PP}(t_1 - t_2). \end{aligned} \tag{20.199}$$

As in formula (20.96), the assumption of vanishing spatial dispersion and anisotropy has been made. This approach leads to a simple relation between

W_2 and the solvent reorganization energy (compare with Eq. (20.113)), namely

$$W_2(t) = \frac{8\pi E_{\rm R}^{\rm sol}}{c_{\rm P}\hbar^2} \int_{t_0}^{t} dt_1 \int_{t_0}^{t_1} dt_2 \tilde{C}_{PP}(t_1 - t_2). \tag{20.200}$$

For further evaluation of W_2 we employ the structure of $\tilde{C}_{PP}(t_1 - t_2)$ as explicitly given by Eqs. (20.97). The time integrations in formula (20.200) can then be performed without any difficulty. This yields

$$W_2(t) = -\frac{2E_{\rm R}^{\rm sol}}{c_{\rm P}\pi\hbar} \int_0^{\infty} \frac{d\omega}{\omega^2}[(\exp(-i\omega t) - 1)(1 + df_{\rm BE}(\omega)) + (\exp(i\omega t) - 1)f_{\rm BE}(\omega) + i\omega t]\frac{\Im\varepsilon(\omega)}{|\varepsilon(\omega)|^2}. \tag{20.201}$$

Chapter 21

Electronic Friction in Molecule–Surface Interactions

Dynamic interactions between molecules and surfaces determine many phenomena of high systematic importance for molecular dynamics and practical relevance for its application to cases of technological interest [750]. These include scattering and diffusion, adsorption and desorption of molecules, besides surface-modifying processes such as corrosion and corrugation.

What impact do nonadiabatic processes have on the behavior of molecules in contact with solid substrates? This question, which is basic for any analysis of molecule–surface interactions in terms of electron dynamics, has been addressed by both experiment (e.g. [751]–[759]) and theory (e.g. [760]–[764]). Among the characteristic features to be explained is the observation that the energy transfer from molecular vibrational degrees of freedom to the surface is highly efficient in the case of a metal substrate while it is strongly suppressed if instead an insulator is used, as demonstrated by measurements involving vibrationally excited nitric oxide (NO) molecules impinging on an Au as opposed to an LiF surface [757, 764, 765].

An appropriate and much studied quantity for assessing the molecule–surface coupling mechanisms is the life time of an excited molecular vibration mode in various vicinities. For an isolated molecule, the elementary deexcitation process is the spontaneous emission of electromagnetic radiation, associated with time constants in the millisecond range. This changes as the molecule approaches a semiconductor or insulator surface. In this case, high-frequency molecular vibrations have been shown to couple to low-frequency phonons of the substrate, resulting in non-radiative energy transfer from adsorbate to surface degrees of freedom, and the lower limit of the characteristic life times sinks into the nanosecond regime (e.g. [762]).

Whenever the substrate is a metal, a variety of new and specific interaction modes appear. The extensively studied $\nu = 1$ state of CO on copper or platinum surfaces ([762], [766]–[769]) provides a prototypical manifestation of these effects. For this case, life times as low as picoseconds have been reported. These observations have been related to nonadiabatic relaxation mechanisms, correlating the vibronic deexcitation of the molecule with an electronic transition within the surface. In particular, a highly characteristic dependence of the deexcitation time on the distance between the molecule and the surface is found. This time is seen to drop drastically to the nanosecond regime and lower as the molecular adsorbate is located at distances ranging between ten and several hundred angstroms from the surface, documenting the onset of a new relaxation channel. In this region, the oscillating dipole associated with the molecular vibration gives rise to the creation of electron-hole pairs in the conduction band of the metallic substrate. This field-induced relaxation process is not influenced by any surface-adsorbate bonding features.

In the close vicinity of the substrate, where the molecule physically attaches to the surface, the electrodynamic deexcitation model overestimates the experimental life times by factors of five to ten ([770]–[772], [762]). One concludes from this failure that the specific features associated with chemisorption or physisorption are of importance in this regime. This situation calls for a comprehensive treatment of both the nuclear and the electronic degrees of freedom involved in the problem. For the particular case of CO on a metal surface, an alternative channel that involves the coupling of vibronic modes of the molecule to surface phonons has been shown to be a rather weak competitor [773].

In the following, we will introduce and discuss methodologies that are suitable to treat the interaction of a molecule with a periodic substrate. This problem differs categorically from that of two finite units in contact with each other, as exemplified by the intermolecular or atom-molecule problems considered so far in this text. In the latter cases, understanding nonadiabatic effects involves generally the notion of enumerable electronic states, and usually a selected number of well-defined relevant states that are coupled by the dynamically evolving nuclear degrees of freedom. As a molecule interacts with a surface, however, a continuum of electronic excitations takes the place of clearly separated electronic excited states. Correspondingly, various numerical techniques discussed in previous chapters for the treatment of nonadiabatic situations are not readily transferable to the problem of molecules in contact with periodic structures or, more generally, with infinitesimally close-lying potential energy surfaces. For instance, the trajectory surface hopping schemes introduced in Chapter 11 are based on the existence of a discrete manifold of potential energy surfaces that are available for trajectory propagation. This type of treatment is obviously not applicable to cases where energy bands are of relevance. Thus, any simulation method that relies on distinct surfaces will be inadequate for the study of dynamic phenomena associated with periodic systems.

A procedure that employs trajectories advancing under the influence of an averaged potential appears more appropriate to address the challenge posed by close-lying potential energy surfaces. Thus, Ehrenfest dynamics may provide a viable starting point for the development of a formalism adjusted to molecule–surface interactions. As mentioned in Chapter 10, the Ehrenfest approximation reaches its limits wherever it leads to unphysical averaging over qualitatively differing processes, such as reflection and adsorption. For potential energy surfaces of similar shape, however, as expected to arise from the one-particle excitation of a strongly delocalized conduction band electron [774] and especially in the extreme case where a set of parallel potential energy surfaces represents the physical situation reasonably well, use of the Ehrenfest methodology appears justified.

This theory may be employed to design an approach that responds to the particular demands of molecular interactions with surfaces. With reference to metal substrates which will be the focus of this chapter, energy transfer from the molecule to the surface results in multiple electron-hole pair creation processes within an energy continuum rather than single, well-characterized electronic excitation events. The less individual and more collective nature of this effect is in many cases adequately described by a *generalized Langevin model* [199, 775] which captures the essential features of the phenomenon in terms of statistical concepts. Novel paradigms that are not present in the theory of intermolecular processes as surveyed in the previous chapters emerge naturally from this treatment, such as *electronic friction* and *fluctuating forces*. This will be outlined in the following.

21.1 Langevin Formulation of Ehrenfest Dynamics

Reverting to the Ehrenfest methodology, we ask for an averaged, or "best", potential that is representative of the continuum of potential energy surfaces due to excitations of conduction band electrons. The nuclear degrees of freedom, represented here by the vibrational coordinates **X**, are described classically, corresponding to a trajectory that evolves under the influence of the effective potential. As in Section 10.1, the electronic wave function is described by the expansion

$$\Psi(\mathbf{x}, \mathbf{X}) = \sum_j c_j(t)\Phi_j(\mathbf{x}, \mathbf{X}), \tag{21.1}$$

with **x** as electronic coordinates. The effective potential is obtained from the expectation value

$$V = \mathbf{c}^\dagger \mathbf{H}_{\mathrm{el}}\mathbf{c}. \tag{21.2}$$

Inserting (21.1) into the TDSE, we find

$$i\hbar\frac{\partial \mathbf{c}}{\partial t} = \mathbf{H}_{\mathrm{el}}\mathbf{c} - i\hbar\sum_k \frac{\partial X_k}{\partial t}\mathbf{A}_k\mathbf{c} \tag{21.3}$$

where $\mathbf{A}_k$ is the usual derivative coupling matrix defined by Eq. (2.10). In order to deduce the force acting on the nuclei, one may use the conservation of the semiclassical total energy $E = T_{\mathrm{kin}} + V$. Setting the total time derivative of E to zero, we arrive at

$$\mu_k\frac{\partial^2 X_k}{\partial t^2} = -\mathbf{c}^\dagger\frac{\partial \mathbf{H}_{\mathrm{el}}}{\partial X_k}\mathbf{c} + \mathbf{c}^\dagger[\mathbf{H}_{\mathrm{el}}, \mathbf{A}_k]\mathbf{c}, \tag{21.4}$$

where μ_k stands for the reduced mass corresponding to the kth vibrational degree of freedom. In the following step, we introduce formally phase space variables for the electronic subsystem by mapping the complex amplitudes c_j on the pair of mutually conjugate coordinates (x_j, p_j), in analogy to Eq. (12.72) [774]:

$$c_j = x_j + ip_j. \tag{21.5}$$

Substituting for **c** in the TDSE (21.3), we botain the two coupled differential equations[1]

$$\hbar \frac{\partial \mathbf{x}}{\partial t} = \mathbf{H}_{\text{el}}\mathbf{p} - \hbar \sum_k \frac{\partial X_k}{\partial t} \mathbf{A}_k \mathbf{x} \tag{21.6}$$

and

$$\hbar \frac{\partial \mathbf{p}}{\partial t} = -\mathbf{H}_{\text{el}}\mathbf{x} - \hbar \sum_k \frac{\partial X_k}{\partial t} \mathbf{A}_k \mathbf{p}. \tag{21.7}$$

Whenever the derivative coupling matrices vanish, these two relations reduce to direct analogs of the classical Hamilton equations for coupled oscillators. Upon inserting Eq. (21.5) into Eq. (21.4), the latter equation turns into

$$\mu_k \frac{\partial^2 X_k}{\partial t^2} = \mathbf{x}^\dagger \left(-\frac{\partial \mathbf{H}_{\text{el}}}{\partial X_k} + [\mathbf{H}_{\text{el}}, \mathbf{A}_k] \right) \mathbf{x} + \mathbf{p}^\dagger \left(-\frac{\partial\, \mathbf{H}_{\text{el}}}{\partial X_k} + [\mathbf{H}_{\text{el}}, \mathbf{A}_k] \right) \mathbf{p}. \tag{21.8}$$

In a subsequent step, we reduce the number of variables by taking into account both the normalization and the overall phase of the electronic wave function. Thus, we select the phase angle obtained from the polar representation of the amplitude c_0 and map the electronic coordinates (x_j, p_j) onto the new pair (y_j, q_j) using the prescription

$$c_j \exp(-i\theta_0) = y_j + iq_j, \quad j \neq 0. \tag{21.9}$$

As q_0 is chosen as zero, y_0 is determined by the normalization constraint. Expressing the product $c_j^* c_0$ through both the old and the new set of variables, one readily shows that y_j and q_j depend on x_j and p_j in the following way:

$$y_j = y_0^{-1}(x_j x_0 + p_j p_0), \tag{21.10}$$

$$q_j = y_0^{-1}(p_j x_0 - x_j p_0), \tag{21.11}$$

and $y_0 = (x_0^2 + p_0^2)^{\frac{1}{2}}$. Differentiating Eqs. (21.10) and (21.11) with respect to time yields, in conjunction with Eqs. (21.6) and (21.7), we get

$$\frac{\partial y_j}{\partial t} = \omega_{j0} q_j - \sum_{k,m} \frac{\partial X_k}{\partial t} \left(A_{jm,k} y_m - A_{0m,k} \frac{q_m q_j}{y_0} \right), \tag{21.12}$$

$$\frac{\partial q_j}{\partial t} = -\omega_{j0} y_j - \sum_{k,m} \frac{\partial X_k}{\partial t} \left(A_{jm,k} q_m - A_{0m,k} \frac{q_m y_j}{y_0} \right). \tag{21.13}$$

[1] Note that the variables **x** and **p** are dimensionless.

In deriving Eqs. (21.12) and (21.13), the adiabatic representation has been adopted, i.e. $\hat{H}_{\rm el}$ has been assumed to be diagonal with elements E_j. Correspondingly, the frequencies ω_{jm} are defined as $(E_j - E_m)/\hbar$. We choose the adiabatic wave functions Φ_j as real, making the matrices $\mathbf{A}_k$ real-valued and antisymmetric. Expressed in the adiabatic frame and by means of the $\{y_j, q_j\}$ set of coordinates, the nuclear equation of motion (21.8) reads

$$\mu_k \frac{\partial^2 X_k}{\partial t^2} = -\sum_j \frac{\partial E_j}{\partial X_k}(y_j^2 + q_j^2) + \sum_{j,m} \hbar\omega_{jm} A_{jm,k}(y_j y_k + q_j q_k). \quad (21.14)$$

Under what conditions can the electron dynamics implicitly contained in Eq. (21.14) and explicitly in Eqs. (21.12) and (21.13) be absorbed into frictional and fluctuating force terms of the Langevin type?[2] By addressing this question, one might be able to replace the detailed dynamic interaction between electronic and nuclear coordinates according to Eq. (21.14) by an effective medium approach which can be characterized as *molecular dynamics with electronic friction*. For the case of linear coupling between the interacting subsystems, such a replacement can be performed exactly. The coupling between nuclear and electronic degrees of freedom in Eqs. (21.12)–(21.14), however, is quadratic. Consequently, Head-Gordon and Tully [774] investigated the question of Langevin substitution on the basis of a linearized version of these equations. Such a version may be generated by stipulating that the values adopted by the variables y_j, q_j $(j > 0)$, are small as compared to y_0:

$$y_0 = \left(1 - \sum_{j>0}(y_j^2 + q_j^2)\right)^{\frac{1}{2}} \cong 1, \quad y_j, q_j \ll 1, \quad j \neq 0. \quad (21.15)$$

To fix the initial values of the electronic degrees of freedom, we specify that only the state $\Phi_0(\mathbf{x}, \mathbf{X})$ is occupied at $t = 0$, such that

$$y_0(0) = 1, y_i(0) = 0 \quad \text{if } i > 0, \quad (21.16)$$

$$q_j(0) = 0 \quad \text{for all } j. \quad (21.17)$$

With these assignments, the condition (21.15) is valid for sufficiently short propagation times or weak nonadiabatic coupling. To simplify Eqs. (21.12)

---

[2] For surveys of generalized Langevin theory see ([776]–[778]).

and (21.13) that govern the electron dynamics, we let those elements of $\mathbf{A}_k$ that involve the state Φ_0, i.e. $A_{0m,k}$ and $A_{j0,k}$, by far outweigh all others, justifying the neglect of the latter. Up to linear order, Eqs. (21.12) and (21.13) then read

$$\frac{\partial y_j}{\partial t} = \omega_{j0} q_j - \sum_{k,m} \frac{\partial X_k}{\partial t} A_{j0,k}, \tag{21.18}$$

$$\frac{\partial q_j}{\partial t} = -\omega_{j0} y_j, \tag{21.19}$$

for $j > 0$. Under the same set of assumptions, the nuclear equation of motion (21.14) simplifies to

$$\mu_k \frac{\partial^2 X_k}{\partial t^2} = -\frac{\partial E}{\partial X_k} + \sum_j 2\hbar \omega_{j0} A_{j0,k} y_j. \tag{21.20}$$

Exercise 21.1. Derive Eqs. (21.18)–(21.20).

Inspecting Eqs. (21.18), (21.19), and (21.20), we note that the variables y_j are easily eliminated by differentiating (21.19) once more with respect to time. This leaves one equation of motion for the electronic and another one for the nuclear degrees of freedom. These read

$$\frac{\partial^2 q_j}{\partial t^2} = -\omega_{j0}^2 q_j + \sum_k \frac{\partial X_k}{\partial t} \omega_{j0} A_{j0,k}, \tag{21.21}$$

$$\mu_k \frac{\partial^2 X_k}{\partial t^2} = -\frac{\partial E_0}{\partial X_k} - \sum_j 2\hbar A_{j0,k} \frac{\partial q_j}{\partial t}. \tag{21.22}$$

In Eq. (21.21), one recognizes a harmonic oscillator differential equation for the electronic coordinates with an added term that accounts for linear coupling to the nuclear subsystem. Superposing a particular solution with the general solution for the homogeneous problem, we find

$$q_j(t) = a \sin(\omega_{j0} t) + b \cos(\omega_{j0} t) + \sum_k \int_0^t dt' A_{j0,k} \frac{\partial X_k}{\partial t'} \sin[\omega_{j0}(t - t')], \tag{21.23}$$

where $a = \omega_{j0}^{-1} \frac{\partial q_j}{\partial t}|_{t=0}$ and $b = q_j(0)$. To obtain the desired Langevin formulation of the nuclear force terms we insert the first time derivative of $q_j(t)$

into Eq. (21.22). Carrying out this substitution, we find the force acting on the nuclei partitioned into three contributions, according to

$$\mu_k \frac{\partial^2 X_k}{\partial t^2} = -\frac{\partial E_0}{\partial X_k} - \sum_m \int_0^t dt' \, \Lambda_{km}(t,t') \frac{\partial X_m}{\partial t} + L_k(t). \qquad (21.24)$$

As can be seen from Eq. (21.24), the nuclear dynamics is determined by two additional forces besides the adiabatic force: one proportional to the nuclear velocities and thus of frictional origin, the other one velocity-independent. The latter is termed *random force*. Its time dependence is found from Eqs. (21.22) and (21.23) to be

$$L_k(t) = 2\hbar \sum_j A_{j0,k}(t)[-\omega_{j0} \sin(\omega_{j0}t) + \cos(\omega_{j0}t)]. \qquad (21.25)$$

The quantity $\Lambda_{mk}(t,t')$ acquires the interpretation of a friction kernel. Explicitly, it is given by

$$\Lambda_{km}(t,t') = 2\hbar \sum_j A_{j0,k}(t) A_{j0,m}(t') \omega_{j0} \cos[\omega_{j0}(t-t')]. \qquad (21.26)$$

Head-Gordon and Tully employed electronic structure theory to arrive at numerical values for the friction term, as described in the following section.

21.2 An *Ab Initio* Model for Electronic Friction

The problem of evaluating the electronic friction matrix (21.26) has been addressed in the framework of Hartree–Fock theory [762, 774], describing the electronic states Φ_j by single determinants. As long as the discussion is confined to metal surfaces in contact with molecules, this approximation is defensible since the relevant electronic excitations involve transitions of conduction band electrons. These particles occupy loosely bound, highly delocalized spin orbitals and are thus unlikely to induce substantial relaxation in the remaining electronic system. We therefore operate with initial and final electronic determinants that differ in one molecular spin orbital (MO) as an occupied orbital $|i\rangle$ is emptied and a virtual orbital $|f\rangle$ populated. Applying the computation rules for matrix elements of single-particle operators acting on determinantal states, as introduced in Section 1.3, we find that the respective derivative coupling term may be expressed as

$$A_{fi,k} = \left\langle \Phi_f \left| \frac{\partial}{\partial X_k} \right| \Phi_i \right\rangle = \left\langle f \left| \frac{\partial}{\partial X_k} \right| i \right\rangle. \qquad (21.27)$$

The MOs $|i\rangle$ and $|f\rangle$ are given in an atomic orbital (AO) basis $\{|\mu\rangle\}$ with expansion coefficients a_i and a_f. The coupling matrix adopts a form conveniently accessible to computation if a matrix $\mathbf{W}_k$ can be found such that

$$A_{fi,k} = \mathbf{a}_f^\dagger \mathbf{W}_k \mathbf{a}_i, \tag{21.28}$$

where $\mathbf{W}_k$ is defined in the AO basis and does not depend on the MOs $\mathbf{a}_f$ and $\mathbf{a}_i$. The matrix $\mathbf{W}_k$ is constructed in the following way [762]: performing the derivative of the MO in (21.27) directly, one finds

$$\left\langle f \left| \frac{\partial}{\partial X_k} \right| i \right\rangle = \mathbf{a}_f^\dagger \frac{\overrightarrow{\partial}}{\partial X_k} \mathbf{S} \mathbf{a}_i + \mathbf{a}_f^\dagger \mathbf{S} \frac{\partial}{\partial X_k} \mathbf{a}_i, \tag{21.29}$$

with $S_{\mu\nu}$ as overlap matrix $\langle \mu|\nu\rangle$. The arrow above the differentiation sign implies here that the derivative is to be taken only of the ket part of $\mathbf{S}$. Comparison between Eq. (21.29) and the desired form (21.28) shows that some manipulation of the second term on the right-hand side of (21.29) is needed. For this purpose, we use the identity

$$\mathbf{H}_{\mathrm{el,AO}} \mathbf{a}_\mathrm{P} = \varepsilon_\mathrm{p} \mathbf{S} \mathbf{a}_\mathrm{p}, \tag{21.30}$$

where the ε_P stand for the MO energies and the symbol $\mathbf{H}_{\mathrm{el,AO}}$ has been used to indicate that the electronic Hamiltonian is expressed in the AO basis. It follows from Eq. (21.30) that

$$\mathbf{a}_f^\dagger \frac{\partial}{\partial X_k} (\mathbf{H}_{\mathrm{el,AO}} \mathbf{a}_i) = \mathbf{a}_f^\dagger \frac{\partial}{\partial X_k} (\varepsilon_i \mathbf{S} \mathbf{a}_i). \tag{21.31}$$

From this statement in conjunction with the counterpart of Eq. (21.30) for $\mathbf{a}_\mathrm{p}^\dagger$ $(\mathbf{a}_\mathrm{p}^\dagger \mathbf{H}_{\mathrm{el,AO}} = \mathbf{a}_\mathrm{p}^\dagger \mathbf{S} \varepsilon_\mathrm{p})$ as well as the orthonormality relation $\mathbf{a}_\mathrm{p}^\dagger \mathbf{S} \mathbf{a}_q = \delta_{pq}$, we obtain

$$\mathbf{a}_f^\dagger \mathbf{S} \frac{\partial}{\partial X_k} \mathbf{a}_i = \frac{1}{\varepsilon_f - \varepsilon_i} \mathbf{a}_f^\dagger \left[\varepsilon_i \frac{\partial}{\partial X_k} \mathbf{S} - \frac{\partial}{\partial X_k} \mathbf{H}_{\mathrm{el,AO}} \right] \mathbf{a}_i. \tag{21.32}$$

Inserting this expression into Eq. (21.29), we write the coupling matrix element as[3]

$$\left\langle f \left| \frac{\partial}{\partial X_k} \right| i \right\rangle = \frac{1}{\varepsilon_f - \varepsilon_i} \mathbf{a}_f^\dagger \left[\varepsilon_i \frac{\overleftarrow{\partial}}{\partial X_k} \mathbf{S} + \varepsilon_f \frac{\overrightarrow{\partial}}{\partial X_k} \mathbf{S} - \frac{\partial}{\partial X_k} \mathbf{H}_{\mathrm{el,AO}} \right] \mathbf{a}_i. \tag{21.33}$$

We notice that the goal of casting the coupling matrix element into the form (21.28) has still not been achieved entirely since in Eq. (21.33), $\mathbf{a}_i$ and $\mathbf{a}_f$

[3]The decomposition $\frac{\partial}{\partial X_k}\mathbf{S} = \frac{\overleftarrow{\partial}}{\partial X_k}\mathbf{S} + \frac{\overrightarrow{\partial}}{\partial X_k}\mathbf{S}$ is used here.

are joined by a matrix that depends on the respective orbital energies ε_i and ε_f, and thus is not independent of the MOs on which it operates. This deficiency, however, can be remedied for orbital energies close to the Fermi level. In this case, the Fermi energy $\varepsilon_{\rm F}$ may give a realistic estimate for both ε_i and ε_f:

$$\varepsilon_i \cong \varepsilon_f \cong \varepsilon_{\rm F}, \tag{21.34}$$

implying

$$A_{fi,k} = \left\langle f \left| \frac{\partial}{\partial X_k} \right| i \right\rangle \approx \frac{1}{\varepsilon_f - \varepsilon_i} \mathbf{a}_f^\dagger \mathbf{G}_k \mathbf{a}_i, \tag{21.35}$$

where use has been made of the definition $\mathbf{G}_k \equiv \varepsilon_F \frac{\partial}{\partial X_k}\mathbf{S} - \frac{\partial}{\partial X_k}\mathbf{H}_{\rm el,AO}$. With this choice, the matrix $\mathbf{W}_k = (\varepsilon_f - \varepsilon_i)^{-1}\mathbf{G}_k$ does not contain anymore the absolute values of the MO energies, but depends on their difference. In the case of vibrational relaxation of a molecule adsorbed to a surface, the value Δ of the vibrational quantum transferred from the molecule to the surface provides an approximation to this difference. Since no band gap exists for a metal surface, it is always possible to identify MOs $|i\rangle$ and $|f\rangle$ such that $\varepsilon_f - \varepsilon_i = \Delta$.

The preceding discussion clarifies that the friction kernel (21.26) is accessible by means of electronic structure theory as one substitutes the result (21.35) for the matrices and implements the prescription

$$\Lambda_{km}(t,t') = 2\sum_{i,f} A_{fi,k}(t)A_{fi,m}(t')(\varepsilon_f - \varepsilon_i)\cos\left[\omega_{fi}(t-t')\right]. \tag{21.36}$$

A compact representation that changes the summation involved in Eq. (21.36) into energy integration is attained by considering the split-time local density of states

$$\mathbf{P}(\varepsilon,t,t') = \sum_{\rm p} \mathbf{a}_{\rm p}(t)\mathbf{a}_{\rm p}^\dagger(t')\delta(\varepsilon - \varepsilon_{\rm p}). \tag{21.37}$$

Employing this expression as well as Eq. (21.35), the friction kernel can be written as [774]:

$$\Lambda_{km}(t,t') = 2\int_{-\infty}^{\varepsilon_{\rm F}} d\varepsilon_i \int_{\varepsilon_{\rm F}}^{\infty} d\varepsilon_f \frac{\mathrm{Tr}[\mathbf{P}(\varepsilon_i,t,t')\mathbf{G}_k(t)\mathbf{P}(\varepsilon_f,t,t')\mathbf{G}_m(t')]}{\varepsilon_f - \varepsilon_i} \times \cos\left[\frac{(\varepsilon_f - \varepsilon_i)(t-t')}{\hbar}\right]. \tag{21.38}$$

A great advantage of this formulation is that it readily allows us to determine the limit of constant coupling, as specified by energy independence of the trace of $(\mathbf{PG}_m\mathbf{PG}_k)$. If this case is realized, one finds

$$\begin{aligned}\Lambda_{km}(t,t') &= 2\mathrm{Tr}[\mathbf{P}(t,t')\mathbf{G}_k(t)\mathbf{P}(t,t')\mathbf{G}_m(t')]\int_{-\infty}^{\varepsilon_\mathrm{F}} d\varepsilon_i \int_{\varepsilon_\mathrm{F}}^{\infty} d\varepsilon_f \\ &\quad\times (\varepsilon_f-\varepsilon_i)^{-1}\cos\left[\frac{(\varepsilon_f-\varepsilon_i)(t-t')}{\hbar}\right] \\ &= 2\mathrm{Tr}[\mathbf{P}(t,t')\mathbf{G}_k(t)\mathbf{P}(t,t')\mathbf{G}_m(t')]\int_0^{\infty} dx\,\cos\left[\frac{(t-t')x}{\hbar}\right] \\ &= 2\pi\hbar\,\mathrm{Tr}[\mathbf{P}(t,t')\mathbf{G}_k(t)\mathbf{P}(t,t')\mathbf{G}_m(t')]\delta(t-t').\end{aligned} \tag{21.39}$$

Exercise 21.2. Prove the result stated in the last line of the chain of equations (21.39). Hint: You may want to use the indefinite integral of the *Ci* function: $\int Ci(z)dz = zCi(z) - \sin(z)$.

Relation (21.39) defines the *Markov limit* for the friction kernel. At this point, we emphasize again that the matrices $\mathbf{G}_k$, $\mathbf{G}_m$ are evaluated in the AO basis. Therefore, the energy dependence of the trace in Eq. (21.38) is determined by the local density of states $\mathbf{P}$. The latter quantity can be assumed to be constant in an energy interval around the Fermi level. For substantially lower energy, the approximation on which Eq. (21.39) is based cannot be justified. For the problem at hand, i.e. molecular adsorption on metal surfaces, however, the regimes $\varepsilon \ll \varepsilon_\mathrm{F}$ and $\varepsilon \gg \varepsilon_\mathrm{F}$ are expected to contribute minimally to the nonadiabatic coupling. Thus, the trace may be modeled according to

$$K_{km}(t,t')\exp\left(-\frac{(\varepsilon_f-\varepsilon_i)^2}{\alpha^2}\right) = \pi\hbar\,\mathrm{Tr}[\mathbf{P}(t,t')\mathbf{G}_k(t)\mathbf{P}(t,t')\mathbf{G}_m(t')], \tag{21.40}$$

with $\mathbf{K}$ as *friction tensor*. The conduction band width of the considered metal surface provides a useful estimate for the parameter α. If Eq. (21.40) is accepted as the form of the trace, the energy integrations involved in formula (21.38) can be carried out analytically to result in

$$\Lambda_{km}(t,t') = 2K_{km}(t,t')\frac{\alpha}{2\hbar\pi^{1/2}}\exp\left(-\frac{\alpha^2(t-t')^2}{4\hbar^2}\right). \tag{21.41}$$

One verifies easily that the Markov limit is recovered as $(\frac{\alpha}{2\hbar} \to \infty)$. In this case, we obtain again the final expression (21.39) of the constant coupling problem, since $\lim(\frac{\alpha}{2\hbar} \to \infty)\Lambda_{km}(t,t') = 2K_{km}(t,t')\delta(t-t')$. This limit is plausibly applicable to the description of nonadiabatic effects assisting the vibrational relaxation of molecules adsorbed to metal substrates, since the vibrational deexcitation energy transferred to the surface is usually a fraction of the conduction band width, the latter being typically on the order of several electron volts. By the same token, a more elaborate approach than Eq. (21.41) is needed if the nuclear kinetic energy reaches or exceeds this margin.

Substituting the friction kernel in the Markov limit for $\Lambda_{km}(t,t')$ in relation (21.24), and observing that $\int_0^t dt'\delta(t-t')f(t') = \frac{1}{2}f(t)$, we arrive at the nuclear equation of motion under the condition of constant coupling. Specifically,

$$\mu_k \frac{\partial^2 X_k}{\partial t^2} = -\frac{\partial E_0}{\partial X_k} - \sum_m K_{km}(t)\frac{\partial X_m}{\partial t} + L_k(t). \tag{21.42}$$

The time dependence of the friction tensor $\mathbf{K}$ arises, just as that of the adiabatic forces, from the temporal changes of the nuclear positions.

21.3 Case Study: Nonadiabatic Effects in the Interaction between the Cu(100) Surface and a CO Molecule

We illustrate the role of electronic friction in molecule–surface processes by the example of CO in contact with Cu(100). Specifically, we focus on mechanisms of vibrational excitation and deexcitation, as well as molecule diffusion after deposition on the surface. In each case, emphasis is placed on the contribution of electronic friction, as it arises from the coupling between vibrational modes of the molecule with the electrons of the surface. As an application of the concepts introduced in the previous two sections, we first discuss the case of vibrational deexcitation of CO attached to Cu(100).

21.3.1 *Vibrational relaxation of CO on the Cu(100) surface: The impact of electronic friction*

The trace in Eq. (21.39) determines both the friction resulting from the electron dynamics and the life time of a vibronic excitation of the molecular adsorbate [762]. For closer examination of this latter aspect, we invoke

Fermi's golden rule that relates the life time τ to the relaxation rate Γ through

$$\Gamma = \frac{1}{\tau} = \frac{2\pi}{\hbar} \sum_f |H_{f,i}|^2 \delta(\varepsilon_f - \varepsilon_i), \tag{21.43}$$

where the Hamiltonian $\hat{H}$ connects the two states characteristic of the molecular adsorption problem discussed, i.e. an initial state involving an excited molecular vibrational mode with the quantum number $\nu = 1$ for the problem of the CO relaxation on metal surfaces, and the electronic ground state of the combined system, while these assignments are reversed in the final state which combines the condition $\nu = 0$ with an electronic excitation. Both the vibronic and electronic parts are written as factors of Born–Oppenheimer products:

$$|\psi_i\rangle = |\nu\rangle|\Phi_i\rangle, \quad |\psi_f\rangle = |\nu'\rangle|\Phi_f\rangle. \tag{21.44}$$

To simplify notation, we restrict the vibrational parameters to a single coordinate which is associated with the internal C-O stretch and the corresponding reduced mass μ. For evaluation of the transition matrix element $H_{f,i}$, two simplifying assumptions may be made, namely [762] (1) the amplitude of the C-O stretch is small by the standard of the equilibrium bond length and (2) the potential energy surfaces corresponding to the electronic states $|\Phi_i\rangle$, $|\Phi_f\rangle$ are parallel, i.e. shifted with respect to each other by the amount of the vibrational quantum Δ, as defined in the context of Eq. (21.35).

Condition (1) justifies computing electronic matrix elements at the CO equilibrium geometry X_0. Since (2) secures that the states $|\nu\rangle$ and $|\nu'\rangle$ are orthogonal, both conditions taken together imply that no electronically diagonal terms (see Eqs. (2.8) and (2.11)) enter the transition matrix elements.

As harmonic oscillator functions are used, the remaining vibrational contribution $\langle\nu'|\frac{\partial}{\partial X}|\nu\rangle(\nu' = \nu - 1)$ can be obtained analytically, and the transition matrix element becomes

$$\begin{aligned} H_{f,i} &= \frac{\hbar^2}{\mu}\left\langle \nu - 1 \left| \left\langle \Phi_f \left| \frac{\partial}{\partial X} \right| \Phi_i \right\rangle \frac{\partial}{\partial X} \right| \nu \right\rangle \\ &= \hbar\left(\Delta\frac{\nu}{2\mu}\right)^{\frac{1}{2}} \left\langle \Phi_f \left| \frac{\partial}{\partial X} \right| \Phi_i \right\rangle |_{X=X_0}, \end{aligned} \tag{21.45}$$

where Δ denotes a single vibrational quantum. Further, the one-determinantal states described in the previous section have been utilized.

As in Eq. (21.27), the initial and final electronic wave functions may be replaced by the two molecular spin orbitals in which they differ. Assume that the final electronic state differs from the ground state by substitution of an MO a for the initial MO i, giving rise to a change of the electronic energy by $\varepsilon_a - \varepsilon_i$.[4] Transfer of the vibrational quantum will then result in a final energy of $\varepsilon_f = \varepsilon_a - \Delta$. Summing over all initial and final MOs under the constraint that the vibrational quantum bridges the gap between their orbital energies leads to

$$\Gamma = \frac{1}{\tau} = \pi\hbar\frac{\Delta}{\mu}\sum_{a,i}\left|\left\langle a\left|\frac{\partial}{\partial X}\right|i\right\rangle\right|^2_{X=X_0}\delta(\varepsilon_i - \varepsilon_a + \Delta), \tag{21.46}$$

where $\nu = 1$ has been assumed. We remodel the square magnitude in this equation by applying the matrix formalism of Eq. (21.35). This approach yields

$$\Gamma = \frac{\pi\hbar}{\mu}\Delta^{-1}\sum_{a,i}(\mathbf{a}_i^{\dagger}\mathbf{G}^{\dagger}\mathbf{a}_a)(\mathbf{a}_a^{\dagger}\mathbf{G}\mathbf{a}_i)\delta(\varepsilon_i - \varepsilon_a + \Delta). \tag{21.47}$$

Carrying out an operation that parallels the step made in going from Eq. (21.37) to Eq. (21.38), we eliminate the molecular orbital coefficients contained in this expression by including the local density of states and find

$$\Gamma = \frac{\pi\hbar}{\mu}\Delta^{-1}\int_0^{\Delta} d\varepsilon\ \mathrm{Tr}[\mathbf{P}(\varepsilon_{\mathrm{F}} - \varepsilon)\mathbf{G}\mathbf{P}(\varepsilon_{\mathrm{F}} + \Delta - \varepsilon)\mathbf{G}]. \tag{21.48}$$

The subsequent move is analogous to that made in Eq. (21.39), as we use the fact that for CO adsorbed on copper or platinum surfaces, as well as other realistic applications of this theory, the interval Δ is small when compared with the scale over which substantial changes of the local density of states are expected to occur as a function of energy. Adopting this approximation, we find Γ from

$$\Gamma = \frac{\pi\hbar}{\mu}\ \mathrm{Tr}[\mathbf{P}(\varepsilon_{\mathrm{F}}^{-})\mathbf{G}\mathbf{P}(\varepsilon_{\mathrm{F}}^{+})\mathbf{G}]. \tag{21.49}$$

The superscripts in the symbols $\varepsilon_{\mathrm{F}}^{-}$ and $\varepsilon_{\mathrm{F}}^{+}$ signify that the Fermi energy has been approached from below in the first and from above in the second

[4]The Coulomb and exchange contributions to the energy difference between two single-determinantal electronic states (see Eq. (1.40)) may be neglected in the case of an infinite system [779].

instance. The similarity of this result with Eq. (21.40) in the Markov limit where $K_{km}(t, t')$ tends towards $\pi\hbar \mathrm{Tr}[\mathbf{PG}_k\mathbf{PG}_m]$ is obvious.

To obtain numerical results for the vibrational relaxation rate and associated life time of CO attached to a Cu(100) surface, Head-Gordon and Tully [762] applied formula (21.49) to a cluster model, simulating the semi-infinite subsystem by a finite aggregation of Cu atoms. A calculated life time of $\tau^{\mathrm{cal}} \approx 1.7$ ps is reported as a typical value resulting from a cluster model that comprises 14 Cu atoms. This finding compares favorably with the measured value of $\tau^{\mathrm{exp}} = 2.0$ ps [780]. The dependence of τ^{cal} on the size of the copper cluster was shown to be relatively weak.

The success of relation (21.49) in the framework of a cluster approach suggests that the periodicity of the substrate is of secondary concern for the problem at hand which, after all, involves a well-defined adsorption site and thus a spatially localized phenomenon. Evidently, the band structure of the molecule–surface composite can in no way be approximated by the molecular orbital energy levels present in an atomic cluster, and condition (21.34) cannot be satisfied if the latter model is adopted. However, the derivation of the relaxation rate, Eq. (21.45) to Eq. (21.48), involved a physically realistic energy difference which ultimately was shown to cancel out. Thus, the discrete energy structure of the cluster as opposed to a semi-infinite system does not pose a significant problem except for the fact that an element of choice enters the formalism with the need for adequate positioning of the Fermi level. Several proposals to treat this issue may be considered, among them the identification of ε_{F} with the HOMO level ε_{H} of the cluster, or with the intermediate position between ε_{H} and the LUMO level ε_{L}, $\frac{1}{2}(\varepsilon_{\mathrm{H}} + \varepsilon_{\mathrm{L}})$. For the larger clusters examined in [762], the former alternative appears preferable since ε_{H} does not deviate substantially from the experimental value of ε_{F}, as identified with the work function of Cu(100) [781]. A similar decision has to be made with respect to the local density of states at the Fermi level over an energy range ΔE. Thus one may define

$$\mathbf{P}(\varepsilon_{\mathrm{F}}) = \frac{1}{\Delta E}(\mathbf{a}_{\mathrm{H}}\mathbf{a}_{\mathrm{H}}^{\dagger} + \mathbf{a}_{\mathrm{L}}\mathbf{a}_{\mathrm{L}}^{\dagger}) \tag{21.50}$$

or

$$\mathbf{P}(\varepsilon_{\mathrm{F}}^{-}) = \frac{2}{\Delta E}(\mathbf{a}_{\mathrm{H}}\mathbf{a}_{\mathrm{H}}^{\dagger}), \quad \mathbf{P}(\varepsilon_{\mathrm{F}}^{+}) = \frac{2}{\Delta E}(\mathbf{a}_{\mathrm{L}}\mathbf{a}_{\mathrm{L}}^{\dagger}). \tag{21.51}$$

At ε_F, the local density of states is naturally determined by the frontier orbitals, represented by $\mathbf{a}_H$ and $\mathbf{a}_L$. If H-1 labels the second highest occupied molecular orbital and L + 1 the second lowest unoccupied molecular orbital, the interval ΔE may be found from

$$\Delta E = (\varepsilon_L - \varepsilon_H) + \frac{1}{2}(\varepsilon_H - \varepsilon_{H-1}) + \frac{1}{2}(\varepsilon_{L+1} - \varepsilon_L). \qquad (21.52)$$

The smearing procedure underlying Eqs. (21.50) and (21.51) is justified for metals with wide flat conduction bands, and thus applicable to Cu. In [762], prescription (21.51) is selected as the form of the local density of states at ε_F since it involves a realization of the quantities $\mathbf{P}(\varepsilon_F^-)$, $\mathbf{P}(\varepsilon_F^+)$ that enter formula (21.49) for the relaxation rate. Physically, the approach (21.51) lends itself directly to the description of the nonadiabatic coupling between occupied and unoccupied cluster orbitals.

This methodology leads to a realistic assessment of the life time of the internal C-O stretch vibrational excitation with $\nu = 1$, as pointed out earlier. The additional vibronic modes of the adsorbed molecule have been treated along the same lines [782]. These are a further single mode of A_1 symmetry, termed *normal frustrated translation*, consisting in the oscillation of the CO molecule as a rigid unit attached to a Cu surface atom. Also, two doubly degenerate modes are distinguished, one associated with the translational motion of CO parallel to the surface (*in-plane frustrated translation*), and the other one with the displacement of the C and the O atom in opposite surface parallel directions (*in-plane frustrated rotation*).

Table 21.1 contains a representation of the four modes.

Table 21.1: The four normal modes for the CO molecules adsorbed to a Cu surface atom.

	Symbol	Description	Degeneracy
(I)	Cu-C-O (↑ above C and O)	In-plane frustrated translation	2
(II)	Cu-C-O (↑ above O, ↓ below C)	In-plane frustrated rotation	2
(III)	Cu←→C-O	Normal frustrated translation	1
(IV)	Cu-C←→O	Internal C-O stretch	1

Using the formalism (21.49), all four normal modes of CO on the copper surface were compared with respect to the vibrational life time τ^{cal} [762]. From this study, the life time associated with the in-plane frustrated rotation mode (II) was found to be on the same order as that of the internal C–O stretch (IV), a few picoseconds. Substantially higher values of τ^{cal}, ranging from about 80 to 100 ps resulted for the two remaining modes which demonstrates that nonadiabatic coupling controls the vibrational deexcitation only in the cases (II) and (IV).

This finding is relevant for understanding electronic friction in problems involving the dynamics of molecules in contact with metal surfaces [783,784]. The basis for the respective computations is the equation of motion (21.42) which incorporates the friction tensor **K**. Diagonalizing **K**, one finds that the three eigenvalues related to the internal C–O stretch and the in-plane frustrated rotation strongly exceed those of the other two vibronic modes. In view of the foregoing discussion of the corresponding life times, this observation is plausible. It is therefore tempting to formulate the friction tensor entirely in terms of the three dominant modes. This succeeds as one adopts the representation

$$\mathbf{K} = \sum_{m=1}^{3} e_m \mathbf{f}_m \mathbf{f}_m^{\dagger}. \tag{21.53}$$

Since the four crucial modes are associated with six independent coordinates, **K** is a 6×6 matrix. From *ab initio* calculations of the friction tensor as a function of the vertical distance of the molecule from the surface, as well as its orientation with respect to the surface, it was found that the tensor elements depend chiefly on three geometric parameters, namely the z-component of the CO center of mass, z_{COM}, and the polar and azimuthal angles Θ and Φ, respectively, that specify the orientation of the molecule [784]. The z-axis is taken here as the surface normal. If one adopts a Cartesian coordinate system in the molecular frame, the displacements in $x-, y-$ and $z-$ direction turn out to be separable to a good approximation. In this system, the $x-$ axis lies in the surface and simultaneously in a plane that contains the $z-$ axis and the internuclear distance of the CO molecule. As the coordinate set $x_{\mathrm{C}}, x_{\mathrm{O}}, y_{\mathrm{C}}, y_{\mathrm{O}}, z_{\mathrm{C}}$, and z_{O} is employed, **K** adopts a block matrix structure, since each eigenvector $\mathbf{f}_m$ reduces to only two non-zero components. Thus, if $\mathbf{f}_1$ corresponds to the first line of the in-plane frustrated rotation mode, involving displacement in $x-$ direction, it may be expressed as

$$\mathbf{f}_1 = (c_1, -\sqrt{1-c_1^2}, 0, 0, 0, 0). \tag{21.54}$$

The friction tensor **K** as given by Eq. (21.53) is therefore block-diagonal in the Cartesian molecular frame. Each of the blocks is written as

$$e_m \begin{pmatrix} c_m^2 & -c_m\sqrt{1-c_m^2} \\ -c_m\sqrt{1-c_m^2} & 1-c_m^2 \end{pmatrix}, \quad m = 1, 2, 3. \tag{21.55}$$

Further, the coefficients c_m and the eigenvalues e_m are seen to depend only on the tilt angle Θ and the center-of-mass distance z_{COM}. They are fitted to match the following functional forms:

$$c_m(\Theta) = c_{1m} + c_{2m}\cos\Theta + c_{3m}\sin^2\Theta, \tag{21.56}$$

$$e_m(\Theta) = (e_{1m} + e_{2m}\cos\Theta + e_{3m}\sin^2\Theta)\exp(-e_{4m}z_{COM}). \tag{21.57}$$

Using the identities (21.53) to (21.57) in conjunction with the equation of motion (21.42), a variety of dynamic interactions between a diatomic molecule and a surface can be explored in terms of electron-hole pair creation. These encompass vibrational excitation or deexcitation upon impact, further scattering from the surface as well as sticking, diffusion, and transient mobility of the molecule on the substrate [783].

To obtain quantitative information about the molecular dynamics of CO on Cu(100) under the influence of electronic damping, a periodic model of the surface has been adopted in the studies reported in [783, 784]. An empirical potential was utilized to simulate the gas-surface interaction, accounting for the first term on the right-hand side of the equation of motion (21.42). Further, a three-layer representation of the substrate was employed, and periodic boundary conditions were implemented parallel to the surface. The effect of the bulk was included by imposing stochastic boundary conditions on the third atomic layer in the vertical direction. This amounts to installing frictional forces as well as Markov random forces $\mathbf{R}(t)$ between the lowest layer of the Cu slab and the adjacent Cu bulk. Under the assumptions that, in the absence of gas molecules, the atoms of the solid are subjected to a harmonic force law, and that only a limited number of surface atoms influence the molecules, one derives the following relation between the time correlation function of the random force $\mathbf{R}(t)$ and the friction kernel $\Lambda(t)$ [199]:

$$\langle \mathbf{R}(t)\mathbf{R}^{\dagger}(t')\rangle = k_B T_s \mathbf{\Lambda}(t,t'). \tag{21.58}$$

In the terminology introduced by Kubo [778], this is the *second fluctuation-dissipation theorem* applied to the boundary between the surface layers and the bulk, resulting in the definition of a *phonon temperature*. The two ingredients of this identity, frictional and random forces, regulate the vibrational flow to and from the bulk. Their equilibration gives rise to a constant temperature T_s. The use of stochastic boundary conditions thus makes it possible to vary the surface temperature systematically and so to study the dynamic gas-substrate interaction as a function of this parameter.

The corresponding relation for the electronic interaction between the gas molecules and the surface defines an electronic temperature T_{el}. With reference to the equation of motion (21.42), one finds that in this case the fluctuation-dissipation theorem takes the following form [785]:

$$\langle \mathbf{L}(t)\mathbf{L}^{\dagger}(t')\rangle = k_B T_{el} \mathbf{K}\delta(t-t'). \tag{21.59}$$

Relations (21.58) and (21.59) may be understood as stating a balance between the energy dissipated due to friction and the energy gained due to the action of the fluctuating force [199].

21.3.2 *Vibrational excitation and hot diffusion*

On the basis of Eqs. (21.58) and (21.59), detailed questions about the temperature dependence of effects arising from inelastic molecule–surface scattering can be answered, and the respective trends for the coupling between molecular modes and surface phonons in the presence of electronic friction can be compared with the case of pure vibrational coupling. To illustrate this notion, we mention results reported in [783] on the vibrational excitation of CO molecules that impinge on the Cu(100) surface at normal incidence, i.e. with the internuclear axis of CO perpendicular to the surface, their initial energy being purely translational. The mean amount of energy acquired by the internal C-O stretching mode due to impact on the surface was monitored in the exit channel of the collision. Figure 21.1 shows this quantity in the surface temperature interval ranging from 100 to 700 K and for various choices of the initial translational energy.

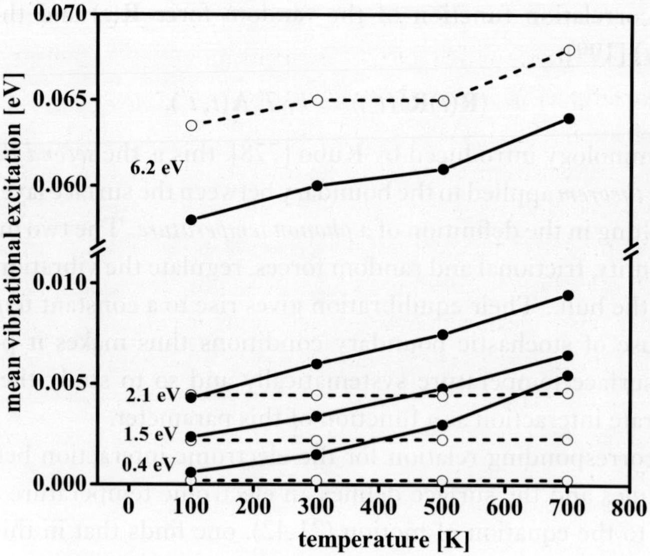

Figure 21.1: Mean vibrational excitation of CO molecules resulting from molecule–surface scattering at normal incidence versus the surface temperature. The initial molecular energy which is purely translational is used as a parameter. The dashed lines refer to the case of exclusively vibrational coupling while the results denoted by solid lines have been obtained by including electronic friction as well. (Reprinted with permission from [783]. Copyright (1998) by the American Institute of Physics.)

For initial energies $E_{\mathrm{in}} \leq 2.1$ eV, the surface temperature is seen to have little influence on the mean vibrational excitation if the effect of electronic friction is neglected. Adding this effect, however, leads to an approximately linear increase of the vibrational excitation with T_{s}. This behavior reflects higher efficiency for energy transfer to the internal C–O stretch from the electron reservoir than from the phonon reservoir of the surface. At the highest incident energy, nonadiabatic interaction with the surface electrons reduces the vibrational excitation content. In this energy regime, the internal C–O stretch excitation due to the impact is found to exceed $k_{\mathrm{B}} T_{\mathrm{s}}$. The coupling to the electron bath, however, moves the energy distribution of the vibrational degrees of freedom more closely to that associated with the surface temperature. In the case considered, this tendency implies a decrease of excitation as compared to the interaction with the phonon system alone.

The nonadiabatic coupling between the nuclear subsystem of the adsorbate and the surface electrons is of relevance for the problem of molecule deposition on substrates and thus affects the dynamics of thin film growth. In this context, the mobility of CO molecules on the Cu(100) surface following their impact has been simulated including or excluding the effect of electronic friction [783]. From simulations of molecular incidence under an angle of 45° and with initial translation energies $E_{in} \leq 1$ eV, it is found that the molecules travel by an average of about 1 Å less far on the surface if allowance is made for electronic friction than in the absence of this effect. To characterize the *hot diffusion* of the molecule after deposition, the root mean square (rms) deviation from its mean displacement along the surface projection of its incident velocity has been plotted in Figure 21.2. Hot diffusion refers to non-ballistic transport, where the diffusing species maintains, for a characteristic period, excess kinetic energy but loses the information about its initial momentum.

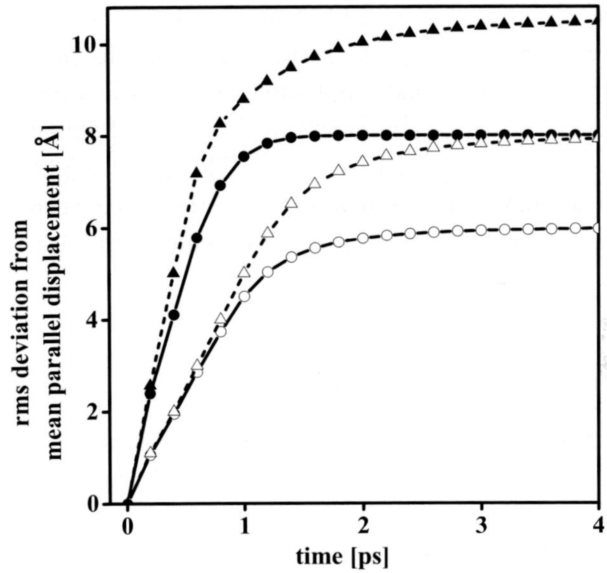

Figure 21.2: Hot diffusion of CO molecules following incidence under 45° on the Cu(100) surface along the [011] azimuth at a surface temperature of 100 K. Shown is the rms deviation from the mean displacement of the molecules along a direction given by the surface projection of the velocity at incidence as a function of time t, where the moment of impact defines $t = 0$. The empty symbols refer to an initial translational energy of $E_{in} = 0.42$ eV, the full symbols to $E_{in} = 0.83$ eV. The results denoted by circles (triangles) include (neglect) the effect of electronic friction. (Reprinted with permission from [783]. Copyright (1998) by the American Institute of Physics.)

Two energies of incidence are compared and results based on a treatment with and without electronic friction are indicated for each of these. From the figure, a significant difference is seen between incorporation and neglect of nonadiabatic coupling to the electronic subsystem of the surface. In the former case, the hot diffusion of the molecules on the surface is markedly damped, and the asymptotic lateral spread of the trapped molecules is reached several picoseconds earlier than in the latter case.

While the coupling to the electron bath strongly influences the vibrational excitation and deexcitation of the CO projectiles as well as their transient mobility, it is of little consequence for the diffusion behavior of the molecules once they have equilibrated on the surface. Computations with and without the electronic friction effect resulted in equal diffusion constants [783]. This indicates that the energy equilibrium is maintained through coupling of the molecules to the surface phonon subsystem alone.

Assessing the dynamics of CO at metal surfaces adequately requires a realistic representation of the respective molecule–surface potential. Improvement of existing model potentials is a topical subject of current research in molecule–surface interactions. Novel developments beyond the potential used in the work of Kindt *et al.* [783] involve addition of dipolar terms to account for lateral long-range forces between co-adsorbed CO molecules, facilitating the study of molecule ensembles at metal surfaces [786].

21.4 Beyond Langevin Theory

Our presentation of electron dynamics in molecule–surface interactions has emphasized the touchstone system CO/Cu(001). Experiments involving other adsorbate and substrate species have added further facets to understanding these effects. As Nienhaus *et al.* [751] were able to show, depositing thermal hydrogen and deuterium atoms on a silver surface generates electron-hole pairs in the metal, giving rise to a *chemicurrent* that was found to pass over a Schottky barrier $\approx$0.5 eV. This and several subsequent measurements ([752]–[755], [787]) gave evidence for the nonadiabatic origin of the observed current and established the detected effect as a widely occurring phenomenon that is not limited to few selected combinations of adsorbate and substrate.

The breakdown of the Born–Oppenheimer approximation in the scattering of NO molecules at high vibrational excitation in contact with the Au(111) surface has been demonstrated by state-to-state molecular beam spectroscopy [759]. To rationalize these findings, an elementary mechanism for the coupling between molecular vibrations and the electronic system of the surface was considered. This model, previously proposed by Gadzuk [788], involves the transition of a surface electron to the oscillating molecule. More specifically, an electron attaches itself to the O atom at the moment of maximum elongation where electron addition to NO is a highly exothermic process [789]. Destabilization of the resulting molecular anion in the course of its contraction or motion away from the surface ultimately causes the electron to return to the metal, inducing vibrational relaxation of the molecule and turning its surplus energy into an excited electron-hole pair. Evidently, one cannot hope to resolve this sequence of events when employing a generalized Langevin approach which allows only for an implicit treatment of the metal electrons. Empirical two-state models, corresponding to the neutral NO molecule and the anion, have been suggested to simulate the nonadiabatic energy transfer from the molecule to the surface in microscopic detail [790, 791].

Tully and coworkers introduced an extension of the trajectory surface hopping procedure (TSH, see Chapter 11) to molecule–metal interaction [764, 789]. Applying this method requires in the first place a discretization of the conduction band energy continuum. In [792] this problem assumes the concrete form of discretizing the *Anderson–Newns Hamiltonian*,

$$\hat{H}_{AN} = \varepsilon_a|a\rangle\langle a| + \int d\varepsilon'\varepsilon'|\varepsilon'\rangle\langle\varepsilon'| + \int d\varepsilon'[V(\varepsilon')|a\rangle\langle\varepsilon'| + V^*(\varepsilon')|\varepsilon'\rangle\langle a|], \tag{21.60}$$

that was proposed by Newns [793] to address the problem of an atom in contact with a surface. The operator $\hat{H}_{\mathrm{AN}}$ represents a many-electron Hamiltonian expressed in terms of single electron states. It is here written for the case of one discrete state $|a\rangle$ coupled to a continuum of states $\{|\varepsilon'\rangle\}$ by an interaction term $V(\varepsilon)$. A prescription for discretizing the Green's functions associated with the Anderson–Newns Hamiltonian by Gaussian quadrature is used to replace the continuum $\{|\varepsilon'\rangle\}$ by a finite set of pseudostates $\{|k\rangle\}$.

This provides the prerequisite for applying the TSH scheme. The particular TSH variant implemented to treat the electron-hole production problem capitalizes on the independent-electron approach implied by the use of H_{AN}. As all electrons evolve independently, their dynamics is governed by single-electron Hamiltonians, and the total energy of the entire system is given at any time of the propagation as the sum over all adiabatic single-electron eigenenergies. The principal virtue of this design is that it allows us to include a large number of possible excitations while the traditional TSH methodology is, by feasibility arguments, restricted to a small selected subset of excited states. The basic tenets of the corresponding *independent electron surface hopping* (IESH) algorithm are made explicit in [764]. The method has been applied to the deexcitation of an NO molecule in an initial vibronic state with $\nu = 15$ and with an initial translational energy of 0.05 eV at a temperature of 300 K, where varying numbers of conduction band orbitals were included. The respective simulations led to a qualitatively correct model, involving the conversion of molecular vibronic energy into electron-hole pair excitations. Further, it was shown that the limit of the electron friction result is reached as the number of contributing conduction band orbitals tends towards infinity.

As compared with Langevin methodology, the IESH technique implies a significant gain in flexibility. While the former accounts for situations characterized by numerous low-energy processes, it does not capture the case of a sizeable amount of energy transferred to a single electron. Both configurations, however, are within the range of IESH. Refining and extending this and other alternatives to the traditional Langevin view [790, 794], and examining their validity by application to an ever larger array of physical systems, is a lively field of present-day experimentation within quantum dynamics.

Epilogue

The theories, methods, and applications of electron dynamics, as summarized in this monograph, present a rich spectrum of very diverse approaches. In concluding, we emphasize that the progress towards solving electron multistate problems in molecular dynamics within the last two or three decades is not a mere reflection of the advances in computer technology during that period. To this date, consistently quantum mechanical approaches, involving, for instance, propagation of nuclear wave packets in a landscape of coupled potential energy surfaces, are extremely demanding in terms of computer time and thus confined to small systems. The development of time-dependent methods that address nonadiabatic molecular dynamics provides an example of creativity stimulated by sparseness of resources. The numerous variations on the theme of nuclear trajectories in conjunction with electrons described by quantum theory, as surveyed in Chapters 8–14 of this text, made it possible to investigate a broad range of physical systems that were out of the reach of any uniformly quantal treatment at the time of the original research. Making a virtue of necessity, they also provided new insight into the semiclassical regime *per se*. Thus, electron nuclear dynamics (Chapter 9) explores the conditions under which the linear quantum mechanical equation of motion can be approximated by nonlinear semiclassical equations in terms of differential geometry. The semiclassical propagator concept turns out to be at the root of both the trajectory surface hopping methodology and direct electron dynamics by use of action-angle variables, as implemented by the classical electron analog theory. In these and numerous other cases, the quest for a practical model of electron dynamics engendered basic research on the transition from the quantum to the classical regime.

When surveying the most recent problems and prospects in electron multistate dynamics, one encounters a perpetually growing, rapidly changing field of research activity. With respect to the directions into which the subject is taken one may notice opposing trends. Among the prevalent goals of many present-day projects is making large systems accessible to existing models, but also describing small many-body systems at high accuracy within a uniformly quantal approach. We single out six areas of current developmental activity in electron dynamics that may be deemed characteristic of the present state of the field:

(1) Braving the challenge of multistate electron processes in large systems, especially biological molecules and atomic as well as molecular clusters and nanostructures.
(2) Exploring nonadiabatic processes in interfaces of hybrid systems of nanotechnological interest.
(3) Shifting the frontiers in the simulation of open systems.
(4) Assessing the impact of nonadiabatic effects on electron transport phenomena.
(5) Extending existing theories of electron dynamics by methods of field theory.
(6) Improving benchmark computations on small molecules by combining nonadiabatic with relativistic approaches.

Commenting on the first of these items, one may accentuate the growing number of studies on photoprocesses in biological systems of large and intermediate size by methods of nonadiabatic molecular dynamics (e.g. [795]). These computations are mostly based on one of the semiclassical procedures described in this monograph, or on a suitable combination of these methods. For instance, a recent study on the reactions involved in the photosynthesis of vitamin D employed linear response TDDFT (Section 16.3.1) in conjunction with the TSH methodology (Sections 11.1 and 11.2) [796]. By the standard of experiment, the nonadiabatic approach was shown to rationalize crucial structural and dynamic features of this process, describing product formation through the non-radiative decay of the intermediate complex previtamin D by passage through a conical intersection.

Current applications of electron multistate dynamics also include complex reactions in nanostructured composites of technological importance. This trend may be illustrated by attempts to unravel the dynamics of photoelectron transfer from a chromophore into the bulk of an adjacent semiconductor. This configuration could be realized by an alizarin molecule attached to TiO_2 [797]. Photonic excitation of the molecule leads to injection of an electron into the conduction band of the semiconductor, which motivates interest in this arrangement for photovoltaics. Electron transfer at the interface between semiconducting materials and organic adsorbates is of major relevance for numerous other areas of nanotechnology, such as molecular electronics [798], photoelectrolysis [799], or lithography [800]. The photoinduced electron dynamics at the interface between the chromophore and the semiconductor involves a competition among several ultrafast electronic processes. Thus, the injected electron relaxes and delocalizes in the conduction band. The relaxing electrons may be back-donated to the chromophore or absorbed by a surrounding electrolyte that restores the charge neutrality of the chromophore. The latter transitions are preferentially undergone by electrons trapped close to the semiconductor surface, and they proceed nonadiabatically, under sizeable energy exchange between electronic and vibrational degrees of freedom. Electron relaxation and delocalization, in contrast, may be governed by adiabatic or nonadiabatic dynamics. Representing the semiconductor by a cluster model, and embedding the TSH procedure into the TDDFT method, Prezhdo *et al.* [797] demonstrated that an adequate model of electron transfer from the chromophore to the bulk requires careful averaging over all of these individual pathways of electron dynamics.

A variety of further examples ([552]– [553]) document that TDDFT supported by TSH has emerged as an efficient procedure for quantum systems of intermediate size. However, it cannot be uniformly applied to describe electron dynamics in complex systems, such as sizeable biological molecules or condensed phases. Those cases require partitioning the system into an open core region to be treated by quantum molecular dynamics, and an environment that is simulated at a different level of theory. The fundamental choices of treating the latter subsystem are by averaging over the environmental degrees of freedom, as realized by Redfield theory (Section 5.6), or allowing for quantum evolution in a classical medium. In the condensed phase, the quantum dynamical process often depends sensitively on the motion of nearby molecules which requires their explicit inclusion into the formalism employed [801]. Substantial progress is presently made in extending the mixed quantum-classical Liouville–von Neumann equation to cases where a few quantum degrees of freedom are coupled to a large reservoir, described as a classical bath. In particular, the quantum-classical limit of quantum correlation functions was shown to yield strategies for calculating transport properties such as rate or diffusion coefficients [801].

Nonadiabatic extensions of electron transport models are of major relevance to molecular electronics as well as spintronics. Consequently, novel time domain propagation methods that allow us to simulate electron as well as spin currents through nanoscale structures attract much current attention. As an example, we mention an equation-of-motion approach to the problem of quantum pumping through interacting quantum dots [802].

Other developmental work aims at increasing the quantum mechanical rigor of existing methods of nonadiabatic dynamics in molecules while simultaneously widening their scope of validity. These activities may be exemplified by pointing at novel trends within MCTDH theory, which was described in Section 15.3 as a time-dependent approach that treats both the electronic and the nuclear subsystem at a quantal level where the nuclear vibrations are represented as distinguishable degrees of freedom. During the past decade, elements of quantum field theory in the framework of second quantization have been imported into the MCTDH scheme ([803–809]). The extended MCTDH method allows us to simulate bosonic or fermionic particles, or mixtures of members of both particle families. Applications of MCTDH theory for identical particles include studying gases of bosonic or fermionic atoms in confining potentials, making the dynamics of interacting many-boson or many-fermion systems accessible to MCTDH analysis. Using this approach, the time behavior of Bose–Einstein condensates [803, 804] or of cold fermionic atoms in two-dimensional harmonic traps [10] has been investigated.

Recently, the second quantization formalism has been applied to the multilayer version of MCTDH theory ([806], see Section 15.3.3). In its fermionic form, this formalism has been used to describe electronic degrees of freedom, taking MCTDH theory from the level of nuclear dynamics in a landscape of coupled potential energy surfaces to that of interacting nuclear and electronic wave packets [806].

Further significant activity aims at highly accurate numerical results from purely quantal approaches involving calculations on small molecules, where the impact of nonadiabatic interactions on various elementary molecular properties is assessed. These include, for instance, the energies of the vibrational spectrum of D_2 [807], or the rovibrational levels [810] and the dissociation energy [811] of molecular hydrogen. These computations allow for a quantitative comparison between the sizes of nonadiabatic corrections and perturbing effects of relativistic or quantum electrodynamical origin. These efforts are aided by computations of the nonadiabatic coupling term (2.10) at increasingly higher levels of quantum chemical theory, reaching from MCSCF and MRCI [812] to equation-of-motion-coupled-cluster (EOM-CC, [813]) procedures.

Electron multistate molecular dynamics as a research subject has not yet crystallized into a set of generally accepted and practiced standard procedures. None of the models discussed in this book has acquired the status of a routinely implemented technology. While the concepts and methods presented here are permanently assessed, modified, and extended, entirely novel methodological ideas and challenging physical applications emerge, revising the map of this field of science, keeping it alive and in constant flux.

landscape of coupled potential energy surfaces to that of interacting nuclear and electronic wave packets [806].

Further significant activity aims at highly accurate numerical results from purely quantal approaches involving calculations on small molecules, where the impact of nonadiabatic interactions on various elementary molecular properties is assessed. These include, for instance, the energies of the vibrational spectrum of D_2 [807], or the rovibrational levels [810] and the dissociation energy [811] of molecular hydrogen. These computations allow for a quantitative comparison between the sizes of nonadiabatic corrections and perturbing effects of relativistic or quantum electrodynamical origin. These efforts are aided by computations of the nonadiabatic coupling term (2.10) at increasingly higher levels of quantum chemical theory, reaching from MCSCF and MRCI [812] to equation-of-motion coupled-cluster (EOM-CC [813]) procedures.

Electron multistate molecular dynamics as a research subject has not yet crystallized into a set of generally accepted and practiced standard procedures. None of the models discussed in this book has acquired the status of a routinely implemented technology. While the concepts and methods presented here are permanently assessed, modified, and extended, entirely novel methodological ideas and challenging physical applications emerge, revising the map of this field of science, keeping it alive and in constant flux.

Bibliography

[1] Erasmus Owen (1564–1616), English alchemist, adventurer, and poet.

[2] E. Deumens, A. Diz, R. Longo, Y. Öhrn, *Rev. Mod. Phys.* 66, 917 (1994).

[3] W. Domcke, G. Stock, *Adv. Chem. Phys.* 100, 1 (1997).

[4] J.C. Tully, in: *Modern Methods for Multidimensional Dynamics Computations in Chemistry*, D.L. Thompson, ed., World Scientific, Singapore (1998).

[5] M.H. Beck, A. Jaeckle, G.A. Worth, H.-D. Meyer, *Phys. Rep.* 324, 1 (2000).

[6] G. Stock, M. Thoss, *Adv. Chem. Phys.* 131, 243 (2005).

[7] J.Z.H. Zhang, *Theory and Application of Quantum Molecular Dynamics*, World Scientific, Singapore (1999).

[8] V. May, O. Kühn, *Charge and Energy Transfer Dynamics in Molecular Systems*, 2nd edition, Wiley-VCH, Weinheim (2004).

[9] H. Domcke, D.R. Yarkony, H. Köppel, eds., *Conical Intersections*, Advanced Series in Physical Chemistry 15, World Scientific, Singapore (2004).

[10] H.-D. Meyer, F. Gatti, G.A. Worth, eds., *Multidimensional Quantum Dynamics*, Wiley, Weinheim (2009).

[11] D. Marx, J. Hutter, *Ab Initio Molecular Dynamics*, Cambridge University Press, Cambridge (2009).

[12] C.A. Ullrich, *Time-dependent Density Functional Theory*, Oxford University Press, Oxford (2012).

[13] M. Born, R. Oppenheimer, *Ann. Phys. Leipzig* 84, 457 (1927).

[14] L.D. Landau, *Phys. Z. Sow.* 1, 46 (1932).

[15] C. Zener, *Proc. Roy. Soc.* A137, 696 (1932).

[16] E.C.G. Stückelberg, *Helv. Phys. Acta* 5, 369 (1932).

[17] E.J. Heller, *J. Chem. Phys.* 62, 1544 (1975).

[18] R. Car, M. Parrinello, *Phys. Rev. Lett.* 55, 2471 (1985).

[19] E.E. Nikitin, *Theory of Elementary Atomic and Molecular Processes in Gases*, Clarendon, Oxford (1974).

[20] D.A. Micha, K. Runge, *Phys. Rev. A* 50, 322 (1994).

[21] H.D. Meyer, W.H. Miller, *J. Chem. Phys.* 70, 3214 (1979).
[22] A. Bjerre, E.E. Nikitin, *Chem. Phys. Lett.* 1, 179 (1967).
[23] G. Granucci, M. Persico, *J. Chem. Phys.* 126, 134114 (2007).
[24] W.H. Miller, *J. Phys. Chem. A* 105, 2942 (2001).
[25] P. Holland, *The Quantum Theory of Motion*, Cambridge University Press, Cambridge (1993).
[26] I. Burghardt, L.S. Cederbaum, *J. Chem. Phys.* 115, 10303 (2001).
[27] P. Hohenberg, W. Kohn, *Phys. Rev.* 136, B864 (1964).
[28] V. Bonacic-Koutecky, R. Mitric, *Chem. Rev.* 105, 11 (2005).
[29] Y.L. Volobuev, M.D. Hack, M.S. Topaler, D.G. Truhlar, *J. Chem. Phys.* 112, 9716 (2000).
[30] C. Zhu, A.W. Jasper, D.G. Truhlar, *J. Chem. Phys.* 120, 5543 (2004).
[31] S.A. Rice, M. Zhao, *Optical Control of Molecular Dynamics*, John Wiley, New York (2000), Section 3.1.2.
[32] M. Shapiro, P. Brumer, *Principles of the Quantum Control of Molecular Processes*, John Wiley, New Jersey (2003).
[33] R.A. Marcus, N. Sutin, *Biochim.Biophys. Acta* 811, 265 (1985).
[34] M. Baer, *Beyond Born–Oppenheimer*, John Wiley, New Jersey (2006).
[35] A. Szabo, N. Ostlund, *Modern Quantum Chemistry*, McGraw-Hill, New York (1982).
[36] I.P. Grant, *Relativistic Quantum Theory of Atoms and Molecules*, Springer Series on Atomic, Optical, and Plasma Physics, Vol. 40, Springer, New York (2007).
[37] T.P. Das, *Relativistic Quantum Mechanics of Electrons*, Harper and Row, New York (1973).
[38] M.R. Pederson, S.N. Khanna, *Phys. Rev. B* 60, 9566 (1999).
[39] M. Tinkham, *Group Theory and Quantum Mechanics*, McGraw Hill, New York (1964).
[40] Y. Öhrn, *Elements of Molecular Symmetry*, John Wiley, New York (2000).
[41] M. Springborg, *Methods for Electronic Structure Calculations*, John Wiley, New York (2000).
[42] M. Klessinger, J. Michel, *Excited States and Photochemistry of Organic Molecules*, 1st edition, VCH, New York (1995).
[43] D.B. Cook, *Handbook of Computational Quantum Chemistry*, Dover Publications, Mineola, New York (2004).
[44] S.S. Shaik, P.C. Hiberty, *A Chemist's Guide to Valence Bond Theory*, Wiley-Interscience, New Jersey (2008).
[45] W.A. Goddard, T.H. Dunning, W.J. Hunt, P.J. Hay, *Acc. Chem. Res.* 6, 368 (1973).
[46] J.B. Foresman, M. Head-Gordon, J.A. Pople, M.J. Frisch, *J. Phys. Chem.* 96, 135 (1992).

[47] B.O. Roos, P.R. Taylor, P.E.M. Siegbahn, *Chem. Phys.* 48, 157 (1980).
[48] K. Hirao, ed., *Recent Advances in Multireference Methods*, World Scientific, Singapore (1999).
[49] J.Čížek, *J. Chem. Phys.* 45, 4256 (1966).
[50] P. Carsky, J. Paldus, J. Pittner, eds., *Recent Advances in Coupled Cluster Methods: Theory and Applications*, Springer, Berlin (2010).
[51] F.A. Evangelista, W.D. Allen, H.F. Schaefer, *J. Chem. Phys.* 125, 154113 (2006).
[52] F.A. Evangelista, W.D. Allen, H.F. Schaefer, *J. Chem. Phys.* 127, 244102 (2007).
[53] D.J. Rowe, *Rev. Mod. Phys.* 40, 153 (1968).
[54] H. Sekino, R.J. Bartlett, *Int. J. Quant. Chem.* 18, 255 (1984).
[55] L. Koziol, S.V. Levchenko, A.I. Krylov, *J. Phys. Chem. A* 110, 2746 (2006).
[56] B.J. Hammond, W.A. Lester, P.J. Reynolds, *Monte Carlo Methods in Ab Initio Quantum Chemistry*, World Scientific, Singapore (1994).
[57] D.R. Yarkony, *J. Phys. Chem. A* 105, 6277 (2001).
[58] R.L. Liboff, *Quantum Mechanics*, 2nd edition, Addison-Wesley, Reading, MA (1992).
[59] I.B. Bersuker, V.Z. Polinger, *Vibronic Interactions in Molecules and Crystals*, Springer, Berlin (1989).
[60] L.I. Schiff, *Quantum Mechanics*, 3rd edition, McGraw-Hill, New York (1955).
[61] M. Born, K. Huang, *Dynamical Theory of Crystal Lattices*, Oxford University Press, New York (1954).
[62] T. Pacher, L.S. Cederbaum, H. Köppel, *Adv. Chem. Phys.* 84, 293 (1993).
[63] A. Bohm, A. Moustafazadeh, H. Koizumi, Q. Niu, J. Zwanziger, *The Geometric Phase in Quantum Systems*, Springer, Berlin (2003).
[64] R. Englman, *The Jahn-Teller Effect in Molecules and Crystals*, Wiley, London (1972).
[65] H.A. Jahn, E. Teller, *Proc. R. Soc. Lond. A* 161, 220 (1937).
[66] I.B. Bersuker, *Jahn-Teller Effect*, Cambridge University Press, Cambridge (2006).
[67] R. Renner, *Z. Phys.* 92, 172 (1934).
[68] M.C.M. O'Brien, C.C. Chancey, *Am. J. Phys.* 61, 688 (1993).
[69] H.C. Longuet-Higgins, *Adv. Spectrosc.* 2, 429 (1961).
[70] M.V. Berry, *Proc. R. Soc. Lond. A* 392, 45 (1984).
[71] Y. Aharonov, D. Bohm, *Phys. Rev.* 115, 485 (1959).
[72] C. Alden Mead, *Chem. Phys.* 49, 23 (1980).
[73] H. Köppel, W. Domcke, in: *Encyclopedia of Computational Chemistry*, P.v.R. Schleyer, eds., Wiley, London (1998).

[74] J. Michl, V. Bonacic-Koutecky, *Electronic Aspects of Organic Photochemistry*, Wiley, New York (1990).
[75] F. Bernardi, M. Olivucci, M.A. Robb, *Chem. Soc. Rev.* 25, 132 (1996).
[76] A. Stolow, *Annu. Rev. Phys. Chem.* 54, 89 (2003).
[77] L.J. Butler, *Annu. Rev. Phys. Chem.* 49, 125 (1998).
[78] D.A. Farrow, W. Qian, E. Smith, D. Jones, *J. Chem. Phys.* 128, 144510 (2008).
[79] C.A. Mead, *J. Chem. Phys.* 72, 3839 (1980).
[80] C.A. Mead, *Rev. Mod. Phys.* 64, 51 (1992).
[81] D.R. Yarkony, in: *Conical Intersections*, H. Domcke, D.R. Yarkony, H. Köppel, eds., World Scientific, Singapore (2004).
[82] D.R. Yarkony, *Rev. Mod. Phys.* 68, 985 (1996).
[83] D.R. Yarkony, *Modern Electronic Structure Theory*, Part I, World Scientific, Singapore (1995).
[84] M.R.C. McDowell, J.P. Coleman, *Introduction to the Theory of Ion–Atom Collisions*, North Holland, Amsterdam (1970).
[85] M. Gell-Mann, M.L. Goldberger, *Phys. Rev.* 91, 398 (1953).
[86] B.A. Lippmann, J.S. Schwinger, *Phys. Rev.* 79, 469 (1950).
[87] R.G. Newton, *Scattering Theory of Waves and Particles*, McGraw-Hill, New York (1966).
[88] M.L. Goldberger, K.M. Watson, *Collision Theory*, Wiley, New York (1967).
[89] N.F. Mott, H.S.W. Massey, *The Theory of Atomic Collisions*, 3rd edition, Clarendon Press, Oxford (1965).
[90] G. Arfken, *Mathematical Methods for Physicists*, 3rd edition, Academic Press Inc., San Diego (1985).
[91] M.S. Child, *Molecular Collision Theory*, Academic Press Inc., London (1974).
[92] M.S. Child, *J. Mol. Spectrosc.* 53, 280 (1974).
[93] G. Herzberg, *Spectra of Diatomic Molecules*, Van Nostrand Reinhold, New York (1950).
[94] M. Erdmann, S. Baumann, S. Gräfe, V. Engel, *Eur. Phys. J. D* 30, 337 (2004).
[95] T.S. Rose, M.J. Rosker, A.H. Zewail, *J. Chem. Phys.* 91, 7415 (1989).
[96] J. Fedor, R. Parajuli, S. Matt-Leubner *et al.*, *Phys. Rev. Lett.* 91, 133401 (2003).
[97] S. Matt-Leubner, J. Fedor, R. Parajuli *et al.*, *Phys. Chem. Chem. Phys.* 7, 1043 (2005).
[98] C.L. Shoemaker, R.E. Wyatt, *Adv. Quantum Chem.* 14, 169 (1981).
[99] R.P. Feynman, *Rev. Mod. Phys.* 20, 367 (1948).
[100] R.P. Feynman, A.R. Hibbs, *Quantum Mechanics and Path Integrals*, McGraw Hill, New York (1965).
[101] L. Schulman, *Techniques and Applications of Path Integration*, Wiley Classics Library, Wiley, New York (1996).

[102] D. Kaiser, *Drawing Theories Apart: The Dispersion of Feynman Diagrams in Postwar Physics*, University of Chicago Press, Chicago (2005).
[103] M. Brack, R.K. Bhaduri, *Semiclassical Physics*, Westview, Boulder (2003).
[104] W. Dittrich, M. Reuter, *Classical and Quantum Dynamics*, 3rd edition, Springer, Berlin (2001).
[105] J.H. Van Vleck, *Proc. Natl. Acad. Sci. USA* 14, 178 (1928).
[106] M. Gutzwiller, *Chaos in Classical and Quantum Mechanics*, Springer, New York (1990).
[107] M. Morse, *The Calculus of Variations in the Large*, American Mathematical Society, Providence (1934).
[108] H. Goldstein, C.P. Poole, J.L. Safko, *Classical Mechanics*, 3rd edition, Addison Wesley, Reading, MA (2002).
[109] G. Wentzel, *Z. Phys.* 38, 518 (1926).
[110] H.A. Kramers, *Z.Phys.* 39, 828 (1926).
[111] L. Brillouin, *Comptes Rendus* 183, 24 (1926).
[112] H. Abramowitz, I.A. Stegun, *Handbook of Mathematical Functions, with Formulas, Graphs, and Mathematical Tables*, Dover, New York (1972).
[113] E.P. Wigner, *Phys. Rev.* 40, 749 (1932).
[114] W.P. Schleich, *Quantum Optics in Phase Space*, Wiley-VCH, Berlin (2001).
[115] R.J. Glauber, *Phys. Rev.* 130, 2529 (1963).
[116] R.J. Glauber, *Phys. Rev.* 131, 2766 (1963).
[117] V. Fock, *Z. Phys.* 75, 622 (1932).
[118] M. Weissbluth, *Photon-Atom Interactions*, Academic Press, Boston (1989).
[119] H. Kleinert, *Path Integrals in Quantum Mechanics, Statistics, Polymer Physics, and Financial Markets*, 3rd edition, World Scientific, New Jersey (2004).
[120] S.T. Ali, J.-P. Antoine, J.-P. Gazeau, *Coherent States, Wavelets and Their Generalizations*, Springer, New York (2000).
[121] G.A. Voth, *J. Phys. Chem.* 97, 8365 (1993).
[122] U. Weiss, *Quantum Dissipative Systems*, 2nd edition, World Scientific, Singapore (1999).
[123] D. Chandler, in: *Classical and Quantum Dynamics in Condensed Phase Simulations*, Proceedings of the International School of Physics, B.J. Berne, G. Ciccotti, D.F. Coker, eds., Lerici, 7–18 July (1997).
[124] D. Chandler, *Introduction to Modern Statistical Mechanics*, 1st edition, Oxford University Press, Oxford (1987).
[125] L. Onsager, *Phys. Rev.* 37, 405 (1931); *Phys. Rev.* 38, 2265 (1931).
[126] P. Haenggi, P. Talkner, M. Borkovec, *Rev. Mod. Phys.* 62, 251 (1990).
[127] J.C. Keck, *Adv. Chem. Phys.* 13, 85 (1967).

[128] G. Mills, G.K. Schenter, D.E. Makarov, H. Jonsson, in: *Classical and Quantum Dynamics in Condensed Phase Simulations*, Proceedings of the International School of Physics, B.J. Berne, G. Cioccotti, D.F. Coker, eds., Lerici, 7–18 July (1997).

[129] G. Voth, D. Chandler, W.H. Miller, *J. Phys. Chem.* 93, 7009 (1989).

[130] C. Kittel, H. Kroemer, *Thermal Physics*, 2nd edition, Freeman, New York (2000).

[131] M.S. Swanson, *Path Integrals and Quantum Processes*, Academic Press, San Diego (1992).

[132] D. Chandler, in: *Liquids, Freezing and the Glass Transition*, Les Houches, Session LI, 1989, J.P. Hansen, D. Levesque, J. Zinn-Justin, eds., Elsevier Science Publishers B.V., New York (1991).

[133] D. Chandler, L.R. Pratt, *J. Chem. Phys.* 65, 2925 (1976).

[134] L.R. Pratt, D. Chandler, *J. Chem. Phys.* 66, 147 (1977).

[135] D. Chandler, P.G. Wolynes, *J. Chem. Phys.* 74, 4078 (1981).

[136] K.S. Schweizer, R.M. Stratt, D. Chandler, P.G.W. Olynes, *J. Chem. Phys.* 75, 1347 (1981).

[137] M. Parrinello, A. Rahman, *J. Chem. Phys.* 80, 860 (1984).

[138] R.W. Hall, B.J. Berne, *J. Chem. Phys.* 860, 3641 (1984).

[139] J. Cao, G.A. Voth, *J. Chem. Phys.* 99, 10070 (1993).

[140] H.F. Trotter, *Proc. Am. Math. Soc.* 10, 545 (1959).

[141] M.P. Allen, D.J. Tildesley, *Computer Simulations of Liquids*, Oxford Science Publications, Clarendon Press, Oxford (1987).

[142] J. Cao, G. Voth, *J. Chem. Phys.* 105, 6856 (1996).

[143] G. Voth, D. Chandler, W.H. Miller, *J. Chem. Phys.* 91, 7749 (1989).

[144] W.H. Miller, *J. Chem. Phys.* 62, 1899 (1975).

[145] P. Pechukas, in: *Dynamics of Molecular Collisions*, Part B, W.H. Miller, ed., Plenum, New York (1976).

[146] J. Cao, C. Minichino, G.A. Voth, *J. Chem. Phys.* 103, 1391 (1995).

[147] M. Schlosshauer, *Decoherence and the Quantum-to-Classical Transition*, Springer, New York (2007).

[148] G. van der Zwan, *Derivation of the Redfield Relaxation Equation*, www.chem.vu.nl/zwan/lectures/redfield.pdf (2001).

[149] A.G. Redfield, *Adv. Magn. Res.* 1, 1 (1965).

[150] G.C. Schatz, M.A. Ratner, *Quantum Mechanics in Chemistry*, Prentice Hall, New Jersey (1993).

[151] J.N. Murrell, S.D. Bosanac, *Introduction to the Theory of Atomic and Molecular Collisions*, John Wiley, Chichester (1989).

[152] G.J. Tawa, S.L. Mielke, D.G. Truhlar, D. Schwenke, *J. Chem. Phys.* 100, 5751 (1994).

[153] T. Wu, T. Ohmura, *Quantum Theory of Scattering*, Prentice Hall, Englewood Cliffs, New Jersey (1962).

[154] E.U. Condon, G.H. Shortley, *The Theory of Atomic Spectra*, Cambridge University Press, Cambridge (1935).

[155] B.R. Johnson, *J. Comp. Phys.* 13, 445 (1973).

[156] F. Mrugala, D. Secrest, *J. Chem. Phys.* 78, 5954 (1983).

[157] D.E. Manolopoulos, *J. Chem. Phys.* 85, 6425 (1986).

[158] D. Secrest, B.R. Johnson, *J. Chem. Phys.* 45, 4556 (1966).

[159] L.M. Delves, *Nucl. Phys.* 9, 391 (1959).

[160] L.M. Delves, *Nucl. Phys.* 20, 275 (1960).

[161] F.T. Smith, *Phys. Rev.* 120, 1058 (1960).

[162] R.T. Pack, G.A. Parker, *J. Chem. Phys.* 87, 3888 (1987).

[163] A.K. Kuppermann, P.G. Hipes, *J. Chem. Phys.* 84, 5962 (1986).

[164] A.T. Le, C.D. Lin, L.F. Errea, L. Mendez, A. Riera, B. Pons, *Phys. Rev. A* 69, 62703 (2004).

[165] M. Baer, ed., *Theory of Chemical Reaction Dynamics*, Vol. I, CRC Press, Boca Raton (1985).

[166] A.M. Arthurs, A. Dalgarno, *Proc. Roy. Soc.* A256, 50 (1960).

[167] J.H. Van Vleck, *Rev. Mod. Phys.* 23, 213 (1951).

[168] A.P.M. Baede, A.M.C. Moutino, A.E. De Vries, J. Los, *Chem. Phys. Lett.* 3, 530 (1969).

[169] R.K.B. Helbing, E.W. Rothe, *J. Chem. Phys.* 51, 1607 (1969).

[170] M. Baer, *Theory of Chemical Reaction Dynamics*, Vol. II, CRC Press, Boca Raton, FL (1985).

[171] G.C. Schatz, *J. Phys. Chem.* 99, 7522 (1995).

[172] T.W.J. Whitley, A.J. Dobbyn, J.N.L. Connor, G.C. Schatz, *Phys. Chem. Chem. Phys.* 2, 549 (2000).

[173] M.A. Alexander, D.E. Manolopoulos, H.-J. Werner, *J. Chem. Phys.* 113, 11084 (2000).

[174] D.M. Newmark, A.M. Wodtke, G.N. Robinson, C.C. Hayden, Y.T. Lee, *J. Chem. Phys.* 82, 3045 (1985).

[175] D.M. Newmark, A.M. Wodtke, G.N. Robinson, *et al. J. Chem. Phys.* 82, 3067 (1985).

[176] D.E. Manolopoulos, *J. Chem. Soc. Faraday Trans.* 93, 673 (1997).

[177] K. Stark, H.J. Werner. *J. Chem. Phys.* 104, 6515 (1996).

[178] D.E. Manolopoulos, K. Stark, H.J. Werner, D.W. Arnold, S.E. Bradforth, D.M. Neumark, *Science* 262, 1852 (1993).

[179] R.T. Skodje, D. Skouteris, D.E. Manolopoulos, S.-H. Lee, F. Dong, K. Liu, *J. Chem. Phys.* 112, 4536 (2000).

[180] F. Rebentrost, W.A. Lester, Jr., *J. Chem. Phys.* 63, 3737 (1975).

[181] F. Rebentrost, W.A. Lester, Jr., *J. Chem. Phys.* 67, 3367 (1977).

[182] J.F. Castillo, D.E. Manolopoulos, K. Stark, H.J. Werner, *J. Chem. Phys.* 104, 6531 (1996).

[183] F.J. Aoiz, L. Banares, J.F. Castillo, *J. Chem. Phys.* 111, 4013 (1999).

[184] G.C. Schatz, P. McCabe, J.N.L. Connor, *Faraday Discuss.* 110, 139 (1998).

[185] R.G. Newton, *Scattering Theory of Waves and Particles*, McGraw-Hill, New York (1966).

[186] W. Kohn, *Phys. Rev.* 74, 1763 (1948).

[187] F. Robicheaux, *J. Chem. Phys.* 91, 708 (1989).

[188] I.C. Percival, *Proc. Phys. Soc.* 76, 206 (1960).

[189] W.H. Miller, *Comment. At. Mol. Phys.* 22, 115 (1988).

[190] J.Z.H. Zhang, S.I. Chu, W.H. Miller, *J. Chem. Phys.* 88, 6233 (1988).

[191] C.W. McCurdy, T.N. Rescigno, B.I. Schneider, *Phys. Rev. A* 36, 2061 (1987).

[192] W.H. Breckenridge, *Adv. Chem. Phys.* 50, 325 (1982).

[193] R. Dueren, V. Lackshewitz, S. Milosevic, *Chem. Phys.* 126, 81 (1988).

[194] P.S. Weiss, J.M. Mestgagh, M.H. Covinsky, B.A. Balko, Y.T. Lee, *Chem. Phys.* 126, 93 (1988).

[195] M. Motzkus, G. Pichler, K.L. Kompa, P. Hering, *Chem. Phys. Lett.* 257, 181 (1996).

[196] P. Botschwina, W. Meyer, I.V. Hertel, W.J. Reiland, *Chem. Phys.* 75, 5438 (1981).

[197] R. de Vivie-Riedle, P. Hering, K.L. Kompa, *Z. Phys.* D 17, 299 (1990).

[198] T.J. Martinez, M. Ben-Nun, R.D. Levine, *J. Phys. Chem. A* 101, 6389 (1997).

[199] J.C. Tully, *J. Chem. Phys.* 73, 1975 (1980).

[200] M. Baer, *Mol. Phys.* 40, 1011 (1980).

[201] P. Halvick, D.G. Truhlar, *J. Chem. Phys.* 96, 2895 (1992).

[202] A. Vibok, G.J. Halasz, T. Vertesi, S. Suhai, M. Baer, J.P. Toennies, *J. Phys. Chem. A* 119, 6588 (2003).

[203] M. Baer, *J. Phys. Chem. A* 105, 2198 (2001).

[204] S.L. Mielke, G.J. Tawa, D.G. Truhlar, *J. Am. Chem. Soc.* 115, 6436 (1993).

[205] Y. Sun, D.J. Kouri, D.G. Truhlar, D.W. Schwenke, *Phys. Rev. A* 41, 4857 (1990).

[206] L. Schlessinger, *Phys. Rev.* 167, 1411 (1969).

[207] G.J. Tawa, S.T. Mielke, D.G. Truhlar, D.W. Schwenke, in: *Advances in Molecular Vibrations and Collision Dynamics*, Volume 2B, J. Bowman, ed., JAI Press, Greenwich, CT (1994).

[208] D.W. Schwenke, K. Haug, M. Zhao, D.G. Truhlar, Y. Sun, J.Z.H. Zhang, D.J. Kouri, *J. Phys. Chem.* 92, 3202 (1988).

[209] S.L. Mielke, G.J. Tawa, D.G. Truhlar, *J. Am. Chem. Soc.* 115, 6436 (1993).

[210] R.S. Friedmann, T.C. Allison, D.G. Truhlar, *Phys. Chem. Chem. Phys.* 1, 1237 (1999).

[211] H. Nakamura, *Nonadiabatic Transition: Concepts, Basic Theory, Applications*, 2nd edition, World Scientific, Singapore (2012).

[212] C. Zhu, H. Kamisaka, H. Nakamura, *J. Chem. Phys.* 115, 11036 (2002).

[213] A. Kuki, *J. Phys. Chem.* 97, 13107 (1993).

[214] R.K. Nesbet, *Variational Principles and Methods in Theoretical Physics and Chemistry*, Cambridge University Press, Cambridge (2003).

[215] R. McWeeny, *Methods of Molecular Quantum Mechanics*, 2nd edition, Academic Press, San Diego (1989).

[216] K. Ohta, *Chem. Phys. Lett.* 329, 248 (2000).

[217] J. Frenkel, *Wave Mechanics: Advanced General Theory*, Clarendon Press, Oxford (1934).

[218] A.D. McLachlan, M.A. Ball, *Rev. Mod. Phys.* 36, 844 (1964).

[219] P. Kramer, M. Saraceno, *Geometry of the Time-Dependent Variational Principle in Quantum Mechanics*, Springer, Berlin (1981).

[220] D. Bohm, D. Pines, *Phys. Rev.* 82, 625 (1951).

[221] M.J. Field, *J. Chem. Phys.* 96, 4583 (1992).

[222] M.J. Field, *J. Chem. Phys.* 172, 83 (1990).

[223] E. Deumens, Y. Öhrn, *J. Chem. Phys.*, 92, 3181 (1988).

[224] M.R. Hoffman, D.J. Fox, J.F. Gaw, *et al.*, *J. Chem. Phys.* 80, 2660 (1984).

[225] R.P. Feynman, *Phys. Rev.* 56, 340 (1939).

[226] H. Hellmann, *Z. Phys.*, 85, 180 (1933).

[227] P. Bendt, A. Zunger, *Phys. Rev. Lett.* 50, 1684 (1983).

[228] G.P. Srivastava, D. Weaire, *Adv. Phys.* 36, 463 (1987).

[229] K. Saita, D.V. Shalashilin, *J. Chem. Phys.* 137, 22A506 (2012).

[230] P. Pulay, *Mol. Phys.* 17, 197 (1969).

[231] G. Pastore, E. Smargiassi, F. Buda, *Phys. Rev. A* 44, 6334 (1991).

[232] D.K. Remler, P.A. Madden, *Mol. Phys.* 70, 921 (1990).

[233] F.A. Bornemann, C. Schütte, *Numer. Math.* 78, 359 (1998).

[234] B. Hartke, E.A. Carter, *Chem. Phys. Lett.* 189, 358 (1992).

[235] B. Hartke, E.A. Carter, *J. Chem. Phys.* 97, 6569 (1992).

[236] M.E. Tuckerman, B.J. Berne, A. Rossi, *J. Chem. Phys.* 94, 1465 (1991).

[237] D.A. Gibson, E.A. Carter, *J. Phys. Chem.* 97, 13429 (1993).

[238] N.L. Doltsinis, D. Marx, *J. Theor. Comp. Chem.* 1, 319 (2002).

[239] D.A. Micha, *J. Phys. Chem. A* 103, 7562 (1999).

[240] W. Fritsch, C.D. Lin, *Phys. Rep.* 202, 1 (1991).

[241] K. Runge, D.A. Micha, E.Q. Feng, *Int. J. Quant. Chem. Symp.* 24, 781 (1990).

[242] K. Runge, D.A. Micha, *Phys. Rev. A* 62, 22703 (2000).

[243] H.F.M. DaCosta, D.A. Micha, K. Runge, *J. Chem. Phys.* 107, 9018 (1997).

[244] K. Runge, D.A. Micha, *Phys. Rev. A* 53, 1388 (1996).
[245] M.B. Shah, H.B. Gilbody, *J. Phys. B* 11, 121 (1978).
[246] W.L. Nutt, R.W. McCullough, K. Brady, M.B. Shah, H.B. Gilbody, *J. Phys. B* 11, 1457 (1978).
[247] L.F. Errea, J.M. Gomez-Llorente, L. Mendez, A. Riera, *J. Phys. B* 20, 6089 (1987).
[248] L.F. Errea, C. Harel, H. Jouin, *Phys. Rev. A* 46, 6617 (1992).
[249] J.M. Maidigan, L. Mendez, B. Pons, A. Riera, *Phys. Rev. A* 46, 5617 (1992).
[250] T.G. Winter, G.J. Hatton, *Phys. Rev. A* 21, 793 (1980).
[251] D.A. Micha, C.D. Stodden, *J. Chem. Phys. A* 105 2890 (2001).
[252] E. Deumens, Y. Öhrn, B. Weiner, *J. Math. Phys.* 32, 1166 (1991).
[253] E. Deumens, Y. Öhrn, *J. Phys. Chem. A* 105, 2660 (2001).
[254] Y. Öhrn, E. Deumens, *Adv. Chem. Phys.* 124 (2002).
[255] E. Deumens, Y. Öhrn, *Int. J. Quant. Chem. Symp.* 23, 31 (1989).
[256] W. Rossmann, *Lie Groups: An Introduction Through Linear Groups*, Oxford University Press, Oxford (2002).
[257] A.M. Perelomov, *Commun. Math. Phys.* 26, 222 (1972).
[258] J. Kucar, H. Meyer, *J. Chem. Phys.* 90, 5566 (1989).
[259] J.B. Delos, *Rev. Mod. Phys.* 53, 287 (1981).
[260] D. Bohm, *Quantum Theory*, Prentice Hall, New York (1951).
[261] J. Morales, A. Diz, E. Deumens, Y. Öhrn, *J. Chem. Phys.* 103, 9968 (1995).
[262] R. Cabrera-Trujillo, J.R. Sabin, Y. Öhrn, E. Deumens, *Phys. Rev. Lett.* 84, 5300 (2000).
[263] R. Cabrera-Trujillo, J.R. Sabin, E. Deumens, Y. Öhrn, *Phys. Rev. A* 66, 22706 (2002).
[264] R. Cabrera-Trujillo, J.R. Sabin, Y. Öhrn, E. Deumens, *Phys. Rev. A* 61, 32719 (2000).
[265] L.I. Schiff, *Phys. Rev.* 103, 443 (1956).
[266] E.A. Mason, J.T. Vanderslice, C.J. Raw, *J. Chem. Phys.* 40, 2153 (1964).
[267] R. Cabrera-Trujillo, Y. Öhrn, E. Deumens, J.R. Sabin, *J. Chem. Phys.* 116, 2783 (2002).
[268] R.S. Gao, L.K. Johnson, G.J. Smith *et al. Phys. Rev. A* 44, 5599 (1991).
[269] M. Kimura, J. Chapman, N.F. Lane, *Phys. Rev. A* 33, 1619 (1986).
[270] R.S. Mulliken, *J. Chem. Phys.* 23, 1833 (1955).
[271] R.S. Mulliken, *J. Chem. Phys.* 23, 1841 (1955).
[272] R.S. Mulliken, *J. Chem. Phys.* 36, 3428 (1962).
[273] R. Cabrera-Trujillo, Y. Öhrn, E. Deumens, J.R. Sabin, *Phys. Rev. A* 62 52714 (2000).

[274] R.F. Bacher, S. Goudsmit, in: *Atomic Energy States*, F.K. Richtmeyer, ed., McGraw-Hill, New York (1932).
[275] S.K. Allison, M. Garcia-Munoz, in: *Atomic and Molecular Processes, Pure and Applied Physics*, Vol. 13, D.R. Bates, ed., Academic Press, New York (1962).
[276] A. Dalgarno, G.W. Griffing, *Proc. Roy. Soc. Lond. Ser. A* 232, 423 (1955).
[277] M. Hedstroem, E. Deumens, Y. Öhrn, *Phys. Rev. A* 57, 2625 (1998).
[278] J.F. Ziegler, J.P. Biersack, U. Littmark, *The Stopping and Range of Ions in Solids*, Pergamon Press, New York (1985).
[279] R. Cabrera-Trujillo, Y. Öhrn, J.R. Sabin, E. Deumens, *Phys. Rev. A* 65, 24901 (2002).
[280] G. Niedner, M. Noll, J. Toennies, C. Schlier, *J. Chem. Phys.* 87, 2686 (1987).
[281] M. Baer, G. Niedner-Schatteburg, J.P. Toennies, *J. Chem. Phys.* 91, 4169 (1989).
[282] A. Blass, E. Deumens, Y. Oehrn, *J. Chem. Phys.* 115, 8366 (2001).
[283] G.M. de Prony, *J. Ec. Polytech. Paris* 1, 24 (1795).
[284] F. Hagelberg, E. Deumens, *Phys. Rev. A* 65, 52505 (2002).
[285] F. Hagelberg, *Int. J. Quantum Chem.* 85, 72 (2001).
[286] E. Deumens, Y. Öhrn, *J. Phys. Chem. A* 105, 2660 (2001).
[287] K. Tsereteli, Y. Yan, J.A. Morales, *Chem. Phys. Lett.* 420, 54 (2006).
[288] J.C. Tully, *J. Chem. Phys.* 93, 1061 (1990).
[289] J.M. Cohen, D.A. Micha, *J. Chem. Phys.* 97, 1038 (1992).
[290] H.D. Meyer, *Chem. Phys.* 82, 199 (1983).
[291] H. Goldstein, C.P. Poole, J.L. Safko, *Classical Mechanics*, Addison-Wesley, Reading, MA, 2001.
[292] W.H. Miller, *J. Chem. Phys.* 53, 3578 (1970).
[293] J.W. Zwanziger, E.R. Grant, G.S. Ezra, *J. Chem. Phys.* 85, 2089 (1986).
[294] J.E. Bayfield, *Quantum Evolution*, John Wiley, New York (1999).
[295] W.H. Miller, *J. Chem. Phys.* 48, 1651 (1968).
[296] J.N.L. Connor, *Chem. Phys. Lett.* 4, 419 (1969).
[297] S.M. Colwell, N.C. Handy, W.H. Miller, *J. Chem. Phys.* 68, 745 (1978).
[298] C. Jaffe, W.P. Reinhardt, *J. Chem. Phys.* 77, 5191 (1982).
[299] T. Uzer, D.W. Nold, R.A. Marcus, *J. Chem. Phys.* 79, 4412 (1983).
[300] H. Köppel, *Chem. Phys.* 77, 359 (1983).
[301] J.C. Tully, R.K. Preston. *J. Chem. Phys.* 55, 562 (1971).
[302] Q. Cui, K. Morokuma, *J. Chem. Phys.* 107, 4951 (1997).
[303] R.K. Preston, J.C. Tully, *J. Chem. Phys.* 54, 4297 (1971).
[304] F.O. Ellison, *J. Am. Chem. Soc.* 85, 3540 (1963).
[305] F.O. Ellison, N.T. Huff, J.C. Patel, *J. Am. Chem. Soc.* 85, 3544 (1963).
[306] K.F. Wong, P.J. Rossky, *J. Phys. Chem. A* 105, 2546 (2001).

[307] F. Webster, P.J. Rossky, R.A. Friesner, *Comput. Phys. Commun.* 63, 494 (1991).

[308] M.J. Bedard-Hearn, R.E. Larsen, B.J. Schwarz, *J. Chem. Phys.* 123, 234106 (2005).

[309] P.V. Parandekar, J.C. Tully, *J. Chem. Phys.* 122, 094102 (2005).

[310] J.C. Tully, *Faraday Discuss.* 110, 407 (1998).

[311] D.F. Coker, L. Xiao, *J. Chem. Phys.* 102, 496 (1994).

[312] M.F. Herman, *J. Chem. Phys.* 76, 2949 (1982).

[313] U. Müller, G. Stock, *J. Chem. Phys.* 107, 6230 (1997).

[314] S. Hammes-Schiffer, J.C. Tully, *J. Chem. Phys.* 101, 4657 (1994).

[315] J.-Y. Fang, S. Hammes-Schiffer, *J. Phys. Chem. A* 103, 9399 (1999).

[316] A. Ferretti, G. Granucci, A. Lami, M. Persico, G. Villani, *J. Chem. Phys.* 104, 5517 (1996).

[317] K.F. Wong, P.J. Rossky, *J. Phys. Chem. A* 105, 2546 (2001).

[318] C. Zhu, H. Nobusada, H. Nakamura, *J. Chem. Phys.* 115, 3031 (2001).

[319] X. Sun, W.H. Miller, *J. Chem. Phys.* 106, 6346 (1997).

[320] T.J. Martinez, M. Ben-Nun, R.D. Levine, *J. Phys. Chem.* 100, 7884 (1995).

[321] M. Ben-Nun, T.J. Martinez, *J. Chem. Phys.* 108, 7244 (1998).

[322] X. Sun, H. Wang, W.H. Miller, *J. Chem. Phys.* 109, 7064 (1998).

[323] W.H. Miller, T.F. George, *J. Chem. Phys.* 56, 5637 (1972).

[324] S. Chapman, *Adv. Chem. Phys.* 82, 2, 423 (1992).

[325] Y.W. Lin, T.F. George, K. Morokuma, *J. Chem. Phys.* 60, 4311 (1974).

[326] Y.W. Lin, T.F. George, K. Morokuma, *Chem. Phys. Lett.* 30, 49 (1975).

[327] T.J. Martinez, M. Ben-Nun, R.D. Levine, *J. Phys. Chem.* 100, 7884 (1994).

[328] M. Ben-Nun, R.D. Levine, *Chem. Phys.* 201, 163 (1995).

[329] M. Chaija, R.D. Levine, *Chem. Phys. Lett.* 304, 385 (1999).

[330] E.J. Heller, *J. Chem. Phys.* 75, 2923 (1981).

[331] W.H. Preuss, S.A. Teukolsky, W.T. Vetterling, B.P. Flannery, *Numerical Recipes*, Cambridge University Press, 2nd edition, Cambridge (1999).

[332] M. Ben-Nun, T.J. Martinez, *Adv. Chem. Phys.* 121, 439 (2002).

[333] M. Ben-Nun, J. Quenneville, T.J. Martinez, *J. Phys. Chem. A* 104, 5161 (2000).

[334] T.J. Martinez, M. Ben-Nun, R.D. Levine, *J. Chem. Phys.* 100, 7884 (1996).

[335] J.H. Barker, R.E. Weston, *J. Chem. Phys.* 65, 1427 (1976).

[336] B.J. Levine, T.J. Martinez, *Ann. Rev. Phys. Chem.* 58, 613 (2007).

[337] M. Ben-Nun, T.J. Martinez, *J. Phys. Chem.* 103, 10517 (1999).

[338] M. Ben-Nun, T.J. Martinez, *J. Phys. Chem. A* 102, 9607 (1998).

[339] M. Ben-Nun, T.J. Martinez, *J. Am. Chem. Soc.* 122, 6299 (2000).

[340] J.D. Coe, T.J. Martinez, *J. Am. Chem. Soc.* 127, 4560 (2005).

[341] J.D. Coe, T.J. Martinez, *J. Phys. Chem. A* 110, 618 (2006).

[342] J.D. Coe, M.T. Ong, B.G. Levine, T.J. Martinez, *J. Phys. Chem. A* 112, 12559 (2008).
[343] M.D. Hack, A.W. Jasper, Y.L. Volobuev, D.W. Schwenke, D.G. Truhlar, *J. Phys. Chem. A* 103, 6309 (1999).
[344] P. Pechukas, *Phys. Rev.* 181, 166 (1969).
[345] P. Pechukas, *Phys. Rev.* 181, 174 (1969).
[346] D.F. Coker, L. Xiao, *J. Chem. Phys.* 102, 496 (1995).
[347] F. Webster, E.T. Wang, P.J. Rossky, R.A. Friesner, *J. Chem. Phys.* 100, 4835 (1994).
[348] R. Marcus, *J. Chem. Phys.* 56, 3548 (1972).
[349] H. Kreek, R. Marcus, *J. Chem. Phys.* 61, 3308 (1974).
[350] M.F. Herman, E. Kluk, *Chem. Phys.* 91, 27 (1984).
[351] G. Stock, M. Thoss, *Phys. Rev. Lett.* 78, 578 (1997).
[352] M. Thoss, G. Stock, *Phys. Rev. A* 59, 64 (1999).
[353] J. Schwinger, in: *Quantum Theory of Angular Momentum*, L.C. Biedenharn, H. Van Dam, eds., Academic Press, New York (1965).
[354] T. Holstein, H. Primakoff, *Phys. Rev.* 58, 1098 (1940).
[355] P. Garbaczewski, *Phys. Rep.* 36, 65 (1978).
[356] W.H. Miller, *J. Phys. Chem. A* 113, 1409 (2009).
[357] M. Thoss, G. Stock, *Adv. Chem. Phys.* 131, 243 (2005).
[358] N. Ananth, C. Venkataraman, W.H. Miller, *J. Chem. Phys.* 127, 084114 (2007).
[359] M. Thoss, W.H. Miller, G. Stock, *J. Chem. Phys.* 112, 10282 (2000).
[360] K.K. Innes, I.G. Ross, W.R. Moomaw, *J. Mol. Spectrosc.* 132, 492 (1988).
[361] C. Woywod, W. Domcke, A. L. Sobolewski, H.-J. Werner, *J. Chem. Phys.* 100, 1400 (1994).
[362] R. Schneider, W. Domcke, H. Köppel, *J. Chem. Phys.* 92, 1045 (1990).
[363] G. Stock, R. Schneider, W. Domcke, *J. Chem. Phys.* 90, 7184 (1989).
[364] G. Stock, W. Domcke, *J. Chem. Phys.* 93, 5496 (1990).
[365] L. Seidner, G. Stock, A.L. Sobolewski, W. Domcke, *J. Chem. Phys.* 96, 5298 (1992).
[366] A.L. Sobolewski, C. Woywod, W. Domcke, *J. Chem. Phys.* 98, 5627 (1993).
[367] G. Stock, C. Woywod, W. Domcke, T. Swinney, B. S. Hudson, *J. Chem. Phys.* 103, 6851 (1995).
[368] S. Krempl, M. Winterstetter, H. Plöhn, W. Domcke, *J. Chem. Phys.* 100, 926 (1994).
[369] S. Krempl, M. Winterstetter, W. Domcke, *J. Chem. Phys.* 102, 6499 (1995).
[370] G. Stock, *J. Chem. Phys.* 103, 2888 (1995).
[371] G. Stock, W. H. Miller, *Chem. Phys. Lett.* 197, 396 (1992).
[372] U. Müller, G. Stock, *J. Chem. Phys.* 107, 6230 (1997).

[373] U. Müller, G. Stock, *J. Chem. Phys.* 111, 77 (1999).

[374] I. Yamazaki, T. Murao, T. Yamanaka, K. Yoshihara, *Faraday Discuss.* 75, 395 (1983).

[375] E. Heller, *J. Chem. Phys.* 68, 2066 (1978).

[376] G. Herzberg, *Molecular Spectra and Molecular Structure III, Electronic Spectra of Polyatomic Molecules*, Van Nostrand, Princeton (1966).

[377] A.R. Walton, D.E. Manolopoulos, *Mol. Phys.* 87, 961 (1996).

[378] G.A. Worth, H.D. Meyer, L. Cederbaum, *J. Chem. Phys.* 109, 3518 (1998).

[379] X. Sun, W.H. Miller, *J. Chem. Phys.* 110, 6635 (1999).

[380] K.G. Kay, *J. Chem. Phys.* 101, 2250 (1994).

[381] M. Herman, *Chem. Phys. Lett.* 275, 445 (1997).

[382] W. Schweizer, *Numerical Quantum Dynamics*, Kluwer Academic Publisher, Dordrecht (2001).

[383] J.D. Doll, T.L. Beck, D.L. Freeman, *J. Chem. Phys.* 89, 5753 (1988).

[384] V.S. Filinov, *Nucl. Phys. B* 271, 717 (1986).

[385] H. Wang, D.E. Manolopoulos, W.H. Miller, *J. Chem. Phys.* 115, 6317 (2001).

[386] H. Wang, X. Sun, W.H. Miller, *J. Chem. Phys.* 108, 9726 (1998).

[387] X. Sun, H. Wang, W.H. Miller, *J. Chem. Phys.* 109, 7064 (1998).

[388] X. Sun, H. Wang, W.H. Miller, *J. Chem. Phys.* 109, 4190 (1998).

[389] N. Makri, K. Thompson, *Chem. Phys. Lett.* 291, 101 (1998).

[390] W.H. Miller, *Faraday Discuss.* 110, 1 (1998).

[391] R. Gelabert, X. Gimenez, M. Thoss, H.B. Wang, W.H. Miller, *J. Phys. Chem.* 114, 2572 (2001).

[392] V.S. Batista, M.T. Zanni, B.J. Greenblatt, D.M. Neumark, W.H. Miller, *J. Chem. Phys.* 110, 3736 (1999).

[393] V. Guallar, V.S. Batista, W.H. Miller, *J. Chem. Phys.* 113, 9510 (2000).

[394] D.E. Skinner, W.H. Miller, *J. Chem. Phys.* 111, 10787 (1999).

[395] M.V. Ovchinnikov, A. Apkarian, G.A. Voth, *J. Chem. Phys.* 114, 7130 (2001).

[396] D. Huber, E.J. Heller, *J. Chem. Phys.* 89, 4753 (1988).

[397] E.J. Heller, *J. Chem. Phys.* 94, 2723 (1991).

[398] R.E. Wyatt, *Quantum Dynamics with Trajectories*, Springer, New York (2005).

[399] L. de Broglie, *Introduction à l'Étude de la Mécanique Ondulatoire*, Hermann et Cie, Paris (1930).

[400] J.S. Bell, *Speakable and Unspeakable in Quantum Mechanics*, Cambridge University Press, Cambridge (1987).

[401] C. Philippides, C. Dewdney, B.J. Hiley, *Nuovo Cimento* 52B, 15 (1979).

[402] R.E. Wyatt, *Chem. Phys. Lett.* 313, 189 (1999).

[403] C.L. Lopreore, R.E. Wyatt, *Phys. Rev. Lett.* 82, 5190 (1999).

[404] R.E. Wyatt, E.R. Bittner, *J. Chem. Phys.* 113, 8888 (2000).

[405] R.E. Wyatt, E.R. Bittner, *J. Chem. Phys.* 113, 8898 (2000).

[406] R.E. Wyatt, D.J. Kouri, D.K. Hoffman, *J. Chem. Phys.* 112, 10730 (2000).
[407] S. Garashchuk, V.A. Rassolov, *Chem. Phys. Lett.* 364, 562 (2002).
[408] B. Poirier, *J. Chem. Phys.* 121, 4501 (2004).
[409] B. Poirier, *J. Chem. Phys.* 128, 164115 (2008).
[410] B. Poirier, *J. Chem. Phys.* 129, 084103 (2008).
[411] D. Babyuk, R.E. Wyatt, *J. Chem. Phys.* 121, 9230 (2004).
[412] D. Babyuk, R.E. Wyatt, *Chem. Phys. Lett.* 400, 145 (2004).
[413] S. Garashchuk, V.A. Rassolov, *J. Chem. Phys.* 121, 8711 (2004).
[414] R.E. Wyatt, C.L. Lopreore, G. Parlant, *J. Chem. Phys.* 114, 5113 (2001).
[415] C.L. Lopreore, R.E. Wyatt, *J. Chem. Phys.* 116, 1228 (2002).
[416] S. Garashchuk, V.A. Rassolov, G.C. Schatz, *J. Chem. Phys.* 124, 244307 (2006).
[417] J.C. Burant, J.C. Tully, *J. Chem. Phys.* 112, 6097 (2000).
[418] S. Garashchuk, V.A. Rassolov, *J. Chem. Phys.* 118, 2482 (2003).
[419] S. Garashchuk, V.A. Rassolov, *Chem. Phys. Lett.* 376, 358 (2003).
[420] S. Garashchuk, V.A. Rassolov, *J. Chem. Phys.* 120, 1181 (2004).
[421] V.A. Rassolov, S. Garashchuk, *J. Chem. Phys.* 120, 6815 (2004).
[422] S. Garashchuk, V.A. Rassolov, G.C. Schatz, *J. Chem. Phys.* 123, 174108 (2005).
[423] N. Zamstein, D.J. Tannor, *J. Chem. Phys.* 137, 22A517 (2012).
[424] N. Zamstein, D.J. Tannor, *J. Chem. Phys.* 137, 22A518 (2012).
[425] Y. Goldfarb, I. Degani, D.J. Tannor, *J. Chem. Phys.* 125, 231103 (2006).
[426] W.A. Frensley, *Rev. Mod. Phys.* 62, 745 (1990).
[427] C.C. Martens, J.Y. Fang, *J. Chem. Phys.* 106 4918 (1997).
[428] I. Burghardt, K.B. Møller, *J. Chem. Phys.* 117 7409 (2002).
[429] I. Burghardt, K.B. Møller, K.H. Hughes, in: *Quantum Dynamics of Complex Molecular Systems*, D.A. Micha, I. Burghardt, eds., Springer, Berlin (2007).
[430] S. Mukamel, *Principles of Nonlinear Optical Spectroscopy*, Oxford University Press, Oxford (1995).
[431] M. Hillery, R.F. O'Connell, M.O. Scully, E.P. Wigner, *Phys. Rep.* 106, 121 (1984).
[432] J.E. Moyal, *Proc. Cambridge Phil. Soc.* 45, 99 (1949).
[433] D. Kosloff, R. Kosloff, *J. Comput. Phys.* 52, 35 (1983).
[434] A. Donoso, C.C. Martens, *J. Phys. Chem. A* 102, 4291 (1998).
[435] A. Donoso, D. Kohen, C.C. Martens, *J. Chem. Phys.* 112, 7345 (2000).
[436] K.H. Hughes, S.M. Parry, I. Burghardt, *J. Chem. Phys.* 130, 054115 (2009).
[437] E. Fick, G. Sauermann, *The Quantum Statistics of Dynamic Processes*, Springer Verlag, Berlin (1990).
[438] L.M. Martyushev, V.D. Selznev, *Phys. Rep.* 426, 1 (2006).
[439] C.D. Levermore, *J. Stat. Phys.* 83, 1021 (1996).

[440] I. Burghardt, L.S. Cederbaum, *J. Chem. Phys.* 115, 10312 (2001).

[441] I. Burghardt, K.B. Møller, G. Parlant, L.S. Cederbaum, E.B. Bittner, *Int. J. Quantum Chem.* 100, 1153 (2004).

[442] Theoretical Chemistry Group Heidelberg, MCTDG References, http://www.pci.uni-heidelberg.de/cms/mctdh.html (2012).

[443] D.J. Tannor, *Introduction to Quantum Mechanics. A Time-Dependent Perspective*, University Science Books, Sausalito, CA (2007).

[444] U. Manthe, in: *Quantum Simulations of Complex Many-Body Systems: From Theory to Algorithms*, Lecture Notes, Vol. 10, J. Grotendorst, D. Marx, A. Muramatsu, eds., Jülich, NIC Series (2002).

[445] J.C. Light, I.P. Hamilton, J.V. Lill, *J. Chem. Phys.* 82, 1400 (1985).

[446] J.C. Light, in: *Time Dependent Quantum Molecular Dynamics*, J. Broeckhove, L. Lathouwers, eds., Springer, Berlin (1992).

[447] D.T. Colbert, W.H. Milller, *J. Chem. Phys.* 96, 1982 (1992).

[448] E.O. Brigham, *The Fast Fourier Transform and its Applications*, Prentice Hall, Englewood Cliffs, NJ (1988).

[449] T.J. Park, J.C. Light, *J. Chem. Phys.* 85, 5870 (1986).

[450] L.F. Shampine, *Numerical Solution of Ordinary Differential Equations*, CRC Press, Boca Raton (1994).

[451] J.A. Fleck, J.R. Morris, M.D. Feit, *Appl. Phys.* 10, 129 (1979).

[452] A. Jeffrey, D. Zwillinger, eds., *Gradshteyn and Ryzhik's Table of Integrals, Sums, and Products*, Academic Press, San Diego, CA (2007).

[453] C.D. Neuhauser, *J. Chem. Phys.* 93, 2611 (1990).

[454] P.A.M. Dirac, *Proc. Cambridge Philos. Soc.* 26, 376 (1930).

[455] A.D. McLachlan, *Mol. Phys.* 8, 39 (1964).

[456] N. Makri, W.H. Miller, *J. Chem. Phys.* 87, 5781 (1987).

[457] H.-D. Meyer, U. Manthe, L.S. Cederbaum, *Chem. Phys. Lett.* 165, 73 (1990).

[458] A.D. Hammerich, U. Manthe, R. Kosloff, H.D. Meyer, L.S. Cederbaum, *J. Chem. Phys.* 101, 5623 (1994).

[459] L. Liu, J.-Y. Fang, H. Guo, *J. Chem. Phys.* 102, 2404 (1995).

[460] B. Pouilly, M. Monnerville, F. Gatti, H.-D. Meyer, *J. Chem. Phys.* 122, 184313 (2005).

[461] G.A. Worth, H.-D. Meyer, L.S. Cederbaum, *J. Chem. Phys.* 105, 4412 (1996).

[462] G.A. Worth, H.-D. Meyer, L.S. Cederbaum, *Chem. Phys. Lett.* 299, 451 (1999).

[463] J.-Y. Fang, H. Guo, *J. Chem. Phys.* 102, 1944 (1995).

[464] A. Borowski, O. Kühn, *Chem. Phys.* 347, 523 (2008).

[465] F. Matzkies, U. Manthe, *J. Chem. Phys.* 110, 88 (1999).

[466] A. Jaeckle, H.-D. Meyer, *J. Chem. Phys.* 109, 2614 (1998).

[467] A. Jaeckle, M.-C. Heitz, H.-D. Meyer, *J. Chem. Phys.* 110, 214 (1999).

[468] R. van Harrevelt, G. Nyman, U. Manthe *J. Chem. Phys.* 126, 084303 (2007).
[469] J.-Y. Fang, H. Guo, *Chem. Phys. Lett.* 235, 341 (1995).
[470] A.P.J. Jansen, H. Burghgraef, *Surf. Sci.* 344, 149 (1995).
[471] R. Milot, A.P.J. Jansen, *J. Chem. Phys.* 109, 1966 (1998).
[472] C. Crespos, H.-D. Meyer, R.C. Mowrey, G.J. Kroes, *J. Chem. Phys.* 124, 074706 (2006).
[473] H.-D. Meyer, U. Manthe, L.S. Cederbaum, *Chem. Phys. Lett.* 165, 73 (1990).
[474] A. Raab, G.A. Worth, H.-D. Meyer, L.S. Cederbaum, *J. Chem. Phys.* 110, 936 (1999).
[475] S. Mahapatra, G.A. Worth, H.D. Meyer, L.S. Cederbaum, H. Köppel, *J. Phys. Chem. A* 105, 5567 (2001).
[476] C. Cattarius, G.A. Worth, H.D. Meyer, L.S. Cederbaum, *J. Chem. Phys.* 115, 2088 (2001) .
[477] H. Köppel, M. Döscher, I. Baldea, H.D. Meyer, P.G. Szalay, *J. Chem. Phys.* 117, 2657 (2002).
[478] E.V. Gromov, A.B. Trofimov, N.M. Vitkovskaja, *J. Chem. Phys.* 121, 4585 (2004).
[479] I. Burghardt, H.D. Meyer, L.S. Cederbaum, *J. Chem. Phys.* 111, 2927 (1999).
[480] G.A. Worth, I. Burghardt, *Chem. Phys. Lett.* 368, 502 (2003).
[481] G.A. Worth, M.A. Robb, I. Burghardt, *Faraday Discuss.* 127, 307 (2004).
[482] M. Barbatti, H. Lischka, *J. Am. Chem. Soc.* 130, 6831 (2008).
[483] B. Cohen, P. Hare, B. Kohler, *J. Am. Chem. Soc.* 125, 13594 (2003).
[484] T. Gustavsson, A. Sharonov, D. Onidas, D. Markovitsi, *Chem. Phys. Lett.* 356, 49 (2002).
[485] S. Perun, A.L. Sobolewski, W. Domcke, *J. Am. Chem. Soc.* 127, 6257 (2004).
[486] R.D. Brown, P.D. Godfrey, D. McNaughton, A.P. Pierlot, *Chem. Phys. Lett.* 156, 61 (1989).
[487] A. Abo-Riziq, L. Grace, E. Nir, M. Kabelac, P. Hobza, M.S. deVries, *Proc. Natl. Acad. Sci. USA* 102, 20 (2005).
[488] A.L. Sobolewski, W. Domcke, C. Hättig, *Proc. Natl. Acad. Sci. USA* 102, 17903 (2005).
[489] A.L. Sobolewski, W. Domcke, *Phys. Chem. Chem. Phys.* 6, 2763 (2004).
[490] Z. Lan, W. Domcke, V. Vallet, A.L. Sobolewski, S. Mahapatra, *J. Chem. Phys.* 122, 224315 (2005).
[491] V. Vallet, Z. Lan, S. Mahapatra, A.L. Sobolewski, W. Domcke, *Faraday Discuss.* 127, 283 (2004).
[492] A.L. Sobolewski, W. Domcke, C. Dedonder-Lardeux, C. Jouvet, *Phys. Chem. Chem. Phys.* 4, 1093 (2002).
[493] H. Köppel, E.V. Gromov, A.B. Trofimov, *Chem. Phys.* 304, 35 (2004).
[494] B. Sellner, M. Barbatti, H. Lischka, *J. Chem. Phys.* 131, 024312 (2009).

[495] M. Barbatti, M. Vazdar, A. Aquino, M. Eckert-Maksic, H. Lischka, *J. Chem. Phys.* 125, 164323 (2006).

[496] V. Poterya, V. Profant, M. Farnik, P. Slavicek, U. Buck, *J. Chem. Phys.* 127, 064307 (2007).

[497] M. Vazdar, M. Eckert-Maksic, M. Barbatti, H. Lischka, *Mol. Phys.* 107, 845 (2009).

[498] S. Faraji, M. Vazdar, V. Sivaranjana Reddy, M. Eckert-Maksic, H. Köppel, *J. Chem. Phys.* 135, 154310 (2011).

[499] W. Kohn, L.J. Sham, *Phys. Rev.* 140, A1133 (1965).

[500] R.G. Parr, W. Yang, *Density Functional Theory of Atoms and Molecules*, 1st edition, Oxford University Press, Oxford (1989).

[501] E. Runge, E.K.U. Gross, *Phys. Rev. Lett.* 52, 12 (1984).

[502] L.H. Thomas, *Proc. Cambridge Phil. Soc.* 23, 542 (1926).

[503] E. Fermi, *Rend. Acad. Lincei* 6, 602 (1927).

[504] J.C. Slater, *Phys. Rev.* 81, 385 (1951).

[505] R. Gáspár, *Acta Phys. Acad. Sci. Hung.* 3, 263 (1954).

[506] P.W. Anderson, *Concepts in Solids*, World Scientific, Singapore (1997), Chapter 2.

[507] M. Levy, *Phys. Rev. A* 26, 1200 (1982).

[508] E.H. Lieb, in: *Physics as Natural Philosophy*, M. Feshbach, A. Shimony, eds., MIT Press, Cambridge, MA (1982).

[509] S.H. Vosko, L. Wilk, M. Nusair, *Can. J. Phys.* 58, 1200 (1980).

[510] J.P. Perdew, A. Zunger, *Phys. Rev. B* 23, 5048 (1981).

[511] L.A. Cole, J.P. Perdew, *Phys. Rev. A* 25, 1265 (1982).

[512] J.P. Perdew, Y. Wang, *Phys. Rev. B* 45, 13244 (1992).

[513] A.D. Becke, *Phys. Rev. A* 38, 3098 (1988).

[514] J.P. Perdew, in: *Electronic Structure of Solids '91*, P. Ziesche, H. Eschrig, eds., Akademie Verlag, Berlin (1991).

[515] J.P. Perdew, Y. Wang, *Phys. Rev. B* 33, 8800 (1986).

[516] A. Becke, *J. Chem. Phys.* 98, 5648 (1993).

[517] C. Lee, W. Yang, R. G. Parr, *Phys. Rev. B* 37, 785 (1988).

[518] R. Gaudoin, K. Burke, *Phys. Rev. Lett.* 93, 173001 (2004).

[519] Publisher's Note, *Phys. Rev. Lett.* 94, 029901 (2005).

[520] O. Gunnarson, B. Lundquist, *Phys. Rev. B* 13, 4274 (1976).

[521] I. Frank, J. Hutter, D. Marx, M. Parrinello, *J. Chem. Phys.* 108, 4060 (1998).

[522] J.P. Perdew, M. Levy, *Phys. Rev. B* 31, 6264 (1985).

[523] A. Goerling, *Phys. Rev. A* 59, 3359 (1999).

[524] A. Goerling, *Phys. Rev. Lett.* 85, 4229 (2000).

[525] A. Theophilou, *J. Phys. C* 12, 5419 (1979).

[526] E.K.U. Gross, L.N. Oliveira, W. Kohn, *Phys. Rev. A* 37, 2805 (1988).

[527] E.K.U. Gross, L.N. Oliveira, W. Kohn, *Phys. Rev. A* 37, 2809 (1988).
[528] L.N. Oliveira, E.K.U. Gross, W. Kohn, *Phys. Rev. A* 37, 2821 (1988).
[529] N.I. Gidopoulos, P.G. Papaconstantinou, E.K.U. Gross, *Phys. Rev. Lett.* 88, 033003 (2002).
[530] R. van Leeuwen, *Phys. Rev. Lett.* 80, 1280 (1998).
[531] R. van Leeuwen, *Int. J. Mod. Phys. A* 15, 1969 (2001).
[532] G. Vignale, *Phys. Rev. A* 77, 062511 (2009).
[533] N.T. Maitra, K. Burke, C. Woodward, *Phys. Rev. Lett.* 89, 023001 (2002).
[534] K. Burke, J. Werschnik, E.K.U. Gross, *J. Chem. Phys.* 123, 062206 (2005).
[535] M. Petersilka, U.J. Gossmann, E.K.U. Gross, *Phys. Rev. Lett.* 76, 1212 (1996).
[536] M.A.L. Marques, E.K.U. Gross, *Annu. Rev. Phys. Chem.* 55, 427 (2004).
[537] Y.M. Niquet, S. Fuchs, X. Gonze, *Phys. Rev. A* 68, 032507 (2003).
[538] M.A.L. Marques, A. Castro, A. Rubio, *J. Chem. Phys.* 115, 3006 (2001).
[539] R.T. Sharp, G.K. Horton, *Phys. Rev.* 90, 317 (1953).
[540] T. Grabo, T. Kreibich, S. Kurth, E.K.U. Gross, in: *Strong Coulomb Correlations in Electronic Structure Calculations: Beyond the Local Density Approximation*, V.I. Anisimov, ed., Gordon and Breach, Amsterdam (2000).
[541] M. Van Faasen, P.L. de Boeij, R. van Leeuwen, J.A. Berger, J.G. Snijders, *Phys. Rev. Lett.* 88, 186401 (2002).
[542] G. Vignale, W. Kohn, *Phys. Rev. Lett.* 77, 2037 (1996).
[543] G. Vignale, C.A. Ullrich, S. Conti. *Phys. Rev. Lett.* 79, 4878 (1997).
[544] G.F. Giuliani, G. Vignale, *Quantum Theory of the Electron Liquid*, Cambridge University Press, Cambridge (2005).
[545] K. Taguchi, J. Haryyama, K. Watanabe, *J. Phys. Soc. Japan* 78, 094707 (2009).
[546] A.D. Becke, K.E. Edgecombe, *J. Chem. Phys.* 92, 5397 (1989).
[547] J.F. Dobson, *J. Chem. Phys.* 98, 8870 (1993).
[548] T. Burnus, M.A.L. Marques, E.K.U. Gross, *Phys. Rev. A* 71, 010501 (2005).
[549] D.B. Chesnut, *J. Phys. Chem. A* 104, 11644 (2000).
[550] A. Savin, R. Nesper, S. Wengert, T.F. Faessler, *Angew. Chem. Int. Ed. Engl.* 36, 1808 (1997).
[551] N. Doltsinis, D. Marx, *Phys. Rev. Lett.* 88, 166402 (2002).
[552] C.F. Craig, W.R. Duncan, O.E. Prezhdo, *Phys. Rev. Lett.* 95, 163001 (2005).
[553] E. Tapavicza, I. Tavernelli, U. Röthlisberger, C. Filippi, M.E. Casida, *J. Chem. Phys.* 129, 124108 (2008).
[554] I. Tavernelli, B.F.E. Chuchod, U. Röthlisberger, *Phys. Rev. A* 81, 052508 (2010).
[555] S.A. Fischer, B.F. Habenicht, A.B. Madrid, W.R. Duncan, O.V. Prezhdo, *J. Chem. Phys.* 134, 024102 (2011).
[556] C. Hu, H. Hirai, O. Sugino, *J. Chem. Phys.* 127, 064103 (2007).

[557] C. Hu, O. Sugino, Y. Tateyama, *J. Chem. Phys.* 131, 114101 (2009).

[558] C. Hu, O. Sugino, H. Hirai, Y. Tateyama, *Phys. Rev. B* 82, 062508 (2010).

[559] I. Tavernelli, E. Tapavicza, U. Röthlisberger, *J. Chem. Phys.* 130, 124107 (2009).

[560] R. Send, F. Furche, *J. Chem. Phys.* 132, 044107 (2010).

[561] J.L. Alonso, X. Andrade, P. Echenique, F. Falceto, D. Prada-Gracia, A. Rubio, *Phys. Rev. Lett.* 101, 096403 (2008).

[562] C. Moss, C.M. Isborn, X. Li, *Phys. Rev.* 80, 024503 (2009).

[563] W.H. Zurek, *Rev. Mod. Phys.* 75, 715 (2003).

[564] W.H. Zurek, *Phys. Today* 44(10), 36 (1991).

[565] E.R. Bittner, P.J. Rossky, *J. Chem. Phys.* 103, 8130 (1995).

[566] G.C. Ghirardi, P. Pearle, A. Rimini, *Phys. Rev. A* 42, 78 (1990).

[567] B.W.N. Lo, R.J.R. Cable, N.R. Carlsen, *Chem. Phys. Lett.* 26, 252 (1974).

[568] A.O. Caldeira, A.J. Leggett, *Ann. Phys.* 149, 374 (1983).

[569] B.J. Schwartz, E.R. Bittner, O.V. Prezhdo, P.J. Rossky, *J. Chem. Phys.* 104, 5942 (1996).

[570] E. Neria, A. Nitzan, *J. Chem. Phys.* 99, 1109 (1993).

[571] O. Prezdho, P. Rossky, *J. Chem. Phys.* 107, 5863 (1997).

[572] E.J. Heller, *J. Chem. Phys.* 75, 2923 (1981).

[573] K.F. Wong, P.J. Rossky, *J. Chem. Phys.* 116, 8429 (2002).

[574] A. Staib, D. Borgis, *J. Chem. Phys.* 103, 2642 (1995).

[575] L. Turi, P.J. Rossky, *J. Chem. Phys.* 120, 3688 (2004).

[576] E. Neria, A. Nitzan, R.N. Barnett, U. Landmann, *Phys. Rev. Lett.* 67, 1011 (1991).

[577] G. Granucci, M. Persico, A. Zoccante, *J. Chem. Phys.* 133, 134111 (2010).

[578] J.C. Thompson, *Electrons in Liquid Ammonia*, Clarendon Press, Oxford (1976).

[579] A.M. Brodsky, A.V. Tsarevsky, *Adv. Chem. Phys.* 44, 483 (1980).

[580] A.M. Kuznetsov, *Charge Transfer in Physics, Chemistry and Biology*, Gordon and Breach, New York (1995).

[581] T. Ichikawa, H. Yoshida, *J. Chem. Phys.* 73, 1540 (1980).

[582] E. Hart, M. Anbar, *The Hydrated Electron*, Wiley, New York (1970).

[583] L. Turi, P.J. Rossky, *Chemical Reviews* 112, 5641 (2012).

[584] E.J. Hart, J.W. Boag, *J. Am. Chem. Soc.* 84, 4090 (1962).

[585] F.H. Long, H. Lu, K.B. Eisenthal, *Phys. Rev. Lett.* 64, 1469 (1990).

[586] F.H. Long, H. Lu, X. Shi, K.B. Eisenthal, *Chem. Phys. Lett.* 185, 47 (1991).

[587] A. Migus, Y. Gauduel, J.L. Martin, A. Antonetti, *Phys. Rev. Lett.* 58, 1559 (1987).

[588] S. Pommeret, A. Antonetti, Y. Gauduel, *J. Am. Chem. Soc.* 113, 9105 (1991).

[589] F.A. Webster, J. Schnitker M.S. Friedrichs, R.A. Friesner, P.J. Rossky, *Phys. Rev. Lett.* 66, 3172 (1991).

[590] T.H. Murphrey, P.J. Rossky, *J. Chem. Phys.* 99, 515 (1993).

[591] J. Schnitker, K. Motakabbir, P.J. Rossky, R.A. Friesner, *Phys. Rev. Lett.* 60, 456 (1988).

[592] D. Eisenberg, W. Kauzmann, *The Structure and Properties of Water*, Oxford University Press, Oxford (1969).

[593] H.J.C. Berendsen, J.P.M. Postma, W.F. van Gunsteren, J. Hermans, in: *Intermolecular Forces*, B. Pullman, ed., Reidel, Dordrecht (1981).

[594] J. Schnitker, P.J. Rossky, *J. Chem. Phys.* 86, 3462 (1987).

[595] J.C. Alfano, P.K. Walhout, Y. Kimura, P.F. Barbara, *J. Chem. Phys.* 98, 5996 (1993).

[596] Y. Kimura, J.C. Alfano, P.W. Walhout, P.F. Barbara, *J. Phys. Chem.* 98, 3450 (1994).

[597] P.J. Reid, C. Silva, P.K. Walhout, P.F. Barbara, *Chem. Phys. Lett.* 228, 658 (1994).

[598] B.J. Schwartz, P.J. Rossky, *J. Chem. Phys.* 101, 6902, 6917 (1994).

[599] B.J. Schwartz, P.J. Rossky *J. Phys. Chem.* 98, 4489 (1994).

[600] B.J. Schwartz, P.J. Rossky, *J. Chem. Phys.* 105, 6997 (1996).

[601] C. Silva, P. Walhout, K. Yokoyama, P.F. Barbara, *Phys. Rev. Lett.* 80, 1086 (1998).

[602] K. Yokoyama, C. Silva, D.H. Son, P.K. Walhout, P. Barbara, *J. Phys. Chem. A* 102, 6957 (1998).

[603] F.A. Webster, E.T. Wang, P.J. Rossky, R.A. Friesner, *J. Chem. Phys.* 100, 4835 (1994).

[604] T.H. Murphrey, P.J. Rossky, *J. Chem. Phys.* 103, 6665 (1995).

[605] J.C. Tully, R.K. Preston, *J. Chem. Phys.* 55, 562 (1971).

[606] C.Y. Yang, K.F. Wong, M.S. Skaf, P.J. Rossky, *J. Chem. Phys.* 114, 3598 (2001).

[607] K.F. Wong, P.J. Rossky, *J. Chem. Phys.* 116, 8418 (2002).

[608] O.V. Prezhdo, P.J. Rossky, *J. Chem. Phys.* 107, 825 (1997).

[609] K.F. Wong, P.J. Rossky, *J. Phys. Chem. A* 105, 2546 (2001).

[610] D. Borgis, P. Rossky, L. Turi, *J. Chem. Phys.* 125, 064501 (2006).

[611] R.E. Larsen, M.J. Bedard-Hearn, B.J. Schwartz, *J. Phys. Chem. B* 110, 20055 (2006).

[612] G. Stock, W. Domcke, in: *Conical Intersections*, W. Domcke, D. Yarkony, H.Köppel, eds., World Scientific, Singapore (2004), Chapter 17.

[613] M.D. Hack, D.G. Truhlar, *J. Phys. Chem. A* 104, 7917 (2000).

[614] H.D. Meyer, W.H. Miller, *J. Chem. Phys.* 70, 3214 (1979).

[615] M.D. Hack, D.G. Truhlar, *J. Chem. Phys.* 114, 2894 (2001).

[616] A.W. Jasper, C. Zhu, S. Nangia, D.G. Truhlar, *Faraday Discuss.* 127, 1 (2004).

[617] C. Zhu, A.W. Jasper, D.G. Truhlar, *J. Chem. Theor. Comp.* 1, 527 (2005).
[618] C. Zhu, S. Nangia, A.W. Jasper, D.G. Truhlar, *J. Chem. Phys.* 121, 7658 (2004).
[619] Y. Elran, P.J. Brumer, *Chem. Phys.* 121, 2673 (2004).
[620] S.C. Cheng, C. Zhu, K.K. Liang, S.H. Lin, D.G. Truhlar, *J. Chem. Phys.* 129, 024112 (2004).
[621] A.W. Jasper, D.G. Truhlar, *J. Chem. Phys.* 122, 44101 (2005).
[622] A.W. Jasper, M.D. Hack, D.G. Truhlar, *J. Chem. Phys.* 115, 1804 (2001).
[623] P.A. Franken, A.E. Hill, C.W. Peters, G. Weinrich, *Phys. Rev. Lett.* 7, 118 (1961).
[624] C.B. Harris, E.P. Ippen, G.A. Mourou, A.H. Zewail, eds., *Ultrashort Phenomena VII*, Springer Verlag, Berlin (1990).
[625] P. Gibbon, *Short Pulse Laser Interactions with Matter: An Introduction*, World Scientific, Singapore (2005).
[626] A. Weiner, *Ultrafast Optics*, Wiley, New Jersey (2009).
[627] V. Sundstroem, ed., *Femtochemistry and Femtobiology*, World Scientific, Singapore (1997).
[628] M. Chergui, ed., *Femtochemistry*, World Scientific, Singapore (1996).
[629] D.A. Wiersma, ed., *Femtosecond Reaction Dynamics*, North Holland, Amsterdam (1994).
[630] A.E. Johnson, W. Jarzeba, G.C. Walker, P.F. Barbara, *Israel. J. Chem.* 33, 199 (1993).
[631] P. Foggi, L. Pettini, I. Santa, R. Righini, S. Califano, *J. Chem. Phys.* 99, 7439 (1995).
[632] T. Baumert, C.R. öttgermann, C. Rothenfusser, P. Thalweiser, V. Weiss, G. Gerber, *Phys. Rev. Lett.* 69, 1512 (1992).
[633] N. Kuthirummal, F.M. Rudakov, C.L. Evans, P.M. Weber, *J. Chem. Phys.* 125, 133307 (2006).
[634] G. Stock, W. Domcke, *Phys. Rev. A* 45, 3032 (1991).
[635] A. Icsevgi, W.E. Lamb, *Phys. Rev.* 185, 517 (1969).
[636] Y.R. Shen, *The Principles of Nonlinear Optics* Wiley, New York (1984).
[637] L. Seidner, G. Stock, W. Domcke, *J. Chem. Phys.* 103, 3998 (1995).
[638] F. Bloch, A. Siegert, *Phys. Rev.* 57, 522 (1940).
[639] W.T. Pollard, S-Y. Lee, R.A. Mathies, *J. Chem. Phys.* 92, 4012 (1990).
[640] S.H. Pine, *Organic Chemistry*, McGraw Hill Book Company, New York (1987).
[641] W.M. Gelbert, K.F. Freed, S.A. Rice, *J. Chem. Phys.* 52, 2460 (1970).
[642] B. Bagchi, G.R. Fleming, D.W. Oxtoby, *J. Chem. Phys.* 78, 7375 (1983).
[643] A. Warshel, Z.T. Chu, J.K. Hwang, *Chem. Phys.* 158, 303 (1991).
[644] L. Seidner, G. Stock, W. Domcke, *Chem. Phys. Lett.* 228, 665 (1994).

[645] A. Bifone, H.J.M. de Groot, F. Buda, *J. Phys. Chem. B* 101, 2954 (1997).

[646] M. Garavelli, P. Celani, F. Bernardi, M.A. Robb, M. Olivucci, *J. Am. Chem. Soc.* 119, 6891 (1997).

[647] S. Hahn, G. Stock, *Phys. Chem. Chem. Phys.* 3, 2331 (2001).

[648] D. Polli, P. Altoe, O. Weingart *et al.*, *Nature* 467, 441 (2010).

[649] Z. Li, J.-Y. Fang, C.C. Martens, *J. Chem. Phys.* 104, 6919 (1996).

[650] M. Hartmann, J. Pittner, V. Bonacic-Koutecky, A. Heidenreich, J. Jortner, *J. Chem. Phys.* 108, 3096 (1998).

[651] M. Hartmann, J. Pittner, V. Bonacic-Koutecky, *J. Chem. Phys.* 114, 2106 (2001).

[652] M. Atos-Abiague, J. Berakdar, *Europhys. Lett.* 69, 277 (2005).

[653] M. Hartmann, J. Pittner, V. Bonacic-Koutecky, *J. Chem. Phys.* 114, 2123 (2001).

[654] S. Vajda, C. Lupulescu, A. Merli *et al. Phys. Rev. Lett.* 89, 213404 (2002).

[655] M. Shapiro, P. Brumer, *Principles of the Quantum Control of Molecular Processes*, John Wiley, New Jersey (2003), Section 1.1.

[656] *ibid.*, Section 2.1.

[657] S.A. Rice, M. Zhao, *Optical Control of Molecular Dynamics*, John Wiley, New York (2000), Section 3.1.2.

[658] R. Kosloff, S.A. Rice, P. Gaspard, S. Tersigni, D.J. Tannor, *Chem. Phys.* 139, 201 (1989).

[659] B. Armstrup, R.J. Carlson, A. Matro, S.A. Rice, *J. Phys. Chem.* 95, 8019 (1991).

[660] B. Hartke, E. Kolba, J. Manz, H.H.R. Schor, *Ber. Bunsengesellschaft Phys. Chem.* 954, 1312 (1990).

[661] M. Shapiro, P. Brumer, *Principles of the Quantum Control of Molecular Processes*, John Wiley, Hoboken (2003), Section 3.1.1.

[662] G.C. Pimentel, J.A. Coonrod, *Opportunities in Chemistry Today and Tomorrow*, National Academy Press, Washington D.C. (1987).

[663] R. Schinke, *Photodissociation Dynamics*, Cambridge University Press, Cambridge (1993).

[664] V.S. Batista, P. Brumer, *J. Phys. Chem. A* 105, 2591 (2001).

[665] F. Shokoohi, S. Hay, C. Wittig, *Chem. Phys. Lett.* 82, 3885 (1985).

[666] W.P. Hess, S.R. Leone, *J. Chem. Phys.* 86, 3773 (1987).

[667] J.F. Black, J.R. Waldeck, R.N. Zare, *J. Chem. Phys.* 92, 3519 (1990).

[668] A. M.O. Morse, K.F. Freed, Y. Band, *J. Chem. Phys.* 70, 3604 (1979).

[669] I. Krylov, R.B. Gerber, *J. Chem. Phys.* 100, 4242 (1994).

[670] E.M. Goldfarb, P.L. Houston, G.S. Ezra, *J. Chem. Phys.* 84, 3120 (1986).

[671] S. Yabushita, K. Morokuma, *Chem. Phys. Lett.* 175, 518 (1990).

[672] Y. Amamtatsu, S. Yabushita, K. Morokuma, *J. Chem. Phys.* 100, 4858 (1991).

[673] I. Nadler, D. Mahgerefteh, H. Reisler, C. Wittig, *J. Chem. Phys.* 82, 3885 (1985).
[674] V.S. Batista, P. Brumer, *J. Chem. Phys.* 114, 10321 (2001).
[675] E.A. Coronado, V.S. Batista, W.H. Miller, *J. Chem. Phys.* 112, 5566 (2000).
[676] R. Mitric, M. Hartmann, J. Pittner, V. Bonacic-Koutecky, *J. Phys. Chem. A* 106, 10477 (2002).
[677] Y. Yan, R.E. Gillilan, R.M. Whitnell, K.R. Wilson, S. Mukamel, *J. Phys. Chem.* 97, 2320 (1993).
[678] Y. Yan, S. Mukamel, *J. Chem. Phys.* 88, 5735 (1988).
[679] J.L. Krause, R.M. Whitnell, K.R. Wilson, Y. Yan, *J. Chem. Phys.* 99, 6562 (1993).
[680] V. Bonacic-Koutecky, R. Mitric, M. Hartmann, J. Pittner, *Int. J. Quant. Chem.* 99, 408 (2004).
[681] R. Mitric, U. Werner, V. Bonacic-Koutecky, *J. Chem. Phys.* 129, 164118 (2008).
[682] R. Mitric, U. Werner, M. Wohlgemut, G. Seifert, V. Bonacic-Koutecky, *J. Phys. Chem. A* 113, 12700 (2009).
[683] M. Shapiro, P. Brumer, *Faraday Discuss. Chem. Soc.* 82, 177 (1987).
[684] C.K. Chan, P. Brumer, M. Shapiro, *J. Chem. Phys.* 94, 2688 (1991).
[685] R. Mitric, J. Petersen, V. Bonacic-Koutecky, *Phys. Rev. A* 79, 053416 (2009).
[686] R. Mitric, J. Petersen, V. Bonacic-Koutecky, in: *Conical Intersections*, Vol.2, H. Köppel, W. Domcke, D. Yarkony, eds., World Scientific, Singapore (2010).
[687] A.M. Kuznetsov, J. Ulstrup, *Electron Transferin Chemistry and Biology. An Introduction to the Theory*, Wiley, Chichester (1999).
[688] H. Taube, *Electron Transfer Reactions in Solution*, Academic Press, New York (1970).
[689] D. Chandler, in: *Classical and Quantum Dynamics in Condensed Phase Simulations*, World Scientific, Singapore (1998), Chapter 2.
[690] P.L. Dutton, K.J. Kaufman, B. Chance, P.M. Rentzepis, *Biophys. J.* 23, 207 (1978).
[691] R.A. Marcus, N. Sutin, in: *Antennas and Reaction Centers of Photosynthetic Bacteria*, M.E. Micel-Beyerle, ed., Springer, New York (1985).
[692] R.A. Kuharski, J.S. Bader, D. Chandler, M. Sprok, M.L. Klein, R.W. Impey, *J. Chem. Phys.* 89, 3248 (1988).
[693] J.S. Bader, R.A. Kuharski, D. Chandler, *J. Chem. Phys.* 93, 230 (1990).
[694] V. May, O. Kühn, *Charge and Energy Transfer Dynamics in Molecular Systems*, 2nd edition, Wiley-VCH, Weinheim (2004), Section 6.2.
[695] *ibid.*
[696] *ibid.*, Section 6.4.1.

[697] A.J. Leggett, S. Chakravarty, A.T. Dorsey, M.P.A. Fisher, A. Garg, W. Zwerger, *Rev. Mod. Phys.* 59, 1 (1987).
[698] N.R. Kestner, J. Logan, J. Jortner, *J. Phys. Chem.* 78, 2148 (1974).
[699] H.J. Kim, J.T. Hynes, *J. Phys. Chem.* 96, 5088 (1992).
[700] J.D. Jackson, *Classical Electrodynamics*, 2nd edition, John Wiley, New York (1962), Section 7.10.
[701] V. May, O. Kühn, *Charge and Energy Transfer Dynamics in Molecular Systems*, Wiley-VCH, 2nd edition, Weinheim (2004), Section 6.9.3.
[702] *ibid.*, Section 5.3.3.
[703] H.J. Kim, J.T. Hynes, *J. Phys. Chem.* 93, 5194 (1990).
[704] I. Rips, J. Jortner, *J. Chem. Phys.* 87, 2090 (1987).
[705] H.J. Kim, J.T. Hynes, *J. Chem. Phys.* 94, 2736 (1990).
[706] J. Blumberger, *Phys. Chem. Chem. Phys.* 10, 5651 (2008).
[707] K. Yoshihara, K. Tominaga, Y. Nagasawa, *Bull. Chem. Soc. Jpn.* 68, 696 (1995).
[708] P.F. Barbara, T.J. Meyer, M.A. Ratner, *J. Phys. Chem.* 100, 13148 (1996).
[709] D. Egorova, A. Kühl, W. Domcke, *Chem. Phys.* 268, 105 (2001).
[710] C.J. Murphy, M.R. Arkin, Y. Jenkins *et al.*, *Science* 262, 1025 (1993).
[711] E.D. Stemp, M.R. Arkin, J.K. Barton, *J. Am. Chem. Soc.* 117, 2375 (1995).
[712] T.J. Kang, W. Jarzeba, P.F. Barbara, T. Fonseca, *Chem. Phys.* 149, 81 (1990).
[713] D. Egorova, M. Thoss, W. Domcke, H. Wang, *J. Chem. Phys.* 119, 2761 (2003).
[714] U. Kleinkathöfer, I. Kondov, M. Schreiber, *Chem. Phys.* 268, 121 (2001).
[715] N.E. Andersen, T. Lian, *Annu. Rev. Phys. Chem.* 56, 491 (2005).
[716] A. Nitzan, M.A. Ratner, *Science* 300, 1384 (2003).
[717] U. Bach, D. Lupo, P. Comte, *Nature* 395, 583 (1998).
[718] M. Grätzel, J.E. Moser, in: *Electron Transfer in Chemistry*, Vol. 5, V. Balzani, ed., Wiley-VCH, Weinheim (2001).
[719] B.L. Tembe, H.L. Friedman, M.D. Newton, *J. Chem. Phys.* 76, 1490 (1982).
[720] J. Logan, M.D. Newton, *J. Chem. Phys.* 78, 4086 (1983).
[721] M. Marchi, D. Chandler, *J. Chem. Phys.* 95, 889 (1991).
[722] N. Metropolis, A.W. Rosenbluth, M.N. Rosenbluth, A.H. Teller, E. Teller, *J. Chem. Phys.* 21, 1087 (1953).
[723] D. Chandler, *Introduction to Modern Statistical Mechanics*, Oxford University Press, Oxford (1987), Section 6.3.
[724] J. Silverman, R.W. Dodson, *J. Phys. Chem.* 56, 846 (1952).
[725] M.D. Newton, *Int. J. Quant. Chem.* 14, 363 (1980).
[726] B.S. Brunschwig, J. Logan, M.D. Newton, N. Sutin, *J. Am. Chem. Soc.* 102, 5798 (1980).
[727] P.G. Wolynes, *J. Chem. Phys.* 87, 6559 (1987).

[728] T. Guarr, E. Buhks, G. McLendon, *J. Am. Chem. Soc.* 105, 3763 (1983).

[729] J. Hudis, R.W. Dodson, *J. Am. Chem. Soc.* 78, 911 (1956).

[730] J.S. Bader, D. Chandler, *Chem. Phys. Lett.* 157, 501 (1989).

[731] P. Siders, R.A. Marcus, *J. Am. Chem. Soc.* 103, 741 (1981).

[732] X. Song, R.A. Marcus, *J. Chem. Phys.* 99, 7768 (1993).

[733] J. Blumberger, M. Sprik, *J. Phys. Chem. B* 108, 6529 (2004).

[734] J. Blumberger, L. Bernasconi, I. Tavernelli, R. Vuilleumier, M. Sprik, *J. Am. Chem. Soc.* 126, 3928 (2004).

[735] J. Blumberger, M. Sprik, *J. Phys. Chem. B* 109, 6793 (2005).

[736] Y. Tateyama, J. Blumberger, M. Sprik, I. Tavernelli, *J. Phys. Chem.* 122, 234505 (2005).

[737] J. Blumberger, I. Tavernelli, M.L. Klein, M. Sprik, *J. Chem. Phys.* 124, 064507 (2006).

[738] M. Sulpizi, S. Raugei, J. VandeVondele, P. Carloni, M. Sprik, *J. Phys. Chem. B* 111, 3969 (2007).

[739] A. Warshel, M. Levitt, *J. Mol. Biol.* 103, 227 (1976).

[740] Y. Zhang, H. Liu, W. Yang, *J. Chem. Phys.* 112, 3483 (2000).

[741] H. Hu, W. Yang, *Annu. Rev. Phys. Chem.* 59, 034105 (2008).

[742] X. Zeng, H. Hu, X. Hu, A.J. Cohen, W. Yang, *J. Chem. Phys.* 128, 124510 (2008).

[743] J. Blumberger, G. Lamoureux, *Mol. Phys.* 106, 1362 (2008).

[744] H. Oberhofer, J. Blumberger, *Angew. Chem. Int. Ed.* 49, 1 (2010).

[745] J.R. Rustad, K.M. Rosso, A.R. Felmy, *J. Chem. Phys.* 120, 7607 (2004).

[746] V. May, O. Kühn, *Charge and Energy Transfer Dynamics in Molecular Systems*, 2nd edition, Wiley-VCH, Weinheim (2004), Section 5.8.

[747] T.M. Chaikin, T.C. Lubensky, *Principles of Condensed Matter Physics*, Cambridge University Press, Cambridge (1995).

[748] D.F. Calef, P.G. Wolynes, *J. Phys. Chem.* 87, 3387 (1983).

[749] B.U. Felderhof, *J. Chem. Phys.* 67, 493 (1977).

[750] G.D. Billing, *Dynamics of Molecule-Surface Interactions*, John Wiley and Sons, New Jersey (2000).

[751] H. Nienhaus, H.S. Bergh, B. Gergen, A. Majumdar, W.H. Weinberg, E.W. McFarland, *Phys. Rev. Lett.* 82, 446 (1999).

[752] B. Gergen, H. Nienhaus, W.H. Weinberg, E.W. McFarland, *Science* 294, 2521 (2001).

[753] G.A. Somorjai, *Catal. Lett.* 101, 1 (2005).

[754] J.Y. Park, G.A. Somorjai, *Chem. Phys. Chem.* 7, 1409 (2006).

[755] B. Mildner, E. Hasselbrink, D. Diesing, *Chem. Phys. Lett.* 432, 133 (2006).

[756] Y. Huang, A.M. Wodtke, H. Hou, C.T. Rettner, D.J. Auerbach, *Phys. Rev. Lett.* 84, 2985 (2005).

[757] J.D. White, J. Chen, D. Matsiev, D.J. Auerbach, A.M. Wodtke, *Nature* 433, 503 (2005).

[758] J.D. White, J. Chen, D. Matsiev, D.J. Auerbach, A.M. Wodtke, *J. Chem. Phys.* 124, 064702 (2006).

[759] Y.H. Huang, C.T. Rettner, D.J. Auerbach, A.M. Wodtke, *Science* 290, 111 (2000).

[760] B.N.J. Persson, *J. Phys. C* 11, 4251 (1978).

[761] H. Metiu, J.W. Gadzuk, *J. Chem. Phys.* 74, 2641 (1981).

[762] M. Head-Gordon, J.C. Tully, *J. Chem. Phys.* 96, 3939 (1992).

[763] S. Roy, N. Shenvi, J.C. Tully, *J. Phys. Chem. C* 113, 16311 (2008).

[764] N. Shenvi, S. Roy, J.C. Tully, *J. Chem. Phys.* 130, 174107 (2009).

[765] Y. Huang, A.M. Wodtke, H. Hou, C.T. Rettner, D.J. Auerbach, *Phys. Rev. Lett.* 84, 2985 (2005).

[766] M. Persson, B. Hellsing, *Phys. Rev. Lett.* 49, 662 (1982).

[767] T.T. Rantala, A. Rosen, *Phys. Rev. B* 34, 837 (1986).

[768] R. Ryberg, *Adv. Chem. Phys.* 76, 1 (1989).

[769] L.F. Sutcu, H.W. White, J.L. Wragg, *Surf. Sci. Lett.* 249, L343 (1991).

[770] B.N.J. Persson, M. Persson, *Surf. Sci.* 97, 609 (1980).

[771] A.G. Eguiluz, *Phys. Rev. B* 30, 4366 (1984).

[772] A. Liebsch, *Phys. Rev. Lett.* 54, 67 (1985).

[773] B.N.J. Persson, M. Persson, *Solid State Commun.* 36, 175 (1980).

[774] M. Head-Gordon, J.C. Tully, *J. Chem. Phys.* 103, 10137 (1995).

[775] S.A. Adelman, J.D. Doll, *J. Chem. Phys.* 64, 2375 (1976).

[776] R. Zwanzig, *Annu. Rev. Phys. Chem.* 16, 67 (1965).

[777] H. Mori, *Progr. Theor. Phys.* 33, 423 (1965).

[778] R. Kubo, *Rep. Progr. Phys.* 29, 255 (1966).

[779] M. Kertész, *Adv. Quantum Chem.* 15, 161 (1982).

[780] M. Morin, N.J. Levinos, A.L. Harris, *J. Chem. Phys.* 96, 3950 (1992).

[781] K. Herman, P.S. Bagus, C.J. Nelin, *Phys. Rev. B* 35, 9467 (1987).

[782] M. Head-Gordon, J.C. Tully, *Phys. Rev. B* 46, 1853 (1992).

[783] J.T. Kindt, J.C. Tully, M. Head-Gordon, M.A. Gomez, *J. Chem. Phys.* 109, 3629 (1998).

[784] J.C. Tully, M. Gomez, M. Head-Gordon, *J. Vac. Sci. Technol.* A77, 1914 (1993).

[785] C. Springer, M. Head-Gordon, *Chem. Phys.* 205, 73 (1996).

[786] M. Ceotto, D. Dell'Angelo, G.F. Tantardini, *J. Chem. Phys.* 133, 054701 (2010).

[787] H. Nienhaus, W.H. Weinberg, E.W. McFarland, *Science* 294, 2521 (2001).

[788] J.W. Gadzuk, *J. Chem. Phys.* 79, 6341 (1983).

[789] S. Roy, N.A. Shenvi, J.C. Tully, *J. Chem. Phys.* 130, 174716 (2009).

[790] S. Li, H. Guo, *J. Chem. Phys.* 117, 4499 (2002).

[791] N.A. Shenvi, S. Roy, P. Parandekar, J.C. Tully, *J. Chem. Phys.* 125, 154703 (2006).

[792] N.A. Shenvi, J.R. Schmidt, S.T. Edwards, J.C. Tully, *Phys. Rev. A* 78, 022502 (2009).

[793] D. Newns, *Phys. Rev.* 178, 1123 (1969).

[794] G. Katz, Y. Zeiri, R. Kosloff, *J. Phys. Chem. B* 109, 18876 (2005).

[795] S. Matsika, P. Krause, *Annu. Rev. Phys. Chem.* 62, 621 (2011).

[796] E. Tapavicza, A.M. Meyer, F. Furche, *Phys. Chem. Chem. Phys.*, 13, 20986 (2011).

[797] O.V. Prezhdo, W.R. Duncan, V.V. Prezhdo, *Prog. Surf. Sci.* 84, 30 (2009).

[798] S.O. Obare, T. Ito, G.J. Meyer, *J. Am. Chem. Soc.* 128, 712 (2006).

[799] N.S. Lewis, *J. Electroanal. Chem.* 508, 1 (2001).

[800] R.D. Sytnik, V.V. Prezhdo, V.Y. Matveenk, G.M. Legoshin, *Glass. Sci. Technol.* 74, 161 (2001).

[801] R. Kapral, *Ann. Rev. Phys. Chem.* 57, 129 (2006).

[802] A. Croy, U. Saalmann, A.R. Hernandez, C.H. Lewenkopf, *Phys. Rev. B* 85, 035309 (2012).

[803] J. Zanghellini, M. Kitzler, C. Fabian, T. Brabec, A. Scrinzi, *Laser Physics* 13, 1064 (2003).

[804] T. Kato, H. Kono, *Chem. Phys. Lett.* 392, 533 (2004).

[805] A.I. Streltsov, O.E. Alon, L.S. Cederbaum, *Phys. Rev. Lett.* 99, 030402 (2007).

[806] H. Wang, M. Thoss, *J. Chem. Phys.* 131, 024114 (2009).

[807] S. Bubin, M. Stanke, L. Adamowicz, *J. Chem. Phys.* 135, 074110 (2011).

[808] O.E. Alon, A.I. Streltsov, L.S. Cederbaum, *Phys. Rev. A* 033613 (2008).

[809] M. Nest, T. Klamroth, P. Saalfrank, *J. Chem. Phys.* 122, 124102 (2005).

[810] K. Pachucki, J. Komasa, *J. Chem. Phys.* 130, 164113 (2009).

[811] K. Piszczatowski, G. Lach, M. Przybytek, J. Komasa, K. Pachucki, B. Jeziorski, *J. Chem. Theory Comp.* 5, 3039 (2009).

[812] H. Lischka, M. Dallos, P.G. Szalay, D.R. Yarkony, R. Shepard, *J. Chem. Phys.* 120, 7322 (2004).

[813] A. Tajti, P.G. Szalay, *J. Chem. Phys.* 131, 124104 (2009).

Index

1,3-butadiene, 433

$\tilde{A}$ continuum, 731, 732
ab initio, xxi, xxxvi, 3, 37, 45, 255, 272, 430, 469, 588, 589, 591, 595, 598, 605, 610, 732, 824, 833
    ab initio electron dynamics, xxxvii
    ab initio molecular dynamics, 305
absorption, 664, 668, 673, 674, 676, 718, 773
    absorption band, 576, 731
    electronic absorption, 433
    linear absorption, 686
    optical absorption, 773
acceptor, xl, 187, 749–751, 753, 754, 757, 758, 760, 763–765, 767–769, 771, 772, 784, 790, 792, 795, 799
action, 132, 136–138, 293, 295, 324, 351, 397, 398, 418, 446, 465, 481, 492, 500–503, 506, 507, 512, 536, 599, 600, 620, 630, 638, 647, 698, 800
    action variables, 391, 393–395
action-angle, 391, 397, 451, 646, 651
    classical action-angle variables, 387
activation, 799, 800
    activation energy, 765, 802
    activation free energy, 793, 797, 798, 800–802
    adiabatic activation energy, 762
    thermal activation, 751, 793
active space, 39
Adams–Bashforth, 557
Adams–Moulton, 557
adenine, 575–577
adiabatic, xxii, xxv, xxviii, xxxi, xxxvi, xxxix, 45, 46, 54, 58, 59, 61, 62, 68, 76, 121, 190, 202, 245, 246, 249, 251, 253, 254, 274, 283, 285, 288, 316, 389, 391, 398, 400, 405, 406, 414, 416, 417, 424, 428, 467, 506, 512, 513, 583, 584, 603, 604, 644, 665, 688, 692, 701, 706, 760, 761, 763, 840, 843
    adiabatic coupling, 205
    adiabatic evolution, 292
    adiabatic frame, 822
    adiabatic motion, 406
    adiabatic phase, 61, 77
    adiabatic potential energy, 63
    adiabatic propagation, 77
    adiabatic representation, 47, 57–60, 82, 822
    adiabatic state, 287, 288, 316, 317
    adiabatic switching, 397, 398
    adiabatic theory, 251
    adiabaticity factor, 287
    crude adiabatic approximation, 56, 57
    crude adiabatic representation, 82
adjoint, 100, 101, 304
adsorbate, xli, 818, 828, 836
    organic adsorbates, 843
adsorption, 384, 579, 817, 819, 827, 829, 831

Aharonov–Bohm effect, 71, 77
molecular Aharonov–Bohm effect, 77
Airy differential equation, 140
Airy function, 140–142
algebraic, 241, 263, 265, 266, 277, 279
algebraic manipulations, 551
algorithm, xxx, xli, 557
integration algorithms, 311
Metropolis algorithm, 795
multistep algorithms, 557
alizarin, 843
alizarin chromophore, 610
alkali, xxx, 78, 321, 700
alkali-halide
alkali-halide chain, 707
non-stoichiometric alkali-halide, 706, 708, 710
non-stoichiometric alkali-halide clusters, xxxix, xl, 663, 700, 736
alkali-halogen, 711
non-stoichiometric alkali-halogen, 716
alkene, 683
allene, 574
amino acids, 576
amino groups, 576
amplitude, 113, 115, 119, 123, 133, 134, 137, 138, 144, 146, 175, 184, 214, 285, 303, 324, 327, 500–503, 505, 515, 567, 569, 570, 574, 615, 616, 718, 728, 821, 829
amplitude density, 277, 278
channel amplitudes, 214
complex-valued amplitudes, 183
electronic transition amplitude, 460
excitation amplitudes, 343
normal mode amplitudes, 361
probability amplitude, 393, 423, 448
scattering amplitude, 238, 364
trajectory amplitudes, 423
transition amplitude, 441
analytical, 593
analytical continuation, 418
angle, 581, 652, 673
azimuthal angles, 833
tilt angle, 834
torsion angle, 683
angular momentum, 215, 225, 234, 247, 257, 377, 441, 657
angular momentum algebra, 236
angular momentum quantum number, 555
angular momentum raising and lowering operators, 235
electronic angular momentum, 247, 248, 261, 273
nuclear angular momentum, 203, 249, 378–380
orbital angular momentum, 104, 203, 214
pseudorotational angular momentum, 379, 380, 398
rotational angular momentum, 379, 380
Schwinger's theory of angular momentum, 451, 452, 454, 455
Smith's grand angular momentum operator, 228
anharmonic, 51, 704
anharmonicity, 378
anion, 839
anisotropy, 781
annihilation operator, 156
antibonding, 13, 14
anticommutator, 519
anti-Hermitian, 100
antisymmetric, 284, 355, 822
approximation, 123, 321, 322, 349, 361, 384, 420, 428, 441, 444, 445, 451, 470, 477–479, 484, 508, 589, 592, 593, 603 623, 625, 626
adiabatic approximation, 600, 603
adiabatic representation, 443
averaged velocity coupling approximation, 515
Becke three-parameter approximation, 595
Born–Huang approximation, 49
Born–Oppenheimer approximation, xx, xxv, xxvi, 5, 45, 46, 49, 54, 63, 71, 86, 312, 325, 342

Condon approximation, 693, 768
dipole approximation, 691
eikonal approximation, xxxi, 144, 332, 338, 354
frozen bond approximation, 708
frozen Gaussian approximation, 349, 421, 629
generalized gradient approximation, 594
harmonic approximation, 457, 756, 757, 764
independent trajectory approximations, 491, 515
IVR approximation, 458
local density approximation, 593
local harmonic approximation, 419–422
non-overlapping pulse approximations, 673
propagating least-squares approximation, 504
random phase approximation, 603
Redfield approximation, 792
rigid rotor approximation, 734
rotating wave approximation, 668, 673, 674, 681, 696, 740
SC-IVR approximation, 462–464
Schiff approximation, 369
semiclassical approximation, 437, 442
slowly varying envelope approximation, 666, 668
stationary phase approximation, 440
sudden approximation, 702
TDH approximation, 567
Van Vleck–Gutzwiller approximation, 462
velocity coupling approximation, 514, 515
WKB approximation, xxix
aqueous electron, 636
arrangement, 223, 230, 239, 258, 727, 843
asymptotic arrangement, 229, 323
final arrangement, 227, 241
initial arrangement, 241
multiple arrangement problem, 242
triatomic arrangement, 229
Arrhenius, 761, 765
Arrhenius activation barrier, 256
Arrhenius factor, 800, 803
asymmetry, xxxii, 79–81, 753
asymptotic region, 105
atomic, xxiii, 15, 256, 331, 368, 369, 406
atomic layer, 834
atomic positions, 740, 744
atomic shell structure, 609
atomic states, 258
atomic basis
infinite atomic basis, 32
atomic orbital (AO), 272, 343, 825
Au, 817
Au(111), 839
autocorrelation, 582
autocorrelation function, 468, 470, 488, 489
averaged velocity coupling approximation (AVCA), 515
avoided crossing, xx, 39, 46, 63, 65, 78, 114, 283, 404–406, 414–416, 420, 426, 428, 462, 463, 465, 584, 622, 702, 703, 706, 707, 760, 761, 762
dual avoided crossing, 414, 416
azine, 576

B3LYP, 595
back action, xxxi, xxxii, xxxvi, 192, 613, 631
electronic back action, 632, 634
backward reaction, 385
bacteriochlorins, 579
Bader, J.S., 491, 803, 804
Baker–Campbell–Hausdorff formula, 156, 808
balance, 835
detailed balance, 413, 766, 767
Ball, M.A., 295
band gap, 826
band structure, 831
Barbara, P.F., 639
barrier, 163, 183, 261, 262, 502, 579, 583, 585, 749, 754, 767
adiabatic barrier, 426, 428
barrier coordinate, 183

barrier partition function, 185
barrier region, 167, 183
barrier top, 163, 169, 172, 184, 185
barrier top partition function, 184
Eckart barrier, 509
free energy barrier, 793, 798, 805
parabolic barrier, 183–186
potential barrier, 182, 499, 751
reaction barrier, 765
base pair, 575, 578
adenine-thymine base pair, 577
DNA base pairs, 579
hydrogen-bonded base pairs, 577
basis, 29, 34, 37, 48, 59, 82, 155, 201, 206, 229, 230, 232, 237, 270, 279, 301, 310, 314, 358, 390, 408, 420, 422, 452, 454, 543, 551, 555–557, 561, 562, 572, 606, 610, 621, 681, 692, 760
$\mathcal{L}^2$ basis, 265
$\mathcal{L}^2$ basis functions, 271
adaptive basis, 423
adiabatic basis, 68, 389, 621
adiabatic basis functions, 62, 202, 229
AO basis, 825
atomic basis, 11, 27, 29, 329, 330, 348
atomic basis function, 330, 331, 350, 351, 359
basis function, xxviii, 49, 59, 205, 232, 235, 257–259, 269, 278, 283, 306, 313, 347, 350, 423–425, 435, 454, 552–555, 561, 681
basis of eigenfunctions, 717
basis of traveling orbitals, 358
basis set, 62, 269, 277, 548, 564, 751
basis set functions, 268
basis state, xxvi, 38, 39, 235, 265, 279, 405, 424, 555
basis vectors, 31, 35, 67, 267, 295
Cartesian basis, 41
channel basis functions, 202, 206
CI basis, 38
complete basis, 31, 34, 47, 88, 202, 218, 558, 606
configuration basis, 31
configuration interaction basis, 37
continuous coherent state basis, 158
diabatic basis, 62, 260, 670
direct product basis, 681, 685
discrete basis, 551, 552
DVR basis, 553, 555
finite basis, 295, 304, 313, 329
grid basis, 552, 553, 556, 560
instantaneous basis state, 441
instantaneous nuclear basis, 439
instantaneous orthonormal basis, 443
momentum basis, 556
non-$\mathcal{L}^2$ basis, 278
non-orthogonal basis functions, 359
parity-adapted basis functions, 234
particle number basis, 154
plane wave basis, 125, 556
pointer basis, 613, 614
product basis, 203
reactant basis, 259
single-valued basis, 75
spatial grid basis, 556
spawned basis functions, 425
time-dependent basis sets, 548
traveling atomic basis functions, 331
traveling atomic orbital basis, 322
truncated basis, 205
variational basis, 551, 552
well-balanced hybrid basis, 279
bath, 192, 466, 471, 620, 625, 751, 775
bath fluctuations, 625
bath trajectories, 631
classical bath, 844
dielectric bath, 775
electron bath, 836, 838
harmonic bath, 479
nuclear bath, 626
vibrational bath, xl, 752, 786, 787, 806
Batista, V.S., 731–733
Beck, M.H., 567
Becke, A., 608
Bell, J.S., 491
benzene, 574
Berry, M.V., 77
Berry's phase, 77

Bessel function, 365
Riccati–Bessel, 106
bichromatic coherent control, 737
bifurcation, 618, 738
wave packet bifurcation, 731
biogenetic, 574
biological, 78, 750, 751, 842, 843
biological complex, 751
biological species, 574
bistable, 163, 164
Bittner, E.R., 616, 623
Bjerre, A., 403
Blass, A., 374
body-fixed, 229, 232, 233, 235, 247–250, 257, 259, 276
body-fixed axis, 247
body-fixed frame, 224
Bohm, A., xxxiv, 304
Bohm, D., 491
Bohm–de Broglie, 492, 498
Bohm–de Broglie theory, 492, 498, 499, 507, 540, 542
nonadiabatic Bohm–de Broglie theory, 512
Bohmian mechanics, xxxv, 491, 492, 512, 539, 540
Bohr–Sommerfeld quantization rules, 144
Boltzmann, 796
bond, 431, 729
antibonding, 580
bond breaking, 39, 700, 706, 709, 711
bond dissociation, 30
bond length, 407
C-N bonds, 576
carbon–carbon double bond, 682
frozen ionic bond, 748
hydrogen bonds, 775
ionic bond, 700, 710
ionic Na-F bond, 706, 709
metallic, 710
multiple bonds, 608
Na-Na bond, 701, 704, 709
non-bonding, 576
olefinic bond, 683
polar, 700
bonding, 13, 14, 289, 290, 525, 683, 706, 794, 818
antibonding, 683
bonding properties, 608
chemical bonding, 608
covalent bonding, 289
molecular bonding properties, 608
boosted self-consistent field equations, 360
Borgis, D., 632
Born approximation, 210–212
distorted-wave Born approximation (DWBA), 212, 213, 285, 288
first Born approximation, 364
single-channel Born approximation, 211
Born–Oppenheimer approximation, xxv, 839
adiabatic Born–Oppenheimer model, 292
Born–Oppenheimer products, 829
Born–Oppenheimer dynamics, xxxvi
BO, 318
Born–Oppenheimer separation, 4
Bose–Einstein, 844
Bose–Einstein distribution, 770, 809
boson, 455, 759
N-boson problem, 452
bosonic, 844
bosonization, 466
bosonization strategy, 465
many-boson, 844
bounce
bounce trajectory, 186
boundary condition, 89, 107, 129, 145, 215, 232, 240, 270, 393, 554, 599, 600, 646, 722
complex boundary conditions, 278
incoming boundary conditions, 98
inelastic scattering boundary conditions, 240
K matrix boundary condition, 110, 111, 219
Neumann boundary condition, 218
outgoing boundary condition, 97, 206, 209
periodic boundary conditions, 640

S-matrix boundary conditions, 109, 110, 275
stochastic boundary conditions, 835
T-matrix boundary conditions, 109
Br, 725
Br_2, 725
Br*, 725
branching plane, 81
branching ratio, 262, 426, 725, 726, 728, 729, 736
branching space, 67
Breit–Wigner phase relation, 117
Breit–Wigner form, 117
bridge, 754, 760
bridge states, 757
Brillouin theorem, 35, 38
broadening, 471
broadened distribution, 471
Brumer, P., xxxix, 715, 729, 731–733
butatriene, 574

c number, 152
C_{2v}, 14, 15, 746
C_{2v} isomer, 707, 738
$\hat{C}_2$, 15
C-O stretch, 829, 832
internal C-O stretch, 833, 836
Campbell–Baker–Hausdorff formula, 346
canonical, 709
canonical ensemble, 169
capture, 370
electron capture, 370
Car–Parrinello, xxxvi, 292
CP, 318
Car–Parrinello dynamics, 318
Car–Parrinello molecular dynamics CPMD, 805
cation, 700
cationic ground state, 666
cationic state, 709
Cauchy principal value, 99
causality, 499, 599, 600, 779, 781
causality dilemma, 600
CEA, 387–389, 391–393, 395, 400, 402, 437, 438, 458
cell, 486, 487
fluid cell, xxxiv, 500, 501, 502, 505, 508, 539
cellular dynamics, 482
center of mass, 50, 202, 225, 232, 323, 376, 378, 427, 581, 724, 833, 834
centroid, 163, 181–183, 185, 800, 802, 803
centroid density, 182
centroid dynamics, 181
centroid expansion, 183
centroid free energy, 173
centroid partition function, 173
centroid summation procedure, 182
centroid variable, 182
reduced centroid density, 173
CH_3I, 338
Chandler, D., 793
channel, 161, 201, 202, 204, 209, 210, 212–214, 220, 228, 232, 241, 242, 262, 323, 361, 366, 426, 509, 707, 719, 723–727, 733, 818
accessible channels, 646
arrangement channel, 226
electronic channels, 245, 285, 511, 736
excitation channels, 706
rearrangement channel, 227
scattering channel, 323
single-channel scattering, 201
transfer channel, 767
chaotic, 397, 398, 472, 489
charge density, 10
charge transfer, 201, 337, 366, 369, 374, 406
charging parameter, 801, 803
Chebyshev, 560, 562
Chebyshev formalism, 563
Chebyshev polynomial expansion, 562
Chebyshev procedure, 564
Chebyshev scheme, 564
chemical dynamics, 59
chemical regime, 381, 383, 384
chemicurrent, 838
chemiluminescence, 749
chemisorption, 818
chlorophyll, 579

chromophore, xxxix, 843
retinal chromophore, 683
CI, 40
circulation, 493, 494
CIS, 38
cis-trans isomerization, xxxix, 86
classical, xix, xxvii–xxix, xxxi, xxxiii, xxxv, xxxvi, 122, 123, 144, 162, 164, 173, 178, 185, 308, 334, 341, 354, 362, 383, 387, 389, 390, 392, 398, 400, 407, 419, 420, 433, 437, 451, 452, 457, 459, 472, 482, 512, 514, 517, 519, 521, 522, 529, 547, 614, 616, 623, 625, 639, 640, 646, 650, 695, 708, 716, 750, 790, 793, 799, 802, 803, 841
classical action, 123, 134, 135, 137, 448, 459, 465
classical action functional, 127, 128
classical action integral, 439
classical continuous variables, 456
classical degrees of freedom, 642
classical limit, xxxii, 137, 151, 170, 171, 185, 350, 356, 393, 512, 515, 520, 541, 693
classical mechanics, xxxiii, 122, 127, 128, 135, 136, 162, 177, 292, 353–355, 386, 387
classical nuclei, xxxi, 308, 373, 381, 614
classical observables, 438
classical observer, 614
classical particle, 123, 127
classical path, 129, 135, 421, 440, 616
classical positions, 409
classical statistics, 178
classical trajectory, 129, 484
classically forbidden, 412, 413, 484
quasi-classical, 146
quasi-classical approximation, 155
uniformly classical, xxxii
uniformly classical model, 383
classical electron, 649
classical electron analog, xxxii, xxxiii, xxxv, xxxviii, 383, 386, 388, 437, 451, 452, 456, 458, 547, 646, 733, 734, 841
CEA, 383, 386
CEA dynamics, 394
CEA translation rules, 397
classical electron analog Hamiltonian, 461
classical Hamilton–Jacobi equation, 122
classical mechanics, 494
classical nuclei, 372
classical particles, 386
classical partition function, 183
classical paths, 128
classically allowed region, 285
Clebsch–Gordan coefficient, 67, 231, 260
close-coupling, 368
close-coupling equation, 205, 220
closed, 534, 537, 538, 542, 613
closed subset, 542
closed system, 613
closed-shell, 24, 306–308, 376, 594
closed-shell system, 10
closure, 104, 237, 546, 769
closure relation, 157
cluster, 321, 482, 587, 831, 843
atomic cluster, 831
copper cluster, 831
molecular clusters, 842
cluster amplitudes, 40
CN, 733–735
CO, xli, 818, 828, 831–833, 835–837
COH_2, 39
coherence, xxxv, xxxviii, 196, 197, 409, 420, 517, 518, 522, 523, 526, 530, 534, 613, 614–616, 619, 645, 693, 792
coherence loss, 620, 631, 632, 637
complete quantum coherence, 640
electronic coherence, 532, 648, 669
loss of coherence, 632
nuclear coherence, 543, 544
quantum coherence, 416, 479
coherent, 154, 155, 615, 651, 659
canonical coherent state, 372
coherent limit, 517, 623
coherent nuclear state, 349
coherent photon ensembles, 154
coherent rovibrational state, 430

coherent state, xxix, xxxii, 154–160, 341, 343, 347, 348, 350, 356, 372, 373, 375, 376, 462, 479
coherent state dynamics, 122
determinantal coherent state, 342, 347
general coherent state, 160
Glauber coherent state, 450
incoherent, 624
incoherent mixture, 659
nuclear coherent state, xxxii, 122, 159, 160, 348, 383, 461, 482
vibrational coherent state, 373
Coker, D.F., 442
Colbert, D.T., 554
collinear, 81, 255
collinear geometry, 261
collision, xxx, 204, 224, 246, 256, 281, 333, 361, 363, 365, 427, 429, 432, 835
atom–diatom collisions, 660
atom–molecule collisions, xix
collinear collision, 222
intermolecular collisions, xix
ion–atom collision, 332, 334
molecular collision theory, 243
proton–hydrogen collisions, 333
reactive collisions, 227
rearrangement collision, 201, 214, 223, 224, 241, 243, 245, 255
commutation relation, 102, 156, 346, 452
commutator, 150–152, 156, 192, 519–521, 780, 804
anticommutation relations, 154, 346
anticommutator, 519, 520, 521
bosonic commutation relations, 455, 759
Fermi anticommutation relations, 344
non-abelian commutator, 254
reversed commutation relations, 235
complete active space multiconfigurational (CASMC), 381
complete active space self-consistent field (CASSCF), 38, 39, 577, 581, 582
completeness, 31, 49, 57, 61, 121, 126, 148, 175, 187, 203, 390, 759
basis set incompleteness, 314
complex energy plane, 116
computation, 824, 825, 833
computational accessibility, xxxvi
computational approaches, xxi, xxviii
computational cost, 564
computational effort, xxvii, 548, 698
computer capacity, xx
non-perturbative computation, 682
condensed phase, xxviii, xl, 644, 844
conduction band, 818–820, 824, 827, 828, 839, 832, 843
conduction band orbitals, 840
conductivity, 749
configuration, xxxix, 31, 32, 38, 82, 227, 257, 272, 303, 324, 393, 406, 426, 429, 434, 565, 579, 580, 585, 672, 680, 706, 743, 749, 765, 796, 797, 843
asymptotic configurations, 229
average electronic configuration, 315
bonding configuration, 290
classical configuration, 616
configuration space, 584, 739, 740, 795
electronic configuration, 245, 324, 327, 341
multiconfigurational, 318
multireference configuration interaction, 581, 585
product configurations, 229, 238
reactant configuration, 259, 800
configuration interaction, 31–33
multireference configuration interaction, 259
truncated configuration interaction, 32
configuration space, 89, 174, 408, 423
configuration state functions (CSFs), 39
conical intersection, xx, xxvii, xxxix, 39, 46, 60, 63, 65, 67, 76, 78–81, 86, 114, 273, 280, 395, 402, 431, 438, 466, 467, 470, 574–577, 579, 580, 582, 585, 613,

660, 690, 700, 707–711, 732, 733, 738, 743, 744, 747, 842
K-fold intersection, 83
conical intersection dynamics, 585
conjugate, 349, 355, 360, 675
complex conjugate, 352, 673, 716
conjugate momentum, 123, 165
conjugate pair, 739
conjugate points, 134, 135, 438
Hermitian conjugation, 521
conservation, 425, 502, 529, 820
conservation of actions, 398
conservation of particle number, 101
conservation of probability, 391
conservative system, 557
energy conservation, 137, 411, 412, 446, 492, 494, 598, 642, 654, 655
momentum conservation, 443
norm conservation, 515, 566
probability conservation, 652
continuity, 137, 492, 495, 541
continuity equation, 498, 502, 539, 544, 545, 604
continuous surface switching (CSS), 648
continuum, xli, 138, 240, 291, 442, 596, 751, 789, 819, 820, 839
continuum mechanics, 605
continuum states, xxx
dielectric continuum, 775, 776, 789, 804
double continuum, 240
control, xxviii, 715, 725, 741, 747, 748
bichromatic coherent control, 731
bichromatic control, 728, 732, 733, 736
coherent bichromatic control, 729
coherent control, xxxix, 715, 725, 729, 730, 736
conversion times, 580
laser control, 718, 719, 726, 730, 736, 738, 742
laser control theory, 726, 730, 736
optimal control, xl, 719, 736, 737, 744
optimal control theory, 738
coordinates, 53, 168, 174–176, 182, 183, 191, 202, 220, 224–227, 319, 333, 355, 361, 377, 392, 421, 438, 439, 449, 458, 459, 462, 466, 479, 480, 506, 518, 534, 536, 550, 553, 555, 568, 582, 724, 734, 739, 740, 745, 746, 751, 759, 763, 764, 775, 779, 794, 822, 829, 833
acceptor equilibrium coordinate, 765
active nuclear coordinate space, 580
adapted coordinates, 224, 225
angular coordinate, 157, 231, 397
Cartesian coordinate system, 333, 833
center-of-mass coordinates, 230
channel coordinate, 214
collective coordinates, 734
complex-valued coordinates, 419
coordinate space, 255, 450, 480, 485
coupling coordinates, 81
Delves coordinates, 255, 257
dimensionless coordinates, 457, 684
electronic coordinates, xxx, 56, 72, 247, 248, 250, 252, 257, 259, 324, 459, 513, 820, 821, 823
environmental coordinates, 791
hyperspherical coordinate, 229
internal coordinates, 79, 81, 202
intersection-adapted coordinates, 79, 80
Jacobi coordinates, 224, 225, 229, 247, 254, 272, 426, 429, 430, 581, 732
mass-scaled coordinates, 226, 228
mass-weighted coordinates, 51
normal coordinates, 53, 70
phase space coordinates, 494, 529, 532, 699, 741
polar coordinates, 68, 157
reaction coordinate, 163, 165, 169, 173, 183, 385, 576, 577, 579, 761–765, 791
rotating coordinate frame, 247
scattering coordinate, 113, 204, 213, 214, 218, 227, 238, 241
torsion coordinate, 684
vibrational coordinate, 230, 373, 471, 757, 758, 785, 820
copper, 818, 830

correlation, xxxvi, 30, 31, 34, 193, 197, 321, 567, 573, 603, 629
antisymmetrized correlation function, 781
correlation energy, 31, 32, 34
correlation function, 629, 713, 782
correlation time, 697, 699, 712
dynamic correlation, 40
electron correlation, 31, 34, 36
local density exchange, 595
quantum correlation functions, 844
short-range correlations, 603
spatio-temporal correlation, 775
static correlation, 30
time correlation function, 166, 472, 479–481, 699, 835
time-dependent correlation function, 781
correlation time, 194, 573
corrosion, 817
corrugation, 817
Coulomb, 18, 594, 830
interelectronic Coulomb term, 30
Coulomb integral, 16
counterpropagation wave method (CPWM), 504
coupled channel
close-coupled channel, 337, 338
coupled-channel equations, 260, 262
coupled-cluster, 40
coupled potential energy surfaces, xix
coupled trajectories, 491
coupling, xx, xxvi, xxxi, xxxviii, xli, 213, 275, 277, 279, 284, 321, 322, 324, 338, 341, 342, 353, 356, 360, 376, 383, 386, 406, 421, 425, 430, 431, 469, 471, 507, 514, 528, 529, 574, 580, 582, 605, 609, 626, 645, 684, 702, 707, 719, 720, 731, 752, 759, 761, 762, 764, 789, 791, 818, 822, 825, 828, 835, 836, 838, 839
adiabatic coupling, 405
coupling between electrons and nuclei, 247, 292
coupling potential, 522
coupling region, 418, 510
coupling strength, 284, 772
derivative coupling, 824
diabatic coupling, 273, 406, 429, 430, 434, 510, 622, 660
electronic coupling, 213, 752, 762, 789
electronic–vibronic coupling, 466, 756, 771, 772
molecule–surface coupling, 817
nonadiabatic coupling, xxxviii, 331, 390, 406, 408, 414, 417, 431, 434, 610, 615, 621, 629, 639, 644, 645, 654, 708, 832, 833, 845
nonadiabatic coupling strength, xxxviii
spin-orbit coupling, 11, 12, 257
spin-orbit-coupled electronic states, 260
strong coupling, 373, 379, 657, 790
strong nonadiabatic coupling, 406
triatomic coupling, 648
vibrational coupling, 835, 836
weak coupling, 471
weak nonadiabatic coupling, 822
covalent, 30, 38, 245, 246, 289, 290, 575
covariant derivative, 60
CPMD, see also Car–Parrinello, 805
Crank–Nicolson, 557, 559
cross section, 118, 119, 241, 262, 334, 362, 689
absorption cross section, 468
charge transfer cross section, 338, 370
differential cross section, 111, 232, 322, 324, 334, 336, 337, 362, 368, 370, 374, 375
elastic cross section, 111
integral cross section, 113, 337
photoabsorption cross section, 607
semiclassical differential cross section, 365
state-to-state cross section, 329
stopping cross section, 371, 372
total cross section, 362, 370

crossing, 46, 65, 286, 288, 289, 418, 428, 524, 525, 532, 576, 577, 579, 580, 583, 624, 660, 730, 760, 762
diabatic crossing, 731, 761, 799
crystal, 593
quartz crystal, 665
CSFs, see configuration state function, 39
CSS, 649
Cu, 832
Cu surface, 832
Cu(100), 828, 831, 834, 835, 837
cumulant, 632, 813
cumulant expansion, 632
cumulant series, 787
current, 537, 544, 609
spin currents, 844
current density, 109, 110, 112
one-particle current density, 110
radial current density, 113–115
scattered current density, 111
curvature, xxxiv, 498, 539
curved metric, xxxii
cyclic, 390, 397
cyclobutene, 433
cytosine, 575, 577–579

D, 750, 753
D_2, 845
D_2O, 638–642, 793, 803
D_{3h} doublet, 65
D_{3h} symmetry, 65
damping, 96, 791
electronic damping, 834
Darwin term, 11
de Broglie, L., xxxiv, 144, 491
de Broglie wave length, 143, 144
local de Broglie wave length, 144
nuclear de Broglie wave lengths, 325
Debye form, 788
decay, xix, 115, 119, 197, 280, 431, 580, 626, 629, 640, 644, 652, 724
decay-of-mixing, 615, 648, 651, 653, 654, 657, 659, 660
nonadiabatic decay, 639
radiationless decay, 580, 684
ultrafast non-radiative decay, 402
decoherence, xxxviii, 197, 412, 534, 538, 613–615, 617–621, 623–626, 629, 631, 635–639, 641, 643, 644, 649, 651, 652, 654, 655, 657, 659
decoherence factor, 643
decoherence force, 651, 654, 655
decoherence function, 631, 632, 637
decoherence time, 620, 623, 624, 626, 629, 631, 634, 635, 639, 641, 657
decoherent force, 656–658
decoherent state, 656–660
direction of decoherence, 655–657
quantum decoherence, 634
state-specific decoherence times, 652
decomposition, 249, 276, 330, 334, 387, 457, 478, 539, 650, 825
decomposition of the total angular momentum, 247
decomposition of unity, 125
Helmholtz decomposition, 812
singular value decomposition, 423
deexcitation, xix, xxxix, 161, 271, 336, 429, 637, 643, 659, 710, 738, 744, 746, 747, 817, 828, 834, 838, 840
deexcitation mechanism, 281
deexcitation time, 818
radiationless deexcitation, 78, 576, 579, 683, 709
ultrafast deexcitation, 575
vertical deexcitation, 701
vibrational deexcitation, 828, 833
deflection angle, 483
deflection function, 365, 367
degeneracy, xxvi, 46, 61, 63, 64, 67–71, 73, 76–79, 82, 84, 256, 376, 386, 395, 637, 682, 794, 795, 799
electronic degeneracies, 76
K-fold degeneracy, 83
near-degeneracy, 395, 682
delay, 118, 666, 668, 676, 679
delay time, 665, 669, 676, 686, 698, 703, 715, 743, 744, 747
optimal delay time, 747
delocalization, 432, 491, 515

delta function, 103, 148, 167, 182, 187, 189, 349, 462, 476, 486, 528, 530, 678, 695, 742, 764, 768, 783, 795, 812
longitudinal delta function, 812
temporal delta functions, 721
Delves coordinate, 257
density, xxxiv, 500, 504, 506, 509, 518, 520, 524, 526, 528, 529, 523, 532, 535, 536, 538, 540, 541, 587–593, 597–599, 601, 603, 604, 606–609, 693, 697, 699
canonical density, 472
charge density, 783
current density, 495, 500, 536, 538, 544, 583, 603, 604
density of states, 770, 786, 789, 813, 826, 831
electronic density, 329, 591, 598, 608
ensemble density, 597
ensemble probability density, 498
excited state density, 595, 698
external charge density, 777
ground state density, xxxvi, 587, 589, 590, 595, 697, 699, 741
ground state electron density, 588, 589
local density, 536, 544
local density of states, 827, 830, 832
local spin density, 594
off-diagonal density, 545
Ohmic spectral density, 773
one-particle density, 370
phase space density, 152, 699, 745
probability density, 498, 499, 534, 544
single-electron density, 587
solvent spectral density, 788
spectral density, 772, 773, 788, 789, 791, 803, 804
time-dependent densities, 609
transition density, 506, 544
vibronic density, 694
density functional theory, see also DFT, xxxvi, 41, 43, 315, 322, 381, 587, 588
density functional methodology, 589
DFT ensemble density, 597
DFT ensemble energy, 597
ensemble DFT, 597
non-local DFT, 595
time-dependent current density functional theory, 603
time-dependent density functional theory, 597, 598
density operator, 10, 304
deoxyribonucleic acid (DNA), 575, 790
dephasing, 631, 651
electronic dephasing, 658
pure dephasing, 632
derivative, 284, 312, 320, 359, 360, 388, 441, 444, 504, 569, 572, 604, 801, 825
energy derivative, 360
first time derivative, 557
functional derivative, 590, 599, 600, 601, 811
second derivative, 297, 488, 627
time derivative, 295, 303, 310, 311, 331, 352, 359, 388, 441, 496, 507, 605, 651, 654, 655, 656, 659, 668, 823
total derivative, 539
total time derivative, 820
derivative coupling, 60, 64
second derivative coupling, 72
desorption, 817
determinant, 303, 306, 346, 441
determinantal state, 329, 347, 824
electronic determinant, 329, 343, 345, 350, 824
excited determinants, 303, 596
excited state determinants, 301
ground state determinant, 301
Jacobian determinant, 134
Kohn–Sham ground state determinant, 610
single determinant, 303, 304, 306, 343, 591, 824
Slater determinant, 301
substitutional determinant, 296, 301
Van Vleck–Gutzwiller determinant, 488
deterministic, 162, 164, 165, 491
Deumens, E., 341
deuterium, 838

DFT, 320, 322, 587, 588, 590, 592–599, 605, 610, 611
diabatic, xl, 45, 58, 61, 62, 202, 204, 205, 246, 251, 253, 254, 257, 273, 274, 282, 283, 289, 316, 388, 389, 405, 424, 426–431, 512, 577, 583, 584, 644, 645, 660, 672, 682, 749, 760–762, 764, 768, 784
    diabatic approximation, 45, 58, 59, 290
    diabatic basis, 688
    diabatic coupling, 58, 163, 245
    diabatic crossing, 245, 407
    diabatic frame, 54, 59, 64, 121
    diabatic potentials, 416
    diabatic representation, 57, 59, 61, 434
    diabatizing, 273
    quasi-diabatic, 756
diatomic, 65, 114, 224, 225, 240, 245, 247, 248, 260, 282, 283, 406, 648, 834
    atom–diatom combinations, 226
    diatom, 225, 230, 232, 406, 407
    diatom center of mass, 225, 238
    diatomic fragment, 407
    diatomic subsystem, 257
diatomics in molecules (DIM), 272, 406
dielectric, 776, 779, 781, 783, 785, 811
    dielectric constant, 782
    dielectric continuum, xl, 798
    dielectric dispersion, 804
    dielectric function, 777, 788, 804
    optical dielectric function, 782
    static dielectric function, 782
differential equation, 196, 212, 223, 229, 236, 246, 253, 262, 309, 317, 325, 328, 332, 352, 422, 423, 459, 484, 667, 823
    coupled differential equations, 204, 251, 354, 691, 821
    coupled ordinary differential equations, 557
    differential equations of first order in time, 307
    first-order differential equations, 571
    inhomogeneous differential equations, 99
    time-dependent coupled differential equations, 566
differential geometry, 841
dimension, xl, 343, 474, 475, 484, 548, 551, 562, 645
    dimensionless, 658, 821
    full-dimensional, 482
    low-dimensional, xxviii
    m-dimensional domain, 474
    multidimensional, 476, 548
    number of dimensions, 516
    reduced dimensionality, 585
dimensional consistence, 387
dimensionality, 473, 474, 565
dipolar
    dipolar approximation, 665
    dipolar interaction, 668, 669, 718, 719
    dipolar interaction operator, 672
    dipolar moment, 336, 718
    dipolar overlap matrix, 722
dipole, 750, 776, 777
    dipole density, 777, 783
    dipole moment, 638
    dipole number density, 776
    dipole–dipole interaction, 812
    oscillating dipoles, 664, 818
Dirac notation, 175
discontinuity, 207, 274, 493
    quantum discontinuities, 404
discontinuity relation, 108
discontinuous, 433, 481, 493, 511, 515, 619, 648, 651
discrete, xxxiii, xxxiv, 114, 174, 291, 299, 552, 771, 819, 839
    discrete degree of freedom, 453
    discretization error, 559
    discretized elements, 501
discrete adiabatic, 622
discrete variable representation, see also DVR
    DVR grid points, 553
    DVR points, 553
disintegration modes, 426

dispersion
spatial dispersion, 779, 781, 814
displacement, 299, 377, 751, 832–834, 837
mean displacement, 837
dissipation, 471, 538
dissipative, xxxviii, 534, 614, 641, 649
dissipated over multiple degrees of freedom, 576
dissociation, 30, 38, 114, 240, 280, 525, 580, 584, 585, 725, 730, 747
diabatic photodissociation, 338
dissociation channels, 718
dissociation continuum, 291
dissociation limit, 38, 40, 585
ICN photodissociation, 730, 732
photodissociation, 715, 718–720, 725–727, 729, 736, 737
distorted waves, 212, 213, 275, 285
distribution, 532, 542, 582, 743, 764, 772
Boltzmann–Maxwell distribution, 763
canonical distribution, 745, 795, 796, 797
charge distribution, 750, 777
intramolecular vibrational redistribution (IVR), 730
phase space distributions, 746, 747
p-like charge distribution, 639
Poisson distribution, 620
thermal distribution, 764, 768, 769
Wigner distribution, 699, 706, 711
divergence, xxxiv, 270, 271, 413, 496, 503, 615, 616, 631
divergence of streamlines, 326, 327
dividing surface, 165, 169, 173, 183
donor, xl, 187, 749–751, 753, 754, 757, 758, 760, 762–765, 767–772, 784–787, 789, 790, 792, 795, 799, 806, 814
donor equilibrium, 780
excited donor, 790
donor–acceptor, 759, 760, 772
donor–acceptor complex, 752
double cone, 79, 81
double slit, 482, 617
double-slit arrangement, 483
driving force, 764, 765, 770, 776, 785–787
DVR, 549, 552–555, 560, 567
dye, 792
dynamic, xxii, xxvii, xxviii, xxxiii, xxxvi, 36, 89, 305, 312, 328, 344, 349, 357, 358, 376, 381, 395, 407, 420, 426, 429, 430, 432, 437, 451, 484, 486, 491, 522, 524, 534, 539, 544, 548, 569, 574, 579, 588, 598, 608, 609, 613, 615, 625, 636, 645, 646, 648, 649, 657, 658, 660, 663, 678, 686, 704, 700, 702, 725, 752, 783, 819, 833, 837, 838, 840, 842, 843, 844
ab initio dynamics, xix, xxii, xxxviii, 36, 615
ab initio quantum dynamics, 702
adiabatic dynamics, 406, 413
Born–Oppenheimer (BO) dynamics, 305, 317, 318, 320, 445
cellular dynamics, 438, 484–486, 488
centroid dynamics, 793
cis-trans photoisomerization dynamics, 683
classical dynamics, 420, 422, 514, 522, 532, 654
classical molecular dynamics, 448, 790
conical intersection dynamics, 690
density functional dynamics, 587
direct electron dynamics, xxi
direct molecular dynamics, 305, 605
dissipative dynamics, 470, 621
dynamic constraint, 721, 722
dynamic electronic variables, 393
dynamics in imaginary time, 163
dynamic many-body problems, 599
dynamic response, 372
Ehrenfest dynamics, xxxvi, 305, 312, 315, 383, 385, 389, 413, 465, 519, 547, 610, 611, 644, 646, 819, 820
electron dynamics, xx, xxi, xxvii, xxviii, xxxvi, xli, 121, 162, 318, 320, 322, 547, 605, 822, 823, 828, 838, 841–843
electron nuclear dynamics (END) theory, 325, 337, 341
electronic multistate dynamics, xxv, xxvii, 613

excited state dynamics, 706
fictitious electron dynamics, 320
fluid dynamics, 494
fragmentation dynamics, 433
Hamiltonian dynamics, 354, 452, 529
hindered dynamics, 583
internal conversion dynamics, 579
laser driven dynamics, 748
linear dynamics, 485
molecular dynamics, xix, xx, xxii, xxv, xxxi, xxxv, xxxix, xli, 320, 404, 438, 492, 526, 610, 617, 621, 631, 663, 711, 803, 817, 822, 834, 841
molecular quantum dynamics, xxviii, 324, 438, 564, 565
molecular reaction dynamics, 332
moment dynamics, 537, 546
multichannel reaction dynamics, 308
multistate dynamics, xxxvi, xl
multistate molecular dynamics, xxxix, 845
multistate nonadiabatic dynamics, 748
nonadiabatic dynamics, xxxix, 386, 414, 419, 457, 464, 466, 512, 515, 529, 609, 619, 644, 665, 706, 711, 733, 747, 843, 844
non-equilibrium dynamics, 534
nonlinear dynamics, 484
nonlinear phase space dynamics, 349
nuclear dynamics, xxi, xxii, xxv, xxvi, xxxvi, xxxvii, 3, 341, 353, 398, 408, 703, 824, 844
nuclear wave packet dynamics, 419
photoinduced dynamics, 580
photoinduced electron dynamics, 843
quantum-classical molecular dynamics, 616
quantum dynamics, xxii, xxx, xxxii–xxxiv, 341, 383, 413, 483, 499, 518, 530, 565, 735, 751, 840
quantum molecular dynamics, xix, xx, xxviii–xxxi, 383, 548, 843
reaction dynamics, xxviii, 471
relaxation dynamics, 641, 644, 790
semiclassical dynamics, 518, 699, 734, 741
semiclassical phase space dynamics, 738
semiclassical quantum dynamics, 395
state-specific molecular dynamics, 321
stimulated emission dynamics, 679
TDHF dynamics, 307, 308, 342, 356, 357
wave packet dynamics, 574, 582
Dyson-type equation, 602

Edgecombe, K.E., 608
effective coupling, 432
effective path method, 383
effective potential, 442
effective screened interaction potential, 363
Ehrenfest, xxxviii, 292, 315–318, 320, 341, 382–384, 386–389, 392, 403, 411, 422, 464, 465, 615, 642, 644–646, 648, 649, 658, 660, 819, 820
Ehrenfest approximation, 325
Ehrenfest limit, xxxviii, 647
Ehrenfest representation, xxxi
Ehrenfest trajectory, 465, 660
Ehrenfest's principle, 313, 316
nonadiabatic Ehrenfest approach, 292
eigenfunction, 96, 101, 102, 104, 123, 124, 153, 155, 169, 189, 203, 205, 215, 218, 231, 232, 235, 313, 316, 372, 452, 455, 597, 643, 727, 735
adiabatic eigenfunctions, 229, 621
angular momentum eigenfunction, 232
bound molecular eigenfunctions, 726
continuum eigenfunction, 95, 97
energy eigenfunctions, xxx
momentum eigenfunction, 450
nuclear eigenfunctions, 768
particle-in-a-box eigenfunctions, 554
position eigenfunction, 450
vibrational eigenfunctions, 582

eigenstate, 101, 124, 155, 275, 596
bound eigenstate, 101
outgoing asymptotic eigenstates, 95
eigenvalue, 22, 123, 124, 135, 157, 170, 189, 194, 215, 219, 316, 391, 394, 397, 399, 454, 551, 559, 562, 563, 597, 742, 746, 833
angular momentum eigenvalues, 398
diabatic eigenvalue, 204
electronic energy eigenvalues, 632
energy eigenvalues, 153, 458, 603, 768
position eigenvalues, 553
semiclassical eigenvalue, 397
semiclassical energy eigenvalues, 399
Eik-TDHF, 332, 333, 336–339
eikonal approximation, 322–324, 326–328
elastic, 203
elastic interactions, 87
scattering processes, 111
electric field, 154, 192, 664, 666–668, 675, 696, 698, 728, 776–778, 783, 784, 811
electric radiation field, 664
electrostatic field, 588
external electric field, 192, 670
incident electric field, 667
macroscopic electric field, 777
periodically changing electric field, 300
time-dependent electric field, 666
electrolyte, 843
electron affinity, 22, 245
electron-hole pair, 819, 818, 839
electron-hole pair excitations, 840
electron hydration, 636
electron liquid, 605
electron localization function, 608
electron nuclear dynamics theory, see also END
asymptotic END wave function, 362, 366
complete electron nuclear dynamics (CEND), 381
complete END theory, 361
electronic END wave function, 363
END cross sections, 368
END equations, 343, 348, 354, 356, 358, 360
END trajectories, 372, 375
linearized END equations, 357
minimal END, 358, 382
electron transfer, xix, xxii, xl, 45, 162, 187, 190, 245, 289, 290, 479, 749–754, 756, 760–762, 764–767, 770–776, 780, 783, 784, 786, 787, 789–794, 797–799, 803, 805, 806, 811, 813, 843
aqueous electron transfer, 805
biological electron transfer, 790
electron transfer rate, 786, 798
intramolecular electron transfer, 763, 789
nonadiabatic electron transfer, 760, 786, 793
photoinduced electron transfer, 790
ultrafast electron transfer, 197, 752, 790, 792
electron translation factors (ETFs), 330, 331, 339, 341, 358
electronic, xix, xxi, xxii, xxv, xxvi, xxviii, xxx–xxxii, xxxvi, xl, xli, 3, 45, 121, 122, 201, 213, 243, 245, 247, 249, 251, 257, 271, 273, 296, 312, 316, 319, 321, 322, 325, 327, 328, 342, 343, 356, 358, 359, 363, 369, 376, 377, 381, 386, 392, 393, 401–404, 409, 410, 439, 447, 451, 457–459, 464, 469, 505, 519, 544, 564, 587, 608, 626, 631, 666, 681, 688, 704, 718, 719, 752, 782, 797, 804, 811, 820, 822–824, 826, 829, 839, 843, 844
adiabatic electronic states, 251, 273
diabatic electronic states, 276
dissociative electronic state, 280
electronic back action, 357
electronic continuum functions, 369
electronic coupling, 187
electronic degrees of freedom, xxi, xxvi, xxxi, xxxii, 45, 121, 190, 202, 224, 230, 243, 246, 249, 251, 261, 282, 310, 318, 320, 341, 343, 377, 379, 383, 386, 400, 402, 438, 451, 457, 459–461, 465, 547, 605, 631, 646, 650, 734, 818, 822, 844

electronic relaxation, 332, 711
electronic state, xxv, xxxviii, 37, 39, 187, 245, 252, 262, 278, 288, 306, 343, 347, 350, 421, 423–425, 426, 573, 574
electronic structure, 318, 319, 321, 575, 589
electronic structure calculations, 590
electronic structure methods, 315
electronic subsystem, 249, 318, 320, 321, 461, 802, 838
electronic time scale, 332
friction kernel, 835
multiple electronic states, 317
orthonormal electronic states, 315
time-dependent electronic states, 392
weak electronic coupling, 187, 190
electronic friction, xli, 817, 820, 824, 828, 833, 835–838
friction kernel, xli, 824, 826, 827
frictional force, 822
electronic Hamiltonian, 9, 16, 825
emission, 337, 668, 673, 674, 676, 679, 773
atomic emission, 336
dipolar emission, 334
emission time, 334
light emission, 333, 336
optical emission, 773
power emission, 334
spontaneous emission, 817
stimulated emission, 671, 676–681, 686, 718
stimulated emission scattering, 676
stimulated emission signal, 688
ultraviolet emission, 665
END, xxxii, xxxvi, 342, 343, 349, 351–353, 356–358, 360–362, 365, 366, 368–370, 372–376, 381–383, 465, 547, 611
energy, xxix, 171, 187, 194, 196, 245, 262, 270, 272, 273, 274, 291, 334, 335, 373, 398, 399, 407, 408, 416, 426, 428, 431, 441, 443, 462, 463, 509, 580, 585, 590, 591, 595, 597, 623, 624, 627, 642, 655, 666, 671, 683, 685, 686, 702, 704, 726, 735, 820, 827, 835, 836, 840
chemical energy regime, 515
classical nuclear energy, 428
classical nuclear kinetic energy, 646
collision energy, 373
configurational energy, 179
correlation energy, 593
diabatic potential energy, 646, 660, 685
dissociation energy, 274
energy barrier, 162, 163
energy barrier top, 162
energy conservation, 187, 336
energy dissipated, 667, 835
energy expectation value, 297, 372, 784
energy functional, 590
energy gap, 406, 431, 435, 639, 640, 679, 695, 704, 709, 741
energy landscapes, 751
energy minimization, 598
exchange-correlation energy, 593, 594
excitation energy, 596, 597, 602, 603, 733
excited state energies, 595–597
excited state reorganization energies, 773
first excited energy, 597
free energy, 173, 783, 786, 789, 795–797, 799, 800
free energy surfaces, 802
instantaneous vibrational energy, 650, 658
ionization energy, 245
kinetic energy, 176, 177, 180, 204, 226, 308, 316, 369, 377, 378, 379, 415, 416, 498, 507, 539, 572, 583, 589, 590, 636, 740, 744, 747, 756, 764, 800
kinetic energy density, 536, 537, 539, 544, 608
kinetic nuclear energy, 412, 447 631
molecular energy levels, 728
nuclear energy loss, 372
nuclear kinetic energy, xxvii, 36, 249, 253, 319, 320, 377, 384, 388, 402, 446, 457, 630, 631
orbital energies, 830

photonic energy, 574
projectile energy, 337, 338
reorganization energy, 765, 772, 773, 776, 785
single-electron eigenenergies, 840
solvent reorganization energy, 785, 815
TDHF energy, 338
total energy, 204, 260, 280, 281, 309, 311, 323, 326, 572, 590, 591, 593, 599, 608, 721
transfer of energy, 271, 280
translational energy, 510, 837, 840
zero-point energy, 373, 547, 768
ensemble, 498–500, 509, 511, 517, 528, 531, 532, 589, 620, 637, 698, 699, 702, 704, 709, 710, 740, 743, 744, 746, 747
canonical ensemble, 174, 795
coherent ensemble, 617
ensemble average, 192, 797
excited state ensemble, 741
quantum-classical ensemble, 632, 635
quantum ensemble, 178
thermal ensemble, 706
trajectory ensemble, 744
ensemble average, 166, 631–633
entropy, 750
envelope function, 667, 696, 698, 741
environment, xxix, 161, 163, 165, 167, 190–195, 197, 538, 541, 613, 614, 625, 635–637, 639, 641, 658, 659, 750, 751, 772, 789, 791, 792, 798
aqueous environment, 793
classical environment, 413, 613, 614, 620, 802
dielectric environment, 780, 786
environmental degrees of freedom, xxix, 189, 191, 192, 632, 637, 649, 843
macroscopic environment, 783
molecular environment, 790, 793
nuclear environment, 615, 631
polar environment, 752, 755, 775, 776, 784, 786, 792, 806, 810
quantum environment, 802
equation of motion (EOM), xxxii, xxxiv, 40, 87, 89, 90, 136, 155, 192, 193, 295, 298, 301, 308, 321, 326, 339, 349, 358, 388, 391, 428, 441, 498, 499, 518, 519, 521, 534, 535, 537, 540, 543, 544, 546, 573, 606, 625, 650, 653, 654, 686, 708, 791, 823, 833–835, 841
classical equation of motion, 151, 308, 393, 410
Ehrenfest equations of motion, 389
Newtonian equation of motion, 539, 702
nonlinear equations of motion, 454
nuclear equation of motion, 822, 823
quantum equations of motion, 614
equation-of-motion-coupled-cluster (EOM-CC), 845
equilibrium, 4, 51, 55, 57, 69, 163, 192–194, 196, 577, 637, 707, 757, 765, 776, 780, 785, 797, 812
C_{2v} equilibrium, 701
C_S equilibrium, 744, 745
equilibration, 636
equilibrium average, 194
equilibrium bond length, 289, 829
equilibrium charge, 795
equilibrium charge state fractions, 371
equilibrium constraint, 197
equilibrium expectation value, 780
equilibrium geometry, 829
equilibrium mole fraction, 169
equilibrium populations, 413
equilibrium states, 162
ground state equilibrium, 638, 704, 709
non-equilibrium, 164, 640, 810
non-equilibrium average, 166
product equilibrium mole fractions, 164
reactant equilibrium mole fraction, 165
thermal equilibrium, 161, 168, 174
equipotential lines, 427
ergodic, 166
ethylene, 433

Euclidean action, 186, 799–801
Euclidean action integral, 176
Euclidean path integral, 177, 178, 180, 184, 802, 798
Euclidean time, 176, 177, 180, 181, 802
Euler angles, 229, 233, 236, 247, 248, 250, 258
Euler–Lagrange equations, 129, 293, 319, 353, 354, 811
Eulerian, 496, 500, 501
  Eulerian frame, 500
evolution, xxxi, xxxiv, xxxviii, 36, 89, 91, 119, 151, 153, 292, 300, 303–305, 322, 329, 343, 386, 404, 408, 410, 412, 424, 429, 443, 461, 484, 498, 500, 501, 505, 509, 517, 525, 531, 545, 557, 568, 572, 573, 575, 577, 582, 583, 617, 623, 645, 647, 649, 651, 657, 663, 665, 669, 689, 703, 706–709, 715, 721, 725, 732–734, 738, 746, 790, 811
  adiabatic evolution, 321, 706
  coherent evolution, 620
  coherent quantum evolution, 631
  decoherent molecular evolution, xxxviii, 615
  dynamic evolution, 342, 377, 426
  equation of evolution, 499
  excited state evolution, 711
  linear evolution, 484
  molecular evolution, xxxii, xxxv, 318, 517, 615, 663
  nonadiabatic evolution, xxxvi, 701
  nonadiabatic molecular evolution, 715
  nonlinear evolution, 484
  quantum evolution, 843
  time evolution, 309, 312, 388
  unitary evolution, 613
  wave function evolution, 723
exact exchange
  (EXX), 603
excess electron, 751–753, 755–757, 761
exchange, 20, 27, 255, 259, 369, 593, 793, 794, 830, 843
  electron exchange, 369, 775, 792
  energy exchange, 165, 515
  exchange amplitude, 78
  exchange contribution, 17, 21
  exchange-correlation, 602
  exchange-correlation functional, 606
  exchange-correlation kernel, 599, 600, 603
  exchange electron, 750
  exchange integral, 17, 594
  exchange of particles, 161
  exchange probabilities, 369
  non-local exchange, 603
  particle exchange, 163
  PGG exchange-correlation kernel, 603
exciplex, 275, 280
  exciplex funnel dynamics, 280, 282
  exciplex levels, 281
  exciplex states, 281
excitation, xxxix, xl, 32, 34, 40, 113, 161, 202, 258, 304, 345, 336, 343, 465, 524, 579, 584, 585, 595, 602, 605, 609, 676, 684, 700, 707, 710, 711, 719, 726–728, 732, 733, 751, 820, 836, 840
  adiabatic excited state, 290
  alternative excitation routes, 729
  bichromatic excitation, 728
  diabatic excited state, 686
  diabatic projection operator, 687
  donor excitation, 790
  doubly excited determinants, 33
  doubly excited states, 35
  electronic excitation, xix, xli, 37, 41, 344, 366, 369, 372, 376, 595, 596, 601, 730, 735, 790, 819, 824, 829
  excitation amplitudes, 345
  excitation energies, 40, 407, 596
  excitation probabilities, 416
  excited electronic state, xxv, 36, 37, 40, 308, 576, 595, 665, 734
  excited level, 676
  excited nonstationary state, 665
  excited state, xxxvi, 4, 37, 38, 256, 300, 416, 417, 430, 453, 466, 525, 576, 580, 595–597, 605, 610, 623, 638–640, 677, 686, 691, 692,

696–698, 702, 703, 709, 710, 719, 720
first excited adiabatic state, 290
first excited energy, 755
first excited state, xxxix, 195, 272, 429, 430, 455, 465, 597, 637, 640–643, 670, 701, 707–709, 712, 755, 761
first excited state population, 196
low-lying excited states, 576, 577, 700, 702
one-particle excitation, 819
optical excitation, 773, 792
optically accessible excited states, 579
photo-excitations, 86
photoexcited state, 683
second excited state, 700, 712
single substitutions, 35, 36
UV excitation, 579
vertical excitation, 585, 701, 772
vibrational excitation, 376, 828, 832, 834–836, 838
excited state trajectory, 743
exit channel, 835
exothermic, 256, 839
expectation value, 11, 21, 25, 152, 174, 215, 301, 319, 356, 381, 387, 431, 432, 456, 630, 645, 686, 687, 721, 738, 740, 779, 812, 820

F-H_2, 246
Fang, J.Y., 517, 520, 521, 525
Faraji, S., 585
fast Fourier transform (FFT), 525, 548
FFT grid procedure, 525
fast variables, 77
FB-IVR, 481–483
Fe, 798, 801, 803
Fe^{2+}, 793, 794
Fe^{2+}+Fe^{3+}, 795, 799
$Fe^{2+}(H_2O)_6$, 793
Fe^{3+}, 793, 794, 805
Fe^{3+}+Fe^{2+}, 795
$Fe^{3+}(H_2O)_6$, 793
femtochemical, 636
femtosecond, xix, xxv, xxxix, 629, 641, 663, 706, 742
femtosecond spectroscopy, 711
femtosecond pump–probe, 665
Fermi, E., 589
Fermi energy, 826, 830
Fermi hole, 8
fermionic, 6, 844
ferroelectricity, 777
ferrous–ferric exchange, xl, 750, 753, 795, 804
aqueous ferrous–ferric exchange, 802–804
Feshbach partitioning, 119
Feshbach resonance, 204, 114
fewest switches, 411–415, 434, 623, 659, 708, 744
fewest switches algorithm, 408, 412, 413, 447, 610
fewest switches hopping algorithm, 435
fewest switches TSH, 610
Feynman, R.P., xxxiii, 122, 183, 615
Feynman representation of quantum mechanics, 437
FFT, 549, 553, 555, 556, 567, 583
FH, 258
fictitious mass, 320, 321
field, 312, 356, 664, 667, 668, 674, 691, 716, 726, 779, 780, 782, 783
dielectric displacement field, 775–779, 812
electromagnetic fields, 715
electrostatic fields, 783
field theory, 842
fluid velocity field, xxxiv
monochromatic fields, 778
perturbing field, 779
probe field, 666, 669, 673
pump field, 666
quantum field theory, 844
weak field limit, 672, 719, 730
Field, M.J., 305, 312, 356
field-induced surface hopping (FISH), 743
field theory, 122
Filinov, V.S., 476, 477

filter, 476, 698
Filinov filtering, 476, 478, 486
filter function, 475, 699
filtering efficiency, 478
filtering technique, 476
spectral filter, 699
flow
compressible flow, 506
Hamiltonian flow, 528
irrotational flow, 495
momentum flow, 539
vibrational flow, 835
fluctuating force, 822
fluctuation, 163, 166, 380, 431, 632, 636, 802
fluctuation-dissipation theorem, 835
mean-square fluctuation, 154
second fluctuation-dissipation theorem, 835
spontaneous fluctuation, 166
structural fluctuations, 636
zero-point fluctuation, 154
fluid, xxxiv, 500, 501, 503, 509, 510
classical fluid, 501
fluid element, 501, 541
probability fluid, 500
quantum fluid, 501, 504
fluorescence, 281
fluorine, 255, 257, 700
flux, 114, 169, 655
electromagnetic flux, 334–336
incoming flux, 111, 114
particle flux, 495
reactive flux, 167, 172
FMS, 429, 431, 432, 434
Fock space, 154, 158, 159
Fock space states, 158
force, 309, 312–314, 316, 319, 320, 338, 341, 357, 359, 384, 440, 447, 464, 541, 545, 615, 635, 655, 657, 748, 820, 824
adiabatic force, 824
adiabatic Hellmann–Feynman forces, 640
delocalized force, 442
dragging force, 331
Ehrenfest force, 317, 318, 320, 465, 642
fluctuating force, xli, 820, 835
frictional, 835
generalized quantum force, 541
Hellmann–Feynman force, 312–314, 317, 633, 634, 637, 639, 643
instantaneous force, 442, 625
linearized quantum force, 516
local force, 3, 443
Markov random forces, 834
mean-field force, 643
nonadiabatic force, 447
nuclear force, 312, 823
Pechukas force, 443, 444, 446, 447, 640
quantum force, 339, 501, 539–541
random force, 835
repulsive force, 367
forward backward-initial variable representation
(FB-IVR), 472, 480
Fourier expansion, 183
Fourier integral, 136
Fourier transform, 299, 304, 335, 400, 401, 468, 602, 604, 669, 778, 781
discrete Fourier transform, 556
fast Fourier transform (FFT), 414, 525, 548, 555
Fourier representation, 627
inverse Fourier transform, 300
Franck–Condon, 631, 768
Franken, P.A., 665
free energy, 784, 786, 789, 794, 803
classical free energy, 802
diabatic free energy, 798
equilibrium free energy, 789
Frenkel variational principle, 295, 306
frequency, 335–337, 372, 373, 376, 400, 607, 631, 635, 665, 666, 668, 686, 691, 728, 740, 761, 762, 768, 772–774, 777, 778, 782, 791, 803, 804
dump laser frequencies, 702
harmonic frequency, 704
high-frequency limit, 782
longitudinal frequency, 788

photon frequency, 607
radiation frequency, 664
resonance frequencies, 689
transition frequency, 668
vibrational frequencies, 761, 762
friction, 48, 835
electronic friction, 822
friction kernel, 828
friction tensor, 828, 833
frozen Gaussian, 635
full multiple spawning (FMS), 426
functional, 41, 263, 460, 587–590, 592, 594, 595, 596, 598, 599, 603, 604, 741, 785, 810, 811
action functional, 122, 129, 132, 438
anharmonic influence functionals, 480
correlation functional, 595
density functional, 606
energy functional, 597
exact functional, 595
exchange-correlation functional, 590
free energy functional, 775, 783, 784, 810–813
hybrid functional, 595, 596
orbital-dependent functional, 603
target functional, 748
total energy functional, 589
variation functional, 722
functional derivative, 19, 41, 42, 313, 319, 601
functional differentiation, 41
second functional derivative, 42
funnel, xx, 275, 280, 580, 707
electronic funnels, 433
funnel states, 281
photochemical funnel, 683, 684
furan, 574

Gadzuk, J.W., 839
Garashchuk, S., 515
gas, 635, 665
free electron gas, 592
gas molecules, 835
gas phase, 755, 786, 789
gas phase Hamiltonians, 786
homogeneous electron gas, 603
uniform electron gas, 609
gas–substrate interaction, 835
gauge, 59, 352
gauge field, 61
gauge function, 294
gauge invariance, 60, 293
gauge symmetries, 61
Gaussian, xxx, 160, 184, 349, 351, 414, 419, 438, 477, 482–484, 486, 488, 524, 531, 542, 546, 574, 582, 619, 630, 634, 691, 713, 739
center of the Gaussian wave packet, 422
dressed with frozen Gaussians, 424
frozen Gaussian, 629, 635
frozen Gaussian approximation, 630
Gaussian basis function, 427
Gaussian-dressed paths, xxxv
Gaussian integration, 190
Gaussian quadrature, 839
Gaussian statistics, 802
Gaussian target, 739
Gaussian wave packet, xxxiii, 420, 483, 484, 508, 509, 511, 516
Gaussian widths, 423, 639
spawned Gaussians, 425
traveling Gaussians, 433
Gell-Mann, M., 96
generalized gradient approximation (GGA), 594
generalized Langevin model, 819
genetic code, 577
geometric phase, 46, 61, 75, 77, 78, 86, 376
complex geometric phase, 76
George, G.F., 418
gerade, 13, 14
GFP chromophore, 610
Ghirardi, G.C., 616
gh-space, 84
gh-plane, 81, 83, 84

Goerling, A., 597
Goldberger, M., 96
golden rule, xxix, xl, 163, 187, 189, 190, 196, 197, 626, 629, 644, 752, 763, 767, 790, 792, 799, 800, 829
Fermi's golden rule, 626, 632
golden rule integral, 764
Gomer–Noyes mechanism, 610
Gossmann, U.J., 603
gradient, 252, 253, 326, 477, 495, 501, 502, 507, 684, 702, 784
excited state gradients, 700
total energy gradient, 353
Green's function, 99, 108, 135, 206–208, 241, 278, 839
asymptotic Green's function, 241
half-integrated Green's function, 278, 279
multichannel Green's function theory, 206
outgoing Green's function, 210, 364
radial Green's function, 107, 206
single-channel Green's function, 207
Green's operator, 229
grid, 487, 504, 549, 550, 559
coordinate grid, 552
grid basis, 552
multidimensional grids, 564
non-evenly spaced grids, 556
Gross, E.K.U., 603
ground state, xix, xxv, xxxvi, 4, 15–17, 20, 21, 25, 30, 32, 35–39, 46, 70, 72, 113, 190, 195, 256, 258, 261, 262, 272, 275, 290, 302, 306, 318, 320, 333, 346, 366, 376, 378, 400, 417, 429–431, 455, 524, 576–578, 580, 581, 585, 588, 595, 598, 637–639, 665, 666, 670, 672, 676, 682–686, 692, 698–701, 704, 706, 707, 709, 720, 726, 727, 730–732, 734, 737, 740, 755, 772, 773, 830
adiabatic ground state, 289, 582, 684, 760
electronic ground state, 457, 465, 576, 577, 583, 585, 609, 610, 719, 732, 829
ground state density, 589–592
ground state equilibrium, 684
ground state population, 624
ground state reactivity, 261
ground state surface, 732, 743–745, 747
group, 346
C_{2v} group, 429
generators, 346
Lie group, 347
group theory, 70
guanine, 575, 577–579
guiding wave, 491, 496, 498, 509
Gutzwiller, M., 134, 135
GVB, 321

H, 259, 333, 336, 337, 366, 407, 512, 581
H^+, 30, 374, 407
H^+ projectile, 334
$H^+ + D_2$, 406
$H^+ + H_2$, 406, 419
H^-, 30
H_2, xxx, 12–14, 30, 32, 33, 38, 40, 259, 260, 272, 274, 281, 365, 373–375, 406, 407, 429–432, 512
H_2 bond length, 431
H_2 dissociation, 30
H_2^+, 336, 406, 407, 753
H_3, 376, 723
H_3^+, 376–380, 406, 407
H_4, 33
H_2O, 14, 15, 637–642, 730, 793, 795, 803, 804
H_2O symmetry, 638
Hamilton equations, 316, 354, 355, 388, 418, 734, 821
Hamilton function, 126, 151, 176, 316, 388–390, 392, 394, 397, 456, 457, 459, 460, 649, 655, 695, 735, 763
adiabatic Hamilton function, 389, 391, 394
diabatic electronic Hamilton function, 394
electronic Hamilton function, 388, 395

time-dependent Hamilton function, 481
Hamilton's equation, 386, 388, 456, 481, 528
Hamilton's principle, 129
Hamilton–Jacobi equation, 136, 137, 143, 324, 494, 512, 514
Hamiltonian, 53, 58, 79, 82, 87, 89–92, 94, 104, 124, 169, 186, 187, 190, 193, 223, 226, 228, 229, 231, 235, 236, 238, 239, 247, 256–258, 268, 275, 293, 294, 296, 323, 363, 389, 397–400, 404, 441, 452, 454–457, 461, 464, 469, 471, 505, 532, 556, 559, 561, 567, 572, 573, 589, 590, 596, 605, 613, 622, 624, 626, 627, 670, 681, 684, 691, 692, 694, 716, 733, 753, 755, 758, 769, 791, 794, 812, 829
acceptor Hamiltonians, 786, 806
adiabatic electronic Hamiltonian, 391
adiabatic Hamiltonian, 229, 389, 396
Anderson–Newns Hamiltonian, 839
asymptotic Hamiltonian, 238, 276, 323
classical Hamiltonian, 698, 699
coupling Hamiltonian, 192, 791
Delves Hamiltonian, 262
diabatic Hamiltonian, 467, 581, 692, 776
distorted-wave Hamiltonian, 275
electron transfer Hamiltonian, 756
electronic Hamiltonian, 23, 35, 47, 62, 259, 306, 315, 387, 408, 444, 458, 645, 753–755
four-mode model Hamiltonian, 471
harmonic oscillator Hamiltonian, 372
image Hamiltonian, 456
interaction Hamiltonian, 195–197, 791
Kohn–Sham Hamiltonian, 610
Landau–Zener–Stückelberg Hamiltonian, 393
many-electron Hamiltonian, 304
mapping Hamiltonian, 460, 465
mean-field Hamiltonian, 567
molecular Hamiltonian, 56
multichannel Hamiltonian, 202
one-electron Hamiltonian, 328
perturbation Hamiltonian, 780, 787
product Hamiltonian, 223
reactant Hamiltonian, 223
space-fixed Hamiltonian, 235
spin-boson Hamiltonian, 756, 759–761, 802, 804
tight binding Hamiltonian, 755
time-dependent electronic Hamiltonian, 440
transfer Hamiltonian, 790
triatomic Hamiltonian, 225
two-state diabatic Hamiltonian, 692
vibrational Hamiltonian, 752
zeroth-order Hamiltonian, 206, 692
Hankel function, 107
harmonic oscillator, 74, 144, 222, 372, 373, 465, 524, 686, 735, 791, 792, 823, 829
displaced harmonic oscillator, 635
harmonic oscillator eigenfunction, 469, 807, 808
harmonic oscillator frequency, 158
harmonic oscillator ground state, 635
harmonic oscillator states, 461
one-dimensional harmonic oscillator, 155
Hartmann, M., 706
Hartree, D.R., 567
Hartree product, 569
Hartree–Fock, see also HF, xxx, xxxii, 16, 23, 29–31, 313, 587, 594, 595, 824
canonical HF equations, 292
Hartree–Fock energy, 23, 25, 34
Hartree–Fock equation, 17, 20, 24, 27, 29, 590, 594
Hartree–Fock ground state, 21, 35–37
Hartree–Fock limit, 29, 31
Hartree–Fock reference determinant, 33, 35
Hartree–Fock theory, 3, 22, 28
post-Hartree–Fock, 30, 33
restricted Hartree–Fock, 23–28, 321, 594, 596

restricted open-shell Hartree–Fock, 321
unrestricted Hartree–Fock, 23, 25–28, 30, 594
He^+ ion, 113
He_2^+, 337
Head-Gordon, M., 822, 824, 831
heat bath, 161, 190
Heaviside factor, 800
Heisenberg picture, 89–91, 123, 175
helical, 238
Heller, E., 349, 468, 483, 484, 629
Hellmann–Feynman force, 407
Hellmann–Feynman theorem, 305, 312, 313, 384
Hellmann–Feynman limit, 313, 314
heme, 579
Herman, M.F., 450
Herman–Kluk, 462, 480, 489, 734
Herman–Kluk factor, 460
Herman–Kluk prefactor, 472
Hermite polynomials, 542
Hermitian, 9, 19, 100, 242, 268, 298, 299, 304, 355, 520–522, 540
anti-Hermitian, 310–312, 515, 568
Hermitian constraint operators, 568
Hessian, 478
heterocyclic aromatic, 579
heterodyne, 671
HF, 292, 301, 303, 314, 315, 319, 322
HgAr, 725
HHD, 723
Hibbs, 184
Hilbert space, xxviii, 292, 296, 349, 452, 549, 619
Hilbert subspace, 292
infinite-dimensional Hilbert space, 549
M-dimensional Hilbert space, 387
spin Hilbert space, 453, 454
two-oscillator Hilbert space, 454
history, 176, 615, 616, 618
arbitrary path, 616
coarse-grained history, 616–618, 620
fine-grained histories, 616, 617, 621
Hohenberg, P., 587, 589, 590
Hohenberg–Kohn
first Hohenberg–Kohn theorem, 595, 598
Hohenberg–Kohn theorem, 587, 589, 595, 598
second Hohenberg–Kohn theorem, 596
hole, 259, 343
single-hole, 570
single-hole functions, 568, 569
Holstein, T., 454
HOMO, 32, 831
homodyne, 669
homogeneous electron gas, 593
hopping, xxxiii, 403, 404, 406–409, 411, 412, 414–417, 419, 420, 424, 433–435, 437, 442, 443, 445, 462, 463, 511, 515, 615, 642–644, 648, 649, 659, 839
hopping criterion, 407
hopping locations, 413
hopping probability, 640, 643, 748
hops between surfaces, 408
horizontal jumps, 413
instantaneous hops, 447
non-vertical hopping, 413
position-conserving jumps, 413
stationary phase surface hopping, 640
Hughes, K.H., 542
Husimi transform, 462
hybrid, 30, 512, 642, 645
hydrated electron, 615, 636–644
hydrated electron deexcitation, 638
hydrated electron relaxation, 638, 641
hydrodynamic, xxxiv, xxxv, 491, 500, 509, 542, 604
classical hydrodynamics, 495, 505, 541, 542
generalized quantum hydrodynamics, 534
hydrodynamic equation, 538, 540, 541, 543
hydrodynamic field, 544
Lagrangian quantum hydrodynamics, 499–501, 503, 505, 508–511, 535
local hydrodynamic fields, 518, 537

nonadiabatic Lagrangian quantum hydrodynamics, 505, 507, 510
quantum hydrodynamic approach, 532
quantum hydrodynamics, xxxiv, 419, 491, 492, 504, 515, 517, 536, 542, 547
hydrogen, 223, 257, 429, 580, 838, 845
hydrogen beam, 371
hydrogen elimination, 580
hydroxylamine, 86
hydroxyphenyl oxazole, 482
hyperangle, 228
hypercube, 474, 475
hyperradius, 227, 228
hyperspherical, 262

I_2, 725
ICN, 731, 732, 736, 737
Icsevgi, A., 667
IESH, 840
imaginary frequency, 53
imaginary time, 136, 163, 171, 799
impact parameter, 335, 362, 365, 367–369, 431, 432
incoherent, 616, 617
incoherent superposition, 534
independent electron surface hopping, 840
independent sets of dynamic variables, 306
infinite-order sudden approximation (IOSA), 373
initial value problem, 89
initial value representation (IVR), 361, 438
inner sphere, 749
in-plane frustrated rotation, 833
instanton, 186
insulator, xli, 817
integrability, 475
integral, 29, 42, 61, 93, 98, 99, 109, 255, 259, 285, 314, 329, 471, 473–476, 478, 479, 485, 493, 502, 529, 549, 593, 619, 629, 640, 664, 697, 716, 764, 768, 772
coupled integral equations, 742
energy integration, 826
F-dimensional integral, 477
IVR integrals, 472
multidimensional integral, 469, 474
overlap integral, 634
phase space integrals, 523
SC-IVR integrals, 473
three-center integrals, 754
two-center integral, 634
two-electron integral, 29, 36
integral equation, 98, 107, 127, 217, 222, 229, 238, 242
Fredholm-type system, 742
interaction, xxii, xxvi, xxviii, xxxi, xxxiv, xxxviii, xl, 32, 224, 227, 246, 259, 263, 271, 272, 274, 288, 296, 304, 323, 331, 338, 361, 363, 366, 367, 376, 377, 411, 412, 415, 431, 460, 470, 472, 506, 512, 576, 580, 588, 604, 605, 607, 610, 613, 614, 636, 640, 659, 670, 672–675, 682, 687, 690, 711, 715, 716, 749–752, 761, 783, 790, 791–793, 817, 818, 828, 836
atom–diatom interactions, 238
bra interaction, 674
exchange interaction, 589
field–matter interactions, 674
gas–surface interaction, 834
H^+–H interaction, 333
interacting electrons, 592, 597
ket interactions, 674
molecule–metal interaction, 839
molecule–surface interaction, xix, xxviii, xli, 817, 819, 838
nonadiabatic interaction, xxxiii, 339, 644, 735
optical interactions, 664, 675, 690, 772, 773
photonic interactions, 663
projectile-target interaction, 369
self-interaction correction, 594
self-interaction effect, 594
self-interaction problem, 595
interaction Hamiltonians, 91
interaction picture, 89–92, 191, 668
interaction picture operators, 91
interelectronic repulsion, 26, 304

interference, xxxiii, xl, 403, 416, 418, 482, 543, 547, 613, 615, 617, 623, 689, 698, 727, 729, 748, 811
interference of trajectories, 532
internal angles, 229
internal degrees of freedom, 202
internal motion, 373
internal nuclear degrees of freedom, 203
intrastate interference effects, 428
particle interference, 499
trajectory interference, 534
internal conversion, xx, 576
ultrafast internal conversion, 574, 579, 683
internal nuclear coordinates, 80
internal quantum numbers, 718
invariant, 454, 589
adiabatic invariance, 397
inverted region, 765, 766
inverted Marcus region, 767
ionization, 369, 700, 701, 711
electron-beam ionization, 114
ionization thresholds, 601
multiphoton ionization, 578, 636
iron, 794, 795, 797, 799, 805
irreducible representation, 14, 346
isomer, 577, 579, 683, 700, 738, 739
C_{2v} isomer, 707, 709, 738, 740, 744, 745
isomeric structures, 578
isomerization, 575, 700, 704, 744, 746, 747
cis-trans isomerization, 433, 682
cis-trans photoisomerization, 663, 683, 684, 687, 690, 711
excited state isomerization, 701, 706
isomerization pathway, 747
isomerization threshold, 740
isomerization time, 690
photoisomerization, 686, 690, 738, 748
isopotomer, 723
isotope effect, 638, 639
isotropic medium, 196
IVR (initial variable representation), 450, 451, 472, 473, 485, 486
IVR-SCF, 473

Jacobian, 131, 184, 441, 449, 459
Jahn–Teller distortion, 68, 73
Jahn–Teller effect, 65, 80, 376
$E \times e$ Jahn–Teller Hamiltonian, 395
Jahn–Teller stabilization energy, 69
Jahn–Teller theorem, 69, 70
Johnson, B.R., 222
jump condition, 108

kernel, 600, 602, 742–744
ket, 348, 350, 359
ket notation, 123
Kindt, J.T., 838
Kluk, E., 450
Kohn, W., 587, 589–591, 593, 604, 667
Kohn anomalies, 269, 270
Kohn integral, 267
Kohn–Sham, 594, 597
auxiliary Kohn–Sham system, 597
generalized Kohn–Sham theory, 603
Kohn–Sham analogous equations, 596
Kohn–Sham determinants, 605, 609, 611
Kohn–Sham energy levels, 596
Kohn–Sham equations, 587, 594, 598, 604
Kohn–Sham ground state determinant, 596
Kohn–Sham orbitals, 606, 607
restricted open-shell Kohn–Sham, 596
time-dependent Kohn–Sham equation, 605
time-dependent Kohn–Sham theory, 606
virtual Kohn–Sham orbitals, 606
Koopmans' theorem, 21, 22
Kosloff, R., 525, 723
Kramer, P., 300
Kramers' degeneracy, 70
Kronecker delta, 242

Krylov space, 561
Kubo, R., 835
Kuehn, O., 806
Kuharski, R.A., 793, 795, 797

$\mathcal{L}^2$ basis, 266
laboratory frame, 363, 368, 373
Lagrange multiplier, 17, 86, 302, 304, 307, 319, 590, 592, 721, 722, 741, 742
Lagrangian, xxxix, 85, 126, 176, 293, 319, 349, 350, 353, 500, 501, 630
classical Lagrangian, 125, 503
generalized Lagrangians, 507
Lagrangian dynamics, 177
Lagrangian frame, 500, 501, 506
quadratic Lagrangian, 131
quantum Lagrangian, 503
Lagrangian minimization, xxxi, 84–86
Lagrangian multiplier, 314
time-dependent Lagrangian multiplier, 293
Laguerre polynomials, 809
Lan, Z., 579
Lanczos, 562
Lanczos recursion procedure, 561
Lanczos recursion scheme, 560
Landau, L.D., xxx, 282, 288
Landau–Zener, 762, 764
Landau–Zener transition probability, 762
Landau–Zener length, 288
Landau–Zener–Stückelberg theory, 246, 660
Langer modification, 394, 395, 458, 461
Langevin, xli, 820, 822, 823, 838, 840
laser, 154, 729
continuous wave laser, 728
laser field, 671, 748
laser frequency, 640, 666
laser intensity, 729
laser ray in the monochromatic limit, 300
optimal laser control, 747
pulsed lasers, 663
ruby laser, 665
ultrashort laser pulse, 524
LCAO, 9, 10, 12, 15, 27, 28
LDA, 593, 595, 603
Lee, C., 595
Legendre polynomials, 104, 113, 203
associated Legendre polynomial, 234
Levi-Civita tensor, 235
Levine, R.D., 419
Li_3, 376
Li_3^+, 376
LiF, 817
life time, xli, 86, 113, 281, 576, 641, 642, 792, 817, 818, 829, 831–833
excited state life times, 641
radiative life time, 281
vibrational life time, 833
Light, T.J., 333, 556, 664, 690, 718
light emission, 333
light wave, 718
linear combinations of atomic orbitals, see also LCAO, 9
linear response, 677, 842
LR, 601
linear response theory, 610, 798
Liouville, 519, 520, 526, 533, 621, 667, 711, 739
classical Liouville equation, 151
Liouville mechanics, 152
Liouville theory, 517, 518
quantum Liouville approach, 541
quantum Liouville equation, 151
semiclassical Liouville approach, 525
semiclassical Liouville approximation, 525
semiclassical Liouville formalism, 518, 524, 532
semiclassical Liouville theory, 526, 528, 532, 692
Liouville–von Neumann equation, xix, xxxiv, xxxviii, 149, 153, 191, 195, 304, 329, 336, 517, 518, 522, 535–537, 543, 545, 546, 614, 620, 621, 631, 632, 641, 649, 663, 667, 691, 693, 694, 844
dissipative Liouville–von Neumann equation, 615

extended Liouville–von Neumann equation, 642
semiclassical Liouville–von Neumann equations, 517, 529
TDHF Liouville–von Neumann equation, 305, 329
Lippmann–Schwinger equation, xxix, 87, 98, 99, 107, 111, 206, 209, 238, 239, 242, 264, 266, 275
modified Lippmann–Schwinger equation, 277
liquid, 636, 639, 640, 665
non-polar liquids, 636
polar liquids, 636
lithography, 843
local, 58, 172, 292, 295, 321, 447, 579, 600, 604
autolocalized state, 636
delocalization, 609, 843
delocalized, 750, 819, 824
electron localization, 608, 609, 636
localization, 446, 447
localized, 187, 188, 308, 450, 608, 637, 702, 756, 760, 794, 795
localized electronic states, 755
localized wave functions, 754
non-local, xxxiv, 229, 441, 537, 599
non-locality, 77, 447, 599, 603, 604, 778, 779
local density approximation, see also LDA, 593
local density of states, 826
local spin density (LSD), 594
log derivative, 218–220, 223, 260
log derivative matrix, 218
log derivative propagation, 217
lone pairs, 608
longitudinal, 604, 657, 782, 783, 812
longitudinal vector potential, 604
Longuet-Higgins, H.C., 77
Lopreore, C.L., 508
Lorentzian, 284
LSC-IVR, 479, 482, 483
LSD, 595
LUMO, 32, 831
LZS, 288–290
macrocycles, 579
magnetic field, 334, 335, 604, 613, 664
magnetic moment, 11
magnetic permeability, 334
malonaldehyde, 433
Manolopoulos, D.E., 221, 469, 488
many-body, 5, 16, 22, 30, 597, 598
many-body effects, 590
many-body perturbation theory, 31, 33, 34
many-electron determinant, 16
many-electron wave function, 40
map, 620
reduction map, 617, 619, 624, 626
mapping, xx, 395, 438, 451–455, 457, 464, 465, 468, 596, 735, 820
mapping rules, 455
operator mapping, 458
SC-IVR mapping, 465
Schwinger mapping formalism, 452
Schwinger mapping procedure, 453, 454
Schwinger mapping rules, 453
Marcus, R.A., 750, 798, 804, 805
Marcus theory, xl, 750, 752, 775, 805
classical Marcus theory, 766
Marcus formula, 766
Markov approximation, 538
Markovian, 791
Markov limit, 827, 828, 831
Martens, C.C., 517, 520, 521, 525
Martinez, T.J., 419
Maslov index, 135, 438, 448
Mason, E.A., 365
Massey parameter, 287–289, 405, 406
matrix, 49, 52, 54, 59, 64, 67, 68, 82, 90, 134, 135, 196, 204, 206, 209, 216, 218, 220, 221, 222, 228, 241, 242, 252, 254, 259, 266, 267, 273, 284, 302, 304, 307, 311, 312, 330–332, 347, 354–356, 357, 359, 360, 409, 420, 423, 428, 429, 445, 448, 454, 477, 479, 488, 506, 520, 544, 549, 551, 552, 562, 569, 619, 622, 640, 641, 648, 681, 721, 760, 822, 824–826, 827, 830
adiabatic potential energy matrix, 252
block matrix, 221, 833

charge density bond order matrix, 10
charge density matrix, 28, 29, 305, 332
classical S-matrix, 387
coupling matrix, 64, 204, 205, 221, 232, 581, 825
density matrix, xix, xxxix, 161, 163, 185, 305, 322, 329, 409, 412, 517, 519, 522, 543, 546, 569, 571, 623, 626, 631, 645, 653, 658, 660, 663, 688, 690, 694, 695, 711, 739
derivative coupling matrix, 48, 63, 71, 255, 820, 821
diabatic coupling matrix, 284
diabatic Hamiltonian matrix, 622
diabatic matrix, 415, 416, 467, 469, 471, 622, 648, 684
diabatic potential energy matrix, 62, 67, 72, 79, 254, 457, 648
eigenvalue matrix, 28
electronic density matrix, 305, 381
Fock matrix, 27–29, 307, 310, 312, 359, 360
idempotent matrices, 551
identity matrix, 306, 307, 348
inelastic T-matrix, 243
K-matrix, 111, 232, 237, 238, 267, 271
metric matrix, 358, 360
monodromy matrix, 133–135, 437, 450, 473, 480, 487, 488
overlap matrix, 9, 27, 29, 266
Pauli matrices, 11, 67
propagator matrix, 222, 452
R-matrix, 217
reactance matrix, 111
reactive T-matrix, 243
rotation matrix, 233, 234, 374
S-matrix, 100–102, 104, 110, 210, 212, 217, 270
T-matrix, 103, 264–266, 276
transition matrix, xxx, 101–103, 209, 239, 240, 263, 265, 266, 268, 285, 768
tridiagonal matrix, 130
unitarity of the *S* matrix, 217
unitary matrix, 19, 62, 312
Wronskian matrix, 208
maximum entropy method, 542
May, V., 806
McLachlan, A.D., 295
MCSCF, 33, 38, 40, 585, 845
MCTDH, 469, 470, 548, 549, 565, 567, 574, 585, 844
mean-field (MF), 16, 384, 385, 642–644
mean-field formalism, 564
mean-field theory, 643
medium, xl, 161, 163, 539, 636, 637, 639, 664, 666–668, 775–777, 784
biological media, xxix
classical medium, 802, 843
condensed media, xxii, 749, 771, 793, 805
continuous dielectric medium, 783
effective medium, 822
gaseous medium, xxxi
isotropic medium, 194
mesh regularization, 504
polar medium, 750, 789
random isotropic media, 665
memory effects, 600
metal, xli, 818, 839
metal electrons, 839
metal surfaces, 824, 827, 829, 838
metallointercalators, 790
metastable, 113–115, 118, 275, 280, 281
methanol, 634, 644
metric, 176, 357, 358
curved metric, 342
flat metric, 342
metric tensor, 357
Mexican hat profile, 69
Meyer, H.D., 386, 387, 567, 646
microreversibility, xxix, 213, 216, 217, 384–386, 413, 646
Miller, W.H., 386, 387, 418, 451, 460, 554, 646
Miller–George theory, 419
Mitric, R., 740
MO, see also molecular orbital, 7–13, 27, 29, 830

mobility, 837
transient mobility, 834, 838
mode, 70, 374, 376, 455, 471, 567, 571, 729, 738, 807, 832
angular vibration mode, 583
bath modes, 471
bending mode, 70, 71, 704
coupling mode, 466, 581, 584, 683
degenerate modes, 832
discrete modes, 771
four-mode approach, 470, 471
non-totally symmetric, 70
non-totally symmetric normal modes, 70
normal mode, 470, 565, 637, 638, 763, 833
nuclear mode, 387, 460, 628, 635, 772
rotational modes, 377, 637
stretching mode, 585
tuning mode, 683, 684, 689
vibrational mode, 585, 613, 641, 663, 763, 807, 810, 828
zero-frequency modes, 53
modulus, 492, 509, 735
wave function modulus, 498, 499, 539
Moebius strip, 77
molecular electronics, 792, 843, 844
molecular fluid, 178
molecular mechanics (MM), 690
molecular orbital (MO), 5, 13, 830, 832
molecule, xxii, xxix, xl, 162, 180, 201, 202, 576, 587, 593, 819
biological molecules, 548, 574, 575, 614
diatomic molecule, 247, 289
H_2 molecule, 256, 275
halogen molecules, 245
linear polyatomic molecule, 71
Møller operators, 101
moment, 518, 535–542, 555
first moment, 545, 546
hierarchy of moments, 518, 535, 538, 541
moment closure, 542
moment coupled equations, 546
moment of zeroth order, 535, 536
quantum-classical moment approach, 537
second moment, 538, 539
transition moments, 545
moment of inertia, 582, 638
torsional moment of inertia, 684
momentum, xxxii, 146, 147, 154, 238, 325, 358, 381, 414, 421, 428, 447, 451, 481, 514, 526, 540, 541, 634, 643, 650, 657, 658, 739
classical momentum, 140
conjugate momenta, 734
electronic momentum, 356
generalized momentum, 391
instantaneous momentum, 702
linear channel momentum, 214
momentum density, 539
momentum field, 539, 545
momentum jump, 482
momentum space, 147–149, 450, 556
momentum transfer, 304, 365
momentum variance, 540
phase space momentum, 518
trajectory momentum, 326, 327, 495
monodromy matrix, 459
monoenergetic, 111
Monte Carlo, 473–475, 477, 753, 793, 795, 803
Monte Carlo error, 475
Monte Carlo integration, 473
Monte Carlo sampling, 423
Monte Carlo trajectory, 796
stationary Monte Carlo, 473
stationary phase Monte Carlo, 475
Morales, J.A., 381
Morse, M., 135
Morse potential, 524
motion in lockstep, 500
Moyal, J.E., 153
Moyal formalism, 153
Moyal function, 154
MRCI, 40, 469, 845
multichannel, xxix, 87, 201, 207, 208, 214, 219, 240, 251, 363

adiabatic multichannel equation, 204, 205
multichannel equation, 205, 210, 284
multichannel scattering, 201, 213, 214, 243
multichannel theory, 206
multiconfiguration, 38, 342, 565, 569
multiconfiguration time-dependent Hartree (MCTDH), xxi, 376, 468, 564, 567
multiconfiguration-self-consistent field (MCSCF), 32
multiconfiguration time-dependent Hartree (MCTDH), xxxv, 376, 548
MCTDH equations, 569, 571, 572, 574
MCTDH wave function, 568, 570
multiconfigurational
complete active space multiconfigurational, 381
multiple independent spawning, 434
MIS, 429, 434
multireference configuration interaction (MRCI), 40
multistate molecular dynamics, xxii
multivalued, 76, 493, 495

N-C cleavage, 585
N-H bond, 580
N-H stretch, 580, 583
N-H stretch coordinate, 581
N-methylethaniminium, 748
N-representable
N-representability, 592
Na, 272–274, 280, 281, 429, 431, 432
Na*, 431, 432
Na^*-H_2, 433
Na*-quenching, 429
$Na^* + H_2$ exciplex, 431
Na^+-F^- bonds, 702
Na_2F, xxxix, 701–707
Na_2F^+, 704
Na_3F, xxxix
Na_3F_2, 706–711, 737, 738, 740, 743–747
$Na_3F_2^*$, 707
Na_nF_m, xxxix
Na_nF_{n-1}, 700–702, 706
Na_nF_{n-2}, 700, 701
Na_4, 321
Na* projectile, 431
Na-H_2, 246, 429, 430
NaI, 114
Nakamura, H., 289
nanodevices, 792
nanoparticle, 792
nanoscience, xli
nanostructures, 587, 842
Ne_2, 114
Ne_2^+, 114
near-crossing, 46
near-infrared, 637
NeExNe, 700, 701
NeExPo, 700, 701, 704, 707, 710
neglect of diatomic differential overlap (NDDO), 306
NeNePo, 700
Neria, A., 635
Newns, D., 839
Newton, M.D., 793
Newton–Raphson method, 86
Newton's law, 357, 512
Newtonian, xxii, xxxi, 312, 540
Newtonian equation, 440, 496, 504, 507, 541
Nienhaus, H., 838
Nikitin, E.E., 403
nilpotent, 346
nitric oxide (NO), 817, 839, 840
nitrogen, 371
Nitzan, A., 635
node, 493, 504, 505
nodal line, 493
nodal point, 495, 504
nodal regions, 493, 504
nonadiabatic, xx, xxvi, xxxi–xxxv, xxxix, xli, 46, 62, 113, 121, 186, 280, 288, 321, 330, 339, 376, 395, 400–404, 413, 419, 441, 457–461, 492, 505, 507, 508, 510–512, 546, 548, 573, 574, 588, 610, 637, 641, 659, 663, 671, 708, 710, 730, 732, 736, 744, 749, 761–763, 794, 817, 819, 828, 842, 844

electronically nonadiabatic, xxx, xxxix, 78
nonadiabatic coupling, 72, 253, 414, 836, 838
nonadiabatic dynamics, 79, 86
nonadiabatic effects, xx, 252, 828
nonadiabatic limit, 762
nonadiabaticity, 64, 71, 600, 752, 756, 789
non-Boltzmann sampling, 796, 797
non-Euclidean, 357
non-interacting systems, 32
nonlinear, xxxiv, 155, 664, 665, 672
nonlinear optical processes, 711
third-order nonlinear processes, 675
non-orthogonal, 157
non-radiative transitions, 711
non-relativistic, 31
normalization, 6, 147, 156, 293, 295, 297, 314, 345, 347, 348, 350, 391, 439, 486, 502, 524, 565, 566, 590, 616, 721, 735, 739, 770, 821
energy normalization, 102, 103
mass normalization, 240
normalization constraint, 293, 294
scattering state normalization, 264
nuclear, xx, xxx, xxxi, xxxiii, xxxviii, xl, 45, 46, 76, 85, 189, 190, 202, 213, 258, 273, 342, 343, 348, 356, 381, 383, 386, 400–404, 419, 420, 433, 439, 451, 457–459, 462, 465, 505, 514, 630, 631, 708, 757, 761, 797, 822, 836, 844
effective nuclear mass, 308
internal nuclear degrees of freedom, 751
nuclear boosts, 341
nuclear centers, 330, 513, 532, 634
nuclear charges, 336
nuclear degrees of freedom, xix, xxvi, xxviii, xxxi, xxxv, xli, 36, 45, 47, 201, 281, 313, 318, 332, 339, 342, 349, 353, 357, 360, 376, 377, 381, 383, 387, 408, 420, 433, 438, 492, 547, 565, 567, 598, 605, 610, 614, 615, 632, 644, 650, 681, 683, 686, 693, 756, 761, 766, 819, 820, 823
splitting of the nuclear wave packet, 361
nucleic acid
nucleic acid bases, 575, 576

observable, xxix, 89–91, 91, 152, 174
momentum observables, 123
occupation number, 154
occupation states, 155, 158
OH, 512
OH^-, 805
Ohmic, 791
Öhrn, Y., 341
on-the-fly, 3, 37, 308, 574, 585, 643, 702, 747
Onsager's regression hypothesis, 166
opacity function, 362
open quantum systems, 197
open system, 161
open-shell, 23, 376, 594
restricted, 23
unrestricted, 23
operator, xxvi, 26, 53, 60, 89–91, 95, 100, 102, 123, 125, 152, 154, 156, 158, 160, 170, 178, 188, 189, 192, 196, 231, 235, 239, 251, 254, 256, 294, 298, 300, 332, 347, 351, 352, 452–454, 457, 480, 481, 518–521, 535, 540, 548, 549, 553, 554, 558–560, 572, 597, 588, 596, 600, 624, 670, 692, 694, 738, 755, 756, 760, 808
adjoint Møller operator, 101
angular momentum operators, 247
annihilation operator, 344, 452, 455, 456, 758
antisymmetrization operator, 328
bounded operator, 619
cluster operators, 40
conjugate operator pairs, 535
conjugation operator, 215
coupling operator, 192, 285, 791
creation operators, 155, 344
density operator, 147, 149, 152, 153, 304, 336, 540, 543, 621, 692, 693
dipolar moment operator, 675
displacement operator, 807
electric dipole operator, 676

electron spin operator, 248
equilibrium statistical operator, 769
evolution operator, 170, 175
exchange operators, 18, 20, 25
Fock operator, 19–22, 24, 26–29, 301, 313, 320, 329, 339, 370
Green's operator, 99, 206, 238, 239
integral operators, 742
ladder operator, 158, 236, 247, 347, 452
Liouville operator, xxxix, 192, 522, 691
matrix operator, 242
mean-field operator, 569
Møller operators, 96, 97, 100, 102
momentum operator, 124, 313, 456, 457
multiplicative operator, 552
normalized operator, 563
number operator, 157
operator equation, 239
operator mapping, 152
operator product, 152, 192, 346, 347, 482, 518, 560
perturbation operator, 299
Poisson bracket operator, 519
population operator, 169, 188
projection operator, 187, 297, 534, 583, 566, 617, 688, 723, 754
reactive $\hat{T}$ operator, 242
reactive transition operator, 239
reduced density operator, 192–194
$\hat{S}$ operator, 95
Schrödinger time evolution operator, 92
shift operators, 806
spin operator, 11, 25, 27, 215, 452–454
split-operator scheme, 582
statistical operator, 175, 180, 188, 193, 518, 780
$\hat{T}$ operator, 103
time evolution operator, 92, 94, 95, 97, 332
time order operator, 93
time-reversal operator, 215
transition operator, 101, 109, 211, 240, 242, 276
two-electron interaction operator, 328
vector operator, 726
optical, 673
linear optical processes, 690
nonlinear optical phenomena, 665
optical control, 715
optically active, 733
optimization, 38, 720, 723, 738
global optimization, 741
orbital, 259, 311, 319, 343, 344, 429, 580, 594, 607, 608, 683, 826
active, 39
antibonding orbital, 576
atomic orbitals, 9, 33, 306, 329, 330, 341, 343, 348
canonical orbital, 33
cluster orbitals, 832
dynamic orbitals, 343, 346, 370
frontier orbitals, 832
highest occupied molecular orbital (HOMO), 32
Kohn–Sham orbitals, 592, 595, 603
molecular orbital, 4, 11, 12, 14, 15, 28, 30, 33, 38, 39, 241, 301–305, 306, 309–311, 314, 319, 328–330, 342, 576, 587, 591, 608, 609, 831, 832
molecular spin orbital, 328, 824, 830
occupied orbital, 34, 303, 307, 310, 345, 606
spatial orbital, 7, 23, 25, 591, 594, 596
spin orbital, 7, 17–24, 32, 35, 328, 343, 824
symmetry-adapted molecular orbitals, 14
time-dependent molecular orbitals, 307, 343
unoccupied orbital, 311, 344, 345
valence orbital, 580
virtual molecular orbital, 306, 310, 312
orbital angular momentum, 70, 74
orbitally degenerate, 70
organic, 577, 682

organism, 575
aquatic organisms, 575
biochemical organisms, 575
orientation, 433, 638, 833
orientation of the molecule, 833
ortho-H_2, 259
ortho-hydrogen, 260
orthogonality, 40, 83, 250
non-orthogonal, 241, 266
orthogonal complement, 294
orthogonal set, 251
orthogonal states, 263
orthogonalization technique, 81
orthonormality, 17, 24, 25, 32, 48, 113, 156, 236, 302, 306, 311, 312, 319, 550–552, 754, 755, 825
oscillation, 416, 418, 477, 478, 623, 704, 832
bosonic oscillator, 452
Brownian oscillators, 625
coupled oscillators, 821
forced oscillation, 299–301
free oscillation, 299, 300, 301
harmonic oscillator, 456
N independent oscillators, 455
N oscillators, 455
oscillating dipole, 664
oscillation amplitudes, 624
oscillation period, 680, 762
oscillator coupled to a bath, 658
oscillatory integrands, 473
Rabi oscillations, 748
shape oscillation, 705
Stückelberg oscillations, 416, 418, 420
two bosonic oscillators, 453
two-dimensional quartic oscillator, 472
zero-point oscillations, xxxv
outer sphere, 750
outgoing wave variation principle (OWVP), 276, 280
oxidation, 750
oxidized complex, 749
oxygen, 371
ozone layer, 575
para-H_2, 259
parabola, 757
acceptor parabola, 765, 767
N-dimensional parabolas, 758
parallel transport, 77
parity, 70, 259
triatomic parity, 260
Park, T.J., 556
Parr, R., 595
partial wave, 104, 203
partial wave analysis, 104
partial wave expansion, 112, 206
partial wave representation, 112
particle, 128, 180, 183, 343, 356, 369, 483, 492, 496, 498, 499–501, 504, 508, 539, 617, 636, 793
classical particle, 128, 186, 315, 406, 623
ensemble of particles, 498
fermionic particles, 844
free particle, 173, 184, 185, 439
guided particle, 498, 499
identical particles, 844
nanoparticle, 792
particle ensemble, 498
particle number state, 156, 158
point particle, 492, 498, 514
single particle, 498, 565–567, 570–574, 592
two-particle interactions, 181
two-particle terms, 588
partition function, xxix, 162, 163, 169, 173–177, 181, 182, 184, 185, 627, 795, 799, 800, 802
canonical partition function, 178
classical partition function, 168, 169, 172, 183
discretized partition function, 179
partition function path integral, 178
quantum partition function, 179
path, 127, 128, 134, 135, 163, 176, 177, 181, 255, 440, 442, 500, 503, 615–618, 620, 745, 797
path expansion, 184
path in complex time, 419
stationary paths, 442

path integral, xxix, 122, 123, 128, 129, 131, 136, 163, 173, 175, 176, 178, 183, 184, 187, 437, 441–443, 447, 460, 615, 616, 793, 794, 799, 800, 804
  closed loop path integral, 177
  Euclidean path integral, xl, 173, 174, 753
  Feynman path integral, 438, 480
  path integration, 185
  real-time path integral, 177
  semiclassical path integral, 440
Pauli exclusion principle, 6, 17
Pauli matrices, 760
Pearle, P., 616
Pechukas, P., 438, 442, 443, 447, 460
Pekar factor, 782, 813
peptide, 683
Perelomov, A.M., 347
periodic, xxx, 265, 268–270, 377, 431, 475, 478, 556, 685, 703, 819, 834
  near-periodic alternation, 704
  periodic bending, 704
  periodic states, 88
  quasi-periodic recurrences, 686
periodic boundary conditions, 9, 834
perturbation, 34, 56, 57, 91, 134, 188, 275, 298, 397, 398, 607, 663, 671, 672, 691, 692, 694, 695, 748, 770, 780, 790, 791
  electric perturbation, 777
  external perturbation, 575, 607
  first-order perturbation theory, 728
  harmonic perturbation, 300
  Liouvillian perturbation formalism, 692
  Møller–Plesset perturbation theory, 36
  non-perturbative, xxxix, 663, 669, 671, 681, 685, 711
  perturbation theory, 23, 54, 91, 626, 717, 718
  perturbative, 663, 669, 671, 672, 680, 711
  perturbative recursion scheme, 675
  Rayleigh–Schrödinger perturbation theory, 34
  second-order perturbation theory, 740
Petersilka, M., 603
pH value, 805
phase, xxxiv, 73, 74, 76, 77, 82, 105, 106, 116, 119, 128, 133, 135, 137, 195, 352, 365, 374, 376, 395, 403, 416, 421, 440, 476, 478, 483, 484, 492, 493, 495, 498, 501, 502, 509, 589, 615, 676, 718, 722, 730, 735, 736, 821
  adapted phase, 505
  asymptotic phase, 718
  centripetal phase, 106
  condensed phases, 635
  dephasing factor, 470
  dephasing mechanisms, 197
  dynamic phase factor, 311, 408
  global phase, 574
  phase coherence, 545, 615
  phase factor, 135, 138, 216, 309, 479, 481, 565, 616, 630, 673, 674
  phase integral, 398
  phase shift, 77, 105, 143, 736
  stationary phase, 138, 189, 440, 441, 443, 447, 460
  stationary phase approximation, 186
  Stückelberg phase, 287
phase space, xxxii, 144, 146, 147, 149, 159, 342, 351, 352, 354, 357, 397, 420, 425, 449, 479, 480, 485, 518, 520, 521, 530, 532, 693, 699, 709, 711, 738, 739, 820
  classical phase space, 349, 518
  generalized END phase space, 356
  generalized phase space, xxxii, 339, 342, 349, 358
  phase space representation, 152
phenol, 579
phonon, 836
  phonon reservoir, 836
  phonon temperature, 835
  surface phonon, 818, 835, 838
photoabsorption, 565
photobiological, xix, xxv
photochemical, xix, xxv, 78
photodetachment, 700
photodissociation, 565

photoexcitation, 433, 466, 468, 580, 638, 731, 732, 736, 751, 790, 804
vertical photoexcitation, 683
photoinduced, xix, 610, 682
photoionization, 402
photolysis, 731
photon, 576, 668
photon echo, 665
photon number, 721
photon number state, 154
photostability, 548, 574–576
photostable organic species, 576
photosynthesis, 749, 751, 842
photovoltaics, 792, 843
physisorption, 818
picosecond, 641, 643, 818, 833, 838
subpicosecond, xx
pilot wave, 495, 498–500
Pines, D., 304
pitch, 79–81
plane wave, 323, 555
outgoing plane wave, 213
plateau regime, 167, 168
platinum, 818, 830
point group, 14, 69, 70
Poisson bracket, 151, 354, 355, 519, 520, 522, 523, 528
Poisson distribution, 372
polar representation, 504
polar solvent, xl
polarizability, 607, 778–782
dynamic polarizability, 607
polarization, 663, 664, 667–674, 678, 680–682, 686, 716, 750, 775–780, 782, 785, 789, 797, 811
electronic polarization, 636, 779, 812
equilibrium polarization, 750, 751, 814
linear polarization, 664, 676, 678
macroscopic polarization, 667
maximum polarization, 797
nonlinear polarization, 664, 666, 669, 711
orientational polarization, 784, 785, 813
polarization field, 668, 673, 784
polarization fluctuations, 814
polarization signal, 673, 676, 681
solvent polarization, 751, 775, 798
third-order polarization, 665, 676, 677, 680
Pollard, W.T., 679
Polli, D., 690
polyatomic, xx, xxvii, 14, 46, 53, 70, 84, 86, 89, 574
polyenes, 682
polymer, 162, 180, 749, 802
classical polymer, 163, 178, 180, 802
classical polymer isomorphism, 178
interacting polymers, 799
polymer loop index, 180
polynomial, 553, 562, 773
Chebyshev, 553
Laguerre, 553
Legendre, 553
polynomial expansion, 560, 562
population, xxxv, xxxix, 48, 196, 317, 370, 372, 384, 385, 409, 411, 420, 426, 428, 465, 509, 511, 515, 517, 518, 522, 525, 532, 543–545, 583, 619, 650, 652, 659–671, 686–690, 693, 696, 725
adiabatic population, 583, 584, 682
diabatic populations, 523, 688
electronic population, 370, 402, 669, 670, 686
Mulliken population analysis, 370
nuclear population, xxxiii, 411
population transfer, 424, 426, 431
porphyrins, 579
position, 147, 421, 423, 425, 442, 456, 457, 464, 487, 502, 512, 582, 594, 634, 643, 739
Dirac position eigenstates, 479
potential, xxxi, 58, 80, 125, 143, 144, 150, 180, 182, 203, 206, 210, 211, 251, 266, 294, 325, 326, 339, 363, 457, 465, 483, 496, 508, 524, 525, 545, 587, 592, 594, 598, 600, 636, 650, 756, 784, 820
adiabatic potential, 50, 64, 65, 69, 283, 288, 289, 585
chemical potential, 592

Coulomb potential, 11, 26, 55, 105, 590
coupling potential, 282, 512, 524, 544, 545, 646, 669, 692
diabatic coupling potential, 58, 65
diabatic potentials, 58, 65, 282, 508, 582
effective potential, 16, 105, 179, 180, 385, 386, 440, 648, 793, 802, 820
Ehrenfest potential, 325, 334
exchange correlation potential, 592–594, 603
EXX potential, 603
molecule–surface potential, 838
Morse potential, 582
orbital-dependent exchange-correlation potentials, 596
pair interaction potentials, 793
quantum potential, xxxiv, 122, 137, 143, 325, 354, 496, 498, 501, 503, 504, 514, 515, 539–541
vector potential, 60, 61, 76, 125, 252, 604, 716
potential energy landscape, xxxiii, xxxiv, 68, 78
potential energy surfaces, xx, xxvi, xxv, xxxi, xxxiii, xxxii, xxxiii, xxxvi, xxxv, xxxvi, xxxviii, xxxix, 3, 36, 45, 46, 50, 51, 53, 57, 61, 68, 71, 78, 82, 83, 85, 114, 121, 225, 245, 256, 259, 261, 308, 317, 322, 338, 342, 376, 378, 384, 402–405, 406, 410,–413, 418, 420–424, 426–430, 445–447, 461, 465, 507–509, 512, 514, 515, 520, 548, 574, 579, 610, 627, 629, 631–633, 635, 640, 643, 647, 648, 676, 679, 688, 697, 698, 701–703, 708, 711, 712, 723, 725, 731–733, 735, 748, 749, 756, 757, 758, 760, 761, 784, 786, 789, 801, 819, 820, 829, 841
adiabatic potential energy surface, xxv, 46, 50, 62, 68–70, 80, 251, 272, 398, 513, 615, 626
average potential energy surface, xxxii, 338, 384, 403, 411, 465, 610, 644
ground state potential energy surface, 281, 720, 723, 727
multiple potential energy surfaces, 186, 213, 263, 517
predictor-corrector, 557
predissociation, 565
electronic predissociation, 113, 114
predissociation dynamics, 113
Preston, R.K., 403, 404, 406, 412, 640
previtamin D, 842
Prezhdo, O., 637, 843
Primakoff, H., 454
principle of minimal action, 123
probability, 123, 124, 146, 148, 149, 174, 197, 213, 232, 261, 262, 287, 288, 329, 362, 366, 370, 390, 393, 405, 410, 417, 418, 422, 439, 442, 483, 498, 499, 509, 510, 534, 580, 583, 608, 616, 620, 623, 640, 652, 659, 688, 720, 725–728, 738, 748, 750, 761, 762
classical particle, 623
conditional pair probability, 608
momentum space probability distributions, 147
probability density, 6, 8, 10
product, 163, 164, 166, 167, 169, 182, 187–190, 245, 262, 272, 426, 429, 686, 749, 750, 770, 790, 799–802, 804, 842
dissociation product, 727, 732
ionized products, 245
product channel, 720
projectile, 202, 331, 332, 334, 336, 337, 365, 366, 367, 369, 370, 371
atomic projectile, 362, 363
CO projectiles, 838
H^+ projectiles, 373
Prony, Gaspard de, 374,
Prony analysis, 376
Prony procedure, 374
propagation, 220, 305, 311, 318, 320, 321, 343, 403, 419, 423, 424, 426, 428, 435, 465
excited state propagation, 321
FMS propagation, 431
frozen Gaussian propagation, 422

wave function propagation, 330
wave packet propagation, 308, 414, 415
propagator, 94, 125–131, 134, 176, 188, 187, 220, 221, 437, 438, 440, 448, 451, 454, 455, 458, 460–462, 469, 472, 480, 487, 557–559, 562, 563, 734
classical propagator, 530
Euclidean propagator, 178
exact propagator, 442
factorization of the propagator, 126
Herman–Kluk propagator, 488, 734
IVR propagator, 449, 480
propagator in coordinate representation, 438
quantum propagator, xxix, 122–124, 126, 175, 439, 448, 640
real-time quantum propagator, 163
reduced propagator, 439–441
SC-IVR propagator, 458, 472
semiclassical propagator, xxxiii, xxxiv, 394, 413, 442, 448, 449
split operator propagation, 559
Van Vleck propagator, xxix, 128, 133–135
proteins, 576
protons, 336, 376
pseudopotential, 637, 640, 754, 755
pseudorotation, 60, 73, 74, 76, 376–378, 381
molecular pseudorotation, 73, 75
pseudorotation angular momentum, 74–76
pseudorotation quantum number, 76
pseudorotational torque, 378
Pulay, P., 315
pulse, 664, 666, 667, 674, 698, 699, 709, 712, 715, 716, 721–723, 725, 741–744
coherent light pulse, 717
delta-shaped pulses, 678, 679
dump pulse, 703, 704, 720, 721, 738, 741–747
electromagnetic pulse, 723–727
Gaussian laser pulses, 686, 687
Gaussian-shaped pulse, 667
ionizing probe pulse, 701, 704, 707
optimum pulse, 721
overlapping pulses, 670, 682
probe pulse, 666–668, 674–677, 679, 681, 686, 691, 698, 700, 704, 707, 709, 713
pulse envelope, 666, 678
pump pulse, 666, 668, 674–677, 679, 682, 684, 687, 698–704, 720, 745, 746
pump–dump, 700, 701
pump–dump experiments, 748
pump–dump signals, 703, 706
pump–dump spectroscopy, 740
pump–probe, xxxix, 78, 663, 665, 666, 672, 674, 700, 701, 702, 706, 747
pump–probe correlation, 699
pump–probe detection, 677
pump–probe experiment, 669, 679, 698, 711
pump–probe signal, 666–668, 676, 686, 689, 690, 698, 704, 705, 706, 708–711, 713
pump–probe spectroscopy, xl, 663, 693, 700, 701, 736, 738
pure state, 146, 147, 149, 534
pyrazine, 466, 467, 469–471, 574, 613
excited state, 469
pyrazine absorption spectrum, 470
pyrrole, 574, 579–581, 585

Q-space, 82, 85
QM/MM, 790, 805
quantum, xxix, xxx, xxxii, xxxiii, 123, 295, 441, 491, 614, 625, 629, 636, 690, 753, 767, 790, 841
canonical quantum theory, 122
discrete quantum numbers, 102
fictitious quantum subsystem, 408
interpretation of quantum theory, 498
primary quantum subsystem, 412, 643

quantum chemical, xxi, 259, 292, 748, 793, 845
quantum chemistry, xxi, xxvii, xxxvi, xli, 805
quantum control, 715
quantum dots, 844
quantum electrodynamical, 845
quantum measurement, 613
quantum pumping, 844
semiempirical quantum method, 805
vibrational quantum, 826
quantum-classical, xxvii, 315, 357, 383, 386, 413, 419, 437, 442, 491, 511, 514, 538, 613–615, 621, 642, 644, 844
quantum-classical hybrid, 419
quantum dynamics, see under dynamics
quantum electrodynamics, 154
quantum fluid, xxxiv
quantum force, 496
quantum interference, 416
quantum mechanical, 307, 308, 332, 341, 356, 376, 381, 383, 386, 387, 391, 392, 395, 397–399, 402, 403, 415, 416, 419, 437, 438, 451, 452, 463, 482, 491, 499, 508, 509, 512, 515, 517, 521, 525–528, 532, 534, 547, 548, 614, 646, 649, 659, 660
quantum mechanics, 387
quantum Monte Carlo, 41
quantum number, 214, 216, 231–233, 237, 238, 257, 260, 261, 278, 281, 323, 327, 362, 555, 583, 616, 719, 726, 727, 733, 736, 829
azimuthal quantum number, 555
collective quantum number, 251
internal quantum numbers, 718, 719
magnetic quantum numbers, 214–217
spin quantum numbers, 257
quantum optics, 154
quantum rate constant, 170
quantum theory, xxvii, xxxi, xxxii, 312, 498, 526, 841
causal interpretation, 491
Copenhagen interpretation, 499
quantum theory of motion, xxxiv, 492, 499, 512, 515, 542
quantum trajectory method (QTM), 499
quasi-bound, 113–115, 119, 262
quasi-classical, 391, 403, 404, 418
Rabi limit, 524
radiation, 334, 576, 673, 682, 716, 773
electromagnetic, 817
electromagnetic radiation, 337, 664, 669, 711, 719, 782
rainbow angle, 362, 367, 374
Raman, C.V.
Raman scattering, 676, 690
stimulated Raman scattering, 665, 671, 676, 679
random, 474, 620, 796
random number, 406, 410, 473, 708, 796
random walk, 795
random phase approximation, 304, 361
Rassolov, V.A., 515
rate, 169, 626, 627, 629, 651, 667, 752, 767, 770, 844
dissipation rate, 668
nonadiabatic transition rate, 629, 644
rate constant, xxix, 162–164, 168, 186, 187, 189, 190, 775, 783, 786, 789, 790, 792, 798, 799, 803, 806
classical full rate constant, 167
classical rate constant, 171
golden rule rate constant, 187
rate equation, 186
rate theory, 136, 161, 162, 195
classical rate theory, 163
reactance, 263
reactant, xxxix, 163–165, 167–169, 182, 187–190, 223, 225, 227, 238, 256, 262, 272, 426, 509, 686, 715, 749, 750, 763, 770, 790, 799–802
reactant equilibrium, 800, 804
reaction, xxx, xxxix, 87, 162, 168, 187, 190, 240, 245, 256, 259–261, 271, 272, 281, 361, 368, 426, 429, 512, 660, 729, 749, 750, 752, 775, 789, 790, 793, 842
back reaction, 767
$F + H_2$ reaction, 255
gas phase reaction, 164

heteronuclear reaction, 750
reaction channels, 384
reaction coordinate, 760, 762
reaction path branching, 547
rearrangement reaction, 246, 366
redox reactions, 751
triatomic reaction, 248
reactive flux, 162, 167
reactivity, 262, 747
realism, 499
rearrangement, xl, 202, 226, 229, 240, 241, 263, 285, 324, 515, 571, 638, 659, 765, 772, 773
atomic rearrangement, 224
rearrangement channel, 238
reciprocity theorem, 217
Redfield equation, 195, 196
Redfield tensor, 195, 196
Redfield theory, xxix, 163, 196, 197, 843
redox, 749, 783, 794, 795, 801
redox pair, 750, 765, 783, 812
reduced mass, 203, 225, 240, 257, 323, 582, 650, 734, 820
three-body reduced mass, 226
reduction map, 620, 622
redundancy, 310, 311, 422, 423
reference system propagator algorithm (RESPA), 321
reflection, 384, 415, 416, 434, 463, 464, 509, 510, 585, 623, 664, 819, 841
reflection probability, 415, 417,418, 461, 463, 464
upper state reflection probability, 434
refraction, 664
refractive index, 667
regularization, 212, 269
relativistic, 842, 845
relax and drive, 332
relaxation, xl, 22, 161, 433, 579, 639, 641, 683, 690, 701, 710, 737, 765, 804, 824, 829, 843
barrier-free relaxation pathways, 576
Debye relaxation time, 788
nonadiabatic relaxation, 818
non-radiative relaxation, 433, 610
nuclear relaxation, 773
relaxation behavior of spin systems, 197
relaxation channel, 818
relaxation rate, 829, 831, 832
relaxation times, 639
solvent relaxation, 790
vibrational relaxation, 826, 828, 839
Renner–Teller, 71
reorganization energy, 789
representation, 45, 70, 88, 93, 98, 103, 122, 123, 132, 152, 155, 160, 175, 178, 182, 224, 232, 234, 238, 240, 260, 264, 310, 312, 316, 331, 332, 342, 347, 383, 389, 417, 427, 450, 454, 457, 472, 477, 519, 536, 538, 547, 552, 567, 570, 579, 582, 583, 608, 618
adiabatic representation, xxvi, 114, 195, 229, 247, 251, 316, 317, 384, 389, 390, 405, 408, 512, 606, 645, 654, 761
body-fixed representation, 230, 232, 235, 247
diabatic representation, xxviii, 249, 253, 282, 388, 392, 402, 408, 428, 439, 457, 505
discrete variable representation (DVR), 548, 553
eikonal representation, xxxi, 292, 322, 324, 339, 353
finite basis representation, 549, 550
forward-backward initial value representation, 479
grid representation, 548, 550, 551, 556
Herman–Kluk representation, 469, 472
infinite product representation, 185
irreducible representation, 70, 346, 429
Jacobi representation, 580
Lehmann representation, 602
momentum representation, 450, 556
path integral representation, 176
phase space representation, 146, 462

polar representation, xxxiv, 492, 493, 506, 513, 821
space-fixed representation, 230, 233
Thouless representation, 342, 343
VB representation, 553
Wigner representation, 519, 536, 693, 697, 741, 745
repulsion, 304, 588, 590
interelectronic repulsion, 589
reservoir, 161, 190, 775
macroscopic reservoir, 773
resolvent, 242
resolvent identity, 98
resonance, 113–116, 118, 281, 698, 733, 750, 795
Feshbach resonances, 113, 280
funnel resonance energies, 282
resonance energy, 119
resonance line shape, 119
shape resonance, 113
response, xix, 607, 625, 637, 664, 777, 779, 797
current–current response function, 604
density–density response function, 601, 607
Kohn–Sham response function, 601, 602
linear response, 588, 601, 603, 672
linear response theory, 604
materials response functions, 742, 744
nonlinear response, 672, 678, 679
restricted open-shell Hartree–Fock (ROHF), 321
restricted open-shell Kohn–Sham (ROKS), 596, 609
reversible work, 794, 797
rhodopsin, xxxix, 683, 690
Riccati equation, 220, 221
Riccati–Bessel function, 107, 207
Riccati–Hankel functions, 106, 107, 208, 214
Riccati–Neumann functions, 106, 108, 207
Rice, S.A., xxxix, 715, 725, 738
Riemann integral, 98
Rimini, A., 616
R-matrix, 219, 220, 223
R-matrix propagation, 220, 223, 280
Robicheaux, F., 271
root search, xxxiii, 438, 442, 447, 449, 471, 485–487
Roothaan, C.C.J., 27, 29
Roothaan equations, 28, 29
Rossky, P.J., 616, 623, 637, 641
rotation, 51, 72–74, 232, 250, 377, 379, 638, 734
frustrated rotation, 833
hindered rotation, 262
rotating frame, 334, 335
rovibrational, 187, 361, 372, 374, 376, 845
rovibrational levels, 280
Runge–Gross theorem, 598, 606
Rustad, J.R., 805
ruthenium, 805

saddle point, 53
Saraceno, M., 300
SC-IVR, 460, 462, 463, 465, 466, 469–473, 476, 479, 481, 482, 613
SC-IVR representation, 460, 479
scattering, xxi, xxix, 78, 87, 102, 202, 204, 205, 214, 223, 224, 226, 232, 249, 263, 274, 276, 308, 323, 332, 336–338, 362, 363, 366, 368, 369, 373, 376, 402, 509, 511, 659, 817, 834, 839
atomic scattering, 332, 333
collinear scattering, 254
elastic scattering, 104, 105, 109, 114, 214
glory scattering, 362
inelastic electron–phonon scattering, 114
inelastic scattering, xxix, 201, 202, 206, 223, 228, 241–243, 369
ion–atom scattering, 337
molecular scattering, xxx, 46, 246, 262, 282, 305, 357, 361, 369, 547
molecule–surface scattering, 836
multichannel scattering, 201
principal-value scattering state, 100, 111

rainbow scattering, 367
reactive scattering, xxix, xxx, 201, 223, 229, 238, 240, 241, 256, 263, 565
resonant scattering, 118
scattering amplitude, 112, 365
scattering angle, 362, 365, 368
scattering function, 102, 203, 209, 213, 263
scattering matrix, 94, 96, 101, 109, 113, 209, 268
scattering operator, 95, 96, 101, 103, 213
scattering plane, 333, 335
scattering solution, 93, 99, 103, 107, 108, 113, 115, 209, 215, 217, 241, 265, 275, 733
state-to-state scattering, 255, 260, 733
s-wave scattering, 267
Schatz, G.C., 262
Schiff, L.I., 364
Schiff base, 433, 748
Schlessinger, J., 276
Schottky, W.H., 838
Schrödinger equation, 31, 214, 215, 217, 598
classical limit Schrödinger equation, 514
hyperspherical Schrödinger equation, 229
stationary Schrödinger equation, xxx, 40, 48–50, 57, 60, 71, 75, 88, 104, 107, 137, 140, 154, 202–204, 231, 236, 242, 249, 250, 268, 276, 292, 318, 320, 322, 324, 363, 405, 591
time-dependent Schrödinger equation, see TDSE, 88
Schrödinger picture, 89, 91, 124, 669
seam, 46, 78–80, 82, 85, 406, 408, 429, 585
crossing seam, 430
intersection seam, 430
one-dimensional seam, 65
seam dimension, 84
seam line, 84
seam space, 81, 82
second harmonic generation, 664
second quantization, 456, 844
Seidner, L., 669, 681
self-consistent, 28, 411, 441, 603, 649, 659, 667
complete active space self-consistent field, 577, 581
multiconfiguration self-consistent field, 585
time-dependent self-consistent field, 564
self-consistent field (SCF), 27, 29, 296, 381
boosted self-consistent field, 357
boosted self-consistent field equations, 359
semiclassical, xix–xxii, xxvii–xxix, xxxii, xxxv, 121, 122, 128, 132, 136–138, 154, 160, 162, 213, 246, 282, 289, 291, 349, 362, 391, 420, 433, 437, 438, 441, 442, 447, 448, 451, 454, 471, 472, 481, 484, 485, 518, 519, 522, 527, 528, 531, 532, 534, 564, 614, 626, 628, 629, 631, 658, 690, 711, 735, 736, 740, 747, 804, 820, 841, 842
nonlinear semiclassical equations, 841
semiclassical phase shift, 365
semiclassical propagation, 442, 471, 515, 747
semiclassical propagator, xxxiii, 123, 134, 135, 437, 438, 451, 453, 475, 486, 487, 841
semiclassical quantization, 395, 397–399, 451
uniform semiclassical approach, 402
semiclassical initial value representation (SC-IVR), 448, 458, 733
separable, 53, 397
non-separable, 397
non-separable degrees of freedom, 403
Sham, L.J., 590, 591, 593
Shapiro, M., xxxix, 715, 729
simple point charge (SPC), 637, 793, 804
Simpson algorithm, 473
simulated annealing, 321

single-channel, 206, 210, 267
    elastic single-channel case, 214
single-excitation configuration interaction (CIS), 38
single-valued, 74, 495
    single-valuedness, 493
singlet states, 585
singularities, 61, 63, 64, 98, 115, 117, 220, 270, 271, 362, 449
size consistency, 32, 33, 40
Slater determinant, 7, 12, 16, 23, 39
slow variables, 77
Sobolewski, A.L., 579
sodium, 700
    excited sodium atoms, 429
solute, 161, 163, 636, 752, 775–778, 780, 783–785, 789, 791, 805, 810–812
    excited solute, 161
    solute field, 784
solute–solvent, 794
solvated electron, 629, 634, 636, 640, 641
    solvated electron relaxation, 638, 641
solvation, 638–640
    first solvation shell, 789, 798, 804
    solvation shell, 794
solvation energy, 750
solvent, xl, 161, 162, 636, 638, 643, 750, 752, 770, 775–778, 783–786, 788, 789, 793, 799, 801, 802, 804, 805, 810, 811, 813
    classical solvent model, 802
    first solvation shells, 775
    free solvent, 780
    macroscopic solvent, 775
    non-polar solvents, 775
    polar solvent, 774, 786, 789, 790
    quantum solvent, 802
    solvent polarizability, 775
    solvent polarization, 780, 810
    solvent relaxation, 804
Song, X., 804
space-fixed, 224, 229, 230, 232, 233, 248, 250, 257
space reflection, 259
spawning, xxxiii, xxxv, 403, 413, 419, 420, 424–428, 433–435, 491, 644
    back-spawning, 424
    full multiple spawning, 426
    multiple independent spawning, 429, 434
    multiple spawning, 423, 426, 427, 429
    spawning threshold, xxxiii, 425
spectroscopic, 78, 636, 663, 672, 680, 681, 701
    NeExPo spectroscopy, 705
    nonlinear spectroscopy, 663
    nonlinear ultrafast spectroscopy, 716
    optical spectroscopy, 690
    pump–probe spectroscopy, 706
    spectroscopic detection, 671
    time-resolved spectroscopy, 678, 709
    ultrafast spectroscopy, 700
spectroscopy
    frequency-resolved spectroscopy, 669, 675
    molecular beam spectroscopy, 839
    nonlinear spectroscopy, 665
    optical femtosecond spectroscopy, xxxix
    pump–probe spectroscopy, 666
    ultrafast optical spectroscopy, 663
spectrum, xxxiii, 76, 336, 564, 622, 686
    absorption spectrum, 468–471, 613
    continuous spectrum, 87, 95, 101, 240, 271
    discrete spectrum, xxviii, 47, 87, 240, 619, 751
    Franck–Condon spectra, 488
    photoabsorption spectra, 606
    pseudospectral, 549, 582
    pseudospectral intensity, 582
    pump–probe spectrum, 682, 686, 689
    stimulated emission spectrum, 679
    UV absorption spectrum, 578
    vibrational spectrum, 769, 845
spin, 7, 8, 10, 11, 13, 23–26, 28, 215, 217, 247, 248, 328, 452, 453, 455, 608, 609
    pseudospin, 759
    spin coordinate, 328
    spin orbitals, 26

spin-orbit splitting, 273
spin singlet, 321
spin contamination, 25, 27
spintronics, 844
square-integrable, 265, 291
$\mathcal{L}^2$ functions, 277
squeezed state, 154
Staib, A., 632
Stark, K., 259
state, 196, 415–418, 426, 427, 450, 453, 543, 546, 573, 577, 580, 584, 585, 598, 613, 615, 617, 637, 647, 656, 659, 696, 700, 702, 719, 723, 760, 822, 829
adiabatic state, xxviii, 415, 443, 463, 626, 628, 762
cis state, 690
coupled electronic states, 507, 518, 542
diabatic state, 463, 464, 670, 799
electronic state, xix, xxi, 323, 407, 409, 412, 420, 428, 429, 442, 465, 431–433, 445, 446, 452, 460–462, 465, 505, 507, 523, 543, 544–546, 605, 615, 634, 637, 650, 656, 659, 663, 666, 682, 687, 692, 698, 702, 730, 731, 732, 735, 741, 748, 756, 759, 760, 768, 771, 772, 818, 824, 829, 830
excited electronic state, 682, 732
excited state, 587, 677, 772, 773, 840
forbidden states, 576
multistate, 320, 659
nuclear coherent state, 419
one-particle states, 313
pure state, xxxv, 518, 538, 541, 542, 651, 659, 739
quantum states, 413, 613, 618, 640, 644
superposition state, 543
target state, 722
stationary, xx, xxviii–xxx, xxxvi, 88, 128, 129, 291, 293, 295, 302, 303, 333, 478, 587, 595, 598
non-stationary, 665, 677
stationarity condition, 280
stationary points, 599
statistical, xxix, 162–164, 169, 174, 190, 819
statistical sampling, 803
statistical weight, 803
statistics, xl, 122, 178
Boltzmann statistics, 413
classical statistics, 146
quantum-statistical, 517, 534
statistical mixture, 146, 518, 534, 614, 619, 649
statistical operator, 534
statistical weights, 534
step function, 165–167, 169, 647
stimulated emission scattering
SE, 676
stimulated Raman scattering
SR, 676
stochastic, xli, 165, 648
stochastic variable, 164
stochastic boundary conditions, 834
Stock, G., 451, 454, 455
Stokes shift, 773
Stückelberg, E.C.G., xxx, 282
substitution, 31, 39, 284, 286, 303, 331, 355, 392, 444, 446, 456, 461, 501, 505, 520, 601, 628, 695, 697, 712, 729, 756, 824, 830
single substitutions, 303
substitutional states, 301
sum over histories, 127
Sun, X., 451, 460
supermolecule, 33
superoperator, 535, 690
dissipative superoperator, 537
superposition, 38, 112, 213, 241, 405, 412, 416, 505, 542, 583, 615, 646, 706, 726, 736, 779
coherent superpositions, 509
quantum superposition, 615, 624, 626, 649
superposition principle, 135
surface, xxii, xxxviii, xl, xli, 574, 580, 582, 584, 585, 605, 615, 616, 617, 629, 631, 634, 635, 643, 677, 679, 685, 757, 761, 817–819, 826, 828, 834, 835, 837, 838
adiabatic surfaces, 617, 762

average surface, xxxviii
copper surface, xli, 833
coupled surfaces, 635
excited state surface, 676, 679, 699, 703, 704, 706, 743
first excited state surface, 706
free energy surfaces, 782, 784–786
ground state surface, 738
metal surface, 826
multiple electronic surfaces, 574
silver surface, 838
single-surface propagation, 647, 648
surface atoms, 834
surface electrons, 836
surface layers, 835
surface temperature, 835–837
susceptibility, 778, 811
linear electric susceptibility, 777
susceptibility tensor, 664
Sutin, N., 750
switching, 659, 660
coherent state switching, 659
continuous surface switching, xxxviii, 644, 648, 649
switching criterion, 405, 408, 411, 659
switching threshold, 411
symmetry, 202, 217, 226, 231, 242, 256, 259, 272, 350, 451, 460, 556, 600, 799
A_1 symmetry, 832
C_{2v} symmetry, 274, 280, 429–431, 638
D_{3h} symmetry, 84
antisymmetrized, 285
inversion symmetry, 665
ungerade symmetry, 672

Tannor, D.J., xxxix, 715, 725, 738
Tannor–Rice, 716, 719, 720, 741
Tannor–Rice optimization, 725
target, xxxix, 202, 331, 333, 336, 337, 363, 365, 369, 371, 740, 744
atomic targets, 366
Hermitian target, 738
hydrogen targets, 336
intermediate target, 743–747
nitrogen target, 371
target state, 459, 468, 721, 733
tautomer, 576
minor tautomer, 577
tautomeric selectivity, 579
Tawa, G.J., 272, 276, 277
TDCDFT (time-dependent current density functional theory), 604, 605
TDDFT (time-dependent density functional theory), 588, 599–601, 604–606, 608–611, 842, 843
TDHF (time-dependent Hartree–Fock), xxxvi, 291, 292, 301, 302, 304, 305, 312, 322, 327–329, 333, 338, 341, 343, 353, 357, 360, 383, 465, 519, 547, 598, 605
TDSE (time-dependent Schrödinger equation), xxxi, 291–293, 315, 320, 327, 328, 349, 351, 388, 405, 409, 420, 435, 441, 445, 456, 484, 491, 492, 495, 496, 498, 500, 501, 505–511, 513, 517, 525, 532, 533, 548, 549, 557, 564, 600, 605, 650, 663, 669, 670, 681, 686, 708, 716, 717, 719, 721, 722, 820, 821
TDVP (time-dependent variation principle), 356
Teller, E., 70
temperature, 161, 320, 635, 658, 702–704, 745, 752, 767, 783, 797, 835
electronic temperature, 320, 835
nuclear temperature, 321
tensor, 353, 355, 812
friction tensor, 827, 833, 834
Levi-Civita tensor, 381
Redfield tensor, 791, 792
thermal, 710
thermal average, 626, 633, 787
thermal rate constants, 752
thermal ensemble, 702
thermal equilibrium, 772
thermally activated, 183
theta function, 740, 781
Thomas, L.H., 589
Thomas–Fermi method, 589
Thoss, M., 451, 454, 455, 469

Thouless parameter, 343, 344, 346, 349, 352–355, 361
Thouless representation, 347, 611
three-center problem, 224, 230
thymine, 575, 577
tight binding, 748, 794, 797, 801
time correlation function, 172
time non-locality, 600
time scale, xx, xxv, xxvi, xl, xxxix, 45, 402, 538, 614, 629, 634, 637, 643, 658, 665, 704, 709, 742, 749, 752, 761, 790, 811
    electronic, 761
    nuclear time scales, 395, 761
time-dependent Hartree (TDH), 564, 565
time-dependent Hartree–Fock, see also TDHF, 305
    first-order, 304
    TDHF equation, 292, 301, 305, 307, 330
    TDHF time evolution equations, 322
time-dependent Hartree–Fock equation, xxxi
time-independent Hartreee-Fock
    linearized TDHF, 361
time-order
    time-ordered succession, 675
    time-ordering prescription, 94
time-reversed, 213, 215, 216
    time reversal, 214
TiO_2, 610, 843
topological effects, 78
torsion, 576, 577
trace, 149, 169, 171, 172, 174, 176, 189, 192, 528, 529, 688, 741, 787, 795, 806, 807, 809, 810, 813, 827
    invariance of the trace, 394
    partial traces, 532
    phase space trace, 523
    time-dependent trace, 799
    trace product formula, 152
trajectory, xix, xxvii, xxix, xxxi, xxxiv, xxxv, 37, 88, 122, 123, 127–129, 134, 135, 162, 165, 177, 181, 255, 308, 315, 327, 332, 357, 361, 365, 367, 376, 384, 387, 392, 397, 398, 402–406, 411, 412, 415, 417, 418, 420, 421, 424–426, 428, 434, 440, 441, 442, 448, 449, 459, 473, 479, 482, 484, 488, 489, 495, 500–502, 504, 509–512, 528, 529, 531–533, 547, 615–617, 623, 629, 630, 633, 634, 640, 643, 645, 649, 651, 658, 659, 698, 699, 702, 706, 708, 734, 735, 743, 802, 819, 796, 820
    average trajectory, 315, 361
    best nuclear trajectory, 384
    chaotic trajectories, 472, 473
    classical trajectory, xxii, xxx, xxxiii, 121, 129, 291, 292, 307, 308, 332, 341, 342, 358, 382, 403, 421, 422, 424, 440, 441, 449, 459, 460, 472, 481, 483, 498, 615, 628, 658, 698, 704
    complex-valued trajectories, 418
    coupled quantum trajectories, 517
    ensemble of trajectories, 398, 526
    fraction of trajectories, 412
    interacting trajectories, xxxv, 531
    mean trajectory, 519
    nearby trajectories, 503
    nonadiabatic trajectories, 645
    phase space trajectory, 528, 529, 534, 738
    quantum trajectory, xxxiv, 122, 491, 495, 500, 503, 504, 512, 515, 518
    quantum trajectory method, 499
    trajectory ensemble, 700, 702, 706, 708
    trajectory interference, 361, 362
    trajectory swarm, 411, 443, 698
    unstable trajectories, 473
    virtual trajectory, 426, 428
trajectory hopping dynamics, 649
trajectory surface hopping (TSH), xxxiii, xxxviii, 289, 403, 404, 413, 438, 446, 615, 640–643, 651, 657, 659, 660, 706, 747, 748, 819, 841
    surface hopping scheme, 322
    TSH simulation, 413
transfer, xl, 507, 676, 749–752, 754, 756, 757, 760, 765, 766, 768, 773, 775, 776, 783, 790, 799

activation-free transfer, 766
barrierless proton transfer, 579
electron transfer, xxviii, xxix, 635, 750, 751
energy transfer, xxxii, 657, 819, 836
excitation transfer, xli
intramolecular electron transfer, 757
nonadiabatic energy transfer, 839
non-radiative energy transfer, 817
photoelectron transfer, 843
population transfer, 534
proton transfer, 578, 579
transfer complex, 763, 767
transfer rate, 762, 764
transformation, 177, 226–228, 249, 250, 253, 255, 342, 389, 397, 454, 478, 479, 504, 549, 551, 556, 668
adiabatic-to-diabatic transformation, 273
basis set transformation, 394
canonical transformation, 392
coordinate transformation, 392, 449
Holstein–Primakoff transformation, 454
linear transformation, 345
unitary transformation, 94, 130, 310, 320, 348, 565, 608
transient, 167, 637
transition, xl, xli, xxv, 32, 48, 71, 114, 123, 147, 163, 164, 167, 169, 187, 190, 194, 223, 226, 239, 240, 258, 263, 280, 281, 287–289, 317, 345, 384, 392, 393, 403, 406, 408, 410–412, 416, 419, 428, 430, 441, 443, 445–448, 452, 457, 466, 469, 472, 506, 509, 518, 521, 534, 541, 550, 574, 576, 604, 613, 637, 638, 643, 676, 677, 683, 693, 698, 703, 706, 708, 715, 720, 730, 749, 750, 762, 773, 783, 793, 796, 797, 799, 801, 824, 841, 843
cis-trans transition, 690
dipolar transition moment, 719
dipole-allowed transition, 730
discrete transitions, 618
electronic transition, xxx, xxxi, xxxviii, 78, 246, 263, 277, 329, 392–394, 403, 404, 410, 509, 510, 515, 517, 588, 675, 762, 767, 818
electronic transition amplitude, 439
Landau–Zener transition, 762
nonadiabatic transition, 282, 407, 416, 425, 431, 632, 639, 640, 641, 644
non-radiative transition, 402, 631, 638
optical transition, 576, 772
photoinduced transition, 690
quantum transition rate, 171, 173
state-to-state transition probabilities, 281, 322
transition matrix, 733, 829
transition probability, 217, 288, 366, 385, 406, 407, 509, 544, 641, 720
transition rate, 184, 185, 190, 510, 511, 626 632, 763
transition state, 743–745, 793, 795–797, 799–801
ultrafast optical transitions, xxxix
vertical transition, 578, 704, 706, 737
transition metal
multiligated transition metal centers, 750
transition metal complexes, 750
transition state approximation, 168, 186
transition state theory, 172, 181
classical transition state approximation, 171
classical transition state theory, 162
quantum transition state approximation, 171
quantum transition state theory, 163, 169, 185
translation, 638
in-plane frustrated translation, 832
translational degree of freedom, 236
transmission, 162, 416, 417, 461, 463, 464, 482, 502, 509, 510, 623, 624
transmission probability, 417, 463, 509, 623
transmittance, 666
transient differential transmittance, 680

transport, 844
activated transport, 164
electron transport, 842, 844
energy transport, 749
non-ballistic transport, 837
transport through molecules, 601
traveling atomic orbitals (TAOs), 330
Trotter theorem, 178
Tsereteli, K., 611
TSH, see also trajectory surface hopping, xxxviii, 374, 406–408, 411–416, 418–420, 424, 428, 434, 442, 447, 610, 744, 839, 840, 842, 843
Tully, J.C., 403, 404, 406–408, 412, 413, 446, 460, 610, 623, 640, 659, 708, 822, 824, 831, 839
Tully canon, 414, 426, 433, 512, 515
tunneling, xxvii, 114, 162, 308, 403, 413, 428, 499, 500, 547, 583, 749, 751, 754, 765, 793, 798, 799, 803, 804, 806
intrastate tunneling, 433
nuclear tunneling, 361, 752, 767
tunneling factor, 767
tunneling splitting, 755, 794, 799
two-state, 64, 273, 282–284, 410, 413, 479, 659, 721, 748, 752, 758, 759, 839
electronic two-state problem, 391–393

uncertainty, 158, 474
minimal uncertainty, 122, 154
position-momentum uncertainty relation, 146, 158
time-energy uncertainty, 658
ungerade, 13, 14
unitarity, 100, 101, 217, 442
quasi-unitarity, 101
unidirectional unitarity, 100
unitarity of the S-matrix, 288
unitary, 102, 552, 559, 560, 562, 631
non-unitary, 558

V-representable, 592
vacuum, 160, 777
valence bond, 33, 272
generalized valence bond theory (GVB), 33
Vallet, V., 579, 580, 582, 585
van der Waals, J.D., 262
Van Vleck, J.H., 129
variance, 158, 166, 474, 541, 632, 739
variation, 129, 132, 143, 144, 264, 266, 277, 293–295, 298, 350–352, 379 566, 569, 570, 584, 590, 593, 607, 617, 689, 721, 722, 742, 752
first variation, 133, 264, 293–295
linear variation, 129
second variation, 129, 134, 135
temporal variation, 306
variation functional, 263
variational, 32, 213, 246, 263, 264, 266, 268, 270, 271, 275, 277, 292, 720, 725, 741, 784
Frenkel variational principle, 302, 566, 569, 570, 572
Kohn variational method, 270
Kohn variational principle, 267, 271
matrix-based variational principle, 270
outgoing wave variation principle, 277
T-matrix variational principle, 264
time-dependent variational principle, 291, 292, 297, 300, 339, 349, 351, 598, 611
time-independent variational theory, 271
variational functional, 295
variational principle, xxx, 263, 291, 295, 300, 590, 596, 597
variational basis (VB), 552
vector, 225, 229, 232, 238, 249, 252, 283, 297, 299, 311, 316, 335, 360, 361, 364, 479, 551, 613, 627, 633, 645, 650, 654, 668, 655, 657, 675, 681, 688, 716
basis vectors, 266, 549, 551
bra vector, 673
eigenvector, 558, 833
Hilbert space vectors, 122
internuclear vector, 247

Jacobi vector, 257, 258
k vector, 304
ket vector, 673
nonadiabatic coupling vector, 407, 411, 412, 446, 606, 645, 657
polarization vector, 811
Poynting vector, 334
vector field, 501, 776, 812
vector operator, 247
vector potential, 604
wave vector, 323, 665, 667, 673–675, 681

Vibok, A., 273
vibration, 162, 372, 377, 638, 761
angular coordinate, 583
coupled vibrational degrees of freedom, 775
molecular vibrations, 839
vibrational degrees of freedom, 372–374, 468, 752, 756, 759, 768, 770, 783, 817, 820, 836, 843
vibrational energy redistribution, 585
vibrational mode, 52, 59, 65, 373, 374, 773
vibrational quantum number, 260
vibrational spectrum, 54

Vignale, G., 604
virtual, 311
virtual orbital, 824
virtual trajectory, 428
vitamin D, 842
Volobuev, Y.L., 644, 645, 647
vortex, 494
quantized hydrodynamic vortex, 494
vortex chirality, 495
vorticity, 494

Walton, A.R., 469, 488
water, 634, 636, 637, 639, 644, 750, 775, 793, 797, 804
deuterated water, 634, 639
water molecule, 640, 793, 797, 802–804

Watson–Crick, 577
Watson–Crick form, 578, 579
Watson–Crick structure, 579

wave, 668
electromagnetic wave, 668
four-wave mixing, 665
p-like waves, 637
three wave mixing, 665
wave equation, 667

wave function, xix, xxx, 47–50, 56–58, 61, 73, 75, 76, 89–91, 95, 96, 98, 104, 115, 116, 127, 139, 174, 188, 202, 204, 209, 216, 220, 226, 233, 239, 250, 251, 259, 260, 265, 275, 285, 293–297, 304, 312, 318, 326, 327, 344, 373, 441, 442, 445, 448, 451, 461, 488, 491, 493, 495, 497–500, 502, 504–506, 509, 517, 540, 556, 560, 564, 567, 572, 573, 587, 589, 590, 618, 668, 671, 672, 677, 681, 706, 723, 734, 748, 756
adiabatic wave functions, 822
auxiliary wave function, 407, 413, 643
Bohm–de Broglie wave function, 495
Born–Oppenheimer wave function, 49
channel wave functions, 241
closed-shell wave function, 309
collapse of the wave function, 620
determinantal wave function, 304
diabatic electronic wave functions, 686
diabatic nuclear wave functions, 253, 254
double-valued electronic wave function, 76
electronic wave function, 59, 75, 76, 82, 144, 256, 305, 306, 311, 315, 325, 342, 381, 387, 395, 443, 451, 459, 627, 644, 649, 708, 753, 768, 820, 821, 830
Gausian wave function, 483
Kohn–Sham wave function, 610
many-body wave function, 595, 597, 605
molecular wave function, xxvi, xxviii, 47, 54, 57, 74, 76, 322, 324, 350, 352, 420, 424, 583, 669, 681, 685, 721–723, 725, 727, 732
multichannel wave function, 363

nuclear wave function, 48, 50, 58, 72, 251, 342, 361, 420, 421, 427, 442, 459, 461, 462, 491, 601, 626, 629, 630, 767
phase space wave functions, 462
reference wave function, 413, 643
single-valued wave function, 76
spawned wave functions, 424
TDHF wave function, 312
wave function following method, 222
WKB wave functions, 285
wave number, 363, 415, 418, 434
complex wave number, 114
wave packet, xxix, xxx, xxxiv, xxxv, 88, 89, 291, 308, 418, 419, 483, 484, 509, 514, 524, 525, 548, 549, 557, 567, 583, 585, 676, 679, 684, 688, 738, 739
bra wave packet, 676, 677
delocalized wave packets, 512
electronic wave packets, xxxi, 78, 845
excited wave packet, 676
Gaussian wave packet, xxxiii, 160, 461
infinitely narrow wave packet, 349
ket wave packet, 677
L^2-integrable wave packets, 87
norm of the wave packet, 558
nuclear wave packet, xxvii, xxxi, xxxiii, xxxiv, 77, 121, 349, 403, 517, 547–549, 565, 687, 841
semiclassical wave packet, 482
splitting of the nuclear wave packet, 491
square-integrable wave packet, xxx
traveling Gaussian wave packets, 421
wave packet propagation, xxx, 547, 548, 558, 563–565, 573, 574, 579, 580, 711, 751
wave packet spreading, 629
wave packet time behavior, 548
Webster, F.A., 640
attractive well, 733
Wentzel–Kramers–Brillouin (WKB), 122
Werner, H.J., 259
Weyl–Wigner correspondence, 152
Wick rotation, 177
Wigner, 146, 479, 518, 696
reduced Wigner functions, 147
Wigner function, xxix, 122, 146–149, 150–154, 740, 747
Wigner transform, 423, 518, 519, 524, 536, 693, 694
Wigner–Moyal, 521
Wigner–Moyal equations, 546
Wigner–Moyal expansion, 519, 520
Wigner–Moyal limit, 521, 522
window, 617, 618, 621, 690
excitation window, 698
probe window function, 699
pump window function, 699
WKB, 142, 144, 285, 286, 326, 465
matching condition, 142
WKB approximation, 139, 143, 324
WKB wave function, 138–140, 142, 144, 145, 285
Wong, K.F., 641
work function, 831
Wronskian, 108, 208, 240
generalized Wronskian, 207
Wyatt, R.E., 499, 508

xenon, 482
Xiao, L., 442, 443

Yan, Y., 738
yield, 748
relative yield, 736
yield optimization, 739

Zener, C., xxx, 282, 288
zeroth-order nuclear wave function, 56
Zeng, X., 805
Zhu, C., 289, 657, 659
Zwanziger, J.W., 395, 397